Lecture Notes in Artificial Intelligence 607

Subseries of Lecture Notes in Computer Science
Edited by J. Siekmann

Lecture Notes in Computer Science
Edited by G. Goos and J. Hartmanis

D. Kapur (Ed.)

Automated Deduction – CADE-11

11th International Conference on
Automated Deduction
Saratoga Springs, NY, USA, June 15-18, 1992
Proceedings

Springer-Verlag
Berlin Heidelberg New York
London Paris Tokyo
Hong Kong Barcelona
Budapest

Series Editor

Jörg Siekmann
University of Saarland
German Research Center for Artificial Intelligence (DFKI)
Stuhlsatzenhausweg 3, W-6600 Saarbrücken 11, FRG

Volume Editor

Deepak Kapur
Institute for Programming and Logics, Dept. of Computer Science
State University of New York, 1400 Washington Av., Albany, NY 12222, USA

CR Subject Classification (1991): I.2.3, F.4.1-2

ISBN 3-540-55602-8 Springer-Verlag Berlin Heidelberg New York
ISBN 0-387-55602-8 Springer-Verlag New York Berlin Heidelberg

Printed in Germany

Typesetting: Camera ready by author/editor
Printing and binding: Druckhaus Beltz, Hemsbach/Bergstr.
45/3140-543210 - Printed on acid-free paper

Preface

This volume contains the papers presented at the Eleventh International Conference on Automated Deduction (CADE-11) held on June 15-18, 1992, in Saratoga Springs, NY, about 20 miles north of Albany, NY.

One hundred thirty-six (136) research papers were submitted for presentation by researchers from nearly 20 countries including Algeria, Australia, Canada, Colombia, France, Germany, Italy, India, Japan, Netherlands, People's Republic of China, Russia, South Korea, Spain, Sweden, Switzerland, Ukraine, United Kingdom, United States. Papers covered many topics including resolution; term-rewriting; natural deduction; theorem proving, in particular in algebra and geometry; parallel theorem provers; unification theory; constraint solving; logic programming; verification; multi-valued, temporal and nonclassical logics; non-monotonic reasoning; planning; proof theory; higher-order logics; and inductive theorem proving.

Each submission was reviewed by at least three program committee members. A large number of papers were found to be of acceptable quality; however because of the short duration of the conference, only 46 papers could be accepted for presentation at the conference. Nearly ninety referees assisted the program committee in the reviewing process. On behalf of the program committee, I thank the referees for their timely reports.

This volume also contains short descriptions of 23 implementations of automated deduction systems. These descriptions were also reviewed by the program committee.

CADE is pleased to announce the establishment of the *Herbrand Award for Distinguished Contributions to Automated Reasoning* to honor an individual or (a group of) individuals for exceptional contributions to the field of automated deduction. Details about this award are given on the next page.

I am pleased to announce that the first Herbrand award will be given at CADE-11 to Dr. Larry Wos. Dr. Wos has made numerous contributions to the field of Automated Reasoning including his seminal papers on the set-of support strategy, paramodulation and demodulation. He has been a principal force behind theorem proving research at Argonne National Laboratory; he is the editor-in-chief of *J. of Automated Reasoning* as well as a founder of the *Association of Automated Reasoning.*

I am grateful to the program committee for their efforts and cooperation in deciding the program as well as other matters related to CADE; to Neil Murray for making the local arrangements for the conference; to the invited speakers Larry Wos (Keynote Address), Raymond Smullyan (Banquet Speaker) and Grigori Mints; to Alan Bundy for organizing a workshop on inductive theorem proving; to the organizers of the tutorials; to Rusty Lusk, his family and Argonne National Laboratory for hosting the program committee meeting; and to Sally Goodall whose meticulous organizational abilities made my work easier.

Finally I am especially indebted to Xumin Nie for maintaining the data base of submissions and doing everything that needed to be done to facilitate my task in organizing the conference.

CADE is partially funded by the Institute for Programming and Logics, the State University of New York, Albany.

Deepak Kapur
Chair, CADE-11

April 1992

Herbrand Award for
Distinguished Contributions to Automated Reasoning

CADE is pleased to announce the establishment of *Herbrand Award for Distinguished Contributions to Automated Reasoning* to honor an individual or (a group of) individuals for exceptional contributions to the field of Automated Deduction.

Nominations for this award can be made at any time to the CADE chair. A nomination should include a letter (up to 2000 words) from a principal nominator describing the nominee's contributions, along with two other letters (up to 2000 words) of endorsement.

The CADE Chair will be responsible for soliciting opinions and evaluations, and carrying out a vote. Nominations will be kept confidential. The nomination of a group of individuals who are collaborators, will be considered as a single nomination.

The program committee, the board of trustees of CADE Inc. (assuming that is in place; otherwise the chairs of last 5 CADEs), and the past award winners will participate in making the selection.

Program Committee

Peter Andrews, Carnegie Mellon University
Wolfgang Bibel, Technische Hochschule Darmstadt
W.W. Bledsoe, University of Texas at Austin
Robert S. Boyer, University of Texas at Austin
Alan Bundy, University of Edinburgh
Edmund Clarke, Carnegie Mellon University
Robert Constable, Cornell University
Ryuzo Hasegawa, ICOT, Japan
Larry Henschen, Northwestern University
Deepak Kapur, University at Albany - SUNY
Claude Kirchner, CRIN and INRIA Lorraine
Kurt Konolige, SRI International
Jean-Louis Lassez, IBM Research, Yorktown Heights
Vladimir Lifschitz, University of Texas at Austin
Donald Loveland, Duke University
Ewing Lusk, Argonne National Laboratory
William McCune, Argonne National Laboratory
Grigori Mints, Stanford University
David Musser, Rensselaer Polytechnic Institute
Hans Jürgen Ohlbach, MPI-I Saarbrücken
David A. Plaisted, University of North Carolina, Chapel Hill
Jörg Siekmann, Universität Saarbrücken
John Slaney, Australian National University
Mark E. Stickel, SRI International

Local Arrangements

Neil V. Murray, University at Albany - SUNY

Referees

R. Amadio	F. Baader	D. Billington
A. Boudet	A. Brodsky	J.R. Burch
H.-J. Burckert	R. Caferra	S.-C. Chou
H. Comon	N. Dershowitz	E. Domenjoud
E. Eder	U. Egly	N. Eisinger
J. Eusterbrock	H. Eveking	W.M. Farmer
M. Fitting	B. Fronhoefer	U. Furbach
H. Ganzinger	R. Girle	G. Grosse
M. Grundy	H. Kirchner	M. Hanus
N. Heintze	M. Hermann	L.M. Hines
S. Hoelldobler	J. Hong	J. Hsiang
J. Jaffar	M. Kaufmann	F. Klay
M. Kohlhase	C. Kreitz	K. Kunen
P. Lescanne	R. Letz	D.E. Long
D. Lugiez	C. Lynch	M. Maher
K. Marriott	L. Maranget	H. Maria
D. Miller	N. Murray	P. Narendran
G. Neugebauer	X. Nie	T. Nipkow
A. Nonnengart	W. Nutt	M. O'Donnell
F. Oppacher	P. Lee	L.C. Paulson
D. Pearce	J. Pelletier	F. Pfenning
U.S. Reddy	D. Reed	J.-L Remy
C. Ringeissen	M. Rusinowitch	V. Saraswat
T. Schaub	H. Schlingleff	M. Schmidt-Schauss
R.C. Sekar	N. Shankar	W. Sieg
G. Smolka	W. Snyder	R. Socher-Ambrosius
J. Stillman	R. Topor	P. Viry
M. Vittek	R. Waldinger	C. Walther
C. Weidenbach	G. Wrightson	W.T. Wu
H. Zhang	X. Zhao	

Table of Contents

Session IV

Session V: Banquet Address

Session VI

Session VII

Session VIII A

Session VIII B

Session IX

Session X: Invited Talk

Session XI

Session XII A

Session XII B

Session XIII

System Abstracts

Problem Sets

The Impossibility of the Automation of Logical Reasoning[1]

Larry Wos

Mathematics and Computer Science Division
Argonne National Laboratory
Argonne, Illinois 60439-4801
wos@mcs.anl.gov
(708) 252-7224

Abstract

Mathematicians and logicians are justly revered for the power of their minds and the depth of their reasoning. Their success in applying logical reasoning to the proof of one significant theorem after another is awe inspiring. And yet, who among these scientists can say precisely how such proofs are found? Therefore, how could anyone have had the audacity to even *consider* the possibility of automating logical reasoning with the goal of assigning to a computer program the task of proving theorems? Indeed, in 1948 the eminent logician Lukasiewicz asserted that a formalized proof cannot be "*discovered* mechanically", but can only be "*checked* mechanically".

Given the importance of his research, it would be erroneous to interpret the Lukasiewicz assertion as evidence of naivete or a lack of foresight. Clearly, therefore, the successful automation of proof finding is out of the question—or so it was thought—especially if the theorem under consideration is interesting and deep. Nevertheless, colleagues, we have beaten the odds, done the impossible, automated reasoning so effectively that our programs have even answered open questions.

Yet skeptics still exist, funding is not abundant, and recognition of our achievements is inappropriately small and not sufficiently widespread. How might we remedy this situation, and how might we prevent people from taking seriously articles entitled "Computers Still Can't Do Beautiful Mathematics" (*New York Times*, July 14, 1991, Week in Review)? Perhaps the following suggestions will provide ammunition for waging the war in which we are

[1]This work was supported by the Applied Mathematical Sciences subprogram of the Office of Energy Research, U.S. Department of Energy, under Contract W-31-109-Eng-38.

obviously engaged, and perhaps my colleagues can use that ammunition to *win the war*. Let me warn you first, however, that these suggestions come from one who lacks a talent for reticence and the avoidance of superlatives.

To me, less than the following understates the case: The degree to which we have automated logical reasoning in less than thirty years is marvelous, stunning, and simply beautiful. For example, Boyer and Moore have produced a splendid program that has verified algorithms in use and all of the details for a new hardware/software system. Chou's program is impressive in its ability to prove one geometry theorem after another. McRobbie and Meyer have a program that has answered open questions in nonstandard logics. Bledsoe's programs have obtained marked success in analysis and proved most useful in Gypsy in the context of correctness proofs for software. McCune and I have answered open questions with his powerful program OTTER, deep questions from a variety of fields of mathematics and logic.

Of course, given a randomly selected theorem from some mathematics book, the probability is small that an existing automated reasoning program can prove it. However, finally, we can cite numerous examples of being presented with interesting theorems, taken from an unfamiliar area, that were proved on the first try by a program. The point is to think of the glass as one fourth full, not three fourths empty.

Given the cited successes and realizing that many others exist, one might wonder why we are insufficiently recognized. The fault, dear Brutus, rests in part with inadequate dissemination of our achievements and in part with too few examples of successes with problems designated by others as "real problems" or Grand Challenges. Of course, some of our work is familiar to other computer scientists, but far wider exposure and recognition is needed, requiring a nontrivial expenditure of time and energy. Precisely how might we proceed?

Campaign 1, graduate departments in universities: We contact such departments—mathematics and logic are particularly amenable—to disseminate our achievements and to accrue appropriate open questions to attack. An obvious mechanism is that of sending letters, papers, and even books.

Campaign 2, the *Journal of Automated Reasoning*: We seek to expand its subscriber list to exceed 1200. Perhaps a chain letter: You subscribe to the *Journal* and add four names under yours on a circulating list, each of whom must subscribe and supply four additional names.

Campaign 3, what are called *real* applications: We attempt to axiomatize aspects of physics, chemistry, biology, and other sciences and—of greater importance—then demonstrate the ability to solve *hard* problems in the corresponding fields. Of course, I am not referring to test problems or prototype problems, but, instead, to problems that the world loves to discuss. I recognize that my poor scholarship may have overlooked examples that already exist; if so, please educate me.

Do I think that such campaigns will be fun? On the contrary, my estimate of the appeal of these campaigns approximates zero. Nevertheless, by winning these campaigns, we win the war.

As we fight this latest war, let us keep in mind the many battles we have won, victories over various aspects of the automation of logical reasoning, symbolized by answers to open questions and the verification of complicated algorithms. Of course, more powerful strategies and more effective inference rules are needed, breakthroughs in implementation are crucial, and—to select one major area not attacked often—the capacity to generate large models and counterexamples efficiently is vital. In addition, perhaps powerful metrics can be discovered, metrics for accurately measuring progress and making sharp comparisons between programs, between inference rules, between strategies, and between paradigms. Despite the given needs and many not cited, we have accomplished much. Indeed, to us as researchers goes the credit for producing this beautiful edifice, this marvel of computer science, this program that functions as an automated reasoning assistant.

We must win the current war and continue to make advances at the current rate, for the importance of automated reasoning to science and to society as a whole cannot be overvalued. In addition, since mathematicians and logicians throughout the world applaud the magic of sitting alone in a room and reading, line by line, a detailed proof of a deep theorem, the field of automated reasoning clearly merits substantial praise, for some of our programs contribute to magic by producing just such an intriguing proof that can be read and fully understood and appreciated.

As we did with the goal of the automation of logical reasoning, we can again do the impossible: Through research and effective campaigning, we can correctly gain for our field wide recognition as one of the most significant and astounding advances of civilization. That we have barely begun is clear; that challenges will always exist is preferred; that automated reasoning offers intrigue, excitement, and beauty will continually be true.

Automatic Proofs in Mathematical Logic and Analysis*

Kurt Ammon
Windmühlenweg 27, W-2000 Hamburg 52
Germany

Abstract. The SHUNYATA program contains heuristics which model reasoning processes of mathematicians in the discovery of proofs. For example, a heuristic constructs sets which form the central "ideas" of proofs. In order to prove the existence of a particular element, another heuristic derives properties of such an element and applies them to its definition. Some heuristics control the application of other heuristics, for example, by time limits which interrupt a heuristic if it achieves no result. Because the heuristics simulate reasoning processes of mathematicians, they produce readable proofs. In experiments, SHUNYATA generated proofs of Banach's fixed point theorem, the Schröder-Bernstein theorem, Gödel's incompleteness theorem, and the Bolzano-Weierstrass theorem. Its architecture and its mode of operation are illustrated in detail by its proof of Heine's theorem which says that every continuous function on a compact set of real numbers is uniformly continuous. Further experiments with a learning procedure in SHUNYATA suggest that an automatic construction of its heuristics on the basis of elementary functions is feasible.

1 Introduction

The main problem in the discovery of many proofs is the construction of a particular set. For example, the intermediate value theorem says that if f is a continuous function on a closed interval $[a, b]$ with $f(a) \leq 0$ and $f(b) \geq 0$, then $f(x) = 0$ holds for some $x \in [a, b]$. In order to prove the theorem, a mathematician may attempt to construct a suitable set and prove that $f(x) = 0$ holds for the supremum x of this set. Because predicates define sets, he may first apply the simple and efficient heuristic to look for a suitable predicate in the theorem. Indeed, the predicate $f(a) \leq 0$ in the theorem defines a suitable set $\{a : f(a) \leq 0\}$ for whose supremum x, $f(x) = 0$ holds. In order to prove the existence of an element $x \in S$ such that a property $P(x)$ holds, a mathematician may apply the heuristic to derive the definition of such an $x \in S$ from the property $P(x)$. An example: If the existence of a solution x of a linear equation $ax = b$ is to be proved, the definition $x = a^{-1}b$ can be derived from $ax = b$. The SHUNYATA program contains heuristics mathematicians use for the discovery of proofs, i.e., it simulates reasoning processes of mathematicians. Preparatory experiments yielded automatic proofs of Banach's fixed point theorem

*This work, in whole or in part, describes components of machines or processes protected by one or more patents or patent applications in Europe, the United States of America, or elsewhere. Further information is available from the author.

(Ammon, 1988a) and the Schröder-Bernstein theorem.[1] In more recent experiments, SHUNYATA generated proofs of Gödel's incompleteness theorem[2] and the Bolzano-Weierstrass theorem.[3] Its architecture and mode of operation are illustrated by its proof of Heine's theorem which says that every continuous function on a compact set of real numbers is uniformly continuous. To my knowledge, these are the first automatic proofs of these theorems in mathematical logic and in standard analysis (see Bledsoe, 1977, p. 31). Section 2 describes heuristics which were applied to generate a proof of Heine's theorem. Sections 3 and 4 give Heine's theorem and a proof of this theorem which was generated by the heuristics. Section 5 discusses the generality of the heuristics and explains their power.

2 Heuristics

The SHUNYATA program contains heuristics which model reasoning processes of mathematicians in the discovery of proofs. We describe important heuristics in SHUNYATA which were applied to the automatic generation of a proof of Heine's theorem.

The input of a heuristic is a *goal*, which may contain a constraint such as a time limit, and a *partial proof*, which is a list of proof steps. A *proof step* contains a proposition or an assumption, and an explanation which gives the heuristic and the preceding proof steps, definitions, and lemmas that were used to derive the proposition. The heuristics produce further proof steps and subgoals. Besides the goal to prove a proposition, some heuristics pursue the goal

$$\textit{prove } x : P(x) \tag{1}$$

to prove a proposition $P(x)$ with a free variable x, i.e., they construct an x such that a given property $P(x)$ holds.

The *proof heuristic* attempts to prove a proposition. If the proposition is not contained in a preceding proof step, it finds out the type of the proposition such as an element relation, a universal or existential proposition, or an inequality and calls special heuristics which process propositions of a special type. If the special heuristics fail, the proof heuristic calls the reduction heuristic and the replacement heuristic.

The *element heuristic* proves element relations $x \in S$. By means of subset relations $T \subseteq S$, it reduces the goal to prove $x \in S$ to the subgoal to prove $x \in T$. The application of the subset relation $T \subseteq S$ to $x \in T$ yields the element relation $x \in S$ to be proved. By means of functions f on a set T into S, it reduces the goal to prove $x \in S$ to the subgoal to prove $x \in T$. The application of the function f to $x \in T$ yields the element relation $x \in S$ to be proved. The element heuristic also processes goals

$$\textit{prove } x : x \in S \tag{2}$$

[1] The Schröder-Bernstein theorem says that two sets M and N are equinumerous if M is equinumerous with a subset of N and N is equinumerous with a subset of M. Kaufmann (1989) checked a proof of this theorem.

[2] Gödel's incompleteness theorem says that every formal number theory contains an undecidable formula, i.e., neither this formula nor its negation can be proved in the theory. Shankar (1987) checked a detailed proof of this theorem.

[3] The Bolzano-Weierstrass theorem says that every infinite bounded set of real numbers has an accumulation point (see Apostol, 1957, p. 43).

to prove an element relation with a free variable x, i.e., to construct an element $x \in S$.

The *universal heuristic* proves universally quantified implications, that is, propositions of the form $\forall x(P(x) \rightarrow Q(x))$. It introduces the assumption that $P(x)$ holds and attempts to derive $Q(x)$. If it succeeds, it applies the rules of implication and generality introduction (see Kleene, 1952, pp. 98–99) which yields the universally quantified implication to be proved.

The *existential heuristic* proves existential propositions $\exists x(x \in S \wedge P(x))$. It searches for elements $a \in S$ in the proof and attempts to prove $P(a)$. If $P(a)$ is true, the application of the rule of existence introduction (see Kleene, 1952, p. 99) to $a \in S \wedge P(a)$ yields the existential proposition to be proved. If an element $a \in S$ cannot be found or $P(a)$ cannot be proved, the existential heuristic calls the concept heuristic which transforms concepts in the proof into a form in which they can be used to find properties of an element $x \in S$ with $P(x)$. Then, it introduces the assumption $x \in S$ and attempts to prove $P(x)$. If subgoals to prove equality or inequality relations $R(x)$ containing the variable x are generated by other heuristics, these relations are regarded as properties of $x \in S$ and are assumed to be true. If a proof of $P(x)$ is achieved, the existential heuristic constructs an $x \in S$ from the relations $R(x)$ by means of simple rules. For example, if a relation $R(x)$ says that x is less than any element of a finite set S of numbers, such a rule defines x as the minimum of S. Then, the existential heuristic proves the relation $R(x)$ on the basis of the definition of $x \in S$. Finally, it applies the rule of existence introduction to $x \in S \wedge P(x)$ which yields the existential proposition to be proved.

The *concept heuristic* searches for terminal concepts in the proof, i.e., concepts that are not applied to the definition of other concepts, and calls the Skolem heuristic and the set heuristic which transform the terminal concepts into a form in which they can be used by other heuristics.

The *Skolem heuristic* replaces existential quantifiers in the consequents of terminal concepts by functions, i.e., it skolemizes these quantifiers.

The *set heuristic* constructs sets and collections of sets. If a terminal concept is a universally quantified implication whose antecedent contains a variable denoting a set or a collection of sets, it constructs examples of such sets or collections of sets. It searches for predicates $P(a, x)$ in preceding proof steps and applies a variable such as a as an index of a set or a collection of sets $\{x : P(a, x)\}$. If a predicate $P(a, x, y)$ contains a free variable y, it is replaced by a new function $f(a)$ which yields a set or a collection of sets $\{x : P(a, x, f(a))\}$.

The *inequality heuristic* proves inequalities. It calls the reduction, transitivity, triangle, and replacement heuristic within time limits. We first describe the transitivity and the triangle heuristic and then the reduction and the replacement heuristic which are also called by the proof heuristic.

The *transitivity heuristic* applies the transitivity of the inequality relation, i.e., $x < y$ and $y < z$ implies $x < z$ for all real numbers x, y, and z. If $x < z$ is an inequality to be proved, the transitivity heuristic searches in the proof for an inequality of the form $x < y$, where the variable y denotes a term, and attempts to prove $y < z$. If $y < z$ is true, it applies the transitivity of the inequality relation to $x < y$ and $y < z$ which yields the inequality $x < z$ to be proved.

The *triangle heuristic* applies the triangle inequality to prove an inequality. The triangle inequality says that $|x - z| \leq |x - y| + |y - z|$ holds for all real numbers x, y, and z. If $|x - z| \leq r$ is an inequality to be proved, the triangle heuristic generates the subgoal to construct a real number y and attempts to prove $|x - y| \leq r/2$ and $|y - z| \leq r/2$. If the inequalities $|x - y| \leq r/2$ and $|y - z| \leq r/2$ are true, the triangle inequality is applied which yields the inequality $|x - z| \leq r$ to be proved.

The *reduction heuristic* matches the consequent $Q(x)$ in a universal proposition $\forall x(P(x) \rightarrow Q(x))$ with a proposition $Q(a)$ to be proved and attempts to prove the instantiated antecedent $P(a)$. If $P(a)$ is true, it applies the universal proposition to $P(a)$ which yields the proposition $Q(a)$ to be proved. The reduction heuristic also processes the goal to prove a proposition $Q(x)$ with a free variable x, i.e., to construct an x such that $Q(x)$ holds. In this case, it reduces the goal to prove $Q(x)$ to the goal to prove $P(x)$, i.e., to construct an x such that $P(x)$ holds.

The *replacement heuristic* changes the representation of a proposition to be proved and attempts to prove the changed proposition. It searches for subterms s of a proposition $P(s)$ to be proved and calls the equation heuristic to prove equations of the form $s = t$, where t is a term to be constructed. Then, it replaces the term s in $P(s)$ by t and generates the subgoal to prove the resulting proposition $P(t)$. Because of $s = t$, the proposition $P(t)$ implies the proposition $P(s)$ to be proved.

The *equation heuristic* proves equations of the form $s = t$, where s is a given term and t is a term to be constructed. It matches an equation $a = b$ in the consequent of a universal proposition with the equation $s = t$ and generates the subgoal to prove the instantiated antecedent of the proposition. The application of the universal proposition to this antecedent yields an equation of the form $s = t$.

The SHUNYATA program contains more sophisticated variants of the heuristics described above. For example, the reduction heuristic applies universal propositions $\forall x, y, ...(P(x, y, ...) \rightarrow Q(x, y, ...))$ containing several variables $x, y, ...$ to reduce the goal to prove a proposition $P(a, b, ...)$ to the subgoal to prove $Q(a, b, ...)$. Furthermore, it contains additional heuristics. For example, a simplification heuristic attempts to simplify propositions in new proof steps. A paraphrase heuristic attempts to represent a proposition in a new proof step more explicitly: It searches for subterms s of the proposition, calls the equation heuristic to prove an equation of the form $s = t$, where t is a term to be constructed, and replaces the term s in the proposition by t. Special techniques improve SHUNYATA's efficiency. For example, a technique deletes proof steps that have been used to achieve preceding subgoals. This entails that SHUNYATA must only process a small number of proofs steps at the same time.

3 Heine's Theorem

Definition 1 A collection C of sets is called an *open covering* of a set S of real numbers if $S \subseteq \bigcup_{A \in C} A$ and all sets $A \in C$ are open, i.e., for all $a \in A$, there is an $r > 0$ such that all real numbers x with $|a - x| < r$ are contained in A (see Apostol, 1957, p. 41 and p. 51). A formalization is:

$$\forall C, S(\mathit{collection}(C, \mathbf{R}) \wedge S \subseteq \mathbf{R} \rightarrow \\ (\mathit{open\text{-}covering}(C, S) \leftrightarrow S \subseteq \textstyle\bigcup_{A \in C} A \wedge \forall A(A \in C \rightarrow \mathit{open}(A))))$$

Definition 2 A set S of real numbers is called *compact* if every open covering of S contains a finite subcollection which also covers S. A formalization is:

$$\forall S(S \subseteq \mathbf{R} \rightarrow \\ (compact(S) \leftrightarrow \forall C(collection(C, \mathbf{R}) \wedge open\text{-}covering(C, S) \rightarrow \\ \exists D(D \subseteq C \wedge finite(D) \wedge S \subseteq \bigcup_{A \in D} A))))$$

Definition 3 A function f on a set S of real numbers with function values in the real numbers is called *continuous* if the following statement holds: For all $a \in S$ and $\epsilon > 0$, there is a $\delta > 0$ such that $|f(a) - f(x)| < \epsilon$ holds for all $x \in S$ with $|a - x| < \delta$. A formalization is:

$$\forall S, f(S \subseteq \mathbf{R} \wedge function(f, S, \mathbf{R}) \rightarrow \\ (continuous(f, S) \leftrightarrow \\ \forall a, \epsilon(a \in S \wedge \epsilon \in \mathbf{R} \wedge 0 < \epsilon \rightarrow \\ \exists \delta(\delta \in \mathbf{R} \wedge 0 < \delta \wedge \\ \forall x(x \in S \wedge |a - x| < \delta \rightarrow |f(a) - f(x)| < \epsilon)))))$$

Definition 4 A function f on a set S of real numbers with function values in the real numbers is called *uniformly continuous* if the following statement holds: For all $\epsilon > 0$, there is a $\delta > 0$ such that $|f(x) - f(y)| < \epsilon$ holds for all $x \in S$ and $y \in S$ with $|x - y| < \delta$. A formalization is:

$$\forall S, f(S \subseteq \mathbf{R}) \wedge function(f, S, \mathbf{R}) \rightarrow \\ (uniformly\text{-}continuous(f, S) \leftrightarrow \\ \forall \epsilon(\epsilon \in \mathbf{R} \wedge 0 < \epsilon \rightarrow \\ \exists \delta(\delta \in \mathbf{R} \wedge 0 < \delta \wedge \\ \forall x, y(x \in S \wedge y \in S \wedge |x - y| < \delta \rightarrow \\ |f(x) - f(y)| < \epsilon)))))$$

Theorem 1 (Heine) Every continuous function on a compact set of real numbers is uniformly continuous (see Apostol, 1957, p. 75). A formalization is:

$$\forall S, f(S \subseteq \mathbf{R} \wedge compact(S) \wedge function(f, S, \mathbf{R}) \wedge continuous(f, S) \rightarrow \\ uniformly\text{-}continuous(f, S))$$

4 Proof

The SHUNYATA program took some four hours to produce a proof of Heine's theorem on an IBM PC AT. It contains some 120 proof steps which are more than 18,000 bytes long. A trace of the subgoals and the proofs steps generated in the course of the proof discovery is some 2,200,000 bytes long. We give the proof in ordinary representation, explain its computational representation, and describe how it was generated by the heuristics in Section 2.

The proof is: We assume that f is a continuous function on a compact set S of real numbers. The definition of continuous functions yields that for all $a \in S$ and $\epsilon > 0$, there is a $\delta > 0$ such that $|f(a) - f(x)| < \epsilon$ for all $x \in S$ with $|a - x| < \delta$. This means that there is a function g on S and the positive real numbers such that if $a \in S$, $x \in S$, $\epsilon > 0$, and $|a - x| < g(a, \epsilon)$, then

$$|f(a) - f(x)| < \epsilon. \tag{3}$$

Let $\epsilon > 0$. We define a function h on S by

$$h(a) = \frac{g(a, \epsilon/2)}{2}, \tag{4}$$

where $a \in S$. For all $a \in S$, we furthermore define sets

$$N(a) = \{x : |a - x| < h(a)\} \tag{5}$$

of the real numbers. Thus, $N(a)$ is the set of all real numbers whose "distance" from a is less than $h(a)$. The sets $N(a)$, where $a \in S$, form an open covering of S because for all $a \in S$, a is contained in $N(a)$ and $N(a)$ is open. Since S is compact, this implies that there is a finite subset T of S such that

$$S \subseteq \bigcup_{a \in T} N(a). \tag{6}$$

We define

$$\delta = \min_{a \in T} h(a). \tag{7}$$

Now, we assume $x \in S$ and $y \in S$ with

$$|x - y| < \delta \tag{8}$$

in order to prove

$$|f(x) - f(y)| < \epsilon. \tag{9}$$

Because of (6), there is an element $a \in T$ such that $x \in N(a)$ holds. Because of (5), we obtain

$$|a - x| < h(a). \tag{10}$$

Because of (4), this means that $|a - x| < g(a, \epsilon/2)$, which implies

$$|f(a) - f(x)| < \frac{\epsilon}{2} \tag{11}$$

according to (3). Because of (7) and (8), we have

$$|x - y| < h(a). \tag{12}$$

The application of the triangle inequality to (10) and (12) yields

$$|a - y| \leq |a - x| + |x - y| < h(a) + h(a) = g(a, \epsilon/2) \tag{13}$$

because of (4). From (13), we obtain

$$|f(a) - f(y)| < \frac{\epsilon}{2}. \tag{14}$$

because of (3). The application of the triangle inequality to (11) and (14) yields

$$|f(x) - f(y)| \leq |f(x) - f(a)| + |f(a) - f(y)| < \epsilon. \tag{15}$$

Thus, the proof of (9) is complete. It follows that the function f is uniformly continuous.

Tables 1 and 2 contain a part of the proof generated by the SHUNYATA program. Each proof step consists of a number, a proposition or an assumption, and an explanation which gives the heuristic and the preceding proof steps, definitions, and lemmas that were used to derive the proposition. For example, step 3 in Table 1 was generated by the concept heuristic by means of step 1 and Definition 2 and step 42 in Table 2 by the equation heuristic by means of step 41 and Lemma 5.

The proof steps 42, 58, and 64–67 in Table 2 require five lemmas and another definition:

1.	*assume* $S, f(S \subseteq \mathbf{R} \wedge compact(S) \wedge$ $function(f, S, \mathbf{R}) \wedge continuous(f, S))$	Universal
2.	*assume* $\epsilon(\epsilon \in \mathbf{R} \wedge 0 < \epsilon)$	Universal
3.	$\forall C(collection(C, \mathbf{R}) \wedge open\text{-}covering(C, S) \rightarrow$ $\exists D(D \subseteq C \wedge finite(D) \wedge S \subseteq \bigcup_{A \in D} A))$	Concept, 1, D2
4.	$\forall a, \epsilon(a \in S \wedge \epsilon \in \mathbf{R} \wedge 0 < \epsilon \rightarrow$ $\exists \delta(\delta \in \mathbf{R} \wedge 0 < \delta \wedge$ $\forall x(x \in S \wedge \|a - x\| < \delta \rightarrow$ $\|f(a) - f(x)\| < \epsilon)))$	Concept, 1, D3
5.	$\exists g(function(g) \wedge$ $\forall a, \epsilon(a \in S \wedge \epsilon \in \mathbf{R} \wedge 0 < \epsilon \rightarrow$ $g(a, \epsilon) \in \mathbf{R} \wedge 0 < g(a, \epsilon) \wedge$ $\forall x(x \in S \wedge \|a - x\| < g(a, \epsilon) \rightarrow$ $\|f(a) - f(x)\| < \epsilon)))$	Skolem, 4
6.	*assume* $g(function(g) \wedge \forall a, \epsilon(a \in S \wedge \ldots))$	Skolem, 5
7.	*assume* $h(function(h, S))$	Set
8.	*assume* $(\forall a(a \in S \rightarrow h(a) \in \mathbf{R} \wedge 0 < h(a))$	Set
9.	*assume* $N(function(N, S) \wedge \forall a(a \in S \rightarrow$	Set
	$N(a) = \{x : x \in \mathbf{R} \wedge \|a - x\| < h(a)\}))$	Set
10.	$collection(N(S), \mathbf{R})$	Set, 9
	...	
39.	$open\text{-}covering(N(S), S)$	Reduction, ..., D1

Table 1: A first part of SHUNYATA's proof of Heine's theorem

Lemma 1 Two is a natural number. A formalization is: $2 \in \mathbf{N}$.

Lemma 2 A fraction of a real number is a real number. A formalization is:

$$\forall x, n(x \in \mathbf{R} \wedge n \in \mathbf{N} \rightarrow x/n \in \mathbf{R})$$

Lemma 3 A fraction of a positive real number is positive. A formalization is:

$$\forall x, n(x \in \mathbf{R} \wedge 0 < x \wedge n \in \mathbf{N} \rightarrow 0 < x/n)$$

Lemma 4 The absolute value of the difference of two real numbers x and y is equal to the absolute value of the difference of y and x. A formalization is:

$$\forall x, y(x \in \mathbf{R} \wedge y \in \mathbf{R} \rightarrow |x - y| = |y - x|)$$

Lemma 5 A finite subset of the image of a function is the image of a finite subset of its domain. A formalization is:

$$\forall f, A, S(A \subseteq f(S) \wedge finite(A) \rightarrow \exists T(T \subseteq S \wedge finite(T) \wedge A = f(T)))$$

Definition 5 Let $(A_i)_{i \in I}$ be an indexed family of sets. Then $\bigcup_{i \in I} A_i$ is the set of all x such that $x \in A_i$ for some $i \in I$. A formalization is:

$$\forall x, A_i, I(x \in \bigcup_{i \in I} A_i \leftrightarrow \exists i(i \in I \wedge x \in A_i))$$

40.	$\exists D(D \subseteq N(S) \wedge \mathit{finite}(D) \wedge S \subseteq \bigcup_{A \in D} A)$	Set, 10, 39, 3
41.	*assume* $D(D \subseteq N(S) \wedge \mathit{finite}(D) \wedge S \subseteq \bigcup_{A \in D} A)$	Set, 40
42.	$\exists T(T \subseteq S \wedge \mathit{finite}(T) \wedge D = N(T))$	Equation, 41, L5
43.	*assume* $T(T \subseteq S \wedge \mathit{finite}(T) \wedge D = N(T))$	Equation, 42
44.	$\bigcup_{A \in D} A = \bigcup_{A \in N(T)} A$	Equation, 43
45.	$S \subseteq \bigcup_{A \in N(T)} A$	Paraphrase, 41, 44
	...	
48.	$S \subseteq \bigcup_{a \in T} N(a)$	Paraphrase, 45, 47
	...	
54.	$S \subseteq \bigcup_{a \in T} \{x : x \in \mathbf{R} \wedge \|a - x\| < h(a)\}$	Paraphrase, 48, 53
55.	*assume* $\delta(\delta \in \mathbf{R} \wedge 0 < \delta)$	Existential
56.	*assume* $x, y(x \in S \wedge y \in S \wedge \|x - y\| < \delta)$	Universal
57.	$x \in \bigcup_{a \in T} \{x : x \in \mathbf{R} \wedge \|a - x\| < h(a)\}$	Element, 54, 56
58.	$\exists a(a \in T \wedge x \in \{x : x \in \mathbf{R} \wedge \|a - x\| < h(a)\})$	Reduction, 57, D5
59.	$x \in \mathbf{R} \wedge \exists a(a \in T \wedge \|a - x\| < h(a))$	Simplification, 58
60.	*assume* $a(a \in T \wedge \|a - x\| < h(a))$	Reduction, 59
61.	$a \in S$	Element, 43, 60
62.	$f(a) \in \mathbf{R}$	Element, 1, 61
63.	$f(x) \in \mathbf{R}$	Element, 1, 56
64.	$\|f(a) - f(x)\| = \|f(x) - f(a)\|$	Equation, 62, 63, L4
65.	$2 \in \mathbf{N}$	Search, L1
66.	$\epsilon/2 \in \mathbf{R}$	Reduction, 2, 65, L2
67.	$0 < \epsilon/2$	Reduction, 2, 65, L3
68.	$h(a) \leq g(a, \epsilon/2)$	Existential, h
69.	$\|a - x\| < g(a, \epsilon/2)$	Transitivity, 60, 68
70.	$\|f(a) - f(x)\| < \epsilon/2$	Reduction, 56, 61, 66, 67, 69, 6
71.	$\|f(x) - f(a)\| < \epsilon/2$	Replacement, 70, 64
72.	$h(a) \leq \frac{g(a,\epsilon/2)}{2}$	Existential, h
73.	$\|a - x\| < \frac{g(a,\epsilon/2)}{2}$	Transitivity, 60, 72
74.	$\delta \leq \frac{g(a,\epsilon/2)}{2}$	Existential, δ
75.	$\|x - y\| < \frac{g(a,\epsilon/2)}{2}$	Transitivity, 56, 74
76.	$a \in \mathbf{R}$	Element, 1, 61
77.	$y \in \mathbf{R}$	Element, 1, 56
78.	$\|a - y\| \leq \|a - x\| + \|x - y\|$	Triangle, 76, 77, 59
79.	$\|a - x\| + \|x - y\| < g(a, \epsilon/2)$	Triangle, 74, 75
80.	$\|a - y\| < g(a, \epsilon/2)$	Triangle, 78, 79
81.	$\|f(a) - f(y)\| < \epsilon/2$	Reduction, 56, 61, 66, 67, 80, 6
82.	$f(y) \in \mathbf{R}$	Element, 1, 56
83.	$\|f(x) - f(y)\| \leq \|f(x) - f(a)\| + \|f(a) - f(y)\|$	Triangle, 63, 82, 62
84.	$\|f(x) - f(a)\| + \|f(a) - f(y)\| < \epsilon$	Triangle, 71, 81
85.	$\|f(x) - f(y)\| < \epsilon$	Triangle, 83, 84

Table 2: A second part of SHUNYATA's proof of Heine's theorem

The proof in Tables 1 and 2 was generated by the heuristics in Section 2 as follows: Because the theorem to be proved is a universally quantified implication, the universal heuristic introduced the assumption in step 1 in Table 1 that its antecedent holds, i.e., that f is a continuous function on a compact set S, and generated the subgoal to prove that f is uniformly continuous. By means of the definition of uniformly continuous functions, the reduction heuristic reduced this subgoal to the subgoal to prove that

$$\begin{aligned}&\exists\,\delta(\delta \in \mathbf{R} \wedge 0 < \delta \wedge \\ &\quad \forall x, y(x \in S \wedge y \in S \wedge |x-y| < \delta \rightarrow |f(x) - f(y)| < \epsilon))\end{aligned} \tag{16}$$

holds for all $\epsilon \in \mathbf{R}$ with $\epsilon > 0$ (see Definition 4). In order to prove this universal proposition, the universal heuristic introduced the assumption in step 2 that $\epsilon > 0$ holds for an $\epsilon \in \mathbf{R}$ and attempted to prove the existential proposition (16). Therefore, the proof heuristic called the existential heuristic which could not find a suitable δ in a preceding proof step. Therefore, it called the concept heuristic which found the terminal concepts *continuous* and *compact* in step 1 and applied their definitions (Definitions 2 and 3) which yielded steps 3 and 4. Because the consequent $\exists\,\delta(\delta \in \mathbf{R} \wedge ...)$ in the definition of continuous functions in step 4 is an existential proposition, the concept heuristic called the Skolem heuristic which replaced the existentially quantified variable δ in step 4 by a function g and produced steps 5 and 6. Then, the concept heuristic called the set heuristic because the antecedent in the definition of compact sets in step 3 contained a collection of sets. In order to construct examples of such collections, the set heuristic searched for predicates in the definition of continuous functions in step 4 and for example found the predicate $|a-x| < \delta$. It used the variable $a \in S$ in this predicate as an index of a collection of sets

$$N(a) = \{x : x \in \mathbf{R} \wedge |a - x| < \delta\}. \tag{17}$$

Because the variable δ in (17) was free, it was replaced by a function $h(a)$ on S such that $h(a) \in \mathbf{R}$ and $h(a) > 0$ hold for all $a \in S$. This yielded steps 7–10 in Table 1. Then, the set heuristic called the proof heuristic to show that $N(S)$ is an open covering of S which required twenty-nine proof steps the last one of which is step 39 in Table 1. Now, the set heuristic applied the definition of compact sets in step 3 to steps 10 and 39 which yielded steps 40 and 41 in Table 2. The paraphrase heuristic attempted to represent the proposition $\bigcup_{A \in D} A$ in step 41 more explicitly. This yielded steps 42–54 in Table 2. Thus, the work of the set heuristic was complete and the control was returned to the existential heuristic.

The existential heuristic introduced a $\delta \in \mathbf{R}$ with $\delta > 0$ in step 55 in Table 2 and generated the subgoal to prove the proposition

$$\forall x, y(x \in S \wedge y \in S \wedge |x-y| < \delta \rightarrow |f(x) - f(y)| < \epsilon) \tag{18}$$

in (16) to find properties of the function h in steps 7 and 8 and the number δ in step 55. Therefore, the universal heuristic introduced two real numbers $x \in S$ and $y \in S$ with $|x-y| < \delta$ in step 56 and generated the subgoal to prove the consequent

$$|f(x) - f(y)| < \epsilon \tag{19}$$

in (18). The inequality heuristic attempted to prove (19) by means of the reduction, transitivity, and replacement heuristic. Because these heuristics failed, it called

1. Triangle heuristic

$$prove\ x : x \in \mathbf{R} \tag{20}$$

Result: $f(a) \in \mathbf{R}$ in step 62

2. Element heuristic with $function(f, S, \mathbf{R})$ in step 1

$$prove\ a : a \in S \tag{21}$$

Result: $a \in S$ in step 61

3. Element heuristic with $T \subseteq S$ in step 43

$$prove\ a : a \in T \tag{22}$$

Result: $a \in T$ in steps 58–60

4. Reduction heuristic with Definition 5

$$prove\ x, A_i : x \in \bigcup_{i \in T} A_i \tag{23}$$

Result: $x \in \bigcup_{a \in T} \{x : x \in \mathbf{R} \wedge |a - x| < h(a)\}$ in step 57

5. Element heuristic with $S \subseteq \bigcup_{a \in T} \{x : x \in \mathbf{R} \wedge |a - x| < h(a)\}$ in step 54

$$prove\ x : x \in S \tag{24}$$

Result: $x \in S$ in step 56

Table 3: How SHUNYATA constructed $f(a) \in \mathbf{R}$ for the triangle heuristic

the triangle heuristic which generated the subgoal (20) in Table 3 to prove the proposition $x \in \mathbf{R}$ with the free variable x, i.e., to construct a real number $x \in \mathbf{R}$ (see Section 2). Table 3 describes how this subgoal was achieved in five stages. Each stage gives a heuristic, a subgoal it generated, and the result produced by the subsequent stages. For example, the triangle heuristic in the first stage generated the subgoal (20) and the second stage provided the result $f(a) \in \mathbf{R}$ in step 62. The element heuristic in the second stage reduced subgoal (20) to the subgoal (21) to construct an element $a \in S$ by means of the fact that f is a function on S into $\mathbf{R}$ in step 1. Another application of the element heuristic in the third stage reduced subgoal (21) to the subgoal (22) to construct an element $a \in T$. Now, the reduction heuristic matched the proposition $a \in T$ in (22) with the proposition $i \in I$ in Definition 5 and used the resulting substitutions a/i and T/I to reduce (22) to the subgoal (23) to construct an x and a family A_i of sets such that $x \in \bigcup_{i \in T} A_i$. Finally, the element heuristic reduced subgoal (23) to the subgoal (24) to construct an $x \in S$ which is contained in step 56 in Table 2. From $x \in S$, the element heuristic and the reduction heuristic produced a real number $f(a) \in \mathbf{R}$ and thus satisfied subgoal (20) which had been generated by the triangle heuristic.

In order to prove the inequality (19) in step 85 in Table 2, the triangle heuristic used the real number $f(a) \in \mathbf{R}$ to generate the subgoals to prove the inequalities

$$|f(x) - f(a)| < \frac{\epsilon}{2} \tag{25}$$

and

$$|f(a) - f(y)| < \frac{\epsilon}{2} \tag{26}$$

in steps 71 and 81. Because the reduction and the transitivity heuristic failed to prove (25) in step 71, the inequality heuristic called the replacement heuristic which for example generated the subgoal

$$\mathit{prove}\ t : |f(x) - f(a)| = t \tag{27}$$

to construct a term t that is equal to the subterm $|f(x)-f(a)|$ of (25). The equation heuristic matched the term $|x - y|$ in Lemma 4 with the term $|f(x) - f(a)|$ in (27) and used the resulting substitutions $f(x)/x$ and $f(a)/y$ to reduce subgoal (27) to the subgoal to prove $f(a) \in \mathbf{R}$ and $f(x) \in \mathbf{R}$ according to the antecedent in Lemma 4. The proposition $f(a) \in \mathbf{R}$ was contained in step 62 and proposition $f(x) \in \mathbf{R}$ in step 63 was proved by the element heuristic. Thus, the equation heuristic generated the equation

$$|f(x) - f(a)| = |f(a) - f(x)| \tag{28}$$

in step 64 which satisfied subgoal (27). Now, the control was returned to the replacement heuristic which applied (28) to transform the subgoal to prove (25) into the subgoal to prove the inequality

$$|f(a) - f(x)| < \frac{\epsilon}{2} \tag{29}$$

in step 70. The reduction heuristic matched the inequality (29) with the inequality $|f(a)-f(x)| < \epsilon$ in step 6 (see step 5) and used the resulting substitutions to reduce the subgoal to prove (29) to the subgoals to prove the propositions $\epsilon/2 \in R$, $0 < \epsilon/2$, and

$$|a - x| < g(a, \epsilon/2) \tag{30}$$

in steps 66, 67, and 69. The reduction heuristic proved the propositions $\epsilon/2 \in R$ and $0 < \epsilon/2$ in steps 66 and 67 by means of Lemmas 2 and 3. Step 65 uses Lemma 1. Because the reduction heuristic failed to prove (30), the inequality heuristic called the transitivity heuristic which used the inequality $|a - x| < h(a)$ in step 57 to transform the subgoal to prove (30) into the subgoal to prove

$$h(a) \leq g(a, \epsilon/2) \tag{31}$$

in step 68. Because (31) contained the function h, the existential heuristic regarded (31) as a property of h and therefore generated step 68. Now, the transitivity, reduction, and replacement heuristic were applied to step 68 which produced the inequality (25) in step 71.

The second inequality (26) was proved by the transitivity, element, triangle and the reduction heuristic in steps 72–81. This yielded another property of h in step 72 and a property of δ in step 74. Then, the triangle heuristic used the inequalities (25) and (29) in steps 71 and 81 to prove the inequality (19) in step 85. Finally, the universal heuristic produced the universal proposition (18) from steps 56 and 85 and returned the control to the existential heuristic.

The proof steps so far generated are valid on the assumption that the function h and the real number δ satisfy the properties in steps 68, 72, and 74. Now, the existential heuristic attempted to derive one of these properties from the remaining ones. Because the property in step 68 could be derived from the property in step 72, it was abandoned. Then, the existential heuristic generated the definition

$$\forall a(a \in S \rightarrow h(a) = \frac{g(a, \epsilon/2)}{2}) \tag{32}$$

of h from the property of h in step 72 and the proposition $a \in S$ in step 61. Furthermore, it generated the definition

$$\delta = \min_{a \in T} \frac{g(a, \epsilon/2)}{2} \tag{33}$$

of δ from the property of δ in step 74 and the propositions $a \in T$ and $\mathit{finite}(T)$ in steps 43 and 60 (see Section 2). It inserted the definition (32) of h into step 7 in Table 1 and called the proof heuristic to prove the proposition in step 8 that $h(a) \in \mathbf{R}$ and $f(a) > 0$ hold for all $a \in S$. Then, it inserted the definition (33) of δ into step 55 in Table 2 and proved the propositions $\delta \in \mathbf{R}$ and $\delta > 0$ in this step. Finally, it proved the properties of h and δ in steps 68, 72, and 74. Finally, the existential heuristic generated the existential proposition (16) and the universal heuristic completed the proof of Heine's theorem.

5 Discussion

The SHUNYATA program models heuristics mathematicians use in the discovery of proofs, i.e., it simulates reasoning processes of mathematicians. If a mathematician attempts to prove Heine's theorem, he may realize that the antecedent in the theorem contains a compact set S and a continuous function f on S (see Section 3). In order to apply the fact that S is compact, he needs an open covering of S according to the definition of compact sets (Definition 2). The definition of continuous functions (Definition 3) implies the existence of a $\delta > 0$ such that $|f(a) - f(x)| < \epsilon$ for all $x \in S$ with $|a - x| < \delta$. The predicate $|a - x| < \delta$ provides a collection of sets $\{x : |a - x| < \delta\}$ which is an open covering of S. Because the δ in these sets depends on $a \in S$, a mathematician may replace it by a function $h(a)$ to be defined subsequently. These reasoning processes are simulated by the concept and the set heuristic in the SHUNYATA program. In order to prove that the function f in Heine's theorem is uniformly continuous (see Definition 4), a mathematician may introduce a $\delta > 0$ and elements $x, y \in S$ with $|x - y| < \delta$ and attempt to prove $|f(a) - f(x)| < \epsilon$ for a given $\epsilon > 0$ in order to find properties of the function h and the number δ. Finally, he uses these properties to define h and δ which completes the proof of Heine's theorem. These reasoning processes are simulated by the existential heuristic in the SHUNYATA program. The fact that SHUNYATA simulates heuristics that have proven effective for mathematicians over the centuries explains its power (see Bledsoe, 1986). Another important advantage of such a program is that mathematicians can easily understand its reasoning processes and its proofs.

A mathematician is able to select the definitions and lemmas needed in a proof from a large set of definitions and lemmas he knows (see Bledsoe, 1986). A knowledge base in the user input of SHUNYATA contains knowledge in elementary mathematics such as the fact that 2 is a natural number (Lemma 1) and definitions and lemmas of a textbook. For example, the knowledge base also includes definitions and lemmas needed in the proof of the Bolzano-Weierstrass theorem such as the definitions of finite and infinite sets (see Apostol, 1957, p. 31). SHUNYATA contains a heuristic that selects the definitions and lemmas needed in a proof from this knowledge base. The selection is controlled by a list of mathematical concepts that occur in subgoals or proof steps previously generated. The representation of the definitions and lemmas in the knowledge base models their ordinary representation in textbooks.

The heuristics in SHUNYATA are general because they are rather elementary. For example, the set heuristic merely searches for sets or collections of sets in the proof which are determined by predicates. In order to prove the existence of an x such that a property $P(x)$ holds, the existential heuristic derives properties of x and uses them to define an x with $P(x)$. The triangle heuristic reduces the goal to prove an inequality $|x - z| < r$ to the subgoals to find a number y and to prove the inequalities $|x - y| < r/2$ and $|y - z| < r/2$. The generality of the heuristics is also illustrated by the number of their successful applications in the proof of Heine's theorem: The reduction heuristic was successfully applied 28 times, the replacement heuristic 16 times, the transitivity heuristic 3 times, and the triangle heuristic 2 times.

Some heuristics in SHUNYATA cannot be applied in all mathematical theories. For example, the triangle heuristic can only be applied in theories in which the triangle inequality holds. Therefore, an extensive automation of theorem proving requires an automatic construction of heuristics such as the triangle heuristic. SHUNYATA contains a learning procedure which constructs concepts and heuristics by composing elementary functions such as the subset relation. For example, the procedure can easily construct the concept of a minimum or a maximum of a finite set of natural numbers (Ammon, 1991). In a more complex experiment, it developed a theorem prover from a proof of the simple theorem that $(x^{-1})^{-1} = x$ holds for all elements x of a group. The theorem prover produced proofs of nine further theorems in group theory such as $x^2 = 1$ implies group commutativity and a proof of SAM's Lemma without any human intervention.[4] The theorem that $x^2 = 1$ implies group commutativity is "the limit of the capability" of the heuristic theorem prover ADEPT developed at MIT in the mid-sixties (Loveland, 1984, p. 13). The Markgraf Karl Refutation Procedure is one of the largest software projects in the history of automatic theorem proving (see Bläsius *et al.*, 1981, p. 516; and Loveland, 1984, p. 22). After some fifteen years of development, Ohlbach and Siekmann (1989, p. 58) give SAM's Lemma as the only "more difficult" theorem that their theorem prover has proved. ADEPT and the Markgraf Karl Refutation Procedure were developed manually. In contrast, SHUNYATA automatically developed the "ideas" for the theorem prover on the basis of elementary functions, implemented them in a program, and applied the resulting theorem prover to the new theorems without any human intervention.

[4] Ammon (1988b) describes an earlier comparable experiment which also applied the learning procedure to a simple proof. The proofs of SAM's Lemma generated in the two experiments are identical.

6 Related Work

Bledsoe's (1979) set variable method constructs sets on the basis of sophisticated rules. The intermediate value theorem says that if f is a continuous function on an interval $[a,b]$ with $f(a) \leq 0$ and $f(b) \geq 0$, then $f(x) = 0$ holds for some $x \in [a,b]$. For example, the application of Bledsoe's method to this theorem produces the set $\{z : z \leq b \wedge f(z) \leq 0\}$. The proof that $f(x) = 0$ holds for the supremum x of this set is straightforward. The set heuristic in SHUNYATA is simpler than Bledsoe's set variable method because it merely searches for a predicate which defines a set. Its application to the intermediate value theorem produces the set given above from the predicate $f(a) \leq 0$ in the theorem. If the set heuristic cannot find a suitable predicate, a composition heuristic constructs predicates by composing elementary functions which represent concepts such as the inequality relation. In the proof of the Bolzano-Weierstrass theorem which says that a bounded infinite set S has an accumulation point (see Apostol 1957, p. 43), the composition heuristic produced the predicate that the set of all $y \in S$ with $y < x$ is finite. The supremum of the set of all numbers x satisfying this predicate is an accumulation point of S. The predicate is the central "idea" of the proof of the Bolzano-Weierstrass theorem (see Section 2). In the proof of Gödel's incompleteness theorem, the composition heuristic produced an undecidable formula, i.e., neither this formula nor its negation can be proved in the theory (see Section 2). The formula is the central "idea" of the proof. These examples show that the composition heuristic can generate the central "ideas" of significant proofs in mathematical logic and analysis. A variant of the composition heuristic can be used as a learning procedure which constructs mathematical concepts and theorem provers (see Section 5).

Bledsoe and Hines (1980) present a resolution-based theorem prover for proving inequalities. A variable elimination method in this theorem prover implicitly uses interpolation axioms such as $\exists x(x \leq a)$. A chaining method implicitly uses the transitivity axiom that the inequalities $x \leq y$ and $y \leq z$ imply $x \leq z$. SHUNYATA does not contain a heuristic which uses interpolation axioms such as $\exists x(x \leq a)$. The existential heuristic in SHUNYATA reduces the goal to prove $\exists x(x \leq a)$ to the subgoal to prove $a \leq a$ by means of an element relation $a \in S$. Bledsoe and Hines' chaining method attempts to unify b and b' if $a < b$ and $b' < c$ hold and applies the resulting unifier to $a < c$. The transitivity heuristic in SHUNYATA reduces the goal to prove $a < c$ to the subgoal to prove $b < c$ if $a < b$ holds. Bledsoe and Hines' theorem prover does not contain methods comparable with the triangle and inequality heuristic in SHUNYATA. The triangle heuristic generates the subgoal to construct a real number which is used as a substituent for the variable y in the triangle inequality (see Table 3). The inequality heuristic controls the reduction, transitivity, replacement, and triangle heuristic by time limits which interrupt a heuristic if it achieves no result.

An important principle of SHUNYATA's heuristics is that their core contains elementary and efficient methods which are applied first. This entails that simple subgoals can be achieved easily and quickly. For example, the existential heuristic uses element relations $a \in S$ in a proof to reduce the goal to prove $\exists x(x \in S \wedge P(x))$ to the subgoal to prove $P(a)$. In the proof of Heine's theorem, this method reduced the goal to prove $\exists x(x \in S \wedge a \in N(x))$ to the subgoal to prove the trivial proposition $a \in N(a)$ by means of the element relation $a \in S$ in a preceding proof step. If this simple and efficient method fails, the existential heuristic applies a more sophisticated method which derives properties of an $x \in S$ with $P(x)$ and constructs such an $x \in S$ from these properties. The learning procedure discussed in Section 5 can automatically modify and extend SHUNYATA's heuristics.

SHUNYATA can easily and quickly prove many theorems which are "more difficult" problems for conventional theorem provers. For example, Ballantyne and Bledsoe's (1977) theorem prover generated a proof of Heine's theorem in nonstandard analysis which simplifies this and many other proofs considerably. SHUNYATA took some three minutes to generate a proof of Heine's theorem in nonstandard analysis. The structure of this proof is simple: Its core was produced by a forward heuristic which merely applies definitions and lemmas to preceding proof steps. McCune's (1990, 1991) theorem prover OTTER 2.2 takes some eight minutes to generate a proof of SAM's Lemma on an IBM PC AT from the axioms for a modular lattice in McCharen *et al.* (1976). This set of axioms is not complete because it does not contain the ordinary lattice operations "join" and "meet". A learning procedure in SHUNYATA automatically developed theorem provers from simple proofs (see Section 5). An optimizing compiler transforms these theorem provers, which are represented by set constructors, into efficient procedural code. A variant of such a theorem prover generated a proof of SAM's Lemma from a complete set of axioms in ordinary representation in less than four minutes. The structure of the proof is simple because it consists of eight equations seven of which simplify the right-hand side of the first equation (see Ammon, 1988b).

7 Conclusion

The SHUNYATA program contains heuristics which simulate reasoning processes of mathematicians. For example, the replacement heuristic changes the representation of a proposition to be proved and attempts to prove the changed proposition. In order to prove the existence of a particular element, the existential heuristic derives properties of such an element and applies them to its definition. Some heuristics control the application of other heuristics, for example, by time limits which interrupt a heuristic if it achieves no result. The core of the heuristics contains simple and efficient methods. For example, the core of the set heuristic constructs sets from predicates in the theorem to be proved. Such sets form the central "ideas" of proofs of Heine's theorem and the intermediate value theorem. If the core of the set heuristic does not provide a suitable set, a composition heuristic contructs predicates by composing elementary functions which represent concepts. It produced the central "ideas" of proofs of Gödel's incompleteness theorem and the Bolzano-Weierstrass theorem. SHUNYATA selects the definitions and lemmas needed in a proof from a knowledge base of definitions and lemmas and produces readable proofs. To my knowledge, it is the first program that automatically constructed proofs of very significant theorems in mathematical logic and standard analysis, for example, Gödel's incompleteness theorem, the Bolzano-Weierstrass theorem, and Heine's theorem. The proofs were generated without any human intervention. A learning procedure in SHUNYATA can automatically change and extend heuristics and develop new heuristics on the basis of elementary functions.

Acknowledgements

The author would like to thank the reviewers for helpful suggestions. Thanks also to Andreas Keller for providing useful comments on an earlier draft. This work was supported in part by the German Science Foundation (DFG).

References

Ammon, K. 1988a. Discovering a proof for the fixed point theorem: A case study. *Proceedings of the Eighth European Conference on Artificial Intelligence*, August 1–5, Munich, Germany, pp. 613–618.

Ammon, K. 1988b. The automatic acquisition of proof methods. *Proceedings of the Seventh National Conference on Artificial Intelligence*, August 21–26, St. Paul, Minnesota, pp. 558–563.

Ammon, K. 1991. Constructing programs from input-output pairs. *Proceedings of the Fifteenth German Workshop on Artificial Intelligence*, September 15–20, Bonn. Berlin: Springer.

Apostol, T. M. 1957. *Mathematical Analysis.* Reading, Mass.: Addison-Wesley.

Ballantyne, A. M., and Bledsoe, W. W. 1977. Automatic proofs of theorems in analysis using nonstandard techniques. *Journal of the Association of Computing Machinery* 24, pp. 353–374.

Bläsius, K., Eisinger, N., Siekmann, J., Smolka, G., Herold, A., and Walther, C. 1981. The Markgraf Karl Refutation Procedure. *Proceedings of the Seventh International Joint Conference on Artificial Intelligence*, August, Vancouver.

Bledsoe, W. W. 1977. Non-resolution theorem proving. Artificial Intelligence 9, pp. 1–35.

Bledsoe, W. W. 1979. A maximal method for set variables in automatic theorem proving. In J. E. Hayes, D. Michie, and L. I. Mikulich, eds., *Machine Intelligence*, Vol. 9. Chichester, England: Ellis Horwood.

Bledsoe, W. W. 1986. Some thoughts on proof discovery. In *Symposium on Logic Programming*, September 22–25, Salt Lake City. IEEE, New York.

Bledsoe, W. W., and Hines, L. M. 1980. Variable elimination and chaining in a resolution-based prover for inequalities. In W. Bibel and R. Kowalski, eds., *5th Conference on Automated Deduction.* Berlin: Springer.

Kaufmann, M. 1989. DEFN-SK: An extension of the Boyer-Moore Theorem Prover to handle first-order quantifiers. Technical Report 43, Computational Logic, Inc., Austin, Texas.

Kleene, S. C. 1952. *Introduction to Metamathematics.* Amsterdam: North-Holland.

Loveland, D. W. 1984. Automated theorem proving: a quarter century review. In W. W. Bledsoe and D. W. Loveland, eds., *Automated Theorem Proving: After 25 Years.* Providence, R.I.: American Mathematical Society.

McCharen, J. D., Overbeek, R. A., and Wos, L. A. 1976. Problems and experiments for and with automated theorem-proving programs. *IEEE Transactions on Computers*, Vol. C-25, No. 8, pp. 773–782.

McCune, W. W. 1990. OTTER 2.0 Users Guide. Report ANL-90/9, Argonne National Laboratory, Argonne, Illinois.

McCune, W. W. 1991. What's new in OTTER 2.2. Report ANL/MCS-TM-153, Argonne National Laboratory, Argonne, Illinois.

Ohlbach, H. J., and Siekmann, J. 1989. The Markgraf Karl Refutation Procedure. University of Kaiserslautern, Department of Computer Science, SEKI Report SR-89-19.

Shankar, N. 1987. Proof checking metamathematics. Technical Report 9, Computational Logic, Inc., Austin, Texas.

Proving Geometry Statements of Constructive Type*

Shang-Ching Chou and Xiao-Shan Gao†

Department of Computer Science
The Wichita State University, Wichita, KS 67208, USA
e-mail: chou@wiz.wsu.ukans.edu

Abstract. This paper presents a method to generate non-degenerate conditions in geometric form for a class of geometry statements of constructive type, called Class C. We prove a mathematical theorem that in the irreducible case, the non-degenerate conditions generated by our method are *sufficient* for a geometry statement in Class C to be valid in *metric geometry*. About 400 among 600 theorems proved by our computer program are in Class C.

Keywords. Geometry theorem proving, Wu's method, non-degenerate condition, generally true, constructive geometry statement, Euclidean geometry, metric geometry, algebraically closed field.

1. Introduction

In the past decade highly successful algebraic methods for mechanical geometry theorem proving have been developed. This began with Wu's pioneering work in 1978 [10]. In the subsequent developments, hundreds of hard theorems in Euclidean and Non–Euclidean geometries have been proved by computer programs based on Wu's method [11], [12], [1], [2]. Inspired by the success of Wu's work, several groups also successfully applied the Gröbner basis (GB) method to prove the same class of geometry statements that Wu's method addresses [7], [8], [9]. In most of the current research, two different but related formulations for geometry statements have been considered.

Formulation (Approach) F1. Introducing parameters and the notion of "generally true" for a geometry statement. The present techniques can prove a statement to be generally true, at the same time giving nondegenerate conditions (usually in algebraic form) automatically.

Formulation (Approach) F2. Giving nondegenerate conditions in geometric form manually (or mechanically) at the beginning as a part of the hypotheses. Then the prover only needs to answer whether the conclusion follows the hypotheses *without adding* any other conditions.

The method originally developed by Wu is for Formulation F1. One of the advantages of Formulation F1 is that non-degenerate conditions can be taken care of without

* The work reported here was supported in part by the NSF Grants CCR–902362 and CCR–917870.

† The current and permanent address: Institute of Systems Science, Academia Sinica, Beijing.

explicitly specifying them. Formulation F1 gives a clear insight into the nature of a geometric statement: if a geometry statement is proved to be generally false, then it cannot be valid no matter how many additional non-degenerate conditions are added so long as the hypotheses keep consistent.

On the other hand, how to translate algebraic nondegenerate conditions into their geometric forms is also an interesting and important topic. This is the aim of this paper. In this paper we presents a method for generating non-degenerate conditions in *geometric form* for a class of geometry statements, called Class C, when using Formulation 1. We prove a mathematical theorem stating that non-degenerate conditions produced by our method are complete in *metric geometry*, i.e., if the statement is not valid under these non-degenerate conditions, then it is generally false.

The main results are based on the unpublished thesis [2]. Though they have been long known to the authors, only the content of Sections 3.1 and 3.2 was published in the book [3] in an implicitly way.

2. Difficulties with Non-Degenerate Conditions

Theorems in geometry textbooks often implicitly assume some non-degenerate conditions that are necessary for the theorems to be valid. Finding all non-degenerate conditions sufficient for a geometry statement to be true may be difficult in many cases. The situation becomes more complicated when we use Wu's method (or the GB method) for proving theorems, because these methods are complete only for metric geometry, not for Euclidean geometry. The following are two examples.

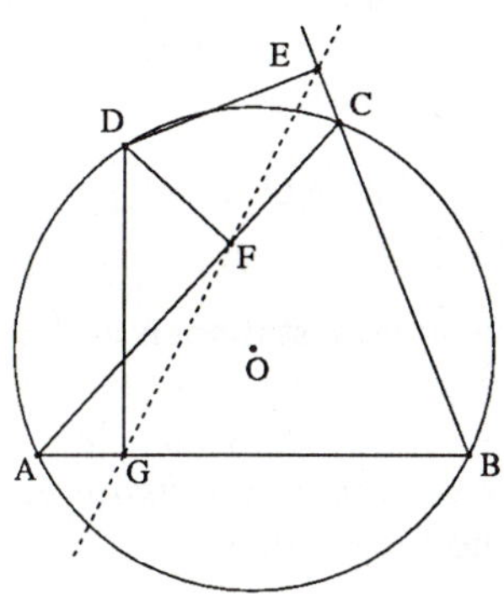

Figure 1

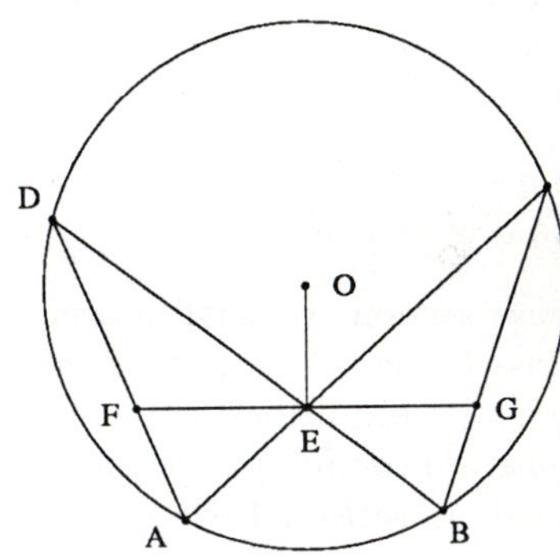

Figure 2

Example (2.1). (Simson's Theorem). Let D be a point on the circumscribed circle (O) of triangle ABC. From D three perpendiculars are drawn to the three sides BC, CA and AB of $\triangle ABC$. Let E, F and G be the three feet respectively. Show that E, F and G are collinear (Figure 1).

The obvious non-degenerate condition for this statement seems to be "A, B and C are not collinear". Indeed, in Euclidean geometry, Simson's theorem is valid under this condition. However, if we try to prove Simson's theorem under this condition with

Wu's method or the GB method according to Formulation F2 (i.e., without adding any other conditions), then the statement cannot be confirmed. The reason for this phenomenon is that Wu's method (or the GB) method is complete only for metric geometry, not for Euclidean geometry. Here we will not give a detailed account of metric geometry (in Wu's sense), the reader can find the discussion in [3], [12]. We only mention that the theory of metric geometry has many models, among which are Euclidean geometry $\mathbf{R}^2$, unordered metric geometries (e.g., complex geometry $\mathbf{C}^2$), etc.

If we want to decide whether Simson's theorem is a theorem of the theory of metric geometry, then the following additional non-degenerate condition is necessary:

(2.1.1) $AB, BC,$ and CA are not isotropic

An isotropic line is a line perpendicular to itself; it does not exist in Euclidean geometry. However, there exist isotropic lines in other models of the theory of metric geometry (e.g., $\mathbf{C}^2$). The result in using Wu's method shows that Simson's theorem can be proved in unordered geometries (i.e., without using the axioms of order) under (2.1.1); however, if we drop (2.1.1) from the hypotheses, then the axioms of order is necessary. Unless we have a clear understanding of the nature of the method and the statements to be proved, it is very hard to come up with non-degenerate conditions such as (2.1.1). This paper gives a satisfactory solution to these problems.

Even in Euclidean geometry, there are some nondegenerate conditions hard to find as illustrated by the following example.

Example (2.2). (the Butterfly Theorem) A, B, C and D are four points on circle (O). E is the intersection of AC and BD. Through E draw a line perpendicular to OE, meeting AD at F and BC at G. Show $FE \equiv GE$ (Figure 2).

Here we need a necessary non-degenerate condition that EO is not perpendicular to AD. The necessity of this condition is hard to perceive.

In the next section, we shall classify a class of geometry statements of constructive type (Class C) and present a mechanical method for producing sufficient number of non-degenerate conditions in geometric form for a statement in this class. Informally, non-degenerate conditions are those for which the corresponding figure must be well-constructed. In section 4, we shall prove our method is complete.

3. A Class of Geometric Statements of Constructive Type

3.1. Definition of Class C

Most theorems in elementary geometry can be described in a constructive way: given a certain arbitrary points, lines, circles and points on these circles and lines, new points, lines and circles are constructed step by step using geometric constructions such as taking the intersection of two lines, an intersection of a line and a circle, or an intersection of two circles. In this subsection, we use the natural language to give a definition of such a statement. In Section 3.3, we will give the precise formula of such a statement using geometric predicates.

First, let us give "circle" a formal definition. A *circle* h is a pair of a point O and a segment (AB): $h = (O, (AB))$, where A and B are two points. Two circles $(O, (AB))$ and $(P, (CD))$ are equal if and only if $O = P$ and congruent(A, B, C, D) (for the definition of "congruent", see 3.2). O is called the center of the circle and (AB) the radius. A point P is on circle $(O, (AB))$ if congruent(O, P, A, B).

Let Π be a finite set of points. We say line l is constructed *directly* from Π if

(i) The line l joins two points A and B in Π. We denote it by $l = L(AB)$; or

(ii) The line l passes through one point C in Π and parallel to a line joining two points A and B in Π. We denote it by $l = P(C, AB)$; or

(iii) The line l passes through one point C in Π and perpendicular to a line joining two points A and B in Π. We denote it by $l = T(C, AB)$; or

(iv) The line l is the perpendicular–bisector of AB with A and B in Π. We denote it by $l = B(AB)$.

A line l constructed directly from Π is *well defined* if the two points A and B mentioned above are distinct.

Likewise, we say a circle $c = (O, (AB))$ is constructed directly from Π if points O, A and B are in Π. The lines and circles constructed directly from Π are said to be *in* Π, for brevity.

Definition. A geometry statement is of constructive type or in Class C if the points, lines, and circles in the statement can be constructed in a definite prescribed manner using the following ten constructions, assuming Π to be the set of points already constructed so far:

Construction 1. Taking an arbitrary point.

Construction 2. Drawing an arbitrary line. This can be reduced to taking two arbitrary points.

Construction 3. Drawing an arbitrary circle. This can be also reduced to taking two or three arbitrary points.

Construction 4. Drawing an arbitrary line passing through a point in Π. This can be reduced to taking an arbitrary point.

Construction 5. Drawing an arbitrary circle knowing its center in Π. This can be also reduced to taking one or two arbitrary points.

Construction 6. Taking an arbitrary point on a line in Π.

Construction 7. Taking an arbitrary point on a circle in Π.

Construction 8. Taking the intersection of two lines in Π.

Construction 9. Taking an intersection of a line and a circle in Π.

Construction 10. Taking an intersection of two circles in Π.

The conclusion is a certain (equality) relation among the points thus constructed.

In the actual prover [2], [6], we have included more constructions such as taking midpoints and constructions involving angle congruence, the radical axis of two circles, taking a point on a circle knowing three points on the circle, etc.

Example (3.1). Simson's theorem can be specified as a statement in Class C by the following *construction sequence*:

Points A, B, and C are arbitrarily chosen,	construction 1
$O = B(AB) \cap B(AC)$,	construction 8
D is on circle $(O, (OA))$,	construction 7
$E = T(D, BC) \cap L(BC)$,	construction 8
$F = T(D, AC) \cap L(AC)$,	construction 8
$G = T(D, AB) \cap L(AB)$,	construction 8

with conclusion collinear(E, F, G).

The Butterfly theorem can be specified as a statement in Class C by the following construction sequence:

O and A are arbitrarily chosen,	construction 1
B is on $(O, (OA))$,	construction 7
C is on $(O, (OA))$,	construction 7
D is on $(O, (OA))$,	construction 7
$E = L(AC) \cap L(BD)$,	construction 8
$F = L(AD) \cap T(E, OE)$,	construction 8
$G = L(EF) \cap L(BC)$,	construction 8

with conclusion midpoint(F, E, G).

In the above examples, we use a *construction sequence* to express a statement in Class C. We will soon present an algorithm for generating non-degenerate conditions for a statement in Class C, knowing its construction sequence. Before presenting the algorithm, we first specify what exact geometric predicates we use.

3.2. The Basic Predicates

In order to describe the logical formula of a statement in Class C, we need four basic (non-logical) predicates: collinear(A, B, C), parallel(A, B, C, D), perpendicular(A, B, C, D), and congruent(A, B, C, D). In our actual prover [2], [3], there are many other predicates. For the complete list of all these predicates and their algebraic equations see pp.97–99 of [3]. We should emphasize that these predicates do include degenerate cases. To be more precise, let $A = (x_1, y_1)$ $B = (x_2, y_2)$, $C = (x_3, y_3)$ and $D = (x_4, y_4)$.

(1) Predicate "collinear(A, B, C)" means that points A, B and C are on the same line; they are not necessarily distinct. Its corresponding algebraic equation is

$$(x_1 - x_2)(y_2 - y_3) - (x_2 - x_3)(y_1 - y_2) = 0.$$

(2) Predicate "parallel(A, B, C, D)" means that

$[(A = B) \vee (C = D) \vee (A, B, C, D$ are on the same line$) \vee (AB \parallel CD)]$. Its algebraic equation is

$$(x_1 - x_2)(y_3 - y_4) - (x_3 - x_4)(y_1 - y_2) = 0.$$

(3) Predicate "perpendicular(A, B, C, D)" means that $[(A = B) \vee (C = D) \vee (AB \perp CD)]$. Its algebraic equation is

$$(x_1 - x_2)(x_3 - x_4) + (y_1 - y_2)(y_3 - y_4) = 0.$$

We define a new predicate isotropic(A, B) to be perpendicular(A, B, A, B), which means $A = B$ or $L(AB)$ is an isotropic line.

(4) Predicate "congruent(A, B, C, D)" means [(isotropic(A, B) $\wedge$ isotropic(C, D)) $\vee$ (AB is congruent to CD)]. Its algebraic equation is

$$(x_1 - x_2)^2 + (y_1 - y_2)^2 - (x_3 - x_4)^2 - (y_3 - y_4)^2 = 0.$$

There are many advantages of using the above predicates. Each of the above predicates corresponds to only one equation, thus its negation corresponds to only one inequation. E.g., $\neg$parallel(A, B, C, D) is "$(A \neq B) \wedge (C \neq D) \wedge (A, B, C, D$ are notcollinear$) \wedge \neg(AB \parallel CD)$". Its corresponding inequation is

$$(x_1 - x_2)(y_3 - y_4) - (x_3 - x_4)(y_1 - y_2) \neq 0,$$

which is *the exact non-degenerate condition we want for intersecting two lines* AB and CD: they have only one common point. Note that this condition implies the condition $(A \neq B \wedge C \neq D)$.

3.3. Mechanical Generation of Non-Degenerate Conditions for Class C

For a statement in Class C, we can generate non-degenerate conditions following the construction sequence step by step. Suppose we have already generated a set of non-degenerate conditions DS under the previous constructions. Let HS be the set of the equation hypotheses under the previous constructions, and Π be the set of points constructed so far. The next construction is one of the ten constructions in Section 3.1. First we add the point(s) to be constructed to the set Π. Since the first five constructions are reduced to taking arbitrary points, nothing is added to HS or DS. Thus we assume the next construction is one of constructions 6–10. We use abbreviations coll(), perp(), para() and cong() for predicates collinear(), perpendicular(), parallel() and congruent(), respectively.

Construction 6. Taking an arbitrary point D on a line l in Π. There are four kinds of lines in Π.
Case 6.1. $l = L(AB)$.
$HS := \{\text{coll}(A, B, D)\} \cup HS$; $DS := \{A \neq B\} \cup DS$.

Case 6.2. $l = P(C, AB)$.
$HS := \{\text{para}(A, B, C, D)\} \cup HS$; $DS := \{A \neq B\} \cup DS$.
Case 6.3. $l = T(C, AB)$.
$HS := \{\text{perp}(A, B, C, D)\} \cup HS$; $DS := \{A \neq B\} \cup DS$.
Case 6.4. $l = B(AB)$.
$HS := \{\text{cong}(A, D, B, D)\} \cup HS$; $DS := \{A \neq B\} \cup DS$.
Construction 7. Taking an arbitrary point A on a circle $(B, (CD))$ in Π.
$HS := \{\text{cong}(A, B, C, D)\} \cup HS$.
Construction 8. Taking the intersection I of two lines in Π.
Since there are four types of lines in Π, there are 10 types of intersections: types LL, LP, LT, LB, PP, PT, PB, TT, TB, and BB.
Let the two lines be given by the following equations:

$$l_1 : a_1x + b_1y + c_1 = 0,$$
$$l_2 : a_2x + b_2y + c_2 = 0.$$

The elegance of our approach is that for all 10 types of intersections, the only non-degenerate condition in algebraic form is $\Delta = a_1b_2 - a_2b_1 \neq 0$ and for different types it has different geometric meanings (non-isotropic, non-perpendicular, etc.)
Case 8.1. Type LL: $I = L(AB) \cap L(CD)$.
$HS := \{\text{coll}(A, B, I), \text{coll}(C, D, I)\} \cup HS$; $DS := \{\neg\text{para}(A, B, C, D)\} \cup DS$.
Note that $\neg\text{para}(A, B, C, D)$ implies $A \neq B$ and $C \neq D$.
Case 8.2. Type LP: $= L(AB) \cap P(E, CD)$.
$HS := \{\text{coll}(A, B, I), \text{para}(C, D, E, I)\} \cup HS$; $DS := \{\neg\text{para}(A, B, C, D)\} \cup DS$. In the special case,

Case 8.2.1. If $B = D$, DS becomes $DS := \{\neg\text{coll}(A, B, C)\} \cup DS$.

Case 8.3. Type LT: $I = L(AB) \cap T(E, CD)$.
$HS := \{\text{coll}(A, B, I), \text{perp}(C, D, E, I)\} \cup HS$; $DS := \{\neg\text{perp}(A, B, C, D)\} \cup DS$. In the special case,
Case 8.3.1. (The foot from E to AB) $A = C$ and $B = D$. $DS := \{\neg\text{isotropic}(A, B)\} \cup DS$.
Case 8.4. Type LB: $I = L(AB) \cap B(CD)$.
$HS := \{\text{coll}(A, B, I), \text{cong}(I, C, I, D)\} \cup HS$; $DS := \{\neg\text{perp}(A, B, C, D)\} \cup DS$. In the special case,

Case 8.4.1. $A = C$ and $B = D$. $DS := \{\neg\text{isotropic}(A, B)\} \cup DS$.[1]

Case 8.5. Type PP: $I = P(E, AB) \cap P(F, CD)$.
$HS := \{\text{para}(A, B, E, I), \text{para}(C, D, F, I)\} \cup HS$; $DS := \{\neg\text{para}(A, B, C, D)\} \cup DS$. In the special case,

Case 8.5.1. $B = D$. $DS := \{\neg\text{coll}(A, B, C)\} \cup DS$.

Case 8.6. Type PT: $I = P(E, AB) \cap T(F, CD)$.
$HS := \{\text{para}(A, B, E, I), \text{perp}(C, D, F, I)\} \cup HS$; $DS := \{\neg\text{perp}(A, B, C, D)\} \cup DS$.

[1] This is one of the two ways to specify the midpoint.

Case 8.7. Type PB: $I = P(E, AB) \cap B(CD)$.
$HS := \{\text{para}(A, B, E, I), \text{cong}(I, C, I, D)\} \cup HS$; $DS := \{\neg\text{perp}(A, B, C, D)\} \cup DS$.
Case 8.8. Type TT: $I = T(E, AB) \cap T(F, CD)$.
$HS := \{\text{perp}(A, B, E, I), \text{perp}(C, D, F, I)\} \cup HS$; $DS := \{\neg\text{para}(A, B, C, D)\} \cup DS$.
In the special case,

Case 8.8.1. $B = D$. $DS := \{\neg\text{coll}(A, B, C)\} \cup DS$.

Case 8.9. Type TB: $I = T(E, AB) \cap B(CD)$.
$HS := \{\text{perp}(A, B, E, I), \text{cong}(I, C, I, D)\} \cup HS$; $DS := \{\neg\text{para}(A, B, C, D)\} \cup DS$.
In the special case,

Case 8.9.1. $B = D$. $DS := \{\neg\text{coll}(A, B, C)\} \cup DS$.

Case 8.10. Type BB: $I = B(AB) \cap B(CD)$.
$HS := \{\text{perp}(I, A, I, B), \text{cong}(I, C, I, D)\} \cup HS$; $DS := \{\neg\text{para}(A, B, C, D)\} \cup DS$. In the special case,

Case 8.10.1. $B = D$. $DS := \{\neg\text{coll}(A, B, C)\} \cup DS$.

Construction 9. Taking an intersection Q of a line and a circle in Π. Let the line be $L(AB)$, or $P(C, AB)$, or $T(C, AB)$, or $B(AB)$, the circle be $(O, (DE))$. $DS := \{\neg\text{isotropic}(A, B)\} \cup DS$.
If $Q = L(AB) \cap (O, (DE))$, then $HS := \{\text{coll}(A, B, Q), \text{cong}(O, Q, D, E)\} \cup HS$.
If $Q = P(C, AB) \cap (O, (DE))$, then $HS := \{\text{para}(A, B, C, Q), \text{cong}(O, Q, D, E)\} \cup HS$.
If $Q = T(C, AB) \cap (O, (DE))$, then $HS := \{\text{perp}(A, B, C, Q), \text{cong}(O, Q, D, E)\} \cup HS$.
If $Q = B(AB) \cap (O, (DE))$, then $HS := \{\text{cong}(Q, A, Q, B), \text{cong}(O, Q, D, E)\} \cup HS$.
Case 9.1. In the special case when one of the intersections, say S, of the circle and the line is already in Π. $DS := \{\neg\text{isotropic}(A, B), S \neq Q\} \cup DS$.
Construction 10. Taking an intersection Q of two circles in Π. Let the two circles be $(O, (AB))$ and $(P, (CD))$.
$HS := \{\text{cong}(O, Q, A, B), \text{cong}(P, Q, C, D)\} \cup HS$; $DS = \{\neg\text{isotropic}(O, P)\} \cup DS$. In the special case,
Case 10.1. One of the intersections is already in Π, say, S. $DS := \{\neg\text{isotropic}(O, P), S \neq Q\} \cup DS$.

Repeating the above steps until every construction is processed, finally we have two parts for the hypotheses: one is $HS = \{H_1, ..., H_r\}$, called the ***equation part*** of the hypotheses; the other is $DS = \{\neg D_1, ..., \neg D_s\}$, called the ***inequation part*** of the hypotheses representing non-degenerate conditions of the statement. Let C be the conclusion of the statement, whose algebraic form is a polynomial equation in the coordinates of the points in Π. Then the exact statement is[1]

$$\forall P \in \Pi(HS \wedge DS \Rightarrow C). \tag{3.2}$$

According to our translation, we can denote a statement S in Class C by (HS, DS, C).

---

1 Depending on the context, HS can also denote the conjunction of its elements, i.e., $HS = H_1 \wedge \cdots \wedge H_r$. The same convention is for DS and other sets of geometric conditions.

3.4. An example

Example (3.4). (the Butterfly theorem in Example 3.1) According to the construction sequence of the Butterfly theorem in Example (3.1), the non-degenerate conditions (the inequation part of the hypotheses) are

$$
\begin{array}{ll}
DS_b = & \\
\neg\text{parallel}(A, C, B, D), & \text{Case 8.1} \\
\neg\text{perpendicular}(A, D, O, E), & \text{Case 8.3} \\
\neg\text{parallel}(E, F, B, C). & \text{Case 8.1}
\end{array}
$$

Here we automatically generate the degenerate condition perpendicular(A, D, O, E).

The equation part of the hypotheses is

$$
\begin{array}{ll}
\text{congruent}(O, A, O, B), & \\
\text{congruent}(O, A, O, C), & \\
\text{congruent}(O, A, O, D), & \\
\text{collinear}(A, E, C), & \\
\text{collinear}(B, E, D), & HS_b \\
\text{perpendicular}(O, E, E, F), & \\
\text{collinear}(E, F, G), & \\
\text{collinear}(F, A, D), & \\
\text{collinear}(G, B, C). &
\end{array}
$$

The exact statement of the Butterfly theorem according to the constructions in (3.1) is:

$$\forall A \cdots \forall G[HS_b \wedge DS_b \Rightarrow \text{midpoint}(F, E, G)]. \tag{3.5}$$

Note that for the same theorem, the construction sequence is usually not unique. Different construction sequences lead to different non-degenerate conditions. However, for all different construction sequences, *the equation part of the hypotheses is always the same.* In this example, it is always HS_b. The reader should try other construction sequences for the Butterfly theorem and Simson's theorem. We have at least 8 essentially different construction sequences for Simson's theorem [5].

The results in Section 4 and 5 show that either (3.5) is a theorem in the theory of metric geometry, or it cannot be a theorem no matter how many additional non-degenerate conditions are added as long as the hypotheses keep consistent.

4. The Completeness of Non-Degenerate Conditions for Metric Geometry

The completeness of our method for generating non-degenerate conditions DS can be stated as the following theorem.

Theorem (4.1). For an irreducible (to be defined later) statement in Class C, our mechanically generated non-degenerate conditions are sufficient for the statement to be valid in the theory of metric geometry. To be more precise, let S = (HS, DS, C)

be a statement in Class C, where $HS = \{H_1, ..., H_r\}$ is the equation part of the hypotheses, $DS = \{\neg D_1, ..., \neg D_s\}$ is the inequation part of the hypotheses, and C is the conclusion. Let Π be the set of all points involved in S. If S is irreducible (to be defined later) and the formula

$$\forall P \in \Pi(HS \wedge DS \Rightarrow C), \tag{4.2}$$

is not valid in a model Ω of the theory of metric geometry whose associated field F_Ω is *algebraically closed*, then (4.2) cannot be a theorem in Ω by adding any set of additional conditions $\neg D_{s+1}, ..., \neg D_t$ *as long as they keep the consistency with the hypotheses*, where each D_i is a geometric condition whose algebraic form is an equation. The consistency means that

$$\forall P \in \Pi(HS \wedge DS \Rightarrow D_i), \tag{4.3}$$

are not valid in Ω, for $i = s + 1, ..., t$. ∎

Now we are proving Theorem (4.1). Our final goal is to prove Theorem (4.8) which is the algebraic form of Theorem (4.1). The proof here was originally in [2].

We use the algebraic approach. Following Hilbert and Wu, we use two kinds of variables: the parameters u_j and the dependent variables x_k. *Our proof also provides a method to choose the parameters u, the dependent variables x, the variable order in x, and a method to decide whether (4.2) is valid in Ω.*

After adopting an appropriate coordinate system, each point P in the statement S corresponds to a pair of coordinates: $P = (x_p, y_p)$. We introduce new parameters u, dependent variables x, and equations according to the steps of constructions. Under the previous constructions, suppose we have already introduced parameters $u_1, ..., u_{j-1}$, dependent variables $x_1, ..., x_{k-1}$, and the equations $h_1 = 0, ..., h_{k-1} = 0$ corresponding to a part of hypotheses $\{H_1, ..., H_{k-1}\}$, and an ascending chain of the form:

$$\begin{array}{l} f_1(u_1, \ldots, u_{j-1}, x_1) \\ f_2(u_1, \ldots, u_{j-1}, x_1, x_2) \\ \ldots \\ f_{k-1}(u_1, \ldots, u_{j-1}, x_1, \ldots, x_{k-1}). \end{array} \tag{4.4}$$

Furthermore, we assume the ascending chain (in week sense) $f_1, ..., f_{k-1}$ is irreducible.[2] Let Π be the set of points constructed so far. First we add the next point(s) to be constructed to Π. Since Constructions 1–5 introduce only arbitrarily chosen points, we only need to assign new parameters to the coordinates of the points. E.g., for construction 1 (taking any point A), we can let $A = (u_j, u_{j+1})$. Thus we assume that the next construction is one of Constructions 6–10.

[2] For the definition of ascending chains, see [11], [3], or [4]. The ascending chain $f_1, ..., f_{k-1}$ is irreducible if $f(u, x_1, ..., x_i)$ is irreducible in the field $\mathbf{Q}(u)[x_1, ..., x_i]/(f_1, ..., f_{i-1})$, for $i = 1, ..., k-1$. Here here $(f_1, ..., f_{i-1})$ is the ideal generated by $f_1, ..., f_{i-1}$.

Construction 6. Taking an arbitrary point D on a line l in Π. Let the corresponding condition in HS be H_k. Let the line equation $h_k = 0$ for l, which is the algebraic form of H_k, be

$$ax + by + c = 0.$$

Here a, b, and c are polynomials in coordinates of the previously constructed points. E.g., if $l = T(C, AB)$ and $C = (x_2, y_2)$, $A = (x_3, y_3)$, $B = (x_4, y_4)$, then the equation is:

$$(x - x_2)(x_3 - x_4) + (y - y_2)(y_3 - y_4) = 0,$$

i.e., $a = x_3 - x_4$, $b = y_3 - y_4$ and $c = -x_2(x_3 - x_4) - y_2(y_3 - y_4)$.

Our first step is to check whether $R_a = prem(a; f_1, ..., f_{k-1})$ and $R_b = prem(b; f_1, ..., f_{k-1})$ are zero. (Here *prem* denotes the successive pseudo remainder of a polynomial by an ascending chain, see [4] for details.)

Case 6.1. $R_a = 0$ and $R_b = 0$. Then the line l is not well defined. We detect the inconsistency of the hypotheses with adding $A \neq B$. In that case, we either can say that the hypotheses do not satisfy the dimensionality constraint required by Formulation F1 (see p.47 [3]), or it is a theorem according to Formulation 2 because of the inconsistency of the hypotheses.

Case 6.2. One of R_a and R_b, say R_b, is zero. We can let $D = (x_k, u_j)$ and have a new equation:

$$f_k = ax_k + bu_j + c = 0,$$

where u_j and x_k are the new parameter and dependent variables introduced. We have a new irreducible ascending chain $f_1, ..., f_k$. Then the condition $a \neq 0$ is equivalent to that the line l is well defined ($A \neq B$).

Case 6.3. Both R_a and R_b are not zero. We can do the same as in case 6.2. The only difference is that the condition $a \neq 0$ is no longer equivalent to $A \neq B$. But $(a \neq 0 \vee b \neq 0)$ is equivalent to $A \neq B$. We will come back to this in the proof of (4.8).

Construction 7. Taking an arbitrary point A on a circle $(B, (CD))$ in Π. Let $A = (x_k, u_j)$, $B = (x_2, y_2)$, $C = (x_3, y_3)$ and $D = (x_4, y_4)$. Then the algebraic form of the corresponding hypothesis H_k in HS is the equation

$$h_k = (x_k - x_2)^2 + (u_j - y_2)^2 - (x_3 - x_4)^2 - (y_3 - y_4)^2 = 0.$$

Our next step is to check whether CD is isotropic, i.e., whether $R = prem((x_3 - x_4)^2 + (y_3 - y_4)^2; f_1, ..., f_{k-1})$ is zero. If $R \neq 0$, then let $f_k = h_k$, and $f_1, ..., f_k$ is irreducible (see [2]). We always assume this is the case.

Construction 8. Taking the intersection I of two lines l_1 and l_2 in Π. We have two corresponding hypotheses H_k and H_{k+1} in HS, whose algebraic forms are two equations for lines l_1 and l_2:

$$h_k = a_1 x + b_1 y + c_1 = 0,$$
$$h_{k+1} = a_2 x + b_2 y + c_2 = 0.$$

First, we check whether $R = prem(\Delta; f_1, ..., f_{k-1})$ is zero, where $\Delta = a_1b_2 - a_2b_1$. Note that $\Delta \neq 0$ is the algebraic form of the non-degenerate condition generated in cases 8.1–8.10 of Section 3.3.

Case 8.1. $R = 0$. Then adding condition $\Delta \neq 0$ causes inconsistency with the previous constructions.

Case 8.2. $R \neq 0$. Letting $I = (x_{k+1}, x_k)$, then we have two new equations:

$$f_k = \Delta x_k + d = 0,$$
$$f_{k+1} = \Delta x_{k+1} + e = 0.$$

where $d = a_1c_2 - a_2c_1$ and $e = b_2c_1 - b_1c_2$. We have an irreducible ascending chain $f_1, ..., f_{k+1}$.

Construction 9. Taking an intersection Q of a line l and a circle c in Π. We have two corresponding hypotheses H_k and H_{k+1} in HS whose corresponding algebraic forms are the equations for the circle c and the line l:

$$h_k = y^2 + x^2 + by + ax + c = 0$$
$$h_{k+1} = b_1y + a_1x + c_1 = 0.$$

First, we check whether $R = prem(a_1^2 + b_1^2; f_1, ..., f_{k-1})$ is zero.

Case 9.1. $R = 0$. Then the hypothesis $HS \wedge DS$ is inconsistent.

Case 9.2. $R \neq 0$. One of $R_a = prem(a_1; f_1, ..., f_{k-1})$, $R_b = prem(b_1; f_1, ..., f_{k-1})$ is zero, say, R_b. (They cannot be both zero, otherwise R would be zero). Then $a_1 \neq 0$ means that the line l is well defined. We introduce two dependent variables x_k, x_{k+1} and let $Q = (x_{k+1}, x_k)$. Eliminating y from equation h_k, we have

$$f_k = (b_1^2 + a_1^2)x_k^2 + (2a_1c_1 + ab_1^2 - a_1bb_1)x_k + (c_1^2 - bb_1c_1 + b_1^2c) = 0,$$
$$f_{k+1} = a_1x_{k+1} + b_1x_k + c_1 = 0.$$

Now we have an ascending chain (in weak sense) $f_1, ..., f_{k+1}$. We can check whether $f_1, ..., f_{k+1}$ is irreducible using the algorithm introduced in [1] and implemented in our prover. *If it is reducible, generally it is still open whether non-degenerate conditions DS are sufficient.* In the statement of Theorem (4.1), we assume $f_1, ..., f_{k+1}$ is irreducible.

Case 9.3. $R \neq 0$, and both R_a and R_b are non-zero. We can do the same as in Case 9.2. The only difference is that $(a_1 \neq 0 \vee b_1 \neq 0)$ is the condition that the line l is well defined. We will come back to this condition in the proof of Theorem (4.8).

Construction 10. Taking an intersection Q of two circles c_1 and c_2 in Π. We have two corresponding hypotheses H_k and H_{k+1} in HS whose corresponding algebraic forms are the equations for the circles c_1 and c_2:

$$h_k = y^2 + x^2 + by + ax + c = 0,$$
$$h_{k+1} = y^2 + x^2 + ey + dx + j = 0.$$

We can replace h_{k+1} by $h_k - h_{k+1}$. Then $h_{k+1} = 0$ is a line equation:

$$h_{k+1} = b_1 y + a_1 x + c_1 = 0,$$

where $a_1 = a - d$, $b_1 = b - e$, $c_1 = c - j$. This is the line l joining the two intersection points of c_1 and c_2, if they intersect (this is generally the case if the field associated with the geometry Ω is algebraically closed.) But line l exists in Ω even if c_1 and c_2 do not have common points in Ω. This is the radical axis of the two circles. Note that l is non-isotropic iff the line joining the two centers is non-isotropic. Now we check whether $R = prem(a_1^2 + b_1^2; f_1, ..., f_{k-1})$ is zero.

Case 10.1. $R = 0$. This means the radical is isotropic, hence the line OP joining the two centers is isotropic and the hypothesis $HS \wedge DS$ is inconsistent.

Case 10.2. $R \neq 0$, and one of $R_a = prem(a_1; f_1, ..., f_{k-1})$ and $R_b = prem(b_1; f_1, ..., f_{k-1})$ is zero. Then we have exactly the same situation as in case 9.2.

Case 10.3. $R \neq 0$, and both R_a and R_b are not zero. Then we have exactly the same situation as in case 9.3.

Repeating this process until we complete all constructions. Finally, we have an irreducible ascending chain:

$$\begin{array}{ll} & f_1(u_1, \ldots, u_d, x_1) \\ & f_2(u_1, \ldots, u_d, x_1, x_2) \\ (4.5) & \qquad\ldots \\ & f_r(u_1, \ldots, u_d, x_1, \ldots, x_r). \end{array}$$

This is the *definition* of a statement in Class C to be *irreducible* in the statement of Theorem (4.1). Now we want to ask whether the formula, i.e., the exact statement of S

$$(4.6) \qquad \forall P \in \Pi[(H_1 \wedge \cdots \wedge H_r \wedge \neg D_1 \wedge \cdots \wedge \neg D_s) \Rightarrow C]$$

is valid in Ω, or in its equivalent algebraic form, whether the formula

$$(4.7) \qquad \forall \in ux[(h_1 = 0 \wedge \cdots \wedge h_r = 0 \wedge d_1 \neq 0 \wedge \cdots \wedge d_s \neq 0) \Rightarrow c = 0]$$

is valid in F_Ω, where $h_1, ..., h_r$, and c are the polynomials corresponding to $H_1, ..., H_r$ and C, respectively, $d_1, ...d_s$ are polynomials corresponding to $D_1, ..., D_s$.

Remark. Each $\neg D_i$ is one of the negations of the four predicates: collinear, parallel, perpendicular, and point equal. The algebraic form for each of the first three is a polynomial equation. For the last one, it is also an equation if we introduce a new variable z (see [5]).

We have the following theorem which is the algebraic form of Theorem (4.1).

Theorem (4.8). For a statement $S = (HS, DS, C)$ in Class C, let the ascending chain $ASC = f_1, ..., f_r$ of the form of (4.5), obtained by the above procedure, be irreducible and $I_i = lc(f_i)$ be the initial of f_i. If F_Ω is algebraically closed, then the following conditions are equivalent:

(1) $prem(c; ASC) = 0$;
(2) The formula

$$\forall ux \in F_\Omega[(f_1 = 0 \wedge \cdots \wedge f_r = 0 \wedge I_1 \neq 0 \wedge \cdots \wedge I_r \neq 0) \Rightarrow c = 0] \tag{4.9}$$

is valid in F_Ω;
(3) Formula (4.7) is valid in F_Ω;
(4) Formula (4.6) is valid in Ω;
(5) $prem(d \cdot c; ASC) = 0$ for any polynomial d with $prem(d; ASC) \neq 0$.

Proof. (1) $\Rightarrow$ (2). Suppose $R = prem(c; ASC) = 0$. Since we have the remainder formula (p.13 of [3]):

$$I_1^{s_1} \cdots I_r^{s_r} c = Q_1 f_1 + \cdots + Q_r f_r + R,$$

and $R = 0$, formula (4.9) is valid.

(2) $\Rightarrow$ (1). Since F_Ω is algebraically closed and the ascending chain ASC is irreducible, (1) follows from (2) by Theorem (3.7) on p.30 in [3].

(2) $\Rightarrow$ (3). Let J be the set $\{I_i \mid I_i$ is not a constant, $i = 1, ..., r\}$. Let $N = \{d_1, ..., d_s\}$. Note that $N \subset J$. We want to show that those I_k in J but not in N can be removed in (4.9). Such an I_k can be only the following three cases:

Case 1: Case 6.3. In this case $I_k = a$. We can let $A = (u_j, x_k)$ and $f_k' = bx_k + au_j + c = 0$. The ascending chain $ASC' = f_1, ..., f_{k-1}, f_k', f_{k+1}, ..., f_r$ is also irreducible. By Lemma (A1.1) [5], $prem(c; ASC) = 0$ if and only if $prem(c; ASC') = 0$. Thus we only need the condition $(a \neq 0 \vee b \neq 0)$. This is equivalent to one of the conditions $d_i \neq 0$.

Case 2: Case 9.3. In this case $I_k = a_1$. Using the same technique as in Case 1, we can come to the conclusion that $I_k \neq 0$ can be replaced by a weaker condition $(a_1 \neq 0 \vee b_1 \neq 0)$, which is implied by $a_1^2 + b_1^2 \neq 0$, i.e., by the condition that line l is non-isotropic.

The details work as follows. Assuming the order $x_{k+1} < x_k$, we have an ascending chain: $ASC' = f_1, ... f_{k-1}, f_k', f_{k+1}', f_{k+2}, ..., f_r$, where

$$f_k' = (b_1^2 + a_1^2)x_{k+1}^2 + (2b_1c_1 - aa_1b_1 + a_1^2 b)x_{k+1} + (c_1^2 - aa_1c_1 + a_1^2 c) = 0,$$
$$f_{k+1}' = b_1 x_k + a_1 x_{k+1} + c_1 = 0.$$

The discriminate of f_k' is $-a^2\delta$ and that of f_k is $-b^2\delta$, where $\delta = 4c_1^2 - 4bb_1c_1 - 4aa_1c_1 + 4b_1^2 c + 4a_1^2 c - a^2 b_1^2 + 2aa_1bb_1 - a_1^2 b^2$. Thus f_k' is irreducible over $F = \mathbf{Q}(u)[x]/(f_1, ..., f_{k-1})$ iff f_k is irreducible over F. Hence f_k' is irreducible. Since $prem(f_i; ASC') = 0$ and $prem(f_i'; ASC) = 0$ (for $i = k, k+1$), $prem(c; ASC) = 0$ iff $prem(c; ASC') = 0$ by Lemma (A1.1) [5]. Thus we only need the condition $(a_1 \neq 0 \vee b_1 \neq 0)$ which is implied by $d_i = a_1^2 + b_1^2 \neq 0$.

Case 3: Case 10.3. The same as in Case 2.

Thus the formula $\forall ux[(f_1 = 0 \wedge \cdots \wedge f_r \wedge d_1 \neq 0 \wedge \cdots \wedge d_s \neq 0) \Rightarrow c = 0]$ is valid. Since $(h_1 = 0 \wedge \cdots \wedge h_r = 0) \Rightarrow (f_1 = 0 \wedge \cdots \wedge f_r = 0)$, (3) follows from (2).

(3) $\Rightarrow$ (2). By the remainder formula, $(f_1 = 0 \wedge \cdots \wedge f_r = 0 \wedge I_1 \neq 0 \wedge \cdots \wedge I_r \neq 0) \Rightarrow h_i = 0$. Also $I_1 \neq 0 \wedge \cdots \wedge I_r \neq 0 \Rightarrow d_i \neq 0$. Thus (4.9) follows from (4.7).

(3) $\Leftrightarrow$ (4). Since (4.7) is the algebraic form of (4.6), (3) and (4) are equivalent.

(1) $\Leftrightarrow$ (5). Since ASC is irreducible, (1) and (5) are equivalent. ∎

Remark. Interpreted in other way, (5) is the completeness of the non-degenerate conditions $\neg D_i$. Suppose $\neg D$ is another non-degenerate condition whose algebraic form is $d \neq 0$. If $(HS \wedge DS) \Rightarrow C$ is not valid in Ω, then $prem(c; ASC) \neq 0$. By (5), $prem(d \cdot c; ASC) \neq 0$, i.e., $(HS \wedge DS \wedge \neg D) \Rightarrow C$ is still not valid.

References

[1] S.C. Chou, "Proving Elementary Geometry Theorems Using Wu's Algorithm", in *Automated Theorem Proving: After 25 years*, Ed. By W.W. Bledsoe and D. Loveland, AMS Contemporary Mathematics Series **29** (1984), 243-286.

[2] S.C. Chou, "Proving and Discovering Theorems in Elementary Geometries Using Wu's Method", PhD Thesis, Department of Mathematics, University of Texas, Austin (1985).

[3] S.C. Chou, *Mechanical Geometry Theorem Proving,* D. Reidel Publishing Company, Dordrecht, Netherlands, 1988.

[4] S. C. Chou and X. S. Gao, "Ritt-Wu's Decomposition Algorithm and Geometry Theorem Proving", in *Proceedings of CADE-10, Lecture Notes in AI*, Vol.449, 207–220, 1990.

[5] S.C. Chou and X.S. Gao, "A Class of Geometry Statements of Constructive Type and Geometry Theorem Proving", TR-89-37, Computer Sciences Department, The University of Texas at Austin, November, 1989.

[6] S.C. Chou, "A Geometry Theorem Prover for Macintoshes", TR-91-8, Computer Sciences Department, The University of Texas at Austin, March, 1991.

[7] S.C. Chou and W.F. Schelter, "Proving Geometry Theorems with Rewrite Rules", *Journal of Automated Reasoning,* **2(4)** (1986), 253–273.

[8] D. Kapur, "Geometry Theorem Proving Using Hilbert's Nullstellensatz", in Proceedings of the 1986 Symposium on Symbolic and Algebraic Computation, 202-208.

[9] B. Kutzler and S. Stifter, "Automated Geometry Theorem Proving Using Buchberger's Algorithm", in Proceedings of the 1986 Symposium on Symbolic and Algebraic Computation, 209-214.

[10] Wu Wen-tsün, "On the Decision Problem and the Mechanization of Theorem Proving in Elementary Geometry", *Scientia Sinica* **21** (1978), 157-179.

[11] Wu Wen-tsün, "Basic Principles of Mechanical Theorem Proving in Geometries", *J. of Sys. Sci. and Math. Sci.* **4(3)**, 1984, 207-235, republished in *Journal of Automated Reasoning* **2(4)** (1986), 221-252.

[12] Wu Wen-tsün, *Basic Principles of Mechanical Theorem Proving in Geometries,* (in Chinese) Peking 1984.

The Central Variable Strategy of Str+ve*

L. M. Hines
Department of Computer Science
University of Texas
Austin, Texas 78759
hines@cs.utexas.edu

Abstract

STR+VE is an upgraded version of our 1980 general inequality theorem prover [6] that contains a restrictive implementation of the transitivity property of inequality and the interpolation property of dense fields. Its underlying system is complete for first order logic. In this paper, we formally present and discuss some of its features. Included is the central variable strategy (not part of the 1980 prover) which gives STR+VE a "large step" capacity.

These features enhance STR+VE and enable it to prove more difficult examples.

1 Introduction

Bledsoe and I introduced a resolution-based prover for general inequalities in [6]. This 1980 program proved some moderately difficult theorems in real analysis, including the continuity of the sum of two continuous functions. Here we present an upgraded version of that prover which we call STR+VE (pronounced strive).

STR+VE includes a *central* variable strategy whereby a selected variable is removed by a series of chaining steps. Linking together these steps gives the prover more overall guidance and control. This strategy along with other enhancements detailed below enable STR+VE to quickly and easily prove theorems such as a first-order version of the intermediate value theorem.[1] Such theorems have been beyond the reach of other automated theorems provers.

In Section 2, we formalize and discuss STR+VE (its domain and its strategy). We state the inference rules in Section 3 including a new form of factoring which has been incorporated into chaining. (On the first reading, one may prefer to skip the subsections of Section 3.1.) We then present the central variable strategy in Section 4. The completeness of STR+VE is discussed in Section 5. Finally, we demonstrate STR+VE on some theorems from real analysis.

*This work supported by National Science Foundation Grants 26-1099-26 and CCR-9101980.

[1]The second-order variable in the least upper bound principle was appropriately instantiated before the theorem was given to the prover. We have named this example IMV.

2 Description

STR+VE is designed for formulas from *dense linear orders without endpoints.* Because arbitrary quantification and uninterpreted function symbols are permitted, first-order logic is contained within these formulas. Also permitted are the function symbols $+$ and $\times$, in order to encode the usual semantics of addition and scalar multiplication.[2]

Such formulas necessarily contain the transitivity property of inequality, as well as the interpolation property of dense orders without endpoints. These properties (in clausal form) contain literals which individually unify with every inequality literal. Resolvents derived from these properties can easily swamp a general purpose prover.[3]

In STR+VE, transitivity and interpolation are built-in as inference rules. Specifically, the chaining inference rule (defined in Section 3) implements the transitivity property. Moreover, chaining is restricted to pairs of terms one of which must be a *shielding term* and the other a non-variable term. A term $f(t_1, t_2, \ldots, t_n)$ is called a *shielding term* if f is neither $+$ nor $\times$, and for some $i \leq n$, t_i is either a variable or contains a variable. This shielding term restriction greatly reduces the prover's search space.

The variable elimination inference rule builds in the interpolation property. It functions as a rewrite rule, deleting any variable which does not occur within any shielding term. Thus, variable elimination directly reduces the number of variables and, consequently, reduces the prover's search space. We say that variables not contained within a shilding term are *eligible.*

Factoring is now incorporated in the chaining rule and is restricted to variants.[4]

The 1980 prover employed a strategy based upon the cooperative nature of variable elimination, chaining and a fast ground-inequality decision procedure. The objective is to transform a *general* clause to a corresponding *ground* clause because such ground clauses can be handled quickly by the ground prover. This objective is attained by: using chaining to remove a variable's shielding terms, thereby moving the variable closer to being eligible; and using variable elimination to remove eligible variables, thereby reducing the number of variables in the clause and bringing the clause is closer to being ground.

This prover's effectiveness was also due to the replacement of the standard resolution rule with the chaining rule. A chaining inference essentially combines two resolution inferences and, subsequently, the prover's basic inferences represent slightly "larger steps" in contrast to a resolution prover.

But, in the 1980 prover, we failed to exploit all of the potential for high-level

---

[2]E.g. $4 \times B$. We use the terminology *scalar multiplication* to indicate multiplication restricted to a numeric constant times an arbitrary term. Also note that neither $+$ nor $\times$ is allowed within any uninterpreted function symbol. For example, the term $f(a+b)$ is not permitted. Such terms can be included by employing an algebraic unifier such as [24].

[3]See [4].

[4]Rabinov established the completeness of resolution with factoring restricted to variants. [19]

guidance. Consequently, it suffers from the problems inherent to other resolution-based or "small step" provers. We address this difficulty in STR+VE by combining chainings to obtain a "larger step".

The central variable strategy links a series of chainings, culminating in a variable elimination (or a chaining which instantiates a variable with a ground term). The implicit goal throughout this chaining series is the eventual removal of the variable. The resolvents in the midst of the series are restricted by this goal: each chaining must remove a shielding term of the variable. These intermediate resolvents cannot be used in any other way and can be discarded once the goal is attained.

The prover systematically removes variables until the clause set is ground inconsistent. Thus, a central search decision is ascertaining the order in which variables are removed. This decision is determined by the order induced by the *dependencies among variables* (see Section 4.2).

3 Inference Rules

3.1 Chaining

Chaining, in STR+VE, is restricted by its term orientation. Specifically, chaining is applied to pairs of unifiable terms, whereas resolution is applied to pairs of unifiable literals.

Neither term may be a variable. Although the use of chaining and variable elimination contains additional restrictions, the prohibition against chaining on variables is the most fundamental one. Not only does the restriction prevent the generation of an uncalculated number of unnecessary resolvents, but it also (along with the term orientation) permits the inclusion of addition and scalar multiplication without extensive use of associative-commutative unification.

We exclude chaining on pairs of ground terms because those deductions are carried out by the ground decision procedure (see Section 3.4). Thus, we require that chaining be applied to two unifiable terms. Neither term may be a variable and one must not be ground. This non-ground term is a *shielding term.*

The 1980 prover used the basic chaining inference rule given below in Section 3.1.1, while STR+VE uses the chaining rule with *equivalence* factoring given in Section 3.1.3. Chaining with equivalence factoring (fe-chaining) was introduced in [12] in order to prove the completeness of a proof system with mandatory variable elimination. Notwithstanding its theoretical roots, this type of chaining is also of practical interest in that it provides additional focus and power to the prover's search strategy. To demonstrate the difference between the fe-chaining rule and the basic chaining rule together with the usual factoring rule, consider the clause set

1. $a < f(x) \quad z < f(w) \quad g(b) < g(z)$
2. $f(y) < b$
3. $b \leq a$

where w, x, y and z are the only variables. We can derive by factoring (from 1) the clause

$$4.\quad a < f(x) \quad g(b) < g(a)$$

and then by basic chaining (from 1 and 2 or from 4 and 2) the clauses

$$\begin{array}{llll} 5. & a < f(x) & z < b & g(b) < g(z) \\ 6. & a < b & z < f(w) & g(b) < g(z) \\ 7. & a < b & & g(b) < g(a) \\ 8. & a < b & z < b & g(b) < g(z). \end{array}$$

By our new fe-chaining, we can not derive 4 or 7, instead we only derive clauses 5, 6 and 8.

3.1.1 Basic chaining

If L is an inequality literal of the form $(a \trianglelefteq b)$[5] then $Right(L) = b$, $Left(L) = a$ and $Op(L) = \trianglelefteq$. If D is a set of inequality literals then $RS(D) = \{Right(L) : L \in D\}$ and $LS(D) = \{Left(L) : L \in D\}$. Define $lt(\trianglelefteq_1, \trianglelefteq_2)$ to be the symbol $<$ if $\trianglelefteq_1$ or $\trianglelefteq_2$ is $<$, and the symbol $\leq$ otherwise.

A *chain resolvent* of clauses C_1 and C_2 with distinct variables is any clause of the form

$$(C_1 \setminus \{L_1\})\sigma \cup (C_2 \setminus \{L_2\})\sigma \cup \{(r \trianglelefteq s)\}\sigma$$

where $L_1 \in C_1$ is of the form $(r \trianglelefteq_1 t_1)$; $L_2 \in C_2$ is of the form $(t_2 \trianglelefteq_2 s)$; σ is the mgu of t_1 and t_2; $t_1\sigma \neq r\sigma$ and $t_2\sigma \neq s\sigma$; either t_1 or t_2 is a shielding term; neither t_1 nor t_2 is a variable; and $\trianglelefteq\ = lt(\trianglelefteq_1, \trianglelefteq_2)$.

For example, given the clauses

$$\begin{array}{llll} 1. & a < f(x) & z < f(w) & g(b) < g(z) \\ 2. & f(y) < b & & \end{array}$$

where w, x, y and z are the only variables, we can derive

$$5.\quad a < f(x) \quad z < b \quad g(b) < g(z)$$

by letting C_1 be clause 1, L_1 be $z < f(w)$ in clause 1, C_2 be clause 2, L_2 be $f(y) < b$ in clause 2 and σ be $\{x/y\}$.

3.1.2 Self-Chaining

A *restricted self-chain resolvent* of clause C is any clause of the form

$$(C \setminus \{L\})\sigma,$$

where SC1: $L \in C$;

[5] We use the symbol $\trianglelefteq$ to stand for either inequality symbol, $\leq$ or $<$.

SC2: σ is the mgu of $Left(L)$ and $Right(L)$;

SC3: either $Left(L)$ or $Right(L)$ is a shielding term;

SC4: neither $Left(L)$ nor $Right(L)$ is a variable; and

SC5: $OP(L)$ is $<$.

For example, given the clause

$$a < f(x) \quad z < f(w) \quad g(b) < g(z)$$

where w, x and z are the only variables, we can derive

$$a < f(x) \quad b < f(w)$$

by letting clause 1 be C, $g(b) < g(z)$ in clause 1 be L and σ be $\{b/z\}$.

Note that we must generalize self-chaining to include addition and scalar multiplication in the same fashion that we generalized basic chaining.

3.1.3 Chaining with Equivalence Factoring

A *fe-chain resolvent* of clauses C_1 and C_2 with distinct variables is any clause of the form

$$\begin{aligned}&(C_1 \setminus D_1)\sigma \cup (C_2 \setminus D_2)\sigma\\ &\cup\{(Left(L_1)\sigma \trianglelefteq Right(L_2)\sigma) : \; L_1 \in D_1 \wedge L_2 \in D_2\\ &\qquad\qquad\qquad\qquad\qquad\qquad\quad \wedge \trianglelefteq = lt(Op(L_1), Op(L_2))\},\end{aligned}$$

where

FE1: $\emptyset \neq D_1 \subseteq C_1$; $\emptyset \neq D_2 \subseteq C_2$;

FE2: σ is the mgu of $RS(D_1) \cup LS(D_2)$;

FE3: $LS(D_2)$ is a singleton set and $RS(D_1)$ is a set of variants terms;

FE4: $RS(D_1)$ or $LS(D_2)$ contains a shielding term;

FE5: no element in $RS(D_1)$ or in $LS(D_2)$ is a variable; and

FE6: $t \notin LS(D_1)\sigma$ and $t \notin RS(D_2)\sigma$, where t is the single element in $RS(D_1)\sigma = LS(D_2)\sigma$.

For example, given the clause set

$$\begin{array}{ll}1. & a < f(x) \quad z < f(w) \quad g(b) < g(z)\\ 2. & f(y) < b\end{array}$$

where w, x, y and z are the only variables, we can derive

$$\begin{array}{llll}5. & a < f(x) & z < b & g(b) < g(z)\\ 6. & a < b & z < f(w) & g(b) < g(z)\\ 8. & a < b & z < b & g(b) < g(z).\end{array}$$

To derive clause 5, D_1 is the set $\{z < f(w)\}$; to derive clause 6, the set $\{a < f(x)\}$; and to derive clause 8, the set $\{a < f(x), z < f(w)\}$. In each derivation, D_2 is the set $\{f(y) < b\}$.

3.2 Variable Elimination

A *variable elimination* resolvent of a clause C is any clause of the form

$$C' \vee \{b_j \trianglelefteq_{i,j} a_i : i = 1, n; j = 1, m\},$$

if C can be written as

$$\{x \trianglelefteq'_i a_i : i = 1, n\} \vee \{b_j \trianglelefteq''_j x : j = 1, m\} \vee C'$$

where x is a variable that does not occur in any a_i, b_j or any literal in C' and where $\trianglelefteq_{i,j}$ is $\leq$ if $\trianglelefteq'$ or $\trianglelefteq''$ is $\leq$ and is $<$ otherwise. Note that if n or m is zero then the resolvent is C'.

For example, given the clause $a \leq z \quad z < x \quad g(b) < z \quad f(x) < a$
with variables, x and z, we derive $a \leq x \quad g(b) < x \quad f(x) < a$
or, given the clause $a \leq z \quad x < z \quad g(b) < z \quad f(x) < a$,
we derive (since $n = 0$) $f(x) < a$.

Previously, we stated that a variable in a clause is eligible if it does not occur in any shielding term in the clause. That is, the variables within shielding terms are "shielded" from variable elimination. Alternatively, we can say that a variable is eligible if its clause can be written in the form of the above disjunction. In the above example, z is eligible, but x is not since $f(x)$ shields it.

3.2.1 Mandatory Variable Elimination

MVE: Variable elimination is applied immediately to any eligible variable.

This requirement is called *mandatory variable elimination* and means that the clause of the eligible variable is *replaced* by the results of variable-eliminating the variable. This requirement is critical. With it, variable elimination decreases the number of variables in the clause set, reducing the prover's search space.

3.3 Equality

Neither chaining nor self-chaining finds a refutation in clause sets such as, $\{c_x \leq d_x,\ d_x \leq c_x,\ f(c_x) < f(d_x)\}$, where x is a variable. To have a refutation, the function f must be *well defined.* We need a clause of the form

$$x_1 < y_1 \vee \ldots \vee x_n < y_n \vee y_1 < x_1 \vee \ldots \vee y_n < x_n \vee f(x_1, \ldots, x_n) \leq f(y_1, \ldots, y_n).$$

"Well-definedness" clauses are not added for Skolem functions (see [7]), so this method adds only a few clauses for most problems.

Nevertheless, we have implemented "well-definedness" as an inference in STR+VE and, thus, we avoid adding any such clauses.

A "well-definedness" resolvent of a clause C is any clause of the form

$$C \setminus \{L\} \cup \{(x_i < y_i) : i = 1, ..., n\} \cup \{(y_i < x_i) : i = 1, ..., n\}$$

where $L \in C$; $Left(L)$ and $Right(L)$ are not unifiable; either $Left(L)$ or $Right(L)$ is a shielding term with a non-Skolem function symbol; neither $Left(L)$ nor $Right(L)$ is a variable; $OP(L)$ is $<$; x_i is the ith argument of $Left(L)$; and y_i is the ith argument of $Right(L)$.

3.4 The Ground Prover

The strategy behind STR+VE presumes a fast ground inequality prover that determines whether a set of clauses is ground inconsistent. The clauses are added incrementally (as they are derived). Consequently, the ground prover must maintain a database which is checked for consistency after each addition. The inconsistent literals in the newly added clause are dropped from the clause and newly entailed equalities are added to the database.

A number of ground inequality packages can be used. These include provers by King [15], Nelson and Oppen [17], and Shostak [22]. Our 1980 prover employed Bledsoe's Sup-Inf prover [2]. In STR+VE, we employed a matrix-based algorithm originally written by Don Simon (described in [9]).

3.5 Addition and Scalar Multiplication

Addition and scalar multiplication are implemented by employing some reduction rules and by generalizing the inference rules given in the preceding subsection.

The reductions include combining addends with common terms, associating addition terms, and distributing scalar multiplication over addition. These rules simplify literals such as $(2 \times a < a)$ to $(a < 0)$ and $(-5 \times c + b < a)$ to $b < a + (5 \times c)$.

To generalize the inference rules, we need to isolate the appropriate term on the inequality's right or left side before applying the rule.

We say that t is a *simple term* if it is a variable or a constant (but not a number), or its topmost function symbol is not $+$ or $\times$. For example, d, $f(x)$ and c are the simple terms in the expression $(5 \times d) + f(x) + (2 \times c))$.

Given a literal L, *Isolate-Left*(L) is the set of fully reduced and semantically equivalent literals, L' where $Left(L')$ is a simple term. Likewise, *Isolate-Right*(L) is defined to "solve" for simple terms on the right side.

For example, *Isolate-Left*$((5 \times d) + f(x) < (2 \times c))$ is $(d < (\frac{2}{5} \times c) + (\frac{-1}{5} \times f(x)))$ and $(f(x) < (2 \times c) + (-1 \times d))$. *Isolate-Right*$((5 \times d) + f(x) < (2 \times c))$ is $((\frac{5}{2} \times d) + (\frac{1}{2} \times f(x)) < c)$.

So, we may define fe-chain resolvents with addition and scalar multiplication of clauses C_1 and C_2 with distinct variables to be any clause of the form

$$\begin{aligned}&(C_1 \setminus E_1)\sigma \cup (C_2 \setminus E_2)\sigma \\ &\cup\{(Left(L_1)\sigma \trianglelefteq Right(L_2)\sigma) : \; L_1 \in D_1 \wedge L_2 \in D_2 \\ &\qquad\qquad\qquad\qquad\qquad\qquad\quad \wedge \trianglelefteq = lt(Op(L_1), Op(L_2))\},\end{aligned}$$

where

FEP1: $\emptyset \neq E_1 \subseteq C_1$; $\emptyset \neq E_2 \subseteq C_2$;

FEP2: D_1 is a set of literals such that for each literal $L \in E_1$ there is a literal $L' \in D_1$ where $L' \in$ *Isolate-Right*(L);

FEP3: D_2 is a set of literals such that for each literal $L \in E_2$ there is a literal $L' \in D_2$ where $L' \in$ *Isolate-Left*(L); and

FEP4: the restrictions FE2 through FE6 of fe-chaining all apply.

4 The Central Variable Strategy

The central variable strategy joins together a series of chainings with the goal of removing a selected variable. Each successive intermediate resolvent can be used only to remove the variable's shielding terms. Thus, the role of these resolvents is "reduced to that of purely temporary scratchpad entities" [21]. The strategy thereby avoids the derivation of redundant or useless resolvents and, hence, provides more overall guidance and control to STR+VE.

The following example demonstrates the power of the strategy.

1. $g(z) \leq d \quad s(z) < a \quad b < f(x) \quad h(x) \leq z$
2. $y \leq g(y)$
3. $a \leq s(w)$
4. $f(v) \leq b$
5. $d < h(u)$

where u, v, w, x, y and z are variables. Initially, four fe-chainings resolvents can be derived (See Appendix 4.3). Twelve more resolvents can be derived from these four and twenty-four more from the twelve. Another round of fe-chaining generates a final twenty-four resolvents. Most of these sixty-four resolvents are useless (tautological, redundant, or derived from such a resolvent). STR+VE's central variable strategy controls the derivation in such a way that these useless resolvents are not generated. In fact, only four resolvents are generated.

The strategy has two components: First, it selects a variable, initiating the series. The selection is based upon *dependencies between variables* [10]. Second, the strategy restricts fe-chaining and self-chaining to the selected variable's leftmost shielding term.

4.1 Justification

Consecutive chainings commute. (See [12] or [20].) Consider a derivation $\mathcal{D}$ of chainings on the shielding terms of the variables x and y that derives a resolvent R. $\mathcal{D}$ can be re-ordered, obtaining a derivation $\mathcal{D}_1$ following by a derivation $\mathcal{D}_2$. $\mathcal{D}_1$ consists only of chainings on the shielding terms of one of the variables and, perhaps, culminates in its variable elimination. Similarly, $\mathcal{D}_2$ is limited to the other variable. Then, following the proof in [12], the combined derivation $\mathcal{D}_1\mathcal{D}_2$ derives a resolvent which subsumes R.

Moreover, each subseries within the series can be re-ordered such that each chaining is on the variable's leftmost shielding term.

Note that we did not state above whether $\mathcal{D}_1$ was limited to x or limited to y; we just limited it to one or the other. But, the selection of which variable is *crucial.* The selection of x represents half of the search space and the selection of y the other half. Pruning one is clearly a major advantage.

STR+VE's dependencies between variables routine partially orders each clause's variables, stating which variable should be selected first, second and so on. Two variables, x and y, are independent (un-ordered), dependant (one goes before the other) or interdependant (equivalent). If x and y are independent, then the clause can be split (à la Davis-Putman [8]). Hence, STR+VE arbitrary selects one. If x depends on y, then STR+VE selects y first. If x and y are interdependent, STR+VE may initiate a derivation with x first and another with y first. (See [9] or [11] for more details.)

4.2 Implementation

We impose on fe-chaining and self-chaining the restrictions given below. These restrictions refer to central variables defined as:

> Every variable in an input clause is considered a *central variable.* In a clause derived by chaining, only the variables in the chained shielding terms are considered central.[6] If none of those variables appear within the clause, then every variable in the clause is considered central. (The terms D_1, D_2, *etc.* refer back to the definition of fe-chaining in Section 3.)
>
> **CV1:** $RS(D_1)$ is either a set of ground terms or a set of shielding terms of a *central* variable in C_1. Likewise, $LS(D_2)$ is either a set of ground terms or a set of shielding terms of a central variable in C_2.
>
> **CV2:** $LS(D_1)$ must contain the leftmost shielding term of a central variable in C_1 and $RS(D_2)$ must contain the leftmost shielding term of

[6] Of course, the variables of any newly derived clause should be renamed apart from the variables of the previous clauses. Thus, the central variables of the parent clauses must be tracked through this renaming.

a central variable in C_2.

CV3: If all the variables in C_1 are central then $RS(D_1)$ must be either a set of ground terms or a set of shielding terms of a dependency-selected variable. Likewise, if all the variables in C_2 are central then $LS(D_2)$ must be either a set of ground terms or a set of shielding terms of a dependency-selected variable.

4.3 Central Variable Strategy Example

Given the five clauses

$$\begin{array}{lllll}
1. & g(z) \leq d & s(z) < a & b < f(x) & h(x) \leq z \\
2. & y \leq g(y) & & & \\
3. & a \leq s(w) & & & \\
4. & f(v) \leq b & & & \\
5. & d < h(u), & & &
\end{array}$$

four fe-chaining resolvents can be derived.

$$\begin{array}{llllll}
6. & z \leq d & s(z) < a & b < f(x) & h(x) \leq z & 1,2 \\
7. & g(z) \leq d & \not a \not< \not a & b < f(x) & h(x) \leq z & 1,3 \\
8. & g(z) \leq d & s(z) < a & \not b \not< \not b & h(x) \leq z & 1,4 \\
9. & g(z) \leq d & s(z) < a & b < f(x) & d < z & 1,5.
\end{array}$$

Each of these resolvents contains three shielding terms and, therefore, three additional resolvents can be derived from each. These twelve resolvents each contain two shielding terms, so we generated in the next round twenty-four resolvents, each containing a single shielding term. The last round then generates twenty-four also.

At the cost of forward subsumption and tautology checking, 15 of these can be discarded. Doing so then prevents the generation of thirty-three of them. This leaves 28 resolvents generated and 13 resolvents retained.

STR+VE's central variable strategy generates only four resolvents. The variable selected in clause 1 is z because x depends on z. Thus, the potential resolvents derivable from 1 and 4 and from 1 and 5 are prohibited, pruning the search space of a total of 32 resolvents. Furthermore, the shielding term $s(z)$ is not the leftmost shielding term of z; so, the chaining 1 and 3 is prohibited (pruning 16 resolvents). We derive only clause 6 given above. Clause 6 is an intermediate clause and z is its central variable. $s(z)$ is z's leftmost shielding term, so we derive only

$$11. \quad z \leq d \quad \not a \not< \not a \quad b < f(x) \quad h(x) \leq z,$$

(pruning 10 resolvents). By mandatory variable elimination, clause 11 is re-written to

$$11. \quad h(x) \leq d \quad b < f(x).$$

z has been removed and we select a new variable. x, the only remaining variable, is selected. $h(x)$ is the leftmost shield, so we derive

$$12. \quad \not{d} \not< \not{d} \quad b < f(x),$$

(pruning 2 resolvents). Clause 12 is also an intermediate clause, x its central variable and $f(x)$ the leftmost shield. We derive

$$13. \quad \square.$$

5 Completeness

Completeness of a chaining and variable elimination system was first established in [7], wherein it is shown that a ground inconsistent set of inequalities can be derived from any inconsistent set of inequalities by means of chaining, self-chaining and variable elimination. Hence, a ground decision procedure completes the refutation.

The proof system described in [7] does not include the important *mandatory* requirement of variable elimination. As noted above, not including this mandatory requirement considerably weakens the power of the prover. Non-mandatory variable elimination *increases* the clause set, rather than *reduces* it.

Subsequently, we established the completeness of another proof system [12] for theorems from the theory of dense total orders without endpoints. That system differs from the one described in [7] in two major ways: It includes mandatory variable elimination and uses term-based equivalence factoring as part of the chaining rule.

STR+VE is based upon this complete system. The axioms of addition and scalar multiplication are also built into STR+VE. So, one may ask whether STR+VE is complete for Abelian groups within dense total orders without endpoints. The answer is not yet known.

5.1 Subsumption and Tautology Deletion

The traditional subsumption rule (i.e., C is subsumed by D if $D\sigma \subseteq C$ for some substitution σ) is used in STR+VE. This traditional subsumption deletion is compatible with the completeness proof.[7]

Tautology deletion, on the other hand, is not compatible. It is provided as an option and is used in proving the theorems in Section 9. However, if fe-chaining is extended as described in [9], tautology deletion might be compatible.

6 Conclusion

STR+VE quickly and easily proves moderately difficult first-order theorems from real analysis. One such example is LIM+ (the sum of limits is equal to the limit

[7] An alternative subsumption rule appears in [10] whereby a clause of the form $a < b$ subsumes $a \le b$ and a clause of the form $a \le c$ subsumes $a \le b \lor b \le c$.

of the sum). As noted in [4], a general purpose prover is inundated with a rapidly multiplying number of resolvents when trying to prove LIM+. Such general purpose provers will likely fail even with pruning or priority schemes (unless these schemes are tailored for the particular theorems being proved). Only model elimination [1] and OTTER [16] has had any success with the simpler versions of LIM+.

By building in the transitivity axiom of inequality via chaining and the interpolation axiom of dense fields via variable elimination, the clauses corresponding to these properties are not explicitly added to the clause set. Not only are the proofs shorter, but chaining and variable elimination are restricted, excluding untold numbers of unproductive and unnecessary resolvents.

The central variable strategy provides more overall guidance. The prover selects a variable and, systematically, removes its shielding terms. The strategy greatly prunes the search space, enabling the prover to find proofs of more difficult examples such as those given in Section 9. STR+VE is the first fully automated theorem prover to prove examples 1 through 4. (IMV has been proved previously only by the prover described in [18] (using an interpretation specifically designed for the problem) and by the prover described in [10].)

7 Related Research

"Built-in" procedures for transitivity and interpolation have previously been suggested and/or used in other provers with varying degrees of success. The *ground* inequality packages of King [15], Nelson and Oppen [17], Shostak [22], and Bledsoe et al. [2] provide alternatives to the ground prover used here. For general inequalities, previous procedures include: the built-in partial ordering of Slagle and Norton [23]; the methods of Hodes [14]; and the restriction intervals method [5].

Some of the concepts employed in STR+VE appeared in the prover described in [10]. Chaining was extended to hyper-chaining, and additional axioms such as the triangle inequality axiom were built-in. Consequently, that prover found proofs to such theorems as the Continuity of the Product and Quotient of Two Continuous Functions and IMV.

Also, in the linked inference principle [25] a series of smaller steps are joined together.

8 Acknowledgements

The research was first suggested to me by Woody Bledsoe. He contributed heavily to the design of our 1980 prover and, consequently, to STR+VE. He continues to be a sounding board with respect to newer ideas that may be incorporated into STR+VE.

He and Laura Buss also proofread earlier drafts of this paper.

9 Examples

STR+VE is written in AKCL (Austin Kyoto Common Lisp). The examples given below were run on a Sun 3-280. Time is given in CPU seconds. The "Input" column contains the number of clauses found by putting the example in CNF and "Res Gen" is the total number of resolvents generated. The "Int Res" is the number of intermediate resolvents generated. In these resolvents, the parents' central variable was not completely removed. Under "Length" is the number of resolvents actually used in the proof.

Once the example has been transformed to negation normal form, STR+VE splits examples which contain a top-level "or". Such happens with the "Subsets have Maximums" example. Thus, we have two lines (one for each case) under that example.

Example	Time	Input	Res Gen	Int Res	Length
Discontinuity	1.2	6	6	0	4
Squeeze	2.9	12	10	4	5
Continuity of Comp.	29.1	10	63	43	17
C&B Subsets - Max	10.0	8	15	11	5
		9	65	40	9
IMV	15.7	15	60	35	10

1. Discontinuity.

$$a < b \ \wedge\ \forall x(x < c \rightarrow f(x) < a) \ \wedge\ \forall x(c < x \rightarrow b < f(x))$$
$$\rightarrow \neg(\exists l \forall \varepsilon > 0 \rightarrow \exists \delta > 0 \wedge \forall y(c \leq y + \delta \wedge y \leq c + \delta \rightarrow l \leq f(y) + \varepsilon \wedge f(y) \leq l + \varepsilon)).$$

2. Squeeze. (suggested by Ken Kunen)

$$[\forall c \forall \varepsilon(0 < \varepsilon \rightarrow \exists \delta(0 < \delta \wedge \forall y(|c - y| < \delta \rightarrow |f(y) + l| \leq \varepsilon)))$$
$$\wedge \forall c \forall \varepsilon(0 < \varepsilon \rightarrow \exists \delta(0 < \delta \wedge \forall y(|c - y| < \delta \rightarrow |g(y) + l| \leq \varepsilon)))]$$
$$\wedge \forall x(f(x) < h(x) < g(x))$$
$$\longrightarrow \forall \varepsilon(0 < \varepsilon \rightarrow \exists \delta(0 < \delta \wedge \forall y(|c - y| < \delta \rightarrow |h(y) + l| \leq \varepsilon)))$$

3. Continuity of Composition.

$$[\forall \varepsilon(0 < \varepsilon \rightarrow \exists \delta(0 < \delta \wedge \forall y(|c - y| < \delta \rightarrow |f(y) + f(c)| \leq \varepsilon)))$$
$$\wedge \forall c \forall \varepsilon(0 < \varepsilon \rightarrow \exists \delta(0 < \delta \wedge \forall y(|f(c) - y| < \delta \rightarrow |g(y) + g(f(c))| \leq \varepsilon)))]$$
$$\longrightarrow \forall \varepsilon(0 < \varepsilon \rightarrow \exists \delta(0 < \delta \wedge \forall y(|c - y| < \delta \rightarrow |g(f(y)) + g(f(c))| \leq \varepsilon)))$$

4. Nonempty, Closed & Bounded Subsets have a Maximum.

$$[1 \leq P(A) \wedge \exists u \forall v(1 \leq P(v) \rightarrow v \leq u) \wedge \forall r(\forall \varepsilon(0 < \varepsilon \rightarrow \exists s 1 \leq P(s) \wedge |s - r| \leq \varepsilon))$$
$$\wedge \exists a(1 \leq P(a)) \wedge \exists u \forall v(1 \leq P(v) \rightarrow v \leq u)$$
$$\rightarrow \exists m(\forall w(1 \leq P(w) \rightarrow w \leq m) \wedge \forall y(\forall z(1 \leq P(z) \rightarrow z \leq y) \rightarrow m \leq y))]$$
$$\longrightarrow \exists s(1 \leq P(s) \wedge \forall x(1 \leq P(x) \rightarrow x \leq s))$$

5. IMV

$$\exists l(a \leq l \leq b \wedge \forall w(a \leq w \leq b \wedge f(w) < 0 \rightarrow w \leq l)$$
$$\wedge \forall y(\forall z(a \leq z \leq b \wedge f(z) < 0 \rightarrow z \leq y) \rightarrow l \leq y))$$
$$\wedge \forall u(a \leq u \leq b$$
$$\rightarrow \forall \varepsilon(0 < \varepsilon \rightarrow \exists \delta(0 < \delta \wedge \forall v(a \leq v \leq b \wedge |u - v| < \delta \rightarrow |f(u) - f(v)| \leq \varepsilon))))$$
$$\wedge a < b \wedge f(a) < 0 < f(b)$$
$$\longrightarrow \exists x(a \leq x \leq b \wedge f(x) = 0)$$

References

[1] Astrachan, O. METEOR: *Model Elimination Theorem Proving for Efficient OR-Parallelism*, Masters Thesis, Computer Science, Duke University, 1989.

[2] Bledsoe, W. W., "The Sup-Inf Method in Presburger Arithmetic". The University of Texas at Austin, Math Department Memo ATP-18. December 1974. Essentially the same as: A new method for proving certain Presburger formulas. Fourth IJCAI, Tbilisi, USSR, September 3-8, 1975.

[3] Bledsoe, W. W., "A Resolution-based Prover for General Inequalities", Technical Report ATP-52, University of Texas, Austin, July 1979.

[4] Bledsoe, W. W., "Using Hyper-Resolution on Problem One of LIM+", Univ of Texas CS Dept Memo, ATP 91, (May 1989).

[5] Bledsoe, W. W., P. Bruell and R. Shostak, "A Prover for General Inequalities", Technical Report ATP-40A, University of Texas, Austin, February 1979. Also, in *IJCAI-79*, Tokyo, Japan, August 1979.

[6] Bledsoe, W. W., and L. M. Hines, "Variable Elimination and Chaining in a Resolution-Base Prover for Inequalities", *Proc. 5th Conference on Automated Deduction*, Les Arcs, France, Springer-Verlag, (July 1980) 70-87.

[7] Bledsoe, W. W., K. Kunen and R. Shostak, "Completeness Results for Inequality Provers", *AI Journal 27*, (1985) 255-288.

[8] Davis, M. and H. Putman, "A Computing Procedure for Quantification Theory", *J. Assoc. Comput. Mach. 7*, (1960) 201-215.

[9] Hines, L. M. and W. W. Bledsoe, "The Str+ve Prover", Technical Report ATP-94, University of Texas at Austin, 1990.

[10] Hines, L. M., "Building In Axioms and Lemmas", Ph.D. Dissertation, University of Texas at Austin, (May 1988).

[11] Hines, L. M., "Hyper-Chaining and Knowledge-Based Theorem Proving", *Proc. 9th International Conference on Automated Deduction*, Argonne, Illinois, Springer-Verlag, (May 1988) 469-486.

[12] Hines, L. M., "Completeness of a Prover for Dense Linear Orders", *Journal of Automated Reasoning*, Vol. 8, 45-75, 1992, Netherlands, Kluwer Academic Pulishers.

[13] Hines, L., "STR+VE⊆: The STR+VE-based Subset Prover", *10th International Conference on Automated Deduction*, July 1990, Kaiserslautern, Germany, Springer-Verlag, 193-206.

[14] Hodes, L., "Solving Problems by Formula Manipulation in Logic and Linear Inequality", *IJCAI-71*, London, 1971, pp 553-559.

[15] King, J. C., *A Program Verifier*, Phd Dissertation, Carnegie-Mellon University, 1969.

[16] McCune, W. W., "Otter 2.0 Users' Guide", Technical Report ANL-90/9, Argonne National Laboratory, Argonne, Illinois.

[17] Nelson, G., and D. Oppen, "A Simplifier Based on Efficient Decision Algorithms", *Proc. Fifth ACM Symp. on Principles of Programming Languages*, 1978.

[18] Nie, X. and D. Plaisted, "A Complete Semantic Back Chaining Proof System", *Proc. 10th Conference on Automated Deduction*, Kaiserslautern, FRG, Springer-Verlag, (July 1990) 16-27.

[19] Rabinov, A., "A Restriction of Factoring in Binary Resolution." *Proc. 9th International Conference of Automated Deduction*, Argonne, Illinois, USA, Springer-Verlag, (May 1988) 582-591.

[20] Richter, M., "Some Reordering Properties for Inequality Proof Trees", *SLNS 171*, (1987) 183-197.

[21] Robinson, J. A., "A Machine-oriented Logic Based on the Resolution Principle", *JACM 12*, 1965, pp 23-41.

[22] Shostak, R., "A Practical Decision Procedure for Arithmetic with Function Symbols", *JACM*, April 1979.

[23] Slagle, J. R., and L. Norton, "Experiments with an Automatic Theorem Prover Having Partial Ordering Rules", *CACM 16*, 1973, pp 683-688.

[24] Stickel, M. A., "A Complete Unification for Associative-Commutative Functions", *IJCAI-75*, Tbilisi, USSR, 1975, pp 71-76.

[25] Wos, L., R. Veroff, B. Smith, and W. McCune. "The Linked Inference Principle, II: The User's Viewpoint." *CADE-7*, Napa, California, (May 1984) 316-332.

Unification in the Union of Disjoint Equational Theories: Combining Decision Procedures

Franz Baader
DFKI
Stuhlsatzenhausweg 3
6600 Saarbrücken 11, Germany
e-mail: baader@dfki.uni-sb.de

Klaus U. Schulz
CIS
University Munich, Leopoldstr. 139
8000 München 40, Germany
e-mail: schulz@cis.uni-muenchen.dbp.de

Abstract

Most of the work on the combination of unification algorithms for the union of disjoint equational theories has been restricted to algorithms which compute finite complete sets of unifiers. Thus the developed combination methods usually cannot be used to combine decision procedures, i.e., algorithms which just decide solvability of unification problems without computing unifiers. In this paper we describe a combination algorithm for decision procedures which works for arbitrary equational theories, provided that solvability of so-called unification problems with constant restrictions—a slight generalization of unification problems with constants—is decidable for these theories. As a consequence of this new method, we can for example show that general A-unifiability, i.e., solvability of A-unification problems with free function symbols, is decidable. Here A stands for the equational theory of one associative function symbol.

Our method can also be used to combine algorithms which compute finite complete sets of unifiers. Manfred Schmidt-Schauß' combination result, the until now most general result in this direction, can be obtained as a consequence of this fact. We also get the new result that unification in the union of disjoint equational theories is finitary, if general unification—i.e., unification of terms with additional free function symbols—is finitary in the single theories.

1 Introduction

E-unification is concerned with solving term equations modulo an equational theory E. The theory is called "unitary" ("finitary") if the solutions of a system of equations can always be represented by one (finitely many) solution(s). Otherwise the theory is of type "infinitary" or "zero" (see e.g., [Si89,JK91] for an introduction to unification theory). Equational theories which are of unification type unitary or finitary play an important rôle in automated theorem provers with "built in" theories (see e.g., [Pl72,St85]), in generalizations of the Knuth-Bendix algorithm (see e.g., [JK86,Ba87]), and in logic programming with equality (see e.g., [JL84]). The reason is that these applications usually require algorithms which compute finite complete sets of unifiers, i.e., finite sets of unifiers from which all unifiers can be generated by instantiation. However, with the recent development of constraint approaches to theorem proving (see e.g., [Bü90]), term rewriting (see e.g., [KK89]), and logic programming (see e.g., [JL87]), the computation of finite complete sets of unifiers is no longer indespensable for these applications. It is enough to decide satisfiability of the constraints, that means e.g., solvability of the unification problems. In the present paper, the design of decision procedures for unification problems will be a major issue.

When considering unification in equational theories one has to be careful with regard to the signature over which the terms of the unification problems can be built. This leads to the distinction between *elementary unification* (where the terms to be unified are built over the signature of the equational theory, i.e., the function symbols occurring in the axioms of the theory), *unification with constants* (where additional free constant symbols may occur), and *general unification* (where additional free function symbols of arbitrary arity may occur).

The following facts show that there really is a difference between the three types of E-unification:

- There exists an equational theory for which elementary unification is decidable, but unification with constants is undecidable (see [Bü86]).
- From the development of the first algorithm for associative-commutative unification (AC-unification) with constants [St75,LS75] it took almost a decade until the termination of an algorithm for general AC-unification was shown by Fages [Fa84].

The second fact shows that the question of how algorithms for elementary unification (or for unification with constants) can be used to get algorithms for general unification is nontrivial. This question is also of practical importance because the applications of theory unification mentioned above usually require algorithms for general unification. For example, in theorem proving with built-in theories free constants and function symbols are often generated during Skolemization.

Even more generally, one often would like to derive algorithms for *unification in the union of disjoint equational theories*, i.e., in the union of several equational theories over disjoint signatures, from unification algorithms in the single theories. That this so-called "combination problem" is important for applications can be illustrated by an example from the area of term rewriting. If we want to compute a canonical term rewriting system for the theory of Boolean rings, we have to use rewriting modulo associativity and commutativity of addition "+" and multiplication "$*$" in Boolean rings. Thus critical pairs are computed with the help of general unification modulo the union of the two disjoint equational theories AC_+ and AC_*.

When considering the combination problem, until now attention was mostly restricted to finitary unifying theories, and by unification algorithm one meant a procedure which computes a finite complete set of unifiers. The problem was first considered in [St75,St81,Fa84,HS87] for the case where several AC-symbols and free symbols may occur in the terms to be unified. More general combination problems were, for example, treated in [Ki85,Ti86,He86,Ye87,BJ89], but the theories considered in these papers always had to satisfy certain restrictions (such as collapse-freeness or regularity) on the syntactic form of their defining identities.

The problem was finally solved in its until now most general form by Schmidt-Schauß [Sc89]. His combination algorithm imposes no restriction on the syntactic form of the identities. The only requirements for a combination of disjoint theories E, F are:

- Unification with constants must be finitary for E and F.
- All constant elimination problems must be finitary solvable in E and F.

A more efficient version of this combination algorithm has been described by Boudet [Bo90].

The method of Schmidt-Schauß can also handle theories which are not finitary. In this case, procedures which enumerate complete sets of unifiers for the single theories can be combined to a procedure enumerating a complete set of unifiers for their union. However, even if unification in the single theories is decidable, this does not show how to get a decision algorithm for unifiability in the combined theory.

The infinitary theory $A = \{f(f(x,y),z) = f(x,f(y,z))\}$ is an example for this case. In 1972, Plotkin [Pl72] has described a procedure which enumerates minimal complete sets of A-unifiers for general A-unification problems, and in 1977 Makanin [Ma77] has shown that A-unification with constants is decidable. But in 1991, decidability of general A-unification was still mentioned as an open problem by Kapur and Narendran [KN91] in their table of known decidability and complexity results for unification. Such a decision procedure could, for example, be useful when building associativity into a theorem prover via constraint resolution; and it could be used to make Plotkin's enumeration procedure terminating for equations having finite complete sets of A-unifiers.

Schmidt-Schauß [Sc89] also treats the problem of how to combine decision procedures. But in this case he needs decision procedures for general unification in the single theories as prerequisites for his combination algorithm. Thus his result cannot be used to solve the open problem of decidability of general A-unification.

The research which will be presented in this paper is based on the ideas of Schmidt-Schauß and Boudet. It was motivated by the question of how to get a decision procedure for general A-unification. However, the results we have obtained are more general. We shall present a method which allows one to decide unifiability in the union of arbitrary disjoint equational theories, provided that solvability of so-called unification problems with constant restrictions—a slight generalization of unification problems with constants—is decidable for the single theories. In addition, our method can also be used to combine algorithms which compute finite complete sets of unifiers.

These main results and some of the interesting consequences will be described in the next section. In Section 3 we shall present the combination algorithm for the decision problem, and describe how it can also be used to generate complete sets of unifiers. Section 4 proves the correctness of the method for the decision problem. The proof for the case of computing complete sets of unifiers can be found in [BS91]. In the fifth section we shall describe conditions under which algorithms for solving unification problems with constant restrictions exist. Some of the consequences mentioned in Section 2 depend on these results.

2 Main results and consequences

As mentioned in the introduction, we have to consider a slight generalization of E-unification problems with constants, so-called E-unification problems with constant restriction, which will be introduced below. Having an algorithm which solves these kind of problems is the only prerequisite necessary for our combination method.

An *E-unification problem with constant restriction* is an ordinary E-unification problem with constants, $\Gamma = \{s_1 \doteq t_1, \ldots, s_n \doteq t_n\}$, where each free constant c occurring in the problem Γ is equipped with a set V_c of variables, namely, the variables in whose image c must not occur. A solution of the problem is an E-unifier σ of Γ such that for all c, x with $x \in V_c$, the constant c does not occur in $x\sigma$. Complete sets of solutions of unification problems with constant restriction are defined as in the case of ordinary unification problems.

It turns out that our combination method does not really need an algorithm which can handle E-unification problems with arbitrary constant restrictions; it is enough to deal with problems with a so-called *linear constant restriction.* Such a restriction is induced by a linear ordering on the variables and free constants as follows: Let X be the set of all variables and C be the set of all free constants occurring in Γ. For a given linear ordering $<$ on $X \cup C$, the sets V_c are defined as $\{x \mid x \text{ is a variable with } x < c\}$.

We are now ready to formulate our first main result, which is concerned with combining decision algorithms. The combination algorithm which is used to establish this result will be described in the next section.

Theorem 2.1 *Let $E_1, \ldots, E_n$ be equational theories over disjoint signatures such that solvability of E_i-unification problems with linear constant restriction is decidable for $i = 1, \ldots, n$. Then unifiability is decidable for the combined theory $E_1 \cup \ldots \cup E_n$.*

By "unifiability" we mean here solvability of elementary unification problems. However, it is easy to see that the result can be lifted to general unification, and to solvability of unification problems with linear constant restriction. The theorem also has several other interesting consequences, which are listed below.

1. *Let E be an equational theory such that solvability of E-unification problems with linear constant restriction is decidable. Then solvability of general E-unification problems is decidable.*
   In fact, for a given set Ω of function symbols we can always build the free theory $F_\Omega := \{f(x_1, \ldots, x_n) = f(x_1, \ldots, x_n) \mid f \text{ is an } n\text{-ary symbol in } \Omega\}$. It is easy to see that F_Ω satisfies the assumption of the theorem; and obviously, any general unification problem modulo E can be seen as an elementary unification modulo $E \cup F_\Omega$ (if Ω contains all the additional free function symbols occurring in the problem).

2. *General A-unifiability is decidable.*
   For A, decidability of unification problems with constant restriction is an easy consequence of a result by Schulz [Sh91] on a generalization of Makanin's procedure. This result shows that it is still decidable whether a given A-unification problem with constants has a solution for which the words substituted for the

variables in the problem are elements of given regular languages over the constants. It is easy to see that problems with constant restriction are a special case of these more generally restricted problems.

3. *General AI-unifiability, where* $AI := A \cup \{f(x,x) = x\}$, *is decidable.*
This was also stated as an open problem in [KN91]. For AI, decidability of unification problems with constant restriction easily follows from the well-known fact (see e.g., [Ho76]) that finitely generated idempotent semigroups are finite.

4. *If solvability of the* E_i*-unification problems with linear constant restriction can be decided by an NP-algorithm, then unifiability in the combined theory is also NP-decidable.*
This fact will become obvious once we have described our combination algorithm. As a consequence one gets simple proofs of Kapur and Narendran's results [KN91] that solvability of general AC- and ACI-unification problems can be decided by NP-algorithms. For these theories, NP-decidability of unification problems with constant restriction can be shown very similarly as in the case of ordinary unification problems with constants.

5. *Let* $E_1, \ldots, E_n$ *be equational theories over disjoint signatures such that solvability of general* E_i*-unification problems is decidable for* $i = 1, \ldots, n$. *Then unifiability is decidable for the combined theory* $E_1 \cup \ldots \cup E_n$. This result, which was first proved by Schmidt-Schauß (see [Sc89], Theorem 10.6), can also be obtained as a corollary to our theorem. In fact, we can show that solvability of E-unification problems with linear constant restriction can be reduced to solvability of general E-unification problems (see Section 5).

The algorithm which will be developed for proving Theorem 2.1 can also be used to compute complete sets of unifiers.

Theorem 2.2 *Let* $E_1, \ldots, E_n$ *be equational theories over disjoint signatures such that all* E_i*-unification problems with linear constant restriction have finite complete set of solutions* ($i = 1, \ldots, n$). *Then the combined theory* $E_1 \cup \ldots \cup E_n$ *is finitary.*

Again, we are talking about elementary unification for the combined theory; but as for the case of the decision problem, the result can easily be lifted to general unification, and to unification problems with linear constant restriction. It should be noted that this result is effective in the sense that we really get an algorithm computing finite complete set of unifiers for the combined theory, provided that for the single theories there exist algorithms computing finite complete sets of solutions of unification problems with linear constant restriction. In the following, we mention two other interesting consequences of the theorem.

6. *Let* $E_1, \ldots, E_n$ *be equational theories over disjoint signatures which are finitary with respect to general unification. Then the combined theory* $E_1 \cup \ldots \cup E_n$ *is finitary.*
In fact, we can show how finite complete sets of unifiers for general E_i-unification problems can be used to construct finite complete sets of solutions for unification problems with linear constant restriction (see Section 5).

7. Algorithms which compute finite complete sets of unifiers for unification with constants, and finite complete sets of constant eliminators can be used to get an algorithm which computes finite complete sets of solutions for unification problems with constant restriction (see Section 5). As a consequence, the *combination result of Schmidt-Schauß* ([Sc89], Corollary 7.14) mentioned in the introduction can also be *obtained as a corollary to Theorem 2.2.*

3 The combination algorithm

For the sake of convenience we shall restrict the presentation to the combination of two theories. The combination of more than two theories can be treated analogously. Before we can start with the description of the algorithm we have to introduce some notation.

Let E_1, E_2 be two equational theories built over the disjoint signatures Ω_1, Ω_2, and let $E = E_1 \cup E_2$ denote their union. Since we are only interested in elementary E-unification, we can restrict our attention to terms built from variables and symbols of $\Omega_1 \cup \Omega_2$. The elements of Ω_1 will be called 1-*symbols* and the elements of Ω_2 2-*symbols*. A term t is called i-*term* iff it is of the form $t = f(t_1, ..., t_n)$ for an i-symbol f $(i = 1, 2)$. A subterm s of a 1-term t is called *alien subterm* of t iff it is a 2-term such that every proper superterm of s in t is a 1-term. Alien subterms of 2-terms are defined analogously. An i-term s is *pure* iff it contains only i-symbols and variables. An equation $s \doteq t$ is pure iff there exists an $i, 1 \leq i \leq 2$, such that s and t are pure i-terms or variables; this equation is then called an i-equation. Please note that according to this definition equations of the form $x \doteq y$ where x and y are variables are both 1- and 2-equations. In the following, the symbols x, y, z, with or without indices, will always stand for variables.

Example 3.1 Let Ω_1 consist of the binary (infix) symbol "$\circ$" and Ω_2 of the unary symbol "h", let $E_1 := \{x \circ (y \circ z) = (x \circ y) \circ z\}$ be the theory which says that "$\circ$" is associative, and let $E_2 := \{h(x) = h(x)\}$ be the free theory for "h".

The term $y \circ h(z \circ h(x))$ is a 1-term which has $h(z \circ h(x))$ as its only alien subterm. The equation $h(x_1) \circ x_2 \doteq y$ is not pure, but it can be replaced by two pure equations as follows. We replace the alien subterm $h(x_1)$ of $h(x_1) \circ x_2$ by a new variable z. This yields the pure equation $z \circ x_2 \doteq y$. In addition, we consider the new equation $z \doteq h(x_1)$. This process of replacing alien subterms by new variables is called *variable abstraction*. It will be the first of the five steps of our combination algorithm.

The input for the *combination algorithm* is an elementary E-unification problem, i.e., a system $\Gamma_0 = \{s_1 \doteq t_1, \ldots, s_n \doteq t_n\}$, where the terms $s_1, \ldots, t_n$ are built from variables and the function symbols occurring in $\Omega_1 \cup \Omega_2$, the signature of $E = E_1 \cup E_2$. The first two steps of the algorithm are deterministic, i.e., they transform the given system of equations into one new system.

Step 1: variable abstraction. Alien subterms are successively replaced by new variables until all terms occurring in the system are pure. To be more precise, assume that $s \doteq t$ or $t \doteq s$ is an equation in the current system, and that s

contains the alien subterm s_1. Let x be a variable not occurring in the current system, and let s' be the term obtained from s by replacing s_1 by x. Then the original equation is replaced by the two equations $s' \doteq t$ and $x \doteq s_1$. This process has to be iterated until all terms occurring in the system are pure. It is easy to see that this can be achieved after finitely many iterations. Now all the terms in the system are pure, but there may still exist non-pure equations, consisting of a 1-term on one side and a 2-term on the other side.

Step 2: split non-pure equations. Each non-pure equations of the form $s \doteq t$ is replaced by two equations $x \doteq s, x \doteq t$ where the x are always new variables.

It is quite obvious that these two steps do not change solvability of the system. The result is a system which consists of pure equations. The third and the fourth step are nondeterministic, i.e., a given system is transformed into finitely many new systems. Here the idea is that the original system is solvable iff at least one of the new systems is solvable.

Step 3: variable identification. Consider all possible partitions of the set of all variables occurring in the system. Each of these partitions yields one of the new systems as follows. The variables in each class of the partition are "identified" with each other by choosing an element of the class as representative, and replacing in the system all occurrences of variables of the class by this representative.

Step 4: choose ordering and theory indices. This step doesn't change a given system, it just adds some information which will be important in the next step. For a given system, consider all possible strict linear orderings $<$ on the variables of the system, and all mappings *ind* from the set of variables into the set of theory indices $\{1, 2\}$. Each pair $(<, ind)$ yields one of the new systems obtained from the given one.

The last step is again deterministic. It splits each of the systems already obtained into a pair of pure systems.

Step 5: split systems. A given system Γ is split into two systems Γ_1 and Γ_2 such that Γ_1 contains only 1-equations and Γ_2 only 2-equations. These systems can now be considered as unification problems with linear constant restriction. In the system Γ_i, the variables with index i are still treated as variables, but the variables with alien index $j \neq i$ are treated as free constants. The linear constant restriction for Γ_i is induced by the linear ordering chosen in the previous step.

The output of the algorithm is thus a finite set of pairs (Γ_1, Γ_2) where the first component Γ_1 is an E_1-unification problem with linear constant restriction, and the second component Γ_2 is an E_2-unification problem with linear constant restriction.

Proposition 3.2 *The input system Γ_0 is solvable if and only if there exists a pair (Γ_1, Γ_2) in the output set such that Γ_1 and Γ_2 are solvable.*

A proof of this proposition is described in the next section. Obviously, if solvability of E_1- and E_2-unification problems with linear constant restrictions is decidable, the proposition implies decidability of elementary E-unifiability, which proves Theorem 2.1.

Example 3.3 We consider the theories E_1 and E_2 of Example 3.1, and the unification problem

$$\{h(x) \circ y = y \circ h(z_1 \circ z_2)\}.$$

Step 1: variable abstraction. This step results in the new system

$$\{x_1 \circ y = y \circ x_2, x_1 = h(x), x_2 = h(x_3), x_3 = z_1 \circ z_2\}.$$

Step 2: split non-pure equations. Since all equations are already pure, nothing is done in this step.

Step 3: variable identification. As an example, we consider the partition where x_1 and x_2 are in one class, and all the other variables are in singleton classes. Choosing x_1 as representative for its class, we obtain the new system

$$\{x_1 \circ y = y \circ x_1, x_1 = h(x), x_1 = h(x_3), x_3 = z_1 \circ z_2\}.$$

Step 4: choose ordering and theory indices. As an example, we take the linear ordering

$$z_1 < z_2 < x_3 < x < x_1 < y,$$

and the theory indices

$$ind(x_1) = ind(x) = ind(z_1) = ind(z_2) = 2 \text{ and } ind(x_3) = ind(y) = 1.$$

Step 5: split systems. On the one hand, we get the system

$$\Gamma_1 = \{x_1 \circ y = y \circ x_1, x_3 = z_1 \circ z_2\}$$

consisting of pure 1-equations. In this system the variables with index 1, i.e., x_3 and y, are still treated as variables, but the variables of index 2, i.e., x_1, z_1 and z_2, are treated as free constants. The linear constant restriction induced by the linear ordering is given by $V_{x_1} = \{x_3\}, V_{z_1} = V_{z_2} = \emptyset$.
On the other hand, we obtain the system

$$\Gamma_2 = \{x_1 = h(x), x_1 = h(x_3)\}$$

consisting of pure 2-equations. Here x and x_1 are treated as variables, and x_3 is treated as free constant. The constant restriction is given by $V_{x_3} = \emptyset$.

This pair (Γ_1, Γ_2) is one element in the set which is the output of the algorithm. It is easy to see that Γ_1 has the solution $\{x_3 \mapsto z_1 \circ z_2, y \mapsto x_1\}$, and Γ_2 has the solution $\{x_1 \mapsto h(x_3), x \mapsto x_3\}$. Consequently, the proposition implies that the original system has a solution.

The combination algorithm can also be used to compute complete sets of unifiers for elementary $(E_1 \cup E_2)$-unification problems, provided that one can compute finite complete sets of solutions for all E_i-unification problems with linear constant restriction $(i = 1, 2)$. The reason is that solutions of the problems Γ_1, Γ_2 in the output of the algorithm can be combined to solutions of the original input system. This *combined solution* is defined inductively over the linear ordering chosen in Step 4 of the algorithm.

Assume that σ_1 is a solution of Γ_1 and σ_2 is a solution of Γ_2. Without loss of generality we may assume that the substitution σ_i maps all variables of index i to terms containing only variables of index $j \neq i$ (which are treated as free constants in Γ_i) or new variables, i.e., variables not occurring in Γ_0, Γ_1, or Γ_2. This can simply be achieved by renaming variables if necessary. First, we define the combined solution σ on the variables occurring in the system obtained after Step 4 of the algorithm. Note that the input system Γ_0 may contain additional variables which have been replaced during the variable identification step.

Let x be the least variable with respect to the linear ordering chosen in Step 4, and let i be its index. Since the solution σ_i of Γ_i satisfies the constant restriction induced by the linear ordering, the term $x\sigma_i$ does not contain any variables of index $j \neq i$ (Recall that these variables are treated as free constants in Γ_i.) Thus we can simply define $x\sigma := x\sigma_i$.

Now let x be an arbitrary variable with index i, and let $y_1, \ldots, y_m$ be the variables with index $j \neq i$ occurring in $x\sigma_i$. Since σ_i satisfies the constant restriction induced by the linear ordering, the variables $y_1, \ldots, y_m$ (which are treated as free constants in Γ_i) have to be smaller than x. That means that $y_1\sigma, \ldots, y_m\sigma$ are already defined. The term $x\sigma$ is now obtained from $x\sigma_i$ by replacing the y_k by $y_k\sigma$ $(k = 1, \ldots, m)$. Because we have assumed that the other variables occurring in $x\sigma_i$ are new variables, we thus have $x\sigma = x\sigma_i\sigma$.

Finally, let x be a variable of the input system which has been replaced by the variable y during the variable identification step. Thus $y\sigma$ is already defined, and we can simply set $x\sigma := y\sigma$. For all variables z not occurring in the input system, or in Γ_1 or Γ_2, we define $z\sigma := z$.

For the above example, the solutions $\sigma_1 = \{x_3 \mapsto z_1 \circ z_2, y \mapsto x_1\}$ and $\sigma_2 = \{x_1 \mapsto h(x_3), x \mapsto x_3\}$ of Γ_1, Γ_2 are combined to $\{z_1 \mapsto z_1, z_2 \mapsto z_2, x_3 \mapsto z_1 \circ z_2, x \mapsto z_1 \circ z_2, x_1 \mapsto h(z_1 \circ z_2), x_2 \mapsto h(z_1 \circ z_2), y \mapsto h(z_1 \circ z_2)\}$.

This construction can now be used to generate complete sets of unifiers for elementary $(E_1 \cup E_2)$-unification problems. For a given input system Γ_0, let $\{(\Gamma_{1,1}, \Gamma_{1,2}), \ldots, (\Gamma_{n,1}, \Gamma_{n,2})\}$ be the output of the combination algorithm. For $i = 1, \ldots, n$ and $j = 1, 2$, let $M_{i,j}$ be a complete set of solutions of the E_i-unification problem with linear constant restriction, $\Gamma_{i,j}$.

Proposition 3.4 *The set of substitutions*

$$\bigcup_{i=1}^{n} \{\sigma \mid \sigma \text{ is the combined solution obtained from } \sigma_1 \in M_{i,1} \text{ and } \sigma_2 \in M_{i,2}\}$$

is a complete set of $(E_1 \cup E_2)$-unifiers of the input system Γ_0.

A proof of this proposition can be found in [BS91]. Obviously, if all the sets $M_{i,j}$ are finite, then the complete set given by the proposition is also finite, which proves Theorem 2.2.

4 Correctness of the combination algorithm

In this section we shall prove Proposition 3.2, which shows that our combination method is correct when applied for the decision problem. The proof for the computation of complete sets of unifiers follows the same lines but is a bit more arduous (see [BS91]). Before we can start with our task, we have to introduce a useful tool, which has first been utilized in connection with the combination problem in [BJ89], namely unfailing completion of the combined theory.

Let E_1, E_2 be equational theories over disjoint signatures Ω_1, Ω_2. We assume that both theories are consistent, that means, they have at least one model of cardinality greater than one, or equivalently, the identity $x =_{E_i} y$ does not hold in either theory. One can now apply unfailing completion (see e.g., [DJ87] for definitions and properties) to the combined theory $E = E_1 \cup E_2$. This yields a possibly infinite ordered-rewriting system R which is confluent and terminating on ground terms. In the following, we shall also apply this system to terms containing variables from a fixed countable set of variables X_0; but this is not a problem because these variables can simply be treated like constants. In particular, this means that the simplification ordering used during the completion must also take care of these additional "constants." It is easy to see that, because the signatures of E_1 and E_2 are disjoint, the system R is the union of two systems R_1 and R_2, where the terms in R_i are built over the signature Ω_i $(i = 1, 2)$. The R_i is just the system which would be obtained by applying unfailing completion to E_i.

Let $T(\Omega_1 \cup \Omega_2, X_0)$ be the set of terms built from function symbols in $\Omega_1 \cup \Omega_2$ and variables in X_0, and let $T_{\downarrow R}$ denote its R-irreducible elements. We consider an arbitrary bijection $\pi : T_{\downarrow R} \longrightarrow Y$ where Y is a set of variables which is disjoint to X_0. This bijection induces mappings π_1, π_2 of terms in $T(\Omega_1 \cup \Omega_2, X_0)$ to terms in $T(\Omega_1 \cup \Omega_2, Y)$ as follows. For variables $x \in X_0$, $x^{\pi_1} := \pi(x)$ (Note that variables are always R-irreducible.) If $t = f(t_1, \ldots, t_n)$ for a 1-symbol f, then $t^{\pi_1} := f(t_1^{\pi_1}, \ldots, t_n^{\pi_1})$. Finally, if t is a 2-term then $t^{\pi_1} := y$ where $y = \pi(s)$ for the unique R-irreducible element s of the $=_E$-class of t. The mapping π_2 is defined analogously. We write these mappings as superscripts to distinguish them from substitutions.

A substitution σ is called R-normalized on a finite set of variables Z iff $z\sigma \in T_{\downarrow R}$ for all variables $z \in Z$. The next lemma will be important in the proof of Proposition 3.2 given below. A proof of the lemma can be found in [BS91].

Lemma 4.1 *Let s, t be pure i-terms or variables, and let σ be a substitution which is R-normalized on the variables occurring in s, t. Then*

$$s\sigma =_E t\sigma \quad \textit{iff} \quad (s\sigma)^{\pi_i} =_{E_i} (t\sigma)^{\pi_i}.$$

Proof of Proposition 3.2

First, we shall show *soundness of the combination algorithm*, that means, we have to demonstrate that Γ_0 is solvable if there exists a pair (Γ_1, Γ_2) in the output set such that Γ_1 and Γ_2 are solvable.

Assume that σ_1 is a solution of Γ_1 and σ_2 is a solution of Γ_2. In the previous section we have already described how these two solutions of the single problems can be combined to a substitution σ, which we have called the combined solution. It remains to be shown that σ is in fact a solution of Γ_0. Obviously, it is sufficient to prove that σ is a solution of the system Γ' which was obtained by Step 4 of the algorithm, and which in Step 5 was split into Γ_1 and Γ_2. Let $s \doteq t$ be an equation in Γ', and assume without loss of generality that this equation was put into Γ_1 in Step 5. Thus we know that $s\sigma_1 =_{E_1} t\sigma_1$. Obviously, this implies $s\sigma_1\sigma =_{E_1} t\sigma_1\sigma$, and thus also $s\sigma_1\sigma =_E t\sigma_1\sigma$. As an easy consequence of the definition of σ, one gets that $\sigma = \sigma_1\sigma$, which finally yields $s\sigma =_E t\sigma$.

In the second part of the proof we have to show *completeness of the combination algorithm*, that means, we have to demonstrate that there exists a pair (Γ_1, Γ_2) in the output set such that Γ_1 and Γ_2 are solvable if Γ_0 is solvable.

Let σ be a solution of Γ_0. Without loss of generality we assume that σ is also a solution of the system obtained after the first two steps of the algorithm, that the set Y_0 of all variables occurring in this system is disjoint to X_0, and that σ is R-normalized on Y_0. The solution σ can be used to find the right alternatives in the nondeterministic steps of the combination algorithm:

- The partition of the set of all variables, which has to be chosen in the third step, is defined as follows. Two variables y and z are in the same class iff $y\sigma = z\sigma$. Obviously, this means that σ is also a solution of the system obtained after the variable identification step corresponding to this partition.

- In the fourth step, the variable y gets index i if $y\sigma$ is an i-term. If $y\sigma$ is itself a variable, y gets index 1 (This is arbitrary, we could have taken index 2 as well.)

- In the fourth step, we also have to choose an appropriate linear ordering on the variables occurring in the system. Consider the strict partial ordering defined by $y < z$ iff $y\sigma$ is a strict subterm of $z\sigma$. We take an arbitrary extension of this partial ordering to a linear ordering on the variables occurring in the system.

The choices we have just described determine a system Γ' in the set of systems obtained after Step 4 of the algorithm, and thus a particular pair of systems (Γ_1, Γ_2) in the output set of the combination algorithm. It remains to be shown that Γ_1, Γ_2 are solvable. In order to define solutions σ_i of these systems, we consider a bijection π from the R-irreducible elements of $T(\Omega_1 \cup \Omega_2, X_0)$ onto a set of variables Y.

This bijection has to satisfy two conditions. First, Y should contain all the variables occurring in Γ'. Since σ is assumed to be R-normalized on Y_0, we have that $y\sigma$ is R-irreducible for all variables y occurring in Γ'. The second condition on π is that $\pi(y\sigma) = y$ for all these variables y. For the satisfiability of these conditions, the variable identification step is important. The reason is that only because of this step we can be sure that Γ' does not contain two different variables y, y' with $y\sigma = y'\sigma$.

As described above, the bijection π induces mappings π_1, π_2. These mappings will now be used to construct the solutions $\sigma_i, i = 1, 2$. The substitution σ_i is defined on the variables y occurring in Γ' by $y\sigma_i := (y\sigma)^{\pi_i}$.

If y is a variable of index $j \neq i$, the term $y\sigma$ is either a variable in X_0 or a j-term. In both cases we get $y\sigma_i = (y\sigma)^{\pi_i} = \pi(y\sigma) = y$. This shows that σ_i really treats the variables of index j as constants.

Now assume that $s \doteq t$ is an equation in Γ_i. Since this equation is also contained in Γ', and since σ solves Γ', we know that $s\sigma =_E t\sigma$. Since σ was assumed to be R-normalized on Y_0, and since $s \doteq t$ is an i-equation, we can apply the lemma to get $(s\sigma)^{\pi_i} =_{E_i} (t\sigma)^{\pi_i}$. Using the definition of σ_i and the fact that $s \doteq t$ is an i-equation, it is easy to see that $(s\sigma)^{\pi_i} = s\sigma_i$ and $(t\sigma)^{\pi_i} = t\sigma_i$. Thus σ_i really solves the equation $s \doteq t$.

It remains to be shown that σ_i satisfies the constant restriction. Assume that x is a variable of index i, and that the variable y of index $j \neq i$ (which is treated as a constant in Γ_i) occurs in $x\sigma_i$. We have to show that x is not an element of V_y, i.e., that $x \not< y$. Recall that $x\sigma_i = (x\sigma)^{\pi_i}$, and that $x\sigma$ is R-irreducible. Thus, since $y \notin X_0$, the occurrence of y in $x\sigma_i$ must come from the occurrence of $y\sigma$ as a subterm of $x\sigma$. Because of the identification step, the fact that x and y are different variables also implies that $x\sigma$ and $y\sigma$ are different terms. Thus $y\sigma$ is a strict subterm of $x\sigma$, which yields $y < x$ because of the way the linear ordering was chosen. □

5 Solving unification problems with constant restriction

In the first part of this section we shall describe how algorithms for general unification can be used to solve unification problems with linear constant restrictions. In the second part, constant elimination algorithms together with algorithms for unification with constants are used to solve unification problems with arbitrary constant restriction. In the following, F is assumed to be an arbitrary consistent equational theory.

Using algorithms for general unification

Let Γ be an F-unification problem with a linear constant restriction, and let $<$ be the linear ordering which induces this restriction. Our goal is to construct a general F-unification problem Γ' such that Γ is solvable iff Γ' is solvable.

In this new system Γ', the free constants of Γ will be treated as variables, i.e., the solutions are allowed to substitute terms for these "constants." For any free constant c of Γ we introduce a new (free) function symbol f_c of arity $|V_c|$. Recall that $V_c = \{x \mid x \text{ is a variable with } x < c\}$ is the set of variables in whose σ-image c must not occur for a solution σ of the problem Γ. The general F-unification problem—in which the free constants of Γ are treated as variables—is now defined as

$$\Gamma' := \Gamma \cup \{c \doteq f_c(x_1, \ldots, x_n) \mid c \text{ is a free constant of } \Gamma \text{ and } V_c = \{x_1, \ldots, x_n\}\}.$$

Proposition 5.1 *The F-unification problem with linear constant restriction, Γ, is solvable iff the general F-unification problem Γ' is solvable.*

The proof of this proposition can be found in [BS91]. Please note that the proposition only holds for unification problems with *linear* constant restriction. The construction we have just described can also be used to get a complete set of solutions of Γ, provided that a complete set of F-unifiers of Γ' can be found.

Let R be the possibly infinite ordered-rewriting system which is obtained when applying unfailing completion to F. We assume that the simplification ordering used during the completion also takes the additional symbols f_c and variables (which are however treated as constants by the ordering) out of a countable set X_0 of new variables into account. This means that we can apply R to terms built out of symbols in the signature of F, the additional symbols f_c, and variables in X_0. Let $T_{\downarrow R}$ be the R-irreducible elements of the set of these terms.

We shall now show how an element σ' of a complete set of F-unifiers of Γ' can be used to define a solution σ of Γ. Without loss of generality we assume that σ' is R-normalized on the variables occurring in Γ'.

Let π be a bijection from $T_{\downarrow R}$ onto a set of variables Y. This bijection has to satisfy two conditions. First, Y should containing all the free constants occurring in Γ (which are treated as variables in Γ'). Since σ' is assumed to be R-normalized on the variables occurring in Γ', we have that $c\sigma'$ is R-irreducible for all free constants c occurring in Γ. The second condition on π is that $\pi(c\sigma') = c$ for all these constants c. This condition is satisfiable because for $c \neq c'$ we have $c\sigma' \neq c'\sigma'$. In fact, since σ' solves Γ', we know that $c\sigma' =_F f_c(x_1\sigma', \ldots, x_n\sigma')$ and $c'\sigma' =_F f_{c'}(x_1\sigma', \ldots, x_n\sigma')$. But this implies that $c\sigma'$ has f_c as root symbol, and $c'\sigma'$ the different symbol $f_{c'}$.

As described in the previous section, the bijection π induces a mapping π_1. To this purpose we treat the symbols of the signature of F as 1-symbols and the symbols f_c as 2-symbols. The mapping π_1 is now used to define our solution σ of Γ. For all variables x occurring in Γ we define $x\sigma := (x\sigma')^{\pi_1}$. The constants c of Γ are really treated as constants by σ, i.e., $c\sigma = c$. However, note that $c = (c\sigma')^{\pi_1}$ holds, anyway.

Proposition 5.2 *Let $C(\Gamma')$ be a complete set of F-unifiers of Γ'. The set $\{\sigma \mid \sigma' \in C(\Gamma')\}$, where σ is constructed out of σ' as described above, is a complete set of solutions of the F-unification problem with linear constant restriction, Γ.*

The proof can be found in [BS91]. Again, the proposition only holds for unification problems with *linear* constant restriction.

Using algorithms for constant elimination and for unification with constants

A *constant elimination problem* in the theory F is a finite set $\{(c_1, t_1), \ldots, (c_n, t_n)\}$ where the c_i's are free constants (i.e., constant symbols not occurring in the signature of F) and the t_i's are terms (built over the signature of F, variables, and free constants). A solution to such a problem is called a *constant eliminator*. It is a substitution σ such that for all $i, 1 \leq i \leq n$, there exists a term t_i' not containing the free constant c_i with $t_i' =_F t_i\sigma$. The notion *complete set of constant eliminators* is defined analogously to the notion complete set of unifiers.

Let Γ be an F-unification problem with *arbitrary* constant restriction. The goal is to construct a complete set of solutions of this problem. In the first step, we just ignore the constant restriction, and solve Γ as an ordinary F-unification problem with constants. Let $C(\Gamma)$ be a complete set of F-unifiers of this problem. In the second step, we define for all unifiers $\sigma \in C(\Gamma)$ a constant elimination problem Δ_σ as follows:

$$\Delta_\sigma := \{(c, x\sigma) \mid c \text{ is a free constant in } \Gamma \text{ and } x \in V_c\}.$$

For all $\sigma \in C(\Gamma)$, let C_σ be a complete set of solutions of the constant elimination problem Δ_σ.

Before we can describe the complete set of solutions of the F-unification problem with constant restriction, Γ, we have to define a slightly modified composition "$\otimes$" of substitutions. Let σ be an element of $C(\Gamma)$, and let τ be a constant eliminator in C_σ. Without loss of generality we assume that τ is the identity on variables not occurring in Δ_σ, and that the terms $y\tau$ for variables y occurring in Δ_σ contain only new variables.

For a given variable x, let $\{c_1, \ldots, c_n\}$ be the set of all constants c_i occurring in Γ such that $x \in V_{c_i}$. If this set is empty (i.e., $n = 0$,) we define $x(\sigma\otimes\tau) := x(\sigma\tau)$. Now assume that $n > 0$. Obviously, we have $\{(c_1, x\sigma), \ldots, (c_n, x\sigma)\} \subseteq \Delta_\sigma$. Since τ is a solution of Δ_σ, there exist terms $s_1, \ldots, s_n$ such that for all $i, 1 \leq i \leq n$, $x\sigma\tau =_F s_i$ and c_i does not occur in s_i. It is easy to see that this also implies the existence of a single term s such $c_1, \ldots, c_n$ do not occur in s and $x\sigma\tau =_F s$. We define $x(\sigma\otimes\tau) := s$.

Proposition 5.3 *The set*

$$\bigcup_{\sigma \in C(\Gamma)} \{\sigma\otimes\tau \mid \tau \in C_\sigma\}$$

is a complete set of solutions of the F-unification problem with constant restriction, Γ.

See [BS91] for a proof.

6 Conclusion

We have presented a new method for treating the problem of unification in the union of disjoint equational theories. Unlike most of the other methods developed for this purpose, it can be used to combine decision procedures as well as procedures computing finite complete sets of unifiers. Applicability of our method depends on a new type of prerequisite, namely on the solvability of unification problems with *linear* constant restrictions. Presupposing the existence of a constant elimination algorithm—as necessary for the method of Schmidt-Schauß—seems to be a stronger requirement. In fact, we have seen that constant elimination procedures can be used to solve unification problems with *arbitrary* constant restrictions. However, it is still an open problem whether there exists an equational theory for which solving unification problems with linear constant restrictions is finitary (or decidable) but solving unification problems with arbitrary constant restrictions is not.

Our main results together with the results described in the previous section show that there is a close correspondence between solving unification problems with linear constant restrictions and solving general unification problems. For a given equational theory, the first kind of problems is decidable (finitary solvable) if and only if the second kind of problems is. As an interesting open problem it remains to be shown whether there exists an equational theory for which unification with constants is decidable (finitary) but general unification—or equivalently, solving unification problems with linear constant restrictions—is not.

To make the presentation and the proof of correctness of the combination method more concise, we did not consider possible optimizations which would rule out certain partitions in Step 3 and certain linear orderings in Step 4 of the algorithm.

References

[BS91] F. Baader, K.U. Schulz, "Unification in the Union of Disjoint Equational Theories: Combining Decision Procedures," DFKI Research Report RR-91-33.

[Ba87] L. Bachmair, *Proof Methods for Equational Theories*, Ph.D. Thesis, Dept. of Comp. Sci., University of Illinois at Urbana-Champaign, 1987.

[Bo90] A. Boudet, "Unification in a Combination of Equational Theories: An Efficient Algorithm," *Proceedings of the 10th International Conference on Automated Deduction, LNCS* **449**, 1990.

[BJ89] A. Boudet, J.P. Jouannaud, M. Schmidt-Schauß, "Unification in Boolean Rings and Abelian Groups," *J. Symbolic Computation* **8**, 1989.

[Bü86] H.-J. Bürckert, "Some Relationships Between Unification, Restricted Unification, and Matching," *Proceedings of the 8th International Conference on Automated Deduction, LNCS* **230**, 1986.

[Bü90] H.-J. Bürckert, "A Resolution Principle for Clauses with Constraints," *Proceedings of the 10th International Conference on Automated Deduction, LNCS* **449**, 1990.

[DJ87] N. Dershowitz, J.P. Jouannaud, "Rewrite Systems," In J. van Leeuwen (editor), Volume B of *Handbook of Theoretical Computer Science*, North-Holland, 1990.

[Fa84] F. Fages, "Associative-Commutative Unification," *Proceedings of the 7th International Conference on Automated Deduction, LNCS* **170**, 1984.

[He86] A. Herold, "Combination of Unification Algorithms," *Proceedings of the 8th International Conference on Automated Deduction, LNCS* **230**, 1986.

[HS87] A. Herold, J.H. Siekmann, "Unification in Abelian Semigroups," *J. Automated Reasoning* **3**, 1987.

[Ho76] J.M. Howie, *An Introduction to Semigroup Theory*, London: Academic Press, 1976.

[JL84] J. Jaffar, J.L. Lassez, M. Maher, "A Theory of Complete Logic Programs with Equality," *J. Logic Programming* **1**, 1984.

[JL87] J. Jaffar, J.L. Lassez, "Constraint Logic Programming," *Proceedings of 14th POPL Conference*, Munich, 1987

[JK86] J.P. Jouannaud, H. Kirchner, "Completion of a Set of Rules Modulo a Set of Equations," *SIAM J. Computing* **15**, 1986.

[JK91] J.P. Jouannaud, C. Kirchner, "Solving Equations in Abstract Algebras: A Rule-Based Survey of Unification," In J.-L. Lassez, G. Plotkin (editors), *Alan Robinson's Anniversary Book*, 1991.

[KN91] D. Kapur, P. Narendran, "Complexity of Unification Problems with Associative-Commutative Operators," Preprint, 1991. To appear in *J. Automated Reasoning*.

[Ki85] C. Kirchner, *Méthodes et Outils de Conception Systématique d'Algorithmes d'Unification dans les Théories equationnelles*, Thèse d'Etat, Univ. Nancy, France, 1985.

[KK89] C. Kirchner, H. Kirchner, "Constrained Equational Reasoning," *Proceedings of SIGSAM 1989 International Symposium on Symbolic and Algebraic Computation*, ACM Press, 1989.

[LS75] M. Livesey, J.H. Siekmann, "Unification of AC-Terms (bags) and ACI-Terms (sets)," Internal Report, University of Essex, 1975, and Technical Report 3-76, Universität Karlsruhe, 1976.

[Ma77] G.S. Makanin, "The Problem of Solvability of Equations in a Free Semigroup," *Mat. USSR Sbornik* **32**, 1977.

[Pl72] G. Plotkin, "Building in Equational Theories," *Machine Intelligence* **7**, 1972.

[Sc89] M. Schmidt-Schauß, "Combination of Unification Algorithms," *J. Symbolic Computation* **8**, 1989.

[Sh91] K.U. Schulz, "Makanin's Algorithm – Two Improvements and a Generalization," CIS-Report 91-39, CIS, University of Munich, 1991.

[Si89] J.H. Siekmann, "Unification Theory: A Survey," in C. Kirchner (ed.), *Special Issue on Unification, Journal of Symbolic Computation* **7**, 1989.

[St75] M. Stickel, "A Complete Unification Algorithm for Associative-Commutative Functions," *Proceedings of the International Joint Conference on Artificial Intelligence*, 1975.

[St81] M.E. Stickel, "A Unification Algorithm for Associative-Commutative Functions," *J. ACM* **28**, 1981.

[St85] M.E. Stickel, "Automated Deduction by Theory Resolution," *J. Automated Reasoning* **1**, 1985.

[Ti86] E. Tiden, "Unification in Combinations of Collapse Free Theories with Disjoint Sets of Function Symbols," *Proceedings of the 8th International Conference on Automated Deduction, LNCS* **230**, 1986.

[Ye87] K. Yelick, "Unification in Combinations of Collapse Free Regular Theories," *J. Symbolic Computation* **3**, 1987.

Reduction and Unification in Lambda Calculi with Subtypes

Tobias Nipkow*
TU München

Zhenyu Qian†
Universität Bremen

Abstract

Reduction, equality and unification are studied for a family of simply typed λ-calculi with subtypes. The subtype relation is required to relate base types only to base types and to satisfy some order-theoretic conditions. Constants are required to have a least type, i.e. "no overloading". We define the usual β and a subtype-dependent η-reduction. These are related to a typed equality relation and shown to be confluent in a certain sense.

A generic algorithm for pre-unification modulo $\beta\eta$-conversion and an arbitrary subtype relation is presented. Furthermore it is shown that unification w.r.t. any subtype relation is universal.

1 Introduction

Subtypes have long been recognized as an important means of succinctly expressing inheritance relations. They are particularly valuable in the area of automated deduction because they can be built into the inference engine by means of order-sorted unification, thus reducing the search space significantly [21]. This has sparked a lot of research on order-sorted equational reasoning for first-order terms (see, for example, [18]).

More recently there has been a drive towards the use of typed λ-calculus for the representation of terms and formulae with bound variables [12, 14]. Again, this is not only conceptually simpler than first-order approaches to bound variables, it can also be automated by means of Huet's higher-order unification algorithm [5] or more recent variations on it [9, 13].

This paper is a first attempt to combine both approaches. A certain amount of work has already gone into combining λ-calculus and subtypes for functional programming languages (see, for example, [11]). Our perspective is different because we are not concerned with type inference in such systems but with equational deduction, especially unification. Consequently we deal with explicitly typed terms, i.e. the Church approach to λ-calculus.

Our combination results are directly applicable to theorem proving systems like TPS [1] and Isabelle [14], both of which are based on higher-order unification. A second application area is the representation of the abstract syntax of programming languages. Subtypes arise naturally because some syntactic classes are contained in others, for example every identifier is an expression. This principle is exploited in systems like Centaur [8] and PSG [19]. PSG is also based on order-sorted unification. The representation of variable bindings in programming languages using typed λ-calculus was first advocated by Huet and Lang [6] and more recently by Pfenning and Elliott [15]. However, there was no theory, let alone a system combining both principles.

*Author's address: Institut für Informatik, TU München, Postfach 20 24 20, W-8000 München 2, Germany. E-mail: `Tobias.Nipkow@Informatik.TU-Muenchen.De`. Research supported by ESPRIT BRA 3245, *Logical Frameworks*, and an SERC Advanced Fellowship.

†Author's address: Universität Bremen, FB Informatik, Postfach 330440, W-2800 Bremen 33, Germany. E-mail: `qian@informatik.uni-bremen.de`. Research partially supported by ESPRIT Basic Research WG *COMPASS* 3264.

1.1 Overview

The research in this paper is carried out in the framework of a general notion of subtype. A subtype relation $\leq$ is required to relate base types only to base types and to satisfy some order-theoretic conditions. The former restriction is a consequence of the intended application to simply-typed λ-terms, the latter a generalization of different particular choices found in the literature [10, 16]. The precise definitions are given in Section 2.

After introducing β and η-reduction on our typed λ-terms, we find that their combination creates a confluence problem (Section 3). This problem is overcome with the introduction of a typed equality of the form $s = t : \tau$ which is closely related to reduction (Section 4). Finally we present a generic pre-unification algorithm for an arbitrary subtype relation $\leq$. Furthermore we show that unification w.r.t. any subtype relation is universal in the sense that unification w.r.t. any other subtype relation can be reduced to it.

2 Terms, Types and Subtypes

Given a set $\mathcal{T}_0$ of *base types* (e.g., *int* and *nat*), the set of (*simple*) *types* is a set $\mathcal{T}$ inductively defined as the smallest set containing $\mathcal{T}_0$ such that $\sigma, \tau \in \mathcal{T}$ implies $\sigma \to \tau \in \mathcal{T}$. The type constructor $\to$ is assumed to associate to the right: $\tau_1 \to \tau_2 \to \tau_3$ abbreviates $\tau_1 \to (\tau_2 \to \tau_3)$. We use a vector notation $\overline{\tau_n} \to \tau$ to denote the type $\tau_1 \to \cdots \to \tau_n \to \tau$ such that $\tau \in \mathcal{T}_0$, unless otherwise stated.

For every type $\tau \in \mathcal{T}$ there exists a set $\mathcal{C}_\tau$ of *constants* such that $\mathcal{C}_{\tau_1} \cap \mathcal{C}_{\tau_2} = \{\}$ if $\tau_1 \neq \tau_2$. Hence we assume that there is no "overloading", i.e. every constant has a least type. Let $\mathcal{C} = \bigcup_{\tau \in \mathcal{T}} \mathcal{C}_\tau$. For every $\tau \in \mathcal{T}$, we assume a countably infinite set $\mathcal{V}_\tau$ of *variables* disjoint from $\mathcal{C}$ such that $\mathcal{V}_{\tau_1} \cap \mathcal{V}_{\tau_2} = \{\}$ if $\tau_1 \neq \tau_2$. Let $\mathcal{V} = \bigcup_{\tau \in \mathcal{T}} \mathcal{V}_\tau$. To indicate that $x \in \mathcal{V}_\tau$ we sometimes write x_τ.

The set $\mathcal{L}(\mathcal{C}, \mathcal{V})$ (short: $\mathcal{L}$) of *λ-terms* (short: *terms*) constructed from $\mathcal{C}$ and $\mathcal{V}$ are defined as usual, i.e. the least set $\mathcal{L}$ such that $\mathcal{C} \cup \mathcal{V} \subseteq \mathcal{L}$, $(u\ v) \in \mathcal{L}$ for $u, v \in \mathcal{L}$ (Application) and $\lambda x.u \in \mathcal{L}$ for $x \in \mathcal{V}, u \in \mathcal{L}$ (Abstraction). We use the following abbreviations: $\lambda \overline{x_n}.s$ stands for $\lambda x_1. \ldots \lambda x_n.s$; $a(\overline{u_n})$ stands for $a(u_1, \ldots, u_n)$ which in turn stands for $(\ldots(a\ u_1)\ldots)u_n$.

A *context* is a term in $\mathcal{L}(\mathcal{C} \cup \{\Box\}, \mathcal{V})$ with a single occurrence of $\Box$. Given a context u, $u[s]$ denotes the term obtained from u by replacing $\Box$ with s.

A variable x in a term is said to occur *bound* if it occurs in some subterm of the form $\lambda x.t$; otherwise it occurs *free*. We assume that no variable is bound more than once in a term and that no variable occurs both bound and free in a term. This assumption is without loss of generality since terms will only be considered modulo α-conversion defined below. The set of all free variables in a syntactic object O is denoted by $\mathcal{FV}(O)$.

Given a relation $\leq$ on types, a term t *has type* τ (*w.r.t.* $\leq$) if $t : \tau$ can be derived by the following well-known rules (see e.g. [10]):

$$\frac{a \in \mathcal{C}_\tau \cup \mathcal{V}_\tau}{a : \tau} \qquad \text{(intro)}$$

$$\frac{x \in \mathcal{V}_\sigma, \quad u : \tau}{\lambda x.u : \sigma \to \tau} \qquad \text{(abs)}$$

$$\frac{u : \sigma \to \tau, \quad v : \sigma}{(u\ v) : \tau} \qquad \text{(app)}$$

$$\frac{u : \sigma, \quad \sigma \leq \tau}{u : \tau} \qquad \text{(coerce)}$$

If we want to emphasize that $t : \tau$ holds w.r.t. some particular relation $\leq$ we write $\vdash_{\leq} t : \tau$.

A term t is *(well-)typed* if $t : \tau$ for some type τ; t is *ill-typed* if it is not well-typed. In this paper we concentrate only on the well-typed terms. So terms are assumed to be typed unless otherwise stated.

We are only interested in a certain subclass of relations $\leq$ which we call subtype relations. The formal definition is an abstraction of two particular subtype relations studied in the literature. The most popular choice is $\preceq$, used, for example, by Cardelli and Mitchell [3, 10], which obeys

$$\sigma \to \tau \preceq \sigma' \to \tau' \iff \sigma' \preceq \sigma \wedge \tau \preceq \tau'. \tag{1}$$

Another particular subtype relation is $\sqsubseteq$ [16] which satisfies

$$\sigma \to \tau \sqsubseteq \sigma' \to \tau' \iff \sigma = \sigma' \wedge \tau \sqsubseteq \tau'. \tag{2}$$

Going from $\preceq$ to $\sqsubseteq$, contravariance in the domain type is replaced by equality. This results in reduced typability: if $\tau \prec \tau'$, then $\vdash_{\preceq} t : \tau' \to \sigma$ implies $\vdash_{\preceq} t : \tau \to \sigma$, whereas $\tau \sqsubset \tau'$ and $\vdash_{\sqsubseteq} t : \tau' \to \sigma$ does not imply $\vdash_{\sqsubseteq} t : \tau \to \sigma$. Although this appears to be a weaker calculus, it will turn out that w.r.t. equality reasoning there is no loss of generality: instead of t one may use $\lambda x_\tau.(t\ x)$, which does have type $\tau \to \sigma$ w.r.t. $\sqsubseteq$. (See Section 5.)

Abstracting from the two special relations $\preceq$ and $\sqsubseteq$ we arrive at the following definition:

Definition 2.1 A *subtype relation* $\leq$ is a partial order on $\mathcal{T}$ which relates function types only to function types (and hence base types only to base types), and satisfies the following properties:

$$\sigma \to \tau \leq \sigma' \to \tau' \implies \sigma' \leq \sigma \wedge \tau \leq \tau' \tag{3}$$

$$\tau \leq \tau' \implies \sigma \to \tau \leq \sigma \to \tau' \tag{4}$$

$$\sigma \to \tau \leq \sigma' \to \tau' \implies \sigma \to \tau \leq \sigma' \to \tau \tag{5}$$

Condition (4) is called *covariance* and is naturally assumed for all known definitions of subtype. *Contravariance*, i.e. the implication

$$\sigma' \leq \sigma \implies \sigma \to \tau \leq \sigma' \to \tau$$

is more contentious and is, for example, not satisfied by $\sqsubseteq$ above. Condition (3) is fairly intuitive, whereas (5) is a technical requirement to make η-reduction work. All three conditions are implied by both (1) and (2).

Throughout this paper we try to be as independent of the special nature of the subtype relation as possible. Unless stated otherwise, $\leq$ should be assumed to be an arbitrary subtype relation. The symbols $\preceq$ and $\sqsubseteq$ denote subtype relations satisfying conditions (1) and (2) respectively. Note that fixing a partial order on the base types uniquely determines the corresponding subtype relations $\preceq$ and $\sqsubseteq$. The relations $\sqsubseteq$ and $\preceq$ are in fact the least and greatest ones among all the subtype relations that agree on the base types:

Lemma 2.2 *If $\leq$, $\sqsubseteq$ and $\preceq$ agree on the base types, then for all types σ and τ*

$$\sigma \sqsubseteq \tau \implies \sigma \leq \tau \quad \textit{and} \quad \sigma \leq \tau \implies \sigma \preceq \tau$$

Proof By induction on the structure of types using properties (4) and (3). □

As a direct consequence we obtain

Corollary 2.3 *If $\leq$, $\sqsubseteq$ and $\preceq$ agree on the base types then*

$$\vdash_{\sqsubseteq} s : \tau \implies \vdash_{\leq} s : \tau \quad \textit{and} \quad \vdash_{\leq} s : \tau \implies \vdash_{\preceq} s : \tau$$

A term may have more than one type because of coercion along the subtype relation. If a term t has a least type τ then we write $t :: \tau$. The above definition of subtype relation, together with the fact that constants cannot be overloaded, implies that every typed term has a least type. This can be shown by a normal form theorem for typing proofs: any derivation in the above system can be transformed such that (coerce) appears only at the root or directly above the second premise of (app). Alternatively, this can be seen as a relation between the above system of inference rules and a slightly modified one where (coerce) has been dropped and (app) has become

$$\frac{u : \sigma \to \tau, \quad v : \sigma', \quad \sigma' \leq \sigma}{(u\ v) : \tau} \qquad \text{(app')}$$

Lemma 2.4 *Let* $\Vdash$ *denote provability in the new typing system. Then* $\vdash_{\leq} s : \tau$ *iff there is a* $\tau' \leq \tau$ *such that* $\Vdash_{\leq} s : \tau'$.

Proof By a transformation of proofs in $\vdash$ into $\Vdash$ using (3) and (4). □

From this lemma it follows directly that each well-typed term has a least type, namely the type computed by $\Vdash$.

A *substitution* $\theta \in \mathcal{SUB}_{\leq}$ is a function $\mathcal{V} \to \mathcal{L}$ such that $\theta(x) \neq x$ holds only for finitely many $x \in \mathcal{V}$, and $\vdash_{\leq} \theta(x_\tau) : \tau$ holds for all $x \in \mathcal{V}$. The *domain* of a substitution θ is defined as $\mathcal{D}om(\theta) = \{x \in \mathcal{V} \mid \theta(x) \neq x\}$. If $\mathcal{D}om(\theta) = \{x_1, \cdots, x_n\}$, then θ can be written as

$$\{x_1 \mapsto \theta(x_1), \cdots, x_n \mapsto \theta(x_n)\},$$

or as $\{\overline{x_n \mapsto \theta(x_n)}\}$ in the vector notation. Substitutions are extended to terms as in the simply typed λ-calculus, where bound variables are renamed to avoid name captures.

It is easy to show that the types of a term are preserved under substitution:

Lemma 2.5 *If* $\theta \in \mathcal{SUB}_{\leq}$ *and* $\vdash_{\leq} u : \tau$ *then* $\vdash_{\leq} \theta(u) : \tau$.

3 Reductions

Any respectable λ-calculus must be equipped with reductions. In our case we can define α and β-reduction in the customary way:

- α*-reduction:* $u[\lambda x_\tau.s] \longrightarrow_\alpha u[\lambda y_\tau.\{x \mapsto y\}s]$ if $y \notin \mathcal{FV}(s)$
- β*-reduction:* $u[(\lambda x.s\ t)] \longrightarrow_\beta u[\{x \mapsto t\}s]$

where the substitution $\{x \mapsto v\}s$ is assumed to happen without name conflict.

The closure of $\longrightarrow_\gamma$, where $\gamma \in \{\alpha, \beta, \eta, \beta\eta\}$, under transitivity, reflexivity and symmetry is denoted by $=_\gamma$ and called γ-conversion. Note that α-conversion does not change the type of bound variables but only their name. Terms are usually considered modulo α-conversion.

It follows from Lemma 2.5 that β-reduction preserves types:

Lemma 3.1 *If* $u \longrightarrow_\beta v$*, then* $\vdash_{\leq} u : \tau$ *implies* $\vdash_{\leq} v : \tau$.

Not so for the usual notion of η-reduction: if $\vdash_{\leq} t : \tau \to \tau'$ and $\sigma < \tau$, then $\vdash_{\leq} \lambda x_\sigma.(t\ x) : \sigma \to \tau'$ but $\vdash_{\leq} t : \sigma \to \tau'$ only if $\leq$ is contravariant! For example w.r.t. $\sqsubseteq$ we find that $t : \sigma \to \tau'$ only if $\sigma = \tau$.

To avoid this loss of types we use the following modified version of

- η*-reduction:* $\vdash_{\leq}\ u[\lambda x_\tau.(t\ x)] \longrightarrow_\eta u[t]$ if $x \notin \mathcal{FV}(t)$ and if $\vdash_{\leq} t : \tau \to \sigma$ for some σ.

The decoration $\vdash_{\leq}$ has become necessary because η-reduction depends on types. Using property (5) of $\leq$ one can show that subject-reduction holds for $\longrightarrow_\eta$.

Lemma 3.2 *If $u \longrightarrow_\eta v$, then $\vdash_{\leq} u : \tau$ implies $\vdash_{\leq} v : \tau$.*

Thus the preservation of types under η-reduction has been built in, but at the expense of simplicity. Fortunately there is a simplified version of η-reduction for $\preceq$:

Lemma 3.3 *If $x \notin \mathcal{FV}(t)$ then $\vdash_{\preceq} u[\lambda x_\tau.(t\ x)] \longrightarrow_\eta u[t]$.*

Substitutivity of η-reduction is not completely trivial: because of the typing condition the rule is not automatically closed under substitution. But since Lemma 2.5 shows that types are preserved under substitution we do in fact have:

Lemma 3.4 *If $\vdash_{\leq} u \longrightarrow_\eta v$ then $\vdash_{\leq} \theta u \longrightarrow_\eta \theta v$ for any substitution θ.*

Let us now turn to termination and confluence.

Strong normalization for the simply typed λ-calculus has the following corollary:

Corollary 3.5 *There are no infinite $\beta\eta$-reductions starting from well-typed terms.*

As in the untyped and simply typed λ-calculus, β-reduction is Church-Rosser.

Theorem 3.6 *β-reduction is Church-Rosser for all terms.*

Of course this implies confluence on well-typed terms for any subtype relation.

Hence each typed term u has a unique (up to α-conversion) β-normal form, denoted by $u{\downarrow_\beta}$. Well-typed β-normal forms must be of the form $\lambda\overline{x_k}.a(\overline{t_n})$ where a is a variable or constant, called the *head*, and each t_i is a well-typed β-normal form. A well-typed β-normal form is called *flexible* if its head is a free variable, *rigid* if not.

Unfortunately, the combination of β- and η-reduction is not necessarily Church-Rosser. If $\sigma \preceq \tau$ we have the following "critical pair" w.r.t. $\preceq$:

$$\lambda y_\tau.y \underset{\eta}{\longleftarrow} \lambda x_\sigma.(\lambda y_\tau.y\ x) \underset{\beta}{\longrightarrow} \lambda x_\sigma.x \tag{6}$$

but $\lambda y_\tau.y \neq_\alpha \lambda x_\sigma.x$ if $\sigma \neq \tau$, and both terms are irreducible.

We return to this problem in the following section. This lack of confluence is also investigated by Curien and Ghelli [4] for the more complex system $F_{\leq}$. They arrive at a confluent system by employing typed reductions of the form $s \longrightarrow t : \tau$.

4 Typed Equality, Reduction and Confluence

So far we have only considered reduction, which already lead to a problem, namely the lack of confluence for $\beta\eta$-reduction. Extending reduction to equality, i.e. considering expansions as well, creates another dubious situation: β- and η-expansion can turn well-typed terms into ill-typed ones because the type of a subterm may increase: if $c :: \tau$, $f :: \tau \to \tau'$ and $\tau < \sigma$, then $(\lambda x_\sigma.x)c$ is a well-typed β-expansion of c, but $f((\lambda x_\sigma.x)c)$ is an ill-typed β-expansion of $f(c)$. This poses the question whether derivations which involve ill-typed terms are semantically sound (ignoring the fact that we haven't mentioned semantics at all). In order to circumvent this awkward point and to see that there is really no problem, we introduce an equality judgement of the form $s = t : \tau$, which means that both s and t are of type τ and are equal on that type.

$$\frac{t : \tau}{t = t : \tau} \qquad \frac{s = t : \tau}{t = s : \tau} \qquad \frac{s = t : \tau \quad t = u : \tau}{s = u : \tau}$$

$$\frac{s_1 = s_2 : \sigma \to \tau \quad t_1 = t_2 : \sigma}{(s_1\ t_1) = (s_2\ t_2) : \tau} \qquad \frac{x \in \mathcal{V}_\sigma \quad s = t : \tau}{\lambda x.s = \lambda x.t : \sigma \to \tau}$$

$$\frac{\lambda x.s : \tau \quad x, y \in \mathcal{V}_\sigma \quad y \notin \mathcal{FV}(s)}{\lambda x.s = \lambda y.\{x \mapsto y\}s : \tau} \qquad \frac{(\lambda x.s\ t) : \tau \quad x \in \mathcal{V}_\sigma \quad t : \sigma}{(\lambda x.s\ t) = \{x \mapsto t\}s : \tau}$$

$$\frac{\lambda x.(t\ x) : \tau \quad t : \tau \quad x \notin \mathcal{FV}(t)}{\lambda x.(t\ x) = t : \tau}$$

The first five rules define a typed equational logic and the last three axiomatize α, β and η-conversion, respectively. Note that these rules are fairly independent of the actual type system used. However we assume in the sequel that the typing judgement of Section 2 is used. This enables us to exploit the close connection with simply-typed terms. In order to indicate which subtype relation is used we write $\vdash_{\leq} s = t : \tau$.

It is a simple exercise to show that $s = t : \tau$ actually implies $s, t : \tau$. It may also be noted that w.r.t. $\preceq$, η-equality (the final inference rule above) does not need the premise $t : \tau$ (compare with Lemma 3.3). As a consequence of Lemma 2.5 it also follows that typed equality is preserved under substitutions:

Lemma 4.1 *If $\theta \in \mathcal{SUB}_{\leq}$ and $\vdash_{\leq} s = t : \tau$ then $\vdash_{\leq} \theta s = \theta t : \tau$*

Although the prospect of deriving equalities via these rules may look daunting, we can show that w.r.t. reduction, types can essentially be ignored. The following easy lemma explains why:

Lemma 4.2 *If $\vdash_{\leq} s : \tau$ and $\vdash_{\leq} s \longrightarrow^*_{\beta\eta} t$ then $\vdash_{\leq} s = t : \tau$.*

In order to formulate the results about equality we need the concept of *untyped* α-conversion:

$$\lambda x_\sigma.s =^\alpha \lambda y_\tau.\{x \mapsto y\}s$$

without the requirement that $\sigma = \tau$.

Theorem 4.3 *If $\vdash_{\leq} s_1, s_2 : \tau$, then the following conditions are equivalent:*

1. $\vdash_{\leq} s_1 = s_2 : \tau$
2. $\vdash_{\leq} s_1 =_{\beta\eta} s_2$
3. $\vdash_{\leq} s_1{\downarrow_{\beta\eta}} =^\alpha s_2{\downarrow_{\beta\eta}}$

Proof The implication $1 \Longrightarrow 2$ is trivial: typed equality is obviously subsumed by (untyped) reduction and expansion.

The implication $2 \Longrightarrow 3$ can be shown by mapping into a simply-typed λ-calculus where all base types are identified. In that calculus the s_i are also well-typed and hence $s_1 =_{\beta\eta} s_2$ implies $s_1{\downarrow_{\beta\eta}} =_\alpha s_2{\downarrow_{\beta\eta}}$. The latter equality implies $\vdash_{\leq} s_1{\downarrow_{\beta\eta}} =^\alpha s_2{\downarrow_{\beta\eta}}$ in the system with subtypes.

For the implication $3 \Longrightarrow 1$ let $t_i = s_i{\downarrow_{\beta\eta}}$. By induction on size of t_1 + size of t_2 we show that $t_1 = t_2 : \tau$. The claim follows from Lemma 4.2. We distinguish two cases.

If t_i is an application, then $t_i = f\ u^i_1\ \dots\ u^i_n$ for some constant $f :: \overline{\tau_n} \to \tau$ and terms $u^i_j : \tau_j$ such that $u^1_j =^\alpha u^2_j$. By induction hypothesis it follows that $u^1_j = u^2_j : \tau_j$. Using the congruence law it follows that $f\ u^1_1\ \dots\ u^1_n = f\ u^2_1\ \dots\ u^2_n : \tau$, i.e. $t_1 = t_2 : \tau$.

If $t_i = \lambda x_{\sigma_i}.u_i$, where $u_i :: \sigma'_i$, then $\tau = \sigma \to \sigma' \geq \sigma_i \to \sigma'_i$. Hence $\sigma \leq \sigma_i$ and $\sigma' \geq \sigma'_i$. Let $u'_i = \{x_{\sigma_i} \mapsto y_\sigma\}u_i$, where y is new. It follows that $u'_i : \sigma'$ and $u'_1 =^\alpha u'_2$. Thus the induction hypothesis implies $u'_1 = u'_2 : \sigma'$. It now follows directly from the proof rules for typed equality that $\lambda x_{\sigma_1}.u_1 = \lambda y_\sigma.(\lambda x_{\sigma_1}.u_1\ y) = \lambda y_\sigma.u'_1 = \lambda y_\sigma.u'_2 = \lambda y_\sigma.(\lambda x_{\sigma_2}.u_2\ y) = \lambda x_{\sigma_2}.u_2 : \tau$. □

The force of this theorem is twofold. It shows that typed equality and $=_{\beta\eta}$ coincide, provided the two terms are well-typed. This is to be expected, because $\beta\eta$-conversion between well-typed terms cannot introduce essential ill-typedness: intermediate derivations involving untyped terms can be eliminated. It is also a solution to the confluence problem raised in the previous section: type labels should be ignored when comparing normal forms. Note, however, that this result applies only to typed terms.

As a simple corollary to Theorem 4.3 we find that if two terms of a common type are equal, they are equal in all common types:

Corollary 4.4 *If $s,t:\tau$ and $s=t:\tau$ for some τ, then $s=t:\sigma$ for all σ such that $s,t:\sigma$.*

In some sense this means that type labels in equalities can be dispensed with.

Theorem 4.3 may create the somewhat misleading impression that confluence of $\beta\eta$-reduction can only be recovered by comparison modulo $=^\alpha$. At least for $\sqsubseteq$ this is not true. The reason is that the "critical pair" (6) is not a critical pair w.r.t. $\sqsubseteq$: the type condition rules out the η-reduction.

By the usual casuistic [2] we can show:

Theorem 4.5 *The reduction $\longrightarrow_\eta$ is confluent for all well-typed terms.*

The proof of this theorem is standard. The correctness of our version relies essentially on Lemma 3.2, subject-reduction for η.

Theorem 4.6 *With respect to $\sqsubseteq$, $\longrightarrow_\beta^*$ and $\longrightarrow_\eta^*$ commute for all well-typed terms: if $s:\tau$, $s\longrightarrow_\beta^* s_1$ and $s\longrightarrow_\eta^* s_2$ then there is a t such that $s_1\longrightarrow_\eta^* t$ and $s_2\longrightarrow_\beta^* t$.*

This theorem is not true for all subtype relations as the critical pair (6) shows.

Theorem 4.7 *With respect to $\sqsubseteq$, $\longrightarrow_{\beta\eta}$ is confluent for all well-typed terms.*

Proof Follows from the two preceding theorems (see [2]). □

With respect to $\sqsubseteq$ this implies that every typed term has a unique (up to $=_\alpha$, not $=^\alpha$!) $\beta\eta$-normal form and, by Theorem 4.3, that typed equality is equivalent to the existence of common $\beta\eta$-normal forms.

5 Unification

The purpose of this section is twofold: to present a unification algorithm for an arbitrary subtype relation, and to show that unification modulo any subtype relation can be reduced to unification modulo any other subtype relation. The latter result is particularly interesting as it shows that w.r.t. equality reasoning contravariance ($\preceq$) can be replaced by equality ($\sqsubseteq$) without loss of generality. In order to relate the different subtype relations ($\leq$, $\sqsubseteq$, $\preceq$, ...) we assume that they all agree on the base types. We begin technicalities by defining what unification is.

Given two terms s and t such that $\vdash_\leq s,t:\tau$, the set of their unifiers is defined as

$$\mathcal{U}_\leq(s =^? t:\tau) \;=\; \{\theta\in\mathcal{SUB}_\leq \mid \vdash_\leq \theta s=\theta t:\tau\}.$$

Using Corollary 4.4 it is easy to see that, if $\vdash_\leq s,t:\sigma$ and $\vdash_\leq s,t:\tau$, then $\mathcal{U}_\leq(s=^? t:\sigma) = \mathcal{U}_\leq(s=^? t:\tau)$. Hence it suffices to talk about $\mathcal{U}_\leq(s=^? t)$, provided s and t have a common type. The remaining standard notions of unification theory [7] carry over unchanged to our order-sorted λ-calculi. Given $\theta,\theta'\in\mathcal{SUB}_\leq$ and $V\subseteq\mathcal{V}$ we write $\vdash_\leq \theta=\theta'\ [V]$ if $\vdash_\leq \theta x=\theta' x:\tau$ for all $x_\tau\in V$, and $\vdash_\leq \theta\leq\theta'\ [V]$ if there is a $\delta\in\mathcal{SUB}_\leq$ such that $\vdash_\leq \theta'=\delta\circ\theta\ [V]$.

It should be pointed out that our notion of unification is slightly more restrictive than necessary: $\mathcal{U}_{\leq}(s =^? t : \tau)$ is only defined if s and t already have the common type τ. This is not a prerequisite for solvability (the problem $x_{int} = 5 : nat$ has the obvious unifier $\{x \mapsto 5\}$ although x_{int} is not of type nat) but simplifies technicalities considerably.

Although we have only formulated unification modulo $\beta\eta$-conversion, we have also looked at unification modulo β-conversion in isolation. For lack of space we only present $\beta\eta$-unification which has the more challenging underlying theory, is simpler to present and is more appropriate for most applications.

Both the formulation of the unification algorithm and the reduction to $\sqsubseteq$ are greatly simplified by the following combination of η-expansion with an adjustment of the types of bound variables. Given a β-normal form $s = \lambda\overline{x_k}.a(\overline{s_m})$ such that $a :: \overline{\tau_{m+n}} \to \tau$, $\vdash_{\leq} s : \sigma$ and $\sigma = \overline{\sigma_{k+n}} \to \sigma'$, where τ and σ' are base types, the *η-adjustment of s with respect to σ* is defined as

$$\eta[s,\sigma] = \lambda\overline{y_{k+n}}.((\lambda\overline{x_k}.a(\overline{\eta[s_m,\tau_m]}))(\eta[y_1,\sigma_1],\ldots,\eta[y_{k+n},\sigma_{k+n}]))\downarrow_\beta$$

where the $y_i :: \sigma_i$ are new variables. For notational convenience the η-adjustment of a variable x is usually abbreviated by x.

The intention of η-adjustment is twofold: to put the term into *η-expanded form* as in [20] and to transform a term typed w.r.t. $\leq$ into a $\beta\eta$-equivalent term of the same type, but now typable w.r.t. $\sqsubseteq$. These two key properties of η-adjustment are:

Lemma 5.1 *If* $\vdash_{\leq} t : \tau$, *then* $\vdash_{\leq} t = \eta[t,\tau] : \tau$ *and* $\vdash_{\sqsubseteq} \eta[t,\tau] : \tau$.

As a consequence, the types of the outermost bound variables of $\eta[t,\tau]$ must reflect the domain type of τ.

Now typed equality can also be determined by α-conversion (not the untyped $=^\alpha$!) of η-adjustments.

Lemma 5.2 *If* $\vdash_{\leq} u_i : \tau$ *then* $\vdash_{\leq} u_1 = u_2 : \tau \iff \eta[u_1\downarrow_\beta,\tau] =_\alpha \eta[u_2\downarrow_\beta,\tau]$.

Proof The proof of $\Leftarrow$ follows directly from Lemma 5.1, since $\vdash_{\leq} u_1 = u_1\downarrow_\beta = \eta[u_1\downarrow_\beta,\tau] = \eta[u_2\downarrow_\beta,\tau] = u_2\downarrow_\beta = u_2 : \tau$.

For the proof of $\Rightarrow$ let $s_i = u_i\downarrow_{\beta\eta}$. By Theorem 4.3.3, $\vdash_{\leq} s_1 =^\alpha s_2$. By the definition of η, we may prove that $\eta[s_1,\tau] =_\alpha \eta[s_2,\tau]$ by an easy induction on the structure of $s_{1/2}$. Since $\eta[u_i\downarrow_\beta,\tau] =_\alpha \eta[s_i,\tau]$, we have $\eta[u_1\downarrow_\beta,\tau] =_\alpha \eta[u_2\downarrow_\beta,\tau]$. □

5.1 A Pre-Unification Algorithm

This section presents a unification algorithm for an arbitrary subtype relation $\leq$. The presentation is inspired by the work of Snyder and Gallier [20]. Following Huet, we only investigate *pre-unification* where a solution is a substitution together with a set of solvable constraints, the so-called flex-flex pairs.

The first step in this algorithm is to η-adjust all unification problems: instead of some set S of typed unification problems we solve $\{\eta[s,\tau] =^? \eta[t,\tau] : \tau \mid (s =^? t : \tau) \in S\}$. The correctness and completeness of this step are a simple consequence of Lemmas 5.1 and 4.1. From now on we can assume that all unification pairs are of the form $\lambda\overline{x_k}.a(\overline{s_m}) =^? \lambda\overline{x_k}.b(\overline{t_n}) : \tau$. This property is kept invariant by the transformation rules below! In fact, τ has served its purpose and will be ignored during unification.

A *unification pair*, denoted by $u =^? v$, is an unordered pair of typed terms. It is called *rigid-rigid* if both u and v are rigid, *flex-rigid* if u is flexible and v rigid, and *flex-flex* if both u and v are flexible. A *system*, often denoted by S, is a finite multiset of unification pairs. A unification pair $F_\tau =^? t \in S$ is said to be *solved in S* if $F \notin \mathcal{FV}(t) \cup \mathcal{FV}(S - \{F =^? t\})$ and $t : \tau$; in this case F is called a *solved variable in S*. A system is said to be *presolved* if every pair in

S is either solved in S or a flex-flex pair where both components have the same least type. If S is presolved, define $\vec{S} = \{F \mapsto t \mid (F =^? t) \in S \text{ is solved in } S\}$. A *pre-unifier* of a system S is a substitution that is more general than some unifier of S. The terminology is justified because if S is presolved, $\vec{S}$ is a pre-unifier of S. Our notion of pre-unifier is more general than Snyder and Gallier's: their pre-unifiers have a simple extension to a unifier, whereas this extension may not be computable for our pre-unifiers. However, if T is presolved, then $\vec{T}$ is also a pre-unifier in the sense of Snyder and Gallier.

The main algorithm is expressed by rewrite rules on systems S. The set of solutions for a system S is given by those normal forms T which are presolved. Each presolved T represents a solution to the initial problem in the usual way [20], i.e. as a substitution $\vec{T}$ and a set of constraints.

The transformation rules below are essentially a reconstruction of Huet's pre-unification algorithm for simply typed terms modulo $\beta\eta$-conversion, with the additional proviso that any substitution must be well-typed. This proviso accounts for the type constraints in Imitation and Projection. In contrast to the simply typed case, flex-flex pairs can be of different type, and yet be unifiable. This requires an extra step to unify the types in flex-flex pairs. Similar adjustments to Huet's algorithm for β-conversion alone lead to a β-unification algorithm with subtypes. The details are not shown because of lack of space.

Trivial pairs are removed:

$$\{u \stackrel{?}{=} u\} \cup S \implies S \tag{Del}$$

Solutions are propagated:

$$\{F \stackrel{?}{=} u\} \cup S \implies \{F \stackrel{?}{=} u\} \cup (\{F \mapsto u\}S){\downarrow_\beta} \tag{Rep}$$

if $F \in \mathcal{FV}(S) - \mathcal{FV}(u)$.

Rigid-rigid pairs are decomposed:

$$\{\lambda\overline{x_k}.a(\overline{u_n})) \stackrel{?}{=} \lambda\overline{x_k}.a(\overline{v_n}))\} \cup S \implies \{\lambda\overline{x_k}.u_i \stackrel{?}{=} \lambda\overline{x_k}.v_i \mid i = 1\ldots n\} \cup S \tag{Dec}$$

if $a \in \{\overline{x_k}\} \cup \mathcal{C}$.

Flex-rigid pairs are processed according to the following generic rule:

$$\{\lambda\overline{x_k}.F(\overline{u_m}) \stackrel{?}{=} \lambda\overline{x_k}.a(\overline{v_n})\} \cup S \implies \{F \stackrel{?}{=} \eta[t, \overline{\sigma_m} \to \sigma], \lambda\overline{x_k}.F(\overline{u_m}) \stackrel{?}{=} \lambda\overline{x_k}.a(\overline{v_n})\} \cup S$$

where $F :: \overline{\sigma_m} \to \sigma$ and $a :: \overline{\tau_n} \to \tau$. The possible values for t are determined by imitation and projection.

Imitation:

$$t = \lambda\overline{y_m}.a(\overline{H_n(\overline{y_m})}) \tag{I}$$

if $a \in \mathcal{C}$, $y_i :: \sigma_i$ for $i = 1\ldots m$, $H_j :: \overline{\sigma_m} \to \tau_j$ for $j = 1\ldots n$, and $\tau \leq \sigma$.

Projection:

$$t = \lambda\overline{y_m}.y_p(\overline{H_l(\overline{y_m})}) \tag{P}$$

if $a \in \{\overline{x_k}\} \cup \mathcal{C}$, $1 \leq p \leq m$, $u_p :: \overline{\gamma_l} \to \gamma$, $\gamma \leq \sigma$, $y_i :: \sigma_i$ for $i = 1\ldots m$, and $H_j :: \overline{\sigma_m} \to \gamma_j$ for $j = 1\ldots l$.

Flex-flex pairs need to have the same least type:

$$\{\lambda\overline{x_k}.F(\overline{u_m}) \stackrel{?}{=} \lambda\overline{x_k}.G(\overline{v_n})\} \cup S \Longrightarrow \{F \stackrel{?}{=} F', G \stackrel{?}{=} G', \lambda\overline{x_k}.F(\overline{u_m}) \stackrel{?}{=} \lambda\overline{x_k}.G(\overline{v_n})\} \cup S \qquad \text{(FF)}$$

where $F :: \overline{\sigma_m} \to \sigma$, $G :: \overline{\tau_n} \to \tau$, $\sigma \neq \tau$, $\gamma \leq \sigma$, $\gamma \leq \tau$, $F' :: \overline{\sigma_m} \to \gamma$, $G' :: \overline{\tau_n} \to \gamma$.

Each application of (I), (P), and (FF) is implicitly followed by applications of (Rep) eliminating F and G.

Theorem 5.3 (Soundness) *If $S \Longrightarrow^* T$, then every unifier of T is also a unifier of S.*

Completeness is proved with the same machinery introduced by Snyder and Gallier. A reduction $\Longrightarrow$ on pairs of substitutions θ and systems S is defined by three clauses:

$$\langle\theta, S\rangle \Longrightarrow \langle\theta, S'\rangle$$

if $S \Longrightarrow S'$ via one of the rules (Del), (Dec), or (Rep).

$$\langle\theta, S\rangle \Longrightarrow \langle\theta \cup \delta, S'\rangle$$

if $S \Longrightarrow S'$ via one of the rules (I) or (P) where $\eta[(\theta F)\downarrow_\beta, \overline{\sigma_m} \to \sigma] = \lambda\overline{y_m}.a(\overline{t_n})$ and $\delta = \{H_i \mapsto \lambda\overline{y_m}.t_i \mid i = 1 \ldots n\}$.

$$\langle\theta, S\rangle \Longrightarrow \langle\theta \cup \delta, S'\rangle$$

if $S \Longrightarrow S'$ via rule (FF) and $\delta = \{F' \mapsto \theta F, G' \mapsto \theta G\}$.

Now it is routine to prove the following lemmas.

Lemma 5.4 *If $\theta \in \mathcal{U}_{\leq}(S)$ and $\langle\theta, S\rangle \Longrightarrow \langle\theta', S'\rangle$, then $S \Longrightarrow S'$, $\theta' \in \mathcal{U}_{\leq}(S')$, and $\vdash_{\leq} \theta = \theta'$ $[\mathcal{FV}(S)]$.*

Lemma 5.5 *If $\langle\theta, S\rangle$ is in normal form, then S is in presolved form.*

Lemma 5.6 *The reduction $\Longrightarrow$ on pairs $\langle\theta, S\rangle$ terminates.*

The completeness theorem itself is a straightforward consequence of these lemmas.

Theorem 5.7 (Completeness) *If $\theta \in \mathcal{U}_{\leq}(S)$, then there exists a presolved T such that $S \Longrightarrow^* T$, $\vdash_{\leq} \vec{T} \leq \theta$ $[\mathcal{FV}(S)]$, and θ is a pre-unifier of all flex-flex pairs in T.*

Note that the final condition (which is missing in [20]) is necessary to ensure that the remaining flex-flex constraints are consistent with the initial solution θ.

5.2 $\mathcal{U}_{\leq}$ is Universal

Having presented a pre-unification algorithm for an arbitrary subtype relation $\leq$, we will now show that $\leq$ can be fixed: unification w.r.t. any subtype relation can be reduced to unification w.r.t. the fixed $\leq$. Which relation is most suitable as a universal relation is not clear in general and may depend on the implementation. The key to connecting two arbitrary subtype relations is to connect each of them with $\sqsubseteq$ by η-adjustment. To simplify matters we only treat unification rather than pre-unification.

Let us define *$\eta_{\sqsubseteq}$-long forms* as β-normal forms $\lambda\overline{x_k}.a(\overline{s_m})$ such that $\vdash_{\sqsubseteq} \lambda\overline{x_k}.a(\overline{s_m}) : \overline{\sigma_k} \to \gamma$ for some base type γ, $a :: \overline{\tau_m} \to \tau$, $x_i :: \sigma_i$, and each s_i is an $\eta_{\sqsubseteq}$-long form with $\vdash_{\sqsubseteq} s_i : \tau_i$. It is easy to check that the terms obtained by η-adjustment are always $\eta_{\sqsubseteq}$-long forms. The uniqueness and the preservation of application of $\eta_{\sqsubseteq}$-long forms can be proved.

Lemma 5.8 *For a term u with $\vdash_{\leq} u : \sigma$, the $\eta_{\sqsubseteq}$-long form s satisfying $\vdash_{\leq} s = u : \sigma$ is unique (up to α-conversion).*

Proof It is easy to check by an induction on the structure of terms that if the above u is an $\eta_{\sqsubseteq}$-long form then it is α-convertible to $\eta[u{\downarrow_\beta}, \sigma]$. Hence, the lemma holds due to Lemmata 4.2 and 5.2. □

Lemma 5.9 *If s and t are $\eta_{\sqsubseteq}$-long forms, so is $(s\ t){\downarrow_\beta}$.*

Proof Follows by an induction on the structure of $\eta_{\sqsubseteq}$-long forms. □

Changing the viewpoint, we can show that application and η-adjustment commute:

Lemma 5.10 *If $\vdash_{\leq} s : \sigma \to \tau$ and $\vdash_{\leq} t : \sigma$ then $\eta[(s\ t){\downarrow_\beta}, \tau] =_\alpha (\eta[s{\downarrow_\beta}, \sigma \to \tau]\ \eta[t{\downarrow_\beta}, \sigma]){\downarrow_\beta}$.*

Proof Follows from Lemma 5.8 and Lemma 5.9. □

Given $\theta \in \mathcal{SUB}_{\leq}$, let us define $\eta[\theta] = \{x \mapsto \eta[(\theta x){\downarrow_\beta}, \tau] \mid x_\tau \in \mathcal{Dom}(\theta)\}$. Lemma 5.1 implies that $\eta[\theta] \in \mathcal{SUB}_{\sqsubseteq}$.

Lemma 5.11 *If $\vdash_{\leq} s : \sigma$ then $\eta[(\theta s){\downarrow_\beta}, \sigma] =_\alpha (\eta[\theta])(\eta[s{\downarrow_\beta}, \sigma]){\downarrow_\beta}$*

Proof Follows from Lemma 5.10. □

Lemma 5.12 *Let $\theta, \theta' \in \mathcal{SUB}_{\leq}$. Then $\vdash_{\sqsubseteq} \eta[\theta'] \circ \eta[\theta] = \eta[\theta' \circ \theta]\ [\mathcal{V}]$.*

Proof Follows from Lemma 5.11 □

Now unifications w.r.t. $\leq$ and $\sqsubseteq$ are related. Firstly, $\sqsubseteq$ is shown universal in the sense that unification w.r.t. an arbitrary subtype relation $\leq$ can be reduced to unification w.r.t. $\sqsubseteq$. Let $\mathcal{CSU}$ denote the complete set of unifiers.

Theorem 5.13 *If $\vdash_{\leq} s, t : \tau$, then $\mathcal{CSU}_{\sqsubseteq}(\eta[s,\tau] =^? \eta[t,\tau] : \tau)$ is a complete set of unifiers of $\vdash_{\leq} s =^? t : \tau$.*

Proof By Corollary 2.3, every element $\theta \in \mathcal{CSU}_{\sqsubseteq}(\eta[s,\tau] =^? \eta[t,\tau] : \tau)\}$ satisfies $\vdash_{\leq} \theta(x) : \sigma$ for every $x_\sigma \in \mathcal{Dom}(\theta)$. Thus, $\theta \in \mathcal{SUB}_{\leq}$. By Lemma 5.1, $\theta \in \mathcal{U}_{\leq}(s =^? t : \tau)$.

To prove completeness, let $\theta \in \mathcal{U}_{\leq}(s =^? t : \tau)$. Then we have

$$
\begin{array}{lll}
 & \vdash_{\leq} \theta s = \theta t : \tau & \\
\Longrightarrow & \eta[(\theta s){\downarrow_\beta}, \tau] =_\alpha \eta[(\theta t){\downarrow_\beta}, \tau] & \text{Lemma 5.2} \\
\Longrightarrow & ((\eta[\theta])(\eta[s{\downarrow_\beta}, \tau])){\downarrow_\beta} =_\alpha ((\eta[\theta])(\eta[t{\downarrow_\beta}, \tau])){\downarrow_\beta} & \text{Lemma 5.11}
\end{array}
$$

and thus $\eta[\theta] \in \mathcal{U}_{\sqsubseteq}(\eta[s,\tau] =^? \eta[t,\tau])$. By Lemma 5.1, $\vdash_{\leq} \theta = \eta[\theta]\ [\mathcal{V}]$. Hence, there is $\theta' \in \mathcal{CSU}_{\sqsubseteq}(\eta[s,\tau] =^? \eta[t,\tau] : \tau)$ with $\vdash_{\leq} \theta' \leq \theta\ [\mathcal{FV}(s,t)]$. □

To go in the inverse direction, it can also be shown that unification modulo $\sqsubseteq$ is reducible to unification modulo an arbitrary subtype relation $\leq$.

Theorem 5.14 *If $\vdash_{\sqsubseteq} s, t : \tau$, then $\{\eta[\theta] \mid \theta \in \mathcal{CSU}_{\leq}(s =^? t : \tau)\}$ is a complete set of unifiers of $\vdash_{\sqsubseteq} s =^? t : \tau$.*

Proof Let $\vdash_{\sqsubseteq} s, t : \tau$ and $\theta \in \mathcal{CSU}_{\leq}(s =^? t : \tau)$. Then $\vdash_{\leq} \theta(s) = \theta(t) : \tau$. By Lemma 5.1, $\vdash_{\leq} \eta[\theta](s) = \eta[\theta](t) : \tau$. By $\Rightarrow$ of Lemma 5.2, $\eta[\eta[\theta](s){\downarrow_\beta}, \tau] =_\alpha \eta[\eta[\theta](t){\downarrow_\beta}, \tau]$. Since, in addition, we have $\vdash_{\sqsubseteq} \eta[\theta](s), \eta[\theta](t) : \tau$, applying $\Leftarrow$ of Lemma 5.2 yields $\vdash_{\sqsubseteq} \eta[\theta](s) = \eta[\theta](t) : \tau$. Hence, $\eta[\theta] \in \mathcal{U}_{\sqsubseteq}(s =^? t : \tau)$.

If $\theta' \in \mathcal{U}_{\sqsubseteq}(s =^? t : \tau)$, we have $\vdash_{\sqsubseteq} \theta' s = \theta' t : \tau$ and $\vdash_{\sqsubseteq} \eta[\theta'] = \theta'\ [\mathcal{V}]$ (Lemma 5.1). By Corollary 2.3 it easily follows that $\vdash_{\leq} \theta' s = \theta' t : \tau$ and hence $\theta' \in \mathcal{U}_{\leq}(s =^? t : \tau)$. Thus there are $\theta \in \mathcal{CSU}_{\leq}(s =^? t : \tau)$ and $\delta \in \mathcal{SUB}_{\leq}$ such that $\vdash_{\leq} \delta \circ \theta = \theta'\ [\mathcal{FV}(s,t)]$. By Lemma 5.2 it follows that $\vdash_{\sqsubseteq} \eta[\delta \circ \theta] = \eta[\theta']\ [\mathcal{FV}(s,t)]$ and thus by Lemma 5.12 $\vdash_{\sqsubseteq} \eta[\delta] \circ \eta[\theta] = \eta[\theta'] = \theta'\ [\mathcal{FV}(s,t)]$, i.e. $\vdash_{\sqsubseteq} \eta[\theta] \leq \theta'\ [\mathcal{FV}(s,t)]$. Hence, the set is complete. □

Now we are ready to relate unification modulo any two arbitrary subtype relations.

Corollary 5.15 *Let $\leq_1$ and $\leq_2$ be two arbitrary subtype relations that agree on base types. If $\vdash_{\leq_1} s,t : \tau$, then $\{\eta[\theta] \mid \theta \in \mathcal{CSU}_{\leq_2}(\eta[s,\tau] =^? \eta[t,\tau] : \tau)\}$ is a complete set of unifiers of $\vdash_{\leq_1} s =^? t : \tau$.*

Proof Follows from Theorems 5.13 and 5.14. □

6 Conclusion

This paper presented a first step on the way to a full combination of order-sorted unification and λ-calculus. The main difficulty in combining subtypes with λ-calculus is the treatment of η-reduction. The main technical challenge that we avoided is overloading. Hence our theory compares with that investigated by Walther [22]. The logical next step is the extension to overloaded constants as presented, for example, by Smolka et al. [18]. Eventually this should lead to a term declaration logic for λ-terms along the lines of Schmidt-Schauß [17].

Acknowledgement

Franz Baader was a very patient reader of an earlier version of this paper which contained many technical inaccuracies. We sincerely thank him for his debugging efforts and apologize for not always having gone as far as he suggested.

References

[1] P. B. Andrews, S. Issar, D. Nesmith, and F. Pfenning. The TPS theorem proving system. In M. Stickel, editor, *Proc. 10th Int. Conf. Automated Deduction*, pages 641–642. LNCS 449, 1990.

[2] H. P. Barendregt. *The Lambda Calculus, its Syntax and Semantics.* North Holland, 2nd edition, 1984.

[3] L. Cardelli. A semantics of multiple inheritance. *Information and Computation*, 76:130–164, 1988.

[4] P.-L. Curien and G. Ghelli. Subtyping + extensionality: Confluence of $\beta\eta$top reduction in $F_{\leq}$. In T. Ito and A. R. Meyer, editors, *Proc. Conf. Theoretical Aspects of Computer Software*, pages 731–749. LNCS 526, 1991.

[5] G. Huet. A unification algorithm for typed λ-calculus. *Theoretical Computer Science*, 1:27–57, 1975.

[6] G. Huet and B. Lang. Proving and applying program transformations expressed with second-order patterns. *Acta Informatica*, 11:31–55, 1978.

[7] J.-P. Jouannaud and C. Kirchner. Solving equations in abstract algebras: A rule-based survey of unification. In J.-L. Lassez and G. Plotkin, editors, *Computational Logic: Essays in Honor of Alan Robinson*, pages 257–321. MIT Press, 1991.

[8] G. Kahn. Natural semantics. In *Proc. 4th Annual Symp. Theoretical Aspects of Computer Science*, pages 22–39. LNCS 247, 1987.

[9] D. Miller. A logic programming language with lambda-abstraction, function variables, and simple unification. In P. Schroeder-Heister, editor, *Extensions of Logic Programming*, pages 253–281. LNCS 475, 1991.

[10] J. C. Mitchell. Coercion and type inference. In *Proc. 11th ACM Symp. Principles of Programming Languages*, pages 175–185, 1984.

[11] J. C. Mitchell. Type inference with simple subtypes. *J. Functional Programming*, 1:245–285, 1991.

[12] G. Nadathur and D. Miller. An overview of λProlog. In R. A. Kowalski and K. A. Bowen, editors, *Proc. 5th Int. Logic Programming Conference*, pages 810–827. MIT Press, 1988.

[13] T. Nipkow. Higher-order unification, polymorphism, and subsorts. In *Proc. 2nd Int. Workshop Conditional and Typed Rewriting Systems*. LNCS 516, 1991.

[14] L. C. Paulson. Isabelle: The next 700 theorem provers. In P. Odifreddi, editor, *Logic and Computer Science*, pages 361–385. Academic Press, 1990.

[15] F. Pfenning and C. Elliott. Higher-order abstract syntax. In *Proc. SIGPLAN '88 Symp. Language Design and Implementation*, pages 199–208. ACM Press, 1988.

[16] Z. Qian. Higher-order order-sorted algebras. In *Proc. 2th Int. Conf. Logic and Algebraic Programming*, pages 86–100. LNCS 463, 1990.

[17] M. Schmidt-Schauß. *Computational Aspects of an Order-Sorted Logic with Term Declarations.* LNCS 395, 1989.

[18] G. Smolka, W. Nutt, J. Goguen, and J. Meseguer. Order-sorted equational computation. In H. Aït-Kaci and M. Nivat, editors, *Resolution of Equations in Algebraic Structures, Volume 2*, pages 297–367. Academic Press, 1989.

[19] G. Snelting. The calculus of context relations. *Acta Informatica*, 28:411–445, 1991.

[20] W. Snyder and J. Gallier. Higher-order unification revisited: Complete sets of transformations. *J. Symbolic Computation*, 8:101–140, 1989.

[21] C. Walther. A mechanical solution of Schubert's Steamroller by many-sorted resolution. *Artificial Intelligence*, 26:217–224, 1985.

[22] C. Walther. Many-sorted unification. *J. ACM*, 35:1–17, 1988.

A Combinatory Logic Approach to Higher-order E-unification (Extended Abstract)

Daniel J. Dougherty
Dept. of Mathematics
Wesleyan University
Middletown, CT 06457 USA
ddougherty@eagle.wesleyan.edu

Patricia Johann
Dept. of Mathematics and Computer Science
Hobart and William Smith Colleges
Geneva, NY 14456 USA
johann@hws.bitnet

Abstract

Let E be a first-order equational theory. A translation of higher-order E-unification problems into a combinatory logic framework is presented and justified. The case in which E admits presentation as a convergent term rewriting system is treated in detail: in this situation, a modification of ordinary narrowing is shown to be a complete method for enumerating higher-order E-unifiers. In fact, we treat a more general problem, in which the types of terms contain type variables.

1 Introduction

Investigation of the interaction between first-order and higher-order equational reasoning has emerged as an active line of research. The collective import of a recent series of papers, originating with [Bre88] and including (among others) [Bar90], [BG91a], [BG91b], [Dou91], [JO91] and [Oka90], is that when various typed λ-calculi are enriched by first-order equational theories, the validity problem is well-behaved, and furthermore that the respective computational approaches to verifying equations (β-reduction and term rewriting) interact in a "modular" fashion.

This paper is concerned with the satisfiability problem in such a combined system, that is, with higher-order E-unification. The main novelty in our approach lies the use of combinatory logic ($\mathcal{CL}$) rather than λ-calculus ($\mathcal{LC}$) as a formalization of higher-order logic. The claim herein is that, as an algebraic treatment of higher-order reasoning, $\mathcal{CL}$ provides a congenial setting for incorporating first-order unification into the higher-order problem, eliminating the complexities (both conceptual and practical) involved with the presence of bound variables.

We provide evidence for this claim by analyzing the situation in which the equational theory E admits a presentation as a convergent (confluent and terminating) term rewriting system. We develop and outline the completeness of an algorithm which is essentially a normalized narrowing algorithm described in terms of transformations on systems, and which enumerates a complete set of higher-order E-unifiers for any system of terms. Future work will explore a combinatory approach to preunifcation and will address specific equational theories such as associativity and commutativity.

It is natural to try to solve higher-order unification problems by passing to $\mathcal{CL}$. Under any of the standard effective translations between $\mathcal{LC}$ and $\mathcal{CL}$, one might translate the relevant $\mathcal{LC}$-terms into $\mathcal{CL}$-terms and attempt to unify these. The fact that the basic $\mathcal{CL}$ axioms admit a convergent reduction relation makes this program particularly appealing, since such theories support narrowing as a unification procedure. Encouraged by the well-known techniques for combining unification algorithms for *first-order* theories ([Sch89], [Bou90]), we can hope for a modular solution to the problem of combining the first-order and higher-order problems.

But the reader familiar with combinatory logic will immediately recognize a difficulty, namely that even when E is empty, the equality generated by basic $\mathcal{CL}$-equations is *not* the equality induced by the translation from $\mathcal{LC}$ to $\mathcal{CL}$. Furthermore, no convergent rewrite system is known for this induced equality.

This difficulty is solved here — generalizing the techniques of [Dou90] — by defining a certain notion of reduction on *systems* of $\mathcal{CL}$ terms. When E has a convergent presentation, this reduction captures the induced equality described above and supports the standard unification strategy of narrowing (and so yields a solution to the unification problem in its usual formalization). The situation turns out to be particularly pleasant when the rewrite system underlying E is left-linear.

Classical higher-order unification concerns unification of terms of the explicitly simply typed λ-calculus; here we treat a more general problem in which the types of terms contain type variables which are eligible for instantiation by our answer substitutions.

Compiling functional programs into combinators has become a standard technique, motivated by the inefficiencies inherent in instantiating terms in the presence of bound variables (see [Pey87] for a discussion); implementations of higher-order unification problems should enjoy similar benefit from the passage to combinators.

Moreover, reasoning about and implementing substitution is simpler in an algebraic setting than in the traditional framework, and the use of combinators spares us the usual manipulation of long η-normal forms of $\mathcal{LC}$-terms. There is more than notational simplicity at stake: as pointed out by Snyder ([Sny90]), the presence of bound variables during paramodulation steps causes significant technical complications.

The incorporation of type-variables represents a significant generalization of the classical problem. It is instructive to see how Huet's classical algorithm ([Hue75]), in particular, the step there called "Projection", fares in the presence of type-variables. Consider a pair $\langle xM_1 \cdots M_r, aN_1 \cdots N_p \rangle$ of terms of the same base-type τ. When some M_i has type $(\tau_1 \to \cdots \to \tau_k \to \tau)$, the ith Projection step introduces the partial substitution $x := \lambda\vec{w}.w_i(h_1\vec{w})\cdots(h_k\vec{w})$, where $\vec{w}$ is a sequence of new variables $w_1, \ldots, w_r$ corresponding to $M_1, \ldots, M_r$, and the h_j are new variables. But suppose the type of M_i can look like $(\tau_1 \to \cdots \to \tau_k \to s)$ where s is a type-variable. In this case there are infinitely many instantiations of the type of M_i corresponding to functions with results of base type, and it is not clear how to account for the infinitely many relevant Projection steps. Put simply, we cannot know how many h_j to use in the binding for x. This point is made in [Nip90], where an (incomplete) extension to Huet's algorithm is presented. Type-variables present no difficulties in the present approach since our algorithm is directed by the shapes of the terms (as usual in algebraic unification) rather than the shape of the types.

Finally, the use of type-variables allows a finite axiomatization of typed combinatory logic, which in turn supports a complete unification procedure with a finitely branching search space.

Pure higher-order unification has found application in automated theorem proving in higher-order logic, specification of higher-order logics, machine learning, type inference in polymorphic lambda calculus, and extensions of logic programming. Unification in the presence of a first-order equational theory E has been surveyed in [Sie89] and [JK92]; the use of narrowing as an algebraic unification procedure originates with Fay ([Fay79]).

The seminal work in higher-order unification is [Hue75]; Gallier and Snyder ([GS89b]) have presented Huet's algorithm in a transformational setting. Snyder ([Sny90]) has given transformations for higher-order unification in the presence of an *arbitrary* equational theory E; Nipkow and Qian ([NQ91]) refined these to allow a modular approach (to pre-unification as well as unification) when a unification algorithm for E is known. The methods in [NQ91] are complete when E satisfies certain (strong) constraints.

There have been attempts to extend classical higher-order unification to allow more flexible typing-schemes. Nipkow ([Nip90]) treats a λ-calculus with type-variables (and a notion of constraints on type-variables), and the procedure given there has been incorporated into the generic theorem prover Isabelle. Elliott ([Ell89]) presents an algorithm for unification in the presence of dependent function types, designed as the basis for a generalization of the programming language λ-Prolog. Each of these algorithms is based on Huet's method; neither of them is a complete unification procedure.

Preliminaries

We will often draw upon classical results about the lambda calculus and combinatory logic (see, for example [HS86]) and use the basic results on the combination of lambda calculus and first-order rewriting ([Bre88], [BG91a], [BG91b], [Dou91]). We will assume familiarity with the use of transformations on systems to study unification ([MM82], [GS89a]).

Fix a set of equations E.

Terms and equalities

In the course of testing equality or unifiability of terms we will find it convenient to introduce constants not occurring in any terms under consideration; this is the motivation for the set *Args* defined below. It will also be convenient to arrange that distinct term-variables do not become identical by virtue of a type-substitution and so we require a precise notion of type-erasure for term-variables.

The *types* are formed by closing a set of *base types* and *type-variables* under the operation $(\alpha_1 \rightarrow \alpha_2)$ for types α_1 and α_2. Base types and type-variables are called *atomic types*.

Fix an infinite well-ordered set of *indeterminates* and an infinite well-ordered set of *parameters*. A *term-variable* is an ordered pair consisting of an indeterminate and a type; a *constant* is an ordered pair consisting of a parameter and a type; an *atom* is either a term-variable or a constant. The *type-erasure* of an atom is the first element of the pair.

Certain constants comprise the *signature* over which our first-order equations are defined. When discussing combinatory logic we assume that the parameters include the symbols I, K, and S. We assume that the set of parameters has a distinguished infinite set *Args* disjoint from $\{I, K, S\}$ and the parameters of the signature.

$\mathcal{LC}$ is the set of explicitly simply typed lambda terms over the atoms excluding I, K, and S; $\mathcal{CL}$ is the set of explicitly simply typed combinatory logic terms over these atoms together with the various I, K, and S typed as usual (the typed I, K, and S are called *redex* atoms). The *support* of a term T, *Supp*(T), is the set of type-variables occurring in T together with the indeterminates occurring among the type-erasures of the atoms in T; a *pure* term is a term in which no constant occurs whose erasure is in *Args*. A *fresh* indeterminate or parameter is one not occurring in any term in the current context; we will often refer to a choice of a term T with fresh variables, by which we mean that *Supp*(T) is disjoint from all type-variables and indeterminates in the current context.

Syntactic equality between terms or types is denoted by $\equiv$. We will not explicitly indicate the types of terms unless it is necessary.

Fix any of the standard $\Lambda : \mathcal{CL} \to \mathcal{LC}$ and $\mathcal{H} : \mathcal{LC} \to \mathcal{CL}$ such that $\Lambda(\mathcal{H}(M)) =_{\beta\eta} M$. Given a first-order equational theory E, define *extensional combinatory E-equality* by $X =_{CE} Y$ iff $\Lambda(X) =_{\beta\eta E} \Lambda(Y)$; it follows that for any $\mathcal{LC}$-terms M and N, $M =_{\beta\eta E} N$ iff $\mathcal{H}(M) =_{CE} \mathcal{H}(N)$.

Substitutions and unification

A *type substitution* is an ordinary algebraic substitution over the algebra of types; a type substitution θ_0 induces a type-shifting mapping on terms in an obvious way, and we shall denote this map by θ_0 as well. A *term substitution* θ_1 is an ordinary (type-preserving) substitution on $\mathcal{LC}$ or $\mathcal{CL}$ terms, as appropriate. A *substitution* θ is a pair consisting of a type substitution θ_0 and a term substitution θ_1; such a pair induces a mapping on $\mathcal{LC}$ and on $\mathcal{CL}$, also denoted θ, by the rule $\theta T \equiv \theta_1(\theta_0 T)$ (application of a substitution to a term, as well as composition of substitutions, will be indicated by juxtaposition). A *pure* substitution is one whose range is a set of pure terms. It will be notationally convenient to allow a term substitution θ_1 to act as the identity on types, so that we may refer to $\theta\alpha$ when α is a type.

Our dual substitutions behave in most ways just as ordinary substitutions (for example, a substitution θ is idempotent iff both θ_0 and θ_1 are idempotent, and if two terms are unifiable they possess a most general unifier). Details for standard unification are worked out in [Dou90].

An instance of the *higher-order E-unification* problem is a pair $\langle M, N\rangle$ of $\mathcal{LC}$-terms; a solution is a substitution θ such that $\theta M =_{\beta\eta E} \theta N$.

Suppose $\mathcal{W}$ is a set of type-variables and indeterminates. Whenever $=_*$ is a notion of equality on terms the notation $\theta =_* \theta'[\mathcal{W}]$ means that

1. for every type-variable $t \in \mathcal{W}$, $\theta_0(t) \equiv \theta'_0(t)$, and
2. for every term-variable x whose erasure is in $\mathcal{W}$, $\theta_1(x) =_* \theta'_1(x)$.

A notion of equality $=_*$ on terms induces an order $\leq_*$ on substitutions defined by: $\theta \leq_* \theta'[\mathcal{W}]$ if there is a substitution η with $\eta\theta =_* \theta'[\mathcal{W}]$.

The transfer to combinatory logic

If σ is an $\mathcal{LC}$-substitution, let the $\mathcal{CL}$-substitution $(\mathcal{H} \circ \sigma)$ be defined by $(\mathcal{H} \circ \sigma)X \equiv (\mathcal{H} \circ \sigma_1)(\sigma_0 X)$; if θ is a $\mathcal{CL}$-substitution, define $(\Lambda \circ \theta)$ analogously. The justification for our strategy of translating the unification problem from $\mathcal{LC}$ to $\mathcal{CL}$ is embodied in the following lemma, which follows from the facts that for any $\mathcal{LC}$-term M and substitution σ, $\mathcal{H}(\sigma M) \equiv (\mathcal{H} \circ \sigma)\mathcal{H}(M)$, and for any $\mathcal{CL}$-term X and substitution θ, $\Lambda(\theta X) \equiv (\Lambda \circ \theta)\Lambda(X)$.

Lemma 1.1 *Let M and N be $\mathcal{LC}$-terms. The $\beta\eta E$-unifiers of M and N are (up to pointwise $\beta\eta$-conversion) those of the form $(\Lambda \circ \theta)$, where θ ranges over the $\mathcal{CL}$-substitutions such that $\theta\mathcal{H}(M) =_{CE} \theta\mathcal{H}(N)$.*

If we define *extensional combinatory E-unification* as the problem of unifying $\mathcal{CL}$ terms with respect to extensional combinatory E-equality, the above discussion shows how a method for extensional combinatory E-unification yields a method for higher-order E-unification as originally presented.

The rest of this paper will be concerned with extensional combinatory E-equality, henceforth *CE-equality*, and extensional combinatory E-unification, henceforth *CE-unification*. The unqualified word "term" will mean "combinatory logic term".

When E is presented by a term rewriting system R we will often abuse notation by writing, e.g., $X =_{CR} Y$ for $X =_{CE} Y$, or by referring to "CR-validity" instead of "CE-validity." When E is empty we will refer to "C-validity" and "C-equality," and write $X =_C Y$.

Systems

A *pair* is either a *term-pair* or a *type-pair*, where a term-pair is a two-element multiset of $\mathcal{CL}$-terms and a type-pair is a two-element multiset of types. A pair is *trivial* if its elements are identical, and *CE-valid* if its elements are CE-equal (it will be convenient to consider trivial type-pairs to be CE-valid).

A *system* is a set of pairs in which no two distinct variables have the same type-erasure; it is *trivial* if each of its pairs is trivial and *CE-valid* if each of its pairs is CE-valid. As is customary, we write $\Gamma, \langle X, Y\rangle$ to abbreviate $\Gamma \cup \{\langle X, Y\rangle\}$. Since this is ambiguous as a decomposition of the system in question (Γ may or may not contain $\langle X, Y\rangle$), we introduce the notation $\Gamma; \langle X, Y\rangle$ to refer to $\Gamma \cup \{\langle X, Y\rangle\}$ with the understanding that $\langle X, Y\rangle$ is *not* a pair in Γ.

A substitution θ is a *unifier* of a system Σ if $\theta\Sigma$ (obtained by applying θ to each type and term occurring in Σ) is trivial. A substitution θ is a *CE-unifier* of a system Σ if $\theta\Sigma$ is CE-valid.

The restriction on type-erasures of the variables in a system is designed to avoid the technical complications which would result if distinct variables could become identical after a type-substitution. The restriction is only a convenience: if Σ is a unification problem presented as a set of pairs in which some indeterminates are given distinct types, successive applications of ordinary syntactic unification on the types assigned to these indeterminates will result in a system with the same CE-unifiers.

Let Σ be a system. If $\langle t, \alpha\rangle$ is a type-pair in Σ and there are no occurences (in type- or term-pairs) of t in Σ other than the one indicated, then t is *solved* in Σ and $\langle t, \alpha\rangle$ is a *solved type-pair*. If $\langle x, X\rangle$ is a term-pair in Σ, x and X have the same type, and there are no occurences of x in Σ other than the one indicated, then x is *solved* in Σ and $\langle x, X\rangle$ is a *solved term-pair.*

If each non-trivial term- or type-pair in Σ is solved, then Σ is a *solved system*, and its pairs determine an idempotent substitution in the usual way. Similarly, any idempotent substitution can be represented as a solved system — if σ is an idempotent substitution, write $[\sigma]$ for any solved system which represents it.

2 Adding algebra to combinatory logic

It will be helpful to have a brief discussion of the combinator-based approach to simple higher-order unification (without first-order equations).

2.1 Transformations for C-validity

Certainly C-equality is decidable; we may simply pass to $\mathcal{LC}$ and use (convergent) $\beta\eta$-reduction to test for equality of terms there. We might hope for a corresponding rewrite relation defined directly over $\mathcal{CL}$-terms, since narrowing is a well-understood unification technique for equational theories admitting a convergent presentation. Unfortunately no such relation is known (the classical *strong reduction* relation in $\mathcal{CL}$, which does capture C-equality, is clearly not suitable as foundation for unification — it is not finitely axiomatizable, and indeed even recognizing the set of rules is non-trivial).

However, there is defined in [Dou90] a well-behaved notion of reduction on *systems* which decides C-equality. C-equality can be obtained from weak equality by the addition of the extensionality rule:

$$\text{Infer } M =_C N \text{ from } Mz =_C Nz, \text{ when } z \text{ is not free in } M \text{ or in } N.$$

Deciding C-equality thus reduces to deciding C-equality between terms whose type is a base type or a type-variable. A weak normal form of such a type has a non-redex atom at the head. But two terms $hM_1 \cdots M_k$ and $h'N_1 \cdots N_{k'}$ are C-equal iff $h \equiv h'$, $k \equiv k'$, and $M_i =_C N_i$ for each i. This motivates the following set of transformations.

Definition 2.1 The set VT consists of the following three reductions:

- WEAK REDUCE
$$\Gamma;\ \langle X, Y\rangle \longrightarrow \Gamma,\ \langle X', Y\rangle,$$
when X weakly reduces to X'.

- ADD ARGUMENT
$$\Gamma;\ \langle X, Y\rangle \longrightarrow \Gamma,\ \langle Xd, Yd\rangle,$$
when X and Y have the same type, at least one of X and Y is of one of the forms: I, K, KA, S, SA, or SAB, and d is built from the first parameter in *Args* not occurring in $\langle X, Y\rangle$, and given the appropriate type.

- DECOMPOSE
$$\Gamma;\ \langle hX_1 \cdots X_k, hY_1 \cdots Y_k\rangle \longrightarrow \Gamma,\ \langle X_1, Y_1\rangle \ldots,\ \langle X_k, Y_k\rangle,$$
when h is a non-redex atom.

Here and elsewhere, we will observe the convention that no transformation is to be done out of a trivial pair. Observe the use of ";" on the left-hand sides of transformations, so that the effect of the transformation is unambiguous, and the use of "," on the right-hand sides, to preclude repetition of identical pairs.

The notation for WEAK REDUCE exploits the fact that pairs are unordered; we intend of course that either element of a pair may be reduced. A similar remark applies in several places below. The use, in ADD ARGUMENT, of new constants rather than new variables will serve to remind us in unification that the new arguments are not part of the original term and should not be instantiated. The neccessity for the restriction on d in ADD ARGUMENT may be seen by considering the non-C-valid pair $\langle Kd, I\rangle$, which can be reduced by an improper application of ADD ARGUMENT to the C-valid pair $\langle Kdd, Id\rangle$.

The key facts (proved in [Dou90]) about VT are the following:

Theorem 2.2 *Let Σ be any system.*

1. *(Soundness) Suppose $\Sigma \longrightarrow \Sigma'$. Then Σ is C-valid if Σ' is C-valid.*
2. *(Sufficiency) Suppose Σ is* VT*-irreducible. Then Σ is C-valid iff it is trivial.*
3. *(Termination) Every sequence of* VT *reductions terminates.*

The combinator-based higher-order unification method can characterized simply: it is the notion of narrowing on systems induced by VT as a reduction relation; Theorem 2.2 is the foundation of the completeness proof. The analogy with rewriting here is not perfect and so the proof of completeness involves some subtleties.

In fact (in the absence of algebraic equations) part (1) of Theorem 2.2 can be strengthened to read "if and only if". Thus VT is the basis for a simple algorithm to decide C-validity between terms.

2.2 Transformations for CR-validity

In the move to higher-order E-unification, the difficult step is that of constructing a set of transformations analogous to VT which capture CE-validity. The rest of this section is devoted to such a construction.

We must first understand how first-order equality, especially when presented in terms of a term rewriting system R, interacts with C-equality. Known results concerning rewriting and the lambda calculus are encouraging, but do not apply directly. If R is convergent, termination of R-weak reduction follows from termination of βR-reduction, but R-weak reduction will not capture CR-equality. Furthermore, we are committed to unification relative to *extensional* combinatory equality, and yet R-reduction and η-reduction will not be jointly confluent in general.

Nevertheless, we are led to consider the following set of transformations obtained from VT by naively incorporating R-reduction.

Definition 2.3 Let R be a first-order rewrite system. The set RVT of transformations is obtained by

- adding to VT the transformation R-REDUCE:

  $$\Gamma;\ \langle X, Y\rangle \longrightarrow \Gamma,\ \langle X', Y\rangle,$$

  where $X \longrightarrow X'$ is an instance of R-reduction, and by

- enforcing the constraint that DECOMPOSE may be applied only to a pair of terms which are R- and weakly irreducible.

The restriction on DECOMPOSE will prevent the unsound step of decomposing a head R-redex.

We might hope for an analogue of Theorem 2.2, but unfortunately, sufficiency can fail:

Example 2.4 Let R consist of the rule $fzz \longrightarrow a$ and let Σ be the system $\langle f(x(SK))(x(KI)), a\rangle$. Then Σ is CR-valid (since $SK =_{CR} KI$) and yet allows no application of a rule from RVT.

On the other hand, the existence of repeated variables in the rewrite rule is the only difficulty:

Theorem 2.5 *Let R be convergent and left-linear. Then the results of Theorem 2.2 hold for RVT reduction.*

Proof. This is a special case of the main result of this section, Theorem 2.8 below. □

An important consequence of Theorem 2.5 is that, when R is left-linear, naively adding R-narrowing to the higher-order narrowing transformations induced by VT allows enumeration of all unifiers.

In order to treat non-left-linear R, we note that for any such R there is a natural *conditional* rewrite system ([BK86]) determining the same equality relation, in which the unconditional parts of the conditional rules are left-linear. We may define such a conditionalization R^L of R by renaming repeated variable occurrences in the left-hand sides of R-rules and recording the necessary constraints among the variables as a collection of equations between variables. In fact, for each rule this collection of constraints is naturally a *system* Δ. If we denote by R^0 the collection of unconditional parts of the new conditional rules (called the *unconditional part* of R^L), we may describe a potential conditional rewrite step in transformational terms as an R^0-step together with a *witnessing system* of conditions $\sigma\Delta$, where Δ is the system of variable-equations associated with the rule and σ is the matching substitution for the R^0-reduction. Such a transformation corresponds to a proper conditional R-step precisely when the witnessing system is CR-valid. Of course, an ordinary R-step corresponds to an R^L-step in which the witnessing system is trivial.

Definition 2.6 Let R, R^L, and R^0 be as in the preceding discussion. The set of transformations R^LVT is obtained by

- adding to VT the transformation R^L-REDUCE:

$$\Gamma;\ \langle X, Y\rangle \longrightarrow \Gamma,\ \langle X', Y\rangle, \sigma\Delta,$$

  where $X \longrightarrow X'$ is an instance of R^0-reduction with associated system of conditions $\sigma\Delta$, and by

- enforcing the constraints that
  - the system of conditions associated with an R^L-REDUCE step is non-trivial only if the redex term X is R- and weakly irreducible, and that

– DECOMPOSE may be applied only to a pair of terms which are R- and weakly irreducible.

The conditions on R^L-REDUCE steps are designed to minimize non-determinism; we will not want to perform an R^L-REDUCE step out of a system when there are R- and weak reductions available. We will treat R^LVT as an extension of RVT by identifying R-REDUCE with "trivially-conditional R^L-REDUCE."

An R^LVT step is *sound* if it is a VT step or it is an R^L-REDUCE step whose associated system of conditions is CR-valid. Note that RVT steps are always sound.

We revisit the earlier example which defeated RVT :

Example 2.7 Let R and Σ be as in Example 2.4. A conditionalization of R has unconditional part $fz'z \longrightarrow a$ and variable condition $z' = z$. Application of R^L-REDUCE yields $\langle a, a\rangle$, $\langle x(SK), x(KI)\rangle$. DECOMPOSE and two applications of ADD ARGUMENT further give $\langle a, a\rangle$, $\langle SKpq, KIpq\rangle$. Several WEAK REDUCE steps yield $\langle a, a\rangle$, $\langle q, q\rangle$, and, anticipating the next lemma, we conclude that Σ was CR-valid. □

The following analogue of Theorem 2.5 embodies the facts about R^LVT critical to our program.

Theorem 2.8 *Let R be convergent.*

1. *Suppose* $\Sigma \longrightarrow \Sigma'$. *Then* Σ *is CR-valid if* Σ' *is CR-valid; when the given transformation is sound then* Σ *is CR-valid iff* Σ' *is CR-valid.*

2. *Suppose* Σ *is irreducible with respect to sound* R^LVT *steps. Then* Σ *is CR-valid iff it is trivial.*

3. *Every sequence of sound* R^LVT *reductions terminates.*

Proof (discussion). The proof of (1) is straightforward.

The proof of (2) essentially consists of demonstrating that if X and Y are CR-equal atomic-type weak normal forms, then either DECOMPOSE or R^L-REDUCE applies out of the pair $\langle X, Y\rangle$.

The proof of (3) is very delicate. By considering the multiset of terms occurring in a system, we may proceed by multiset induction on the following relation, derived by examining the effects of the various R^LVT steps on individual terms: $X \mapsto Y$ if Y is the result of (i) adding an argument, (ii) extracting a subterm not containing the head-symbol of X, (iii) a weak reduction, or (iv) the unconditional part of a sound R^L-reduction.

Observe that adding an argument may induce weak or R^L-reductions, extracting a subterm may increase the length of the type, and both weak and R^L-reduction can increase the size of the term. So it seems that no simple argument is available based on an ordering of terms (with respect to size, type, or R- and weak reduction size).

A further difficulty is that linearization can transform a terminating rewrite system into a non-terminating one. For example, if R has the single rule $fx(gx) \longrightarrow fxx$, then R is terminating but the linearization $fx(gy) \longrightarrow fxx$ is not (the term $f(ga)(ga)$ reduces to itself). The restriction to sound linearized steps is, in itself, not enough; if we add the rule $gx \longrightarrow x$ to the system just presented, then $f(ga)(ga)$ reduces to itself by a sound step. So the restriction to irreducible terms in (properly conditional) R^L-steps is important.

The proof of termination is an adaptation of the technique of *logical relations*, a fundamental tool in the study of the lambda-calculus (see [Mit90] for an introduction and references). □

3 Transformations for higher-order E-unification

As suggested in the previous section, we may generate unification transformations from R^LVT in precisely the same way that narrowing transformations are derived from a convergent term rewriting system.

Definition 3.1 Let R be a convergent term rewriting system, and R^0 the unconditional part of a conditionalization of R. The set R^LUT is obtained by adding the following transformations to those for standard unification:

- *Weak Narrow:*
$$\Gamma;\ \langle X, Y\rangle \Longrightarrow [\mu],\ \mu\Gamma,\ \langle \mu X^*, \mu Y\rangle\,,$$
where there exists a non-variable ocurrence d of X and a weak reduction rule $S \longrightarrow T$ with fresh variables such that μ is a most general unifier of X/d and S, and $X^* \equiv X[d \rightarrow T]$.

- *Add Argument:*
$$\Gamma;\ \langle X, Y\rangle \Longrightarrow [\mu],\ \mu\Gamma,\ \langle(\mu X)d, (\mu Y)d\rangle\,,$$
where $\mu \equiv (\pi \rightarrow \pi')$ is a most general type-unifier of the set consisting of the type of X, the type of Y, and (just in case these are each atomic types) the type $(s \rightarrow t)$, for fresh type-variables s and t, and where d is built from the first fresh parameter in *Args*, given type π.

- *Split:*
$$\Gamma;\ \langle xX_1\cdots X_n, hZ_1\cdots Z_mY_1\cdots Y_n\rangle \Longrightarrow$$
$$[\mu],\ \mu\Gamma,\ \langle z_1, \mu Z_1\rangle, \ldots, \langle z_m, \mu Z_m\rangle, \langle \mu X_1, \mu Y_1\rangle, \ldots, \langle \mu X_n, \mu Y_n\rangle\,,$$
where $m, n \geq 0$, $x \in Vars$, h is a pure atom, each z_i is a fresh indeterminate given the same type as Z_i, $1 \leq i \leq m$, and μ is a most general unifier of x and $hz_1\cdots z_m$.

- R^L*-Narrow*
$$\Gamma;\ \langle X, Y\rangle \Longrightarrow [\mu], \mu\Gamma,\ \langle \mu X^*, \mu Y\rangle, \mu\Delta,$$
when there exists a non-variable occurrence d of X and a rule $S \longrightarrow T$ in R^0 with fresh variables and associated system of conditions Δ such that μ is a most general unifier of X/d and S, and $X^* \equiv X[d \leftarrow T]$.

We adopt the convention that no R^LUT transformation is to be done out of a solved or trivial pair. This respects the intuition that the solved part of a system is merely a record of an answer substitution being constructed.

The R^LUT transformations are sound:

Lemma 3.2 (Soundness) *If θ is a pure CR-unifier of Σ' and $\Sigma \Longrightarrow \Sigma'$ via an R^LUT step, then $\theta\Sigma$ is CR-valid.*

Proof. Use the notation of Definition 3.1. Our hypothesis entails that $\theta[\mu]$ is CR-valid, so $\mu \leq_{CR} \theta$, and hence $\theta\mu =_{CR} \theta$. It follows that $\theta\mu\Gamma =_{CR} \theta\Gamma$, and so we need only show that θ CR-unifies the "redex pair" of the transformation.

When the transformation is *Weak Narrow* or R^L-*Narrow*, the argument is exactly as for first-order narrowing.

When the transformation is *Add Argument* we want to see that $\theta X =_{CR} \theta Y$. But $\theta\mu(Xd) =_{CR} \theta\mu(Yd)$, that is, $(\theta X)(\theta d) =_{CR} (\theta Y)(\theta d)$ and we may invoke the extensionality rule since θ is pure and so θd is guaranteed to be new to $\langle \theta X, \theta Y \rangle$.

In the case of *Split*, the fact that $\theta X_i =_{CR} \theta\mu X_i =_{CR} \theta\mu Y_i =_{CR} \theta Y_i$ for $1 \leq i \leq n$ implies that we need only argue that $\theta \langle x, hZ_1 \cdots Z_m \rangle$ is CR-valid. We compute: $\mu x \equiv \mu h z_1 \cdots z_m$ by definition of μ, so $\theta x =_{CR} \theta\mu x \equiv \theta\mu(hz_1 \cdots z_m) =_{CR} \theta(hz_1 \cdots z_m)$, but our hypothesis implies that for each i, $\theta z_i =_{CR} \theta Z_i$. □

We now address completeness. The Lifting Lemma below is the main tool for proving that for any system Σ, the set R^LUT of transformations can enumerate a complete set of CR-unifiers for Σ when R is convergent. For its statement and proof we require the following notion:

Definition 3.3 Let R be a term rewriting system. A $\mathcal{CL}$ term X is a *CR-normal form* if there exists a long βR-normal form M such that $X \equiv \mathcal{H}(M)$.

Say that an idempotent substitution θ is a *normalized CR-unifier* of system Σ if $D\theta_0$ and the type-erasures of the terms in $D\theta_1$ are contained in *Supp*(Σ), $\theta\Sigma$ is CR-valid, and for each unsolved variable x of Σ, θx is CR-normal. Write NCRU(Σ) for the set of *normalized CR-unifiers of* Σ.

Lemma 3.4 (Lifting Lemma) *Let $\theta \in NCRU(\Sigma)$ and let $\langle X, Y \rangle$ be an unsolved pair in Σ. If*

$$\theta\Sigma \longrightarrow \Pi$$

is a sound R^LVT step out of $\theta \langle X, Y \rangle$, then there exists a Σ' and θ' with

$$\Sigma \Longrightarrow \Sigma'$$

such that

1. *$\theta' \equiv \theta[Supp(\Sigma)]$,*
2. *$\theta'\Sigma'$ and Π differ only by trivial pairs, and*
3. *$\theta' \in NCRU(\Sigma')$.*

Proof (discussion). In outline, the proof is a standard lifting-lemma construction, but there are subtleties in the use of normalized substitutions. We need to know that CR-normal forms are weakly irreducible and R-irreducible, unique in their CR-equivalence classes, and closed under the formation of subterms. All but the last assertion can be derived from classical facts about normal forms in $\mathcal{LC}$ and $\mathcal{CL}$. Closure under subterm, however, seems to require a complete reconstruction of the classical theory of *strong reduction* ([CF58], [Hin67], [Ler67]) in the presence of R-reduction to show that a generalization of Curry's normal form theorem and its converse hold, i.e., that the classes of CR-normal forms and terms which are R-strongly irreducible are the same. □

3.1 Refinements

Before proving that the R^LUT transformations can enumerate a complete set of CR-unifiers for convergent R and arbitrary systems of $\mathcal{CL}$ terms, we indicate a way to decrease the non-determinism inherent in our transformation-based CR-unification method.

Definition 3.5 A system is *semi-simple* if every term occurring has a variable at the head and is in R- and weak normal form.

Lemma 3.6 *Any sequence of* WEAK REDUCE, ADD ARGUMENT, *and R-reductions applied to a system will terminate in a semi-simple system with the same CR-unifiers.*

We will see that it suffices to apply R^LVT transformations only to semi-simple systems. Reducing a system to a semi-simple one recalls the SIMPL phase of Huet's classical higher-order unification algorithm. In the present setting, we may view such a reduction as the normalization phase of a normalized narrowing algorithm ([Ret87]).

3.2 The Algorithm and Its Completeness

Definition 3.7 The non-deterministic algorithm $\mathcal{RU}$ is the following process:
Repeatedly:

1. Reduce the system to a semi-simple system via R^LVT steps and then apply some R^LUT transformation out of a non-trivial unsolved pair.
2. If at any point the system is syntactically unifiable by a pure substitution then return a most general unifier of the system without transforming the system (but do not necessarily halt).

It follows from Lemmas 3.2 and 3.6 that if Algorithm $\mathcal{RU}$ is run on an initial system Σ and returns a substitution θ, then θ is a CR-unifier of Σ. The main result of this section is a converse.

Example 3.8 Consider the rewrite system and conditionalization of Example 2.4, together with the system Σ comprising the pair

$$\langle f(ux)(u(SKx)), a\rangle\,,$$

where u has type $(r \to r) \to 0$ and x has type $r \to r$. One application of R^L-*Narrow* yields

$$\langle y, ux\rangle\,,\ \langle z, u(SKx)\rangle\ \langle ux, u(SKx)\rangle\,,$$

(here and below we suppress writing trivial pairs). Term Decomposition gives

$$\langle y, ux\rangle\,,\ \langle z, u(SKx)\rangle\,,\ \langle x, (SKx)\rangle\,.$$

After an application of *Add Argument* we have

$$\langle y, ux\rangle\,,\ \langle z, u(SKx)\rangle\,,\ \langle xp, (SKx)p\rangle\,.$$

Two WEAK REDUCE steps further give

$$\langle y, ux\rangle\ ,\ \langle z, u(SKx)\rangle\ ,\ \langle xp, p\rangle\ ,$$

and one *Weak Narrow* using $Sz_1z_2z_3 \rightarrow z_1z_3(z_2z_3)$ and pair $\langle xp, p\rangle$ yields

$$\langle y, u(Sz_1z_2)\rangle\ ,\ \langle z, u(SK(Sz_1z_2))\rangle\ ,\ \langle x, Sz_1z_2\rangle\ ,\ \langle z_1p(z_2p), p\rangle\ ,\ \langle z_3, p\rangle\ .$$

Another *Weak Narrow*, this time employing $Kw_1w_2 \rightarrow w_1$ and pair $\langle z_1p(z_2p), p\rangle$ leaves us with

$$\langle y, u(SKz_2)\rangle\ ,\ \langle z, u(SK(SKz_2))\rangle\ ,\ \langle x, SKz_2\rangle\ ,\ \langle z_1, K\rangle\ ,\ \langle w_1, p\rangle\ ,\ \langle w_2, z_2p\rangle\ ,\ \langle z_3, p\rangle\ .$$

This system is solved, so we extract the CR-unifier θ where θ_0 is the identity and $\theta_1 = \{x \mapsto SKz_2\}$.

Theorem 3.9 (Completeness) *Let R be convergent, and let θ be a pure CR-unifier of Σ. Then there is a computation of Algorithm $\mathcal{RU}$ on Σ producing a pure CR-unifier σ of Σ with $\sigma \leq_{CR} \theta[Supp(\Sigma)]$.*

Proof (discussion). Without loss of generality we may assume $\theta \in NCRU(\Sigma)$. By Theorem 2.8 there is a sequence of sound R^LVT transformations taking the system $\theta\Sigma$ to a trivial pair, and the proof inductively lifts these to R^LUT steps out of Σ.

The induced R^LUT steps are indeed $\mathcal{RU}$ steps, and and the substitution σ computed by the sequence of liftings can be seen to satisfy $\sigma \leq_{CR} \theta[Supp(\Sigma)]$. □

4 Conclusion

We have defended the claim that the formulation of higher-order logic using combinators allows routine algebraic techniques to be applied to higher-order unification and leads to a smooth interaction between these methods and the standard tools of first-order unification.

The ordinary higher-order unification problem submits to a narrowing algorithm in the combinator setting. For higher-order E-unification in the situation that E itself allows narrowing, the combined problem is solved by combining these algorithms (in a naive way for the left-linear case, and by passing to conditional rewriting otherwise). The result is a complete enumeration procedure with a finitely-branching search space; we expect that this technique will yield an alternative approach to that in [Sny90] to higher-order E-unification under arbitrary E.

In light of the theoretical intractability of general higher-order unification and the practical utility of *pre-unification*, a crucial line of investigation is now an analysis and implementation of higher-order E-preunification in the combinator framework, and the development of tools to control the redundancy which is inherent (as shown by Huet) in any complete enumeration method for higher-order unification. It will also be important to focus on particular standard equational theories such as associativity and commutativity; our conjecture is that the known algorithms for such situations can be readily adapted to the algebraic higher-order context.

References

[Bar90] F. Barbanera. Adding algebraic rewriting to the Calculus of Constructions: strong normalization preserved. *Extended Abstracts, The Second International Workshop on Conditional and Typed Rewriting Systems*, Center for Pattern Recognition and Machine Intelligence, 1990.

[BG91a] V. Breazu-Tannen and J. Gallier. Polymorphic rewriting conserves algebraic strong normalization. To appear in *Theoretical Computer Science.*

[BG91b] V. Breazu-Tannen and J. Gallier. Polymorphic rewriting conserves algebraic confluence. To appear in *Information and Computation.*

[BK86] J. A. Bergstra and J. W. Klop. Conditional rewrite rules: confluence and termination. *Journal of Computer and System Sciences* 32, pp. 322–362, 1986.

[Bou90] A. Boudet. Unification in a combination of equational theories: an efficient algorithm. In *Proceedings of the Tenth Conference on Automated Deduction*, Springer-Verlag LNAI 449, pp. 292–307, 1990.

[Bre88] V. Breazu-Tannen. Combining algebra and higher-order types. In *Proceedings of the Third Annual IEEE Symposium on Logic in Computer Science*, IEEE Press, pp. 82–90, 1988.

[CF58] H. B. Curry, R. Feys. *Combinatory Logic, Vol. I*, North-Holland, 1958.

[Dou91] D. J. Dougherty. Adding algebra to the untyped lambda calculus. In *Proceedings, Fourth International Conference on Rewriting Techniques and Applications*, Springer-Verlag LNCS 488, pp. 37–48, 1991. To appear, *Information and Computation.*

[Dou90] D. J. Dougherty. Higher-order unification via combinators. Preprint, 1990. To appear, *Theoretical Computer Science.*

[Ell89] C. Elliott. Higher-order unification with dependent function types. In *Proceedings of the Third International Conference on Rewriting Techniques and Applications*, Springer-Verlag LNCS 355, pp. 121–136, 1989.

[Fay79] M. Fay. First order unification in an equational theory. *Proceedings of the Fourth Workshop on Automated Deduction*, 1979.

[GS89a] J. H. Gallier and W. Snyder. Complete sets of transformations for general *E*-unification. *Theoretical Computer Science* 67, pp. 203–260, 1989.

[GS89b] J. H. Gallier and W. Snyder. Higher-order unification revisited: complete sets of transformations. *Journal of Symbolic Computation* 8, pp. 101–140, 1989.

[Hin67] R. Hindley. Axioms for strong reduction in combinatory logic. *Journal of Symbolic Computation* 32, pp. 224–236, 1967.

[HS86] J. R. Hindley and J. P. Seldin. *Introduction to Combinators and λ-Calculus*, Cambridge University Press, 1986.

[Hue75] G. Huet. A unification algorithm for typed λ-calculus. *Theoretical Computer Science* 1, pp. 27–57, 1975.

[JK92] J.-P. Jouannaud and C. Kirchner. Solving equations in abstract algebras: A rule-based study of unification. In *Computational Logic: Essays in Honour of Alan Robinson*, ed. J. Lassez and G. Plotkin, MIT Press. To appear.

[JO91] J.-P. Jouannaud and M. Okada. A computation Model for executable higher-order algebraic specification languages. In *Proceedings of the Sixth Annual IEEE Symposium on Logic in Computer Science*, IEEE Press, pp. 350–361, 1991.

[Joh91] P. Johann. *Complete Sets of Transformations for Unification Problems.* Dissertation, Wesleyan University, 1991.

[Ler67] B. Lercher. Strong reduction and normal form in combinatory logic. *Journal of Symbolic Logic* 2, pp. 213–223, 1967.

[Mit90] J. Mitchell. Type systems for programming languages. In *Handbook of Theoretical Computer Science, Volume B*, ed. J. van Leeuwen, MIT Press/Elsevier, pp.365–458, 1990.

[MM82] A. Martelli and U. Montanari. An efficient unification algorithm. *ACM Transactions on Programming Languages and Systems* 4, pp. 258–282, 1982.

[Nip90] T. Nipkow. Higher-order unification, polymorphism, and subsorts. *Extended Abstracts, The Second International Workshop on Conditional and Typed Rewriting Systems*, Center for Pattern Recognition and Machine Intelligence, 1990.

[NQ91] T. Nipkow and Z. Qian. Modular Higher-order E-unification. In *Proceedings, Fourth International Conference on Rewriting Techniques and Applications*, Springer-Verlag LNCS 488, pp. 200–214, 1991.

[Oka90] M. Okada. Strong normalizability for the combined system of the typed lambda calculus and an arbitrary convergent rewrite system. In *Proceedings, ISSAC 89*, 1989.

[Pey87] S. L. Peyton-Jones. *The Implementation of Functional Programming Languages.* Prentice-Hall, 1987.

[Ret87] P. Réty. Improving basic narrowing techniques. In *Proceedings of the Second International Conference on Rewriting Techniques and Applications*, 1987.

[Sch89] M. Schmidt-Schauss. Unification in a combination of arbitrary disjoint equational theories. *Journal of Symbolic Computation* 8, pp. 51–99, 1989.

[Sie89] J. Siekmann. Unification theory. *Journal of Symbolic Computation* 7, pp. 207–274, 1989.

[Sny90] W. Snyder. Higher-order E-unification. In *Proceedings of the Tenth Conference on Automated Deduction*, 1990.

Cycle Unification

Wolfgang Bibel[1], Steffen Hölldobler[1], Jörg Würtz[2]

[1] Fachgruppe Intellektik, Fachbereich Informatik, Technische Hochschule Darmstadt, Alexanderstraße 10, 6100 Darmstadt, Germany
[2] Deutsches Forschungszentrum für Künstliche Intelligenz, Stuhlsatzenhausweg 3, 6600 Saarbrücken 11, Germany

Abstract. Two-literal clauses of the form $L \leftarrow R$ occur quite frequently in logic programs, deductive databases, and – disguised as an equation – in term rewriting systems. These clauses define a cycle if the atoms L and R are weakly unifiable, ie. if L unifies with a new variant of R. The obvious problem with cycles is to control the number of iterations through the cycle. In this paper we consider the cycle unification problem of unifying two literals G and F modulo a cycle. We review the state of the art of cycle unification and give some new results for a special type of cycles called matching cycles, ie. cycles $L \leftarrow R$ for which there exists a substitution σ such that $\sigma L = R$ or $L = \sigma R$. Altogether, these results show how the deductive process can be efficiently controlled for special classes of cycles without losing completeness.

1 Introduction

It is the foremost goal of the research in the field of automated deduction to develop general *and* adequate proof methods and techniques for the logics under consideration. It is comparatively easy to invent a general proof method, but it is much more difficult to develop a general *and* adequate proof technique. For example, the resolution principle [15] and the connection method [1] are general proof methods for first-order logic. But are they adequate? What is the meaning of adequateness in the first place? Roughly speaking, we will consider a technique as being adequate if it solves simpler problems faster than more difficult ones. We illustrate the notion of adequateness by a problem, where the known general proof techniques face difficulties whereas trained humans seem to be able to solve it quite reasonably.

For this purpose, consider the following set of clauses in Prolog-like notation which is taken from [14] and was originally studied by Łucasiewicz.

$$\begin{array}{lll} & \leftarrow Pi(iab, i(ibc, iac)). & \text{G} \\ Pw & \leftarrow Pv,\ Pivw. & \text{MP} \\ Pi(i(ixy, z), i(izx, iux)). & & \text{A} \end{array}$$

The terms represent implicational formulas, ie. iab encodes $a \rightarrow b$ and P asserts the derivability of its argument. Thus, the second clause represents modus ponens. It contains several *cycles* [2] defined by the connections between the atom Pw and the atoms $Pivw$ and Pv. The clause MP can be applied to itself and this may lead to an exponential growth of the search space. The obvious problem is to control the self-applicability of MP while retaining completeness. Łukasiewicz has found a 29 step proof. He must have exercised a good control over MP! Quintus PROLOG on a Sun SPARC station 2 did not find a proof in several days. Nearly all existing automatic theorem provers cannot solve this problem as well since they are not able to exercise a good control over MP. E. Lusk[3] reports that the parallel version of Otter at Argonne is able to obtain a hyperresolution proof with about 150 proof-steps while generating 6.5 million clauses in about half an hour during the search for it. Their prover does not have a good control over MP as well. It solves the problem by sheer power.

In [3] it was conjectured that a problem like the Łucasiewicz-formula could be solved in less than a second by way of a technique called *cycle unification*. At present this conjecture remains a challenge since the Łucasiewicz-formula is a particularly difficult instance of a class of formulas which could eventually be treated by cycle unification. In this paper we make a first step towards this goal by restricting our attention to the special case of formulas with exactly one cycle. In fact, we even focus our analysis on the simple class of two-literal clauses of the form $Pl_1 \ldots l_n \leftarrow Pr_1 \ldots r_n$ which consists of nothing but a single cycle. This additional restriction simplifies the discussion without loss of generality of the method.

Such a two-literal clause is usually embedded in the context of some larger formula, or set of clauses. Again for simplicity of the discussion and without loss of generality, we restrict the treatment to the case of two additional clauses, namely a goal clause – referred to as *(calling) goal* – of the form $\leftarrow Ps_1 \ldots s_n$, which calls the cycle, and a fact – called *(terminating) fact* – of the form $Pt_1 \ldots t_n$, which terminates the cycle. In our restricted case a *cycle unification problem* is then the following one:

> Is there a substitution σ such that $\sigma Ps_1 \ldots s_n$ is a logical consequence of $Pl_1 \ldots l_n \leftarrow Pr_1 \ldots r_n$ and $Pt_1 \ldots t_n$?

If such a substitution σ exists, then σ is said to be a *solution* for the cycle unification problem. For more general cases, cycle unification can be defined in an analogue way.

In order to be able to control a cycle we have to answer the following questions. Is cycle unification decidable? How many independent most general solutions has a cycle unification problem? Does there exist a unification algorithm which enumerates a minimal and complete set of solutions for a cycle unification problem? Answers to these questions may help to increase the power of automated theorem provers significantly. For example, if a cycle is embedded in a larger formula and it can be determined that the corresponding cycle unification problem is unsolvable, then the clauses defining the cycle can be eliminated from the formula. If a minimal and complete set Σ of solutions for a cycle unification problem exists and can be enumerated, then any other solution is subsumed by a solution in Σ and need not to be considered. If Σ is finite, then this may prune a potentially infinite search space to a finite one. But theorem proving is not the only task which may benefit from cycle unification. There are a variety of applications for cycle unification such as intelligent backtracking, deductive databases, program transformation, and termination proofs for logic programs, to mention just a few.

Although cycle unification is of significant importance for the field of automated deduction, it has received surprisingly little attention in the literature. Function-free cycle unification problems, ie. cycle unification problems defined over variables and

[3] private communications

constants only, occur mainly in deductive databases and it can be shown that under certain conditions these problems do not give rise to infinite computations (cf. [10]). In [13] the number of iterations through a cycle can be limited via a user-defined parameter. L. Vielle shows that certain cycle unification problems can be solved by generalization and subsumption [20]. There, after several iterations through a cycle, subterms occurring in a goal are replaced by variables. Subsumption techniques may now be applied to terminate otherwise infinite derivations. The technique is shown to be complete. Unfortunately, answers to a generalized goal need not to be answers to the initial goal.

M. Schmidt–Schauß [16] has shown that cycle unification is decidable provided that the goal and the fact are ground, ie. they do not contain variable occurrences. Independently, P. Devienne [7] has given a more general result for cycle unification problems with linear goals and facts, ie. each variable occurs at most once in the goal and the fact. He uses essentially the same ideas as Schmidt–Schauß, but a very special technique based on directed weighted graphs. Devienne's results were used by De Schreye et al. [17] to decide whether cycles admit non-terminating queries to deductive systems. Another approach has been taken by H.J. Ohlbach [11] who represented sets of terms by so-called abstraction trees which may compress the search space. Moreover, abstraction trees can be used to compile two-literal clauses and in certain cases a finite abstraction tree can represent infinitely many solutions of a cycle unification problem [12].

This paper is a first step towards a theoretical foundation of cycle unification. After some preliminary notes on definitions and notations we formally define cycle unification in Section 3. In Section 4 we define various new classes of restricted cycle unification problems with increasing complexity and show that their unification problems are decidable, determine the unification types, and develop unification algorithms. The paper concludes with a summary of the results on cycle unification and an outline of future work.

2 Definitions and Notations

Our definitions and notations follow those suggested in [6]. Throughout this paper capital letters such as P, Q, ... denote *predicate symbols*, small letters such as a, b, ... denote *constants*, f, g, ... *function symbols*, and x, y, ... *variables*. A *term* is either a variable or of the form $f(t_1, \ldots, t_n)$, where t_1, ..., t_n are terms. s, t, ... denote terms. An *atom* is of the form $P(t_1, \ldots, t_n)$. Let X be an atom or a term. $\mathcal{V}ar(X)$ denotes the set of variables occurring in X. X is called *ground* iff X does not contain any variable. X is called *linear* iff every variable occurs at most once in X. By X^k we denote the syntactic object where each variable occurring in X has the index k attached to it.

A *substitution* is a mapping from the set of variables into the set of terms which is equal to the identity almost everywhere. Hence, it can be represented as a finite set of pairs $\{x_1 \mapsto t_1, \ldots, x_n \mapsto t_n\}$, $x_i \neq t_i$, $1 \leq i \leq n$. Substitutions are denoted by small greek letters such as σ, θ, ... The identity substitution is called ε. $\sigma t = \sigma(t)$ if t is a variable and $\sigma t = f(\sigma t_1, \ldots, \sigma t_n)$ if $t = f(t_1, \ldots, t_n)$. $\mathcal{D}om(\sigma) = \{x \mid x \text{ is a variable and } \sigma x \neq x\}$ is the *domain* of σ. $\mathcal{VR}an(\sigma) = \bigcup_{x \in \mathcal{D}om(\sigma)} \mathcal{V}ar(\sigma x)$ is the *variable range* of σ.

The *composition* $\sigma\tau$ of two substitutions σ and τ is defined by $(\sigma\tau)x = \sigma(\tau x)$. The *restriction* of the substitution σ to the set V of variables is defined by $\sigma|_V\, x = \sigma x$ if $x \in V$ and $\sigma|_V\, x = x$ otherwise. A substitution σ is called *variable-pure* if $\{\sigma x \mid x \in \mathcal{D}om(\sigma)\}$ only consists of variables. A *renaming* is a variable-pure substitution σ such that $\sigma x = \sigma y$ implies $x = y$ for $x,\ y \in \mathcal{D}om(\sigma)$.

If W is a set of variables, then $\sigma = \tau\ [W]$ iff $\forall x \in W : \sigma x = \tau x$. A substitution σ is called *more general* than a substitution τ on W, $\sigma \lesssim \tau\ [W]$, iff there exists a substitution ρ such that $\rho\sigma = \tau\ [W]$. Two substitutions σ and τ are called *independent* on W iff $\sigma \not\lesssim \tau\ [W]$ and $\tau \not\lesssim \sigma\ [W]$.

σ is called a *unifier* for t and t' iff $\mathcal{D}om(\sigma) \subseteq \mathcal{V}ar(t) \cup \mathcal{V}ar(t')$ and $\sigma t = \sigma t'$. A unifier σ of t and t' is called *most general unifier* iff $\sigma \lesssim \tau\ [\mathcal{V}ar(t) \cup \mathcal{V}ar(t')]$ for all unifiers τ of t and t'. σ is called a *matcher* for t and t' iff $\mathcal{D}om(\sigma) \subseteq \mathcal{V}ar(t')$ and $t = \sigma t'$ holds. A matcher σ of t and t' is called a *most general matcher* iff $\sigma \lesssim \tau\ [\mathcal{V}ar(t')]$ for all matchers τ of t and t'. The definitions above can be extended to atoms, equations, and sets of equations in the obvious way.

3 Cycle Unification

$C = \{L \leftarrow R\}$ is called a *cyclic theory*, or *cycle* for short, if the atoms L and R are weakly unifiable, ie. there exist two substitutions σ and σ' such that $\sigma L = \sigma' R$ [8]. Let G and F be two atoms such that $\mathcal{V}ar(G) \cap \mathcal{V}ar(F) = \emptyset$. A *cycle unification problem* $\langle G \xrightarrow[C]{\circ} F\rangle$ (or $\langle G \xrightarrow{\circ} F\rangle_C$) is the problem whether there exists a substitution σ such that σG is a logical consequence of F and C. A substitution σ is a *solution* for the cycle unification problem if $\mathcal{D}om(\sigma) \subseteq \mathcal{V}ar(G)$ and σG is a logical consequence of F and C.[4]

Since solutions to cycle unification problems are substitutions, the notions of more general, independent, etc. substitutions can be extended to more general, independent, etc. solutions of cycle unification problems in the obvious way. Let $\mathcal{C} = \langle G \xrightarrow[C]{\circ} F\rangle$ be a cycle unification problem. A set Σ of substitutions is a *complete* set of solutions for $\mathcal{C}$ iff each substitution in Σ is a solution for $\mathcal{C}$ and for each solution θ for $\mathcal{C}$ we find a substitution σ in Σ such that $\sigma \lesssim \theta[\mathcal{V}ar(G)]$. A complete set Σ of solutions for $\mathcal{C}$ is said to be *minimal* iff for all $\sigma,\ \theta \in \Sigma$ we find that $\sigma \lesssim \theta[\mathcal{V}ar(G)]$ implies $\sigma = \theta$.

As a first example consider the problem

$$\langle Pa \xrightarrow{\circ} Pfffa\rangle_{\{Px \leftarrow Pfx\}}.$$

The empty substitution ε is the only most general solution for this problem. However, there may be more than one solution as the example

$$\langle Pxy \xrightarrow{\circ} Pab\rangle_{\{Pvw \leftarrow Pwv\}}$$

shows. This problem has the two independent most general solutions $\{x \mapsto a,\ y \mapsto b\}$ and $\{x \mapsto b,\ y \mapsto a\}$. But, there may be even infinitely many independent most

[4] A cycle unification problem should not be confused with a theory unification problem $\langle G =_C F\rangle$, ie. the problem whether there exists a substitution σ such that $\sigma G =_C \sigma F$ [1, 19].

general solutions. As an example consider

$$\langle Px \xrightarrow{\rho} Pa \rangle_{\{Pfy \leftarrow Py\}}.$$

This problem has the most general solutions $\{x \mapsto a\}$, $\{x \mapsto fa\}$, $\{x \mapsto ffa\}$, ...

For a cycle unification problem $\langle G \xrightarrow{\rho} F \rangle_{\{L \leftarrow R\}}$ to be solvable, the atoms F and G must be of the form $P(t_1, \ldots, t_n)$ and $P(s_1, \ldots, s_n)$, respectively. Since L and R are weakly unifiable, their predicate symbols must also be identical, ie. L and R must be of the form $P'(l_1, \ldots, l_n)$ and $P'(r_1, \ldots, r_n)$, respectively. In the sequel we will only consider cycle unification problems of this form. Furthermore, as the case $P \neq P'$ is trivial, we assume $P = P'$.

To solve a cycle unification problem $\langle G \xrightarrow[c]{\rho} F \rangle$ we have to find a substitution which either unifies G and F or unifies – viz. simultaneously unifies each equation in –

$$\mathcal{C}^k = \mathcal{N} \cup \mathcal{Y}^k \cup \mathcal{X}^k,$$

where

$\mathcal{N} = \{s_1 \doteq l_1^1, \ \ldots, \ s_n \doteq l_n^1\}$ is the set of *entry equations*,

$\mathcal{Y}^k = \{r_1^i \doteq l_1^{i+1}, \ \ldots, \ r_n^i \doteq l_n^{i+1} \mid 1 \leq i \leq k\}$
is the set of *cycle equations* for k iterations through the cycle, and

$\mathcal{X}^k = \{r_1^{k+1} \doteq t_1, \ \ldots, \ r_n^{k+1} \doteq t_n\}$
is the set of *exit equations* after k iterations through the cycle.

The following proposition is an immediate consequence of the completeness and soundness of the connection method [1] or SLD-resolution, eg. [9]. Due to lack of space we had to omit the proof of this proposition and all further theorems. They can be found in detail in [4].

Proposition 1. *σ is a solution for $\langle G \xrightarrow[c]{\rho} F \rangle$ iff there exists a substitution θ such that θ unifies G and F and $\sigma = \theta|_{\mathcal{V}ar(G)}$ or there exists a natural number k such that θ unifies $\mathcal{C}^k$ and $\sigma = \theta|_{\mathcal{V}ar(G)}$.*

Throughout the paper τ will denote the most general unifier of G and F restricted to $\mathcal{V}ar(G)$, if it exist. Similarly, τ_k will denote the most general unifier of $\mathcal{C}^k$ restricted to $\mathcal{V}ar(G)$, if it exists.

In order to be able control a cycle, we are interested in the answer to three basic questions. Is cycle unification decidable? How many independent most general solutions has a cycle unification problem? Does there exist a unification algorithm which enumerates a minimal and complete set of solutions for a cycle unification problem?

Following [18], we define the type of a cycle unification problem as follows. A cycle unification problem is of type *unitary* iff there exists a single most general solution, *finitary* iff there exist finitely many most general solutions, and *infinitary* iff there exist infinitely many most general solutions.

4 Matching Cycles ($\mathcal{C}_m$)

A cycle $\{L \leftarrow R\}$ is called *left matching* and *right matching* iff there is a substitution σ such that $\sigma L = R$ and $L = \sigma R$, respectively. We assume σ to be the most general matcher. Throughout the paper we will consider only right matching cycles. The corresponding results for left matching cycles can be obtained analogously. The class of cycle unification problems with matching cycles is denoted by $\mathcal{C}_m$.

Matching cycles show some interesting properties. The substitution σ_i which is the most general unifier of R^i and L^{i+1} in the i-th iteration through the cycle can easily be obtained from the substitution σ which matches L against R as follows. Let

$$P_\sigma = \{(x, \sigma x) \mid x \in \mathcal{D}om(\sigma)\} \cup \{(x, x) \mid x \in \mathcal{V}ar(R) \setminus \mathcal{D}om(\sigma)\}.$$

Then,

$$\sigma_i = \{x^i \mapsto t^{i+1} \mid (x, t) \in P_\sigma\}.$$

For example, the cycle $\{Pgz,ga,z \leftarrow Pxyz\}$ is right matching with most general matcher $\sigma = \{x \mapsto gz,\ y \mapsto ga\}$ and we obtain $P_\sigma = \{(x, gz),\ (y, ga),\ (z, z)\}$ and $\sigma_i = \{x^i \mapsto gz^{i+1},\ y^i \mapsto ga,\ z^i \mapsto z^{i+1}\}$. Furthermore, since $\mathcal{D}om(\sigma_i) \cap \mathcal{D}om(\sigma_{i+1}) = \emptyset$ the most general solution of the cycle equations $\mathcal{Y}^k$ is simply the composition $\rho = \sigma_k \cdots \sigma_1$. The solution τ_k for C^k can now be obtained by simultaneously unifying the atoms ρG, ρL^1 and ρR^{k+1}, ρF and restricting their most general unifier to $\mathcal{V}ar(G)$, if it exists. Since $\rho G = G$, $\rho R^{k+1} = R^{k+1}$, and $\rho F = F$, the interesting bindings in ρ are those for the variables occurring in L^1.

We will now consider two classes of matching cycle unification problems defined by the matching substitution σ. These are the classes of variable-pure matching cycles and non-recursive matching cycles.

4.1 Variable-pure Matching Cycles ($\mathcal{C}_{vp}$)

A variable-pure matching cycle $L \leftarrow R$ is a matching cycle, where the matching substitution σ is variable-pure. Hence, σ must be of the form

$$\{x_1 \mapsto y_1,\ \ldots,\ x_l \mapsto y_l\},$$

where $x_j \in \mathcal{V}ar(R)$, $y_j \in \mathcal{V}ar(L)$, $1 \leq j \leq l$, and, thus, P_σ is a set of variable-pairs.

A set of variable-pairs P recursively defines a *sequence* of variables as follows. If $(x_1, x_2) \in P$, then P defines the sequence $\langle x_1, x_2 \rangle$. If $\langle x_1, \ldots, x_l \rangle$ is a sequence defined by P and $(x_l, x_{l+1}) \in P$, then $\langle x_1, \ldots, x_l, x_{l+1} \rangle$ is a sequence defined by P. A sequence $\langle x_1, \ldots, x_l \rangle$ is called *linear* iff $x_i \neq x_j$, $1 \leq i, j \leq l$, $i \neq j$. A sequence $\rho = \langle x_1, \ldots, x_l \rangle$ *contains* (or is a *subsequence* of) sequence π iff there exists an $i \geq 1$ and an $j \leq l$ such that $\pi = \langle x_i, \ldots, x_j \rangle$. A sequence π is called *maximal* iff there is no longer sequence containing π. A sequence $\pi = \langle x_1, \ldots, x_l \rangle$ is called a (*cyclic*) *permutation* iff $\langle x_1, \ldots, x_{l-1} \rangle$ is linear and $x_1 = x_l$. A sequence $\langle x_1, \ldots, x_l \rangle$ is called a (*cyclic*) *permutation with linear entry-sequence* iff $\langle x_1, \ldots, x_{l-1} \rangle$ is maximal and linear and there exists an j, $1 < j < l$, such that $x_l = x_j$. It is obvious that a permutation with linear entry-sequence can be divided into the linear entry-sequence and the permutation. Finally, a sequence $\langle x_1, \ldots, x_{l+1} \rangle$ has *length* l.

Let $C = \{L \leftarrow R\}$ be a variable-pure matching cycle with matching substitution σ. C is a *linear sequence* iff P_σ defines a single maximal and linear sequence. A cycle C is a *permutation* iff P_σ defines a permutation such that the variables occurring in it are equal to the variables occurring in P_σ. C is a *permutation with linear entry-sequence* iff P_σ defines precisely one permutation with linear entry-sequence.

In the following paragraphs we will formally investigate the properties of the classes of cycles defining a linear sequence ($\mathcal{C}_{ls}$), a permutation ($\mathcal{C}_p$), and a permutation with linear entry-sequence ($\mathcal{C}_{pls}$). As a combination we obtain results for the class of variable-pure matching cycles ($\mathcal{C}_{vp}$).

Linear Sequences ($\mathcal{C}_{ls}$). In this section we consider variable-pure matching cycles $\{L \leftarrow R\}$ with matching substitution σ where P_σ defines a single, maximal linear sequence. Such a cycle is contained in the unification problem

$$\langle Pw_1w_2w_3 \overset{\circ}{\longrightarrow} Pfa,bc\rangle_{\{Pfy,zv \leftarrow Pfx,yz\}},$$

where $\sigma = \{x \mapsto y,\ y \mapsto z,\ z \mapsto v\}$ is the matching substitution. P_σ defines the linear sequence $\langle x, y, z, v\rangle$ with length $l = 3$ and $\sigma_i = \{x^i \mapsto y^{i+1},\ y^i \mapsto z^{i+1},\ z^i \mapsto v^{i+1}\}$. As mentioned before, the solution for the cycle-equations $\mathcal{Y}^k$ is $\sigma_k \cdots \sigma_1$ and we are only interested in the restriction of $\sigma_k \cdots \sigma_1$ to $\mathcal{V}ar(L^1)$. Thus, we compute

$$\sigma_1|_{\{y^1,\ z^1,\ v^1\}} = \{y^1 \mapsto z^2,\ z^1 \mapsto v^2\}$$

and

$$\sigma_2\sigma_1|_{\{y^1,\ z^1,\ v^1\}} = \{y^1 \mapsto v^3,\ z^1 \mapsto v^2\}.$$

Since $\mathcal{D}om(\sigma_i)$, $i \geq 1$, does not contain any (superscribed) variable v, we find that for all $k > 2$

$$\sigma_k \cdots \sigma_1|_{\{y^1,\ z^1,\ v^1\}} = \sigma_2\sigma_1|_{\{y^1,\ z^1,\ v^1\}}$$

holds. One should observe that (i) a (superscribed) variable v is never in the domain of a most general solution for the exit equations and (ii) more and more variables in $\mathcal{V}ar(L^1)$ are mapped on a (superscribed) variable v until eventually all variables in $\mathcal{V}ar(L^1)$ are mapped on (superscribed) v. This is not only true in the example but holds for all solvable linear sequences. Intuitively, the more iterations through the cycle are considered the less is the influence of the exit equations on the solution of the cycle unification problem $\mathcal{C}^k$. In the example we obtain the solutions $\tau_0 = \{w_1 \mapsto fb,\ w_2 \mapsto c\}$ for $\mathcal{C}^0$, $\tau_1 = \{w_1 \mapsto fc\}$ for $\mathcal{C}^1$, $\tau_2 = \{w_1 \mapsto fv^3\}$ for $\mathcal{C}^2$, $\tau_3 = \tau_2$ for $\mathcal{C}^3$, etc.

One should observe that in the general case of a linear sequence of length l

$$\tau_{l-1} = \tau_{l+i} \text{ for } 0 \leq i$$

holds. Thus, for cycle unification problems defining linear sequences with length l we have only to consider τ, ie. the restriction of the most general unifier of G and F to $\mathcal{V}ar(G)$, if it exists, and the first $l-1$ iterations through the cycle to obtain all possible most general solutions for a cycle unification problem in the class $\mathcal{C}_{ls}$. In the example, $\tau_{l-1} = \tau_2$ is more general τ_1 and we conjecture that in the general case $\tau_{l-1} \preceq \tau_{l-i}$, $0 < i \leq l$, holds. Furthermore, in the example τ_2 is even more general than the most general unifier $\tau = \{w_1 \mapsto fa,\ w_2 \mapsto b,\ w_3 \mapsto c\}$ of $Pw_1w_2w_3$ and Pfa,bc, but this is not the case in general.

Conversely, if neither G and F are unifiable nor any of the sets C_i, $0 \leq i < l$, is solvable, then the cycle unification problem is unsolvable.

Permutations (C_p). We recall that a sequence $\langle x_1, \ldots, x_{l+1} \rangle$ is a permutation iff $\langle x_1, \ldots, x_l \rangle$ is linear and $x_1 = x_{l+1}$. The cycle unification problem

$$\langle Pxy \xrightarrow{\Omega} Pab \rangle_{\{Pvw \leftarrow Pwv\}}$$

defines the permutation $\langle w, v, w \rangle$ of length $l = 2$. If we solve

$$C^0 = \{x \doteq v^1,\ y \doteq w^1,\ w^1 \doteq a,\ v^1 \doteq b\},$$

we obtain the solution $\tau_0 = \{x \mapsto b,\ y \mapsto a\}$. Considering one iteration through the cycle we have to solve

$$C^1 = \{x \doteq v^1,\ y \doteq w^1,\ w^1 \doteq v^2,\ v^1 \doteq w^2,\ w^2 \doteq a,\ v^2 \doteq b\},$$

which results in $\tau_1 = \{x \mapsto a,\ y \mapsto b\}$. Two iterations through the cycle and solving

$$C^2 = \{x \doteq v^1,\ y \doteq w^1,\ w^1 \doteq v^2,\ v^1 \doteq w^2,\ w^2 \doteq v^3,\ v^2 \doteq w^3,\ w^3 \doteq a,\ v^3 \doteq b\}$$

yields $\{x \mapsto b,\ y \mapsto a\} = \tau_0$. Thus, we periodically return to previously computed solutions. In general, we have only to consider the unifier of G and F, if it exists, and finitely many iterations through the cycle to obtain all possible most general solutions for a cycle unification problem in the class of permutations. Conversely, if neither G and F are unifiable nor any one of the sets C^i, $0 \leq i < l$, is solvable, then the cycle unification problem is unsolvable.

Permutations with Linear Entry-sequence (C_{pls}). We recall that a permutation with linear entry-sequence has the form

$$\langle x_1, \ldots, x_l, x_{l+1}, \ldots, x_{l+m}, x_{l+1} \rangle$$

in which $\langle x_1, \ldots, x_{l+1} \rangle$ is a linear (entry-) sequence and $\langle x_{l+1}, \ldots, x_{l+m}, x_{l+1} \rangle$ is a permutation. Since the permutations with linear entry-sequence can be splitted into these two parts, their behaviour is determined as a combination of these parts. After l–1 iterations through the cycle the variables occurring in L^1 depend only on the (superscribed) variables $x_{l+1}, \ldots, x_{l+m}$. The values for $x_{l+1}, \ldots, x_{l+m}$ are solely determined by the permutation $\langle x_{l+1}, \ldots, x_{l+m}, x_{l+1} \rangle$ and the exit equations. As before, we have only to consider finitely many – viz. $m-1$ – further iterations through the cycle to obtain all possible most general solutions for a cycle unification problem in the class C_{pls}. Conversely, if neither G and F are unifiable nor any one of the sets C^i, $0 \leq i < l+m-1$, is solvable, then the cycle unification problem is unsolvable. We will give an example of C_{pls} at the end of Section 4.1.

Variable-pure Matching Cycles (C_{vp}). Variable-pure matching cycles define combinations of permutations, permutations with linear entry-sequence, and maximal linear sequences. Let a variable-pure matching cycle define p permutations

$$\langle x_{1,i},\ \ldots,\ x_{m_i,i},\ x_{1,i} \rangle,\ 1 \leq i \leq p,$$

pl permutations with linear entry-sequence

$$\langle y_{1,i},\ \ldots,\ y_{l_i,i},\ y_{l_i+1,i},\ \ldots,\ y_{l_i+n_i,i},\ y_{l_i+1,i} \rangle,\ 1 \leq i \leq pl,$$

and l maximal linear sequences

$$\langle z_{1,i}, \ldots, z_{\tilde{l}_i+1,i}\rangle,\ 1 \leq i \leq l.$$

Let $M = \max(1, l_1, \ldots, l_{pl}, \tilde{l}_1, \ldots, \tilde{l}_l)$ and $N = \operatorname{lcm}(1, m_1, \ldots, m_p, n_1, \ldots, n_{pl})$, where lcm denotes the least common multiple.

As in the previous subsection we find that after M–1 iterations through the cycle the variables occurring in L^1 depend only on the (superscribed) variables occurring in the permutations – ie. in $\{x_{1,i}, \ldots, x_{m_i,i} \mid 1 \leq i \leq p\}$.[5] – and that the values for the (superscribed) variables $x_{1,i}, \ldots, x_{m_i,i}$, $1 \leq i \leq p$, are solely determined by the permutations $\langle x_{1,i}, \ldots, x_{m_i,i}, x_{1,i}\rangle$, $1 \leq i \leq p$, and the exit equations. As before, we have only to consider finitely many – viz. $N-1$ – further iterations through the cycle to obtain all possible most general solutions for a cycle unification problem in the class C_{vp}. More formally, we can show that for all $i \geq 0$ and $k > 0$ the claims

$$C^{M-1+i} \text{ is solvable iff } C^{M-1+i+k\cdot N} \text{ is solvable}$$

and

$$\tau_{M-1+i} \stackrel{\bullet}{\leq} \tau_{M-1+i+k\cdot N}$$

hold. One should observe that this result subsumes the result of linear sequences (where $N = 1$), of permutations (where $M = 1$) and of permutations with linear entry-sequence.

Let $\langle G \stackrel{\Omega}{\longrightarrow} F\rangle_{\{L \leftarrow R\}}$ be a cycle unification problem. The steps in Figure 1 define a cycle unification algorithm for variable-pure matching cycles. Algorithms for C_{ls}, C_p, and C_{pls} are special cases. To illustrate the algorithm consider the cycle unification problem

$$\langle Pu_1u_2u_3u_4u_5 \stackrel{\Omega}{\longrightarrow} Pabcde\rangle_{\{Pyzvzv \leftarrow Pxyzvw\}}.$$

We obtain the following steps.

1. $Pu_1u_2u_3u_4u_5$ and $Pabcde$ are unifiable with most general unifier $\tau = \{u_1 \mapsto a,\ u_2 \mapsto b,\ u_3 \mapsto c,\ u_4 \mapsto d,\ u_5 \mapsto e\}$.
2. $\sigma = \{x \mapsto y,\ y \mapsto z,\ z \mapsto v,\ v \mapsto z,\ w \mapsto v\}$ is a most general right matcher of $Pxyzvw$ against $Pyzvzv$. P_σ defines two permutations with linear entry-sequence, viz. $\langle x, y, z, v, z\rangle$ and $\langle w, v, z, v\rangle$. Hence, $M = 2$, $N = 2$ and the problem is in C_{vp}.
3. $\tau_0 = \{u_1 \mapsto b,\ u_2 \mapsto c,\ u_3 \mapsto d,\ u_4 \mapsto c,\ u_5 \mapsto d\}$,
   $\tau_1 = \{u_1 \mapsto c,\ u_2 \mapsto d,\ u_3 \mapsto c,\ u_4 \mapsto d,\ u_5 \mapsto c\}$,
   $\tau_2 = \{u_1 \mapsto d,\ u_2 \mapsto c,\ u_3 \mapsto d,\ u_4 \mapsto c,\ u_5 \mapsto d\}$
   are the most general solutions obtained by solving C^0, C^1, and C^2, restricted to $\{u_1, u_2, u_3, u_4, u_5\}$, respectively.
4. We obtain the set $\{\tau, \tau_0, \tau_1, \tau_2\}$ as a minimal and complete set of solutions.

[5] Note, $\{\langle y_{l_i+1,i}, \ldots, y_{l_i+n_i,i}, y_{l_i+1,i}\rangle \mid 1 \leq i \leq pl\} \subseteq \{\langle x_{1,i}, \ldots, x_{m_i,i}, x_{1,i}\rangle \mid 1 \leq i \leq p\}$.

1. **If G and F are unifiable, then compute τ as the most general unifier for G and F restricted to the variables in G.**
2. **If $\langle G \twoheadrightarrow F\rangle_{\{L\leftarrow R\}} \in \mathcal{C}_{vp}$, then compute the lengths $l_1, \ldots, l_i$ of all defined maximal linear sequences/linear entry-sequences and the lengths $m_1, \ldots, m_j$ of all defined permutations. Let $M = \max(1, l_1, \ldots, l_i)$ and $N = \mathrm{lcm}(1, m_1, \ldots, m_j)$.**
3. **If $\mathcal{C}^k$ is solvable, then compute τ_k as the most general unifier for $\mathcal{C}^k$, restricted to the variables occurring in G, $0 \leq k \leq M+N-2$.**
4. **Let Σ be the set of solutions obtained in steps (1) and (3). If $\Sigma = \emptyset$, the problem is unsolvable. Otherwise, iteratively eliminate a substitution α if the current set of solutions contains another substitution δ with $\delta \preceq \alpha\ [\mathcal{V}ar(G)]$. The obtained set is a minimal and complete set of solutions for the cycle unification problem $\langle G \twoheadrightarrow F\rangle_{\{L\leftarrow R\}}$.**

Fig. 1. A Unification Algorithm for $\mathcal{C}_{vp}$.

Theorem 2. *Let C be a variable-pure matching cycle.*

(a) $\langle G \xrightarrow[C]{\ \ } F\rangle$ *is decidable.*

(b) $\langle G \xrightarrow[C]{\ \ } F\rangle$ *is finitary.*

(c) *There exists an algorithm computing a minimal and complete set of solutions for* $\langle G \xrightarrow[C]{\ \ } F\rangle$.

One should observe that this result holds for $\mathcal{C}_{ls}$, $\mathcal{C}_p$, and $\mathcal{C}_{pls}$ as well because they are subsets of $\mathcal{C}_{vp}$.

4.2 Non-recursive Matching Cycles ($\mathcal{C}_{nr}$)

In the preceding subsection we have considered only variable-pure matching cycles. We will now lift this restriction by considering matching cycles in which the matching substitution may bind variables to terms including function-symbols and constants. As an example consider the cycle unification problem

$$\langle Py \twoheadrightarrow Pa\rangle_{\{Pfx\leftarrow Px\}},$$

which has an infinite, complete and minimal set $\{\{y \mapsto a\}, \{y \mapsto fa\}, \{y \mapsto ffa\}, \ldots\}$ of solutions. Since we do not yet know how to control such cycles, we restrict ourselves and exclude bindings like $x \mapsto fx$, as it is contained in the matcher of the example. A variable x is called *recursive* iff there exists an i, $i > 0$, such that $x \mapsto t \in \sigma^i$, $x \in \mathcal{V}ar(t)$, and $t \neq x$.[6] Because $t \neq x$, variables occurring in permutations are not recursive. However, the variable x in the example above

[6] By σ^i we denote the i–fold composition of σ with itself, ie. $\sigma^1 = \sigma$ and $\sigma^i = \sigma(\sigma^{i-1})$.

is recursive. It is easy to verify that it is decidable whether σ contains recursive variables by considering σ^j, $1 \leq j \leq k$, where k is the number of bindings in σ. A cycle $\{L \leftarrow R\}$ is called *non-recursive matching* iff the domain of the matcher does not contain any recursive variable. The class of non-recursive matching cycle unification problems is denoted by $\mathcal{C}_{nr}$.

One should observe that if σ is a matching substitution for a cycle in $\mathcal{C}_{nr}$, then P_σ may contain pairs (x,t), where t is not a variable. To be able to deal with those pairs, we have to extend the definition of sequences. A set of variable-term pairs P defines a *sequence* as follows. If $(x,t) \in P$ and t is ground, then P defines $\langle x,t\rangle$. If $(x,t) \in P$ and $\mathcal{V}ar(t) = \{x_1,\ldots,x_n\}$, then P defines $\langle x,x_1\rangle$, $\ldots$, $\langle x,x_n\rangle$. Let $\langle x_1,\ldots,x_n\rangle$ be a sequence defined by P and $(x_n,t) \in P$. If t is ground, then P defines $\langle x_1,\ldots,x_n,t\rangle$. If $\mathcal{V}ar(t) = \{y_1,\ldots,y_n\}$, then P defines $\langle x_1,\ldots,x_n,y_1\rangle$, $\ldots$, $\langle x_1,\ldots,x_n,y_n\rangle$. A sequence $\pi = \langle x_1,\ldots,x_l,t\rangle$ is called *linear* iff each variable occurs at most once in π. One should observe that since non-recursive cycles do not contain recursive variables, a (superscribed) variable x occurring in a permutation cannot be mapped on a term t such that $x \in \mathcal{V}ar(t)$ and $t \neq x$. Thus, permutations are constructed from pairs of variables in P_σ only.

It is easy to see, that we can allow linear sequences ending in a ground term without changing our previous results. Let a non-recursive matching cycle define p permutations, pl permutations with linear entry-sequence, and l maximal linear sequences possibly ending in a ground term. Furthermore, let M and N be defined as in the previous subsection. As before we obtain

$$\mathcal{C}^{M-1+i} \text{ is solvable iff } \mathcal{C}^{M-1+i+k\cdot N} \text{ is solvable}$$

and

$$\tau_{M-1+i} \stackrel{\triangleleft}{=} \tau_{M-1+i+k\cdot N}$$

for $i \geq 0$ and $k > 0$. Thus, the results for variable-pure cycle unification problems can be generalized to non-recursive matching cycle unification problems.

The unification algorithm for non-recursive matching cycles is analogous to the algorithm for variable-pure matching cycles but with the extended definition of sequences. As an example consider the cycle unification problem

$$\langle Pu_1u_2u_3u_4 \xrightarrow{\Omega} Pabcd\rangle_{\{Pfyz,vwz \leftarrow Pxyzw\}}.$$

An application of the algorithm yields the following results.

1. $Pu_1u_2u_3u_4$ and $Pabcd$ are unifiable with the most general unifier $\tau = \{u_1 \mapsto a,\ u_2 \mapsto b,\ u_3 \mapsto c,\ u_4 \mapsto d\}$.

2. $\sigma = \{x \mapsto fyz,\ z \mapsto w,\ w \mapsto z,\ y \mapsto v\}$ is a most general right matcher of $Pxyzw$ against $Pfyz,vwz$. P_σ defines the sequences $\langle x,y,v\rangle$ and $\langle x,z,w,z\rangle$ such that $M=2$, $N=2$ and the problem is in $\mathcal{C}_{nr}$ because the cycle's matcher does not contain any recursive variable.

3. $\tau_0 = \{u_1 \mapsto fbc,\ u_2 \mapsto v^1,\ u_3 \mapsto d,\ u_4 \mapsto c\}$,
   $\tau_1 = \{u_1 \mapsto fv^2d,\ u_3 \mapsto c,\ u_4 \mapsto d\}$
   $\tau_2 = \{u_1 \mapsto fv^2c,\ u_3 \mapsto d,\ u_4 \mapsto c\}$
   are the most general solutions obtained by solving $\mathcal{C}^0, \mathcal{C}^1$, $\mathcal{C}^2$, restricted to $\{u_1,\ u_2,\ u_3,\ u_4\}$, respectively.

4. We obtain the set $\{\tau, \tau_1, \tau_2\}$ as a minimal and complete set of solutions.

Theorem 3. *Statements (a), (b), and (c) of Theorem 2 hold also for non-recursive matching cycles.*

5 Summary and Future Work

In this paper we have formally defined cycle unification (for a restricted class of formulas). We have considered various classes of cycle unification problems with increasing complexity, have shown that they are decidable and finitary, and have specified a minimal and complete unification algorithm for these classes. Table 1 gives an overview of these results as well as of previous work. In each row we state the decidability and the unification type for a particular class of cycle unification problems, indicate whether there exists an algorithm to compute a minimal and complete set of solutions, and provide the reference if there exists one. $\mathcal{C}$ denotes the class of unrestricted cycle unification problems. In $\mathcal{C}_l$ and $\mathcal{C}_g$ goals and facts are restricted to be linear and ground, respectively. $\mathcal{C}_u$ denotes the class of *unifying* cycles, ie. cycles $L \leftarrow R$, for which L and R are unifiable. This class has recently been investigated in [22]. For the definition of the more complicated classes $\mathcal{C}_m$ and $\mathcal{C}_{nr}$ the reader is referred to Section 4. The various classes are related as shown in Figure 2.

Class	Decidability	Type	Algorithm	References
$\mathcal{C}$	open	infinitary	open	
$\mathcal{C}_l$	decidable	infinitary	open	[7]
$\mathcal{C}_g$	decidable	unitary	yes	[16]
$\mathcal{C}_m$	open	infinitary	open	
$\mathcal{C}_{nr}$	decidable	finitary	yes	this paper
$\mathcal{C}_u$	decidable	finitary	yes	[22]

Table 1. Properties of cycle unification classes.

One might think that our results for the class $\mathcal{C}_{nr}$ may easily be extended for cycle unification problems $\langle G \xrightarrow{\sigma} F\rangle_{\{L\leftarrow R\}}$ such that there exist two non-recursive substitutions σ and τ with $\sigma L = \tau R$. But this is not true. For the cycle $\{Px \leftarrow Pfx\}$ we find $\{x \mapsto fy\}Px = \{x \mapsto y\}Pfx$. Yet, as shown in Subsection 4.2 there are infinitely many independent solutions for this problem. In the case $\{Pfx,z \leftarrow Pyfx\}$ we find $\{z \mapsto fx\}Pfx,z = \{y \mapsto fx\}Pyfx$ and one might expect that only a single instance of the cycle is needed. However, as shown in [22], the latter example belongs to a class of *unifying* cycles, ie. cycles $L \leftarrow R$ for which there exists a substitution σ with $\sigma L = \sigma R$, which need one cycle iteration, ie. two instances of the clause.

One of the major open problems in our restricted context is the question whether $\mathcal{C}$ is decidable. $\mathcal{C}_l$, $\mathcal{C}_u$, and $\mathcal{C}_{nr}$ are decidable. However, there are several results which point into the opposite direction for the case of $\mathcal{C}$. In [5] it is shown that

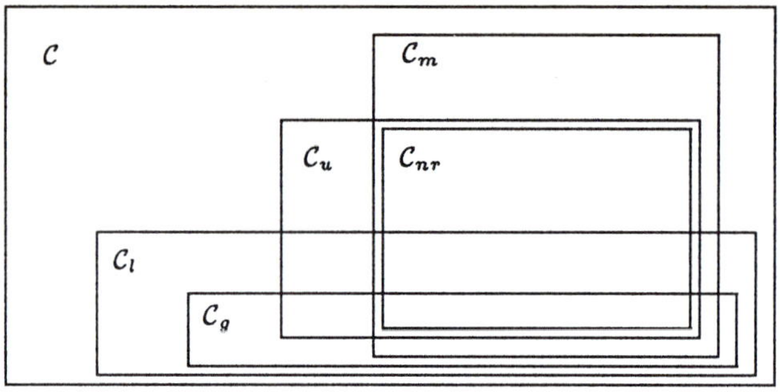

Fig. 2. The relation between the classes $\mathcal{C}$, $\mathcal{C}_l$, $\mathcal{C}_g$, $\mathcal{C}_m$, $\mathcal{C}_{nr}$, and $\mathcal{C}_u$.

the termination of a one rule term rewriting system, where rewriting may occur at proper subterms, is undecidable. Similarly, we know from [16] that the class of Horn clauses consisting of two clauses of the form $L \leftarrow R$ and two ground unit clauses is undecidable. It is, however, not obvious, how these result could be adapted to cycle unification problems.

In the future we intend to develop heuristics to control further classes of cycle unification problems. We are looking for a well–founded ordering based on a measure of complexity for the instances of the cycle in order to apply an idea similar to the one contained in [16]. Certain cycles $L \leftarrow R$ cause some of the terms occurring in L and R to grow or shrink monotonically at each iteration of the cycle. If there were an upper bound for these terms defined by G or F, then one would be able to decide the cycle unification problem $\langle G \xrightarrow{\Omega} F \rangle_{\{L \leftarrow R\}}$. For illustration of this idea consider the cycle unification problem

$$\langle Pffx,x \xrightarrow{\Omega} Pufu \rangle_{\{Pfffy,fz \leftarrow Pfy,z\}}.$$

The i-th instance of the right-hand side of the cycle $Pfffy,fz \leftarrow ePfy,z$ is matched against the $i+1$-st instance of the left-hand side by $\sigma_i = \{y^i \doteq ffy^{i+1},\ z^i \doteq fz^{i+1}\}$. We observe that the depth of y and z decreases with each iteration through the cycle and the goal as well as the fact define upper bounds because the multiple occurrences of x and u correlate y and z via the entry- and exit equations. In [4] we have exploited this insight for the computation of the number k of iterations through the cycle to obtain a solution. For the example we obtain $k = 2$ and the solution $\tau_2 = \{x \mapsto f^5 y^3\}$.

As shown in Section 3, there might be infinitely many independent solutions of a cycle unification problem. Hence, we need a compact representation of infinitely many terms for such cases and intend to use the one suggested in [11].

In order to solve the Łukasiewicz formula mentioned in the introduction as a challenge problem, the results obtained so far and in the future need to be generalized to the case of more than one interacting cycles. [3] contains first ideas how this might be achieved. Altogether there is quite a bit of work ahead of us until this challenge problem might be solved in less than a second as thought possible in [3].

As an example of an application of cycle unification in logic programming mentioned in the introduction consider the cycle $max(X,Y,Z)\ :-\ max(Y,X,Z)$ in PROLOG notation. It is contained in a clause set representing SAM's lemma [21]. The cycle expresses the commutativity of the first two arguments of the maximum-predicate. This clause may be used in a PROLOG program which computes the maximum of two numbers.

$$\begin{aligned} max(X,Y,Y) &:- X =< Y. \\ max(X,Y,Z) &:- max(Y,X,Z). \end{aligned}$$

If we ask the query $?-\ max(2,3,V)$, PROLOG yields the desired result and V is bound to 3. But if we ask for all solutions, the program does not terminate because the search space is infinite. Similarly, negative queries handled by negation-as-failure may not produce the expected results because of possibly infinite evaluation trees; for example consider the query $?-\ \sim max(2,3,2)$[7] leading to the subgoal $?-\ max(2,3,2)$.

With the tools of this paper it can be seen, however, that the cycle consists of a permutations of length 1 and a permutation of length 2. Therefore, it is sufficient to consider at most one self-application of the cycle to obtain all correct answers. As a result, the infinite search space collapses to a trivial one and the computation of the two program clauses becomes roughly equivalent with the following program.

$$\begin{aligned} max(X,Y,X) &:- X > Y. \\ max(X,Y,Y) &:- X =< Y. \end{aligned}$$

No query asked to this program will give rise to a non-terminating computation because all queries have a finite evaluation tree.

Acknowledgement: The third author was partially supported by the ESPRIT project MEDLAR and the Stadt Dreieich. We would like to thank Franz Baader and two anonymous referees. Their very valuable comments led to a significant improvement of the paper.

References

1. W. Bibel. *Automated Theorem Proving.* Vieweg Verlag, Braunschweig, 2 edition, 1987.
2. W. Bibel. Advanced topics in automated deduction. In R. Nossum, editor, *Fundamentals of Artificial Intelligence II*, pages 41–59. Springer, LNCS *345*, 1988.
3. W. Bibel. Perspectives on automated deduction. In R. S. Boyer, editor, *Automated Reasoning: Essays in Honor of Woody Bledsoe*, pages 77–104. Kluwer Academic, Utrecht, 1991.
4. W. Bibel, S. Hölldobler, and J. Würtz. Cycle unification. Technical Report AIDA-91-15, FG Intellektik, FB Informatik, TH Darmstadt, 1991.
5. M. Dauchet. Simulation of a Turing machine by a left-linear rewrite rule. In *Proceedings of the Conference on Rewriting Techniques and Applications*, pages 109–120. Springer, LNCS *355*, 1989.
6. N. Dershowitz and J.-P. Jouannaud. Notations for rewriting. *EATCS Bulletin*, 43:162–172, 1991.

[7] $\sim$ denotes the negation-as-failure.

7. P. Devienne. Weighted graphs: A tool for studying the halting problem and time complexity in term rewriting systems and logic programming. *Journal of Theoretical Computer Science*, 75:157–215, 1990.
8. E. Eder. Properties of substitutions and unifications. *Journal of Symbolic Computation*, 1:31–46, 1985.
9. J. W. Lloyd. *Foundations of Logic Programming*. Springer, 1984.
10. J. Minker and J.-M. Nicolas. On recursive axioms in deductive databases. *Information Systems*, 8(1):1–13, 1983.
11. H. J. Ohlbach. Abstraction tree indexing for terms. In L. C. Aiello, editor, *Proceedings of the European Conference on Artificial Intelligence*, pages 479–484, 1990.
12. H. J. Ohlbach. Compilation of recursive two-literal clauses into unification algorithms. In P. Jorrand and V. Sgurev, editors, *Proceedings of the AIMSA*, pages 13–22, 1990.
13. H. J. Ohlbach and G. Wrightson. Solving a problem in relevance logic with an automated theorem prover. In R. E. Shostak, editor, *Proceedings of the Conference on Automated Deduction*, pages 496–508. Springer, LNCS *170*, 1984.
14. F. Pfenning. Single axioms in the implicational propositional calculus. In E. Lusk and R. Overbeek, editors, *Proceedings of the Conference on Automated Deduction*, pages 710–713. Springer, LNCS *310*, 1988.
15. J. A. Robinson. A machine-oriented logic based on the resolution principle. *Journal of the ACM*, 12:23–41, 1965.
16. M. Schmidt-Schauß. Implication of clauses is undecidable. *Journal of Theoretical Computer Science*, 59:287–296, 1988.
17. D. De Schreye, K. Verschaetse, and M. Bruynooghe. A practical technique for detecting non-terminating queries for a restricted class of Horn clauses, using directed, weighted graphs. In *Proceedings of the International Conference on Logic Programming*, pages 649–663, 1990.
18. J. H. Siekmann. Unification theory. *Journal of Symbolic Computation*, 7:207 – 274, 1989.
19. M. E. Stickel. Automated deduction by theory resolution. *Journal of Automated Reasonsing*, 1:333–356, 1985.
20. L. Vielle. Recursive query processing: The power of logic. Technical Report TR-KB-17, European Computer-Industry Research Center, 1987.
21. L. Wos. The problem of finding a strategy to control binary paramodulation. *Journal of Automated Reasonsing*, pages 101–107, 1988.
22. J. Würtz. Unifying cycles. Technical report, Deutsches Forschungszentrum für Künstliche Intelligenz, 1992. To appear.

A Parallel Completion Procedure for Term Rewriting Systems [1]

Katherine A. Yelick
University of California at Berkeley

Stephen J. Garland
MIT and The Hebrew University

Abstract. We present a parallel completion procedure for term rewriting systems. Despite an extensive literature concerning the well-known sequential Knuth-Bendix completion procedure, little attention has been devoted to designing parallel completion procedures. Because naive parallelizations of sequential procedures lead to over-synchronization and poor performance, we employ a transition-based approach that enables more effective parallelizations. The approach begins with a formulation of the completion procedure as a set of transitions (in the style of Bachmair, Dershowitz, and Hsiang) and proceeds to a highly tuned parallel implementation that runs on a shared memory multiprocessor. The implementation performs well on a number of standard examples.

1 Introduction

We describe a parallel completion procedure for term rewriting systems. A sequential completion procedure was formulated first by Knuth and Bendix [13]. Extensions, modifications, and applications to algebra, theorem proving, and data type induction are described by Buchberger [3] and Dershowitz [5].

Performance is an important factor that limits the applicability of completion procedures, and of term rewriting systems in general. We show how parallelism can lead to significantly better performance. Opportunities for parallelism abound, because completion is not inherently sequential. But straightforward parallelizations of the Knuth-Bendix procedure perform poorly. Careful algorithm and data structure design is needed, as in sequential completion procedures, to avoid superfluous work. How parallel tasks are scheduled must be tuned, because the order in which steps are performed plays a crucial role in performance.

This paper is divided as follows. Section 2 defines the completion problem. Section 3 describes the issues that arise in finding good parallel solutions. Section 4 presents transition axioms for a completion procedure using the inference rules of Bachmair, Dershowitz, and Hsiang [2]. Section 5 transforms these axioms into ones suitable for parallel implementation. Section 6 describes the implementation. Section 7 describes its performance. Sections 8 and 9 describe related work and summarize our results.

2 The Completion Problem

We assume a familiarity with the notions of terms and substitutions. An equation is an unordered pair of terms, written $(s \doteq t)$. We write $(s \doteq t)$ when we want to

---

[1] Both authors were supported in part by the Advanced Research Projects Agency of the Department of Defense, monitored by the Office of Naval Research under contract N00014-89-J-1988, by the National Science Foundation under grant CCR-8910848, and by the Digital Equipment Corporation. K. Yelick was supported in part by AT&T and by the University of California at Berkeley. S. Garland was supported in part by a Fulbright Lectureship.

Authors' addresses: K. A. Yelick, Computer Science Division, University of California, Berkeley, CA 94720. e-mail: yelick@cs.berkeley.edu. S. J. Garland, MIT Laboratory for Computer Science, 545 Technology Square, Cambridge, MA 02139. e-mail garland@lcs.mit.edu.

distinguish an orientation; i.e., $(s \doteq t)$ is either $(s \doteq t)$ or $(t \doteq s)$.

A rewrite rule is an ordered pair of terms, written $s \to t$. A term rewriting system R is a set of rewrite rules, which defines a relation $\to_R$ on terms such that $s \to_R t$ (*s rewrites to t*) if and only if there is a rule $l \to r$ in R and a substitution σ such that s contains σl as a subterm (l is said to *match* this subterm) and t is formed by replacing the occurrence of σl by σr. Let $\to_R^+$, $\to_R^*$, and $\leftrightarrow_R^*$ be the transitive, reflexive transitive, and reflexive symmetric transitive closures of $\to_R$. Since $\leftrightarrow_R^*$ is symmetric, it is well-defined even when R is replaced by a set of equations E. The relation $\leftrightarrow_E^*$ is the *equational theory of* E and is also denoted by E^*; i.e., an equation $(s \doteq t)$ is in E^* if and only if $s \leftrightarrow_E^* t$.

R is *confluent* if, whenever $r \to_R^* s$ and $r \to_R^* t$, there is a term u such that $s \to_R^* u$ and $t \to_R^* u$. R is *noetherian* if $\to_R^+$ contains no infinite chains. R is *convergent* if it is both confluent and noetherian. If R is convergent, then for any term t there exists a unique irreducible (i.e., unrewritable) term $t\downarrow_R$ (called the *normal form* of t) such that $t \to_R^* t\downarrow_R$. When R is simply noetherian, we write $t\downarrow_R$ to mean some one of the possibly many normal forms of t. When R is convergent, $s \leftrightarrow_R^* t$ if and only if $s\downarrow_R = t\downarrow_R$.

Some applications of term rewriting require convergent systems; others attempt to establish convergence by creating new rules in a process known as *completion*. Such applications raise the following related problems.

Definition. Given a set of equations E, the *finite completion problem* is to find a convergent rewriting system R such that $\leftrightarrow_R^*$ and $\leftrightarrow_E^*$ are the same relation.

For some some sets of equations E, no finite R exists that solves the finite completion problem. In such cases, it is sometimes useful to find approximations R_i to E, in the following sense.

Definition. Given a set of equations E, the *completion problem* is to produce a (possibly infinite) sequence of rewriting systems $R_0, R_1, \ldots$ such that each R_i is noetherian, $\to_{R_i}^*$ is contained in $\leftrightarrow_E^*$, and, for any equation $(s \doteq t)$ in E^*, there exists i such that for all $j > i$, s and t have unique normal forms in R_j, and $s\downarrow_{R_j} = t\downarrow_{R_j}$.

If the finite completion problem can be solved for E, the resulting rewriting system provides a decision procedure for E^*. If the completion problem can be solved, the resulting sequence of rewriting systems provides a semidecision procedure. However, there may not be a solution to the completion problem. For example, if E consists of a single commutativity axiom, no noetherian rewriting system has the same equational theory as E.

More general formulations of the completion problem increase the number of sets of equations for which solutions exist. Some generalizations allow function symbols in R that are not in E, requiring only that $\leftrightarrow_R^*$ be a conservative extension of $\leftrightarrow_E^*$. Others include completion modulo equations (such as commutativity) [17]. Although these generalizations fall in the class addressed by our approach, we do not consider them in this paper.

Solving the completion problem involves generating additional rules, if necessary, to ensure confluence. Such rules can be found using *unification* and *critical-pairing*. Two terms s and t are *unifiable* if there is a substitution σ (called their *unifier*) such that $\sigma s = \sigma t$. If $\sigma s = \sigma t$, and for all other unifiers σ', there is a substitution τ

such that $\sigma' = \sigma \circ \tau$, then σ is a *most general unifier* of s and t. If two terms are unifiable, then they have a most general unifier. A substitution τ is a *renaming* if, for all variables v in the domain of τ, $\tau(v)$ is a variable. Most general unifiers are unique up to renaming, i.e., up to composition with renaming substitutions.

Let r_1 and r_2 be the rewrite rules $s \rightarrow t$ and $l \rightarrow r$. Assume r_1 and r_2 have no variables in common. Then $(s' \doteq t')$ is a *critical pair* between r_1 and r_2 if some σ unifies l with a non-variable subterm of s, s' is formed from σs by replacing the subterm σl by σr, and t' is σt. Let $crit(r_1, r_2)$ be the set of critical pairs between r_1 and r_2, with r_2's variables renamed, if necessary, to avoid conflicts with r_1's. Let $crit\_all(R)$ be $\bigcup\{crit(r_1, r_2) : r_1, r_2 \in R\}$. Note that $crit\_all(R)$ contains $crit(r, r)$ for all r in R. Both *crit* and *crit_all* are unique up to variable renaming.

Solving the completion problem also involves establishing that a rewriting system is noetherian, which itself is an undecidable problem. The most common approach to proving that a system is noetherian is to use a *reduction ordering* on terms, i.e., a monotonic well-founded ordering that is stable under substitution [4]. If $>$ is a reduction ordering and $l > r$ for every rule $l \rightarrow r$ in R, then R is noetherian. Completion procedures generally employ a fixed reduction ordering and halt with failure if this ordering is not powerful enough to orient some equations that arise during completion. This prompts the following definition.

Definition. Given a set of equations E and a reduction ordering $>$, a *completion procedure* produces a possibly infinite sequence $R = \langle R_0, R_1, \ldots \rangle$ of rewriting systems such that:

1. Each R_i is provably noetherian; i.e., $l > r$ for all $l \rightarrow r$ in R_i.
2. Each R_i is consistent with E; i.e., $\leftrightarrow^*_{R_i}$ is contained in $\leftrightarrow^*_E$.
3. If some R_i solves the finite completion problem, then the procedure halts with success, and this R_i is the last element in R.
4. If no R_i solves the finite completion problem, then either (a) the procedure halts with failure or (b) R solves the general completion problem, i.e., R is infinite, and, for any equation $s \leftrightarrow^*_E t$, $s\downarrow_{R_i} = t\downarrow_{R_i}$ for all sufficiently large i.

The first two conditions are *safety properties*, which prevent completion procedures from producing rewriting systems that do not terminate, or that incorrectly reduce two terms to a common normal form. The last two conditions are *liveness properties*, which require completion procedures to achieve certain results.

Condition (4a) classifies trivial procedures, which always halt with failure, as completion procedures. Completion procedures typically differ on the set of inputs for which they fail, and their ability to resist failure is one of the qualities by which they are judged. Completion procedures can be made *failure resistant* by allowing the reduction ordering to be enlarged in restricted ways [8]. Completion procedures can be made *unfailing* by leaving some equations unordered and restricting the application of rewrite rules [2]. Because these generalizations do not raise interesting new questions concerning parallelism, we do not consider them in this paper.

3 Design Issues for Parallel Completion Procedures

The completion problem provides many opportunities for parallelism. At the same time, it presents many pitfalls. Ideally, a parallel completion procedure running on n processors should be close to n times faster than a well-tuned sequential completion procedure running on a single processor. Such speed-ups, however, may be difficult to attain. Because of synchronization overhead, tasks cannot be too fine-grained and parallel data structures must minimize synchronization bottlenecks. Because all processors must be kept equally busy to achieve maximum performance, tasks cannot be so coarse-grained as to prevent effective load-balancing. And because highly parallel processes can exhibit nondeterministic behavior, care must be taken to maintain correctness while optimizing performance.

We explore these issues by discussing various ways to parallelize the sequential Knuth-Bendix completion procedure [13]. This procedure has two alternating phases: *internormalization*, which rewrites and eliminates equations and rules, and *critical pairing*, which creates new rules. Although the requirements for completion procedures do not mention internormalization, experience with sequential completion procedures has shown that it is essential for good performance.

3.1 Granularity

The simplest approach to designing a parallel completion procedure is to keep the overall sequential structure of the Knuth-Bendix procedure, but to use fine-grained parallelism for low level computations. For example, we can use theoretically efficient parallel procedures to determine if two terms match during internormalization. Yet this helps little in practice: even with lightweight threads, terms must be huge—with hundreds or thousands of operators—before the gains outweigh the overhead of starting and synchronizing parallel tasks. Although very large terms arise in some applications, they are not the norm.

At a higher level of granularity, we can parallelize the operations of rewriting and computing critical pairs, for example, by attempting to match and unify different subterms in parallel. But this also requires very large terms to justify the overhead.

At better levels of granularity, we can (and do) attempt to rewrite a given term by several rewrite rules, or to compute critical pairs between different rules in parallel.

3.2 Load Balancing

A further way to parallelize the sequential completion procedure is to perform the two processes of internormalization and critical pairing in parallel, thereby creating a two-stage pipeline with a feedback loop from the critical pairing process to the internormalization process. Additional parallelism can be used within the two stages to try to balance the pipeline. Implementing this design taught two lessons that led us to abandon developing parallel programs from conventional sequential ones.

The first lesson is that the amount of time spent in normalization far exceeds the time spent in critical pairing. The magnitude of the difference—a factor of 20 is not unusual—limits the potential speedup of the pipeline to 5%. Furthermore, on a six-processor machine, dedicating one sixth of the processing power to 5% of the work is not a good use of resources.

The second lesson is that, even with many processors, performance instability between the two stages makes it difficult to balance the pipeline: an expensive critical pairing stage that generates many new rules is generally followed by an expensive internormalization stage. On typical iterations, the ratio of the time spent internormalizing to the time spent critical pairing varied between 1/2 and 79. We can address these specific performance problems, for example, by having the critical pairing stage work on more rules to make it relatively more expensive, by further optimizing internormalization to make it less expensive, or by dividing the stages into parallel subtasks and dynamically allocating additional processors. But these solutions avoid the real problem: there is a synchronization point after each iteration of the pipeline, when each stage gets new rules from the other. The situation is even worse without the pipeline, there being two synchronization points per iteration, one after internormalization and one after critical pairing.

Unnecessary synchronization points are artifacts of basing parallel programs on sequential programs. Hence we use a different, transition-based approach [23] that allows parallelism both between and within critical pairing and internormalization.

3.3 Correctness

Achieving performance while maintaining correctness requires care in choosing algorithms and data structures. For example, if we attempt to rewrite both sides of an equation in parallel, then failed rewrites must not modify the terms being rewritten, and successful simultaneous rewrites must not interfere with one another.

When internormalizing in parallel, we must be careful to prevent two rules that are the same up to the names of their variables from reducing one another to trivial rules, i.e., to rules with identical left and right sides.

4 High-level Transition Axioms for Completion

We base the design of our parallel completion procedure not on traditional sequential procedures, but on a reformulation of the original Knuth-Bendix procedure [13] by Bachmair, Dershowitz and Hsiang [2] as a set of nondeterministically-applied transition axioms. Figure 1 presents transition axioms for a completion procedure similar to the inference rules in [2]. The state consists of a set of equations E and a set of rewrite rules R. Initially, E holds the user's input and R is empty. (Our formulation differs slightly from that in [2], which assumes that neither E nor R contains elements that differ only by renamings. Such an assumption is difficult to implement.) The procedure stops if and when all guards are false, e.g., because the *fail* transition sets both E and R to *ordering_failure*.

The transition axioms preserve the invariant that each rule $s \rightarrow t$ in R is ordered with respect to $>$, i.e., $s > t$. An important property of reduction orderings is that if $s \rightarrow t$ and r both satisfy this invariant, and if r rewrites t to t', then $s \rightarrow t'$ also satisfies the invariant; hence *right_reduce* preserves the invariant. Because *left_reduce* may not satisfy the invariant, and because it may reduce a rule to a triviality, it turns a rewritten rule into an equation.

The requirement $\neg age\_prevents(s \rightarrow t, r)$ in the guard for *left_reduce* prevents the completion procedure from removing all instances of a redundant rewrite rule. As noted earlier, when two rules are renamings of one another, either can reduce the other to a triviality. The details of this problem are not important for this paper, but the solution affects our presentation. We associate an *age* with each rewrite rule and define $age\_prevents(r_1, r_2)$ to be true if r_1 is older than r_2 and the left sides of r_1 and r_2 are renamings of each other. Using the age of rules is mentioned in [2], and the validity of this solution was confirmed by Dershowitz [6].

Any procedure that performs a fair interleaving of these actions solves the completion problem. CP fairness ensures both that every equation appearing in E is

State Components

E : set[equation] + ordering_failure
R : set[rule] + ordering_failure

Initially

E = user input
R = $\emptyset$

Transition Axioms

simplify % Apply one rewrite step to either side of an equation
$(s \doteq t) \in E \;\&\; t \to_R t' \Rightarrow$
$E := E - (s \doteq t) + (s \doteq t')$

delete % Delete trivial equation
$(s \doteq s) \in E \Rightarrow$
$E := E - (s \doteq s)$

orient % Convert equation into rewrite rule using reduction ordering
$(s \doteq t) \in E \;\&\; s > t \Rightarrow$
$E := E - (s \doteq t) \;\&\; R := R + (s \to t)$

right_reduce % Apply one rewrite step to right side of a rule
$(s \to t) \in R \;\&\; t \to_R t' \Rightarrow$
$R := R - (s \to t) + (s \to t')$

left_reduce % Apply one rewrite step to left side of a rule
$(s \to t) \in R \;\&\; r \in R \;\&\; s \to_{\{r\}} s' \;\&\; \neg\text{age\_prevents}(s \to t, r) \Rightarrow$
$R := R - (s \to t) \;\&\; E := E + (s' \doteq t)$

deduce % Add one critical pair between rules in R to E
$(s \doteq t) \in \text{crit\_all}(R) \Rightarrow$
$E := E + (s \doteq t)$

fail % Halt if there is an unorderable, nontrivial, normalized equation
$(s \doteq t) \in E \;\&\; s \neq t \;\&\; s\downarrow_R = s \;\&\; t\downarrow_R = t \;\&\; s \not> t \;\&\; t \not> s \Rightarrow$
$E, R :=$ ordering_failure

Liveness

CP fairness: for any nonterminating execution $(E_0, R_0), (E_1, R_1), \ldots$ and any i, $\bigcap_{j>i} E_j = \emptyset$ and, if $e \in \bigcap_{j>i}\text{crit\_all}(R_j)$, then some renaming of e is in some E_k.

CP termination: if $(E_0, R_0), (E_1, R_1), \ldots$ is nonterminating, then no R_i solves the finite completion problem for E_0.

Figure 1: High-level transition axioms for completion

eventually ordered, simplified, or deleted, and also that every critical pair is eventually added to E. CP termination ensures that the completion procedure terminates if and when it solves the finite completion problem.

The completion procedure can fail if there is a nontrivial equation e in E that can be neither ordered nor rewritten, since no fair execution can leave e in E forever. Most implementations, however, will run indefinitely with an unorderable equation as long as there is other work to do. These procedures are not CP fair, but this is an academic point: finite machine resources will eventually take care of termination.

5 Implementable Transition Axioms for Completion

The transition axioms in Figure 1 are not appropriate for direct implementation. To make them implementable, we must find ways to ensure liveness, balance their granularity, and simplify their guards. At the same time, we seek to enhance performance by avoiding repetitive work. As in good sequential completion procedures, we want to avoid normalizing an equation or rule multiple times by the same set of rules, to avoid computing critical pairs for a given pair of rules more than once, and to avoid computing critical pairs between unnormalized rules.

5.1 Liveness

In sequential completion procedures, liveness is generally established by proving that appropriate invariants are true at different points of control. For example, nested loops might normalize all equations with respect to all rules, and invariants about which equations are normalized with respect to which rules may depend on the control point within each loop.

In our parallel completion procedure, we establish liveness instead by dividing the state (i.e., the sets of equations and rules) into components that satisfy appropriate invariants, e.g., about which equations and rules have been normalized with respect to which other rules, and which rules have had their critical pairs computed. A significant part of the design effort for parallel completion involves defining appropriate state components and invariants.

5.2 Granularity

Most transitions in Figure 1 perform a single rewrite step or compute a single critical pair. They incur considerable overhead, and are too fine-grained to be useful on a small number of processors. Therefore, we seek to increase their granularity.

Since actions that are too coarse-grained result in insufficient parallelism, we avoid both extremes by defining transitions that apply a single rule, at most once, to (at most) every rule or equation in the system, or that apply all rules as many times as possible to a single equation or rule. We avoid transitions that apply all rules to all equations.

5.3 Simple Guards

We simplify the guards for the transitions by turning them into inexpensive tests for the existence of an element in a state component or into tests involving the value of a scalar variable. The guards in Figure 1 involve more costly tests: a term being in normal form, an equation being orderable, or an equation being a critical pair between two rules. Fortunately, the same technique that helps with liveness can be

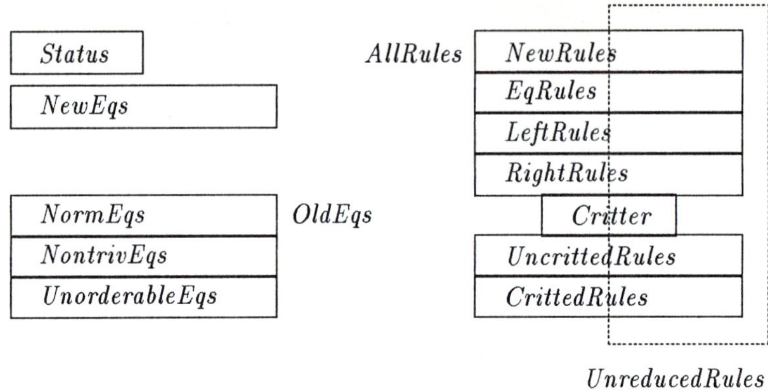

Initially, *NewEqs* = user input, *Status* = *running*, others = $\emptyset$.

Figure 2: Venn diagram of state components for completion

used to simplify guards: we put equations and rules into different state components, and we define invariants for these components that match the guards of Figure 1.

Figure 2 shows our state components. All but one is implemented as a *queue*; the queues *OldEqs*, *AllRules*, and *UnreducedRules* share elements with other queues, but the rest are mutually disjoint. The queues can simply be semiqueues, which ensure only that anything enqueued will eventually be dequeued. We use FIFO queues. The "usual" path that data (i.e., rules and equations) take in Figure 2 is from *NewEqs* to *OldEqs* to *AllRules* to *CrittedRules*. A final state component, *Status*, can have one of two values: *running* or *ordering_failure*.

The state components that contain equations have the following invariants. *OldEqs* contains the same elements as *NormEqs*, *NontrivEqs*, and *UnorderableEqs*. All equations in *OldEqs* are normalized with respect to *AllRules* $-$ *NewRules*. No equation in *NontrivEqs* or *UnorderableEqs* is trivial. No equation in *UnorderableEqs* can be ordered by the reduction ordering.

The state components that contain rewrite rules have the following invariants. *UnreducedRules* is a subset of *AllRules*, which has the same elements as the seven other queues shown in Figure 2. If $s \rightarrow t$ is in *AllRules*, then $s > t$. The left side of every rule in *AllRules* is normalized with respect to *AllRules* $-$ *NewRules* $-$ *EqRules*. The right side of every rule in *AllRules* $-$ *UnreducedRules* is normalized with respect to *RightRules*, *Critter*, *UncrittedRules*, and *CrittedRules*. *Critter* contains at most one rule; all critical pairs between this rule and itself have been added to *NewEqs* (but may no longer be there). For any r_1 and r_2 in *UncrittedRules* or *CrittedRules*, $crit(r_1, r_2)$ has been added to *NewEqs*. All critical pairs between the rule in *Critter* and those in *CrittedRules* have been added to *NewEqs*. (*UncrittedRules* is so-called because critical pairs have not yet been computed between its rules and *Critter*.)

Figures 3 and 4 contain directly implementable transition axioms for completion. These transitions move rules from one queue to another, maintaining the above invariants. In their descriptions, $rewritable(Q, r)$ contains the elements of Q that can be rewritten using r; $rewrites(Q, r)$ contains these elements, each rewritten once by r; $left\_reducible(Q, r)$ contains those $s \rightarrow t$ in Q such that $\neg age\_prevents(s \rightarrow t, r)$ and s can be rewritten by r; $left\_reduced(Q, r)$ contains these rules, with s rewritten

Transition Axioms

```
normalize_eqn
    e = head(NewEqs) ⇒
        NewEqs := NewEqs − e &
        NormEqs := NormEqs + e ↓AR &
        OldEqs := OldEqs + e ↓AR
filter_eqn
    (s ≐ t) = head(NormEqs) ⇒
        NormEqs := NormEqs − (s ≐ t) &
        if (s ≠ t) then NontrivEqs := NontrivEqs + (s ≐ t)
orient_eqn
    (s ≐ t) = head(NontrivEqs) ⇒
        NontrivEqs := NontrivEqs − (s ≐ t) &
        if (s > t) then NewRules := NewRules + (s → t)
                & AllRules := AllRules + (s → t)
                & OldEqs := OldEqs − (s ≐ t)
            elseif (t > s) then NewRules := NewRules + (t → s)
                & AllRules := AllRules + (t → s)
                & OldEqs := OldEqs − (s ≐ t)
            else UnorderableEqs := UnorderableEqs + (s ≐ t)
back_simplify
    r = head(NewRules) ⇒
        NewRules := NewRules − r &
        EqRules := EqRules + r &
        NewEqs := NewEqs + rewrites(OldEqs, r) &
        foreach X in (NormEqs, NontrivEqs, UnorderableEqs, OldEqs)
            X := X − rewritable(X, r)
left_reduce
    r = head(EqRules) ⇒
        EqRules := EqRules − r &
        LeftRules := LeftRules + r &
        NewEqs := NewEqs + left_reduced(AllRules, r) &
        foreach X in (NewRules, EqRules, LeftRules, RightRules, Critter,
                UncrittedRules, CrittedRules, UnreducedRules, AllRules)
            X := X − left_reducible(X, r)
right_reduce
    r = smallest(LeftRules) ⇒
        LeftRules := LeftRules − r &
        RightRules := RightRules + r &
        UnreducedRules := UnreducedRules + right_reduced(AllRules, r)
                − right_reducible(UnreducedRules, r) &
        foreach X in (NewRules, EqRules, LeftRules, RightRules, Critter,
                UncrittedRules, CrittedRules, AllRules)
            X := X − right_reducible(X, r) + right_reduced(X, r)
```

Figure 3: Directly implementable transitions, Part I

right_normalize
  $(s \to t) = \text{head}(\text{UnreducedRules}) \Rightarrow$
    UnreducedRules := UnreducedRules $- (s \to t)$ &
    foreach X **in** (NewRules, EqRules, LeftRules, RightRules,
      Critter, UncrittedRules, CrittedRules, AllRules)
      if $(s \to t) \in X$ **then** $X := X - (s \to t) + (s \to t \downarrow_{\text{AllRules}})$
add_critical$_1$
  $r_1 = \text{head}(\text{Critter})$ & $r_2 = \text{head}(\text{UncrittedRules}) \Rightarrow$
    UncrittedRules := UncrittedRules $- r_2$ &
    CrittedRules := CrittedRules $+ r_2$ &
    NewEqs := NewEqs $\cup$ crit(r_1, r_2)
add_critical$_2$
  UncrittedRules $= \emptyset$ & $r = \text{head}(\text{RightRules}) \Rightarrow$
    UncrittedRules := Critter $\cup$ CrittedRules &
    CrittedRules := $\emptyset$ &
    Critter := $\{r\}$ &
    RightRules := RightRules $- r$ &
    NewEqs := NewEqs $\cup$ crit(r, r)
fail
  UnorderableEqs $\neq \emptyset \Rightarrow$
    Status := ordering_failure

Liveness

Weak fairness: *fail* must be executed if its guard remains true.

Figure 4: Directly implementable transitions, Part II

once by r. *right_reducible*(Q, r) contains those $s \to t$ in Q such that t can be rewritten by r; *right_reduced*(Q, r) contains these rules, with t rewritten once by r.

Note that when *orient_eqn* creates a new rule, the equations in *OldEqs* must be renormalized. *Back_simplify* moves an equation rewritten by the new rule to *NewEqs*, where it will be renormalized by *normalize_eqn*. *Left_reduce* moves rewritten rules back to *NewEqs*, where they can be deleted or oriented in the other direction. *Right_reduce* puts rewritten rules into *UnreducedRules*, where they will be renormalized by *right_normalize*.

The liveness property in Figure 4 is *weak fairness* between *fail* and all other actions. Thus, an unorderable equation may exist in *UnorderableEqs* for a long time before the completion process fails, but unless the equation is moved or deleted, the process must eventually halt with failure. Note that there is no fairness requirement between any of the other actions.[2]

The axioms in Figures 3 and 4 can be viewed as an implementation of Figure 1, and standard refinement mapping techniques [1] applied to prove correctness. In these methods, the set of possible executions of the implementation are shown to be a subset of the executions of the specification, with a refinement mapping used to associate states of one with states of the other.

[2] Weak fairness is stronger than necessary, because the guard on *fail* may be continuously true even though *UnorderableEqs* has no persistent element. Moreover, the fairness condition may be dropped entirely if the semidecision procedure property is not desired.

5.4 Designing for Performance

Among the lessons learned from sequential implementations of completion are that internormalization enhances performance, but redundant normalization is inefficient. The *UnreducedRules* queue helps balance these conflicting requirements. When the right side of a rule is rewritten, it may no longer be in normal form with respect to the other rules; but equations and other rules that were previously unrewritable by this rule remain unrewritable. Hence we implement a "reminder" that the rule must be right-normalized by placing it in *UnreducedRules*, while at the same time leaving it in the queue to which it belonged so as not to lose the information contained in the invariant for that queue.

Although we designed our state components and transitions to avoid extra normalizations and critical pairing, the design does not prevent this from happening when objects become outdated. This trade-off keeps communication overhead low.

We implement *RightRules* as a priority queue, which is a special case of a semiqueue in which the "smallest" element is dequeued first. We choose this implementation because rules dequeued from *RightRules* will be used to compute critical pairs, and sequential implementations have demonstrated the practical importance of computing critical pairs between small rules before larger ones.

6 The Implementation

A two-step implementation based on the transition axioms in Figures 3 and 4 follows the transition-based approach described in [23]. First, we implement each transition axiom by a transition procedure. Our approach requires that these transition procedures appear atomic with respect to one another: any concurrent execution of the procedures must be equivalent to a sequential execution in which each procedure invocation takes a step specified by some transition axiom. Second, we implement a scheduler to execute the transition procedures in parallel. The scheduler, which is a parallel program that runs on all processors, guarantees liveness; being application-specific, it gives far better performance than would be achieved by allowing the operating system to schedule tasks. The weak liveness property in Figure 4 gives us the freedom to choose a scheduling order that performs well.

An easy way to make the transition procedures appear atomic is to make them atomic, e.g., to have them run within critical regions. But this leads to a completely sequential, albeit nondeterministic, completion procedure. Thus, significant algorithm and data structure design goes into implementing the transition procedures so that they appear atomic when they are actually highly concurrent.

Instead of describing our concurrent data structures here, we describe some of their properties and the kinds of parallelism they admit. We lock sparingly; the few critical regions are typically only a few instructions long. For example, each queue has two locks—one on the pointer to its head, the other on its pointer to the tail; each lock is held only long enough to increment its pointer. The queues contain pointers to rewrite rules and equations; the extra level of indirection allows rules and equations to be in several queues at once, as indicated by the overlapping regions in the Venn diagram of Figure 2.

Rewrite rules are modified in place, for example, by the *right_normalize* procedure, which must replace a rule by a right-normalized rule in each of seven queues. This procedure is implemented by modifying the rule in *UnreducedRules*, which is a shared copy of the rules in the other queues.

Rewrite rules are modified without locking. The code relies on the underlying shared memory, which guarantees that if a memory location is read at the same time it is written, then the read will observe either the value before or the value after the write—it will not observe an intermediate nonsensical value. A rewrite rule is represented by two locations, and simultaneous rewrites to the two terms are allowed. If two simultaneous rewrites are done to the same term, one may be lost. However, this loss does not affect correctness. It may affect performance in that the lost rewrite may have to be redone, but lost rewrites rarely happen in practice, and the effect on performance is not significant.

The scheduler that runs on each processor consists of a loop that repeatedly invokes transition procedures. Most of the design decisions in the scheduler are related to performance tuning, and are discussed in Section 7 along with the performance results. The problem of detecting termination within the scheduler is nontrivial, because it requires taking a global snapshot of the system to determine whether all guards are simultaneously false. If a process examines the queues in turn to see whether they are empty, by the time it finishes new elements may have been enqueued. Therefore, an agreement algorithm must be run between the schedulers on the separate processors to ensure that all continue running until they agree there is no work to do.

7 Scheduling and Performance

Experiments show that the best scheduler executes the transition procedures in a loop with the order given by Figures 3 and 4. This scheduler repeatedly invokes the same transition procedure as long as possible, then moves on to the next. The only exceptions are the *add_critical* procedures, which it executes only once before repeating the other procedures. This choice was motivated by experience with sequential completion procedures: performance, both in time and space, is better if equations and rules are kept in normal form. Thus, critical pair computations are stopped as soon as renormalization is required.

Figure 5 shows the performance of this scheduler on a Firefly with six CVAX processors when completing typical sets of equations taken from the term rewriting literature. The first column shows the number of milliseconds taken by the program running on one processor. The other columns show the relative performance with more processors. These figures were obtained by averaging five executions for each example; they show that speedups are better for the larger examples.

The first example generates a complete set of rewrite rules for group theory. The next provides some simple axioms for arithmetic, which *fib4* extends to compute the fourth Fibonacci number. Completing *homomorphism* establishes the fact that a map from one group into another that preserves group multiplication also preserves inverses and the identity. Completing *group56* produces a complete presentation of a group of order 56. The last example is due to Ursula Martin.

The time required to complete *group56* depends dramatically on the order in which critical pairs are computed, because one particular critical pair eliminates most of the other rules [16]. A round-robin scheduler, which executes each transition procedure once before going on to the next, seems to exhibit super-linear speed-ups for this example, gaining a factor of eight with two processors and fifteen with six.

	1 (ms)	1 : 2	1 : 3	1 : 4	1 : 5	1 : 6
group	1718	2.0	2.4	2.1	4.1	2.9
arith	4666	1.9	2.7	3.4	3.3	2.9
fib4	6603	1.8	3.1	2.9	2.2	3.5
homomorphism	11980	2.0	2.8	2.5	4.2	4.2
group56	18796	2.2	2.3	3.1	3.7	3.8
domino	44162	2.0	2.9	3.7	3.8	5.1

Figure 5: Transition-based completion using the best scheduler

However, the round-robin scheduler takes more time in all cases than the scheduler we use. Hence looking at speedups without considering absolute performance is a very bad way to tune parallel programs. Furthermore, leaving scheduling to the operating system is likely to produce terrible results.

Performance depends not just on the way the completion procedure schedules tasks, but also on the way the operating system allocates processes to processors. The times in Figure 5 were obtained using a system scheduler that does not move processes from one processor to another, thereby preserving cache context. When processes are allowed to move, performance is significantly worse: the speed-ups with six processors were about 65% of those in Figure 5.

Our parallel completion procedure is consistently faster, even when running on one processor, than the completion procedure provided by the Larch Prover [12]. While this comparison is somewhat unfair to the Larch Prover, which has more functionality, the comparison does show that our parallel implementation is fast enough to be of practical use.

These performance results are encouraging. For the larger examples, performance continues to improve as processors are added. This does not guarantee scalability beyond a small number of processors, but neither does it provide evidence against scalability.

8 Related Work

Dershowitz and Lindenstrauss [7] describe an abstract concurrent machine for rewriting a term by a set a rules, and they use it to compare various rewriting strategies. The parallelism in their model is very fine grained. They associate a process with each operator in a term and try to minimize the normalization time for that term. This is quite different from our implementation, which emphasizes larger grained parallelism; it allows parallel rewriting during internormalization, but does not intentionally schedule multiple rewrites of the same term.

Some sequential completion procedures [11, 14, 15] also divide the state into components, either to achieve better performance or to facilitate reasoning about correctness. The data structures used for our state components differ from these, however, because they are concurrent and have more control information to allow for the additional executions that come from parallelism.

Slaney and Lusk [20] describe a parallel closure computation, which divides the problem of completing $R = \{r_1, \ldots, r_n\}$ into tasks that compute critical pairs between r_i and $\{r_1, \ldots, r_i\}$, adding any results not subsumed by R to a new list S, and a single task that moves rules from S to R, incrementing n and rechecking the rest of S for subsumption. Because this method fails to address issues (such as internormalization) that affect the total amount of work performed, any speed-ups it exhibits are not very meaningful. For example, if two critical pairing tasks generate members s_1 and s_2 of S such that s_2 subsumes s_1, but not *vice versa*, a task will be spawned to compute critical pairs between s_1 and all members of R, without there being any way to abort such an expensive task when it is found to be superfluous.

A completion procedure that has been parallelized successfully is Buchberger's algorithm for computing Gröbner Bases [3]. Like us, Ponder observed performance problems when using a straightforward parallelization of Buchberger's algorithm [18]. Vidal [22] redesigned the data structures and rearranged the top-level procedure to get superior speed-ups. His procedure does internormalization in a separate sequential phase. It is therefore simpler than our procedure, but less efficient, since unreduced polynomials may cause extraneous computation.

9 Summary and Conclusions

Completion is a complicated problem, and the design of parallel completion procedures differs in nontrivial ways from the design of sequential procedures. The nondeterministic description of completion in terms of transition axioms given by Bachmair *et al* [2] is a good starting point for a parallel design, particularly because it avoids some of the unnecessary serialization inherent in parallelizing a conventional sequential program. But there is a large step between their description and the directly implementable transition axioms that describe our procedure. In particular, the liveness requirement (CP fairness) must be encoded into data structures without creating too many serialization points.

Our design technique appears promising as a means of parallelizing other applications that are highly irregular in their data structures, control structures, and communication patterns. Examples of problems that have been difficult to parallelize by other means include circuit verification and simulation, certain sparse matrix problems, and many symbolic applications.

Our implementation runs on a shared memory multiprocessor and performs well when completing typical sets of equations, generally achieving speedups of $4n/5$ or better when run with n processors. We are currently porting our implementation to an eight-processor Sequent, to test the scalability of our implementation.

We believe the completion procedure will scale to more processors as long as there is sufficient data to keep the transition procedures busy. Because completion often generates hundreds of intermediate rewrite rules, and because of the combinatorial nature of applying these rules, an implementation for upwards of a hundred processors should be interesting. There does not appear to be sufficient parallelism, however, to support tens of thousands of processors. To utilize that many processors, one would need to parallelize the processes of rewriting, matching, and unification.

References

[1] M. Abadi and L. Lamport. The existence of refinement mappings. Research Report 29, Digital Equipment Corporation Systems Research Center, Palo Alto, CA, 1988.

[2] L. Bachmair, N. Dershowitz, and J. Hsiang. Orderings for equational proofs. In *Proceedings of the Symposium on Logic in Computer Science*, pages 346–357. IEEE, 1986.

[3] B. Buchberger. History and basic features of the critical pair/completion procedure. *Journal of Symbolic Computation*, 3(1&2):3–38, February/April 1987.

[4] N. Dershowitz. Orderings for term-rewriting systems. *Theoretical Computer Science*, 17:279–301, 1982.

[5] N. Dershowitz. Completion and its applications. In H. Ait-Kaci and M. Nivat, editors, *Resolution of Equations in Algebraic Structures*, volume II: Rewriting Techniques, pages 31–86. Academic Press, New York, 1989.

[6] N. Dershowitz, 1990. Private communication.

[7] N. Dershowitz and N. Lindenstrauss. An abstract machine for concurrent term rewriting. In *Proceedings of the Second International Conference on Algebraic and Logic Programming, Berlin.* LNCS, October 1990.

[8] D. Detlefs and R. Forgaard. A procedure for automatically proving termination of a set of rewrite rules. In *Proceedings of the First International Conference on Rewriting Techniques and Applications, Dijon, France*, pages 255–270. LNCS 202, May 1985.

[9] C. Dwork, P. C. Kanellakis, and J. C. Mitchell. On the sequential nature of unification. *Journal of Logic Programming*, 1:35–50, June 1984.

[10] C. Dwork, P. C. Kanellakis, and L. Stockmeyer. Parallel algorithms for term matching. *SIAM Journal of Computing*, 17(4):711–731, August 1988.

[11] S. J. Garland and J. V. Guttag. A guide to LP, the Larch Prover. Technical Report 82, Digital Equipment Corporation Systems Research Center, Palo Alto, CA, 1991.

[12] S. J. Garland, J. V. Guttag, and J. L. Horning. Debugging Larch Shared Language specifications. Technical Report 60, Digital Equipment Corporation Systems Research Center, Palo Alto, CA, 1990.

[13] D. E. Knuth and P. B. Bendix. Simple word problems in universal algebras. In J. Leech, editor, *Computational Problems in Abstract Algebra*, pages 263–297. Pergamon, Oxford, 1970.

[14] P. Lescanne. Completion procedures as transition rules + control. In *TAPSOFT '89*, pages 28–41. LNCS 351, 1989.

[15] P. Lescanne. Orme, an implementation of completion procedures as sets of transition rules. In M. Stickel, editor, *Proceedings of the 10th International Conference on Automated Deduction*, pages 661–662. LNCS 449, 1990.

[16] U. Martin. Doing algebra with REVE. Technical report, University of Manchester, Manchester, England, 1986.

[17] G. E. Peterson and M. E. Stickel. Complete sets of reductions for some equational theories. *J. ACM*, 28(2):233–264, Apr. 1981.

[18] C. Ponder. Evaluation of performance enhancements in algebraic manipulation systems. Technical Report UCB/CSD-88/438, Computer Science Division, University of California, Berkeley, CA 94720, 1988.

[19] R. Ramesh and I. Ramakrishnan. Optimal speedups for parallel pattern matching in trees. In *Proceedings of the 2nd International Conference on Rewriting Techniques and Applications, Bordeaux, France*, pages 274–285. LNCS 256, May 1987.

[20] J. K. Slaney and E. W. Lusk. Parallelizing the closure computation in automated deduction. In *Proceedings of the 10th International Conference on Automated Deduction*, pages 28–29. LNCS 449, 1990.

[21] R. M. Verma and I. Ramakrishnan. Tight complexity bounds for term matching problems. *Information and Computation*, 1990.

[22] J.-P. Vidal. The computation of Gröbner bases on shared memory multiprocessors. Technical Report CMU-CS-90-163, School of Computer Science, Carnegie Mellon University, Pittsburg, PA, 1990.

[23] K. A. Yelick. *Using Abstraction in Explicitly Parallel Programs.* PhD thesis, MIT Laboratory for Computer Science, Cambridge, MA 02139, December 1990. Also appeared as MIT/LCS/TR-507, July 1991.

Grammar Rewriting

David McAllester

MIT Artificial Intelligence Laboratory
545 Technology Square
Cambridge Mass. 02139

Abstract. We present a term rewriting procedure based on congruence closure that can be used with arbitrary equational theories. This procedure is motivated by the pragmatic need to handle equational theories where confluence can not be achieved. The procedure uses context free grammars to represent equivalence classes of terms. The procedure rewrites grammars rather than terms and uses congruence closure to maintain certain congruence properties of the grammar. Grammars provide concise representations of large term sets. Infinite term sets can be represented with finite grammars and exponentially large term sets can be represented with linear sized grammars. Although the procedure is primarily intended for use in nonconfluent theories, it also provides a new kind of confluence that can be used to give canonical rewriting systems for theories that are difficult to handle in other ways.

1 Introduction

In most practical applications of term rewriting systems, such as verifications done with the Boyer-Moore prover [3], confluence can not be achieved. In such cases attempts to prove true equations often fail. In this paper a new term rewriting procedure is described that is intended to improve the success rate for proof attempts in nonconfluent theories. In cases where normal term rewriting fails, success can sometimes be achieved by generating a set of normal forms rather than an individual normal form. This is done using a term ordering where a single term can have many normal forms all of which are minimal under that ordering. Somewhat surprisingly, canonicalization under such weak term orderings is possible. The canonical form of a given term is taken to be the *set* of all normal forms.

The main problem with using sets as canonical forms is that the sets involved can be quite large. In fact, to improve the success rate in nonconfluent theories we would like the canonical sets to be as large as possible. Fortunately, large sets can be compactly represented with context free grammars. A finite grammar can represent an infinite set of terms. More importantly, very large finite term sets can be represented by compact grammars. Consider the equational theory consisting of the single commutativity axiom of the form $x+y=y+x$. In this theory the equivalence class of a sum of n constants contains order 2^n terms. However, a grammar for generating this class contains only order n productions. As another example consider the theory consisting of a single associativity axiom $x+(y+z)=(x+y)+z$. In this theory the size of the equivalence class of a sum of n constants is the Catalan number of n — a hyperexponential function. However the equivalence class of a sum of n constants can be generated by a grammar with order n^3 productions. In the equational theory that contains both associativity and commutativity the grammar for generating the equivalence class of a sum of n constants grows exponentially in n. However, the grammar is still vastly smaller than the equivalence class itself.

Using grammars to represent equivalence classes involves the well known congruence closure procedure. Congruence closure is an efficient algorithm for determining the consequences of a finite set of ground equations [8], [13], [11], [5]. A finite set of ground equations can be converted to a grammar that encodes the congruence relation on terms implicit in the equations. Congruence closure can be viewed as an algorithm for converting ground equations to grammars. The rewriting procedure described here operates on grammars — a grammar is repeatedly rewritten to incorporate new ground equations.

The rewriting procedure described here is analogous to ordered rewriting systems [2], [6], [9], [12]. It rewrites (representations of equivalence classes of) ground terms using a set of unordered equations. This rewriting is done in the presence of a well founded order on ground terms and the rewriting process is guaranteed to terminate. However, unlike previous ordered rewriting systems, the order on ground terms is not assumed to be total. The commutativity equation $x + y = y + x$ can be handled by placing *both* sides of the equation into the representation of the set being rewritten. Thus the grammar rewriting procedure described here provides a way of handling non-orientable equations that is different from both ordered rewriting and from the use of special unification algorithms [7].

The grammar rewriting technique described here is somewhat similar to a congruence closure based rewriting technique for nonground rewrite rules described by Chew [4]. However, Chew's procedure requires that the set of rewrite rules be nonoverlapping, i.e., that there are no critical pairs. Chew's procedure also essentially requires that the normal form of a term be a single term rather than a set represented by a grammar.

The full version of this paper (availible form the author) shows how grammar rewriting can be used to give a canonical rewriting system for idempotent semigroups. It has been shown that there is no canonical term or word rewriting system for this theory [1]. Unfortunately, the canonical grammar rewriting system for idempotent semigroups is highly exponential and is of theoretical interest only.

The grammar rewriting procedure described here has been incorporated into the Ontic verification system which is under continued development by the author, Robert Givan, Carl Witty, and Kevin Zalondek. Experimentation with the procedure in Ontic is currently under way.

2 Congruence Grammars

This section reviews basic properties of congruence grammars. A congruence grammar is formally equivalent to a deterministic bottom-up finite state tree automaton [14] [10]. The grammar view used here seems more natural than the tree automaton view in the term rewriting domain.

Congruence grammars provide a way of compactly representing equivalence classes of first order terms. Each equivalence class is represented by a nonterminal symbol of the grammar which generates the elements of the class. Equivalence classes are always disjoint sets. This observation motivates the restriction that no two productions of a congruence grammar can have the same right hand side. For example, we can not have $X \rightarrow a$ and $Y \rightarrow a$ where X and Y are distinct nonterminal symbols.

> **Definition:** A *Congruence Grammar* is a set of productions of the form $X \rightarrow f(Y_1, \cdots, Y_n)$, where f is an n-ary function symbol, each Y_i is a nonterminal symbol, and where no two productions have the same right hand side.

We use the standard definition of a first order term where we assume an infinite set of function symbols of each arity (number of arguments). Constant symbols are treated as function symbols of no arguments. So, for example, the production $X \rightarrow a$ is a production of the above form where a is function of no arguments. The above definition allows us to prove that distinct nonterminal symbols generate disjoint classes. The following lemma can be proved by induction on the size of terms.

> **Lemma:** A given term is generated by at most one nonterminal of a congruence grammar.

Example, Representing an Infinite Class: Consider the equivalence class of the constant symbol a under the single equation $a = f(a)$. This infinite equivalence class is a context free language generated by the two productions $X \rightarrow f(X)$ and $X \rightarrow a$.
Example, An Equivalence Class under Commutativity: Consider n distinct constants $a_1, a_2, \cdots, a_n$ and define the term t_i to be the sum of the first i constants associated to the left, i.e., t_1 is a_1, t_2 is $a_1 + a_2$, t_3 is $(a_1 + a_2) + a_3$, and so on. Note that for $1 < i \leq n$ we have that t_i is $t_{i-1} + a_i$. Suppose that $+$ is commutative but not associative. In this case the equivalence class of the term t_n is generated by the nonterminal X_n in the grammar containing the $2(n-1)$ productions of the form

$$X_i \rightarrow (a_i + X_{i-1})$$

$$X_i \rightarrow (X_{i-1} + a_i)$$

together with the production

$$X_1 \rightarrow a_1.$$

The 2^{n-1} terms in this equivalence class are generated by a grammar with $2n - 1$ productions.
Example, An AC Equivalence Class: Consider n constants $a_1, a_2, \cdots, a_n$ and suppose that $+$ is both associative and commutative. Let S be any non-empty subset of these constants. We define t_S to be the term that is the sum of the elements of S. For each non-empty subset S of the constants we introduce a nonterminal symbol X_S. Now consider the grammar which contains all productions of the form

$$X_S \rightarrow (X_W + X_V)$$

where W and V are disjoint and S is $W \cup V$ plus all productions of the form $X_{\{a_i\}} \rightarrow a_i$. The equivalence class of t_U is generated by the nonterminal X_U of this grammar. More generally, the nonterminal X_S generates the equivalence class of t_S.

The grammar that generates the equivalence class of t_U grows exponentially in size of U — the grammar contains order 3^n productions. This fact can be derived by observing that each production $X_S \rightarrow (X_W + X_V)$ classifies each element of U in one of three ways — either as a member of W, or a member of V, or as a member of neither. There are 3^n ways of classifying n constants into three groups. The number of productions in this grammar should be contrasted with the number of terms in the equivalence class. The grammar for the set of terms equivalent to a sum of n variables contains $O(3^n)$ productions while the equivalence class contains $O(4^n(n+1)!)$ different terms.[1] When a term ordering is used the grammar that generates the set of normal forms may contain fewer productions.

3 Congruence Grammars and Ground Equations

A congruence grammar is a representation of an equivalence relation on terms. It turns out that those relations which can be represented by finite congruence grammars are exactly those relations which are the deductive closure of a finite set of ground equations. This section states some basic results relating congruence grammars and ground equations.

First, we define the equivalence relation on (all) ground terms represented by a congruence grammar. This is done by defining an interning operation on ground terms. The word "interning" is commonly used to describe the way strings are mapped to variables in most programming languages. Here we are mapping semantically equivalent terms to the same internal data structure. We assume a given congruence grammar and define an interning operation such that for any term t, the result of interning t, denoted $I[t]$, is a nonterminal of the grammar such that $I[t]$ generates t. If there is no nonterminal of the grammar which generates t then the grammar is extended with new productions in such a way that the desired nonterminal is created.

Procedure for Interning a term t:

1. Let $s_1, \cdots, s_n$ be the (possibly empty) sequence of immediate subterms of t, i.e., the terms such that t is $f(s_1, \cdots, s_n)$.
2. Let $Y_1, \cdots, Y_n$ be the result of recursively interning the terms $s_1, \cdots, s_n$.
3. Let $X \rightarrow f(Y_1, \cdots, Y_n)$ be the production whose right hand side is $f(Y_1, \cdots, Y_n)$. If there is no such production then create one with a new nonterminal X and add it to the grammar.
4. Return the nonterminal symbol X.

Since no two productions share the same right hand side the productions can be stored in a hash table indexed by their right hand sides. Throughout this section we assume the presence of such a hash table and assume that hash table operations can be performed in unit time. Under these assumptions the above intern procedure runs in linear time in the size of the given term.

Note that the interning procedure has a "side effect" — it can add productions to the grammar. This ensures that for any sequence of interning operations the

[1] The order of the number of terms in the equivalence class is derived from Sterling's approximation applied to a formula involving Catalan numbers.

nonterminal symbol $I[t]$ that results from interning t will be the same for each repeated interning of t even if the original grammar did not contain any nonterminal that generated t. Although a purely functional treatment of interning is possible, the functional treatment is more complex. In spite of the implementation based on side effects, the intern function is best viewed as a function from terms to nonterminals where the mapping from terms to nonterminals is completely determined by the initial grammar. In fact, the initial congruence grammar determines an equivalence relation on terms such that two terms s and t are equivalent if $I[s]$ equals $I[t]$. For example, suppose the initial grammar contains the two productions $X \rightarrow f(X)$ and $X \rightarrow a$ and let $f^n(a)$ be an abbreviation for the term $f(f(\cdots f(a)))$ with n occurrences of f. In this case we have that $I[f^n(a)]$ equals X for any term of the form $f^n(a)$. This implies that $I[g(f^n(a))]$ will be equal to $I[g(f^m(a))]$ for any nonnegative integers n and m. Computing $I[g(f^n(a))]$ gives a nonterminal symbol Y such that the grammar has been extended to include the production $Y \rightarrow g(X)$. Although interning can add productions to the grammar, it does not change the equivalence relation on terms defined by the grammar. Different grammars define different intern mappings and we will write $I_G[s]$ to mean the result of interning the term s with respect to the grammar G.

> **Theorem:** For any finite set H of equations between ground terms there exists a finite congruence grammar $G_r(H)$ such that that for any two ground terms s and t we have that $H \models s = t$ if and only if $I_{G_r(H)}[s]$ equals $I_{G_r(H)}[t]$. Furthermore, assuming that hash table operations can be done in unit time, the grammar $G_r(H)$ can be computed from H in $n \log n$ time where n is the written length of H.

The proof of this theorem is based on well known algorithms for congruence closure [8], [13], [11], [5]. A procedure for incrementally incorporating new equations directly into a congruence grammar is given in section 6. A nonoptimal algorithm for converting a set of equations H to a congruence grammar $G_r(H)$ can be described as follows. For each term s occuring in H we introduce a nonterminal symbol X_s.[2] For each constant symbol a appearing in H we construct the production $X_a \rightarrow a$. For each term $f(w_1, \cdots w_k)$ occuring in H we add the production $X_{f(w_1, \cdots w_k)} \rightarrow f(X_{w_1}, \cdots, X_{w_n})$. Now we process the equations in H while maintaining a union-find structure on nonterminal symbols. For each equation $s = w$ in H we call union on the nonterminals X_s and X_w. After processing all equations we select a canonical representative from each equivalence class of nonterminal symbols and replace every nonterminal in the grammar by its canonical representative. The resulting grammar need not be a congruence grammar because it is now possible that two distinct productions have the same right hand side. Any time we have two productions $X \rightarrow f(W_1, \cdots W_n)$ and $Y \rightarrow f(W_1, \cdots W_n)$ with the same right hand side we uniformly replace all occurances of X in the grammar by Y. Such replacement is continued until we have a congruence grammar.

The transformation from equations to grammars has an inverse — one can transform a grammar back into a set of equations. For each nonterminal symbol X we

[2] (We say that s occurs in H if s is either one side of an equation in H *or* s is a subterm of a term in an equation in H.

construct a new constant symbol c_X. The constants of the form c_X will be called *internal* constants to distinguish them from the other (external) constants. A term that does not contain any of these internal constants will be called an external term. We now have the following definition and lemma.

Definition: If G is a congruence grammar we define $E_q(G)$ to be a set of equations of the form $c_X = f(c_{Y_1}, \cdots, c_{Y_n})$ where G contains the production $X \to f(Y_1, \cdots, Y_n)$.

For example, if G contains the two productions $X \to f(X)$ and $X \to a$, then $E_q(G)$ contains the two equations $c_X = f(c_X)$ and $c_X = a$.

Lemma: For any congruence grammar G, and any two external ground terms s and t, we have that $I_G[s]$ equals $I_G[t]$ if and only if $E_q(G) \models s = t$.

Proof Sketch: One can show by induction on the size of a term s that $E_q(G) \models s = c_{I_G[s]}$. Then if $I_G[s] = I_G[t] = X$ we have $E_q(G) \models s = c_X$ and $E_q(G) \models t = c_X$ so $E_q(G) \models s = t$. To show the converse we add a production $X \to c_X$ for each nonterminal X. These productions do not alter the intern function on external terms. We then consider the congruence relation (on both internal and external terms) defined by the intern function for this extended grammar. This congruence relation provides a semantic model of $E_q(G)$. So if $I_G[s] \neq I_G[t]$ then $E_q(G) \not\models s = t$. ∎

The operation E_q introduces internal constants into the equation set. These internal constants are irrelevant to the equivalence relation on external terms defined by the equation set. One can define the operation G_r so that it eliminates internal constants from the grammar. The equation sets constructed by E_q are useful for analysis and conceptual definitions but they are never actually computed. All computation is done directly on grammars.

4 Grammar Rewriting

This section defines the basic concepts of grammar rewriting.

Definition: A grammar rewriting system is a pair $<E, w>$ where E is a set of equations between first order terms (usually containing variables) and w is a weight function which assigns a positive integer to every ground term.

The weight function induces a well founded order on ground terms by setting $s < t$ if and only if $w[s] < w[t]$. Unlike the orderings used in traditional ordered rewriting systems, e.g., [9], the weight orderings used in grammar rewriting are not total on ground terms. In term rewriting one is interesting in simplifying one particular term. Grammar rewriting is also focused on a particular term. However, in grammar rewriting this term is represented by a nonterminal symbol of a grammar.

Definition: We define the "one step" rewrite relation $\mapsto_E$ on ground terms so that $\sigma[s] \mapsto_E \sigma[t]$ provided either $s = t \in E$ or $t = s \in E$, every free variable of t appears in s, and σ is a ground substitution.

For example, if E is the set $\{g(x, f(x)) = c\}$ then we have that $g(h(a),\ f(h(a))) \mapsto_E c$ but we do not have $c \mapsto_E g(h(a),\ f(h(a)))$.

Definition: A *grammar term* is a pair $<X,\ G>$ where G is a finite congruence grammar and X is a nonterminal that appears in G.

Definition: We say that a term s is a minimal representative of a grammar term $<X,\ G>$, with respect to a weight function w, if s is generated by X under G and no other term generated by X is smaller than s according to w.

Definition: We define the one step rewrite relation $\mapsto_{<E,\ w>}$ on grammar terms so that

$$<X,\ G> \mapsto_{<E,\ w>} <X',\ G_r(E_q(G) \cup \{u = v\})>$$

provided u is a subterm of some minimal representative of $<X,\ G>$ under the weight function w, $u \mapsto_E v$, $I_G[u] \neq I_G[v]$, and X' is the nonterminal of $G_r(E_q(G) \cup \{u = v\})$ that generates the terms generated by X under G.

The one-step grammar rewriting operation defined above corresponds to selecting a term minimally generated by X, rewriting that term according to some equation in E, and then modifying the grammar so that the result of the rewrite is included in the language generated by X. A procedure for efficiently computing $G_r(E_q(G) \cup \{w = u\})$ from G and the equation $w = u$ is given in section 6. This procedure does not construct an equation set — it performs a direct operation on the grammar.

The definition of the rewrite relation on grammars requires that the new equation being incorporated into the grammar is indeed new, i.e., it can not be an equation that is already implied by $E_q(G)$. This allows for the existence of normal forms as defined below.

Definition: We say that a grammar term $<X,\ G>$ is *in normal form* if there is no grammar term $<X',\ G'>$ such that $<X,\ G> \mapsto_{<E,\ w>} <X',\ G'>$.

As an example of rewriting, let C be the equation set consisting of the single commutative law $x+y = y+x$. Let 1 be the weight function that assigns every term the weight 1 (this weight function corresponds to the empty ordering on terms). Let G be the grammar consisting of the productions $X_1 \rightarrow a_1 + X_2$, $X_2 \rightarrow a_2 + X_3$, $\cdots$, $X_{n-1} \rightarrow a_{n-1} + X_n$, $X_n \rightarrow a_n$. In other words, G is the grammar generated by interning the term $a_1 + (a_2 + (\cdots + a_n))$ starting with the empty grammar. In the grammar rewriting system $<C,\ 1>$ we have that the grammar term $<X_1,\ G>$ rewrites to a normal form $<X_1,\ G'>$ where G' is the grammar consisting of the productions of the form

$$X_i \rightarrow a_i + X_{i+1}$$
$$X_i \rightarrow X_{i+1} + a_i$$

together with the production

$$X_n \rightarrow a_n.$$

A similar example can be given for the equation set AC consisting of the associative and commutative laws.

Certain restrictions on the weight function w can be used to ensure that grammar rewriting always terminates.

Definition: A grammar rewriting system $<E, w>$ is called *terminating* if there are no infinite rewrite chains of the form

$$<X_1, G_1> \mapsto_{<E, w>} <X_2, G_2> \mapsto_{<E, w>} <X_3, G_3> \mapsto_{<E, w>} \cdots$$

Definition: A weight function w will be called a *polynomial weight function* if for each function symbol f of n arguments there exists a polynomial $P_f(x_1, \cdots, x_n)$ in n variables with coefficients greater than or equal to 1, where each x_i appears in at least one term, where there is a constant term of at least 1, and such that for any ground terms $s_1, \cdots, s_n$ we have $w[f(s_1, \cdots, s_n)] = P_f(w[s_1], \cdots, w[s_n])$.

Orderings based on polynomial weight functions are well known in the term rewriting literature. For a given finite set of function and constant symbols, and a given weight k, there are only finitely many terms that can be constructed from those symbols that have weight less than or equal to k. Because grammar rewriting with a finite equational theory can not generate infinitely many constant or function symbols, and because matching is restricted to minimal weight terms, we have the following well foundedness lemma for grammar rewriting.

Lemma: If w is a polynomial weight function and E is any finite set of equations then $<E, w>$ is terminating.

We use $\mapsto^*_{<E, w>}$ to denote the reflexive transitive closure of the relation $\mapsto_{<E, w>}$.

Definition: Let $I_0[s]$ be the grammar term that results from interning s relative to the empty grammar. If $I_0[s] \mapsto^*_{<E, w>} <X, G>$ and $<X, G>$ is in normal form, then $<X, G>$ is called a *normal form of s*.

Definition: For any ground term s and set of equations E we define $|s|_E$ to be the set of terms that can be proven to be equal to s using the equations in E.

Definition: A grammar rewriting system $<E, w>$ is called *complete* if for each ground term s, and any normal form $<X, G>$ of s, every minimal member of $|s|_E$ is generated by X under G.

Definition: A grammar rewriting system is called *canonical* if it is terminating and complete.

A canonical grammar rewriting system $<E, w>$ provides a decision procedure for the equational theory E.

Theorem: Let $<E, w>$ be a canonical grammar rewriting system and let s and t be any two ground terms. Let $<X, G>$ and $<X', G'>$ be normal forms for s and t respectively under the rewriting system $<E, w>$. Let G'' be the grammar that encodes all equivalences encoded in G or G', i.e., G'' is $G_r(E_q(G) \cup E_q(G'))$. We have that $E \models s = t$ if and only if $I_{G''}[s] = I_{G''}[t]$.

This lemma follows from the invariant that for any ground term u if $I_0[u] \mapsto^*_{<E, w>} <X, G>$ then X generates u under G and if $<X, G>$ is a normal form of u then X generates all minimal elements of $|u|_E$.

5 Examples

We let w be the simple polynomial weight function such that for every term $f(s_1, \cdots, s_n)$ we have that $w[f(s_1, \cdots, s_n)] = w[s_1] + \cdots + w[s_n] + 1$. Let A and C be the two equation sets $\{x + (y + z) = (x + y) + z\}$ and $\{x + y = y + x\}$ respectively. Let AC be $A \cup C$. One can easily verify that the systems A, C, and AC are all canonical under this ordering.

Let w be any polynomial weight function satisfying $w[s + t] = 2w[s] + w[t] + 1$. Under this weight function we have that $w[s + t] < w[t + s]$ provided $w[s] < w[t]$ and we have $w[s + (t + u)] < w[(s + t) + u]$ for any terms s, t, and u. The equation set $AC \cup \{x + (y + z) = y + (x + z)\}$ forms a canonical system where every term normalizes to a grammar term whose minimal representatives are the weight sorted permutations of the addends under a standard parenthesization. If the addends are linearly ordered by weight then there is only one minimal representative.

We now consider Abelian groups. Let w be any order satisfying $w[s+t] = w[s] + w[t] + 1$ and $w[-s] = 2w[s]$. Under this ordering we have that $w[(-s) + (-t)] < w[-(s + t)]$. The equation set $AC \cup \{-(x + y) = (-x) + (-y),\ \ x + (-x) = 0,\ \ x + 0 = x\}$ is a canonical system under this ordering. Every term normalizes to a grammar whose minimal representatives form an AC equivalence class. Refinements of the ordering can give canonical systems which generate smaller grammars.

Let E be $AC \cup \{f(x+x) = 1\}$. This is given in [9] as an example of a theory that can not be handled by ordered term rewriting without special unification procedures. However, this equation set is canonical under grammar rewriting using the simple ordering given in the first example of this section.

The theory of idempotent semigroups is the theory defined by the associativity axiom $x + (y + z) = (x + y) + z$ and the idempotence axiom $x + x = x$. The theory idempotent semigroups has the property that for any finite set of constants the set of terms that can be constructed from those constants and the operation $+$ fall into only a finite number of equivalence classes. Such a theory is called *locally finite*. A theory is called *regular* if every equation has the property that every variable appearing in one side of the equation also apears on the other side of the equation. In the full paper it is shown that any locally finite regular theory is canonical under fair grammar rewriting relative to the empty order on terms. Hence grammar rewriting gives a canonical rewriting system for the theory of idempotent semigroups. It has been shown that there is no canonical term or word rewriting system for the theory of idempotent semigroups [1].

6 Incorporating Equations into Grammars

This section gives an algorithm for computing $G_r(E_q(G) \cup \{s = t\})$ from the grammar G and the equation $s = t$. The grammar $G_r(E_q(G) \cup \{s = t\})$ is directly computed from G by incrementally adding and removing productions. The algorithm given here is quite similar to the congruence closure procedure described in [11]. The algorithm has been reformulated here to operate on grammars and optimized to run in order $n \log n$ time (under the assumption that hash table operations take unit time).[3]

[3] A somewhat different $n \log n$ algorithm for congruence closure is described in [5].

This procedure uses the internal constants of the form c_Z that are associated with the nonterminals of the grammar. The procedure maintains an equivalence relation represented by three sets of equations. First, the procedure maintains a congruence grammar. Second, the procedure maintains a queue of equations of the form $c_Z = c_Y$. Third, an additional set of equations between constants of the form c_Z is maintained in a union-find structure on these constants. The equivalence relation determined by these three sets of equations is maintained as a fixed invariant of the procedure.

As mentioned above, equations between constant symbols can be handled by the well known union-find procedure. Consider a set of equations $a_1 = b_1$, $a_2 = b_2$, $\cdots$, $a_n = b_n$. This set of equations can be represented in a union-find structure by executing **union**(a_1, b_1), **union**(a_2, b_2), $\cdots$, **union**(a_n, b_n). To see if an equation $c = d$ follows from the given equations one computes **find**(c) and **find**(d). The equation $c = d$ is provable if and only if these two find operations return the same value.

> **Definition:** A constant c_X is said to be *dead* if **find**(c_X) is some constant other than c_X. A constant that is not dead is said to be *alive*, i.e., a constant c_Y is alive if **find**(c_Y) equals c_Y.

Procedure for computing $G_r(E_q(G) \cup \{s = t\})$:

1. Let Z and W be the nonterminals $I_G[s]$ and $I_G[t]$.
2. Initialize S to be a queue containing the single equation $c_Z = c_W$.
3. While the queue S is not empty do the following.
   (a) Remove an equation $c_Z = c_W$ from the queue S.
   (b) Let c_X be **find**(c_Z) and let c_Y be **find**(c_W).
   (c) If c_X is the same symbol as c_Y then do nothing, otherwise:
   (d) Call union(c_X, c_Y).
   (e) Swap the roles of X and Y if necessary so that c_X is dead and c_Y is still alive.
   (f) Let $\mathcal{P}$ be the set of all productions involving X.
   (g) Remove all the productions in $\mathcal{P}$ from the grammar.
   (h) For each production $Z \rightarrow f(W_1, \cdots W_n)$ in $\mathcal{P}$ do the following.
      i. Let $c_{Z'}$ be **find**(c_Z) and for each W_i let $c_{W_i'}$ be **find**(c_{W_i})
      ii. If there is no production whose right hand side is $f(W_1', \cdots, W_n')$ then add the production $Z' \rightarrow f(W_1', \cdots, W_n')$.
      iii. If there is already a production $U \rightarrow f(W_1', \cdots, W_n')$ where U is different from Z' then add the equation $c_U = c_{Z'}$ to the queue S.

In the above procedure we assume that the find operation is such that **find**(c_X) is a canonical member of the equivalence class of c_X relative to the equivalence relation encoded in the union-find structure. In other words, **find**(c_X) is a constant c_Y such that **find**(c_Y) equals c_Y. We also assume that when two equivalence classes are merged with a union operation the canonical representative of the resulting equivalence class is selected to be one of the two previous canonical representatives.

The procedure maintains the invariant that no two productions in the grammar have the same right hand side, i.e., the grammar is always a congruence grammar. Furthermore, the equivalence relation encoded in the equations in the grammar, the queue, and the union-find procedure is maintained as a fixed invariant. To check this one must check that every added equation is derivable from previous equations and that every equation removed in step g is derivable from previous equations plus those equations added in step h. The procedure also maintains the invariant that every nonterminal in every production is "alive", i.e., if X appears in a production of the grammar then $\texttt{find}(c_X)$ equals c_X. Every nontrivial execution of the main loop reduces the number of living nonterminal symbols, so the procedure must terminate. Furthermore, when the procedure terminates the equational theory enoded in the union-find structure can be dropped without altering the induced equivalence relation on external terms. To prove this consider an extended grammar that includes all productions of the form $X \rightarrow c_Y$ where X is a living nonterminal and $\texttt{find}(c_Y)$ is c_X. The equation set of the extended grammar encodes the equivalence relation of the original grammar plus the equivalences in the union find structure. However, this extended grammar encodes the same intern function on external terms as the unextended grammar.

If we assume a bound on the number of arguments taken by function symbols, e.g., no function takes more than three arguments, and assume that hash table operations can be performed in unit time, then under an appropriate implementation of union-find it can be shown that the above procedure terminates in order $n \log n$ time where n is the number of productions in the original grammar. In practice the incorporation of a single new equation into a large grammar requires the manipulation of only a small subset of the grammar.

7 A Grammar Rewriting Algorithm

In this section we consider a fixed but arbitrary grammar rewriting system $<E, w>$, where w is a polynomial weight function, and give a procedure for computing all the ways in which a given grammar term can be rewritten under $<E, w>$. The procedure is incremental so that if $<X, G> \mapsto_{<E, w>} <X', G'>$ then the set of possible ways of rewriting $<X', G'>$ can be computed incrementally from the set of possible ways of rewriting $<X, G>$. Incremental procedures can be defined by "inference rules" that are run in an incremental forward chaining manner. We first give rules for deriving the weight of nonterminal symbols as defined below.

Definition: For each nonterminal Y of a congruence grammar the weight of Y, denoted $w[Y]$, is minimum weight of all terms generated by Y.

The following inference rule can be used to propagate bounds on weights.

$$\frac{\begin{array}{l} w[X_1] \leq w_i \\ \vdots \\ w[X_n] \leq w_n \\ Z \rightarrow f(X_1, \cdots, X_n) \end{array}}{w[Z] \leq P_f(w_1, \cdots, w_n)}$$

For constant symbols (functions of no arguments) the above inference rule can be used to generate a weight bound directly from a production of the form $Z \rightarrow c$.

Weight bounds generated by constants can then be propagated to other symbols. One can show that, for any given nonterminal Y, the tightest bound that can be derived for $w[Y]$ using the above rule is in fact the weight of Y as defined above. In practice, simply running this rule over the entire grammar until the tightest bounds are derived seems to be an acceptable incremental algorithm for computing the weight of nonterminals. However, the theoretical worst case behavior of this this algorithm is quite bad. An $n \log n$ algorithm can be derived by placing derived bounds on a priority queue and processing the tightest bounds first.

Procedure to compute the weight of all nonterminals:

1. Initialize S to be the priority queue containing all pairs of the form $<Y, w>$ where the grammar contains the production $Y \rightarrow a$ where a is a constant and w is the weight of a.
2. Until S is empty, or until every nonterminal has been assigned a weight, do the following.
   (a) Remove a pair $<Y, w>$ from S such that w is the minimum weight of all pairs on S.
   (b) If Y has already been assigned a weight do nothing. Otherwise:
   (c) Assign Y the weight w.
   (d) For each production $Z \rightarrow f(W_1, \cdots, W_n)$ such that some W_i is Y and such that each W_i has been assigned a weight w_i, add the pair $<Z, P_f(w_1, \cdots, w_n)>$ to the queue S.

Each pair added to the queue has a larger weight than the last pair removed from the queue. This implies that the pairs removed from the queue have monotonically increasing weight. This, plus the assumed properties of the polynomial weights, implies that for each nonterminal symbol Y the weight assigned to Y is the minimum over all the productions from Y of the weight computed from that production. One can check that by selecting productions that minimize weight that if Y has been assigned weight w then Y generates a term of weight w. Furthermore, one can prove by induction on the weight of a term s that the weight of s is at least as large as the weight assigned to $I[s]$, the symbol that generates s. Since the procedure only assigns a single weight to each nonterminal symbol the number of pairs placed on the priority queue is linear in the size of the grammar. Order n insertions and deletions from a priority queue can be done in $n \log n$ time. Assuming a bound on the number of arguments taken by any function symbol, the other operations in this procedure can be performed in order n time so the total running time is order $n \log n$.
The next step is to identify those nonterminals in the grammar that generate subterms of minimal representatives of $<X, G>$.

Definition: A production $Z \rightarrow f(X_1, \cdots, X_n)$ is *minimal* if $w[Z] = P_f(w[X_1], \cdots, w[X_n])$.

Lemma: If s is a minimal representative of the grammar term $<X, G>$ then s is generated by X in that subset of G which consists of just the minimal productions of G.

Definition: We say that a nonterminal symbol Y is a *minimal subterm nonterminal* of a grammar term $<X, G>$ if either Y is X (X is a minimal subterm nonterminal) or G contains a minimal production $W \rightarrow f(Z_1, \cdots, Z_n)$ where W is a minimal subterm nonterminal and some Z_i is Y.

As an example consider the grammar term $<X, G>$ where G consists of the five productions

$$X \rightarrow f(Y), \ Y \rightarrow g(Z), \ Z \rightarrow a$$

$$Z \rightarrow h(W), \ W \rightarrow f(Z).$$

The production $Z \rightarrow h(W)$ is not minimal — it does not provide a smallest term generated by Z. However, all other productions are minimal, including the production $W \rightarrow f(Z)$ which provides a smallest term generated by W. However, the nonterminal W is not a minimal subterm nonterminal — it does not generate a subterm of a minimal representative of $<X, G>$.
The procedure for constructing matches uses a simple extension of the grammar.

Definition: An *extended congruence grammar* is a congruence grammar such that, for each nonterminal X appearing in a production of the grammar, the grammar also contains the production $X \rightarrow c_X$.

We define a match to be a triple, **match**$[Y, s, \sigma]$, where Y is a nonterminal, s is a term, and σ is a substitution such that $I[\sigma[s]]$ is Y. Matches can be computed by starting with "basic matches" and creating new matches according to inference rules that generate new matches from old matches. For each minimal subterm nonterminal Y, and for each variable x occurring in E, we create the basic match **match**$[Y, x, \{x \mapsto c_Y\}]$. To minimize the number of basic matches it is important to rename the variables in equations in E to minimize the total number of variables in E. Typically there will be no more than 3 or 4 variables in E. We also create basic matches for constant symbols that appear in E. For each constant a appearing in E we create the basic match **match**$[I_G[a], a, \emptyset]$ where $\emptyset$ is the empty substitution. More complicated matches can be constructed using the following rule.

$$\frac{\begin{array}{l}\mathtt{match}[Y_1, s_1, \tau_1] \\ \vdots \\ \mathtt{match}[Y_n, s_n, \tau_n] \\ Z \rightarrow f(Y_1, \cdots, Y_n)\end{array}}{\mathtt{match}[Z, f(s_1, \cdots, s_n), \sigma]}$$

The substitutions are represented by finite lists of variable-value pairs. In all the substitutions constructed by the matching procedure the values assigned to variables are always internal constants. The above rule only applies when the substitutions τ_i agree on all shared variables — if τ_i and τ_j both contain a pair assigning a value to the variable x then they must both assign x to the same internal constant. The substitution σ is simply the union of the τ_i's. The above rule is also restricted to the case where Z is a minimal subterm nonterminal, the production $Z \rightarrow f(Y_1, \cdots, Y_n)$ is minimal, and $f(s_1, \cdots, s_n)$ is a term occurring in E.

The restrictions on the above inference rule are "nonmonotonic". The addition of a new production can cause other productions to go from being minimal to being nonminimal. When matches are computed incrementally as new productions are added it is possible that previously computed matches become "obsolete" in the sense that they were computed from productions that have now become nonminimal.[4] The failure to remove matches that are obsolete due to productions that are no longer minimal will cause minor overgeneration of matches. The time required to detect and remove these obsolete matches is probably greater than the time taken to perform any extra rewrites they cause. Rewrites generated by these obsolete matches are semantically sound and do no harm.

Finally, we construct "equate relations" of the form $c_Z \mapsto_E v$ where c_Z is an internal constant and v is a ground term involving internal constants. More specifically, if we derive $\mathtt{match}[Z,\ s,\ \sigma]$ and E contains either $s = t$ or $t = s$ where every free variable in t occurs in s, then we can derive $c_Z \mapsto_E \sigma[t]$. The set of derived equate relations of the form $c_z \mapsto_E v$ provide all the possible ways of rewriting the grammar. For each equate relation $c_Z \mapsto_E v$, the grammar can be rewritten by using the procedure of section 6 to equate c_Z and v.

8 Summary

Grammar rewriting is motivated by the desire to increase the success rate of attempts to prove equations in nonconfluent rewrite systems. Intuitively, the rewrite process generates a set of terms rather than an individual term, and by generating two sets of terms, rather than two classical normal forms, we increase the likelihood of proving the desired equation. Although grammar rewriting is primarily motivated by the need to handle nonconfluent theories, it also provides a new kind of canonical rewriting system. Under grammar rewriting there exist finite canonical systems for equational theories, such as idempotent semigroups, that do not have finite canonical systems under traditional notions of rewriting.

9 Acknowledgements

I would like to thank Robert Givan and Carl Witty for their help in the implementation of grammar rewriting in the Ontic theorem proving system. Support for the laboratory's artificial intelligence research is provided in part by the Advanced Research Projects Agency of the Department of Defense under Office of Naval Research contract N00014-85-K-0124.

References

1. Franz Baader. Rewrite systems for varieties of semigroups. In *CADE-10, LNAI 449*, pages 381–395. Springer-Verlag, 1990.

[4] Matches can also become obsolete if nonterminals or constant symbols involved in the match become "dead" due to generated equations between nonterminals. Matches that are obsolete due to references to dead nonterminals or dead constants are easily detected and eliminated.

2. L. Bachmair, N. Dershowitz, and D. Plaisted. Completion without failure. In *Proc. Col. on Resolulution of Equations in Algebraic Structures*, 1987.
3. Robert S. Boyer and J Struther Moore. *A Computational Logic*. ACM Monograph Series. Academic Press, 1979.
4. Leslie P. Chew. An improved algorithm for computing with equations. In *FOCS80*, pages 108–117. IEEE Computer Society Press, 1980.
5. Peter J. Downey, Ravi Sethi, and Robert E. Tarjan. Variations on the common subexpression problem. *JACM*, 27(4):758–771, October 1980.
6. J. Hsaing and M. Rusinowitch. On word problems in equational theories. In *ICALP-87, LNCS 267*, pages 54–71. Springer-Verlag, 1987.
7. J. P. Jouannaud and H. Kirchner. Completion of a set of rules modulo a set of equations. *SIAM Journal of Computing*, 15:1155–1194, 1986.
8. Dexter C. Kozen. Complexity of finitely presented algebras. In *Proceedings of the Ninth Annual ACM Symposium on the Theory of Compututation*, pages 164–177, 1977.
9. Ursula Martin and Tobias Nipkow. Ordered rewriting and confluence. In *CADE-10, LNAI 449*, pages 365–380. Springer-Verlag, 1990.
10. J. Mezei and J. B. Wright. Algebraic automata and context free sets. *Information and Control*, 11:3–29, 1965.
11. Greg Nelson and Derek Oppen. Fast decision procedures based on congruence closure. *JACM*, 27(2):356–364, April 1980.
12. Gerald E. Peterson. Complete sets of reductions with constraints. In *CADE-10, LNAI 449*, pages 381–395. Springer-Verlag, 1990.
13. R. Shostak. An algorithm for reasoning about equality. *Comm. ACM.*, 21(2):583–585, July 1978.
14. W. Thomas. Automata on infinite objects. In *Handbook of Theoretical Computer Science, Volume B, Formal Methods and Semantics*, pages 133–164. MIT Press, 1990.

This article was processed using the LaTeX macro package with LLNCS style

Polynomial Interpretations and the Complexity of Algorithms

Adam Cichon Pierre Lescanne

CRIN & INRIA-Lorraine

BP 101, 54602 Villers-lès-Nancy, France

Abstract. The ability to use a polynomial iterpretation to prove termination of a rewrite system naturally prompts the question as to what restriction on complexity this imposes. The main result of this paper is that a polynomial interpretation termination proof of a rewrite system $\mathcal{R}$ which computes a number theoretic function implies a polynomial bound on that function's rate of growth.

1 Introduction

The ability to use a polynomial iterpretation to prove termination of a rewrite system, a method due to Lankford [4, 5], naturally prompts the question as to what restriction on complexity this imposes. Various claims and conjectures have been made, significant amongst which is:

> *Of course, polynomial interpretations do not suffice in general, since they give a polynomial upper bound on the complexity of the computations by $\mathcal{R}$, interpreted as a program computing over integers, whereas arbitrary recursive functions can be defined by term rewriting systems.*
>
> HUET AND OPPEN, 1980

At first sight, it would appear that this claim, which appears in [3], has, once and for all, been refuted by Lautemann [6] and Geupel [2], who show that polynomial interpretation admits doubly exponential derivation lengths. Lautemann and Geupel give the following system of rewrites for the number theoretic squaring function, together with a polynomial interpretation which proves termination:

Rewrite Rules	*Polynomial Interpretation*
$x + 0 \rightarrow x$	$[0] = 2$
$x + s(y) \rightarrow s(x + y)$	$[s](X) = X + 1$
$d(0) \rightarrow 0$	$[+](X, Y) = X + 2Y$
$d(s(x)) \rightarrow s(s(d(x)))$	$[d](X) = 3X$
$q(0) \rightarrow 0$	$[q](X) = X^3$
$q(s(x)) \rightarrow s(q(x) + d(x))$	

They then consider terms of the form $q^{(n)}(s(s(0)))$. These terms satisfy:

$$q^{(n)}(s(s(0))) \to_{\mathcal{R}}^{+} s^{2^{2^n}}(0)$$

Since no rule increases the number of outermost occurences of s by more than one, the length of this derivation must be at least doubly exponential. Lautemann and Geupel then go on to prove:

Theorem (Geupel [2], Lautemann [6])
If $\mathcal{R}$ has a polynomial interpretation then there is a constant c such that, for any terms t, s,

$$\text{if } t \to_{\mathcal{R}}^{n} s \text{ then } n \leq 2^{2^{c.\mathcal{L}(t)}}$$

where $\mathcal{L}(t)$ is the number of symbols in t and $t \to_{\mathcal{R}}^{n} s$ means that t rewrites to s in n steps under the rules in $\mathcal{R}$.

An examination of these results shows that they do not directly address the question implicit in the remark of Huet and Oppen. In the example given above, while there are terms which admit doubly exponential derivation lengths, the rewrite system itself is intended for the calculation of the squaring function over the integers. If we therefore only consider terms of the form $q(s^n 0)$ we see that the lengths of derivations are indeed bounded by a polynomial in n. We can note also that the size of the output term is similarly bounded by a polynomial in n.

One might ask whether this is always the case concerning the bounds on derivation lengths. The answer is no – and we shall give a counterexample in the text. On the other hand, we shall show that the size of the output term *is* always bounded by a polynomial in the size of the input term. Furthermore, we shall exhibit a curious relationship between the input-output relation and derivation lengths, namely that the "longer" the derivation the "smaller" the output. More precisely, we prove that if a rewrite system $\mathcal{R}$, which admits a polynomial interpretation termination proof, computes a number theoretic function F in a time which is polynomial in the input size then there is also a polynomial bound on the size of the output. But if $\mathcal{R}$ computes F in exponential time, then the size of the output is bounded by a *linear* polynomial.

We have not come across any examples of non-trivial functions F with this last property, see example 1. This leads us to suggest that if $\mathcal{R}$ computes F on input $s^n 0$ in exponential time then $\mathcal{R}$ is a "bad" choice of rewrite system for F and as a consequence Huet and Oppen's statement holds true for "good" rewrite systems.

2 Preliminaries

The term rewriting notations used in this paper are based on those advocated by N.Dershowitz and J-P.Jouannaud in [1]. We shall use in particular:

$\overset{+}{\to}$ to denote the *transitive closure* of the binary relation $\to$.

$\overset{!}{\to}$ to denote *normalisation*. That is, $t \overset{!}{\to} s$ means that t rewrites to s and s is in normal form.

$\xrightarrow{n}$ where $t \xrightarrow{n} s$ indicates that there exists a rewrite sequence from t to s whose length is n.

2.1 The {0,s}-Constructor Discipline

The main emphasis in this paper is on the use of term rewriting systems for defining *number theoretic functions*, that is, functions:$\mathbf{N}^k \mapsto \mathbf{N}$. We therefore restrict our attention to term rewriting systems which are assumed to have the following properties:

1. $\mathcal{R}$ is a finite set of rewrite rules over the set $\mathcal{T}(\mathcal{F}, \mathcal{X})$ of terms.
2. The set $\mathcal{F}$ of function symbols is finite, contains only one constant (0-ary function symbol), 0, and contains the unary function symbol s.
3. $\mathcal{R}$ is terminating and confluent, and the set

$$\{0, s(0), \ldots, \underbrace{s(s(\ldots s(s(0))\ldots))}_{n\ times}, \ldots\}$$

   is precisely the set of normal forms for the ground terms in $\mathcal{T}$. This means that for every function symbol f in $\mathcal{F}$, which is not 0 or s, there is an f-eliminating rule in $\mathcal{R}$.

Henceforth we write $s^{(n)}(0)$, and frequently $s^n 0$, for $\underbrace{s(s(\ldots s(s(0))\ldots))}_{n\ times}$.

The term $s^n 0$ represents the numeral n. A function symbol f in $\mathcal{F}$ represents the number theoretic function $\{f\}$ if

$$f(s^{m_1}0, \ldots, s^{m_k}0) \xrightarrow{!}_{\mathcal{R}} s^m 0$$
$$\text{if and only if}$$
$$\{f\}(m_1, \ldots, m_k) = m.$$

It is in this sense, therefore, that we say $\mathcal{R}$ *computes* $\{f\}$.

2.2 Polynomial Interpretation Termination Proofs

A *polynomial interpretation termination proof* for a rewrite system $\mathcal{R}$ over a set $\mathcal{T}(\mathcal{F}, \mathcal{X})$ of terms is achieved by assigning to each function symbol f in $\mathcal{F}$ a polynomial with integer coefficients. We denote this polynomial by $[f]$. If f is n-ary then $[f]$ must be a polynomial in n variables. The polynomial $[f]$ must satisfy the monotonicity condition:

$$x < y \Longrightarrow [f](\cdots x \cdots) < [f](\cdots y \cdots)$$

and must ensure that terms are mapped into nonnegative integers only. Each rule must be reducing with respect to the its interpretation, that is

$$l \rightarrow r \in \mathcal{R} \Longrightarrow [r] < [l]$$

for all values of variables greater than the minimum of the interpretations of the ground terms.

Remark 1 *Whenever $\mathcal{R}$ has a polynomial interpretation termination proof the following useful inequality arises:*

$$\text{if } t \to_{\mathcal{R}}^{+} s \text{ then } [s] < [t].$$

3 Exponential Functions and Polynomial Interpretations

In this section we shall show that functions of exponential growth cannot be computed by rewrite systems with polynomial interpretation termination proofs. The proof is achieved by first establishing a relationship between the rate of growth of a function and the length of its computation and then using the theorem of Lautemann and Geupel.

Definition 1 *We define the height, $|t|$, of a term t in $T(F,X)$ as follows:*

$$|t| = \begin{cases} 0 & \text{when } t \text{ is a constant or a variable,} \\ \max_{i\in 1..n}\{|t_i|\}+1 & \text{when } t = f_k(\vec{t}). \end{cases}$$

where $\vec{t} = t_1, \ldots, t_n$.

Definition 2 *Suppose that $\mathcal{R}$ is a finite set of rewrite rules $\{l(\vec{x}) \to r(\vec{x})\}$. Let*

$$M_{\mathcal{R}} = \max_{l(\vec{x})\to r(\vec{x})\in\mathcal{R}} \{|r(\vec{x})|\}$$

Definition 3 *Proper subtraction, $\dot{-}$, is defined:*

$$x \dot{-} y = \begin{cases} 0 & \text{when } x < y, \\ x - y & \text{otherwise.} \end{cases}$$

The following theorem connects the *input, output* behaviour of term rewriting systems with derivation lengths.

Theorem 1

$$\text{If } t \to_{\mathcal{R}}^{1} s \text{ then } |s| \dot{-} |t| \le M_{\mathcal{R}}.$$

and hence

$$\text{If } t \to_{\mathcal{R}}^{n} s \text{ then } n \ge \frac{|s| \dot{-} |t|}{M_{\mathcal{R}}}$$

Proof: *The proof is given by lemmas 1, 2, 3 below.* □

Lemma 1 *For any substitution $\vec{d}\ (= d_1, \ldots, d_n)$ for $\vec{x}\ (= x_1, \ldots, x_n)$,*

$$|t(\vec{d})| \le |t(\vec{x})| + \max_{i\in 1..n}\{|d_i|\}.$$

Proof: *The proof is by induction on the term tree for $t(\vec{x})$. When t is a constant or*

a variable, the result is trivial.
If $t(\vec{x}) = f(t_1(\vec{x}), \ldots, t_m(\vec{x}))$ *then*

$$\begin{aligned} |t(\vec{d})| &= \max_{j\in 1..m}\{|t_j(\vec{d})|\}+1 \\ &\underset{IH}{\leq} \max_{j\in 1..m}\{|t_j(\vec{x})|+\max_{i\in 1..n}\{|d_i|\}\}+1 \\ &\leq \max_{j\in 1..m}\{|t_j(\vec{x})|\}+1+\max_{i\in 1..n}\{|d_i|\} \\ &= |t(\vec{x})|+\max_{i\in 1..n}\{|d_i|\}. \end{aligned}$$

□

Lemma 2 *For any substitution* $\vec{d}$ *for* $\vec{x}$,

$$|r_i(\vec{d})| \leq |l_i(\vec{d})| + M_{\mathcal{R}}$$

Proof: *By lemma 1,*

$$\begin{aligned} |r_i(\vec{d})| &\leq |r_i(\vec{x})|+\max_{i\in 1..n}\{|d_i|\} \\ &\leq M_{\mathcal{R}}+\max_{i\in 1..n}\{|d_i|\} \\ &\leq M_{\mathcal{R}}+|l_i(\vec{d})|. \end{aligned}$$

□

The next lemma generalises lemma 2 to the case where a one step rewrite occurs by applying a rule in $\mathcal{R}$ to a proper subterm of a term t. We shall use the notation $t[u]$ to denote a term t with u as a proper subterm.

Lemma 3 *For any substitution* $\vec{d}$ *for* $\vec{x}$,

$$|t[r_i(\vec{d})]| \dot{-} |t[l_i(\vec{d})]| \leq M_{\mathcal{R}}.$$

Proof: *This is now a straightforward induction over the term tree for* t. □

We can apply the results above to obtain a connection between the input-output behaviour and the lengths of derivations of a function $\{f\}$ computed by a rewrite system $\mathcal{R}$. The computation of $\{f\}(m_1, \ldots, m_k)$ is achieved by normalising $f(s^{m_1}(0), \ldots, s^{m_k}(0))$ under the rules in $\mathcal{R}$.
Suppose that $\{f\}(m_1, \ldots, m_k) = m$ so that

$$f(s^{m_1}(0), \ldots, s^{m_k}(0)) \to^! s^m(0).$$

By theorem 1, if n is the number of steps to normalisation, we obtain

$$n \geq \frac{|s^m(0)| \dot{-} |f(s^{m_1}(0), \ldots, s^{m_k}(0))|}{M_{\mathcal{R}}}.$$

Now, since both

$$|\, s^m(0)\,| = m = \{f\}(m_1, \ldots, m_k)$$

and

$$|\, f(s^{m_1}(0), \ldots, s^{m_k}(0))\,| = \max_{i \in 1..k}\{m_i\} + 1$$

we therefore have

$$n \geq \frac{\{f\}(m_1, \ldots, m_k) \;\dot{-}\; \max_{i\in 1..k}\{m_i\} + 1}{M_{\mathcal{R}}}$$

In particular, if $\{f\}$ is an exponential function, then the number of steps to normalisation is at least exponential in the height of the starting term.

We now obtain:

Theorem 2 *If $\mathcal{R}$ computes the number theoretic function $\{f\}$ and $\{f\}$ has exponential growth then $\mathcal{R}$ has no polynomial interpretation termination proof.*
Proof: *Without serious loss of generality and with considerably greater clarity we give the proof in the case where*

$$\{f\}(x) = 2^x.$$

For any j, consider the term $f^{(j)}(0)$. Then

$$f^{(j)}(0) \rightarrow^n_{\mathcal{R}} s^m(0).$$

We write 2_k to denote $2^{2^{\cdot^{\cdot^{2}}}}\Big\}$ k times. We have

$$m = 2_{j-1}$$

and, by theorem 1,

$$n \geq \frac{2_{j-1} - j + 1}{M_{\mathcal{R}}}.$$

Now, if R had a polynomial interpretation termination proof then by the theorem of Lautemann and Geupel we would have

$$\begin{aligned} n &\leq 2^{2^{c\mathcal{L}(f^{(j)}(0))}} \\ &= 2^{2^{cj}} \end{aligned}$$

so that

$$\frac{2_{j-1} - j + 1}{M_{\mathcal{R}}} \leq 2^{2^{cj}}, \textit{for all } j$$

which is clearly impossible. □

Theorem 2 shows that functions with at least exponential growth cannot be computed by rewrite systems with polynomial interpretation termination proof.

In the next section we shall sharpen this and show that a polynomial interpretation termination proof of a rewrite system $\mathcal{R}$ which computes a number theoretic function implies a polynomial bound on that function's rate of growth.

4 Polynomial Interpretation Termination Proof Implies A Polynomial Bound On $\{f\}$

This section is devoted to the proof of the following theorem:

Theorem 3 *Suppose that $\mathcal{R}$ is a rewrite system which computes the function $\{f\}$ and that $\mathcal{R}$ has a polynomial interpretation termination proof. Then, according to the interpretation of s, we have the following:*

1. *If $[s](X) = X + q$, where q is a constant ≥ 1, then $\{f\}$ is bounded by a polynomial function P i.e. $\{f\}(\vec{x}) \leq P(\vec{x})$, for all $\vec{x}$.*

2. *If $[s](X) = aX^p + Q(X)$, where degree $Q < p$ and either $a \geq 1$ and $p > 1$ or $a > 1$ and $p \geq 1$, then $\{f\}$ is bounded by a linear polynomial.*

4.1 s has a linear interpretation

In this section we consider the case when the constructor s has a linear interpretation, namely

$$[s](X) = aX + q \text{ where } a(\geq 1) \text{ and } q(\geq 1) \text{ are constants.}$$

Without loss of generality, we suppose that f is a monadic function. Assume that $[f](X)$ is a polynomial of degree d with leading coefficient c and $[0] = b$.

There are two cases to consider:

Case 1 $a = 1$.

We have

$$[s^n(0)] = b + nq$$

Therefore

$$[f(s^n(0))] = [f](b + nq)$$

Since $[f]$ is a polynomial, so clearly is $\lambda n.[f](b + nq)$. From the inequality

$$[f(s^n(0))] > [s^{\{f\}(n)}(0)]$$

which is a consequence of remark 1 and $f(s^n(0)) \rightarrow^+_{\mathcal{R}} s^{\{f\}(n)}(0)$, it follows that $\{f\}$ is polynomially bounded.

Case 2 $a > 1$.

We now have

$$[s^n(0)] = ba^n + R(a) \text{ with } \textit{degree } R < n$$

Therefore

$$[f(s^n(0))] = [f](ba^n + R(a)) \sim cb^d a^{nd}$$

On the other hand

$$[s^{\{f\}(n)}(0)] \sim ba^{\{f\}(n)}$$

From the inequality

$$[f(s^n(0))] > [s^{\{f\}(n)}(0)]$$

one gets the following inequality on the leading terms:

$$cb^d a^{nd} \geq ba^{\{f\}(n)}$$

which implies

$$n \geq \frac{1}{d}\{f\}(n) + constant$$

This means that in this case the function $\{f\}$ is bounded by a linear polynomial.

An example of such a situation is the following rewrite system which has exponential derivation lengths:

Example 1

$$\begin{array}{rcl} f(0, y) & \rightarrow & y \\ f(s(x), y) & \rightarrow & f(x, f(x, y)) \end{array}$$

Its termination can be proved by the polynomial interpretation

$$\begin{array}{rcl} [0] & = & 2 \\ [s](X) & = & 2X + 1 \\ [f](X, Y) & = & X + Y \end{array}$$

and indeed $\{f\}(m, n) = n$.

4.2 When the constructor s has a non linear interpretation

In this section we consider the case when the constructor s has a non linear interpretation, namely

$$[s](X) = aX^p + Q(X) \text{ with } p \geq 2 \text{ and } degree\ Q < p$$

Again, we suppose that f is a monadic function and that $[f](X)$ is a polynomial of degree d with leading coefficient c and $[0] = b$. One has

$$[s^n(0)] = a^n b^{p^n} + R(b) \text{ with } d^\circ R < p^n$$

Therefore

$$[f(s^n(0))] = [f](a^n b^{p^n} + R(b)) \sim ca^{dn} b^{dp^n}$$

On the other hand

$$[s^{\{f\}(n)}(0)] \sim a^{\{f\}(n)} b^{p^{\{f\}(n)}}$$

Once more, from the inequality

$$[f(s^n(0))] > [s^{\{f\}(n)}(0)]$$

we get the following inequality on the leading terms:

$$ca^{dn} b^{dp^n} \geq a^{\{f\}(n)} b^{p^{\{f\}(n)}}$$

which implies

$$n + log_p d \geq \{f\}(n)$$

if one considers only the exponents of b. This means that in this case the function $\{f\}$ is less than linear. A natural example of such a situation is the system:

Example 2

$$\begin{array}{rcl} half(s(s(x))) & \rightarrow & s(half(x)) \\ half(s(0)) & \rightarrow & 0 \\ half(0) & \rightarrow & 0 \end{array}$$

Its termination can be proved by the polynomial interpretation

$$\begin{array}{rcl} [s](X) & = & X^2 \\ [half](X) & = & X^d \end{array}$$

and indeed $\{half\}(n) = \lfloor \frac{n}{2} \rfloor$ *is less than* n.

5 Practical issues

This result has interesting practical consequences. First it sets a strict limit on the possible interpretations of the constructors, namely essentially $[s](X) = X + c$ and $[0] = 2$ and no specific limit for the defined functions. On the other hand, we can guess from the function $\{f\}$ whether the rewrite system that defines f has a polynomial interpretation proof of termination. Such a proof exists only if $\{f\}$ has polynomial growth. For instance, a system that defines the factorial or the exponential cannot be proved to terminate using polynomial interpretations.

References

[1] N. Dershowitz and J-P. Jouannaud. Notations for rewriting, 1991. LaTeXscript.

[2] O. Geupel. Terminationbeweise bei Termersetzungssytem, 1988. Diplomarbeit.

[3] G. Huet and D.C. Oppen. Equations and rewrite rules: A survey. In R. Book, editor, *Formal Language Theory: Perspectives and Open Problems*, pages 349–405. Academic Press, 1980.

[4] D. Lankford. Canonical algebraic simplification in computational logic. Report ATP–25, University of Texas, 1975.

[5] D. Lankford. On proving term rewriting systems are noetherian. Report MTP-3, Louisiana Tech. University, 1979.

[6] C. Lautemann. A note on polynomial interpretation. In *Bulletin of the European Association for Theoretical Computer Science*, volume 4, pages 129–131, October 1988.

Uniform Traversal Combinators: Definition, Use and Properties *

Leonidas Fegaras[1], *Tim Sheard*[2], *and David Stemple*[1]

[1] Department of Computer and Information Science, University of Massachusetts, Amherst, MA 01003. (*email:* {fegaras,stemple}@cs.umass.edu)
[2] Department of Computer Science and Engineering, Oregon Graduate Institute, Beaverton, OR 97006. (*email:* sheard@cse.ogi.edu)

Abstract

In this paper we explore ways of capturing well-formed patterns of recursion in the form of generic reductions. These reductions, called *uniform traversal combinators*, can substantially help the theorem proving process by eliminating the need for induction and can also be an aid in achieving effective program synthesis.

1 Introduction

Recursive structures, such as lists and trees, can be defined inductively in most functional languages [6]. The recursive types of these structures can be formalized using axiom sets generated automatically from their type definition, which are basically equivalent to Hoare's axioms for recursive data structures [5]. Programs that operate on instances of these types can be expressed as recursive functions in a pure applicative language. Theorems about these functions can be proved using induction principles on the structure of the parameter types of these functions. The Boyer-Moore theorem prover [3], for example, proves theorems about recursive functions mechanically by using axioms, definitions, and previously proved theorems, along with powerful induction mechanisms on recursive structures. Program synthesis needs techniques similar to those used in theorem proving. However it is more difficult, partially because induction methods cannot be applied directly for synthesizing the recursive definition of a function.

Proving theorems about computations over recursive structures can be made easier by requiring that functions be expressed in stereotyped ways. One kind of stereotyping is the systematic use of higher order functions that carry out all the traversal of recursive structures. Such traversal functions can capture common patterns of recursion that occur often during programming and, therefore, minimize the explicit use of recursion, which now becomes encapsulated by these functions. The well-known map function that applies a function to each element of a list is an example of a higher order function that encapsulates a traversal. By proving theorems about such traversal functions, some properties of functions using them can be proven by shallower reasoning than would be required

* This paper is based on work supported by the National Science Foundation under grants IRI-8606424 and IRI 8822121 and by the Office of Naval Research University Research Initiative contract, number N00014-86-K-0764

if the traversals were not "pre-analyzed" in isolation. In particular, the use of induction proofs can be substantially diminished.

Reductions [12, 11] are convenient abstractions for expressing manipulations of bulk data types represented by recursive structures. They accumulate results as they traverse a structure and can be used for more computations than are expressible using a mapping traversal. Reductions tailored to particular recursive types can be generated automatically by a compiler by examining the type details. Reductions over lists and finite sets are expressive enough to directly simulate all primitive recursive functions [7]. The work reported here extends these reductions to cover a larger set of recursion patterns and is motivated by the desire to use them as an aid both in theorem proving and program synthesis.

In this paper we explore a broad class of traversal functions and prove their fundamental properties. We introduce a family of generic functions, called **traversal combinators**, that capture a large family of type-safe primitive recursive functions. Most functions expressed as recursive programs where only one parameter becomes smaller at each recursive call are members of this family. This restriction excludes some valid functions, such as structural equalities and ordering, because they require their two input structures to be traversed simultaneously.

Our generic functions are combinators as each takes functions as inputs and return a new function as output (the one that performs the actual reduction). The most important contribution of this paper is our treatment of the class of traversal combinators resulting from restricting their input functions to be themselves traversal combinators. We call these functions **uniform traversal combinators**. This offers a disciplined and uniform treatment of functions. This uniformity introduces some nice properties, such as these combinators being closed under composition, that aid in theorem proving and program synthesis. In order to prove equality theorems it was necessary to extend our language to include structural equality as a special primitive. Programs expressed in this algebra can be tested for functional equivalence in a systematic and complete way, based on the fact that there is a unique way for expressing a function as a traversal.

Our algebra is at least equivalent to the first order logic. It cannot capture some interesting functions, such as transitive closure and integer exponentiation. Nevertheless, our system can express and prove complex theorems. We envision a system where all theorems expressible in our algebra are proved using the efficient algorithms presented in this paper, while the rest are tested by a theorem prover based on heuristics, such as the Boyer-Moore theorem prover.

2 Related Work

The work reported here is a new method for theorem proving with structural induction. Even though there is some research on analyzing the properties of some highly stereotyped recursions similar to our traversals, there is no work reported on defining a systematic method for applying these properties to theorem proving.

The most influential work on realizing the importance of capturing recursions into a few powerful patterns was by Richard Bird [1]. Even though his work was focused on specific types, such as lists, it suggested ways of extending these methods to other types. The work of Grant Malcolm on homomorphisms [8] generalized these methods for all types. His paper introduced a generic form of homomorphisms, very similar to our traversal combinators, stated without proof the promotion theorem for any homomorphism, and used it for proving some general properties of recursive types. Another work with similar definitions of recursive patterns, called iterative functions, was by Bohm and Berarducci [2], based on the second order lambda calculus. The most complete work on analyzing the properties of stereotyped recursions was by Meijer, Fokkinga, and Paterson [9]. They presented four classes of generic functions: catamorphisms (similar to our traversal combinators), anamorphisms, hylomorphisms, and paramorphisms. They proved a large number of generic theorems for each class, such as the fusion law (similar to our promotion theorem) and the uniqueness property for catamorphisms (for proving equalities of two functions).

3 Definitions

The language used in this and subsequent sections is ADABTPL [4], which is a strongly-typed functional programming language [6]. All types described in this paper are **canonical types**. The set of canonical types is a restricted subset of all the types that can be expressed in ADABTPL. They are constructed exclusively using 1) parameterization, 2) recursion, 3) the singleton type constructor, 4) the tuple type constructor, and 5) the union type constructor. For purposes of explanation, in the definition of a type T we restrict recursion to be a direct reference to type T, not to any type expression $t(T)$ that depends on T. For example, we do not permit the definition of the part-subpart tree structure where a subpart consists of a list of parts. In Section 8 we will remove this restriction to allow any recursively defined type.

An example of a tuple type definition is:

```
person = struct make_person ( name: string, address: string );
```

The tuple value constructor here is `make_person` of type `[string,string]->person`, that is, it takes two strings as input and returns a new object of type `person` as output. Examples of polymorphic union types are lists and trees (they have one type parameter **alpha** that can be instantiated to any canonical type):

```
list(alpha) = union
    ( null: singleton nil,
      consp: struct cons ( head: alpha, tail: list(alpha) ) );
tree(alpha) = union
    ( emptytree: singleton empty,
      fulltree: struct node
                ( info: alpha, left: tree(alpha), right: tree(alpha) ) );
```

The singleton type constructor creates a type with only one value. The value is constructed by the nullary constructor function named following `singleton`, such as `nil` and `empty`. Lists and trees have the following constructors:

```
nil: []->list(alpha)
```

```
cons: [alpha,list(alpha)]->list(alpha)
empty: []->tree(alpha)
node: [alpha,tree(alpha),tree(alpha)]->tree(alpha)
```

In general, any canonical type T has a number of constructors $C_1, \ldots, C_n$. More specifically, a tuple type has only one constructor whose input types are not recursive. Each alternative of a union type has one constructor and therefore this union type has all these constructors. We assume that any union type has at least one constructor with no recursive input types. Nullary constructors of type $C_i : () \rightarrow T$ are considered to be constant values of type T. For example, `nil` is a constructor of type `[]->list(alpha)` but it is used as a constant of type `list(alpha)`. To make our notation simpler, we assume that each constructor C_i has the variables of type T separated from the other variables: we write $C_i(\overline{x_i}, \overline{y_i})$ to indicate that the variables $\overline{x_i} = x^i_1, \ldots, x^i_{i_r}$ are of any type other than T and the variables $\overline{y_i} = y^i_1, \ldots, y^i_{i_s}$ are of type T.

One generic function over lists is the list reduction function `tc_list(fnil,fcons)`:

```
function(alpha,beta)
   tc_list ( fnil:  []->beta,
             fcons: [alpha,list(alpha),beta]->beta ): [list(alpha)]->beta;
[x]->case x
     { nil -> fnil();
       cons(a,r) -> fcons(a,r,tc_list(fnil,fcons)(r)) };
```

where $[x_1, \ldots, x_n] \rightarrow \text{exp}$ is a lambda abstraction with variables $x_1, \ldots, x_n$ and body exp (expressed as $\lambda x_1 \ldots \lambda x_n.\text{exp}$ in lambda calculus) and `case` matches `x` with one of the list constructor patterns. The list of variables following the keyword `function`, that is, `(alpha,beta)` in our example, denotes free type variables of a polymorphic function definition. Note that the body of `tc_list` is a lambda abstraction, that is, this function returns a closure. In other words, since the inputs are functions this is a combinator. For this reason, applying `tc_list` of the functions `f1` and `f2` to a list `l` is written as `tc_list(f1,f2)(l)`. For example, the list length function is computed by `tc_list([]->0,[?,?,i]->i+1)`, where ? is a don't-care parameter. The list append function `append(x,y)` is computed by `tc_list([]->y,[a,?,l]->cons(a,l))(x)`, while the list reverse is computed by `tc_list([]->nil,[a,?,l]->append(l,cons(a,nil)))`.

Reductions can be generalized to cover all canonical types. We call these generalized reductions *traversal combinators*. They are combinators because they accept functions as inputs, such as `fnil` and `fcons` in `tc_list`, and return a new function as output. The 'traversal' part of the name is justified because the output function of the combinator traverses the hierarchical structure of its input object.

Definition 1 (Traversal combinator) *Let T be a canonical type with constructors $C_i(\overline{x_i}, \overline{y_i})$. A traversal combinator $\mathcal{H}_T(f_1, \ldots, f_n) : T \rightarrow b$, where b is any type, is defined as follows:*

$[x] \rightarrow$ *case* x

$\{ \quad \ldots$

$\quad C_i(\overline{x_i}, \overline{y_i}) \;\rightarrow\; f_i(\overline{x_i}, \overline{y_i}, \mathcal{H}_T(f_1, \ldots, f_n)(y^i_1), \ldots, \mathcal{H}_T(f_1, \ldots, f_n)(y^i_{i_r}));$

$\quad \ldots$

$\}$

Variables z_k^i in $f_i(\overline{x_i}, \overline{y_i}, \overline{z_i})$, where $z_k^i = \mathcal{H}_T(f_1, \ldots, f_n)(y_k^i)$, are called **accumulative result variables** because they accumulate the results of the recursive calls to $\mathcal{H}_T(f_1, \ldots, f_n)$, while variables from $\overline{x_i}$ and $\overline{y_i}$ are called non-accumulative result variables. Note that if $C_i(\overline{x_i}, \overline{y_i})$ has k variables of a type other than T (these are the $\overline{x_i}$ variables) and m variables of type T (these are the $\overline{y_i}$ variables), then f_i has $k + 2m$ variables: $k + m$ non-accumulative and m accumulative result variables. For example, in the expression `tc_list([]->y,[a,l,r]->cons(a,r))(x)` variable `r` is accumulative while `a`, `l` and `x` are not.

From Definition 1 we can see that if f is the traversal combinator $\mathcal{H}_T(f_1, \ldots, f_n)$ then:

$$f(C_i(\overline{x_i}, \overline{y_i})) = f_i(\overline{x_i}, \overline{y_i}, f(y_1^i), \ldots, f(y_{i_r}^i))$$

In addition, if $\forall i: \ f_i(\overline{x_i}, \overline{y_i}, \overline{z_i}) = C_i(\overline{x_i}, \overline{z_i})$ then $f = \lambda x.x$; that is, f is the identity function for T.

In ADABTPL, a traversal combinator $\mathcal{H}_T$ is written as `tc_T`. For example, the integer type `int` has two constructors: `zero: int` and `succ: [int]->int`. The traversal combinator over integers is `tc_int(fz,fs)`:

```
function(beta) tc_int ( fz: []->beta, fs: [int,beta]->beta ): [int]->beta;
[x]->case x
     { zero -> fz();
       succ(i) -> fs(i,tc_int(fz,fs)(i)) };
```

If `beta=int` then this combinator can simulate all primitive recursive functions for integers [10]. For example, `tc_int([]->y,[?,i]->succ(i))(x)` computes $x + y$;
`tc_int([]->zero,[?,i]->i+y)(x)` computes $x * y$;
`tc_int([]->true,[?,i]->(i=false))(x)` computes the predicate $even(x)$;
`tc_int([]->true,[i,r]->(i=y) or r)(x)` computes $x \leq y$;
`tc_int([]->zero,[i,r]->succ(if (i=y) then zero else r))(x)` computes $x - y$;
and `tc_int([]->succ,[?,f]->tc_int([]->f(succ(zero)),[?,i]->f(i)))(m)(n)` computes the Ackermann function Ack(m,n) [2].

The boolean type is just the union of the singletons `true` and `false`. The boolean traversal combinator `tc_boolean` is the thinly disguised if-then-else function:

```
function(beta) tc_boolean
   ( ftrue: []->beta, ffalse: []->beta ) : [boolean]->beta;
[x]->case x { true -> ftrue(); false -> ffalse() };
```

A theorem, very useful for proving properties about traversal combinators is the promotion theorem [8] (or fusion law [9]). It states the condition for the composition of a function with a traversal combinator to be a traversal combinator too. It says that the composition of a function g with a traversal combinator $\mathcal{H}_T(f_1, \ldots, f_n)$ is also a traversal combinator if the composition of g with each f_i promotes the g call only to the accumulative result variables of f_i. This theorem is used in Section 4 as a reduction rule for composing traversal combinators.

Theorem 1 (Promotion theorem)
if $\forall i \forall \overline{x_i} \forall \overline{y_i} \forall \overline{z_i}: \ g(f_i(\overline{x_i}, \overline{y_i}, \overline{z_i})) = \phi_i(\overline{x_i}, \overline{y_i}, g(z_1^i), \ldots, g(z_{i_r}^i))$ *then*

$$g \circ \mathcal{H}_T(f_1, \ldots, f_n) = \mathcal{H}_T(\phi_1, \ldots, \phi_n)$$

Proof by structural induction: We will prove that $\forall x: \; g(f(x)) = \phi(x)$, where $f = \mathcal{H}_T(f_1,\ldots,f_n)$ and $\phi = \mathcal{H}_T(\phi_1,\ldots,\phi_n)$. If x is a construction $C_i(\overline{x_i})$ (that is, C_i has no arguments of type T), then $g(f(C_i(\overline{x_i}))) = g(f_i(\overline{x_i})) = \phi_i(\overline{x_i}) = \phi(C_i(\overline{x_i}))$. Let $x = C_i(\overline{x_i},\overline{y_i})$. We assume the theorem is true for all $y : T$ that are subtrees of the tree x. Then $g(f(x)) = g(f_i(\overline{x_i},\overline{y_i},f(y^i_1),\ldots,f(y^i_{i_r}))) = \phi_i(\overline{x_i},\overline{y_i},g(f(y^i_1)),\ldots,g(f(y^i_{i_r})))$. But each y^i_k is a subtree of x and thus $g(f(y^i_k)) = \phi(y^i_k)$. Therefore, $\phi_i(\overline{x_i},\overline{y_i},g(f(y^i_1)),\ldots,g(f(y^i_{i_r}))) = \phi_i(\overline{x_i},\overline{y_i},\phi(y^i_1),\ldots,\phi(y^i_{i_r})) = \phi(C_i(\overline{x_i},\overline{y_i})) = \phi(x)$.□

For example, for lists we have:

$$\left.\begin{array}{ll} g(f_1()) & = \phi_1() \\ \forall a \forall l \forall s : g(f_2(a,l,s)) & = \phi_2(a,l,g(s)) \end{array}\right\} \Rightarrow g \circ \text{tc\_list}(f_1,f_2) = \text{tc\_list}(\phi_1,\phi_2)$$

The promotion theorem for integers is:

$$\left.\begin{array}{ll} g(f_1()) & = \phi_1() \\ \forall i \forall r : g(f_2(i,r)) & = \phi_2(i,g(r)) \end{array}\right\} \Rightarrow g \circ \text{tc\_int}(f_1,f_2) = \text{tc\_int}(\phi_1,\phi_2)$$

The following corollary says that there is a unique way for expressing a function as a traversal combinator [9]. It is used in Section 5 for testing the functional equality of two traversal combinators:

Corollary 1 (Uniqueness property)

$$\forall i \forall \overline{x_i} \forall \overline{y_i} : \; g(C_i(\overline{x_i},\overline{y_i})) = \phi_i(\overline{x_i},\overline{y_i},g(y^i_1),\ldots,g(y^i_{i_r})) \Longleftrightarrow g = \mathcal{H}_T(\phi_1,\ldots,\phi_n)$$

Proof: $\Rightarrow$: From the promotion theorem with $f_i(\overline{x_i},\overline{y_i},\overline{z_i}) = C_i(\overline{x_i},\overline{z_i})$.
$\Leftarrow$: From Definition 1. □

Structural equalities for all but the trivial canonical types cannot be captured as traversal combinators. We need to define a special combinator instead. In the following definition of structural equality we do not separate the arguments of C_i of type T from the others in order to make the notation more readable:

Definition 2 (Structural Equality) *Let $T(\overline{\alpha})$ be a canonical type, where $\overline{\alpha}$ is a sequence of type parameters $\alpha_1,\ldots,\alpha_r$, $r \geq 0$. A structural equality $\mathcal{EQ}_T(\varepsilon_1,\ldots,\varepsilon_r) : T(\overline{\alpha}) \times T(\overline{\alpha}) \rightarrow$ boolean, where $\varepsilon_i : \; \alpha_i \times \alpha_i \rightarrow$ boolean, over the canonical type $T(\overline{\alpha})$ is defined as:*

$$\begin{array}{l} [x,y] \rightarrow \mathit{case}\; x,\; y \\ \qquad \{ \;\; \ldots \\ \qquad\quad C_i(x_1,\ldots,x_{k_i}),\; C_i(y_1,\ldots,y_{k_i}) \;\rightarrow\; \bigwedge_{j=1}^{k_i} \mathcal{R}_{i,j}(x_j,y_j); \\ \qquad\quad \ldots \\ \qquad\quad \mathit{other} \;\rightarrow\; \mathit{false} \\ \qquad \} \end{array}$$

where each x_j and y_j are of type $T_{i,j}(\overline{\alpha})$ and

$$\mathcal{R}_{i,j} = \left\{ \begin{array}{ll} \varepsilon_s & \mathit{if}\; \exists s : \; T_{i,j}(\overline{\alpha}) = \alpha_s \\ \mathcal{EQ}_{T_{i,j}}(\varepsilon_1,\ldots,\varepsilon_r) & \mathit{otherwise} \end{array}\right.$$

In ADABTPL, a structural equality $\mathcal{EQ}_T$ is written as equal_T. For example, the list equality has only one parameter ea that corresponds to the type parameter alpha:

```
function(alpha) equal_list
   ( ea: [alpha,alpha]->boolean ) : [list(alpha),list(alpha)]->boolean;
[x,y]->case x, y
       { nil, nil -> true;
         cons(a,l), cons(b,r) -> ea(a,b) and equal_list(ea)(l,r);
         other -> false };
```

We will now enhance the definition of structural equality $\mathcal{EQ}_T$ in such a way that the composition algorithm in the next section is true. This is done by taking the two inputs of type T of the output function of $\mathcal{EQ}_T$, such as the two lists of equal_list, as nullary functions and by adding a continuation that maps the result of the equality to a type β (typically this continuation is expressed as a tc_boolean combinator). The reason behind this enhancement is that when we compose a traversal combinator with an equality combinator we want to yield another equality combinator so that our language is closed under composition. This is achieved with the extra continuation ϕ.

Definition 3 (Equality Combinator) *Let T be a canonical type with structural equality $\mathcal{EQ}_T$, f and g functions of type $() \rightarrow T$, and ϕ a function of type boolean $\rightarrow \beta$. The equality combinator for T is $\mathcal{E}_T$: $() \rightarrow \beta$, defined as:*

$$\mathcal{E}_T(f, h, \phi) = \phi(\mathcal{EQ}_T(\varepsilon_1, \ldots, \varepsilon_r)(f, h))$$

Parameters ε_k are instantiated to equalities whenever the type parameters of T are instantiated to types. From now on we will ignore them because they do not affect our analysis. We will assume that they are hidden in $\mathcal{EQ}_T$. The promotion theorem for equality combinators is simply:

$$g \circ \mathcal{E}_T(f, h, \phi) = \mathcal{E}_T(f, h, g \circ \phi)$$

In ADABTPL, an equality combinator $\mathcal{E}_T$ is written as eq_T. For example:

```
function(beta) eq_list
  ( f: []->list(int), h: []->list(int), c: [boolean]->beta ) : beta;
c(equal_list(equal_int)(f(),h()));
```

4 Uniform Traversal Combinators

A very interesting class of traversal combinators results from restricting all functions f_i in $\mathcal{H}_T(f_1, \ldots, f_n)$ to be traversal combinators themselves. This restriction is very important because any theorem or algorithm that refers to such combinators can also work on each f_i recursively. This strict discipline of the function form offers us a more uniform treatment of functions. Functions have now a tree-like form, where each traversal combinator $\mathcal{H}_T(f_1, \ldots, f_n)$ is a node and each f_i is a child. This structured view of functions is aimed at simplifying theorem proving and facilitating program synthesis.

Definition 4 (Uniform Traversal combinator) *A uniform traversal combinator from T to b, where T and b are canonical types and all variables z and $\overline{z}$ are bounded variables, is one of the following:*

- **projection***: a lambda abstraction* $\lambda\overline{x}.\ z$*, where* z *is a variable of type* b*;*
- **construction***: an expression* $\lambda\overline{x}.\ C(h_1(\overline{z}),\ldots,h_s(\overline{z}))$ *where* C *is a constructor of* b *and each* h_i *is a uniform traversal combinator;*
- **traversal***: a traversal combinator* $\lambda\overline{x}.\ \mathcal{H}_T(f_1,\ldots,f_n)(z)$ *from* T *to* b *where each* f_i *is a uniform traversal combinator and variable* z *is a non-accumulative result variable;*
- **equality***: an equality combinator* $\lambda\overline{x}.\ \mathcal{E}_T(f,h,\phi)$*, where* f*,* h*, and* ϕ *are uniform traversal combinators.*

For example, integer multiplication is computed by the uniform traversal combinator:

```
[x,y]->tc_int([]->zero,[?,i]->tc_int([]->i,[?,j]->succ(j))(y))(x)
```

The function `reverse(x)` computed by:

```
[x]->tc_list([]->nil,[a,?,l]->tc_list([]->cons(a,nil),
                                       [c,?,s]->cons(c,s))(l))(x)
```

is not a uniform traversal combinator, because the inner `tc_list` is on `l` which is an accumulative result variable. The following is a uniform traversal combinator that when applied over a list of lists x it returns the list of lengths of x:

```
[x]->tc_list([]->nil,[a,?,r]->cons(tc_list([]->zero,
                                           [?,?,i]->succ(i))(a),r))(x)
```

Here the inner `tc_list` is over `a`, which is not an accumulative result variable. Integer subtraction `x-y` can be computed by the following uniform traversal combinator:

```
[x,y]->tc_int([]->zero,
              [i,r]->succ(eq_int([]->i,[]->y,
                                 [z]->tc_boolean([]->zero,[]->r)(z))))(x)
```

Two points need to be clear in the definition of uniform traversal combinators. First, traversals are over variables, not over expressions. In addition, this definition does not say that the composition of uniform traversal combinators is also a uniform traversal combinator. But we will prove next that this is true for any such composition. Second, the variable z of a traversal must not be an accumulative result variable. This is a necessary condition for having these combinators closed under composition. The intuition behind this is that we cannot traverse the values that are accumulated during the traversal of a structure (but we can pass them as whole values to constructions or to equalities). This restriction is very important because it substantially limits the expressiveness of our algebra (it makes our language deterministic logspace instead of polynomial time). Even though there is an alternative way of expressing the reverse function, there are some interesting functions, such as the transitive closure and the integer exponentiation, that cannot be captured in our algebra.

4.1 Composition of Uniform Traversal Combinators

Suppose that we have the composition $g(h(\overline{x}))$, where g and h are uniform traversal combinators. We can synthesize the traversal ϕ equal to $g \circ h$ by applying the promotion theorem to break it down into simpler cases. These cases are also made simpler by applying the promotion theorem again. The following constructive proof composes any uniform traversal combinators.

Theorem 2 (Composition of Uniform Traversal Combinators)
The composition of uniform traversal combinators is a uniform traversal combinator.

Constructive proof: First we will present the algorithm for composing combinators and then we will prove that the algorithm is correct. We denote $\Phi(g, [h_1, \ldots, h_r], \rho)$ the application of the function g over $h_1 \ldots h_r$, where all g, $h_1, \ldots, h_r$ are uniform traversal combinators. Variable ρ contains bindings from combinators to combinators. To find this composition, the g call is pushed inside the expressions h_i by applying the promotion theorem. The promotion theorem says that for $f = \mathcal{H}_T(f_1, \ldots, f_n)$: $g \circ f = \mathcal{H}_T(\phi_1, \ldots, \phi_n)$ if $g(f_i(\overline{x_i}, \overline{y_i}, \overline{z_i})) = \phi_i(\overline{x_i}, \overline{y_i}, g(z^i_1), \ldots, g(z^i_{i_r}))$. If $w^i_k = g(z^i_k)$ then $\phi_i(\overline{x_i}, \overline{y_i}, w^i_1, \ldots, w^i_{i_r})$ is the composition $g \circ f_i$, provided that all references to z^i_k are eliminated (as it will be proved below). For that reason we pass a binding list ρ to Φ that contains the bindings $g(z^i_k) = w^i_k$ and $z^i_k = f(y^i_k)$ (from Definition 1). The algorithm consists of the following rules:

Algorithm 1 (Composition Algorithm)

$$\begin{array}{ll}
\Phi(\lambda\overline{x}.x_k, [\ldots, h_i, \ldots], \rho) & \rightarrow h_k \\
\Phi(g, [\lambda\overline{x}.x_i], \rho) & \rightarrow \text{if } \rho \vdash g(x_i)/e \text{ then } \lambda\overline{x}.e \text{ else } \lambda\overline{x}.g(x_i) \\
\Phi(\lambda\overline{x}.C(\ldots, e_i(\overline{x}), \ldots), [\ldots, h_i, \ldots], \rho) & \rightarrow \lambda\overline{x}.C(\ldots, \Phi(e_i, [\ldots, h_i, \ldots], \rho), \ldots) \\
\Phi(\lambda\overline{x}.\mathcal{H}_T(\ldots, f_i, \ldots)(z), [C_i(\overline{u}, \overline{w})], \rho) & \rightarrow \Phi(f_i, [\overline{u}, \overline{w}, \ldots, \Phi(\lambda\overline{x}.\mathcal{H}_T(\ldots, f_i, \ldots)(z), [w_k], \rho), \ldots], \rho) \\
\Phi(\lambda\overline{x}.\mathcal{E}_T(f, h, \phi), [\ldots, h_i, \ldots], \rho) & \rightarrow \lambda\overline{x}.\mathcal{E}_T(\Phi(\lambda\overline{x}.f, [\ldots, h_i, \ldots], \rho), \Phi(\lambda\overline{x}.h, [\ldots, h_i, \ldots], \rho), \phi) \\
\Phi(g, [\lambda\overline{x}.\mathcal{E}_T(f, h, \phi)], \rho) & \rightarrow \lambda\overline{x}.\mathcal{E}_T(f, h, \Phi(g, [\phi], \rho)) \\
\Phi(g, [\lambda\overline{x}.\mathcal{H}_T(\ldots, f_i, \ldots)(z)], \rho) & \rightarrow \lambda\overline{x}.\mathcal{H}_T(\ldots, \Phi(g, [\lambda\overline{x_i}\lambda\overline{y_i}\lambda\overline{w_i}.f_i(\overline{x_i}, \overline{y_i}, \overline{z_i})], \rho'), \ldots)(z) \\
 & \quad \text{where } \rho' = \rho[\ldots, g(z^i_k)/w^i_k, z^i_k/\mathcal{H}_T(\ldots, f_i, \ldots)(y^i_k), \ldots]
\end{array}$$

Expression $\rho[u/w]$ extends ρ with the binding from the combinator u to the combinator w, while expression $\rho \vdash u/w$ returns true if there is a binding in ρ from u to w. The last rule comes from the promotion theorem. It renames the variables of f_i from $\overline{w_i}$ to $\overline{z_i}$ but it binds each w^i_k in $\overline{w_i}$ to $g(z^i_k)$. So if later a call $g(y^i_k)$ is found, it is replaced with w^i_k.

It is obvious that the above reductions are correct and they always terminate, because one of the two parameters of Φ becomes smaller in each recursive call to Φ. The resulting expression has the form of a uniform traversal combinator but possibly with some variables unbound. These are the z^i_k variables from the last rule. The only thing that remains to prove is that there are no such unbound variables z^i_k at the end. But the last rule applies whenever g is a traversal. In that case Φ is called recursively with g as the first argument. This is true for the fourth and sixth rule too. This recursion terminates whenever we find a call $\Phi(g, [h], \rho)$ that does not match these recursive rules. Therefore, this call must match the second rule. Then, if x_i in $\lambda\overline{x}.x_i$ is one of the z^i_k in ρ then $g(x_i)$ is replaced with the variable w^i_k. If z^i_k does not appear in a call to g, then z^i_k is replaced with $f(y^i_k)$, where $f = \mathcal{H}_T(\ldots, f_i, \ldots)$. The only other place that z^i_k could appear is in z in the last rule. But this is not permitted because traversals cannot be done over accumulative result variables. □

For example, suppose that we want to find a traversal combinator tc_list(h1,h2)(x) equivalent to the composition length(append(x,y)), where append and length are:

```
append(x,y) = tc_list([]->y,[a,?,s]->cons(a,s))(x)
length(x) = tc_list([]->zero,[?,?,i]->succ(i))(x)
```

We apply the promotion theorem with g =length:

```
1) h1() = length(y) = tc_list([]->zero,[?,?,i]->succ(i))(y)
2) h2(a,?,length(s)) = length(cons(a,s)) = succ(length(s))
     => h2(?,?,u)=succ(u)   where u=length(s)
```

Therefore, the composition length(append(x,y)) is:

```
tc_list([]->tc_list([]->zero,[?,?,i]->succ(i))(y),[?,?,u]->succ(u))(x)
```

Let length(x)+length(y) be equal to tc_list(h1,h2)(x), where

```
x+y = tc_int([]->y,[?,j]->succ(j))(x)
```

We apply the promotion theorem for lists with

```
g(x) = tc_int([]->tc_list([]->zero,[?,?,i]->succ(i))(y),[?,j]->succ(j))(x)
```

to compose g(length(x)) = tc_list(h1,h2)(x):

```
1) h1() = g(zero) = tc_list([]->zero,[?,?,i]->succ(i))(y)
2) h2(?,?,g(i)) = g(succ(i)) = succ(g(i))
     => h2(?,?,w)=succ(w)    where w=g(i)
```

Therefore, length(x)+length(y) is:

```
tc_list([]->tc_list([]->zero,[?,?,i]->succ(i))(y),[?,?,w]->succ(w))(x)
```

From these two examples we can see that

```
length(append(x,y)) = length(x)+length(y)
```

This is an example of a theorem proved without using induction explicitly. Here testing the equality of length(append(x,y)) and length(x)+length(y) was trivial. In Section 5 we will present a complete method for testing functional equalities.

We will compose now g(f(x)), where:

```
f(x) = tc_list([]->nil,[a,?,r]->cons(tc_list([]->zero,
                                    [?,?,i]->succ(i))(a),r))(x)
g(y) = tc_list([]->zero,[b,?,j]->tc_int([]->j,[?,k]->succ(k))(b))(y)
```

(if x is a list of lists then f(x) returns the list of lengths of x and if y is a list of integers then g(y) returns the sum of all these integers). The promotion theorem for g(f(x)) equal to tc_list(f1,f2)(x) gives:

```
1) f1() = g(nil) = zero
2) f2(a,?,g(r)) = g(cons(tc_list([]->zero,[?,?,i]->succ(i))(a),r))
       = tc_int([]->g(r),[?,k]->succ(k))
         (tc_list([]->zero,[?,?,i]->succ(i))(a))
```

Let `h(x) = tc_int([]->u,[?,k]->succ(k))(x)`, where `u=g(r)`, and `h(tc_list([]->zero,[?,?,i]->succ(i))(a)) = tc_list(h1,h2)(a)`, then:

```
1) h1() = u
2) h2(?,?,h(i)) = succ(h(i)) => h2(?,?,w)=succ(w)
```

Therefore, `g(f(x))` is

```
tc_list([]->zero,[a,?,u]->tc_list([]->u,[?,?,w]->succ(w))(a))(x)
```

5 Equality of Uniform Traversal Combinators

The following algorithm tests whether any two uniform traversal combinators compute the same function. It is based on the uniqueness property that says that there is a unique way for expressing a function as a traversal combinator. We will make use of this algorithm in Section 6 for proving equality theorems.

Theorem 3 (Equality theorem) *Let $\lambda\overline{x}.f(\overline{x})$ and $\lambda\overline{x}.g(\overline{x})$ be two uniform traversal combinators. Then $\forall\overline{x}:\ f(\overline{x})=g(\overline{x})$ iff $\mathcal{E}(\lambda\overline{x}.f(\overline{x}),\lambda\overline{x}.g(\overline{x}),[])=$ true, where $\mathcal{E}$ is defined as: (for purposes of explanation we ignore variable renaming)*

Algorithm 2 (Equality of Combinators)

$\mathcal{E}(\lambda\overline{x}.z,\lambda\overline{x}.z,\rho) \rightarrow \text{true}$

$\mathcal{E}(g,\lambda\overline{x}.z,\rho) \rightarrow \rho\vdash z/g$

$\mathcal{E}(\lambda\overline{x}.C(\ldots,e_i(\overline{z}),\ldots),\lambda\overline{x}.C(\ldots,h_i(\overline{z}),\ldots),\rho) \rightarrow \forall i:\ \mathcal{E}(e_i,h_i,\rho)$

$\mathcal{E}(\lambda\overline{x}.C_1(\ldots),\lambda\overline{x}.C_2(\ldots),\rho) \rightarrow \text{false}$

$\mathcal{E}(g,\mathcal{E}_T(f,h,\phi),\rho) \rightarrow \text{if } \mathcal{E}(f,h,\rho) \text{ then } \mathcal{E}(g,\phi(\text{true}),\rho) \text{ else } \mathcal{E}(g,\phi(\text{false}),\rho)$

$\mathcal{E}(\lambda\overline{x}.g(z),\lambda\overline{x}.\mathcal{H}_T(\ldots,\phi_i,\ldots)(z),\rho) \rightarrow \forall i:\ \mathcal{E}(g(C_i(\overline{x_i},\overline{y_i})),\lambda\overline{x_i}\lambda\overline{y_i}.\phi_i(\overline{x_i},\overline{y_i},\overline{z_i}),\rho')$ where $\rho'=\rho[\ldots,z_k^i/g(y_k^i),\ldots]$

Proof: From the uniqueness property we have:

$$\forall i\forall\overline{x_i}\forall\overline{y_i}:\ g(C_i(\overline{x_i},\overline{y_i}))=\phi_i(\overline{x_i},\overline{y_i},g(y_1^i),\ldots,g(y_{i_r}^i))\ \Leftrightarrow\ g=\mathcal{H}_T(\phi_1,\ldots,\phi_n)$$

That is, g is equal to $\mathcal{H}_T(\ldots,\phi_i,\ldots)$ if and only if for all i: $g(C_i(\overline{x_i},\overline{y_i}))$ is equal to $\phi_i(\overline{x_i},\overline{y_i},\overline{z_i})$, where $z_k^i=g(y_k^i)$ (this is the last rule). In that case, ρ is extended to include all bindings $z_k^i=g(y_k^i)$ so that if later we need to test whether $z_k^i=g(y_k^i)$ then this is true (this is the second rule). □

For example, suppose that we want to prove the commutativity law for integer addition `x+y==y+x`, where `==` is the structural equality `equal_int` for integers. This is expressed as:

```
tc_int([]->y,[?,i]->succ(i))(x) == tc_int([]->x,[?,j]->succ(j))(y)
```

Let `g(y)=tc_int([]->y,[?,i]->succ(i))(x)` then we apply the uniqueness property for the equality `g(y)==tc_int([]->x,[?,j]->succ(j))(y)`:

```
1) y=zero:    (g(zero)==x) = (tc_int([]->zero,[?,i]->succ(i))(x)==x) = true
2) y=succ(j): (g(succ(j))==succ(g(j))) =
              (tc_int([]->succ(j),[?,i]->succ(i))(x)==succ(g(j)))
```

Let `f(x)=succ(g(j)))`. We apply the uniqueness property again:

```
1) x=zero:    (f(zero)==succ(j)) = (succ(j)==succ(j)) = true
2) x=succ(i): (f(succ(i))==succ(f(i))) = (succ(f(i))==succ(f(i))) = true
```

6 Theorem Proving

The algorithms for composing uniform traversal combinators and for testing their functional equalities can be applied for proving theorems about uniform traversal combinators. Suppose that we want to prove that an expression $e(\overline{x})$ is always true. First we need to find all uniform traversal combinators associated with each function call in e. Then we use the composition algorithm to synthesize the uniform traversal combinator $f(\overline{x})$ equivalent to $e(\overline{x})$. Then we are left with the simpler task of proving whether $f(\overline{x})$ is always true (a tautology).

Algorithm 3 (Tautology) *Let $f(\overline{x})$ be a uniform traversal combinator of type boolean. Then $\forall\overline{x}: \ f(\overline{x}) = c$, where c is either true or false, iff $\mathcal{T}(f(\overline{x}), c)$, where $\mathcal{T}$ is defined as:*

$$\begin{array}{lll}
\mathcal{T}(c,c) & \rightarrow \mathit{true} & \\
\mathcal{T}(\lambda\overline{x}\lambda\overline{y}.y_i, c) & \rightarrow \mathit{true} & (y_i \textit{ is an accumulative result variable}) \\
\mathcal{T}(\lambda\overline{x}.\mathcal{H}_T(f_1,\ldots,f_n)(z), c) & \rightarrow \forall i: \ \mathcal{T}(f_i, c) & \\
\mathcal{T}(\lambda\overline{x}.\mathcal{E}_T(f,h,\phi), c) & \rightarrow \mathit{if}\ \mathcal{E}(f,h,[])\ \mathit{then}\ \mathcal{T}(\phi(\mathit{true}),c)\ \mathit{else}\ \mathcal{T}(\phi(\mathit{false}),\neg c) & \\
\mathit{otherwise} & \rightarrow \mathit{false} &
\end{array}$$

Suppose that we want to prove the associativity law for integer multiplication, that is: `(x*y)*z==x*(y*z)`, where multiplication is computed by:

```
x*y = tc_int([]->zero,[?,i]->tc_int([]->i,[?,j]->succ(j))(y))(x)
```

We start by composing `(x*y)*z`. We apply the promotion theorem with

```
g(u) = u*z = tc_int([]->zero,[?,i]->tc_int([]->i,[?,j]->succ(j))(z))(u)
```

Let `g(tc_int([]->zero,[?,i]->tc_int([]->i,[?,j]->succ(j))(y))(x))` be equal to `tc_int(f1,f2)(x)`. Then from the promotion theorem we have:

```
1) f1() = g(zero) = zero
2) f2(?,g(i)) = g(tc_int([]->i,[?,j]->succ(j))(y)) = tc_int(h1,h2)(y)
```

Let `m=g(i)`. We apply the promotion theorem again:

```
1) h1() = g(i) = m
2) h2(?,g(j)) = g(succ(j)) = tc_int([]->g(j),[?,k]->succ(k))(z)
     => h2(?,w) = tc_int([]->w,[?,k]->succ(k))(z)
```

Therefore, (x*y)*z is

```
tc_int([]->zero,
       [?,m]->tc_int([]->m,[?,w]->tc_int([]->w,[?,k]->succ(k))(z))(y))(x)
```

We will compose now x*(y*z):

```
x*(y*z) = tc_int([]->zero,[?,i]->tc_int([]->i,[?,j]->succ(j))(y*z))(x)
```

Let g(u) = u+i = tc_int([]->i,[?,j]->succ(j))(u). Expression x*(y*z) becomes:

```
g(y*z) = tc_int(f1,f2)(y)
       = tc_int([]->zero,[?,u]->tc_int([]->u,[?,w]->succ(w))(z))(y)
```

and the promotion theorem gives:

```
1) f1() = g(zero) = i
2) f2(?,g(u)) = g(tc_int([]->u,[?,w]->succ(w))(z)) = tc_int(h1,h2)(z)
```

Let k=g(u). We apply the promotion theorem again:

```
1) h1() = g(u) = k
2) h2(?,g(w)) = g(succ(w)) = succ(g(w))  => h2(?,m) = succ(m)
```

Therefore, x*(y*z) is

```
tc_int([]->zero,
       [?,i]->tc_int([]->i,[?,k]->tc_int([]->k,[?,m]->succ(m))(z))(y))(x)
```

Finally, we can see that (x*y)*z is equal to x*(y*z).

7 Program Synthesis

Let g and ϕ be uniform traversal combinators. The following algorithm synthesizes all uniform traversal combinators f such that $g(f(\overline{x})) = \phi(\overline{x})$ is true. We express that as: $f \in \mathcal{S}(g, \phi)$, that is, $\mathcal{S}(g, \phi)$ returns the set of uniform traversal combinators f that satisfy $g(f(\overline{x})) = \phi(\overline{x})$. If ϕ is a traversal, then from the promotion theorem we have $\forall i: \; g(f_i(\overline{x_i}, \overline{y_i}, \overline{z_i})) = \phi_i(\overline{x_i}, \overline{y_i}, g(z_1^i), \ldots, g(z_r^i))$. If all such f_i are found, then f is $\mathcal{H}_T(f_1, \ldots, f_s)$. But $\phi_i(\overline{x_i}, \overline{y_i}, g(z_1^i), \ldots, g(z_r^i))$ is the composition of the uniform traversal combinators ϕ_i and g and thus it can be derived using the composition algorithm. Therefore, we need to synthesize all f_i that satisfy the equation $g(f_i(\overline{x_i}, \overline{y_i}, \overline{z_i}))$ = a known traversal. This is achieved by calling the same algorithm for synthesizing combinators recursively. The detailed algorithm is the following:

Algorithm 4 (Synthesis algorithm)

$$
\begin{array}{ll}
\mathcal{S}(\lambda y.y, \phi) & \to \{\phi\} \\
\mathcal{S}(\lambda y.C(\ldots), \lambda \overline{z}.z_i) & \to \mathcal{S}(\lambda y.C(\ldots), \lambda \overline{x_i} \lambda \overline{y_i}.C(\overline{x_i}, \overline{y_i})) \\
\mathcal{S}(\lambda y.C_1(\ldots), C_2(\ldots)) & \to \emptyset \\
\mathcal{S}(\lambda y.C(\ldots, g_i, \ldots), C(\ldots, h_i, \ldots)) & \to mgu(\ldots, \mathcal{S}(g_i, h_i), \ldots) \\
\mathcal{S}(\mathcal{H}_T(\ldots), \lambda \overline{z}.z_i) & \to \mathcal{S}(\mathcal{H}_T(\ldots), \lambda \overline{z}.\mathcal{H}_T(\ldots, C_i, \ldots)) \\
\mathcal{S}(g, \mathcal{H}_T(\ldots, \lambda \overline{x_i} \lambda \overline{y_i} \lambda \overline{z_i}.\phi_i(\overline{x_i}, \overline{y_i}, \overline{z_i}), \ldots)) & \to \{\mathcal{H}_T(\ldots, f_i, \ldots) / \\
 & \quad f_i \in \mathcal{S}(g, \Phi(\phi_i, [\overline{x_i}, \overline{y_i}, \ldots, g(z_k^i), \ldots], []))\} \\
\mathcal{S}(\mathcal{H}_T(\ldots), C(\ldots)) & \to \bigcup_i \mathcal{S}(\Phi(\mathcal{H}_T(\ldots), C_i, []), C(\ldots)) \\
 & \quad \textit{where } C_i \textit{ is a constructor of } T
\end{array}
$$

where mgu($g_1, \ldots, g_n$) returns the most general unifier for all g_k. The last rule tries every constructor C_i of T as the output value of $f(\overline{x})$

For example, suppose that there is a uniform traversal combinator g that satisfies:

```
length(g(x,y)) = tc_list([]->tc_list([]->zero,[?,?,j]->succ(j))(y),
                         [?,?,i]->succ(i))(x)
```

where `length` is `tc_list([]->zero,[?,?,i]->succ(i))`. We want to find every g that satisfies the above equation. Let g be `tc_list(g1,g2)(x)`. From the promotion theorem we have:

```
1) length(g1) = tc_list([]->zero,[?,?,j]->succ(j))(y)
   let g1=tc_list(h1,h2)(y) then
   1.1) length(h1) = zero
          []->zero is the only component of length that returns zero => h1 = nil
   1.2) length(h2(c,s,j)) = succ(length(j))
          [?,?,i]->succ(i) is the only component of length that returns succ.
          Let h2(c,s,j)=cons(x1,x2) then length(cons(x1,x2))=succ(length(j))
          => succ(length(x2))=succ(length(j)) => h2(c,s,j) = cons(x1,j)
2) length(g2(b,r,i)) = succ(length(i)) => g2(b,r,i) = cons(x3,i)
```

Therefore, `g(x,y)` is:

```
tc_list([]->tc_list(nil,[?,?,j]->cons(x1,j))(y),[?,?,i]->cons(x3,i))(x)
```

where `x1` and `x3` are universally quantified variables that yield a class of solutions for `g(x,y)`.

8 Model Extensions

We can extend the definition of canonical types to include all recursively defined types. In that case, if a type T has a constructor $C(\ldots, x_i, \ldots)$, where x_i is of type $t(T)$, that is a type expression that refers to T, then instead of calling $\mathcal{H}_T$ recursively in the definition of $\mathcal{H}_T$ we call $\mathcal{M}_{T'}(\mathcal{H}_T)$, where $T'(\alpha) = t(\alpha)$, α is a type parameter, and $\mathcal{M}_{T'}$ is the generic map over T' [12]. A generic map of a canonical type $T(\alpha_1, \ldots, \alpha_n)$ is the function $\mathcal{M}_T(m_1, \ldots, m_n)$ that assigns a mapping m_i to each type parameter α_i. All theorems and algorithms for combinators are still valid because any generic map $\mathcal{M}_T(m_1, \ldots, m_n)$ is a uniform traversal combinator if all m_i are uniform traversal combinators. For example, the map over lists is `map_list(m)` and it is equal to `tc_list([]->nil,[a,?,r]->cons(m(a),r))`.

We can also extend our model to include higher order expressions: Let $T = T_1 \times \ldots \times T_n \rightarrow T_0$ be a function type. The uniform traversal combinator for T is $\mathcal{H}_T(c, r_1, \ldots, r_n) : T \rightarrow \beta$, where $c : T_0 \rightarrow \beta$ and $r_i : \rightarrow T_i$, and it is defined as $\lambda f.\ c(f(r_1(), \ldots, r_n()))$. Again variable f is not permitted to be an accumulative result variable. The promotion theorem for T is simply $g \circ \mathcal{H}_T(c, r_1, \ldots, r_n) = \mathcal{H}_T(g \circ c, r_1, \ldots, r_n)$. The composition algorithm is still correct because if h is a uniform traversal combinator so is $\mathcal{H}_T(c, r_1, \ldots, r_n)(h)$. That way we can prove high-order theorems, that is, for any function of a specific type.

In ADABTPL, a combinator for the type $T = T_1 \times \ldots \times T_n \rightarrow T_0$ is written as `tc_T1x...xTn_T0`. For example, the following proves that the list map distributes over append:

```
map_list(f)(append(x,y)) = append(map_list(f)(x),map_list(f)(y))
```

where `map_list(f)=tc_list([]->nil,[b,?,r]->tc_a_b([z]->cons(z,r),[]->b)(f))`. We apply the promotion theorem for `g(append(x,y))=tc_list(f1,f2)(x)`, where `g(x)=map_list(f)(x)`:

```
f1() = g(y) = map_list(f)(y)
f2(b,?,g(r)) = g(cons(b,r)) = tc_a_b([z]->cons(z,g(r)),[]->b)(f)
  => f2(b,?,u) = tc_a_b([z]->cons(z,u),[]->b)(f))
```

Therefore, `map_list(f)(append(x,y))` is

```
tc_list([]->map_list(f)(y),[b,?,u]->tc_a_b([z]->cons(z,u),[]->b)(f))(x)
```

which is the same with `append(map_list(f)(x),map_list(f)(y))`.

9 Acknowledgments

We are grateful to Neil Immerman and Sushant Patnaik for their help in analyzing the expressiveness of our language.

References

1. R. S. Bird. The Promotion and Accumulation Strategies in Transformational Programming. *ACM Transactions on Programming Languages and Systems*, 6(4):487–504, October 1984.
2. C. Bohm and A. Berarducci. Automatic Synthesis of Typed Λ-Programs on Term Algebras. *Theoretical Computer Science*, 39:135–154, 1985.
3. R. S. Boyer and J. S. Moore. *A Computational Logic*. Academic Press, New York, 1979.
4. L. Fegaras, T. Sheard, and D. Stemple. The ADABTPL Type System. In *Proceedings of the Second International Workshop on Database Programming Languages, Salishan, Oregon*, pages 243–254, 1989.
5. C. A. Hoare. Recursive Data Structures. *Journal of the ACM*, 4(2):105–132, June 1975.
6. P. Hudak. Conception, Evolution, and Application of Functional Programming Languages. *ACM Computing Surveys*, 21(3):359–411, September 1989.
7. N. Immerman, S. Patnaik, and D. Stemple. The Expressiveness of a Family of Finite Set Languages. *Proceedings of the Tenth ACM Symposium on Principles of Database Systems, Denver, Colorado*, pages 37–52, May 1991.
8. G. Malcolm. Homomorphisms and Promotability. In *Mathematics of Program Construction*, pages 335-347. Springer-Verlag, June 1989.
9. E. Meijer, M. Fokkinga, and R. Paterson. Functional Programming with Bananas, Lenses, Envelopes and Barbed Wire. In *fifth ACM Conference on Functional Programming Languages and Computer Architecture, Cambridge Massachusetts*, pages 124–144, August 1991.
10. H. Rogers. *Theory of Recursive Functions and Effective Computability*. McGraw-Hill, 1967.
11. T. Sheard. Generalized Recursive Structure Combinators. Technical report, University of Massachusetts at Amherst, 1989. COINS Technical Report 89-26.
12. T. Sheard. Automatic Generation and Use of Abstract Structure Operators. *ACM Transactions on Programming Languages and Systems*, 19(4):531–557, October 1991.

This article was processed using the LaTeX macro package with LMAMULT style

Sorted Unification Using Set Constraints

Tomás E. Uribe *

Dept. of Computer Science and Beckman Institute, University of Illinois,
405 North Mathews Avenue, Urbana, Illinois, 61801, USA

Abstract. This paper describes a new representation for sortal constraints and a unification algorithm for the corresponding constrained terms. Variables range over sets of terms described by *systems of set constraints* that can express limited inter-variable dependencies. These sets of terms are more general than regular tree languages, but are still closed under intersection. The new unification algorithm shows sorted unification to be decidable for a broad class of sorted signatures, which we call *semi-linear*, and, more generally, for sort theories with a least Herbrand model that can be represented using the new constraints. A finite representation of a complete set of well-sorted unifiers can always be found, even in those cases where this set is infinite.

1 Introduction

Sorts are widely used to add more "structure" to first order logic, improving the efficiency of deductive systems [Cohn, 1989]. Increasingly, sorted logic is being viewed as an instance of constraint logic [Comon and Delor, 1991; Frisch, 1991]. This is the approach adopted in this paper: Instead of defining "well-sorted" terms and substitutions, we define a class of monadic constraints where variables range over sets of ground terms described by *systems of set constraints*. An algorithm that finds the solutions to a set of equations between constrained terms is then presented. The sets of ground terms over which variables can range are more general than regular tree languages, but are still closed under intersection.

Nearly all specifications of order-sorted logics impose restrictions on the sorted signature in order to ensure the decidability of unification. Implicitly, these restrictions have been such that sorts are at most as general as regular tree languages. Schmidt-Schauß [1989] considered the specification of sorted signatures through the more general mechanism of *term declarations*, but left open the problem of unification under *linear* signatures, where no dependencies between variables in the term declarations can be enforced. Linear signatures also correspond to regular tree languages, so that our algorithm shows unification to be decidable not only for these but a more general class of signatures, which we call *semi-linear*, where limited inter-variable dependencies in the term declarations are allowed. In general, if we are given a sort theory whose least Herbrand model can be represented with the new constraints, then sorted unification under that theory can be carried out (see Section 3).

Our constraints are closely related to the extended *membership constraints* presented by Comon and Delor [1991] in the case where *tree automata with equality*

* Current address: Apartado Aéreo 100677, Bogotá, Colombia. Phone: (1-571) 213-7160.

tests [Bogaert and Tison, 1992] are used. Instead of referring to the normalization and determinization of these automata, we present alternate algorithms based on transforming systems of set constraints. The results in this paper are theoretical in nature, and the problem of finding efficient implementations for the algorithms presented here is left open. Such implementations could increase the overall efficiency of deductive systems by expanding the range of sort theories that the unifier can reason about, and by allowing, in some cases, the finite representation of otherwise infinite sets of answers.

This paper summarizes a portion of [Uribe, 1991], where all complete proofs can be found. Section 2 presents the preliminary definitions and introduces the new constraints, together with a solved form for them. Section 3 briefly describes the relationship among these constraints, sort theories and sorted signatures. Section 4 presents an algorithm to rewrite constraints into solved form. Section 5 presents the actual unification algorithm, and Section 6 discusses related work.

2 Set Expressions and Constraints

We are given a countably infinite set of variables $\mathcal{V}$ and a ranked set of function symbols $\mathcal{F}$, from which terms are constructed in the usual way.[2] For any syntactic object P, $Var(P)$ denotes the set of variables occurring in P. A term t is *ground* if $Var(t) = \emptyset$. A substitution θ is written as $\{x_1 \mapsto t_1, \ldots, x_n \mapsto t_n\}$, where each variable x_i is mapped to t_i, $x_i \neq x_j$ for $i \neq j$, and $t_i \neq x_i$ for all i. The term t_i is denoted by $x_i\theta$. The set of all variables x such that $x \neq x\theta$ is denoted by $Dom(\theta)$, and the set of all terms t such that $t = x\theta$ for $x \in Dom(\theta)$ is denoted by $Rng(\theta)$. Given substitutions θ and Ω, the substitution obtained by composing Ω onto θ is denoted by $\theta \cdot \Omega$; for all $x \in \mathcal{V}$, $x(\theta \cdot \Omega) = (x\theta)\Omega$. A substitution θ is *idempotent* if $\theta = \theta \cdot \theta$. Given a set of equations between terms $S = \{s_1 = t_1, \ldots, s_n = t_n\}$, a substitution θ is a *unifier* for S if $s_1\theta = t_i\theta$ for all i, $1 \leq i \leq n$. A substitution θ is *more general* than a substitution ρ if $\rho = \theta \cdot \gamma$ for some substitution γ.

We now define *systems of set constraints* (SSC's). These are based on the normal form for collections of set equations used in [Heintze and Jaffar, 1990], from which most of the notation in this section is in fact borrowed. However, new definitions and algorithms have to be used: Even though the syntax is simpler, we include limited inter-variable dependencies that allow SSC's to describe non-regular tree languages. The basic building blocks for SSC's are *set expressions*:

Definition 1 (Set Expressions). A *set expression* is (i) a term, (ii) one of the symbols "$\top$" and "$\bot$", or (iii) $(t \sqcup s)$ or $(t \sqcap s)$, where s and t are set expressions.

We do *not* allow "$\top$", "$\bot$", or any set expression that contains "$\sqcap$" or "$\sqcup$" to be nested inside a function symbol. We treat "$\sqcup$" and "$\sqcap$" as the idempotent, associative and commutative operators they will represent, so that a set expression of the form $(e_1 \sqcap (e_2 \sqcap \ldots (e_{n-1} \sqcap e_n) \ldots))$ can be written as $e_1 \sqcap \ldots \sqcap e_n$. A set expression is *atomic* if it is "$\bot$", "$\top$", or a term. A set expression is in *disjunctive normal form* (DNF) if it is of the form $e_1 \sqcup \ldots \sqcup e_m$, where each e_i is of the form $a_1 \sqcap \ldots \sqcap a_n$ for atomic set expressions $a_1, \ldots, a_n$. The result of rewriting a set expression *exp* into DNF is denoted by $DNF(exp)$.

[2] $\mathcal{F}$ should contain at least one constant symbol, so that the *Herbrand Universe* — the set of all ground terms — is not empty.

Definition 2 (Set Equations and Constraints). A *set equation* is a pair of set expressions $\{E_1, E_2\}$, written as $E_1 \equiv E_2$. A *set constraint* is an *ordered* set equation $X \equiv E$, where X is a variable and E is a set expression, written as $\langle X \equiv E\rangle$. The variable X is said to be *constrained* by E.

Set constraints will be written between brackets, as in $\langle X \equiv (Y \sqcap Z) \sqcup f(a)\rangle$.

Definition 3 (SSC's). A *system of set constraints* (SSC) is a finite set C of set constraints, $\{\langle Y_1 \equiv E_1\rangle, \ldots, \langle Y_n \equiv E_n\rangle\}$, such that $Y_i \neq Y_j$ if $i \neq j$. The set of variables $\{Y_1, \ldots, Y_n\}$ is the *domain* of C, denoted by $Dom(C)$, and the set of set expressions $\{E_1, \ldots, E_n\}$ is the *range* of C, denoted by $Rng(C)$. For each $Y_i \in Dom(C)$, the set expression E_i is denoted by $C(Y_i)$.

For our purposes, we will assume that $Var(Rng(C)) = Dom(C)$ for all SSC's C.

Definition 4 (Interpretations). An *interpretation* I defined on a finite set of variables $V \subset \mathcal{V}$ is a mapping from the variables in V into sets of ground terms. This mapping is extended to set expressions as follows:

1. $I(\top)$ is the Herbrand Universe; $I(\bot)$ is the empty set, $\emptyset$.
2. If X is a variable in V, then $I(X)$ is given by the original interpretation. If X is a variable not in V, then $I(X) \stackrel{\text{def}}{=} \emptyset$.
3. $I(e_1 \sqcap e_2) \stackrel{\text{def}}{=} I(e_1) \cap I(e_2)$; $I(e_1 \sqcup e_2) \stackrel{\text{def}}{=} I(e_1) \cup I(e_2)$.
4. Finally, if t is an arbitrary non-variable term, $I(t) \stackrel{\text{def}}{=} \{t\theta \mid \theta$ is a substitution with $X\theta \in I(X)$ for all $X \in Var(t)\}$. That is, $I(t)$ contains all those terms that can be obtained by replacing all occurrences of each variable X in t by an element of $I(X)$. Note that $I(t) = \emptyset$ if $I(X) = \emptyset$ for some $X \in Var(t)$.

The last condition ensures that equal variables within $f(t_1, \ldots, t_n)$ correspond to the same term, and is the main difference between these set constraints and Heintze and Jaffar's [1990] normal forms.

Two interpretations I and I' are *equivalent* over a set of variables V, written as $I =_V I'$, if for all $X \in V$, $I(X) = I'(X)$. A partial order on interpretations can be defined in the usual way, where $I \leq I'$ if for all $X \in \mathcal{V}$, $I(X) \subseteq I'(X)$.

Definition 5 (Models). Given an SSC C, an interpretation I is a *model* of C if $I(X) = I(C(X))$ for all $X \in Dom(C)$.

Two SSC's C and C' are *equivalent* over a set of variables V if for any model I of C, there exists a model I' of C' such that $I =_V I'$, and vice-versa. In this case, we write $C \Longleftrightarrow_V C'$. Given an SSC C and a set equation $E_1 \equiv E_2$, we write $C \models E_1 \equiv E_2$ if for any model I of C, $I(E_1) = I(E_2)$.

In general, SSC's need not have unique models over their constrained variables (though they always have a model). However, there is a simple sufficient condition under which an SSC has a unique model:

Definition 6 (Cycles). A *cycle* in an SSC C is a sequence of variables in $Dom(C)$, $X_1, \ldots, X_n$, such that $X_1 = X_n$ and for each i, $1 \leq i < n$, $C(X_i)$ contains an un-nested occurrence of X_{i+1}.[3]

[3] A variable is *un-nested* in a set expression if it is not a subterm of any other term.

Intuitively, a cycle means that a variable is defined directly in terms of itself, with no mediating function symbol. For example, the SSC $\{\langle X \equiv X \sqcap (a \sqcup b)\rangle\}$ has the simplest possible kind of cycle, while $\{\langle X \equiv f(X)\rangle\}$ does *not* contain a cycle.

Proposition 7. *Let C be an* SSC *with no cycles. Then for any two models I and I' of C, $I =_{Var(C)} I'$. Therefore, the model I of C such that $I(X) = \emptyset$ for all $X \notin Dom(C)$ is the* least *model of C.*

Definition 8 (Constrained Expressions). For any SSC C with no cycles, we denote the least model of C by M_C. A set expression E *constrained* by C, written as E/C, is said to stand for the set of terms $M_C(E)$. If $t \in M_C(E)$, we say that t is an *instance* of E/C.

All the SSC's we consider will have no cycles, so M_C will always be well-defined. Note that if $C_1 \Longleftrightarrow_V C_2$ and $Var(E) \subseteq V$, then $M_{C_1}(E) = M_{C_2}(E)$.

Given an arbitrary SSC C and $X_1, X_2 \in Dom(C)$, the question of whether the intersection of $M_C(X_1)$ and $M_C(X_2)$ is empty is undecidable. This follows from the undecidability of sorted unification for general sorted signatures with term declarations [Schmidt-Schauß, 1989], given the equivalence between SSC's and sorted signatures described in Section 3. This is why we will consider only *shallow* SSC's, where the only nested terms in $Rng(C)$ are variables:

Definition 9 (Shallow SSC's). An SSC C is *shallow* if the only terms in $Rng(C)$ are variables, constants, or of the form $f(V_1, \ldots, V_n)$ for an n-ary function f and variables $V_1, \ldots, V_n$, $n \geq 1$.

Definition 10 (Solved Form). An SSC C is in *solved form* if C has no cycles and for all $X \in Dom(C)$, $C(X)$ is "$\bot$", "$\top$", or $e_1 \sqcup \ldots \sqcup e_m$, $m \geq 1$, where $e_i \neq e_j$ for all $i \neq j$ and each e_i is a variable, a constant, or $f(X_1, \ldots, X_n)$ for variables $X_1, \ldots, X_n$, $n \geq 1$.

Note that all SSC's in solved form are shallow. As we show in Section 4, any shallow SSC C can be rewritten into an SSC in solved form, C', such that $C \Longleftrightarrow_{Dom(C)} C'$.

Example 1. $C = \{\langle Y \equiv X\rangle, \langle X \equiv f(X,X) \sqcup a\rangle\}$ is a shallow SSC in solved form; $M_C(Y) = M_C(X)$ is the infinite set of terms $\{t_0, t_1, t_2, \ldots\}$, where $t_0 = a$ and $t_n = f(t_{n-1}, t_{n-1})$, for $n \geq 1$. This is not a regular tree language. The set of ground instances of the constrained term $g(Y)/C$ is

$$M_C(g(Y)) = \{g(a),\ g(f(a,a)),\ g(f(f(a,a), f(a,a))),\ \ldots\}.$$

3 Sorted Signatures, Sort Theories and Constraints

Sorted logics can be specified using a *sorted signature* that describes well-sorted and ill-sorted terms [Schmidt-Schauß, 1989], or through the use of monadic constraints that refer to a background *sort theory* [Frisch, 1991].[4] For the purposes of this paper, sorted signatures can be described in terms of corresponding sort theories: A *sort theory* is a finite set of first order sentences containing only unary *sort predicates*.

[4] See [Cohn, 1989] for an overview of other dimensions among which sorted logics can vary. The above are the only two of interest to us now.

If equational theories are not included, a sorted signature Σ is equivalent to a sort theory $\mathcal{A}_\Sigma$ that can be written as a collection of Horn clauses. $\mathcal{A}_\Sigma$ is the signature's *relativization*, as defined by Schmidt-Schauß [1989]. For each sort symbol S in Σ, $\mathcal{A}_\Sigma$ contains a sort predicate with the same name. The sets of ground terms of a sort S as given by Σ correspond to the interpretation of S in the least Herbrand model of $\mathcal{A}_\Sigma$; that is, a ground term g is said to be of sort S in Σ if, and only if, $\mathcal{A}_\Sigma \models S(g)$. In both formulations, we write $x{:}S$ to indicate that x is a variable of sort S.

An *elementary* sorted signature[5] is equivalent to a set of Horn clauses of the form $S_1(x) \rightarrow S_2(x)$ for each subsort declaration $S_1 \sqsubseteq S_2$, and of the form $S_1(x_1) \wedge \ldots \wedge S_n(x_n) \rightarrow S_0(f(x_1, \ldots, x_n))$, where $x_i \neq x_j$, for each function declaration $f(S_1 \times \ldots \times S_n){:}S_0$. As first pointed out by Comon [1989], these signatures are also equivalent to finite frontier-to-root tree automata: The sets of ground terms that belong to each sort are regular tree languages. This is not always the case if we allow arbitrary *term declarations* as presented by Schmidt-Shauß [1989]. A term declaration $S_0[t]$ is equivalent to a Horn clause of the form $S_1(x_1) \wedge \ldots \wedge S_n(x_n) \rightarrow S_0(t)$, where $x_1, \ldots, x_n$ are the different variables in t and $S_1, \ldots, S_n$ are their respective sorts.

In the general case, sorted unification under term declarations is undecidable. Usually, only elementary signatures are considered, and their *regularity* — that every term have a unique least sort — is often required to ensure that there is always a finite number of maximally general unifiers.[6] These two restrictions are not needed if SSC's can be used: The infinite sets of unifiers can be represented finitely. We do make some general assumptions about sort theories (and hence signatures): The subsort relation should be a partial order, and for each sort S, there should be a ground term g of sort S, which ensures that sorts are not empty. Under these conditions, unification under a sorted signature is operationally equivalent to unification under the corresponding sort theory: The well-sorted substitutions and the generality ordering between them are the same [Uribe, 1991].

Definition 11. Let Σ be a sorted signature. An SSCC_Σ is *equivalent* to Σ if for each sort S in Σ there is a variable $\Upsilon_S \in Dom(C_\Sigma)$ such that the set of ground terms of sort S is equal to $M_{C_\Sigma}(\Upsilon_S)$.

If Σ can be rewritten as a Horn clause sort theory, then the model M_{C_Σ} corresponds to the least Herbrand model $\mathcal{M}$ of Σ, in the sense that $M_{C_\Sigma}(\Upsilon_S)$ is equal to the interpretation of the predicate S under $\mathcal{M}$.

If an SSC C_Σ is equivalent to a sorted signature Σ, then a sorted term t can be translated into a constrained term t'/C as follows: Let t' be the result of replacing each variable $x{:}S$ in t by a new variable N_x, and let C be the union of C_Σ and the set constraints $\{\langle N_x \equiv \Upsilon_S\rangle \mid x{:}S \in t\}$. The set of "well-sorted ground instances" of t will be exactly equal to $M_C(t')$. A unification problem given Σ can be rewritten as the problem of finding the solutions to a set of equations between terms constrained by SSC's, by translating in this way the terms in the original equation set. Thus,

[5] One with only subsort and function declarations, in Schmidt-Schauß's terms; these are often simply called "sorted signatures."

[6] Regularity is in general an undecidable property of signatures with term declarations [Schmidt-Schauß, 1989]. A consequence of the results presented here is that it is decidable for semi-linear signatures.

the algorithms of the following sections show that sorted unification under a class of signatures is decidable if these signatures can be represented as equivalent shallow SSC's.

Definition 12 (Semi-linear). A term t is *linear* if each variable in t occurs exactly once in t. A term t is *semi-linear* if each occurrence of a given variable in t has the same prefix.[7] A sorted signature is (semi-) linear if t is (semi-) linear for all term declarations $S[t]$. An SSC C is (semi-) linear if all terms appearing in $Rng(C)$ are (semi-) linear.

If Σ is a finite *semi-linear* sorted signature (or a corresponding sort theory) such that the subsort relation is a partial order, then one can easily obtain an SSC C_Σ in solved form that is equivalent to Σ in the sense of Definition 11 [Uribe, 1991].

Declarations (Σ)	Axioms ($\mathcal{A}_\Sigma$)
$\text{EVEN} \sqsubseteq \text{INT}$	$\forall x\ Even(x) \rightarrow Int(x)$
$\text{ODD} \sqsubseteq \text{INT}$	$\forall x\ Odd(x) \rightarrow Int(x)$
$\text{SQUARE} \sqsubseteq \text{INT}$	$\forall x\ Square(x) \rightarrow Int(x)$
$\text{EVEN}[0]$	$Even(0)$
$\text{EVEN}[s(x{:}\text{ODD})]$	$\forall x\ Odd(x) \rightarrow Even(s(x))$
$\text{ODD}[s(x{:}\text{EVEN})]$	$\forall x\ Even(x) \rightarrow Odd(s(x))$
$\text{EVEN}[sum(x{:}\text{EVEN}, y{:}\text{EVEN})]$	$\forall x \forall y\ Even(x) \wedge Even(y) \rightarrow Even(sum(x,y))$
$\text{SQUARE}[mult(x{:}\text{INT}, x{:}\text{INT})]$	$\forall x\ Int(x) \rightarrow Square(mult(x,x))$

Table 1. Sorted signature declarations and corresponding sort theory axioms.

Example 2. Table 1 shows a set of subsort and term declarations Σ together with a corresponding sort theory. Note that Σ is semi-linear: The only term declaration with repeated variables is $\text{SQUARE}[mult(x{:}\text{ODD}, x{:}\text{ODD})]$. An SSC in solved form equivalent to Σ is:

$$C_\Sigma = \left\{ \begin{array}{cc} \langle \Upsilon_{int} \equiv \Upsilon_{even} \sqcup \Upsilon_{odd} \sqcup s(\Upsilon_{int}) \rangle, & \langle \Upsilon_{square} \equiv mult(\Upsilon_{int}, \Upsilon_{int}) \rangle, \\ \langle \Upsilon_{even} \equiv 0 \sqcup s(\Upsilon_{odd}) \sqcup sum(X_e, Y_e) \rangle, & \langle \Upsilon_{odd} \equiv s(\Upsilon_{even}) \rangle, \\ \langle X_e \equiv \Upsilon_{even} \rangle, & \langle Y_e \equiv \Upsilon_{even} \rangle \end{array} \right\}$$

A linear SSC with no cycles is equivalent to a term grammar: $M_C(X)$ is a regular tree language for each $X \in Dom(C)$. Linear signatures can be expressed as equivalent linear SSC's. Any linear SSC with no cycles can be rewritten into an SSC in solved form, equivalent to C over $Dom(C)$, that is also linear. In this case, simpler versions of the algorithms presented here can be used.

4 Simplifying Constraints

A crucial element in most constraint-solving procedures is the ability to rewrite constraints into a solved form. In a constraint logic framework, the only operational requirement is that we be able to check whether a conjunction of constraints has

[7] For example, $f(g(x), g(x), y)$ is semi-linear, but $f(h(x, g(x)), y, h(x, g(x)))$ is not.

a (non-empty) solution (see [Jaffar and Lassez, 1987]). Rewriting SSC's into solved form (see Definition 10) is part of the process by which we do this.

We now present an algorithm that rewrites a shallow SSC into an equivalent one in solved form. This is an extension of the corresponding simplification procedure presented by Heintze and Jaffar [1990]. The main difficulty appears when simplifying subexpressions of the form $f(X_1, \ldots, X_n) \sqcap f(Y_1, \ldots, Y_n)$. Since we are expressing inter-variable dependencies within terms, we cannot do a simple pairwise matching of variables; instead, a form of unification must be used.

We begin with a number of transformations on an SSC C. Each replaces a set expression E in a set constraint $\langle X \equiv E \sqcup exp \rangle \in C$ by a new set expression and possibly adds new set constraints over new variables:[8]

S1: Replace $f(X_1, \ldots, X_n) \sqcap g(Y_1, \ldots, Y_m)$ by $\perp$ if $f \neq g$, for $m, n \geq 0$.

S2: Replace $(exp \sqcap \perp)$ by $\perp$, $(exp \sqcap \top)$ by exp, $(exp \sqcup \perp)$ by exp, and $(exp \sqcup \top)$ by $\top$, where exp is any set expression. Replace $(t \sqcap t)$ by t, and $(t \sqcup t)$ by t, for any term t.

S3: Replace $V \sqcap exp$ by $DNF(C(V) \sqcap exp)$, where $V \in Dom(C)$ and exp contains no un-nested occurrences of V.

S4: Replace $f(X_1, \ldots, X_n) \sqcap f(Y_1, \ldots, Y_n)$ by $f(Z_1, \ldots, Z_n)$, provided that:
(a) $Z_i = Z_j$ if, and only if, $X_i = X_j$ or $Y_i = Y_j$, $1 \leq i, j \leq n$, and
(b) $C \models Z_i \equiv \bigcap_{V \in S} V$, for each i, $1 \leq i \leq n$, where

$$S = \{X_j \mid Z_j = Z_i, 1 \leq j \leq n\} \cup \{Y_j \mid Z_j = Z_i, 1 \leq j \leq n\}.$$

In order to satisfy (b), new set constraints of the form $\langle Z_i \equiv V_1 \sqcap \ldots \sqcap V_n \rangle$ can be added to C, where Z_i is a new variable and each $V_i \in Dom(C)$.

Proposition 13. *If C' is obtained by applying any of the above transformations to an* SSC *C, then $C \Longleftrightarrow_{Dom(C)} C'$.*

To ensure termination, the actual simplification algorithm needs to keep track of the variables introduced by transformation S4. Some of these new variables will represent the intersection of variables in $Dom(C)$, while others will be "renamings" of these, needed to avoid introducing new inter-variable dependencies. To do this, we maintain a set that defines a function from variables into subsets of $Dom(C)$, indicating when a variable stands for the intersection of the variables in a subset of $Dom(C)$; Int is a set of pairs of the form $\langle V, \{V_1, \ldots, V_n\} \rangle$, where $Int(V) = S$ if $\langle V, S \rangle \in Int$ and $\{V\}$ otherwise. We also need to maintain Rep, a one-to-one mapping from subsets of $Dom(C)$ into the unique variable that represents its intersection (if such a variable exists); $Rep(S) = V$ if $\langle S, V \rangle \in Rep$.[9] We now define the following variant of transformation S4, as applied to an SSC C and the sets Int and Rep:

[8] In this and the following algorithms we assume that all set expressions in $Rng(C)$ are in disjunctive normal form. In practice, rewriting set expressions into DNF can be delayed until strictly necessary, to increase efficiency.

[9] This is an extension of the labeling scheme used by Heintze and Jaffar [1990], where variables are subscripted with the set of variables whose intersection they represent.

Input: A shallow SSC C with no cycles.
Output: An SSC in solved form.
Let $Int = \emptyset$, and let $Rep = \{\langle\{X\}, X\rangle \mid X \in Dom(C)\}$.
Exhaustively apply transformation S3.
WHILE any of transformations S1, S2 or S4′ apply, DO:
    Apply one of transformations S1, S2, or S4′.
    IF transformation S4′ is applied adding new elements to Rep,
        THEN exhaustively apply transformation S3.

Fig. 1. Simplification algorithm

S4′:

Consider an occurrence of $f(X_1, \ldots, X_n) \sqcap f(Y_1, \ldots, Y_n)$.
Let $t_1 = f(X_1, \ldots, X_n)$ and $t_2 = f(Y_1, \ldots, Y_n)$.
Let π be a renaming substitution such that $Var(t_2\pi)$ is a set of new variables.
Let θ be a most general unifier for $\{t_1 = t_2\pi\}$ such that $Var(Rng(\theta)) \subseteq Var(t_2\pi)$.
FOR EACH variable V in $Var(f(X_1, \ldots, X_n)\theta)$, DO:

- Let $S_V := \bigcup_{X \in \{X_i \mid X_i\theta = V\} \cup \{Y_i \mid Y_i\pi\theta = V\}} Int(X)$.
- If there exists a variable Z such that $Rep(S_V) = Z$, add the set constraint $\langle V \equiv Z\rangle$; otherwise, add the set constraint $\langle V \equiv V_1 \sqcap \ldots \sqcap V_m\rangle$, where $\{V_1, \ldots, V_m\} = S_V$, and let $Rep := Rep \cup \langle S_V, V\rangle$.
- Let $Int := Int \cup \{\langle V, S_V\rangle\}$.

Replace $t_1 \sqcap t_2$ by $t_1\theta$.

Proposition 14. *Given that $Int(V) = \{X_1, \ldots, X_n\}$ only if $C \models V \equiv X_1 \sqcap \ldots \sqcap X_n$ and that $Rep(\{X_1, \ldots, X_n\}) = V$ only if $C \models V \equiv X_1 \sqcap \ldots \sqcap X_n$, then transformation S4′ is equivalent to transformation S4. Furthermore, the above hypotheses will still hold for C' and the new Int and Rep after transformation S4′ is applied.*

The simplification algorithm can now be obtained by initializing *Int* and *Rep* and applying transformations S1–S3 and S4′ as shown in Figure 1. Transformations S1–S3 do not change *Int* and *Rep*. If the input is a shallow SSC with no cycles, then the simplification algorithm of Figure 1 terminates, and the output SSC is in solved form. Note that if C has no cycles then transformation S3 can only be applied a finite number of times to C; and if C contains no cycles, then the resulting C' after applying any of transformations S1–S4 contains no cycles either.

We can show that at any point during the simplification algorithm, if $Int(X) = \{Y_1, \ldots, Y_n\}$ or $Rep(\{Y_1, \ldots, Y_n\}) = X$, then $C \models X \equiv Y_1 \sqcap \ldots \sqcap Y_n$. Together with Propositions 13 and 14, this means that if C is the input SSC and C' is the output SSC (which is in solved form), then $M_C(X) = M_{C'}(X)$ for all $X \in Dom(C)$.

Recall that we allow $M_C(X) = \emptyset$ for a variable X constrained by an SSC C. Since we will use SSC's to represent non-empty solutions to sets of equations between constrained terms, we will need to be able to identify these *empty variables* once solved form is obtained.

Input: An SSC C in solved form. Let all variables in $Dom(C)$ be unmarked. While any of the following applies for an unmarked variable $X \in Dom(C)$, do:

- If $\langle X \equiv \top \sqcup exp \rangle \in C$, mark X.
- If $\langle X \equiv k \sqcup exp \rangle \in C$ for some constant k, mark X.
- If $\langle X \equiv Y \sqcup exp \rangle \in C$ and Y is a marked variable, mark X.
- If $\langle X \equiv f(Y_1, \ldots, Y_n) \rangle \in C$ and each Y_i is a marked variable, mark X.

At the end, a variable $X \in Dom(C)$ is unmarked if, and only if, $M_C(X) = \emptyset$.

Fig. 2. Identifying empty variables

Figure 2 presents an algorithm that identifies all the empty variables in a solved form SSC C, again adapted from the corresponding algorithm given by Heintze and Jaffar [1990]. At the end of the algorithm, $X \in Dom(C)$ is empty in C if, and only if, X is unmarked.[10]

5 The Unification Algorithm

Definition 15 (Constrained Equation Sets). A *constrained equation set* is a finite set S of equations between terms together with an SSC C such that $Var(S) \subseteq Dom(C)$, and is written as S/C. A substitution θ *solves* an equation set S/C if (i) for all equations $t_1 = t_2$ in S, $t_1\theta = t_2\theta$, and (ii) $X\theta \in M_C(X)$ for all $X \in Dom(\theta)$. Equivalently, we say that θ is a *solution* of S/C.

The unification problem is to find all the substitutions (if any) that solve a constrained equation set S/C. The algorithm presented here is based on a well-known approach to sorted unification. Given S/C, an unsorted most general unifier ψ for S is computed (ignoring C), and is then *weakened*: Constraints are found for the variables in $Rng(\psi)$ such that that for all $X \in Dom(\psi)$, $X\psi$ ranges over a subset of $M_C(X)$. Furthermore, the new constraints should represent all the solutions to the original constrained equation set S/C. Section 5.1 presents a procedure that will do this, by solving sets of *membership constraints*. The actual unification algorithm is then presented in Section 5.2.

5.1 Solving Membership Constraints

Given a substitution $\{X_1 \mapsto t_1, \ldots, X_n \mapsto t_n\}$ and an SSC in solved form C, where $\{X_1, \ldots, X_n\} \subseteq Dom(C)$ and $Var(\{t_i\}) \cap Dom(C) = \emptyset$, we want to find all the substitutions ρ such that $t_i\rho \in M_C(X_i)$ for all i, $1 \le i \le n$. This process is equivalent to weakening an unsorted substitution so as to make it "well-sorted." The problem can be formulated in a more general way, as follows:

[10] Once empty variables have been identified, C can be rewritten into an equivalent SSC, C', such that that no empty variables appear in $Rng(C')$; this algorithm is straightforward (see [Uribe, 1991]), so it is not presented here.

Definition 16 (Membership Constraints). Given an SSC C, a *membership constraint* is an expression $t \Subset E$, for a term t and a set expression E. A substitution ρ *solves* a set of membership constraints $\{t_1 \Subset E_1, \ldots, t_n \Subset E_n\}$ if for all i, $1 \leq i \leq n$, $t_i\rho \in M_C(E_i)$.[11]

The following algorithm finds the solutions to a set of membership constraints given a *solved form* SSC C and *shallow* set expressions E_i. The solutions are represented using substitutions θ_j and SSC's C_j such that $M_{C_j}(t_i\theta_j) \subseteq M_C(E_i)$ for each membership constraint $t_i \Subset E_i$. In general, there is no single pair $\langle \theta, C' \rangle$ that can capture all possible solutions; however, we will always be able to find a *finite* set of such pairs such that every solution is an instance of one of them.

We assume that C contains no occurrences of "$\perp$": After using the algorithm in Section 4 to identify empty variables, it is simple to obtain an SSC equivalent to C such that for all empty variables X, $C(X) = \perp$ and X appears nowhere else in $Rng(C)$. If any of the variables X for an initial membership constraint $t \Subset X$ is empty in C, then there are no (non-empty) solutions. Otherwise, all the set constraints of the form $\langle X \equiv \perp \rangle$ can be removed from C.

The algorithm maintains a set $\mathcal{S}$ of *partial solutions*, each of which is a pair $\langle \theta_i, U_i \rangle$ for a set of membership constraints U_i and a substitution θ_i. Part 1 of the algorithm applies transformations to a set $\mathcal{S}$ of such pairs. Each transformation acts upon a pair $\langle \theta, U \rangle$ in $\mathcal{S}$. The substitution θ is not changed unless otherwise noted:

W1: $U = U_o \cup \{f(t_1, \ldots, t_n) \Subset g(X_1, \ldots, X_m) \sqcup exp\} \Longrightarrow U_o \cup \{f(t_1, \ldots, t_n) \Subset exp\}$ where exp is not empty, $f \neq g$, $m, n \geq 0$. Otherwise, remove $\langle \theta, U \rangle$ from $\mathcal{S}$.
W2: $U = U_o \cup \{f(t_1, \ldots, t_n) \Subset X \sqcup exp\} \Longrightarrow U_o \cup \{f(t_1, \ldots, t_n) \Subset C(X) \sqcup exp\}$.
W3: Consider a pair $\langle \theta, U \rangle$ where $U = U_o \cup \{f(t_1, \ldots, t_n) \Subset f(X_1, \ldots, X_n)\}$, for $n \geq 2$.

Let Ω be an idempotent most general unifier for the equation set $\{t_i = t_j \mid X_i = X_j\}$.
If this unification fails, remove $\langle \theta, U \rangle$ from $\mathcal{S}$. Otherwise:
Let U_o' be the result of applying Ω to
the left hand side of each membership constraint in U_o.
Replace U by $U_o' \cup \{t_1\Omega \Subset X_1, \ldots, t_n\Omega \Subset X_n\}$ and replace θ by $\theta \cdot \Omega$.

W4: If $U = U_o \cup \{f(t_1, \ldots, t_n) \Subset e_1 \sqcup \ldots \sqcup e_m\}$ for $n > 1, m > 1$, where each e_i is of the form $f(Y_1, \ldots, Y_n)$, replace the pair $\langle \theta, U \rangle \in \mathcal{S}$ by the pairs $\langle \theta, U_o \cup \{f(t_1, \ldots, t_n) \Subset e_1\} \rangle$, $\ldots$, $\langle \theta, U_o \cup \{f(t_1, \ldots, t_n) \Subset e_m\} \rangle$.
W5: $U = U_o \cup \{f(t) \Subset f(X_1) \sqcup \ldots \sqcup f(X_n)\} \Longrightarrow U_o \cup \{t \Subset X_1 \sqcup \ldots \sqcup X_n\}$.
W6: $U = U_o \cup \{c \Subset e_1 \sqcup \ldots \sqcup e_n\} \Longrightarrow U_o$ for a constant c, if $e_i = c$ for some i, $1 \leq i \leq n$; if c is a constant and $e_i \neq c$ for all i, remove $\langle \theta, U \rangle$ from $\mathcal{S}$.
W7: $U = U_o \cup \{t \Subset \top \sqcup exp\} \Longrightarrow U_o \cup \{X_1 \Subset \top, \ldots, X_n \Subset \top\}$, where t is any non-variable term and $\{X_1, \ldots X_n\} = Var(t)$.

Figure 3 shows the weakening algorithm, which always terminates with a finite number of solutions. All the elements of each set of membership constraints U_i at

[11] In Comon and Delor's [1991] notation, a set of membership constraints would be the conjunction $t_1 \in S_1 \wedge \ldots \wedge t_n \in S_n$, where each S_i is the tree language $M_C(E_i)$.

Input: A set of membership constraints $U = \{t_i \Subset X_i\}$ and an SSC C in solved form, where $Var(\{t_i\})$ and $Var(C)$ are disjoint.
Output: A set of pairs $\langle\theta_i, C_i\rangle$, for SSC's C_i and substitutions θ_i.

Part 1: Let $\mathcal{S} = \{\langle\{\}, U\rangle\}$. Exhaustively apply transformations W1–W7 to $\mathcal{S}$.

Part 2: For each $\langle\theta_i, U_i\rangle \in \mathcal{S}$ do:
Let C_i be the union of C and the set constraints
$\langle X \equiv exp_1 \sqcap \ldots \sqcap exp_n\rangle$, for each $X \in Var(U)$, where
$\{X \Subset exp_1, \ldots, X \Subset exp_n\}$ are the membership constraints for X in U_i.

Return $\{\langle\theta_i, C_i\rangle\}$.

Fig. 3. Weakening algorithm

the end of Part 1 are of the form $X \Subset exp$, and the SSC C_i constructed from U_i in Part 2 is a shallow SSC with no cycles.

We must note that the substitution component of the partial solutions can be replaced by a set of equations to which transformations analogous to those in a normal unification algorithm are applied. The "variable isolation" transformation would also instantiate the terms in the left hand side of the membership constraints, while a new version of transformation W3 simply adds the appropriate equations to this set. No additional control is necessary to ensure the termination of this extended set of transformations — see [Uribe, 1991] — so that the algorithm of Figure 3 is a particular version of that more general one.

We now present the main steps in a proof of the correctness and completeness of the weakening algorithm:

Definition 17 (Solving Partial Solutions). A substitution ρ *solves* a partial solution $\langle\theta, U\rangle$ if (i) $t_i\rho \in M_C(E_i)$ for all membership constraints $t_i \Subset E_i$ in U, and (ii) for all variables X, $X\rho = (X\theta)\rho$.

It is simple, though tedious, to show that a substitution ρ solves an element of a set of partial solutions S if, and only if, ρ solves an element of the set S' resulting from applying any of transformations W1–W7 to S. Thus, a substitution ρ solves the set of membership constraints input to the algorithm of Figure 3 if, and only if, ρ solves some partial solution $\langle\theta_i, U_i\rangle$ in $\mathcal{S}$ at the end of Part 1. It is also easy to verify that each C_i obtained in Part 2 is a shallow SSC with no cycles. The following theorem puts everything together:

Theorem 18. *A substitution ρ solves the set $\{t_1 \Subset X_i, \ldots, t_n \Subset X_n\}$ input to the algorithm of Figure 3 if, and only if, for some output pair $\langle\theta_i, C_i\rangle$, $X\rho \in M_{C_i}(X\theta_i)$ for all $X \in Var(\{t_1, \ldots, t_n\})$.*

That is, all possible solutions to the input are represented by the output, and all the solutions represented by the output are correct: $\bigcup_i M_{C_i}(t\theta_i)$ is exactly equal

Input: A constrained equation set S/C, where C is in solved form.
Output: A set of pairs $\langle\theta_i, C_i\rangle$ where θ_i is a substitution and C_i is an SSC in solved form.

Step 1: If S is not unifiable, return $\emptyset$. Otherwise: Let Γ be an idempotent most general unifier for S. Let Θ be obtained from Γ by renaming $Var(Rng(\Gamma))$ into new variables, so that $Var(Rng(\Theta))$ and $Var(C)$ are disjoint.
$\Theta = \{X_1 \mapsto t_1, \ldots, X_n \mapsto t_n\}$, where $X_i \in Var(S)$.
Step 2: Let $U = \{t_i \Subset X_i\}$, and let $\mathcal{T} = \{\langle\theta_i, C_i\rangle\}$ be the result of solving the set of membership constraints U given C, using the algorithm of Figure 3.
Step 3: Rewrite the SSC's in $\mathcal{T}$ into solved form, using the algorithm of Figure 1. If $M_{C_i}(X) = \emptyset$ for some $X \in Var(Rng(\Theta))$, remove $\langle\theta_i, C_i\rangle$ from $\mathcal{T}$.
Step 4: For each pair in $\mathcal{T}$, replace θ_i by $\theta_i' = \Theta \cdot \theta_i$ and return the new $\mathcal{T}$.

Fig. 4. Unification algorithm for constrained equation sets

to the set of all the instances of t that can belong to $M_C(exp)$ for each input membership constraint $t \Subset exp$. Note that the SSC's in the output are not necessarily in solved form — but they are always shallow. Therefore, the simplification algorithm of Section 4 and the algorithm that identifies empty variables (in Figure 2) have to be used to check whether the constrained variables output by the weakening algorithm are empty. This is included in the unification algorithm of the following section, which integrates all the algorithms presented so far.

5.2 Unifying Constrained Terms

Given a constrained equation set S/C, we want to characterize all the substitutions ρ such that $t_1\rho = t_2\rho$ for all equations $t_1 = t_2 \in S$ and $X\rho \in M_C(X)$ for all $X \in Var(S)$. Figure 4 describes the unification algorithm that does this.

Since we assume that C is in solved form and $Var(S) \subseteq Dom(C)$, the input for the weakening algorithm in Step 2 is in the required form. Thus, the solving procedure terminates, and the SSC's C_i at the end of Step 2 are all shallow. Therefore, Step 3 also terminates. The algorithm is *correct* and *complete*, as follows:

Theorem 19. *A substitution ρ solves the constrained equation set S/C input to the unification algorithm of Figure 4 if, and only if, for some pair $\langle\theta_i', C_i\rangle$ in the output, $X\rho \in M_{C_i}(X\theta_i')$ for all $X \in Var(S)$.*

Note that if $\emptyset$ is returned, then the input equation set has no solutions.

The following very simple example is presented by Frisch [1991] to show that a finite set of terms can have an infinite number of incomparable maximally general well-sorted unifiers. Given the sort theory

$$\Sigma = \{\forall x\ \mathrm{T}(i(x)) \rightarrow \mathrm{T}(i(s(x))),\ \mathrm{T}(i(a))\},$$

the two terms $z{:}\mathrm{T}$ and $i(s(y))$ have an infinite number of most general well-sorted unifiers, none of which is more general (with respect to Σ) than the others:

$$\left\{ \begin{array}{ll} \{y \mapsto a,\ z{:}\mathrm{T} \mapsto i(s(a))\}\,, & \{y \mapsto s(a),\ z{:}\mathrm{T} \mapsto i(s(s(a)))\}\,, \ldots \\ \{y \mapsto s * n(a),\ z{:}\mathrm{T} \mapsto i(s(s * n(a)))\}\,, & \ldots \end{array} \right\}$$

The following SSC in solved form represents the least model of Σ:

$$C_\Sigma = \{\, \langle \Upsilon_{Top} \equiv \top \rangle,\, \langle \Upsilon_T \equiv i(V) \rangle,\, \langle V \equiv a \sqcup s(V) \rangle \,\}$$

Unifying $i(s(y))$ and z under the constraint $C_\Sigma \cup \{\langle y \equiv \Upsilon_{Top} \rangle,\ \langle z \equiv \Upsilon_T \rangle\}$ results in $\Theta = \{y \mapsto N, z \mapsto i(s(N))\}$, where the new constraint is $C_\Sigma \cup \{\langle N \equiv V \rangle\}$. Thus, an infinite number of unifiers is represented using a single constrained substitution.

6 Related Work

Constraint logic has been extensively studied, particularly the case of constraint logic programming (CLP), where the unification procedure is replaced by one that determines the solvability of constraints over a given structure [Jaffar and Lassez, 1987]. Burckert [1990] presents a general resolution principle for constrained clauses. Chen and Hsiang [1991] propose a method for finitely representing, and unifying, structured infinite sets of terms, and show how these *recurrence domains* can be incorporated into the CLP framework. These domains are neither more nor less general than the class of tree languages generated by shallow SSC's.

The *substitutional framework* for sorted deduction presented by Frisch [1991] provides the theoretical basis for building hybrid deductive systems where monadic constraints are imposed on variables, and where a constrained variable is treated as schematic for the set of ground terms known to satisfy the constraint. If a sort theory has a least Herbrand model that can be represented using SSC's, it follows from Frisch's [1991] results that the correctness and completeness of a variety of deductive systems are preserved if SSC's are incorporated as constraints and the ordinary unification procedure is replaced by the unification algorithm for constrained terms presented here [Uribe, 1991].

Shallow SSC's are the "grammar-like" counterpart of a subclass of the *tree automata with equality tests* presented by Bogaert and Tison [1992]. These extended tree automata are used by Comon and Delor [1991] to generalize a simplification algorithm for arbitrary equational formulae with membership constraints; this algorithm was initially presented using regular tree languages as constraints. Our unification algorithm can be seen as a instance of this procedure, using a different representation for constraints and where negation is not allowed. The languages generated by SSC's correspond to Bogaert and Tison's [1992] $Rec_=$ class (i.e. are recognized by tree automata whose tests contain no inequalities). Comon and Delor point out that their basic algorithm can be generalized to any class of tree languages that has certain closure and decidability properties.

Similarly, Cohn and Frisch [1992] present a method for solving sorted unification problems under an arbitrary sort theory, provided an oracle that answers certain questions about the theory.[12] The unification algorithm we present can be seen as

[12] The properties for classes of languages identified by Comon and Delor [1991] ensure that these questions can be effectively answered for the corresponding sort theories.

an effective version of this procedure for those cases where the sort theory has a least model that can be described using a shallow SSC.

Algorithms for solving general systems of set constraints (that do not express inter-variable dependencies) are presented in [Heintze and Jaffar, 1991] and [Aiken and Wimmers, 1991]. Heintze and Jaffar [1990] present an algorithm that approximates the least model of a logic program by ignoring inter-variable dependencies. The approximation is represented as a term grammar — that is, as a linear, shallow SSC. Therefore, this approximation algorithm can be used to "pre-process" a sort theory, in those cases where the approximation is exact, so that the algorithms we present here (in a more specialized version) can be used.

The problem of finding similar approximation algorithms that express inter-variable dependencies is left open; so is that of finding efficient implementations of the algorithms presented here. Future prospects also include extending the results to allow for more expressive set constraints, particularly those that contain inequations as well as equations between sets.

Acknowledgements

Thanks to Anthony Cohn, Alan Frisch, Peter Haddawy, David Page, Christian Prehofer and Richard Scherl, with whom I had useful discussions about this work, as well as the anonymous referees for their comments. Special thanks to David Page, Richard Scherl and the Universidad de los Andes Department of Mathematics for their assistance in the completion of this paper. This research was partially supported by NASA under grant number NAG 1-613.

References

[Aiken and Wimmers, 1991] A. Aiken and E. L. Wimmers. Solving systems of set constraints. Research report 8257 (#75451), IBM Almaden Research Center, 1991.

[Bogaert and Tison, 1992] B. Bogaert and S. Tison. Equality and disequality constraints on direct subterms in tree automata. In *Proc. Symposium on Theoretical Aspects of Computer Science*, pages 161–171. Springer-Verlag, 1992.

[Burckert, 1990] H-J. Burckert. A resolution principle for clauses with constraints. In *Proc. 10th International Conference on Automated Deduction*, pages 178–192. Springer-Verlag, July 1990.

[Chen and Hsiang, 1991] H. Chen and J. Hsiang. Recurrence domains: Their unification and application to logic programming. Technical report, State University of New York at Stony Brook, July 1991. Short version appears as *Logic Programming with Recurrence Domains*, Proc. ICALP 91.

[Cohn and Frisch, 1992] A. G. Cohn and A. M. Frisch. An abstract view of sorted unification. In *Proc. 11th International Conference on Automated Deduction*. Springer-Verlag, 1992. To appear.

[Cohn, 1989] A. G. Cohn. Taxonomic reasoning with many-sorted logics. *Artificial Intelligence Review*, 3:89–128, 1989.

[Comon and Delor, 1991] H. Comon and C. Delor. Equational formulae with membership constraints. Rapport de Recherche 649, LRI, Université de Paris Sud, Orsay, France, March 1991. Revised version of *Equational Formulas in Order-sorted Algebras*, Proc. ICALP 90.

[Comon, 1989] H. Comon. Inductive proofs by specifications transformation. In *Proc. Rewriting Techniques and Applications*, pages 76–91. Springer-Verlag, April 1989.

[Frisch, 1991] A. M. Frisch. The substitutional framework for sorted deduction: Fundamental results on hybrid reasoning. *Artificial Intelligence*, 49:161–198, 1991.

[Heintze and Jaffar, 1990] N. Heintze and J. Jaffar. A finite presentation theorem for approximating logic programs. Research report 16089 (#71415), IBM T.J. Watson Research Center, September 1990. Preliminary version appears in *Proc. 17th ACM Symposium on Principles of Programming Languages*, 1990.

[Heintze and Jaffar, 1991] N. Heintze and J. Jaffar. A decision procedure for a class of herbrand set constraints. Technical Report CMU-CS-91-110, Carnegie Mellon University, February 1991. Abstract appears in *Proc. of the 5th Annual IEEE Symposium on Logic in Computer Science*, 1990.

[Jaffar and Lassez, 1987] J. Jaffar and J. Lassez. Constraint logic programming. In *Proc. of the 14th ACM Principles of Programming Languages Conference*, pages 111–119, Munich, January 1987.

[Schmidt-Schauß, 1989] M. Schmidt-Schauß. *Computational Aspects of an Order-Sorted Logic with Term Declarations.* Springer-Verlag, New York, 1989.

[Uribe, 1991] T. E. Uribe. Sorted unification and the solution of semi-linear membership constraints. Technical Report UIUCDCS-R-91-1720, University of Illinois, Urbana, Illinois, December 1991.

This article was processed using the LaTeX macro package with LLNCS style

An Abstract View of Sorted Unification

Alan M. Frisch[1] and Anthony G. Cohn[2]

[1] Dept. of Computer Science and Beckman Institute, University of Illinois, 405 North Mathews Avenue, Urbana, Illinois, 61801, USA

[2] Division of Artificial Intelligence, School of Computer Studies University of Leeds, Leeds, LS2 9JT, England

Abstract. The study of sorted unification has progressed by developing algorithms for more and more general languages. This paper addresses the question of what can be said about sorted unification independent of the sorted language being used. This is done by abstracting away from the particulars of sorted languages and formulating a set of transformation rules for solving sorted unification problems in general. Strategies for controlling these transformation rules are examined to see which prevent a certain kind of infinite execution.

1 Introduction

The study of sorted unification has progressed by developing algorithms for more and more general languages. Reiter [1977] considered the simple case of unifying atoms in a language all of whose function symbols had zero arity. Later, Walther [1988] allowed arbitrary function symbols, but required functions to have a monomorphic sort behavior. Following this, Schmidt-Schauss [1989] developed algorithms that can handle polymorphic (overloaded) functions provided their sort behavior is described by a finite set of term declarations. Many other algorithms have also been developed in this line of research [Frisch, 1991; Jouannaud and Kirchner, 1991; Smolka *et al.*, 1989], each making some assumptions about the sorted language. This progression leads us to ask what can be said about sorted unification independent of the sorted language being used. This paper addresses this question by examining what happens when we remove all restrictions imposed on the problem by the choice of language.

We are able to address this question by abstracting away from the particulars of sorted languages and considering only the one important feature that matters for sorted unification—that each sorted language identifies a set of substitutions that it considers to be well-sorted. We assume that the set meets certain requirements—ones that are met by every known sorted language. Indeed, it is hard to see how anything that reasonably could be called a sorted language would fail to meet the requirements. We consider only sorted unification problems with uninterpreted function symbols, that is, with an empty equational theory.

This paper presents a procedure, called the Generalized Sorted Unification Procedure (GSUP), for solving all sorted unification problems within the abstract characterization. The GSUP is based on the now-common view of unification as equation solving as introduced by Herbrand and brought to prominence by Martelli and Montanari [1982]. Procedures based on this view non-deterministically apply a set of transformations to an equational formula (a formula whose atoms are all equations) until the formula is in a form where its solutions can be read off. The GSUP employs a standard set of six equation-solving transformations for ordinary unification together with a single transformation, called *weakening*, for handling sorts. The weakening transformation is uncomputable in general. Indeed, it will be seen that sorted unification problems can be solved if, and only if, weakening is computable. Thus the unification algorithms presented by others can be seen as identifying restrictions that are sufficient to ensure the computability of weakening.

Every execution of the GSUP can be thought of as a derivation of one equational formula from another using the transformations as inference rules. We distinguish two kinds of infinite derivations—those that are infinitely broad, and those that are infinitely deep. Though infinitely broad derivations cannot in general be avoided, we identify a non-deterministic control strategy that guarantees that no derivation is infinitely deep. We show that some, but not all, control strategies that guarantee the finite depth of derivations in certain sorted logics, admit infinite depth derivations in general.

In summary, the goal of this paper is to reveal what is intrinsic to the computation of sorted unifiers without regard to any particular sorted language. This is done by abstracting away from the particulars of sorted languages and formulating a correct procedure for the abstraction.

2 Sorted Substitutions and Unifiers

The notion of sorted unification rests on the definition of sorted substitution, which varies across the many different sorted languages in the literature. Nevertheless, all of the definitions posses certain abstract properties, which is all that is required to define and prove correct our abstract sorted unification procedure.

In all sorted languages, whether or not a substitution is well sorted depends upon what is known about the sorts. The source of this information about sorts varies from language to language. In numerous sorted languages [Schmidt-Schauß, 1989; Walther, 1988; Smolka *et al.*, 1989; Jouannaud and Kirchner, 1991], information about sorts is encoded in a sort signature, which is part of the definition of the language. The signature usually defines an ordering on the set of sort symbols and may also contain declarations defining the sortal behavior of the function symbols of the language. Languages of this kind vary considerably in terms of the declarations that they permit. In other sorted languages [Reiter, 1977; Frisch, 1991] information about the sorts is encoded by a set of sentences of the language itself. Such a set is often called a *sort theory*. Sentences of the sort theory are constructed like ordinary logical sentences except that they contain no ordinary predicate symbols; in their place are sort symbols acting as monadic predicate symbols. A language may place restrictions on what can be expressed in a sort theory. Yet another approach that could be taken is to interpret sort symbols with respect to an interpretation or set of interpretations, as is usually done for constraints in constraint logic [Höhfeld and Smolka, 1988; Jaffar and Lassez, 1987; Bürckert, 1990].

To summarize, whether or not a substitution is well-sorted depends upon sort information, which comes from either a signature, theory, interpretation, or set of interpretations. In this paper we abstract away from the source of the information and just speak of the information coming from a *sort definition*, which is usually denoted by Σ.

The sorted terms to be unified are just like ordinary logical terms except that sorted variables are used instead of ordinary variables. A sorted variable is a pair, $x{:}\tau$, where x is a variable name and τ is a sort symbol. Typographically, sort symbols are written entirely in small capitals as such: MAMMAL. Throughout this paper τ and ω stand for arbitrary sort symbols.

We shall not be concerned with what method is used to define which substitutions are well sorted. We shall abstract this away by assuming that some set of substitutions has been defined as well sorted relative to a sort definition Σ. Such substitutions are said to be ***Σ-well sorted*** and are called ***Σ-substitutions.***

We require that the set of Σ-substitutions satisfies three conditions:

1. The set of Σ-substitutions is closed under composition.
2. Every renaming substitution that maps each variable to a variable of the same sort is a Σ-substitution. This includes the identity substitution.
3. A substitution θ is Σ-well sorted if, and only if, $\{x{:}\tau \mapsto (x{:}\tau)\theta\}$ is Σ-well sorted for every variable $x{:}\tau$.

Requiring closure under composition is equivalent to requiring the instantiation ordering to be transitive. Together, the first two requirements entail that variants of Σ-substitutions are also Σ-substitutions and that terms that are variants are instances of each other. As a special case, the requirements entail that the instantiation ordering is reflexive. The third requirement means that well sortedness can be determined by examining what a substitution does to each variable independently. Every known definition of sorted substitution meets these three requirements; doing otherwise appears nonsensical.

2.1 A Particular Sorted Language

Though the ideas of this paper are developed for the abstract definition of sorted substitutions just presented, it will prove useful to have a concrete definition in hand for presenting examples. Because of the generality of the approach we shall consider the case where the sort definition is encoded in a sort theory.

As previously stated, the sentences of the sort theory are constructed like ordinary logical sentences except that they contain no ordinary predicate symbols; in their place are sort symbols acting as monadic predicate symbols. The sentences may contain function symbols that are drawn from the same set as those that may occur in the terms to be unified. Σ_1 is an example of a sort theory and is used in examples later in the paper.

$$
\begin{aligned}
\Sigma_1 = \{ & \forall x, y\ \textsc{odd}(x) \wedge \textsc{odd}(y) \rightarrow \textsc{evn}(sum(x,y)), \\
& \forall x, y\ \textsc{evn}(x) \wedge \textsc{evn}(y) \rightarrow \textsc{evn}(sum(x,y)), \\
& \forall x, y\ \textsc{evn}(x) \wedge \textsc{odd}(y) \rightarrow \textsc{odd}(sum(x,y)), \\
& \forall x, y\ \textsc{odd}(x) \wedge \textsc{evn}(y) \rightarrow \textsc{odd}(sum(x,y)), \\
& \forall x\ \textsc{odd}(x) \vee \textsc{evn}(x) \rightarrow \textsc{int}(x), \\
& \forall x\ \textsc{evn}(x) \rightarrow \neg\textsc{odd}(x), \\
& \textsc{evn}(two)\}
\end{aligned}
$$

Roughly speaking, a substitution is well sorted relative to a sort theory if it maps each variable to a term that satisfies the sort associated with the variable. More precisely, a substitution θ is well sorted relative to a sort theory Σ if, and only if, it maps each variable $x{:}\tau$, to a term t such that $\Sigma \models \forall x_1, \ldots x_n\ \tau(t)$, where $x_1, \ldots, x_n$ are the freely-occurring variables of t. To illustrate the definition, $\{z{:}\text{EVN} \mapsto sum(two, y{:}\text{EVN})\}$ is well sorted with respect to Σ_1 but $\{z{:}\text{EVN} \mapsto sum(x, y{:}\text{ODD})\}$ is not.

Two special cases of this definition are worth noting. If θ is well sorted relative to Σ and maps $x{:}\tau$ to a ground term t, then it must be that $\Sigma \models \tau(t)$. In other words, Σ must entail that t is of sort τ. If θ maps $x{:}\tau$ to a variable $y{:}\omega$ then it must be that $\Sigma \models \forall y{:}\omega\ \tau(y)$. That is, Σ must entail that ω is a subset of τ.

One can easily see that this definition of Σ-substitution meets the second and third requirements of a sort definition. Frisch [1991] has proved that the definition meets the first requirement, that the set of Σ-substitutions is closed under composition.

The reader should bear in mind that this definition of well-sorted substitution is used only in the examples of this paper.

2.2 Sorted Unification

Given two terms, s and t, the sorted unification problem is that of finding a characterization of the set containing every Σ-substitution θ such that $s\theta = t\theta$. Alternatively, we say that θ is a Σ-solution of the equation $s = t$. In some studies the set of solutions to an equation is characterized by a set of substitutions, called unifiers. In other studies, the solutions are characterized by a formula in a solved form. Solved forms are defined so that it is easy to determine if a formula in solved form has any solutions. Often solved forms are defined so that a set of unifiers characterizing all solutions can be read off directly from a formula in solved form.

The emphasis of this study is on computing a solved form formula that characterizes the Σ-solutions to a sorted unification problem. Though unifiers are only of secondary concern, a set of unifiers can be read off directly from a formula in our solved form. Many automated deduction techniques traditionally formulated in terms of unifiers have, in recent years, been reformulated in terms of constraints, which are often maintained in solved form. Thus, our approach is in keeping with this trend.

To solve a sorted unification problem with two terms, s and t, the GSUP starts with the formula $s = t$ and repeatedly applies transformations rules until a formula in solved form is obtained. The formulas manipulated by the GSUP are possibly infinite disjunctions. Each disjunct is either the atomic formula *False* or an existentially quantified finite conjunction of equations between terms. Thus, a formula is a disjunction of a possibly infinite number of disjuncts, each of which is either *False* or of the form

$$\exists \vec{z}\ x_1 = t_1 \wedge \cdots \wedge x_n = t_n \tag{1}$$

From this point on "conjunction" shall refer only to a conjunction of equations, "disjunct" shall refer only to *False* or an existentially quantified conjunction, and "disjunction" shall refer only to a disjunction of such disjuncts. We shall consider the conjunction and disjunction operators to apply to the set of their constituent formulas. Thus these operators are commutative and associative, so the order of their constituents is irrelevant. This treatment of the operators also allows conjunctions and disjunctions of zero, one, or more constituents. Thus a single equation is the trivial case of a conjunction of only one equation and a single conjunction is the trivial case of a disjunction of only one disjunct.

Throughout this paper the following symbols are used. t is an arbitrary term and f is an arbitrary function or predicate symbol, possibly of arity 0. $f(t_1,\ldots,t_n)$ is an arbitrary non-variable term. If $n = 0$, this is just an individual symbol and the parentheses are omitted. U, V, and W are conjunctions and Γ and Λ are disjunctions.

We speak of a substitution solving or Σ-solving a formula in the following way:

Definition 1 (Solves and Σ-Solves). Let θ be a substitution (Σ-substitution). Then, θ *solves* (*Σ-solves*) $t_1 = t_2$ if, and only if, $t_1\theta = t_2\theta$. θ solves (Σ-solves) a conjunction if, and only if, it solves (Σ-solves) every conjunct and it solves (Σ-solves) a disjunction if, and only if, it solves (Σ-solves) at least one of the disjuncts. No substitution solves (Σ-solves) *False*. Finally, θ solves (Σ-solves) $\exists x\ \phi$ if, and only if, some substitution, θ', solves (Σ-solves) ϕ and $v\theta = v\theta'$ for every variable v except, possibly, x.

Notice that the empty conjunction is solved by every substitution and that the empty disjunction is solved by no substitution.

Two equational formulas are said to be equivalent if they have the same Σ-solutions.

The unification procedure operates by repeatedly transforming a formula until it reaches a particular form called *solved form* at which point a maximally general solution to the conjunction can be read off directly. The usual definition of solved form is extended to handle solving equations between sorted terms relative to a sort definition Σ.

Definition 2 (Σ-Solved Form). An equation is Σ-well sorted if it is of the form $x{:}\tau = t$, and $\{x{:}\tau \mapsto t\}$ is Σ-well sorted. An occurrence of an equation in a conjunction is in *solved form* if the equation is of the form $x{:}\tau = t$ and $x{:}\tau$ does not occur in t or in any other occurrence of an equation in the conjunction. If, in addition, the equation is Σ-well sorted then it is in Σ-solved form. A conjunction or disjunction is in solved (Σ-solved) form if every equation it contains is in solved (Σ-solved) form.

It is easy to read off a complete set of Σ-unifiers from a disjunction in Σ-solved form. Each disjunct other than *False* is of the form shown in (1) and yields a single unifier

$$\{x_1 \mapsto t_1, \cdots, x_n \mapsto t_n\}.$$

Example 1. The equation $z{:}\text{EVN} = sum(v{:}\text{INT}, w{:}\text{INT})$ is in solved form, but not in Σ_1-solved form. In fact, it is not equivalent to any equation or disjunct in Σ_1-solved form, though it is equivalent to the following disjunction in Σ_1-solved form:

$$(\exists x{:}\text{EVN}, y{:}\text{EVN}\ z{:}\text{EVN} = sum(x{:}\text{EVN}, y{:}\text{EVN}) \wedge v{:}\text{INT} = x{:}\text{EVN} \wedge w{:}\text{INT} = y{:}\text{EVN}) \vee$$
$$(\exists x{:}\text{ODD}, y{:}\text{ODD}\ z{:}\text{EVN} = sum(x{:}\text{ODD}, y{:}\text{ODD}) \wedge v{:}\text{INT} = x{:}\text{ODD} \wedge w{:}\text{INT} = y{:}\text{ODD})$$

Notice that the two disjuncts are incomparable in that each has Σ_1-solutions that the other doesn't.

Example 2. (from [Smolka *et al.*, 1989]) Consider the following sort theory:

$$\begin{aligned}\Sigma_2 = \{&\mathrm{A}(c),\\ &\mathrm{B}(c),\\ &\forall x\ \mathrm{A}(x) \to \mathrm{A}(f(x)),\\ &\forall x\ \mathrm{B}(x) \to \mathrm{B}(f(x))\}\end{aligned}$$

The equation $x{:}\mathrm{A} = y{:}\mathrm{B}$ is not in Σ_2-solved form. In fact, it is not equivalent to any equation, disjunct or finite disjunction in Σ_2-solved form, though it is equivalent to the following infinite disjunction in Σ_2-solved form:

$$(x{:}\mathrm{A} = c \wedge y{:}\mathrm{B} = c) \vee (x{:}\mathrm{A} = f(c) \wedge y{:}\mathrm{B} = f(c)) \vee \cdots$$

3 The GSUP Transformation Rules

Figure 1 shows the seven transformations used by the GSUP. Each transformation may be matched against any subformula of a disjunction. Since conjunctions are taken to be sets, transformations may match a conjunct in any position. For example, the left-hand-side of variable isolation matches any disjunct that contains an equation of the form $x{:}\tau = t$, where x is any variable, τ is any sort symbol, and t is any term. The matched disjunct might have only one equation—in which case U is empty—and it might have no existential quantifiers. The presence of the existential quantifiers on the left-hand-side of transformations 6 and 7 indicates that the transformations are matched to an *entire* disjunct, not just some of the conjuncts within it.

Notice that a special case of the decomposition transformation is that if f is of arity zero then the transformation simply deletes the equation $f = f$. Also notice that the substitution applied in variable isolation may be ill-sorted.

The first six transformations are standard transformations for ordinary unification. Weakening is the only transformation that takes sorts into account. If an equation of the form $x{:}\tau = t$ is not Σ-well sorted then weakening is applicable. This provision ensures that weakening is not a no-op. The effect of weakening is to introduce a formula, called a *weakening formula*, that explicitly represents the conditions needed to make $x{:}\tau = t$ Σ-well sorted. More precisely, weakening must find a characterization of every Σ-substitution θ such that $x{:}\tau = t\theta$ is Σ-well sorted. Since the set of all such Σ-substitutions may have multiple incomparable maximal elements, it must be characterized, in general, by a disjunction.

We stipulate that immediately after rewriting a formula each transformation implicitly transforms the resulting formula into an equivalent one of the proper disjunctive form. Specifically, match failure and occurs check failure may produce a disjunct that contains multiple equations, one of which is *False*. Such disjuncts are replaced with *False*. For tidyness, all vacuous quantifiers are removed as part of each transformation. A formula that results from weakening is not necessarily a disjunction of existentially-quantified conjunctions of equations. To get it in this

1. **Transposition**
   $f(t_1, \ldots, t_n) = x{:}\tau \Longrightarrow x{:}\tau = f(t_1, \ldots, t_n)$
2. **Identity Elimination**
   $U \wedge x{:}\tau = x{:}\tau \Longrightarrow U$
3. **Decomposition**
   $f(t_1, \ldots, t_n) = f(t'_1, \ldots, t'_n) \Longrightarrow t_1 = t'_1 \cdots \wedge t_n = t'_n$
4. **Match Failure**
   $U \wedge f(t_1, \ldots, t_n) = g(t'_1, \ldots, t'_m) \Longrightarrow False$
   provided f and g are distinct.
5. **Occurs Check Failure**
   $U \wedge x{:}\tau = f(t_1, ..., t_n) \Longrightarrow False$
   provided $x{:}\tau$ occurs in $f(t_1, ..., t_n)$.
6. **Variable Isolation**
   $\exists \vec{z}\, U \wedge x{:}\tau = t \Longrightarrow \exists \vec{z}\, U\{x{:}\tau \mapsto t\} \wedge x{:}\tau = t$,
   provided $x{:}\tau$ occurs in U but not in t.
7. **Weakening**
   $\exists \vec{z}\, U \wedge x{:}\tau = t \Longrightarrow \exists \vec{z}\, U \wedge x{:}\tau = t \wedge \Lambda$
   where
   - Λ is a Σ-solved disjunction such that θ Σ-solves Λ if, and only if, $x{:}\tau = (t\theta)$ is Σ-well sorted, and
   - every free variable in Λ occurs in t,

   provided $x{:}\tau = t$ is not Σ-well sorted.

Fig. 1. Transformations for Sorted Unification

form, $\wedge$ must be distributed over $\vee$ and each existential quantifiers must be pulled to the front of the disjunct in which it occurs—renaming variables where necessary.

By distributing $\wedge$ over $\vee$, the overall effect of weakening is to replace a single disjunct with a disjunction that has one disjunct for each disjunct of the weakening formula. This is the only transformation that can increase the number of disjuncts in a disjunction. Finally, observe that if there is no substitution θ such that $x{:}\tau = t\theta$ is Σ-well sorted then the weakening equation must be *False* (or an equivalent formula) and the effect of weakening is to eliminate the disjunct containing $x{:}\tau = t$.

The problem of finding a weakening formula is, in general, uncomputable. It is easy to see that this is a consequence of placing so few constraints on the set of well-sorted substitutions; the set need not even be recursively enumerable. Schmidt-Schauss [1989] has shown that the problem of determining whether an equation has a Σ-well sorted solution is undecidable even if the sort definition is given by a *finite* signature with term declarations. It is easy to show that the problem of computing a weakening formula can be reduced to the sorted unification problem with a single equation between a variable and a term. Thus, sorted unification is computable only if sorted weakenings are; so it is not unreasonable for our procedure to depend upon the computation of weakening formulas.

Example 3. Consider using the GSUP transformations on the equation $z{:}\mathrm{EVN} = sum(v{:}\mathrm{INT}, w{:}\mathrm{INT})$ from Example 1. Weakening is the only applicable transformation. Using the following weakening formula, which has two disjuncts,

$$(\exists x{:}\mathrm{EVN}, y{:}\mathrm{EVN}\ v{:}\mathrm{INT} = x{:}\mathrm{EVN} \wedge w{:}\mathrm{INT} = y{:}\mathrm{EVN}) \vee$$
$$(\exists x{:}\mathrm{ODD}, y{:}\mathrm{ODD}\ v{:}\mathrm{INT} = x{:}\mathrm{ODD} \wedge w{:}\mathrm{INT} = y{:}\mathrm{ODD})$$

weakening yields an equational formula that also has two disjuncts:

$$(\exists x{:}\mathrm{EVN}, y{:}\mathrm{EVN}\ z{:}\mathrm{EVN} = sum(v{:}\mathrm{INT}, w{:}\mathrm{INT}) \wedge v{:}\mathrm{INT} = x{:}\mathrm{EVN} \wedge w{:}\mathrm{INT} = y{:}\mathrm{EVN}) \vee$$
$$(\exists x{:}\mathrm{ODD}, y{:}\mathrm{ODD}\ z{:}\mathrm{EVN} = sum(v{:}\mathrm{INT}, w{:}\mathrm{INT}) \wedge v{:}\mathrm{INT} = x{:}\mathrm{ODD} \wedge w{:}\mathrm{INT} = y{:}\mathrm{ODD})$$

Applying variable isolation four times—to the second and third conjuncts of each disjunct—results in:

$$(\exists x{:}\mathrm{EVN}, y{:}\mathrm{EVN}\ z{:}\mathrm{EVN} = sum(x{:}\mathrm{EVN}, y{:}\mathrm{EVN}) \wedge v{:}\mathrm{INT} = x{:}\mathrm{EVN} \wedge w{:}\mathrm{INT} = y{:}\mathrm{EVN}) \vee$$
$$(\exists x{:}\mathrm{ODD}, y{:}\mathrm{ODD}\ z{:}\mathrm{EVN} = sum(x{:}\mathrm{ODD}, y{:}\mathrm{ODD}) \wedge v{:}\mathrm{INT} = x{:}\mathrm{ODD} \wedge w{:}\mathrm{INT} = y{:}\mathrm{ODD})$$

Notice that this is the Σ_1-solved form from Example 1 and that no transformations can be applied to this formula.

Example 4. Consider using the GSUP transformations on the equation $x{:}\mathrm{A} = y{:}\mathrm{B}$ from Example 2. Weakening is the only applicable transformation. Every weakening formula for this equation has an infinite set of disjuncts. Using the following weakening formula

$$y{:}\mathrm{B} = c \vee y{:}\mathrm{B} = f(c) \vee \cdots$$

weakening yields a disjunction with an infinite number of disjuncts:

$$(x{:}\mathrm{A} = y{:}\mathrm{B} \wedge y{:}\mathrm{B} = c) \vee (x{:}\mathrm{A} = y{:}\mathrm{B} \wedge y{:}\mathrm{B} = f(c)) \vee \cdots$$

Variable isolation can now be applied once to each disjunct, resulting in:

$$(x{:}\mathrm{A} = c \wedge y{:}\mathrm{B} = c) \vee (x{:}\mathrm{A} = f(c) \wedge y{:}\mathrm{B} = f(c)) \vee \ldots$$

Notice that this is the Σ_2-solved form from Example 2 and that no transformations can be applied to this formula.

One can easily see that no GSUP transformation applies to a disjunction in Σ-solved form. In the next section, we will prove that the converse holds even when certain control restrictions are placed on the transformations. We now establish that the GSUP transformations preserve the set of Σ-solutions. Thus, if these transformations are repeatedly applied to a disjunction until none apply, the resulting disjunction is in Σ-solved form and has has precisely the same Σ-solutions as the original disjunction.

Theorem 3 (Preservation Theorem). *A substitution Σ-solves a disjunction if, and only if, it Σ-solves the disjunction after applying any of the transformations of the Generalized Sorted Unification Procedure.*

Proof. Results on unsorted unification tell us that the first six transformations preserve the set of solutions, and hence they preserve the set of Σ-solutions. Now for the weakening transformation we show that a Σ-substitution σ Σ-solves $U \wedge x{:}\tau = t$ if, and only if, it Σ-solves $U \wedge x{:}\tau = t \wedge \Lambda$. The if direction is obvious. For the other direction assume σ Σ-solves $U \wedge x{:}\tau = t$. By the third requirement on Σ-substitutions (see Section 2), $\{x{:}\tau \mapsto (x{:}\tau)\sigma\}$ is a Σ-substitution. Since $(x{:}\tau)\sigma$ and $t\sigma$ are identical, $\{x{:}\tau \mapsto t\sigma\}$ is also a Σ-substitution. Therefore, by the definition of Λ, σ Σ-solves Λ. ∎

4 The GSUP Control Strategy

There are two ways that an execution of GSUP rule application can be infinite. An execution can be infinitely broad in that it computes an infinite disjunction, and it can be infinitely deep in that an infinite sequence of transformations is applied to a single disjunct. Though by necessity some executions of the GSUP must be infinitely broad, this section develops a flexible control scheme for rule application that admits no infinitely deep executions.

Observe that the application of a transformation to a disjunction affects only a single disjunct, and it does so independently of the other disjuncts present. The affected disjunct is replaced with one or more disjuncts. This observation allows us to define what we shall call a *derivation tree.* A derivation tree is a possibly infinitary tree in which each node is labeled by a disjunct. Furthermore, for every interior node there must be a GSUP transformation that rewrites the node's label to the disjunction whose disjuncts are precisely the labels of the node's immediate descendants. A derivation tree is said to be *complete* if each of its leaf nodes is labeled by a disjunct to which no transformations apply.

It should be clear that we can view any execution of GSUP rule applications as a process of growing a derivation tree and that the process must halt as soon as it grows a complete derivation tree. We can now be precise and define an execution to be finitely broad if its derivation tree is finitary and to be finitely deep if every path in its derivation tree is finite. A great deal of our effort in developing the GSUP has been devoted to formulating a restriction on the ordering of the transformation rules that guarantees finite depth derivation trees but imposes as little control as possible.

The presentation shall be eased by considering a macro transformation called *weakening/isolation.* Applying this macro consists of applying weakening followed immediately by applying variable isolation, where possible, to each equation introduced by weakening.

First, let us demonstrate that an unconstrained application of the GSUP transformation rules can result in an infinite depth derivation. In this example weakening/isolation alters some well-sorted equations in the conjunct, making them ill-sorted. When weakening is applied to these ill-sorted equations, the cycle begins again.

Example 5. This example is based on the sort definition given by the following sort theory:

$$\begin{aligned}\Sigma_2 = \{&\forall x, y\ \mathrm{B}(x) \rightarrow \mathrm{A}(f(g(y, f(y), x))),\\ &\forall x, y\ \mathrm{A}(x) \rightarrow \mathrm{B}(h(g(y, x, h(y)))),\\ &\forall x\ \mathrm{T}(x),\\ &\mathrm{A}(a),\\ &\mathrm{B}(b)\}\end{aligned}$$

Throughout this example we do not display the sort of any variable whose sort is T. Consider an execution of the GSUP starting with a single conjunction:

$$\begin{array}{ll} x{:}\mathrm{A} = f(y) \wedge & (1)\\ w{:}\mathrm{B} = h(z) \wedge & (2)\\ y = z & (3)\end{array}$$

Applying weakening/isolation to (2) adds (4) to the conjunction and modifies (2) and (3):

$$\begin{array}{ll} \exists z', x'{:}\mathrm{A} & \\ \quad x{:}\mathrm{A} = f(y) \wedge & (1)\\ \quad w{:}\mathrm{B} = h(g(z', x'{:}\mathrm{A}, h(z'))) \wedge & (2)\\ \quad y = g(z', x'{:}\mathrm{A}, h(z')) \wedge & (3)\\ \quad z = g(z', x'{:}\mathrm{A}, h(z')) & (4)\end{array}$$

Then applying weakening/isolation to (1) adds (5) to the conjunction and modifies (1) and (3):

$$\begin{array}{ll} \exists z', x'{:}\mathrm{A}, y', w'{:}\mathrm{B} & \\ \quad x{:}\mathrm{A} = f(g(y', f(y'), w'{:}\mathrm{B})) \wedge & (1)\\ \quad w{:}\mathrm{B} = h(g(z', x'{:}\mathrm{A}, h(z'))) \wedge & (2)\\ \quad g(y', f(y'), w'{:}\mathrm{B}) = g(z', x'{:}\mathrm{A}, h(z')) \wedge & (3)\\ \quad z = g(z', x'{:}\mathrm{A}, h(z')) \wedge & (4)\\ \quad y = g(y', f(y'), w'{:}\mathrm{B}) \wedge & (5)\end{array}$$

Decomposing (3) replaces it with the conjunction:

$$\begin{array}{ll} x'{:}\mathrm{A} = f(y') \wedge & (3.1)\\ w'{:}\mathrm{B} = h(z') \wedge & (3.2)\\ y' = z' & (3.3)\end{array}$$

Now notice that (3.1), (3.2) and (3.3) are alphabetic variants of (1), (2) and (3). This entire cycle of transformations can begin again.

We now define a syntactic feature, called *safeness*, and use it to formulate a restriction on the application of weakening.

Definition 4 (Oriented Equations and Safeness). Let U be a conjunction of equations. Then an equation $x{:}\tau = t$ occurring in U is *oriented* in U if, and only if, t does not contain x and every other occurrence (if any) of $x{:}\tau$ in U is on the right hand side of an equation oriented in U. A variable occurring in U is *safe* in U if, and only if, it is solved in U or has no occurrence in U except in the right hand sides of the oriented equations of U.

As an example, in

$$x{:}\tau = f(x{:}\tau) \wedge f(y{:}\tau) = g(x{:}\tau) \wedge y{:}\tau = a \wedge z{:}\tau = g(v{:}\tau) \wedge w{:}\tau = g(z{:}\tau)$$

the oriented equations are $z{:}\tau = g(y{:}\tau)$ and $w{:}\tau = g(z{:}\tau)$, and the safe variables are $w{:}\tau$ and $v{:}\tau$.

Observe that every solved equation in a conjunction is oriented.

The GSUP Control Strategy.

- Use transformations 1 through 6 and weakening/isolation.
- Repeatedly apply any transformation to any part of the formula until no transformations apply but:
  - Never apply weakening to $\exists\vec{z}\, U \wedge x{:}\tau = t$ unless every variable in t is safe in $U \wedge x{:}\tau = t$.

Example 6. Consider the unification problem from Example 5. In the initial conjunction all equations are oriented, and all variables except y are safe. Thus, as before, weakening/isolation can be applied to (2). In the resulting conjunction, all equations are oriented and, again, all variables except y are safe. Therefore, unlike the previous example, weakening/isolation cannot be applied to (1) since its right-hand side contains an unsafe variable. The only applicable transformation is the application of variable isolation to (3), which modifies (1) and results in:

$$\begin{array}{ll} \exists z', x'{:}\mathrm{A} & \\ \quad x{:}\mathrm{A} = f(g(z', x'{:}\mathrm{A}, h(z'))) \wedge & (1) \\ \quad w{:}\mathrm{B} = h(g(z', x'{:}\mathrm{A}, h(z'))) \wedge & (2) \\ \quad y = g(z', x'{:}\mathrm{A}, h(z')) \wedge & (3) \\ \quad z = g(z', x'{:}\mathrm{A}, h(z')) & (4) \end{array}$$

All of the equations are now solved and thus all of the variables are safe. The only transformation now applicable is the application of weakening/isolation to (1). No well-sorted substitution applied to the right hand side of equation (1) will make the equation well-sorted, so the weakening formula is *False*. Consequently, the result of this transformation is *False*.

Before proving that the Generalized Sorted Unification Procedure is guaranteed to halt, we prove that it will never halt too soon, that is, it will never halt before transforming the disjunction into Σ-solved form.

Theorem 5 (GSUP Solved Form Theorem). *When the GSUP terminates, the resulting disjunction is in Σ-solved form.*

Proof. Assume that the disjunction contains an equation, E, that is not in Σ-solved form. If every equation in the disjunction is in solved form, then every equation is oriented in the conjunction and therefore every variable is safe in the disjunction. Thus E meets the provisions on weakening/isolation and thus the transformation can be applied to E. Otherwise, some equation in the disjunction is not solved and therefore one of the transformations of ordinary unification can be applied to it. (The Solved Form Theorem for unsorted unification tells us this.) ■

The safeness provision on the weakening transformation guarantees that variable isolation is never applied to an equation whose left hand side is an existentially quantified variable. This guarantee coupled with the fact that isolated variables stay isolated assures us that variable isolation will only be applied a finite number of times, at most once for every free variable in the disjunction.

Lemma 6 (Isolation of Variables). *The GSUP will never apply variable isolation to an equation whose left hand side is an existentially quantified variable.*

Proof. (Sketch) If a variable is safe in a conjunction, it remains safe after any of the transformations are performed. Variable isolation is never applied to an equation whose left hand side is a safe variable. Every existentially quantified variable introduced by weakening/isolation is safe. ∎

Theorem 7 (GSUP Finite Depth Theorem). *Every derivation tree for the GSUP has finite depth.*

Proof. We specify a well-founded ordering over the set of all disjuncts and show that the disjunct at any node in a derivation tree is less than the disjunct at the node's parent. The ordering is obtained by associating a quadruple $\langle v_1, v_2, v_3, v_4 \rangle$ of natural numbers with every disjunct, where

v_1 is the number of unsolved free variables,
v_2 is the number of occurrences of function symbols (including rank zero ones) in equations of the form $t = t'$ where t and t' are not variables,
v_3 is the number of equations of the form $t = x{:}\tau$ or $x{:}\tau = x{:}\tau$, where t is not a variable, and
v_4 is the number of equations of the form $x{:}\tau = t$ that are not well sorted.

We say that one disjunct is less than another if the quadruple of the first lexicographically precedes the quadruple of the second. By inspection one can see that any disjunct is reduced in this ordering when any of the *applicable* transformations of the GSUP are applied to it. Notice that, by the preceding lemma, variable isolation reduces v_1. ∎

Corollary 8 (Termination of GSUP). *Given an oracle for computing weakening formulas, the GSUP always terminates if the oracle always returns a finite formula.*

From this corollary and the observation that the problem of computing weakening formulas can be reduced to the sorted unification problem, it follows that sorted unification is computable if, and only if, weakening formulas are computable. This has never before been shown.

5 Other Control Strategies

Sorted unification algorithms often employ a control strategy that ensures that variable isolation is applied only to well-sorted equations. This strategy is captured most generally by the following control strategy:

The Clean Isolation Control Strategy.

- Use transformations 1 through 6 and weakening/isolation.
- Repeatedly apply any transformation to any part of the formula until no transformations apply but:
  - Never apply variable isolation to an equation that is not well-sorted.

In particular, these algorithms often are obtained by modifying an ordinary unification algorithm to weaken a term t immediately before it is substituted for a variable $x{:}\tau$. (See [Frisch, 1991] and [Walther, 1988] for examples of such algorithms.) In other words, a substitution θ is applied to t to ensure that $\{x{:}\tau \mapsto t\theta\}$ is well-sorted. If there is no such substitution, failure is reported.[3]

These algorithms can be thought of as an uncontrolled execution of transformations 1 through 5 and a single macro transformation, which first applies weakening/isolation to an equation and then applies variable isolation to all immediate descendants of that equation. Obviously, this control strategy, which we call the weaken-when-isolating strategy, is a specialization of the clean isolation control strategy.

Though the clean isolation control strategy, and hence the weaken-when-isolating strategy, have been proven to terminate in several sorted unification algorithms for particular sorted logics, these strategies are not guaranteed to terminate in general. This can be seen by observing that the non-terminating execution of Example 5 does in fact follow the clean isolation control strategy and the weaken-when-isolating strategy.

Another common control strategy is to compute an ordinary unifier and then weaken it to obtain well-sorted unifiers. (See [Smolka *et al.*, 1989] and [Uribe, 1992] for examples of this strategy.) This strategy can be formulated as:

The Weaken Last Control Strategy.

- Use transformations 1 through 6 and weakening/isolation.
- Repeatedly apply any transformation except weakening/isolation to any part of the formula until none of those transformations apply.
- Then repeatedly apply weakening/isolation to any part of the formula until it no longer applies.

Observe that after transformations 1 through 6 have been applied till termination, the resulting formula is in solved form and thus all its variables are safe. Since solved form formulas remain in solved form after the application of any transformation, all variables remain safe. Thus, all subsequent applications of weakening/isolation satisfy the safeness provision of the GSUP control strategy. Consequently, every execution of the weaken last control strategy is an execution of the GSUP control strategy[4] and therefore the weaken last strategy is guaranteed to terminate.

For certain sorted logics an unconstrained control strategy can be shown to terminate. For example, Schmidt-Schauss [1989] shows termination for a logic with polymorphic term declarations and a sort structure that is a semi-lattice. Jouannaud and Kirchner [1991] claim termination for a logic with regular signatures and function declarations.

In summary, the GSUP control strategy and its specialization, the weaken last control strategy, always generate finite depth derivation trees. With only certain

[3] Some, but not all, authors are careful to point out that this weakening must be considered even when there are no occurrences of $x{:}\tau$ to get modified by the application of $\{x{:}\tau \mapsto t\theta\}$.

[4] Though not vice-versa, as illustrated by Example 6.

sort definitions are the uncontrolled strategy and the clean isolation control strategy guaranteed to generate only finite depth derivation trees. An open problem is to identify the conditions necessary to guarantee finiteness for these two control strategies.

6 Related Work

Prior to this work, the most general formulation of a sorted unification procedure was the set of GSOUP transformation rules formulated by Schmidt-Schuass [1989]. Though he assumes that the sort definition is specified by a *finite* signature with arbitrary term declarations, it appears that the restriction to finiteness is not essential for the approach. One can easily show that every sort definition can be produced by a possibly infinite signature with term declarations.

Whereas we consider our transformations to be chosen with don't-care nondeterminism, Schmidt-Schauss considers his transformations to be chosen with don't-know nondeterminism. Thus, our Preservation Theorem is much stronger than his corresponding completeness result. Furthermore, we show how our don't-care nondeterminism can be controlled to avoid the generation of infinite paths in the derivation tree; Schmidt-Schuass does not address this issue.

An exciting recent trend in the study of sorted unification has been the attempt to formulate expressive representations of sorts for which minimal complete sets of unifiers are always finite and computable. For example, Uribe[1992] represents each sort, considered as a set of terms, as a shallow system of set constraints, which is essentially a grammar for a class of term languages. From the viewpoint of our work, his results can be seen as identifying conditions under which sorted unification—and, in particular, weakening—is decidable. For instance, since the class of languages that his grammars can represent is closed under intersection, he is requiring that the ordering of sorts has unique greatest lower bounds. Other work within this trend has represented sorts by regular tree languages [Comon and Delor, 1991], by tree automata with equality constraints [Bogaert and Tison, 1992], and by ρ-terms [Chen and Hsiang, 1991].

7 Conclusion

This paper has demonstrated how the details of particular sorted languages can be abstracted away from the study of sorted unification. A set of transformation rules for solving sorted unification problems in the general case was presented and proved correct. The paper identified two ways that the transformations could be applied infinitely. Though infinitely broad executions cannot be avoided, we presented a control strategy that admits no infinitely deep executions. Certain common strategies were shown to be special cases of our control strategy. Other common control strategies known to prevent infinitely deep derivations in special cases were shown to admit such derivations in the more general setting.

Acknowledgments

We thank Joxan Jaffar, Michael Maher, Kim Marriott, Nachum Dershowitz, Tomás Uribe, and Claude Kirchner for stimulating and helpful discussions. The comments of several anonymous referees greatly helped to improve the paper. This work has been partially supported by a visiting fellowship awarded to Alan Frisch by the U.K. Science and Engineering Research Council under grant GR/G/24231.

References

[Bürckert, 1990] Hans-Jürgen Bürckert. *A Resolution Principle for Clauses with Constraints.* Research Report RR-90-02, German Research Center for Artificial Intelligence, March 1990.

[Bogaert and Tison, 1992] B. Bogaert and S. Tison. Equality and disequality constraints on brother terms in tree automata. In A. Finkel, editor, *Proc. of the Ninth Symposium on Theoretical Aspects of Computer Science*, Paris, Springer-Verlag, 1992.

[Chen and Hsiang, 1991] H. Chen and J. Hsiang. Logic programming with recurrence domains. In *Proc. of the Eighteenth International Colloquium on Automata, Languages and Computation*, pages 20–34, Madrid, Spain, Springer-Verlag, July 1991.

[Comon and Delor, 1991] H. Comon and C. Delor. *Equational Formulae with Membership Constraints.* Rapport de Recherche 649, LRI, Université de Paris Sud, Orsay, France, March 1991.

[Frisch, 1991] A. M. Frisch. The substitutional framework for sorted deduction: Fundamental results on hybrid reasoning. *Artificial Intelligence*, 49:161–198, 1991.

[Höhfeld and Smolka, 1988] M. Höhfeld and G. Smolka. *Definite Relations over Constraint Languages.* Lilog-Report 53, IBM Deutschland, October 1988.

[Jaffar and Lassez, 1987] J. Jaffar and J. Lassez. Constraint logic programming. In *Proc. of the* 14th *ACM Principles of Programming Languages Conference*, pages 111–119, Munich, January 1987.

[Jouannaud and Kirchner, 1991] J.-P. Jouannaud and C. Kirchner. Solving equations in abstract algebras: A rule-based survey of unification. In J.-L. Lassez and G. Plotkin, editors, *Computational Logic: Essays in Honor of Alan Robinson*, chapter 8, pages 257–321, MIT Press, Cambridge, MA, 1991.

[Martelli and Montanari, 1982] Alberto Martelli and Ugo Montanari. An efficient unification algorithm. *ACM Transactions on Programming Languages and Systems*, 4(2):258–282, April 1982.

[Reiter, 1977] Raymond Reiter. *An Approach to Deductive Question-Answering.* BBN Technical Report 3649, Bolt Beranek and Newman, Inc., 1977.

[Schmidt-Schauß, 1989] M. Schmidt-Schauß. *Computational Aspects of an Order-Sorted Logic with Term Declarations.* Volume 395 of *Lecture Notes in Computer Science*, Springer-Verlag, New York, NY, 1989.

[Smolka *et al.*, 1989] Gert Smolka, Werner Nutt, Joseph A. Goguen, and José Meseguer. Order-sorted equational computation. In Hassan Aït-Kaci and Maurice Nivat, editors, *Resolution of Equations in Algebraic Structures, Volume 2, Rewriting Techniques*, chapter 10, pages 297–367, Academic Press, New York, 1989.

[Uribe, 1992] T. E. Uribe. Sorted unification using set constraints. In D. Kapur, editor, *Proceedings of the Eleventh International Conference on Automated Deduction*, Saratoga Srpings, NY, June 1992.

[Walther, 1988] Christoph Walther. Many-sorted unification. *Journal of the ACM*, 35(1):1–17, January 1988.

Unification in Order-Sorted Algebras with Overloading

Alexandre Boudet*

LRI, CNRS URA 410
Université Paris-Sud, Centre d'Orsay
91405 Orsay Cedex

Abstract

We present an algorithm for unification in the combination of a theory Th_1 and one of its *overloaded extensions* Th_2 in the order-sorted framework. This problem is a particular combination problem where the signatures are not disjoint. A major consequence is that an equality proof between two pure terms in Th_1 may need the use of an axiom of Th_2. This makes the usual combination techniques incomplete, in particular the solving of pure equations in the theory to which they belong. To solve the problem, we need a *separated normal form* as well as a *complete set of normalizing substitutions*.

1 Introduction

To unify two terms s and t in a term algebra is to find a substitution σ that makes the formula $s\sigma \stackrel{?}{=} t\sigma$ valid when the predicate $\stackrel{?}{=}$ is interpreted as the syntactic equality (syntactic unification), or an equational theory (semantic unification). Equivalently, one can transform a unification problem using some solution-preserving rules until a solved form is obtained from which all such substitutions are built in a straightforward manner [13, 9]. Unification is a fundamental mechanism in logic programming, term rewriting, automated theorem proving, and artificial intelligence in general. This is so because it is the unification process that makes the applicable instances of the inference rules appear, as in Robinson's resolution rule [14].

Order-sorted algebras have been introduced by Goguen [8] in order to enrich the pure equational logic in which such features as partial operators or overloading cannot be expressed. Equational deduction in order-sorted algebras has been studied in *e.g.* [16, 12].

In this paper, we study semantic unification in order-sorted algebras with overloaded operators as a particular case of unification in a combination of equational

*This research was supported in part by GRECO Programmation CNRS and ESPRIT Working Group COMPASS.

theories over non-disjoint signatures. This has already been partially done by Kirchner [11] who gives an algorithm essentially based on the nondeterministic guess of the least sort of the values of the variables by a solution. Kirchner's algorithm is not complete because it doesn't take into account a mechanism that we describe here. Roughly, a theory Th_2 is an *overloaded extension* of a theory Th_1 if the function symbols operate on smaller sorts and have more properties in Th_2 than in Th_1. Overloaded extensions are used in, *e.g.*, the OBJ specification language [8], in which an operator may be declared with the theory $A \cup AC$ as in figure 1. The main difficulty is that a pure term in Th_1 (*i.e.*, having no subterm in Th_2) can be Th_1-equal to a non pure term in which some operators have the properties of Th_2. This makes incomplete the solving of a pure problem in the theory to which it belongs. We clarify this on an example: consider the specification of figure 1 and the equation $x + a \stackrel{?}{=} y + b$. Just choosing the sorts of x and y and then using the standard combination techniques [10, 15, 1] won't yield all the solutions. In particular, the problem $x + a \stackrel{?}{=} y + b \ \wedge \ x \in \underline{s_1} \ \wedge \ x \notin \underline{s_2} \ \wedge \ y \in \underline{s_1} \ \wedge \ y \notin \underline{s_2}$ has a solution even if the $+$ symbols in both sides of the equation are only associative since they have an argument which is in in $\underline{s_1}$ but not in $\underline{s_2}$. Indeed, the substitution $\sigma = \{x \mapsto c + b, y \mapsto c + a\}$ is a solution since $c + b$ and $c + a$ are in $\underline{s_1}$ and not in $\underline{s_2}$ and we have the following proof:

$$(x + a)\sigma \equiv (c + b) + a \underset{Th_1}{\longleftrightarrow} c + (b + a) \underset{Th_2}{\longleftrightarrow} c + (a + b) \underset{Th_1}{\longleftrightarrow} (c + a) + b \equiv (y + b)\sigma$$

sorts : $\underline{s_1}, \underline{s_2}$
subsorts : $\underline{s_2} < \underline{s_1}$

Th_1			Th_2		
$+$	$: \underline{s_1} \times \underline{s_1} \rightarrow \underline{s_1}$	$[assoc]$	$+$	$: \underline{s_2} \times \underline{s_2} \rightarrow \underline{s_2}$	$[assoc - comm]$
a	$: \ \rightarrow \underline{s_2}$		a	$: \ \rightarrow \underline{s_2}$	
b	$: \ \rightarrow \underline{s_2}$		b	$: \ \rightarrow \underline{s_2}$	
c	$: \ \rightarrow \underline{s_1}$				

Figure 1: The theory $A \cup AC$. The associative theory Th_1 of $+$ in $\underline{s_1}$ and its overloaded extension, the associative-commutative theory Th_2 of $+$ in $\underline{s_2}$.

Another new result is that our algorithm terminates which was left open in Kirchner's paper. Termination does not require an elaborated proof as in the case of assiciative-commutative unification [6, 2], due to our assumptions on the sort structure and the choice *a priori* of the theory in which each variable may be instantiated.

2 Definitions

2.1 Order-Sorted Signatures and Tree Automata

Let $\mathcal{S}$ be a finite, non-empty set of *sort symbols*, and $\leq_{\mathcal{S}}$ a partial ordering on $\mathcal{S}$. The sort symbols in $\mathcal{S}$ will be denoted by underlined letters $\underline{s_1}, \ldots, \underline{s_n}, \ldots$ Let $\mathcal{F}$ be a finite set of *function symbols*, every symbol $f \in \mathcal{F}$ being equipped with an integer *arity* $a(f)$, and a *typing function* $\tau(f) \in 2^{\mathcal{S}^{a(f)+1}}$. If $(\underline{s_1}, \ldots, \underline{s_{a(f)}}, \underline{s}) \in \tau(f)$, we will write

$$f: \underline{s_1} \times \cdots \times \underline{s_{a(f)}} \rightarrow \underline{s}$$

$\mathcal{X}$ is a countable set of variables, $\mathcal{T}(\mathcal{F}, \mathcal{X})$ is the free $\mathcal{F}$-algebra over $\mathcal{X}$, while $\mathcal{T}(\mathcal{F})$ denotes the algebra of *ground terms, i.e.*, of the terms without any variables. The application of the substitution σ to a term t, and the composition of two substitutions σ and θ are denoted, using postfix notation, by $t\sigma$ and $\sigma\theta$ respectively. The set of equations corresponding to a substitution σ is $\sigma_=$. The set of ground substitutions is denoted by $Gsubst(\mathcal{F})$. We will use $\equiv$ for syntactic equality. For any syntactic object o, $V(o)$ denotes the set of variables occurring in o. Our notations are consistent with [5], for instance, $t|_p$ denotes the subterm of t at position p and $t[u]_p$ denotes the term t in which the subterm at position p is replaced by u.

Definition 1 *An* order-sorted signature *is a 4-uple* $\Sigma = (\mathcal{S}, \mathcal{F}, \leq_{\mathcal{S}}, \tau)$.

Comon [3, 4] pointed out that an order-sorted signature is a bottom-up finite tree automaton A_Σ where the states are the sorts of $\mathcal{S}$ and for every pair (s,s') such that $\underline{s} <_{\mathcal{S}} \underline{s'}$ there is an ϵ-transition $\underline{s}(t) \rightarrow \underline{s'}(t)$; for every $f \in \mathcal{F}$, for every $(\underline{s_1}, \ldots, \underline{s_{a(f)}}, \underline{s}) \in \tau(f)$ there is a transition $f(\underline{s_1}(t_1), \ldots, \underline{s_{a(f)}}(t_{a(f)})) \rightarrow \underline{s}(f(t_1, \ldots, t_{a(f)}))$; the accepting states are the sorts of $\mathcal{S}$; the alphabet is the set $\mathcal{F}$ of function symbols.

$\mathcal{T}_\Sigma$ is the regular tree language recognized by the automaton A_Σ. The *terms of sort* $\underline{s} \in \mathcal{S}$ are the terms of $\mathcal{T}(\mathcal{F})$ that are recognized by the automaton $A_{\underline{s}}$ obtained from A_Σ by replacing the set $\mathcal{S}$ of final states by the singleton $\{\underline{s}\}$. We denote by $\mathcal{T}_{\underline{s}}$ the regular tree language of ground terms of sort $\underline{s}$. Note that we drop the usual notion of *well-sorted terms* and *well-sorted substitutions*, as used in [8, 16, 12, 11, 7], all the sort information being expressed by the means of *sort constraints* involving usual terms and *sort expressions*.

2.2 Sort Expressions and Sort Constraints

Definition 2 *A* sort expression *is a propositional formula over* $\mathcal{S} \cup \{\top_{\mathcal{S}}, \bot_{\mathcal{S}}\}$, *built up using the connectives* $\wedge$, $\vee$ *and* $\neg$. *The set of sort expressions is* $\mathcal{SE}$ [1].

The semantics of sort expressions is natural in terms of intersection, union and complement of regular tree languages:

- $\mathcal{T}_{\top_{\mathcal{S}}} = \mathcal{T}(\mathcal{F})$, $\mathcal{T}_{\bot_{\mathcal{S}}} = \emptyset$
- $\mathcal{T}_{\underline{s_1} \wedge \underline{s_2}}$ (resp. $\mathcal{T}_{\underline{s_1} \vee \underline{s_2}}$) is the language $\mathcal{T}_{\underline{s_1}} \cap \mathcal{T}_{\underline{s_2}}$ (resp. $\mathcal{T}_{\underline{s_1}} \cup \mathcal{T}_{\underline{s_2}}$)
- $\mathcal{T}_{\neg \underline{s_1}}$ is the complement in $\mathcal{T}(\mathcal{F})$ of the language $\mathcal{T}_{\underline{s_1}}$

[1] In [4], Comon also uses the construction $f(\underline{s_1}, \ldots, \underline{s_n})$ which is not needed here and which allows to create new sorts like succ(Nat).

$$
\begin{array}{rcl}
f(t_1,\ldots,t_n) \in \underline{s} & \Rightarrow & \bigvee_{(\underline{s_1},\ldots,\underline{s_n}) \in f^{-1}(\underline{s})} t_1 \in \underline{s_1} \wedge \cdots \wedge t_n \in \underline{s_n} \\
t \in \underline{s_1} \wedge t \in \underline{s_2} & \Rightarrow & t \in (\underline{s_1} \wedge \underline{s_2}) \\
t \in \underline{s_1} \vee t \in \underline{s_2} & \Rightarrow & t \in (\underline{s_1} \vee \underline{s_2}) \\
\neg t \in \underline{s} & \Rightarrow & t \in \neg \underline{s} \\
t \in \underline{s} & \Rightarrow & T \quad \text{if } \mathcal{T}_{\underline{s}} = \mathcal{T}(\mathcal{F}) \\
t \in \underline{s} & \Rightarrow & F \quad \text{if } \mathcal{T}_{\underline{s}} = \emptyset \\
\neg(\phi \wedge \psi) & \Rightarrow & \neg\phi \vee \neg\psi \\
\neg(\phi \vee \psi) & \Rightarrow & \neg\phi \wedge \neg\psi \\
\phi \wedge T & \Rightarrow & \phi \\
\phi \wedge F & \Rightarrow & F \\
\phi \vee T & \Rightarrow & T \\
\phi \vee F & \Rightarrow & \phi
\end{array}
$$

Figure 2: Normalization of sort constraints

For any sort expression $\underline{s}$, we can build the usual way a bottom-up finite tree automaton that recognizes the language $\mathcal{T}_{\underline{s}}$.

Comon [4] defines the following operation on sort expressions: for every $f \in \mathcal{F}$, of arity n, for every sort expression $\underline{s}$, $f^{-1}(\underline{s})$ is a finite set of n-uples of sort expressions such that $f(t_1,\ldots,t_n) \in \mathcal{T}_{\underline{s}}$ iff there exists $(\underline{s_1},\ldots,\underline{s_n}) \in f^{-1}(\underline{s})$ such that $t_i \in \mathcal{T}_{\underline{s_i}}$ for $1 \leq i \leq n$. We use $f^{-1}(\underline{s})$ only for the normalization of the sort constraints and we refer the reader to [4] its construction.

Definition 3 *An* atomic sort constraint *is either T or F or a formula $t \in \underline{s}$ or $t \notin \underline{s}$ where t is a term of $\mathcal{T}(\mathcal{F},\mathcal{X})$ and $\underline{s}$ is a sort expression. If ϕ and ψ are sort constraints, then $\phi \wedge \psi$, $\phi \vee \psi$ and $\neg\phi$ are sort constraints. We write $t \notin \underline{s}$ for $\neg t \in \underline{s}$.*

The semantics of sort constraints is given by interpreting the unary predicates $\cdot \in \underline{s}$ as the membership of $\mathcal{T}_{\underline{s}}$. Note that we have negation for free in the sort constraints, since the regular languages are closed under boolean operations.

2.3 Normalization of the Sort Constraints

The "sorts as constraints" approach allows to solve the sort constraints separately. We will assume in the following that at every moment the sort constraints are normalized using the set of rules of figure 2. These rules are to be applied modulo the associativity-commutativity and idempotence of the connectives $\wedge$ and $\vee$.

2.4 Order-Sorted Specifications

Definition 4 *An* equational axiom *or* identity *is of the form* $\phi: \ t_1 = t_2$ *where* ϕ *is a sort constraint of the form* $x_1 \in \underline{s_1} \ \wedge \cdots \wedge \ x_n \in \underline{s_n}$ *with* $\underline{s_1}, \ldots, \underline{s_n} \in \mathcal{SE}$*, and* t_1 *and* t_2 *are terms of* $\mathcal{T}(\mathcal{F}, \mathcal{X})$*. We make no distinction between the axioms* $\phi: \ t_1 = t_2$ *and* $\phi: \ t_2 = t_1$*. Given a set* E *of equational axioms over an order-sorted signature* Σ*, the pair* $Th = (\Sigma, E)$ *is called an* order-sorted specification *(or* theory*).*

An equational axiom $\phi: \ t_1 = t_2$ represents the set of the ground equations $t_1\sigma = t_2\sigma$ where σ is a ground solution of ϕ.

Definition 5 *Given an order-sorted specification* $Th = (\Sigma, E)$ *and* $t_1, t_2 \in \mathcal{T}(\mathcal{F})$*, we write* $t_1 \underset{Th}{\longleftrightarrow} t_2$ *if there exists an axiom* $x_1 \in \underline{s_1} \ \wedge \cdots \wedge \ x_n \in \underline{s_n} : \ l = r$ *in* E*, a position* $p \in \mathcal{D}om(t_1)$ *and a ground substitution* σ *such that* $t_1|_p = l\sigma$ *and* $t_2 = t_1[r\sigma]_p$ *and for all* $i \in [1..n]$*,* $x_i\sigma \in \mathcal{T}_{\underline{s_i}}$*.* $t_1 \underset{Th}{\longleftrightarrow} t_2$ *is called a* one-step equational proof*. The reflexive, transitive closure of* $\underset{Th}{\longleftrightarrow}$ *is* $\overset{*}{\underset{Th}{\longleftrightarrow}}$*.*

2.5 Unification Problems

Definition 6 *A* unification problem *is a first order predicate calculus formula built up using the connectives* $\wedge$ *and* $\vee$*, the existensial quantifier* $\exists$ *and where the atoms are of one of the following forms:*

- *T (the trivial unification problem), or F (the unsolvable unification problem)*
- *an equation* $t_1 \overset{?}{=} t_2$ *with* $t_1, t_2 \in \mathcal{T}(\mathcal{F}, \mathcal{X})$
- *an atomic sort constraint* $t \in \underline{s}$*, or* $t \notin \underline{s}$ *where* $\underline{s}$ *is a sort expression*

Definition 7 *The set* $Sol_{Th}(P)$ *of the (ground) Th-solutions of a unification problem is inductively defined by*

- $Sol_{Th}(T) = Gsubst(\mathcal{F})$
- $Sol_{Th}(F) = \emptyset$
- $Sol_{Th}(t_1 \overset{?}{=} t_2) = \{\sigma \in Gsubst(\mathcal{F}) \mid t_1\sigma \overset{*}{\underset{Th}{\longleftrightarrow}} t_2\sigma\}$
- $Sol_{Th}(t \in \underline{s}) = \{\sigma \in Gsubst(\mathcal{F}) \mid t\sigma \in \mathcal{T}_{\underline{s}}\}$
- $Sol_{Th}(t \notin \underline{s}) = \{\sigma \in Gsubst(\mathcal{F}) \mid t\sigma \in \mathcal{T}_{\neg\underline{s}}\}$

- $Sol_{Th}(P_1 \wedge P_2) = \{\sigma \in Gsubst(\mathcal{F}) \mid \sigma \in Sol_{Th}(P_1) \cap Sol_{Th}(P_2)\}$
- $Sol_{Th}(P_1 \vee P_2) = \{\sigma \in Gsubst(\mathcal{F}) \mid \sigma \in Sol_{Th}(P_1) \cup Sol_{Th}(P_2)\}$
- $Sol_{Th}((\exists x)\ P) = \{\sigma \in Gsubst(\mathcal{F}) \mid \exists t \in \mathcal{T}(\mathcal{F}), \sigma \in Sol_{Th}(P\{x \mapsto t\})\}$

2.6 Unifiers and Solved Forms

Definition 8 *A* constrained substitution *is a pair (σ, ϕ) where σ is a substitution and ϕ a sort constraint.*
A unifier *of a unification problem P is a constrained substitution (σ, ϕ) such that if θ is a ground substitution of the form $\sigma\sigma'$ where σ' is a solution of ϕ, then θ is a solution of P. If in addition every solution of P is of the above form, then (σ, ϕ) is called a* most general unifier *of P.*

Some unification problems represent their own most general unifier:

Definition 9 *A unification problem P is in a solved form if $P \equiv T$ or $P \equiv F$, or*

$$P \equiv (\exists y_1, \ldots, y_p)\ x_1 \stackrel{?}{=} t_1 \ \wedge \cdots \wedge \ x_n \stackrel{?}{=} t_n \ \wedge \ z_1 \in \underline{s_1} \ \wedge \cdots \wedge \ z_m \in \underline{s_m}$$

where

- *x_i is free for $1 \leq i \leq n$*
- *$x_i \neq x_j$ for $1 \leq i, j \leq n$ and $i \neq j$*
- *$x_i \notin \mathcal{V}ar(t_j)$ for $1 \leq i, j \leq n$*
- *$\underline{s_i} \in \mathcal{SE}$ for $1 \leq i \leq m$*
- *z_i is free or there exists $j \in [1..n]$ such that $z_i \in V(t_j)$ for $1 \leq i \leq m$*

Lemma 1 *If $P \equiv (\exists y_1, \ldots, y_p)\ x_1 \stackrel{?}{=} t_1 \wedge \cdots \wedge x_n \stackrel{?}{=} t_n \wedge z_1 \in \underline{s_1} \wedge \cdots \wedge z_m \in \underline{s_m}$ is in a solved form, then $(\{x_1 \mapsto t_1, \ldots, x_n \mapsto t_n\}, z_1 \in \underline{s_1} \ \wedge \cdots \wedge \ z_m \in \underline{s_m})$ is a most general unifier of P.*

2.7 Overloaded extensions

We will consider unification problems in the combination of a theory Th_1 and one of its *overloaded extensions* Th_2 as defined below:

Definition 10 *Let $\mathcal{S}$ be a set of sorts, and $\leq_{\mathcal{S}}$ a partial ordering on $\mathcal{S}$.*
Let $\Sigma_1 = (\mathcal{S}, \leq_{\mathcal{S}}, \mathcal{F}_1, \tau_1)$ and $\Sigma_2 = (\mathcal{S}, \leq_{\mathcal{S}}, \mathcal{F}_2, \tau_2)$.
A theory $Th_2 = (\Sigma_2, E_2)$ is an overloaded extension *of a theory $Th_1 = (\Sigma_1, E_1)$ if $\mathcal{T}_{\Sigma_2} \subseteq \mathcal{T}_{\Sigma_1}$ and $\forall s, t \in \mathcal{T}_{\Sigma_2}$ $\quad s \xleftrightarrow[Th_1]{*} t \Rightarrow s \xleftrightarrow[Th_2]{*} t$.*

To solve the problem, we need the following extra assumptions:

1. The theory $Th_1 \cup Th_2$ is a *simple theory, i.e.*, there is no proof $t_1 \xleftrightarrow[Th_1 \cup Th_2]{*} t_2$ where t_1 is a proper subterm of t_2. This assumption makes the *occur-check* rule complete.

2. The axioms of Th_1 (resp. Th_2) apply only to terms of $\mathcal{T}_{\Sigma_1}$ (resp. $\mathcal{T}_{\Sigma_2}$). This corresponds to the usual assumption that the equations involve well-formed terms.

3. The axioms are *theory preserving, i.e.*, if t is a term of $\mathcal{T}_{\Sigma_2}$ (resp. of $\mathcal{T}_{\Sigma_1} \setminus \mathcal{T}_{\Sigma_2}$), then so is every term u s.t. $t \xleftrightarrow[Th_1 \cup Th_2]{*} u$. This corresponds to considering *collapse-free* theories in the one-sorted case: there is no equality proof between two terms of different theories.

4. $Th_1 \cup Th_2$ has a *separated normal form*, and *complete sets of normalizing substitutions*, these two notions will be introduced in section 3.

The two theories of figure 1 meet all these requirements, and all the examples in the following will use the specification of figure 1.

Under the above conditions, we can restrict our attention to the simple sort structure $\mathcal{T}_{\perp_S} \subseteq \mathcal{T}_{\Sigma_2} \subseteq \mathcal{T}_{\Sigma_1} \subseteq \mathcal{T}_{\top_S}$.

Notation In the following, $\underline{\eta_0}$ denotes the sort expression corresponding to the set $\mathcal{T}_{\top_S} \setminus \mathcal{T}_{\Sigma_1}$; $\underline{\eta_1}$ corresponds to $\mathcal{T}_{\Sigma_1} \setminus \mathcal{T}_{\Sigma_2}$; $\underline{\eta_2}$ corresponds to $\mathcal{T}_{\Sigma_2}$. The sort structure may of course be more complex inside Th_1 and Th_2. In the case of the theory $A \cup AC$ of figure 1, $\underline{\eta_0}$ corresponds to $\neg \underline{s_1}$, $\underline{\eta_1}$ to $\underline{s_1} \wedge \neg \underline{s_2}$ and $\underline{\eta_2}$ to $\underline{s_2}$. Note that $\underline{\eta_0}$, $\underline{\eta_1}$ and $\underline{\eta_2}$ form a partition of $\mathcal{T}(\mathcal{F}_1 \cup \mathcal{F}_2)$ and that no axioms can apply to the terms of $\underline{\eta_0}$, hence Th_0-equality is just syntactic equality. Syntactic unification has to be used to solve problems of the form $s \stackrel{?}{=} t \wedge \phi$ where s and t are both pure in Th_0 under ϕ (purity will be defined in the next section).

3 Separating the Proofs

In this section, we show how to reduce our problem to the case of disjoint signatures.

Definition 11 *A subterm u is an* alien subterm *of a term $t[u]_p$* at position *p if $t \notin \mathcal{T}_{\Sigma_i}$ and u is a maximal subterm of t such that $u \in \mathcal{T}_{\Sigma_i}$, for $i = 1$ or $i = 2$. A non-variable subterm u is an* alien subterm *of a term $t[u]_p$* at position *p* under *ϕ if for every solution σ of ϕ, $u\sigma$ is an alien subterm of $t\sigma$ at position p. A term t is pure (under ϕ) in Th_0 (resp. Th_1, Th_2) if it has no alien subterms (under ϕ), and $t \in \mathcal{T}_{\top_S} \setminus \mathcal{T}_{\Sigma_1}$ (resp. $\mathcal{T}_{\Sigma_1} \setminus \mathcal{T}_{\Sigma_2}$, $\mathcal{T}_{\Sigma_2}$).*

Example 1
The term $(x + y) + z$ is pure in Th_1 under $\phi \equiv x \in (\underline{s_1} \wedge \neg \underline{s_2}) \wedge y \in \underline{s_2} \wedge z \in \underline{s_2}$, while the term $x + (y + z)$ is not. Note that the two terms are equal in Th_1: this example shows how an application of an axiom of Th_1 to a pure term in Th_1 may create new alien subterms.

The previous example illustrates our main problem: solving a pure problem in Th_1 may not be complete since a proof between two pure terms of Th_1 may use some applications of axioms of Th_2. To solve this problem, we will define a normal form for the terms such that every proof between two terms in normal form is an instance of a proof (maybe of length zero) in Th_1. For this purpose, we distinguish some one-step proofs as the one shown in the example, where the application of an axiom of Th_1 creates a pure term in Th_2.

Definition 12 *Let $\phi : l = r$ be an axiom of Th_1. A* critical scheme *is an axiom (valid in Th_1) $\psi : l = r$ such that $Sol(\psi) \subseteq Sol(\phi)$, and every solution of ψ is a solution of $l \in \underline{\eta_1}$ and of $r \in \underline{\eta_1}$, and l or r is not pure in Th_1 under ψ. A critical scheme $\psi : l = r$ is* orientable *if l or r is pure in Th_1 under ψ. An* oriented critical scheme *is a rewrite rule $\psi : l \rightarrow r$ where $\psi : l = r$ is an orientable critical scheme and l is pure in Th_1 under ψ. The rewrite system containing all the oriented critical schemes is denoted by CS, and the corresponding rewrite relation by $\xrightarrow[CS]{}$.*

Example 2 *In $A \cup AC$, the axiom $x \in \underline{s_1} \wedge y \in \underline{s_2} \wedge z \in \underline{s_2} : (x+y)+z = x+(y+z)$ is a critical scheme. This axiom applies to the term $(c+b)+a$, yielding the term $c+(b+a)$ where $b+a$ is a new alien subterm in Th_2 with an associative-commutative $+$. This is due to the existence of an overlap between the right-hand side of the critical scheme and the axiom of commutativity.*

Actually, in $A \cup AC$, there are only two critical schemes and both are orientable. The rewrite system CS contains the two following rules:
$x \in (\underline{s_1} \wedge \neg \underline{s_2}) \wedge y \in \underline{s_2} \wedge z \in \underline{s_2} : (x+y)+z \rightarrow x+(y+z)$
$x \in \underline{s_2} \wedge y \in \underline{s_2} \wedge z \in (\underline{s_1} \wedge \neg \underline{s_2}) : x+(y+z) \rightarrow (x+y)+z.$

Assume now that Th_1 has the axiom
$x \in \underline{s_1} \wedge y \in \underline{s_1} \wedge z \in \underline{s_1} \wedge t \in \underline{s_1} : (x+y)+(z+t) = (x+z)+(y+t)$
and that $+ : \underline{s_2} \times \underline{s_2} \rightarrow \underline{s_2}$ belongs to $\mathcal{F}_2$. Now the critical scheme
$x \in \underline{s_2} \wedge y \in \underline{s_2} \wedge z \in (\underline{s_1} \wedge \neg \underline{s_2}) \wedge t \in \underline{s_2} : (x+y)+(z+t) = (x+z)+(y+t)$
is not orientable because both sides of the equation have an alien subterm ($x+y$ in the lhs and $y+t$ in the rhs).

Definition 13 *A proof $u_0 \xleftrightarrow[Th_1 \cup Th_2]{*} u_n$ is* separated *if no one-step proof $u_i \xleftrightarrow[Th_1]{} u_{i+1}$ can be performed by applying a critical scheme instead of the original axiom. Such a separated proof is denoted by $u_0 \xleftrightarrow[Th_1 \oplus Th_2]{*} u_n$.*

Example 3 *The proof $(c+a)+b \xleftrightarrow[Th_1]{} c+(a+b) \xleftrightarrow[Th_2]{} c+(b+a) \xleftrightarrow[Th_1]{} (c+b)+a$ is not separated because the first and last step can be performed with the critical scheme $x \in (\underline{s_1} \wedge \neg \underline{s_2}) \wedge y \in \underline{s_2} \wedge z \in \underline{s_2} : (x+y)+z = x+(y+z)$. Note that in the case of disjoint signatures, the relations $\xleftrightarrow[Th_1 \cup Th_2]{*}$ and $\xleftrightarrow[Th_1 \oplus Th_2]{*}$ are the same.*

The importance of the relation $\xleftrightarrow[Th_1 \oplus Th_2]{*}$ is illustrated by the following lemma:

Lemma 2 *Let $P \equiv s \stackrel{?}{=} t \wedge \phi$ be a unification problem such that s and t are pure in Th_1 under ϕ. Assume that θ is a $Th_1 \oplus Th_2$-solution of P.*
Then there exists a substitution σ such that $\forall x \in V(P)$ $x\sigma \overset{*}{\underset{Th_2}{\longleftrightarrow}} x\theta$ and σ is a Th_1 solution of P.

Proof: Let us distinguish a representative in every class for the equivalence $\overset{*}{\underset{Th_2}{\longleftrightarrow}}$ in η_2, the representative of the class of a term t being denoted by $\overline{t}$. Let $\overline{\theta}$ be the substitution θ where all the alien subterms have been replaced by their representatives. Since s and t are pure under ϕ and θ is a solution of ϕ, the alien subterms in $s\theta$ (and $t\theta$) are all in the substitution part (*i.e.*, at positions introduced by θ), hence, $\overline{s\theta} \equiv s\overline{\theta}$ (and $\overline{t\theta} \equiv t\overline{\theta}$). Consider the proof

$$s\theta \equiv u_0 \overset{*}{\underset{Th_1 \oplus Th_2}{\longleftrightarrow}} u_1 \overset{*}{\underset{Th_1 \oplus Th_2}{\longleftrightarrow}} \cdots \overset{*}{\underset{Th_1 \oplus Th_2}{\longleftrightarrow}} u_n \equiv t\theta$$

We show that this proof lifts to a proof of the form $\overline{s\theta} \overset{*}{\underset{Th_1}{\longleftrightarrow}} \overline{t\theta}$. Indeed, every step $u_i \overset{*}{\underset{Th_1}{\longleftrightarrow}} u_{i+1}$ in the proof, corresponds to a one-step proof $\overline{u_i} \overset{*}{\underset{Th_1}{\longleftrightarrow}} \overline{u_{i+1}}$ because (i) the proof being separated, the axiom applies above the alien subterms, and (ii) two equal alien subterms in u_i are replaced by the same representative in $\overline{u_i}$. Moreover, for every step $u_i \overset{*}{\underset{Th_2}{\longleftrightarrow}} u_{i+1}$ using an axiom of Th_2, we have $\overline{u_i} \equiv \overline{u_{i+1}}$ since the axiom has been applied to an alien subterm which does not change its representative. Let $\sigma \equiv \overline{\theta}$. We have $x\sigma \overset{*}{\underset{Th_2}{\longleftrightarrow}} x\theta$ $\forall x \in V(P)$, and $s\sigma \equiv \overline{s\theta} \overset{*}{\underset{Th_1}{\longleftrightarrow}} \overline{t\theta} \equiv t\sigma$. □

The lemma 2 shows that solving a pure equation in Th_1 is complete with respect to the relation $\overset{*}{\underset{Th_1 \oplus Th_2}{\longleftrightarrow}}$. We are now going to reduce $Th_1 \cup Th_2$-unification to $Th_1 \oplus Th_2$-unification. The idea is the following: first, we define a normal form for terms such that two terms in normal form are $Th_1 \cup Th_2$-equal if and only if they are $Th_1 \oplus Th_2$-equal. Second we define a complete set Σ of *normalizing substitutions* for a unification problem P such that every (ground) $Th_1 \cup Th_2$-solution of P is of the form $\sigma\sigma'$ for some $\sigma \in \Sigma$, where σ' is a $Th_1 \oplus Th_2$-solution of the normal form of $P\sigma$.

Definition 14 *A* separated normal form $(\cdot\downarrow)$ *is a mapping from $\mathcal{T}(\mathcal{F}_1 \cup \mathcal{F}_2)$ into itself such that $t \overset{*}{\underset{Th_1 \cup Th_2}{\longleftrightarrow}} t\downarrow$ for every $t \in \mathcal{T}(\mathcal{F}_1 \cup \mathcal{F}_2)$ and for $t_1, t_2 \in \mathcal{T}(\mathcal{F}_1 \cup \mathcal{F}_2)$, $t_1 \overset{*}{\underset{Th_1 \cup Th2}{\longleftrightarrow}} t_2$ implies $t_1\downarrow \overset{*}{\underset{Th_1 \oplus Th2}{\longleftrightarrow}} t_2\downarrow$.*
Let $\mathcal{V} \subset \mathcal{X}$ be a finite set of variables. We write $s \overset{*}{\longleftrightarrow}_\phi t$ if for every solution σ of ϕ, $s\sigma \overset{*}{\longleftrightarrow} t\sigma$. A separated normal form $(\cdot\downarrow_\phi)$ under a sort constraint ϕ *is a mapping from $\mathcal{T}(\mathcal{F}_1 \cup \mathcal{F}_2, \mathcal{V})$ into itself such that for $t_1, t_2 \in \mathcal{T}(\mathcal{F}_1 \cup \mathcal{F}_2, \mathcal{V})$, $t_1 \overset{*}{\underset{Th_1 \cup Th2}{\longleftrightarrow}}_\phi t_1\downarrow_\phi$ and $t_1 \overset{*}{\underset{Th_1 \cup Th_2}{\longleftrightarrow}}_\phi t_2$ implies $t_1\downarrow_\phi \overset{*}{\underset{Th_1 \oplus Th_2}{\longleftrightarrow}}_\phi t_2\downarrow_\phi$.*

Definition 15 *Let $\mathcal{V} \subset \mathcal{X}$ be a finite set of variables. A* complete set of normalizing substitutions (CSNS) for $\mathcal{V}$ *is a set Σ of constrained substitutions such that for every ground substitution θ with $Dom(\theta) \subseteq \mathcal{V}$ there exist a constrained substitution (σ, ϕ) and a ground substitution σ' such that*

- *σ' is a solution of ϕ*
- $\forall v \in \mathcal{V} \;\; v\theta \overset{*}{\underset{Th_1 \cup Th_2}{\longleftrightarrow}} v\sigma\sigma'$
- $\forall t \in \mathcal{T}(\mathcal{F}_1 \cup \mathcal{F}_2, \mathcal{V}) \;\; (t\sigma)\downarrow_\phi \sigma'$ *is in a separated normal form.*

The next theorem is our main result. It shows that every $Th_1 \cup Th_2$-solution of an equation $t_1 \overset{?}{=} t_2$ is a $Th_1 \oplus Th_2$-solution of the equation to which a normalizing substitution has been applied. This reduces $Th_1 \cup Th_2$-unification to $Th_1 \oplus Th_2$-unification (that is to the case of disjoint signatures), provided that a separated normal form and a complete set of normalizing substitutions are known.

Theorem 1 *Let $P \equiv t_1 \overset{?}{=} t_2$ be a unification problem. Let Σ be a complete set of normalizing substitutions for $V(P)$. Assume θ is a (ground) $Th_1 \cup Th_2$-solution of P. Then there exist a constrained substitution $(\sigma, \phi) \in \Sigma$ and a ground substitution σ' such that σ' is a $Th_1 \oplus Th_2$-solution of $(P\sigma)\downarrow_\phi \wedge \phi$.*

Proof: By definition of a complete set of normalizing substitutions, there exist $(\sigma, \phi) \in \Sigma$ and a solution σ' of ϕ such that $x\theta \overset{*}{\underset{Th_1 \cup Th_2}{\longleftrightarrow}} x\sigma\sigma'$ for every $x \in V(P)$ and $(t_1\sigma\downarrow_\phi)\sigma'$ and $(t_2\sigma\downarrow_\phi)\sigma'$ are in a separated normal form.
Hence, by definition of a separated normal form, $(t_1\sigma\downarrow_\phi)\sigma' \overset{*}{\underset{Th_1 \oplus Th_2}{\longleftrightarrow}} (t_2\sigma\downarrow_\phi)\sigma'$, that is σ' is also a $Th_1 \oplus Th_2$-solution of $(P\sigma)\downarrow_\phi$. □

4 The Rules

Figure 3 gives the rules for unification in $Th_1 \cup Th_2$. We assume that at every step, the problem is a disjunction of problems of the form $P \equiv (\exists y_1, \ldots, y_p) \;\; P_= \wedge P_C$ where $P_=$ is a conjunction of equations and P_C is a sort constraint. The rules apply to the members of such a disjunction. P_i is the conjunction of the equations that are pure in Th_i under P_C, for $i = 0, 1, 2$ (Th_0 being the free theory on $\underline{\eta_0}$).

These rules are very similar to the rules for associative-commutative unification given in [2]. The main difference is the rule **Choose-NS** which chooses a theory for each variable and a normalizing substitution. Note that **Solve-P_1** is not complete unless **Choose-NS** has previously been applied, hence the rules must be applied with the given control, while the sort constraints are normalized at every step using the rules of figure 2. With this control, termination is easy to prove. The only non-straightforward step is step 3. It is enough to notice that a variable x occurring in an equation $x \overset{?}{=} t$ in a solved subproblem P_i cannot appear in P_j for $j > i$ since P_j is pure in Th_j. Hence, solving a subproblem P_j and applying **Coalesce** to an equation

Figure 3: The Algorithm

1. Apply **Choose-NS** with a CSNS for $V(P_=)$.
2. Apply **Variable Abstraction** as long as possible.
3. Apply as long as possible, with decreasing priority **Theory-Clash**, **Check***, **Coalesce**, **Solve-P_0**, **Solve-P_1**, **Solve-P_2**.
4. Apply as long as possible **Replacement**, **EQE$_=$** and **EQE$_C$**.

Choose-NS
$P_= \;\Rightarrow\; \bigvee_{(\sigma,\phi)\in\Sigma}\; (\exists y_1,\ldots,y_p)\; (P_=\sigma)\downarrow_\phi \wedge \sigma_= \wedge \phi$
if Σ is a CSNS for $V(P_=)$ and $y_1,\ldots,y_p$ are the new variables introduced by σ

Variable Abstraction
$t[u]_p \stackrel{?}{=} t' \wedge P_C \;\Rightarrow\; (\exists x)\; t[x]_p \stackrel{?}{=} t' \wedge x \stackrel{?}{=} u$
if u is an alien subterm of $t[u]_p$ at position p under P_C

Coalesce
$(\exists y_1,\ldots,y_p)\; x \stackrel{?}{=} y \wedge P \;\Rightarrow\; (\exists y_1,\ldots,y_p)\; x \stackrel{?}{=} y \wedge P\{x \mapsto y\}$
if $x, y \in V(P)$ and x is existentially quantified or y is free

Solve-P_i
$P_i \;\Rightarrow\; P_i'$
if $0 \leq i \leq 2$ and P_i is not solved, and P_i' is a solved form of P_i

Theory-Clash
$x \stackrel{?}{=} t_1 \wedge x \stackrel{?}{=} t_2 \;\Rightarrow\; F$
if t_1 is pure in Th_i and t_2 is pure in Th_j with $i \neq j$

Check*
$x_1 \stackrel{?}{=} t_1[x_2] \wedge x_2 \stackrel{?}{=} t_2[x_3] \wedge \cdots \wedge x_n \stackrel{?}{=} t_n[x_1] \;\Rightarrow\; F$
if one $t_i \notin \mathcal{X}$ for $1 \leq i \leq n$

Replacement
$x \stackrel{?}{=} s \wedge P \;\Rightarrow\; x \stackrel{?}{=} s \wedge P\{x \mapsto s\}$
if $x \in V(P)$

EQE$_=$
$(\exists x)\; x \stackrel{?}{=} t \wedge P \;\Rightarrow\; P$
if x does not occur in P

EQE$_C$
$(\exists x)\; x \in \underline{s} \wedge P \;\Rightarrow\; P$
if x does not occur in P

$x \stackrel{?}{=} y$ of the solved form cannot make a subproblem P_i unsolved for $i < j$, and, with our control, the rule **Solve-P_i** will be applied at most once to each subproblem.

Theorem 2 *The rules of figure 3 terminate with the given control. They preserve the sets of solutions, and the irreducible problems are solved forms.*

5 Application to $A \cup AC$

5.1 Computation of Separated Normal Forms

Since two terms in separated normal form which are equal in the theory $Th_1 \oplus Th_2$ can (by definition) be proved equal without applying a critical scheme, the natural idea is to orient the critical schemes so as to make the new alien subterms appear. The following lemma gives a method for computing a separated normal form when orienting the critical schemes does not change the theory.

Lemma 3 *Assume that all the critical schemes are orientable and that the relation*

$$(\underset{Th_1 \oplus Th_2}{\longleftrightarrow} \cup \underset{CS}{\longrightarrow})$$

is confluent and that the relation

$$(\underset{Th_1 \oplus Th_2}{\overset{*}{\longleftrightarrow}} \underset{CS}{\longrightarrow})$$

is terminating. Then the terms in normal form for $(\underset{Th_1 \oplus Th_2}{\overset{*}{\longleftrightarrow}} \underset{CS}{\longrightarrow})$ *are separated normal forms.*

In other words, if orienting the critical schemes does not change the theory, and if their application modulo $Th_1 \oplus Th_2$ terminates, then two terms to which no critical scheme can be applied are equal in $Th1 \cup Th_2$ if and only if they are equal in $Th_1 \oplus Th_2$. The theory $A \cup AC$ meets the conditions of the lemma. Moreover, since the critical schemes are left linear, it is enough to apply as long as possible the critical schemes modulo only Th_1.

5.2 Computation of a CSNS

We show now how to compute a complete set of normalizing substitutions in the theory $A \cup AC$. Without loss of generality, we restrict our attention to the constrained substitutions (σ, ϕ) such that σ is in separated normal form under ϕ. We will also take advantage of the fact that a term in $\underline{\eta_2}$ is always in a separated normal form, and that the critical schemes may only apply above the alien subterms (*i.e.*, the alien subterms are in the substitution), hence a normalizing substitution of the form

$$(\{x \mapsto t[x_1 + x_2], \ldots\}, x_1 \in \underline{\eta_2} \ \wedge \ x_2 \in \underline{\eta_2} \ \wedge \ \cdots)$$

may be simplified, yielding

$$(\{x \mapsto t[x_1], \ldots\}, x_1 \in \underline{\eta_2} \ \wedge \ \cdots)$$

Lemma 4 *Assume that the critical schemes have no multiple occurrences of the variables in $\underline{\eta_2}$ in their left-hand sides. Let t be a term such that $V(t) = \{x\}$, in a separated normal form under the trivial sort constraint T. Let t' be a ground term in a separated normal form. Assume that a critical scheme applies to $t\{x \mapsto t'\}$. Then, either $t' \in \underline{\eta_2}$, or t' is an instance of a non-variable proper subterm of a left-hand side of a critical scheme.*

The previous lemma indicates that besides the constrained substitutions

$$(\{x \mapsto x'\}, x' \in \underline{\eta_0}) \tag{1}$$

$$(\{x \mapsto x'\}, x' \in \underline{\eta_1}) \tag{2}$$

$$(\{x \mapsto x'\}, x' \in \underline{\eta_2}) \tag{3}$$

a CSNS for $\{x\}$ must contain the two constrained substitutions corresponding to the proper non-variable subterms of the left-hand sides of the critical schemes of $A \cup AC$:

$$(\{x \mapsto x_1 + x_2\}, x_1 \in \underline{\eta_1} \wedge x_2 \in \underline{\eta_2}) \tag{4}$$

$$(\{x \mapsto y_1 + y_2\}, y_1 \in \underline{\eta_2} \wedge y_2 \in \underline{\eta_1}) \tag{5}$$

Now, the values of x_1 and y_2 may themselves be instances of a proper subterm of a critical scheme. Hence, we need to "apply" each of the two above constrained substitutions to x_1 and y_2. We get the four following constrained substitutions

$$(\{x \mapsto x_1' + x_2' + x_2\}, x_1' \in \underline{\eta_1} \wedge x_2' \in \underline{\eta_2} \wedge x_2 \in \underline{\eta_2}) \tag{6}$$

$$(\{x \mapsto y_1' + y_2' + x_2\}, y_1' \in \underline{\eta_2} \wedge y_2' \in \underline{\eta_1} \wedge x_2 \in \underline{\eta_2}) \tag{7}$$

$$(\{x \mapsto y_1 + x_1' + x_2'\}, y_1 \in \underline{\eta_2} \wedge x_1' \in \underline{\eta_1} \wedge x_2' \in \underline{\eta_2}) \tag{8}$$

$$(\{x \mapsto y_1 + y_1' + y_2'\}, y_1 \in \underline{\eta_2} \wedge y_1' \in \underline{\eta_2} \wedge y_2' \in \underline{\eta_1}) \tag{9}$$

The constrained substitutions (6) and (9) may be simplifyed, yielding renamings of (4) and (5) respectively. The constrained substitutions (7) and (8) differ only by a renaming. At this point, the reader can check that no new simplified constrained substitution can be generated by repeating the process, hence

$$\begin{array}{ll} \{ & (\{x \mapsto x'\}, x' \in \underline{\eta_0}), (\{x \mapsto x'\}, x' \in \underline{\eta_1}), (\{x \mapsto x'\}, x' \in \underline{\eta_2}), \\ & (\{x \mapsto x_1 + x_2\}, x_1 \in \underline{\eta_1} \wedge x_2 \in \underline{\eta_2}), \\ & (\{x \mapsto y_1 + y_2\}, y_1 \in \underline{\eta_2} \wedge y_2 \in \underline{\eta_1}), \\ & (\{x \mapsto z_1 + z_2 + z_3\}, z_1 \in \underline{\eta_2} \wedge z_2 \in \underline{\eta_1} \wedge z_3 \in \underline{\eta_2}) \ \} \end{array}$$

is a complete set of normalizing substitutions for $\{x\}$ in $A \cup AC$.
It it easy to show that

$$\{(\sigma_1 \cdots \sigma_n, \phi_1 \wedge \cdots \wedge \phi_n) \mid (\sigma_i, \phi_i) \in CSNS\{x_i\}\}$$

is a CSNS for $\{x_1, \ldots, x_n\}$.

The solution of the problem $x + a \stackrel{?}{=} y + b \wedge x \in \underline{\eta_1} \wedge y \in \underline{\eta_1}$ shown in the introduction is computed by our algorithm by choosing the constrained substitution

$$(\{x \mapsto x_1 + x_2, y \mapsto y_1 + y_2\}, x_1 \in \underline{\eta_1} \wedge x_2 \in \underline{\eta_2} \wedge y_1 \in \underline{\eta_1} \wedge y_2 \in \underline{\eta_2}) \in CSNS(\{x, y\})$$

6 Conclusion

The study of the interactions of a theory and one of its overloaded extensions at the proof level has made us discover a mechanism (unnoticed so far) that plays a key role in presence of overloaded operators. As far as we know, this is the first solution to a problem of unification in a combination of equational theories over non-disjoint signatures. Separated normal forms should be used not only for unification but also for matching as well as for deciding equality in $Th_1 \cup Th_2$.

The method that we use here does not work for all theories. To apply our method, we first need to orient all the critical schemes. This will not be possible if Th_1 has an axiom like

$$(x + y) + (z + t) = (x + z) + (y + t)$$

Second, we need to compute a finite complete set of normalizing substitutions, and such a set does not always exist.

Our algorithm trivially applies to the combination of Th_1 and one of its overloaded extensions Th_2 whenever Th_1 is a *shallow theory* that is a theory presented by a set of axioms where all the variables are at depth at most 1. Indeed, in this case, there are no critical schemes, and it is just enough to guess nondeterministically the theory of each variable, and to apply the standard combination techniques. The method that we use here for computing a complete set of normalizing substitutions does not work in general. For instance, if Th_1 is the left distributivity, the process of instantiating a variable introduced by a normalizing substitution by a normalizing substitution will yield infinitely many incomparable constrained substitutions, even if a CSNS is finite.

Note that when Th_1 is a shallow theory (or whenever there are no critical schemes), our algorithm reduces to Kirchner's which is complete in this case. For instance, if Th_1 is the commutative theory of $+$ on $\underline{s_1}$ and Th_2 is the associative-commutative theory of $+$ on $\underline{s_2}$ with $\underline{s_2} < \underline{s_1}$, then the step 1 of our algorithm reduces to the choice of a theory for each variable, which is the base of Kirchner's method.

The algorithm might extend to non-regular or collapsing theories, but it will not be enough to adapt the techniques of [1, 15] for solving theory conflicts and cycles: a term of $\underline{\eta_1}$ may be equal to a term of $\underline{\eta_2}$, violating one of our main assumptions.

Acknowledgement: I thank J.-P. Jouannaud and H. Comon for many helpful discussions.

References

[1] Alexandre Boudet. Unification in combination of equational theories : an efficient algorithm. In *Proc. 10th Conf. on Automated Deduction, Kaiserslautern, LNCS 449*. Springer-Verlag, July 1990.

[2] Alexandre Boudet, Evelyne Contejean, and Hervé Devie. A new AC-unification algorithm with a new algorithm for solving diophantine equations. In *Proc. 5th IEEE Symp. Logic in Computer Science, Philadelphia*, June 1990.

[3] Hubert Comon. Inductive proofs by specifications transformation. In *Proc. 3rd Rewriting Techniques and Applications 89, Chapel Hill, LNCS 355*, pages 76–91. Springer-Verlag, April 1989.

[4] Hubert Comon. Equational formulas in order-sorted algebras. In *Proc. ICALP, Warwick*. Springer-Verlag, July 1990.

[5] Nachum Dershowitz and Jean-Pierre Jouannaud. Rewrite systems. In J. van Leeuwen, editor, *Handbook of Theoretical Computer Science*, volume B, pages 243–309. North-Holland, 1990.

[6] François Fages. Associative-commutative unification. *J. Symbolic Computation*, 3(3), June 1987.

[7] A. Frisch. A general framework for sorted deduction: Fundamental results for hybrid reasoning. In *Conf. on Principles of Knowledge Representation and Reasoning*. Morgan Kaufmann, 1989.

[8] K. Futatsugi, Joseph Goguen, Jean-Pierre Jouannaud, and J. Meseguer. Principles of OBJ2. In *Proc. 12th ACM Symp. Principles of Programming Languages, New Orleans*, 1985.

[9] Jean-Pierre Jouannaud and Claude Kirchner. Solving equations in abstract algebras: A rule-based survey of unification. In Jean-Louis Lassez and Gordon Plotkin, editors, *Computational Logic: Essays in Honor of Alan Robinson*. MIT-Press, 1991.

[10] Claude Kirchner. *Méthodes et Outils de Conception Systématique d'Algorithmes d'Unification dans les Théories equationnelles.* Thèse d'Etat, Univ. Nancy, France, 1985.

[11] Claude Kirchner. Order-sorted equational unification. In *Proc. 5th Int. Conference on Logic Programming, Seattle*, August 1988.

[12] Claude Kirchner, Hélène Kirchner, and José Meseguer. Operational semantics of OBJ-3. In *Proc. 15th ICALP, Tampere, LNCS 317*. Springer-Verlag, July 1988.

[13] A. Martelli and U. Montanari. An efficient unification algorithm. *ACM Transactions on Programming Languages and Systems*, 4(2):258–282, 1982.

[14] J. A. Robinson. A machine-oriented logic based on the resolution principle. *Journal of the ACM*, 12(1):23–41, 1965.

[15] M. Schmidt-Schauss. Unification in a combination of arbitrary disjoint equational theories. *J. Symbolic Computation*, 1990. Special issue on Unification.

[16] G. Smolka, W. Nutt, J. A. Goguen, and J. Meseguer. Order-Sorted Equational Computation. In Hassan Aït-Kaci and Maurice Nivat, editors, *Resolution of Equations in Algebraic Structures, Volume 2, Rewriting Techniques*, chapter 10, pages 297–367. Academic Press, New York, N.Y., 1989.

Puzzles and Paradoxes

Raymond M. Smullyan

Department of Philosophy
Indiana University
Bloomington, Indiana

Abstract

This will be a discussion of some new variants of known puzzles and paradoxes, with perhaps a bit of magic thrown in.

Experiments in Automated Deduction with Condensed Detachment*

William McCune and Larry Wos

Mathematics and Computer Science Division
Argonne National Laboratory
Argonne, Illinois 60439-4801
U.S.A.

Abstract

This paper contains the results of experiments with several search strategies on 112 problems involving condensed detachment. The problems are taken from nine different logic calculi: three versions of the two-valued sentential calculus, the many-valued sentential calculus, the implicational calculus, the equivalential calculus, the R calculus, the left group calculus, and the right group calculus. Each problem was given to the theorem prover OTTER and was run with at least three strategies: (1) a basic strategy, (2) a strategy with a more refined method for selecting clauses on which to focus, and (3) a strategy that uses the refined selection mechanism and deletes deduced formulas according to some simple rules. Two new features of OTTER are also presented: the refined method for selecting the next formula on which to focus, and a method for controlling memory usage.

1 Introduction

The aim of this paper is to examine the role of strategy in the study of logic calculi with condensed detachment. We present results of experiments with the theorem-proving program OTTER on 112 problems, all of which contain the axiom (or, from another point of view, inference rule) condensed detachment.

All of the problems concern axiomatizations of various logic calculi, including the two-valued sentential calculus and two of its variations, the many-valued sentential calculus, the implicational fragment of sentential calculus, equivalential calculus, and three subsystems of the equivalential calculus: the R calculus, the left group calculus, and the right group calculus. The problems should also serve well as test problems for evaluating other search strategies and other theorem-proving programs.

We have experimented extensively with most of the problems, and we have developed specialized strategies for particular logic calculi. For the experiments presented in this paper, however, we sought strategies that perform well on all of the problems.

*This work was supported by the Applied Mathematical Sciences subprogram of the Office of Energy Research, U.S. Department of Energy, under Contract W-31-109-Eng-38.

Aside from a default basic strategy, we experimented with a guidance strategy and a deletion strategy. The guidance strategy, which we call the ratio strategy (Section 1.2.1), combines best-first search with breadth-first search when selecting the next formula on which to focus. The deletion strategy (Section 1.2.2) causes derived formulas that are instances of simple patterns to be deleted.

1.1 Condensed Detachment

All of the problems use C. A. Meredith's *condensed detachment* [6, 14], a rule of inference that combines detachment (modus ponens) and instantiation. Let C be the binary operation of concern. If $C(\alpha, \beta)$ and γ are both theorems (renamed so that they have no variables in common), and if α and γ unify with most general unifier σ, then $\beta\sigma$ is deduced by condensed detachment. The formula $C(\alpha, \beta)$ is the *major premise*, and γ is the *minor premise*. The binary operation is usually interpreted as "implies", "equivalent", or some variation of the group operation, depending on the calculus.

The logic calculi can be studied as first-order theories by a trivial transformation [5]. First, a unary predicate P, interpreted as "is a theorem" or "is the group identity", is introduced. Then, each axiom of the calculus is preceded by P, with its variables universally quantified. Finally, condensed detachment becomes an axiom of the theory.

$$\forall x \forall y (P(C(x, y)) \ \&\ P(x) \ \ \rightarrow \ \ P(y)).$$

An application of hyperresolution with the axiom condensed detachment corresponds directly to an application of the inference rule condensed detachment. Although we used hyperresolution exclusively for the experiments presented in this paper, any inference rule for first-order logic is applicable.

The AN calculus, which is a variation of the two-valued sentential calculus, has a binary operation o, which can be interpreted as disjunction, and a unary operation n, which can be interpreted as negation. For the AN calculus, the following variation of condensed detachment is used:

$$\forall x \forall y (P(o(n(x), y)) \ \&\ P(x) \ \ \rightarrow \ \ P(y)).$$

The study of logic calculi with condensed detachment has been one of the first and most successful applications of automated theorem proving. Original research has been conducted and open questions have been answered by relying heavily on automated theorem proving programs [3, 16, 15, 4, 22, 18, 21, 11, 10].

1.2 OTTER and Simple Strategies

OTTER [9] is a resolution/paramodulation theorem-proving program for first-order logic with equality. Its basic algorithm, restricted to hyperresolution with condensed detachment, is shown in Figure 1.

1.2.1 Selecting the Given Clause

Our default strategy for selecting the given clause in Step 1 of the basic algorithm has traditionally been to select a clause with the fewest symbols; if there is more

Start with *sos* list containing all axioms and with *usable* list containing the axiom for condensed detachment.
Loop:
1. G = select-given-clause(*sos*);
2. move G from *sos* to *usable*;
3. apply condensed detachment as much as possible, with G as one premise, taking the other premise from *usable*; append to *sos* the results that are not subsumed by anything in *sos* or *usable*;

end loop.

Figure 1: OTTER's Basic Algorithm with Condensed Detachment

than one clause of minimum length, the first of those is selected. We call the default strategy "selecting the smallest clause as given". However, some problems in the logic calculi yield quickly to a breadth-first search, which is accomplished by selecting the first clause in the *sos* list as the given clause. The method we use for most of the experiments presented in this paper combines those two methods. In every fourth iteration of the loop, the first clause is selected, and in the remaining iterations, the first clause of minimum length is selected. We call this refined method "selecting the given clause with ratio 3". The refinement allows large clauses to enter the search while the focus remains mainly on small clauses. It is similar to a selection strategy used by J. Kalman in one of his early programs [3].

1.2.2 Deleting Derived Formulas

In the equivalential calculus, the R calculus, and the left and right group calculi (all of which have binary operator e), we found that formulas containing subformulas that are instances of $e(x, x)$ are generally not as useful or as powerful as formulas without such instances. Searches in which those formulas are deleted are generally more effective, although they can result in longer proofs. The strategy also applies to the implicational calculi by deleting deduced formulas with instances of $i(x, x)$, although it appears to be less effective there. The strategy applies to the AN calculus, in which the binary operation is disjunction, by deleting formulas with instances of $o(n(x), x)$.

In the calculi with unary operation n, meaning negation, we found that deduced formulas containing instances of $n(n(x))$ caused redundancy in the search spaces and that deleting those formulas generally improved the searches. We also ran experiments deleting formulas with instances of $n(n(n(x)))$.

We used demodulation of derived formulas to implement the deletion strategy. When the strategy was in use, demodulation usually accounted for between one third and one half of the CPU time.

1.2.3 Controlling Memory Usage

We limited the OTTER jobs to 12 Mbytes of memory, in which OTTER can store roughly 20,000 formulas. Even with the deletion strategy of the preceding subsection,

OTTER quickly fills 12 Mbytes. The list *sos* typically grows much faster than does the number of given clauses that are removed from it. Thus, most formulas in *sos* never enter the search, and memory is wasted.

Our current solution to that problem is the following. When one third of available memory has been filled, we impose a limit on the number of symbols in deduced clauses. The limit, say n, is such that 5% of all formulas in *sos* have $\leq n$ symbols. Every tenth iteration of the main loop after the initial limit has been set, calculate a prospective new limit n' in the same way. If $n' < n$, then the limit is reset to n'. We arrived at the values 1/3 and 5% by trial and error. Although this method is incomplete, its use with condensed detachment problems typically does not have a great effect on the sequence of given clauses, or therefore, on the search. We have not experimented heavily with this method on other problems.

1.3 The Experiments

We ran all of the experiments on SPARCstation 1+ computers with 16 megabytes of main memory. In that environment, OTTER can infer several thousand formulas per second, most of which are deleted because they are subsumed by existing formulas or by the deletion strategy. (Back subsumption, in which newly kept formulas cause the deletion of weaker existing formulas, was not used.)

Here is an example of the way in which problems are presented. Given the equivalential calculus formulas

(EC-1)	$e(e(e(x,y),e(z,x)),e(y,z))$
(EC-4)	$e(e(x,y),e(y,x))$
(EC-5)	$e(e(e(x,y),z),e(x,e(y,z)))$

the problem (EC-4,EC-5 ⇒ EC-1) is to find a refutation of the following set of clauses. Symbols x, y, and z are variables, and a, b, and c are Skolem constants.

$\neg P(e(x,y)) \mid \neg P(x) \mid P(y)$.	% Condensed Detachment
$P(e(e(x,y),e(y,x)))$.	% EC-4
$P(e(e(e(x,y),z),e(x,e(y,z))))$.	% EC-5
$\neg P(e(e(e(a,b),e(c,a)),e(b,c)))$.	% Denial of EC-1

Each problem was run with several strategies with a time limit of four hours each. In the tables that follow, "Fail" indicates that no proof was found within four hours, and "*" indicates that no proof is possible, because the goal would be deleted by the deletion strategy. All of the times are given in seconds. The strategies are the following.

Basic. The smallest formula is selected as given.

Ratio. Given clauses are selected with ratio 3.

R-e. Given clauses are selected with ratio 3, and deduced clauses containing an instance of $e(x,x)$ as a subformula are deleted.

R-i. Given clauses are selected with ratio 3, and deduced clauses containing an instance of $i(x,x)$ as a subformula are deleted.

R-nn. Given clauses are selected with ratio 3, and deduced clauses containing an instance of $n(n(x))$ are deleted.

R-nnn. Given clauses are selected with ratio 3, and deduced clauses containing an instance of $n(n(n(x)))$ are deleted.

R-i-nn. Given clauses are selected with ratio 3, and deduced clauses containing an instance of $i(x, x)$ as a subformula or an instance of $n(n(x))$ are deleted.

R-i-nnn. Given clauses are selected with ratio 3, and deduced clauses containing an instance of $i(x, x)$ as a subformula or an instance of $n(n(n(x)))$ are deleted.

R-o-nn. Given clauses are selected with ratio 3, and deduced clauses containing an instance of $o(n(x), x)$ as a subformula or an instance of $n(n(x))$ are deleted.

R-o-nnn. Given clauses are selected with ratio 3, and deduced clauses containing an instance of $o(n(x), x)$ as a subformula or an instance of $n(n(n(x)))$ are deleted.

We present here a sequence of problems that reflects a wide range of difficulty and that roughly follows the historical development of the individual calculi.

2 Two-Valued Sentential Calculi

We experimented with three versions of two-valued sentential calculus: (1) the CN calculus, with operators intended to mean implication and negation, (2) the C0 calculus, with implication and falsehood, and (3) the AN calculus, with disjunction and negation. If appropriate definitions are added for the missing operators, each version is equivalent to the classical propositional calculus.

2.1 The Implication/Negation Two-Valued Sentential Calculus (CN)

Each of the following formulas holds in the two-valued sentential calculus (CN). The numbering of the formulas is from [7, p. 42–51].

(CN-1)	$i(i(x, y), i(i(y, z), i(x, z)))$
(CN-2)	$i(i(n(x), x), x)$
(CN-3)	$i(x, i(n(x), y))$
(CN-16)	$i(x, x)$
(CN-18)	$i(x, i(y, x))$
(CN-19)	$i(i(i(x, y), z), i(y, z))$
(CN-20)	$i(x, i(i(x, y), y))$
(CN-21)	$i(i(x, i(y, z)), i(y, i(x, z)))$
(CN-22)	$i(i(x, y), i(i(z, x), i(z, y)))$
(CN-24)	$i(i(i(x, y), x), x)$
(CN-30)	$i(i(x, i(x, y)), i(x, y))$
(CN-35)	$i(i(x, i(y, z)), i(i(x, y), i(x, z)))$
(CN-37)	$i(i(i(x, y), z), i(n(x), z))$
(CN-39)	$i(n(n(x)), x)$

(CN-40)	$i(x,n(n(x)))$
(CN-46)	$i(i(x,y),i(n(y),n(x)))$
(CN-49)	$i(i(n(x),n(y)),i(y,x))$
(CN-54)	$i(i(x,y),i(i(n(x),y),y))$
(CN-59)	$i(i(n(x),z),i(i(y,z),i(i(x,y),z)))$
(CN-60)	$i(i(x,i(n(y),z)),i(x,i(i(u,z),i(i(y,u),z))))$
(CN-CAM)	$i(i(i(i(i(x,y),i(n(z),n(u))),z),v),i(i(v,x),i(u,x)))$

According to Lukasiewicz [8, p. 136], the first axiom system for the two-valued sentential calculus was {CN-18,CN-21,CN-35,CN-39,CN-40,CN-46} and was due to Frege. We use that as our starting point. Lukasiewicz showed that CN-21 depends on the remaining axioms of Frege's system (Problem 1, Table 1). Another early

Table 1: CN Calculus, Frege and Hilbert Systems

#	Theorem	Basic	Ratio	R-i-nn	R-i-nnn	R-nn	R-nnn
1	CN-18,CN-35,CN-39,CN-40,CN-46 ⇒ CN-21	Fail	246	16	176	26	254
2	CN-3,CN-18,CN-21,CN-22,CN-54 ⇒ CN-30	3	5	2	2	6	6
3	CN-3,CN-18,CN-21,CN-22,CN-54 ⇒ CN-35	Fail	Fail	8005	6371	3657	3864
4	CN-3,CN-18,CN-21,CN-22,CN-54 ⇒ CN-39	<1	9	*	8	*	11
5	CN-3,CN-18,CN-21,CN-22,CN-54 ⇒ CN-40	14	10	*	34	*	13
6	CN-3,CN-18,CN-21,CN-22,CN-54 ⇒ CN-46	7366	1534	1467	1509	2857	2401

axiomatization of CN was due to Hilbert [8, p. 136]: {CN-3,CN-18,CN-21,CN-22,CN-30,CN-54}. Lukasiewicz showed that CN-30 is not necessary (Problem 2, Table 1). Problems 3–6, Table 1, are to derive Frege's simplified system from Hilbert's simplified system.

Lukasiewicz axiomatized CN with {CN-1,CN-2,CN-3} [7]. Other axiom systems for CN are {CN-18,CN-35,CN-49} (Church [1]), {CN-19,CN-37,CN-59} (Lukasiewicz [7]), {CN-19,CN-37,CN-60} (Wos [19]), and {CN-CAM} (C. A. Meredith [12]). Problems 7–24, Table 2, are to start with {CN-1,CN-2,CN-3} and derive formulas in the other axiomatizations. Problems 25–36, Table 3, are to derive Lukasiewicz's system {CN-1,CN-2,CN-3} from the other systems.

2.2 The Implication/Falsehood Two-Valued Sentential Calculus (C0)

Each of the following formulas holds in the C0 calculus:

(C0-1)	$i(i(x,y),i(i(y,z),i(x,z)))$
(C0-2)	$i(x,i(y,x))$
(C0-3)	$i(i(i(x,y),x),x)$
(C0-4)	$i(F,x)$
(C0-5)	$i(i(i(x,F),F),x)$
(C0-6)	$i(i(x,i(y,z)),i(i(x,y),i(x,z)))$
(C0-CAM)	$i(i(i(i(i(x,y),i(z,F)),u),v),i(i(v,x),i(z,x)))$

Each of the sets {C0-1,C0-2,C0-3,C0-4} (Tarski-Bernays, according to [12]), {C0-2,C0-5,C0-6} (Church [1]), and {C0-CAM} (C. A. Meredith [12]) axiomatizes the

Table 2: CN Calculus, Starting with {CN-1,CN-2,CN-3}

#	Theorem	Basic	Ratio	R-i-nn	R-i-nnn	R-nn	R-nnn
7	CN-1,CN-2,CN-3 ⇒ CN-16	<1	<1	<1	<1	<1	<1
8	CN-1,CN-2,CN-3 ⇒ CN-18	60	7	5	5	10	9
9	CN-1,CN-2,CN-3 ⇒ CN-19	60	7	5	5	10	9
10	CN-1,CN-2,CN-3 ⇒ CN-20	89	147	23	28	65	157
11	CN-1,CN-2,CN-3 ⇒ CN-21	104	148	23	28	66	158
12	CN-1,CN-2,CN-3 ⇒ CN-22	105	589	74	184	177	595
13	CN-1,CN-2,CN-3 ⇒ CN-24	105	71	31	40	40	86
14	CN-1,CN-2,CN-3 ⇒ CN-30	109	71	32	45	40	86
15	CN-1,CN-2,CN-3 ⇒ CN-35	Fail	Fail	Fail	Fail	Fail	Fail
16	CN-1,CN-2,CN-3 ⇒ CN-37	105	33	31	40	34	40
17	CN-1,CN-2,CN-3 ⇒ CN-39	104	31	*	37	*	41
18	CN-1,CN-2,CN-3 ⇒ CN-40	106	32	*	38	*	42
19	CN-1,CN-2,CN-3 ⇒ CN-46	1021	1262	423	1434	470	1378
20	CN-1,CN-2,CN-3 ⇒ CN-49	260	73	31	36	45	88
21	CN-1,CN-2,CN-3 ⇒ CN-54	1195	1608	447	1552	509	1763
22	CN-1,CN-2,CN-3 ⇒ CN-59	Fail	Fail	Fail	Fail	Fail	Fail
23	CN-1,CN-2,CN-3 ⇒ CN-60	Fail	Fail	Fail	Fail	Fail	Fail
24	CN-1,CN-2,CN-3 ⇒ CN-CAM	Fail	Fail	Fail	Fail	Fail	Fail

Table 3: CN Calculus, Deriving {CN-1,CN-2,CN-3}

#	Theorem	Basic	Ratio	R-i-nn	R-i-nnn	R-nn	R-nnn
25	CN-18,CN-35,CN-49 ⇒ CN-1	Fail	1083	89	531	91	1137
26	CN-18,CN-35,CN-49 ⇒ CN-2	4	3	8	11	1	4
27	CN-18,CN-35,CN-49 ⇒ CN-3	3	1	3	3	12	2
28	CN-19,CN-37,CN-59 ⇒ CN-1	6038	303	89	245	99	286
29	CN-19,CN-37,CN-59 ⇒ CN-2	622	359	3107	6800	257	592
30	CN-19,CN-37,CN-59 ⇒ CN-3	161	12	5	12	6	15
31	CN-19,CN-37,CN-60 ⇒ CN-1	5611	515	493	682	480	702
32	CN-19,CN-37,CN-60 ⇒ CN-2	753	546	Fail	Fail	511	755
33	CN-19,CN-37,CN-60 ⇒ CN-3	239	224	345	337	329	332
34	CN-CAM ⇒ CN-1	Fail	Fail	Fail	Fail	Fail	Fail
35	CN-CAM ⇒ CN-2	Fail	Fail	Fail	Fail	Fail	Fail
36	CN-CAM ⇒ CN-3	6	10	Fail	Fail	14	14

C0 calculus. Problems 37–49, Table 4, involve deriving each axiom system from the others.

Table 4: The C0 Calculus

#	Theorem	Basic	Ratio	R-i
37	C0-2,C0-5,C0-6 ⇒ C0-1	419	72	54
38	C0-2,C0-5,C0-6 ⇒ C0-3	337	103	98
39	C0-2,C0-5,C0-6 ⇒ C0-4	<1	<1	<1
40	C0-2,C0-5,C0-6 ⇒ C0-CAM	Fail	Fail	Fail
41	C0-1,C0-2,C0-3,C0-4 ⇒ C0-5	38	7	7
42	C0-1,C0-2,C0-3,C0-4 ⇒ C0-6	1251	953	1010
43	C0-1,C0-2,C0-3,C0-4 ⇒ C0-CAM	Fail	Fail	Fail
44	C0-CAM ⇒ C0-1	Fail	Fail	Fail
45	C0-CAM ⇒ C0-2	<1	<1	<1
46	C0-CAM ⇒ C0-3	13	24	30
47	C0-CAM ⇒ C0-4	<1	<1	<1
48	C0-CAM ⇒ C0-5	5	9	12
49	C0-CAM ⇒ C0-6	Fail	Fail	Fail

2.3 The Disjunction/Negation Two-Valued Sentential Calculus (AN)

Each of the following formulas holds in the AN calculus:

(AN-1) $o(n(o(n(y),z)),o(n(o(x,y)),o(x,z)))$
(AN-2) $o(n(o(x,y)),o(y,x))$
(AN-3) $o(n(x),o(y,x))$
(AN-4) $o(n(o(x,x)),x)$
(AN-CAM) $o(n(o(n(o(n(x),y)),o(z,o(u,v)))),o(n(o(n(u),x)),o(z,o(v,x))))$

Each of the sets {AN-1,AN-2,AN-3,AN-4} (Whitehead-Russell, according to [12]) and {AN-CAM} (C. A. Meredith [12]) axiomatizes the AN calculus. Problems 50–54, Table 5, are to derive each system from the other. Recall that the clause form of condensed detachment for the AN calculus is $\neg P(o(n(x),y)) \mid \neg P(x) \mid P(y)$.

Table 5: The AN Calculus

#	Theorem	Basic	Ratio	R-o-nn	R-o-nnn	R-nn	R-nnn
50	AN-1,AN-2,AN-3,AN-4 ⇒ AN-CAM	Fail	Fail	Fail	Fail	Fail	Fail
51	AN-CAM ⇒ AN-1	Fail	Fail	Fail	Fail	516	Fail
52	AN-CAM ⇒ AN-2	3472	Fail	Fail	Fail	449	5365
53	AN-CAM ⇒ AN-3	34	58	133	78	137	78
54	AN-CAM ⇒ AN-4	11447	Fail	Fail	Fail	2657	Fail

3 The Many-Valued Sentential Calculus (MV)

Each of the following formulas holds in the many-valued sentential calculus:

(MV-1) $i(x,i(y,x))$
(MV-2) $i(i(x,y),i(i(y,z),i(x,z)))$
(MV-3) $i(i(i(x,y),y),i(i(y,x),x))$
(MV-4) $i(i(i(x,y),i(y,x)),i(y,x))$
(MV-5) $i(i(n(x),n(y)),i(y,x))$
(MV-24) $i(n(n(x)),x)$
(MV-25) $i(i(x,y),i(i(z,x),i(z,y)))$
(MV-29) $i(x,n(n(x)))$
(MV-33) $i(i(n(x),y),i(n(y),x))$
(MV-36) $i(i(x,y),i(n(y),n(x)))$
(MV-39) $i(n(i(x,y)),n(y))$
(MV-50) $i(n(x),i(y,n(i(y,x))))$

Łukasiewicz defined the many-valued sentential calculus $L_{\aleph_0}$ and conjectured that it is axiomatized by {MV-1,MV-2,MV-3,MV-4,MV-5} [8]. Wajsberg proved the conjecture, and C. A. Meredith later proved MV-4 dependent on the remaining axioms [8, p. 144]. Problems 55–62, Table 6, are to prove MV-4 and several other formulas from {MV-1,MV-2,MV-3,MV-5}. (Problem 55 has been called "Luka5" by members of the Argonne group.)

Table 6: The MV Calculus

#	Theorem	Basic	Ratio	R-i-nn	R-i-nnn	R-nn	R-nnn
55	MV-1,MV-2,MV-3,MV-5 ⇒ MV-4	Fail	Fail	Fail	Fail	Fail	Fail
56	MV-1,MV-2,MV-3,MV-5 ⇒ MV-24	3	2	*	8	*	2
57	MV-1,MV-2,MV-3,MV-5 ⇒ MV-25	4475	8	5	5	9	9
58	MV-1,MV-2,MV-3,MV-5 ⇒ MV-29	3	2	*	8	*	2
59	MV-1,MV-2,MV-3,MV-5 ⇒ MV-33	Fail	2036	1468	2665	1827	3955
60	MV-1,MV-2,MV-3,MV-5 ⇒ MV-36	Fail	2035	3138	2664	3812	3955
61	MV-1,MV-2,MV-3,MV-5 ⇒ MV-39	7	17	675	25	628	16
62	MV-1,MV-2,MV-3,MV-5 ⇒ MV-50	Fail	2041	3151	2674	3825	3964

4 The Implicational Propositional Calculus (IC)

The implicational propositional calculus (IC) is the part of the sentential calculus in which the negation operation does not occur. Each of the following formulas holds in IC:

(IC-1) $i(x,x)$
(IC-2) $i(x,i(y,x))$
(IC-3) $i(i(i(x,y),x),x)$
(IC-4) $i(i(x,y),i(i(y,z),i(x,z)))$
(IC-5) $i(x,i(i(x,y),y))$
(IC-JL) $i(i(i(x,y),z),i(i(z,x),i(u,x)))$

Each of the sets {IC-2,IC-3,IC-4} (Tarski-Bernays, according to [8, p. 296]) and {IC-JL} (Łukasiewicz [8, p. 295]) axiomatizes IC. Problems 63–68, Table 7, are to derive each system from the other.

Table 7: The Implicational Propositional Calculus

#	Theorem	Basic	Ratio	R-e
63	IC-2,IC-3,IC-4 ⇒ IC-JL	50	101	100
64	IC-JL ⇒ IC-1	8	<1	27
65	IC-JL ⇒ IC-2	8	<1	<1
66	IC-JL ⇒ IC-3	32	47	26
67	IC-JL ⇒ IC-4	7933	13985	12224
68	IC-JL ⇒ IC-5	2172	3753	2715

5 Equivalential and Group Calculi

The equivalential and group calculi have one binary operator, e. In the equivalential calculus (EC) [8], $e(\alpha, \beta)$ is normally interpreted as equivalence of α and β; however, it can also be interpreted as the group operation $\alpha\beta$ in Boolean groups (groups in which the square of every element is the identity). Under the group interpretation, the theorems of EC are exactly the formulas that are equal to the group identity in Boolean groups.

The theorems of the R calculus [13] are exactly the formulas equal to the identity in Abelian groups when $e(\alpha, \beta)$ is interpreted as $\alpha\beta^{-1}$. There is also an L calculus, whose theorems are equal to the identity when $e(\alpha, \beta)$ is interpreted as $\alpha^{-1}\beta$. We have not experimented with the L calculus, but for completeness, we list here YOL, the shortest single axiom for the L calculus [16]. No other of length 11 exists.

(YOL) $\quad e(e(x,y),e(e(e(z,y),x),z))$

The theorems of the left group (LG) calculus [2] are exactly the formulas equal to the identity in (general) groups when $e(\alpha, \beta)$ is interpreted as $\alpha^{-1}\beta$. Similarly, the theorems of the right group (RG) calculus [2] are exactly the formulas equal to the identity in (general) groups when $e(\alpha, \beta)$ is interpreted as $\alpha\beta^{-1}$.

The following relationships exist between the equivalential and group calculi:

LG theorems ⊂ L theorems ⊂ EC theorems.
RG theorems ⊂ R theorems ⊂ EC theorems.

5.1 The Equivalential Calculus (EC)

The following formulas hold in the equivalential calculus:

(EC-1) $\quad e(e(e(x,y),e(z,x)),e(y,z))$
(EC-2) $\quad e(e(x,e(y,z)),e(e(x,y),z))$
(EC-4) $\quad e(e(x,y),e(y,x))$
(EC-5) $\quad e(e(e(x,y),z),e(x,e(y,z)))$

According to Łukasiewicz [8, p. 252], the first axiomatization of EC was {EC-1,EC-2}, due to Leśniewski. Soon after, Wajsberg produced others, including {EC-4,EC-5}. Problems 69 and 70, Table 8, are to derive the Leśniewski system from the Wajsberg system.

Each of the following formulas is a single axiom for EC, in roughly the order in which they were discovered. None shorter exists, nor does there exist any other of length 11.

(YQL)	$e(e(x,y),e(e(z,y),e(x,z)))$	Łukasiewicz
(YQF)	$e(e(x,y),e(e(x,z),e(z,y)))$	Łukasiewicz
(YQJ)	$e(e(x,y),e(e(z,x),e(y,z)))$	Łukasiewicz
(UM)	$e(e(e(x,y),z),e(y,e(z,x)))$	Meredith
(XGF)	$e(x,e(e(y,e(x,z)),e(z,y)))$	Meredith
(WN)	$e(e(x,e(y,z)),e(z,e(x,y)))$	Meredith
(YRM)	$e(e(x,y),e(z,e(e(y,z),x)))$	Meredith
(YRO)	$e(e(x,y),e(z,e(e(z,y),x)))$	Meredith
(PYO)	$e(e(e(x,e(y,z)),z),e(y,x))$	Meredith
(PYM)	$e(e(e(x,e(y,z)),y),e(z,x))$	Meredith
(XGK)	$e(x,e(e(y,e(z,x)),e(z,y)))$	Kalman
(XHK)	$e(x,e(e(y,z),e(e(x,z),y)))$	Winker
(XHN)	$e(x,e(e(y,z),e(e(z,x),y)))$	Winker

Problems 71–84, Table 8, are to start with each single axiom and derive the system that precedes it.

Table 8: EC

#	Theorem	Basic	Ratio	R-e
69	EC-4,EC-5 ⇒ EC-1	244	366	279
70	EC-4,EC-5 ⇒ EC-2	<1	<1	<1
71	YQL ⇒ EC-4	<1	<1	<1
72	YQL ⇒ EC-5	23	2	2
73	YQF ⇒ YQL	2	5	2
74	YQJ ⇒ YQF	34	54	33
75	UM ⇒ YQJ	558	1074	159
76	XGF ⇒ UM	<1	<1	<1
77	WN ⇒ XGF	98	164	85
78	YRM ⇒ WN	326	474	425
79	YRO ⇒ YRM	188	250	151
80	PYO ⇒ YRO	281	592	516
81	PYM ⇒ PYO	245	449	352
82	XGK ⇒ PYM	Fail	Fail	499
83	XHK ⇒ XGK	Fail	Fail	886
84	XHN ⇒ XHK	750	1690	484

5.2 The R Calculus (R)

Each of the following formulas is a single axiom for the R calculus:

(QYF)	$e(e(e(x,y),e(x,z)),e(z,y))$	Meredith
(YQM)	$e(e(x,y),e(e(z,y),e(z,x)))$	Meredith
(WO)	$e(e(x,e(y,z)),e(z,e(y,x)))$	Meredith
(XGJ)	$e(x,e(e(y,e(z,x)),e(y,z)))$	Winker

Problems 85–88, Table 9, are to show the four formulas equivalent in a circular manner.

Table 9: R Calculus

#	Theorem	Basic	Ratio	R-e
85	YQM $\Rightarrow$ QYF	<1	<1	<1
86	WO $\Rightarrow$ YQM	21	11	5
87	XGJ $\Rightarrow$ WO	Fail	Fail	362
88	QYF $\Rightarrow$ XGJ	41	42	21

5.3 The Left Group Calculus (LG)

Kalman's axiomatization of the LG calculus is {LG-1,LG-2,LG-3,LG-4,LG-5} [2].

(LG-1) $e(e(e(x,e(e(y,y),x)),z),z)$
(LG-2) $e(e(e(e(e(x,y),e(x,z)),e(y,z)),u),u)$
(LG-3) $e(e(e(e(e(e(x,y),e(x,z)),u),e(e(y,z),u)),v),v)$
(LG-4) $e(e(e(e(x,y),z),u),e(e(e(x,v),z),e(e(y,v),u)))$
(LG-5) $e(e(e(x,e(e(y,x),z)),e(e(u,x),v)),e(e(e(e(x,y),u),z),v))$

(P-1) $e(e(e(x,y),z),e(e(u,y),e(e(x,u),z)))$
(P-4) $e(x,e(e(e(e(y,z),e(y,u)),e(z,u)),x))$
(Q-1) $e(x,e(e(y,z),e(e(z,y),x)))$
(Q-2) $e(e(x,y),e(e(z,x),e(z,y)))$
(Q-3) $e(e(e(x,y),e(e(y,x),z)),z)$
(Q-4) $e(e(e(x,y),e(x,z)),e(y,z))$
(LG-27-1690) $e(e(e(e(x,y),z),e(e(u,v),e(e(e(w,v),e(w,u)),s))),e(z,e(e(y,x),s)))$

With great assistance from OTTER, McCune later showed that each of the sets {LG-2,LG-3}, {LG-2,P-1}, {LG-2,P-4}, {LG-2,Q-1,Q-2}, {P-1,Q-3}, {P-4,Q-3}, {Q-1,Q-2,Q-3}, {Q-1,Q-3,Q-4}, and {LG-27-1690} also axiomatizes the LG calculus [11, 10]. Problems 89–101, Table 10, roughly parallel the discovery of the new axiom systems for the LG calculus.

5.4 The RG Calculus (RG)

Kalman's axiomatization of the RG calculus is {LG-1′,LG-2′,LG-3′,LG-4′,LG-5′} [2].

(LG-1′) $e(x,e(x,e(e(y,e(z,z)),y)))$
(LG-2′) $e(x,e(x,e(e(y,z),e(e(y,u),e(z,u)))))$
(LG-3′) $e(x,e(x,e(e(y,e(z,u)),e(y,e(e(z,v),e(u,v))))))$
(LG-4′) $e(e(e(x,e(y,z)),e(u,e(y,v))),e(x,e(u,e(z,v))))$
(LG-5′) $e(e(x,e(y,e(z,e(u,v)))),e(e(x,e(v,z)),e(e(y,e(v,u)),v)))$

(Q-2′) $e(e(e(x,y),e(z,y)),e(x,z))$

Table 10: LG Calculus

#	Theorem	Basic	Ratio	R-e
89	LG-2,LG-3,LG-4 ⇒ LG-1	115	115	*
90	LG-2,LG-3 ⇒ LG-4	222	6	2
91	LG-3 ⇒ LG-4	109	8	9
92	LG-2,LG-3 ⇒ LG-5	Fail	Fail	845
93	LG-2,P-1 ⇒ LG-3	Fail	Fail	536
94	LG-2,P-4 ⇒ P-1	<1	<1	<1
95	LG-2,Q-1,Q-2 ⇒ P-1	16	30	5
96	P-1,Q-3 ⇒ LG-2	211	295	15
97	P-4,Q-3 ⇒ P-1	<1	<1	1
98	Q-1,Q-2,Q-3 ⇒ LG-2	10905	Fail	7617
99	Q-1,Q-4 ⇒ Q-2	<1	<1	<1
100	LG-27-1690 ⇒ P-1	<1	<1	<1
101	LG-27-1690 ⇒ Q-3	<1	2	3

(Q-3′) $e(x, e(e(x, e(y, z)), e(z, y)))$
(Q-4′) $e(e(x, y), e(e(x, z), e(y, z)))$

With great assistance from OTTER, McCune later showed that each of the pairs {Q-2′,Q-3′} and {Q-3′,Q-4′} axiomatizes the RG calculus and that each of the following formulas is a single axiom for the RG calculus [11, 10]:

(LG-2′) $e(x, e(x, e(e(y, z), e(e(y, u), e(z, u)))))$
(S-2′) $e(e(x, e(y, z)), e(x, e(e(y, u), e(z, u))))$
(S-3′) $e(x, e(x, e(e(e(y, z), e(u, z)), e(y, u))))$
(S-4′) $e(e(x, e(y, z)), e(e(x, e(u, z)), e(y, u)))$
(P-4′) $e(e(x, e(e(y, z), e(e(y, u), e(z, u)))), x)$
(S-6′) $e(e(x, e(e(e(y, z), e(u, z)), e(y, u))), x)$

Problems 102–112, Table 11, roughly parallel the discovery of the new axiom systems for the RG calculus.

6 Summary

We have presented 112 condensed detachment problems that offer a large range of difficulty to automated theorem-proving programs, and we have shown how OTTER, using several simple strategies (see Section 1.3), performs on those problems.

For the equivalential, R, RG, and LG calculi (the problems with functor e), strategy R-e wins on nearly all problems. We note that the deletion in strategy R-e prevents proofs in problems 89 and 102. For the CN, C0, and MV calculi, no clear overall winner was found. For the AN calculus problems, strategy R-nn performed best. For the IC problems, the basic strategy performed best. Although we did not run experiments using deletion while selecting the smallest clause as given, we can compare the performance of basic and ratio strategies (both without deletion). No

Table 11: RG Calculus

#	Theorem	Basic	Ratio	R-e
102	LG-2′ ⇒ LG-1′	130	133	*
103	LG-2′ ⇒ LG-3′	Fail	Fail	104
104	LG-2′ ⇒ LG-4′	Fail	889	62
105	LG-2′ ⇒ LG-5′	Fail	Fail	809
106	Q-2′,Q-3′ ⇒ LG-2′	9609	Fail	5634
107	Q-3′,Q-4′ ⇒ Q-2′	<1	<1	<1
108	S-2′ ⇒ LG-2′	757	1495	136
109	S-3′ ⇒ LG-2′	91	142	40
110	S-4′ ⇒ LG-2′	5837	Fail	Fail[a]
111	P-4′ ⇒ LG-2′	<1	<1	<1
112	S-6′ ⇒ LG-2′	120	143	19

[a]The deletion strategy eliminates *all* interesting paths.

clear winner was found, but the ratio strategy performed slightly better than the basic strategy overall. The results of the experiments reinforce our long-held position that a single strategy cannot be effective on a wide range of problems.

Several of the problems have been particularly challenging for us. Problem 67, posed as a challenge problem in [17] and called "imp4" by members of the Argonne group, was the first truly difficult condensed detachment theorem proved by OTTER. It has been used extensively as a benchmark for parallel deduction programs. Problem 34, to derive CN-1 from CN-CAM, has resisted all of our attempts at automated proofs. (One attempt generated 1.4 billion formulas and consumed 17 CPU days on a Solbourne 5e/900 computer.) Problem 55, to show the dependence of MV-4 in Lukasiewicz's system for $L_{\aleph_0}$, has also resisted all of our attempts. (One attempt generated 983 million formulas.) We have, however, found many proofs for Problem 55 using OTTER in various proof-checking modes [20]. OTTER's search for a proof for Problem 44, to derive C01 from C0-CAM, is impeded by the memory control feature. The weight limit is lowered, either too much or too soon, which causes key formulas to be discarded. OTTER has found a proof in about seven hours with a strategy similar to the basic strategy but with a constant weight limit of 18 instead of the memory control feature. We have not obtained proofs for problems 24, 40, 43, and 50, which are to derive complicated single axioms, because our strategies are biased towards finding simple formulas. The remaining problems for which the tables list complete failure have yielded to specialized strategies.

References

[1] A. Church. *Introduction to Mathematical Logic*, volume I. Princeton University Press, 1956.

[2] J. A. Kalman. Axiomatizations of logics with values in groups. *Journal of the London Math. Society*, 2(14):193–199, 1975.

[3] J. A. Kalman. Computer studies of $T_{\rightarrow} - W - I$. *Relevance Logic Newsletter*, 1:181–188, 1976.

[4] J. A. Kalman. A shortest single axiom for the classical equivalential calculus. *Notre Dame Journal of Formal Logic*, 19:141–144, 1978.

[5] J. A. Kalman. Condensed detachment as a rule of inference. *Studia Logica*, LXII(4):443–451, 1983.

[6] E. J. Lemmon, C. A. Meredith, D. Meredith, A. N. Prior, and I. Thomas. Calculi of pure strict implication. Technical report, Canterbury University College, Christchurch, 1957. Reprinted in *Philosophical Logic*, Reidel, 1970.

[7] J. Lukasiewicz. *Elements of Mathematical Logic*. Pergamon Press, 1963. English translation of the second edition (1958) of *Elementy logiki matematycznej*, PWN, Warsaw.

[8] J. Lukasiewicz. *Selected Works*. North-Holland, 1970. Edited by L. Borkowski.

[9] W. McCune. OTTER 2.0 Users Guide. Tech. Report ANL-90/9, Argonne National Laboratory, Argonne, IL, March 1990.

[10] W. McCune. Automated discovery of new axiomatizations of the left group and right group calculi. Preprint MCS-P220-0391, Mathematics and Computer Science Division, Argonne National Laboratory, Argonne, IL, 1991.

[11] W. McCune. Single axioms for the left group and right group calculi. Preprint MCS-P219-0391, Mathematics and Computer Science Division, Argonne National Laboratory, Argonne, IL, 1991.

[12] C. A. Meredith. Single axioms for the systems (C,N), (C,0), and (A,N) of the two-valued propositional calculus. *Journal of Computing Systems*, 1:155–164, 1953.

[13] C. A. Meredith and A. N. Prior. Equational logic. *Notre Dame Journal of Formal Logic*, 9:212–226, 1968.

[14] D. Meredith. In memoriam Carew Arthur Meredith (1904–1976). *Notre Dame Journal of Formal Logic*, 18:513–516, 1977.

[15] J. G. Peterson. Shortest single axioms for the classical equivalential calculus. *Notre Dame Journal of Formal Logic*, 17(2):267–271, 1976.

[16] J. G. Peterson. The possible shortest single axioms for EC-tautologies. Report 105, Dept. of Mathematics, University of Auckland, 1977.

[17] F. Pfenning. Single axioms in the implicational propositional calculus. In E. Lusk and R. Overbeek, editors, *Proceedings of the 9th International Conference on Automated Deduction, Lecture Notes in Computer Science, Vol. 310*, pages 710–713, New York, 1988. Springer-Verlag.

[18] L. Wos. Meeting the challenge of fifty years of logic. *Journal of Automated Reasoning*, 6(2):213–232, 1990.

[19] L. Wos. Automated reasoning and Bledsoe's dream for the field. In R. S. Boyer, editor, *Automated Reasoning: Essays in Honor of Woody Bledsoe*, chapter 15, pages 297–342. Kluwer Academic Publishers, 1991.

[20] L. Wos and W. McCune. The application of automated reasoning to proof translation and to finding proofs with specified properties: A case study in many-valued sentential calculus. Tech. Report ANL-91/19, Argonne National Laboratory, Argonne, IL, 1991. In preparation.

[21] L. Wos, S. Winker, W. McCune, R. Overbeek, E. Lusk, R. Stevens, and R. Butler. Automated reasoning contributes to mathematics and logic. In M. Stickel, editor, *Proceedings of the 10th International Conference on Automated Deduction, Lecture Notes in Artificial Intelligence, Vol. 449*, pages 485–499, New York, July 1990. Springer-Verlag.

[22] L. Wos, S. Winker, B. Smith, R. Veroff, and L. Henschen. A new use of an automated reasoning assistant: Open questions in equivalential calculus and the study of infinite domains. *Artificial Intelligence*, 22:303–356, 1984.

Caching and Lemmaizing in Model Elimination Theorem Provers*

Owen L. Astrachan[1] and Mark E. Stickel[2]

[1] Department of Computer Science, Duke University, Durham, NC 27706, ola@cs.duke.edu
[2] Artificial Intelligence Center, SRI International, Menlo Park, CA 94025, stickel@ai.sri.com

Abstract. Theorem provers based on model elimination have exhibited extremely high inference rates but have lacked a redundancy control mechanism such as subsumption. In this paper we report on work done to modify a model elimination theorem prover using two techniques, *caching* and *lemmaizing*, that have reduced by more than an order of magnitude the time required to find proofs of several problems and that have enabled the prover to prove theorems previously unobtainable by top-down model elimination theorem provers.

1 Introduction

Model Elimination (*ME*) [17, 19] is a complete inference procedure for the first-order predicate calculus. It is the method underlying the Prolog Technology Theorem Prover (*PTTP*) [28, 29], the *SETHEO* prover [16], and several or-parallel theorem provers [26, 9, 2]. The use of model elimination, an input proof procedure, has enabled *ME*-based provers to draw on techniques developed by the logic programming community (hence the name *PTTP*) that enable the implementations to be very efficient in the use of space, to have a high inference rate, and to be readily parallelized. *METEOR* is a high-performance implementation of *ME* written in C that runs under the UNIX operating system. It compiles clauses into a data structure that is then "interpreted" at run-time by a uniprocessor, a multiprocessor, or a network of uniprocessors. *METEOR* and *PTTP* perform exactly the same number of inferences in solving the problems reported in [29] when *METEOR* is run using inference count as the depth measure (see Section 3).

The high inference rate and modest storage requirements of *PTTP* and *METEOR* make them attractive inference engines useful for seeking shallow proofs. In some domains, the high inference rate may overcome the lack of redundancy control and permit the discovery of hard theorems with deep proofs. For example, using *METEOR* we are able to find proofs of two problems [2] from a set of real-analysis challenge problems [7] that are difficult if not unobtainable for *OTTER* [21], a prover which employs both subsumption and a notion of best-first search. The proof of the third challenge problem from this set is too deep for *METEOR* to obtain without

* This research was supported by the National Science Foundation under Grant CCR-8922330. The views and conclusions contained herein are those of the authors and should not be interpreted as necessarily representing the official policies, either expressed or implied, of the National Science Foundation or the United States government.

some modification. Parallel implementations of model elimination theorem provers have resulted in very high performance provers [26, 9, 2], though they have not produced a proof of a theorem infeasible to obtain by running the prover on a single processor.

In general, the lack of both a redundancy-control mechanism such as subsumption and a best-first search methodology are severe impediments to finding deep proofs. Many theorems obtainable by *OTTER* cannot be proven in our systems because the size of the search space and the lack of redundancy control quickly overwhelm the high inference rate. In this paper we report on the modification of the search mechanism used in *METEOR* by the addition of *caching*, which replaces search, and *lemmaizing*, which augments search. Our goal has been to implement these modifications with minimal degradation of the high inference rate. These modifications have enabled *METEOR* to prove theorems not previously obtainable by top-down *ME* provers and can reduce by more than an order of magnitude the time required to find proofs of some "difficult" theorems. We use the two-dimensional grid in Fig. 1 to categorize our approaches.

	replace search	augment search
discovery cost	caching	
other cost	heuristic caching	lemmaizing

Fig. 1. changing the search mechanism.

In a broad sense, the cost referred to in Fig. 1 is a measure of the computational resources used to find a solution of a goal. More precisely, *METEOR* employs an iterated form of depth-first search called iterative deepening [30, 15] in which the maximum depth of search is bounded during each iteration. This bound limits the computational resources available to solve a goal; the resources actually used to find a solution of a goal constitute the discovery cost of the solution. Details of the search mechanism and the depth bounds used in *METEOR* are given in Section 3.

In our terminology, *caching* refers to a mechanism that optionally replaces the normal search mechanism at a lower computational cost, but yields essentially identical results to search. Cached goals are solved by lookup instead of search. Proofs will be found with the same cost bound as when caching is not used, and no more inferences will be performed with caching than without (in practice, many fewer inferences are required when caching is used). Caching reduces the number of inferences because replacing search for solutions of previously seen (cached) goals by lookup avoids repeating inferences on "failure branches" of the search tree explored during the search for the cached solution; lookup ideally will return each distinct solution only once, whereas search may repeatedly generate the same solution, and fewer more general solutions may be retrieved from the cache instead of many more specific ones. Whether there is a net performance gain depends on how efficiently the cache is implemented, i.e., what are the relative costs of search and cache lookup.

Caching charges discovery cost to reproduce essential features of search, e.g., the same solutions with the same cost. Charging discovery cost is not the only option, however. Charging some cost other than discovery cost leads to what we call *heuristic caching*, which is identical to caching in concept and implementation except for the cost charged for looked-up solutions. The guarantee that caching will not result in an increase in the number of inferences is absent for heuristic caching, but in some domains heuristic caching can be extremely successful.

The objective of caching is to make effective use of results discovered in past search. Caching simply stores results of past searches and replaces future searches by cache lookup. Another way to use results discovered in past searches is to record some seemingly useful solutions as *lemmas* and use them in future inferences in the same way as input clauses are used. Note that lemmas, unlike caching, can introduce substantial additional redundancy in the search space, since theorems can then be proved both either entirely from the input clauses as before or by use of lemmas. Lemmas (as stored solutions) can be beneficial only if less is charged for their use than for their solution from the input clauses and, even then, a lemma must actually be used in the proof of the top goal for there to be any reduction in the total size of the search space. We use the word *lemma-izing* or *lemmaizing*[3] to refer to this mechanism that augments the search by introducing lemmas that are treated as input clauses .

This paper is organized as follows: In Section 2 we briefly describe the model elimination proof procedure. In Section 3 we outline the search mechanism used in *METEOR*. Our modification to this mechanism are described in Section 4 and Section 5. Results generated using these modifications are given in Section 6. Related work is outlined in Section 7 and conclusions presented in Section 8. A more detailed account of the work presented here can be found in [3].

2 Model Elimination

In this section we give a description of the *ME extension* and *reduction* inference rules and other *ME* terminology sufficient for an understanding of the remaining sections. We assume familiarity with terminology of resolution proof procedures, e.g., *terms*, *atomic formulas (atoms)*, *literals*, *clauses* and *unification*. For a description of these, see [19], which also gives a complete description of the model elimination procedure. We use Prolog notation in which variables are represented by capital letters and functions, constants and predicates are represented by lowercase letters.

The *ME* proof procedure uses a kind of annotated clause called a *chain*. Roughly, the annotations in a chain record previous inferences that have been made in the current sequence of inference steps and identify information that can be used as the proof is expanded. Literals in a chain are either *B-literals* or *A-literals*. An A-literal has been used in an extension operation (and is thus in some sense an ancestor literal) and may participate in the *ME* reduction operation. Initially all literals are *B-literals*.

The *ME* procedure begins with some designated input clause as the initial chain. The leftmost literal in this chain is unified with a complementary literal of an input clause. The leftmost literal in the chain is designated as an A-literal (ancestor literal),

[3] Although the juxtaposition of vowels in this word may be inharmonious, recall memoizing [22] used to mean essentially what we call caching.

the other literals (if any) in the input clause are added to the front of the chain, and the unifying substitution is applied. This is the *ME* extension operation. It is the same as the Prolog inference operation except that it retains the unified literal as an A-literal, which may then be used in subsequent *ME* reduction operations. A-literals appear in brackets in the following descriptions. We assume that all chains and clauses are renamed apart so that they are variable disjoint as necessary. Formally we have

Definition 2.1 *Given chain C_1 of the form l_1C_0 with leftmost B-literal l_1 and input clause C_2 with literal l_2 complementary to l_1 such that the atoms of l_1 and l_2 are unifiable with most general unifier (mgu) θ, the ME extension operation of C_1 with C_2 on l_2 yields the chain $\{(C_2 - l_2)[l_1]C_0\}\theta$ where $[l_1]$ is an* A-literal *and the literals in $(C_2 - l_2)$ may be reordered. We use the notation* extend$(C_1, l_1, C_2, l_2, \theta)$ *to denote the result of the extension.*

Example: If $C_1 = q(f(X), Y)[r(Y, Z)]p(X, Z)$ and $C_2 = \neg q(f(a), c)\neg r(c, b)$ (note that C_1 has one A-literal and two B-literals) then C_1 extended with C_2 on $\neg q(f(a), c)$ is $\neg r(c, b)[q(f(a), c)][r(c, Z)]p(a, Z)$.

Definition 2.2 *Given chain C_1 of the form l_1C_0 with leftmost B-literal l_1 and complementary A-literal l_2 such that the atoms of l_1 and l_2 are unifiable with mgu θ, the ME reduction operation yields chain $C_0\theta$. We use the notation* reduce(C_1, l_1, l_2, θ) *to denote the result of the reduction.*

Example: In the resultant chain above, $\neg r(c, b)[q(f(a), c)][r(c, Z)]p(a, Z)$, the (only possible) reduction operation yields $[q(f(a), c)][r(c, b)]p(a, b)$.

Note that both the reduction operation and extension with a unit clause (in which no literals are added to the chain) can make the leftmost literal of the chain an A-literal. As the *ME* inference rules require the leftmost literal to be a B-literal, any leftmost A-literals are removed after the extension and reduction operations are performed. This is the *contraction* operation as defined in [19]. In the chain used in the examples above, the chain $[q(f(a), c)][r(c, b)]p(a, b)$ is contracted to the chain $p(a, b)$. In practice this operation is incorporated into the extension and reduction operations. The A-literals that are removed by contraction represent solved goals or lemmas.

3 Search Mechanism

Although *ME* is a complete proof procedure in that there is always an *ME* derivation of the empty chain from an unsatisfiable set of input clauses, a complete search strategy must be employed to ensure that such a derivation is found. Prolog, for example, uses unbounded depth-first search and may fail to find a proof because of infinite branches in the search tree.

Rather than employ a breadth-first strategy with its exponential storage requirements, *METEOR* and *PTTP* use iterative deepening [30, 15] to ensure completeness of the search strategy. Rather than storing intermediate results as is done in breadth-first search, results are recomputed at each stage of the iterative deepening search.

We impose a cost bound on prospective proofs. Our cost bounds are not bounds on the entire search space (except implicitly), but rather only on each portion of it that forms a single (partial) proof. Thus, for example, a bound on the number of inference steps in a proof is a cost measure, but a bound on the total number of inferences performed in the process of finding the proof, including those on failing branches of the search space, is not. In *PTTP*, proofs of minimal length (in number of *ME* inferences) are found since the total number of nodes in a proof tree is bounded during each stage of iterative deepening with successively higher resource limits. We call this depth measure ***inference depth*** or D_{inf}. Alternatively, bounding the depth of the proof tree yields what we call ***A-literal depth*** or D_{Alit} since the depth of a proof tree is the number of A-literals present in the chain that represents the current state of the deduction. This metric is the default depth measure used in *SETHEO* and was used in one of the earliest implementations of *ME* [12]. Neither of these measures is clearly superior to the other in that there seems to be no *a priori* method for determining which measure yields a proof more quickly for any particular theorem.

When bounded search is used, each goal has an associated cost bound (derived from the current global bound) that must not be exceeded during an attempt to solve the goal. We use the notation $\langle \mathcal{G}, n \rangle$ to refer to a single literal goal $\mathcal{G}$ with associated cost bound n. The search mechanism employed in *METEOR* is described in Fig. 2.

```
boolean
Solve(chain C,resource n)
  if C is the empty chain then
        return TRUE
  goal G ← leftmost literal in C
  R ← A-literals of C potentially unifiable with complement of G (for reductions)
  E ← input clauses with literals potentially unifiable with complement of G (for extensions)
            /* try to solve ⟨G,n⟩ */
  for each l_R in R do
        n_new ← resources available if reduction made
        if n_new ≥ 0 and l_R and complement of G unify with mgu θ then
              if Solve(reduce(C, G, l_R,θ),n_new) then
                    return TRUE
  endfor (reduction)
  for each clause C in E with literal l_C do
        n_new ← resources available if extension of C with C is made
        if n_new ≥ 0 and and l_C and complement of G unify with mgu θ then
              if Solve(extend(C,G,C,l_C,θ), n_new) then
                    return TRUE
  endfor (extension)
  return FALSE
```

Fig. 2. the search procedure.

Several optimizations can be applied in the search mechanism without affecting its completeness [29]. Many of these optimizations are implicit in the definition of an acceptable chain and the accepting transformation that is applied to chains in the original presentation of *ME* [17, 18, 19]. The most effective of these tends to be the *identical ancestor pruning rule.* Before any reductions or extensions are attempted, the A-literals in the chain to the right of $\mathcal{G}$ are examined to see if any are identical to $\mathcal{G}$. If this is the case, **Solve** returns FALSE; it is not necessary to solve a goal in the context of a previous attempt to solve the same goal. This pruning rule is highly effective but its use is limited when caching is employed, in a manner described later.

4 Caching

By *caching* we mean the use of a device (the *cache*) that on occasion replaces the regular search mechanism and yields substantially identical results to search. This means that solutions should be retrieved from the cache only when it is known that the cache contains complete information, i.e., it contains all solutions that would be generated by search. When the cache is complete in this sense its use can replace the normal search mechanism. To this end, the cache consists of two logical parts: the *cache directory*, which stores information about which goals have solutions stored in the cache, and the *cache store*, which contains the solutions. When a goal and its associated resource bound are submitted to the **Solve** procedure (see Fig. 2), the directory is consulted and the cache store used, if possible, before line 4. If the cache store is used, procedure **Solve** is exited (with success or failure) before lines 4–14 are executed. If the cache store is not used, lines 4–14 are executed as in the regular search procedure.

The caching method we describe here is applicable only to cases of model elimination in which the reduction operation is not used; this includes problems expressed in Horn clauses. For such problems all solutions of the pair $g_1 = \langle\mathcal{G},n\rangle$ are also solutions of the pair $g_2 = \langle\mathcal{G},m\rangle$ if $n \leq m$ since the sequence of inferences that solve $\mathcal{G}$ in g_1 will also solve $\mathcal{G}$ in g_2 provided the identical-ancestor pruning rule is partially disabled. If during the search for solutions of some goal $\langle\mathcal{G},n\rangle$ a branch of the search tree below $\mathcal{G}$ is pruned using identical-ancestor pruning with an ancestor $A_{\mathcal{G}}$ of $\mathcal{G}$, a solution might be missed that would be found in another context in which $A_{\mathcal{G}}$ did not appear as an ancestor. To prevent such inconsistencies, and to avoid the need for storing an environment of ancestor literals, identical-ancestor pruning of subgoals of cacheable goals is disabled. More precisely, a goal cannot be pruned by any ancestor of a cacheable goal. Pruning is permitted if the pruning goal is not being stored in the cache, or its descendants are not.

For non-Horn problems the sequence of deductions that solve $\mathcal{G}$ in g_1 may include reductions with ancestors of $\mathcal{G}$. This same sequence of deductions will solve $\mathcal{G}$ in g_2 only if the same reductions are possible, i.e., only if the necessary A-literals are present in the chain. Thus caching with non-Horn problems would seem to require that a cached solution contain some record of the A-literals present when the solution is generated.

Although a modification of *ME* has been proposed [24] that can decrease the number of A-literals that are stored with a cached solution (at the expense of potentially longer proofs), and a method of generalized A-literals is addressed in [1], in this paper we do not address the issue of caching with non-Horn problems. We do, however, note several successful applications of lemmaizing to non-Horn problems in Section 6.

4.1 The Cache Mechanism

The cache store contains all the cached solutions. A cached solution consists of a substitution instance of the subgoal and the resource bound used in obtaining the solution as defined in Definition 4.1.

Definition 4.1 *A cached solution is a pair* $\langle \mathcal{G}', n_{\mathcal{G}'} \rangle$ *where* $\mathcal{G}'$ *is* $\mathcal{G}\theta$ *for some goal* $\mathcal{G}$*,* θ *is the composition of unifiers used in solving* $\mathcal{G}$*, and* $n_{\mathcal{G}'}$ *is the measure of the resources used in producing* $\mathcal{G}'$*.*

A cached solution stores only the instantiation $\mathcal{G}'$—nothing to identify the goal $\mathcal{G}$ it was used to solve. Thus, the cache store contains solutions (provable literals) divorced from the goals during whose proof they were found.

The cache directory, which is consulted to determine if the cache store of solutions should be used, consists of cache templates defined in Definition 4.2.

Definition 4.2 *A cache template or template is a triple* $\langle \mathcal{G}, m, m_S \rangle$ *that indicates that the cache is m-complete for goal* $\mathcal{G}$*. If* $m_S \leq m$*, then* m_S *is the smallest resource needed to solve* $\mathcal{G}$*; if* $m_S > m$*, then* $\mathcal{G}$ *has no solution with cost* $\leq m$*.*

The templates in the cache directory are used to determine when the solutions in the cache store include a complete set of solutions for any particular goal.

Definition 4.3 *A cache is complete for* $\langle \mathcal{G}, n \rangle$ *if all solutions of (sub)goal* $\mathcal{G}$ *that can be obtained with a resource bound of at most n are in the cache store. In this case we say that the cache is n-complete for* $\mathcal{G}$*.*

Given a goal pair $\langle \mathcal{G}, n \rangle$, the cache directory is searched to see if $\mathcal{G}$ appears as the first component of a template (there may be more than one applicable template if we are using template-subsumption, see Section 4.2). If a template is found and it indicates that the cache is m-complete for $\mathcal{G}$ with $m \geq n$ then the cache can be used in lieu of the regular search mechanism. As an optimized special case, we note that if $m \geq n$ and $m_S > n$ then there are no solutions bounded by n, so further cache lookup to find solutions is unnecessary. This use of the cache directory to indicate failure corresponds to the *failure cache* outlined in [11]; when templates are used in this way we call them *failure templates*.

If the cache is complete for a goal, solutions to the goal can be found by retrieving from the cache store all cached solutions that are instances of the goal. If subsumption of cached solutions is employed, however, unifiable solutions rather than instances must be retrieved from the cache store. When caching is used, the modifications indicated in Fig. 3 are made to the search routine of Fig. 2.

In its implementation in *METEOR*, the code in Fig. 3 is guarded by a statement that enables cache use only when n, the resource available, is above some user-specified threshold value. In the current implementation, the same threshold is used to guard both solution storage and template retrieval. Because the identical-ancestor pruning rule is so effective and must be partially disabled when caching is used and because the normal inference rate in *METEOR* is high enough that using the cache for shallow searches is counterproductive, cache use is prevented when the global cost bound is below the threshold. We present data in Section 6 showing how different threshold values affect the performance of the prover; in general, low thresholds severely degrade cache performance.

```
       /* added before line 4 in Fig. 2 */
[3.1]  ⟨G,m,m_S⟩ ← template corresponding to ⟨G,n⟩
        /* more than one template may be applicable
         if template-subsumption is being used */

[3.2]  if m ≥ n then
[3.3]   if n ≥ m_S then
[3.4]    return CacheSolve(C,n)
[3.5]   else
[3.6]    return FALSE
       /* if we reach here then use regular
        search mechanism */
```

```
boolean
CacheSolve(C,n)
     {G is the leftmost literal in C}
[1]  L ← all solution pairs ⟨G',n_G'⟩ such that G'
      is potentially unifiable with G and
      such that n_G' ≤ n

[2]  for each ⟨G',n_G'⟩ in L do
[3]   if G and G' unify with mgu θ then
[4]     n_new ← n − n_G'
[5]     if Solve(extend(C,G,G', G',θ), n_new) then
[6]      return TRUE
     end for
[7]  return FALSE
```

Fig. 3. using the cache.

4.2 Storing Templates and Solutions

To conserve cache storage and to minimize the effective branching factor of the search space, a solution subsumed by an entry already in the cache store is not entered in the store. Subsumption of solution pairs is defined in Definition 4.4.

Definition 4.4 *If* $S_1 = \langle \mathcal{H}, n_{\mathcal{H}} \rangle$ *and* $S_2 = \langle \mathcal{G}, n_{\mathcal{G}} \rangle$ *are solution pairs then the pair* S_1 <u>subsumes</u> *the pair* S_2 *if and only if* $\mathcal{H}$ *subsumes* $\mathcal{G}$ *(there exists a substitution* σ *with* $\mathcal{H}\sigma = \mathcal{G}$*) and* $n_{\mathcal{H}} \leq n_{\mathcal{G}}$.

If the resource bound $n_{\mathcal{H}}$ of a pair $\langle \mathcal{H}, n_{\mathcal{H}} \rangle$ is greater than the resource bound $n_{\mathcal{G}}$ of a pair $\langle \mathcal{G}, n_{\mathcal{G}} \rangle$ then both pairs must be stored in the cache even if $\mathcal{H}$ subsumes $\mathcal{G}$ since the solution $\mathcal{G}$ might be usable when $\mathcal{H}$ is not due to its lower resource requirements.

Since the cache replaces search with a (it is hoped) more efficient mechanism, and since previously unseen goals cannot use the cache, it is worth investigating methods that allow the search for solutions of an unseen goal to be replaced with a cache lookup. This is the motivation for the concept of *template-subsumption* (defined in Definition 4.5): to allow the cache to be used when a specific goal is encountered for the first time with a given resource bound.

When a goal pair $\langle \mathcal{G}, n \rangle$ is encountered for the first time, it is possible that the cache is m-complete for a more general goal with $m \geq n$. In this case the cache may be used instead of the regular search mechanism; we say that the pair $\langle \mathcal{G}, n \rangle$ is *template-subsumed* as defined in Definition 4.5.

Definition 4.5 *If cache template* $T = \langle \mathcal{H}, m, m_S \rangle$ *and goal pair* $G = \langle \mathcal{G}, n \rangle$ *then* T <u>template-subsumes</u> G *if and only if* $\mathcal{H}$ *subsumes* $\mathcal{G}$ *and* $m \geq n$.

If such a goal pair $\langle \mathcal{G}, n \rangle$ is template-subsumed by a cache template $\langle \mathcal{H}, m, m_S \rangle$, then since the cache is m-complete for the more general goal $\mathcal{H}$ all solutions of $\mathcal{H}$ are stored in the cache. These solutions are a superset of the solutions of $\mathcal{G}$ given that $\mathcal{H}$ subsumes $\mathcal{G}$ and that the resource bounds m and n satisfy the constraints of Definition 4.5. Thus the cache is m-complete for the goal pair as well and the cache replaces search.

The use of template-subsumption in the cache is a parameter that the user of the system can set. Results indicate that the more frequent cache access enabled by template subsumption more than compensates for the increased lookup time of a subsuming template.

4.3 Heuristic Caching

We have investigated an alternative to caching in which a cost other than discovery cost is incurred when a cached solution is used. We call the method ***heuristic caching*** and, as shown in Section 6, we have found a domain in which its use yields substantial performance gains over the normal caching mechanism.

When a solution is retrieved from the cache, a cost is incurred as shown on line 4 of Fig. 3. When caching recreates the search space this cost reflects the cost used in creating the solution. Consider charging some other cost, e.g., a cost less than that used to discover the solution. In such a case a solution may be used whose discovery cost exceeds the current cost bound. Using such a solution in effect permits a search beyond that constrained by the current cost bound. Of course, charging less than the discovery cost can also permit many deep but fruitless paths to be searched as well. In general, it is difficult to identify those solutions that should be stored with a cost less than the discovery cost. In certain domains, however, it may be possible to treat all solutions uniformly and realize a substantial performance gain over caching. We report on such a domain in Section 6.

5 Lemmaizing

The difficulties of caching for non-Horn problems and the potential for large caches for deep Horn problems have led us to investigate alternatives to caching that can decrease the storage requirements inherent in caching all solutions and that can allow high-cost proofs to be discovered with low cost bounds. In this section we investigate such an approach we have called *lemmaizing*.

Lemmaizing differs from heuristic caching in that not all solutions are stored for a given goal, but only some (it is hoped relevant) solutions are stored thus augmenting rather than replacing search. Since lemmas are not needed for completeness, we may impose syntactic and semantic criteria in deciding which lemmas to retain. The idea is to store lemmas that are used to eliminate repeated subdeductions. In this sense the use of lemmas allows us to combine an aspect of bottom-up reasoning with the top-down reasoning in *METEOR*. By imposing strict criteria on lemmas we retain a complete inference theorem that will hopefully allow us to prove theorems otherwise unobtainable. In one of the first implementations of *ME* [12], this kind of lemma use was explored in an ad-hoc manner. No notion of lemma cost was explored and lemmas were not found to be useful in general since the potential for a shorter proof was not realized due to the increased branching factor induced by allowing lemmas as alternatives for extension.

5.1 Lemma Storage and Retrieval

If solutions used in the proof of the top goal incurred a smaller cost than other solutions, the proof might be found more quickly. The intuition here is that these repeated solutions function as *lemmas*; they reflect useful information to be used without requiring that the information be rederived each time it is used. Identifying useful lemmas is a non-trivial task and an important one in automated reasoning systems [6]. We are only beginning to explore this use of stored solutions and report on several successful uses of lemmaizing in Section 6. We have used several syntactic and semantic criteria in determining what lemmas to store. The primary criterion we have employed is to limit the nesting depth of terms that may appear in lemmas.

In some domains (e.g., group theory) demodulators may be used to rewrite solutions to a canonical form. This reduces redundancy since subsumption checks permit rewritten solutions to be discarded that might otherwise appear (redundantly) in the lemma store. Demodulation also permits terms that might violate a syntactic criterion such as nesting depth to be rewritten to a form that does not violate the criterion. Although demodulation and resolution may not, in general, result in a complete proof procedure, we can impose any restrictions on lemmas (including demodulating them) and retain completeness in *ME*. We have experimented with demodulating lemmas in group and ring problems by including a complete set of rewrite rules as demodulators as well as using demodulators generated during the search to rewrite all lemmas; we report on several successful applications in this area in Section 6. Details of the rewriting techniques employed can be found in [3].

6 Results

In this section we include results for a variety of problems we have run using the caching and lemmaizing methods outlined in the previous sections. Detailed accounts of the implementation of these methods can be found in [1, 3]. Rather than give an exhaustive set of results based, for example, on the problems reported in [29]; we include problems that have been historically difficult for *ME* based provers. All the results in this section are based on running an unoptimized version of *METEOR* (in the sense that the compiler debug rather than optimize flags were set) on a Sun SPARC-station 2 with 64 megabytes of memory.

We have used heuristic caching in proving SAM's lemma [35] and achieved spectacular results for top-down or *ME* theorem provers. Although *OTTER* solves the same formulation of SAM's lemma in about seven seconds, it has been an intractable problem for provers not employing some form of redundancy control. In the formulation we use, the input clauses for this problem are from the domain of function free Datalog problems. For this and other Datalog-like problems storing all solutions but charging unit retrieval cost seems a promising method.

We have also experimented with lemmas in several group theory problems. By imposing limits on the nesting depth of function symbols that appear in lemma terms (and by use of demodulation) we have been able to prove both the commutator problem and the theorem that if $x^2 = x$ in a ring then the ring is commutative [35] whose proofs have, heretofore, been unobtainable by top-down *ME* theorem provers.

We also note a successful proof of the intermediate value theorem of calculus (as formulated in [34]). This non-Horn problem is proved using lemmaizing and retaining all lemmas with a nesting depth of less than six. To our knowledge, this problem has been beyond the capabilities of linear provers.

In both Tables 1 and 2 each cell of the table indicates the number of seconds needed to find the proof (the top number) and the number of successful inferences made.

In Table 1 the columns indicate (respectively) the problem name, statistics using the default (D_{inf}) search strategy, using failure templates only, the best caching results over several runs (see Table 2), using all solutions with function symbols nested at most one as lemmas with unit cost, and rewriting these lemmas using using a complete set of reductions as well as demodulators generated during the proof. Note that use of failure templates only fails to overcome the disadvantage of having to partially disable identical-ancestor pruning and thus requires more inferences than the default strategy; full caching overcomes this disadvantage and shows substantial net gains over the default strategy.

Table 1. results of caching.

problem	D_{inf}	fail. temp.	cache	unit lemma	demod.
wos10	13.06	26.51	3.72	6.40	2.73
	78,669	129,643	10,714	19,562	7,979
wos 1	19.64	35.96	6.39	316	0.48
	139,068	223,455	10,551	1,221,686	1,273
wos21	283.43	840.2	85.51	584	39.73
	2,200,583	5,397,293	368,426	2,134,087	132,307
wos15	13,841		1,356	29.8	4.37
	91,879,275		5,399,388	104,883	15,701
sam	10^{14}‡		280.37	40.83†	
	$5(10^{17})$‡		948,444	155,480	
wos22	11,388		1,565		
	71,143,961		7,217,820		

†heuristic caching
‡projected measure

In Table 2 results are given using different cache thresholds, i.e., varying the minimum level at which the cache is consulted rather than relying solely on the normal search mechanism. Although the user can set the threshold, the default threshold used in *METEOR* is five.

Table 2. using different cache thresholds.

	Cache Threshold Level						
problem	**1**	**2**	**3**	**4**	**5**	**6**	**7**
wos10	4.87	4.66	4.12	3.72	4.45	6.54	8.83
	8,482	8,482	8,482	10,714	17,563	31,698	46,561
wos 1	11.57	9.43	7.12	6.39	6.68	7.96	13.11
	8,499	8,499	8,499	10,551	14,136	23,059	49,582
wos21	492	483	358	226	133	96	86
	133,536	133,536	139,843	156,766	203,618	292,732	368,426
wos15	31,822	31,043	16,203	5,720	2,141	1,515	1,356
	1,507,114	1,507,114	1,509,839	1,721,907	2,413,466	3,765,384	5,399,388
sam†	42	42	41	41	43	52	
	126,650	126,650	127,451	130,328	156,433	242,347	
wos22	36,078	35,009	17,288	6,075	2,504	1,686	1,565
	1,846,619	1,846,619	1,921,009	2,280,075	3,231,997	4,942,331	7,217,820

†heuristic caching

Table 3 shows results from running several group theory problems using demodulators generated during the proof in addition to using the standard demodulators for free groups. No back demodulation was employed and all terms appearing in lemmas were restricted by limiting the level to which function symbols could be nested.

Table 3. using demodulation with lemma generation.

	Dynamic Demodulation				
problem	**time (secs)**	**inferences**	**# proof steps proof depth**	**# stored lemmas**	**lemmas generated**
wos10	2.73	7,979	7/8	38	7,101
wos 1	0.48	1,273	7/7	10	1,585
wos21	39.73	132,307	9/9	33	92,640
wos15	4.37	15,701	7/9	21	12,872
commutator	430.7	1,281,052	7/9	417	611,148
$x^2 = x$ ring	1,495	4,763,795	5/10	170	2,349,653

In Table 4 results are given for lemmaizing with several non-Horn problems. The results are given as number of seconds and number of inferences needed to solve the problems. The problem labeled ivt is the intermediate value theorem. As noted above, this problem has been beyond the range of top-down, linear provers. It is proved by the *STR+VE* prover [8] and the *HD-PROVER* in [34]. With the addition of several non-automatically constructed rewrite rules it is proved by the prover in [25]. The problem labeled nonobv is a problem given in [23] and subsequently cited in [20]. We should note that when D_{Alit} is used as the depth measure *METEOR* solves this problem in under one second. The problem labeled salt is Lewis Carrol's salt and mustard logic puzzle.

METEOR and *PTTP* are attractive inference engines because of minimal memory requirements and because of the high inference rate. One of the potential drawbacks incurred by caching and lemmaizing is the (potentially large) size of the cache and lemma store. Table 5 shows the memory requirements for the problems reported in this section. For problems that are solved quickly, the requirements are quite modest

Table 4. non-Horn problems.

problem	**normal search**	**lemmaizing**
ivt D_{Alit}		915 3,216,208
nonobv D_{inf}	657 6,526,914	1.42 12,071
salt D_{Alit}	60.3 660,774	35.7 309,434

Table 5. memory requirements.

problem	**# sol's**	**# trie nodes**	**size (Mbytes)**
wos10	473	1,300	0.05
wos1	2,301	7,269	0.38
wos21	8,867	28,428	1.5
sam†	143	407	0.013
wos15	55,076	176,037	10.2
wos22	79,871	243,227	15.38
commutator‡	417	1,094	0.48
$x^2 = x$ ring‡	170	469	0.19
ivt‡	91	289	0.14
nonobv‡	31	80	0.002

†heuristic caching
‡lemmaizing

which makes caching and lemmaizing attractive methods for this class of problem.

For hard problems, the memory requirements needed to store all solutions as is done in caching makes lemmaizing an attractive alternative. Of course lemmaizing requires the identification of useful lemmas which is a difficult task itself.

7 Related Work

Much work has been done in the area of query optimization for deductive databases [4]. This work tends to focus on reducing redundant (recursive) derivations by program transformation techniques [5], by introducing a control language [13, 14], and by run-time analysis [32]. In general, these techniques are designed to work with function-free, Horn (Datalog) programs. As our results with SAM's lemma indicate, caching and heuristic caching can work well for this class of problem. The framework of SLD-AL resolution [33] is closely related to our framework, the concept of a lemma in SLD-AL resolution corresponds exactly to (and is antedated by) the use of lemmas in model elimination; the QSQR implementation [32] of SLD-AL resolution also uses iterative deepening. The OLDT resolution procedure [31] is very closely related and involves an iterative deepening search of Datalog programs. As database optimizations, these methods concentrate on reducing redundancy when all solutions to a goal are desired; in a theorem proving context we (usually) only search for one proof.

Extension tables as used in [10] are closely related to the OLDT procedure. Although an outline of an iterative deepening prover is given there, no empirical data is given and it appears that the method has not yet been implemented. Plaisted [24] has implemented a theorem prover in which solved goals are stored although no notion of cache completeness is used. Although he reports some favorable results, caching in his prover could lead to longer proofs, did not work for the same class of problems we report on here, and did not admit proofs for problems previously unprovable in his system. In fairness, his implementation was not optimized and access to the store of solved goals can be particularly slow in the system employed in the prover.

Elkan [11] reports on the idea of caching to reduce redundancy in a resolution based prover for Horn problems, but only reports on the use of caching to solve one problem. His prover has been used in the realm of explanation based learning and the use of the prover with what we call lemmaizing is reported in [27].

8 Conclusions

We have outlined two modifications to the *ME* search mechanism used in *METEOR*. These modifications, caching and lemmaizing, have enabled *METEOR* to prove theorems previously unobtainable by top-down model elimination theorem provers and have reduced by more than an order of magnitude the time required to prove some typically difficult theorems.

Other work in this area has focused almost exclusively on lemmaizing. We have studied, and shown to be successful, a method (caching) that consciously replaces search rather than augmenting it. In our implementation we have succeeded not only in reducing the number of inferences (which is easy and guaranteed for exhaustive searches), but in reducing the time required to find reduced-inference proofs, which is not so easy. The volume of data caching and lemmaizing uses demands indexing schemes unnecessary for ordinary *ME*; adding and using such schemes in an already fast theorem prover indicates not only the promise of the methods, but the versatility of our prover.

In the future we hope to address optimizations to the caching mechanism that will increase its efficiency both in terms of storage requirements and in its redundancy reducing capabilities thus permitting caching to be applicable to a larger class of problem. We also plan to investigate methods for identifying useful lemmas that will allow us to combine aspects of bottom-up reasoning with the goal-directedness of top-down provers in solving both Horn and non-Horn problems.

References

1. O.L. Astrachan. *Investigations in Theorem Proving based on Model Elimination.* PhD thesis, Duke University, 1992. (expected).
2. O.L. Astrachan and D.W. Loveland. METEORs: High performance theorem provers using model elimination. In R.S. Boyer, editor, *Automated Reasoning: Essays in Honor of Woody Bledsoe.* Kluwer Academic Publishers, 1991.
3. O.L Astrachan and M.E. Stickel. Caching and lemmaizing in model elimination theorem provers. Technical Report 513, SRI International, Menlo Park, CA, December 1991.
4. F. Bancilhon and F. Ramakrishnan. Performance evaluation of data intensive logic programs. In J. Minker, editor, *Foundations of Deductive Databases and Logic Programming*, chapter 12, pages 439–517. Morgan Kaufmann, 1988.
5. C. Beeri and R. Ramakrishnan. On the power of magic. In *Proceedings of the 6th Symposium on Principles of Database Systems*, pages 269–283, 1987.
6. W.W. Bledsoe. Some thoughts on proof discovery. In *Proceedings of the IEEE Symposium on Logic Programming*, pages 2–10, 1986.
7. W.W. Bledsoe. Challenge problems in elementary calculus. *Journal of Automated Reasoning*, 6(3):341–359, 1990.
8. W.W. Bledsoe and L. Hines. Variable elimination and chaining in a resolution-based prover for inequalities. In *Proceedings of the Fifth Conference on Automated Deduction*, pages 281–292. Springer-Verlag, 1980.
9. S. Bose, E. Clarke, D.E. Long, and S. Michaylov. Parthenon: A parallel theorem prover for non-Horn clauses. In *Proceedings of the Symposium on Logic in Computer Science*, 1989.
10. S.W. Dietrich. Extension tables: Memo relations in logic programming. In *Proceedings of the IEEE Symposium on Logic Programming*, pages 264–272, 1987.
11. C. Elkan. A conspiratorial and caching and/or tree searcher for theorem-proving. In *Proceedings of the Eleventh International Joint Conference on Artificial Intelligence*, 1989.
12. S. Fleisig, D. Loveland, A. Smiley, and D. Yarmash. An implementation of the model elimination proof procedure. *Journal of the Association for Computing Machinery*, 21:124–139, January 1974.
13. A.R. Helm. Detecting and eliminating redundant derivations in logic knowledge bases. In W. Kim, J.-M. Nicolas, and S. Nishio, editors, *Deductive and Object-Oriented Databases*, pages 145–161. Elsevier Science Publishers, 1990.

14. A.R. Helm. On the elimination of redundant derivations during execution. In S. Debray and M. Hermenegildo, editors, *Proceedings of the North American Conference on Logic Programming*, pages 551–568, 1990.
15. R.E. Korf. Depth-first iterative deepening: An optimal admissible tree search. *Artificial Intelligence*, 27:97–109, 1985.
16. R. Letz, S. Bayerl, J. Schumann, and W. Bibel. SETHEO—a high-performance theorem prover. (to appear).
17. D.W. Loveland. Mechanical theorem proving by model elimination. *Journal of the Association for Computing Machinery*, 15(2):236–251, April 1968.
18. D.W. Loveland. A simplified format for the model elimination procedure. *Journal of the Association for Computing Machinery*, 16(3):349–363, July 1969.
19. D.W. Loveland. *Automated Theorem Proving: A Logical Basis*. North-Holland, 1978.
20. R. Manthey and F. Bry. SATCHMO: a theorem prover implemented in Prolog. In *Proceedings of the Ninth International Conference on Automated Deduction*, pages 415–434. Springer-Verlag, 1988.
21. W. McCune. *OTTER 2.0 Users Guide*. Argonne National Laboratory, March 1990.
22. D. Michie. Memo functions and machine learning. *Nature*, 218:19–22, April 1968.
23. F.J. Pelletier and P. Rudnicki. Non-obviousness. *AAR Newsletter*, (6):4–5, 1986.
24. D. Plaisted. A sequent style model elimination strategy and a positive refinement. *Journal of Automated Reasoning*, 6(4), 1990.
25. D. Plaisted and S.-J. Lee. Inference by clause linking. Technical Report TR90-022, University of North Carolina, Department of Computer Science. Chapel Hill, NC, 1990.
26. J. Schumann and R. Letz. PARTHEO: A high performance parallel theorem prover. In *Proceedings of the Tenth International Conference on Automated Deduction*, pages 40–56, 1990.
27. A. Segre and D. Scharstein. Practical caching for definite-clause theorem proving. [draft], September 1991.
28. M.E. Stickel. A Prolog technology theorem prover. *New Generation Computing*, 2(4):371–383, 1984.
29. M.E. Stickel. A Prolog technology theorem prover: Implementation by an extended Prolog compiler. *Journal of Automated Reasoning*, 4:343–380, 1988.
30. M.E. Stickel and W.M. Tyson. An analysis of consecutively bounded depth-first search with applications in automated deduction. In *Proceedings of the Ninth International Joint Conference on Artificial Intelligence*, pages 1073–1075, August 1985.
31. H. Tamaki and T. Sato. OLD resolution with tabulation. In *Proceedings of the Third International Conference on Logic Programming*, 1986.
32. L. Vieille. Recursive axioms in deductive databases: the query/subquery approach. In *Proceedings of the 1st International Conference on Expert Database Systems*, pages 179–193, 1986.
33. L. Vieille. Recursive query processing: The power of logic. *Theoretical Computer Science*, 69(1):1–53, 1989.
34. T.C. Wang and W.W. Bledsoe. Hierarchical deduction. *Journal of Automated Reasoning*, 3:35–77, 1987.
35. L. Wos. *Automated Reasoning: 33 Basic Research Problems*. Prentice Hall, 1988.

LIM+ Challenge Problems by RUE Hyper-Resolution

Vincent J. Digricoli
Eugene Kochendorfer

Department of Computer Science, Fordham University
Bronx, New York 10458

Abstract. Our purpose is to discuss a series of experiments proposed by Bledsoe [1] using a general purpose theorem prover, complete for first order logic, based on the resolution principle of Robinson but enhanced by an extended definition of resolution which incorporates the axioms of equality.

1 Introduction

As an inference system we use RUE resolution with denotes resolution by unification and equality. We resolve either by complete unification of complementary literals or by partial unification of literals resolving to inequalities. This inference rule implicitly applies the properties of equality as defined by the axioms of equality: transitivity, commutivity, reflexivity, substitution in predicates & functions, and requires no use of paramodulation. Refutations by RUE resolution, based on disagreement sets, have an abbreviated form much shorter than equivalent refutations with the equality axioms. Furthermore, this inference principle creates a new environment for the definition of heuristics facilitating proof search. A full treatment of RUE resolution with an analysis of prior experiments performed is found in [3,4].

The RUE theorem prover is now being completely recast and expanded to introduce diverse restriction strategies and our first step has been to implement the complete form of RUE Hyper-resolution which has been applied in five new experiments proposed by Bledsoe as LIM+ challenge problems which he describes as follows:

Acknowledgment: National Science Foundation Grant CCR-9024953 supports this research.

Lim+ Challenge Problems:

"The following is a list of challenge problems for automated provers. They are based on the theorem in calculus that the sum of two continuous functions is continuous. These problems are graded so that Problems 1 & 2 are easiest and 4 & 5 the most difficult.... Problems 1 and 2 can be proved by some of the current provers, but even they are difficult for provers that are complete for first order logic. Problems 3, 4 and 5 are harder, but hopefully can also be proved by the best current provers."

We have successfully performed the first four experiments with RUE hyper-resolution and are on the verge of completing problem 5 which we have proven in a reduced form with a 24 step hyper-refutation.

Performance Table

Problem	Proof Length	Total Unif. in search	CPU Time
1.	12 steps	3954	61.96 sec
2.	13 "	2738	64.06 "
3.	19 "	6851	81.61 "
4.	26 "	5399	138.50 "
(reduced) 5.	24 "	6335	82.80 "

Each of the above experiments deals with the same theorem made progessively more difficult by modifying the input set to be further removed from the theorem's conclusion. The experiments were performed on a 33 mHerz 486 Intel CPU and the system is now being ported to an IBM RISC workstation with an expected substantial reduction of cpu time.

In the previous 20 experiments performed with RUE resolution in the fields of boolean algebra, group theory and ring theory, the input sets were composed soley of unit clauses in the equality predicate and the longest theorem was an 18 step RUE refutation proving the associativity of logical-or in boolean algebra. This theorem is difficult for a human to prove and the computer generated proof in [4] improves upon the human proof published by Birkhoff [8].

The Lim+ set, particulary in problems 4 and 5, involves longer refutatons than previously attempted, with larger input sets including non-unit clauses and predicates other than equality. The resolvents produced are more complex employing ten distinct function symbols and nesting arguments eight levels deep. In this paper we present and analyze the 24 step RUE hyper-refutation for problem 5 describing the heuristics applied to find the proof. We have modified the Bledsoe input set for Problem 5 by introducing the axioms: Min(X,Y)<=X, Min(X,Y)<=Y, X~<=Half(Z) v Y~<=Half(Z) v X+Y<=Z, which Bledsoe derives from more elementary axioms. We are currently working on the full form of Problem 5.

2. RUE Hyper-resolution Applied to Problem 5

We wish to prove with automated reasoning the theorem: "The sum of continuous functions is a continuous function" and use the hyper-resolution input set: S = E U N, where E is the electron set of positive clauses and N, the nucleus set of non-positive clauses. A clause with one or more negated literals is non-positive. The axioms used and transcription to clause form is taken from Bledsoe [1].

E: Electrons

```
1.  (X+Y)+Z = X+(Y+Z)      (associativity)
2.  X+(Y+Z) = (X+Y)+Z      (assoc.variant)
3.  X+Y = Y+X              (commutivity)
4.  |X+Y| <= |X|+|Y|
5.  X<=0  v  |J(X)+(-A)| <= X
    (X>0 -->  |J(X)+(-A)| <= X,
     given X>0, let J(X) select a point p
     in the interval |p-A|<=X)
6.  Min(X,Y) <= X
7.  Min(X,Y) <= Y
8.  X<=Y  v  Y<=X
```

N: Nuclei

```
 1. U~<=V  v  V~<=W  v  U<=W     (transitivity of <=)

 2. |V+(-A)|~<=D(U)  v  |G(V)+(-G(A))|<=U  v  U<=0

    (U>0,|V-A|<=D(U) --> |G(V)-G(A)|<=U,
     U is epsilon, D(U)is delta,
     G(V) is continuous at point A)
```

```
 3. |V+(-A)|~<=C(U)   v   |F(V)+(-F(A))|<=U   v   U<=0
    (U>0,|V-A|<=C(U) --> |F(V)-F(A)| <=U,
     F(V) is continuous at point A)

 4. U~<=Half(W) v V~<=Half(W) v U+V<=W

 5. Negated Theorem:
    | [F(J(U))+G(J(U))] + [(-F(A))+(-G(A))] | ~<= E  v  U<=0

    (U>0 --> |[F(J(U))+G(J(U))] - [F(A)+G(A)]| > E,
     where U is delta and E, epsilon,
     J(U) selects a point in the interval |J(U)-A)|<=U)

 6. U~<=V  v  Min(U,V)=U

 7. D(U)~<=0  v  U<=0    (U>0 --> D(U)>0)
    (given epsilon>0 there exists a delta>0)

 8.  V~<=U  v  Min(U,V)=V

 9. C(U)~<=0  v  U<=0    (U>0 --> C(U)>0)
    (given epsilon>0 there exists a delta>0)

10. Half(U)~<=0  v  U<=0   (U>0 --> Half(U)>0)

11. E~<=0    (epsilon>0)
```

In the above, X,Y,Z,U,V,W are variables and A,E are constants. The input set has 19 clauses with 10 distinct function symbols (|,C,D,F,G,+,unary -,J, Half and Min) and two predicates (=,<=).

A hyper-resolution clash is defined as a series of successive resolution steps: { N, E_1, E_2, ..., E_K }, where we resolve the nucleus N with the electron E_1 to produce a resolvent which is resolved with E_2, and so on until the resolution with E_K produces a positive clause, a new electron called the clash resolvent. The original N set is stable and does not expand while the E set grows as new clash resolvents are added to it. The complete formal description of hyperresolution is found in [9] and in RUE form in [10].

The RUE hyper-refutation for this theorem is 24 steps, producing 13 clash resolvents; 10 of the 24 steps involve equality wherein the RUE inference rule generates inequalities implicitly applying the axioms of equality. Hence, this is a good hybrid proof using standard resolution with the mgu and disagreement sets under RUE theory.

The most complex resolvent produced in the refutation is a four literal resolvent with 10 distinct function symbols, arguments being nested 8 levels deep, with Min appearing 6 times and Half, 7 times, with a clause length of 178 characters, which forces the theorem prover to use relaxed complexity bounds making it more difficult to find the proof.

We now state the hyper-refutation which proves the theorem. We denote a nucleus by a negative index to N and electrons by a positive index to E:

```
{-1 5 6}      E15:  Min(X,Y)<=0   v   |J(Min(X,Y))+(-A)| <=X

{-1 5 7}      E16:  Min(X,Y)<=0   v   |J(Min(X,Y))+(-A)| <=Y

{-2 15}       E25:  Min(D(X),Y)<=0   v
                    |G(J(MIN(D(X),Y)))+(-G(A)))|<=X   v   X<=0

{-3 16}       E35:  Min(X,C(Y))<=0   v
                    |F(J(MIN(X,C(Y))))+(-F(A)))|<=Y   v   Y<=0

{-4 35 25}
              E67:  Min(D(H(X)),Y)<=0   v
                    Min(Z,C(H(X)))<=0   v   Half(X)<=0   v
|F(J(Min(Z,C(H(X)))+(-F(A))|+|G(J(Min(D(H(X)),Y)+(-G(A))|<=X

{-1 4 67}
              E125: | [F(J(Min(Z,C(H(X)))+(-F(A))] +
                      [G(J(Min(D(H(X)),Y)+(-G(A))] | <=X   v
              Min(D(H(X),Y)<=0 v Min(Z,C(H(X)))<=0 v H(X)<=0

{-5 125 1 s1 2 s2 3}

                E133: H(E)<=0   v   Min(D(H(E)),C(H(E)))<=0

{-6 8}          E137: X<=Y   v   Min(Y,X)=Y

{-7 133 137}    E191: C(H(E))<=D(H(E))   v   H(E)<=0

{-8 191}        E217: H(E)<=0 v Min(D(H(E)),C(H(E))) = C(H(E))

{-9 133 217}    E307: H(E)<=0

{-10 307}       E317: E<=0

{-11 317}       empty clause.
```

Let us examine the central clash of the proof:

```
                 E133: {-5 125 1 s1 2 s2 3}

(-5) | [F(J(U)+G(J(U))] + [(F(A)+(-G(A))] | ~<= E  v  U<=0
     (neg.th.)
   |
   |--(125)| [F(J(M(Z,C(H(X)))+(-F(A))] +
   |         [G(J(M(D(H(X)),Y))))+(-G(A)))] |<=X  v
   |         M(D(H(X)),Y)<=0 v M(Z,C(H(X)))<=0  v H(X)<=0
   |
   |         subst: M(Z,C(H(X)))/U, E/X
   |
[F(J(M(Z,C(H(E)))))+(-F(A))]+[G(J(M(D(H(E)),Y))))+(-G(A)))]
  ~=
[F(J(M(Z,C(H(E))))))+G(J((M(Z,C(H(E)))))))] + [-F(A)+(-G(A))]
  v  M(D(H(E)),Y)<=0  v  H(E)<=0  v  M(Z,C(H(E)))<=0

(let us abbreviate the above notation)
   [FJ+-FA]+[GJ+-GA] ~=[FJ+GJ]+[-FA+-GA] v M(D(H(E)),Y)<=0...
                       |
                       |--(1) (X+Y)+Z = X+(Y+Z)
                       |   sub: FJ/X,-FA/Y,(GJ+-GA)/Y
                       |
FJ+[-F(A)+(GJ+-G(A)] ~=[FJ+GJ]+[-FA+-GA] v M(D(H(E)),V)<=0..
                       |
                 (symmetry)
[FJ+GJ]+ [-FA+-GA] ~=FJ+[-FA+(GJ+-GA)] v M(D(H(E)),V)<=0 ...
                       |
                       |--(1)  (X+Y)+Z = X+(Y+Z)
                       |       sub: FJ/X,GJ/Y,(-FA+-GA)/Z
                       |
FJ+[GJ+[-FA+-GA]] ~=FJ+[-FA+[GJ+-GA]] v M(D(H(E)),V)<=0 v...
                       |
             (descent to lower level inequality)
                       |
   GJ+[-FA+-GA] ~= -FA+[GJ+-GA] v  M(D(H(E)),V)<=0  v ...
                       |
                       |--(2)  X+(Y+Z) = (X+Y)+Z
                       |       sub: GJ/X,-FA/Y,-GA/Z
                       |
   [GJ+-FA]+-GA ~= -FA+[GJ+-GA] v M(D(H(E)),V)<=0  v ...
                       |
                 (symmetry)
      -FA+[GJ+-GA] ~=[GJ+-FA]+-GA v M(D(H(E)),V)<=0 v ...
                       |
                       |--(2)  X+(Y+Z) = (X+Y)+Z
                       |       sub: -FA/X,GJ/Y,-GA/Z
                       |
      [-FA+GJ]+-GA ~=[GJ+-FA]+-GA v M(D(H(E)),V)<=0 v ...
                       |
```

```
           (descent to lower level inequality)
         [-FA+GJ] ~= [GJ+-FA]  v  M(D(H(E)),V)<=0  v ...
                              |
                (resume full notation)
                              |
-F(A)+G(J(M(D(H(E)),V))) ~= G(J(M(Z,C(H(E)))))+(-F(A))  v
                          |   M(D(H(E)),V)<=0  v  H(E)<=0 v
                          |   M(Z,C(H(E)))<=0
                          |
                          |--(3)  X+Y = Y+X
                          |       sub:-F(A)/X,
                          |       G(J(M(D(H(E)),V)))/Y,
                          |       D(H(E))/Z, C(H(E))/V
                          |
      E133: H(E)<=0  v  M(D(H(E)),C(H(E)))<=0     (merging)
```

In the above clash, in the first RUE inference we resolve to the topmost disagreement which is crucial. In subsequent steps a reordering of terms occurs by applying associative axioms four times and then finally commutivity.

Two other clashes resolve to inequalities:

```
       {-7 133, 137}:

       (-7)  D(U) ~<= 0  v  U<=0
                    |
                    |--(133)  M(D(H(E)),C(H(E)))<=0 v H(E)<=0
                    |
M(D(H(E)),C(H(E))) ~= D(U)  v H(E)<=0  v  U<=0
                    |
                    |--(137)  M(Y,X)=Y   V  X<=Y
                    |   sub:  D(H(E))/Y, C(H(E))/X, H(E)/U
                    |
     E191: C(H(E))<=D(H(E))  v  H(E)<=0
```

and {-9: 133, 217}:

```
     (-9) C(U) ~<= 0  v  U<=0
                |
                |--(133)  M(D(H(E)),C(H(E)))<=0 v H(E)<=0
                |
M(D(H(E)),C(H(E))) ~= C(U) v H(E)<=0 v U<=0
                |
                |--(217) M(D(H(E)),C(H(E)))=C(H(E)) v H(E)<=0
                |        sub: H(E)/U
                |
   E307: H(E) <= 0  (merging).
```

The above in two further steps is reduced to the empty clause since we are given E>0. Note that paramodulation is effectively incorporated into RUE resolution by the use of inequalities.

3. The Management of Heuristic Search

It is well known that despite the completeness of resolution it cannot be used efficiently with level saturation even under a restriction strategy. A heuristic ordering must be imposed on the search bringing into the foreground what is deemed relevant and doable in acceptable computer time. The effective use of a theorem prover requires human participation to create and adapt heuristics to a specific problem in the context of a particular inference system. The heuristics employed in our experiments are derived from the particular character of the RUE inference principle and from the structure of hyper-resolution.

In forming a substitution, the theorem prover scans potentially complementary literals, left to right, unifying at every point of disagreement but it passes over non-unifiable terms to compute a partial unifier when the mgu does not exist. If there is an mgu, the empty set becomes the disagreement set and standard resolution is applied. The first six clashes are based on the mgu but at clash:

E133: {-5 125, 1, s1, 2, s2, 3}

partial unification comes into play and inequalities appear in resolvents. In this clash, s1 denotes the symmetry interchange of the arguments of an inequality before applying electron 1 and similarly for s2.

The proof of the completeness of RUE resolution in open form states that a composite substitution exists which underlies a refutation but it is not specified. An RUE unifier is defined which is conjectured to satisfy completeness. Even use of the mgu (when it exists) in certain cases can forestall completeness in RUE resolution. This complex issue is discussed in [3]. What is important is that empirically we have found that the partial unifier is an effective heuristic choice leading to refutations.

In respect to choice of disagreement set in applying the RUE inference rule, we descend to the lowest level (innermost) disagreement not containing an irreducible inequality. An inequality is irreducible if we cannot deduce the corresponding equality, e.g., F(a)~=G(a) is irreducible since we cannot deduce F(a)=G(a) from the input set since we are only given that F and G are continuous at a and not equal. This heuristic selection of a disagreement set has proven effective in all experiments performed together with the partial unifier previously discussed.

At the end of each run, the theorem prover tabulates a list of inequalities present in resolvents from which the user selects what are judged to be irreducible inequalities. The latter are given to the theorem prover as input in the next run so that no resolvents are formed with such inequalities. This sometimes directs the theorem prover to use the topmost disagreement set in place of the innermost. This occurred in resolving nucleus (-5) with electron E125 in clash {-5 125 1 s1 2 s2 3} and enabled the theorem prover to then apply the associativity axioms to the topmost inequality to reorder terms and arrive at the correct refutation path.

The refutation search was pruned by a fairly standard use of complexity bounds in which resolvents are discarded if they exceed limits in respect to depth of function nesting, too many literals, too many distinct variables, excessive use of individual function symbols or excessively long clash development.

We now consider heuristics derived from the structure of hyper-resolution. An E-unsatisfiable input set has an RUE hyper-refutation composed of clashes in the form:

$$\{N: E_1,\ldots,E_k\} \text{ ---> new electron}$$

The final clash generates the empty clause which necessarily implies that the input set contains a nucleus composed only of negated literals, which resolves with unit electrons to generate the empty clause.

The N set does not change during the proof search as only new electrons are generated and added to the E set. Intermediate resolvents within a clash are discarded. The hyper-resolution algorithm simply repeatedly sweeps the stable N set applying old and new electrons until the empty clause is derived. A complete sweep of N producing no new electrons indicates that the input set is satisfiable and terminates execution. At the end of a computer run, a tabulation is given the user of the generated E set which is a measure of the success of the search.

The user may then specify for each nucleus a set of E-ranges: (r_1, r_1'), ..., (r_n, r_n') specifying that in the next run the nucleus is to be acted upon only with electrons within these ranges (the range number is simply a subscript into the array of electrons comprising the E set). This gives the user direct control to either constrain or expand the search for clashes pertaining to a nucleus.

Furthermore, the order of nuclei in the N set is important since some nuclei are used at the beginning of a refutation to generate electrons for other nuclei later in the refutation. The input order of the nuclei should approximate the order in which nuclei are used in the refutation and user should experiment with this order when the proof search stalls. It is also worthwhile to expand the N set at the beginning of a run by duplicating certain nuclei so that a sweep of the N set strategically emphasizes certain nuclei to exploit new electrons as they are generated.

4. Conclusion

In [1] Bledsoe presents a paramodulation proof of this theorem, using an input set where the axioms we have used: Min(X,Y)<=X, Min(X,Y)<=Y and X~<=H(Z) v Y~=H(Z) v X+Y<=Z, are not present but instead are derived from more elementary axioms. He also presents a much longer proof with the equality axioms which shows the importance of using paramodulation to reduce the length of equality proofs. RUE and paramodulation refutations are basically the same length and a major improvement over explicit use of the equality axioms. The refutations stated by Bledsoe are not hyper-refutations and are stated as being hand derived. We are continuing our experiments and should have the RUE proof for Problem 5 (unreduced) available for conference presentation.

We have currently embarked on a two year program of development to completely recast the RUE theorem prover to include diverse restriction strategies such as Lock resolution, Set of Support, Linear resolution and Unit resolution, all of which must be adapted to the RUE context. We expect to perform many more experiments, testing various restriction strategies and developing further the body of RUE heuristics for proof search. We will also introduce a new user interface to facilitate use of the theorem prover by other researchers.

Appendix

We state the complete computer generated proof including steps within clashes. We have as the input set:

E: (set of electrons)

```
 1. (X+Y)+Z = X+(Y+Z)     (associativity)
 2. X+(Y+Z) = (X+Y)+Z
 3. X+Y = Y+X             (commutivity)
 4  |X+Y|<=|X|+|Y|
 5. X<=0  v  |J(X)+(-A)|<=X
 6. M(X,Y)<=X
 7. M(X,Y)<=Y
 8. X<=Y  v  Y<=X
```

N: (set of nuclei)

```
 1. U~<=V  v  V~<=W  v  U<=W    (transitivity)
 2. |V+(-A)|~<=D(U)  v  |G(V)+(-G(A))|<=U  v  U<=0
 3. |V+(-A)|~<=C(U)  v  |F(V)+(-F(A))|<=U  v  U<=0
 4. U~<=H(W)  v  V~<=H(W)  v  U+V<=W
 5. (|(F(J(U))+G(J(U)))+(-F(A)+(-G(A)))|~<=E  v  U<=0
 6. U~<=V  v  M(U,V)=U
 7. D(U)~<=0  v  U<=0
 8. V~<=U  v  M(U,V)=V
 9. C(U)~<=0  v  U<=0
10. H(U)~<=0  v  U<=0
11. E~<=0
```

Refutation:

The substitution applied is the mgu or mgpu and we rename variables in resolvent electrons to X,Y,Z. The parent clauses of a resolvent are stated to the right with indices to the literals resolved upon.

```
R1:  X<=W  v  |J(X)+(-A)|<W  v  X<0  [-1(1),5(2)]

E15: M(X,Y)<=0  v  |J(M(X,Y)+(-(A))|<=X  [R1(1),6(1)]

R2:  X<=W  v  |J(X)+(-A)|<W  v  X<0  [-1(1),5(2)]

E16: M(X,Y)<=0  v  |J(M(X,Y)+(-(A))|<=Y  [R2(1),7(1)]

E25: M(D(U),Y)<=0  v  |G(J(M(D(U),Y)))+(-G(A)|<=U  v  U<=0
renamed:
     M(D(X),Y)<=0  v  |G(J(M(D(X),Y)))+(-G(A)|<=X  v  X<=0
     [-2(1),15(2)]
```

```
E35: M(X,C(U))<=0  v  |F(J(M(X,C(U))))+(-F(A))|<U  v  U<0
renamed:
     M(X,C(Y))<=0  v  |F(J(M(X,C(Y))))+(-F(A))|<Y  v  Y<0
     [-3(1),16(2)]

R3:  V~<=H(W)  v  M(X,C(H(W)))<=0  v
     |F(J(M(X,C(H(W)))))+(-F(A))|+V<=W  v  H(W)<=0
     [-4(1),35(2)]

E67: M(D(H(W)),Y)<=0 v M(X,C(H(W)))<=0 v H(W)<=0 v
|F(J(M(X,C(H(W)))))+(-F(A))|+|G(J(M(D(H(W),Y)))+(-G(A))|<=W
renamed:
     M(D(H(X)),Y)<=0 v M(Z,C(H(X)))<=0 v H(X)<=0 v
|F(J(M(Z,C(H(X)))))+(-F(A))|+|G(J(M(D(H(X),Y)))+(-G(A))|<=X
     [R3(1),25(2)]

R4:  |X|+|Y|<=W  v  |X+Y|<W  [-1(1),4(1)]

E125: M(D(H(X)),Y)<=0  v  M(Z,C(H(X)))<=0  v  H(X)<=0  v
 |(F(J(M(Z,C(H(X)))))+(-F(A)))+(G(J(M(D(H(X),Y)))+(-G(A)))<=X
  [R4(1),67(4)]

R5:(F(J(M(Z,C(H(E)))))+(-F(A)))+(G(J(M(D(H(E)),V))))+(-G(A))))
 ~=(F(J(M(Z,C(H(E))))))+G(J((M(Z,C(H(E)))))))+(-F(A)+(-G(A)))
  v M(D(H(E)),Y)<=0 v H(E)<=0 v M(U,C(H(E)))<=0 (renamed)
  [-5(1),125(4)]

R6: F(J(M(Z,C(H(E)))))+ (-F(A)+(G(J(M(D(H(E),Y))))+(-G(A)))
   ~=(F(J(M(Z,C(H(E)))))+G(J(M(Z,C(H(E))))))+ (-FA+(-G(A)))
    v M(D(H(E)),V)<=0  v  H(E)<=0  v  M(Z,C(H(E)))<=0
    [R5(1),1(1)]

R7: (F(J(M(Z,C(H(E)))))+G(J(M(Z,C(H(E))))))+ (-FA+(-G(A)))
   ~= F(J(M(Z,C(H(E)))))+(-F(A)+(G(J(M(D(H(E),Y))))+(-G(A)))
    v M(D(H(E)),V)<=0  v  H(E)<=0  v  M(Z,C(H(E)))<=0
    [symmetry of inequality]

R8:    G(J(M(Z,(C(H(E)))))+(-F(A)+(-G(A)))
   ~= -F(A)+(G(J(M(D(H(E)),V)))+(-G(A))))
      v M(D(H(E)),V)<=0  v  H(E)<=0  v  M(Z,C(H(E)))<=0
      [R7(1), 1(1)]

R9:  (G(J(M(Z,C(H(E)))))+(-F(A)))+ (-G(A)))
  ~= -F(A)+ (G(J(M(D(H(E)),V)))+(-G(A)))
      v M(D(H(E)),V)<=0  v  H(E)<=0  v  M(Z,C(H(E)))<=0
      [R8(1), 2(1)]

R10:    -F(A)+ (G(J(M(D(H(E)),V)))+(-G(A)))
     ~= (G(J(M(Z,C(H(E)))))+(-F(A)))+ (-G(A)))
      v M(D(H(E)),V)<=0  v  H(E)<=0  v  M(Z,C(H(E)))<=0
      [symmetry of inequality]
```

```
R11: -F(A)+G(J(M(D(H(E)),V))) ~= G(J(M(Z,C(H(E)))))+(-F(A))
      v M(D(H(E)),V)<=0  v  H(E)<=0  v  M(Z,C(H(E)))<=0
      [R10(1), 2(1)]

E133: H(E)<=0  v  M(D(H(E)),C(H(E)))<=0
      [R11(1),3(1)]

E137: V<=U  v  M(U,V)=U  renamed to  X<=Y  v  M(Y,X)=Y
      [-6(1), 8(1)]

R12: M(D(H(E)),C(H(E)))~=D(U)  v  H(E)<=0  v  U<=0
     [-7(1), 133(2)]

E191: C(H(E))<=D(H(E))  v  H(E)<=0
      [R12(1),137(2)]

E217: H(E)<=0  v  M(D(H(E),C(H(E)))=C(H(E))
      [-8(1),191(1)]

R13: M(D(H(E)),C(H(E)))~=C(U)  v  H(E)<=0  v  U<=0
     [-9(1),133(2)]

E307: H(E)<=0
      [R13(1),217(2)]

E317: E<=0
      [-10(1),307(1)]

E318: empty clause
      [-11(1),317(1)]
```

The anomaly that exists in theorem proving is that a proof which may be difficult for a human may not be difficult for a computer and conversely. Here the proof of continuity of the sum of continuous functions is easy for a human since the proof structure is intuitively evident from the definition of continuity, yet the proof is difficult for a computer which relies on syntactic search in place of intuition. In contrast, a human finds it difficult to prove associativity of the logical-or [4] in boolean algebra where there is no direct intuitive approach but the computer finds it a shorter and less difficult proof than the above.

References

1. W.W.Bledsoe:Challenge Problems in Elementary Calculus. Jn. Automated Reasoning,Vol.6,No.3,Sept1990,341-359.

2. J.A.Robinson:Automatic Deduction with Hyperresolution. Int.Journal of Computational Math,1965,227-234.

3. V.J.Digricoli,M.C.Harrison:Equality-Based Binary Resolution.Journal ACM,v33,n2,Apr1986,253-289.

4 V.J.Digricoli:The Management of Heuristic Search in Boolean Exp's with RUE Resolution.IJCAI-85,1154-1161.

5. V.J.Digricoli,J.Lu,V.Subrahmanian:And-Or Graphs Applied to RUE Resolution.IJCAI-89,354-358.

6. L.A.Wos,R.A.Overbeek,L.Henschen:Hyperparamodulation- a Refinement of Paramodulation.CADE-5,1980,208-219.

7. J.B.Morris:E-Resolution:an Extension of Resolution to Include the Equality Relation.IJCAI-69,287-294.

8. G.G.Birkhoff:Distributive Postulates for Systems Like Boolean Algebra.Trans Am Math Society,v60,July-Dec1946.

9. C.Chang,R.Lee:Symbolic Logic and Mechanical Theorem Proving.Academic Press, 1989.

10. V.J.Digricoli:Resolution by Unification and Equality. Dissertation Courant Institute,February 1983.

Computing Prime Implicates Incrementally

Peter Jackson

McDonnell Douglas Research Laboratories
St. Louis, Missouri, USA

Abstract. We describe an algorithm (called PIGLET) which takes a theory already in prime implicate form and computes the prime implicates of that theory extended by a single clause. Then we compare PIGLET's performance with that of an alternative algorithm for this incremental computation (called IPIA). We critique some optimizations that have been proposed for IPIA, and show that two of them interact to render the modified method incomplete. Finally, we present data which show that PIGLET outperforms IPIA, in that it generates a smaller search space and takes less CPU time. Its superiority is evident both when updating a randomly-generated theory by a random clause and when we generalize the two programs for the task of updating by more than one clause. Its high performance is mainly due to heuristics which capitalize on merges to promote early backward subsumption and pre-empt forward subsumption.

1 Introduction

In the 1950s, Quine [13, 14, 15] discussed the problem of devising a general mechanical procedure for reducing any formula to its simplest equivalent. This problem is relevant to a number of applications in electrical engineering and computer science, such as the minimization of Boolean functions [1, 10, 17], and also to applications in automated reasoning and artificial intelligence, such as truth maintenance and belief revision [6, 12, 16]. Quine proposed a solution based on the developed Disjunctive Normal Form, but noted that the procedure was 'laborious.' Assuming $P \neq NP$, the problem is now considered to be intractable in general, since a worst case solution requires resources which are exponential in the number of distinct propositional variables occuring in the formula.

In his original papers, Quine used the term 'prime implicant of a formula Φ' to refer to 'a fundamental formula that logically implies Φ but ceases to when deprived of any one literal' ([15], pp. 164-165). Formulas were sums of products of literals, *i.e.*, they were in Disjunctive Normal Form (DNF), and were shown to be logically equivalent to the alternation of their prime implicants. Since then, others [16] have used the term to refer to the dual notion for the Conjunctive Normal Form (CNF), where formulas are products of sums of literals and a set of clauses is logically equivalent to the conjunction of its prime implicates.

Our terminology and notation shall be as follows. The term *theory* will refer to any finite set of formulas in Conjunctive Normal Form, constructed over a finite alphabet, but not necessarily closed under deduction. The letter T will stand for such a theory, the letter C will stand for a disjunction of literals (*i.e.*, a clause), while the lower case $\pm c$ will stand for a literal of indifferent sign.

Definition. A disjunction C is a *prime implicate* of a theory T iff

(i) C is not a tautology;
(ii) $T \models C$; and
(iii) for all literals $\pm c$ in C, $T \not\models C'$ where C' is C less $\pm c$.

If C and C' are disjunctions, then C *subsumes* C' iff $C \models C'$.

(The dual notion for the Disjunctive Normal Form can be defined as follows. A conjunction C is a *prime implicant* of a theory T iff (i) C is not a contradiction; (ii) $C \models T$; and (iii) for all literals $\pm c$ in C, $C' \not\models T$ where C' is C less $\pm c$. If C and C' are two conjunctions, then C *subsumes* C' iff $C' \models C$.) ||

There are a number of algorithms in the literature for putting a theory, T, in prime implicate form (*e.g.*, [8, 17, 18]). Here we present two novel algorithms, one (called PIG) for the full conversion from T to its prime implicates, and one (called PIGLET) for the special case of extending a theory already in prime implicate form. Thus given the set of prime implicates of T, $\mathbf{PI}(T)$, and an *input clause* C, PIGLET computes the prime implicates of $\mathbf{PI}(T) \cup \{C\}$. In another paper [5], we show that the PIG algorithm's mean performance on random problems is generally better than [17] and [18], and comparable with [8]. Here we consider the incremental computation, for which [8] is not well-suited.

Prime implicate programs, such as those mentioned above, have obvious applications in artificial intelligence, *e.g.*, when using an ATMS to perform abduction. If C is a fault symptom, then the minimal supports of $\neg C$ w.r.t. some theory T constitute parsimonious explanations of C in terms of the knowledge in T, and prime implicates play a major role in the computation of minimal supports [16]. A number of recent proposals for belief revision also involve the use of prime implicates (*e.g.*, [6], [12]), in both the specification and the implementation of revision functions.

In this paper, we present data which demonstrate that PIGLET outperforms the best alternative algorithm in the literature, due to Kean & Tsiknis, called IPIA [9]. We also critique four optimizations which have been proposed for IPIA and show that two of these interact to render the method incomplete. Kean & Tsiknis showed that the incremental computation has the same complexity as the full transformation to prime implicate form, and is therefore intractable in the worst case. However, they argue that the mean complexity of the incremental algorithm is less than the full computation. One could also argue that the worst case may not be the general case for many applications in automated reasoning.

2 The Incremental Prime Implicate/Implicant Algorithm (IPIA)

IPIA is based on Tison's [18] method for generating the prime implicates of a set of clauses T. One key notion is that of a *biform variable*: a variable that occurs both positively and negatively in T. Another is what Kean and Tsiknis call *consensus*: a restricted kind of resolution in which we restrict our attention to those resolvents of clauses in T which are not tautologies. (Quine and others use the term 'consensus' somewhat differently to a describe an operation that is essentially the dual of resolution in DNF. We follow Kean's terminology here to facilitate comparison with [9], though all examples are given in CNF.)

Tison's method can be described by the following loop. For each biform variable c and for every pair of clauses C_i, C_j in T, add the consensus of C_i, C_j on c to T, if such a consensus exists, and delete every subsumed clause from T. At termination, T will contain all the prime implicates of the original set of clauses.

The IPIA algorithm of Kean & Tsiknis resembles Tison's method, except that it stores the new implicates of $\mathbf{PI}(T) \cup \{C\}$ in a set S. It turns out that the incremental computation admits of the following simplifications: (i) we need only perform consensus with respect to biform variables of T occuring in the input clause; and (ii) we need only perform consensus between clauses from the sets S and T, but not within the same set. Various optimizations have been proposed for IPIA, which we discuss in the following section.

The IPIA Algorithm

Given a theory T in prime implicate form, and an input clause C, IPIA performs the following sequence of steps, and returns the set $S \cup T$.

1. $S := \{C\}$;
   delete any subsumed clauses from the set $S \cup T$;
   if C is deleted, then halt.
2. For each biform variable c occuring in C do
   for each $C_i \in S$ and $C_j \in T$ such that C_i, C_j have consensus on c do
   let C_k be the consensus of C_i and C_j on c;
   $S := S \cup \{C_k\}$;
   delete any subsumed clauses from the set $S \cup T$. *(End of Algorithm)*

It can be shown that, on termination of the algorithm, $S \cup T$ will contain all and only the prime implicates of $T \cup \{C\}$.

3 Optimized IPIA

Kean & Tsiknis propose four optimizations of IPIA, which we shall discuss in the order in which they appear in Section 7 of their paper [9]. Some additional notation will be required. If $C = \{\pm c_1, ..., \pm c_n\}$ is an input clause for theory T, then

$[C] = \{c_i \mid \pm c_i \in C$ and c_i is a biform variable w.r.t. $T\}$,

Also, let $C_i|_x C_j$ denote the consensus of C_i and C_j on variable x. We shall omit the consensus variable when it is not germane to the discussion.

(1) *Root optimization.* If $C|_x C_i$ subsumes C for some $x \in [C]$, then terminate the search for consensi on x and make $C|_x C_i$ the root of a new consensus tree. Obviously, any further consensi involving C will be subsumable by consensi involving $C|_x C_i$. However, the problem of finding such $C|_x C_i$ quickly is not addressed by the optimized algorithm.

(2) *Single biform selection.* When looking for consensi with C on the ith biform variable, x_i, only consider candidates in T which have $\pm x_i$ as the only literal which is complementary

to literals in C. More than one complementary literal will not generate a consensus but a tautology.

(3) *The history restriction.* For each clause in S (*i.e.*, for C and for each clause derived by consensus), define its *history* as follows:

$$history(C) = \emptyset;$$
$$history(C_i|_x C_j) = history(C_i) \cup \{x\}.$$

Thus the history of a clause contains all the biform variables that have figured in its generation. The claim is that, when looking for consensi with a clause C_i, we need only consider clauses C_j in T such that $history(C_i) \cap C_j = \emptyset$. A non-null intersection means that the consensus would reintroduce variables from C_i's history unnecessarily.

(4) *Local subsumption check.* If $C_i|_x C_j$ subsumes C_i, then $C_i|_x C_j$ subsumes all consensi $C_i|_x C_k$ formed so far such that $j \neq k$. If $C_i|_x C_j$ subsumes C_j, then $C_i|_x C_j$ subsumes all consensi $C_k|_y C_j$ formed so far with any $C_k \in S$ such that $i \neq k$, and any biform variable y. Early elimination of these subsumed clauses is obviously advantageous.

The trouble is that (3) and (4) interact to cause incompleteness, as can be illustrated by the following simple example.

Example 1. Let $T = \{\{b, \neg d, f\}, \{\neg b, \neg d, \neg f\}\}$ and let $C = \{b, e, \neg d, \neg f\}$. So $[C] = \{b, f\}$, and we can build the consensus tree shown in Figure 1(i). But (3) disallows the final step, since it reintroduces the variable b which appears in the history of $\{e, \neg d, \neg f\}$. This would cause no problems if we could form the consensus

$$\{b, e, \neg d, \neg f\}|_f\{b, \neg d, f\}.$$

$\{e, \neg d, \neg f\}$ subsumes $\{b, e, \neg d, \neg f\}$, however, so the latter has already been deleted, courtesy of (4).

Had we performed

$$\{b, e, \neg d, \neg f\}|_f\{b, \neg d, f\}$$

before performing

$$\{b, e, \neg d, \neg f\}|_b\{\neg b, \neg d, \neg f\},$$

we would still be in trouble, because $\{b, e, \neg d, \neg f\}$ would still be deleted, and (3) would prevent the consensus

$$\{b, e, \neg d\}|_b\{\neg b, \neg d, \neg f\}$$

from generating the clause $\{e, \neg d, \neg f\}$, since this step reintroduces the variable f, which appears in the history of $\{b, e, \neg d\}$ — see Figure 1(ii).

These two heuristics therefore combine to cause incompleteness since, in our example, we will miss generating one or other of the prime implicates $\{b, e, \neg d\}$ and $\{e, \neg d, \neg f\}$, depending upon the order in which we perform consensi. ||

In Section 5, we shall see that (4) contributes surprisingly little to IPIA's performance on randomly-generated problems.

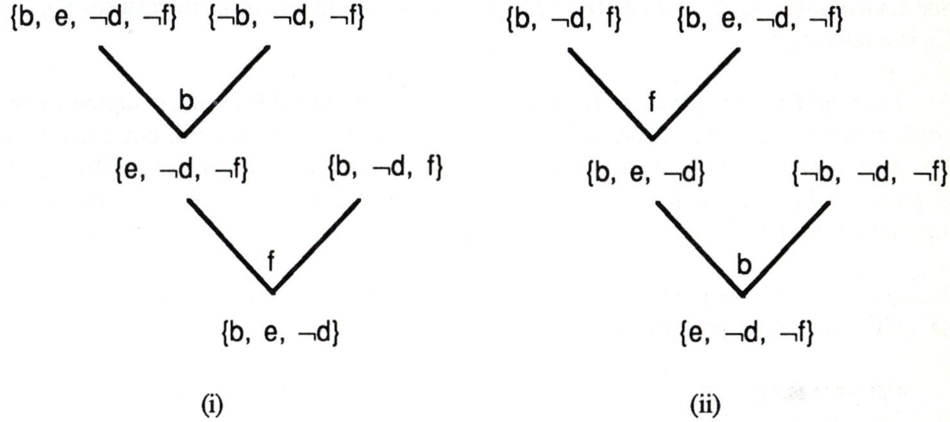

Figure 1. Two consensus trees for Example 1. Trees (i) and (ii) show different orders for performing consensi among clauses, each of which results in incompleteness (see text).

The remainder of the paper has the following plan. Section 4 introduces a novel algorithm (called PIG, for Prime Implicate Generator) which also resembles Tison's method, except that it computes prime implicates using a particular resolution strategy which concentrates on finding merges. Then we outline a specialization of PIG, called PIGLET, for the incremental computation. In Section 5, we compare PIGLET with IPIA on computations in which we update randomly-generated theories with one or more randomly-generated input clauses. Section 6 discusses the results of this experiment, considers related and future avenues of research, and serves as a summary and conclusion.

4 The Prime Implicate Generator (PIG)

To describe the PIG algorithm in detail, we shall need the following notation. Let T be a set of clauses, $\{C_1, ..., C_n\}$. $LHS(C_i) \subseteq C_i$ denotes the set of positive literals of C_i, and $RHS(C_i) \subseteq C_i$ denotes the set of negative literals of C_i. $|C_i|$ denotes the cardinality of C_i, and $C_i|C_j$ the consensus of C_i and C_j. $COMPLEMENTS(C_i)$ is based on the set $COMPS(C_i)$, which is given by

$$\{C_j \in T \mid |LHS(C_i) \cap RHS(C_j)| = 1 \text{ or } |RHS(C_i) \cap LHS(C_j)| = 1, \text{ but not both}\}.$$

Intuitively, the $COMPS$ of a clause are those other clauses that could resolve with it without generating a tautology, as in the 'single biform selection' optimization of Kean & Tsiknis. We also use the following optimizations when generating the $COMPLEMENTS$.

(1) *Avoid duplicate steps.* Since we compute the *COMPLEMENTS* of clauses of T in a particular order, we can arrange things so that $C_j \notin COMPLEMENTS(C_i)$ if we already have $C_i \in COMPLEMENTS(C_j)$ to avoid performing the same consensus twice.

(2) *Preempt forward subsumption.* Let $<C_j>_i$ denote C_j less those merge literals which it shares with C_i. Considering any C_j in $COMPS(C_i)$, we can disbar from the set of complements all those C_k in $COMPS(C_i)$ such that $<C_j>_i \subseteq <C_k>_i$. The complements that remain make up $COMPLEMENTS(C_i)$. We assume that none of the clauses C_i, C_j, C_k is a tautology.

The need for (1) is obvious. It should also be obvious that this cannot compromise the completeness of the PIG algorithm. (2) is perhaps less straightforward, but Example 2 illustrates how such C_k lead to consensi with C_i that will be forward subsumed by $C_i|C_j$. A proof to the effect that (2) preserves the completeness of PIG can be found in the Appendix (Theorem 1).

Example 2. Let $C_i = \neg p \vee q$, and let $C_j = \neg p \vee \neg q \vee r$ and $C_k = \neg q \vee r \vee t$ be members of $COMPS\ (C_i)$. If we perform the two resolution steps

$$C_i|C_j = \neg p \vee r$$
$$C_i|C_k = \neg p \vee r \vee t$$

we see that the first resolvent subsumes the second. However, if we first apply the preempt forward subsumption heuristic, we find that

$$<C_j>_i = \{\neg q, r\}$$
$$<C_k>_i = \{\neg q, r, t\}$$

so $<C_j>_i \subseteq <C_k>_i$ and we omit C_k from $COMPS\ (C_i)$. ||

The PIG Algorithm

1. $S := \{C_i \in T \mid COMPLEMENTS(C_i) \neq \emptyset\}$.

2. For each $C_i \in S$, and for every $C_j \in COMPLEMENTS(C_i)$,

$$COST(C_i, C_j) = (|C_i| + |C_j|) - (|LHS(C_i) \cap LHS(C_j)| + |RHS(C_i) \cap RHS(C_j)|).$$

$$COST^*(C_i) = min[COST(C_i, C_j)], \qquad \text{for } C_j \in COMPLEMENTS(C_i).$$

3. Sort S so that $C_i < C_k$ only if $COST^*(C_i) < COST^*(C_k)$.

4. If S is empty, then halt and return T, else for the next $C_i \in S$,

4.1. Remove C_i from S.

4.2. If C_i has been deleted, then go to 4.

4.3. If C_i's best (*i.e.*, lowest cost) complement, C_j, has been deleted, then recompute $COST^*(C_i)$, reschedule C_i, and go to 4.

4.4. Compute the consensus $C_i|C_j$ and remove C_j from the complements of C_i.
4.5. If $C_i|C_j$ is the empty clause, then halt and return $\{\perp\}$.
4.6. Perform both forward and backward subsumption on $C_i|C_j$ with respect to the clauses in T.
4.7. If $C_i|C_j$ survives forward subsumption, then add it to T and sort-insert it into S using the cost function of Step 2.
4.8. Reschedule C_i only if it has complements left.
4.9. Go to 4. *(End of Algorithm)*

The algorithm assumes that tautologies and subsumed clauses have been deleted from T; this could easily be incorporated as a preliminary step. The cost function in Step 2 biases the search towards consensi that generate *merges*, *i.e.*, resolutions involving pairs of clauses that contain literals of the same sign in addition to complementary literals. The resulting merges contain fewer literals than non-merge resolvents derived from parents of similar size. These resolvents have both less chance of being subsumed and more chance of subsuming existing clauses before the latter generate more resolvents. Needless to say, the empty clause subsumes all other clauses, and its generation signifies that T is inconsistent.

A correctness proof for PIG is outlined in the Appendix (Theorem 2).

In order to compare the our resolution-based approach with that of IPIA, we first derived a version of PIG (called PIGLET), which is specialized for the incremental computation, but which handles any number of new clauses. Since not every compound proposition can be represented by a single clause, we deemed this generalization to be advantageous for applications in automated reasoning.

PIGLET can be outlined as follows. Let T be the set of prime implicates of the original theory and let A be the *set* of new clauses to be incorporated in T. Then PIGLET is simply a restriction of PIG in which:

(1) we never resolve any two clauses in T;
(2) if A is a singleton, then we never resolve any two consequences of $T \cup A$;
(3) if A is not a singleton, then we give top priority to resolving clauses in A with each other, *i.e.*, clauses in A have a cost of zero while they have *COMPLEMENTS* in A; and
(4) for any clause C_i, $COMPLEMENTS(C_i)$ is constructed over $T \cup A$, instead of over T.

The PIGLET Algorithm

1. $S := \{C_i \in A \mid COMPLEMENTS(C_i) \neq \emptyset\}$.
2. For each $C_i \in S$, and for every $C_j \in COMPLEMENTS(C_i)$,

$$COST(C_i, C_j) = \text{if } C_j \in A, \text{ then } 0 \text{ else } (|C_i| + |C_j|) - (|LHS(C_i) \cap LHS(C_j)| + |RHS(C_i) \cap RHS(C_j)|).$$

$$COST^*(C_i) = min[COST(C_i, C_j)], \qquad \text{for } C_j \in COMPLEMENTS(C_i).$$

3 & 4. As in PIG. *(End of Algorithm)*

A correctness proof for PIGLET is outlined in the Appendix (Theorem 3).

5 Experimental Procedure and Results

We implemented PIGLET and the IPIA algorithm, both with and without the four optimizations suggested in [9]. We noted in Section 3 that two of the optimizations (the *history restriction* and *local subsumption checking*) interact to render Optimized IPIA incomplete, while *root optimization* is not effective unless guided by heuristics. The data reported in this paper was derived using *single biform selection* and *local subsumption checking*. However, we also implemented a variant that used the *history restriction* in place of *local subsumption checking*, and reran some of the data. The results of this exercise are discussed below.

In fact, we implemented a straightforward generalization of IPIA which, like PIGLET, can accommodate any number of new clauses. The only modification is to Step 2, where we must allow ourselves to compute the consensus of two clauses in S, if and only if S is not a singleton intially, although we still refrain from resolving any pair of clauses in the original set T. This modification has no effect upon the performance of the program when updating by a single clause.

Both algorithms were implemented in an object-oriented extension to Lisp, called SLOOP [4]. They share much code, including many classes and methods. Clauses, theories and atoms are all objects; typical operations perform indexing, subsumption checking, searching for resolution steps, *etc.* The programs are not built for speed, and would benefit from reimplementation in another language. Our main focus is upon the size of the search space generated by each algorithm.

In comparing the programs, the measure that we shall use is the Total Number of Nodes expanded (hereafter TNN), where each node represents a consensus step. We use TNN rather than penetrance because it is easy to see that neither algorithm is admissible. Consequently, it is not possible to compute the smallest consensus tree using the programs themselves, and it is infeasible to do this by hand for large theories.

In order to effect an impartial comparison, a program was written which generated propositional calculus theories in CNF using pseudo-random numbers. (The same program is briefly described in [8].) The parameters to the program were as follows: (i) the number of clauses to appear in the theory; (ii) the maximum number of literals in such clauses; and (iii) the size of the alphabet from which literals should be formed.

In addition, the theories generated were subject to the following constraints. Unit clauses were outlawed, since their presence can lead to too many clauses being subsumed (making the theory too 'easy'), or the theory being inconsistent (in which case its sole prime implicate is $\bot$). Theories contained no tautologies, and no clauses contained redundant literals.

We compared the two programs on three classes of theories, updated by one, three and five clauses. These clauses were guaranteed not to be subsumed by existing clauses, since this might have made the incremental computation too easy. In the handful of cases where the new clauses turned out to be inconsistent with the theory, a new theory and a new set of updating clauses were generated.

Each class contained 20 theories with a maximum clause size of 5 literals. The A classes were constructed over an alphabet of 8 propositional letters, the B classes over an alphabet of 10, and the C classes over an alphabet of 12. Each of the classes A, B and C contained as many clauses as they contained propositional letters, *i.e.*, 8, 10 and 12 respectively. This was to ensure that the combinatorial effect of increasing alphabet size was not unduly restricted by the size of the theory. (Table 1 shows that theory size *per se* has no reliable effect upon the size of the search space generated by either program.) Finally, we generated an additional pair of theory classes, D1 and E1, with a theory size and alphabet size of 14 and 16 respectively, in order to confirm a trend in the data when updating by a single clause. Thus we generated 220 random theories in all.

We defer a discussion of the appropriateness of using random problems for this comparison until Section 6.

Class	PIs (mean)	PIGLET (mean)	IPIA (mean)	PIGLET (sd)	IPIA (sd)
A1	17.0 (19.5)	11.7	17.6	7.83	13.30
A3	16.8 (20.6)	37.7	86.6	6.99	71.25
A5	17.5 (28.3)	117.1	359.8	116.26	340.85
B1	28.0 (32.4)	22.6	36.4	22.31	34.42
B3	24.5 (44.6)	168.6	429.0	193.96	599.91
B5	30.4 (44.4)	209.3	738.4	181.52	646.66
C1	44.3 (51.5)	54.5	73.1	49.06	68.23
C3	55.4 (84.2)	438.5	1152.0	397.50	1241.00
C5	40.9 (85.5)	635.1	1886.9	500.99	1593.70
D1	80.1 (82.2)	103.0	152.4	136.57	215.47
E1	159.5 (159.5)	307.6	541.9	360.84	751.02

Table 1. Data summary of the experiment: means (to 1 d.p.) and standard deviations (to 2 d.p.) of TNN (Total Number of Nodes expanded) for PIGLET and IPIA for each class of theory. The number in the name of the theory class indicates the number of new clauses to be incorporated. The PIs column shows the mean number of prime implicates of the original theory, while the number in parentheses gives the mean number of prime implicates of the augmented theory.

Theories were put in prime implicate form initially using the MM program [8]. The data for the incremental computation are given in Table 1; the version of IPIA used throughout is the one with 'local subsumption checking.' The data are displayed graphically in Figures 2 and 3, where means atop columns are rounded to the nearest whole number.

Figure 2 shows that PIGLET outperforms IPIA in a remarkably regular fashion when updating by a single clause. The ratio between the size of the search spaces that they generate is about 2:3, regardless of alphabet size. The standard deviation data indicates that PIGLET's performance is also much less variable than IPIA's. Figure 3 shows that PIGLET's advantage grows as we increase the number of new clauses. These data do not show an exponential trend as in Figure 2, but the ratio is larger, about 1:3.

The version of IPIA that incorporated the 'history restriction' instead of the 'local subsumption checking' performed so similarly to the latter that it is not worth reproducing the data. We did not rerun the modified program on the whole data, just on classes B1, B2, C1 and C2 in order to check the behaviour on updating by both a single clause and more than one clause in two classes. The total number of nodes expanded diverged from the original run on just 2 out of the 80 theories, resulting in moderate savings.

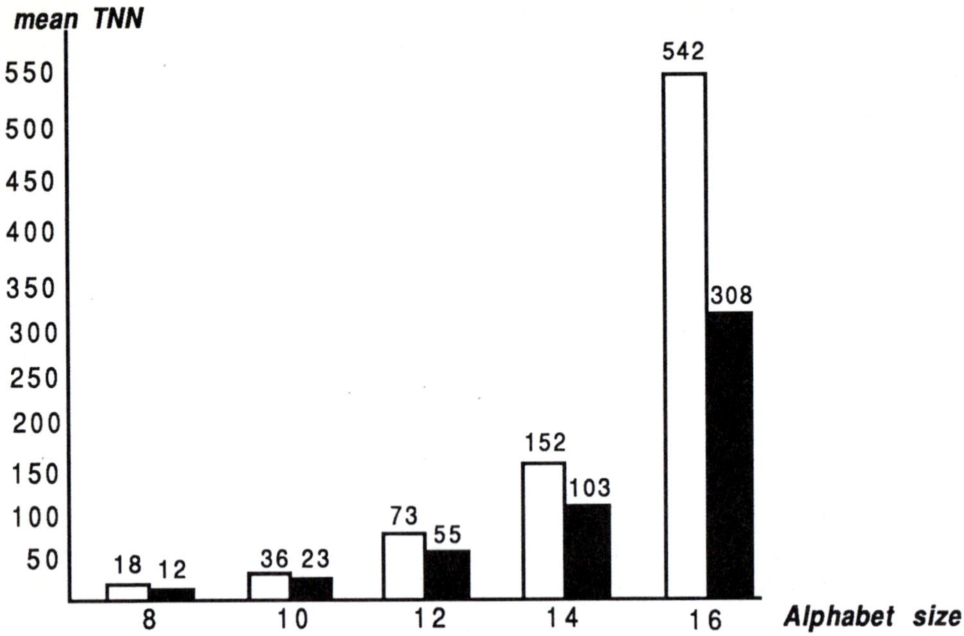

Figure 2. Bar chart plotting mean number of nodes expanded by IPIA (light bars) and PIGLET (dark bars) against alphabet size (classes A1, B1, C1, D1, and E1).

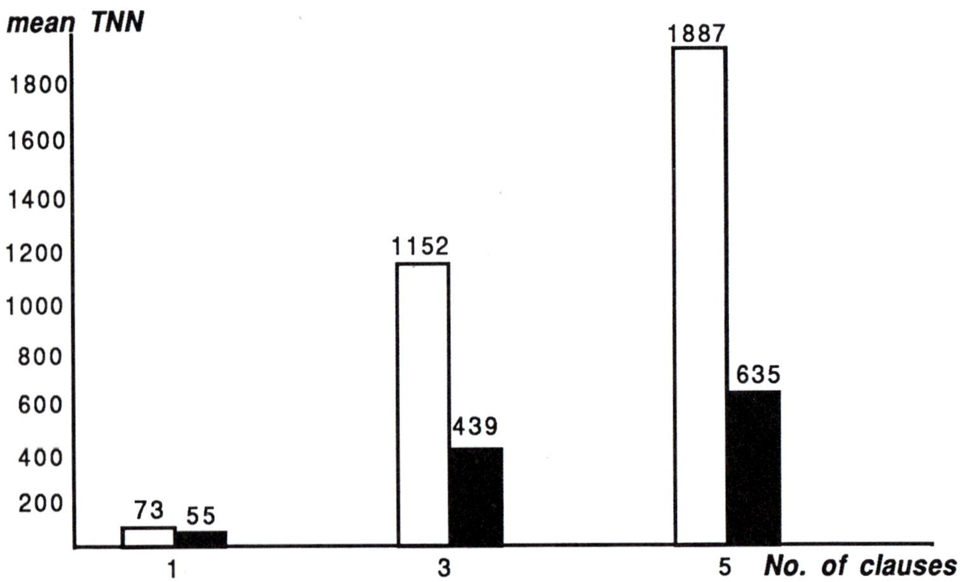

Figure 3. Bar chart plotting mean number of nodes expanded by IPIA (light bars) and PIGLET (dark bars) against the number of new clauses for Class C.

Timing data were collected using the LISP function (TIME <*function call*>) available on the Symbolics. This is a fairly blunt instrument, so we shall not reproduce the data here. However, it is good enough to expose gross disparities in performance. PIGLET ran faster than IPIA by a factor of about 2. This shows that PIGLET's heuristics for narrowing the search space do not consume inordinate amounts of CPU time.

6 Summary, Future Research and Related Work

Kean & Tsiknis note ([9], Section 7) that there seems to be no heuristic for a resolution or consensus based system that is guaranteed to prevent the generation of clauses that will be forward subsumed. The more prime implicates there are, the more opportunities there will be for resolution between them, and the more candidate generation and subsumption checking will have to be done to ensure completeness. IPIA avoids the general form of this problem when updating by a single clause because potential resolution steps are restricted to clauses that come from different pools. We do not need to resolve 'old' clauses, *i.e.*, clauses from the original theory, with each other, because they are already in prime implicate form, and we do not need to resolve 'new' clauses with each other, because doing so only results in tautologies, or existing clauses, or clauses that will be forward subsumed. However, when updating by multiple clauses, part of this advantage is lost: we may now need to resolve two new clauses to compute a prime implicate.

Our data show that PIGLET's heuristic use of merges conveys significant advantages in the incremental computation, regardless of how many clauses one is updating by. The ordering on potential resolution steps induced by the *COST** function promotes early backward subsumption which can only serve to narrow the search space. The forward pruning of potential resolution steps implemented by the *COMPLEMENTS* function limits forward subsumption, and therefore helps to cut down the final bushiness of the search space. This is a very substantial saving, gained at a relatively low computational cost. In [5], we report that the *COMPLEMENTS* heuristic cuts PIG's search space by an order of magnitude on the full prime implicate computation. We have data which show that, if you run PIGLET without this refinement, it still outperforms IPIA when updating by more than one clause, but is outperformed by IPIA when updating by a single clause.

It is worth mentioning that PIG and PIGLET are both 'anytime algorithms' [2], in the sense that they can be halted at any time and produce an approximate answer which improves with processing time. Thus we could use the cost function to limit search, *e.g.*, by only scheduling resolution steps with $COST(C_i, C_j) \leq k$, for some arbitrary cut-off k. This does not guarantee that we will then compute all prime implicates of size less than or equal to k, since some of these may have derivations containing steps whose cost is greater than k. However, the programs do tend to generate smaller and therefore more informative implicates first, even though implicates are not actually generated in strict order of increasing size. Partial runs of the IPIA program are less predictable, since the algorithm does not specify any heuristic means for preferring smaller resolvents, and it is not clear how the program could be enhanced, given that the algorithm's control structure is based on a loop through the biform variables.

It is generally acknowledged that random problems involving prime implicants tend to be harder than the problems normally encountered in logic minimization (see e.g., [1], p.173). However, the relation between random problems and 'typical' automated reasoning tasks is not clear. In certain applications, such as qualitative simulation, domain complexity may render the problem as difficult as the worst case. In other applications, such as abduction or belief revision, the worst case may be encountered less frequently. Experience so far using PIGLET in the context of an abduction engine called PABLO [7] suggests that Horn theories containing diagnostic knowledge and case data can be updated in a fraction of the time taken to process smaller random problems. But further study is required before we can be more precise concerning both the amount of savings and those properties of knowledge bases which bring savings about.

Another area for future research lies in the relation between resolution and connection-based methods for computing prime implicates. We have already shown [4] that resolution-based methods can approach the efficiency of connection-based methods if guided by good heuristics. Meanwhile, a synthesis between resolution and connection-based methods has been proposed [3] to derive an inference rule, called *consolution*, which appears to combine the virtues of these two approaches. Specifically, it shows how resolution can be seen as a path-checking algorithm in the sense of the connection calculus. Understanding more about this relationship will help us to make more insightful comparisons between prime implicate algorithms than are possible at present.

Computing prime implicates by consolution would combine some of the efficiencies of the connection method, as already achieved by the MM algorithm [8], with some of the conveniences of resolution supplied by PIG and PIGLET, such as the anytime character of the computation and the ability to use lemmata. In fact, the algorithms for consolution and MM are quite similar in some respects and even share some heuristics. The main differences between them are that (i) MM follows the connection calculus in failing to make use of lemmata, and (ii) MM incorporates a very effective ordering heuristic for path construction. If we now view PIGLET and IPIA as path-checking algorithms, it seems clear that PIGLET's opportunistic ordering of clauses based on the distribution of merge literals also acts as a path construction heuristic which is more effective than IPIA's control loop with optimizations.

Finally, there is a large body of related work in the Electrical Engineering literature that ought to be mentioned. In the Introduction, we noted that the problem of computing prime implicants/implicates efficiently is related to the problem of Boolean minimization. However, in the latter we seek not only to generate the prime implicants of a logic function but also to extract a *minimum prime cover* of that function. Roughly speaking, a minimum prime cover of a function f is a disjunction of prime implicants of f which expresses f but which ceases to do so if a single disjunct is deleted. Such a cover is sometimes called an *irredundant disjunctive form* for f, and it is not necessarily unique (see [11], p.130).

There are two approaches to Boolean minimization which stand a chance of being efficient. One generates all the prime implicants using a minterm expansion method such as [10] and then selects a 'near minimum' prime cover heuristically, since the cover extraction problem is known to be NP-complete. The other, more recent, approach tries simultaneously to identify and select (not necessarily prime) implicants for a (not necessarily minimum) cover (see e.g., literature review in [1]).

Modern algorithms such as ESPRESSO [1] use an n-dimensional cube representation for a minimization problem of n input variables, and the heuristics they employ are all predicated upon this representation. Consequently it is difficult to compare such techniques with approaches based on either resolution or the connection calculus, although this might be an avenue worth exploring. However, unlike the heuristics used in PIGLET, such methods will typically compromise completeness. In logic minimization, loss of completeness results in suboptimal designs which use more devices or gates than necessary. In an automated reasoning application such as abduction, the consequence may be missed explanations of data, or explanations which are not as parsimonious as they might be. Deciding when and how to forgo completeness is a domain-dependent issue, influenced by the impact of suboptimal solutions. The present paper confines its attention to ensuring that the basic underlying computation is both correct and as efficient as we can make it.

Acknowledgements

I am indebted to a number of anonymous reviewers who provided useful comments and constructive suggestions which improved the quality of this paper. This research was supported by the McDonnell Douglas Independent Research and Development Program.

References

1. R. H. Brayton, G. D. Hachtel, C. T. McMullen, A. L. Sangiovanni-Vincentelli: *Logic Minimization Algorithms for VLSI Synthesis*. Boston, MA: Kluwer-Academic, 1984.
2. T. Dean, M. Boddy: An analysis of time-dependent planning. *7th AAAI*, 49-54, 1988.
3. E. Eder: Consolution and its relation with resolution. *12th IJCAI*, 132-136, 1991.
4. P. Jackson: *The SLOOP Manual.* Artificial Intelligence Applications Institute, Edinburgh University, 1987.
5. P. Jackson: Computing prime implicates. *Proceedings 20th Annual Computer Science Conference*, 65-72, New York: ACM Press, 1992.
6. P. Jackson: Computing minimal refutations. *Proceedings of 3rd Scandinavian Conference on Artificial Intelligence*, 107-118, Amsterdam: IOS Press, 1991.
7. P. Jackson: Possibilistic prime implicates and their use in abduction. *Proceedings of AAAI-91 Workshop on Abduction*, 44-50, 1991.
8. P. Jackson, J. Pais: Computing prime implicants. *Proceedings of 10th International Conference on Automated Deduction*, 543-557, 1990.
9. A. Kean, Tsiknis, G: An incremental method for generating prime implicants/ implicates. *J. Symb. Comp.*, 9, 185-206, 1990.
10. E. L. McCluskey Jr: Minimization of Boolean functions. *Bell Systems Technical Journal*, 35, 1417-1444, 1956.
11. S. Muroga: *Logic Design and Switching Theory.* New York, NY: Wiley, 1979.
12. J. Pais, P. Jackson: Partial monotonicity and a new version of the Ramsey Test. To appear in *Studia Logica.*
13. W. V. O. Quine: The problem of simplifying truth functions. *American Mathematical Monthly*, 59, 521-531, 1952.
14. W. V. O. Quine: A way to simplify truth functions. *American Mathematical Monthly*, 62, 627-631, 1955.
15. W. V. O. Quine: On cores and prime implicants of truth functions. *American Mathematical Monthly*, 66, 1959.
16. R. Reiter, J. de Kleer: Foundations of assumption-based truth maintenance systems: Preliminary report. *6th National Conference on Artificial Intelligence*, 183-188, 1987.
17. J. R. Slagle, C.-L. Chang, R. C. T. Lee: A new algorithm for generating prime implicants. *IEEE Transactions on Computers*, C-19(4), 304-310, 1970.

18. P. Tison: Generalized consensus theory and application to the minimization of boolean functions. *IEEE Transactions on Electronic Computers*, EC-16(4), 446-456, 1986.

Appendix

To simplify both the statement of the following theorems and their proofs, it is convenient to allow any non-tautologous clause to subsume a tautology. Thus we allow that C_i subsumes C_j iff $C_i \subseteq C_j$ or C_j is a tautology and C_i is not. Since we discard both subsumed clauses and tautologies in computing prime implicates, this useful fiction will not affect the correctness of the PIG and PIGLET algorithms.

Theorem 1. Let C_i, C_j, C_k be clauses, such that C_i will resolve with C_j, and C_i will resolve with C_k, where none of C_i, C_j, C_k is a tautology. Then $C_i|_xC_j$ subsumes $C_i|_yC_k$ if $<C_j>_i \subseteq <C_k>_i$.

Proof. We proceed by case analysis. There are two cases to consider: the one where $x = y$, and the one where $x \neq y$. As in the text, we use $\pm x$ to denote a literal of indifferent sign.

If $x = y$, then we have

$$C_i|_xC_j = \{c \in C_i \mid c \neq \pm x\} \cup \{c \in C_j \mid c \neq \pm x\}$$
$$C_i|_yC_k = \{c \in C_i \mid c \neq \pm x\} \cup \{c \in C_k \mid c \neq \pm x\}$$

so $C_i|_xC_j$ subsumes $C_i|_yC_k$ if

$$(\{c \in C_j \mid c \neq \pm x\} - \{c \in C_i \mid c \neq \pm x\}) \subseteq (\{c \in C_k \mid c \neq \pm x\} - \{c \in C_i \mid c \neq \pm x\}),$$

i.e., if

$$<C_j>_i \subseteq <C_k>_i.$$

If $x \neq y$, then either $\pm x \in <C_k>_i$, in which case $C_i|_yC_k$ will be a tautology, or $\pm x \notin <C_k>_i$, in which case $<C_j>_i \subseteq <C_k>_i$ will not hold. ||

Only proof outlines are included for Theorems 2 and 3, as (i) the full proofs are quite lengthy, and (ii) the proofs rest for the most part on previously published results.

Theorem 2. Given a propositional theory, T, the PIG algorithm computes all and only the prime implicates of T.

Proof. Our proof rests entirely upon Theorem 1 and the soundness and completeness of general resolution for consequence-finding. Apart from the 'preempt forward subsumption' heuristic, which deletes only subsumed clauses, the algorithm resolves any two clauses obtained or derived from T that will resolve. The *COST* function merely determines the order in which resolutions occur to promote early backward subsumption. Subsumption deletion does not affect the completeness of unrestricted resolution. ||

<u>Theorem 3</u>. Given a propositional theory, T, and a set of clauses, C, the PIGLET algorithm computes all and only the prime implicates of $T \cup C$.

Proof. There are two cases to consider: (1) the case where C is a singleton; and (2) the case where C contains more than one clause.

(1) The proof rests upon Theorem 1, and the following observations, which have their counterparts in Lemmas 5.1 and 5.3 of Kean & Tsiknis (1990) for the computation of prime *implicants*. Firstly, we need never resolve any pair of clauses in T, since T is already in prime implicate form, and so any such resolvent will be subsumed by a clause in T. Secondly, the resolvent of any pair of clauses from

$$S = \mathbf{Cn}(T \cup C) - (T \cup C)$$

will be subsumed by a clause in S.

(2) As in case (1), except that we are forced to resolve clauses in S for the sake of completeness. ||

Linear-Input Subset Analysis

Geoff Sutcliffe
Dep't of Computer Science, The University of Western Australia
Nedlands, Western Australia, 6009. Email : geoff@cs.uwa.oz.au

Abstract. There are syntactically identifiable situations in which reduction does not occur in chain format linear deduction systems, i.e. situations in which linear-input subdeductions are performed. Three methods of detecting these situations are described in this paper. The first method (Horn subset analysis) focuses on Horn input chains while the second (LISS analysis) and third (LISL analysis) are successive generalisations of the first method. A significant benefit that may be derived from detecting linear-input subdeductions is the applicability of a truth value deletion strategy in such subdeductions. The completeness of the deletion strategy is proved, and its efficacy indicated.

1. Introduction

The exponential size of the search space of the resolution procedure necessitates the use of refinements which restrict the search space. Many refinements of the resolution procedure have been developed, one category of which is the linear refinements. The earliest linear deduction systems were the R3 refinement [Luckham, 1970], s-linear resolution [Loveland, 1970], and the strategy of preference of a 'new' conjunction [Zamov, 1969]. After these initial three systems, two streams of development are evident. One stream developed refinements based on the merging restriction [Andrews, 1968], whilst the other developed the chain format systems. The results presented in this paper are for chain format linear deduction systems.

The first chain format systems developed were the Model Elimination procedure [Loveland, 1969] and Linear resolution with Selection function (SL-resolution) [Kowalski, 1971]. Subsequent chain format systems include Ordered Linear deduction [Chang, 1973], the Graph Construction procedure [Shostak, 1976], Selective Linear Model deduction [Brown, 1974], and Linear resolution with Unrestricted Selection based on Trees (LUST)-resolution [Minker 1982]. There have also been many implementations of chain format systems, e.g. [Fleisig, 1974], [Stickel, 1986], [Sutcliffe, 1990], [Tarver, 1990]. The results presented in this paper are not immediately applicable to the Graph Construction procedure and Selective Linear Model deduction, due to their mechanisms for reusing portions of deductions. However, a simple modification to these deduction systems (as discussed in the conclusion) makes the results applicable.

Linear-input deduction is a refinement of linear deduction which does not permit any form of ancestor resolution, i.e. linear deduction using only the extension operation. Linear-input deduction is complete for sets of Horn clauses, but is not complete for sets of non-Horn clauses. [Ringwood, 1988] provides an interesting synopsis and references for the history of linear-input deduction systems. The use of reduction makes linear deduction complete for sets of non-Horn clauses. There are syntactically identifiable situations in which reduction does not occur in linear deduction systems, i.e. situations in which linear-input subdeductions are performed, and three methods of analysing sets of input chains have been developed for detecting these situations. The first method focuses on Horn input chains while the second and third are successive generalisations of the first method. Other work [Wakayama, 1990] has also noted that ancestor resolution and factoring are not always necessary for obtaining a refutation when the input set is non-Horn. The methods of analysis presented here are, however, more general than Wakayama's, which is restricted to entire input sets.

The detection of linear-input subdeductions is useful for (and was largely motivated by) the imposition of truth value deletion in linear deduction systems. The use of semantic information appears to have the potential to significantly improve the performance of deduction systems. This has been noted in the literature, e.g. "An emphasis on semantics rather than on syntax has far greater potential for producing a dramatic impact on the power of automated reasoning programs" [Wos, 1988, p. 257] (Research towards a "semantically orientated strategy" is Problem 5 in [Wos, 1988]), and "... if searches in symbolic computation are not to fall prey to combinatorial explosion, they must incorporate domain-specific knowledge in such a way so as to give direction to the search." [McRobbie, 1988, p. 198]. Up to this point, truth value deletion has been considered incompatible with linear deduction. The isolation of linear-input subdeductions, through linear-input analysis, makes it possible to use truth value deletion in linear deductions. This is described in section 3 of this paper.

2. Linear-Input Analysis

2.1 Horn Subset Analysis

It has been noted that "...in many proofs, most of the input clauses are Horn clauses..." [Plaisted, 1982, p. 231]. Examination of linear refutations of some input sets reveals that once the positive B-literal of a Horn input chain has been extended against, no reductions are performed until that B-literal (in the guise of an A-literal) is truncated. Horn subset analysis detects these subdeductions.

The *input predicate set* of a set of input chains contains the predicate structures[1] that appear in the input set. To detect situations in which reduction does not occur in a linear deduction from a negative top chain in an input set (lemma 1.1 relies on a negative top chain), the *Horn subset* of the input predicate set is extracted. A predicate structure is in the Horn subset iff (i) it does not occur positively in a non-Horn input chain, and (ii) for every Horn input chain in which the predicate structure occurs positively, every predicate structure in the chain is in the Horn subset.

Example

The Horn subset of `{~r~p~q, ~pq, p~q, pq, r~t~s, t~u, u, s}`, with `~r~p~q` as the top chain, is `{r, t, s, u}`.

[1]A predicate structure is the predicate symbol and arity of a literal.

The Horn subset divides the input chains into two groups, based on whether or not all literals in the chain have predicate structures that are in the Horn subset. Any predicate structures, literals, or input chains which only contain predicate structures that are in the Horn subset, are called *Horn subset objects*.

In a linear deduction from a negative top chain no reduction against Horn subset literals is performed, and once a Horn subset B-literal is selected, no reduction against literals rightwards from the selected B-literal is performed until that B-literal (in the guise of an A-literal) is truncated. Further, only the positive B-literal of a Horn subset input chain is ever resolved against in an extension operation. These properties are now proved. (Concepts similar to those used here were informally introduced in [Sutcliffe, 1989].)

Lemma 1.1
In a linear deduction from a negative top chain : (i) no positive Horn subset A- or B-literal occurs in a centre chain. (ii) negative B-literals in Horn subset input chains are never resolved against in extension operations. (iii) no reduction against Horn subset A- or B-literals is performed.

The proof of part (i) is by contradiction. If a positive Horn subset A- or B-literal occurs in a centre chain then the A-literal immediately to its left must be a Horn subset A-literal, as its complement originates from the same input chain as the first literal. Further, the Horn subset A-literal to the left must be positive, for otherwise the first literal occurs positively in a non-Horn input chain. Iteratively, all the A-literals to the left of a positive Horn subset A- or B-literal must be positive. However, the leftmost A-literal in the centre chain must be negative as the top chain is negative. Contradiction. Hence : (i) no positive Horn subset A- or B-literal occurs in a centre chain. (ii) as there can only be negative Horn subset B-literals in a centre chain, negative B-literals in Horn subset input chains can never be resolved against in extension operations. (iii) as complementary Horn subset A- and B-literals cannot occur in any centre chain, no reduction against such literals is performed. **QED**

Theorem 1.2
In a linear deduction from a negative top chain : (i) every A- and B-literal to the right of a Horn subset A-literal in a center chain, is also a Horn subset literal. (ii) no reduction against A- and B-literals rightwards from a Horn subset A-literal is performed. (iii) once a Horn subset B-literal is selected, no reduction against literals rightwards from the selected B-literal is performed until that B-literal (in the guise of an A-literal) is truncated.

By lemma 1.1, once a Horn subset B-literal in a centre chain has been selected, it is necessarily extended against. From the definition of the Horn subset, the B-literals added to the centre chain in the extension are Horn subset B-literals. Therefore : (i) iteratively, every A- and B-literal to the right of the selected Horn subset B-literal (now an A-literal) is a Horn subset literal. (ii) by lemma 1.1, no reduction against A- and B-literals rightwards from a Horn subset A-literal is performed. (iii) the structure of chain format linear deductions and part (ii) ensure that no reduction against literals rightwards from the selected B-literal is performed until that B-literal (in the guise of an A-literal) is truncated. **QED**

2.2 Linear-Input Subsets for Literal Structures

Horn subset analysis focuses on Horn input chains, and does not provide adequate analysis for input chains which are non-Horn but are Horn in a renaming of the input set. Many results based on the polarity of literals can be generalised to be based on a division of the literal structures[2] that appear in the input set, e.g. P_1 resolution [Robinson, 1965] generalises to P_P resolution [Meltzer, 1966], hyper-resolution [Robinson, 1965] generalises to AM-clashes [Slagle, 1967]. Similarly, Horn subset analysis generalises to results for non-Horn chains, in the form of Linear-Input Subset for literal Structures (LISS) analysis. The generalisation from Horn subsets to LISSs comes at the cost of a more complex analysis. Rather than a direct examination of the input set, LISS analysis requires examination of an abstraction of the deduction search tree.

For an input chain, the corresponding *chain structure set* contains the literal structures that occur in the chain. To detect situations in which reduction does not occur in a deduction from a chosen top chain, the *linear-input subset* of the literal *structures* that occur in the input set is extracted. This is done by building an *extension tree* whose nodes are literal structures. The extension tree has a mythical root whose offspring are the elements of the chain structure set corresponding to the chosen top chain. For each non-root literal structure in an extension tree, its offspring are those literal structures that (i) are in chain structure sets that contain a literal structure complementary to the parent literal structure, (ii) are not the complementary literal structure, and (iii) do not have themselves as ancestors in the extension tree unless, between the offspring and the ancestor, there exists a literal structure which does not have itself as an ancestor above the offspring's ancestor. A literal structure is in the LISS iff for every occurrence in the extension tree (i) it is not complementary to an ancestor, and (ii) all of its descendants are in the LISS.

Example

The first few level of the LISS tree for `{r~p~q, ~pq, p~q, pq, ~r~t~s, tu, ~u, s}`, with `r~p~q` as the top chain, are :

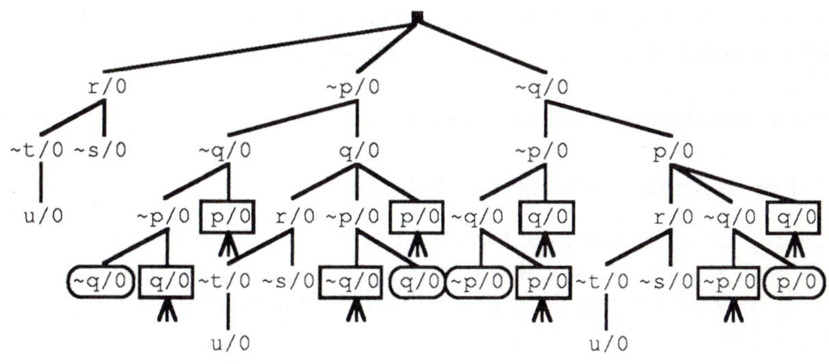

---

[2]A literal structure is the sign, predicate symbol and arity of a literal.

Circled literal structures are leaves of the tree, as dictated by item (iii) in the definition of these trees. Boxes literal structures are complementary to an ancestor. The lower levels of the tree reveal no new information. The LISS is thus {`r, ~t, u, ~s`}. No Horn subset exists for this top chain, as it is non-negative. With `~r~t~s` as the top chain, the Horn subset is {`s`} and the LISS is {`~t, u, ~s`}. Some of the LISS tree

Any literal structures or literals which only contain literal structures that are in the LISS, are called *LISS objects*.

In a linear deduction from a chosen top chain, no reduction against LISS B-literals is performed, and once a LISS B-literal is selected, no reduction against literals rightwards from the selected B-literal is performed until that literal (in the guise of an A-literal) is truncated. These properties are now proved.

Lemma 2.1
In a linear deduction from a chosen top chain no reduction against LISS A- or B-literals is performed.

The root to tip sequence of literal structures in a branch of the extension tree corresponds to possible left to right sequences of A- and B-literal structures in centre chains of a deduction from the chosen top chain. Each node corresponds to a possible A-literal in a centre chain, and literal structures that are lower in the branch correspond to possible B-literals to the left of that A-literal in a centre chain. Therefore : (i) no LISS B-literal in a centre chain has a structure complementary to an A-literal to its left (LISS definition part (i)), and no reduction against LISS B-literals is performed. (ii) no LISS A-literal has a B-literal with a complementary structure to its right (LISS definition part (ii)), and no reduction against LISS A-literals is performed. **QED**

Theorem 2.2
In a linear deduction from a chosen top chain : (i) every A- and B-literal to the right of a LISS A-literal in a center chain, is also a LISS literal. (ii) no reduction against A- or B-literals rightwards from a LISS A-literal is performed. (iii) once a LISS B-literal is selected, no reduction against literals rightwards from the selected B-literal is performed until that B-literal (in the guise of an A-literal) is truncated.

The proof is analogous to that of theorem 1.2.

2.3 Linear-Input Subsets for Literals

In building the extension tree, LISS analysis makes the assumption that every pair of literals with complementary literal structures can unify. A more accurate analysis is possible, by working directly with the literals in the input set. Linear-Input Subset for Literals (LISL) analysis does this.

To detect situations in which reduction does not occur in a deduction from a chosen top chain, the *linear-input subset* of the *literals* in the input set is extracted. This is done by building an extension tree whose nodes are literals from the input set. The method used is similar to that for LISS analysis. For each non-root parent literal in a LISL extension tree, its offspring are those literals that (i) are in chains that contain a literal complementarily unifiable with the parent literal, (ii) are not the complementarily unifiable literal, and (iii) do not have themselves as ancestors in the extension tree unless, between the offspring and the ancestor, there exists a literal which does not have itself as an ancestor above the offspring's ancestor. A literal structure is in the LISL iff for every occurrence in the extension tree (i) it is not complementarily unifiable with an ancestor, and (ii) all of its descendants are in the LISL. Note that although the extension tree uses unifiability, unification is never consummated.

Example

The first few levels of the LISL tree for {`r~p(a)~q, ~p(a)q, p(a)~q, p(a)q, ~r~t~s, tu, ~u, s~p(b), p(b)`}, with `r~p(a)~q` as the top chain, are :

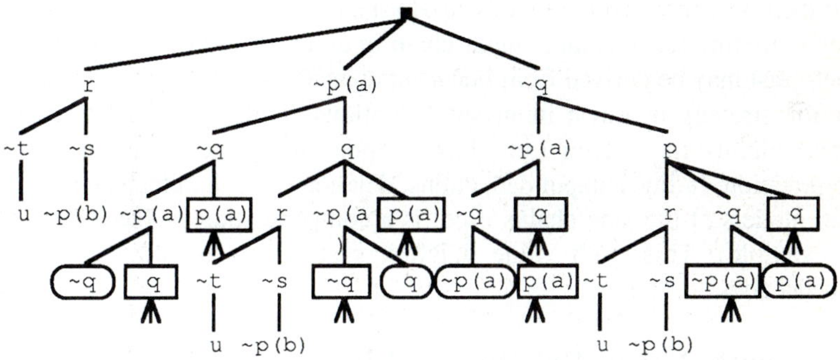

Circled literal structures are leaves of the tree, as dictated by item (iii) in the definition of these trees. Boxes literal structures are complementarily unifiable with an ancestor. The lower levels of the tree produce no new information. The LISL is thus {r^1, $\sim t^5$, u^6, $\sim s^5$, $\sim p(b)^9$} (where the superscripts indicate the chain number that the literal is in). The LISS is {`~t, u`}.

The results and theorem proofs for LISL analysis are analogous to those for LISS analysis. The results are simply stated here.

Lemma 3.1

In a linear deduction from a chosen top chain no reduction against LISL A- or B-literals is performed.

Theorem 3.2
In a linear deduction from a chosen top chain : (i) every A- and B-literal to the right of a LISL A-literal in a center chain, is also a LISL literal. (ii) No reduction against A- or B-literals rightwards from a LISL A-literal is performed. (iii) Once a LISL B-literal is selected, no reduction against A- and B-literals rightwards from the selected B-literal is performed until that B-literal (in the guise of an A-literal) is truncated.

2.4 Taking Advantage of Linear-Input Subsets

The above results show that once a Horn subset/LISS/LISL B-literal (henceforth, Horn subset/LISS/LISL objects will be referred to generically as *linear-input objects*) has been selected, a linear deduction system starts a linear-input subdeduction. The subdeduction end when that B-literal (in the guise of an A-literal) is truncated. The selected B-literal is called the *top literal* of the subdeduction.

In a linear-input subdeduction the reduction operation can be explicitly ignored, so that no effort is expended attempting to find A-literals to reduce against. If Horn subset analysis is used then only the positive literals of Horn subset input chains need ever be considered when searching for a suitable input chain in an extension operation. A more significant benefit that may be derived from linear-input analysis is the completeness of a truth value deletion strategy in linear-input subdeductions. Truth value deletion strategies reject clauses/chains in a deduction, based upon their interpretation value in a given interpretation. In linear-input deductions, deletion of center chains which are not FALSE in all models of the side chains used, is a complete strategy (see [Bundy, 1983, p 147] for example). This truth value deletion system can be transferred to linear-input subdeductions.

3. Truth Value Deletion in Linear Deduction Systems

Def'n : A *side chain model* of a deduction is an interpretation that is a model of the side parent chains used in the deduction.

Def'n : The *rightwards subchain* of a literal in a center chain consists of the literal and all literals to its right.

A truth value deletion system which requires all rightwards subchains of the top literal in a linear-input subdeduction to be interpreted as FALSE, in all side chain models of the subdeduction, is complete. The system arises from the following results.

Lemma 4.1
In a linear refutation in which only extension and truncation operations are performed, all centre chains are interpreted as FALSE in all side chain models of the refutation. (In determining the interpretation value of a centre chain, only B-literals are considered.)

In this situation a linear refutation is reduced to a linear-input refutation. The lemma then follows directly from [Bundy, 1983, Thm 2, p 147]. **QED**

Lemma 4.2
In a linear-input subdeduction, all rightwards subchains of the top literal are interpreted as FALSE in all side chain models of the subdeduction. (In determining the interpretation value of a rightwards subchain, only B-literals are considered.)

Let $C_1R_1,...,C_{n-1}R_{n-1},C_n$ be the centre chains of a linear-input subdeduction, so that R_1 is the top literal, each R_i is a rightwards subchain of the top literal, and R_{n-1} is the top literal in the guise of an A-literal. Theorems 1.2, 2.2 and 3.2 show that no literal in any C_i is used in deduction operations that affect the R_i. Therefore there is a refutation from the top chain R_1 using only extension and truncation operations. By lemma 4.1, all the R_i are interpreted as FALSE in all side chain models of the refutation. Thus in the linear-input subdeduction all the R_i are interpreted as FALSE in all side chain models of the subdeduction. **QED**

Lemma 4.3
In a linear-input subdeduction, all rightwards subchains of the top literal are simultaneously interpreted as FALSE in all side chain models of the subdeduction. (In determining the interpretation value of a rightwards subchain, only B-literals are considered.)

Consider a ground universe instance of a linear-input subdeduction. By lemma 4.2 all rightwards subchains of the top literal are (because they are ground) simultaneously interpreted as FALSE in all side chain models of the ground subdeduction. Side chain models of the original non-ground subdeduction are also side chain models of the ground subdeduction. All rightwards subchains of the top literal in the original subdeduction therefore simultaneously have ground universe instances that are interpreted as FALSE in all side chain models of the original subdeduction. Thus, in a linear-input subdeduction, all rightwards subchains of the top literal are simultaneously interpreted as FALSE in all side chain models of the subdeduction. **QED**

Theorem 4.4
In a linear-input subdeduction, all rightwards subchains of the top literal are simultaneously interpreted as FALSE in all side chain models of the subdeduction. (In determining the interpretation value of a rightwards subchain, all literals, including A-literals, are considered.)

In a linear-input subdeduction, all A-literals in rightwards subchains of the top literal also occur as B-literals in ancestor rightwards subchains. The ancestor rightwards subchains are simultaneously subject to the truth value restriction. Therefore lemma 4.3 can be extended to include A-literals when determining interpretation values. **QED**

Theorem 4 establishes the completeness of a truth value deletion system for chain format linear deduction systems when a linear-input subdeduction is being performed. The deletion system is called the *rightwards subchain* truth value deletion system. The imposition of the rightwards subchain system relies on being able to predict which input chains can be used as side chains of each linear-input subdeduction, so that side chain models can be supplied. This property is called *side chain predictability*. If Horn subset analysis is used then side chain predictability is available immediately, as only non-negative Horn subset input chains are used as side parent chains. For LISS analysis, side chain predictability is obtained by inspection of the extension tree. For a linear-input subdeduction from a given top literal, the input chains that may be used as side parent chains are those that were used in building extension subtrees rooted at the LISS element corresponding to the top literal. Thus a set of possible side parent chains is associated with each LISS element. There are then two options for building side chain models :

- Different side chain models may be built for each LISS element, based upon the associated input chains. Although this may require significant effort, there may be some benefit in constructing models that are local to linear-input subdeductions, as the models need only reflect truth value information relevant to the subdeductions. As is noted in [Plaisted, 1982, p 238], "This is interesting because it corresponds to the fact that in the human theorem proving process attention is given to various specialized models at various stages of the proof".
- Side chain models of the union of the sets associated with the LISS elements may be built. This approach is only possible if the union is a proper subset of the input set. If all input chains are possible side parent chains, the LISS subset may be reduced by adding a new condition for membership of the LISS subset : (iii) the chosen top chain is not used in forming any descendant of the literal structure. If this condition is added, then the top chain of the deduction cannot be a side parent chain in any linear-input subdeduction, and models of the union are possible. If this latter approach only excludes the top chain from being a side parent chain then an additional truth value restriction, that requires the top chain of the deduction to be interpreted as FALSE in the side chain models, may be imposed. This is possible because at least one instance of one input chain used in a refutation must be interpreted as FALSE in every truth value interpretation of an unsatisfiable input set.

Side chain predictability for LISL analysis is the same as for LISS analysis.

4. Conclusion

LISS and LISL analysis have been found to be significantly more effective than Horn subset analysis for finding linear-input subdeductions. An advantage of the generalisations is that no restrictions are placed on the nature of the top chains of deductions. LISS analysis and the truth value deletion system have been implemented as part of the Semantically Guided Linear Deduction (SGLD) system [Sutcliffe, 1992]. The generation of the LISS is a simple iterative task, and is done before deductions are started. Truth value deletion is then applied as appropriate.

Following are some performance figures, for SGLD, which indicate the efficacy of the linear-input analysis and truth value deletion. Results are given for each of SGLD's search strategies, and for two initial bounds for the consecutively bounded depth first search used. The results are in the form <total number of deduction operations> : <time taken in seconds>. The column labelled LIA indicates whether or not linear-input analysis and truth value deletion have been used.

The first example is a program verification problem, similar to the shortburst and burstall problems from Reboh [1972]. This problem is, however, non-Horn. The truth value interpretation used, for the rows where LIA is 'Yes', knows the structure of the program's state space.

Search Style	LIA	Initial Depth Bound	
		Top chain length	6
Literal Selected	No	177 : 3	84 : 2
	Yes	75 : 1	22 : <1
Literal Ordered	No	199 : 3	100 : 2
	Yes	95 : 2	38 : 1
Cell Selected	No	307 : 5	127 : 2
	Yes	156 : 3	55 : 1
Cell Ordered	No	307 : 5	127 : 2
	Yes	156 : 3	55 : 1

Table 1. Results for the Program Verification Problem

The second example is the second group theory problem in [Chang, 1970] - In an associative system with an identity element, if the square of every element is the identity, then the system is commutative. The truth value interpretation used is an Abelian group of four elements (including the identity), which conforms to the hypotheses of the problem. Here the linear-input analysis finds that the entire deduction must be linear-input, so that truth value deletion is applied throughout. As the input set is Horn, this is not surprising. However, it does indicate that linear-input analysis encompasses detection of Horn-ness.

Search Style	LIA	Initial Depth Bound	
		Top chain length	7
Literal Selected	No	1413 : 23	1227 : 20
	Yes	725 : 13	555 : 10
Literal Ordered	No	1389 : 23	1203 : 21
	Yes	734 : 13	564 : 10
Cell Selected	No	2890 : 50	2487 : 44
	Yes	1623 : 30	1236 : 23
Cell Ordered	No	2890 : 51	2487 : 44
	Yes	1623 : 30	1236 : 23

Table 2. Results for the Group Theory Problem

The final example is Schubert's Steamroller problem. The truth value interpretation used is aware of sorts, knows that only animals eat, and that only two animals can be compared in size.

Search Style	LIA	Initial Depth Bound	
		Top chain length	12
Literal Selected	No	15669 : 287	1162 : 22
	Yes	3680 : 51	1343 : 22
Literal Ordered	No	15681 : 300	1005 : 19
	Yes	3728 : 52	1358 : 22
Cell Selected	No	60038 : 1128	61247 : 1512
	Yes	11030 : 230	4638 : 105
Cell Ordered	No	59229 : 1112	32634 : 786
	Yes	10973 : 230	5051 : 113

Table 3. Results for Schubert's Steamroller Problem

In all except for two combinations of strategy, the search space of SGLD is reduced (in some cases dramatically) by the use of linear-input analysis and truth value deletion. In most cases, the time taken is also reduced, in the best case by almost an order of magnitude. The user specified initial bounds (the righthand columns of results) are the values for which SGLD gives the best results without linear-input analysis and truth value deletion. This column illustrates that linear-input analysis and truth value deletion are of utility, even with a careful choice of deduction system parameters. The overall utility of the truth value deletion depends mainly on the manner in which the interpretation used is stored and manipulated. The implementation used in SGLD is moderately efficient.

The combination of linear-input analysis and truth value deletion clearly has the potential to be of benefit in chain format linear deduction systems. In some aspects it is analogous to the Simplified Problem Reduction Format's truth value deletion, the success of which "seems to have something to do with the fact that Horn clauses are common in typical problems." [Plaisted, 1982, p 238] The system presented here has, however, the potential for greater success as a more general notion than Horn-ness is used to determine when truth value deletion is applicable.

In the introduction it was noted that linear-input analysis is not immediately applicable in the Graph Construction procedure and Selective Linear Model deduction. The problem in these two systems is that reduction against C-literals (confusingly called A-literals in SLM) can be performed within, what would otherwise be, linear-input subdeductions. This added possibility is easily dealt with, in one of two ways. Firstly, all linear-input C-literals are inserted at the left most end of center chains, indicating that they are logical consequences of the side parent chains that participated in their production. Therefore the C-literals are TRUE in all models of such side parent chains. Lemma 4.1 is easily extended to cover reduction against such C-literals, and the truth value deletion system is still applicable. The second approach is to add such C-literals to the input set as unit chains, rather than inserting them at the left most end of the centre chain. If this option is chosen, then no reduction against linear-input C-literals can be performed, and all the results of this paper apply. The option of adding unit input chains also has other advantages, as the unit chains can be used throughout deductions.

References

Andrews P.B., Resolution with Merging, In *Journal of the ACM* 15(3), ACM Press, New York, NY, (1968), 367-381.

Brown F.M., SLM, Internal Memo #72, Department of Artificial Intelligence, University of Edinburgh, Edinburgh, Scotland, (1974).

Bundy A., *The Computer Modelling of Mathematical Reasoning*, Academic Press, London, England, (1983).

Chang C-L., The Unit Proof and the Input Proof in Theorem Proving, In *Journal of the ACM* 17(4), ACM Press, New York, NY, (1970), 698-707.

Chang C-L., and Lee R.C-T., *Symbolic Logic and Mechanical Theorem Proving*, Academic Press, New York, NY, (1973).

Fleisig S., Loveland D.W., Smiley A.K., and Yarmush D.L., An Implementation of the Model Elimination Proof Procedure, In *Journal of the ACM* 21(1), ACM Press, New York, NY, (1974), 124-139.

Kowalski R., and Kuehner D., Linear Resolution with Selection Function, In *Artificial Intellience* 2, Elsevier Science, Amsterdam, The Netherlands, (1971), 227-260.

Loveland D.W., A Simplified Format for the Model Elimination Theorem-Proving Procedure, In *Journal of the ACM* 16(3), ACM Press, New York, NY, (1969), 349-363.

Loveland D.W., A Linear Format for Resolution, In Laudet M. et al. (Ed.), *Proceedings of the IRIA Symposium on Automatic Demonstration* (Versailles, France, 1968), Springer-Verlag, New York, NY, (1970), 147-162.

Luckham D., Refinement Theorems in Resolution Theory, In Laudet M. et al. (Ed.), *Proceedings of the Symposium on Automatic Demonstration* (Versailles, France, 1968), Springer-Verlag, New York, NY, (1970), 163-190.

McRobbie M.A., Meyer R.K., and Thistlewaite P.B., Towards Efficient "Knowledge-Based" Automated Theorem Proving for Non-Standard Logics, In Lusk E., Overbeek R. (Ed.), *Proceedings of the 9th International Conference on Automated Deduction* (Argonne, IL, 1988), (Goos G., Hartmanis J. (Ed.), *Lecture Notes in Computer Science 310*), Springer-Verlag, New York, NY, (1988), 197-217.

Meltzer B., Theorem-proving for computers: Some results on resolution and renaming, In *The Computer Journal* 8, The Britsh Computer Society, London, England, (1966), 341-343.

Minker J., and Zanon G., An Extension to Linear Resolution with Selection Function, In *Information Processing Letters* 14(4), Elsevier Science, Amsterdam, The Netherlands, (1982), 191-194.

Plaisted D.A., A Simplified Problem Reduction Format, In *Artificial Intelligence* 18, Elsevier Science, Amsterdam, The Netherlands, (1982), 227-261.

Reboh R., Raphael B., Yates R.A., Kling R.E., and Velarde C., Study of automatic theorem proving programs, Technical Note 72, Artificial Intelligence Center, SRI International, Menlo Park, CA, (1972).

Ringwood G.A., SLD: A Folk Acronym, In Moss C. (Ed.), *Logic Programming Newsletter* 2(1), Association for Logic Programming, London, England, (1988), 5-7.

Robinson J.A., Automatic Deduction with Hyper-resolution, In *International Journal of Computer Mathematics* 1, Gordon and Breach, London, England, (1965), 227-234.

Shostak R.E., Refutation Graphs, In *Artificial Intelligence* 7, Elsevier Science, Amsterdam, The Netherlands, (1976), 51-64.

Slagle J.R., Automatic Theorem Proving with Renamable and Semantic Resolution, In *Journal of the ACM* 14(4), ACM Press, New York, NY, (1967), 687-697.

Stickel M.E., A Prolog Technology Theorem Prover: Implementation by an Extended Prolog Compiler, In Siekmann J.H. (Ed.), *Proceedings of the 8th International Conference on Automated Deduction* (Oxford, England, 1986), (Goos G., Hartmanis J. (Ed.), *Lecture Notes in Computer Science 230*), Springer-Verlag, New York, NY, (1986), 573-587.

Sutcliffe G., Complete Linear Derivation Systems for General Clauses, In Wos L. (Ed.), *Association for Automated Reasoning Newsletter* (13), Association for Automated Reasoning, Argonne, Il, (1989), 3-4.

Sutcliffe G., A General Clause Theorem Prover, In Stickel M.E. (Ed.), *Proceedings of the 10th International Conference on Automated Deduction* (Kaiserslautern, Germany, 1990), (Siekmann J.H. (Ed.), *Lecture Notes in Artificial Intelligence 449*), Springer-Verlag, New York, NY, (1990), 675-676.

Sutcliffe G., The Semantically Guided Linear Deduction System, In Kapur, D. (Ed.), *Proceedings of the 11th International Conference on Automated Deduction* (Saratoga Springs, NY, 1992), Springer-Verlag, New York, NY, (1992).

Tarver M., An Examination of the Prolog Technology Theorem Prover, In Stickel M. (Ed.), *Proceedings of the 10th International Conference on Automated Deduction* (Kaiserslautern, Germany, 1990), (Siekmann J.H. (Ed.), *Lecture Notes in Artificial Intelligence 449*), Springer-Verlag, New York, NY, (1990), 322-335.

Wakayama T., and Payne T.H., Case-Free Programs: An Abstraction of Definite Horn Programs, In Stickel M. (Ed.), *Proceedings of the 10th International Conference on Automated Deduction* (Kaiserslautern, Germany, 1990), (Siekmann J.H. (Ed.), *Lecture Notes in Artificial Intelligence 449*), Springer-Verlag, New York, NY, (1990), 87-101.

Wos L., *Automated Reasoning - 33 Basic Research Problems*, Prentice-Hall, Englewood Cliffs, New Jersey, (1988).

Zamov N.K., and Sharonov V.I., On a class of strategies which can be used to prove theorems by the resolution principle, In (In Russian) (Ed.), *Issled, po konstruktivnoye matematikye i matematicheskoie logikye* III(16), National Lending Library Russian Translating Program 5857, Boston Spa, Yorkshire, (1969), 54-64.

Theoretical study of symmetries in propositional calculus and applications*

Belaid Benhamou and Lakhdar Sais

L.I.U.P. - Université de Provence,
3,Place Victor Hugo - F13331 Marseille cedex 3, France
e-mail : Siegel@frccup51.bitnet

Abstract. Many propositional calculus problems (for example the Ramsey or the pigeon hole problems) can quite naturally be represented by a small set of first order logical clauses which becomes a very large set of propositional clauses when we substitute the variables by the constants of the domain. In many cases, the set of clauses contains several symmetries i.e. the set of clauses remains invariant under a permutation of variable names. We will show how we can shorten the proofs of such problems. We present an algorithm which detects the symmetries and explain how the symmetries are introduced and used in the following methods: Slri, Davis and Putnam and Semantic Evaluation. With symmetries we have got good results on many known problems such pigeon hole, Schur's lemma, Ramsey, the eight queen etc. The most interesting one is that we have been able to prove for the first time the unsatisfiability of the Ramsey problem for 17 vertices and 3 colors.

1 Introduction

The subject of this paper is the study of symmetries in propositional calculus in order to make the automated deduction algorithms more efficient. Using resolution as a base proof system for the propositional calculus, Tseitin[16] has shown how certain tricky mathematical arguments can be considered as short proofs for some propositional tautologies representing mathematical statements. He suggested to increase resolution by the principle of extension which consists of introducing auxiliary variables to represent intermediate formulas so that the length of a proof can be significantly reduced by manipulating these variables instead of the formulas that they stand for. On the other hand, Krishnamurthy[10,9] proposed the principle of symmetries which allows to recognize that a tautology remains invariant under certain permutation of variable names and uses this information to avoid repeated independent derivations of intermediate formulas that are permutations of other ones. Let us consider the logical formulas associated with concrete problem. In most cases, we use first order logic to express such a formula. In other words, we use the usual mathematical language. From this set of first order logic clauses we can obtain propositional clauses by substituting the variables by constants of the domain. Thus, the

* This work has been suported by the PRC-GDR Intelligence Artificielle.

previous set of clauses becomes a large set for many problems and contains several symmetries. The proprety of symmetry can lead to a shorter proof for the problem. However, the current methods of theorem proving do not take advantage of symmetry properties, and in[9] there is no algorithm neither to search for symmetries nor to use them in a formal way. The purpose of our study is the detection and the use of symmetries. First we describe a method which computes the symmetry on a given set of clauses. The algorithm is given with more details in[2,1]. Second we explain how symmetries are introduced and combined with different automated deduction algorithms like SL-Resolution[8], the Davis and Putnam[6] procedure,and Semantic Evaluation[13]. Finally we apply SLRI[5] and the Semantic Evaluation methods with and without symmetry to some classical problems, such pigeon hole [3,4], Schur's lemma[15] and Ramsey problems [13] and the comparison shows the possibility to make shorter proofs for hard propositional tautologies and reduce significantly the complexity of resolution in many cases. One of the most intresting results, is that we have been able to prove the unsatisfiability of Ramsey problem with 17 vertices and 3 colors, which had not been shown by a theorem proving program before.

2 Definitions and notations

We shall assume that the reader is familiar with the propositional calculus. For a formal description of the calculus, see e.g.[12]. Let V be the set of propositional variables called only variables. Variables will be distinguished from literals, which are variables with an assigned parity 1 or 0 {which means TRUE or FALSE, respectively} this distinction will be ignored whenever it is convenient but not confusing. For a propositional variable p, there are two literals: p the positive literal and $\neg p$ the negative one.

A clause is a disjunction of literals $(p_1, p_2, \ldots, p_n)$ such that no literal appears more than once and is denoted by $p_1 \vee p_2 \vee \ldots \vee p_n$. A system S of clauses is a conjunction of clauses. In other words we say that S is in the conjunctive normal form (CNF).

A truth assignment to a system of clauses S is a map I from the set of variables in S to the set {TRUE, FALSE}. If $I[p]$ is the value for the positive literal p then $I[\neg p] = 1 - I[p]$. The value of a clause $p_1 \vee p_2 \vee \ldots \vee p_n$ in I, is the maximum value of its literals in I. By convention we define the value of the empty clause ($n = 0$) to be FALSE. The value $I[S]$ of the system of clauses is TRUE if the value of each clause of S is TRUE, FALSE otherwise. We say that a system of clauses S is satisfiable if there exists some truth assignments in which S takes the value TRUE, it is unsatisfiable otherwise. In the first case I is called a model of S. Let us remark that a system which contain the empty clause is unsatisfiable.

It is well-known that for every propositional formula F there exists a formula F' in conjunctive normal form(CNF) such that the length of F' is at most 3 times as long as the formula F and F' is satisfiable iff F is satisfiable. In the following we will assume that the formulas are given in a conjunctive normal form.

3 Symmetries

First of all, let us define the concepts of permutations and symmetry, and prove significant properties that will enable us to improve the proof algorithms. For more details we refer the reader to [2].

A bijective map $\sigma : V \rightarrow V$ is called a permutation of variables. If S is a set of clauses, c a clause of S and σ a permutation of variables occuring in S, then $\sigma(c)$ is the clause obtained by applying σ to each variable of c and $\sigma(S) = \{\sigma(c)/c \in S\}$.

Definition 1. A set P of literals is called *complete* if $\forall \ell \in P$, $\neg \ell \in P$

Definition 2. Let P be a complete set of literals and S a set of clauses of which all literals are in P.
A permutation σ defined on P ($\sigma : P \rightarrow P$) is called a *symmetry* of S if it satisfies the following conditions:

1. $\forall \ell \in P, \sigma(\neg \ell) = \neg \sigma(\ell)$
2. $\sigma(S) = S$

Definition 3. Two literals (variables) ℓ and ℓ' are symmetrical in S notation ($\ell \sim \ell'$) if there exists a symmetry σ of S such that $\sigma(\ell) = \ell'$. A set $\{\ell, \ell_1, \ell_2, \ldots, \ell_n\}$ of literals is called a cycle of symmetry in S if there exist a symmetry σ defined on S, such that $\sigma(\ell) = \ell_1, \sigma(\ell_1) = \ell_2, \ldots \sigma(\ell_{n-1}) = \ell_n, \sigma(\ell_n) = \ell$.

Remark. All the literals in a cycle of symmetry are symmetrical two by two.

Example 1. Let S be the following set of clauses :
$S : \{a \vee \neg b, c\}$
and σ the map defined on the complete set P of literals occuring in S:

$$\begin{aligned} \sigma :\ & P \longrightarrow P \\ a & \longrightarrow \sigma(a) = \neg b \\ \neg a & \longrightarrow \sigma(\neg a) = b \\ b & \longrightarrow \sigma(b) = \neg a \\ \neg b & \longrightarrow \sigma(\neg b) = a \\ c & \longrightarrow \sigma(c) = c \\ \neg c & \longrightarrow \sigma(\neg c) = \neg c \end{aligned}$$

σ is a symmetry of S , a and $\neg b$ are symmetrical in S ($a \sim \neg b$).
$\sigma(S) = \{\neg b \vee a, c\} = S$.

Remark. $\sigma(S)$ is the set of clauses obtained from S by exchanging the literal positions in the same clause, or the order of the clauses in S. Generally, the two previous operations can be applied simultaneously.
- The identity map is a symmetry.
- The inverse map of a symmetry is also a symmetry.
- The composition of symmetries is a symmetry.

Definition 4. Let P be a complete set of literals, σ a symmetry, I a truth assignment of P and S a set of clauses then
- I/σ is the truth assignment obtained by substituting every literal ℓ in I by $\sigma(\ell)$.
- S/σ is the set of clauses obtained with substituting every literal ℓ in S by $\sigma(\ell)$.

4 Symmetry properties

If I is a model of S and σ a symmetry, we can get another model of S by applying σ on the variables which appear in I. In the following propositions, we assume that σ is a symmetry of the system of clauses S.

Proposition 5. *I is a model of S iff I/σ is a model of S .*

Proof. see Cf[2].

Proposition 6. *Let ℓ be a literal, such that $\ell' = \sigma(\ell)$ and $I' = I/\sigma$.
If I is such that $I[\ell]$=1, then I' is such that $I'[\ell'] = 1$*

Proof. According to prposition 5.
If I[ℓ]=1 then I/σ is a model of ℓ/σ, thus $I'[\ell'] = 1$.

Proposition 7. *If ℓ has the value true in a model of S, then $\sigma(\ell)$ will have the value true in a model of S.*

Proof. Direct consequence of the proposition 6.

Theorem 8. *Let ℓ and ℓ' be two literals of S.
if $\ell \sim \ell'$ in S, then ℓ has a model in S iff ℓ' has a model in S .*

Proof. $\Rightarrow$)
If $\ell \sim \ell'$ in S then there exists σ a symmetry of S such that $\sigma(\ell) = \ell'$
If I is a model of S then I/σ is a model of $S/\sigma = S$ (proposition 5.)
If $I[\ell] = 1$ then $I/\sigma[\ell']$=1(proposition 6.)
$\Leftarrow$)
Let $\ell = \sigma^{-1}(\ell')$, then the proof is identical.

It is very important to understand the difference between symmetrical and equivalent literals. Equivalent literals have exactly the same models. But if ℓ and ℓ' are symmetrical then : ℓ has a model in S iff ℓ' has a model in S, and the two models are generally different, so it is easier to find symmetrical literals than equivalent ones. Note that theorem 8. expresses an important property that we will use to make prune the resolution trees.
If ℓ has no model in S and $\ell \sim \ell'$, then ℓ' will have no model in S, thus we prune the branch which corresponds to the assignment of ℓ' in the resolution tree. Therefore, if there is n symmetrical literals we can cut $n-1$ branches in the resolution tree.

Proposition 9 (necessary condition of symmetries). *If two literals (variables) ℓ and ℓ' are symmetrical in a set of clauses S then the number of occurences of the variable ℓ in S is the same as the number of occurences of ℓ' and there must be a correspondence between the length of clauses in which ℓ occurs and the length of clauses in which ℓ' occurs.
i.e
$\ell \sim \ell'$ then,*

1. *occurence_number(ℓ)=occurence_number(ℓ'); and,*

2. *occurence_number(¬ℓ)=occurence_number(¬ℓ'); and,*
3. $\exists c \in S$ *such that* $|\, c\,|=n$ *and* $\ell \in c \Leftrightarrow \exists \acute{c} \in S$ *such that* $|\, \acute{c}\,|=n$ *and* $\ell' \in \acute{c}$.

Proof. 1) Suppose that he first condition is not satisfied:
Let ℓ and ℓ' be two symmetrical literals in S, and suppose that occurence_number(ℓ) $\neq$ occurence_number(ℓ'). Let occurence_number(ℓ) $= n$, occurence_number(ℓ') $= m$, and $n \neq m$. Then there are m clauses which contain ℓ' in the set S and n clauses which contain ℓ, after all permutation of ℓ and ℓ' the new set S' will have m clauses with the literal ℓ and n clauses with the literal ℓ', but $n \neq m$ then $S \neq S'$. Thus it does not exist a symmetry σ of S, such that $\sigma(\ell) = \ell'$ or $\sigma(\ell') = \ell$, therefore ℓ and ℓ' are not symmetrical. This gives a contradiction with the previous hypothesis(QED). The arguments for $\neg\ell$ are similar.
2) If the third condition is not satisfied:
then $\exists c \in S$ such that $|\, c\,|= n$ and $\ell \in c$ and $\nexists c' \in S$ such that $|\, c'\,|= n$ and $\ell' \in c'$. After all permutation of ℓ and ℓ' the new system S' will have a clause of length n in which ℓ' occurs. Hence, $S \neq S'$, thus it does not exist a symmetry σ of S, such that $\sigma(\ell) = \ell'$ or $\sigma(\ell') = \ell$, therefore ℓ and ℓ' are not symmetrical. This gives a contradiction with the previous hypothesis(QED).

5 Symmetry detection method

Let S be a set of clauses. Finding a symmetry in S is equivalent to finding a permutation σ which keeps S invariant. There can exist many symmetries, but at each resolution step we are interested in only one of them. For example, in the SLRI method (C. Cubbada and M.D. Mouseigne[5]) for a refuting literal ℓ, it is more interesting to get the symmetry which has more literals occuring in the same clause as ℓ. But in Davis and Putnam or in Semantic Evaluation procedure it is better to compute the wider cycle of a given literal ℓ i.e., we try to find a set of literals $\{\ell, \ell_1, \ell_2, \ldots, \ell_n\}$ such that $\sigma(\ell) = \ell_1, \sigma(\ell_1) = \ell_2, \ldots \sigma(\ell_{n-1}) = \ell_n, \sigma(\ell_n) = \ell$ with σ a symmetry defined on S.

It is well known that each permutation can be expressed as a product of elementary permutations called transpositions. Thus we write a permutation β as a product in the form $\beta = (x_1, y_1)(x_2, y_2)\ldots(x_n, y_n), x_i \neq y_i, \forall i \in \{1..n\}$. In order to check if two literals x_0 and y_0 are symmetrical in S we compute a symmetry σ of S, such that $\sigma(x_0) = y_0$. To achieve this, we try to express σ as a product of the form $\sigma = (x_0, y_0)\sigma'$, with σ' a sub-permutation of σ that we shall define as we do the other transpositions. Therefore, to substitute x_0 to y_0, we replace it by y_0 in each clause in which it appears. Thus we relate clauses in which x_0 appears to the clauses in which y_0 does.

To do this, clauses and other variables will be related (taking into account the previous necessary conditions of symmetry: proposition 9.). For instance let us consider the two following clauses: $c = x_0 \vee x_1 \vee x_2 \ldots \vee x_n$, $c' = y_0 \vee y_1 \vee y_2 \ldots \vee y_n$ and try to relate them. After replacing x_0 by y_0 we process the other variables in c, thus each variable $x_i, i \in \{1..n\}$ must be linked to a variable $y_j, j \in \{1..n\}$ and at each step we add the transposition (x_i, y_j) to σ.

If it is possible to substitute all the variables, we say that c is transformed into c' with σ i.e. $\sigma(c) = c'$. If all clauses containing x_0 have been related, then we consider the next transposition and we apply the same process. When there are no more transpositions, the σ, found is a symmetry of S.

The other possibility is when we try the transposition (x_i, y_j) we can not put in relationship a clause in which x_i occurs. Thus, if $(x_i, y_j) = (x_0, y_0)$ then the symmetry σ does not exist, otherwise we bactrack and try to link x_i to another variable either in the same clause in which y_j occurs or we try to relate c to another clause.

Remark that the symmetry between x_{i-1} and y_{j-1} depends of the one between x_i and x_j i.e. if $\sigma = (x_0, y_0) \ldots (x_{i-1}, y_{j-1})(x_i, x_j) \ldots (x_k, y_k)$, then $(x_0 \sim y_0) \Longrightarrow (x_{i-1} \sim y_{j-1}) \Longrightarrow (x_i \sim y_j) \Longrightarrow (x_k \sim y_k)$. In the actual implementation we fix a level of bactracking in order not to spend much time in failures. Thus we take another variable to substitute x_i iff the current number of backtraks is not greater than the fixed level. Otherwise we reject the symmetry, since we are interested only in the ones which do not take more time than resolution. In practice our algorithm compute all symmetries in less than twenty bactracks, and when we go over this number generally the symmetry does not exist. In many cases we find the symmetry with no bactraks.

The problem is now to find a clever strategy (or a choice function) for substituting the variables in order to build σ efficiently.

Strategy
It is clear that the efficiency of the algorithm depends of the ordering of the variables to be permuted; then we propose the following strategy:

1. First consider the variables which can only be substituted by one element.
2. Begin the processing with the variable of which we compute the cycle of symmetry.
3. Each clause we begin to relate must be achieved as soon as possible. In other words, variables in this clause are in priority in order to backtrack quickly in case of wrong choice.
4. Identity is applied for the main variables, when there is no new transpositions to try.

For more details about this method and its implementation we refer the reader to [2].

6 Advantage of symmetries

The SL-Resolution algorithm for the propositional calculus is derived from J.A. Robinson[14], D.W. Loveland[11], R.A. Kowalski[8]. In the case of refutations occuring throughout the algorithm, cylinders without solutions may be stored and used later(B-ancestors). In the case that the algorithm fails to find a refutation tree, the algorithm SLRI of C.M.D. Cubadda[5] saves certain information from the failure tree: the so called I-ancestors (I means: impasse). The set of clauses of the instance F satisfied by the set I of I-ancestors retained at a node is eliminated for a while. C.M.D. Cubadda proved that their algorithm never takes more time and that there is strict inclusion of calculation trees. The SL-Resolution tests a clause c added to a consistent instance and retains no solution in the case of consistency. SLRI takes every instance and, like Davis and Putnam procedure, gives a cylinder of solutions in the case of consistency.

6.1 Introducing symmetries in the SLRI method

Using the result of theorem 8., we can improve the method of Semantic Evaluation as well as the SLRI. Instead of refuting each literal separately, it is possible to refute simultaneously symmetrical literals. Thus, let C=$\{\ell_1, \ell_2, \ldots, \ell_{i-1}, \ell_i, \ldots, \ell_n\}$ be a clause in a system of clauses S. Suppose that the literals $\ell_1, \ell_2, \ldots, \ell_{i-1}$ have been refuted and that ℓ_i is the literal we try to refute. If we prove that ℓ_i is symmetrical with one of the previous literals then it will be directly refuted by the symmetry property.
Indeed, if $\ell_i \sim \ell_j$ in S with $1 \leq j \leq i-1$ then [ℓ_i has a model in S if and only if ℓ_j has a model in S]. As ℓj has no model in S ℓ_i has no model in S, and ℓ_i is refuted. Thus we can cut the branch which coresponds to ℓ_i assignment in the resolution tree. To optimize the use of symmetries, we try to get more symmetrical literals in the same clause, in order to cut a large part of the resolution tree.

Example 2. Let us consider the pigeon-hole problem (put n pigeons in $n-1$ holes such that each pigeon-hole holds at most one pigeon).
For instance let n=3 (3 pigeons and 2 holes), this problem is described with the following set of clauses:

$C_1 : p_1(1) \vee p_1(2)$
$C_2 : p_2(1) \vee p_2(2)$
$C_3 : p_3(1) \vee p_3(2)$
$C_4 : \neg p_1(1) \vee \neg p_2(1)$
$C_5 : \neg p_1(1) \vee \neg p_3(1)$
$C_8 : \neg p_2(1) \vee \neg p_3(1)$
$C_6 : \neg p_1(2) \vee \neg p_2(2)$
$C_7 : \neg p_1(2) \vee \neg p_3(2)$
$C_9 : \neg p_2(2) \vee \neg p_3(2)$

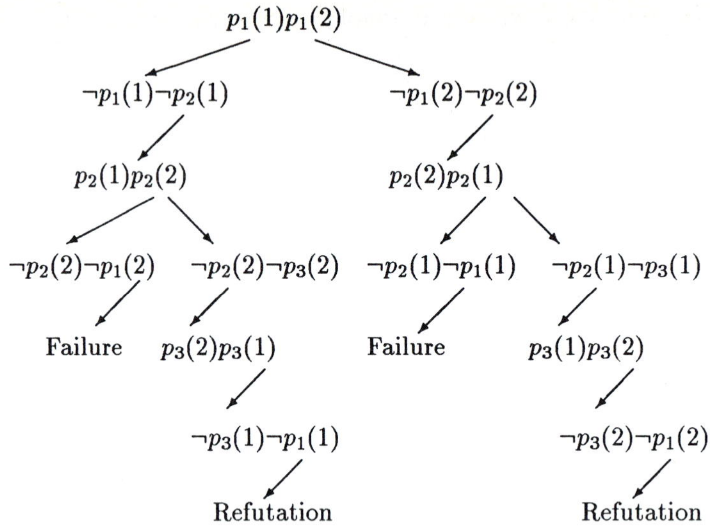

Fig. 1. The resolution tree of SLRI method

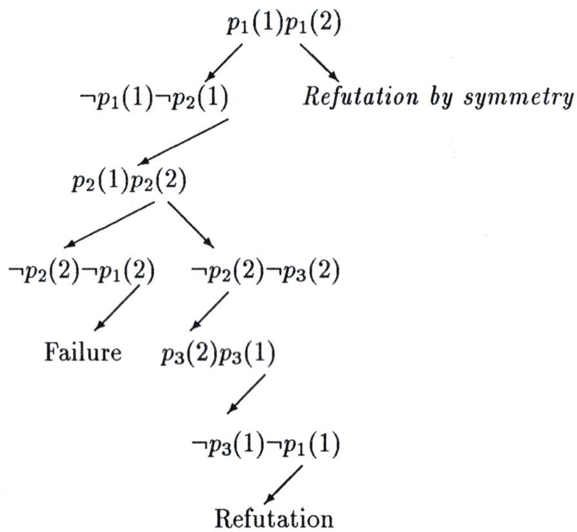

Fig. 2. The resolution tree of the SLRI associated with symmetry

It is clear that the subtree built from $p_1(1)$ in the resolution tree of SLRI (Fig.1) is symmetrical to the one built from $p_1(2)$. This due to the fact that the two literals $p_1(1)$ and $p_1(2)$ are symmetrical in S. Therefore, in the second resolution tree (Fig.2) the sub-tree of $p_1(2)$ disappears, because $p_1(2)$ is refuted directly according to its symmetry with $p_1(1)$. When the number of pigeons grows, symmetries are used at different levels of resolution, thus we cut more than one branch in the resolution tree; the complexity of resolution is linear in number of steps. Let us remark that when the number of pigeons is greater than 8, the resolution of this problem with the SLRI method without symmetry becomes impossible. The Pigeon-hole problem is studied with more details in the next section in order to show the influence of the symmetry property in SLRI method.

6.2 Symmetries in the Semantic Evaluation method

Definition 10. Let S be a system of clauses and $p_1, p_2, \ldots, p_n$ distinct literals which occur in S such that the set $\{p_1, p_2, \ldots, p_n\}$ does not contain both a literal and its opposite.
$Tp_1, p_2, \ldots, p_n(S)$: The set of clauses obtained from S by removing all the clauses containing one of the literals $p_1, p_2, \ldots, p_n$ and all the ocurences of $\neg p_1, \neg p_2, \ldots, \neg p_n$ in the other clauses.
T"$p_1, p_2, \ldots, p_n(S)$ is the set $Tp_1, p_2, \ldots, p_n(S)$ simplified by subsumption : i.e if $Tp_1, p_2, \ldots, p_n(S)$ contains both a clause c and a sub-clause of c then the clause c does not stand in T"$p_1, p_2, \ldots, p_n(S)$.

The QUINE algorithm has a semantic aspect, and checks recursively the satisfiability of a system of clauses.
Let S be a system of clauses.

Satisfiable(S):
if S=ϕ
then S is satisfiable.
else if S contains an empty clause
    then S is insatisfiable
    else begin
        choose arbitrarily one litteral p which ocurs in S
        if T"p(S) is satisfiable
        then S is satisfiable
        else if T"¬p(S) is satisfiable
            then S is satisfiable
            else S is not satisfiable
        end

Many improvements have been added to the basic algorithm (QUINE). The Davis and Putnam procedure is obtained by introducing the two following heuristics :
The first one consists in assigning the monoliterals in priority. The second one consists in assigning the pure literals in priority. Indeed, Davis and Putnam uses the following property :

If p is a pure literal in S, then S is satisfiable iff T"$p(S)$ is satisfiable. Assigning a pure literal allows to make a cut in the proof tree. Generalizing this property leads to the theorem of model partitioning(Oxusoff and Rauzy[13]) which allows to cut more branches in the proof tree when a pure literal has been assigned a value.

The method of Semantic Evaluation comes from the previous property. It is more efficient than the two pevious algorithms.

6.3 Introducing symmetries

It is also possible to introduce symmetries in the Semantic Evaluation method. We will use the same property as above (if the literals ℓ and ℓ' are symmetrical in S, then ℓ has a model in S iff ℓ' has a model in S).

At each step of derivation we assign to a literal the value TRUE or FALSE. If it generates the empty clause, then we insert in the model which is being built the opposite of that literal together with all the other opposites of its symmetrical literals.

If ℓ is a literal in a system S such that T"$\ell(S) \models \square$ ($\square$: is the notation of the empty clause and $\models$ the logical implication symbol). ℓ has no model in S then for all model I of S $I[\neg\ell]$=1 , if $\ell \sim \ell'$ then ℓ' has no model in S (theorem 8.), thus for all model I of S $I[\neg\ell'] = 1$. Now we can cut in the resolution tree the branch which correspond to the assignment of ℓ' and insert $\neg\ell$ and $\neg\ell'$ in the model which is being built.

As a consequence,, the system of clauses is significantly simplified, and many branches of the resolution tree are cut simultaneously, because when backtrak occurs we do not take care of negation assignments of symmetrical literals. Thus for each symmetrical literal we cut a branch in the resolution tree. Then with n symmetrical literals we can make $n-1$ cuts.

To implement this new algorithm, we need a procedure which, for a given literal, computes a cycle of its symmetrical literals which generate the empty clause, such a cycle is obtained by puting in a relationship literals in a same clause. If this cycle exists, then we backtrack immediately, because we know that there exists no model at the current level. Otherwise we compute the larger cycle (which contains more symmetrical literals) in order to get quickly the empty clause or a model of the system of clauses.

7 Application

7.1 SLRI application

The two graphs below illustrate the results of the pigeon-hole problem solved by the SLRI method,with and without taking advantage of symmetries.

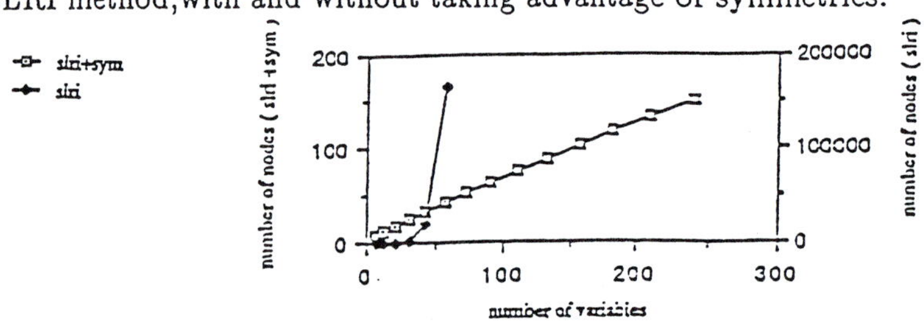

Fig. 3. The pigeon-hole problem: comparison of the number of nodes.

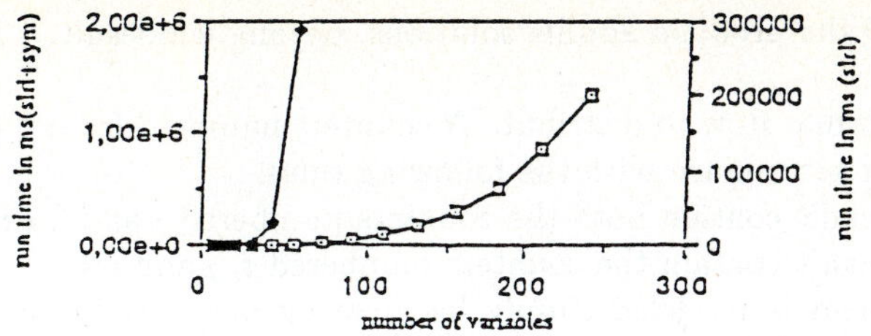

Fig. 4. The pigeon-hole problem: comparison of run times.

Interpretation. Fig.3 shows that the complexity of the SLRI with symmetry detection method is linear in number of elementary steps; whereas SLRI without symmetry can not solve the problem when the number of pigeons is greater than 8. The complexity of resolution is exponential.
Fig 4. shows that the CPU time is no more exponential, but becomes proportionnal to n^2.

The statement of the Ramsey problem with 17 vertices and 3 colors is: "Color a complete graph with 17 vertices with 3 colors, such that there is no monochrome triangle". In the SLRI method we use symmetry on literals which occur in the same clause. Unfortunately such symmetries do not exist at different levels of resolution in the Ramsey problem. Morever, there are symmetrical literals in different clauses at all levels of resolution; therefore the method of Semantic Evaluation can give better results. In that case, we need a good heuristic to decide which is the literal to assign next, in order to keep the symmetries in the next resolution levels.

7.2 Semantic Evaluation applications

In order to check the practical efficiency of our algorithm, we present some results for the Semantic Evaluation method with and without taking advantage of the symmetry property on some known benchmarks. For comparison we present in program Eval_Sem results obtained with the Semantic Evaluation method and in program Eval_Sem+Sym we give the correspondent results of the method with advantage of symmetries.

7.2.1 Description of the benchmarks

- Queens. Placing N queens in $N \times N$ chessboard such that there is no couple of queens attacking each other.

- Erdös's theorem. Find the permutation σ of N first numbers such that for each 4-tuple $1 \leq i < j < k < l \leq$ N none of the two relations $\sigma(i) < \sigma(j) < \sigma(k) < \sigma(l)$ and $\sigma(l) < \sigma(k) < \sigma(j) < \sigma(i)$ is verified.
  This problem is modeled by creating for each couple (i,j) a variable $f_{i,j}$ which means $\sigma(i) < \sigma(j)$. The rules express the associativity of the relation <, and prohibit the misplaced 4-tuples.

For $N \leq 9$ the problem admits solutions, beyong it doesn't.

- Schur's lemma: How to distribute N counters numbred from 1 to N into 3 boxes A, B, C in accordance with the following rules:
  1) A box can't contain both the counters numbered i and $2 * i$
  2) A box can't contain the counters numbered i, j and $i + j$
  This problem is modeled simply by creating one variable by counter and by box.For $N \leq 13$ the problem admits solutions, beyong it doesn't.

- Ramsey problem's: Color the edges of a complete graph on N vertices with three different colors such that no monochromatic triangle appears.
  For $N \leq 16$ the problem admits solutions, beyong it doesn't.

7.3 Results

Table 1. shows some results of the Schur Lemma, Ramsey problems, etc. The most satisfying result is that we have proved for the first time the unsatisfiability of the Ramsey problem with 17 vertices and 3 colors. This result is given by our algorithm, implemented in PASCAL, in 30 min of CPU on a SUN4/110. The Semantic Evaluation algorithm without symmetry property, has run 15 h CPU and more than 1400000 steps on a HP 9000/350 without success.

Table 1. Schur's Lemma, Ramsey's problem, etc

Problems	Size formula		Eval_Sem		Eval_Sem+Sym	
	Clauses	Variables	Steps	Times	Steps	Times
Queens 8	736	64	138	1.5"	64	0.35"
Queens 10	1480	100	395	6.43"	197	2.04"
Erdös 9	420	36	35	0.35"	35	0.36"
Erdös 10	660	45	1070	15"	914	8.21"
Schur13	178	39	684	3.3"	42	0.05"
Schur14	203	42	2061	11.2"	249	1.33"
Ramsey14	1456	273	273	4.7"	217	2.76"
Ramsey15	1785	315	4078	1'.30"	723	50.73"
Ramsey16	2160	360	4094	1'.29"	2957	1'.25"
Ramsey17	2584	408		-	27000	30'

Table 2. contains some results of the pigeon-hole problem described above. In this problem there are symmetries at each level. Thus SLRI and Semantic Evaluation with symmetry detection have a linear resolution complexity. Without symmetry property both methods fail to find the proof when the number of pigeons is greater than 10.

Table 2. Pigeon hole problems

Number of pigeons	Size formula		Eval_Sem+Sym	
	Clauses	Variables	Steps	Times
14	1197	182	193	4.8"
16	1816	240	253	7.73"
18	2619	306	321	13.50"
20	3630	380	397	22.36"
22	4873	462	481	37.11"
24	6372	552	573	55.58"
26	8150	650	673	1'36"
28	10234	756	781	2'2"
30	12645	870	897	3'4"

8 Conclusion

Many problems can not be solved with the classical resolution methods when they involve more than a certain number of variables. However, most of these problems include symmetries; the efficiency of the resolution methods can be significantly increased by taking advantage of this property.

We have proven several results, and implemented the corresponding algorithms for two resolution methods. We have obtained satisfactory CPU times for the pigeon-hole problem; we have also been able to solve the Ramsey problem with 17 vertices and 3 colors, which has always been impossible before.

We intend to use the symmetries in order to compute only the models which are not symmetrical to a model which has already been found during resolution, thus, a problem will be solved when all the basic models will be found; all the other ones can be deduced from these basic models by applying the symmetries. In particular, this method can be applied when considering how to get the two solutions which are not isomorphic for the Ramsey problem with 16 vertices and 3 colors[7]. Promising results are expected in a near future.

We are also working on what we call the "Strongly symmetrical cycles", i.e: for any order of the literals, a cycle is still a cycle. This will be useful when dealing with statement like "N literals are true among the literals of the clause c".

References

1. B. Benhamou and L. Sais. Etude des symétries en calcul propositionnel. Master's thesis, GIA Luminy (Marseille), 1990.
2. B. Benhamou and L. Sais. Etude des symétries en calcul propositionnel. Technical Report 1, Université de provence, 1991.
3. W. Bibel. Short proofs of the pigeon hole formulas based on the connection method. *Automated reasoning*, (6):287–297, 1990.
4. S. Cook. A short proof of the pigeon hole principle using extended resolution. *SIGACT News*, (8):28–32, oct.-dec. 1976.
5. C. Cubbada and M. D. Mouseigne. *Variantes de l'algorithme de SL-Résolution avec retenue d'information.* PhD thesis, GIA Luminy (Marseille), 1988.

6. M. Davis and H. Putnam. A computing procedure for quatification theory. *JACM*, (7):201–215, 1960.
7. J. Kalbfleisch and R. Stanton. On the maximal triangle-free edge-chromatic graphs in three colors. *combinatorial theory*, (5):9–20, 1969.
8. R. Kowalski and D. Kuehner. Linear resolution with selection function. *Artificial Intelligence*, (2):227–260, 1971.
9. B. Krishnamurty. Short proofs for tricky formulas. *Acta informatica*, (22):253–275, 1985.
10. B. Krishnamurty and R. Moll. Examples of hard tautologies in the propositionnal calculus, 1981.
11. D. W. Loveland. A linear format for resolution. In Springer, editor, *Lecture notes in computer science*, number 125, 1970.
12. R. Lyndon. *Notes of logic.* Van Nostrand Mathematical Studies, 1964.
13. L. Oxusoff and A. Rauzy. *L'évaluation sémantique en calcul propositionnel.* PhD thesis, GIA - Luminy (Marseille), 1989.
14. J. A. Robenson. Teorem proving on computer. *JACM*, pages 163–174, 1963.
15. I. SCHUR. Uber die kongruenz $x^m + y^m = z^m mod(p)$. *J ber Deutsch Verein*, (25):114–116, 1916.
16. G. S. Tseitin. On the complexity of derivation in propositional calculus. In *Structures in the constructive Mathematics and Mathematical logic*, pages 115–125. H.A.O Shsenko, 1968.

Difference Matching*

David Basin
Max-Planck-Institut für Informatik
Im Stadtwald, Saarbrücken, Germany

Toby Walsh
Department of AI, Edinburgh University
Edinburgh, Scotland

Abstract

Difference matching is a generalization of first-order matching where terms are made identical both by variable instantiation and by structure hiding. After matching, the hidden structure may be removed by a type of controlled rewriting, called rippling, that leaves the rest of the term unaltered. Rippling has proved highly successful in inductive theorem proving. Difference matching allows us to use rippling in other contexts, e.g., equational, inequational, and propositional reasoning. We present a difference matching algorithm, its properties, several applications, and suggest extensions.

1 Introduction

A central problem in theorem proving is showing that one equality follows from others or more generally that one formula is entailed by others. Many techniques have been developed for this purpose, for example, the use of canonical sets of rewrite rules and resolution theorem proving. Within inductive theorem proving, this problem arises in the proof that the induction conclusion follows from the induction hypothesis. Boyer and Moore, in their theorem prover NQTHM [4, 5], approach this problem essentially by normalizing both the induction hypothesis and conclusion using the same set of rewrite rules. Their approach works remarkably well; however, their normalization procedure (actually, a combination of procedures) is complex, contains a large amount of heuristic information, and is difficult to detach from their theorem prover itself. Motivated by their success, Bundy suggested in [7] an alternative approach based not on exhaustive heuristic based normalization but instead upon the application of structure preserving rewrite rules; he called this rewriting *rippling*.[1]

The ideas behind rippling are fairly straightforward. In an inductive proof, the induction conclusion is an image of the induction hypotheses except for the appearance of certain function symbols applied to the induction variable in the conclusion.

*The first author was supported, while at Edinburgh, by SERC grant GR/F/71799, the second by a SERC PostDoctoral Fellowship. We would like to thank the other members of the Edinbrugh Mathematical Reasoning Group for their feedback on this project.

[1]The name rippling comes from *rippling-out* a term coined by Aubin [1], a student of Boyer and Moore's, during his study of generalization in inductive theorem proving. It is based on an observation that one can iteratively unfold (as in [10]) recursive functions in the induction conclusion, preserving the structure of the induction hypothesis while unfolding.

We call these function symbols *wave-fronts* (terminology and notation will be formally defined in Section 2). The rest of the induction conclusion, which is an exact image of the induction hypothesis is called the *skeleton*. For example, if we wish to prove that a proposition $P(x)$ is true for all natural numbers, we assume it is true for n and attempt to show it is true for $s(n)$. That is, we show $P(s(n))$ follows from the hypothesis $P(n)$. The hypothesis and the conclusion are identical except for the successor function applied to the induction variable n. We mark this wave-front by placing a box around it, and underlining the subterm contained in the skeleton, $P(\boxed{s(\underline{n})})$. Rippling then applies just those rewrite rules, called *wave-rules*, which move the difference out of the way leaving behind the skeleton. In their simplest form, wave-rules are rewrite rules of the form:

$$\alpha(\boxed{\beta(\underline{\gamma})}) \quad \Rightarrow \quad \boxed{\rho(\underline{\alpha(\gamma)})}$$

By design, the skeleton $\alpha(\gamma)$ remains unaltered by their application. We eventually hope to rewrite the conclusion, $P(\boxed{s(\underline{n})})$ using such wave-rules into some function of $P(n)$; that is, into $\boxed{f(\underline{P(n)})}$ (f may be the identity) and then call upon the induction hypothesis.

Rippling, and its extensions (e.g., rippling in other directions than out, and more general forms of wave-rules) are presented in [8] and are employed in the Oyster-Clam system. Many of the same ideas were developed independently by Hutter [14], from ideas in [7], and employed in the INKA system. Both systems have enjoyed a high degree of success. This success seems to stem from three properties of rippling (a more detailed analysis of these properties can be found in [9]):

1. Rippling involves little search. In rippling, the wave-fronts in the pattern must correspond to wave-fronts in the instance. This leads to a very controlled application of rewrite rules that gives very low branching rates.

2. Rippling terminates. Rippling always makes progress moving wave-fronts in a desired direction; hence termination is guaranteed, even when applying rewrite rules that would normally, without wave-front annotation, lead to loops.

3. Rippling applies only "good" rewrites. As wave-rules are structure preserving, if rippling terminates successfully, the hypothesis can be used to prove or simplify the conclusion.

From these properties, we can see that rippling is an attractive kind of rewriting. Inductive theorem proving is well suited for rippling because, in the induction step, there is a natural annotation of the conclusion so that its skeleton is identical to the induction hypothesis.[2] However, there is no reason why rippling cannot be used as a general rewrite procedure when one term has the same skeleton as another; its only precondition is that it is applied to a term with wave-front annotations.

In this report we introduce difference matching, a procedure that supports the general application of rippling in theorem proving and term simplification. A difference matcher takes as inputs two terms (or formulas) s and t. It returns s annotated

[2] This is for constructor style induction schemas; for destructor schemas one could ripple on hypotheses or move the wave-fronts into the conclusion. See [9].

with wave-fronts, and a set of substitutions such that the skeleton of the annotated term equals t under substitution. Although difference matching generalizes first-order matching, it is much more than matching. It is an attempt to make two terms identical not just by variable instantiation, but also by structure hiding; the hidden structure is the part of the term within the wave-front that serves to direct rippling.

A brief example will illustrate the spirit of how difference matching and rippling can be combined. The example is simple enough to describe quickly (more complex examples are given in the experience section) and has the virtue of illustrating an interesting special case of difference matching where there are no match variables to be instantiated. Unlike first-order matching, which degenerates to syntactic identity, difference matching remains non trivial and interesting in the absence of match variables. We will, however, give examples of difference matches involving variable instantiation in Sections 3 and 4.

Suppose we wish to demonstrate that

$$s(y) + x < s(y) * s(y) \tag{1}$$

follows from

$$x < y * y. \tag{2}$$

Here x and y are skolem constants and s is the successor function. An informal proof might first add $s(y)$ to each side of Equation 2 and then y to the righthand side.

$$\begin{aligned} s(y) + x &< s(y) + y * y \\ &< s(y) + y * y + y \\ &= s(y) * s(y) \end{aligned}$$

Hence Equation 1 follows. Automating such a proof would seem to require search to select which terms to add. It would also require reasoning about the properties of addition, multiplication, and basic monotonicity and inequality reasoning.

An alternative is to use difference matching and rippling. That is, to find an annotation of Equation 1 so that its skeleton is the same as Equation 2; afterwards, rippling will hopefully reduce the equation to its skeleton which is our given hypothesis.

We shall need some basic wave-rules that relate plus, times, and less-than.

$$\boxed{Z + \underline{X}} < \boxed{Z + \underline{Y}} \Rightarrow X < Y \tag{W_1}$$

$$X < \boxed{\underline{Y} + Z} \Rightarrow X < Y \tag{W_2}$$

$$\boxed{s(\underline{X})} * Z \Rightarrow \boxed{\underline{X * Z} + Z} \tag{W_3}$$

$$Z * \boxed{s(\underline{X})} \Rightarrow \boxed{Z + \underline{Z * X}} \tag{W_4}$$

Note that $\Rightarrow$ stands for rewriting as opposed to implication. Since we will reason backwards, a wave-rule $\phi \Rightarrow \psi$ is justified by an implication $\psi \rightarrow \phi$; that is, if we can prove ψ, then we can also prove ϕ. For example, W_2 is a valid rule only in the direction shown. These wave-rules could be supplied as ordinary rewrite rules and have wave-front annotation automatically added (as in Edinburgh Clam system).[3] Wave-rules can also be automatically synthesized from function definitions (as in the INKA system and also in Clam).

Difference matching Equation 1 with 2 will annotate the former as

$$\boxed{s(y)+\underline{x}} < \boxed{s(\underline{y})} * \boxed{s(\underline{y})}$$

and rippling will apply wave-rules W_4, W_1, W_3, and W_2 to the above which results in the following inequalities.

$$\begin{array}{rcl}
\boxed{s(y)+\underline{x}} & < & \boxed{s(y)+\underline{\boxed{s(\underline{y})}*y}} \\
x & < & \boxed{s(\underline{y})} * y \\
x & < & \boxed{\underline{y*y}+y} \\
x & < & y*y
\end{array}$$

Since the final inequality is Equation 2, our hypothesis, the proof is completed. Note that explosive search was not required as rippling provided the key ideas needed to transform the conclusion into its skeleton, the hypothesis. More complex examples of this kind of rewriting are found in Section 4.

The idea behind the combination of difference matching and rippling is rather general. The combination captures a basic problem solving strategy of finding differences or mismatches between terms and working to eliminate those differences. Difference identification and reduction are also central themes in the research of [3, 11, 16]. In their work, a partial unification results in a special kind of resolution step (E and RUE-resolution) where the failure to completely unify gives rise to new inequalities that represent the differences between the two terms. This leads to a controlled application of equality reasoning where paramodulation is only used when needed. This is analogous to our use of difference matching and rippling as a means of controlling the use of general rewrite rules. A different strategy for removing differences between two terms is to seek a generalization for the two; there may also be connections between difference matching (and its extensions) to anti-unification[18], although we have not explored this direction.

The remainder of our paper is organized as follows. In Section 2, we present definitions and necessary background. In Section 3, we present the formal properties required of a difference match and give an algorithm which returns exactly the matches with these properties. We analyze some other properties of our algorithm, and prove its correctness. In Section 4, we report on experience using difference matching in finding the sums of series. These results are interesting not only as exercises in equality reasoning, but also in demonstrating the use of difference matching and rippling in solving non-inductive problems. In the final section, we sketch extensions and draw conclusions.

---

[3] Automatically annotating rewrite rules as wave-rules can be seen as a special case of difference unification (see Section 5) between the right and left hand sides of the rewrite rule with universally quantified variables treated as constants. Clam implements such "wave-rule parsing" by simply enumerating annotations and comparing resulting skeletons.

2 Terms and Annotation

Terms

The algorithm we present acts on both terms and formulas. To facilitate this we work in a simple first-order sorted framework (as in [13]) in which sorts can be used to distinguish between such syntactic categories.

A *sort* is the name of a set of syntactic entities called *constants*. For $\mathcal{S}$ a set of sorts, *types* are specified by finite sequences of members of $\mathcal{S}$ which we shall write as $S_1 \times .... \times S_n \rightarrow S$ where the S_i range over $\mathcal{S}$ (notational conventions are given at end of this section). A *signature* Σ *over* $\mathcal{S}$ consists of a set Σ of operators and a typing function, *Ty*, from operators to types. For the remainder of this paper we shall assume that we are given a set of sorts $\mathcal{S}$ and a signature Σ over $\mathcal{S}$.

Object level variables of sort s are members of V_S, where V_S is one of a S-indexed family of sets. We let V denote the union of these sets. *Object level terms of sort s* are members of the set $\mathcal{T}_S(\Sigma, \mathcal{V})$ which is the smallest set of terms that includes $\mathcal{V}_S$ and for all $f \in \Sigma$, if $Ty(f) = S_1 \times ... \times S_n \rightarrow S$ then $f(t_1, ..., t_n)$ is in $\mathcal{T}_S(\Sigma, \mathcal{V})$ whenever $t_i \in \mathcal{T}_{S_i}(\Sigma, \mathcal{V})$. The *terms over* Σ are members of $\mathcal{T}(\Sigma, \mathcal{V})$, which is the S-indexed union of the $\mathcal{T}_S(\Sigma, \mathcal{V})$. Whenever possible, we will omit sorts and types when these are immaterial or context makes our meaning clear.

Let $\mathcal{M}_S$ be a denumerable set of meta-variables of sort S and $\mathcal{M}$ the union of these sets. These variables may be thought of (and are sometimes referred to in the literature) as *match-variables* or *logic-variables*. A *meta-level term (of sort S)* is a member of $\mathcal{T}(\Sigma, \mathcal{V} \cup \mathcal{M})$ (respectively $\mathcal{T}_S(\Sigma, \mathcal{V} \cup \mathcal{M})$). We shall call both object and meta-level terms simply *terms* when there is no danger of confusion. A *ground meta-level term* is one containing no meta-variables, although it may contain members of $\mathcal{V}$, and is ground from the standpoint of the matching algorithm that we will define. Indeed, in defining difference matching and analyzing its correctness, we will simplify matters by treating members of $\mathcal{V}_S$ as if they were constants of sort S. Given a meta-level term t, *Mvars*(t) denotes the members of $\mathcal{M}$ that occur in t.

Annotation

An *object-level W-term* is a member of $\mathcal{T}(\Sigma, \mathcal{V})$ that may contain contain *wave-fronts*. A wave-front is a term t with a proper sub-term t' deleted. The deleted subterm may itself contain wave-fronts. We represent it by an annotation of t which encloses t in a box and underlines the sub-term t'. We let $\mathcal{WT}(\Sigma, \mathcal{V})$ represent the set of such W-terms. For example, we can annotate $f(a, b)$ as $\boxed{f(\underline{a}, b)}$. A meta-level W-term is defined analogously to object-level W-terms except over $\mathcal{T}(\Sigma, \mathcal{V} \cup \mathcal{M})$.

There are various alternative ways W-terms may be represented on paper or in a computer. We shall display them using the new Edinburgh "box-and-hole" notation[9]; however, the older Edinburgh "box" notation[8] and Hutter's representation of his C-terms[14] are other possible options. All that we require of any implementation is that we can tell which parts of the term are "deleted" by a wave-front and hence do not contribute to the skeleton. To this end, if r is a W-term with subterm t, let the predicate *InWave*(r, t) hold when a distinguished occurrence of t (which may be given by a term address) is completely within a wave-font within r.

$HdInWave(r)$ hold if r is a function application and the leading function symbol is within a wave-front. For example, if r is $\boxed{f(\underline{a}, b)}$ then $HdInWave(r)$ and $InWave(r, b)$ both hold and neither $HdInWave(a)$ nor $InWave(r, a)$ hold. In the remainder of this paper we will not distinguish between annotations that appear different but are indistinguishable to these predicates. E.g., $\boxed{s(s(\underline{0}))}$ and $\boxed{s(\boxed{s(\underline{0})})}$. This is sensible as in systems like Clam and INKA it is possible to represent internally both annotated terms and wave-rules in a normal form where each wave-front has an immediate subterm deleted (e.g., the second term in the above example) and there is no loss of generality in rippling with such a representation. Our algorithm, given in the next section returns terms annotated this way.

We now define two functions from W-terms into terms. The first, *Skel*, returns the skeleton, that is the unannotated part of an annotated term. It is defined recursively by cases. Since the definition of W-terms disallows an annotated constant or variable to itself be a W-term (since there are no proper subterms to be "deleted") we must have in the base case that $Skel(X) = X$ when X a constant, variable, or meta-variable. In the step case, if the leading function is not in a wave-front (that is, if $\neg HdInWave(f(t_1, \ldots, t_n))$) we have

$$Skel(f(t_1, \ldots, t_n)) = f(Skel(t_1), \ldots, Skel(t_n));$$

alternatively if $HdInWave(f(t_1, \ldots, t_n))$, it must be the case that for some $j \in \{1..n\}$ $\neg InWave(f(t_1, \ldots, t_n), t_j)$, and then

$$Skel(f(t_1, \ldots, t_n)) = Skel(t_j).$$

For example,

$$Skel(f(\boxed{f(\underline{a}, b)}, b)) = Skel(\boxed{f(\underline{f(a, b)}, b)}) = f(a, b)$$

Skeleton preserving rewrite rules are called *wave-rules* by Bundy and *C-equations* by Hutter.

The second function we define, *Erase*, simply removes wave-front annotation but leaves the term otherwise unchanged. E.g., $Erase(f(\boxed{f(\underline{a}, b)}, b)) = f(f(a, b), b)$. For a W-term t, we call the term computed by $Erase(t)$ the *body* of t and the part returned by $Skel(t)$ its *skeleton*. Note that, since wavefronts are deleted, the skeleton may not well-typed. For the remainder of the paper we shall restrict ourselves to W-terms t whose skeletons are themselves well-typed terms.

A substitution is a sort respecting partial function of type $\mathcal{M} \to \mathcal{T}(\Sigma, \mathcal{V})$ that takes values on only finitely many members of $\mathcal{M}$. Substitutions are extended to terms in the standard way. If σ is a substitution, let $Dom(\sigma)$ be those elements of $\mathcal{M}$ for which σ takes a value and $\sigma(\alpha)$ the value of σ on α. We shall represent a substitution σ by a finite set of pairs $\{\langle\alpha_1, t_1\rangle, \ldots \langle\alpha_n, t_n\rangle\}$ where $t_i \in \mathcal{T}_{S_i}(\Sigma, \mathcal{V})$, $\alpha_i \in \mathcal{M}_{S_i}$. We combine substitutions σ_1 and σ_2 using the operator $\cup$ which takes the set-theoretic union of the sets representing the two substitutions. This returns a well-defined substitution provided that

$$\forall \alpha \in (Dom(\sigma_1) \cap Dom(\sigma_2)).\, \sigma_1(\alpha) = \sigma_2(\alpha)$$

For brevity, we will write this condition as $compatible(\sigma_1, \sigma_2)$. We shall also write *Sub* to represent the set of substitutions.

Notation

We will use the following notational conventions. We shall let S range over sorts; τ will range over types; σ will range over substitutions; $a, b, c, \ldots$ will range over (possibly annotated) constants; $f, g, h, \ldots$ will range over (possibly annotated) members of Σ; $x, y, z, \ldots$ will range over (possibly annotated) members of Var; Greek letters such as $\alpha, \beta, \gamma, \ldots$ will range over members of $\mathcal{M}$; $r, s, t, \ldots$ will range over both (possibly annotated) terms and meta-level terms. These variables may be sub or super-scripted. The empty set symbol, $\emptyset$, shall denote both the empty set and the substitution with empty domain.

3 Difference Matching

The Algorithm

The input to a difference matcher is a term s, possibly containing meta-variables, called the *pattern*, and a term t called the *instance*. The difference matcher returns annotated terms r, and substitutions σ such that r is an annotation of s and the skeleton of r matches t under substitution σ. As the output to a difference matcher is not always unique, we shall give an an algorithm for difference matching in a logic programming like language that computes terms satisfying the relation of type $\mathcal{T}_S(\Sigma, \mathcal{V} \cup \mathcal{M}) \times \mathcal{T}_S(\Sigma, \mathcal{V}) \times \mathcal{WT}_S(\Sigma, \mathcal{V}) \times Sub$. An implementation of this relation,

$$dm(s, t, r, \sigma)$$

should satisfy the property P_1,

$$Erase(r) = s,$$

the property P_2,

$$\sigma(Skel(r))) = t,$$

and the property P_3,

$$Dom(\sigma) = Mvars(Skel(r)).$$

The first property insists that r is simply an annotation of s. The second insists that the skeleton of r equals t under substitution σ. The last property enforces a kind of minimality on substitutions and allows us to ignore those with "extraneous" assignments. As shorthand, we will let $MatchRel(s, t, r, \sigma)$ represent the conjunction of these three properties, $P_1 \wedge P_2 \wedge P_3$.

The following is a specification of a relation dm that has these properties.

% Difference Matching
% Clause 1, Meta-Variable:
$dm(\alpha, t, \alpha, \{\langle\alpha, t\rangle\}) \Leftarrow Ty(\alpha) = Ty(t).$

% Clause 2, Constant (or member of $\mathcal{V}$):
$dm(a, a, a, \emptyset).$

% Clause 3, equal outer functors ($j > 0$):
$dm(f(s_1, \ldots, s_j), f(t_1, \ldots, t_j), f(r_1, \ldots, r_j), \sigma) \Leftarrow$
$\quad \forall i \in \{1..j\}.\, dm(s_i, t_i, r_i, \sigma_i),$
$\quad$ *if* $\forall i, i' \in \{1..j\}.\, compatible(\sigma_i, \sigma_{i'})$ *then* $\sigma = \cup_{i=1}^{j} \sigma_i.$

% Clause 4, (possibly) different outer functors ($j > 0$)
$dm(f_s(s_1, \ldots, s_j), f_t(t_1, \ldots, t_k), \boxed{f_s(s_1, \ldots, s_{i-1}, \underline{r_i}, s_{i+1} \ldots s_j)}, \sigma) \Leftarrow$
$\quad \exists i \in \{1..j\}.\, dm(s_i, f_t(t_1, \ldots, t_k), r_i, \sigma).$

The program notation we use is similar to that of Prolog. We have used some syntactic sugar such as ellipsis and bounded universal and existential quantification which have the obvious "intended meaning" and can be coded in Prolog in a straightforward way. When translated into a Prolog program (which we have done), the algorithm may be executed in in mode(+,+,?,?). We shall call the first two arguments, s and t, the *inputs* and those pairs of r and σ that satisfy dm the *outputs*.

As an example of difference matching, the following table contains the outputs which difference match with the pattern $x + 1 < (\alpha + 1) * (\alpha + 1)$ and the instance $x < (y + 1) * (y + 1)$.

Annotations	*Substitutions*
$\boxed{\underline{x} + 1} < (\alpha + 1) * (\alpha + 1)$	$\{\langle\alpha, y\rangle\}$
$\boxed{\underline{x} + 1} < \boxed{(\underline{\alpha} + 1)} * \boxed{(\underline{\alpha} + 1)}$	$\{\langle\alpha, y + 1\rangle\}$
$\boxed{\underline{x} + 1} < \boxed{(\underline{\alpha} + 1) * (\alpha + 1)}$	$\{\langle\alpha, (y + 1) * (y + 1)\rangle\}$
$\boxed{\underline{x} + 1} < \boxed{(\alpha + 1) * (\underline{\alpha} + 1)}$	$\{\langle\alpha, (y + 1) * (y + 1)\rangle\}$

Properties

Several properties of dm are immediately apparent. First, the dm program can be decomposed into definitions of several logical relations. Clauses 2 and 3 define the relationship of syntactic identity. Adding Clause 1 defines the relationship of (sorted) first-order matching. That is, if inputs s and t first-order match, i.e. there is a substitution σ such that $\sigma(s) = t$, then there is a unique r and σ that satisfies the relation defined by the first three clauses. Adding Clause 4 instead of Clause 1, yields an algorithm for ground difference matching, illustrated in the introduction.

The union of the four clauses defines a relationship more general than first-order matching. For example, when the pattern s contains meta-variables of the same sort as s, a difference match is always possible (and one for each such meta-variable). For instance, if $Ty(f) = Ty(g) = S_1 \to S_1$ and $Ty(\alpha) = Ty(b) = S_1$ then

$$dm(f(\alpha), g(b), \boxed{f(\underline{\alpha})}, \{\langle \alpha, g(b) \rangle\}).$$

Note, by the way, that skeleton of the annotated output term is always well-typed as it is identical to the instance.

Also unlike standard matching, for a given input, there may be, in the worst case, exponentially many (in the size of the pattern) outputs that satisfy the difference matching relation. However, such bad behavior is unusual; in practice there are only a few successful matches. But because there are exponentially many possible ways of annotating a term, difference matching appears more difficult than first order matching. The complexity of the algorithm given, in mode(+,+,− ,−) is exponential because of the j-ary branching in Clause 4. Note that it is linear in mode(+,+,+,+), hence given a pair of inputs, finding a pair of outputs which satisfy the difference matching relation is in NP since we can check if a guessed annotation satisfies *MatchRel* in time linear in the size of the pattern. Of course, there may be faster implementations of difference matching than the one we have given; we have not analyzed the problem's exact complexity.

What is less obvious is that dm returns all and only all those matches that satisfy P_1, P_2, and P_3.

Lemma 1 *Soundness for dm:* $dm(s,t,r,\sigma) \Rightarrow \text{MatchRel}(s,t,r,\sigma)$.

Proof: By structural induction on the pattern s. In the base case, s is either a meta-variable or a constant and *MatchRel* is obviously satisfied by the only applicable clauses, 1 and 2 respectively.

In the step case we have $s = f_s(s_1, \ldots, s_j)$. Suppose Clause 3 applies. Then $t = f_s(t_1, \ldots, t_j)$ and $r = f_s(r_1, \ldots r_j)$. Moreover for $i \in \{1..j\}$, $dm(s_i, t_i, r_i, \sigma_i)$ and by the induction hypothesis *MatchRel* holds of these smaller instances. So

$$Erase(r) = Erase(f_s(r_1, \ldots, r_j)) = f_s(Erase(r_1), \ldots, Erase(r_j)) = f_s(s_1, \ldots, s_j) = s.$$

Also by the induction hypothesis $\sigma_i(Skel(r_i)) = t_i$, and, by Clause 3, $\sigma = \cup_i \sigma_i$. Hence

$$\sigma(Skel(r)) = \sigma(Skel(f_s(r_1, \ldots, r_j)) = f_s(\sigma_1(Skel(r_1)), \ldots \sigma_j(Skel(r_j))) = t.$$

Finally since the $Mvars(Skel(r)) = \cup_i Mvars(Skel(r_i))$, then $Dom(\sigma) = Mvars(Skel(r))$.

Alternatively, suppose that Clause 4 applies, so $t = f_t(t_1, \ldots, t_k)$. Now there is some $i \in \{1..j\}$ and some r' where $dm(s_i, f_t(t_1, \ldots, t_k), r', \sigma)$. Now by the induction hypothesis, $Erase(r') = s_i$, so we have

$$\begin{aligned} Erase(r) &= Erase(\boxed{f_s(s_1, \ldots, s_{i-1}, \underline{r'}, s_{i+1} \ldots s_j)}) \\ &= Erase(\boxed{f_s(s_1, \ldots, \underline{s_i}, \ldots, s_j)}) \\ &= f_s(s_1, \ldots, s_i, \ldots, s_j) = s \end{aligned}$$

Similarly, P_2 holds because by the induction hypothesis $\sigma(Skel(r')) = f_t(t_1, \ldots, t_k)$, so

$$\sigma(Skel(r)) = \sigma(Skel(\boxed{f_s(s_1, \ldots, s_{i-1}, \underline{r'}, s_{i+1} \ldots, s_j)})) = \sigma(Skel(r')) = t.$$

Moreover, as $Skel(r') = Skel(r)$ and $Dom(\sigma) = Mvars(Skel(r')))$, then $Dom(\sigma) = Mvars(Skel(r))$. □

To prove the converse, we use a few facts which may be easily verified by the diligent reader.

Fact 1 *If* $s = f_s(s_1, \ldots, s_j)$, $t = f_t(t_1, \ldots, t_k)$, $r = f_r(r_1, \ldots, r_l)$, $\mathrm{MatchRel}(s, t, r, \sigma)$, *and* $\neg\mathrm{HdInWave}(r)$, *then* $f_s = f_t = f_r$, $j = k = l$, $\forall i \in \{1..j\}. \mathrm{MatchRel}(s_i, t_i, r_i, \sigma_i)$, $\sigma = \cup_{i=1}^{j} \sigma_i$ *and* $\mathrm{Dom}(\sigma_i) = \mathrm{Mvars}(\mathrm{Skel}(r_i))$.

Fact 2 *If* $s = f_s(s_1, \ldots, s_j)$, $t = f_t(t_1, \ldots, t_k)$, $r = f_r(r_1, \ldots, r_l)$, $\mathrm{MatchRel}(s, t, r, \sigma)$, *and* $\mathrm{HdInWave}(r)$, *then* $f_s = f_r$, $j = l$, *and* $\exists i \in \{1..j\}. \mathrm{MatchRel}(s_i, t, r_i, \sigma) \wedge \forall k \in \{1..j\}. k \neq i \Rightarrow \mathrm{InWave}(r, r_k)$.

Lemma 2 *Completeness for dm:* $\mathrm{MatchRel}(s, t, r, \sigma) \Rightarrow dm(s, t, r, \sigma)$.

Proof: Proof by structural induction on s. There are two base-cases. In the first, s is a meta-variable α of the same sort as t. So by P_1, r must also be α and, by P_2 and P_3, σ must be the substitution $\{\langle\alpha, t\rangle\}$. But then we have Clause 1 of *dm* satisfied. In the second base-case, s is a constant a or the same sort as t. By P_1, $t = a$ and, by P_3, since $Var(t) = \emptyset$, we have $\sigma = \emptyset$. So Clause 2 of *dm* is satisfied.

In the step case we must have $s = f_s(s_1, \ldots, s_j)$, $t = f_t(t_1, \ldots, t_k)$ and $r = f_r(r_1, \ldots, r_l)$. But by P_1 we must have that the bodies of s and r are the same so f_r equals f_s and $j = l$. Now f_r may or may not be within a wave-front and we split on these two cases. In the first case, $\neg HdInWave(r)$, so we may apply Fact 1 and conclude that $\forall i \in \{1, ..j\}. MatchRel(s_i, t_i, r_i, \sigma_i)$. From the induction hypothesis it follows that $dm(s_i, t_i, r_i, \sigma_i)$ and since, by Fact 1, $\sigma = \cup\sigma_i$ we conclude $dm(s, t, r, \sigma)$ using Clause 3. In the second case, $HdInWave(r)$, we have by P_1 that the body of r and s are identical and, by Fact 2, $\exists i \in \{1..j\}. \forall k \in \{1..j\}. k \neq i \Rightarrow InWave(r, r_k)$. So r must look like

$$r = \boxed{f_s(s_1, \ldots, s_{i-1}, \underline{r_i}, s_{i+1} \ldots s_j)}.$$

Moreover, $MatchRel(s_i, t, r_i, \sigma)$ holds, so by the induction hypothesis $dm(s_i, t, r_i, \sigma)$. Hence $dm(s, t, r, \sigma)$ by Clause 4. □

4 Experience

We have explored the use of difference matching and rippling in several domains: equational reasoning arising in hardware verification, summing series, and calculating products, derivatives and integrals. This section will focus on examples in summing series.

Although an inductive proof can be used to show that a sum (or a product) has a certain closed form, we have been investigating *non-inductive* methods for *discovering* the sum. For example, one method for summing a series is to manipulate the sum into some function of known standard results. This is analogous to the situation in inductive theorem proving where we try to manipulate the induction conclusion into some function of the induction hypothesis. Consider, for example, how we might solve

$$s_n = \sum_{i=0}^{n} a * b^{i+1}. \tag{3}$$

In our work, we will simplify sums by difference matching against standard results like:

$$\sum_{I=0}^{N} C = (N+1) * C \tag{4}$$

$$\sum_{I=0}^{N} I = \frac{N(N-1)}{2} \tag{5}$$

$$\sum_{I=0}^{N} C^I = \frac{C^{N+1}-1}{C-1} \qquad C \neq 1 \tag{6}$$

$$= N+1 \qquad C = 1$$

where C is a constant. Additionally, we will need various wave-rules for manipulating series and algebraic expressions. Some of these rules include:

$$\sum_{I=A}^{B} \boxed{(\underline{U} + V)} \Rightarrow \boxed{\underline{\sum_{I=A}^{B} U} + \sum_{I=A}^{B} V}$$

$$\sum_{I=A}^{B} \boxed{C * \underline{U}} \Rightarrow \boxed{C * \underline{\sum_{I=A}^{B} U}}$$

$$X^{\boxed{\underline{Y}+1}} \Rightarrow \boxed{X * \underline{X^Y}}$$

Again where C is a constant. We begin by difference matching our goal (3) against known standard results. In this case, we can successfully difference match and ripple against all three of the standard results given above. However, there only exist wave-rules for moving the wave-front annotations returned by difference matching against (6):

$$\sum_{i=0}^{n} \boxed{a * \underline{b^{\boxed{\underline{i}+1}}}}$$

Of course, we are not always so lucky; sometimes, we can successfully difference match against several different standard results. This introduces an element of search (although in practice the search space is rather small).

Rippling can now be used to move these wave-fronts out of the way:

$$\begin{aligned} s_n &= \sum_{i=0}^{n} \boxed{a * \underline{b^{\boxed{\underline{i}+1}}}} \\ &= \sum_{i=0}^{n} \boxed{a * b * \underline{b^i}} \\ &= \boxed{a * \underline{\sum_{i=0}^{n} \boxed{b * \underline{b^i}}}} \\ &= \boxed{a * b * \underline{\sum_{i=0}^{n} b^i}} \end{aligned}$$

Finally, we match with the closed form solution (6):

$$s_n = \begin{cases} a * b * \frac{b^{n+1}-1}{b-1} & \mathit{if}\ b \neq 1 \\ a * b * (n+1) & \mathit{if}\ b = 1 \end{cases}$$

Difference matching and rippling play important roles in several other methods we have investigated for summing series. For example, another method using difference matching and rippling *perturbates* the sum by one term. This is closely related to induction. Consider, for example, trying to sum the geometric progression of Equation 6 from first principles.

$$s_n = \sum_{i=0}^{n} b^i$$

We begin by perturbing this sum by one term:

$$s_{n+1} = s_n + b^{n+1}$$

Now, we can also strip off not the last term but the first term:

$$s_{n+1} = b^0 + \sum_{i-0}^{n} b^{i+1}$$

Combining the last two equations, we get:

$$s_n + b^{n+1} = b^0 + \sum_{i=0}^{n} b^{i+1}$$

This is nearly an equation in s_n. If we difference match the sum in the righthand side of the equation against s_n, we get the wave-fronts:

$$\sum_{i=0}^{n} b^{\boxed{\underline{i}+1}}$$

If we ripple these wave-fronts out of the way, we will have an equation just in s_n which we can algebraically solve:

$$\begin{aligned} s_n + b^{n+1} &= b^0 + \sum_{i=0}^{n} b^{\boxed{\underline{i}+1}} \\ &= b^0 + \sum_{i=0}^{n} \boxed{b * \underline{b^i}} \\ &= b^0 + \boxed{b * \underline{\textstyle\sum_{i=0}^{n} b^i}} \\ &= b^0 + \boxed{b * \underline{s_n}} \end{aligned}$$

Thus,

$$s_n + b^{n+1} = b^0 + b * s_n$$

Solving this last equation (using the collect and isolate methods developed in the algebraic problem solver PRESS [6]) gives:

$$s_n = \frac{b^{n+1} - 1}{b - 1} \qquad b \neq 1$$

The case for $b = 1$ is trivial.

These and other methods based on difference matching for summing series have been implemented in the Oyster-Clam system [19]. However, a couple slight extensions of the basic difference matching algorithm were required. First, representing equations involving summation requires an extension of term syntax as summation is actually a functional, one of whose arguments is a function of the summation index. We shall not go into details here (see [2]), but representing patterns involving sums uses higher-order match variables in an essentially trivial way[4] and we can make direct use of our difference matcher to find substitution instances for such expressions. Another complication was that in some of these examples we were interested matches where the target, not the pattern, was annotated. This too was trivial and suggests a number of useful extensions of difference matching (see next section).

The methods implemented successfully sum a large number of different series. For example, they can tackle most of those problems give in the introductory chapters of a freshman text like *Concrete Mathematics*[12] (*e.g.* $\sum i^m$, $\sum (i+1)*a^i$, $\sum \sin(i*\theta)$). We can also solve those sums reported by Hutter in [14]. Note that, unlike Hutter, we are discovering the closed form for the sums as opposed to verifying the answer using induction.

5 Conclusion and Future Work

This paper presents a difference matching algorithm and proves various properties it possesses (like soundness and completeness). Difference matching appears to be a

[4] Even more trivial than Miller and Nipkow's higher-order patterns[15, 17] as each match variable within the body of the sum is a function precisely of the single binding variable representing the summation index.

promising means of marking differences between terms such that they can be made similar through rippling or some other means of selective simplification. It appears to be a key idea enabling the use of rippling as a controlled means of rewriting in a variety of inductive and non-inductive domains.

There are a number of directions for extensions that we have begun to consider. Below we list several of the more promising directions.

Unification: Matching can be extended by allowing match variables in both the pattern and instance. Another interesting "unification" extension is to return annotations of both the pattern and the instance. This would enable rippling to simplify both hypothesis and conclusion. As previously indicated, a difference unification algorithm can also serve as a wave-rule parser.

Equational Matching: In the previous section we required that matching respect bound variables but other extensions include full second and higher-order matching as well as first-order equational matching.

Annotated Substitutions: Difference matching could also be extended to return annotated substitutions. This will generate a much larger set of matches (when the inputs have at least one solution, they will have infinitely many) so there is a control problem of picking or efficiently enumerating useful matches. However, this extension is simple to implement and involves only an extension to the base-case of the difference matcher.

General Wave Annotations: Currently we allow no more than one sub-term to be deleted within a wave-front; however, returning "multi-wave annotations" (see [8]) requires this restriction to be weakened.

References

[1] R. Aubin. Some generalization heuristics in proofs by induction. In G. Huet and G. Kahn, editors, *Actes du Colloque Construction: Amelioration et verification de Programmes.* Institut de recherche d'informatique et d'automatique, 1975.

[2] David Basin, Alan Bundy, and Toby Walsh. Difference Matching and Rippling. In preparation.

[3] Karl Hans Bläsius and Jörg H. Siekmann. Partial unification for graph based equational reasoning. In *9th International Conference On Automated Deduction*, pages 397 – 414, Argonne, Illinois, 1988. Springer-Verlag.

[4] Robert S. Boyer and J. Strother Moore. *A Computational Logic.* Academic Press, 1979.

[5] Robert S. Boyer and J. Strother Moore. *A Computational Logic Handbook.* Academic Press, 1988. Perspectives in Computing, Vol 23.

[6] Alan Bundy. *The Computer Modelling of Mathematical Reasoning.* Academic Press, 1983.

[7] Alan Bundy. The use of explicit plans to guide inductive proofs. In *9th International Conference On Automated Deduction*, pages 111–120, Argonne, Illinois, 1988.

[8] Alan Bundy, Frank van Harmelen, Alan Smaill, and Andrew Ireland. Extensions to the rippling-out tactic for guiding inductive proofs. In M.E. Stickel, editor, *10th International Conference on Automated Deduction*, pages 132–146. Springer-Verlag, 1990. Lecture Notes in Artificial Intelligence No. 449.

[9] A. Bundy, A. Stevens, F. van Harmelen, A. Ireland, and A. Smaill. Rippling: A heuristic for guiding inductive proofs. Research Paper 567, Dept. of Artificial Intelligence, Edinburgh, 1991. Submitted to Artificial Intelligence.

[10] R.M. Burstall and J. Darlington. A transformation system for developing recursive programs. *Journal of the Association for Computing Machinery*, 24(1):44–67, 1977.

[11] Vincent J Digricoli. The management of heuristic search in boolean experiments with RUE resolution. In *9th IJCAI*, pages 1154 – 1161, Los Angeles, California, 1985.

[12] R.L. Graham, D.E. Knuth, and O. Patashnik. *Concrete Mathematics.* Addison-Wesley, 1989.

[13] Gérard Huet and D.C. Oppen. Equations and rewrite rules: a survey. In R. Book, editor, *Formal Languages: Perspectives and Open Problems.* Academic Press, 1980.

[14] D. Hutter. Guiding inductive proofs. In M.E. Stickel, editor, *10th International Conference on Automated Deduction*, pages 147–161. Springer-Verlag, 1990. Lecture Notes in Artificial Intelligence No. 449.

[15] Dale Miller. A logic programming language with lambda-abstraction, function variables, and simple unification. Technical Report ECS-LFCS-01-159, University of Edinburgh, LFCS, May 1991.

[16] J. Morris. E-resolution: an extension of resolution to include the equality relation. In *Proceedings of the IJCAI-69*, 1969.

[17] Tobias Nipkow. Higher-order critical pairs. In *Symposium on Logic in Computer Science*, 1991.

[18] Gordon D. Plotkin. A note on inductive generalization. *Machine Intelligence,* 5:153-163, 1970.

[19] Toby Walsh, Alex Nunes, and Alan Bundy. The use of proof plans to sum series. In D. Kapur, editor, *11th International Conference on Automated Deduction.* Springer-Verlag, 1992.

Using Middle-Out Reasoning to Control the Synthesis of Tail-Recursive Programs[1]

Jane Hesketh, Alan Bundy and Alan Smaill

Address: Dept of Artificial Intelligence, University of Edinburgh, 80, South Bridge, Edinburgh, Scotland.
Telephone: +44 31 650 {2718,2716,2710}
Email: {hesketh, bundy, a.smaill}%ed.ac.uk@nsfnet-relay.ac.uk

Abstract

We describe a novel technique for the automatic synthesis of tail-recursive programs. The technique is to specify the required program using the standard equations and then synthesise the tail-recursive program using the proofs as programs technique. This requires the specification to be proved realisable in a constructive logic. Restrictions on the form of the proof ensure that the synthesised program is tail-recursive.

The automatic search for a synthesis proof is controlled by proof plans, which are descriptions of the high-level structure of proofs of this kind. We have extended the known proof plans for inductive proofs by adding a new form of generalisation and by making greater use of middle-out reasoning. In middle-out reasoning we postpone decisions in the early part of the proof by the use of meta-variables which are instantiated, by unification, during later parts of the proof. Higher order unification is required, since these meta-variables can represent higher order objects.

The program synthesised is automatically verified to ensure that it satisfies its specification. This type of verification is contrasted with template-based transformation approaches which require proofs that the general transformations described by the templates preserve equivalence.

The technique described is more general than template-based approaches, since it is not tied to program patterns which must be specified in advance. Detailed information about proof structure enables it to use a wider repertoire of rewritings in a more goal-directed way than comparable transformational techniques.

1 Introduction

Consider the following two definitions[2] of procedures for reversing lists (:: is infix *cons*):

[1]The research reported in this paper was supported by SERC grant GR/E/71799, and an SERC Senior Fellowship to the second author. We are grateful for feedback on this paper from Andrew Ireland and Toby Walsh, and for conversations with Dale Miller and other members of the Mathematical Reasoning Group at Edinburgh.

[2]Notice that taking such definitions as equations, as we do, expresses more than the corresponding functional program, which uses the equations' computational content reading from left to right only. This distinction is significant in that the equations constitute a statement about the program at the specification level, not just the program itself. Although currently we rely on the recursive definitions to guide the initial formulation of the problem for the planner, in principle our system could work with a specification which was not completely and purely equationally specified.

$$\begin{array}{rcl|rcl} rev_n(nil) & = & nil & rev_t(l) & = & rev_2(l, nil) \\ rev_n(h :: t) & = & append(rev_n(t), h :: nil) & rev_2(nil, a) & = & a \\ & & & rev_2(h :: t, a) & = & rev_2(t, h :: a) \end{array}$$

The auxiliary procedure, rev_2, is *tail recursive*. That is, recursive calls to rev_2 occur only as the outermost function of the procedure body. The *accumulator* argument, a, is used to build up the output as the recursion is entered, so that nothing remains to be done as the recursion exits. This has important consequences for the efficiency of rev_2 and, hence, rev_t. It is not necessary to maintain a stack of recursive calls during its implementation, which cuts down considerably on the space requirements of a procedure call. Thus, tail recursive procedures can be compiled into iterative ones. In contrast, we will call rev_n a *naïve* procedure. It is the obvious definition of list reversal, and so easy for programmers to discover, but is much less space efficient.

In many cases it is possible to transform naïve procedures into equivalent tail recursive ones. For instance, any linearly recursive procedure can be transformed into a tail-recursive one [11], although the general process for doing this creates rather convoluted and sub-optimal tail-recursive procedures. Various attempts have been made to automate tail recursive transformation, *e.g.* [6], thus freeing the programmer from the burden of discovering the more efficient, but more complex, definition. Another approach, described in [8], uses general second-order templates of transformations, but is inevitably limited to such cases as have been anticipated and proved to preserve equivalence. In this paper we consider a novel technique for automating the synthesis of tail recursive procedures based on the *proofs as programs* technique.

Proofs as programs is a technique for synthesising computer programs from proofs that their specifications are realisable. Suppose $spec(inputs, output)$ is a logical relation between the inputs and outputs of a program. We prove the *specification theorem*:

$$\forall inputs, \exists output.\ spec(inputs, output)$$

in a constructive logic. Intuitively, this proof must show how given any combination of inputs, an output can be constructed that meets the specification. This construction can be extracted from the proof and expressed as the required program. The use of a constructive logic excludes 'pure existence' proofs, where the existence of an output is proved without a suitable construction being exhibited. By using an automatic theorem prover to prove the specification theorem we can automate the process of program synthesis.

Different proofs of the same specification theorem yield different procedures meeting the same specification. For instance, from a specification of list reversal we can synthesise either the naïve or the tail recursive procedure, according to the proof we find. We have identified a characterisation of proofs that synthesise tail recursive procedures, following [10]. The essence is that the witnesses of the two existential quantifiers, one in the induction hypothesis and one in the induction conclusion, should be identical. The *witness* of an existential quantifier is the object which is the evidence of the existence asserted. If we restrict our theorem prover to proofs of this form then we can guarantee to synthesise only tail recursive functions. These witnesses give the value of the function. Equality between them means that the function does not change value as the recursive call is exited.

A key idea is that we can use the naïve equational definition as a specification of the tail recursive program. If $f_n(inputs)$ is the naïve definition, then $output = f_n(inputs)$ can serve as $spec(inputs, output)$. We can prevent the trivial solution to:

$$\forall inputs, \exists output.\ output = f_n(inputs)$$

in which $f_n(inputs)$ is substituted for *output* by insisting on a proof where f_n does not appear as part of the witness.

Achieving proofs with the desired form automatically, requires some insight into the proof process. The synthesis of our equivalent to rev_2 involves a generalised version of the specification theorem which is then be shown to be imply the original. This step guarantees that the synthesised function satisfies the original specification. The exact nature of the generalised theorem is not known initially, but its overall form can be described. Middle-out reasoning is used to identify its components and these define the synthesised function. This reasoning is guided to satisfy the criterion that the existential witnesses for the output are identical in induction hypothesis and conclusion All this is implemented using a proof planning system.

2 Tail-Recursive Reverse Example

To illustrate this process, consider the synthesis of tail recursive reverse. Using the naïve reverse to specify the required procedure gives the specification theorem:

$$\vdash \forall x, \exists z.\ z = rev_n(x)$$

Firstly, we generalise this theorem to:

$$\vdash \forall x, \forall a, \exists z.\ z = append(rev_n(x), a) \tag{1}$$

which leaves us with the obligation to show that the generalised theorem entails the original one:

$$\forall x, \forall a, \exists z.\ z = append(rev_n(x), a) \quad \vdash \quad \forall x, \exists z.\ z = rev_n(x)$$

This generalisation is an example of an *eureka step*. It seems to come 'out of the blue' with no apparent motivation. A major problem of automation of tail recursive transformation is to find techniques for calculating such eureka steps.

To prove (1), the generalised theorem, we use simple list induction on x.

2.1 The Step Case

The step case of this induction is:

$$\forall a, \exists z.\ z = append(rev_n(t), a) \quad \vdash \quad \forall a, \exists z.\ z = append(rev_n(h :: t), a)$$

We rewrite the induction conclusion using first the definition of rev_n:

$$\forall a, \exists z.\ z = append(rev_n(t), a) \quad \vdash \quad \forall a, \exists z.\ z = append(append(rev_n(t), h :: nil), a)$$

and then the associativity of *append*:

$$\forall a, \exists z.\ z = append(rev_n(t), a) \quad \vdash \quad \forall a, \exists z.\ z = append(rev_n(t), append(h :: nil, a))$$

which simplifies using the definition of *append* to:

$$\forall a, \exists z.\ z = append(rev_n(t), a) \ \vdash\ \forall a, \exists z.\ z = append(rev_n(t), h :: a)$$

The step case can now be finished off by stripping off the universal quantifiers and instantiating the a in the induction hypothesis to $h :: a$ in the induction conclusion. The induction hypothesis and induction conclusion are then identical. Note that the existentially quantified variables, z, in the induction hypothesis and the induction conclusion are equal. This equality ensures that a tail recursive procedure will be synthesised by the proof.

2.2 The Base Case

The base case is:

$$\vdash\ \forall a, \exists z.\ z = append(rev_n(nil), a)$$

The definitions of rev_n and *append* simplify this to

$$\vdash\ \forall a, \exists z.\ z = a$$

which can be proved by instantiating z to a.

2.3 The Justification

It only remains to prove the original theorem from the generalised one.

$$\forall x, \forall a, \exists z.\ z = append(rev_n(x), a) \ \vdash\ \forall x, \exists z.\ z = rev_n(x)$$

Stripping the two $\forall x$ quantifiers while identifying the xs, gives:

$$\forall a, \exists z.\ z = append(rev_n(x), a) \ \vdash\ \exists z.\ z = rev_n(x)$$

Instantiating a to nil gives:

$$\exists z.\ z = append(rev_n(x), nil) \ \vdash\ \exists z.\ z = rev_n(x)$$

The hypothesis now simplifies to the conclusion.

The proof is now complete. Analysis of the proof gives the required tail recursive definition of rev_t.

$$\begin{aligned} rev_t(l) &= rev_2(l, nil) \\ rev_2(nil, a) &= a \\ rev_2(h :: t, a) &= rev_2(t, h :: a) \end{aligned}$$

3 Proofs as Programs

We have implemented the proofs as programs technique in the OYSTER system, [4]. OYSTER is a interactive theorem prover for intuitionalist type theory, a higher order, constructive, typed logic based on Martin Löf Type Theory. It is a Prolog reimplementation of the Nuprl system. One advantage of using this logic for program synthesis is that every rule of inference of the logic has an associated program construction rule, so that the synthesised program is built as a side effect of constructing the proof, and no post-analysis of the proof is required to extract it.

The technical basis of this program construction process is that every statement of the logic has the form $a \in A$, with three possible readings:

1. a is an object of type A;
2. a is a proof of proposition A;
3. a is a program meeting the specification A. (a is also called the *extract term.*)

The first two readings were discovered by Curry and Howard, and are called the 'Curry-Howard isomorphism' and the 'propositions as types principle'. The third reading extends this to the 'proofs as programs principle'

Our synthesis technique exploits the ambiguity provided by these three readings. We treat the input and output variables as objects belonging to some type, *e.g.* lists of natural numbers, and the specification as a proposition over these objects. We prove the specification suppressing the details of its extract term. When the proof is complete this extract term is revealed and interpreted as a functional program.

Following Nuprl, OYSTER provides the facility for writing tactics. These are programs (in OYSTER's case Prolog programs) which apply rules of inference of the logic. Tactics can embody some heuristic ideas about how the proof should proceed, and thus be used to guide the theorem prover. The rest of this paper is concerned with the nature of the heuristics we have developed to guide inductive proofs, in particular tail recursive transformations, with how these heuristics are embodied in the OYSTER system, and with how they can automate the discovery of eureka steps.

4 Proof Plans

The recent work of the Mathematical Reasoning Group at Edinburgh derives from the Boyer-Moore, [1], characterisation of proofs as a combination of special purpose components, such as symbolic evaluation and induction, which act in concert. We have reconstructed and extended the Boyer-Moore components and implemented them as OYSTER tactics. Moreover, we use an AI plan formation program, CLAM, to link these components together, [2]. Each tactic is specified by a *method* using a meta-logic. The pair of tactic and method is called a *proof plan.* CLAM operates on a meta-level representation of the proof tree, and uses meta-level reasoning on the methods to select a combination of tactics customised to the current theorem. This combination can then be executed in the object-level logic to produce a proof.

In particular, the meta-level representation of the proof can contain meta-variables. These meta-variables act as place holders for expressions to whose precise identity we are not yet ready to commit. They are introduced during the meta-level application of rules of inference which at the object-level require some commitment, and are later instantiated by unification during the planning process. Meta-variables can be of higher-order type, so higher order unification is required to instantiate them. Meta-variables can be introduced, for instance, during the stripping of an existential quantifier on the right hand side of a sequent or a universal quantifier on the left hand side of a sequent, or during the generalisation of a theorem. At the object-level these proof steps require a commitment to some particular expression, *e.g.* as the witness of a quantifier. We call this use of meta-variables *middle-out reasoning*, because it allows us to turn the search space inside out — doing the middle of the proof first and the beginning later.

The proof plans technique is easily adapted to deal with a variety of types of proof using a body of general-purpose proof plans. Methods describe the preconditions under which their tactics are applicable and the effect of their tactics on given input sequents. A simple, iterative-deepening search strategy is used by the planner. No combinatorial explosion is encountered due to the small size of the search space defined by the proof plans. The object-level search space, on the other hand, is huge, *cf.* [2]. Some standard proof plans are:

induction : selection and application of an appropriate induction rule;
symbolic evaluation : simplification of expressions using definition unfolding;
rippling : rewriting the induction conclusion to make it resemble the induction hypothesis;
tautology : propositional tautology checking plus simple equality manipulation;
existential : middle-out reasoning to postpone selecting an existential witness;
fertilization use of the induction hypothesis to prove the induction conclusion.

By using these proof plans in appropriate combinations, CIAM is capable of proving a sizeable body of theorems from the Boyer-Moore corpus. As in the Boyer-Moore theorem-prover, theorems which have already been proved may be made available as lemmas for later proofs, but this increases the size of the search space.

The key proof plan is rippling. It is also the only one that is not self-explanatory. We outline it below.

In the step case of an induction proof we prove the induction conclusion from the induction hypothesis. These two formulae are very similar. We call their points of difference, *wave fronts*. Wave fronts are expressions with holes in them. We indicate them by putting them in boxes, *e.g.*

$$\forall a, \exists z.\ z = append(rev_n(t), a) \quad \vdash \quad \forall a, \exists z.\ z = append(rev_n(\boxed{h :: \underline{t}}), a)$$

The role of rippling is to move these wave fronts from their innermost position around the induction variable to somewhere where they will not block the matching of induction hypothesis to induction conclusion. There are two such places: (1) surrounding the entire induction conclusion, and (2) surrounding an universally quantified variable, *e.g.* an accumulator. This movement is effected by rewrite rules called *wave rules*. There are two kinds corresponding to the two directions of movement: longitudinal wave rules move the wave fronts outwards, and transverse wave rules move the wave fronts sideways. To move wave fronts to target (1) involves purely longitudinal wave rules, but to move them to target (2) involves longitudinal wave rules followed by transverse wave rules, and sometimes followed by longitudinal wave rules applied backwards. Examples of longitudinal wave rules are:

$$append(X, \boxed{append(\underline{Y}, Z)}) \Rightarrow \boxed{append(\underline{append(X,Y)}, Z)} \qquad (2)$$

$$rev_n(\boxed{X :: \underline{Y}}) \Rightarrow \boxed{append(\underline{rev_n(Y)}, X :: nil)} \qquad (3)$$

$$append(\boxed{X :: \underline{Y}}, Z) \Rightarrow \boxed{X :: \underline{append(Y, Z)}}$$

and examples of transverse wave rules are:

$$append(\boxed{append(X, Y)}, Z) \Rightarrow append(X, \boxed{append(Y, Z)}) \qquad (4)$$

$$\boxed{s(X)} + Y \Rightarrow X + \boxed{s(Y)}$$

X, Y and Z in these rules are *meta-variables*, *i.e.* these rewrite rules are really rule schemata. We adopt the convention that upper case letters represent meta-level variables and lower case letters represent object-level variables. Using meta-variables facilitates the use of unification at the meta-level during middle-out reasoning.

The rewriting of:

$$\forall a, \exists z.\ z = append(rev_n(\boxed{h :: t}), a)$$

to:

$$\forall a, \exists z.\ z = append(\boxed{append(rev_n(t), h :: nil)}, a)$$

using longitudinal wave rule (3) is an example of rippling outwards, and the application of transverse wave rule (4) to this to produce:

$$\forall a, \exists z.\ z = append(rev_n(t), \boxed{append(h :: nil, a)})$$

is an example of rippling sideways. For more details see [5].

5 Proof Plans for Tail-Recursive Synthesis

For tail-recursive synthesis, the existing proof plans were adapted, mostly simply to inhibit inadvertent application in the presence of meta-variables. The rippling proof plan was significantly changed, and a new generalisation proof plan was added to create and introduce the generalised theorem. These will now be described in detail.

5.1 Tail-Recursive Generalisation

We add a new proof plan for tail-recursive generalisation. Our overall strategy is to generalise in order to be able to construct a proof with the special characteristic of not altering the existential witnesses in the induction step case. In particular, the generalisation introduces an accumulator. Suppose $f_n(inputs)$ is the naïve procedure and $f_t(inputs)$ is the tail-recursive procedure. We will synthesise a procedure $f_2(inputs, accumulator)$ and define $f_t(inputs) = f_2(inputs, a_0)$, for some particular value, a_0, of the accumulator. We, therefore, want to generalise the specification, $outputs = f_n(inputs)$, of f_t to a specification, $outputs = g(f_n(inputs), accumulator)$, of f_2. and recover the value a_0 from the justification branch of this generalised proof. Since we don't know what value of g to use to wrap $f_t(inputs)$ and *accumulator* together, we will use a meta-variable, which will be instantiated by later stages of the proof, *i.e.* the new specification will be: $outputs = G(f_n(inputs), accumulator)$.

5.1.1 Preconditions

The proof plan applies when the goal is of the form:

$$\forall x, \forall \underline{y}, \exists z.\ z = f_n(x, \underline{y})$$

where $\underline{y}$ represents zero or more additional parameters.

This is a universally quantified expression consisting of an equality between an existentially quantified variable and an arbitrary term containing any of the universally quantified variables but not the existentially quantified one.

If middle-out reasoning is already taking place on the goal, initiating further middle-out reasoning is likely to be unwieldy or explosive, so a check is made that it does not currently contain any meta-variables. Specifically, since the generalised goal has the same form as the ungeneralised one, the lack of such a check could result in generalisations of generalisations, etc.

5.1.2 Effects and Tactic

The first output sequent of the tactic is the generalisation of the original specification theorem, *i.e.*

$$\vdash\ \ \forall x, \forall \underline{y}, \forall a, \exists z.\ z = G(f_n(x, \underline{y}), a)$$

where G is a meta-variable. The second output sequent is the justification proof branch: *i.e.* that the generalised theorem implies the original one.

$$\forall x, \forall \underline{y}, \forall a, \exists z.\ z = G(f_n(x, \underline{y}), a)\ \ \vdash\ \ \forall x, \forall \underline{y}, \exists z.\ z = f_n(x, \underline{y})$$

5.2 Longitudinal Wave Proof Plan

Rippling is applied during the step case of inductive proofs, after wave fronts have been inserted by the induction method. It consists of repeated, but selective, rewriting with longitudinal and transverse wave rules. Rippling works by calling two sub-proof-plans: longitudinal wave and transverse wave. Our versions of these are identical to the standard one, except that they take account of the possibility that the input sequent could contain meta-variables.

The longitudinal wave plan takes as input a sequent such as the generalisation:

$$\vdash \forall x, \forall \underline{y}, \forall a, \exists z.\ z = G(f_n(\boxed{c(\boxed{x})}, \underline{y}), a)$$

5.2.1 Preconditions

Assume that the longitudinal wave rule applied to this is[3]:

$$\phi(\boxed{\kappa(\boxed{X})}, \underline{Y}) \Rightarrow \boxed{\kappa'(\boxed{\phi(X, \underline{Y})}, \underline{Y})}$$

Wave rules like this are selected and tested in turn. In standard CLAM the left-hand-side of the rule would be tested to see if it unified as a whole with some subterm

[3]This is a simplification, see [5] for general form of wave rules.

of the sequent. Expressions are annotated with marks to indicate wave fronts. The standard CLAM unification algorithm aligns these marks and hence the wave fronts in goal and rule. In our middle-out reasoning system, separate unifications are performed successively on each of the following expression pairs, and progressive instantiation takes place.

- the terms contained by the respective wave fronts, x and X
- the smallest terms containing the wave fronts, $c(x)$ and $\kappa(x)$
- the term to be rewritten and the whole left-hand side of the rule, $f_n(c(x), \underline{y})$ and $\phi(\kappa(x), \underline{Y})$.

The unifications are performed with the wave front marks removed. Wave fronts in goal and rule are still aligned by the first two stages of our three stage unification process.

5.2.2 Effects and Tactic

These are exactly as in the standard proof plan, the application of the lemma or definition is computed and planning continues. The tactic records what is to be applied, in which direction, and the position of the relevant subterm.

5.3 Transverse Wave Proof Plan

This incorporates changes to permit meta-variables as described for the longitudinal wave proof plan. It takes as input a sequent conclusion such as:

$$\vdash \quad \forall \underline{y}, \forall a, \exists z.\ z = G(\boxed{c'(\underline{f_n(x, \underline{y})}\boxed{, \underline{y})}}, a)$$

5.3.1 Preconditions

Assume a transverse wave rule:

$$\phi(\boxed{\kappa'(\underline{X}}\boxed{, \underline{Y})}, A) \rightarrow \phi(X, \boxed{\kappa''(\underline{Y},}\underline{A}\boxed{)})$$

The positions of the wave fronts before and *after* the use of a lemma are used to ensure that the effect of the lemma is a ripple sideways, as intended.

Essentially everything proceeds as for the longitudinal wave proof plan, except for an extra unification stage. We insist that the term at the position around which the wave front is moved is unified with the last universally quantified variable in the induction conclusion, *i.e.* the accumulator added by the generalisation.

5.3.2 Effects and Tactic

Again, these are just the same as in standard CLAM.

6 Tail-Recursive Reverse Example Revisited

We now repeat the example of §2, but this time with annotations to explain how our proof plans are able to find this proof with very little search. Note particularly how meta-variables are used to postpone the commitment to existential witnesses and generalisations. As before, the specification goal is:

$$\vdash \ \forall x, \exists z.\ z = rev_n(x)$$

The only proof plans whose preconditions may be satisfied are induction and tail-recursive generalisation. After some search induction fails the tail-recursive restriction about existential witnesses. The result of generalisation is a new sequent:

$$\vdash \ \forall x, \forall a, \exists z.\ z = G(rev_n(x), a)$$

and a further justification sub-goal to show that this entails the original theorem.

6.1 Proving the Generalisation Using Induction

The only proof plans which might conceivably apply to this are existential, tail-recursive generalisation and induction. The first two are barred since the meta-sequent already contains a meta-variable. Induction is unaffected by the meta-variable. The preconditions of the induction strategy proof plan look ahead to try to find an induction rule that will allow rippling to proceed ([3] explains this look-ahead process). This look-ahead suggests a simple list induction on x.

6.1.1 The Step Case

$$\forall a, \exists z.\ z = G(rev_n(t), a) \ \vdash \ \forall a, \exists z.\ z = G(rev_n(\boxed{h :: \underline{t}}), a)$$

The only applicable proof plan for this meta-sequent is rippling. Again the existential and tail-recursive generalisation's preconditions would fail due to the presence of the meta-variable.

The rippling proof plan could select several wave rules, depending on which were available. It is always debatable, when discussing wave rules, which ones should be assumed to be available. There are broadly three options:

- Minimal - Only necessary definitions, anything else to be created.
- Average - Definitions, along with some collection of lemmas. Specifically *not* just such lemmas as will make for the proof in hand easy.
- Maximal - everything, no matter how trivial, which is true for the theory.

The last of these is impossible. The first is an interesting but not particularly realistic case. What mathematician or automatic theorem prover would we expect to derive everything from first principles all the time? So we have taken the middle way and provided an average collection of rules: 20 longitudinal rules and 12 transverse ones. Below we explain what choices they present to the system reasoning middle-out. From this it should be clear that they do not make the task artificially simple.

In our example the only possible match, which obeys the constraints introduced by rippling, is, (3), the definition of rev_n. This applies, producing a new sequent:

$$\forall a, \exists z.\ z = G(rev_n(t), a) \ \vdash \ \forall a, \exists z.\ z = G(\boxed{append(\underline{rev_n(t)}, h :: nil)}, a)$$

This is submitted afresh to the planner, and exactly the same proof plans apply as for its predecessor — namely only rippling. Each wave-rule is considered in turn. For each, as described before, there is progressive unification of the terms within the wave front, the wave front term, and the whole left-hand-side of the rule with whatever subterm of the conclusion it will match. This is vital not only to achieve the effect we want, but also to control the higher order unification process. At any point, one of these unifications may fail, and the planner will backtrack to get the next wave-rule.

Here either (2) or (4) could apply. Consider the branch of the search space in which (4) is applied. This is a sideways ripple towards the accumulator a. The unification instantiates G to $\lambda u \lambda v.append(u, v)$, so, after β-reduction, the result of the rule application is:

$$\forall a, \exists z.\ z = append(rev_n(t), a) \ \vdash\ \forall a, \exists z.\ z = append(rev_n(t), \boxed{append(h :: nil, \boxed{a})})$$

Rippling finishes by applying symbolic evaluation to the wave front, which gives:

$$\forall a, \exists z.\ z = append(rev_n(t), a) \ \vdash\ \forall a, \exists z.\ z = append(rev_n(t), \boxed{h :: \boxed{a}})$$

Now fertilisation can apply, matching the induction hypothesis and induction conclusion and finishing the step case. a in the induction hypothesis is instantiated to $h :: a$ from the induction conclusion.

6.1.2 The Base Case

The step case work has instantiated G, so the meta-sequent is

$$\vdash\ \forall a, \exists z.\ z = append(rev_n(nil), a)$$

The existential proof plan could apply, and would lead to the introduction of $append(rev_n(nil), a)$ for z. This would be cumbersome, but not wrong. Symbolic evaluation applies twice, using the base definitions of rev_n and $append$:

$$\vdash\ \forall a, \exists z.\ z = a$$

Now, the existential proof plan is the only one which can apply, it introduces a and a meta-variable Z for z. The tautology proof plan now applies, instantiating Z to a, and completing the base case.

6.2 The Justification

Now that G has been instantiated, the justification sub-goal is to prove:

$$\forall x, \forall a, \exists z.\ z = append(rev_n(x), a) \ \vdash\ \forall x, \exists z.\ z = rev_n(x)$$

First, we assume that any universally quantified variables in the conclusion should be identified with their counterparts in the generalisation hypothesis. So each of these is introduced, and echoed in the hypothesis.

$$\forall a, \exists z.\ z = append(rev_n(x), a) \ \vdash\ \exists z.\ z = rev_n(x)$$

An instantiation, a_0, of a is chosen such that the z in the hypothesis is equal to the z in the conclusion, *i.e.* $append(rev_n(x), a_0) = rev_n(x)$.

We can postpone the choice of a_0 by middle-out reasoning. This requires the universal proof plan, a dual of the existential proof plan, which works on universal quantifiers in the hypothesis. A meta-variable, A, is inserted for a, and the symbolic evaluation proof plan is applied to that hypothesis alone, *i.e.* to the expression

$$\exists z.\ z = append(rev_n(x), A)$$

This instantiates A to nil and reduces the hypothesis to:

$$\exists z.\ z = rev_n(x)$$

to which fertilization applies, finishing the justification and the whole proof.

7 Using Higher Order Unification

As noted above, higher-order unification is required to implement middle-out reasoning in our higher-order logic. Unfortunately, higher-order unification is undecidable; it may fail to terminate when two expressions are not unifiable, and it may produce an unlimited number of unifiers when they are. If we only test for unifiability we can obtain better behaviour. Higher order unifiability is semi-decidable. Therefore, we have implemented the higher order unifiability algorithm invented by Huet, [7], which yields unifiers if there are any. This algorithm uses a simply typed lambda calculus, which has proved adequate in practice. However, since our logic includes dependent types, we intend to extend our algorithm to Pym's version for dependent types, [9], in the near future.

Higher order unification may provide more than one unification, possibly infinitely many. Not all of these will be sensible for our problem. Below we explain how a choice is made.

7.1 Interfacing Middle-Out Reasoning and Higher Order Unification

All meta-variables are treated as variables in the higher order unification. Additionally, any universally quantified variables in lemmas or hypotheses are treated as variables. Anything else is deemed a constant. Some pre-processing is necessary:

1. Copies are made of both the terms to be compared and their contexts, which are used for type-checking. The context of a term to be unified is the sequent or theorem from which it originates.
2. The types of the two terms to be compared are tested to ensure they are the same. If not, the algorithm fails immediately.
3. The copies of the terms are normalised by β-reduction.
4. The types of all the symbols in the ground copies of the terms are guessed and recorded.
5. The expressions to be compared are further normalised by η-conversion.
6. All the arguments throughout the term are checked to ensure that they have the correct types to suit the functions they are in.

In practice, it is usually possible to reduce the problem to a sequence of higher-order matches rather than unification. Matching is much more controllable.

7.2 Inadmissible Unifications

Unification suggests instantiations for meta-variables, some of which may be legal but not sensible. Suppose, for example, we have the following sequent:

$$\forall a, \exists z.\ z = G(rev_n(t), a) \quad \vdash \quad \forall a, \exists z.\ z = G(append(rev_n(t), h :: nil), a)$$

where G is a meta-variable to be instantiated through the demands of the subsequent proof. We expect to instantiate G by considering unifications of

$$append(append(rev_n(t), h :: nil), a)$$

with

$$G(append(rev_n(t), h :: nil), a)$$

Higher order unification suggests four unifiers for G:

- $\lambda u \lambda v.append(u, v)$
- $\lambda u \lambda v.append(u, a)$
- $\lambda u \lambda v.append(append(rev_n(t), h :: nil), v)$
- $\lambda u \lambda v.append(append(rev_n(t), h :: nil), a)$

At this point in the proof, each of them would be a valid object to use for G in terms of OYSTER's logic. The last three are not sensible choices, because they build universally quantified or free variables into the identity of the function, G. As G is a function whose identity must be describable externally to the proof, and so should not be dependent on the names of variables used during proof, such instantiations are not appropriate. Effectively, they are barred by *temporal scoping* - a notion suggested by Dale Miller. For this reason, we discard them.

8 Comparison with Related Techniques

Other research has also tackled this problem, and it is related to work on program transformation. There are two major approaches. One is to develop general templates which have an input part, an output part and conditions as in [8], all described in terms of shared variable components. If a program matches the input part and the conditions on the components can be satisfied, the instantiation of the output part yields a tail-recursive version. The other approach, as taken in [6], is to apply a sequence of transformations to the initial program, for example using definitions and known properties of functions. Although this has an outwardly similar appearance to the work reported here, and comparable search problems, there are some notable differences, as will be explained below.

Template-based approaches are static, in the sense that they must declare the form of their transformation in advance. This reduces their potential to handle new problems. Dynamic techniques, such as the one described here and in [6] are more flexible, and able to use whatever information is available at the time.

Darlington's system does not ensure that optimisation has actually occurred during the transformation. In the proofs as programs characterisation, tail-recursiveness becomes the goal which drives the process. The requirement that the proof structure satisfy the criterion regarding the identity of the existential witnesses guides the strategy which controls the search for a synthesis proof.

The work described in this paper is completely automated. In contrast, the account given in [8] is theoretical. Their work concentrates on the preservation of equivalence by the technique without detailed examination of the search problems arising from the need to identify the values of higher-order variables present in the output program, but not the input. This identification is non-trivial in general. Darlington's approach is automated to a considerable degree, but full automation was never a goal of that system. Lacking the tools of the CLAM approach, such as the monitoring of wavefronts, it has less ability to characterise the types of operation, and so needs more guidance. Explicit instructions to unfold using definitions and then use an associativity lemma provide such guidance. The various kinds of wave rules we have described are defined syntactically according to their effect on the conclusion of an inductive proof. Longitudinal waves are not only the recursive steps of definitions, but anything which can achieve the same effect of moving a wavefront upwards in the term structure. Transverse waves are not just derived from associative functions, they too are syntactically defined to include any lemmas which enable a wavefront to be moved sideways. In this sense the proof plans system is more general. It is, however, worth pointing out that Darlington's system is capable of other forms of optimisation than tail-recursiveness, which have not so far been implemented with proof plans.

A significant difference between the work described in this paper and the transformational approaches is that here we prove that the synthesised program satisfies the original specification. Each of these other techniques carries the burden of guaranteeing that their processes of transformation are equivalence-preserving. A disadvantage for our approach is that time must be spent proving this satisfaction on each occasion, but this is compensated for by the fact that any constraints required are those particular to this proof. We do not have to prove that the general process is equivalence preserving, as must be done by computational induction for each higher order template in [8].

9 Results and Conclusions

We have described a novel approach to the transformation of naïve procedures into tail-recursive ones. We specify the tail-recursive procedure using the naïve definition and synthesise it using the proofs as programs technique embodied in the OYSTER proof development system. Restrictions on the form of the proof ensure that the resulting extract term is bound to be tail-recursive. To control the search involved in this proof we have used the proof plans technique as embodied in the CLAM proof planner. This reduces a huge object-level search space to a small meta-level search space, which can be successfully searched by a weak general-purpose strategy: iterative-deepening.

We have introduced a new form of generalisation, which is required for tail-recursive transformation, and extended the middle-out reasoning capabilities of CLAM by allowing the use of meta-variables during generalisation. This has necessitated the improvement of some of the existing CLAM proof plans so that they can cope with meta-variables more successfully.

The approach has been successfully tested on the synthesis of procedures for the summation of an arbitrary function from 0 to an arbitrary value, *times*, *total*, *greatest* and *length*. In each case a simple and natural tail-recursive procedure is synthesised. We plan to continue testing. Initial results suggest that this is a powerful new technique which extends existing ones.

Further work needs to be done to specify all the restrictions which are appropriate to the choice of unification given our understanding of the aims of the task. Work is in progress to improve the flexibility of the unification process so that composite functions can be identified, and to extend Huet's algorithm to dependent types.

References

[1] R.S. Boyer and J.S. Moore. *A Computational Logic*. Academic Press, 1979. ACM monograph series.

[2] A. Bundy, F. van Harmelen, J. Hesketh, and A. Smaill. Experiments with proof plans for induction. *Journal of Automated Reasoning*, 7:303–324, 1991. Earlier version available from Edinburgh as DAI Research Paper No 413.

[3] A. Bundy, F. van Harmelen, J. Hesketh, A. Smaill, and A. Stevens. A rational reconstruction and extension of recursion analysis. In N.S. Sridharan, editor, *Proceedings of the Eleventh International Joint Conference on Artificial Intelligence*, pages 359–365, Morgan Kaufmann, 1989. Also available from Edinburgh as DAI Research Paper 419.

[4] A. Bundy, F. van Harmelen, C. Horn, and A. Smaill. The Oyster-Clam system. In M.E. Stickel, editor, *10th International Conference on Automated Deduction*, pages 647–648, Springer-Verlag, 1990. Lecture Notes in Artificial Intelligence No. 449. Also available from Edinburgh as DAI Research Paper 507.

[5] A. Bundy, F. van Harmelen, A. Smaill, and A. Ireland. Extensions to the rippling-out tactic for guiding inductive proofs. In M.E. Stickel, editor, *10th International Conference on Automated Deduction*, pages 132–146, Springer-Verlag, 1990. Lecture Notes in Artificial Intelligence No. 449. Also available from Edinburgh as DAI Research Paper 459.

[6] J. Darlington. An experimental program transformation and synthesis system. *Artificial Intelligence*, 16(3):1–46, August 1981.

[7] G. Huet. A unification algorithm for lambda calculus. *Theoretical Computer Science*, 1:27–57, 1975.

[8] G. Huet and B. Lang. Proving and applying program transformation expressed with second order patterns. *Acta Informatica*, 11:31–55, 1978.

[9] D.J. Pym. *Proofs, search and computation in general logic*. PhD thesis, University of Edinburgh, 1990. Available as LFCS report ECS-LFCS-90-125.

[10] S.S. Wainer. Programs from proofs. 1989. Seminar given at the Department of Artificial Intelligence, Edinburgh.

[11] ÅWikström. *Functional Programming Using Standard ML*. Prentice Hall, 1987.

The Use of Proof Plans to Sum Series *

Toby Walsh   Alex Nunes   Alan Bundy

Department of AI, Edinburgh University

Abstract

We describe a program for finding closed form solutions to finite sums. The program was built to test the applicability of the *proof planning* search control technique in a domain of mathematics outwith induction. This experiment was successful. The series summing program extends previous work in this area and was built in a short time just by providing new series summing methods to our existing inductive theorem proving system *CIAM*.

One surprising discovery was the usefulness of the *ripple* tactic in summing series. Rippling is the key tactic for controlling inductive proofs, and was previously thought to be specialised to such proofs. However, it turns out to be the key sub-tactic used by all the main tactics for summing series. The only change required was that it had to be supplemented by a *difference matching* algorithm to set up some initial meta-level annotations to guide the rippling process. In inductive proofs these annotations are provided by the application of mathematical induction. This evidence suggests that rippling, supplemented by difference matching, will find wide application in controlling mathematical proofs.

1 Introduction

In [2] we introduced *proof planning*, a new technique for controlling the search for a proof by using the common structure of a family of similar proofs as a guide. The application of proof planning to the control of inductive proofs is described in [3] whilst *rippling*, a key tactic in inductive proofs, is described in [4]. Proof planning has been implemented in the *Oyster/CIAM* system [3].

This paper explores the usefulness of proof planning, in general, and rippling, in particular, in a non-inductive domain: the discovery of closed form solutions to finite series. We describe several methods for summing series and show how they can be represented in the proof plans formalism. They have all been implemented in the *Oyster/CIAM* system and tested on a wide range of series problems. Most of these methods make use of rippling as a key submethod. In order to use rippling in non-inductive domains it is necessary to supplement it with a special matching algorithm called *difference matching* (see [5] in this volume for details).

The mathematical problem we address is to derive closed form solutions for finite

*The research reported in this paper was supported by SERC grant GR/F/71799, a SERC PostDoctoral Fellowship to the first author and a SERC Senior Fellowship to the third author. We would like to thank the other members of the mathematical reasoning group for their feedback on this project.

series like:

$$\sum_{i=0}^{n} s(i).a^i$$

where a is a constant and s is the successor function, *i.e.* $s(i) = i + 1$. By 'derive a closed form solution' we mean find an expression, equal to the finite sum, that is free of the summation operator. In this example, for $a \neq 1$, such a closed form solution would be:

$$\frac{s(n).a^{s(n)}}{a-1} - \frac{a^{s(n)}-1}{(a-1)^2}$$

There has been limited research into this domain. Some researchers have tackled the topic as a verification problem, [9, 7]; both these teams use mathematical induction to prove that a series is equal to a user supplied closed form. In the work reported below, the closed form solution to a series is simultaneously synthesised and verified. None of our solution methods uses induction for either the synthesis or the verification task.

Another approach is to use a decision procedure, like Gosper's algorithm, [8], to compute closed form solutions. Such decision procedures have the drawback of being "black-boxes" of only being applicable to a narrow class of series. The work reported here is applicable to a much wider class of series and the solutions produced can be understood by mathematicians. Indeed our solution methods are modelled on those used by mathematicians.

2 Proof Planning

A brief description of the ideas and concepts involved in proof planning follows in order to set the background for what is to come. A more detailed account is given in [2, 3].

The notion of explicit *proof plans* as a technique for guiding an automatic theorem prover in its search for a proof by mathematical induction originated in a project to develop automated search in the *Oyster* program synthesis system, a reimplementation in Prolog of the Cornell Nuprl system, [6]. *Oyster* is an interactive proof editor for a logic based on Martin-Löf's Intuitionistic Type Theory. Following LCF, the search for a proof in *Oyster* can be guided by programs called *tactics*. *Oyster*'s tactics for inductive proofs are written in Prolog; they are based on and extend ideas in the Boyer-Moore theorem prover, Nqthm, [1].

To control the application of tactics, the *CLAM* plan formation program analyses the current theorem and constructs a special-purpose super-tactic to prove it. To enable *CLAM* to do this, every tactic is (partially) specified by giving preconditions for its attempted application and some of the effects of its successful application. This partial specification is called a *method*. Methods are described in a *meta-logic*, whose domain of discourse is mathematical expressions and which describes syntactic properties of these expressions, *e.g.* the number and location of particular subexpressions.

3 The Ripple Tactic

Since the *ripple* tactic plays a key role both in inductive proofs and in summing series, we will illustrate the proof planning technique by describing it. This description will be necessarily brief and superficial. For more details see [4].

Rippling is used during the step case of an inductive proof. Its job is to rewrite the induction conclusion into a form in which it contains one or more subexpressions which match the induction hypothesis. This enables the next tactic, *fertilize*, to use the induction hypothesis to prove the induction conclusion.

To visualise how rippling works consider the following analogy. Some mountains are reflected in a loch[1] in the valley below them. Someone throws a stone into the loch, disturbing the reflection. The waves from the impact ripple outwards to the shore of the loch leaving the reflection undisturbed again. The mountains are the induction hypothesis, the reflection is the induction conclusion and the wave-fronts are those parts of the induction conclusion which differ from the induction hypothesis. The rippling is the selective application of rewrite rules of a suitable form to move the wave-fronts out of the way.

Induction conclusions are necessarily similar to their induction hypotheses except for the addition of some subexpressions, called *wave-fronts*, which are provided by the form of induction rule used. For instance, in the standard inductive proof of the associativity of $+$, the induction hypothesis is:

$$x + (y + z) \quad = \quad (x + y) + z$$

and the induction conclusion is:

$$\boxed{s(\underline{x})} + (y + z) \quad = \quad (\boxed{s(\underline{x})} + y) + z \tag{1}$$

The wave-fronts are those subexpressions enclosed in boxes less the subexpressions that are underlined. In general, wave-fronts are terms with one or more *wave-holes* in them. The part of the induction conclusion that is similar to the induction hypothesis is called the *skeleton.*

To move these wave fronts outwards the ripple tactic applies *wave-rules.* These are rewrite rules with the property that the left and right hand sides are identical except for the addition of different wave fronts on each side. Furthermore, more of the skeleton is in the wave-hole(s) on the right hand side than it is on the left hand side. The wave-front on the right hand side can be empty. For instance, the recursive definition of $+$ and the substitution law for s provide two wave-rules.

$$\boxed{s(\underline{U})} + V \quad \Rightarrow \quad \boxed{s(\underline{U + V})} \tag{2}$$

$$\boxed{s(\underline{U})} = \boxed{s(\underline{V})} \quad \Rightarrow \quad U = V \tag{3}$$

Wave-rule (2) can be applied to each side of the induction conclusion (1). This causes both the wave-fronts to ripple outwards.

$$\boxed{s(\underline{x + (y + z)})} = \boxed{s(\underline{(x + y)})} + z,$$

Wave-rule (2) can be applied again to the right hand side of the equality to produce:

$$\boxed{s(\underline{x + (y + z)})} = \boxed{s(\underline{(x + y) + z})}.$$

At this point, the two wave-fronts can be eliminated by applying wave-rule (3) to give:

$$x + (y + z) \quad = \quad (x + y) + z$$

[1] The Scottish word for lake.

which is identical to the induction hypothesis. The *fertilize* tactic is then used to prove (trivially) the induction conclusion from the induction hypothesis. The proof of the step case is then complete.

The reasons for the success of rippling are:

- It involves little or no search since the wave-fronts in the goal must correspond to wave-fronts in the wave-rule. The consequence of this is a very controlled application of rewrite rules which in practice means very low branching rates, typically one choice or none at all.
- It terminates. Rippling always makes progress moving wave-fronts in some direction; hence termination is guaranteed, even when applying rewrite rules that would normally, without wave-front annotation, lead to loops.
- It applies only "good" rewrites. As wave-rules are skeleton preserving, if rippling terminates successfully, the hypothesis can be used to prove or simplify the conclusion.

4 Difference Matching

A precondition of rippling is that wave-front annotations have been placed in the formula to be rippled. In inductive proofs these are provided by the induction rule in a natural way. In this paper we observe that rippling can be used by a wide variety of theorem proving tactics provided wave-front annotations can be provided. In particular, rippling is useful for rewriting a goal formula so that it contains a subexpression that matches a hypothesis formula. Many proofs have hypotheses and goals with shared structure. We conjecture that rippling will prove useful for putting these goals into a form in which the hypotheses can be used to prove them. This paper provides supportive evidence for our conjecture.

To annotate the goal formula with wave-fronts we use a *difference matcher*. The difference matcher takes the goal, G, and hypothesis, H, as inputs. It returns G', a copy of G annotated with wave fronts, and substitutions, σ, such that the skeleton of G' equals H under substitution σ. Although difference matching generalises first-order matching, it is not just matching. It is an attempt to make two expressions identical by both variable instantiation and structure hiding; the hidden structure is the wave-front. Further details of an algorithm for difference matching can be found in this volume [5].

5 Methods for Summing Series

We now explain our methods for summing series. They are called: *standard form*, *perturbate*, *conjugate*, *telescope* and *closed form*. The first four are substantive methods whilst the last is just a simplifier and a checker that the solution is in closed form. Each of the first four method makes significant use of rippling augmented with difference matching.

In the rest of the paper, we will adopt the following conventions: The letters i, j and k will be used for indices of summation, *i.e.* bound variables of type natural number. The letters l, m and n will stand for constants of type natural number, *i.e.* they will not depend on any indices of summation. These will typically be used for the bounds of summation. The letters a, b, c and d will stand for constants and variables of type real, *i.e.* they will not depend on any indices of summation. The letters u, v, w, x, y and z will stand for terms of type real, *i.e.* they may depend on indices of summation. In $\sum_{i=0}^{n} a.x$, for instance, x may depend on i, but a and n do not.

5.1 Standard Form

The *standard form* method is the backbone of our methods. It does not find closed form solutions from first principles but tries to reduce the current problem to one which has already been solved. We illustrate this with an example.

Consider the finite sum $\sum_{i=0}^{n} b.i + c$. We can use the *standard form* method to break this into two sub-problems which match previously solved ones, namely the following standard forms:

$$\sum_{i=0}^{n} i = \frac{n.(n-1)}{2} \tag{4}$$

$$\sum_{i=0}^{n} a = s(n).a \tag{5}$$

This will be done by using difference matching to annotate the sum with wave-fronts and then rippling to reduce it to the sub-problems. For the rippling, *standard form* will use the following wave-rules:

$$\sum_{i=m}^{n} \boxed{\underline{x} + \underline{y}} \Rightarrow \boxed{\underline{\sum_{i=m}^{n} x} + \underline{\sum_{i=m}^{n} y}} \tag{6}$$

$$\sum_{i=m}^{n} \boxed{a.\underline{x}} \Rightarrow \boxed{a.\underline{\sum_{i=m}^{n} x}} \tag{7}$$

Note that wave-rule (6) contains two wave-holes, one on the x and one on the y. *CIAM* automatically creates the two weakened versions of this wave-rule which just contain one wave-hole, *eg* :

$$\sum_{i=m}^{n} \boxed{\underline{x} + y} \Rightarrow \boxed{\underline{\sum_{i=m}^{n} x} + \sum_{i=m}^{n} y} \tag{8}$$

Note that wave-rule (7) requires a meta-level condition that a does not contain i. Such meta-level conditions are readily handled by the proof planning mechanism.

First $\sum_{i=0}^{n} b.i + c$ is annotated with wave-fronts by difference matching it with the standard form (4). This gives the annotated sum:

$$\sum_{i=0}^{n} \boxed{b.\underline{i} + c}$$

Rippling with wave-rule (8) gives:

$$\boxed{\underline{\sum_{i=0}^{n} \boxed{b.\underline{i}}} + \sum_{i=0}^{n} c}$$

and then with wave-rule (7) gives:

$$\boxed{b.\underline{\sum_{i=0}^{n} i} + \sum_{i=0}^{n} c}$$

This is then be fertilized with standard form (4), to give:

$$b.\frac{n.(n-1)}{2}+\sum_{i=0}^{n} c$$

Since a summation sign is still present, the current problem is difference matched with standard form (5) to give the new annotated sum:

$$\boxed{b.\frac{n.(n-1)}{2}+\underline{\sum_{i=0}^{n} c}}$$

Since this is already fully rippled it is immediately fertilized with (5) to give:

$$b.\frac{n.(n-1)}{2}+s(n).c$$

which is in closed form, as required.

The *standard form* method can be summarised as follows:

- Find a standard form which difference matches with the current problem and add this as a hypothesis.
- Use difference matching to annotate the current problem with wave-fronts.
- Ripple these wave-fronts outwards.
- Fertilize with the hypothesis.

5.2 Perturbate

The *perturbate* method's proof strategy has many similarities to induction. From the usual recursive definition of a sum we have the following equation:

$$\sum_{i=m}^{s(n)} u_i = \sum_{i=m}^{n} u_i + u_{s(n)}$$

Alternatively, we can strip off terms from the other end to derive the following equation:

$$\sum_{i=m}^{s(n)} u_i = u_m + \sum_{i=m}^{n} u_{s(i)}$$

Combining these two equations gives:

$$\sum_{i=m}^{n} u_i + u_{s(n)} = u_m + \sum_{i=m}^{n} u_{s(i)} \tag{9}$$

The idea of *perturbate* is to rewrite $\sum_{i=m}^{n} u_{s(i)}$ into a function of $\sum_{i=m}^{n} u_i$, say $f(\sum_{i=m}^{n} u_i)$ using rippling. Therefore all occurrences of s(i) in $\sum_{i=m}^{n} u_{s(i)}$ are annotated with wave-fronts which are then rippled outwards, *i.e.* equation (9) is annotated to:

$$\sum_{i=m}^{n} u_i + u_{s(n)} = u_m + \sum_{i=m}^{n} u_{\boxed{s(\underline{i})}} \tag{10}$$

We will call this equation the *perturbation equation*. The wave-fronts in the perturbation equation are rippled outwards until it is in the form:

$$\sum_{i=m}^{n} u_i + u_{s(n)} \quad = \quad u_m + \boxed{f(\underline{\textstyle\sum_{i=m}^{n} u_i})}$$

This equation is then solved for $\sum_{i=m}^{n} u_i$ using the equation solving tactics of PRESS, [10]. There is a possibility of failure since the unknown, $\sum_{i=m}^{n} u_i$, sometimes cancels out.

To illustrate *perturbate* consider the example sum:

$$\sum_{i=0}^{n} i.a^i$$

Now, by the perturbation equation, (10), we have:

$$\sum_{i=0}^{n} i.a^i + s(n).a^{s(n)} \quad = \quad 0.a^0 + \sum_{i=0}^{n} \boxed{s(\underline{i})}.a^{\boxed{s(\underline{i})}}$$

To ripple this we need the wave-rules (8), (7) and:

$$\boxed{s(\underline{x})}.y \quad \Rightarrow \quad \boxed{\underline{x.y} + y} \qquad (11)$$

$$x^{\boxed{s(\underline{y})}} \quad \Rightarrow \quad \boxed{x.\underline{x^y}} \qquad (12)$$

Rippling first with wave-rule (11) gives:

$$\sum_{i=0}^{n} i.a^i + s(n).a^{s(n)} \quad = \quad 0.a^0 + \sum_{i=0}^{n} \boxed{\underline{i.a^{\boxed{s(\underline{i})}}} + a^{s(i)}}$$

then with wave-rule (8) gives:

$$\sum_{i=0}^{n} i.a^i + s(n).a^{s(n)} \quad = \quad 0.a^0 + \boxed{\underline{\textstyle\sum_{i=0}^{n} i.a^{\boxed{s(\underline{i})}}} + \textstyle\sum_{i=0}^{n} a^{s(i)}}$$

then with wave-rule (12) gives:

$$\sum_{i=0}^{n} i.a^i + s(n).a^{s(n)} \quad = \quad 0.a^0 + \boxed{\underline{\textstyle\sum_{i=0}^{n} i.\boxed{a.\underline{a^i}}} + \textstyle\sum_{i=0}^{n} a^{s(i)}}$$

and finally with wave-rule (7) gives:

$$\sum_{i=0}^{n} i.a^i + s(n).a^{s(n)} \quad = \quad 0.a^0 + \boxed{a.\underline{\textstyle\sum_{i=0}^{n} i.a^i} + \textstyle\sum_{i=0}^{n} a^{s(i)}}$$

This equation can be solved for $\sum_{i=0}^{n} i.a^i$ using PRESS's methods (provided $a \neq 1$) giving an equation for $\sum_{i=0}^{n} i.a^i$ in terms of $\sum_{i=0}^{n} a^{i+1}$. The standard from method is then called to replace $\sum_{i=0}^{n} a^{i+1}$ by a closed form expression. This gives:

$$\sum_{i=0}^{n} i.a^i = \frac{a^{s(n)} - a.\frac{a^{s(n)}-1}{a-1}}{a-1}$$

The *perturbate* method can be summarised as follows:

- Instantiate the perturbation equation to the current series.
- Ripple the wave fronts on the right hand side of the equation outwards.
- Solve the resulting equation, using PRESS tactics, treating the current series as the unknown.

The *perturbate* method can be generalised so that it uses more complex forms of perturbation equation based on more complex forms of the recursive definition of summation. For instance, it could use the following two step perturbation equation:

$$\sum_{i=m}^{n} u_i + u_{s(n)} + u_{s(s(n))} = u_m + u_{s(m)} + \sum_{i=m}^{n} u_{s(s(i))}$$

This is useful for series like:

$$\sum_{i=0}^{n} (-1)^i.\frac{1}{2^i}$$

Analogously to mutual recursion, we can also perform mutual perturbations. This is useful for series like $\sum_{i=0}^{n} \sin(i.\theta)$. These generalisations have yet to be implemented. However, we do not envisage any significant difficulties in extending *perturbate* in these ways.

5.3 Conjugate

The *conjugate* method transforms the finite sum of a term into the finite sum of its conjugate, in the hope that it will be easier to find a closed form solution to the conjugate sum than to the original sum. The conjugate can be one of several second order operations, *e.g.* the differential or integral of the original term, or the mapping of a trigonometric series onto the real or imaginary part of a complex series. Thus the *conjugate* method is a generic one covering a wide range of transformations.

The general idea can be understood as follows. Suppose we want to find a closed form for $\sum u$. Let F be a second-order function with an inverse, F^{-1}. That is, there is an equation of the form:

$$F(F^{-1}(v)) = v \tag{13}$$

Let us also assume that there exists a wave-rule which will ripple the function F through the summation operator.

$$\sum \boxed{F(\underline{v})} \Rightarrow \boxed{F(\underline{\sum v})} \tag{14}$$

Thus, combining these two equations we have

$$\begin{aligned}\sum u &= \sum \boxed{F(\underline{F^{-1}(u)})} \\ &= \boxed{F(\underline{\sum F^{-1}(u)})}\end{aligned}$$

This new expression looks syntactically more complicated than the original but often $F^{-1}(u)$ simplifies to some expression u', whereby $\sum u'$ is easier to sum than $\sum u$.

In order to prevent *conjugate* being universally applicable or looping, it is necessary to impose a constraint on it. We have adopted the constraint of a heuristic postcondition that u' must have a lower complexity than u. Complexity is measured using a simple Knuth-Bendix term order.

For example, consider the sum mentioned in the introduction,

$$\sum_{i=0}^{n} s(i).a^{i}$$

where $a \neq 1$. Let F be the differentiation operator and F^{-1} be integration. Now differentiation ripples through summation:

$$\sum \boxed{\frac{\mathrm{d}\underline{u}}{\mathrm{d}x}} \Rightarrow \boxed{\frac{\mathrm{d}\underline{\sum u}}{\mathrm{d}x}} \tag{15}$$

And integrating $s(i).a^{i}$ with respect to the free variable a gives $a^{s(i)}$. Constants of integration can be safely ignored since they will disappear on differentiation. Since $a^{s(i)}$ is simpler than $s(i).a^{i}$ in our Knuth-Bendix order, *conjugate* can proceed. It rewrites the sum to:

$$\sum_{i=0}^{n} \frac{\mathrm{d}a^{s(i)}}{\mathrm{d}a}$$

Wavefront annotations are added by difference matching against the standard result for the sum of a geometric series,

$$\sum_{j=0}^{n} b^{j}$$

This gives:

$$\sum_{i=0}^{n} \boxed{\frac{\mathrm{d}\underline{a^{\boxed{s(\underline{i})}}}}{\mathrm{d}a}}$$

Rippling first with wave-rule (15) gives:

$$\boxed{\frac{\mathrm{d}}{\mathrm{d}a}\underline{\sum_{i=0}^{n} a^{\boxed{s(\underline{i})}}}}$$

And then with wave-rule (12) gives:

$$\boxed{\frac{\mathrm{d}}{\mathrm{d}a}\underline{\sum_{i=0}^{n} \boxed{a.\underline{a^{i}}}}}$$

And finally with wave-rule (7) gives:

$$\boxed{\frac{\mathrm{d}}{\mathrm{d}a}(a.\underline{\textstyle\sum_{i=0}^{n} a^{i}})}$$

This is then fertilized with the standard form for a geometric series, to give:

$$\frac{\mathrm{d}}{\mathrm{d}a}(a.\frac{a^{s(n)}-1}{a-1})$$

The *closed form* method (described in §5.5) then differentiates this expression giving the final answer:

$$\sum_{i=0}^{n} s(i).a^{i} \quad = \quad \frac{s(n).a^{s(n)}}{a-1} - \frac{a^{s(n)}-1}{(a-1)^{2}}$$

The *conjugate* method can be summarised as follows:

- Find a second order operator, F, that ripples through summation.
- Apply F^{-1} to the series term, u, and simplify the result to u'.
- If u' is simpler in the Knuth-Bendix ordering than u then sum the series $\sum u'$ giving an answer v'.
- Simplify $F(v')$ and return this as the final result.

A major use of the *conjugate* method is for summing trigonometric series by transforming them into the real or imaginary parts of exponential series. Consider, for example:

$$\sum_{i=0}^{n} \sin(i.\theta)$$

This is solved by *conjugate* by rewriting it into:

$$\sum_{i=0}^{n} \mathrm{Im}(e^{\sqrt{-1}.i.\theta})$$

This series can then be summed by difference matching against the standard form for a geometric series and rippling. Other series which can be solved in a similar way include $\sum\cos(i.\theta)$, $\sum\sin(i.\theta).\cos(i.\theta)$, $\sum\sin^{2}(i.\theta)$ and $\sum\cos^{2}(i.\theta)$

5.4 Telescope

The *telescope* method is based on the idea that if one part of the term in the series can be cancelled against part of the next term then the sum can be collapsed like a folding "telescope" into a hopefully simpler problem. The version of *telescope* described here concentrates on a restriction of this strategy in which consecutive terms of the series cancel each other out totally.

To give a more rigorous description of this technique we introduce the upper difference operator:

$$\triangle\, u_i \quad = \quad u_{i+1} - u_i$$

This operator has the useful property for summing series that:

$$\begin{aligned}\sum_{i=m}^{n} \triangle\, u_i &= (u_{s(n)} - u_n) + (u_n - u_{n-1}) + \ldots \\ &\quad \ldots + (u_{m+2} - u_{s(m)}) + (u_{s(m)} - u_m) \\ &= u_{s(n)} - u_m \end{aligned} \tag{16}$$

We call this the *telescope equation.* It can be used by *telescope* provided that the terms, v_i, of the series being summed, $\sum v_i$, can be rewritten into the form of an upper difference, $\triangle\, u_i$. The telescope equation, (16), can then be used to reduce the series to $u_{s(n)} - u_m$. At the moment, upper differences are supplied by the user. Recently, however, we have proposed a higher order procedure for discovering upper differences automatically.

To illustrate the *telescope* method consider:

$$\sum_{i=0}^{n} \binom{i}{m}$$

Where:

$$\binom{n}{m} = \frac{n!}{m!(n-m)!}$$

Using the identity:

$$\binom{s(n)}{s(m)} = \binom{n}{m} + \binom{n}{s(m)}$$

We get:

$$\begin{aligned}\binom{i}{m} &= \binom{i+1}{s(m)} - \binom{i}{s(m)} \\ &= \triangle \binom{i}{s(m)}\end{aligned}$$

The series is therefore rewritten by *telescope* into the sum of an upper difference:

$$\sum_{i=0}^{n} \triangle \binom{i}{s(m)}$$

Using (16) as a standard form this is rewritten by the *standard form* method into:

$$\binom{s(n)}{s(m)} - \binom{0}{s(m)}$$

The *telescope* method can be summarised as follows:

- Express the series term as an upper difference.
- Instantiate the telescope equation with this upper difference version of the series.

The *telescope* method can sum a wide variety of series including any series like $\sum i^3$ which is polynomial in the index of summation.

5.5 Closed Form

The *closed form* method terminates all our proof plans by checking that any solutions derived are in closed form. It uses the following definition of closed formedness:

> **Definition 1 (Closed formedness)** : *An expression, exp is a closed form iff it is of the general form:*

$$
\begin{aligned}
exp &:= constant \mid var \mid s(exp) \mid exp + exp \mid exp - exp \mid -exp \mid \\
&\quad exp.exp \mid \frac{exp}{exp} \mid exp^{exp} \mid \ln(exp) \mid \log_{exp}(exp) \mid \\
&\quad \sin(exp) \mid \cos(exp) \mid if(test, exp, exp) \\
constant &:= 0 \mid e \\
var &:= \textit{universally quantified variables} \\
test &:= exp > exp \mid exp < exp \mid exp = exp \mid exp \geq exp \mid exp \leq exp
\end{aligned}
$$

This definition could be easily extended to include a larger set of constants, functions and tests (one obvious extension would be the factorial if one were to reason with products). Its most significant feature is what it leaves out, *i.e.* summation operators, but also the differential, integral, real and imaginary operators.

Before checking that solutions are in closed form, *closed form* simplifies the solution. This has the effect of eliminating any functions that lie outside the closed form grammar and which can be simply eliminated by evaluation. Note that the check for closed formedness is essentially a meta-level operation, *i.e.* it is couched in terms of the syntax of the expression rather than its semantics. This is easily handled by the meta-logical language of the proof plan methods. It shows that some kind of meta-level reasoning is essential in this domain.

6 Implementation and Results

The five series summing methods described above have been implemented as methods in the *CIAM* system and tested successfully on a range of examples.

The problem of summing a series is represented as a logical theorem, *i.e.* to find a closed form for the series $\sum_{i=m}^{n} u_i$ we get *CIAM* to plan the proof of the theorem:

$$\forall m{:}nat\,.\,\forall n{:}nat\,.\,\exists S.\, S = \sum_{i=m}^{n} u_i \tag{17}$$

As yet, we have not written the tactics necessary to execute the plans in *Oyster*; that is, we only build plans and not their corresponding object-level proofs. However, since the preconditions to our methods are *complete* specifications of the methods' applicability, the successful execution of any plan is guaranteed. Indeed, the mapping from plans to their corresponding proofs is purely mechanical.

For reasons of simplicity, the summation operator is represented in the meta-level using a *pseudo* first order term, $sum(i, 0, n, u_i)$. All manipulations of such terms are checked to see that they are valid (*eg* that a bound variable is not being instantiated in an unsound way). This guarantees soundness. We perform simple first order matching on this representation; this looses us completeness since we can only perform imitation (and not arbitrary higher order unification). So far, however, this incompleteness has not proved a significant problem since our manipulations have only required this very restricted form of matching; our methods have failed to find closed form solutions but not because of this incompleteness. We eventually intend to move to a full higher order representation.

In planning a proof, *CIAM* uses the methods in the following order: *closed form* is considered first as it is the only terminating method; *standard form* is considered second as it finishes many of the proofs begun by the other methods; *conjugate* and *telescope* are considered next, in that order; and *perturbate* is considered last as (like the induction method in inductive proof planning) it is nearly always applicable and thus a strategy of last resort. Note that theorem (17) admits a trivial solution in which the witness to S is $\sum_{i=m}^{n} u_i$. However, this trivial solution is not found by *CIAM* because its plans for summing series can only terminate with the *closed form* method, which insists that S is in closed form.

These methods are successful at summing a large number of series with little search. Rippling does need to perform more search than in inductive theorem proving. This is mostly a consequence of the greater number of difference matches possible compared with the (usually) sole induction hypothesis in inductive theorem proving. Rippling is still, however, very controlled as the absence of suitable wave-rules usually terminates unsuccessful branches of the search quickly. Additionally, we are currently developing heuristics for selecting between difference matches which should help to eliminate some of this search.

Among the series that have been summed in this way are those shown in table 1. Problems 8 to 10 are of particular interest as they fall outside the range of Gosper's algorithm, and have not, as far as we are aware, been automatically synthesised before.

No	Problem	Closed Form	Main Method Used
1	$\sum i$	$\frac{n.s(n)}{2}$	*telescope*
2	$\sum i^2$	$\frac{2.n^3+3.n^2+n}{6}$	*telescope*
3	$\sum i + i^2$	1)+2)	*standard form*
4	$\sum a^i$	$\frac{a^{s(n)}-1}{a-1}$	*perturbate*
5	$\sum i.a^i$	$\frac{s(n).a^{s(n)}-a.\frac{a^{s(n)}-1}{a-1}}{a-1}$	*perturbate*
6	$\sum (i+1).a^i$	$\frac{s(n).a^{s(n)}}{a-1} - \frac{a^{s(n)}-1}{(a-1)^2}$	*conjugate*
7	$\sum \frac{1}{i.(i+1)}$	$\frac{n}{s(n)}$	*telescope*
8	$\sum F_i$	$F_{n+2} - 1$	*telescope*
9	$\sum \sin(i.\theta)$	$\frac{(\cos\theta-1).\sin(s(n).\theta)-\sin\theta(\cos(s(n).\theta)-1)}{(\cos\theta-1)^2+\sin^2\theta}$	*conjugate*
10	$\sum \cos(i.\theta)$	$\frac{(\cos\theta-1).(\cos(s(n).\theta)-1)+\sin\theta.\sin(s(n).\theta)}{(\cos\theta-1)^2+\sin^2\theta}$	*conjugate*
11	$\sum \binom{m+i}{i}$	$\binom{m+s(n)}{n}$	*telescope*
12	$\sum \binom{s(i)}{s(m)}$	$\binom{s(n)}{s(s(m))} + \binom{s(n)}{s(m)}$	*standard form*

All sums are from 0 *to* n, $a \neq 1$, F_i *is the ith Fibonacci number, and* $\cos(\theta) \neq 1$. *As well as the main method listed, each problem required the use of difference matching, followed by rippling and fertilization. Each plan was terminated by the* closed form *method.*

Table 1: Some Series Summed by Our System

7 Related Work

Previous work on summing series falls into two camps: verification and decision procedures.

Verification

Sometimes we are given a closed form solution and a series, and we verify that they are equal, usually by mathematical induction. This approach has been adopted by Hutter, [9], and by Clarke and Zhao, [7]. Hutter has used the INKA inductive theorem prover system to verify sums and to prove properties of them, *e.g.*

$$\sum_{i=1}^{n} i + \sum_{i=1}^{n} i = n.s(n) \qquad\qquad \sum_{i=1}^{n} i^3 = \sum_{i=1}^{n} i \,.\, \sum_{i=1}^{n} i$$

Clarke and Zhao have used their Analytica theorem prover, built on top of Mathematica, to prove[2]:

$$\sum_{i=0}^{n} \frac{2^i}{1+a^{2^i}} = \frac{1}{a-1} + \frac{2^{s(n)}}{1-a^{2^{s(n)}}}$$

They also sum series using Gosper's algorithm and the built-in Mathematica simplifier.

In principle, it should be possible to adapt these verification methods to the synthesis of closed form solutions by proving theorems of the form (17). However, existing inductive theorem provers are weak at proving theorems containing existential quantifiers like (17). Moreover, like the methods described in this paper, the methods used by humans and reported in mathematics textbooks often do not use induction.

Decision Procedures

Decision procedures for summation are implemented in general-purpose computer algebra systems like MACSYMA, MAPLE, Mathematica and REDUCE. Such decision procedures are restricted to certain narrow classes of series. For instance, Gosper's algorithm, probably the best decision procedure for summation [8], is restricted to series where the ratio of consecutive partial sums is a rational function. Our technique is not restricted in this way and several of the series listed in table 1 fall outside this class. Another advantage is that these method can be extended to return answers which are not, strictly speaking, closed form (*eg* they can transform certain sums into functions of the Harmonic numbers, $H_n = \sum_{i=1}^{n} \frac{1}{i}$). Additionally, theses methods could equally well be used to reason about infinite absolutely convergent series. Although some such series can sometimes be summed by a decision procedure by considering the limit of a finite series (*eg* $\lim_{n\to\infty} \sum_{i=0}^{n} \frac{1}{i^2}$), many cannot as they have no finite closed form (*eg* $\sum_{i=0}^{n} \frac{1}{i!}$).

Another disadvantage of the decision procedure approach is that they are 'black-boxes', providing no rational explanation of the answers they come up with. Our technique produces proofs which are similar in structure to the methods used by humans and reported in mathematics textbooks. They are, therefore, intelligible to mathematicians.

---

[2]Note that this result is incorrect in the case $a = 1$. This error appears to be due to unsoundness in the Mathematica simplifier.

8 Conclusions

Our research in the domain of summing series has shown that the proof planning search control technique is applicable not just to inductive proofs but also to a non-inductive domain. Indeed, some of the tactics developed specifically for inductive proofs are applicable to summing series. In particular, rippling, already shown to be the key tactic for inductive proofs, turns out, remarkably, to be the key tactic in this new domain. It is used as the main sub-tactic by all the major series summing tactics. Outside inductive proofs, rippling must be supplemented by difference matching [5] to set up the initial wave-fronts. With this addition, we predict that rippling will be widely applicable in automated theorem proving. There is room for further extensions to the tactics described to attack a greater class of series. New series summing tactics could be constructed in the same vein.

The proof planning based technique we have described for summing series extends previous techniques in this area. It can be used to synthesise solutions rather than just verify them. It is not restricted to a small class of series. It was designed and built within the space of a few months as an MSc project. It was a simple matter to adapt our existing programs for inductive proofs to this new domain. Most of the methods we have developed for summing series can be readily adapted to closely related tasks *e.g.* finding closed form solutions to products and integrals. The above observations provide evidence for the general applicability of the proof planning formalism in controlling mathematical proofs.

References

[1] R.S. Boyer and J.S. Moore. *A Computational Logic.* Academic Press, 1979.

[2] A. Bundy. The use of explicit plans to guide inductive proofs. In R. Lusk and R. Overbeek, editors, *9th Conference on Automated Deduction*, pages 111–120, Springer-Verlag, 1988.

[3] A. Bundy, F. van Harmelen, C. Horn, and A. Smaill. The Oyster-Clam system. In M.E. Stickel, editor, *10th International Conference on Automated Deduction*, pages 647–648, Springer-Verlag, 1990.

[3] A. Bundy, F. van Harmelen, J. Hesketh, and A. Smaill. Experiments with proof plans for induction. *Journal of Automated Reasoning*, 7:303–324, 1991.

[4] A. Bundy, F. van Harmelen, A. Smaill, and A. Ireland. Extensions to the rippling-out tactic for guiding inductive proofs. In M.E. Stickel, editor, *10th International Conference on Automated Deduction*, pages 132–146, Springer-Verlag, 1990.

[5] D. Basin and T. Walsh. *Difference Matching.* In D. Kapur, editor, *11th International Conference on Automated Deduction*, Springer-Verlag, 1992.

[6] R.L. Constable, S.F. Allen, H.M. Bromley, et al. *Implementing Mathematics with the Nuprl Proof Development System.* Prentice Hall, 1986.

[7] E. Clarke and X. Zhao. *Analytica - A Theorem Prover for Mathematica.* Technical Report, Carnegie Mellon University, 1991.

[8] R.W. Gosper. Indefinite hypergeometric sums in MACSYMA. In *Proc. MACSYMA Users Conference*, pages 237–252, 1977.

[9] D. Hutter. Guiding inductive proofs. In M.E. Stickel, editor, *10th International Conference on Automated Deduction*, pages 147–161, Springer-Verlag, 1990.

[10] L. Sterling, A. Bundy, L. Byrd, R. O'Keefe, and B. Silver. Solving symbolic equations with PRESS. *J. Symbolic Computation*, 7:71–84, 1989.

Disproving Conjectures

Martin Protzen, TH Darmstadt
Germany[1]

Abstract: A calculus to disprove universally quantified conjectures is presented. It's soundness and completeness are verified. Some strategies and heuristics are presented that yield an effective prover based on the calculus. The prover has been integrated into the induction theorem proving system INKA and has proven successful for disproving conjectures, in particular of those synthesized by the system.

1 Introduction

The degree of mechanization of theorem provers directly depends on the provers' ability to generate useful hypotheses to support reasoning on a given problem. The quality of the hypotheses depends on the systems' knowledge of the domain of interest. In general this knowledge is not enough to avoid generation of false hypotheses. In these cases, time is wasted as provers generally are designed to prove true conjectures. Thus all strategies and heuristics do not apply and the search space grows. An example for methods which generate hypotheses is the termination procedure found in [Walther 88]. Here, in order to show that an algorithm gcd given by

$$DEF_{gcd} = \{ \ \forall x,y: x{=}y \rightarrow gcd(x\ y) = x,\ \forall x,y: x \neq y \wedge x = 0 \rightarrow gcd(x\ y) = y,$$
$$\forall x,y: x{\neq}y \wedge y{=}0 \rightarrow gcd(x\ y) = x,$$
$$\forall x,y: x{\neq}y \wedge x{\neq}0 \wedge y{\neq}0 \rightarrow gcd(x\ y) = gcd(remainder(x\ y)\ remainder(y\ x)) \ \},$$

terminates, the hypothesis

(*) $\forall$ x,y:number $x \neq y \wedge x \neq 0 \wedge y \neq 0 \rightarrow (x \geq y \wedge y \neq 0) \vee (x \neq 0 \wedge y = 0)$

is synthesized.

This hypothesis expresses the fact that if the first argument remainder(x y) in the recursive call of gcd could be proven smaller than the first parameter x of gcd, the termination of gcd could be established. The synthesized conjecture cannot be proven: consider x=1 and y=2. Nevertheless, a theorem proving system would try to prove formula (*) using induction. After falsifying the first hypothesis the method generates other hypotheses which finally prove termination of gcd.

A second example of a method that generates hypotheses is generalization. There the proposition $\forall$k,l: list length(append(k l)) = length(append(l k)) is (over-)generalized to the conjecture $\forall$k,l: list append(k l) = append(l k). For a detailed discussion of generalization and overgeneralization refer to [Aubin 76, Boyer&Moore 79, Hummel 90].

Other examples of methods that generate hypotheses include the computation of induction axioms proposed in [Walther 91] where induction axioms are compared using first order formulas, and the synthesis procedure for skolem-functions described in [Biundo 89].

These examples demonstrate cases where the above mentioned methods fail. Thus it would be advantageous if these hypotheses were rejected without spending too much effort trying to prove them. In this paper we will demonstrate a method which overcomes this problem for most false hypotheses.

[1]Author's address: Technische Hochschule Darmstadt, FB Informatik, Alexanderstr. 10, D-6100 Darmstadt, Germany, Phone: +49 6151 16-4494, E-mail: martin@inferenzsysteme.informatik.th-darmstadt.de

Our approach closely follows a common technique of mathematics: when trying to invalidate a conjecture one searches for a counterexample. So, in order to disprove a conjecture $\forall x^* \varphi$, the method we propose tries to find a substitution for the variables x^* such that $\exists x^* \neg \varphi \Leftrightarrow \neg \forall x^* \varphi$ is proved.
The idea of our approach is as follows: consider the formula

$\exists$ n, m: number plus(n+1 m) = 1.

In order to find a pair of numbers n,m which satisfies the formula we propagate a restriction to the term plus(n+1 m) to it's subterms, i.e. to have the value 1. The definition of plus provides us with two cases, the first one $\forall$x,y: x = 0 → plus(x y) = y demands that the first argument of plus equals 0, which yields the formula

$\exists$ n, m: number n+1 = 0 $\wedge$ m = 1

in our problem. But this formula is not true. So we try the second case $\forall$x,y: x≠0 → plus(x y) = plus(x-1 y) +1 which yields

$\exists$ n, m: number n+1 ≠ 0 $\wedge$ plus(n+1-1 m)+1 = 1.

This can be simplified to

$\exists$ n, m: number plus(n m) = 0.

Again we propagate restrictions to subterms, this time using the first case of plus yielding

$\exists$ n, m: number n = 0 $\wedge$ m = 0 .

and have now found sufficient conditions for the variables n, m such that the initial formula becomes true.

This simple example shows some similarities to narrowing [Hullot 80, Hölldobler 89]. But, of course, it is not narrowing. The main differences are that narrowing uses unification of the left hand side l of an reduction l → r and an occurrence $t \backslash \pi$ in the term under consideration t, whereas our method only uses a matcher of l and t, i.e. it only instantiates variables of l and does not replace proper subterms of t. So we only have to deal with a much smaller search space.

Since a formula is valid iff it's negation is unsatisfiable the method could be used as a theorem prover by exhaustive search for counterexamples. But clearly the method may not terminate and hence is not a decision procedure. On the other hand every theorem prover could substitute our method.

The proposed method overcomes an often criticized shortcoming of conventional theorem provers, i.e. provers not based on completion techniques, which is claimed to be a main advantage of completion based proof procedures: the ability to detect false conjectures. Still one advantage of completion based procedures remains because they integrate proving and disproving into one approach where in the case of conventional provers two different techniques have to be used.

After giving a formal framework, we will introduce a calculus for proving existentially quantified conjectures in section 3. Section 4 gives an overview of the strategies and heuristics applied to obtain an efficient procedure from the calculus. Ideas to extend the method beyond a fast checker for counterexamples are given in section 5, whereas section 6 demonstrates the method for a more complex example.

2 Formal Preliminaries

To have a formal framework for the objects we are concerned with we borrow some notations from [Walther 91] :

We assume a non-empty set $\mathcal{S}$ of *sort symbols* which are names for the various domains under consideration. Given an *$\mathcal{S}$-sorted signature* $\Sigma=(\Sigma_{w,s})_{w\in\mathcal{S}^*, s\in\mathcal{S}}$, $f\in\Sigma_{w,s}$ is a *function symbol* of *rank* ws, *arity* w and of *sort* s, and f is called *reflexive* iff s occurs in w and *irreflexive* otherwise. For a function symbol $f\in\Sigma_{w_1\ldots w_n,s}$ the set of *reflexive positions* rPos(f) is defined as $\{i \mid w_i = s\}$. Each $\mathcal{S}$-sorted signature is separated into a *constructor part* Σ^c, a *selector part* Σ^s and a *defined part* Σ^d. The elements of Σ^c are called *constructors*, elements of Σ^s are called *selectors* and elements of Σ^d are called *defined symbols*.

For each $s \in \mathcal{S}$, $\mathcal{V}_s$ is the set of all *variables* of sort s, $\mathcal{T}(\Sigma,\mathcal{V})_s$ denotes the set of all (*well formed*) *terms* (over Σ and $\mathcal{V}$) of sort s, and $\mathcal{T}(\Sigma,\mathcal{V})$ is the set of all terms (over Σ and $\mathcal{V}$). $\mathcal{V}(o)$ denotes the set of all free variables in an arbitrary object o. $\mathcal{F}(\Sigma,\mathcal{V})$ is the set of all (*well formed*) *formulas* (over Σ and $\mathcal{V}$) and $t_1 \equiv t_2$ (assuming $t_1,t_2 \in \mathcal{T}(\Sigma,\mathcal{V})_s$) denotes an equation. Equations and negated equations are called literals. All substitutions are assumed to be sort preserving, i.e. σ: $\mathcal{V}_s \to \mathcal{T}(\Sigma,\mathcal{V})_s$, and therefore $\sigma(\mathcal{T}(\Sigma,\mathcal{V})_s) \subset \mathcal{T}(\Sigma,\mathcal{V})_s$ when σ is extended as an *endomorphism* for mapping terms. Given a sequence $x^* = x_1 \ldots x_{|w|} \in \mathcal{V}_w$, $\sigma(x^*)$ denotes the corresponding sequence $\sigma(x^*)=\sigma(x_1)\ldots\sigma(x_{|w|}) \in \mathcal{T}(\Sigma,\mathcal{V})_w$. SUB(X, Y) is the set of all substitutions with domain X and codomain Y. We shall also apply substitutions to quantifier-free formulas.

A Σ-algebra A is called *standard* iff the Σ^c-reduct of A is an *initial* Σ^c-algebra. We may restrict ourselves to the set $\mathcal{T}(\Sigma^c)$ of constructor ground terms as the carrier of a standard algebra. So a standard Σ-algebra A represents a mapping $\mathcal{T}(\Sigma) \to \mathcal{T}(\Sigma^c)$, which "evaluates" each ground term t to a constructor ground term A(t). Since we do not need variable assignments for closed formulas, we use standard Σ-algebras also as standard interpretations and standard models for *closed* formulas.

A *specification* S is a triple $(\mathcal{S},\Sigma,\Phi)$, where $\mathcal{S}$ is a set of sort symbols, Σ is an $\mathcal{S}$-sorted signature and Φ is a finite set of closed formulas, called the *axioms* of S. The set of sort symbols $\mathcal{S}$ stores the names for the data structures known to the system, the signature Σ stores the names of the algorithms, and Φ represents the data base of the system.

A Σ-algebra M is a standard model *of the specification* S iff M is a standard model of Φ. S is *admissible* iff S has a standard model. One may consider the standard model M as an *interpreter* for the data structures and algorithms in the data base, and admissibility guarantees that there is *exactly one* such interpreter. For an admissible specification S, the *theory* Th(S) of S is defined as $Th(S) = \{\varphi \in \mathcal{F}(\Sigma,\mathcal{V}) \mid \varphi$ is closed and $M \models \varphi\}$, where M is a standard model of S. Hence each admissible specification *specifies a theory* which can be thought of as the *implicit* knowledge of the system (whereas the set Φ of axioms represents the system's *explicit* knowledge). So if we wonder whether a certain statement φ about the data structures and algorithms holds true, we ask whether φ is in the theory, i.e. $\varphi \in Th(S)$.

Given an admissible specification S, we *extend* S by a new data structure, by a new algorithm, or by a new lemma defined on the data structures and algorithms already in S. But we demand, that these extensions satisfy certain requirements to guarantee that the resulting specification S' is admissible.

We extend a specification $S=(\mathcal{S},\Sigma,\Phi)$ with a data structure D for a new sort symbol s yielding $S'=(\mathcal{S}',\Sigma',\Phi')$, by extending $\mathcal{S}$ with s, Σ with the constructors and selectors of the new data structure and Φ by the *representation formulas* REP_s for s. These are a set of closed Σ'-formulas axiomatizing properties of the new data structure: (1) different constructors denote different objects, (2) constructors are injective, (3) each object can be denoted as an application of some constructor to its selectors, (4) each selector is inverse to the constructor it belongs to and (5) each selector returns an arbitrary but fixed term ("default term") if applied to a constructor it does not belong to.

The structure number, e.g., uses the constructors 0 and s and the selector p and is defined by REP_{number} =

{ (1) $\forall$n:number $\neg$ 0≡s(n), (2) $\forall$ n_1,n_2:number $s(n_1) \equiv s(n_2) \to n_1 \equiv n_2$,
(3) $\forall$n:number n≡0 $\vee$ n≡s(p(n)), (4) $\forall$n:number p(s(n))≡n, (5) p(0)≡0 }

and the structure list is defined by REP_{list} =

{ (1) $\forall$ n:number$\forall$ l:list $\neg$ empty≡add(n l), (3) $\forall$ l:list l≡empty $\vee$ l≡add(head(l) tail(l)),
(2) $\forall n_1,n_2$:number $\forall l_1,l_2$:list add($n_1 l_1$)≡add($n_2 l_2$) $\rightarrow n_1 \equiv n_2 \wedge l_1 \equiv l_2$,
(4a) $\forall$ n:number$\forall$ l:list head(add(n l))≡n, (5a) head(empty)≡0,
(4b) $\forall$ n:number$\forall$ l:list tail(add(n l)≡l, (5b) tail(empty)≡empty }.

We may also extend a specification S by a new algorithm: For a specification S'= ($\mathcal{S}'$,Σ',Φ'), the set of *definition formulas* DEF_f given as DEF_f ={ $\forall$x*:w $\varphi_1 \rightarrow$ fx*≡r_1, ..., $\forall$x*:w $\varphi_k \rightarrow$ fx*≡r_k } defines an *algorithm* for *f* iff each φ_i is a quantifier-free Σ'-formula. The expressions "$\varphi_i \rightarrow$ fx*≡r_i" are the *cases* of f, φ_i is the *condition* and r_i is the *result* of case i. An algorithm for f is *recursive* iff at least one case i of f is *recursive*, i.e. φ_i or r_i contains a term of form *f(t*)*. For instance, the algorithm for plus defined by DEF_{plus} = { $\forall$x,y:number x≡0 $\rightarrow$ plus(x y) ≡ y, $\forall$x,y:number $\neg$ x≡0 $\rightarrow$ plus(x y) ≡ s(plus(p(x) y)) } is a recursive algorithm.

Given an algorithm for f, a formula φ_f can be uniformly derived from f, such that $\varphi_f \in$ Th(S) entails, that if two conditions are satisfied for a given input, then the results agree (*case-functionality*), and also that there is *at least* one condition, which is satisfied for a given input (*case-completeness*). Also a *termination hypothesis* φ_{term} (eventually) can be derived from f, such that $\varphi_{term} \in$ Th(S) entails that f *terminates* for each input.

If $\varphi_f \in$ Th(S) and $\varphi_{term} \in$ Th(S) can be verified, we extend the specification S=($\mathcal{S}$,Σ,Φ) by the algorithm f and obtain the specification S'=($\mathcal{S}$, Σ∪{f}, Φ ∪ DEF_f).

If φ∈ Th(S) can be verified, S'=($\mathcal{S}$,Σ,Φ∪{φ}) is an extension of S=($\mathcal{S}$,Σ,Φ) by the lemma φ. Obviously, we do *not* extend the *theory* of the specification, because Th(S)=Th(S'). That means nothing "new" is defined, when extending an admissible specification by a lemma. We only extend the set of axioms, by making some implicit knowledge explicit with *shifting* a formula from the *theory* of the given specification to the *axioms* of the extended specification, where it can be used for subsequent deductions.

3 The Calculus

In this section a method is proposed which computes a variable assignment - if such an assignment exists - for an unquantified formula φ such that φ under this assignment is evaluated to true. If successful, the proposition [∃ x*:w φ] ∈ Th(S) is proved.

The proposition φ ∈ Th(S) holds iff there is an assignment of ground constructor terms a* to the variables x* of φ such that under this assignment φ is valid in the standard model M of S. Stipulating σ={x*/ a*} yields M ⊨ σ(φ) which can be expressed as σ(φ) ∈ Th(S). Thus to prove the proposition φ valid it suffices to find a ground constructor substitution σ (where domain(σ) ⊂ $\mathcal{V}$(φ)) such that σ(φ) ∈ Th(S). We then speak of "φ being satisfiable with σ". The codomain of the substitution can be extended to $\mathcal{T}(\Sigma, \mathcal{V})$ without harm because the existence of a substitution σ'∈ SUB($\mathcal{V}$,$\mathcal{T}(\Sigma, \mathcal{V})$) satisfying φ implies the existence of a ground constructor substitution σ∈ SUB($\mathcal{V}$, $\mathcal{T}(\Sigma^c)$) which also will satisfy φ.

As we will not allow universal quantifiers in φ, we do not need any quantifiers anymore. In the sequel we regard every variable to be implicitly existentially quantified. Without loosing generality we can assume φ in disjunctive normal form, thus every formula can be seen as a disjunction of conjunctions of literals.

After these preliminaries we now define the terms *problem* and *satisfiability of problems*:

Definition 3-1: A *problem* is defined as a finite set of literals. $\mathcal{P}(\Sigma, \mathcal{V})$ is the set of all problems over Σ and $\mathcal{V}$. A problem[2] P={L_1,...,L_n} stands as an abbreviation for the formula $\bigwedge P = L_1 \wedge \ldots \wedge L_n$, where the empty problem stands for $\bigwedge \emptyset = \text{TRUE}$. In the sequel we occasionally will use problems instead of formulas. ♦

Definition 3-2: A literal L is called *satisfiable with substitution* $\sigma \in \text{SUB}(\mathcal{V}, \mathcal{T}(\Sigma,\mathcal{V}))$, iff $\forall x^*{:}w\ \sigma(L) \in \text{Th}(S)$ where $\mathcal{V}(x^*) = \mathcal{V}(\sigma(L))$. L is called *satisfiable*, iff there is a substitution σ, such that L is satisfiable with σ, and *unsatisfiable* otherwise. A problem P is called *satisfiable with substitution* $\sigma \in \text{SUB}(\mathcal{V}, \mathcal{T}(\Sigma,\mathcal{V}))$, iff every literal $L \in P$ is satisfiable with σ. P is called *satisfiable*, iff there is a substitution σ, such that P is satisfiable with σ, and *unsatisfiable* otherwise. ♦

Subsequently we will use ΔL as a shorthand for a possibly negated literal L, if a statement applies to both the negated and unnegated literal. Furthermore the equality predicate is regarded symmetric such that there is no difference between literals $L = \Delta\, q \equiv r$ and $L' = \Delta\, r \equiv q$.

3.1 Base Rules

To mechanize the test for satisfiability of problems a calculus $\mathfrak{R}$ is now introduced. Our approach closely follows the problem reduction approach [Nilsson 80]: each rule of the calculus reduces a problem P to a problem Q, which - hopefully - is easier to satisfy. During this reduction all necessary bindings to variables are stored into a substitution λ. The reduction stops, if the empty problem has been reached. The resulting substitution then satisfies the original problem. We start our presentation by introducing a minimal set of rules and proving a soundness as well as a completeness theorem for this set. Based on these definitions we extend the set of rules to allow the development of a more efficient algorithm from the calculus.

Each rule has the form $P \vdash Q, \lambda$ where P, Q are problems and λ is a substitution. A rule should be read as follows: if Q is satisfiable with a substitution σ, then P is satisfiable with the substitution $\sigma \circ \lambda$, where $\sigma \circ \lambda$ stands for functional composition of σ and λ. Using the rule system we now try to find a sequence of rule applications, which reduces the initial problem P to the empty problem ∅. Composition of all involved substitutions yields a substitution which satisfies P.

As we will motivate each rule by an example, we will use the data structures number and list and the binary function plus defined in section 2.

Definition 3-3: calculus $\mathfrak{R}$

(1) Language: $\mathcal{P}(\Sigma,\mathcal{V}) \to \mathcal{P}(\Sigma,\mathcal{V}) \times \text{SUB}(\mathcal{V}, \mathcal{T}(\Sigma,\mathcal{V}))$

in the sequel all terms are elements of $\mathcal{T}(\Sigma,\mathcal{V})$, unless otherwise stated.

(2) Rule Templates:

(a) Functionality: Terms having the same leading symbol are equal if all arguments are pairwise equal. If the leading symbol is a constructor symbol, this is a necessary condition for equality because constructors are injective.

Rule: $P \cup \{f s_1 \ldots s_n \equiv f t_1 \ldots t_n\} \vdash P \cup \{s_1 \equiv t_1, \ldots, s_n \equiv t_n\}, \varepsilon$

Example: Let L = plus(s(x) y) ≡ plus(s(y) x). Each solution of {s(x) ≡ s(y), y ≡ x} also solves L. Each substitution satisfies L' = 0 ≡ 0.

(b) Injectivity: Two terms having the same leading symbol are different only if at least one pair of subterms is different. If the leading symbol is a constructor, this condition is

[2]Note the difference between problems and clauses: problems stand for conjunctions of literals, whereas clauses represent disjunctions.

sufficient for the terms being different.
Rule: $P \cup \{\neg\, fs_1...s_n \equiv ft_1...t_n\} \vdash P \cup \{\neg\, s_i \equiv t_i\}, \varepsilon$ for some $i \in \{1...n\}$ and $f \in \Sigma^c$.
Example: Each solution of $L = \neg\, x \equiv y$, solves $L' = \neg\, s(x) \equiv s(y)$ as well as $L'' = \neg\, add(x\ u) \equiv add(y\ v)$.

(c) Deletion rule for inequalities: Constructor terms having different leading symbols are different.
Rule: $P \cup \{\neg\, fs^* \equiv gt^*\} \vdash P, \varepsilon$ where $f, g \in \Sigma^c$ and $f \neq g$.
Example: Every substitution satisfies $L = \neg\, 0 \equiv s(y)$ and in particular ε does.

(d) Paramodulation: Let $L = fs^* \equiv q$, where f is a defined symbol. Then the definition formulas in DEF_f can be applied to fs^*. Let δ denote parameter substitution, i.e. δ applied to the formal parameters x^* in the definition formulas of f yields the actual parameters s^*. Then L is satisfiable, iff the instantiated conditions $\delta(\varphi_i)$ of a definition formula $[\forall \ldots \varphi_i \rightarrow fx^* \equiv r_i]$ and $\delta(r_i) \equiv q$ can be satisfied with the same substitution, i.e. the result term of the definition formula and the term q can be unified.

This rule is central to the calculus. $L = fs^* \equiv q$ denotes a constraint for the term fs^*, having the same value as q. Using the paramodulation rule, constraints for the subterms s_i of fs^*, i.e. the actual parameters of f, can be deduced.
Rule: $P \cup \{\Delta\, fs^* \equiv t\} \vdash P \cup \delta(\varphi_i) \cup \{\Delta\, \delta(r_i) \equiv t\}, \varepsilon$ where $[\forall \ldots \varphi_i \rightarrow fx^* \equiv r_i] \in DEF_f$ and $\delta \in SUB(\mathcal{V}, \mathcal{T}(\Sigma, \mathcal{V}))$ such that $\delta(fx^*) = fs^*$.
Example: Let $L = plus(0\ z) \equiv 0$ and DEF_{plus} as above. Before using the first formula in DEF_{plus} we have to replace the formal parameters x and y by the actual parameters 0 and z, yielding $P' = \{0 \equiv 0, z \equiv 0\}$. Each solution of P' also solves L and P' is "easier" to satisfy than L because it does not contain the symbol plus.

(e) Instantiation: As every satisfiable literal $L = \Delta\, v \equiv t$ where v is a variable also is satisfiable with a ground constructor substitution σ (i.e. $\sigma \in SUB(\mathcal{V}, \mathcal{T}(\Sigma^c))$), there is a constructor c such that $\sigma(v) = cq^*$ where $q^* \in \mathcal{T}(\Sigma^c)$. So there is an instantiation $\sigma' = \{v / cu^*\}$ for v (where the u_i are variables not used before), such that L is satisfiable iff $\sigma'(L)$ is. The only remaining problem is to find the "right" constructor c. This is easy if t has a constructor as leading symbol. For obvious reasons σ' has to be applied to all other literals of the actual problem as well.
Rule: $P \cup \{\Delta\, v \equiv t\} \vdash \lambda(P) \cup \{\Delta\, \lambda(v) \equiv \lambda(t)\}, \lambda$ where $c \in \Sigma^c$, $v \in \mathcal{V}$, $\lambda = \{v / cu^*\}$ and u_i new variables.
Example: Let $P_1 = \{x \equiv s(y)\}$. Instantiation of x by $\tau = \{x / s(u)\}$ yields $P_1' = \{s(u) \equiv s(y)\}$. Applying the functionality rule (a) yields $\{u \equiv y\}$ which is satisfiable with every substitution which unifies u and y.

The solution of $P_2 = \{\neg\, x \equiv s(y)\}$ is very similar: x can either be instantiated by $\tau = \{x / 0\}$ or by $\tau' = \{x / s(u)\}$. In the first case the resulting literal $\neg\, 0 \equiv s(y)$ can be deleted using rule (c), whereas in the second case the satisfiability of $P_2' = \{\neg\, u \equiv y\}$ has to be shown.

Obviously this method is not efficient if a variable and an arbitrary term have to be equated, because the constructor c has to be guessed. For this reason later on we will introduce a rule which unifies a term and a variable without referring to the leading symbol.

If a term has a selector as leading symbol we need to know which constructor the argument of the selector has as leading symbol. We then can make use of one of the representation formulas concerning selectors. In the sequel the symbol $b^{i,j}$ denotes the jth selector of the ith

constructor c^i.

(f) reduce selector term: This rule handles the cases where a selector is applied to the constructor it belongs to.

<u>Rule</u>: $P \cup \{\Delta\, b^{i,j}(q) \equiv r\} \vdash P \cup \{q \equiv c^i(u^*), \Delta\, u_j \equiv r\}$, ε where u_k are new variables.

<u>Example</u>: Let L = tail(l) ≡ t. Each solution of P_1 = {l ≡ add(u v), v ≡ t} also solves L_1. L_2 = head(l) ≡ n is solved by every solution of P_2 = {l ≡ add(u v), u ≡ n}.

(g) eliminate selector: This rule handles the cases where a selector is applied to a term having a constructor not belonging to the selector as leading symbol.

<u>Rule</u>: $P \cup \{\Delta\, b^{i,j}(q) \equiv r\} \vdash P \cup \{q \equiv c^h(u^*), \Delta\ \nabla s^{i,j} \equiv r\}$, ε where $i \neq h$, u_k are new variables, and $\nabla s^{i,j}$ is the default term of the rangesort of the selector $b^{i,j}$.

<u>Example</u>: Let L_1 = head(l) ≡ n. Each solution of P_1 = {l ≡ empty, n ≡ 0} also solves L_1.

This completes the rules of the base calculus. Now we define **(3) deductions in the calculus:**

For problems P, P' and substitutions σ, σ' a *deduction* of (P', σ') from (P, σ) is defined as a sequence of pairs $(P_0, \sigma_0)\ldots(P_n, \sigma_n)$ where $(P_0, \sigma_0) = (P, \sigma)$ and $(P_n, \sigma_n) = (P', \sigma')$, if $\text{domain}(\sigma) \cap \mathcal{V}(P) = \varnothing$ and either

(0) $(P_0, \sigma_0) = (P_n, \sigma_n)$ (deduction of length 0)

or (n) there is a rule $P_i \vdash P_{i+1}, \lambda_{i+1}$ and $\sigma_{i+1} = \lambda_{i+1} \circ \sigma_i$ for all $i \in \{0, \ldots, n-1\}$.

In overloading $\vdash$ we use $(P, \sigma) \vdash (P', \sigma')$ as a shorthand for "there is a deduction of (P', σ') from (P, σ)". If the substitutions involved in a deduction $(P_0, \sigma_0) \vdash (P_1, \sigma_2) \vdash \ldots \vdash (P_n, \sigma_n)$ are not needed, we will use $P_0 \vdash P_1 \vdash \ldots \vdash P_n$. $\vdash_n$ refers to the length of a deduction explicitly.

We call a problem P *solvable with substitution* σ, iff $(P, \varepsilon) \vdash (\varnothing, \sigma)$, σ is called a *solution* of P. $P \vdash \varnothing$ denotes solvability of P (with arbitrary σ), $P \not\vdash \varnothing$ denotes unsolvability, i.e. for all $\sigma \in SUB(\mathcal{V}, \mathcal{T}(\Sigma,\mathcal{V}))$ $(P, \varepsilon) \not\vdash (\varnothing, \sigma)$ holds. ♦

Our calculus correlates the semantic notions "satisfiable" and "binding" to the syntactic or deductive notions "solvable" and "solution" respectively. Of course, the *soundness* of the calculus is demanded, i.e. if we state the solvability of a problem and deduce a solution σ for it, the problem must be satisfiable with σ. Also *completeness* of the calculus is required, i.e. if a problem is satisfiable then a solution should be deduced.

Thus we prove **Theorem 3-1 (soundness):** Let P be a problem and σ a substitution where $(P, \varepsilon) \vdash (\varnothing, \sigma)$. Then P is satisfiable with σ.

<u>Sketch of proof</u>: We show $\sigma(P) \in Th(S)$ by induction over the length of the deduction $(P, \varepsilon) \vdash_n (\varnothing, \sigma)$. For deductions of length 0 we have $P = \varnothing$ and $\sigma = \varepsilon$, thus $\sigma(P) = \varnothing \in Th(S)$ holds by definition. For deductions of length n+1 we separate the first step: $(P, \varepsilon) \vdash_1 (Q, \lambda) \vdash_n (\varnothing, \sigma)$ where $P \vdash Q, \lambda$. We have to prove that there is a substitution τ and a deduction $(Q, \varepsilon) \vdash_n (\varnothing, \tau)$ such that $\sigma = \tau \circ \lambda$. $\tau(Q) \in Th(S)$ holds by induction hypothesis. Now for each rule $P \vdash Q, \lambda$ of the calculus it is shown that $\tau(Q) \in Th(S)$ implies $\tau(\lambda(P)) \in Th(S)$. Thus P is satisfiable with $\sigma = \tau \circ \lambda$. For the details of the proof refer to [Protzen 91]. ♦

Theorem 3-2: The calculus is complete, i. e. for each satisfiable problem P there exists a substitution σ such that $(P, \varepsilon) \vdash (\varnothing, \sigma)$ holds.

<u>Sketch of proof</u>: We define a well founded order $>_{\#}$ and show that for every satisfiable problem P there exists a rule $P \vdash Q, \lambda$ in the calculus, such that Q is still satisfiable and P

$>_\#$ Q holds. As P is satisfiable there is also a ground constructor substitution σ satisfying P. Thus $\sigma(P)$ is ground. We define a cost function $\#_T(t)$ on ground terms t giving an upper estimate of the number of steps needed for the evaluation of t. In the same manner we define cost functions $\#_L(L)$ and $\#_P(P)$ for ground literals L respectively ground problems P: $\#_P(P)$ is defined as the costs of the "most expensive" literal in P. Furthermore we define N(P) for ground problems as the number of literals with maximal costs. For (non-ground) problems Q we define V(Q) as the number of variables v which appear in a literal of the form $\Delta\ v \equiv t$ as top level symbols. Now we are able to define a well-founded order $>_\#$ over pairs (σ, P) of ground constructor substitutions σ and problems P as the lexicographic combination of the orders defined by $\#_P(\sigma(P))$, $N(\sigma(P))$ and V(P). Finally we prove by induction over $>_\#$, that for each (with σ) satisfiable problem $P \neq \emptyset$ there exists a substitution σ' and a rule $P \vdash Q, \lambda$ in the calculus, where (1) $\sigma'|_{\mathcal{V}(P)} = \sigma$, (2) $\sigma'(Q)$ is ground, (3) $\sigma'(Q) \in Th(S)$ and (4) $(\sigma, P) >_\# (\sigma', Q)$. As the pairs (σ', Q) where $Q = \emptyset$ are the only minimal elements with respect to $>_\#$, it is proven that for each satisfiable problem P there exists a deduction $P \vdash \emptyset$. For the details of the proof see [Protzen 91]. ♦

3.2 Extending the Calculus

After having proven the base calculus correct and complete we will extend the calculus by some additional rules, which allow us to develop a more powerful method. If the additional rules are correct by themselves, the extended calculus will be correct as well.

As stated above, instantiation rule (e) is particularly inefficient. The following rule equates the terms q, r of a literal $L = q \equiv r$ in *one* step:

(h) unification: A literal $L = q \equiv r$ is trivially satisfiable if q and r are unifiable, as for every unifier σ of q and r $\sigma(q) = \sigma(r)$ holds. The unifier has to be applied to all other literals in the problem, i.e. in the sequel the instantiated problem $\sigma(P)$ has to be inspected. Because $\sigma(L) \in Th(S)$, it is sufficient to inspect $\sigma(P \setminus L)$.

Rule: $P \cup \{q \equiv r\} \vdash \lambda(P), \lambda$ where λ is most general unifier of q and r.

Example: Let P={s(u)≡plus(u v), plus(u x)≡plus(0 v)}. All substitutions satisfying P have to satisfy L=plus(u x)≡plus(0 v). λ={u/0, x/v} is (most general) unifier of plus(u x) and plus(0 v) and thus satisfies L. So, if σ satisfies λ(P')={s(0)≡plus(0 v)}, then $\sigma \circ \lambda$ will satisfy P.

As all terms have a unique representation as constructor terms the following definition allows to formulate another deletion rule:

Definition 3-4: For terms $q \in \mathcal{T}(\Sigma,\mathcal{V})$ we define the set of *strict subterms*

$$\text{as}\quad S(q) = \begin{cases} \bigcup_{i \in rPos(c)} (\{q_i\} \cup S(q_i)), & \text{where } q = cq^* \text{ and } c \in \Sigma^c. \\ \emptyset & \text{else.} \end{cases}$$ ♦

Example: add(n append(u v)) and append(u v) are strict subterms of q = add(m add(n append(u v))), but u and v are not. m is not a strict subterm as it is not on a reflexive position.

For all $t \in S(q)$ $t \equiv q \notin Th(S)$ holds, because assuming the contrary would entail the existence of terms $t', q' \in \mathcal{T}(\Sigma^c)$ where $t \equiv t'$, $q \equiv q'$, $t' \equiv q' \in Th(S)$ and $t' \neq q'$, which contradicts the representation formulas. Note that this only holds for *strict* subterms, not for ordinary subterms.

Examples: $x \equiv s(x) \notin Th(S)$, whereas $plus(0\ 0) \equiv 0 \in Th(S)$.

This allows the following deletion rule:
(i) deletion rule for strict subterms: If one of the terms q, r of a literal L = q ≡ r is a strict subterm of the other, L is not satisfiable. But the negated literal ¬ L is satisfiable with every substitution.
Rule: $P \cup \{\neg q \equiv r\} \vdash P, \ \varepsilon$ where $q \in S(r)$.
Example: No substitution satisfies L = s(0) ≡ 0, but every substitution satisfies L' = ¬ s(0) ≡ 0. By the same argument ¬ k ≡ add(plus(x y) k) is satisfiable with every substitution.

3.3 Trivial Unsatisfiable Problems

Up to this point we only have discussed rules which allow the deduction of the empty problem. On the other hand there are literals and problems which are not satisfiable with any substitution. If any of these problems occur, the deduction can be aborted, as it will not be successful. The following definition gives syntactical criteria for such problems:

Definition 3-5 (trivial unsatisfiable literals and problems)
Literals L of the form

$L = fs^* \equiv gt^*$ where $f, g \in \Sigma^c$ and $f \neq g$,
$L = q \equiv r$ where $q \in S(r)$ or
$L = \neg q \equiv q$

are called *trivial unsatisfiable literals*.

A problem P is called a *trivial unsatisfiable problem*, iff P contains

a trivial unsatisfiable literal L or
two literals $L_1 = q \equiv r$ and $L_2 = \neg q \equiv r$ or
two literals $L_1 = q \equiv c_1t^*$ and $L_2 = q \equiv c_2t^*$ where $c_1 \neq c_2$ and $c_1, c_2 \in \Sigma^c$. ♦

Theorem 3-3: Trivial unsatisfiable problems are not satisfiable. ♦

Examples: No substitution satisfies the following literals and problems

0 ≡ s(q),	add(0 q) ≡ empty,	q ≡ add(n q),
plus(x y) ≡ s(plus(x y)),		¬ plus(q r) ≡ plus(q r),
{ k ≡ l, ¬ k ≡ l },	{ q ≡ 0, q ≡ s(r) },	{ l ≡ empty, l ≡ add(u v) }.

Remark: The three cases in the definition of trivial unsatisfiable literals correspond to the deletion rules (c), (i) respectively (h) of our calculus where the considered literals appear in their negated form. These rules delete a "trivial satisfiable" literal from the problem (and extend the satisfying substitution if necessary). This duality stems from the fact that for closed formulas φ the negation ¬ φ of a valid formula φ ∈ Th(S) is not valid.

4 Control Strategies

Based on the calculus defined in the last section we will now develop a method which proves the satisfiability of formulas. It has to be checked whether there is a sequence of rule applications which deduces the empty problem (∅, σ) from the initial problem (P, ε). If such a deduction exists the method will terminate and the composition of all intermediate instantiations σ will be the result.

In some situations it is possible that a rule is applied which directs the deduction into a "dead end", i.e. from a satisfiable problem P a problem Q is deduced which is not satisfiable. For this reason in each situation all applicable rules have to be regarded, otherwise the method would not be complete. But sometimes a rule can be selected which is "safe", i.e. will not lead into a dead end. So it is advantageous to apply this rule first.

4.1 Structure of Control

We will use an OR-tree to organize the search. The root of the tree is labelled with the initial formula φ. Without loosing generality we can assume φ in disjunctive normal form. The direct successor nodes of the root are labelled with the single parts of the disjunction φ. All these parts are conjunctions of literals, i.e. they correspond to the problems defined in the last section. In the sequel we will not distinguish between nodes of the tree and their labels.

A leaf P of the tree can be expanded by selecting a literal L from P and applying all possible rules to $P = \{L\} \cup P'$. Each rule application results in a problem Q_i, which will be a direct successor of P in the tree.

If the empty problem is a leaf of the tree the method immediately stops and returns the composition of all instantiations on the path from the root to the empty problem as solution. If a leaf contains a trivial unsatisfiable literal, then this leaf will be labelled *closed*. Non-leaf nodes will be labelled closed, if all their successors are. Nodes not labelled closed are called *open* nodes. A formula φ is unsatisfiable, iff the root of the corresponding tree has been labelled closed. If it can be guaranteed that each node will be expanded during the method (such methods are called *fair*) then a solution will be found, if one exists. If there is no solution the method will stop when the root has been labelled closed, or it will not stop at all, i.e. there will always be open nodes in the tree. Such infinitely expandable OR-trees exist, otherwise there would be a decision procedure for satisfiable problems which contradicts the undecidability of the class of universal formulas of arithmetic.

Crucial for the effectiveness of the method is how the two indeterminisms are resolved: (1) which literal to select from a problem (AND-indeterminism) and (2) which rule to apply to this literal (OR-indeterminism).

<u>Example</u>: Let $P = \{\neg x \equiv y, s(x) \equiv s(s(y))\}$. First we have to select a literal from P (AND-indeterminism). Choosing $\neg x \equiv y$ leaves only the instantiation rule (e) to apply, with the two alternatives to instantiate x by 0 or by s(u) (OR-indeterminism). The first case yields the unsatisfiable problem $\{\neg 0 \equiv y, s(0) \equiv s(s(y))\}$, i.e. we entered a "dead end". The second case yields the satisfiable problem $\{\neg s(u) \equiv y, s(s(u)) \equiv s(s(y))\}$. Entering a dead end could have been avoided, if we initially had selected the second literal $s(x) \equiv s(s(y))$ in P: here only the functionality rule is applicable yielding $\{\neg x \equiv y, x \equiv s(y)\}$. Here it is again advantageous to select the second literal and to apply the unification rule yielding $\{\neg s(y) \equiv y\}$. Now the empty problem can be deduced using deletion rule (i).

4.2 Resolving Indeterminisms

In this section we will state principles for selecting a literal from a problem. To obtain a method as powerful as possible the search space has to be reduced as far as possible without eliminating those parts which contain solutions. This is done by delaying decisions as far as possible, thus at first selecting only those rules which will not lead into a dead end.

Principle 1: detect trivial unsatisfiable literals immediately

Theorem 3-3 gives criteria for unsatisfiable literals (and problems). From those problems the empty problem can never be deduced, so we label these problems closed. If the newly labelled problem was the last open successor of a currently unlabelled node, this node can be labelled closed as well.

Principle 2: use deletion rules
This reduces the size of problems by eliminating irrelevant information.

Principle 3: use explicit information
The first thing we do is to evaluate all terms symbolically whenever possible, i.e. we will replace plus(0 x) by x and p(s(u)) by u. This does not reflect a rule of our calculus but this does not matter as we are only replacing one representative of a term by a simpler one, thus the value of the term remains equal under any interpretation.

If there is a literal $L = fs^* \equiv ft^*$ in P where f is a constructor, then all literals $L_i = s_i \equiv t_i$ have to be satisfiable (with the same substitution). Satisfiability of $P' = \{L_i \mid \forall\ i = 1\ldots n\}$ is necessary and sufficient for the satisfiability of P. Using the functionality rule splits the condition "L satisfiable" into several simpler but equivalent conditions "L_i satisfiable".

Another case of explicit information are literals $L = v \equiv t$ where the variable v does not occur in t. Here the unification rule (h) can be applied using the substitution $\tau = \{v / t\}$. If P is satisfiable with σ, i.e. we are not yet in a dead end, $\sigma(v) \equiv \sigma(t) \in Th(S)$ holds. For all variables $x \neq v$ we have $\sigma(\tau(x)) = \sigma(x)$ and for v we have $\sigma(\tau(v)) = \sigma(t) \equiv \sigma(v) \in Th(S)$. Thus $\sigma(\tau(q)) \equiv \sigma(q) \in Th(S)$ holds for all $q \in \mathcal{T}(\Sigma, \mathcal{V})$ and that is why applying the unification rule to literals $L=v\equiv t$ never leads into dead ends under the conditions mentioned above.

In the general case $L = q \equiv r \in P$ where q, r are not variables, applying the unification rule possibly leads into dead ends, i.e. there is a mgu τ of q and r such that $\tau(P)$ is not satisfiable although P is: let P = {plus(x y) ≡ plus(u v), u ≡ s(x)}. $\tau = \{x / u, y / v\}$ unifies plus(x y) and plus(u v), but τ(P) = {plus(u v) ≡ plus(u v), u ≡ s(u)} is not satisfiable.

Alternately applying functionality and unification rule propagates variable bindings which necessarily hold to other literals of the problem. This can reduce the search space drastically, and as applying these two rules - under the limitations stated above - will never lead into an dead end, it is advantageous to apply these rules whenever possible.
Example: let P = {add(x append(m n)) ≡ add(0 n), x ≡ s(y)}. Using functionality yields {x ≡ 0, append(m n) ≡ n, x ≡ s(y)}, applying unification to x ≡ 0 yields {append(m n) ≡ n, 0 ≡ s(y)} which is trivially unsatisfiable. Thus we could prove unsatisfiability without any search.

Principle 4: select literals to which only one rule can be applied
Now we inspect all literals of the actual problem $P = P' \cup \{L\}$ how many successors they have, i.e. which rules are applicable and do not lead to trivial unsatisfiable successor problems. If there is a literal which allows only one rule application, selecting this literal will not lead into a dead end as we did not make a real decision. This principle does not affect possible solutions but reduces the search space.

These four principles are applied whenever possible, not necessarily in the order listed here. Afterwards all possibilities to delay decisions are exhausted and a literal has to be selected without appealing to postulated principles.

Resolving the OR-indeterminism is much harder since strategies determining which rule to apply depend on the selected literal. As statements about most advantageous rule applications are only possible based on concrete literals we only formulate a very general principle:

Principle 5: use "simple" rules prior to "complex" rules
Obviously rules generating "simple" problems should be preferred to rules generating more complex problems. E.g. using the paramodulation rule a rule is simple if it is based on a

non-recursive definition formula, complex else.

Example: using the first definition formula [$\forall$ x, y:number $x \equiv 0 \rightarrow plus(x\ y) \equiv y$] of plus results in general in simpler successor nodes than using the second definition formula [$\forall$ x, y:number $x \equiv s(p(x)) \rightarrow plus(x\ y) \equiv s(plus(p(x)\ y))$].

For the same reason a rule which succeeds in equating the leading symbols of a literal $L = fs^* \equiv gt^*$ will be preferred to a rule which does not.

4.3 Termination of the Method

The method described in this section will not necessarily terminate, since we can not do better than a semi-decision procedure. Otherwise the class of universal formulas of arithmetic would be decidable.

For this reason we have to equip the method with termination conditions. This can be done by limiting the depth of the OR-tree or the length of deductions. Another possibility is to limit the application of certain rules, which stops indefinite deductions caused by recursively defined symbols (viz. section 5).

As we do not know the actual length of a deduction $(P, \varepsilon) \vdash (\emptyset, \sigma)$, these limitations could cut off branches of the search tree which contain solutions, so the method would not be complete. This can be overcome by iterative deepening: first limit the length of deductions to N steps, then - if not successful - to 2N and so on, ... This again will result in a complete method, but termination can not be guaranteed.

5 Extensions of the Method

This section is devoted to extensions of the method which extend it beyond the usage as a fast checker for counterexamples.

5.1 Incorporating Inductive Reasoning

In some non-terminating deductions using the paramodulation rule certain literals respectively instances of them appear again and again.

Let us look at some examples first. We will restrict ourselves to problems containing only one literal. The examples will use the algorithms diff and power2 which are defined by

$DEF_{diff} = \{$ $\forall n, m: n \equiv 0 \rightarrow diff(n\ m) \equiv 0$, $\forall n, m: \neg n \equiv 0 \wedge m \equiv 0 \rightarrow diff(n\ m) \equiv n$, $\forall n, m: n \equiv s(p(n)) \wedge m \equiv s(p(m)) \rightarrow diff(n\ m) \equiv diff(p(n)\ p(m))\}$ resp.

$DEF_{power2} = \{$ $\forall n: n \equiv 0 \rightarrow power2(n) \equiv s(0)$, $\forall n: n \equiv s(p(n)) \rightarrow power2(n) \equiv double(power2(p(n)))\}$.

Example 1: Let $L_1 = \neg\, diff(x\ x) \equiv 0$. Using the three instances of paramodulation rule corresponding to the three definition formulas of diff we obtain the following successors:

A. $L_1 \vdash \{x \equiv 0, \neg x \equiv 0\}$

B. $L_1 \vdash \{x \equiv 0, x \equiv s(p(x)), \neg 0 \equiv 0\}$

C. $L_1 \vdash \{x \equiv s(p(x)), x \equiv s(p(x)), \neg\, diff(p(x)\ p(x)) \equiv 0\}$

The first two alternatives are not satisfiable, the third allows application of the paramodulation rule for diff again yielding another three successors. Again both the first two alternatives are not satisfiable whereas the third contains the literal $L_1'' = \neg\, diff(p(p(x)))\ p(p(x)) \equiv 0$.

L_1' as well as L_1'' are instances of L_1. Obviously each problem deduced from L_1 will either not be satisfiable or will contain an instance of L_1.

Example 2: Let L_3 = power2(n) ≡ 0. Using paramodulation on power2 we obtain:

A. $L_3 \vdash$ {n ≡ 0, s(0) ≡ 0}

B. $L_3 \vdash$ {n ≡ s(p(n)), double(power2(p(n))) ≡ 0} = P

Successor A. is trivially unsatisfiable. Applying paramodulation on double to the second literal of B. yields two successors:

B_1. P $\vdash$ {n ≡ s(p(n)), power2(p(n)) ≡ 0, 0 ≡0}

B_2. P $\vdash$ {n ≡ s(p(n)), power2(p(n))≡s(p(power2(p(n)))), s(s(double(p(power2(p(n))))))≡0}.

This time B_2 is unsatisfiable, and B_1 contains an instance of L_3.

The assumption that all literals which have instances of themselves in deduced problems are not satisfiable, does not hold true as the following example shows: let L_4 = diff(x x) ≡ 0. Again paramodulation yields three successors:

A. $L_4 \vdash$ {x ≡ 0, x ≡ 0}

B. $L_4 \vdash$ {x ≡ 0, x ≡ s(p(x)), 0 ≡ 0}

C. $L_4 \vdash$ {x ≡ s(p(x)), x ≡ s(p(x)), diff(p(x) p(x)) ≡ 0}

Now alternative C. does contain an instance of L_4, but this time A. is satisfiable and so is L_4. Obviously any substitution will satisfy L_4.

The following observations can be made: here successors corresponding to recursive definition formulas lead to problems containing an instance of the initial literal. In those cases where successors corresponding to non-recursive definition formulas are not satisfiable, the negation of the formula corresponding to the initial literal is an inductive consequence of the axioms, i.e. can only be proven using inductive reasoning. Thus it seems reasonable to integrate techniques into the method which allow inductive reasoning. Clearly this should be done using information obtained by other modules of the prover. We do not follow this approach here as it contradicts the application we intended, i.e. as a fast checker for counterexamples.

5.2 Allowing Universal Quantifications

We can weaken the restriction of the method to formulas of the form ∃... φ such that it is able to handle formulas of the form ∃...∀... φ. This done by inventing a new type of object in the calculus: variable-constants which will be denoted by capital letters. Thus an equation x ≡ Y will stand for the formula ∃x ∀y x ≡ y. For these new objects the calculus has to be adapted:

Literals of the form L = X ≡ X and L = ¬ X ≡ ct* (where c is a constructor symbol and X is a strict subterm of ct*) can be deleted, whereas literals of the form L = ¬ X ≡ X, L = Δ X ≡ Y, L = X ≡ ct* or L = ¬ X ≡ ct* (where X is not a subterm of ct*) are unsatisfiable. The calculus is not complete anymore as for literals L = ¬ X ≡ ct* with X subterm but not strict subterm of ct* in general neither satisfiability nor unsatisfiability can be shown. For details refer to [Protzen 91].

6 Demonstration of the Method

We will now demonstrate the method by a more complex example. We will underline selected literals and indicate rules and instantiations used. There foo(n) indicates the application of the paramodulation rule corresponding to the nth definition formula of foo. Sometimes we condense several steps in the calculus into one.

Example: The subsequent conjecture is synthesized by the termination procedure (viz. section 1)

$$\forall\ x,y{:}number\ \neg\, x{\equiv}y \wedge \neg\, x{\equiv}0 \wedge \neg\, y{\equiv}0 \rightarrow (\neg\, lt(x\ y) \wedge \neg\, y{\equiv}0) \vee (\neg\, x{\equiv}0 \wedge y{\equiv}0).$$

(where lt denotes the "less than"-relation), which is simplified to

$$\forall\ x,y{:}number\quad x \equiv y \vee x \equiv 0 \vee y \equiv 0 \vee \neg\, lt(x\ y)\ .$$

This can uniformly be transformed to the problem P =

$\{\neg\, x \equiv y, \underline{\neg\, x \equiv 0, \neg\, y \equiv 0}, lt(x\ y)\}$

instantiate by $\{x / s(u), y / s(v)\}$ $\{\underline{\neg\, s(u) \equiv s(v), \neg\, s(u) \equiv 0, \neg\, s(v) \equiv 0, lt(s(u)\ s(v))}\}$

functionality + delete + evaluate $\{\neg\, u \equiv v, \underline{lt(u\ v)}\}$

lt(1) $\{\neg\, u \equiv v, \underline{u \equiv 0}, \neg\, v \equiv 0\}$

unify $\{u / 0\}$ $\{\neg\, 0 \equiv v, \neg\, v \equiv 0\}$

instantiate $\{v / s(z)\}$ $\{\underline{\neg\, 0 \equiv s(z)}\}$

delete $\varnothing$.

Thus $\sigma = \{x / s(0), y / s(s(z))\}$ satisfies P.

Example: An conjecture obtained during recursion elimination [Walther 88] is

$$\forall\ x, y{:}\ number\quad y \equiv s(p(y)) \wedge p(y) \equiv s(p(p(y))) \wedge \neg\, remainder(x\ y) \equiv 0$$
$$\rightarrow \neg\, (p(y) \equiv s(p(p(y))) \wedge p(p(y)) \equiv s(p(p(p(y)))) \wedge$$
$$remainder(x\ p(y)) \equiv 0 \wedge gt(x\ p(y)))$$

which uniformly can be transformed to

$\{\underline{y \equiv s(p(y))}, \underline{p(y) \equiv s(p(p(y)))}, \neg\, remainder(x\ y) \equiv 0,$
$\underline{p(p(y)) \equiv s(p(p(p(y))))}, remainder(x\ p(y)) \equiv 0, gt(x\ p(y))\}$

unify by $\{y / s(u), u / s(v), v / s(w)\}$ + evaluate + delete

(*) $\{\neg\, remainder(x\ s(s(s(w)))){\equiv}0, remainder(x\ s(s(w))){\equiv}0, gt(x\ s(s(w)))\}$

remainder[3](2) (as remainder(1) immediately leads to an unsatisfiable problem)

$\{\underline{\neg\, s(s(s(w))) \equiv 0}, gt(s(s(s(w)))\ x), \underline{\neg\, x \equiv 0}, remainder(x\ s(s(w))) \equiv 0,$
$gt(x\ s(s(w)))\}$

unify $\{x / s(z)\}$ + evaluate + delete

$\{gt(s(s(w))\ z), remainder(s(z)\ s(s(w))) \equiv 0, gt(z\ s(w))\}$

This problem again is unsatisfiable which will be detected either by a loop detecting facility or by a criterion suggested in section 5. After having backtracked we continue with problem (*), this time using remainder(3).

$\{\underline{\neg\, gt(s(s(s(w)))\ x)}, \underline{\neg\, s(s(s(w))) \equiv 0},$
$\neg\, remainder(diff(x\ s(s(s(w))))\ s(s(s(w))) \equiv 0,$
$remainder(x\ s(s(w))) \equiv 0, gt(x\ s(s(w)))\}$

three applications of gt(3) and of the instantiation rule (resulting in $\{x / s(s(s(z)))\}$) and evaluation yield

$\{\underline{\neg\, gt(w\ z)}, \neg\, remainder(diff(z\ w)\ s(s(s(w)))) \equiv 0,$
$remainder(s(s(s(z)))\ s(s(w))) \equiv 0, gt(s(z)\ w)\}$

gt(1) $\{\underline{w = 0}, \neg\, remainder(diff(z\ w)\ s(s(s(w)))) \equiv 0,$
$remainder(s(s(s(z)))\ s(s(w))) \equiv 0, gt(s(z)\ w)\}$

unify $\{w/0\}$ $\{\neg\, remainder(\underline{diff(z\ 0)}\ s(s(s(0)))) \equiv 0, remainder(s(s(s(z)))\ s(s(0))) \equiv 0,$
$\underline{gt(s(z)\ 0)}\}$

evaluate $\{\underline{\neg\, remainder(z\ s(s(s(0)))) \equiv 0}, remainder(s(s(s(z)))\ s(s(0))) \equiv 0\}$

remainder(2) $\{gt(s(s(s(0)))\ z), \underline{\neg\, z \equiv 0}, remainder(s(s(s(z)))\ s(s(0))) \equiv 0\}$

[3]remainder is defined by $DEF_{remainder}{=}\{\forall\ x,y{:}\ y{=}0 \rightarrow remainder(x\ y){=}0,\ \forall\ x,y{:}\ gt(y\ x) \rightarrow remainder(x\ y){=}x,$ $\forall\ x,y{:}\ \neg\, y{=}0 \wedge \neg\, gt(y\ x) \rightarrow remainder(x\ y){=}\ remainder\ (diff(x\ y)\ y)\}$

instantiate {z/s(n)} {gt(s(s(s(0))) s(n)), ¬ s(n) ≡ 0, remainder(s(s(s(s(n)))) s(s(0))) ≡ 0}
evaluate, delete {gt(s(s(0)) n), remainder(s(s(s(s(n)))) s(s(0))) ≡ 0}
gt(2) {¬ s(s(0)) ≡ 0, n ≡ 0, remainder(s(s(s(s(n)))) s(s(0))) ≡ 0}
unify {n/0}, delete {remainder(s(s(s(s(0)))) s(s(0))) ≡ 0}
evaluate {0 ≡ 0}
delete ∅.

Thus we have the instantiations {y/s(u), u/s(v), v/s(w), w/0, x/s(s(s(z))), z/s(n), n/0} and so P is satisfiable with σ = {x / s(s(s(s(0)))), y / s(s(s(0)))}.

7 Conclusion

The method has been integrated into the INKA-system [Biundo et al. 86]. Experience with the method has shown that it meets our expectations. This is in particular true for conjectures synthesized by the prover itself, viz. recursion elimination, termination hypotheses [Walther 88], generalization hypotheses [Hummel 90], containment formulas [Walther 91].

The method is less successful if the input problem is not satisfiable as the criteria suggested in section 5 are not yet integrated. In these cases the method only terminates because search depth is limited.

Further developments will have to concentrate on the following topics:
- invention of the criteria of section 5 (detect unsatisfiable problems).
- improvement of the heuristics.
- improvement of the backtrack strategy ("intelligent backtracking"), e.g. store problems which have been shown unsatisfiable if it is highly probable that they will repeatedly occur.
- usage of lemmata for the expansion of the tree and simplification of problems.

Acknowledgements

I thank Christoph Walther who convinced me to write this paper and Dieter Hutter for many helpful discussions during the preparation of this work.

Literature

R. Aubin, Mechanizing Structural Induction, Ph. D. Thesis, University of Edinburgh, 1976

S. Biundo, B. Hummel, D. Hutter, C. Walther, The Karlsruhe Induction Theorem Proving System, Proc. CADE-8, Oxford, Springer LNCS, vol. 230, 1986

S. Biundo, Automatische Synthese rekursiver Algorithmen als Beweisverfahren, Dissertation, Universität Karlsruhe, 1990

R. S. Boyer & J S. Moore, A Computational Logic, Academic Press, 1979

S. Hölldobler, Foundations of Equational Logic Programming, Springer LNCS, vol. 353, 1989

J. M. Hullot, Canonical Forms and Unification, Proc. CADE-5, Les Arcs, 1980

B. Hummel, Generierung von Induktionsformeln und Generalisierung beim automatischen Beweisen mit vollständiger Induktion, Dissertation, Universität Karlsruhe, 1990

N. J. Nilsson, Principles of Artificial Intelligence, Tioga Pub. Comp., 1980

M. Protzen, Disproving Conjectures, Internal Report, TH Darmstadt, 1991

Chr. Walther, Argument-Bounded Algorithms as a Basis for Automated Termination Proofs, Proc. CADE-9, Argonne 1988

Chr. Walther, Computing Induction Axioms, Internal Report, TH Darmstadt, 1991

An Interval-based Temporal Logic in a Multivalued Setting

Mathias Bauer
German Research Center for Artificial Intelligence (DFKI)
Stuhlsatzenhausweg 3,
W – 6600 Saarbrücken, FRG
bauer@dfki.uni-sb.de

Abstract

We describe the embedding of the semantic notions and modal operators of a first-order temporal logic based on time intervals in a multivalued setting. Truth values will be realized as functions from time intervals to "ordinary" truth values like t and f. The main emphasis lies on the realization of the various modal operators contained in the temporal logic as operations on the functional truth values. We show that it is possible to obtain an efficient system sufficient for tasks in the area of diagnostic reasoning.

1 Introduction

Reasoning about a changing world requires mechanisms going beyond the scope of classical predicate logic. Thus, several temporal logics have been proposed to solve problems in areas like hardware and software verification (e.g., [13]), planning ([2]), reasoning about actions (e.g., [9], [15]), and plan recognition (e.g., [4]).

Among these logics those equipped with operators supporting a compositional approach (e.g., [14], [10]) gained particular importance because they allow some kind of modular reasoning in which formulas can be combined to statements about temporally more complex situations. The operator enabling this kind of reasoning is often referred to as *chop* ($\mathcal{C}$). The semantics of logics containing it is usually based on the notion of *intervals* as sequences of states in contrast to logics like *tense logic* (cf. [5]) based on time *points* and Allen's temporal logic with intervals that are not built up from single states (cf. [1]).

Unfortunately, decision procedures for propositional temporal logics as described in [14] are non-elementary, i.e., of exponential height in the nesting depth of the chop operator. If the demand for completeness is relaxed, however, it is possible to implement inference systems of clearly smaller complexity for tasks in the field of diagnostic reasoning (e.g., [9], [4]).

The system MVL described in [6] suggests itself as a basis for such an implementation as it provides a proof system to which modal operators like *chop* can be added (cf. [8]). The crucial idea of MVL (which stands for "Multivalued Logics") is to keep the inference machine and the "bookkeeping" about truth values separated from each other. Thus, exchanging the set of possible truth values while retaining the prover results in a system for a totally different logic.

The aim of this paper is to describe the theoretical foundations of an implementation of FTL, a first-order version of the temporal logic introduced in [14], in the MVL setting and its realization, possible applications, and limitations. The reasons for embedding this logic in MVL are twofold: On the one hand, we will see how MVL generalizes the usual concepts of Kripke-style modal operators and sets of truth values underlying a certain logic (cf. section 3.3). On the other hand, this enables us to give an efficient implementation of a restricted inference machine for FTL (cf. section 3.4).

Sections 2 and 3.1 will introduce syntax and semantics of our temporal logic and the foundations of the MVL system, resp. In sections 3.2 through 3.4, we will describe how to embed our interval-based logic and its modal operators in MVL. As truth values we will use functions from time intervals to "ordinary" values including t and f and demonstrate how complex modal operators can be realized as operations on these functions. We will also address the problem of how to represent the set of intervals and the truth functions so that an effective computation is possible. In section 3.5, the computational overhead caused by the modal operators is shown to play a minor role concerning the complexity of the whole system. Finally, we will consider limitations and possible applications of the resulting system.

2 The Temporal Logic FTL

The temporal logic FTL ("First-order Temporal Logic") presented in this section essentially corresponds to the extension of the system PTL(U,X,C) as described in [14] to first-order logic.

2.1 Syntax

Given a denumerable set X of variables and a signature S, the set of formulas of FTL comprises T, F, and the usual set of first-order formulas with quantifiers $\forall$ and $\exists$ and the connectives $\neg, \wedge, \vee, \rightarrow$, and $\leftrightarrow$ over S and X. Besides, it contains all formulas of the form $\bigcirc p$ (*"next"*), $(p\ \mathcal{U}\ q)$ (*"until"*), and $(p\ \mathcal{C}\ q)$ (*"chop"*). By Φ_0 we denote the set of atomic first-order formulas. The versions of *next* and *chop* presented here are often referred to as "strong next" and "strong chop".

Before presenting the formal semantics of FTL, we give an intuitive description of the meaning of the modal formulas introduced above: We want to consider a formula $\bigcirc p$ true in an interval σ if p is true if we consider the situation one state later, i.e., if p is true in the interval obtained from σ by removing its first state.

We say $(p\ \mathcal{U}\ q)$ holds in an interval σ if q holds sometime within σ and p holds *all of the time before* within σ.

The chop operator $\mathcal{C}$ provides a possibility to *compose* two formulas p and q by concatenating the intervals in which they hold. Considered differently, chop allows to *split up* an interval σ in which $(p\ \mathcal{C}\ q)$ holds into two subintervals σ_1 and σ_2 where p and q hold, resp.

2.2 Semantics of FTL

Definition 2.1 Any subset of Φ_0 is called a *state*. Let Σ be the power set 2^{Φ_0} of Φ_0. Then the elements of $\mathcal{I} = \Sigma^+ \cup \Sigma^\omega$ are called *intervals*.

The idea behind this definition is that a state contains just those atomic formulas true at a certain moment in time. Intervals are non-empty sequences of states and thus model the truth or falsity of formulas over time.
We need some operations on intervals:

Definition 2.2 Let $\sigma, \sigma_1, \sigma_2$ be intervals. Then the *length* of σ is defined by

$$|\sigma| = \begin{cases} \omega, & \text{if } \sigma \text{ is infinite} \\ n, & \text{if } \sigma = \langle S_0, ..., S_n \rangle \end{cases}$$

The *composition* of σ_1 and σ_2 is

$$\sigma_1 \oplus \sigma_2 = \begin{cases} \sigma_1, & \text{if } |\sigma_1| = \omega \\ \langle S_0, ..., S_n, S_{n+1}, ... \rangle, & \text{if } \sigma_1 = \langle S_0, ..., S_n \rangle \text{ and } \sigma_2 = \langle S_n, S_{n+1}, ... \rangle \end{cases}$$

The *nth suffix* of $\sigma = \langle S_0, ... \rangle$ is $\sigma^{(n)} = \langle S_n, ... \rangle$.

Relating FTL to classical modal logics with Kripke-style semantics, we can regard intervals as possible worlds. The accessibility relation between worlds can be described by terminating subintervals defined below:

Definition 2.3 Let σ_1, σ_2 be intervals. Then we define $R_t \subseteq \mathcal{I} \times \mathcal{I}$ by

$$\sigma_1 \; R_t \; \sigma_2 \; :\Longleftrightarrow \; \sigma_1 = \sigma_2^{(1)},$$

and call the first suffix σ_1 of σ_2 *the first terminating subinterval* of σ_2.
We denote the transitive closure of R_t by R_t^+ and the reflexive, transitive closure of R_t by $\overline{R}_t$.
If $\sigma_1 \overline{R}_t \sigma_2$, then there is a unique interval σ' such that $\sigma_2 = \sigma' \oplus \sigma_1$. This interval is the *complement* of σ_1 in σ_2 and is denoted by $compl_{\sigma_2}(\sigma_1)$.

So, a world σ_1 is accessible from σ if σ_1 is a terminating subinterval of σ. Now we are ready to define the notion of satisfiability.

Definition 2.4 Let D denote the non-empty set over which we interpret the logical variables of X. Then $\rho : X \to D$ is called an *assignment*. Let $\sigma = \langle S_0, ... \rangle \in \mathcal{I}$ be an interval, $x \in X$ a variable, $P \in \Phi_0$ an atomic first-order formula, p, q FTL formulas. Then we have

$$\begin{array}{lll}
\sigma \models T & & \\
\sigma \not\models F & & \\
\sigma \models P & \Longleftrightarrow & P \in S_0 \\
\sigma \models \neg p & \Longleftrightarrow & \sigma \not\models p \\
\sigma \models p \wedge q & \Longleftrightarrow & \sigma \models p \text{ and } \sigma \models q \\
\sigma \models \exists x.p & \Longleftrightarrow & \text{there is an assignment } \rho \text{ such that } \sigma \models p^x_{\rho(x)} \\
\sigma \models \bigcirc p & \Longleftrightarrow & \sigma^{(1)} \models p \\
\sigma \models (p \; \mathcal{U} \; q) & \Longleftrightarrow & \text{there is } \sigma'', \sigma'' \overline{R}_t \sigma, \text{ such that } \sigma'' \models q \text{ and for all } \sigma' : \\
 & & \text{if } \sigma' \overline{R}_t \sigma \text{ and } \sigma'' R_t^+ \sigma', \text{ then } \sigma' \models p \\
\sigma \models (p \; \mathcal{C} \; q) & \Longleftrightarrow & \text{there are } \sigma', \sigma'' \text{ such that } |\sigma'| < \omega, \sigma' \models p, \sigma'' \models q, \\
 & & \text{and } \sigma = \sigma' \oplus \sigma''
\end{array}$$

Concerning the other connectives and the universal quantifier, we use the usual recursive definitions.

It should be noted that the truth value of a non-modal formula only depends on the first state of an interval.

On the basis of the operators defined so far, we can derive other modalities useful in temporal reasoning. Examples are

$$\Diamond p \;:\equiv\; (T\,\mathcal{U}\,p)$$

and its dual

$$\Box p \;:\equiv\; \neg\Diamond\neg p\,.$$

These operators are called *sometimes* and *always*. In "classical" modal logics, they correspond to the modalities *possibly* and *necessarily*. Introducing the abbreviation

$$empty \;:\equiv\; \neg\bigcirc T$$

to denote the end of an interval, we can derive the so-called *weak* versions of chop and next:

$$\begin{aligned}(p\;\mathcal{C}\;q) \;&:\equiv\; ((p\;\mathcal{C}\;q)\vee(p\wedge\Box\neg empty\,))\\ \odot p \;&:\equiv\; empty\vee\bigcirc p\,.\end{aligned}$$

3 Review of MVL and the Embedding of FTL

The MVL system by Ginsberg is an attempt to capture many sorts of reasoning within the field of artificial intelligence in a uniform framework (cf. [6]). The basic idea is to split up inference into two parts: One in which the actual process of reasoning takes place – realized by a theorem prover – and one in which some kind of "bookkeeping" of the results obtained from the inference machine is done.

As an example, one might imagine a system for probabilistic reasoning where the bookkeeping consists of combining the numerical values assigned to the formulas used and pruning formulas whose probability is below a certain threshold. Other examples given by Ginsberg are ATMS and default reasoning systems.

Ginsberg formalizes the bookkeeping part of reasoning by attaching two kinds of labels to each bit of information: one describing the amount of knowledge available about a certain statement and one indicating the degree of certainty about its validity. On the basis of these labels, sets of truth values can be given the internal structure of a so-called *bilattice*, which is advantageous with regard to several aspects:

1. **Modularity.** It is possible to develop theorem provers suitable for many different object logics independent of the actual choice of underlying truth values, since the definition of a bilattice forms a unique interface to the bookkeeping part of the inference machine. Selecting a new set of possible truth values yields a reasoner for a totally different logic although the original prover is further used (cf. [7] and [8]).

2. **Efficiency.** As we will see, it is possible to exploit the additional information represented in the bilattice structure during the inference process to render it more efficient.

3. **Modal operators.** It is easy to introduce new modal operators into a given logic, as they can essentially be expressed using primitive operations on the elements of a bilattice. Besides forming the basis for efficient implementations, this is also interesting from a theoretical point of view, as this approach generalizes both the classical concept of Kripke-style modal operators and Moore's autoepistemic operator L (cf. [12]). Thus, we are able to introduce modal operators of arity > 1 (in fact, the FTL operators $\mathcal{U}$ and $\mathcal{C}$ exceed Kripke's approach) and to compare different modal logics within a single uniform framework.

In section 3.1, we will describe the formal basis for the truth values to be chosen and the way in which the closure of a certain set of propositions is computed using this basis. Section 3.2 shows how functions can serve as truth values in this sense and applies these results to FTL. In sections 3.3 and 3.4, we consider the MVL concept of modal operators as described in [8] and its application to our temporal logic before we finally give some complexity results in 3.5.

3.1 Mathematical Preliminaries of MVL

The fundamental notion in connection with MVL truth values is that of a *bilattice* defined below:[1]

Definition 3.1 ([6]) A *bilattice* is a sextuple $(B, \wedge, \vee, \cdot, +, \neg)$ such that

1. $(B, \wedge, \vee)$ and $(B, \cdot, +)$ are both complete lattices.
2. $\neg : B \rightarrow B$ is a mapping with
   (a) $\neg^2 = 1$, and
   (b) $\neg$ is a lattice homomorphism from $(B, \wedge, \vee)$ to $(B, \vee, \wedge)$ and from $(B, \cdot, +)$ to itself.

If the operations $\wedge, \vee, +$, and $\cdot$ distribute with respect to each other, the bilattice is called *distributive.* If only $\wedge, \vee$ and $\cdot, +$ each distribute with respect to each other, it is called *t-distributive* and *k-distributive*, resp.

The elements of a bilattice can be considered as truth values if its operations are interpreted in the following way: The two pairs of operations $\wedge, \vee$ and $\cdot, +$ each induce a partial order on the elements of B, denoted by $\leq_t$ and $\leq_k$, resp. If x and y are elements of B with $x \leq_t y$, we interpret this by saying that y represents a truth value that is "nearer to truth" than the one represented by x. In other words, a formula assigned the truth value y is considered "more true" than one assigned x. An example

[1]The presentation of the mathematical foundations of MVL in this paper has to concentrate on the most essential topics. For an elaborated description, the reader is referred to [6].

for this ordering is $f \leq_t t$. Thus, $\leq_t$ represents the degree of certainty about the validity of a certain statement.

If on the other hand $x \leq_k y$, we say that y stands for a greater amount of knowledge about a certain fact than x. If we allow for some truth value u (*unknown*), we have $u \leq_k f$ and $u \leq_k t$, whereas t and f are incomparable with respect to $\leq_k$.

Completing the truth values used so far with another element denoted by $\perp$ which stands for "both t and f", we obtain the smallest non-trivial bilattice F representing the set of truth values used in first-order logic in MVL. Figure 1 shows this bilattice where $\leq_t$ increases from left to right and $\leq_k$ from bottom to top. The role of the $\neg$ operation is to invert the $\leq_t$ order while retaining $\leq_k$. We now relate the bilattice operations $\wedge, \vee$, and $\neg$ to the interpretation of the elements of the bilattice as truth values. As the notation indicates, there is a strong similarity between these operations in a bilattice and their syntactical counterparts within logic. Considering the bilattice F for example, we get $t \wedge f = f$, $t \vee f = t$, $\neg t = f$ etc., just as expected from traditional logic.[2]

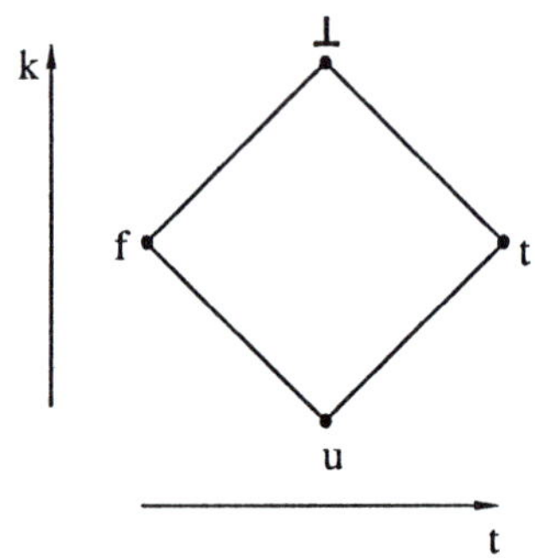

Figure 1: The bilattice F for first-order logic.

Definition 3.2 ([6]) Given some logical language $\mathcal{L}$ and a bilattice B, a *truth assignment* is a mapping $\phi : \mathcal{L} \rightarrow B$.

In MVL, such a mapping corresponds to a declarative database. In conventional logics, inferencing a formula p from a set S of axioms consists in checking if p is a member of the deductive closure of S. The multivalued counterpart for this process is to compute the truth value of p in the *closure* $cl(\phi)$ of a truth assignment ϕ.

Definition 3.3 ([6]) A truth assignment ϕ is *closed* if for all $p, q, p_1, p_2, \ldots \in \mathcal{L}$

1. $\phi(\bigwedge_i p_i) \geq_k \bigwedge_i \phi(p_i)$,
2. $\phi(\neg p) = \neg\phi(p)$, and
3. if $p \models q$, then $\phi(q) \geq_t \phi(p)$.

[2] $\wedge$ yields the greatest lower bound glb_t of its arguments w.r.t. $\leq_t$ and $\vee$ the least upper bound lub_t.

This means that – as already mentioned – the behaviour of the bilattice operation $\neg$ perfectly corresponds to negation in logic (2.), and that a formula q can't be "less true" than a formula p entailing it (3.). The content of clause 1. becomes clear when considering a formula $p \wedge \neg p$ where both conjuncts are assigned u but their conjunction obviously should be assigned f.

According to the "classical" approach to logic where the closure of some set T of sentences is defined to be the intersection of all deductively closed sets containing T, we define the closure $cl(\phi)$ of a truth assignment ϕ by:

Definition 3.4 ([6]) $cl(\phi) = \prod \{\psi \mid \psi \geq_k \phi$ and ψ is closed$\}$, where ϕ and ψ are compared pointwise.

To give a *constructive* account of closure, we will restrict ourselves here to the case of so-called *canonically grounded* bilattices where each element x can be expressed uniquely by a sum $x = g_t(x) + g_f(x)$, the *t-grounding* and *f-grounding* of x, resp. They correspond to the primitive bits of information x is composed of.[3] The bilattice F and the one used for FTL are both canonically grounded.

Now let p and q be sentences of our logical language such that $p \models q$ and $\phi(p) = x$. According to definition 3.3, we have $cl(\phi)(q) \geq_t x$ and thus $cl(\phi)(q) \geq_k g_t(x)$ (since $x = g_t(x) + g_f(x) \geq_k g_t(x)$). So the contribution of p to q's truth value is $g_t(\phi(p))$. If there are many sentences entailing q, this knowledge has to be accumulated by summing over all t-groundings.

For a set $S \subseteq \mathcal{L}$ of sentences we define $\phi(S) := \bigwedge_{p \in S} \phi(p)$. If $p \in \mathcal{L}$, we denote by $\pi_+(p)$ and $\pi_-(p)$ the sets of all subsets of $\mathcal{L}$ entailing p and $\neg p$, resp.:

$$\pi_+(p) = \{S \mid S \models p\}, \quad \pi_-(p) = \{S \mid S \models \neg p\}.$$

Now assume we are given a truth assignment ϕ with $\phi(\neg p) = \neg\phi(p)$.[4] Then we have

Theorem 3.1 ([6]) The closure of ϕ is given by

$$cl(\phi)(p) = \sum_{S \in \pi_+(p)} [\phi(S) \vee u] + \sum_{S \in \pi_-(p)} [\neg\phi(S) \wedge u]$$

$$= \sum_{S \in \pi_+(p)} g_t[\phi(S)] + \sum_{S \in \pi_-(p)} \neg g_t[\phi(S)]$$

This result implies a method to effectively compute the closure by steadily pruning formulas from the search space whose truth value is $<_k$ than the truth value already accumulated during the previous steps of the proof, since they can't make a real contribution to $cl(\phi)(p)$. Furthermore, the summation over $\pi_-(p)$ may be left out if we are only interested in the *truth* of p, i.e., if we want to show $cl(\phi)(p) \geq_k t$.

[3] In the bilattices considered here, $g_t(x) = x \vee u$, $g_f(x) = x \wedge u$. E.g., $g_t(\bot) = t$, $g_f(\bot) = f$.

[4] If ϕ doesn't meet this property, it is easy to compute a corresponding $\phi_\neg$ that is $\neg$-*closed*.

3.2 Functional Truth Values

For some applications it is not sufficient to use some kind of "atomic" truth values like t and f. Instead, it might be convenient to employ *mappings* from a given set to some bilattice as truth values. Ginsberg describes this for a simple temporal logic with truth values $g : I\!N \to F$ from the set of time points to the classical truth values (cf. [8], [9]). After describing the principal concept of functional truth values, we will see in this section how this technique can be applied to FTL.

Given a set S and a bilattice B, the set B^S of functions from S to B obviously inherits the bilattice property from B if the operations $\wedge, \vee, \cdot, +, \neg$ are computed pointwise.

Taking S to be the set of time intervals represented by $I\!N^2$, and B to be the first-order bilattice F, we obtain the new bilattice of truth values of FTL, denoted by B_I. Intervals are represented as pairs of natural numbers $i = (a, l)$, where a is the first state and l is the number of states of i. Thus, the elements of B_I are *total* functions over $I\!N^2$.

Some practical problems arise with this approach: How to represent the infinitely large set $I\!N^2$ and the functions $g : I\!N^2 \to F$?

We begin with the second question and assume the general case of a bilattice B^S. If we can put some order on the set S, it is possible to make the representation of the functions g more compact by only listing those points of S explicitly where the value of g changes and assuming g to be constant between two such so-called *exception points*. As in general there will be no natural total order available for S, we organize S as a directed acyclic graph (DAG) with an induced partial order $\preceq$:

$$s_1 \preceq s_2 \ :\Longleftrightarrow \ \text{there is a path from } s_1 \text{ to } s_2 \text{ in } S.$$

To effectively represent this DAG, it is sufficient to have its root and a function computing the common successors of a given pair of points.

The required structure for the interval DAG D_I can be extracted from definition 2.4. As our time begins with state 0, we define the root of D_I to be $(0,1)$, i.e., the interval consisting only of state 0. Examining those intervals sharing some common properties yields four classes:

1. Intervals of the form $(a, 1)$ consisting of a single state. They correspond to the embedding of time *points* into intervals. Within D_I, they yield a path $(0,1) \to (1,1) \to (2,1) \to \ldots$ called the *α-branch* of D_I.

2. Intervals *beginning with the same state*, as a non-modal formula is valid in (a, l) iff it is in (a, l') for any $l' > 0$. For each $a \in I\!N$, these form a path $(a,1) \to (a,2) \to (a,3) \to \ldots$, the *$\beta_a$-branch* of D_I.

3. Intervals *ending in the same state*. They all are terminating subintervals of a common greater interval and form pathes $(0,l) \to (1,l-1) \to \ldots \to (l-1,1)$. A sequence of this form is called the *γ_l-branch* of D_I as for all $(a', l') \in \gamma_l$: $a' + l' = l$. By $\gamma_{a,l}$ we denote that part of γ_{a+l} beginning with (a, l).

4. "Empty" intervals $(a, 0)$. They are only mentioned for completeness as any formula is considered *unknown* in such an interval. In the following, we won't regard this kind of intervals anymore.

Figure 2 shows a part of the resulting graph D_I.

This structure of D_I allows to pass truth information along the edges to represent all kinds of relations between intervals described in definition 2.4. We now demonstrate how the concepts developed so far can be applied.

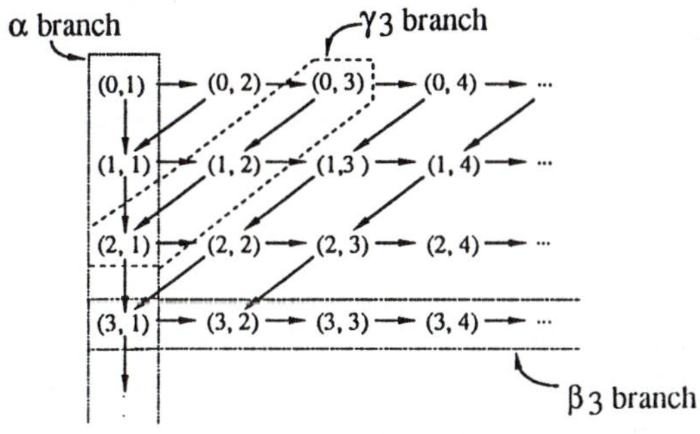

Figure 2: The DAG D_I of intervals.

Example 3.1 Let p be a first-order formula *true* in state 2 and *unknown* everywhere else. According to definition 2.4, p is true in *all* intervals beginning with state 2. This can be expressed by assigning p a truth function $g_1 : I\!N^2 \rightarrow F$ with

$$g_1((a,l)) = \begin{cases} t, & a = 2, l \geq 1 \\ u, & otherwise \end{cases}$$

Listing g_1's exception points and their respective truth values yields $\langle (0,1)-u,\ (2,1)-t,\ (3,1)-u \rangle$. In this case, we use inheritance along the β_2 branch of D_I and the fact that the *interval* $(2,1)$ of the α branch corresponds to *state* 2.

If we want to express that an FTL formula q is true in *all* states of an interval, e.g., $(5,3)$, we use a truth function g_2 with

$$g_2(w) = \begin{cases} t, & w \in \{(5,3),(6,2),(7,1)\} \\ u, & otherwise \end{cases}$$

The exception point list of g_2 is $\langle (0,1)-u,\ (5,3)-t,\ (5,4)-u \rangle$, where t is passed along the $\gamma_{5,3}$ branch of D_I.

3.3 Realization of Modal Operators

As already mentioned in section 3.1, the bilattice operations like "$\wedge$" play two distinct roles: Besides being a function on elements of a bilattice, they occur as binary operators in our logical language. In [8], Ginsberg generalizes this view to arbitrary operations:

"...bilattice operations can be viewed *in general* as establishing semantic meanings for their syntactic counterparts. These syntactic counterparts are generally referred to as *modal operators*."

Definition 3.5 ([8]) Given a bilattice B, any n-ary function $g : B^n \rightarrow B$ is a *modal operator.*

Usually, modal operators are given a semantics using Kripke's approach of possible worlds (cf.[11]). Given an accessibility relation r among worlds and a modal operator Ω, where Ωp is intended to be valid in a world w if p holds in all worlds w' accessible from w, we can define the semantics of this operator by conceptually introducing a function Γ that takes a formula p and a world w and returns the truth value of p in w.[5] Then we have

$$\Gamma(\Omega p, w) = \bigwedge_{\substack{w' \\ r(w,w')}} \Gamma(p, w').$$

How can this be related to the results of the previous section? Taking time intervals as possible worlds and the subinterval relationship $\overline{R}_t$ as accessibility relation,[6] we get for a formula Ωp, where p is assigned a truth function $g : I\!N^2 \rightarrow F$:

$$\Omega(g)(w) = \bigwedge_{w' \, \overline{R}_t \, w} g(w') \qquad (w \in I\!N^2)$$

In this equation, Ω is a bilattice operator.
There exists a distinction between two kinds of modal operators: The so-called *non-deductive* operators, e.g., Moore's autoepistemic operator L (cf. [12]), do not respect the bilattice operations $\cdot$ and $+$. These operators are usually not given a possible-worlds semantics, but are characterized by some functional relationship between the truth values of their "input" formulas and their results. The class of *deductive* modal operators is characterized by the classical Kripke semantics and comprises operators like those of necessity and possibility. When computing the closure $cl(\phi)$ of a truth assignment ϕ, the following inequality holds for deductive operators Ω:

$$cl(\phi)(\Omega(p_1, ..., p_n)) \geq_k \Omega(cl(\phi)(p_1), ..., cl(\phi)(p_n)),$$

while there is a strict equality for the non-deductive ones.

Proposition 3.1 ([8]) Any modal operator on a distributive bilattice B that can be written in terms of $\wedge, \neg, +$, constant functions and projections is deductive.

Projections are functions $\pi_{w,w'}$ indexed by two possible worlds w, w' that – when supplied with a truth function g and a world w'' – return the value of g at w' if $w = w''$ and u otherwise; i.e., such a projection exactly represents the property of accessibility of w' from w.

[5] Ω corresponds to the classical "necessity" operator.
[6] I.E., w' is accessible from w iff $w'\overline{R}_t w$.

3.4 FTL Modal Operators

In the following, we will describe how the basic modal operators of FTL introduced in section 2 can be expressed as bilattice operations on B_I.
Let $g, g_1, g_2 \in B_I$ be truth functions $\mathbb{N}^2 \to F$, and $w, w', w'' \in \mathbb{N}^2$ pairs of natural numbers representing intervals.
Then we can express the *next* operator $\bigcirc$ by

$$next(g)(w) \;=\; g(w') \qquad (1)$$

where w' is uniquely determined by $w' \, R_t \, w$. This corresponds to the intuitive semantics of *next*: To see if $\bigcirc p$ is valid in an interval w, check p's truth value in its first terminating subinterval w'. Obviously, the result of *next* is itself an element of B_I.

With the other basic operators, things become slightly more complicated. Consider the operator *until*. Translating its semantics given in definition 2.4 into the MVL formalism yields the following: Given two truth functions $g_1, g_2 \in B_I$, we can determine the value of $until(g_1, g_2)$ at w by checking if there is some terminating subinterval w'' such that $g_2(w'') = t$ and for all subintervals w' of w *before* w'': $g_1(w') = t$. We get

$$until(g_1, g_2)(w) = \bigvee_{w''\overline{R}_t w} \Big[\bigwedge_{\substack{w'\overline{R}_t w \\ w'' R_t^+ w'}} g_1(w') \wedge g_2(w'') \Big]. \qquad (2)$$

Remark: In (2), lub_t and glb_t are guaranteed to exist even in the infinite case because – according to definition 3.1 – $(B_I, \wedge, \vee)$ is a *complete* lattice. In contrast to traditional logic, we can therefore express existential and universal propositions about worlds by disjunctions and conjunctions, resp.

For the *chop* operator, the translation is similar: Check if there is a terminating subinterval w' of w such that $g_2(w') = t$ and for the complement w'' of w' in w: $g_1(w'') = t$. Again, we get a disjunction

$$chop(g_1, g_2)(w) \;=\; \bigvee_{w'\overline{R}_t w} [g_1(compl_w(w')) \wedge g_2(w')]. \qquad (3)$$

Here, the above remark is also valid yielding the correctness of (3). As a consequence of (1), (2), (3), and proposition 3.1, we get the following result:

Corollary 3.1 The modal operators *next* ($\bigcirc$), *until* ($\mathcal{U}$), and *chop* ($\mathcal{C}$) are deductive.

From the equations listed above, an actual implementation of these operators can be easily derived. For this, notice that

1. $(a', l') \; R_t \; (a, l) \iff l > 1, a' = a + 1$, and $l' = l - 1$.
2. $(a', l') \; \overline{R}_t \; (a, l) \iff a + l = a' + l', l' > 0$, and $a' \geq a$.
3. $(a', l') \; R_t^+ \; (a, l) \iff a + l = a' + l', l' > 0$, and $a' > a$.
4. $compl_{(a,l)}((a + i, l - i)) = (a, i + 1)$.

So we get

$$next(g)((a,l)) = g((a+1, l-1)) \tag{4}$$

and

$$until(g_1,g_2)((a,l)) = \bigvee_{i=0}^{l-1} [\bigwedge_{j=0}^{i-1} g_1((a+j, l-j)) \wedge g_2((a+i, l-i))]. \tag{5}$$

The implementation of *chop* is done by

$$chop(g_1,g_2)((a,l)) = \bigvee_{i=0}^{l-1} [g_1((a,i+1)) \wedge g_2((a+i, l-i))]. \tag{6}$$

To complete this section, we will consider the derived modal operators *always* ($\Box$) and *sometimes* ($\Diamond$). As mentioned in section 2, *sometimes* can be obtained by

$$\Diamond p :\equiv (T \; \mathcal{U} \; p).$$

Denoting the truth functions assigned to T and p by g_T and g_p, resp. (i.e., $\forall w \in \mathbb{N}^2 : g_T(w) = t$), and inserting them into (2), we immediately get

$$\begin{aligned} sometimes(g_p)(w) &= until(g_T, g_p)(w) \\ &= \bigvee_{w''\overline{R}_t w} [\bigwedge_{\substack{w'\overline{R}_t w \\ w'' R_t^+ w'}} t \wedge g_p(w'')] \quad \text{(cf. def. of } g_T) \\ &= \bigvee_{w''\overline{R}_t w} g_p(w'') \end{aligned} \tag{7}$$

Taking advantage of the duality of *sometimes* and *always*, it is possible to derive its realization by

$$\begin{aligned} always(g)(w) &= \neg sometimes(\neg g)(w) \\ &= \neg \bigvee_{w'\overline{R}_t w} \neg g(w') \\ &= \bigwedge_{w'\overline{R}_t w} g(w') \end{aligned} \tag{8}$$

In general, formulas like (2) and (3) cannot be effectively computed as they may contain infinitely large disjunctions and conjunctions. The representation of the set of intervals in a DAG and of the truth functions by only listing their exception points, however, admits these computations for many cases.

Example 3.2 In example 3.1, we assigned a truth function g_2 to a formula q with the intention to express that q is true during the whole interval $(5,3)$. Applying *always* to this function yields $always(g_2) = g_2$, i.e., g_2 in fact formalized what we intended. Consider another formula r with truth function g_3 where

$$g_3(w) = \begin{cases} t, & w \in \{(7,2),(8,1)\} \\ u, & otherwise \end{cases}$$

i.e., r is true in the whole interval $(7,2)$. Then the truth value of $(q \mathcal{C} r)$ is a function $chop(g_2, g_3) = g_4$ with

$$g_4(w) = \begin{cases} t, & w \in \{(5,4),(6,3),(7,2),(8,1)\} \\ u, & otherwise \end{cases}$$

that is represented using the exception points $\langle (0,1) - u,\ (5,4) - t,\ (5,5) - u \rangle$. Applying *next* to g_3, i.e., computing the truth value of $\bigcirc r$, yields $next(g_3) = g_5$ with

$$g_5(w) = \begin{cases} t, & w \in \{(6,3),(7,2)\} \\ u, & otherwise \end{cases}$$

We have a slightly different view on modal operators than Ginsberg has. According to his understanding of modal operators, *next* should *modify* the truth value of r by pushing it one step into the future, whereas the above result is the truth function obtained from the query "In which intervals is $\bigcirc r$ true?"

3.5 Complexity Considerations

For the case of *chop*, we will describe an algorithm for (6) exploiting the sparsity of exception points of truth functions and compute its complexity.

Let $g_1, g_2 \in B_I$, $exc(g_1), exc(g_2)$ their respective sets of exception points with $|exc(g_1)| = n_1$ and $|exc(g_2)| = n_2$. As indicated in equation (6), to compute the value of $chop(g_1, g_2)$ at (a, l), we have to consider each possible splitting of (a, l) into two subintervals (a_1, l_1) and (a_2, l_2) such that $(a, l) = (a_1, l_1) \oplus (a_2, l_2)$ and combine the respective values of g_1 and g_2 using "$\wedge$". All these intermediate results are put together in one disjunction yielding the final value at (a, l). In a first step, we have to complete both sets in the following way:

1. Each $(a_2, l_2) \in exc(g_2)$ can be combined with $(a_1, l_1) \in exc(g_1)$ iff $a_1 + l_1 - 1 = a_2$, i.e., iff $(a_1, l_1) \in \gamma_{a_2+1}$. So, we need all points on γ_{a_2+1} where g_1 changes its value. Apart from the points of $\gamma_{a_2+1} \cap exc(g_1)$, these are all those elements (a_1, l_1) of $exc(g_1)$ passing their truth values into γ_{a_2+1} by inheritance. We have to collect their projections $(a_1, a_2 - a_1 + 1)$ onto γ_{a_2+1}. In the worst case, we therefore have to consider each member of $exc(g_1)$ for each of $exc(g_2)$ and obtain a complexity of $O(n_1 \cdot n_2)$.

2. In the other direction, any $(a_1, l_1) \in exc(g_1)$ is combinable with any $(a_2, l_2) \in exc(g_2)$ where $a_1 + l_1 - 1 = a_2$, i.e., with all members of β_{a_2}. Just as in 1., we have to compute the projections of all points (a_2', l_2') of $exc(g_2)$ influencing the truth values along β_{a_2}. These are the points $(a_2, a_2' + l_2' - a_2)$. The complexity of this step is again $O(n_1 \cdot n_2)$.

3. Combining the points obtained from the two previous steps according to (6) yields the same complexity again.

So, the computational overhead caused by *chop* is merely $O(n_1 \cdot n_2)$. For *until*, the process is similar and takes the same time, whereas *next* can be implemented to consume linear time.

4 Applications

One possible application for a system as the one described is in the field of plan recognition. Assume we are given some observed actions $a_1(t_1)$ and $a_2(t_2)$ with their actual parameters t_i and exact temporal information about their occurrences and some plan hypotheses P_1 and P_2 written as FTL formulas, where

$$\begin{aligned} \forall x_1, x_2, x_3 . P_1(x_1, x_2, x_3) &\equiv (a_1(x_1) \wedge \Diamond (a_2(x_2) \; C \; a_3(x_3))) \\ \forall x_1, x_2, x_3 . P_2(x_1, x_2, x_3) &\equiv (a_1(x_1) \wedge \bigcirc a_2(x_2) \wedge \bigcirc\bigcirc a_3(x_3)) \end{aligned}$$

Using ordinary deduction, we can infer which of P_1 and P_2 is *not* a valid hypothesis for an explanation of the observed action sequence. Assume $a_1(t_1)$ is observed in state 5, i.e., in the interval $(5, 1)$, and $a_2(t_2)$ in state 10. Trying to derive $\neg P_1$ from this database yields a truth function g_{P_1} with value t everywhere except for all intervals $(5, l)$ of β_5 with $l \geq 6$, where g_1 yields u. The reason for this result is the fact that any interval in which P_1 could hold has to begin with state 5 and include at least state 10 where $a_2(t_2)$ takes place. The derivation of $\neg P_2$, however, yields a constant truth function g_{P_2} with value t. The result of these inferences is that P_1 is consistent with the observations and thus a valid hypothesis for the observed action sequence – in contrast to P_2 that is not.

If later on action $a_3(t_3)$ is observed – e.g., at state 12 – we can even derive P_1 with a truth function g'_{P_1} that yields t for all $(5, l)$ with $l \geq 8$ as now each interval satisfying P_1 starts at state 5 and includes at least state 12.

In [4], a similar approach to plan recognition with a temporal modal logic is described.

5 Conclusions

We introduced a modal temporal logic FTL based on work described in [14] and the basic concepts of Ginsberg's MVL presented in [6]. The main emphasis lay on the translation of FTL into the MVL formalism, where the choice of functions $\mathbb{N}^2 \rightarrow F$ as truth values – as counterparts for its interval-based semantics – and the implementation of FTL's modal operators as functions over these truth values played a central role. We finally showed that it is even possible to give efficient implementations for these concepts by exploiting some constraints on the structure of truth values.

As expected, this efficiency is not for free. For example, truth functions changing their value infinitely often (e.g., from state to state) can't be represented using the methods described.

Another – perhaps even more serious – drawback lies in the limitation of possible inferences caused by MVL. To reason about a formula Ωp containing a modal operator Ω, p is required to have a *concrete truth value* that can serve as input for the *function* Ω. Axiom schemata like $\bigcirc A \rightarrow \Diamond A$ are conceptually not supported. Thus, the applicability of inference systems based on MVL is limited to cases of *diagnostic* reasoning, where a set of observations with their actual truth values is given. In such a situation, tasks like temporal projection are also solvable by using modal operators that "push" certain truth values into the future. Examples are reasoning about actions (e.g., Ginsberg's treatment of the Yale Shooting Problem (cf. [9] and [3])), plan recognition as described in [4], and all kinds of fault diagnosis. For such tasks, a more powerful - and less efficient - prover is generally not needed, but can be replaced by an inference system as the one described above.

Acknowledgements

I'd like to thank Gaby Merziger and Dietmar Dengler for fruitful discussions and Jana Köhler for her comments on an earlier version of this paper.

References

[1] J. Allen. Towards a General Theory of Action and Time. *Artif. Intell.*, 23:123–154, 1984.

[2] J. Allen. Planning as Temporal Reasoning. In *Proc. of the 2nd International Conference on Principles of Knowledge Representation and Reasoning*, 1991.

[3] A. Baker and M. Ginsberg. Temporal Projection and Explanation. In *Proc. of the 11th International Joint Conference on AI*, 1989.

[4] M. Bauer, S. Biundo, D. Dengler, M. Hecking, J. Köhler, and G. Merziger. Integrated Plan Generation and Recognition – a Logic-based Approach. In W. Brauer, editor, *4. Internationaler GI-Kongreß Wissensbasierte Systeme.* Springer, 1991.

[5] J. Burgess. Basic Tense Logic. In D. Gabbay and F. Guenthner, editors, *Handbook of Philosophical Logic, Vol. II*, pages 89–133, 1984.

[6] M. Ginsberg. Multivalued Logics: A Uniform Approach to Inference in AI. *Computational Intelligence*, 4:265–316, 1988.

[7] M. Ginsberg. A Circumscriptive Theorem Prover. *Artif. Intell.*, 39:209–230, 1989.

[8] M. Ginsberg. Bilattices and Modal Operators. *J. of Logic and Computation*, 1, 1990.

[9] M. Ginsberg. Computational Considerations in Reasoning about Action. In *Proc. of the 2nd International Conference on Principles of Knowledge Representation and Reasoning*, 1991.

[10] R. Hale. Temporal Logic Programming. In A. Galton, editor, *Temporal Logics and Their Applications*, pages 91–119. Academic Press, 1987.

[11] S. Kripke. Semantical Considerations on Modal Logic. *Acta Philosophica Fennica*, 16:83–94, 1963.

[12] R. Moore. Semantical Considerations on Nonmonotonic Logic. *Artif. Intell.*, 25:75–94, 1985.

[13] B. Moszkowski. A Temporal Logic for Multi-level Reasoning about Hardware. *Computer*, 18:10–19, 1985.

[14] R. Rosner and A. Pnueli. A Choppy Logic. In *Proc. of the Symposium on Logic in Computer Science*, pages 306–313, June 1986.

[15] Y. Shoham. Time for Action: On the Relation between Time, Knowledge and Action. In *Proc. of the 11th International Joint Conference on AI*, pages 954–960, 1989.

A Normal Form for First-Order Temporal Formulae

Michael Fisher*

Department of Computer Science,
University of Manchester
Manchester, U.K.

EMAIL: michael@cs.man.ac.uk

Abstract

In this paper a normal form for formulae of a first-order temporal logic is described. This normal form, called First-Order Separated Normal Form (SNF_f), forms the basis of both a temporal resolution method [5] and a family of executable temporal logics [2]. A first-order temporal logic, based on a linear discrete model structure, is introduced and the procedure for transforming an arbitrary formula of this logic to SNF_f is described. The transformation process not only preserves satisfiability but also ensures that any model of the transformed formula is a model of the original one. These properties ensure that the transformation into SNF_f has applications in both theorem proving and execution.

1 Introduction

One of the barriers to the development of clausal proof methods for modal and temporal logics has been the difficulties encountered in producing a normal form for such formulae. The main problem is that quantifiers and modal or temporal operators cannot be arbitrarily interchanged. For example, $\Diamond\forall x.\, p(x)$ is not, in general, equivalent to $\forall x.\, \Diamond p(x)$.

The technique of *renaming* has been used to preserve the structure of classical formulae and to improve the efficiency of deduction processes which apply to formulae in clausal form [7, 3]. It can be seen that, by renaming selected subformulae, modal and temporal formulae can be transformed whilst retaining their structure. For example, $\Diamond\forall x.\, p(x)$ can be transformed into the 'equivalent' formula, $\Diamond q \wedge \Box(q \Leftrightarrow \forall x.\, p(x))$, where x is a new predicate symbol. In this paper, this observation is used and extended to provide a transformation process, based on renaming, for transforming temporal formulae into a normal form. By using an appropriate normal form, new (renaming) subformulae, such as $\Box(q \Leftrightarrow \forall x.\, p(x))$, can also be transformed into the normal form.

The normal form used in this paper, called First-Order Separated Normal Form (SNF_f), is derived from the Propositional Separated Normal Form (SNF) used in [5]. Formulae in SNF_f are represented using a minimal set of temporal operators, in a form to which both resolution and execution techniques can be applied.

In §2 the temporal logic that will be used is described, with the syntax and semantics of the first-order temporal logic (called FML) being given in §2.1 and §2.2 respectively. In §3,

*This work was partially supported both by ESPRIT under Basic Research Action 3096 (SPEC), and by SERC under Research Grant GR/H/18449.

the normal form used for formulae of FML is outlined, reviewing propositional SNF in §3.1, and defining first-order SNF (SNF_f) in §3.2.

In §4, the procedure for transforming an arbitrary formula of FML into SNF_f is described. This procedure includes renaming transformations to remove quantifiers from inside temporal contexts and renaming transformations to eliminate certain temporal operators. The justifications for this second type of transformation are given elsewhere [5, 4] as they are analogous to the transformations used in the propositional case.

The applications of this transformation into SNF_f, particularly the temporal resolution procedure described in [5] and the direct execution of temporal formulae, are outlined in §5. Finally, in §6, the main results are summarised and further work in this area is outlined.

2 A First-Order Linear Temporal Logic

The first-order temporal logic, based on discrete, linear models with finite past and infinite future, is called FML, and its syntax and semantics are outlined below. Note that, FML incorporates the *constant domain* assumption with regard to quantification over temporal formulae.

2.1 Syntax

Well-formed formulae of FML (WFF_f) are generated from the following symbols.

- A set, $\mathcal{L}_p$, of *predicate symbols* represented by strings of lowercase alphabetic characters.

  Associated with each predicate symbol, p, is a non-negative integer, *arity(p)*. (Note that predicates with arity 0 are termed *propositions*.)

- Classical connectives, $\neg$, $\vee$, $\wedge$, $\Rightarrow$, and $\Leftrightarrow$.

- Future-time temporal operators, including unary operators $\bigcirc$, $\Diamond$ and $\Box$, and binary operators $\mathcal{U}$ and $\mathcal{W}$.

- Past-time temporal operators, including unary operators ⭘, ●, ⧫ and ■, and binary operators $\mathcal{S}$ and $\mathcal{Z}$.

- A set, $\mathcal{L}_v$, of *variable symbols*, x, y, z, etc.

- A set, $\mathcal{L}_c$, of *constant symbols*, a, b, c, etc.

- A set, $\mathcal{L}_f$, of *function symbols*, f, g, h, etc.

  Associated with each function symbol, f, is an arity, given by *arity(f)*.

- Quantifiers, $\forall$ and $\exists$.

- '(' and ')' which are, as usual, used to avoid ambiguity.

The set of *terms*, $\mathcal{L}_t$, contains those strings that are either constants or variables, or are generated by application of function symbols to other elements of $\mathcal{L}_t$. $\mathcal{L}_t$ is defined as follows.

1. Both $\mathcal{L}_v$ and $\mathcal{L}_c$ are subsets of $\mathcal{L}_t$.

2. If $t_1, \ldots, t_n$ are in $\mathcal{L}_t$, and f is a function symbol of arity n, then $f(t_1, \ldots, t_n)$ is in $\mathcal{L}_t$.

The set of well-formed formulae of FML (WFF_f) is defined as follows.

1. If $t_1, \ldots, t_n$ are in $\mathcal{L}_t$, and p is a predicate symbol of arity n, then $p(t_1, \ldots, t_n)$ is in WFF_f.
2. if A and B are in WFF_f, then the following are in WFF_f

$$\begin{array}{llllll} \neg A & A \vee B & A \wedge B & A \Rightarrow B & (A) & \\ \Diamond A & \Box A & A\,\mathcal{U}\,B & A\,\mathcal{W}\,B & \bigcirc A & \\ \blacklozenge A & \blacksquare A & A\,\mathcal{S}\,B & A\,\mathcal{Z}\,B & \boldsymbol{\bigcirc} A & \bullet A \end{array}$$

3. If A is in WFF_f and v is in $\mathcal{L}_v$, then $\exists v.\, A$ and $\forall v.\, A$ are both in WFF_f.

Sub-classifications of WFF_f are defined as follows. A *literal* is defined as either a predicate applied to an appropriate term, or the negation of such a predicate.
A *State-formula* is either a literal or a non-temporal combination of other state-formulae.
Future-time formulae (non-strict) are defined as follows

- if A is a state-formula, then A is a future-time formula,
- if A and B are future-time formulae, then $\neg A$, $A \wedge B$, $A \vee B$, $A \Rightarrow B$, $\exists v.\, A$, $\forall v.\, A$, $A\,\mathcal{U}\,B$, $A\,\mathcal{W}\,B$, $\bigcirc A$, $\Diamond A$, and $\Box B$ are all future-time formulae.

Strict past-time formulae are defined as follows

- if A and B are either state-formulae or strict past-time formulae, then $\boldsymbol{\bigcirc} A$, $\bullet B$, $A\,\mathcal{S}\,B$, $A\,\mathcal{Z}\,B$, $\blacklozenge A$, and $\blacksquare B$ are all strict past-time formulae,
- if A and B are strict past-time formulae, then $\neg A$, $A \wedge B$, $A \vee B$, $A \Rightarrow B$, $\exists v.\, A$ and $\forall v.\, A$ are all strict past-time formulae.

2.2 Semantics

The basic models of FML are discrete, linear structures with finite past and infinite future. To this structure a domain, $\mathcal{D}$, and mappings from elements of the language to denotations are added. Thus the full model structure for FML is

$$\mathcal{M} = \langle \sigma, \mathcal{D}, \pi_c, \pi_f, \pi_p \rangle$$

where

- σ, a sequence of states $s_0, s_1, s_2, s_3, \ldots$,
- $\mathcal{D}$ is the object-level domain,
- π_c is a map from $\mathcal{L}_c$ to $\mathcal{D}$,
- π_f is a map from $\mathcal{L}_f$ to $\mathcal{D}^n \to \mathcal{D}$, where n is the arity of f, and,
- π_p is a map from $\mathbf{N} \times \mathcal{L}_p$ to $\mathcal{D}^n \to \{T, F\}$.

Thus, for a particular state s, and a particular predicate p of arity n, $\pi_p(s,p)$ represents a map from n-tuples of elements of $\mathcal{D}$ to T or F. Note that the *constant domain* assumption is used, i.e. that $\mathcal{D}$ is constant for every state.

Next, *variable assignments* and *term assignments* are defined. A variable assignment is a mapping from $\mathcal{L}_v$ to elements of $\mathcal{D}$. Given a variable assignment, V, and the valuation functions, π_c and π_f, associated with a particular model structure, a term assignment $\tau_{v\pi}$ is a mapping from $\mathcal{L}_t$ to $\mathcal{D}$, defined inductively as follows.

- **if** $c \in \mathcal{L}_c$ **then** $\tau_{v\pi}(c) = \pi_c(c)$
- **if** $f \in \mathcal{L}_f$ **then** $\tau_{v\pi}(f(t_1,\dots,t_n)) = \pi_f(f)(\tau_{v\pi}(t_1),\dots,\tau_{v\pi}(t_n))$ [*where arity*(*f*) = *n*]
- **if** $v \in \mathcal{L}_v$ **then** $\tau_{v\pi}(c) = V(c)$
- **if** $p \in \mathcal{L}_p$ **then** $\tau_{v\pi}(p(t_1,\dots,t_n)) = p(\tau_{v\pi}(t_1),\dots,\tau_{v\pi}(t_n))$ [*where arity*(*p*) = *n*]
- **if** $t = \text{OP}\,t'$ **then** $\tau_{v\pi}(t) = \text{OP}\,\tau_{v\pi}(t')$ [*where* OP $\in \{\neg, \bigcirc, \bullet, \odot, \Diamond, \Box, \blacklozenge, \blacksquare\}$]
- **if** $t = t'\,\text{OP}\,t''$ **then** $\tau_{v\pi}(t) = \tau_{v\pi}(t')\,\text{OP}\,\tau_{v\pi}(t'')$ [*where* OP $\in \{\wedge, \vee, \Rightarrow, \mathcal{S}, \mathcal{Z}, \mathcal{U}, \mathcal{W}\}$]

The semantics of a well-formed formula is given with respect to a model structure, a state at which the temporal formula is to be interpreted, and a variable assignment. The satisfaction relation, $\models$, relates such tuples to well-formed formulae.

The semantics of a predicate is given by the truth value of the predicate application as defined in the model:

$$\langle \mathcal{M}, s_i, V\rangle \models p(x_1,\dots,x_n) \text{ iff } \pi_p(i,p)(\tau_{v\pi}(x_1),\dots,\tau_{v\pi}(x_n)) = T.$$

Two special predicate symbols are identified. In every model, and at every state, s_i, $\pi_p(i, \mathit{true})$ is a function that returns T no matter what its arguments are. Similarly, $\pi_p(i, \mathit{false})$ returns F. When *true* and *false* have arity 0, **true** and **false**, respectively, will be used.

The semantics of the standard propositional connectives is as in classical logic, e.g.,

$$\langle \mathcal{M}, i, V\rangle \models \varphi \vee \psi \text{ iff } \langle \mathcal{M}, i, V\rangle \models \varphi \text{ or } \langle \mathcal{M}, i, V\rangle \models \psi$$

The semantics of the unary future-time temporal operators is defined as follows

$$\begin{array}{lll} \langle \mathcal{M}, s_i, V\rangle \models \bigcirc\varphi & \text{iff} & \langle \mathcal{M}, s_{i+1}, V\rangle \models \varphi \\ \langle \mathcal{M}, s_i, V\rangle \models \Diamond\varphi & \text{iff} & \text{there exists a } j \ge i \text{ such that } \langle \mathcal{M}, s_j, V\rangle \models \varphi \\ \langle \mathcal{M}, s_i, V\rangle \models \Box\varphi & \text{iff} & \text{for all } j \ge i \text{ then } \langle \mathcal{M}, s_j, V\rangle \models \varphi. \end{array}$$

The informal semantics of these operators is as follows: $\bigcirc\varphi$ means that φ must be true in the *next* state; $\Diamond\varphi$ means that φ must be true at *some* state in the future; $\Box\varphi$ means that φ must be true at *all* states in the future.

The two binary future-time temporal operators are interpreted as follows

$$\begin{array}{lll} \langle \mathcal{M}, s_i, V\rangle \models \varphi\,\mathcal{U}\,\psi & \text{iff} & \text{there exists a } k \ge i \text{ such that } \langle \mathcal{M}, s_k, V\rangle \models \psi \\ & & \text{and for all } i \le j < k \text{ then } \langle \mathcal{M}, s_j, V\rangle \models \varphi \\ \langle \mathcal{M}, s_i, V\rangle \models \varphi\,\mathcal{W}\,\psi & \text{iff} & \text{for all } j \ge i \text{ then } \langle \mathcal{M}, s_j, V\rangle \models \varphi \\ & & \text{or } \langle \mathcal{M}, s_i, V\rangle \models \varphi\,\mathcal{U}\,\psi \end{array}$$

If past-time temporal formulae are interpreted at a particular state, s_i, then states with indices less than i are 'in the past' of the state s_i. The semantics of unary past-time operators is given as follows:

$$\begin{array}{lll}
\langle \mathcal{M}, s_i, V\rangle \models & \bullet\varphi & \text{iff } i = 0 \text{ or } \langle \mathcal{M}, s_{i-1}, V\rangle \models \varphi \\
\langle \mathcal{M}, s_i, V\rangle \models & \circ\varphi & \text{iff } i > 0 \text{ and } \langle \mathcal{M}, s_{i-1}, V\rangle \models \varphi \\
\langle \mathcal{M}, s_i, V\rangle \models & \blacklozenge\varphi & \text{iff there exists } j \text{ such that } 0 \leq j < i \text{ and } \langle \mathcal{M}, s_j, V\rangle \models \varphi \\
\langle \mathcal{M}, s_i, V\rangle \models & \blacksquare\varphi & \text{iff for all } j \text{ such that } 0 \leq j < i \text{ then } \langle \mathcal{M}, s_i, V\rangle \models \varphi
\end{array}$$

Note that, in contrast to the future-time operators, the $\blacklozenge$ and $\blacksquare$ operators are interpreted as being *strict*, i.e. the current index is not included in the definition. Also, as there is a unique start state, termed the *beginning of time*, two different last-time operators are used. The difference between '$\circ$' and '$\bullet$' is that for any formula φ, $\circ\varphi$ is false at the beginning of time, while $\bullet\varphi$ is true at the beginning of time. In particular, $\bullet$**false** is only true when interpreted at the beginning of time; otherwise it is false. Note that the equivalence, $\circ\neg\varphi \equiv \neg\bullet\varphi$, relates these operators.

The semantics for the binary past-time operators $\mathcal{S}$ and $\mathcal{Z}$ relates to that for $\mathcal{U}$ and $\mathcal{W}$ just as the unary past-time operators related to the unary future-time operators. Finally, the semantics of quantifiers is defined as follows.

$$\begin{array}{lll}
\langle \mathcal{M}, s, V\rangle \models \forall x.\ \varphi & \text{iff} & \text{for all } d \in \mathcal{D}.\ \langle \mathcal{M}, s, V \dagger [x \mapsto d]\rangle \models \varphi \\
\langle \mathcal{M}, s, V\rangle \models \exists x.\ \varphi & \text{iff} & \text{there exists } d \in \mathcal{D}. \text{ such that } \langle \mathcal{M}, s, V \dagger [x \mapsto d]\rangle \models \varphi
\end{array}$$

As the interpretation consists of a triple, comprising model, state, and assignment components, a well-formed formula, A, is *satisfied* in a particular model, $\mathcal{M}$, at the beginning of time, s_0, under a particular variable assignment, V, if $\langle \mathcal{M}, s_0, V\rangle \models A$. A formula is *satisfiable* if it is satisfied in some model, and for some variable assignment, at the beginning of time. Note that the more standard definition of satisfiability, where the formula is interpreted in *some* state, can be translated to the definition given above by using the following equivalence.

$$\exists \mathcal{M}.\ \exists V.\ \exists s.\ \langle \mathcal{M}, s, V\rangle \models A \qquad \text{if, and only if,} \qquad \exists \mathcal{M}.\ \exists V.\ \langle \mathcal{M}, s_0, V\rangle \models \Diamond A$$

In this paper, closed formulae, i.e. formulae containing no free variables, will mainly be used. In this case, the empty mapping, [ ], is used as the initial variable assignment.

3 A Normal Form for Temporal Formulae

The approach outlined in this paper is to rewrite arbitrary temporal formulae into *First-Order Separated Normal Form* (SNF$_f$). Before defining SNF$_f$, the similar, but simpler, propositional normal form will be reviewed.

3.1 Propositional Separated Normal Form

A temporal formula in the basic (propositional) Separated Normal Form (SNF) described in [5] is of the form

$$\Box \bigwedge_{i=1}^{n} (P_i \Rightarrow F_i).$$

Here, each P_i is a *strict* past-time temporal formula and each F_i is a (non-strict) future-time formula. Each of the '$P_i \Rightarrow F_i$' (called *rules*) is further restricted to be of one the following

$$\begin{array}{lcll}
\bullet\,\textbf{false} & \Rightarrow & \bigvee_{k=1}^{r} l_k & \text{(an } \textit{initial} \ \Box\text{-rule)} \\
\circ \bigwedge_{j=1}^{m} l_j & \Rightarrow & \bigvee_{k=1}^{r} l_k & \text{(a } \textit{global} \ \Box\text{-rule)} \\
\bullet\,\textbf{false} & \Rightarrow & \Diamond l & \text{(an } \textit{initial} \ \Diamond\text{-rule)} \\
\circ \bigwedge_{j=1}^{m} l_j & \Rightarrow & \Diamond l & \text{(a } \textit{global} \ \Diamond\text{-rule)}
\end{array}$$

where each l_j, l_k, or l is a literal. In previous papers, we have shown how propositional temporal formulae can be rewritten into this form [4, 5].

3.2 First-Order Separated Normal Form

In defining first-order SNF (SNF_f), the basic rule form is again required to be '$P_i \Rightarrow F_i$'. However, now the restrictions on P_i and F_i are slightly different:

- each P_i is now a *non-strict* past-time temporal formula;
- each F_i is again a (non-strict) future-time formula.

This restriction would not make any difference in propositional SNF as, in the propositional case, state-formulae on the left-hand side of the implication can always be moved across to the right. However, as will be seen later, the ability to be able to refer to state-formulae on the left-hand side of this implication is required for SNF_f.

The general form of a formula in SNF_f is

$$\Box \bigwedge_{i=1}^{n} \forall \bar{x}_i.\ (\forall \bar{y}_i.\ P_i(\bar{x}_i, \bar{y}_i)) \Rightarrow (\exists \bar{z}_i.\ F_i(\bar{x}_i, \bar{z}_i))$$

with the above restrictions of P_i and F_i. Here, '$\bar{x}_i$' represents a vector of variables, $x_{i_1}, x_{i_2}, \ldots, x_{i_m}$, so $\forall \bar{x}_i$ represents

$$\forall x_{i_1}.\ \forall x_{i_2}.\ \ldots\ \forall x_{i_m}$$

and $P_i(\bar{x}_i, \bar{y}_i)$ represents

$$P_i(x_{i_1}, x_{i_2}, \ldots, x_{i_m}, y_{i_1}, y_{i_2}, \ldots, y_{i_k})\,.$$

Now, in SNF$_f$, each rule is further restricted to be one of the following.

$$\forall \bar{x}.\ \ [(\forall \bar{y}.\ \bullet\mathbf{false} \wedge \bigwedge_{b=1}^{h} l_b(\bar{x},\bar{y})) \Rightarrow \exists \bar{z}.\ \bigvee_{j=1}^{r} m_j(\bar{x},\bar{z})] \qquad \text{(an } \textit{initial} \ \square\text{-rule)}$$

$$\forall \bar{x}.\ \ [(\forall \bar{y}.\ (\bigcirc \bigwedge_{a=1}^{g} k_a(\bar{x},\bar{y})) \wedge \bigwedge_{b=1}^{h} l_b(\bar{x},\bar{y})) \Rightarrow \exists \bar{z}.\ \bigvee_{j=1}^{r} m_j(\bar{x},\bar{z})] \qquad \text{(a } \textit{global} \ \square\text{-rule)}$$

$$\forall \bar{x}.\ \ [(\forall \bar{y}.\ \bullet\mathbf{false} \wedge \bigwedge_{b=1}^{h} l_b(\bar{x},\bar{y})) \Rightarrow \exists \bar{z}.\ \Diamond l(\bar{x},\bar{z})] \qquad \text{(an } \textit{initial} \ \Diamond\text{-rule)}$$

$$\forall \bar{x}.\ \ [(\forall \bar{y}.\ (\bigcirc \bigwedge_{a=1}^{g} k_a(\bar{x},\bar{y})) \wedge \bigwedge_{b=1}^{h} l_b(\bar{x},\bar{y})) \Rightarrow \exists \bar{z}.\ \Diamond l(\bar{x},\bar{z})] \qquad \text{(a } \textit{global} \ \Diamond\text{-rule)}$$

where each k_a, l_b, m_j or l is a literal.

4 Rewriting Formulae into SNF$_f$

In this section, it will be shown how arbitrary well-formed formulae of FML can be transformed, using satisfiability preserving transformations, into SNF$_f$. For simplicity, this translation process will be split into two phases. The first transforms an arbitrary formula from WFF$_f$ into rule form, i.e.,

$$\square \bigwedge_{i=1}^{n} \forall \bar{x}_i.\ Q_i(\bar{x}_i) \Rightarrow G_i(\bar{x}_i)$$

where $Q_i(\bar{x}_i)$ is a past-time formula and $G_i(\bar{x}_i)$ is a future-time formula. The second phase translates a formula in this rule form into one in SNF$_f$.

Rather than giving the full set of transformation rules, only the transformations that are most fundamental to the production of SNF$_f$ will be presented individually, together with an overview of the translation process. The proofs of the satisfiability preserving properties of the transformations used to remove temporal operators can be found in [4], while the correctness of the new transformations introduced in this paper follows from the theorem given below.

Theorem 1 *Given a formula, $\mathcal{A}(B(\bar{x}))$, where $\mathcal{A}$ is a context*[1] *and B is a formula with free variables, $\bar{x}$, then let N be the formula*

$$\mathcal{A}(b(\bar{x})) \wedge \square \forall \bar{y}.\ (b(\bar{y}) \Leftrightarrow B(\bar{y}))$$

where 'b' is a new predicate symbol (i.e. b has not occurred in any previous formula). Given the above constraints, both the following statements hold.

1. *If $\mathcal{A}(B(\bar{x})$ is satisfiable, then N is satisfiable, i.e.*

   if $\exists \mathcal{M}.\exists V.\ \langle \mathcal{M}, s_0, V\rangle \models \mathcal{A}(B(\bar{x})$ *then* $\exists \mathcal{M}'.\exists V'.\ \langle \mathcal{M}', s_0, V'\rangle \models N$.

[1] A context is taken to be a function over WFF$_f$ that embeds its argument at a given position in an enclosing formula. Various contexts will be used to represent subclasses of WFF$_f$.

2. *Any model for N is a model for* $\mathcal{A}(B(\bar{x}))$, *i.e.*

$$\textit{if} \quad \exists \mathcal{M}''. \exists V''. \langle \mathcal{M}'', s_0, V'' \rangle \models N \qquad \textit{then} \quad \langle \mathcal{M}'', s_0, V'' \rangle \models \mathcal{A}(B(\bar{x}))$$

The transformations required for rewriting a formula into SNF_f will now be introduced.

4.1 Generating Rule Form

An arbitrary temporal formula, A, is first transformed as follows.

$$A \longrightarrow \Box \forall \bar{x}. (\bullet \mathbf{false} \Rightarrow A)$$

This is a sound transformation as long as none of the variables in $\bar{x}$ occur in A.

Example The formula $\exists w. \Diamond \blacksquare p(w)$, can be replaced by

$$\Box \forall \bar{x}. (\bullet \mathbf{false} \Rightarrow \exists w. \Diamond \blacksquare p(w)) \tag{1}$$

Since satisfiability is 'anchored' at s_0, and since $\bullet$**false** can only be satisfied at s_0 then, to satisfy $\langle \mathcal{M}, s_0, V \rangle \models \Box \forall \bar{x}. (\bullet \mathbf{false} \Rightarrow A)$, A must be satisfied at s_0.

Next, all the past-time operators are removed from the right-hand side of the formula. This is achieved by renaming the past-time formulae with new predicates and adding the definition of these new predicates to the original formula. Using $\mathcal{F}$ to represent any arbitrary non-strict future-time context, and $B(\bar{v})$ to represent a past-time formula with free variables $\bar{v}$, the following transformation is applied.

$$\Box \forall \bar{x}. P(\bar{x}) \Rightarrow \mathcal{F}(B(\bar{v})) \longrightarrow \left\{ \begin{array}{llcl} & \Box \forall \bar{x}. & P(\bar{x}) & \Rightarrow & \mathcal{F}(b(\bar{v})) \\ \wedge & \Box \forall \bar{y}. & B(\bar{y}) & \Rightarrow & b(\bar{y}) \\ \wedge & \Box \forall \bar{y}. & \neg B(\bar{y}) & \Rightarrow & \neg b(\bar{y}) \end{array} \right\}$$

Example Given the formula (1) above, $\blacksquare p(w)$ can be replaced by a new predicate $q(w)$, defined in such a way that $q(y)$ is satisfied if, and only if, $\blacksquare p(y)$ is satisfied. Thus the new set of rules generated by this transformation is

$$\begin{array}{llcl} & \Box \forall \bar{x}. & \bullet \mathbf{false} & \Rightarrow & (\exists w. \Diamond q(w)) \\ \wedge & \Box \forall y. & \blacksquare p(y) & \Rightarrow & q(y) \\ \wedge & \Box \forall y. & \neg \blacksquare p(y) & \Rightarrow & \neg q(y) \end{array}$$

Thus, this transformation preserves satisfiability as each subformula (possibly containing free variables) is replaced by a new predicate (applied to those free variables) whose satisfiability is directly linked to that of the replaced subformula. For example, the defining formulae for the new predicate q above combine to give

$$\Box \forall y. (\blacksquare p(y) \Leftrightarrow q(y))$$

Thus, the value of $q(y)$ is always the same as that of $\blacksquare p(y)$. In particular, at the moment when $\blacksquare p(w)$ would have been evaluated, $q(w)$ has exactly the same value.

Similarly, any future-time temporal sub-formulae can be removed from the left-hand side of the formula, by using the renaming transformation:

$$\Box\,\forall\bar{x}.\ \mathcal{P}(A(\bar{v})) \Rightarrow F(\bar{x}) \longrightarrow \left\{\begin{array}{rlrcl} & \Box\,\forall\bar{x}. & \mathcal{P}(a(\bar{v})) & \Rightarrow & F(\bar{x}) \\ \wedge & \Box\,\forall\bar{y}. & a(\bar{y}) & \Rightarrow & A(\bar{y}) \\ \wedge & \Box\,\forall\bar{y}. & \neg a(\bar{y}) & \Rightarrow & \neg A(\bar{y}) \end{array}\right\}$$

4.2 From Rule Form to SNF_f

After applying the above transformations, the remaining formulae are in rule form, i.e.

$$\Box\forall\bar{x}.\ (P(\bar{x}) \Rightarrow F(\bar{x})),$$

where $P(\bar{x})$ is a (non-strict) past-time formula and $F(\bar{x})$ is a (non-strict) future-time formula. The translation of this form into SNF_f involves removing all past-time temporal operators (apart from one level of last-time operators) from $P(\bar{x})$, and removing all future-time operators (apart from, at most, one '$\Diamond$' operator) from the $F(\bar{x})$, re-ordering the quantifiers to be in the form required for SNF_f, and rewriting the remaining formulae into the correct conjunctive or disjunctive form (as described in §3.2).

A procedure for carrying out this translation from rule form to SNF_f will now be described. Note that several of the transformations used, namely those involved in the removal of temporal operators, were introduced previously [5, 4]. The procedure described is certainly not the most efficient method of translating to SNF_f, however it has the advantage of being reasonably simple to explain.

Remove all embedded quantifiers from the right-hand side. Here, any embedded quantifiers are removed from the right-hand side of any rule (i.e., from $F(\bar{x})$). Thus, the following two transformations are used to ensure that either no quantifiers appear on the right-hand side of a rule, or that they only occur at the top level, e.g. $\Box\,\forall\bar{x}.\ P(\bar{x}) \Rightarrow (\forall w.\ F(\bar{x}, w))$.

$$\Box\,\forall\bar{x}.\ P(\bar{x}) \Rightarrow \mathcal{F}(\forall w.\ B(\bar{v}, w)) \longrightarrow \left\{\begin{array}{rlrcl} & \Box\,\forall\bar{x}. & P(\bar{x}) & \Rightarrow & \mathcal{F}(b(\bar{v})) \\ \wedge & \Box\,\forall\bar{y}. & b(\bar{y}) & \Rightarrow & \forall w.\ B(\bar{y}, w) \\ \wedge & \Box\,\forall\bar{y}. & \neg b(\bar{y}) & \Rightarrow & \exists w.\ \neg B(\bar{y}, w) \end{array}\right\}$$

$$\Box\,\forall\bar{x}.\ P(\bar{x}) \Rightarrow \mathcal{F}(\exists w.\ B(\bar{v}, w)) \longrightarrow \left\{\begin{array}{rlrcl} & \Box\,\forall\bar{x}. & P(\bar{x}) & \Rightarrow & \mathcal{F}(b(\bar{v})) \\ \wedge & \Box\,\forall\bar{y}. & b(\bar{y}) & \Rightarrow & \exists w.\ B(\bar{y}, w) \\ \wedge & \Box\,\forall\bar{y}. & \neg b(\bar{y}) & \Rightarrow & \forall w.\ \neg B(\bar{y}, w) \end{array}\right\}$$

Example Given the formula $\Box\,\forall\bar{x}.\ P(\bar{x}) \Rightarrow \Diamond(\forall z.\ \Box q(\bar{x}, z))$, the above transformations can be applied to give

$$\begin{array}{rlrcl} & \Box\,\forall\bar{x}. & P(\bar{x}) & \Rightarrow & \Diamond c(\bar{x}) \\ \wedge & \Box\,\forall\bar{y}. & c(\bar{y}) & \Rightarrow & \forall z.\ \Box q(\bar{y}, z) \\ \wedge & \Box\,\forall\bar{y}. & \neg c(\bar{y}) & \Rightarrow & \exists z.\ \neg\Box q(\bar{y}, z) \end{array}$$

Again, a subformula is replaced by a new predicate whose value mirrors that of the subformula. Thus, the predicate $c(\bar{y})$ above is satisfied if, and only if, $\forall z.\ \Box q(\bar{y}, z))$ is satisfied.

Remove all embedded quantifiers from the left-hand side. Here, a similar set of transformations is applied in order to ensure that there are no embedded quantifiers on the left-hand side of a rule, i.e., if a quantifier such as $\forall y$ occurs it will occur as

$$\Box \forall \bar{x}. (\forall y. P(\bar{x}, y)) \Rightarrow F(\bar{x})$$

The transformation rules used in this process are the duals of those given above. For example, embedded existential quantifiers can be removed using the following rule (where $\mathcal{P}$ is a general past-time context).

$$\Box \forall \bar{x}. \mathcal{P}(\exists w. B(\bar{v}, w)) \Rightarrow F(\bar{x}) \longrightarrow \left\{ \begin{array}{lllll} & \Box \forall \bar{x}. & \mathcal{P}(b(\bar{v})) & \Rightarrow & F(\bar{x}) \\ \wedge & \Box \forall \bar{y}. & \exists w. B(\bar{y}, w) & \Rightarrow & b(\bar{y}) \\ \wedge & \Box \forall \bar{y}. & \forall w. \neg B(\bar{y}, w) & \Rightarrow & \neg b(\bar{y}) \end{array} \right\}$$

Now that quantifiers only appear at the 'outer' levels of each rule, transformations aimed at removing various temporal operators can be applied. Initially, implications in either side of the rule are expanded into their definitions in terms of negation and disjunction, and all negations are pushed through the formula so that they apply only to predicates. Next transformations which remove certain temporal operators (those not required for SNF_f) and replace them by new predicates are applied. As these transformations are the obvious first order extensions of those described in earlier papers [5, 4], only a brief outline of their use will be given, and any justification for their correctness will be omitted.

Removing Past-time Temporal Operators. The operators $\blacksquare$ and $\blacklozenge$ can be rewritten in terms of the $\mathcal{Z}$ and $\mathcal{S}$ operators respectively. Next, the $\mathcal{Z}$ and $\mathcal{S}$ operators, and any multiple occurrences of last-time operators (represented below by $\mathcal{L}$), are removed. The following transformations are used for this purpose. (Note that, here, $\mathcal{P}$ represents a general (non-strict) past-time context, and that $A \Leftrightarrow B$ is used as an abbreviation for $(A \Rightarrow B) \wedge (\neg A \Rightarrow \neg B)$.)

$$\Box \forall \bar{x}. \mathcal{P}(A(\bar{v})\, \mathcal{S}\, B(\bar{v})) \Rightarrow F(\bar{x}) \longrightarrow \left\{ \begin{array}{llll} & \Box \forall \bar{x}. \mathcal{P}(s(\bar{v})) & \Rightarrow & F(\bar{x}) \\ \wedge & \Box \forall \bar{y}. \mathbf{O}(B(\bar{y}) \vee (A(\bar{y}) \wedge s(\bar{y}))) & \Leftrightarrow & s(\bar{y}) \end{array} \right\}$$

$$\Box \forall \bar{x}. \mathcal{P}(A(\bar{v})\, \mathcal{Z}\, B(\bar{v})) \Rightarrow F(\bar{x}) \longrightarrow \left\{ \begin{array}{llll} & \Box \forall \bar{x}. \mathcal{P}(z(\bar{v})) & \Rightarrow & F(\bar{x}) \\ \wedge & \Box \forall \bar{y}. \bullet(B(\bar{y}) \vee (A(\bar{y}) \wedge z(\bar{y}))) & \Leftrightarrow & z(\bar{y}) \end{array} \right\}$$

$$\Box \forall \bar{x}. \mathcal{P}(\mathcal{L}(A(\bar{v}))) \Rightarrow F(\bar{x}) \longrightarrow \left\{ \begin{array}{llll} & \Box \forall \bar{x}. \mathcal{P}(l(\bar{v})) & \Rightarrow & F(\bar{x}) \\ \wedge & \Box \forall \bar{y}. \mathcal{L}(A(\bar{y})) & \Leftrightarrow & l(\bar{y}) \end{array} \right\}$$

Removing Future-time Temporal Operators With the future-time operators, all but the '$\Diamond$' temporal operator are removed and the transformations must ensure that each $\Diamond$ only applies to a literal. The $\Box$ operator can be reduced to $\mathcal{W}$, so the following transformations must be used to remove the unwanted temporal operators.

$$\Box\,\forall\bar{x}.\,P(\bar{x}) \Rightarrow \mathcal{F}(A(\bar{v})\,\mathcal{W}\,B(\bar{v})) \longrightarrow \left\{\begin{array}{l} \Box\,\forall\bar{x}.\,P(\bar{x}) \;\Rightarrow\; \mathcal{F}(B(\bar{v}) \vee (A(\bar{v}) \wedge w(\bar{v}))) \\ \wedge \\ \Box\,\forall\bar{y}.\,\bullet w(\bar{y}) \;\Leftrightarrow\; B(\bar{y}) \vee (A(\bar{y}) \wedge w(\bar{y})) \end{array}\right\}$$

$$\Box\,\forall\bar{x}.\,P(\bar{x}) \Rightarrow \mathcal{F}(A(\bar{v})\,\mathcal{U}\,B(\bar{v})) \longrightarrow \left\{\begin{array}{l} \Box\,\forall\bar{x}.\,P(\bar{x}) \;\Rightarrow \\ \qquad \mathcal{F}((B(\bar{v}) \vee (A(\bar{v}) \wedge u(\bar{v}))) \wedge \Diamond B(\bar{v})) \\ \wedge \\ \Box\,\forall\bar{y}.\,\bullet u(\bar{y}) \;\Leftrightarrow\; B(\bar{y}) \vee (A(\bar{y}) \wedge u(\bar{y})) \end{array}\right\}$$

$$\Box\,\forall\bar{x}.\,P(\bar{x}) \Rightarrow \mathcal{F}(\bigcirc A(\bar{v})) \longrightarrow \left\{\begin{array}{lcl} \Box\,\forall\bar{x}.\,P(\bar{x}) & \Rightarrow & \mathcal{F}(n(\bar{v})) \\ \wedge\;\; \Box\,\forall\bar{y}.\,\bullet n(\bar{y}) & \Leftrightarrow & A(\bar{y}) \end{array}\right\}$$

$$\Box\,\forall\bar{x}.\,P(\bar{x}) \Rightarrow \mathcal{F}(\Diamond A(\bar{v})) \longrightarrow \left\{\begin{array}{lcl} \Box\,\forall\bar{x}.\,P(\bar{x}) & \Rightarrow & \mathcal{F}(f(\bar{v})) \\ \wedge\;\; \Box\,\forall\bar{y}.\,f(\bar{y}) & \Leftrightarrow & A(\bar{y}) \end{array}\right\}$$

This last transformation is used only when a '$\Diamond$' operator is not applied to a literal. Note that, the $\mathcal{S}$, $\mathcal{Z}$, $\mathcal{W}$, and $\mathcal{U}$ operators are effectively replaced by their fixpoint definitions [8].

Having removed all the unnecessary temporal operators, the last major transformation that needs to be carried out is to remove all existential quantifiers from the past-time component of each rule and all universal quantifiers from the future-time component of each rule. This is because, though the quantifiers now only appear at the outer level of the formula, each rule is required to be of the form

$$\Box\,\forall\bar{x}.\,(\forall\bar{y}.\,P(\bar{x},\bar{y})) \Rightarrow (\exists\bar{z}.\,F(\bar{x},\bar{z}))$$

Removing Existential Quantifiers from Past-time Components. To remove existential quantifiers from the past-time component of a rule, several renamings are applied. Given a rule of the form

$$\Box\,\forall\bar{x}.\,(\exists y.\,P(\bar{x},y)) \Rightarrow F(\bar{x})$$

a new predicate is introduced to replace the past-time formula $P(\bar{x},y)$. Thus, $P(\bar{x},y)$ is replaced by $q(\bar{x},y)$, and the following new rules are added, ensuring that $q(\bar{x},y)$ is satisfied if, and only if, $P(\bar{x},y)$ is.

$$\begin{array}{llcl} \Box\,\forall\bar{v}.\,\forall w. & [P(\bar{v},w) & \Rightarrow & q(\bar{v},w)] \\ \Box\,\forall\bar{v}.\,\forall w. & [\neg P(\bar{v},w) & \Rightarrow & \neg q(\bar{v},w)] \end{array}$$

Thus, the original rule becomes

$$\Box\,\forall\bar{x}.\,(\exists y.\,q(\bar{x},y)) \Rightarrow F(\bar{x})$$

Now, another new predicate '$e(\bar{x})$' is introduced in order to replace '$\exists y.\,q(\bar{x},y)$'. Thus, the original formula becomes $\Box\,\forall\bar{x}.\,e(\bar{x}) \Rightarrow F(\bar{x})$ and the following new rules, defining the value of the predicate e, are added.

$$\begin{array}{llcl} \Box\,\forall\bar{v}. & [e(\bar{v}) & \Rightarrow & (\exists y.\,q(\bar{v},y))] \\ \Box\,\forall\bar{v}. & [\neg e(\bar{v}) & \Rightarrow & (\forall y.\,\neg q(\bar{v},y))] \end{array}$$

Note that though a new universal quantifier is introduced, it appears at the outer level of the future-time component (such universal quantifiers are removed below). Each renaming operation has introduced a new predicate whose value is linked directly to the subformula it replaced.

Removing Universal Quantifiers from Future-time Components. Given a rule of the form

$$\square \forall \bar{x}.\, P(\bar{x}) \Rightarrow (\forall y.\, F(\bar{x}, y))$$

the fact that the predicate *true* is satisfied for any arguments is used and the above rule is rewritten to

$$\square \forall \bar{x}.\, \forall y.\, (P(\bar{x}) \wedge \mathit{true}(y)) \Rightarrow F(\bar{x}, y)\,.$$

This is a sound non-temporal transformation and ensures that the quantifiers are in the positions specified by SNF_f.

Rewriting Non-Temporal Formulae The rules can now be rewritten using classical transformation rules into the appropriate SNF_f form. This simply involves rewriting the formulae into the appropriate conjunctive or disjunctive forms.

4.3 Example

To show how the transformation proceeds, an example of a formula in rule form will be given and its transformation into SNF_f will be followed.

- The original formula is as follows and is in the 'past implies future' form.

$$\square \forall x.\, [\,\bullet \exists z.\, c(z,x)] \Rightarrow [a(x)\, \mathcal{W}\, (\forall y.\, b(x,y))]$$

- First, the embedded quantifier is removed from the future-time component, generating the following rules.

$$\begin{array}{llrcl} 1. & \square \forall x. & [\,\bullet \exists z.\, c(z,x)] & \Rightarrow & a(x)\, \mathcal{W}\, q(x) \\ 2. & \square \forall w. & q(w) & \Rightarrow & \forall y.\, b(w,y) \\ 3. & \square \forall w. & \neg q(w) & \Rightarrow & \exists y.\, \neg b(w,y) \end{array}$$

- Similarly, the embedded quantifier is removed from the past-time component, giving the following.

$$\begin{array}{llrcl} 1. & \square \forall x. & \bullet e(x) & \Rightarrow & a(x)\, \mathcal{W}\, q(x) \\ 4. & \square \forall w. & e(w) & \Rightarrow & \exists z.\, c(z,w) \\ 5. & \square \forall w. & \neg e(w) & \Rightarrow & \forall z.\, \neg c(z,w) \end{array}$$

- All the quantifiers occur at the outer level of each component, so now the '$\mathcal{W}$' operator is replaced by its definition, giving the following.

$$\begin{array}{llrcl} 1. & \square \forall x. & \bullet e(x) & \Rightarrow & q(x) \vee (a(x) \wedge u(x)) \\ 6. & \square \forall y. & \bullet u(y) & \Rightarrow & q(y) \vee (a(y) \wedge u(y)) \end{array}$$

- Next, the universal quantifiers that appear in the future-time components of rules 2 and 5 are removed, giving the following replacements for these rules.

$$\begin{array}{lllcl} 2. & \Box \forall w.\, \forall y. & q(w) \wedge true(y) & \Rightarrow & b(w,y) \\ 5. & \Box \forall w.\, \forall z. & \neg e(w) \wedge true(z) & \Rightarrow & \neg c(z,w) \end{array}$$

- Finally, rules 1 and 6 are split to ensure that the future-time component of each rule is a disjunction of literals. This splits rule 1 into new rules 1a and 1b, and splits rule 6 into the new rules 6a and 6b. Thus, the final set of SNF_f rules produced from the original formula is

$$\begin{array}{lllcl} 1a. & \Box \forall x. & \bigcirc e(x) & \Rightarrow & q(x) \vee a(x) \\ 1b. & \Box \forall x. & \bigcirc e(x) & \Rightarrow & q(x) \vee u(x) \\ 2. & \Box \forall w.\, \forall y. & q(w) \wedge true(y) & \Rightarrow & b(w,y) \\ 3. & \Box \forall w. & \neg q(w) & \Rightarrow & \exists y.\, \neg b(w,y) \\ 4. & \Box \forall w. & e(w) & \Rightarrow & \exists z.\, c(z,w) \\ 5. & \Box \forall w.\, \forall z. & \neg e(w) \wedge true(z) & \Rightarrow & \neg c(z,w) \\ 6a. & \Box \forall y. & \bigcirc u(y) & \Rightarrow & q(y) \vee a(y) \\ 6b. & \Box \forall y. & \bigcirc u(y) & \Rightarrow & q(y) \vee u(y) \end{array}$$

4.4 Correctness of Transformations

Assuming that the transformation procedure is characterised by $\tau\colon \mathrm{WFF}_f \mapsto \mathrm{WFF}_f$, where for any formula A, $\tau(A)$ is in SNF_f, the following results can be given.

Lemma 1 *Given a formula A in* WFF_f*, if A is satisfiable, so is* $\tau(A)$.

Lemma 2 *Given a formula A in* WFF_f*, if* $\tau(A)$ *is satisfied in a model* $\mathcal{M}$*, then A is also satisfied in* $\mathcal{M}$.

These results follow immediately from Theorem 1 and the corresponding properties of the transformations not based upon renaming [4].

4.5 Complexity

Although a thorough investigation into the complexity of the transformation to SNF_f is part of our future work (see §6), coarse complexity bounds can be given by recognising that the renaming transformations used to remove both quantifiers and temporal operators introduce at most five new rules.

5 Applications

Two applications immediately follow from Lemmas 1 and 2. The first is a temporal resolution method that can be applied to formulae in SNF_f, which relies on Lemma 1. The second is the direct execution of temporal formulae in SNF_f, which relies on both Lemmas.

5.1 Temporal Resolution

Given a formula in SNF$_f$, i.e., $\Box\,\forall\bar{x}.\ (\forall\bar{y}.\ P(\bar{x},\bar{y})) \Rightarrow (\exists\bar{z}.\ F(\bar{x},\bar{z}))$ skolemization can be applied to give a universally quantified formula of the form

$$\Box\,\forall\bar{x}.\ P'(\bar{x}) \Rightarrow F'(\bar{x})$$

where P' and F' contain no quantifiers. To this type of formula, a variety of resolution methods can be applied, ranging from non-clausal resolution [1] to translation methods [6]. However, the particular motivation for developing this normal form was to provide the basis of a first-order extension of the resolution method described in [5].

This resolution process applies to sets of formulae in SNF and is split two components: the application of non-temporal (classical) resolution to $\Box$-rules; and the application of a temporal resolution rule. Temporal resolution is applied to a $\Diamond$-rule containing a future-time component such as $\Diamond q$. In this case, the rule can be resolved with a set of $\Box$-rules, S, which, when satisfied, imply that q will never be satisfied (i.e. $\Box\neg q$). As $\Diamond q$ guarantees that q must be satisfied at some stage in the future, the resolution rule derives the constraint that the formulae in S can never be satisfied while $\Diamond q$ is outstanding.

This temporal resolution rule encodes a form of induction over temporal sequences as hidden $\Box$-formulae must be found from the set of rules in order that they may be resolved with a complementary $\Diamond$-formula. The actual search for the formulae on which to apply the temporal resolution rule can be implemented using graph-theoretic techniques [5].

Further work on the first-order version of this method is under way, as are investigations into the first-order analogue of the graph-theoretic methods that are used to implement temporal resolution in the propositional case.

5.2 Execution of Temporal Formulae

Having produced a formula in SNF$_f$ which is satisfiable, if and only if, the original formula is satisfiable, this formula can be executed, producing a model consistent with either formula. In particular, METATEM [2], a system for executing temporal formulae, accepts formulae in SNF$_f$. The formulae executed within METATEM are represented using *implicit* quantification. As the general form of a formula in SNF$_f$ is

$$\Box\,\forall\bar{x}.\ (\forall\bar{y}.\ P(\bar{x},\bar{y})) \Rightarrow (\exists\bar{z}.\ F(\bar{x},\bar{z}))$$

then this can be represented in the form expected by METATEM by eliminating the quantifiers. The assumptions made by the METATEM system are that

- variables that appear only on the left of the '$\Rightarrow$' are universally quantified
- variables that appear only on the right of the '$\Rightarrow$' are existentially quantified
- variables that appear on both sides '$\Rightarrow$' are universally quantified at the outer level

For example,

$$\bigcirc test(x,y) \wedge run(x,z) \Rightarrow \Diamond terminate(a,w,y)$$

where w, x, y, and z are variables, and a is a ground term, represents the formula

$$\Box\,\forall y.\ [\forall x.\ \forall z.\ (\bigcirc test(x,y) \wedge run(x,z))] \Rightarrow [\exists w.\ \Diamond terminate(a,w,y)]$$

Thus, given a formula in the appropriate form, METATEM will attempt to generate a model for this formula. (For more details of this approach to executable temporal logics, see [2].)

6 Further Work

In this paper a normal form for a first-order temporal logic has been described. The procedure by which arbitrary formulae are transformed into this normal form has been outlined and justified. There are several obvious areas yet to be investigated. For example, more work is required on refining the complexity bounds for the transformation process. Related to this is the question of whether, by using more specific renaming techniques (such as using '$\Rightarrow$' rather than '$\Leftrightarrow$' in certain cases [7]), these bounds can be reduced.

The transformations described in this paper have been applied to a specific temporal logic. One question that remains is whether similar transformations can be applied to related logics, for example 'standard' modal logics, and logics that do not have constant domains.

Work on extending the resolution method described in [5] to this first-order framework is in progress, as is the development of an implementation of this procedure.

7 Acknowledgements

The author wishes to thank Philippe Noël, Rajeev Goré, and the anonymous referees for their helpful comments on an earlier version of this paper.

References

[1] M. Abadi and Z. Manna. Nonclausal Deduction in First-Order Temporal Logic. *ACM Journal*, 37(2):279–317, April 1990.

[2] H. Barringer, M. Fisher, D. Gabbay, G. Gough, and R. Owens. METATEM: A Framework for Programming in Temporal Logic. In *REX Workshop on Stepwise Refinement of Distributed Systems: Models, Formalisms, Correctness*, Mook, Netherlands, June 1989. (Published in *Lecture Notes in Computer Science*, volume 430, Springer Verlag).

[3] T. Boy de-la Tour and G. Chaminade. The Use of Renaming to Improve the Efficiency of Clausal Theorem Proving. In Ph. Jorrand and V. Sgurev, editors, *Artificial Intelligence IV: Methodology, Systems, Applications*, pages 3–12. Elsevier Science Publishers B.V. (North-Holland), 1990.

[4] M. Fisher and P. Noël. Transformation and Synthesis in METATEM – Part I: Propositional METATEM. Technical Report UMCS-92-2-1, Department of Computer Science, University of Manchester, Oxford Road, Manchester M13 9PL, U.K., February 1992.

[5] Michael Fisher. A Resolution Method for Temporal Logic. In J. Mylopoulos and R. Reiter, editors, *Proceedings of the International Joint Conference on Artificial Intelligence (IJCAI)*, Sydney, Australia, August 1991. Morgan Kaufman.

[6] Andreas Nonnengart. Resolution for First-Order Linear Temporal Logics. (Draft), 1991.

[7] D. A. Plaisted and S. A. Greenbaum. A Structure-Preserving Clause Form Translation. *Journal of Symbolic Computation*, 2(3):293–304, September 1986.

[8] Pierre Wolper. Temporal Logic Can Be More Expressive. *Information and Control*, 56, 1983.

Semantic Entailment in Non Classical Logics Based on Proofs Found in Classical Logic

Ricardo CAFERRA   Stéphane DEMRI

LIFIA-IMAG
46, Av. Félix Viallet, 38031 Grenoble Cedex, France

Abstract

A particular way of relating logics, specially useful in the framework of automated theorem proving is proposed. From the definition of the semantics of a logic (called *source logic* and abbreviated henceforth SL) in another logic (called *target logic* and abbreviated henceforth TL or TLS), we translate formulas of SL into TL using known techniques. Then we show how to partially translate proofs found in TL into SL. More precisely, the main theoretical result of the paper is a theorem establishing that for a class of non-classical logics - taking first-order sorted logic with equality as target logic - given a formula f in SL, it is possible from a proof P of f (obtained in TL) to backward translate into SL some (sometimes all) formulas in P. This set of backward translated formulas are proved to be semantically related each other and to define a partial consequence relation in SL. We get therefore an entailment sequence for f *in SL.* Our approach is applicable to source logics either without "computationally interesting" proof systems or without proof systems at all. One running example is fully treated. We compare the results of our method with the ones of a specialized tableaux-based theorem prover for the logic S4(p). Some hints of future work are given.

1 Introduction

This paper deals with the usefulness of relating logics in the field of non classical theorem proving. Relating logics has been profitably used in pure logical studies, and in theorem proving [15, 16, 14, 8] (in [12] proposals in this direction are also made). The present work attacks an aspect of relating logics particularly important in theorem proving, which remains non addressed by the existing approaches. Specifically, the key notion for relating logics in theorem proving is the translation from a *source logic* (SL) to a *target logic* (TL). SL is the logic in which we have expressed our problem, in which we want to prove theorems and for which we do not have a good theorem prover or a theorem prover at all (sometimes - for logics semantically defined- we do not even have a proof system). TL is a logic for which a great deal of results is known and for which we have good theorem provers with good complete strategies. Roughly speaking the approaches in [15, 16, 14, 8] deal with the problem of *proving in TL* formulas *stated in SL*. But the point is that when a user expresses in a logic SL his problems, he very likely expects to handle them in the *same logic* SL and *not* in a target logic. We propose a solution to this problem for a *class* of non classical logics. Traditionally there are two main approaches to the notion of *general* (or *abstract*) logics:

- (i) a *model theoretical approach*, i.e. a logic is a means of characterizing structures. What is emphasized is the *satisfaction* relation (see for ex. [6]).
- (ii) a *proof theoretical approach*, i.e. a logic is a means of doing proofs. What is emphasized is the *consequence relation* (see for ex. [1]) or as called for example in [11] the *entailment relation*.

More recently a very general unifying approach (in the framework of category theory) has been introduced [11] in which

> *"the concept of logic as a harmonious relationship between entailment and satisfaction is particularly simple and, once the observation is made, seems the obvious thing to do" ([11], p.3).*

Meseguer also advocates in his paper the need of a great flexibility about what counts as a "proof" (specially for computer science applications). Basically the approaches (i) and (ii) consider a logic as a couple $(\Sigma, Entail)$ where Σ is a class of languages and *Entail* is a satisfaction relation in the view (i) and a consequence relation in the view (ii) (for properties required on these relations see [6, 1, 11]).
The view in [11] covers (i) and (ii). The notion of *logical system* $\mathcal{S}$ is the most general one in [11] (we present only the main features given in [11] and we omit the technical details).
A *logical system* $\mathcal{S}$ is a couple $(\mathcal{L}, \mathcal{P})$ where $\mathcal{L}$ is a *logic* and $\mathcal{P}$ a *proof calculus* (roughly speaking the specification of inference rules generating a consequence relation). A *logic* is a 3-uple $\mathcal{L} = (\Sigma, \models, \vdash)$ where Σ is a class of languages , $\vdash$ specifies a consequence relation and $\models$ specifies a satisfaction relation. The relations $\vdash$ and $\models$ are required to satisfy a *soundness* condition : if $\Gamma \vdash \varphi$ then $\Gamma \models \varphi$ (where Γ is a set of wffs and φ a wff belonging to a language of Σ). A logic is *complete* if in addition, if $\Gamma \models \varphi$ then $\Gamma \vdash \varphi$. In [11] this is presented with the category theory language. The categorical approach and the traditional one appear to have the same expressive power and a one-to-one terminological correpondence can be established for all interesting notions (see for example [10]). The problem we are interested in is :

Given some premisses and a conclusion in a given logic the source logic (SL) and assuming that :

1. *a translation to a target logical system (TLS) is possible*
2. *a proof in a proof system of TLS can be found for formulas translated from SL*

Is it possible to present in SL a sequence of entailments corresponding (in some sense to be precised) to the proof found in TLS? It should be noticed that we do not require in principle that there is a proof system for SL.
The main theoretical achievement of the present paper is a theorem that is in some sense, stronger than the conjecture in [4]. This theorem establishes that for formulas in a class of non classical logics (including at least S4 and S5) it is possible to give

a sequence of semantic entailments in these logics based on proofs found in classical logic. The semantic entailments define a consequence relation compatible with a possibly existing one. *The omitted proofs in the following sections can be found in [3].*

2 From SL to TLS and back

In the next section we use ideas and techniques inspired by [15, 16, 14, 8].

2.1 From SL to TLS

We consider a transformation ψ (see in [14] the similar notion of *morphism*) between SL and TLS. We shall consider in the rest of the paper that TLS has the expressive power of first-order logic with equality and that the associated proof system is resolution (plus paramodulation). Therefore we shall speak indistinctly of clauses or formulas and deduction or resolution steps in TLS (and so on). In order to prove the main theoretical result of this paper (theorem 2), we require ψ to satisfy the following three properties. Let $\mathcal{I}$ be an SL-interpretation, f and g be SL-formulas, ψ_F be the formula transformation and ψ^{-1} be the interpretation inverse transformation.

- (I) if $\mathcal{I} \models_{SL}$ f then $\psi(\mathcal{I}$, f) $\models_{TLS}$ ψ_F(f) $\wedge$ A, where A is a set of axioms associated to ψ (see the *specification morphism* in [14]) (*Soundness*)
- (II) $\mathcal{I} \models_{SL}$ f iff $\psi^{-1}(\psi(\mathcal{I}$,g)) $\models_{SL}$ f (*Interpretation Projection*)
- (III) if $\mathcal{I} \models_{SL}$ f and $\psi(\mathcal{I}$,f) $\models_{TLS}$ G $\wedge$ A with possibly G = ψ_F(g) then $\psi^{-1}(\psi(\mathcal{I}$,f)) $\models_{SL}$ g - constants in G are in ψ_F(f) - (*Weak Completeness*)

The condition (I) is defined in [14]. The condition (III) is a variant of the completeness condition. Moreover, the condition (II) can be viewed as a useful condition in order to define a partial inverse formula transformation for a *class* of source logics. We introduce a normalization function Norm for the TLS-formulas. This operator transforms a formula f into an equivalent one which can be backward translated. We define a inverse formula transformation ϕ which is partial because its domain is not all the clauses but only the normalized ones. In relation with the transformation ψ we assume also that $\models_{TLS} A \Rightarrow (Norm(F) \Leftrightarrow F)$ for every TLS-formula F (*normalization condition*).
For example, if the set A is composed of equations then the function Norm might be a procedure of normalization for terms. In the context of the resolution calculus in TLS, we introduce in the following definitions syntactic properties about clauses.

Definition: Let f denote a SL-formula. We note,
- $C_1 \wedge C_2 ... \wedge C_n$ the clausal form of ψ_F(f),
- A the set of axioms in clausal form associated to ψ.

Let C be a TLS-clause deduced from $\{C_1, C_2 ..., C_n\} \cup$ A by resolution (with paramodulation).
A literal L is *an et-literal* ("easily translatable literal") *for f* iff the predicate symbol in L corresponds to a predicate symbol in f.
A clause C is *an et-clause* ("easily translatable clause") *for f* iff every literal of C is *an et-literal for f* $\diamond$

In the next proposition, F $=_{renaming}$ G means that G is F up to the renaming of some variables of F.

Proposition 1 **If** *the two conditions below hold,*

- *C is an et-clause for a SL-formula f*
- *$Norm(\psi_F(\phi(Norm(C)))) =_{renaming} Norm(C)$ (this condition expresses the fact that the class of normalized TLS-formulas is closed under the function composition $\psi_F\phi$)*

Then $\models_{TLS} A \Rightarrow (\psi_F(\phi(Norm(C))) \Leftrightarrow Norm(C))$.

The main purpose of the conjecture in [4] is to relate semantically an initial SL-formula and the backward translation of some clauses obtained in TLS. The next theorem is a weaker version of this conjecture ((Hyp 1) is added). It should be also noticed that the condition we require to the partial inverse formula transformation ϕ is that Norm($\psi_F(\phi(Norm(C)))$) $=_{renaming}$ Norm(C) for every et-clause C.

2.2 From TLS to SL

Theorem 1 *Let A denote the set of axioms associated to ψ but not necessarily in clausal form. Let C be an et-clause for the SL-formula f.*
If $Norm(\psi_F(\phi(Norm(C)))) =_{renaming} Norm(C)$ **and** $\models_{TLS} \psi_F(f) \wedge A \Rightarrow C$ *(Hyp1)* **then** $f \models_{SL} \phi(Norm(C))$.

Proof: The interpretation I satisfies the SL-formula f (Hyp2).
(i) $\psi(I,f) \models_{TLS} \psi_F(f) \wedge A$ (from (I) Section 2.1)
(ii) $\psi(I,f) \models_{TLS} A \Rightarrow (Norm(C) \Leftrightarrow C)$ (*normalization condition*)
(iii) $\psi(I,f) \models_{TLS}$ C (from (Hyp1) and (i))
(iv) $\psi(I,f) \models_{TLS} Norm(C)$ (from (i), (ii) and (iii))
(v) $\psi(I,f) \models_{TLS} A \Rightarrow (\psi_F(\phi(Norm(C))) \Leftrightarrow Norm(C))$ (from Proposition 1)
(vi) $\psi(I,f) \models_{TLS} \psi_F(\phi(Norm(C))) \Leftrightarrow Norm(C)$ (from (i) and (v))
(vii) $\psi(I,f) \models_{TLS} \psi_F(\phi(Norm(C)))$ (from (iv) and (vi))
(viii) $\psi^{-1}(\psi(I,f)) \models_{TLS} \phi(Norm(C))$ (from (III) Section 2.1)
(ix) $I \models_{SL} \phi(Norm(C))$ (from (II) Section 2.1 and (viii))
The above deduction is valid for every SL-interpretation I, so f $\models_{SL} \phi(Norm(C))$ □

We want to backward translate the proofs in TLS into sequences of entailments of formulas in SL. More precisely, we are looking for *semantical links* between formulas of SL corresponding (not necessarily one-to-one) to deduction steps in TL. We shall note $\mathcal{R}$ the *resolution operator* which takes as parameters two clauses and returns the set of all its possible resolvents, $\mathcal{PAR}$ and $\mathcal{F}$ are similar operators for the paramodulation and factorization rules respectively.

Definition: Let ψ be a transformation from SL to TLS and f a SL-formula.
We note,
- $C_1 \wedge C_2 ... \wedge C_n$ the clausal form of ψ_F(f),
- A the set of axioms in clausal form associated to ψ , let us suppose B $\subseteq$ A.

The partial inverse formula transformation ϕ has *semantic links of degree one* iff, for all the et-clauses c_1, c_2 and c_3,

if $c_3 \in \mathcal{R}(c_1, c_2)$ then $\phi(Norm(c_1) \wedge Norm(c_2)) \models_{SL} \phi(Norm(c_3))$ and
if $c_2 \in \mathcal{F}(c_1)$ then $\phi(Norm(c_1)) \models_{SL} \phi(Norm(c_2))$.
Similarly, the partial inverse formula transformation has *semantic links of degree two in B* iff, if the conditions below hold :
- c is an et-clause for f deduced from $\{C_1, .., C_n\} \cup$ B.
- $c' \in \mathcal{PAR}$(b,c).
Then c' is an et-clause for f, ϕ(Norm(c')) exists and $\phi(Norm(c)) \models_{SL} \phi(Norm(c'))$
$\diamond$

At this point we have all the notions and auxiliary results needed to state the main theoretical contribution of this paper. It identifies sufficient conditions to build a sequence of semantic entailments in the *source logic* SL based on a proof obtained in TLS.

Theorem 2 *Let ψ be a transformation between SL and TLS and f a SL-formula. We note,*
- $C_1 \wedge C_2 ... \wedge C_n$ *the clausal form of* $\psi_F(f)$,
- *A the set of axioms associated in clausal form to ψ, let us suppose $B \subseteq A$.*

It is assumed that $\psi_F(f)$ is equivalent to its clausal form and that ϕ has semantic links of first and second degree in B. Let $(l_1, ..., l_m)$ be a deduction in TLS such that:

1. $l_m \notin B$.
2. *For $1 \le i \le m$, if $l_i \notin B$ then l_i is an et-clause for f*
3. *For $1 \le i \le m$, one of the conditions below holds*
   (a) $l_i \in B \cup \{C_1, C_2 ..., C_n\}$
   (b) *there exists j such that $1 \le j < i$, $l_j \notin B$ and $l_i \in \mathcal{F}(l_j)$*
   (c) *there exists j and k such that $1 \le j \le k < i$, $l_k, l_j \notin B$ and $l_i \in \mathcal{R}(l_j, l_k)$ with $\models_{SL} \phi(Norm(l_j) \wedge Norm(l_k)) \Leftrightarrow \phi(Norm(l_j)) \wedge \phi(Norm(l_k))$ (*)*
   (d) *there exists j such that $1 \le j \le k < i$, $l_j \notin B$, $l_k \in B$ and $l_i \in \mathcal{PAR}(l_j, l_k)$*

Then *there exists a sequence $(f_0, ..., f_u)$ such that for $0 \le i \le u-1$, $f_i \models_{SL} f_{i+1}$ where f_0 =f and $f_u = \phi(Norm(l_m))$* (entailment condition).

Proof: By induction on the number of steps s of the TLS-deduction.

Base case : s = 1
Necessarily $l_1 \in \{C_1, C_2, ..., C_n\}$. But $\models_{TLS} \psi_F(f) \wedge A \Rightarrow l_1$ because $\models_{TLS} \psi_F(f) \Leftrightarrow C_1 \wedge .. \wedge C_n$. According to Theorem 1, f $\models_{SL} \phi(Norm(l_1))$.

Induction step
Let $(l_1, ..., l_{k+1})$ be a deduction satisfying the initial conditions (we call the l_i's "lemmas"). We distinguish different cases according to the way the lemma l_{k+1} is deduced.

Case 1 : $l_{k+1} \in \{C_1, C_2 ..., C_n\}$
As for the base case f $\models_{SL} \phi(Norm(l_{k+1}))$ (u = 1).

Case 2 : There exists j such that $1 \leq j < k+1$ and $l_{k+1} \in \mathcal{F}(l_j)$
There exists a sequence of formulas $(p_0, ..., p_{u(j)})$ such that for $0 \leq s \leq u(j) - 1$, $p_s \models_{SL} p_{s+1}$ where p_0 = f and $p_{u(j)} = \phi(Norm(l_j))$. We build the sequence $(f_0, ..., f_{u(j)+1})$ in the following way :

- for $0 \leq s \leq u(j), f_s = p_s$ and $f_{u(j)+1} = \phi(Norm(l_{k+1}))$

Since ϕ has semantic links of degree one in B, then $\phi(\text{Norm}(l_j)) \models_{SL} \phi(\text{Norm}(l_{k+1}))$. Obviously this sequence satisfies the required conditions.

Case 3 : There exist i and j such that $1 \leq i \leq j < k+1$ and $l_{k+1} \in \mathcal{R}(l_i, l_j)$
There exists a sequence of formulas $(p_0, ..., p_{u(i)})$ such that for $0 \leq s \leq u(i) - 1$, $p_s \models_{SL} p_{s+1}$ where p_0 = f and $p_{u(i)} = \phi(Norm(l_i))$. Similarly there exists a sequence of formulas $(q_0, ..., q_{u(j)})$ such that for $0 \leq s \leq u(j) - 1$, $q_s \models_{SL} q_{s+1}$ where q_0 = f and $q_{u(j)} = \phi(Norm(l_j))$. We prove the case when u(i)≤u(j) (the symmetrical case is immediate). We build the sequence $(f_0, ..., f_{u(j)+1})$ in the following way :

- $f_0 = f$, for $1 \leq s \leq u(i), f_s = p_s \wedge q_s$,
- for $u(i) + 1 \leq s \leq$ u(j), $f_s = p_{u(i)} \wedge q_s$ $(f_{u(j)} = \phi(Norm(C_i)) \wedge \phi(Norm(C_j))$,
- $f_{u(j)+1} = \phi(Norm(l_{k+1}))$

By the induction hypothesis, for $0 \leq k \leq u(j), f_k \models_{SL} f_{k+1}$, and $f_{u(j)} \models_{SL} f_{u(j)+1}$ and as ϕ has semantic links of degree one and the condition (*) holds.

Case 4 : There exist i and j such that 1 ≤ i ≤ j < k+1, $l_i \in$ B and $l_{k+1} \in \mathcal{PAR}(l_i, l_j)$
There exists a sequence of formulas $(p_0, ..., p_{u(j)})$ such that for $0 \leq t \leq u(j) - 1$, $p_t \models_{SL} p_{t+1}$ where p_0 = f and $p_{u(j)} = \phi(Norm(l_j))$. Since ϕ has semantic links of degree two in B, $\phi(Norm(l_j)) \models_{SL} \phi(Norm(l_{k+1}))$. We build the sequence $(l_0, .., l_{u(j)+1})$ in the following way:

- $l_0 = f$, for $1 \leq t \leq u(j)$, $l_t = p_t$, $l_{u(j)+1} = \phi(Norm(l_{k+1}))$.

Obviously this sequence satisfies the required conditions. □

Remarks :

- In principle Theorem 2 can be applied to the "usual" non-classical logics (S5, S4, ..).
- The sequence which is built in the proof is not necessarily the shortest one. The length of the sequence may be reduced, for instance by deleting the formulas appearing twice. The size of the formulas can also be reduced.
- If we consider that the steps of a proof "keep trace" of the strategy used to obtain them, we can see our method as translating into SL proofs *together* with the strategy used in TLS. In this sense our approach is more general than the one which transfers strategies from TLS into SL and attempt to prove completeness for them (for a non trivial example see [4]).
- The condition (*) could be considered as too restrictive nevertheless different non trivial examples in different logics (for instance see Section 3.6) satisfies this hypothesis.

The following is a corollary of Theorem 2.

Corollary 1 *If the conditions of Theorem 2 are satisfied then* $f \models_{SL} \phi(Norm(l_m))$

This corollary should be compared with Theorem 1. The next two lemmas state that the defined semantic entailment has the properties one reasonably expects. We first recall that a *consequence relation* is reflexive, monotonic (this condition is not required in [1]) and transitive (see [17]).

Lemma 1 *Let E be the following set of wffs :* $\{\varphi \mid$ *there exists a TLS-formula u such that* $\varphi = \phi(Norm(u))\}$. **If** *for* $\varphi_1, \varphi_2 \in E$, *we define* ($\varphi_1 \vdash \varphi_2$ *iff* $\varphi_1 \models_{SL} \varphi_2$) **then** $\vdash$ *is a consequence relation in E.*

Proof: Trivial from the properties of $\models_{SL}$. □

Lemma 2 *Let* $SL = (\Sigma, \models, \vdash_1)$ *be a complete source logic. The consequence relation* $\vdash$ *defined in Lemma 1 is compatible with* $\vdash_1$, *that is to say,* $\vdash \subseteq \vdash_1$.

Proof: Obvious by using the definition of $\vdash$ and the completeness of SL. □

3 One Particular Logic : the Multiagent Logic S4(p)

The notion of interpretation we consider for TLS is the algebraic one, see for example [14]. We adopt the usual definitions of models (reflexive and transitive accessibility relations) and satisfiability relations for S4(p). We work on the propositional level to illustrate our method. Similar results can be surely obtained for the first-order case (at least with constant world domains) on which we are working (see Section 4).

3.1 Definition of the Transformation ψ

We distinguish the formula transformation from the interpretation transformation.

Formula Transformation
The formalism in the definition of the transformation is simpler than the one defined in [14] since we should use only an underlying fragment of the *Context Logic*. In order to compute the translation of the S4(p)-formula f , ψ_F(f), we introduce a ternary function such that ψ_F(f) = ψ_F(f,Init-World,1) (no confusion is possible between the unary and the ternary function). The second argument is a *context term* and the third one is a negation flag. ψ_F is defined inductively as follows :

1. ψ_F(P, C, neg) = P(C) where P is a propositional variable

2. $\psi_F(\Box_{agent}$ f, C, neg) = **if** (neg = 1) **then** ∀ X:'A,W$\rightarrow^{rt}$W' $\psi_F(f, apply_1(apply_2($X, agent), C)) **else** $\neg\psi_F(f, apply_1(apply_2($Sk, agent), C)) with Sk a new constant of sort 'A,W$\rightarrow^{rt}$W' ($\psi_F(\Diamond_{agent}$ f, C, neg) has a dual definition)

3. ψ_F(f $\triangle$ g, C, neg) = ψ_F(f, C, neg) $\triangle$ ψ_F(g, C, neg) where $\triangle \in \{\vee, \wedge\}$

4. $\psi_F(\neg$ f, C, neg) = $\neg\psi_F$(f, C, (1 - neg))

Remarks :

- The TLS-formula ψ_F(f,Init-World,1) has only quantifications with "universal force".
- The third argument could be deleted in ψ_F, if the initial formula is in *negative normal form.*
- The *Strong Skolemisation*, term mainly used in [14], considers only the *domain* variables preceding a possibility operator, when a Skolem function is introduced for that operator. For the propositional case, the absence of domain allows us to introduce only Skolem constants. The soundness of the simplification for S4(p) is proved in [14]. This principle cannot be applied for all modal logics (symmetrical accessibility relations could introduce additional difficulties).

We present a list of axioms which should be satisfied by every translated S4(p)-interpretations in order to verify the soundness condition (I) Section 2.1. The below axioms are related to the ones proposed in [14].

List of axioms

A1 ∀ x,y : 'W$\rightarrow^{rt}$W' ∀ w:World $apply_1(x, apply_1(y, w)) = apply_1(o_1(y, x), w)$
A2 ∀ x,y:'A,W$\rightarrow^{rt}$W' a:Agent $apply_2(o_2(x, y), a) = o_1(apply_2(x, a), apply_2(y, a))$
A3 ∀ a:Agent $apply_2$(IDL,a) = ID
A4 ∀ x:'W$\rightarrow^{rt}$W' $o_1(x, ID) = x$
A5 ∀ x:'W$\rightarrow^{rt}$W' $o_1(ID, x) = x$
A6 ∀ x,y,z:'W$\rightarrow^{rt}$W' $o_1(x, o_1(y, z)) = o_1(o_1(x, y), z)$
A7 ∀ w:World $apply_1$(ID,w) = w
A8 ∀ x,y:'W$\rightarrow^{rt}$W' (∀ w:World $apply_1$(x,w) = $apply_1$(y,w)) ⇒ (x = y)
The axioms A4 and A5 are not necessary since we have introduced the axiom A8.

The definitions of the interpretation transformation and the one for the inverse interpretation transformation are inspired from [14] and they are detailed in [3].

3.2 Some Properties of the Transformation

Proposition 2 *The transformation* ψ *from SL to TLS preserves models, i.e. if* $I \models_{S4(p)} f$ *then* $\psi(I,f) \models_{TLS} \psi_F(f)$. *Moreover every formula* A_i *for* $1 \le i \le 8$, *is satisfied by* $\psi(I,f)$.

Proposition 3 *Let f be an SL-formula and G be a TLS-formula such that possibly* $G = \psi_F(g)$ *- the constants in G are in* $\psi_F(f)$. *If* $I \models_{SL} f$ *and* $\psi(\mathcal{I},f) \models_{TLS} G$ *then* $\psi^{-1}(\psi(\mathcal{I},f)) \models_{SL} g$.

Proposition 4 *The transformation* ψ *verifies the* interpretation projection *condition.*

Proof: By construction $\psi^{-1}(\psi(\text{I,f})) = \text{I}$, so the interpretation projection condition immediately holds. □

Proposition 5 *Let f be a S4(p)-formula. We note,*

- $C_1 \wedge C_2 ... \wedge C_n$ *the clausal form of* $\psi_F(f)$,
- $B = \{A_1,...,A_7\}$ *a set of axioms associated to* ψ

Every et-clause C for f deduced from $\{C_1, ..., C_n, A_1,...,A_7\}$ *satisfies* $f \models_{S4(p)} \phi(Norm(C))$.

3.3 Definition of the Normalization Function

In order to apply the results of the previous section we first define a normalization function Norm, and we prove afterwards that ϕ has semantic links of degree one and two in $\{A_i, 1 \leq i \leq 7\}$. In the sequel, we shall note indifferently 't o_i s' for 'o_i(t,s)' and 't o_i s o_i u' for 'o_i(t,o_i(s,u))' . We propose a set of rewriting rules, which is straightly computed from the set of axioms of the *specification transformation*:

R1.1	$apply_2(t_1 o_2 t_2, t_3) \rightarrow apply_2(t_1, t_3) o_1 apply_2(t_2, t_3)$
R1.2	$apply_2(IDL, a) \rightarrow$ ID
R2.1	$apply_1(t_2, apply_1(t_1, t_3)) \rightarrow apply_1(t_1 o_1 t_2, t_3)$
R2.2	$(t_1 o_1 t_2) o_1 t_3 \rightarrow t_1 o_1 (t_2 o_1 t_3)$
R2.3	ID o_1 x $\rightarrow$ x
R2.4	x o_1 ID $\rightarrow$ x
R3.1	$apply_1$(ID, Init-World) $\rightarrow$ Init-World

We note $\rightarrow_R^*$ the function which applies the rule R as much as possible.
For R $\in$ {R1.1, R1.2, R2.1+R2.2, R2.3+R2.4, R3.1} $\rightarrow_R^*$ is effectively a total function. This is due to the fact that each system R terminates and is confluent. We define the normalization function Norm as the following composition of functions : $\rightarrow^*_{R3.1} \circ \rightarrow^*_{R2.3+R2.4} \circ \rightarrow^*_{R2.1+R2.2} \circ \rightarrow^*_{R1.2} \circ \rightarrow^*_{R1.1}$.

Proposition 6 *For every et-clause C,* $\models_{TLS} A \Rightarrow (C \Leftrightarrow Norm(C))$.

This is a direct consequence of the fact that the rewriting rules are built from the axioms associated to ψ.

3.4 The Partial Inverse Formula Transformation ϕ

The proposed backward transformation is a modification of the one defined in [4]. In the present work, the translation integrates the agents in the syntax, and no more in the definition of sorts. To translate a clause we capture the underlying semantics of the transition from one world to another by introducing a modal operator.

Backward transformation for clauses

The principle of the backward translation consists in gathering the literals which have some *context* in common due to the normal form of translated clauses. Indeed the inverse partial formula transformation is basically defined for the normalized clauses and then for the conjunction of normalized clauses. From the last section, it could be deduced that every normalized clause C has the format:
$L_1 \vee ... \vee L_N$ where the L_i's are normalized literals.

Every normalized literal L_i has the following form :
$L_i = s_i P_i(C_i)$ with $s_i \in \{\Lambda, \neg\}$ (Λ denotes the empty string) and P_i is an unary predicate symbol.
The normalized terms (the C_i's) have the following form :
C_i = Init-World or $C_i = apply_1(t^i_1 o_1 ... o_1 t^i_{f(i)}$, Init-World).
The terms t^j_i have the following form :
$t^i_j = apply_2(x^j_i, A^j_i)$ with A^j_i, a constant of sort Agent and x^j_i a constant or a variable of sort 'A,W$\rightarrow^{rt}$W'. We note $[t_1, ..., t_n]$ the *context* term $apply_1(t_1 o_1 ... o_1 t_n$,Init-World). First of all we algorithmically define auxiliary functions which deal with syntactic transformations.

- generate-op(t) := **if** t' is a variable **then** $\Box_A$ **else** $\Diamond_A$
  % t is a term of the form $apply_2(t', A)$ where t' is a term of sort 'A,W$\rightarrow^{rt}$W', A a constant of sort Agent%

- local-context($(t_1, ..., t_n)$) :=
  % the t_i's have the form $apply_2(t', A)$ and '.' is the concatenation operator%
  if n = 1 **then** generate-op(t_1) **else** generate-op(t_1) . local-context($(t_2, ..., t_n)$)

- $\beta([t_1, ..., t_n]$, m) := **if** n = m **then** Init-World **else** $[t_{m+1}, ..., t_n]$
  % m is an integer, the t_i's have the form $apply_2(t', A)$, $0 \leq m \leq n$ %
  We extend β to the literals, the clauses and the sets of clauses.

- $\phi_1(L)$:= **if** C = Init-World **then** P **else** generate-op(t_1) . ϕ_1 (sP(β(C,1)))
  % L is a literal with the format described above, L = sP(C) %

Definition of ϕ
Let C be a clause $L_1 \vee ... \vee L_n$. We define a partition of the set of literals belonging to C, i.e. we define the set of classes $\{C_i, 1 \leq i \leq k\}$. Moreover the classes satisfy the following properties. For $1 \leq j \leq$ k, there exists α_j such that

1. % $L_i = s_i P_{Li}(apply_1(apply_2(t^{Li}_1, a^{Li}_1) o_1 ... o_1 apply_2(t^{Li}_{ni}, a^{Li}_{ni})$,Init-World)) or $L_i = s_i P_{Li}$(Init-World) %
   For all P,Q $\in C_j$, for all $1 \leq i \leq \alpha_j$, $t^Q_i = t^P_i$, and $a^Q_i = a^P_i$, and $t^Q_{\alpha j+1} \neq t^P_{\alpha_j+1}$
2. For all P $\in C_j$ and $Q \notin C_j, \forall\ 1 \leq i \leq \alpha_j, t^Q_i \neq t^P_i$
3. By convention if α_j = -1 then, for all L $\in C_j$, L = sP(Init-World) .

It is easy to verify that the decomposition in classes is unique. The definition of ϕ(C) uses the function α which takes as argument a set of literals S and which returns a modal formula :
$\alpha(\{l_1, ..., l_p\})$:=
Case :% $l_i = s_i P_i(C_i)$ %
p = 1 : $\phi_1(l_1)$
α_C = -1 : $s_1 P_1 \vee ... \vee s_p P_p$
otherwise local-context($\{t^{l1}_1, ..., t^{l1}_{\alpha S}\}$) . $\phi(\{L | \exists l_i \in$ C such that L = $s_i P_i(\beta(C_i, \alpha_S))\})$.
We can now define $\phi(C) = \alpha(C_1) \vee .. \vee \alpha(C_k)$.

Backward transformation for conjunctions of clauses
We extend the partial inverse formula transformation to the conjunctions of et-clauses. From the idea of "partition" in the definition for clauses, we define some "superclasses" whose elements are sets of et-clauses from different partition of clauses. Indeed, we propose an algorithm to compute $\phi(C_1 \wedge ... \wedge C_N)$ where the C_i's are et-clauses. As defined above, we associate to each clause C_i, a partition $\{c_i^1, ..., c_i^{ui}\}$ and to each class C, the integer α_C and the sequence $(t_1^C, ..., t_{\alpha C}^C)$. For C = c_i^j we also note the sequence $(t_1^{(i,j)}, ..., t_{\alpha(i,j)}^{(i,j)})$.
The point is now to compute some possible superclasses belonging to the power set of the set $\{c_i^j$, for $1 \leq i \leq N$, $1 \leq j \leq u_i\}$. A superclass SC = $\{c_{i1}^{j1}, ..., c_{in}^{jn}\}$ has the following properties :

- for $1 \leq a < b \leq n$, $i_a \neq i_b$ (otherwise $c_{ia}^{ja} \bigcup c_{ia}^{jb}$ is a class)
- there exists E_{max} optimal, with $0 \leq E_{max} \leq$ Min($\{\alpha(i_k, j_k)$, $1 \leq k \leq n\}$) such that for $1 \leq f \leq E_{max}$,
  - for $1 \leq a \leq b \leq n$, $t_f^{ia,ja} = t_f^{ib,jb}$ or,
  - $t_f^{ia,ja}$ and $t_f^{ib,jb}$ have the same agent and both contain a variable (necessarily different, since they belong to different clauses). Moreover, there exists δ such that $f < \delta \leq E_{max}$ and $t_\delta^{ia,ja} = t_\delta^{ib,jb}$.

We note,
ϕ'(SC) = local-context$((t_1^{(i1,j1)}, ..., t_{Emax}^{(i1,j1)}))$. $\bigwedge \{\phi(\beta(c_{ik}^{jk}, E_{max}))$, $1 \leq k \leq n\}$.
Finally, $\phi(C_1 \wedge ... \wedge C_N) = \bigvee \{ [\bigwedge \{ \alpha(c_j^{ij}), 1 \leq j \leq N \}]^*, 1 \leq i_1 \leq u_1, ..., 1 \leq i_N \leq u_N\}$. Moreover,
$[\alpha(c_1) \wedge ... \wedge \alpha(c_l)]^*$:= **if** there exists $\{n1, ..., nk\} \subseteq [1,..., l]$ such that $\{c_{n1}, ..., c_{nk}\}$= U is a superclass **then** $\alpha(c_{n1}) \wedge ... \wedge \alpha(c_{nk})$ is replaced by ϕ'(U). Only maximal superclasses are considered and few replacements are possible.
Otherwise $\alpha(c_1) \wedge ... \wedge \alpha(c_l)$.

Remarks :

- For N = 1, we obtain the restricted definition for clauses.
- The "prefix-stable" property is implicitly used to guarantee existence and unicity of the decompositions [13].
- If no superclass is available in $\phi(C_1 \wedge ... \wedge C_N)$ we can also note it $\phi(C_1) \wedge ... \wedge \phi(C_N)$

Lemma 3 *Let C_1, C_2 be two et-clauses for the SL-formula f. The backward transformation satisfies the two following properties:*

(a) $\phi(Norm(C_1) \wedge Norm(C_2)) \models_{SL} \phi(Norm(C_1)) \wedge \phi(Norm(C_2))$

(b) It is decidable whether $\models_{SL} \phi(Norm(C_1) \wedge Norm(C_2)) \Leftrightarrow \phi(Norm(C_1)) \wedge \phi(Norm(C_2))$.

3.5 Results About the Semantic Entailment

Proposition 7 *If C is a normalized TLS-clause - Norm(C) = C - then there exists a way of introducing new constants for* ψ_F *such that* $Norm(\psi_F(\phi(C))) =_{renaming} C$.

Proposition 8 *The transformation* ϕ *has semantic links of degree one.*

Proposition 9 *The transformation* ϕ *has semantic links of degree two in* $\{A_i,\ 1 \leq i \leq 7\}$.

The transformation ψ, the partial inverse formula transformation ϕ , and the normalized function Norm, satisfy the conditions of Theorem 2, so we effectively have for S4(p) a sequence of semantic entailments for a class of proofs from TLS.

3.6 Comparison with a S4-theorem Prover : a Running Example

We illustrate our method with the McCarthy's Wise Man Puzzle (see for ex. [14]) and we compare the proof using our approach with the one obtained with a tableaux-based prover. In order to axiomatize this puzzle in S4(p), the three wise men are A,B,C and C is the wisest. The propositional variable P_i designates the fact that the wise man i has a white spot on his forehead. At least one of them has a white spot and everyone knows that everybody else knows that his colleagues know this.
A1 $\Box_i\Box_j\Box_k(P_A \vee P_B \vee P_C)$ for i,j,k $\in$ {A,B,C} and {i,j,k}={A,B,C}

The three men can see each other and they know this. Whenever one of them has a white or black spot he knows that his colleagues knows this from each other.
A2 $\Box_i(\neg P_i \Rightarrow \Box_j\neg P_i)$ for i,j $\in$ {A,B,C} and i$\neq$j.
A3 $\Box_i\Box_j(\neg P_i \Rightarrow \Box_k\neg P_i)$ for i,j,k $\in$ {A,B,C} and {i,j,k} = {A,B,C}
A4 $\Box_i\Box_j(\neg P_i \Rightarrow \Box_k\neg P_j)$ for i,j,k $\in$ {A,B,C} and {i,j,k} = {A,B,C}

C knows that B does not know that the colour of his spot and C knows that B knows that A does not know the colour of his spot.
A5 $\Box_C\neg\Box_B P_B$ A6 $\Box_C\Box_B\neg\Box_A P_A$

From A1 to A6 we obtain 26 axioms from which we want to deduce $\Box_C P_C$. To do so, within our inference laboratory ATINF [2, 5] we transform the S4(p)-problem into an TLS problem by translating the 26 axioms and $\neg\Box_C P_C$ to TLS. The shortest proof which has been built with the resolution-based theorem prover of ATINF is showed in Figure 1. The proof was obtained in less than 1 second *cpu* on a 8Mb SUN4 workstation. In order to appreciate experimentally the usefulness of the proposed method we compare it with a tableaux-based prover. With the tableaux-based theorem prover of ATINF parametrized for the multiagent logic S4(p) (which uses mainly the method in [7]) the proof was obtained in about 20 seconds cpu. A lot of choices appeared for the application of π-rules (rules related to the possibility operator) and almost 50 % of the attempted choices were useless.

Proof of the "Wise Man Puzzle" in TLS (Figure 1)

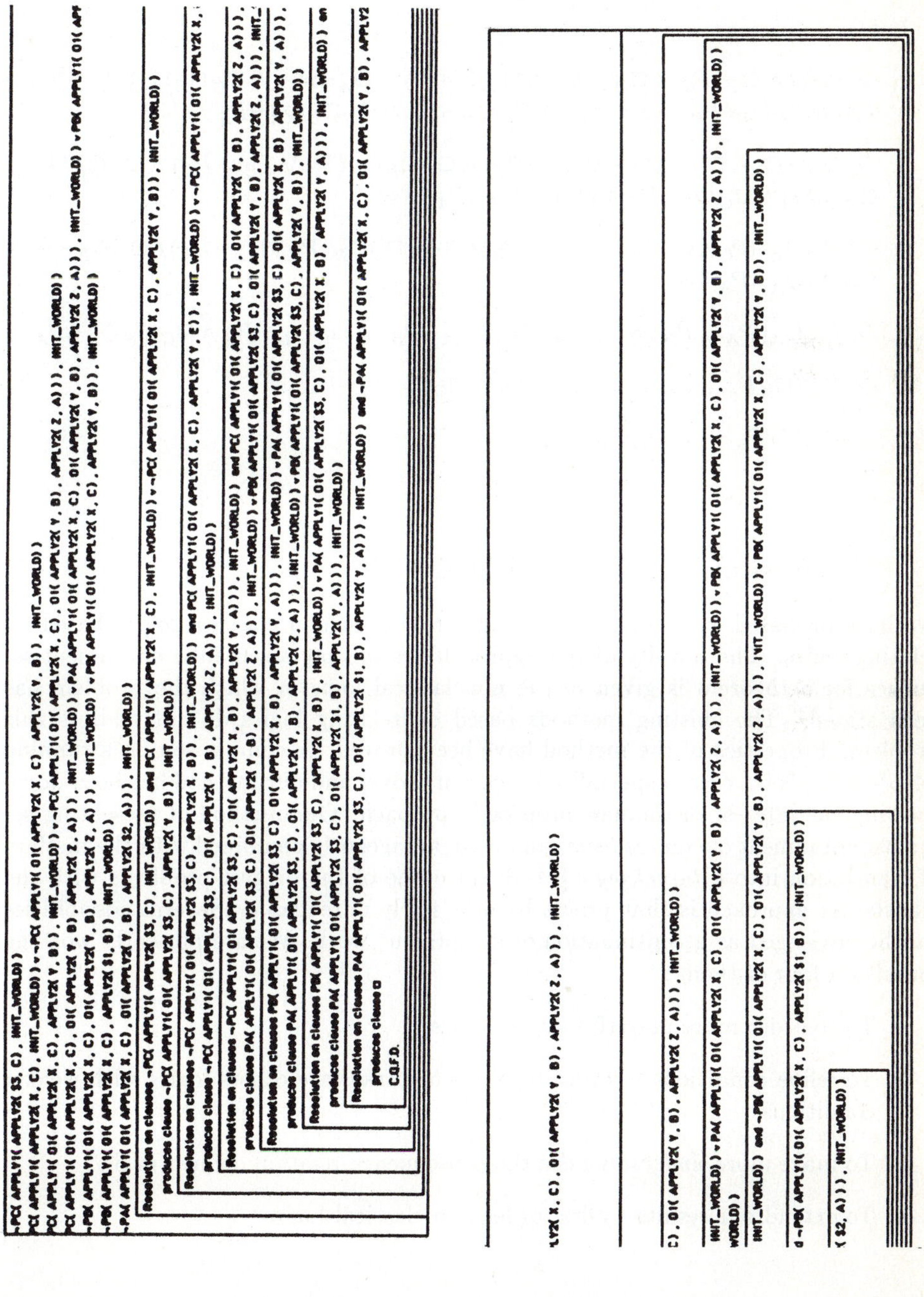

The proof in the multiagent logic T needed about 60 seconds cpu. A lot of choices were attempted but only few were fruitful. This could be compared with Konolige's solution [9]. The sequence of semantic entailments in Figure 2 is built on the same bases as the proof of Theorem 2. We note f the conjunction of the 26 axioms plus the negation of the conclusion.

Sequence of semantic entailments for the "Wise Man Puzzle" in SL (Figure 2)

1. f
2. $\Diamond_C \neg P_C \wedge \Box_C(P_C \vee \Box_B \neg P_C) \wedge \Box_C\Box_B(P_C \vee \Box_A \neg P_C) \wedge \Box_C\Box_B\Box_A(P_A \vee P_B \vee P_C) \wedge \Box_C\Box_B(P_B \vee \Box_A \neg P_B) \wedge \Box_C\Diamond_B \neg P_B \wedge \Box_C\Box_B\Diamond_A \neg P_A$
3. $\Diamond_C\Box_B \neg P_C \wedge \Box_C\Box_B(P_C \vee \Box_A \neg P_C) \wedge \Box_C\Box_B\Box_A(P_A \vee P_B \vee P_C) \wedge \Box_C\Box_B(P_B \vee \Box_A \neg P_B) \wedge \Box_C\Diamond_B \neg P_B \wedge \Box_C\Box_B\Diamond_A \neg P_A$
4. $\Diamond_C\Box_B\Box_A \neg P_C \wedge \Box_C\Box_B\Box_A(P_A \vee P_B \vee P_C) \wedge \Box_C\Box_B(P_B \vee \Box_A \neg P_B) \wedge \Box_C\Diamond_B \neg P_B \wedge \Box_C\Box_B\Diamond_A \neg P_A$
5. $\Diamond_C\Box_B\Box_A(P_A \vee P_B) \wedge \Box_C\Box_B(P_B \vee \Box_A \neg P_B) \wedge \Box_C\Diamond_B \neg P_B \wedge \Box_C\Box_B\Diamond_A \neg P_A$
6. $\Diamond_C\Box_B(\Box_A P_A \vee P_B) \wedge \Box_C\Diamond_B \neg P_B \wedge \Box_C\Box_B\Diamond_A \neg P_A$
7. $\Diamond_C\Diamond_B\Box_A P_A \wedge \Box_C\Box_B\Diamond_A \neg P_A$
8. $\perp$

4 Conclusion and Future Work

We have proposed a method for theorem proving in non classical logics based on relating logics. The novelty of the approach lies in that a semantic entailment sequence for a theorem is given *in* the nonclassical logic in which the problem has been stated. The existing methods based on relating logics do not address this problem. Properties of the method have been proved. A comparison on a running S4(p)-example between a specialized theorem prover and our method has been done, showing the interest of the new proposed approach. Our method of presenting semantic entailment on *source logic* can be advantageously combined with the method of translation into a *target logic* [14, 8] since one of the biggest problems with the translation approach is that proofs become fairly unreadable. Several possibilities can be envisaged as a continuation of the present work. We are presently working mainly on four of them :

1. To consider more powerful target logics
2. To refine conditions in order to backward translate others logics with weaker conditions
3. To make more *understandable* the consequence relation defined in SL
4. To extend our results to first-order non classical logics

References

[1] A. Avron. Simple Consequence Relations. *Information and Computation*, 92:105–139, 1991.

[2] T. Boy de la Tour, R. Caferra, and G. Chaminade. Some tools for an Inference Laboratory (ATINF). In *CADE-9*, pages 744–745. Springer-Verlag, LNCS 310, 1988.

[3] R. Caferra and S. Demri. Semantic entailment in non classical logics based on proofs found in classical logic, 1992. Extended version to appear.

[4] R. Caferra, S. Demri, and M. Herment. Logic morphisms as a framework for the backward transfer of lemmas and strategies in some modal and epistemic logics. In *AAAI-9*, pages 421–426. AAAI, MIT Press, July 1991.

[5] R. Caferra, M. Herment, and N. Zabel. User-oriented theorem proving with the ATINF graphic proof editor. In *FAIR' 91*, pages 2–10. Springer-Verlag, LNAI 535, 1991.

[6] H. D. Ebbinghaus. Extended logics : the general framework. In J. Barwise and Feferman S., editors, *Model theoretic logics*, pages 25–76. Springer-Verlag, 1985.

[7] M. C. Fitting. *Proof methods for modal and intuitionistic logics*. D. Reidel Publishing Co., 1983.

[8] A. Herzig. *Raisonnement automatique en logique modale et algorithmes d'unification*. PhD thesis, Université Paul Sabatier, Toulouse, July 1989.

[9] K. Konolige. *A deduction model of belief*. Pitman, 1986.

[10] C.R. Mann. Equivalence of deduction in proof theory and free cartesian closed categories. *Journal of Symbolic Logic*, 39:380–381, 1974.

[11] J. Meseguer. General logic. In H-D Ebbinghaus, editor, *Logic Colloquium '87*, pages 275–330. North-Holland, 1987.

[12] C. Morgan. Methods for automated theorem proving in non classical logics. *IEEE Transactions on Computers*, 25(8):852–862, August 1976.

[13] H.J. Ohlbach. *A resolution calculus for modal logics*. PhD thesis, FB Informatik Univ. of Kaiserslautern, 1988.

[14] H.J. Ohlbach. Context Logic. Technical report, FB Informatik Univ. of Kaiserslautern, 1989.

[15] E. Orlowska. Resolution systems and their applications I. *Fundamenta Informaticae*, 3:253–268, 1979.

[16] E. Orlowska. Resolution systems and their applications II. *Fundamenta Informaticae*, 3:333–362, 1980.

[17] D. Scott. Completeness and axiomatizability in many-valued logic. In L. Henkin et al., editor, *Tarski Symposium*, pages 411–435, 1974.

Embedding Negation as Failure into a Model Generation Theorem Prover

Katsumi Inoue Miyuki Koshimura* Ryuzo Hasegawa
Institute for New Generation Computer Technology
1-4-28 Mita, Minato-ku, Tokyo 108, Japan
inoue@icot.or.jp, koshi@icot.or.jp, hasegawa@icot.or.jp

Abstract. Here, for the first time, we give an implementation which computes *answer sets* of every class of (function-free) logic programs and deductive databases containing both negation as failure and classical negation. The proposal is based on bottom-up, incremental, backtrack-free computation of the minimal models of positive disjunctive programs, together with integrity constraints over beliefs and disbeliefs. Our translation method not only provides a simple fixpoint characterization of answer sets, but also is very helpful to understand under what conditions each model is "stable" or "unstable". The procedure has been implemented on top of the model generation theorem prover MGTP on a parallel inference machine, and has been applied to a legal reasoning system.

1 Introduction

This paper presents a novel and simple procedure which computes the models of logic programs containing *negation-as-failure* formulas. In traditional top-down proof procedures such as SLDNF-resolution [14], *not* P succeeds if there is no top-down proof of P; the meaning of negation as failure is only procedural. On the other hand, in recent theories of logic programming and deductive databases, declarative semantics have been given to extensions of logic programs, where the negation-as-failure operator is considered to be a *nonmonotonic* modal operator [6, 7, 13]. In particular, logic programs or deductive databases containing both negation as failure (*not*) and classical negation ($\neg$) can be interpreted as Reiter's default theories [19] or disjunctive default theories [9]. With these new semantics, logic programming can be used as a powerful knowledge representation tool, whose applications contain reasoning with incomplete knowledge [7, 8], expression of "don't-care" nondeterminism [20], exception handling [12], default reasoning and abduction [10].

However, for these extended classes of logic programs, the top-down approach cannot be used for computation because there is no local property in evaluating programs. For example, there has been *no* top-down proof procedure which is sound with respect to the *stable model semantics* [6] for general logic programs. Thus, we need *bottom-up* computation for correct evaluation of negation-as-failure formulas. This area is progressing, and there have been some proposals for computing stable models of general logic programs [20, 2, 22]. Unfortunately, these previous approaches are only applicable to a simple (variable-free, disjunction-free) class of programs.

*Presently at: Toshiba Information Systems, Japan.

We show a bottom-up computation of *answer sets* for any class of function-free logic programs, including the *extended disjunctive databases* proposed by Gelfond and Lifschitz [8] the proof procedure of which has not been found. Bottom-up computation reasons forwards, starting from unconditional literals, accumulating proved literals, and outputting models of programs. In evaluating *not* P in a bottom-up manner, it is necessary to interpret *not* P with respect to a fixpoint of computation because even if P is not currently proved, P might be proved in subsequent inferences. We thus come up with a completely different way of thinking for *not*: when we have to evaluate *not* P in a current world, or a partial model, instead of computing "negation by failure to prove P", we split the world into two: (1) the world where P is assumed not to hold, and (2) the world where it is necessary that P holds. Each negation-as-failure formula *not* P is thus translated into negative and positive literals with a modality expressing belief, i.e., "disbelieve P" and "believe P".

We thus provide a translation from any logic program (with negation as failure) into a *positive disjunctive program* (without negation as failure) [16] of which a *model generation theorem prover*, like SATCHMO [15] or the MGTP [5], can compute the minimal models. Some pruning rules with respect to "believed" or "disbelieved" literals are expressed as integrity constraints that are dealt with by using object-level schemata on the MGTP [11]. The MGTP then finds *all* answer sets *incrementally*, *without backtracking*, and *in parallel*. The proposed method is surprisingly simple and does not increase the computational complexity of the problem more than computation of the minimal models of positive disjunctive programs. The procedure has been implemented on top of the MGTP on a parallel inference machine, and has been applied to a legal reasoning system [17]. While we use the MGTP to generate minimal models in this paper, the proposed translation method may be linked with other methods to compute models [4] or a fixpoint construction like [18].

2 Positive Disjunctive Programs

This section shows how the MGTP [5] computes the minimal models of a *positive disjunctive program*, that is, a disjunctive database [8] which contains neither negation as failure nor classical negation. The MGTP can deal with this class of programs, and in later sections every other extended program can be shown to be translated to a positive disjunctive program.

2.1 Minimal Models

A *positive disjunctive program* [16] is a set of rules of the form:

$$A_1 \mid \ldots \mid A_l \leftarrow A_{l+1}, \ldots, A_m , \tag{1}$$

where $m \geq l \geq 0$ and each A_i is an atom. According to [8], we use the connective "|" instead of "$\vee$" although each A_i ($l \geq i \geq 1$) is a disjunct of the consequent of the rule. When $l = 0$, a rule of the form:

$$\leftarrow A_1, \ldots, A_m$$

is called an *integrity constraint.* [1]

[1] We allow for integrity constraints, that is, rules with empty consequents, in every class of logic programs and deductive databases. While this form of rules is not excluded by Gelfond and Lifschitz's definitions [6, 7, 8], the corresponding semantics are not explicitly described.

The meaning of a positive disjunctive program Σ can be given by the *minimal models* of Σ [16]. We represent the semantics in a similar way to the definition given by Gelfond and Lifschitz [8], as follows. A rule containing variables stands for the set of its ground instances. We denote by $\mathcal{L}$ the set of ground literals in the language. An *answer set* of Σ is any minimal subset S of $\mathcal{L}$ satisfying the conditions:

1. For any ground instance $A_1 \mid \ldots \mid A_l \leftarrow A_{l+1}, \ldots, A_m$ ($l \geq 1$) of any rule of Σ, if $A_{l+1}, \ldots, A_m \in S$, then for some i ($1 \leq i \leq l$), $A_i \in S$;

2. For any ground instance $\leftarrow A_1, \ldots, A_m$ of any integrity constraint of Σ, if $A_1, \ldots, A_m \in S$, then $S = \mathcal{L}$.

We say Σ is *contradictory* if it has the answer set $\mathcal{L}$. It is easy to see that a contradictory program has the unique answer set $\mathcal{L}$. Unless a program Σ is contradictory, any answer set of Σ is a set of ground atoms, and the set of answer sets of Σ is equivalent to the set of minimal models of the program when each rule of the form (1) in Σ is identified with a clause (in the sense of classical logic) of the form:

$$A_1 \vee \ldots \vee A_l \vee \neg A_{l+1} \vee \ldots \vee \neg A_m .$$

2.2 MGTP

The answer sets or the minimal models of positive disjunctive programs can be computed by using the MGTP [5]. The MGTP is a parallel and refined version of SATCHMO [15], which is a bottom-up model generation theorem prover that uses hyperresolution and case-splitting on non-unit derived clauses. In order to emphasize that the MGTP computes the models in a bottom-up manner, we express each rule of the form (1) in a positive disjunctive program as

$$A_{l+1}, \ldots, A_m \rightarrow A_1 \mid \ldots \mid A_l . \tag{2}$$

Given a positive disjunctive program Σ, the MGTP extends *model candidates* as follows. A model candidate is a subset of $\mathcal{L}$ and the initial set $\mathcal{S}_0$ of model candidates is given as $\{\emptyset\}$. Let $\mathcal{S}$ be a set of model candidates in any stage. The MGTP applies the following operation to $\mathcal{S}$ as long as possible:

For any $S \in \mathcal{S}$ and any ground instance of any rule in Σ of the form (2), if $A_{l+1}, \ldots, A_m \in S$ and $A_1, \ldots, A_l \notin S$, then remove S from $\mathcal{S}$, and add $S \cup \{A_i\}$ to $\mathcal{S}$ for every $i = 1, \ldots, l$ ($l \geq 0$).

If the MGTP cannot apply any operation to $\mathcal{S}$, it stops and returns $\mathcal{S}$.

In the above operation of the MGTP, we can deal with variables more elegantly. Instead of using ground instances of the rules, for each rule with variables in the form (2) we can obtain a substitution σ such that $A_{l+1}\sigma, \ldots, A_m\sigma$ ($l \geq 0$) is satisfied in a model candidate S. We call the process of obtaining such a substitution σ a *conjunctive matching* of the antecedent literals against elements in S. Note that this process does not need full unification if the *range-restrictedness* condition is imposed on the rules [15] . A rule is said to be range-restricted if every variable in the rule has at least one occurrence in its antecedents. It is sufficient to consider one-way unification, i.e., matching, instead of full unification with occurs check since every model candidate constructed by the MGTP in such a case contain only ground atoms. This is also a nice property for the implementation of the MGTP in KL1 (the kernel language for parallel inference machines developed at ICOT), because KL1 head unification is simply matching. The MGTP also improves the efficiency by removing redundant conjunctive matching with a ***ramified-stack*** algorithm [5].

When there are multiple rules whose antecedents are exactly the same, we want to avoid the same conjunctive matching more than once. Also, in the conversion techniques we present in later sections, we will use those rules whose consequents are disjunctions of conjunctions. For these purposes, the MGTP allows rules of the form:

$$A_{l+1}, \ldots, A_m \rightarrow A_{1,1}, \ldots, A_{1,k_1} \mid \ldots \mid A_{l,1}, \ldots, A_{l,k_l}\,, \tag{3}$$

where $m \geq l \geq 0$, $k_i \geq 1$ $(1 \leq i \leq l)$, and each A_i or $A_{i,j}$ is an atom. Each $A_{i,1}, \ldots, A_{i,k_i}$ represents a conjunction of atoms. We call a rule of this form (3) an *MGTP rule*. In summary, the MGTP operations are formally defined as follows. Let $\mathcal{S}$ be a set of model candidates in some stage, and S any element of $\mathcal{S}$.

1. **(Model candidate extension)** If there is an MGTP rule of the form

   $$A_{l+1}, \ldots, A_m \rightarrow A_{1,1}, \ldots, A_{1,k_1} \mid \ldots \mid A_{l,1}, \ldots, A_{l,k_l} \quad (l \geq 1)$$

   and a substitution σ such that $A_{l+1}\sigma, \ldots, A_m\sigma \in S$, and it does not hold that $A_{i,1}\sigma, \ldots, A_{i,k_i}\sigma \in S$ for any $i = 1, \ldots, l$, then remove S from $\mathcal{S}$, and add $S \cup \{A_{i,1}\sigma, \ldots, A_{i,k_i}\sigma\}$ to $\mathcal{S}$ for all $i = 1, \ldots, l$;

2. **(Model candidate rejection)** If there is an MGTP rule of the form

   $$A_1, \ldots, A_m \rightarrow$$

   and a substitution σ such that $A_1\sigma, \ldots, A_m\sigma \in S$, then remove S from $\mathcal{S}$.

Given a set Σ of MGTP rules and an initial set $\mathcal{S}_0$ of model candidates (usually $\mathcal{S}_0 = \{\emptyset\}$), let $\mathcal{M}_\Sigma(\mathcal{S}_0)$ be the set of model candidates closed under the above two operations. Now, let us denote by $min(\mathcal{S})$ the set of minimal (in the sense of set inclusion of literals) model candidates of $\mathcal{S}$. Then, the output of the MGTP is characterized by the fixpoint operator $MGTP$:

$$MGTP(\Sigma, \mathcal{S}_0) = min(\mathcal{M}_\Sigma(\mathcal{S}_0))\,.$$

In the following, we assume that function symbols in the language are only constants and that the number of constants is finite. When a set Σ of MGTP rules satisfies these assumptions as well as the range-restrictedness, we say Σ is *(finitely) groundable* [1]. It is easy to see that a finitely groundable program has a finite number of answer sets. For any finitely groundable set Σ of MGTP rules, the following properties hold.

Proposition 2.1 If Σ is not contradictory, $MGTP(\Sigma, \{\emptyset\})$ is equivalent to the answer sets of Σ.

Corollary 2.2 $MGTP(\Sigma, \{\emptyset\}) = \emptyset$ if and only if Σ is contradictory.

It is guaranteed that a set of clauses has at least one minimal model if it is satisfiable [1]. The next is the basic property of the MGTP as a theorem prover.

Corollary 2.3 Suppose that every rule in Σ can be identified with the corresponding clause. $MGTP(\Sigma, \{\emptyset\}) = \emptyset$ if and only if Σ is unsatisfiable.

3 General Logic Programs

This section presents how to compute the answer sets of a *general logic program* [6], that is, a logic program which contains negation-as-failure formulas but does not contain classical negation. While this class of logic programs is an instance of the more general class of disjunctive databases [8] introduced in the next section, here we shall first explain the basic idea of bottom-up computation of negation as failure.

A *general logic program* is a set of rules of the form:

$$A_l \leftarrow A_{l+1}, \ldots, A_m, \mathit{not}\, A_{m+1}, \ldots, \mathit{not}\, A_n \tag{4}$$

where, $n \geq m \geq l \geq 0$, $1 \geq l \geq 0$, and each A_i is an atom. The meaning of a general logic program is given by the *stable model semantics* [6]. The semantics again treats a rule with variables as shorthand for the set of its ground instances. Let Π be a general logic program without variables. For any set of literals $S \subseteq \mathcal{L}$, let $\mathit{reduct}(\Pi, S)$ be the set of rules without *not* obtained from Π by removing

1. every rule containing a formula $\mathit{not}\, A$ in its body with $A \in S$, and
2. every negation-as-failure formula $\mathit{not}\, A$ in the bodies of the remaining rules.

Then, S is an *answer set* (or *stable model*) of Π if S is the answer set of $\mathit{reduct}(\Pi, S)$. Note that since $\mathit{reduct}(\Pi, S)$ is a program without containing negation as failure, its answer set is already defined by its minimal model in Section 2.1. Moreover, for any set S of literals, the answer set of $\mathit{reduct}(\Pi, S)$ is uniquely determined as there is no disjunction in any rule of $\mathit{reduct}(\Pi, S)$, while Π may have multiple answer sets.

Since in any general logic program Π we allow for *integrity constraints*, i.e., rules without heads (expressed by $l = 0$ for the form (4)), Π may be *contradictory*, that is, Π may have the answer set $\mathcal{L}$. Again, unless Π is contradictory, every stable model of Π is a set of ground atoms. Unlike positive disjunctive programs, a general logic program may not have any answer set. We say Π is *incoherent* if it has no answer set. Thus, any program is either a *consistent* program (which has consistent answer sets), contradictory program, or incoherent program [10].

Notice that the above definition of stable models is not constructive; S is defined by using itself in a way that a negation-as-failure formula $\mathit{not}\, P$ is true if P is not true in S. This S can be considered as a *guess* of a possible answer set. If S coincides with the smallest set of atoms deductively closed under the rules of Π, then the guess is correct so that it is acceptable as an answer set. Hence, the most direct way to compute the stable models of Π is to generate all possible guesses and then test if each guess is correct. This method is too explosive to realize because we have to generate and test $2^{|\mathcal{A}|}$ sets of atoms, where $\mathcal{A}$ is the set of ground atoms.

We thus make the number of guesses as few as possible. Let Π_P be the set of rules in Π which do not contain *not*, and Π_N the rest of the rules in Π. If Π has consistent stable models, then every stable model S should contain the least model M of Π_P. To compute the rest of the atoms in S, we make guesses as to whether each atom P appearing as $\mathit{not}\, P$ in Π_N is present. These guesses are delayed as long as possible: if there is a rule in which all antecedents that do not contain *not* are satisfied by a set S' of atoms such that $S' \supseteq M$, then we make a guess with respect to every $\mathit{not}\, P$ in the antecedents, and extend S' either by its consequent together with the guess that all such P's will not be present or by a guess that one of such a P will be derived.

Now, we are ready to compute the stable models of Π by using the MGTP. Based on the above discussion, we translate each rule in Π of the form (4):

$$A_l \leftarrow A_{l+1}, \ldots, A_m, \, not\, A_{m+1}, \ldots, not\, A_n$$

to the following MGTP rule:

$$A_{l+1}, \ldots, A_m \rightarrow \neg\mathsf{K}A_{m+1}, \ldots, \neg\mathsf{K}A_n, \, A_l \,|\, \mathsf{K}A_{m+1} \,|\, \ldots \,|\, \mathsf{K}A_n \,. \tag{5}$$

The intuitive meaning of this MGTP rule is as follows. $\mathsf{K}A$ is a guess that A should hold, or A is *believed*, and $\neg\mathsf{K}A$ is a guess that A should not hold, or A is *disbelieved*. In other words, $\mathsf{K}A$ ($\neg\mathsf{K}A$) imposes the condition that A must hold (A must not hold). For any MGTP rule of the form (5), if a model candidate S' satisfies $A_{l+1}, \ldots, A_m$, then S' is split into $n - m + l$ ($n \geq m \geq 0$, $0 \leq l \leq 1$) model candidates.

We might relate the symbol K introduced in literals $\mathsf{K}A$, $\neg\mathsf{K}A$ with the modal operator K in an epistemic logic. We call these literals K-*literals*, and other literals without the symbol K *objective literals*. However, we avoid defining the symbol K either as a new connective or as a modal operator since we would like to remain within the MGTP calculus, or positive disjunctive programs, so that both positive and negative K-literals can be dealt with as atoms. To do so, we need to give the conditions which K-literals should satisfy. The following two schemata are thus introduced to reject model candidates when their guesses turn out to be wrong:

- If some ground instance of some atom A holds in S' and is not believed in S', then reject S'.

$$\neg\mathsf{K}A, \, A \rightarrow \quad \text{for every atom } A\,. \tag{6}$$

- If some ground instance of some atom A is believed in S' and is not believed in S', then reject S'.

$$\neg\mathsf{K}A, \, \mathsf{K}A \rightarrow \quad \text{for every atom } A\,. \tag{7}$$

Given a general logic program Π, we denote by $tr_1(\Pi)$ the set of rules consisting of the two schemata (6) and (7), and the MGTP rules obtained by replacing each rule (4) of Π with a rule (5).

The MGTP then computes the fixpoint $MGTP(tr_1(\Pi), \{\emptyset\})$ of model candidates. Each model candidate output by the MGTP contains K-literals as well as objective literals. Next is the condition that all of guesses made so far in a model candidate are correct. Let be $S' \in MGTP(tr_1(\Pi), \{\emptyset\})$.

- If any ground instance A of any atom is believed in S', then it must be true in S'.

$$\text{For every ground atom } A, \text{ if } \mathsf{K}A \in S', \text{ then } A \in S'\,. \tag{8}$$

We call condition (8) the **T**-*condition*. It is named after axiom **T** in modal logic. That is, in the fixpoint each K-literal satisfies the condition if the model candidate corresponds to a stable model. Note that the **T**-condition is used as a test and cannot be a schema because our K-literals are merely guesses and we should confirm that A is actually derived from the program. Therefore, we *cannot* write the condition as:

$$\mathsf{K}A \rightarrow A \quad \text{for every atom } A\,.$$

Now, for each model candidate S' computed by the MGTP, we denote by *objective*(S') the set of literals obtained from S' by removing all K-literals. The following two theorems guarantee that the above computation by the MGTP is sound and complete with respect to the stable model semantics.

Theorem 3.1 (1) If a model candidate S' in $MGTP(tr_1(\Pi), \{\emptyset\}) \neq \emptyset$ satisfies the **T**-condition (8), then $S = objective(S')$ is a stable model of Π.
(2) If $S \neq \mathcal{L}$ is a stable model of Π, then there is a model candidate S' in $MGTP(tr_1(\Pi), \{\emptyset\})$ such that $S = objective(S')$ and S' satisfies the **T**-condition (8).

Theorem 3.2 (1) $MGTP(tr_1(\Pi), \{\emptyset\}) = \emptyset$ if and only if Π is contradictory.
(2) Π is incoherent if and only if $MGTP(tr_1(\Pi), \{\emptyset\}) \neq \emptyset$ and there is no model candidate which satisfies the **T**-condition (8).

In order that each MGTP rule of the form (5) may be *range-restricted*, in the original rule of the form (4) from which the MGTP rule is translated, every variable has at least one occurrence in its antecedents that do not contain *not*. This restriction is as natural as the range-restrictedness in positive disjunctive programs, and can be satisfied in most AI applications. At least, rules can be easily converted in order to satisfy this kind of range-restrictedness. The MGTP gives high inference rates for range-restricted rules by avoiding computation relative to their useless ground instances.

Example 3.3 Let the general logic program Π_1 consist of the following two rules:

$$P \leftarrow not\, Q\,,$$
$$Q \leftarrow not\, R\,.$$

These two rules are translated to the following MGTP rules:

$$\rightarrow \neg \mathsf{K}Q, P \,|\, \mathsf{K}Q\,,$$
$$\rightarrow \neg \mathsf{K}R, Q \,|\, \mathsf{K}R\,.$$

Then, $tr_1(\Pi_1)$ consists of these two rules and the two schemata (6) and (7). Now, let us see how the MGTP computes the stable models of Π_1. We start from the initial model candidates $\mathcal{S}_0 = \{\emptyset\}$.

1. $\mathcal{S}_1 = \{\, \{\neg \mathsf{K}Q, P\}, \{\mathsf{K}Q\}\,\}$ by extending $\mathcal{S}_0$ with the first MGTP rule.

2. $\mathcal{S}_2 = \{\, S_1, S_2, S_3, S_4\,\}$, where $S_1 = \{\neg \mathsf{K}Q, P, \neg \mathsf{K}R, Q\}$, $S_2 = \{\neg \mathsf{K}Q, P, \mathsf{K}R\}$, $S_3 = \{\mathsf{K}Q, \neg \mathsf{K}R, Q\}$, and $S_4 = \{\mathsf{K}Q, \mathsf{K}R\}$, by extending $\mathcal{S}_1$ with the second MGTP rule.

3. $\mathcal{S}_3 = \{\, S_2, S_3, S_4\,\}$ by rejecting S_1 with the schema (6).

4. No operation is applicable to $\mathcal{S}_3$. Hence, $MGTP(tr_1(\Pi_1), \mathcal{S}_0) = \mathcal{S}_3$.

5. In $\mathcal{S}_3$, only S_3 satisfies the **T**-condition (8). Hence, $objective(S_3) = \{Q\}$ is the unique stable model of Π_1 by Theorem 3.1.

Note that when MGTP operations can be applied to model candidates with multiple rules, these rules can be processed *in any order*. Furthermore, computation is *incremental*. Consider, for example, that only the first rule of Π_1 is given at first. In this case, $\mathcal{S}_1$ is the output of the MGTP, in which only $\{\neg \mathsf{K}Q, P\}$ satisfies the **T**-condition

(8), showing that $\{P\}$ is the stable model of the current program. Then, suppose that the second rule of Π_1 is added to the program that contains only the first rule. This time we can see that the MGTP outputs $\mathcal{S}_3$ (so that the stable model is again $\{Q\}$) by using $\mathcal{S}_1$ as the initial model candidates.

Example 3.4 Let the general logic program Π_2 consist of the following four rules:

$$\begin{aligned} &R \leftarrow not\, R\,, \\ &R \leftarrow Q\,, \\ &P \leftarrow not\, Q\,, \\ &Q \leftarrow not\, P\,. \end{aligned}$$

These rules are translated to the following MGTP rules:

$$\begin{aligned} &\rightarrow \neg\mathsf{K}R, R \,|\, \mathsf{K}R\,, \\ &Q \rightarrow R\,, \\ &\rightarrow \neg\mathsf{K}Q, P \,|\, \mathsf{K}Q\,, \\ &\rightarrow \neg\mathsf{K}P, Q \,|\, \mathsf{K}P\,. \end{aligned}$$

In this example, the first MGTP rule can be further reduced to

$$\rightarrow \mathsf{K}R\,,$$

if we prune the first disjunct by the schema (6). Therefore, the rule has computationally the same effect as the integrity constraint:

$$\leftarrow not\, R\,.$$

This integrity constraint says that every answer set has to contain R: namely, *R should be derived.* Now, it is easy to see that $MGTP(tr_1(\Pi_2), \{\emptyset\}) = \{S_5, S_6, S_7\}$, where $S_5 = \{\mathsf{K}R, \neg\mathsf{K}Q, P, \mathsf{K}P\}$, $S_6 = \{\mathsf{K}R, \mathsf{K}Q, \neg\mathsf{K}P, Q, R\}$, and $S_7 = \{\mathsf{K}R, \mathsf{K}Q, \mathsf{K}P\}$. The only model candidate that satisfies the **T**-condition (8) is S_6, showing that $\{Q, R\}$ is the unique stable model of Π_2. Note that the top-down procedure proposed by Eshghi and Kowalski is not sound [3, pp.251] because it has a top-down proof of P. In our case, we can easily check that $objective(S_5) = \{P\}$ is not a stable model because S_5 contains $\mathsf{K}R$ but does not contain R.

4 Extended Disjunctive Databases

An *extended disjunctive database* [8] is a set of rules of the form:

$$L_1 \,|\, \ldots \,|\, L_l \leftarrow L_{l+1}, \ldots, L_m, not\, L_{m+1}, \ldots, not\, L_n \tag{9}$$

where $n \geq m \geq l \geq 0$ and each L_i is a literal. In particular, when an extended disjunctive database is a set or rules each of whose consequent consists of at most one literal:

$$L_l \leftarrow L_{l+1}, \ldots, L_m, not\, L_{m+1}, \ldots, not\, L_n$$

($n \geq m \geq l \geq 0$, $1 \geq l \geq 0$), it is called an *extended logic program* [7].

Both extended logic programs and extended disjunctive databases allow for *classical negation* as well as negation as failure. Answer sets of extended disjunctive databases are defined as generalizations of both minimal models of positive disjunctive programs and stable models of general logic programs, as follows.

Let Σ be any extended disjunctive database without *not*. An *answer set* of Σ is any minimal subset S of $\mathcal{L}$ satisfying the conditions:

1. For any ground instance $L_1 \mid \ldots \mid L_l \leftarrow L_{l+1}, \ldots, L_m$ $(l \geq 1)$ of any rule of Σ, if $L_{l+1}, \ldots, L_m \in S$, then for some i $(1 \leq i \leq l)$, $L_i \in S$;

2. For any ground instance $\leftarrow L_1, \ldots, L_m$ of any integrity constraint of Σ, if $L_1, \ldots, L_m \in S$, then $S = \mathcal{L}$;

3. If for some ground atom A, $A \in S$ and $\neg A \in S$, then $S = \mathcal{L}$.

Now, let Π be any extended disjunctive database without variable. For any set $S \subseteq \mathcal{L}$, let *reduct*(Π, S) be the set of rules without *not* obtained from Π by removing

1. every rule containing a formula *not* L in its body with $L \in S$, and

2. every negation-as-failure formula *not* L in the bodies of the remaining rules.

Then, S is an *answer set* of Π if S is one of the answer sets of *reduct*(Π, S).

In the same way as general or extended logic programs [10], an extended disjunctive database is classified as either a *consistent*, *contradictory*, or *coherent* database.

The effect of introducing classical negation into programs appears in the third condition of the above definition. Intuitively speaking, each answer set is a possible set of beliefs: each literal in an answer set can be considered to be true in the belief set. Since positive and negative literals have the same status, the result of negation by failure to prove an atom A does not mean that its negation $\neg A$ is true. If an answer set contains neither A nor $\neg A$, the truth value of A is *unknown* in the belief set. Thus, the answer set semantics can provide for indefinite answers in answering queries, and such unknown information can be referred to in an extended disjunctive database. By using two kinds of negation, we can easily represent incomplete knowledge [7], exceptions [12], closed world assumptions [7, 10], defaults and hypotheses [10].

To compute the answer sets of an extended disjunctive database Π, we translate each rule in Π of the form (9):

$$L_1 \mid \ldots \mid L_l \leftarrow L_{l+1}, \ldots, L_m, \mathit{not}\, L_{m+1}, \ldots, \mathit{not}\, L_n$$

to the following MGTP rule:

$$L_{l+1}, \ldots, L_m \rightarrow$$
$$\neg \mathsf{K} L_{m+1}, \ldots, \neg \mathsf{K} L_n, L_1 \mid \ldots \mid \neg \mathsf{K} L_{m+1}, \ldots, \neg \mathsf{K} L_n, L_l \mid \mathsf{K} L_{m+1} \mid \ldots \mid \mathsf{K} L_n . \qquad (10)$$

The next five schemata are introduced to reject model candidates containing wrong guesses. In below, when L is a literal, $\overline{L}$ is the literal complementary to L: for instance, when A is an atom, $\overline{A} = \neg A$ and $\overline{\neg A} = A$.

$$\neg \mathsf{K} L, L \rightarrow \quad \text{for every literal } L \in \mathcal{L}. \qquad (11)$$
$$\neg \mathsf{K} L, \mathsf{K} L \rightarrow \quad \text{for every literal } L \in \mathcal{L}. \qquad (12)$$
$$L, \overline{L} \rightarrow \quad \text{for every literal } L \in \mathcal{L}. \qquad (13)$$
$$\mathsf{K} L, \overline{L} \rightarrow \quad \text{for every literal } L \in \mathcal{L}. \qquad (14)$$
$$\mathsf{K} L, \mathsf{K} \overline{L} \rightarrow \quad \text{for every literal } L \in \mathcal{L}. \qquad (15)$$

In the above schemata, (11) and (12) correspond to the two schemata (6) and (7) for general logic programs. The schema (13) is introduced to reflect the third condition of the definition of answer sets. Namely, if for some literal L and some substitution σ, a model candidate S' contains both $L\sigma$ and $\overline{L\sigma}$, then S' is rejected. The last two schemata, (14) and (15), are used for similar purposes. Now, we denote by $tr_2(\Pi)$ the set of rules consisting of the five schemata, (11), (12), (13), (14) and (15), and the MGTP rules obtained by replacing each rule (9) of Π with a rule (10).

Finally, the **T**-condition is as follows. Let be $S' \in MGTP(tr_2(\Pi), \{\emptyset\})$.

$$\text{For every ground literal } L \in \mathcal{L}, \text{ if } \mathsf{K}L \in S', \text{ then } L \in S'. \tag{16}$$

Theorem 4.1 Suppose that Π is consistent. Then,

$$min(\{\ objective(S') \mid S' \in MGTP(tr_2(\Pi), \{\emptyset\}),\ S' \text{ satisfies the } \mathbf{T}\text{-condition (16)}\ \})$$

is equivalent to the answer sets of Π.

When the program Π is not consistent, unlike general logic programs, we cannot identify whether Π is contradictory or incoherent. Here, we have weaker results.

Proposition 4.2 (1) If Π is contradictory, then $MGTP(tr_2(\Pi), \{\emptyset\}) = \emptyset$.
(2) If $MGTP(tr_2(\Pi), \{\emptyset\}) \neq \emptyset$ and there is no model candidate satisfying the **T**-condition (16), then Π is incoherent.

From the above result, when $MGTP(tr_2(\Pi), \{\emptyset\}) = \emptyset$, we can see that Π is either contradictory or incoherent. [2] Anyway, since Π is inconsistent, if our goal is to compute consistent answer sets, this distinction is not very important. Nevertheless, we could use the following property, which was first given by [10].

Proposition 4.3 Let Π_P be the set of rules in Π not containing *not*. Π is contradictory if and only if Π_P is contradictory.

Since the MGTP can easily compute the answer sets of Π_P by using the schema (13), we can determine whether the whole database Π is contradictory or not.

Example 4.4 Let us verify Theorem 4.1 in the example of the extended disjunctive database Π_3, which is a slightly modified version of "broken-hand" example of Gelfond et al. [9]:

$$\begin{aligned}
&Lh\text{-}Usable(x) \leftarrow Person(x),\ not\ Ab1(x)\,,\\
&Rh\text{-}Usable(x) \leftarrow Person(x),\ not\ Ab2(x)\,,\\
&Ab1(x) \leftarrow \neg Lh\text{-}Usable(x)\,,\\
&Ab2(x) \leftarrow \neg Rh\text{-}Usable(x)\,,\\
&Person(A) \leftarrow,\\
&\neg Lh\text{-}Usable(A) \mid \neg Rh\text{-}Usable(A) \leftarrow\ .
\end{aligned}$$

Of these, the first two rules are translated to the following MGTP rules:

$$\begin{aligned}
&Person(x) \rightarrow \neg \mathsf{K}Ab1(x), Lh\text{-}Usable(x) \mid \mathsf{K}Ab1(x)\,,\\
&Person(x) \rightarrow \neg \mathsf{K}Ab2(x), Rh\text{-}Usable(x) \mid \mathsf{K}Ab2(x)\,.
\end{aligned}$$

[2] The reason why the MGTP cannot output any model candidate with respect to $tr_2(\Pi)$ for an incoherent database Π is that the fourth and fifth schemata (14), (15) force the MGTP to prune model candidates that may either result in an inconsistent answer set $\mathcal{L}$ or violate the **T**-condition (16). Therefore, if these two schemata could be removed from the translated program, some model candidates would remain in the final output for the incoherent database. However, such a translation would reduce the efficiency of computation. See also the proof of Theorem 4.1 in Appendix A.

It is easy to see that $MGTP(tr_2(\Pi_3), \{\emptyset\})$ contains the following two model candidates that satisfy the **T**-condition (16):

$$\{ Person(A),\ \neg Lh\text{-}Usable(A),\ Ab1(A),\ \mathsf{K}Ab1(A),\ \neg\mathsf{K}Ab2(A),\ Rh\text{-}Usable(A) \},$$
$$\{ Person(A),\ \neg Rh\text{-}Usable(A),\ Ab2(A),\ \mathsf{K}Ab2(A),\ \neg\mathsf{K}Ab1(A),\ Lh\text{-}Usable(A) \}.$$

Removing all the K-literals from these model candidates, we can get the two desired answer sets of Π_3.

5 Implementation of Schemata

5.1 Schemata on the MGTP

So far, we have represented schemata to reject model candidates that contain wrong guesses. To implement these schemata, we can simply use object-level schemata on top of the MGTP in the same way as an implementation of tableaux provers of modal logics by [11]. For an atom A, the negative literal $\neg A$ can be expresses as `-A`, and for a literal L, the K-literal $\mathsf{K}L$ can be expressed as `k(L)` in KL1, where neither "`-`" nor "`k`" appears elsewhere in the program as a predicate symbol. The following is an example of expression of the five schemata for extended disjunctive databases.

```
(11)    -k(L), L --> false.
(13)    -A, A --> false.
(14)    k(A), -A --> false.
        k(-A), A --> false.
(15)    k(-A), k(A) --> false.
```

Note that the schema (12) for extended disjunctive databases can be omitted because it is an instance of the rule (13) above. Also, the two schemata (6) and (7) for general logic programs can be expressed by the above rules (11) and (13).

By using object-level schemata on the MGTP, we can use the MGTP without any change. This has a great advantage that the inference rules (logic) and the inference engine (control) can be clearly separated. However, to improve the efficiency, we can consider another method as shown next.

5.2 Restriction of Model Candidate Extensions

In Section 2.2, we have defined the model candidate extension operation for an MGTP rule of the form (3). This operation does not distinguish K-literals from objective literals. However, to improve the efficiency, we can incorporate the schemata into the operation so that extensions should be avoided if the resultant model candidates are to be pruned immediately. For example, consider the MGTP rule (10) obtained by translating from a rule (9) in an extended disjunctive database:

$$L_{l+1}, \ldots, L_m \rightarrow$$
$$\neg\mathsf{K}L_{m+1}, \ldots, \neg\mathsf{K}L_n, L_1 \mid \ldots \mid \neg\mathsf{K}L_{m+1}, \ldots, \neg\mathsf{K}L_n, L_l \mid \mathsf{K}L_{m+1} \mid \ldots \mid \mathsf{K}L_n,$$

where $n \geq m \geq l \geq 0$. For this rule and a model candidate $S' \in \mathcal{S}$, the model candidate extension by the MGTP works as follows:

> **If** for some substitution σ, $L_{l+1}\sigma, \ldots, L_m\sigma \in S'$,
> it does not hold that $\neg\mathsf{K}L_{m+1}\sigma, \ldots, \neg\mathsf{K}L_n\sigma, L_i\sigma \in S'$ for any $i = 1, \ldots, l$,
> and $\mathsf{K}L_j\sigma \notin S'$ for any $j = m+1, \ldots, n$,

then remove S' from $\mathcal{S}$,
add $S' \cup \{\neg \mathsf{K} L_{m+1}\sigma, \ldots, \neg \mathsf{K} L_n\sigma, L_i\sigma\}$ to $\mathcal{S}$ for all $i = 1, \ldots, l$,
and add $S' \cup \{\mathsf{K} L_j\sigma\}$ to $\mathcal{S}$ for all $j = m+1, \ldots, n$.

On the other hand, a "forward-checking" version is as follows:

If for some substitution σ, $L_{l+1}\sigma, \ldots, L_m\sigma \in S'$,
$L_i\sigma \notin S'$ for any $i = 1, \ldots, l$,
and $\mathsf{K} L_j\sigma, L_j\sigma \notin S'$ for any $j = m+1, \ldots, n$,
then remove S' from $\mathcal{S}$,
add $S' \cup \{\neg \mathsf{K} L_{m+1}\sigma, \ldots, \neg \mathsf{K} L_n\sigma, L_i\sigma\}$ to $\mathcal{S}$ for each i $(1 \leq i \leq l)$ such that
$\neg \mathsf{K} L_i\sigma, \overline{L_i\sigma}, \mathsf{K}\overline{L_i\sigma} \notin S'$ and that $L_i\sigma \neq L_j\sigma$ for any $j = m+1, \ldots, n$,
and add $S' \cup \{\mathsf{K} L_j\sigma\}$ to $\mathcal{S}$ for each j $(m+1 \leq j \leq n)$ such that
$\neg \mathsf{K} L_j\sigma, \overline{L_j\sigma}, \mathsf{K}\overline{L_j\sigma} \notin S'$.

The five schemata for extended disjunctive databases are completely incorporated into the above new model candidate extension operation, which checks whether each new objective or K-literals can be safely added or not to S'. [3] At the expense of these extra checking, we can reduce the number of model candidate extensions. In practice, the cost of generating or extending model candidates and keeping them is much more expensive than the cost of these extra checking. Moreover, we can dispense with conjunctive matching of antecedents of the schemata, which is always tried against any model candidate even if it is not to be pruned. Notice also that in the above new operation, a model candidate S' will never be extended if it contains a literal $\overline{L_i\sigma}$ for any $i = 1, \ldots, l$ or any $i = m+1, \ldots, n$. This is because, if S' contains an $L_i\sigma$ $(1 \leq i \leq l)$, then it can not be extended in such a way that new objective literals are added as fewer as possible, and if S' contains an $L_j\sigma$ $(m+1 \leq j \leq n)$, then its extension will either be pruned or never increase a new objective literal.

6 Discussion

In this section, we compare the proposed method to other approaches to evaluate logic programs containing negation-as-failure formulas. The proposed method has several computational advantages: in a word, it can find *all* answer sets for *every* class of groundable logic programs or disjunctive databases, *incrementally*, *without backtracking*, and *in parallel*. We shall examine these characteristics as follows.

1. *Finding all answer sets.* This fact means that the proposed method is *complete* with respect to the answer set semantics. This is due to the fact that the MGTP [5] can find every minimal model of a positive disjunctive program. For positive disjunctive programs, bottom-up computation has recently been recognized to be more useful than top-down computation, and there has been some other methods to compute minimal models [4] or to characterize fixpoint computation [18]. Therefore, our translation method may be linked with those methods as well as the bottom-up SATCHMO [15] prover. Moreover, our method is *sound* with respect to the answer set semantics, again due to bottom-up computation. Top-down computation, on the other hand, can never guarantee the soundness even for general logic programs as shown by [3].

[3] A forward-checking mechanism will work automatically by using a new version of the MGTP (the "lazy" MGTP), which does not extend model candidates to be pruned by some integrity constraints.

2. *Applicable to every class of logic programs or deductive databases.* Several procedures have been proposed to compute the stable models of general logic programs [20, 2, 22]. However, these procedures deal only with the propositional case and none of them can be extended to allow for disjunctive databases because they are based on TMS-like algorithms. Our procedure, on the other hand, can deal with both variables and disjunctions for range-restricted rules very elegantly. Note also that our proposed method does not increase the computational complexity of the problem more than computation of the minimal models of positive disjunctive programs; the size of the translated MGTP rules is the same as the size of the original rules, and the disjuncts introduced by the translation would be of the same size of the positive literals in the heads of the positive disjunctive database if each negation-as-failure formula $not\,A$ were replaced with a negative literal $\neg A$ in the sense of classical clausal logic.

3. *Incremental, backtrack-free computation.* Since we keep K-literals in each model candidate, when new rules are added to the database, the previous set of model candidates can be used as the input to the next computation. Procedures given by [20, 22] cannot be used incrementally. Eshghi's [2] proposal may be used incrementally, but it requires an exponential-time algorithm to convert its data structures into the stable models, which is much more complicated than our use of the **T**-condition (8). Furthermore, by means of case-splitting of the MGTP, a model candidate is split into multiple model candidates, without future backtracking. We thus need not enumerate rules for their applications to model candidate extensions.

4. *Parallel implementation.* Our method is also the first attempt to compute answer sets in parallel. The procedure has been implemented on a distributed-memory multiprocessor machine, Multi-PSI, developed at ICOT. The translation is especially suitable for OR-parallelism because for each negation-as-failure formula we will make guesses to believe or disbelieve it. Multiple model candidates are thus taken as the source for exploiting OR-parallelism of the MGTP.

7 Application to Legal Reasoning

The proposed method has been applied to a legal reasoning system [17]. We can see some advantages of the proposed method from the viewpoint of this application.

It has been recognized that to use two kinds of negation, negation as failure and classical negation, are very powerful in order to represent knowledge of legislation in logic programs [12, 21]. By [21], a *primary* fact (whose proof has to be demonstrated) can be represented by a literal, while a *secondary* fact L (for which proof to the contrary must not be given) as a negation-as-failure formula $not\,\overline{L}$. Usually, in a legal reasoning system, a set of ground facts and a set or general rules or norms are given, and the goal is to obtain the possible *interpretations* containing judicial precedents. However, such an inference is plausible as it is often necessary to reason with incomplete information. Therefore, the system should create explanations or justifications why the conclusions have been derived under some legal concept. This kind of processes sometimes reflects antagonistic arguments by a jury in a court.

For the above purposes, our computation is extremely desirable. Bottom-up computation constructs the model candidates, each of which corresponds to a possible interpretation. In each model candidate, a negative K-literal $\neg \mathsf{K}\overline{L}$ represents an absence of the contrary to a secondary fact $\overline{L}$. Thus, if a proof for L could be given against $\neg\mathsf{K}L$, then the corresponding argument would be rebuted. On the other hand, since the objective literals in a model candidate that does not satisfy the T-condition is not an answer set, the model candidate can be understood as a weak argument. In this case, though, we can see that if only a literal L could be established for each positive K-literal $\mathsf{K}L$ that has not been justified in the model candidate, such an argument might become valid. Hence, those extra information represented by K-literals in model candidates can play an important role in legal reasoning.

8 Conclusion

In this paper, we have presented a novel technique to compute answer sets of logic programs or disjunctive databases. The technique is simply based on a bottom-up model generation method for positive disjunctive databases, together with integrity constraints over K-literals expressed by object-level schemata on the MGTP. The proposed translation is also very simple and does not increase the program size. Moreover, the method has been implemented on a parallel inference machine.

We should comment, though, that while our results have a useful application to legal reasoning, the general question of how to avoid combinatorial explosion in constructing model candidates still needs to be investigated. From our experience in testing our procedure for some kinds of nonmonotonic reasoning and planning, it is necessary for some applications to have a query-answering mechanism in addition to computation of the model candidates. We also expect that for different semantics of logic programs or disjunctive databases from the answer set semantics, it may be possible to have different translations into MGTP rules containing K-literals together with different integrity constraints. These issues will be discussed in a separate paper.

A Appendix: Proofs

Here, we give a sketch of the proof of Theorem 4.1. Since Theorem 3.1 is an instance of Theorem 4.1, we omit its proof. We will also omit proofs of other theorems and propositions as they can be proved more easily. In the following, Π is a finitely groundable extended disjunctive database. We can further assume without loss of generality that Π is a set of ground rule of the form (9). We use the following notations throughout this appendix. For each rule R in Π of the form (9):

$$R \;=\; L_1 \,|\, \ldots \,|\, L_l \leftarrow L_{l+1},\, \ldots,\, L_m,\, not\, L_{m+1},\, \ldots,\, not\, L_n\,,$$

we define the following:

$$\begin{aligned}
tr(R) &= L_{l+1},\ldots,L_m \rightarrow \neg\mathsf{K}L_{m+1},\ldots,\neg\mathsf{K}L_n,\, L_1 \,|\, \ldots \\
&\qquad \ldots \,|\, \neg\mathsf{K}L_{m+1},\ldots,\neg\mathsf{K}L_n,\, L_l \,|\, \mathsf{K}L_{m+1} \,|\, \ldots \,|\, \mathsf{K}L_n\,, \\
red(R) &= L_{l+1},\ldots,L_m \rightarrow L_1 \,|\, \ldots \,|\, L_l\,, \\
dis(R) &= \{L_1,\ldots,L_l\}\,, \\
pos(R) &= \{L_{l+1},\ldots,L_m\}\,, \\
neg(R) &= \{L_{m+1},\ldots,L_n\}\,, \\
\mathsf{K}neg(R) &= \{\mathsf{K}L_{m+1},\ldots,\mathsf{K}L_n\}\,, \\
\neg\mathsf{K}neg(R) &= \{\neg\mathsf{K}L_{m+1},\ldots,\neg\mathsf{K}L_n\}\,.
\end{aligned}$$

Given Π and a set of literals $S \subseteq \mathcal{L}$, we define:

$$reduct{*}(\Pi, S) = \{\, red(R) \mid R \in \Pi,\ neg(R) \cap S = \emptyset \,\} \cup \{\, \text{the schema (13)} \,\}.$$

Lemma A.1 Let be $S' \in MGTP(tr_2(\Pi), \{\emptyset\})$. If $S = objective(S')$ is a minimal set such that S' satisfies the **T**-condition (16), then $S \in MGTP(reduct{*}(\Pi, S), \{\emptyset\})$.

Proof: For every rule $tr(R)$ in $tr_2(\Pi)$, it holds that, if $pos(R) \subseteq S'$, then either $\mathsf{K}neg(R) \cap S' \neq \emptyset$, or $dis(R) \cap S' \neq \emptyset$ and $\neg\mathsf{K}neg(R) \subseteq S'$. Since S' was not pruned by the schema (11) and satisfies the **T**-condition (16), it holds that for every rule $tr(R)$ in $tr_2(\Pi)$, if $pos(R) \subseteq S'$, then either $neg(R) \cap S' \neq \emptyset$, or $dis(R) \cap S' \neq \emptyset$ and $neg(R) \cap S' = \emptyset$. Now, for each rule $tr(R)$ in $tr_2(\Pi)$, there is a rule R in Π. Therefore, for every rule R in Π, if $pos(R) \subseteq S$, then either $neg(R) \cap S \neq \emptyset$, or $dis(R) \cap S \neq \emptyset$ and $neg(R) \cap S = \emptyset$. Because of the minimality of S and the closedness of the MGTP operations, $S \in MGTP(reduct{*}(\Pi, S), \{\emptyset\})$. □

Note that by the above proof, only the two schema, (11) and (13), and the **T**-condition (16) are necessary to guarantee the soundness of computation; other schemata are used for obtaining the efficiency of computation.

Lemma A.2 If $S \in MGTP(reduct{*}(\Pi, S), \{\emptyset\})$, then S is a minimal set satisfying the condition: S is the objective literals of some model candidate S' in $MGTP(tr_2(\Pi), \{\emptyset\})$ such that S' satisfies the **T**-condition (16).

Proof: For every rule R in $reduct{*}(\Pi, S)$, it holds that, if $pos(R) \subseteq S$, then $dis(R) \cap S \neq \emptyset$. We then see that for every rule R in Π, if $pos(R) \subseteq S$, then either $neg(R) \cap S \neq \emptyset$, or $dis(R) \cap S \neq \emptyset$ and $neg(R) \cap S = \emptyset$. Now, for each rule R in Π, there is a rule $tr(R)$ in $tr_2(\Pi)$. Since $MGTP(reduct{*}(\Pi, S), \{\emptyset\})$ is closed under the MGTP operations, S is a minimal set satisfying the condition: there exists a model candidate S' in $MGTP(tr_2(\Pi), \{\emptyset\})$ such that $S = objective(S')$ and that for any rule $tr(R)$ in $tr_2(\Pi)$, it holds that, if $pos(R) \subseteq S'$, then either $\mathsf{K}neg(R) \cap S' \neq \emptyset$, or $dis(R) \cap S' \neq \emptyset$ and $\neg\mathsf{K}neg(R) \subseteq S'$. It is easy to check that this S' is not pruned by the schemata, and that S' satisfies the **T**-condition (16). □

Theorem 4.1 Suppose that Π is consistent. The answer sets of Π is equivalent to:

$$min(\{\, objective(S') \mid S' \in MGTP(tr_2(\Pi), \{\emptyset\}),\ S' \text{ satisfies the } \mathbf{T}\text{-condition (16)} \,\}).$$

Proof: A set $S \subseteq \mathcal{L}$ is an answer set of Π if S is an answer set of $reduct(\Pi, S)$.

$\Leftrightarrow S \in MGTP(reduct{*}(\Pi, S), \{\emptyset\})$ (by a variant of Proposition 2.1).

$\Leftrightarrow S$ is a minimal set satisfying the condition: S is the objective literals of some model candidate S' in $MGTP(tr_2(\Pi), \{\emptyset\})$ such that S' satisfies the **T**-condition (16) (by Lemmas A.1 and A.2). □

References

[1] G. Bossu and P. Siegel, Saturation, nonmonotonic reasoning, and the closed-world assumption, *Artificial Intelligence* **25** (1985) 23–67.

[2] K. Eshghi, Computing stable models by using the ATMS, in: *Proceedings of AAAI-90*, Boston, MA (1990) 272–277.

[3] K. Eshghi and R.A. Kowalski, Abduction compared with negation by failure, in: *Proceedings of the Sixth International Conference on Logic Programming*, Lisbon, Portugal (1989) 234–254.

[4] J.A. Fernández and J. Minker, Bottom-up evaluation of hierarchical disjunctive deductive databases, in: *Proceedings of the Eighth International Conference on Logic Programming*, Paris, France (1991) 660–675.

[5] H. Fujita and R. Hasegawa, A model generation theorem prover in KL1 using a ramified-stack algorithm, in: *Proceedings of the Eighth International Conference on Logic Programming*, Paris, France (1991) 535–548.

[6] M. Gelfond and V. Lifschitz, The stable model semantics for logic programming, in: *Proceedings of the Fifth International Conference and Symposium on Logic Programming*, Seattle, WA (1988) 1070–1080.

[7] M. Gelfond and V. Lifschitz, Logic programs with classical negation, in: *Proceedings of the Seventh International Conference on Logic Programming*, Jerusalem, Israel (1990) 579–597.

[8] M. Gelfond and V. Lifschitz, Classical negation in logic programs and disjunctive databases, *New Generation Computing* **9** (1991) 365–385.

[9] M. Gelfond, V. Lifschitz, H. Przymusińska and M. Truszczyński, Disjunctive defaults, in: *Proceedings of the Second International Conference on Principles of Knowledge Representation and Reasoning*, Cambridge, MA (1991) 230–237.

[10] K. Inoue, Extended logic programs with default assumptions, in: *Proceedings of the Eighth International Conference on Logic Programming*, Paris, France (1991) 490–504.

[11] M. Koshimura and R. Hasegawa, Modal propositional tableaux in a model generation theorem prover, in: *Proceedings of the Logic Programming Conference '91*, Tokyo, Japan (1991) 43–52 (in Japanese).

[12] R.A. Kowalski and F. Sadri, Logic programs with exceptions, in: *Proceedings of the Seventh International Conference on Logic Programming*, Jerusalem, Israel (1990) 598–613.

[13] V. Lifschitz, Nonmonotonic databases and epistemic queries, in: *Proceedings of the Twelfth International Joint Conference on Artificial Intelligence*, Sydney, Australia (1991) 381–386.

[14] J.W. Lloyd, *Foundations of Logic Programming* (Springer-Verlag, First Edition, 1984).

[15] R. Manthey and F. Bry, SATCHMO: a theorem prover implemented in Prolog, in: *Proceedings of the Ninth International Conference on Automated Deduction*, Lecture Notes in Computer Science **310**, Springer-Verlag (1988) 415–434.

[16] J. Minker, On indefinite databases and the closed world assumption, in: *Proceedings of the Sixth International Conference on Automated Deduction*, Lecture Notes in Computer Science **138**, Springer-Verlag (1982) 292–308.

[17] K. Nitta, Y. Ohtake, S. Maeda, M. Ono, H. Ohsaki and K. Sakane, HELIC-II: a legal reasoning system on the parallel inference machine, to appear in: *International Conference on Fifth Generation Computer Systems 1992*, Tokyo, Japan, June 1992.

[18] D.W. Reed, D.W. Loveland and B.T. Smith, An alternative characterization of disjunctive logic programs, in: *Proceedings of the 1991 Internatinonal Logic Programming Symposium*, San Diego, CA (1991) 54–68.

[19] R. Reiter, A logic for default reasoning, *Artificial Intelligence* **13** (1980) 81–132.

[20] D. Saccà and C. Zaniolo, Stable models and non-determinism in logic programs with negation, in: *Proceedings of the Ninth ACM SIGACT-SIGMOD-SIGART Symposium on Principles of Database Systems*, Nashville, TN (1990) 205–229.

[21] G. Sartor, The structure of norm conditions and nonmonotonic reasoning in law, in: *Proceedings of the Third International Conference on Artificial Intelligence and Law*, Oxford, England (1991) 155–164.

[22] K. Satoh and N. Iwayama, Computing abduction by using the TMS, in: *Proceedings of the Eighth International Conference on Logic Programming*, Paris, France (1991) 505–518.

Automated Correctness Proofs of Machine Code Programs for a Commercial Microprocessor[1]

Robert S. Boyer and *Yuan Yu*

Computer Sciences and Mathematics Departments
University of Texas at Austin
Austin, Texas 78712
telephone: (512) 471-9745
email: boyer@cs.utexas.edu or yuan@cs.utexas.edu

Abstract. We have formally specified a substantial subset of the MC68020, a widely used microprocessor built by Motorola, within the mathematical logic of the automated reasoning system Nqthm, i.e., the Boyer-Moore Theorem Prover [4]. Using this MC68020 specification, we have mechanically checked the correctness of MC68020 machine code programs for Euclid's GCD, Hoare's Quick Sort, binary search, and other well-known algorithms. The machine code for these examples was generated using the Gnu C and the Verdix Ada compilers. We have developed an extensive library of proven lemmas to facilitate automated reasoning about machine code programs. We describe a two stage methodology we use to do our machine code proofs.

Key words. Automated reasoning, Nqthm, Boyer-Moore Theorem Prover, formal program verification, object code, Gnu, C, Ada.

1 Introduction

One of the main reasons for our lack of confidence in computing systems is the lack of mathematical theories to forecast accurately the behavior of computing systems. The idea of providing a rigorous mathematical basis for programming dates back to the very beginning of computing. In the classic papers of von Neumann and Goldstine [7], which introduced the first "von Neumann machine," they described how to prove the correctness of machine code programs. Fifteen machine code programs are there specified, coded, and proved correct. Later, methods for proving the correctness of programs written in higher level programming languages were put forward by McCarthy, Floyd, Hoare and others.

A formal program verification is a mathematical proof that a program executed according to a certain model of computation meets some specification. Such a proof involves a formal specification of the computing model on which the program is intended to execute and a formal reasoning system on which the proof is based. Once a program is formally verified, people may place a very high degree of confidence in the correctness of the program. Because correctness proofs can be extremely tedious, it seems to be difficult for humans to check (most importantly, correctly) all the proof details. Automated reasoning systems have been successfully used to assist humans to check the correctness of some computer programs (see, for example, the survey paper [3] and the more recent, large scale efforts [9,1]).

We have recently used the automated reasoning system Nqthm [4] to define formally a mathematical specification for the widely used Motorola MC68020 microprocessor and to verify mechanically the correctness of machine code generated

[1]The work described here was supported in part by NSF Grant MIP-9017499.

by high-level programming language compilers for that microprocessor. We believe our work differs from almost all the other research in formal program verification, which almost exclusively has focused on proving the correctness of programs written in high level programming languages, and has not undertaken the goal of assuring that the software correctly executes upon hardware that is commonly used. In only a very few cases does research on formal program correctness come close to the hardware level, and, as far as we are aware, when it does, the hardware is either "on paper" or very novel. For earlier examples of Nqthm-checked program proofs that are rooted in an interpreter semantics approach, see the exemplary work of Bevier (on Kit), Hunt (on FM8502), Moore (on Piton), and Young (on Micro-Gypsy), all described in [1]. For related work, see also [6].

There are several motivations that led us to study program verification at the machine code level.

- Formal specification and verification at the processor level, e.g., for a compiler correctness proof, is ultimately a necessary ingredient in formal correctness proofs of programs if we take as our goal ensuring that programs are executed correctly on a particular processor.
- Some of the most sensitive programs in the world are currently "verified" at the machine code level anyway, even though written in higher level programming languages—compilers are simply not trusted by those to whom security matters most.
    - Many high level programming languages, such as those typically used in industrial practice, are not precisely specified. It is not easy, or even possible, to give the semantics of some programming features. For example, the `volatile` construct in C.
    - Some "industrial strength" compilers produce erroneous code. But given the sheer size of these compilers, it is unlikely that any will be proved correct in the foreseeable future.
- Programs written in high level languages may need to have assembly code embedded in them in order to communicate with the outside world. No high level language specification can make clear the semantics of the embedded assembly instructions. Furthermore, any sort of real-time analysis is (currently) best done at the machine code level.

It is our belief that success in verifying machine code programs will be a step towards incorporating formal verification into the "real" world of programming.

To familiarize the reader with the features of Nqthm, Section 2 is dedicated to a brief overview of the Nqthm logic and its theorem prover. In Section 3, we briefly describe our formal specification of the user instruction set of the MC68020. For more details of the specification, we refer the reader to the complete documentation of our specification in [5]. In Section 4, we discuss how we prove the correctness of object code generated from C and Ada programs.

2 The Automated Reasoning System Nqthm

The automated reasoning system Nqthm, also known as "the Boyer-Moore Theorem Prover," is a Common Lisp program for proving mathematical theorems. In the twelve years since *A Computational Logic* [2] was published, Nqthm has been used to specify and check the correctness of numerous theorems from many areas of mathematics and computing. An extensive partial listing may be found in [4, pages 5–9]. See also [1]. This section contains an extremely brief overview of Nqthm that may suffice for understanding the rest of this paper. For a thorough and precise description of the Nqthm logic, we refer the reader to the rigorous treatment in [4], especially Chapter 4, in which the logic is precisely defined.

The logic of Nqthm is a quantifier-free first order logic with equality. The syntax is similar to that of Pure Lisp or the lambda calculus, e.g., prefix notation with parentheses. Constants in the logic are functions with no arguments. The basic theory includes axioms defining (1) the Boolean constants `(TRUE)` and `(FALSE)`, abbreviated as `T` and `F`, respectively; (2) the equality function `(EQUAL X Y)`, axiomatized to return `T` or `F` according to whether `X` is `Y`; (3) the if-then-else function `(IF X Y Z)`, axiomatized to return `Z` if `X` is `F` and `Y` otherwise; and (4) the Boolean arithmetic functions `AND`, `OR`, `NOT`, `IMPLIES`, and `IFF`.

The logic of Nqthm contains two "extension" principles under which the user can introduce new concepts into the logic with the guarantee of consistency. The *Shell Principle* allows the user to add axioms introducing "new" inductively defined "abstract data types." Nonnegative integers, ordered pairs, e.g., `(cons 1 2)`, and symbols, e.g., `'running`, are axiomatized in the logic by adding shells. The *Definition Principle* allows the user to define new functions in the logic. Among the functions in the logic are `PLUS`, `DIFFERENCE`, `LESSP`, `TIMES`, `QUOTIENT`, and `REMAINDER`, which correspond to $+, -, <, *, /$, and mod, respectively.

Nqthm is a mechanization of this logic. It takes as input a term in the logic, and repeatedly transforms it in an effort to reduce it to non-`F`. Many heuristics and decision procedures are implemented as part of the transformation mechanism.

The commands to the theorem prover include those for defining new functions, proving lemmas, and adding shells, etc. Two commands are the most often used. (1) `(defn` *fn-name* `(`*arg1 ... argn*`)` *body*`)` defines a new function *fn-name* with *n* arguments *arg1,..., argn* and definition *body*. (2) `(prove-lemma` *lemma-name* `(`*lemma-type*`)` *statement*`)` initiates a proof attempt for conjecture *statement*. If the proof succeeds, the lemma is stored as a rule with type *lemma-type*, under the given name *lemma-name*. The behavior of the prover is influenced profoundly by the proof of appropriate lemmas.

3 The MC68020 Instruction Set Specification

We have formalized most of the user programming model of the MC68020 microprocessor. The formal specification is intended to reflect as closely as possible the "user's manual view" of the MC68020 [10]. We, at the present time, have avoided considering the supervisor level of the MC68020. Any exception caused by user programs simply halts our formalized machine. Before presenting our formal specification, we first give an informal description of the user programming model of the MC68020 and our formalism.

3.1 Formalizing the MC68020 Instruction Set Architecture

We have followed the state transition approach to formalizing the MC68020 microprocessor. We define the MC68020 as an abstract machine and the MC68020 instructions as operations on the states of the abstract machine. We specify the "semantics" of this abstract machine as a function in the Nqthm logic in the most

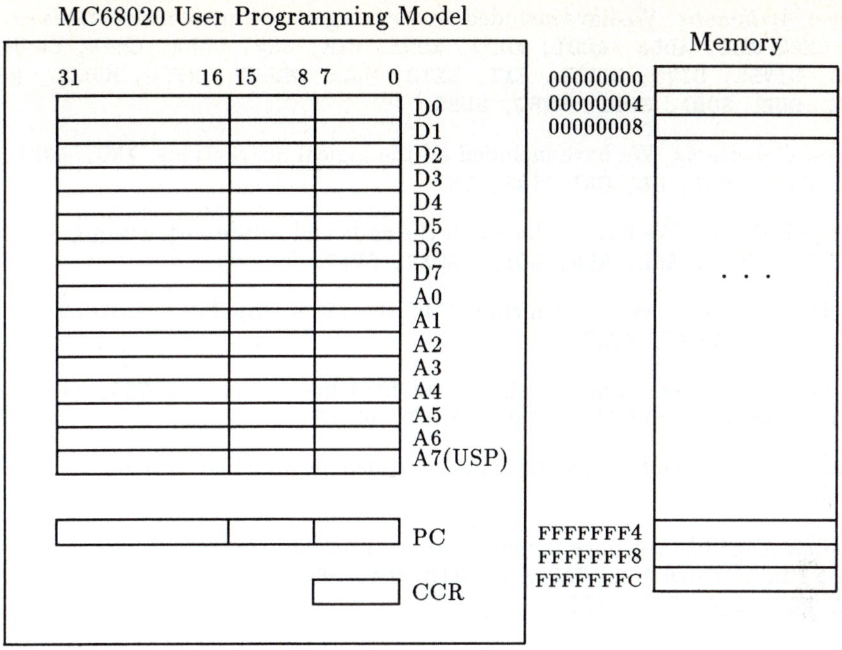

Figure 1: The User Visible Machine State

straightforward way: fetch the current instruction in the current state, decode the instruction, perform the operation, and return a new machine state suitably altered.

Figure 1 provides a two dimensional picture of the user "programming model" for the MC68020, as described in [10]. This model has 16 32-bit general-purpose registers (8 data registers, D0-D7, and 8 address registers, A0-A7), a 32-bit program counter PC, and an 8-bit condition code register, CCR. The address register A7 is also used as the user stack pointer (USP). The 5 least significant bits in CCR are condition codes for carry, overflow, zero, negative, and extend. This "model" is the only part of the state of an MC68020 that a user program can read or write under our formal semantics. Not present in this model are such arcane actualities as the instruction cache, memory management, and the supervisor stack.

In our formalization, we have focused exclusively on the user available instructions of the MC68020 instruction set. Our specification consists of about 80% of all the user available instructions. Most of the instructions we have left unspecified have some *undefined* effects on the machine state. For example, some of the condition codes of the instruction `CMP2` are described as *undefined* in [10]. We have deliberately excluded these instructions in our specification. Fortunately, these instructions constitute only a small portion of the instruction set, and most of them are rarely used. We summarize below those instructions formalized.

The instructions of the MC68020 instruction set are classified into ten categories according to their functions [10].

1. *Data Movement.* We have included all the data movement instructions: `EXG`, `LEA`, `LINK`, `MOVE`, `MOVEA`, `MOVEM`, `MOVEP`, `MOVEQ`, `PEA`.

2. *Integer Arithmetic.* We have included all the integer arithmetic instructions except `CMP2`: `ADD, ADDA, ADDI, ADDQ, ADDX, CLR, CMP, CMPA, CMPI, CMPM, DIVS, DIVSL, DIVU, DIVUL, EXT, EXTB, MULS, MULSL, MULU, MULUL, NEG, NEGX, SUB, SUBA, SUBI, SUBQ, SUBX.`

3. *Logical Operations.* We have included all the logical instructions: `AND, ANDI, EOR, EORI, NOT, OR, ORI, TAS, TST`

4. *Shift and Rotate.* We have included all the shift and rotate instructions: `ASL, ASR, LSL, LSR, ROL, ROR, ROXL, ROXR, SWAP.`

5. *Bit Manipulation.* We have included all the bit manipulation instructions: `BCHG, BCLR, BSET, BTST.`

6. *Bit Field.* We have included all the bit field instructions: `BFCHG, BFCLR, BFEXTS, BFEXTU, BFFFO, BFINS, BFSET, BFTST.`

7. *Binary coded decimal.* None of the binary coded decimal instructions has been considered.

8. *Program Control.* We have included all the program control instructions except a pair of instructions `CALLM` and `RTM`: `Bcc, DBcc, Scc, BRA, BSR, JMP, JSR, NOP, RTD, RTR, RTS.`

9. *System Control.* Only 5 of the 21 system control instructions are formalized: `ANDI to CCR, EORI to CCR, MOVE from CCR, MOVE to CCR, ORI to CCR.`

10. *Multiprocessor.* None of the multiprocessor instructions have been considered.

We have formalized all eighteen MC68020 addressing modes. An addressing mode can specify a constant that is the operand, a register that contains the operand, or a location in memory where the operand is stored. For a complete description of the MC68020 addressing modes, we refer the reader to Motorola's MC68020 user's manual [10].

3.2 The Formal Instruction-Level Specification

Before we present some details of the formal specification, we first formally define the user visible state and its internal representation.

3.2.1 The User Visible State

The only type of object manipulated at the instruction level is the *bit vector*, which is represented as a nonnegative integer in our specification. For example, the value of the program counter is represented as a nonnegative integer with range between 0 and $2^{32}-1$, inclusive. Each of the operations on bit vectors can then be formalized as some sort of operation on nonnegative integers. Here are a few basic operations on bit vectors and their definitions in the Nqthm logic.

```
(defn head (x n)
  (remainder x (exp 2 n)))

(defn tail (x n)
```

```
  (quotient x (exp 2 n)))

(defn bits (x i j)
  (head (tail x i) (plus 1 (difference j i))))

(defn app (n x y)
  (plus (head x n) (times y (exp 2 n))))
```

Intuitively, **head** returns the bit vector of the first **n** bits of **x**; **tail** returns the bit vector obtained by discarding the first **n** bits of **x**; **bits** returns the bit vector consisting of bits **i** through **j** of **x**; **app** returns the bit vector obtained by concatenating **x** and **y**.

A user visible state is represented as a list of length five, e.g.,

```
(list status regs pc ccr mem)
```

where the content of each of the five fields has the following interpretation:

- **status** is the machine status word, which is either the symbol **'running** or some error message if an exception occurs. This status field is not actually present in any MC68020. Rather, it is the artifice of our state formalization by which we indicate that an actual error has arisen or that an aspect of the MC68020 not defined in our formalization has been encountered during execution.
- **regs** is the register file, which is represented as a list of 16 nonnegative integers.
- **pc** is the program counter, which is represented as a nonnegative integer.
- **ccr** is the condition code register, which is represented as a nonnegative integer.
- **mem** is the memory, which is represented as a pair of binary trees. A binary representation for memory provides some efficiency for simulating MC68020 instructions. One of the binary trees is a formalization of memory protection—one may specify that any byte of memory is **'ram**, **'rom**, or **'unavailable**; the other binary tree holds the data, i.e., the actual bytes stored. As discussed elsewhere in this paper, we use the notion of read-only memory to deal with the issue of cache consistency. We also believe that it is unrealistic to assert the correctness of machine code programs without carefully characterizing which parts of memory are read and written—few MC68020 chips are connected to a full 4 gigabytes of RAM. Memory protection issues are not specified in [10].

mc-status, **mc-rfile**, **mc-pc**, **mc-ccr** and **mc-mem** are accessors to the machine status word, the register file, the program counter, the condition codes and the memory, respectively.

3.2.2 The Specification

The top-level loop of our specification is defined by a pair of functions, the "single-stepper" **stepi** and the "stepper" **stepn**:

```
(defn stepi (s)                        ; the single stepper.
  (if (evenp (mc-pc s))
      (if (pc-word-readp (mc-pc s) (mc-mem s))
```

```
        (execute-ins (current-ins (mc-pc s) s)
                     (update-pc (add (l) (mc-pc s) (wsz)) s))
      (halt (pc-signal) s))
   (halt (pc-odd-signal) s)))

(defn stepn (s n)                    ; executes n instructions.
  (if (or (mc-haltp s) (zerop n))
      s
    (stepn (stepi s) (sub1 n))))
```

stepi calls **execute-ins** to compute the new machine state from the current state **s** by executing the current instruction if the program counter is aligned on a word boundary, as required by the MC68020, and points to readable memory. **stepn** simply executes **n** instructions by calling the single stepper **stepi**.

Roughly speaking, **execute-ins** decodes the current instruction according to the opcode and jumps to the specification of the instruction identified. We formalize each individual instruction with one main function and a few auxiliary functions, which calculate the effective addresses, fetch the operands, perform the specific operation, update the condition codes, and store the results.

Altogether, our formal specification is about 128,000 bytes long, which takes up approximately 80 pages of text. It consists of 569 function definitions. About two thirds of the specification is devoted to the formalization of individual instructions.

In addition to using Nqthm to prove general theorems about the correctness of MC68020 programs under the semantics provided by **stepn**, as we discuss in subsequent sections, it is noteworthy that it is actually possible, within Nqthm, to "run" **stepn** on concrete data. That is, Nqthm together with **stepn** provides a simulator for the MC68020, albeit one that requires approximately 1,000,000 Sun-3 (MC68020) instructions to simulate a single MC68020 instruction. We mention this simulation possibility only to emphasize the important point: our "semantics" for the MC68020 is an *operational* semantics in the strictest sense of the word. There are several advantages to having such an operational characterization of the semantics of our computational model:

- It is possible to "test" the specification's correctness by executing it on specific data and comparing the result with the behavior of an actual MC68020. (While testing does not find all bugs, it does find some!)
- By giving the MC68020 semantics entirely with definitions instead of with an *ad hoc* collection of axioms, we are guaranteed that the specification is consistent, relative to the consistency of elementary number theory.

Rather than describing the details of any particular instruction specification, we instead focus on some of the interesting issues that have come up in the specification.

Cache Consistency. The MC68020 has an on-chip instruction cache, and a write operation does not invalidate or modify the corresponding entry in the instruction cache. Rather than formalizing the details of the MC68020 cache (which usually changes from MC680x0 processor to processor), we have adopted, for the time being, the strategy of *requiring* that instruction fetches be from *read-only* parts of the memory, and therefore, if the instruction cache is entirely valid at the beginning of the execution, it will remain valid all throughout the execution.

Evaluation Order We found some MC68020 instructions are sensitive to internal evaluation order. For instance, the MOVE instruction has two effective address calculations. Because of the side effect of effective address calculation, it is necessary to know which address is calculated first. This information is not specified in the Motorola literature, but by speaking with Motorola engineer Jim Eifert in April 1990, we learned that it is an internal Motorola policy that the source effective address is always calculated first.

Condition Code Computation. Ideally, we would specify the condition codes in a way most natural to the "user." But in order to assure full compliance with the MC68020 specification [10], we have followed the syntactical definition described in Table 3-11 of [10]. For instance, we define the carry bit of the SUB instruction as follows:

```
(defn sub-c (n sopd dopd)
  (let ((result (sub n sopd dopd)))
    (b-or (b-or (b-and (bitn sopd (sub1 n))
                       (b-not (bitn dopd (sub1 n))))
                (b-and (bitn result (sub1 n))
                       (b-not (bitn dopd (sub1 n)))))
          (b-and (bitn sopd (sub1 n)) (bitn result (sub1 n))))))
```

To paraphrase this, the carry bit is set to $(Sm \wedge \overline{Dm}) \vee (Rm \wedge \overline{Dm}) \vee (Sm \wedge Rm)$, where Sm, Dm, and Rm denote the most significant bit of source, destination and result, respectively. This characterization is perhaps not the way the user views the carry bit of a SUB (subtraction) instruction! One of the problems we have to deal with in the verification phase is to eliminate these "semantic gaps."

Effective Address Calculation. The MC68020 provides a very rich set of addressing modes. The definition of effective address calculation is rather complicated and required great pain to formalize completely.

4 Machine Code Verification

Among the possible applications of the MC68020 formal specification, we are currently primarily concerned with studying the verification of specific machine code programs. To date we have successfully verified many small machine code programs generated from their C and Ada counterparts with the Gnu C compiler and the Verdix Ada compiler. In this section, we will briefly report our work in this direction.

4.1 A Library of Lemmas

The development of lemmas is the key to success in any use of an interactive theorem proving system, certainly of Nqthm. Lemmas are saved as derived inference rules that affect the future behavior of the system. The quality of the lemmas often determines the success of the entire proof effort. Our approach to developing a lemma library can be roughly viewed as "bottom-up." We carefully study each of the concepts involved, in the hope of proving a set of lemmas that fully characterizes these concepts. In general, the library is intended to be the mechanization of a basic theory of formal reasoning about machine code programs which will have utility in the verification of many different programs.

We have invested a vast amount of time creating our lemma database, probably the most time spent in the entire project. Currently, our library of lemmas is about 180,000 bytes, or 100 pages, long. Our experiments with the library, the topic of the next section, have been very satisfactory. Next, we briefly review some of the important issues we have dealt with in our development of the library.

4.1.1 Arithmetic

All the bit vector operations are defined with nonnegative integer arithmetic; hence theorems about bit vectors are merely theorems about nonnegative integer arithmetic. We have focused on reasoning about these functions: **plus**, **difference**, **times**, **remainder**, **quotient** and **exp**. During the development, we have been greatly benefited from an integer library developed at Computational Logic, Inc.

Due to the fixed size of operations at the MC68020 machine level, it is inevitable that we study modulo arithmetic. Our purpose here is to establish a set of proof rules to support modulo arithmetic reasoning at a relatively high level. For example, addition is defined here as $(x + y) \bmod 2^n$:

```
(defn add (n x y) (head (plus x y) n))
```

One of the rules for modulo addition is associativity:

```
(prove-lemma add-associativity (rewrite)
     (equal (add n (add n x y) z) (add n x (add n y z))))
```

4.1.2 Alternative Interpretations

We followed Motorola's description of the MC68020 very literally while writing our formal specification. But it is sometimes the case that the descriptions Motorola provides are not the most useful mathematical characterizations of operations to use when it is time to prove the correctness of particular machine code programs. An important type of lemma in our library is the kind which expresses in a more useful or intuitive fashion the semantics of a machine code operation. For example, we have previously discussed the rather syntactic formulation of the changes to the condition codes given in the Motorola manual. The following lemma establishes, roughly speaking, that the carry bit after a subtraction instruction is set iff $y < x$.

```
(prove-lemma sub-bcs&cc (rewrite)
    (implies (and (nat-rangep x n)
                  (nat-rangep y n)
                  (not (zerop n)))
             (equal (bcs (sub-c n x y))
                    (if (lessp (nat-view y) (nat-view x))
                        1 0))))
```

In a similar vein, the following lemma characterizes in arithmetic terms (using exponentiation) the effects of arithmetic shifting, which is defined in the specification by "bit movement." Roughly, this lemma says that shifting x left s bits equals multiplying x by 2^s, if there is no overflow.

```
(prove-lemma asl-int (rewrite)
     (implies (and (nat-rangep x n)
                   (int-rangep (nat-to-int x n) (difference n s)))
              (equal (nat-to-int (asl n x s) n)
                     (itimes (nat-to-int x n) (exp 2 s)))))
```

4.1.3 Memory Management

Memory management is probably the most difficult part of the library. It mainly concerns general theorems about the machine state and its components. For example, the following lemma "tells" the prover how to read the byte at memory location x.

```
(prove-lemma byte-read-write (rewrite)
     (equal (byte-read x (byte-write v y mem))
            (if (mod32-eq x y)
                (if (nat-rangep v 8) (fix v) (head v 8))
              (byte-read x mem))))
```

Roughly, this says that the result of reading at location x after writing v at location y is either v or the previous contents of x, according to whether x is y or not.

4.2 Correctness Proofs

We turn now to the most interesting part of our project—the correctness proof of object code generated from higher level languages. In this section, we will explain our approach with the correctness proof of a machine code program that computes the greatest common divisor of two nonnegative integers by Euclid's algorithm. To obtain our machine code program, we start with the following C program.

```
int gcd(int a, int b)
{
  while (a != 0){
    if (b == 0) return (a);
    if (a > b)
      a = a % b;
    else b = b % a;
  };
  return (b);
}
```

We next run this C program through the Gnu C compiler, load the object code into memory, and use the Gnu debugger to obtain the machine code both in symbolic format (for human consumption, only) and in numeric format (for Nqthm's consumption). The symbolic format is:

```
gcd:            linkw a6,#0
                moveml d2-d3,sp@-
                movel a6@(8),d2
                movel a6@(12),d3
gcd+16:         tstl d2
                beq 0x22f6 <gcd+48>
                tstl d3
```

```
                bne 0x22e2 <gcd+28>
                movel d2,d0
                bra 0x22f8 <gcd+50>
gcd+28:         cmpl d2,d3
                bge 0x22ee <gcd+40>
                divsll d3,d0,d2
                movel d0,d2
                bra 0x22d6 <gcd+16>
gcd+40:         divsll d2,d0,d3
                movel d0,d3
                bra 0x22d6 <gcd+16>
gcd+48:         movel d3,d0
gcd+50:         moveml a6@(-8),d2-d3
                unlk a6
                rts
```

The numeric format, expressed as a list of nonnegative integers, is given by this Nqthm function:

```
(defn gcd-code ()
  '(78    86    0     0     72    231   48    0
    36    46    0     8     38    46    0     12
    74    130   103   28    74    131   102   4
    32    2     96    22    182   130   108   8
    76    67    40    0     36    0     96    232
    76    66    56    0     38    0     96    224
    32    3     76    238   0     12    255   248
    78    94    78    117))
```

4.2.1 The Correctness Statement

The correctness statement for the foregoing GCD program should fully characterize the effects of the execution of the program on the machine state. The most important requirement of the correctness statement is that it be "context-free" and "universally" applicable. In our formalism, the correctness of a machine code program means:

- The execution terminates, and the new machine state is "normal," e.g., no read or write to unavailable memory occurred, no illegal instruction was executed.
- The program counter is set to the "right" location.
- The correct results are stored in the right place.
- The register file is properly managed, e.g., A7, the User Stack Pointer, is set to the right location, and some registers used as temporary storage are restored to their original values.
- The program only accesses and changes the intended portion of memory.

In our example, the correctness of our GCD program is given by the following three theorems, which formalize exactly what we have described above.

```
(prove-lemma gcd-correctness (rewrite)
   (implies (gcd-statep s a b)
            (and (equal (mc-status (stepn s (gcd-t a b))) 'running)
                 (equal (mc-pc (stepn s (gcd-t a b))) (rts-addr s))
                 (equal (iread-dn 32 0 (stepn s (gcd-t a b))) (gcd a b))
                 (equal (read-an 32 6 (stepn s (gcd-t a b)))
                        (read-an 32 6 s))
                 (equal (read-an 32 7 (stepn s (gcd-t a b)))
                        (add 32 (read-an 32 7 s) 4)))))

(prove-lemma gcd-rfile (rewrite)
   (implies (and (gcd-statep s a b)
                 (d2-7a2-5p rn)
                 (leq oplen 32))
            (equal (read-rn oplen rn (mc-rfile (stepn s (gcd-t a b))))
                   (read-rn oplen rn (mc-rfile s)))))

(prove-lemma gcd-mem (rewrite)
   (implies (and (gcd-statep s a b)
                 (disjoint x k (sub 32 12 (read-sp s)) 24))
            (equal (read-mem x (mc-mem (stepn s (gcd-t a b))) k)
                   (read-mem x (mc-mem s) k))))
```

`(gcd-statep s a b)`, which is the hypothesis that specifies the assumptions on the initial state, asserts, roughly speaking:

- The machine state s is in the user mode.
- The program counter of s is even.
- The 60 consecutive bytes in the memory of s, starting from the address pointed to by the program counter of s, store the GCD program given above by `(gcd-code)`.
- There is "enough" space on the stack.
- The integers a and b are on the stack, and both are nonnegative.

Informally, the theorem `gcd-correctness` states that if s is as characterized by `(gcd-statep s a b)`, then there is an integer n, given by the expression `(gcd-t a b)`, which tells us how many instructions to run the MC68020 starting with s before the GCD program returns, such that after running s for n steps the resulting state s' has these properties:

- s' is still `'running`, i.e., no errors occurred.
- The pc of s' points to the return address on the top of the stack in s.
- Register D0 of s' contains `(gcd a b)`, the greatest common divisor of a and b.
- Register A6, which is used by the LINK instruction, is unchanged.
- Register A7, the stack pointer, has been incremented by 4.

The theorem **gcd-rfile** further asserts that all of the registers of s' have the same values as those of s, except D0, D1, A0, A1, A6, and A7. Finally, the theorem **gcd-mem** asserts that every memory location of s' has the same value as it did in s, except for the 12 bytes on either side of the stack pointer of s. (**read-rn** and **read-mem** are the primary functions for reading the contents of the registers and memory.)

The completeness of detail that is necessary when proving the correctness of machine code programs is especially obvious when one verifies recursive machine code programs, such as we have done with Quick Sort.

4.2.2 The Proof

In our approach to the verification of specific machine code programs, there are always two independent phases: one deals with the correctness of the underlying algorithm and the other deals with the correctness of its implementation. Success in separating the two issues and tackling each of them in isolation makes the correctness proof easier. Therefore, our correctness proofs are always divided into two steps:

1. We formalize the underlying algorithm as a function in the Nqthm logic and prove the equivalence of the algorithm with the result of running the MC68020 specification on the given machine code. What we establish in this step is that the implementation does implement the algorithm. Note that this says nothing about the correctness of the algorithm.

2. We prove that the algorithm, formalized as an Nqthm function, is correct. Note we do not need to deal with any MC68020 related specifics in this step. So, we can focus completely on the mathematical properties of the algorithm.

Thus, for the previous example, we formalize the Euclid's algorithm in Nqthm as follows:

```
(defn gcd (a b)
  (if (zerop a) (fix b)
  (if (zerop b) a
  (if (lessp b a) (gcd (remainder a b) b)
                  (gcd a (remainder b a))))))
```

The first step is to prove that the functional behavior of the machine code is equivalent to the above function. The second step is to prove that the above function does indeed compute the greatest common divisor of a and b, which is proved by the following two theorems:

```
; (gcd a b) is a common divisor of a and b.
(prove-lemma gcd-is-cd (rewrite)
             (and (equal (remainder a (gcd a b)) 0)
                  (equal (remainder b (gcd a b)) 0)))

; (gcd a b) is the greatest, i.e., it divides any common divisor of a
; and b.
(prove-lemma gcd-the-greatest (rewrite)
             (implies (and (not (zerop a))
                           (not (zerop b))
                           (equal (remainder a x) 0)
```

```
            (equal (remainder b x) 0))
        (not (lessp (gcd a b) x))))
```

4.2.3 Timing Analysis for GCD

The function `gcd-t`, which was used above in the theorem `gcd-correctness`, returns the exact number of MC68020 instructions executed by the GCD program. The definition of `gcd-t` is:

```
(defn gcd-t1 (a b)
  (if (zerop a) 6
  (if (zerop b) 9
  (if (lessp b a) (plus 9 (gcd-t1 (remainder a b) b))
                  (plus 9 (gcd-t1 a (remainder b a)))))))

(defn gcd-t (a b) (plus 4 (gcd-t1 a b)))
```

Using the definition of `gcd-t`, we have mechanically proved that the number of instructions executed by the GCD program is at most 598.

To study the real-time bounds of programs, we need to incorporate time information for each individual instruction, which seems to us a quite natural extension to our specification.

5 Conclusions

Our experience with machine code verification has been encouraging. We have managed to verify, with Nqthm, the object code produced by the Gnu C compiler for some of the C functions in Kernighan and Ritchie's book [8], such as binary search, Quick Sort, and also some C library functions, e.g., strlen and strcpy. We have also mechanically verified the object code produced by the Verdix Ada compiler for an integer square root algorithm. This success leads us to believe that it is completely feasible to verify the object code of moderate pieces of software in a reasonable time span. To scale up, some automated tools must be developed to assist the user in the proof. At the present time, we have not implemented any such tools. But we have been considering the implementation of something like a verification condition generator for machine code reasoning.

Our MC68020 specification has been greatly influenced by the need for reasoning about machine code. This consideration has complicated, to some extent, our specification task.

Building the library of lemmas consumes most of our energy. It is still under development as we are increasing our ability to handle more and more programming language constructs and data types. It is interesting to see whether we can apply the mathematics so developed to another computer architecture, say, RISC; we believe we can. To ensure that any change is indeed an improvement, we have followed the "proveall" discipline described in [4], i.e., the practice of making sure that after we make changes to our specification or lemmas, we can still prove the most important of our previous results.

Much can be said about future directions for such research. To specify supervisor mode, especially interrupt handling, is both desirable and challenging, but we have not yet attended to it. The formal specification we have already developed will let us investigate: (a) the correctness of some moderate sized piece of software that is in critical use; (b) the real-time execution bounds of programs; (c) the correctness of high level programming language compilers; and (d) the correctness of some lower level software, e.g., software for cache and memory management. We believe that success in any of these directions would be a major contribution to formal reasoning.

It can be expected that the full details of this project, including the complete MC68020 specification, lemma library, and programs proved will appear in subsequent publications, for example in the Ph. D. dissertation of Yuan Yu.

6 Acknowledgements

We would like to thank Bill Bevier, Don Good, Warren Hunt, Matt Kaufmann, J Moore, and Bill Schelter for their many constructive suggestions and discussions. Special thanks to Fay Goytowski for her meticulous reading of the MC68020 specification, which revealed a dozen or so errors. The general style of Nqthm formalization used in this MC68020 specification is the product of over a decade of study by the authors of Nqthm and their students. Especially influential was the FM8502 and Piton work [1]. The development of Nqthm was primarily supported by NSF, ONR, and DARPA.

References

[1] William Bevier, Warren Hunt, J Strother Moore, and William Young. Special issue on system verification. *Journal of Automated Reasoning*, 5(4), 1989.

[2] Robert S. Boyer and J Strother Moore. *A Computational Logic*. Academic Press, New York, 1979.

[3] Robert S. Boyer and J Strother Moore. Program verification. *Journal of Automated Reasoning*, 1(1):17–23, 1985.

[4] Robert S. Boyer and J Strother Moore. *A Computational Logic Handbook*. Academic Press, 1988.

[5] Robert S. Boyer and Yuan Yu. A formal specification of some user mode instructions for the Motorola 68020. Technical Report TR-92-04, Computer Sciences Department, University of Texas at Austin, 1992.

[6] Jeffrey V. Cook. Verification of the C/30 microcode using the state delta verification system (SDVS). In *13th National Computer Security Conference*, volume 1, pages 20–31, 1990.

[7] Herman H. Goldstine and John von Neumann. Planning and coding problems for an electronic computing instrument. In *John von Neumann, Collected Works*, volume V, pages 34–235. Pergamon Press, Oxford, 1961.

[8] Brian W. Kernighan and Dennis M. Ritchie. *The C Programming Language, Second Edition*. Prentice Hall, Englewood Cliff, New Jersey, 1988.

[9] J Strother Moore. Piton: A verified assembly-level language. Technical Report CLI-22, Computational Logic, Inc., Austin, Tx, June 1988.

[10] Motorola, Inc. *MC68020 32-bit Microprocessor User's Manual*. Prentice Hall, New Jersey, 1989.

Proving the Chinese Remainder Theorem by the Cover Set Induction*

Hantao Zhang, Xin Hua
Department of Computer Science
The University of Iowa
Iowa City, IA 52242, U.S.A.
{hzhang, xin}@cs.uiowa.edu

Abstract

An experiment of the cover set induction principle in *RRL* is presented with a proof of the Chinese Remainder theorem. To the best of our knowledge, this is the first machine proof of the theorem. The proof itself can be viewed as the correctness proof of a program which computes the least positive simultaneous solution of n congruence equations. We also discussed the problems involved in proving the theorem: designs of good specifications and induction schemes and control of rewriting.

1 Introduction

In the recent years, algebraic specification has become a popular language for formally specifying system designs such as abstract data types, algorithms, digital circuits [4]. An algebraic specification usually consists of a signature whose functions are recursively defined by a set of equations. To reason about properties of an algebraic specification, mathematical induction is required. The structural induction principle of Burstall [3] and the inductive completion procedure originally proposed by Musser [11] are the two well-known approaches for proving inductive properties of algebraic specification.

In the theorem prover *RRL* [10], we implemented two methods based on the approach of inductive completion [8], [9]. We also developed a generalization of Burstall's structural induction principle based on the concept of *cover set* for automated induction in algebraic specification [20]. Using the cover set method, hundreds of theorems about lists, sequences, natural numbers and integers have been proved. The cover set method has been used to verify the correctness of various sorting algorithms and the longest common subsequence algorithm (all the algorithms written in algebraic specification) [14]. The method has also been used to verify circuits designs [6] and has been combined with the Floyd-Hoare approach to verify the correctness of algorithms written in imperative languages [15]. The method has also been used to prove several interesting theorems in mathematics: the unique prime factorization theorem in number theory [20], Ramsey's theorem in graph theory [16], and Lagrange's theorem in group theory [14].

The cover set method is closely related to the Boyer-Moore's logic [2]. Recently, we conducted a comparison study of the cover set method and the Boyer-Moore theorem prover [1, 16] by comparing

*Partially supported by the National Science Foundation Grants no. CCR-9009414 and INT-9016100.

proofs of Ramsey's theorem in graph theory [16]. Even though the two methods use different formal languages, the proofs in the two methods are very similar. Despite of the close relationship between the Boyer-Moore's computational logic and the algebraic specification on which the cover set induction method is based, it is still a challenge to reproduce the computer proofs of many interesting problems done by the Boyer-Moore theorem prover, because a problem presented in one approach may not be a problem of other approaches and the same problem may need different solutions.

To further identify the problems of automated induction involved in dealing with algebraic specification and to further improve the cover set method, we took a more challenging step in the fall of 1991: Instead of reproducing Boyer-Moore's proof of an interesting problem, we decided to attack a famous theorem whose computer proof was unknown. The Chinese remainder theorem in number theory was chosen for this purpose.

The Chinese Remainder Theorem: *Let $m_1, m_2, ..., m_n$ be pairwise relatively prime positive integers. Then the system of congruences*

$$S(n): \begin{cases} x \equiv a_1 \pmod{m_1}, \\ x \equiv a_2 \pmod{m_2}, \\ \quad \cdots \\ x \equiv a_n \pmod{m_n}, \end{cases}$$

has a unique solution modulo $M_n = m_1 m_2 ... m_n$.

A typical proof of this theorem consists of two parts:

- **Existence:** There exists a simultaneous solution to $S(n)$.
- **Uniqueness:** For any solutions x_1 and x_2 of $S(n)$, $x_1 \equiv x_2 \pmod{M_n}$.

The uniqueness proof presented no serious difficulty to the cover set method. The existence proof of $S(n)$ can be recursively reduced to the existence proof of $S(2)$, which, however, needs some intricate construction and is not easily formalized; because of the space restriction, we have to omit the existence proof of $S(2)$ in this paper (see [17]).

Instead, we will pay more attention to general problems and issues involved in proving the Chinese remainder theorem. We will present (in section 2) the computer proof of the Chinese remainder theorem, assuming that the solution of $S(2)$ is known and some related lemmas are true. In *RRL*, however, every lemma must be proved to be true first, before it can be used to prove other lemmas. We will first outline a hand proof of the theorem, then formulate it in the language of the algebraic specification, and finally give the proof produced by *RRL*. In section 3, we present several problems encounted in proving the Chinese remainder theorem and our solutions.

2 A Proof of the Chinese Remainder Theorem

In this section, the following are assumed to be true in the proof of the Chinese remainder theorem; their proofs can be found in the extended version of this paper [17]:

Lemma 0: Let m_1, m_2 be relatively prime positive integers. For any a_1 and a_2, there exists a number $s_2 = sol_2(m_1, a_1, m_2, a_2)$ such that (i) $s_2 \equiv a_1 \pmod{m_1}$, (ii) $s_2 \equiv a_2 \pmod{m_2}$ and (iii) $s_2 < m_1 m_2$.

Lemma 1: If x, y, z are pairwise relatively prime, so are x and yz.

Lemma 2: If x, y are relatively prime and $x|z$, $y|z$, then $xy|z$.[1]

Assuming the above lemmas are true, a typical proof of the Chinese remainder theorem goes as follows.

For the existence part, we construct by induction a simultaneous solution s_n to $S(n)$, the system of n congruences. The basis case when $n = 1$ is trivial. By the induction hypothesis, the system $S(n-1)$ has a simultaneous solution, say s_{n-1}. By Lemma 1 and a simple induction, $m_1 m_2 ... m_{n-1}$ and m_n are relatively prime. We thus convert the n-congruence problem into a two-congruence problem:

$$x \equiv s_{n-1} \pmod{M_{n-1}},$$
$$x \equiv a_n \pmod{m_n},$$

where $M_{n-1} = m_1 m_2 ... m_{n-1}$. Let $s_n = sol_2(m_n, a_n, M_{n-1}, s_{n-1})$, which satisfies Lemma 0. It is easy to verify that $s_n = a_i \pmod{m_i}$ for all $1 \leq i \leq n$.

For the uniqueness part, we show that any two solutions are congruent modulo M_n. Let x_1 and x_2 both be simultaneous solutions to $S(n)$. Then, for each k, $x_1 \equiv x_2 \equiv a_k \pmod{m_k}$, so that $m_k|(x_1 - x_2)$. By Lemmas 1 and 2, we see that $M_n|(x_1 - x_2)$. Therefore, $x_1 \equiv x_2 \pmod{M_n}$. This shows that the simultaneous solution of $S(n)$ is unique modulo M_n. □

2.1 Formulation of the Theorem in RRL

In this section, we define in *RRL* various predicates and functions to formulate the Chinese remainder theorem. Let $ma = \{\langle m_i, a_i \rangle \mid 1 \leq i \leq n\}$ be the data of $S(n)$, and suppose that predicates *allpositive*, *allprime*2 and *allcongruent* and the function *products*1 have been properly defined such that

- *allpositive*(ma) if and only if $m_i > 0$ for $1 \leq i \leq n$.
- *allprime*2(ma) if and only if $m_1, ..., m_n$ are pairwise relatively prime.
- *allcongruent*(x, ma) if and only if $x \equiv a_i \pmod{m_i}$ for $1 \leq i \leq n$.
- *products*1(ma) $= m_1 m_2 \cdots m_n$.

In *RRL*, the numbers may be constructed by the functions `0` and `suc` (successor) and the lists are constructed by `null` (the empty list) and `cons`. The pairs are constructed by the function `mkpair` with the properties that `first(mkpair(x, y)) = x` and `second(mkpair(x, y)) = y`. That is, the function `first` (`second`) returns the first (second) element of a pair. The functions `0`, `suc`, `null`, `cons` and `mkpair` are called *constructors*. Other functions can be defined by equations with respect to constructors. For instance, the predicate `allcongruent` can be defined in *RRL* as follows:

```
[allcongruent : num, list -> bool] /* 'list' is a list of integer pairs. */
 allcongruent(x, null) := true
 allcongruent(x, cons(y, z)) :=
             allcongruent(x, z) and (rem(x, first(y)) = rem(second(y), first(y)))
```

[1] $x|z$ reads as x *divides* z, or equivalently, the remainder of z by x is 0.

where $rem(x, y)$ returns the remainder of x by y.

The definition of *allprime2* needs the auxiliary function *prime2list*:

```
[prime2list : num, list -> bool]
 prime2list(x, null) := true
 prime2list(x, cons(y, z)) := prime2(x, first(y)) and prime2list(x, z)

[allprime2 : list -> bool]
 allprime2(null) := true
 allprime2(cons(y, z)) := prime2list(first(y), z) and allprime2(z)
```

where $prime2(x_1, x_2)$ is true if and only if x_1 and x_2 are relatively prime, and $prime2list(x, xl)$ is true if and only if x is relatively prime with the first element of every pair in the list xl.

The reader should have no difficulty to provide the definitions of `allpositive` and `products1`.

The Chinese remainder theorem is stated as:

$$\forall Y(allpositive(Y) \wedge allprime2(Y)) \Rightarrow \\ (\exists x\ allcongruent(x, Y) \wedge \\ \forall x_1 \forall x_2((allcongruent(x_1, Y) \wedge allcongruent(x_2, Y)) \Rightarrow (rem(x_1 - x_2, products1(Y)) = 0))).$$

Because the rewriting approach to algebraic specification cannot handle quantifiers, to get rid of the quantifiers in the above formula, we may skolemize it by introducing the skolem function $soln(Y)$ for the existentially quantified variable x. The result is the following quantifier-free formula (assuming all the variables are universally quantified):

$$(allpositive(Y) \wedge allprime2(Y)) \Rightarrow \\ (allcongruent(soln(Y), Y) \wedge \\ ((allcongruent(x_1, Y) \wedge allcongruent(x_2, Y)) \Rightarrow (rem(x_1 - x_2, products1(Y)) = 0))).$$

The above quantifier-free formula is easily converted into formulas accepted by *RRL* by substituting `and` for $\wedge$, `not` for $\neg$, and `P if C` for $C \Rightarrow P$. This is, there are two goals for *RRL* prove:

Goal 1 (existence): `allcongruent(soln(y), y) if allpositive(y) and allprime2(y)`

Goal 2 (uniqueness):
```
rem(x1 - x2, products1(y)) == 0 if
         allpositive(y) and allprime2(y) and
         allcongruent(x1, y) and allcongruent(x2, y)
```

Before we start the proof process, we have to provide an explicit definition of $soln(ma)$. The proof of the Chinese remainder theorem in the previous section provides us a clue on how to construct $soln(ma)$. Let $s_2 = sol_2(m_1, a_1, m_2, a_2)$ be the least positive solution of two congruences, such that $s_2 = a_1 \pmod{m_1}$ and $s_2 = a_2 \pmod{m_2}$, where m_1 and m_2 are relatively prime positive numbers. We can define the function *soln* in *RRL* as follows:

```
[soln : list -> num]
 soln(null) := 0
 soln(cons(x, z)) := sol2(first(x), second(x), products1(z), soln(z))
```

The argument *ma* of *soln*(*ma*) is assumed to be a list of pairs of integers. If *ma* satisfies the conditions of the Chinese remainder theorem, i.e., *allprime*2(*ma*) and *allpositive*(*ma*), we can expect that *soln*(*ma*) will return a simultaneous solution of the n congruences, as ensured by the previous proof of the theorem. The mechanical proof of Goal 1 will confirm that our expectation is right. Since our theorem prover cannot construct automatically the function *soln*(*ma*), this kind of proofs is termed as "proof verification" in the literature.

2.2 Use of the Cover Set Method in *RRL*

The general steps to use the cover set induction method in *RRL* are as follows:

1. Input the definitions of functions (in form of quantifier-free equations or conditional equations) as well as the arities of functions.
2. Declare the constructors of each new sort.
3. Make rewrite rules from the definitions by orienting them from left to right.
4. Test the completeness of the functions with respect to constructors.
5. For each lemma to be proved, apply a mixture of the following techniques:
   (a) **simplification**: Simplifying a formula by "contextual rewriting" [20] and the consistency checking procedure.
   (b) **case-split**: Transforming the formula $P(cond(X, Y, Z))$ into $P(Y)$ **if** X and $P(Z)$ **if** $\neg X$.
   (c) **generalization**: Either replacing some common subterms in a formula by new variables or eliminating irrelevant premises in the formula.
   (d) **induction**: Using the cover set induction method reported in [20].
6. After each lemma is proved, it is made into a rewrite rule and is stored in the system for proving other lemmas.

The cover set induction is based on the concept of *cover set*, which is a finite set of terms covering all possible values that a data type can have. This cover set is used to generate an induction scheme for proving formulas by induction. Different cover sets are used for the same data type based on the way the data type is used in function definitions.

The soundness of the cover set induction replies on (3) and (4) above: the well-founded induction is ensured by the termination of rewrite rules, and the completeness of a cover set is ensured by the completeness of that function [14]. For the Chinese remainder theorem, *RRL* can automatically prove that all the rewrite rules made from the definitions are terminating, and most of the definitions are complete. For those functions whose completeness cannot be automatically proved, *RRL* can still prove them with the user's assistance.

2.3 The Proof of Goal 1

Let us illustrate the use of the cover set method by the proof of Goal 1. Assuming that the equations defining `allcongruent` have been made into terminating rewrite rules, the function `allcongruent` has been tested to be complete, and some auxiliary lemmas have been proved, Goal 1 is proved by *RRL* as follows:

```
RRL-> prove  allcongruent(soln(Y), Y) if allpositive(Y) and allprime2(Y)

Let P(Y) be [main] allcongruent(soln(Y), Y) if (allprime2(Y)) and (allpositive(Y))

The induction will be done on Y in allcongruent(soln(Y), Y), and will follow the scheme:
    [#1] P(null)
    [#2] P(cons(y1, z)) if  { P(z) }

By rule [228],       /* [228] is the first equation of the definition of allcongruent. */
    [#1] allcongruent(soln(null), null) if (allpositive(null)) and (allprime2(null))
    is reduced to true.

Conjecture [#2] is split into:
 [#2.1] allcongruent(soln(cons(y, z)), cons(y, z)) if
          allprime2(cons(y, z)) and allpositive(cons(y, z)) and not(allprime2(z))
 [#2.2] allcongruent(soln(cons(y, z)), cons(y, z)) if
          allprime2(cons(y, z)) and allpositive(cons(y, z)) and not(allpositive(z))
 [#2.3] allcongruent(soln(cons(y, z)), cons(y, z)) if
          allprime2(cons(y, z)) and allpositive(cons(y, z)) and allcongruent(soln(z), z)
 ... ...
All subgoals of [main] are proved, hence
    [main] allcongruent(soln(Y), Y) == true if (allprime2(Y)) and (allpositive(Y))
    is an inductive theorem.
```

The induction scheme used to prove Goal 1 is derived from the definition of **allcongruent** and is identical to that by the structural induction [3]. The basic case **[#1]** is easily proved by the definition of **allcongruent**.

The inductive case **[#2]** is split into **[#2.1]**, **[#2.2]** and **[#2.3]**, of which the first two correspond to the cases when the premises of **P(z)** are false and the last one corresponds to the case when the premises of **P(z)** are all true. If we view **[#2.1]**, **[#2.2]** and **[#2.3]** as clauses, then the splitting is the same as converting **[#2]** into clauses.

The proofs of **[#2.1]** and **[#2.2]** are fairly easy: **[#2.1]** is proved by the contradiction between **allprime2(cons(y, z))** and **not(allprime2(z))**, and **[#2.2]** is proved by the contradiction between **allpositive(cons(y, z))** and **not(allpositive(Z))**.

The proof of **[#2.3]** needs the following five previously proved lemmas:

```
 [215] rem(sol2(xm1, xa1, xm2, xa2), xm1) ---> rem(xa1, xm1)  if
             { not((0 = xm1)), not((0 = xm2)), prime2(xm1, xm2) }
 [230] (products1(y) = 0) ---> false  if   { allpositive(y) }
 [231] prime2(x, products1(y)) ---> true  if
             { allprime2(y), allpositive(y), not((0 = x)), prime2list(x, y) }
 [235] allcongruent(u, z) ---> true  if
             { allpositive(z), allprime2(z), allcongruent(rem(u, products1(z)), z) }
 [236] rem(sol2(xm1, xa1, products1(z), soln(z)), products1(z)) ---> soln(z)  if
             { allpositive(z), allprime2(z), not((0 = xm1)), prime2list(xm1, z) }
```

By the definition of **soln**, **allcongruent(soln(cons(y, z)), cons(y, z))** in **[#2.3]** is reduced to **allcongruent(sol2(first(y), second(y), products1(z), soln(z)), cons(y, z))**. Let S2 be the shorthand for **sol2(first(y), second(y), products1(z), soln(z))**. Then the above formula is further reduced, by the definition of **allcongruent**, to

allcongruent(S2, z) and (rem(second(y), first(y)) = rem(S2, first(y))).

The first part is reduced to true by rules [235], [236] and the induction hypothesis (i.e., **allcongruent(soln(z), z)**). The second part of the conjunction is reduced to true by rules [215], [230] and [231]. Note that the premises of [#2.3] imply **allprime2(z)** and **allpositive(z)**.

Let us explain briefly how the above five lemmas are obtained. Rule [215] is part (i) of Lemma 0, rule [230] can be proved by an easy induction, and rule [231] is a corollary of Lemma 1 (i.e., if x, y, z are pairwise relatively prime, so are x and yz. *RRL* has proved Lemmas 0 & 1 prior to this proof.)

Rule [235] is derived from **rem(x, y) + y * div(x, y) = x** and

```
[234] allcongruent(v + (u * products1(z)), z) ---> true  if
                         allprime2(z), allpositive(z), allcongruent(v, z)
```

Rule [236] is derived from parts (ii) and (iii) of Lemma 0. That is, *RRL* first proved

```
[232] soln(y) < products1(y) ---> true if (allpositive(y) and allprime2(y))
```

by part (iii) of Lemma 0 and a simple induction. Then

```
rem(sol2(xm1, xa1, products1(z), soln(z)), products1(z))
      = rem(soln(z), products1(z))          /* by part (ii) of Lemma 0 */
      = soln(z)                             /* by rem(x, y) = x if x < y and [232]. */
```

It is interesting to note that rule [232] tells us that $soln(y)$ returns the least positive solution of $S(n)$. This property is not required in the original theorem but needed in the current machine proof.

2.4 The Proof of Goal 2

The induction scheme for proving Goal 2 (uniqueness) is the following:

```
RRL-> prove rem((x1 - x2), products1(Y)) == 0 if
              (allpositive(Y) and (allprime2(Y) and
              (allcongruent(x1, Y) and allcongruent(x2, Y))))

Let P(Y) be [main] rem((x1 - x2), products1(Y)) == 0 if
                      (allprime2(Y) and (allpositive(Y)) and
                      (allcongruent(x1, Y)) and (allcongruent(x2, Y))

The induction will be done on Y in allcongruent(x1, Y), and will follow the scheme:
    [#1] P(null)
    [#2] P(cons(y1, z)) if  { P(z) }
 ... ...
```

The following four lemmas are proved before the proof of Goal 2.

```
  [98] rem(x, suc(0)) ---> 0
 [133] rem(y, (x1 * x2)) ---> 0 if { (rem(y, x1) = 0), (rem(y, x2) = 0), prime2(x1, x2) }
 [144] rem((x1 - x2), y) ---> 0 if { (rem(x2, y) = rem(x1, y)) }
 [231] prime2(x, products1(y)) ---> true  if
             { allprime2(y), allpositive(y), not((0 = x)), prime2list(x, y) }
```

Rule [133] is Lemma 2. By rule [98] and the definition of *products1*, the basis case [#1] is reduced to true. Subgoal [#2] is split into five conditional equations, the first four of which can be easily proved by the contradiction among the premises of each equation. With the help of rules [133], [144] and [231], the fifth equation is reduced to true.

3 Problems and Issues

3.1 Choosing Good Specification

One of the major difficulties of using the cover set method is to prepare a good algebraic specification of the problems. Sometimes, it is not a trivial job to formally specify an idea in the language of algebraic specification; sometimes there are several specifications available for the same idea. In [5], Gramlich discussed the case of defining the predicate which asserts a list is sorted by different ways when verifying different sorting algorithms. However, there are no general rules to guide the user to design a good specification. In the following, we hope our experiments can provide some insight to this problem. For the Chinese remainder theorem, we spent at least one third of the total time (about one month) in formally specifying the problem.

The proof of Lemma 0 (see section 2) involves the use of the following two sets:

$$\begin{aligned} E &= \{\langle x,y\rangle \mid 0 \leq x < m_1, 0 \leq y < m_2\}, \\ R &= \{\langle rem(z,m_1), rem(z,m_2)\rangle \mid 0 \leq z < m_1 m_2\}. \end{aligned}$$

Let $b_1 = rem(a_1,m_1)$ and $b_2 = rem(a_2,m_2)$. If $\langle b_1,b_2\rangle \in R$, that is, if there exists z such that $\langle rem(z,m_1), rem(z,m_2)\rangle = \langle b_1,b_2\rangle$, then z is the solution we are looking for, because $rem(z,m_1) = b_1 = rem(a_1,m_1)$, $rem(z,m_2) = b_2 = rem(a_2,m_2)$ and $z < m_1 m_2$.

In an algebraic specification, a set can be specified in different ways. In this experiment, a set is represented by a list containing no duplicates. The following function *enum* is defined to generate the set E, while the auxiliary function $enum1(n,m)$ generates the set $\{\langle n,y\rangle \mid 0 \leq y < m\}$.

```
[enum1 : num, num -> list]
 enum1(xm, 0) :=  null
 enum1(xm, suc(xn)) := cons(mkpair(xm, xn), enum1(xm, xn))

[enum : num, num -> list]
 enum(0, xm) := null
 enum(suc(xn), xm) := append(enum1(xn, xm), enum(xn, xm))
```

For instance, $enum(3,2)$ returns the list $(\langle 2,1\rangle\langle 2,0\rangle\langle 1,1\rangle\langle 1,0\rangle\langle 0,1\rangle\langle 0,0\rangle)$.

However, it is not easy to specify the set R. By trial-and-error, we finally decided to use the following specification: Let $rempair(x_1,m_1,x_2,m_2)$ denote the pair of the remainder of x_1 divided by m_1 and that of x_2 by m_2. The function **rempairs** is defined to generate the set R.

```
[rempair : num, num, num -> pair]
 rempair(x1, xm1, x2 xm2) := mkpair(rem(x1, xm1), rem(x2, xm2))

[rempairs : num, num, num -> list]
 rempairs(0, xm1, xm2) := null
 rempairs(suc(xn), xm1, xm2) := cons(rempair(xn, xm1, xn, xm2), rempairs(xn, xm1, xm2))
```

For instance, $rempairs(6,3,2)$ generates the list $(\langle 2,1\rangle\langle 1,0\rangle\langle 0,1\rangle\langle 2,0\rangle\langle 1,1\rangle\langle 0,0\rangle)$. It is easy to see that $rempairs(6,3,2)$ and $enum(3,2)$ generate the same set.

Now it is clear that the set E can be represented by `enum`(m_1, m_2), and the set R can represented by `rempairs(m1 * m2, m1, m2)` in *RRL*. However, if we know that there exists $\langle c_1, c_2\rangle \in R$ such that $\langle c_1, c_2\rangle = \langle b_1, b_2\rangle$, how do we find the value z such that $rem(z, m_1) = c_1$ and $rem(z, m_2) = c_2$? Note that the core of the proof of Lemma 0 is to show that the set E and R are equivalent. For this reason, it is not convenient for us to work with the set R':[2]

$$R' = \{\langle z, rem(z,m_1), rem(z,m_2)\rangle \mid 0 \le z < m_1m_2\}.$$

Our solution to the above question is the function **index**:

```
[index : univ, list -> num]
 index(x, null) := 0
 index(x, cons(y, xl)) := cond((x = y), length(xl), index(x, xl))
```

Intuitively, `index(x, xl)` returns the position of `x` in the list `xl` (the position is counted from right to left, with the position of the rightmost element being 0). For instance, if $R = (...\langle 2,3\rangle\langle 0,0\rangle)$, $index(\langle 2,3\rangle, R)$ returns 1.

Let us look at another example. In the textbooks, when we say "two numbers are relatively prime", we assume that the gcd of these two numbers is 1. That is, the predicate $prime2(x,y)$ is expressed as $gcd(x,y) = 1$. We found that this definition of $prime2(x,y)$ is not easy for a theorem prover to handle. Instead, we use the fact that two integers are relatively prime if and only if they share no prime factors. That is, let $primefacs(x)$ denote the multi-set of prime factors of x, then $prime2(x,y)$ if and only if $primefacs(x) \cap primefacs(y) = \emptyset$. Using this definition for $prime2(x,y)$, the proofs of Lemmas 1 and 2 (see section 2) can be easily mechanized.

3.2 Designing Good Induction Scheme

The following lemma is the so-called "pigeon-hole theorem" in set theory, which plays an important role in the equivalence proof of the two sets E and R given in the beginning of section 3.1.

Lemma 3: For any two sets x and y, if $x \subseteq y$ and $|x| = |y|$, then $y \subseteq x$.

Proof. Let the statement in the pigeon-hole theorem be denoted as $P(V,U)$. A natural induction scheme for $P(V,U)$ will be based on the size of U, and will have the form:

- **Basis case:** $\forall U \forall V[|U| = 0 \Rightarrow P(V,U)]$.
- **Inductive case:** $\forall U \forall V[(\forall U'\forall V'(|U'| < |U|) \wedge P(V',U')) \Rightarrow P(V,U)]$.

The basis case is trivial because $|U| = |V| = 0$ implies $U = V = \emptyset$. The inductive case goes as follows: Suppose $U = \{x\} \cup W$, where $x \notin W$. Then $U \subseteq V$ implies $x \in V$ and $W \subseteq (V - \{x\})$. It is easy to see that $|W| = |U| - 1$ and $|V - \{x\}| = |V| - 1$, so we instantiate U' by W and V' by $V - \{x\}$ in the induction hypothesis. That is, we assume $P(V - \{x\}, W)$ is true, or $(V - \{x\}) \subseteq W$. It is easy to show that $V \subseteq (\{x\} \cup W)$ if and only if $(V - \{x\}) \subseteq W$. □

[2] If R' is used, we have to define another function which eliminates the first element from each triple of R' when comparing with the set E.

The formulation of this theorem in *RRL* is easy. In *RRL*, a set is represented by a list of distinct elements. Suppose the lists are constructed by the constructors **null** and **cons**. Besides the usual functions *member* and *delete* on lists (*member*(x, xl) is true if and only if x is in xl, and *delete*(x, xl) deletes one x from xl if x is in xl), the following functions are used in this example:

```
[length : list -> num]          /* length(xl) is the number of elements of xl. */
 length(null) := 0
 length(cons(x, y)) := suc(length(y))

[distinct : list -> bool]       /* distinct(xl) iff each element of xl is distinct. */
 distinct(null) := true
 distinct(cons(x, y)) := not(member(x, y)) and distinct(y)

[subset : list, list -> bool]  /* subset(u, w) iff u is a subset of w. */
 subset(null, w) := true
 subset(cons(x, y), w) := member(x, w) and subset(y, w)
```

The pigeon-hole theorem is input to *RRL* as the following:

```
RRL-> prove subset(v, u) if distinct(u) and distinct(u) and subset(u, v) and
                                   (length(u) = length(v))
```

Prior to proving this theorem, the following lemmas are proved first in *RRL*:

```
Lemma 1.  distinct(delete(x, xl)) ---> true if  distinct(xl)
Lemma 2.  length(delete(x, xl)) ---> sub1(length(xl))  if  member(x, xl)
Lemma 3.  subset(xl, cons(x, yl)) ---> subset(delete(x, xl), yl)  if  distinct(xl)
```

These lemmas are used explicitly or implicitly in one way or another in the previous hand proof of the pigeon-hole theorem.

To prove a theorem by induction, an induction scheme must be formed in the way that has a good chance of success. In the current implementation of the cover set principle, an induction scheme is chosen heuristically from the definition of a function appearing in the conjecture (this idea is learned from the Boyer-Moore theorem prover). In other words, different definitions may give different induction schemes. While this heuristic is very useful to prove many theorems, because of its heuristic nature, some schemes do not lead to a successful proof. This problem is manifested in the proof of the pigeon-hole theorem.

The following is quoted from the transcript of *RRL*:

```
Let P(u) be [main] subset(v, u)  if  (distinct(u)) and (distinct(v)) and
                                     (subset(u, v)) and (length(v) = length(u))
The induction will be done on u in subset(u, v), and will follow the scheme:
    [#1] P(null)
    [#2] P(cons(x, y)) if  { P(y) }
```

The message indicates that the induction scheme is derived from the definition of ***subset*** and the induction will done on the variable u. That is, subgoals **[#1]** and **[#2]** correspond to the first and second equations in the definition of ***subset***. Note that P(y) in **[#2]** comes from the right-hand side of the second equation because that equation expresses a recursive definition (i.e., the term ***subset***(y, w)). Unfortunately, this induction scheme does not lead to a successful proof.

If we change the heuristics of *RRL* such that the term *length*(u) (or *distinct*(u)) is chosen, then the induction scheme will be the same as above. On the other hand, if one of the other three terms, i.e., *subset*(v, u), *length*(v) and *distinct*(v), is chosen, then the induction scheme will be based on v (the theorem will be denoted as $P(v)$). Again, none of these induction schemes leads to a successful proof.

From the previous hand proof, we know the induction scheme that really works is the following (assuming the theorem is denoted by $P(v, u)$):

```
[#1] P(v, null)
[#2] P(v, cons(x, w)) if  { P(delete(x, v), w) }
```

That is, the proper instance of the induction hypothesis should be

```
subset(delete(x, v), w)  if  distinct(w) and distinct(delete(x, v)) and
               subset(w, delete(x, v)) and (length(delete(x, v)) = length(w))
```

From the above discussions, we know that no matter how we change heuristics in *RRL*, the above desired scheme cannot be automatically generated. When heuristics fail, the guidance of the user is essential to the success. Inspired by the hint facilities provided by the Boyer-Moore theorem prover, in order to prove this lemma as well as other lemmas, we added into *RRL* some facilities through a proof manager called FRI [7] or through the hint mechanism. Currently, the user can use a different induction scheme by choosing a function name; this function may not exist in the current conjecture, but must exist in the system and must satisfy certain completeness property [14]. As we have said before, the definition of that function decides the induction scheme the user intends to choose. Thus, the user may define a *dummy* function to generate any desired induction scheme.

For the problem in the above example, we can define the *dummy* function as follows:

```
[dummy : list, list  -> list]
 dummy(v, null) := null
 dummy(v, cons(x, w)) := dummy(delete(x, v), w)
```

Note that the right-hand side of the second equation specifies the format of induction hypothesis.

The following script of *RRL* shows how to use *dummy* through the hint mechanism:

```
RRL-> prove subset(v, u)  if  distinct(u) and distinct(v) and
                subset(u, v) and (length(v) = length(u)) and hint(induc(dummy(v, u)))

Proving subset(v, u)  if  distinct(u) and distinct(v) and
                             subset(u, v) and (length(v) = length(u))
        by hint(induc(dummy(v, u)))

Let P(v, u) be [main] subset(v, u)  if  distinct(u) and distinct(v) and
                               subset(u, v) and (length(v) = length(u))

The induction will be done on v, u in dummy(v, u), and will follow the scheme:
    [#1] P(v, null)
    [#2] P(v, cons(x, w)) if  { P(delete(x, v), w) }
```

After $P(v, u)$ is proved, it is made into a rewrite rule (without the hint term) and stored for proving other lemmas.

Besides the hint mechanism described above and the proof manager described in [7], we also studied the way how induction hypotheses are handled in the inductive completion approach and developed a new method to handle induction hypotheses in the cover set method. With the new method being implemented, *RRL* automatically proved the pigeon-hole lemma; see [18] for details.

3.3 Control of Rewriting

The implementation of the cover set method in *RRL* has been supported by a set of auxiliary proof techniques, among which term rewriting is the most important one. In *RRL*, every (conditional) equation is converted into a (conditional) rewrite rule and these rules perform the so-called "contextual rewriting" [19]. We illustrate the use of contextual rewriting by a simple example.

Example 3.1 Suppose we are given a rewrite rule $\mathbf{r}: rem(u * v, u) \rightarrow 0$ **if** $(u \neq 0)$ and a clause

$$\mathbf{c}: (x * y \neq y * z) \vee (rem((y * z), x) = 0) \vee (x = 0)$$

to be simplified. When we try to rewrite $(rem((y * z), x) = 0)$ in $\mathbf{c}$, we take the negation of the rest literals of $\mathbf{c}$, i.e., $C = \{(x * y = y * z), (x \neq 0)\}$, as the *context*. At first, $(rem((y * z), x) = 0)$ is reformulated by C to $rem((x * y), x) = 0$, on which the left-hand side of $\mathbf{r}$ applies with the matching substitution $\sigma = \{u \leftarrow x, v \leftarrow y\}$. Because the condition of $\sigma\mathbf{r}$ is $x \neq 0$, which is reformulated by C to *true*, the term $rem((x * y), x)$ is replaced by the right-hand side of $\sigma\mathbf{r}$, i.e., 0. In short, the literal $(rem((y * z), x) = 0)$ is simplified by $\mathbf{r}$ with C to true. This process of simplification is called "contextual rewriting". □

Contextual rewriting is a generalization of conditional rewriting and cross-fertilization of the Boyer-Moore theorem prover [2] and has been formally studied in [19]. For resolution-based theorem proving, it has been shown that contextual rewriting is a very powerful simplification rule which subsumes tautology deletion, subsumption and demodulation altogether. We think that one of the major factors to the efficiency of the cover set method (see [16] for a comparison study) is that we always keep the rewrite rules which are simplified by other rules in the system.

Like any kind of term rewriting, the following properties are desirable: (a) **soundness**: Terms (or clauses) are only rewritten to logically equivalent ones. (b) **termination**: No infinite rewritings are possible. (c) **confluence**: The order of rewritings is indifferent.

The soundness of contextual rewriting has been proved in [19]. However, termination and confluence of contextual rewriting cannot be guaranteed in the proof of the Chinese remainder theorem. In the following, we briefly discuss our solutions to this problem.

There are two sources of nontermination in contextual rewriting: (i) the right-hand side of a rewrite rule is not smaller than its left-hand side; (ii) the condition of a rewrite rule is not smaller than the left-hand side of its head.

For case (i), the equation $delete(x_1, delete(x_2, y)) = delete(x_2, delete(x_1, y))$ is an example. This equation can be proved to be true by *RRL* and is needed in the proofs of several other lemmas. In general, the techniques developed for unfailure completion procedures can be adopted to handle this problem. We have chosen a variant of the technique called "constraint rewriting" proposed in [12] in our experiment.

For case (ii), the rewrite rule [235] given in section 2.3 is an example:

```
[235] allcongruent(u, z) ---> true  if
        { allpositive(z), allprime2(z), allcongruent(rem(u, products1(z)), z) }
```

When this rule applies to the term `allcongruent(a, b)`, during the process of establishing the truth of the condition of `[235]`, the same rule, i.e., `[235]` itself, can be used to rewrite the term `allcongruent(rem(a, products1(b)), b)`. Similarly, the same rule will be used to rewrite `allcongruent(rem(rem(a, products1(b)), products1(b)), b)`, etc. This will cause an infinite loop.

To solve the above problem, we restrict the ***depth*** of simplification[3] to a small number, say 2. For instance, suppose we are given the following rules:

$$\begin{aligned} \mathbf{r}_1: &\quad a_0 \rightarrow b_0 \textbf{ if } a_1 = b_1 \\ \mathbf{r}_2: &\quad a_1 \rightarrow b_1 \textbf{ if } a_2 = b_2 \\ \mathbf{r}_3: &\quad a_2 \rightarrow b_2 \textbf{ if } a_3 = b_3 \end{aligned}$$

To simplify the clause $a_0 = b_0 \vee a_2 \neq b_2$ to ***true*** by contextual rewriting, a depth 2 simplification is needed. To simplify $a_0 = b_0 \vee a_3 \neq b_3$ to ***true***, a depth 3 simplification is needed. If the depth is limited to 2, then $a_0 = b_0 \vee a_3 \neq b_3$ cannot be simplified to ***true***. To compensate the loss of the power by depth restriction, the user can use ***bridge lemmas***. For this example, a bridge lemma will be $\mathbf{r}_4 : a_1 \rightarrow b_1$ **if** $a_3 = b_3$. When this bridge lemma is added (after it is proved to be true), the clause $a_0 = b_0 \vee a_3 \neq b_3$ can be simplified to ***true*** by a depth 2 simplification.

Currently *RRL* does not support automatic techniques to solve the confluence problem of contextual rewriting. The user can ***freeze*** (or ***disenable***) some rewrite rules to avoid undesirable rewriting or tells *RRL* explicitly through an interactive interface [7] to use a particular rule. Note that the facility of freezing or unfreezing rewrite rules was added to *RRL* during the experiment of proving the Chinese remainder theorem.

4 Conclusions

Because of space restriction, we have only partially presented a computer-verified proof of the Chinese remainder theorem. To the best of our knowledge, this is the first machine proof of the theorem. Our experiment with the cover set method has revealed several interesting problems, such as how to choose right specifications, how to choose induction schemes, how to handle the problem related to rewriting.

#Tokens	2664
#Definitions	31
#Lemmas	118
Replay Time	51 seconds

Table 1: The proof statistics of RRL.

Table 1 presents some statistics of the proof. The number of tokens is the total number of identifiers in the input file. The number of lemmas includes the final goals of the theorem. The replay time is the total time to execute all the input commands after a complete input file is

[3]Suppose initially the *depth* of the simplification is 0. If the current simplification is at *depth* d, and a conditional rewrite rule is applied, then the simplification of the conditions of that rule is at *depth* $d+1$.

available. The time is measured in AKCL on a Sun Sparc station 1 with 16 megabytes of main memory. In fact, we did not start from the scratch when working on this theorem. We started with a *RRL* input file which leads to a proof of the unique prime factorization theorem [20]. A lot of effort has been saved by using many lemmas in that file. These lemmas are all counted in Table 1.

We wish that our presentation allows other theorem provers to reproduce the proof. We think that the proof complexity of this theorem is comparable to that of Wilson's theorem in number theory which has been proved by the Boyer-Moore theorem prover [13]. We like to point out that the objective of this experiment is not to prove this difficult and well-known theorem with extreme rigor, but to provide insights on how to develop techniques for proof construction and presentation in an algebraic specification. That is, if other theorem provers can prove the Chinese remainder theorem, we are particularly interested in whether the proofs are easy (to construct), convincing, readable and useful. Because algebraic specification is very popular among computer scientists ands very suitable for inductive proofs (it is a multiple-sort language), we believe that our method is accessible to a wide range of audience.

We may view our proof of the Chinese remainder theorem as an example of program verification. The function *soln* we constructed in section 2 returns the least positive solution of n congruence equations. This function can be considered as a program (in equational form) which calls a number of other functions such as *sol*2, etc. Our proof guarantees the correctness of this program. We plan to study the correctness proofs of the algorithms which are based on the Chinese remainder theorem and produce these proofs in *RRL*.

The work reported here is a case study in a long term project for developing techniques of automated induction in *RRL*. The project is driven by a number of case studies from hardware and software verification, in a hope to identify problems and issues that have to be addressed in the further development of *RRL*. The proof of the Chinese remainder theorem is a milestone of the progresses in this project.

Acknowledgement: We would like to thank Deepak Kapur for his support and enthusiasm to the work reported here. We also thank anonymous referees whose comments helped us to restructure the paper.

References

[1] Basin, D., Kaufmann, M.: (1990) The Boyer-Moore prover and Nuprl: an experimental comparison. In: Proc. of the BRA Logical Frameworks Workshop.

[2] Boyer, R.S. and Moore, J S.: (1979) A computational logic. New York: Academic Press.

[3] Burstall, R.: (1969) Proving properties of programs by structural induction. Computer Journal 12(1), 41-48.

[4] Ehrig, H., Mahr, B.: (1985) Fundamentals of Algebraic Specification. Vol. 1&2. Springer - Verlag.

[5] Gramlich, B.: (1990) Completion based inductive theorem proving: A case study in verifying sorting algorithms. SEKI-Report SR-90-04, FB Informatik, Universität Kaiserslautern.

[6] Hua, X., Zhang, H.: (1992) Axiomatic semantics of a hardware specification language. Proc. of the Second IEEE Great Lakes Symposium on VLSI, Kalamazoo, MI, February 1992.

[7] Hua, X., Zhang, H.: (1992) FRI: Failure Resistant Induction in RRL. See this volume.

[8] Jouannaud, J.-P., and Kounalis, E.: (1986) Proofs by induction in equational theories without constructors. In: Proc. of Logic in Computer Science Conference, Cambridge MA.

[9] Kapur, D., Narendran, P., and Zhang, H.: (1986) Proof by induction using test sets. Proc. of 8th Intl Conf. on Automated Deduction (CADE-8), Oxford, U.K.

[10] Kapur, D., Zhang, H.: (1989) An overview of RRL: Rewrite Rule Laboratory. *Proc. of the third International Conference on Rewriting Techniques and its Applications* (RTA-89), Chapel Hill, NC, Springer Verlag LNCS 355, 513-529.

[11] Musser, D.R.: (1980) On proving inductive properties of abstract data types. In *Proceedings 7th ACM Symp. on Principles of Programming Languages*, pages 154–162. ACM. Las Vegas, NE.

[12] Peterson, G.L.: (1990) Complete sets of reductions with constraints. Proc. of 10th International Conference on Automated Deduction. Lecture Notes in Artificial Intelligence Vol. 449. Springer-Verlag, Berlin, pp. 381-395

[13] Russinoff, D.M.: (1985) An experiment with the Boyer–Moore theorem prover: a proof of Wilson's theorem. *J. of Automated Reasoning* **1**: 121-139

[14] Zhang, H.: (1992) Automated induction in algebraic specification. Research monograph (under preparation).

[15] Zhang, H., Guha, A., Hua, X.: (1991) Using algebraic specifications in Floyd-Hoare assertions. In: Rus, T. (ed.): Proc. of Second International Conference on Algebraic Methodology and Software Technology, Iowa City, Iowa.

[16] Zhang, H., Hua, X.: (1992) Proving Ramsey theorem by cover set induction: a case and comparison study. Presented at the Second International Symposium on Artificial Intelligence and Mathematics. Fort Lauderdale, Florida.

[17] Zhang, H., Hua, X.: (1992) A computer proof of the Chinese remainder theorem. Unpublished manuscript.

[18] Zhang, H., Hua, X., Kapur, D.: (1992) Use of induction hypotheses in the cover set induction: a case and comparison study. Unpublished manuscript.

[19] Zhang, H., Kapur, D.: First-order logic theorem proving using conditional rewrite rules. In: Lusk, E., Overbeek, R., (eds.): Proc. of 9th international conference on automated deduction. Lecture Notes in Computer Science 310, Springer 1988, pp. 1-20

[20] Zhang, H. and Kapur, D., and Krishnamoorthy, M.S.: (1988) A Mechanizable Induction Principle for Equational Specifications. Proc. of *Ninth International Conference on Automated Deduction (CADE-9)*, Argonne, IL, May 1988. Springer-Verlag LNCS 310, pp. 250-265.

Automatic Program Optimization Through Proof Transformation[1]

Peter Madden
Department of Artificial Intelligence, University of Edinburgh,
80 South Bridge, Edinburgh EH1 1HN, Scotland.
Tel: 031-650-2722, E-mail: madden@uk.ac.ed.aisb

1 Introduction

This paper concerns *the automatic transformation of programs by transforming their synthesis proofs.* This forms part of an ongoing research effort at the University of Edinburgh Department of Artificial Intelligence concerned with tackling the demands on complexity, reliability and quantity of software by tackling the problems of automated synthesis, verification and transformation of programs.

The author has implemented a program specialization and optimization system: the OYSTER *meta-level proof transformation system* (henceforth the OMTS). The main applications of the proof transformations are the *optimization of recursive programs* and the adaptation, or *specialization*, of programs to special situations. The transformations are correctness guaranteed and automatic.

In [12], we described in outline the rudiments of the specialization application of the OMTS which adapts programs to special situations, *specialization*, by transforming constructive synthesis proofs. We compared this with the original implementation [10, 9].[2]

In this paper we shall concentrate on a broader application of proof transformation: the optimization of recursive programs by transforming the induction schema employed within the corresponding synthesis proofs. This is a novel approach to program optimization. Key questions which we address are how program synthesis proofs can be exploited in order to transform programs and what advantages this approach has over the more traditional approach to program transformation where transformation rules are applied directly to the source program in order to construct the target program.

1.1 Transformation of Inductive Synthesis Proofs

In **Fig.** 1 we schematically depict an example source to target meta-level proof transformation: a single source inductive synthesis proof is depicted on the *left hand side* of the diagram: The proof yields a complex source algorithm which recurses with *exponential* behaviour (*exp*) due to the fact that a particular induction – *course_of_values* – is employed during the synthesis (the term *ext.* represents the program extraction process). The target proof is represented on the *right hand side*: the proof is not interactively synthesized from the specification, as may be the source, but rather *automatically* constructed by the application of meta-level operators which map and then transform portions of the object-level source proof. In particular, the source *course_of_values* induction is transformed into the more efficient *stepwise* target induction, thus yielding a target extract algorithm that recurses on its data-structure in more efficient *linear* fashion. Synthesis proofs contain more information than programs since not all the information required to complete a proof is concerned with execution. This extra information is used to control the proof transformations and largely accounts for the systems cotrol and automatability.

---

[1]This research was supported by SERC grant GR/F/71799, a SERC studentship and a SERC Postdoctoral Fellowship to the author. I would like to thank Prof. Alan Bundy and Dr Alan Smaill for supervision regarding the research documented in this paper.

[2]A detailed exposition of the author's specialization transformations is included in [13].

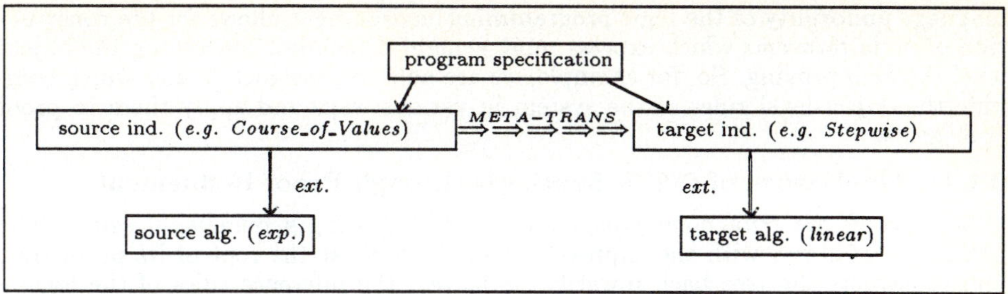

Figure 1: Recursion optimization through induction schema transformation

In §2 we provide a background to the OMTS by discussing the duality between proofs and programs (in particular the duality between induction and recursion), the OYSTERsystem and the typical structure of inductive proofs. In §3 we provide a brief overview of the OMTS DESIGN (details may be obtained from [13]). In §4 we describe some relevant existing techniques for program transformation – in particular the *fold/unfold* technique and the use of *dependency graph* analysis for *tupling program transformations.* In §5 we explain, through example, the methodology of the OMTS. In §6 we compare the properties of the OMTS with existing program transformation systems and highlight the advantages of the former. Finally, in §7, we provide a summary.

2 Background: The Duality Between Programs and Proofs

Constructive logic allows us to correlate computation with logical inference. This is because proofs of propositions in such a logic require us to construct objects, such as functions and sets, in a similar way that programs require that actual objects are constructed in the course of computing a procedure. Historically, this duality is accounted for by the *Curry-Howard isomorphism* which draws a duality between the inference rules and the functional terms of the λ-calculus [7, 11].

Such considerations allow us to correlate each proof of a proposition with a specific λ-term, λ-terms with programs, and the proposition with a specification of the program. Hence different constructive proofs of the same proposition correspond to different ways of computing a specific program specification.

2.1 The OYSTER System

The OYSTER system is an implementation of a constructive type theory which is based on Martin-Löf type theory, [14].OYSTER is written in Quintus Prolog, and run at the Prolog prompt level, so it is controlled by using Prolog predicates as commands. Proof tactics can be built as Prolog programs, incorporating OYSTER commands (which are simply Prolog predicates).

The main benefit of using type theory is that, recalling the previous section, it nicely combines typing properties with the properties of constructivism, such that we can both correlate the propositions of the λ-calculus with specifications of programs and correlate the proofs of the propositions with how the specification is computed. The main benefit of using a sequent calculus notation, as opposed to that of any of the numerous natural deduction systems, is that at any stage (node) during a proof development, all the dependencies (assumptions and hypotheses) required to complete that proof stage are explicitly presented within a *hypothesis list.* A sequent is of the form [HYPOTHESES] $\vdash$ [CONCLUSION], where, in the course of proving the conclusion, refinements may either act upon the hypotheses (so called *elim* refinements) or act upon the conclusion (so called *intro* refinements).

A major motivation behind the development of the OYSTER system is that the language uniformity of the logic programming environment allows for the construction of *meta-theorems* which express more general principles, concerning the object level theorem proving. So, for example, we are able to construct *tactics* which combine the object-level rules of the system in various ways and apply them to proof (sub)goals.

2.1.1 The Nature of OYSTER Synthesis through Proof Refinement

OYSTER proofs are *refinement proofs*, and are edited using a *refinement editor*. The OYSTER proof starts with the expression to be proved at the root of its proof tree, and constructs the tree back towards the leaves: the inference rules of the logic – *refinement rules* – are applied in reverse to a goal, to reduce, or *refine*, it to a set of sub-goals which, in turn, require proving in order to complete the overall proof.

Any proof is *complete* when the proof tree has been sufficiently developed *backwards* such that every leaf node can be proved without producing any further sub-goals. We refer to such proofs as being *goal-directed.* The refinement editor allows proof trees to be *traversed*, and refinement rules (or combinations thereof called proof tactics) to be applied to chosen nodes. The end-nodes, or leaves, of a proof will always correspond either to axiomatic equalities, well-formedness goals or unification.

At any stage during the development of a proof it is possible to access the *extract term* of the proof constructed so far. Each construct in the extract term corresponds to a proof construct. As such, the extract term reflects the algorithmic ideas behind the proof of the theorem. The extract programs consist of λ-calculus function terms, $\lambda(x, f_x)$ where f is the computed function and f_x the output when f is applied to input x.

Within type theory, each mathematical sentence, or proposition, is considered as a type, the elements of which are proofs of that sentence. A *type*, by definition, is a term which can be *inhabited* by other terms, or, equivalently, all types can have members. The existence of an extract term, corresponding to a particular proposition, is evidence that the proposition's type is inhabited, and this is equivalent to the proposition being constructively proved. All constructs of a completed proof that have an associated extract term of computational significance are collectively referred to as the *synthesis component* of the proof. The *verification component* of a proof is not used in executing the extract term, but ensures that the extract term satisfies the specification. Hence, although all components of a proof's extract term correspond directly to components in the proof, the relation is not bi-directional.

2.2 The Induction-Recursion Duality

OYSTER provides primitive recursion schemas for the basic types: integers, natural numbers and lists. The recursion schemas enable one to define recursive functions through case analyses, where the cases are determined by the structure of the type; and apply induction as an inference (refinement) rule, thus enabling one to synthesize the dual recursion in the extract program.

Employing any of the induction schemas in a (synthesis) proof will induce the corresponding, or *dual*, recursion schema in the extract algorithm. So, for example, *stepwise* recursion over the natural numbers is synthesized by applying *stepwise* induction, conventionally represented thus (where s is the successor (constructor) function):

$$\frac{\vdash A(0) \qquad \forall y : nat.\, A(y) \vdash A(s(y))}{\vdash \forall x : nat.\, A(x)}.$$

This states that A holds of any natural number, x, iff one can establish that A holds of 0 (the base case), and that, assuming A holds of some natural number y, that A holds of $s(y)$ (the step case).

Stepwise induction on the naturals, integers and on lists constitute the *primitive induction schemas*, and are built into the OYSTER system.Employing such induction as an inference rule will split the proof into the corresponding cases. Each case will have a corresponding proof and extract component. The structure of the program extracted from the complete proof will mirror that of the (instantiated) dual induction schema. Indeed, a strong heuristic that applies to synthesis through inductive theorem proving is that the behaviour of the induction variable should mirror that of the recursive terms in the function's definition.

More sophisticated induction schemas can be established by performing higher order proofs that appeal to the primitive schemas in order to justify the sophisticated scheme. An example of a non-primitive scheme is *course_of_values* induction. As with the primitive schemas, *course_of_values* recursion over the natural numbers is synthesized by applying *course_of_values* induction. This is conventionally represented thus:

$$\frac{\forall x : nat,\ \forall y : nat.\ ((y < z) \rightarrow A(y)) \vdash A(z)}{\vdash \forall x : nat.\ A(x)}.$$

This states that A holds of any natural number, x, iff one can establish that A holds of any natural number, z, assuming that A holds of any natural number, y, less than z.

Employing *course_of_values* induction as an inference rule does not automatically split the proof into a separate base and step case. Rather, the resulting subgoal represents the original proof tree with the induction hypothesis, $(y < z) \rightarrow A(y)$, entered into the proof as a new assumption (which tacitly includes the assumption that the hypothesis itself has a proof). The onus for splitting the proof into various cases, as defined by the function being synthesized, then lies with the user.

2.3 Exploiting Inductive *Proof Plans*

Practically all inductive proofs follow the same *general* strategy. This is one main reason why inductive proofs are a good candidate for transformation – due to the *similarity in form* of inductive proofs they afford us *general* mechanisms for optimizing recursive algorithms. That is, there is a typical *proof plan* for inductive proofs [2, 3].

In **fig.** 2 we have represented the key decisions and choice commitments made during a typical inductive proof. These will involve applying one of the OYSTER induction rules and then witnessing the existential quantifier, using $\exists - intro$, at each of the induction cases (where, as indicated in **fig.** 2, the application of the *intro* rules are specific to inductive *synthesis* proofs). Finally, we must verify that the instantiated schema will yield a recursive schema that will compute the input-output relation specified in the main conjecture represented in **fig.** 2 by $\forall x$ input $\exists y$ output spec(input,output). Note that we have indicated, within dashed boxes, that, following the witnessing steps of the (outermost) induction, we may wish to perform a further *nested* induction. These will take the same format as the outermost induction.

The verification stages will nearly always involve a process whereby formulae are "unpacked" - or *unfolded* - by replacing terms by suitably instantiated definitions. The proliferation of this process such that recursive terms are gradually removed from the recursive branches – by the repeated unpacking of induction terms – is part of the (heuristic) process known as *rippling-out* (following [1]). Simple examples of this would be the application of the following rewrites:

$$\begin{aligned} s(x) + y &\Rightarrow s(x + y), \\ append(e :: l_1, l_2) &\Rightarrow e :: append(l_1, l_2), \end{aligned}$$

where, in each case, the terms $x+y$ and $append(l_1, l_2)$ would unify with the respective induction hypotheses.

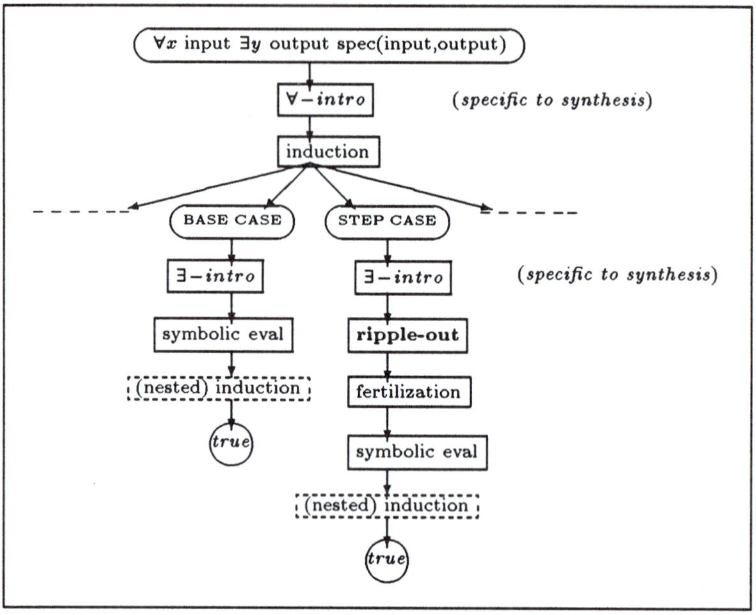

Figure 2: A General (Typical) Inductive Proof Plan (Strategy)

The goal of rippling-out is to reduce the induction step case to terms which can be unified with those in the induction hypothesis, or those in subsequent derivations of the induction hypothesis. This unification process, performed at the induction step, has been coined *fertilization*. The rationale behind fertilization rests, again, on the induction-recursion duality: by ensuring that we include the induction hypothesis in the construction of the induction step, we ensure that recursion is built into the λ-function being synthesized.

The rippling-out stages of the proof will contain an account of the dependencies between facts involved in the computation of the recursive branch of the function being synthesized. This *dependency information* is not usually available for inspection in traditional program code, and is exploited for the purposes of program through proof transformation.

The proof is completed when all the terms in the induction cases can be reduced to tautologies. Generally, this is done through the unification of terms in the sub-goals with those in any of the hypotheses made earlier in the proof, and by straightforward symbolic evaluation.

3 A Brief Overview of the OMTS Design

Proof trees are internally represented within OYSTER as quite complex Prolog data-structures. To avoid computational effort being expended on attempting to access individual semantic units the OYSTER representations of the proof trees are processed, by abstraction, into more accessible list structures called *rule-trees*. The information contained in a typical rule-tree consists of, some of the assumptions (hypotheses) made during the proof, the branching structure of the proof, the rules applied along with any corresponding arguments, and an account of the dependencies between facts in the proof. The dependency information consists of inter-relations between (sub)goals, and inter-relations between (sub)goals and assumptions (hypotheses).

The dependency information has to be abstracted from the rule-tree, which, in effect, retains a record of the inter-relations between *proof hypotheses* and sub-goals. So, recalling the Curry-Howard isomorphism, the rule trees contain an account of the dependencies between facts involved in the computation of the λ-function constructed by the corresponding proof.

Each rule-tree is akin to a proof plan which combines a number of proof tactics and/or rules into a large tactic such that a complete proof can be (re)produced from the plan. The rule-trees also serve as skeletons of proofs in which the inference rules are recorded, but not the formulae to which they are applied. The nesting pattern of the rule-tree list structure mirrors the branching pattern of the corresponding proof.

The rule-tree, and sub-trees thereof, are also akin to large *proof tactics* in that they consist of arbitrary combinations of inference rules and proof tactics, such as quantifier elimination, case analysis application, and induction, by means of the pre-defined tactical *then*.

A source proof is transformed by the application of *transformation tactics* to the source rule-tree. Hence the the transformation tactics perform transformations on proof tactics (viz., the OMTS rule-trees). They are called transformation *tactics* because it is their function to develop the target proof by modifying the source rule-tree according to certain pre (and post) conditions (see below). Using the transformation tactics, (sub)branches of the source proof can be accessed and the appropriate transformations made. The resulting target rule tree can then be applied, to the target specification, at the OYSTER object-level.

4 Relevant Work: Fold/Unfold and Tupling

Tupling is an important means of linearizing exponential procedures. It works by grouping together, in a single recursive tuple function, the separate recursive expressions in the source procedure. So, with $i \geq 2$ the conditions for tupling are as follows:

• **pre-condition:** There exist two or more *recursive calls*, $f(n), ..., f(n-i)$, which share some *common recursion variable(s)* in a function definition.
• **post-condition:** There exists a fixed sized tuple - the *eureka tuple* - within which common subsidiary recursive calls arising from the execution of each of $f(n), ..., f(n-i)$ can be merged.

The provision of the fixed sized tuple constitutes the *eureka step* for program transformation by tupling. In the majority of systems that employ the tupling technique, such as Darlington's pioneering NPL program transformation system [4], considerable user interaction is required to achieve the eureka step, and the subsequent folding step (see below).

The transformation process starts with the *source Fibonacci* proof of the previous section, duplicated below:

$$
\begin{array}{lll}
(1) & fib(0) & = 1; \\
(2) & fib(1) & = 1; \\
(3) & fib(n+2) & = fib(n+1) + fib(n).
\end{array}
$$

In order to calculate $fib(n)$ one must first calculate $fib(n-1)$ and $fib(n-2)$. Each of these sub-goals leads to another two recursive calls on fib and so on. In short the computational tree is exponential where the number of recursive calls on fib approaches 2^n.

The next step is the eureka step and consists of defining the desired optimization in terms of the "course of values" definition. This is done by the introduction of the auxiliary function fib_{tup} (thus satisfying condition 2 above):

$$(4) \quad fib_{tup}(n) \;=\; \langle fib(n+1), fib(n)\rangle.$$

This auxiliary function acts as a tuple – the *eureka tuple* – which, in effect, replaces the source recursion schema with a target schema which combines identical recursive calls. So Darlington's strategy is motivated by the observation that significant optimization of a (declarative) program generally implies the use of a new recursion schema. This process depends on the user providing the requisite definition of the *eureka* tuple fib_{tup}.

The system proceeds to evaluate the recursive branches of the auxiliary function, given the original equations and the instantiated base cases, by employing the *fold/unfold* strategy [8]. Armed with the original equations, it is a simple matter for the system to evaluate the base case for fib_{tup} given the left hand side of the equation, $fib_{tup}(0)$,

(5) $fib_{tup}(0) \;=\; \langle 1, 1\rangle.$

By introducing a *where* clause the system produces a definition of Fibonacci in terms of the auxiliary function fib_{tup}:

(6) $fib(n+2) \;=\; (u1+u2)$, where $\langle u1, u2\rangle = fib_{tup}(n)$.

Forced folding then comes into play for the optimization of $fib_{tup}(n+1)$: given the instantiated left hand side of the recursive step, unfolding produces the equation:

(7) $fib_{tup}(n+1) \;=\; \langle fib(n+1)+fib(n), fib(n+1)\rangle.$

The system then attempts to fold this equation with

(8) $fib_{tup}(n) \;=\; \langle fib(n+1), fib(n)\rangle,$

but fails since there is no direct match between the two. By observing that all the components necessary to match equation (8) are present within equation (7), the system *forces* the match by rearranging equation (7) to the following:

(9) $fib_{tup}(n+1) \;=\; \langle u1+u2, u1\rangle$, where $\langle u1, u2\rangle = \langle fib(n+1), fib(n)\rangle$

This now easily folds with (8) yielding the desired optimized function definition:

(10) $fib_{tup}(n+1) \;=\; \langle u1+u2, u1\rangle$, where $\langle u1, u2\rangle = fib_{tup}(n)$

In general, the tupling technique transformations allow for the "collapsing" of less efficient recursion schemas into more efficient ones. In this way subsidiary recursive calls can be merged into a single step in the optimized definition.

The *eureka step*, and the subsequent fold step(s) required to introduce a recursion into the target, present the greatest obstacles to automation. Until recently, these steps required considerable user intervention.

4.1 Exploiting Dependency graphs in Existing Systems

By an analysis of *symbolic dependency graphs*, based on [15], [5] describes an automatic procedure for finding a pair of *matching tuples* by the unfolding of selected calls to the source program, and then using matching as a means of testing for successful folding.

A dependency graph, DG, is a representation of a particular function call's evaluation tree which shows the calling structure of the subsidiary recursive calls. A *symbolic* DG is based on function calls which are potentially infinite in size. The initial portion of the symbolic DG for *Fibonacci* is shown below in **fig.** 3. The multiple evocations of subsidiary calls, the redundancy pattern, is exhibited by more than one arrow directed at any particular node. The main idea taken from [15] is that:

> An appropriate eureka tuple can be found if and only if there exists a *progressive sequence* of *cuts* that *match* one another, in the function's dependency graph.

A *cut* is defined as a subset of nodes across a dependency graph that when removed will divide the graph into two disconnected halfs. A *progressive sequence* of cuts is a sequence of cuts ordered according to size (i.e., according to the number of nodes in the subset). A pair of cuts *match* if a consistent substitution can be obtained when each function call of the first cut is matched with the corresponding function call of the second cut (formal definitions are given in [5]).

The finding of an appropriate *eureka tuple* depends on the notion of a continuous sequence of cuts. This is defined by Chin as follows:

> A *continuous* sequence of cuts, $cut_1, cut_2, ..., cut_N$, is a successive series of cuts which starts with the root node as its first cut. This sequence successively

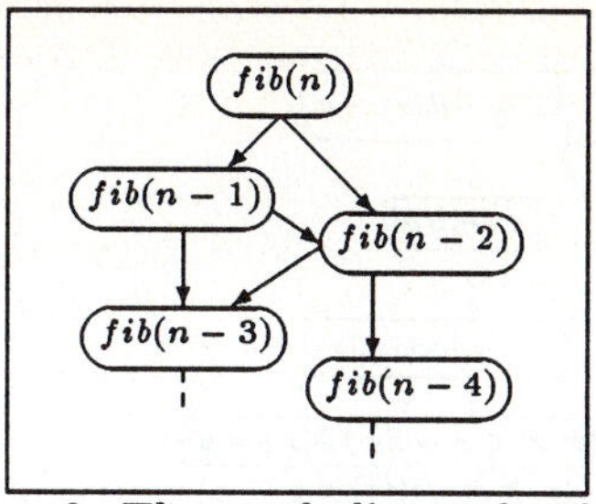

Figure 2: The symbolic DG for $fib(n)$

> **obtains the next cut by giving up a subset of nodes... from the *topmost set* of the current cut in order to acquire the children for the next cut.**

The topmost set of a cut is defined as a set of nodes whose ancestors are not present in the cut itself.

Returning to the example and starting with the main function call, Chin's analysis replaces $fib(n)$, the first cut, with its two subsidiary calls, $\langle fib(n-1), fib(n-2)\rangle$. This gives us the second cut. The analysis then proceeds by unfolding only that call in a cut which is *not* a subsidiary call of the other call, i.e., the topmost item. So, since the function call $fib(n-2)$ is a subsidiary call of $fib(n-1)$, only $fib(n-1)$ is unfolded. This gives the third cut, $\langle fib(n-2), fib(n-3)\rangle$. The third cut matches the second cut, thus providing the analysis with a matching tuple.

So, the fold/unfold steps required for the tupling transformation are achieved by locating a pair of matching tuples by the unfolding of appropriately selected calls and then using matching as a means of testing for successful folding. The main difference between Chin and Darlington's systems is that the use of such selection ordering allows for a considerable degree of automation, since once this analysis succeeds the main task of the tupling transformation – finding a successful fold – will have been achieved.

5 OMTS Proof Transformation

Chin's automation of Darlington's *fold/unfold* strategy hinges on the production of a continuous sequence of cuts. This in turn hinges on producing an appropriate *ordering* for selecting nodes to unfold during the analysis. So the production of (symbolic) DGs, and their subsequent analysis, are an essential ingredient of *automating* the tupling technique. This is not the case where synthesis proofs are concerned: the rippling out stages of an inductive proof will display all the (dependency) information required for the automatic construction of target tuples:

5.1 Explanation by Example

Remaining with the *Fibonacci* example, we can fill in the details of the schematic proof plan of **Fig.**2 to depict both an outline of the source *course_of_values* proof for synthesizing *Fibonacci*, **Fig.**4(a), and an outline of the target *stepwise* proof for synthesizing *Fibonacci*, **Fig.**4(b). For the sake of clarity, we omit many of the type checking and elimination rules. We use the arcs, M1 to M5, that pass from **Fig.**4(a) to **Fig.**4(b) to depict those (sub)structures of the source proof which are used to develop the target proof. These "mappings" will be explained in §5.1.2.

We shall first briefly describe the nature of the source proof (i.e. **Fig.**4(a)). The specification, fib_{spec}, for computing the Fibonacci numbers :

$$fib_{spec}: \ \forall x{:}nat, \exists y{:}nat.\ fib(x) = y,$$

where fib is defined through the use of the following lemmas:

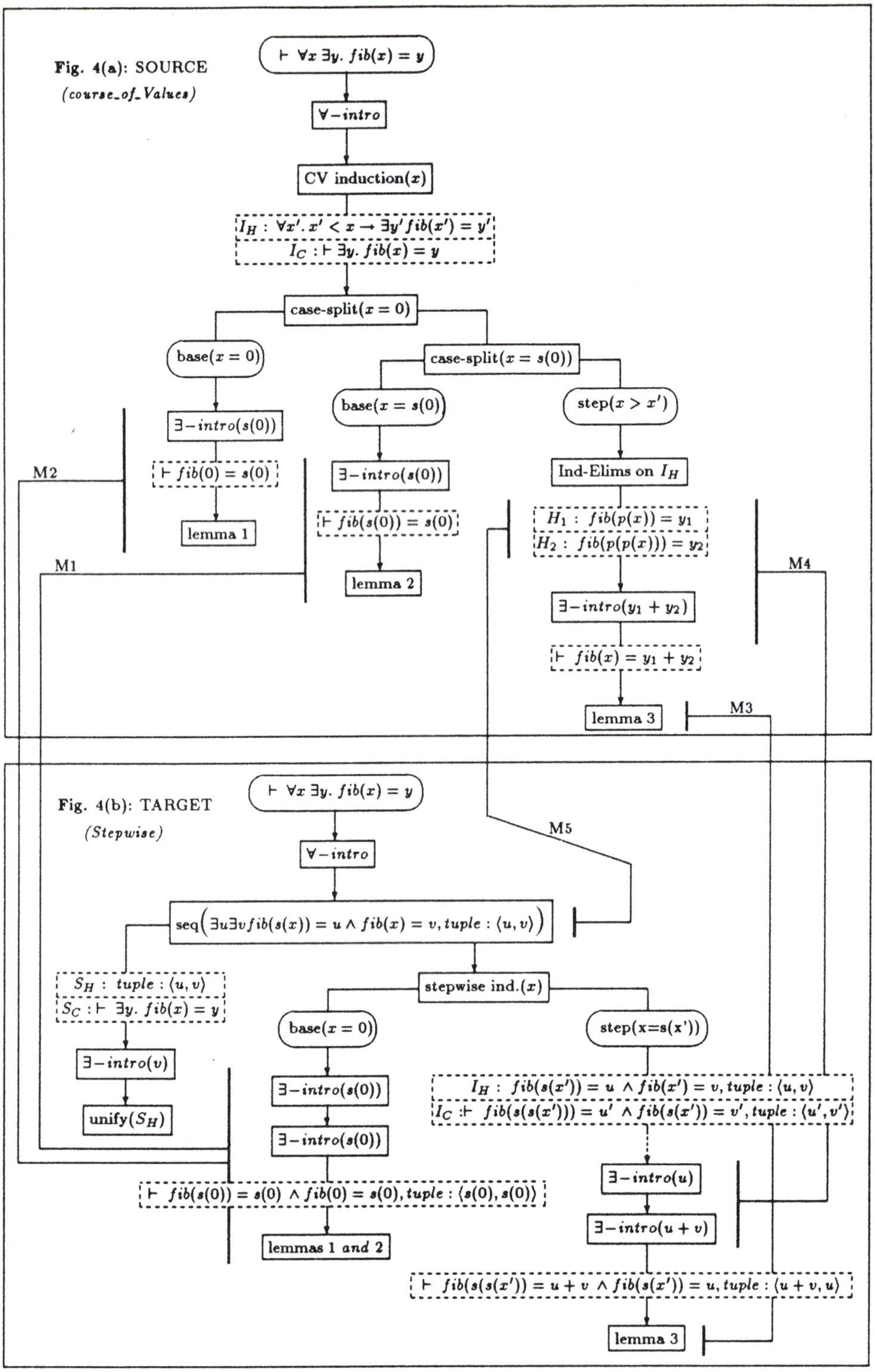

Figure 4: Schematic Representation of Source to Target Proof Mappings for *Fibonacci*

$$
\begin{array}{lll}
fib\_lemma\ 1: & fib(0) & = s(0); \\
fib\_lemma\ 2: & fib(s(0)) & = s(0); \\
fib\_lemma\ 3: & \multicolumn{2}{l}{\forall x{:}nat,\ \exists y{:}nat,\ \exists z{:}nat\ .\,(x = 0 \rightarrow void) \wedge (x = s(0) \rightarrow void)} \\
 & \multicolumn{2}{l}{\wedge\ fib(p(x)) = y \wedge fib(p(p(x))) = z\ \rightarrow\ fib(x) = y + z,}
\end{array}
$$

where p is the *predecessor* function defined by induction over the naturals such that $fib(x-1) \equiv fib(p(x))$ and $fib(x-2) \equiv fib(p(p(x)))$. p is usefully employed as a destructor function of a function's data-structure (as opposed to using the canonical successor function, s, to build constructor definitions).

The third lemma, *fib_lemma* 3, defines the recursive case and is naturally a course_of_values definition. Hence, by the induction-recursion duality, fib_{spec} is most naturally proved by *course_of_values* induction.

Basically the proof requires an initial application of the $\forall - intro$ refinement.[3] This has the effect of *intro*ducing the universal quantifiers, followed by course of values induction on x (CV induction(x)).

The cases of the induction schema are then satisfied by setting up a nested case analysis structure by performing two case-split refinements, where the second case-split is nested within the first. The outermost case split corresponds to $x = 0 \vee x \neq 0$, and the innermost case to split to $x = s(0) \vee x \neq s(0)$.

By having the case splits nested in this way, we cover all the conditions specified in the course of values definition. By using the $\exists{-}intro(W)$ rule, a suitable witness, W, is introduced at each case, and then verification is performed by appealing to (unfolding with) the relevant lemma (with various well-formedness goals being satisfied along the way). Within the dashed-boxes we have included key hypotheses and (sub)goals (conclusions): the application of *course_of_values* induction yields the induction hypothesis, I_H, and the induction conclusion, I_C. At the two base cases, B1 and B2, we provide, in both cases, a witness of $s(0)$. At the step case, S, we perform a series of eliminations (Ind-Elims) on I_H which yield the hypotheses H_1 and H_2. H_1 and H_2 provide outputs for $fib(p(x))$, namely y_1, and for $fib(p(p(x)))$, namely y_2. Recursion is then built into the function being constructed by using H_1 and H_2 to provide a witness for the step case $fib(x)$, namely $y_1 + y_2$.

Upon completion of the synthesis component of the target proof, verification is performed by mapping across and applying precisely the same lemmas that are used for the source proof verification.

5.1.1 Proof Transformation Conditions

The OMTS transformations are akin to tactic transformations guided in part by whether or not certain syntactic properties are true of the source rule-trees. Such syntactic properties function as transformation tactic pre-conditions. We can also predict the probable outcome of the application of a transformation tactic in terms of syntactic properties of the target rule-tree. A source to target transformation will be deemed successful if the target rule tree satisfies the post-conditions.[4]

The pre- and post-conditions for the induction schema transformations of an exponential procedure to a linear procedure include amongst their pre-conditions that the dominant induction in the proof is a *course_of_values* induction. Amongst the post-conditions will be the presence of a *stepwise* construct in the target rule-tree.

The conditions also include those for proof tupling which, analogous to the program tupling conditions (§4), can be stated thus:

- **pre-condition:** There exist two or more induction terms, $f'(n), ..., f'(n-i)$, which share some *common induction variable(s)* in a function definition (where $i \geq 2$).

---

[3] Since OYSTER proofs are *goal-directed*, then rules such as $\forall - introduction$ have the *quantifier stripping* effect usually associated with $\forall - elimination$ in forwards proof systems.

[4] If the source rule-tree satisfies the pre-conditions then only in exceptional cases will a complete target rule tree be produced which violates the post-conditions.

• **post-condition:** There must be present(constructed) a fixed sized tuple - the *eureka tuple* - within which common subsidiary function calls arising from the unfoldings of each of $f'(n), ..., f'(n-i)$ are merged.

We refer to the use of the tupling technique within proof transformations as *proof tupling.*

5.1.2 How the OMTS constructs the Target

We now provide a description of the key stages of the source to target proof transformation. We shall refer to the arc labels, M1 - M5, of **Fig.** 4 when ever appropriate.

The first step is to map across the source main goal to form an identical target main goal (thus ensuring that the target retains the same functionality as the source). The subsequent $\forall-intro$ application is also mapped across.

• *Tuple Construction*: The next step involves mapping (sub)structures from the source in order to construct an appropriate tuple (the *eureka step*) for the target definition. This is depicted by arc M5. Functions which fit the following schema:

$$f(x) \Leftarrow \textbf{if } b(x) \textbf{ then } c(x) \textbf{ else } h(x, f(d_1(x)), f(d_2(x))).$$

can be linearized by tabulating the members of the set of subsidiary calls between the main function call, $f(x)$, and that call in the body, $f(\delta^j(x))$ which takes the maximum number of applications of the *common descent function*, δ, to the recursive argument x (see [6] for details). A common descent function is one in which both d_1 and d_2 may be defined. This means each descent function is related to each other through δ in that each is cashed out in terms of applying δ a certain number of times, i.e., $d_1 = \delta^i$ and $d_2 = \delta^j$, where δ^n is to be interpreted as the application of δ n times.

In the case of *Fibonacci* the common descent function is the pedecessor function, p. By noting the maximum number, Φ, of applications of p to the induction argument x, required for the construction of the course_of_values induction step, the OMTS can determine the requisite size and contents of the target tuple (namely Φ and $fib(p^1(x), ..., p^\Phi(x))$. This gives $\langle fib(p(x)), fib(p(p(x)))\rangle$ as the target tuple definition in terms of the source definition. This information is provided by the hypotheses, H_1 and H_2, resulting from eliminations on the induction hypothesis I_H. H_2 provides us with a value of $\Phi = 2$. The OMTS must enter such a target tuple definition as a new goal to be proved *before* the application of the target stepwise induction. Hence, to allow the subsequent constructor version of stepwise induction to proceed smoothly, the target tuple definition is converted into the following constructor form (where $tuple : \langle u, v\rangle$ denotes a tuple of objects, u and v, each inhabiting the natural number type):

$$\exists u \exists v fib(s(x)) = u \wedge fib(x) = v, tuple : \langle u, v\rangle$$

This is *sequenced* (seq) into the target proof thus producing two new sub-goals: the first is the main goal to be proved along with the sequenced definition as an additional hypothesis (S_H in **Fig.**4). The second sub-goal requires constructing the tuple specified by the sequenced goal.

Note that the *eureka step* of the program tupling transformations is automatically achieved through exploiting information contained within the rippling-out stages of the source proof.

In general , if we schematicaly represent the recursive step of a bi-linear function, f, thus:

$$h(x, f(\delta^i(x)), f(\delta^\Phi(x))),$$

then the *eureka tuple* for computing $f(x)$ is obtained by creating exactly Φ existentially quantified variables, where each quantifier ranges over one of the tuple components. Each quantified variable is then witnessed, respectively, by each member of the progression:

$$f(\delta^1(x)), f(\delta^2(x)), f(\delta^3(x)), ..., f(\delta^n(x)), ..., f(\delta^\Phi(x)),$$

to produce the tuple:

$$f_{tup}(x) = \langle f(\delta^1(x)), f(\delta^2(x)), f(\delta^3(x)), ..., f(\delta^n(x)), ..., f(\delta^\Phi(x))\rangle,$$

where the value of the nth member will be a subsidiary call on f such that the value of its recursive argument is n less than that in the main call, such that, for example:[5]

$$f(\delta^n(x)) = f(x - n).$$

• *Instantiating the Target Induction*: The system then selects stepwise induction, which has a more efficient computation rule than course_of_values, and applies it to the above target tuple definition. This results in the following cases:

Induction hypotheses:	$\exists u : nat\, \exists v : nat.fib(s(x')) = u \wedge fib(x') = v$
Induction base:	$\vdash \exists u' : nat\, \exists\, v' : nat.fib(s(0)) = u' \wedge fib(0) = v'$
Induction step:	$\vdash \exists u' : nat\, \exists\, v' : nat.fib(s(s(x'))) = u' \wedge fib(s(x')) = v'$

The task now is for the system to supply witnesses (instantiations) for the existential quantifiers at each of the cases. The *base case* witnessing is quite straightforward and can be mapped almost directly from the source (arcs M1 and M2 of **Fig.**4): the system splits the target base formula into its conjuncts, and then applies the same $\exists - intro$ witnessing steps as were applied in order to satisfy the source base cases. So with both u' and v' instantiated to $s(0)$, this provides the requisite target tuple, $\langle s(0), s(0)\rangle$, for the recursive base case. The *step case* witnessing is achieved, like the initial *eureka step*, through the exploitation of dependency information contained in the induction step branch of the source proof. We require existential witnesses for both u' and v' at the target recursive step. Since $fib(p(x')) = v'$, then we can obtain a witness for v' simply by introducing the value for $fib(p(x'))$ provided by the target induction hypothesis, namely u. A witness for u' is obtained by observing the dependent subsidiary calls required for evaluating $fib(x)$ in the source induction step. The required target calls are determined by, *firstly*, observing the source subsidiary calls obtained through the elimination applications in the source induction step ($fib(p(x'))$ and $fib(p(p(x')))$), and, *secondly*, by noting the function applied to these two subsidiary calls to provide an output for $fib(x)$ at the source induction step (+). The abstraction of this information is denoted by arc M 3 of **Fig.**4. Hence the OMTS can determine, basically through a consistent substitution of $s(s(x))$ for x, that since $u' = s(s(x))$ then a witness for u' is provided by adding the values for $fib(s(x'))$ and $fib(x')$. These are provided by the target induction hypothesis, giving a step case witness for u' of $u + v$. This provides the requisite target tuple, $\langle u + v, u\rangle$, where v provides the output for $fib(x)$ in the main goal (i.e. hypothesis S_H in **fig.** 4(b)). So, in general, providing witnesses for the target induction step involves, (i), *substituting* the *target induction hypothesis* tuple components for those in the *source induction step* in order to satisfy the first tuple component, and, (ii), a direct 1-1 mapping of the first component, $fib(x - 1)$, of the *target induction hypothesis* to satisfy the second tuple component.

Each of the target witnesses must then be verified to produce the correct output for the respective branch of recursive function being synthesized. This is achieved through mapping and then applying the lemmas used in the source proof.

5.1.3 Further Applications of the OMTS

The explanation of the proof tupling procedures is very much a simplification concerning the workings of, and programs transformed by, the proof transformation system. For example, the system is not limited to optimizing only bi-linear functions. Furthermore, proofs containing *nested* inductions yield function definitions with nested recursion schemas. The OMTS is capable of transforming such proofs

[5] If we are dealing with constructor definitions, such that the common descent function, δ, is the successor function, s, then $f(\delta^n(x)) = f(x + n)$.

to target proofs which combine, through proof tupling, the nested inductions into a single induction thereby yielding a target program with a single recursion. Finally, programs synthesized through stepwise induction may be transformed to programs synthesized through *divide and conquer* induction. In this way the system transforms linear recursion to logarithmic recursion. These and other applications, and a far more thorough and detailed account of the proof transformations, especially of the flexibility of control that the induction schemas give us over the resulting recursive structures *and* of the automatic construction of the auxiliary function definition used to define the target tuple, can be found in [13].

6 Comparisons with Program Tupling Transformations

We now discuss the differences, and advantages, that the OMTS approach to optimization has on the exploitation of dependency information, correctness, automatability and control and search issues.

6.1 The Reduced Workload Regarding Dependency Analyses

The rule-tree abstractions are designed to preserve precisely that information which proofs contain in addition to the programs that they synthesize: a description of the task being performed; a verification of the method; and an account of the dependencies between facts involved in the computation.

The fact that they contain an account of the dependencies between the facts involved in the computation means that, in order to construct an appropriate recursive definition for the target, we do not have to appeal to complex tuple formation procedures that construct *symbolic dependency graphs* and subsequently analyse these in order to form the requisite *Eureka* tuple.

Recall, from §4.1, that previous tuple analysis procedures involve constructing a representation of a function's evaluation tree (a symbolic dependency graph) which shows the calling structure of the subsidiary recursive calls, and then analysing the recursive calls to find the matching tuples. Such an analysis tells us two things: firstly, *the number* of subsidiary calls of the main function calls required to form the tuple (i.e., the determination of the tuple size); and secondly, *which* subsidiary calls are to be tabulated. An advantage of *proof tupling* is that *both* of these things, required for the tuple formation, are contained in the source proof, and preserved in the source proof rule tree abstraction. This means that they can readily be abstracted from the proof and exploited for the construction of the target tuple *without* any additional dependency graph construction and analysis procedures.

An interesting point to note is that Chin's analysis of unfolding cuts, followed by matching (or unification) is very similar to the process of rippling out followed by fertilization during synthesis. The analogy nicely illustrates the fact that the information required by Chin's tuple analysis is (explicitly) present within our synthesis proofs.

6.2 Correctness

More recent incarnations of the fold/unfold strategy have been shown to be correctness guaranteed for specified classes of functions (*cf.* [16] and [5]). However, each extension to the class of functions requires a corresponding extension to the correctness procedures, and this leads to a considerable work overhead (proportional to the range of transformations – or *generality* – of the system).

This is not a problem regarding the OMTS, and any future extensions thereof. We can summarize what has been said regarding the correctness of the OMTS transformations as follows: Extract programs are correct with respect to the complete specifications of the synthesis proofs from which they are extracted. Given the respective specifications, the target and source rule-trees contain all the information required to construct the respective proofs. The respective specifications are, in the case of optimization, the same. Therefore, correctness of the target proof *and* of the source to target transformation is ensured. Hence the correctness of *all* terminating transformations is ensured *without* having to additionally provide, or extend, any correctness criteria, or proof, each time we extend the range of programs to which the transformations are applicable.

6.3 Search

Since we can regard the rule-trees, together with pre- and post-conditions, as proof plans then a general advantage of performing *tactic transformations* – i.e., meta-level transformations on the object-level tactics – is that the transformation space is equivalent to a planning search space which is far smaller than the object-level search space.

Furthermore, the portions of the source proof that are accessed for the analysis correspond to specific semantic units: the specification, the application of induction, the induction base and step cases, the unfolding step, and the witnessing rule. These are clearly represented as distinct sub-lists within the rule-tree abstractions, and the OMTS knows precisely where to look in order to access any of the aforementioned units. So, unlike program tupling, the OMTS proof tupling optimizations do not require the construction of a (potentially infinite) dependency graph, nor does it require any procedures for searching the dependency graph in order to find a *matching tuple*.

The motivation that drives the tuple formation procedures also has a considerable bearing on the amount of search involved during tupling transformations: Darlington's NLP, and Chin's HOPE^{+}, tuple analysis is motivated by the desire to find a tuple which can be used for *folding*. This involves quite extensive search in order to find a successful fold. In the original NLP system, the search for a fold requires considerable user interaction. In more recent systems considerable effort is required to automate the process of forced folding whereby equations must be re-arranged such that a match can be found, and recursion introduced into the definition of target program.

The OMTS analysis, on the other hand, is motivated by the desire to find witnesses for the tuple components at the induction step of a synthesis *proof*. Once this has been achieved then the proof is completed in much the same way as any inductive synthesis proof: by a process of *rippling* until all terms in the conclusion match terms in the proof hypotheses. The iterative application of rippling, at least on available evidence, is far easier to automate. It has been done for the current OMTS system *and* within the context of automatic *proof plan* application through the automation of the *rippling out process* [2].

7 Summary

We described the fundamentals of a working synthesis proof transformation system. The novel aspect of this research is that program optimization is achieved through the transformation of *synthesis* proofs. In particular, recursive programs are optimized by transforming inductive synthesis proofs. Techniques from the field of program transformation may be used to transform the computational content of a proof. An important technique for transforming exponential behaviour into linear behaviour is *tupling*. The OMTS, unlike other existing transformation systems, performs this technique on (synthesis) proofs. The system satisfies the desirable properties for a transformation system of correctness, generality, automatability and the means to guide search through the transformation space.

The benefits of the proof transformation approach include the fact that extra information contained in the proofs, but not programs, can be exploited to automatically guide the transformations. In particular, dependency information abstracted from the source proof guides the transformations without the need for any extensive dependency graph analysis.

The source and target programs of traditional program transformation systems do not have a formal specification present. This means there is no immediate means of checking that the target program meets the desired operational criteria. Regarding proof transformation, all transformed programs are correct with respect to their specifications, and we ensure that the target computes the same specified input/output relation as the source (only more efficiently).

An important commitment regarding the recursive behaviour of an extract program is the choice of induction schemata (and how the cases are satisfied). By exploiting the common structure of OYSTER inductive synthesis proofs we can transform the induction schema employed in a proof yielding an inefficient program into a schema such that the new target proof yields a more efficient program. Transformation is achieved through the application of *proof transformation tactics* to internal representations of the OYSTER proofs. Since we can provide a general proof plan for inductive (synthesis) proofs, then we can build general transformation tactics for optimizing the recursive programs that they synthesize.

References

[1] R. Aubin. Some generalization heuristics in proofs by induction. In G. Huet and G. Kahn, editors, *Actes du Colloque Construction: Amélioration et vérification de Programmes*. Institut de recherche d'informatique et d'automatique, 1975.

[2] A. Bundy, F. van Harmelen, J. Hesketh, and A. Smaill. Experiments with proof plans for induction. *Journal of Automated Reasoning*, 7:303–324, 1991.

[3] A. Bundy, van Harmelen. F., C. Horn, and A. Smaill. The Oyster-Clam system. In M.E. Stickel, editor, *10th International Conference on Automated Deduction*, pages 647–648. Springer-Verlag, 1990. Lecture Notes in AI No.449.

[4] R.M. Burstall and J. Darlington. A transformation system for developing recursive programs. *Journal of the Association for Computing Machinery*, 24(1):44–67, 1977.

[5] W. N. Chin. *Automatic Methods for Program Transformation*. PhD thesis, University of London (Imperial College), 1990.

[6] N. H. Cohen. Eliminating redundant recursive calls. volume 5 No. 3, pages 265–299. 1983.

[7] H.B. Curry and R. Feys. *Combinatory Logic*. North-Holland, 1958.

[8] J. Darlington. *A Semantic Approach to Automatic Program Improvement*. PhD thesis, Dept. of Artificial Intelligence, Edinburgh, 1972.

[9] C. A. Goad. Computational uses of the manipulation of formal proofs. Technical report, Stanford University, 1980. STAN-CS-80-819.

[10] C. A. Goad. Proofs as descriptions of computation. In *Lecture Notes in Computer Science*. Academic Press, 1980.

[11] W.A. Howard. The formulae-as-types notion of construction. In J.P. Seldin and J.R. Hindley, editors, *To H.B. Curry; Essays on Combinatory Logic, Lambda Calculus and Formalism*, pages 479–490. Academic Press, 1980.

[12] P. Madden. The specialization and transformation of constructive existence proofs. In N.S. Sridharan, editor, *Proceedings of the Eleventh International Joint Conference on Artificial Intelligence*. Morgan Kaufmann, 1989. Also available from Edinburgh as DAI Research Paper 416.

[13] P. Madden. *Automated Program Transformation Through Proof Transformation*. PhD thesis, University of Edinburgh, 1991.

[14] Per Martin-Löf. Constructive mathematics and computer programming. In *6th International Congress for Logic, Methodology and Philos ophy of Science*, pages 153–175, Hanover, August 1979. North Holland, Amsterdam. 1982.

[15] A. Pettorossi. A powerful strategy for deriving programs by transformation. In *ACM Lisp and Functional Programming Conference*, pages 405–426, 1984.

[16] H. Tamaki and T. Sato. A transformation system for logic programs that preserves equivalence. Technical Report TR-018, ICOT, 1984.

Proof Search Theory and Practice in the (former) USSR
(Tentative)

Grigori Mints

Department of Philosophy
Stanford University
Stanford, California

Basic Paramodulation and Superposition

Leo Bachmair* Harald Ganzinger†
Christopher Lynch‡ Wayne Snyder§

March 15, 1992

Abstract

We introduce a class of restrictions for the ordered paramodulation and superposition calculi (inspired by the *basic* strategy for narrowing), in which paramodulation inferences are forbidden at terms introduced by substitutions from previous inference steps. These refinements are compatible with standard ordering restrictions and are complete without paramodulation into variables or using functional reflexivity axioms. We prove refutational completeness in the context of deletion rules, such as simplification by rewriting (demodulation) and subsumption, and of techniques for eliminating redundant inferences. Finally, we discuss experimental data obtained from a modification of Otter.

1 Introduction

Paramodulation is a refutational theorem proving method for first-order logic with equality, originally presented in [17] and refined in various ways since that time (e.g., [1, 7, 13, 15, 18, 24]). For instance, paramodulation is refutationally complete if inferences are not allowed at variable positions, and also if various ordering restrictions are imposed. In addition, techniques for simplifying and removing redundant clauses have been developed. In this paper we strengthen the former refinements significantly by extending the principle of *basic* narrowing to paramodulation. In essence, basic paramodulation forbids paramodulation at terms introduced by substitutions in earlier inferences. We show that this strategy is compatible with ordering restrictions and can also be applied to the superposition calculus. It turns out that stronger restrictions on forbidden terms can be imposed if the ordering restrictions are weakened. We will examine this tradeoff from several points of view and discuss three specific inference systems, called *basic superposition*, *basic ordered paramodulation*, and *basic paramodulation*. We also describe a technique, called *variable*

*Department of Computer Science, SUNY at Stony Brook, Stony Brook, NY 11794, U.S.A., leo@sbcs.sunysb.edu; partially supported by NSF Grant No. CCR-8901322.

†Max-Planck-Institut für Informatik, Im Stadtwald, D-W-6600 Saarbrücken, Germany, hg@cs.uni-sb.de

‡Computer Science Department, Boston University, 111 Cummington St., Boston, MA 02215, U.S.A., lynch@cs.bu.edu; partially supported by NSF grants CCR-8901647 and CCR-8814339.

§ibid, snyder@cs.bu.edu; partially supported by NSF Grant No. CCR-8910268.

abstraction, for propagating forbidden terms around a clause, and explore the role of simplification rules such as demodulation, subsumption and blocking.

To illustrate the basic ideas, let us consider the paramodulation inference

$$\frac{\rightarrow Q(ga), f(hz,z) \approx gz \qquad P(f(x,gy)) \rightarrow k(x,gy) \approx hy}{P(ggy) \rightarrow Q(ga), k(hgy,gy) \approx hy}$$

and possible further paramodulations into its conclusion. Using boxes to indicate subterms at which further paramodulations are forbidden, we obtain the following representation of the conclusion

$$P(gg\boxed{y}) \rightarrow Q(ga), k(hg\boxed{y}, g\boxed{y}) \approx h\boxed{y}$$

(we always disallow paramodulation into variables). Basic superposition forbids inferences at any term introduced as part of the substitution,

$$P(g\boxed{gy}) \rightarrow Q(ga), k(\boxed{hgy}, g\boxed{y}) \approx h\boxed{y},$$

while basic ordered paramodulation also forbids inferences at any term positioned below a former paramodulation inference,

$$P(\boxed{ggy}) \rightarrow Q(ga), k(\boxed{hgy}, g\boxed{y}) \approx h\boxed{y}.$$

An alternate strategy with weaker ordering constraints also forbids inferences at any term introduced by the left premise,

$$P(\boxed{ggy}) \rightarrow Q(\boxed{ga}), k(\boxed{hgy}, g\boxed{y}) \approx h\boxed{y}.$$

Finally, restrictions on a term t can be propagated to other occurrences of t,

$$P(\boxed{ggy}) \rightarrow Q(\boxed{ga}), k(\boxed{hgy}, \boxed{gy}) \approx h\boxed{y}.$$

We have not indicated any applicable ordering constraints in this example, but as we shall see below, the ordering constraints gradually become weaker as the "basic" constraints get stronger.

These strategies can be implemented easily either by using a simple marking strategy (with a Boolean flag indicating forbidden terms) or, alternately, by directly implementing the formalism of closures (i.e., pairs of clauses and substitutions) in which we describe our inference systems.

In some ways, the techniques presented here are reminiscent of the *set-of-support* strategy for resolution, which directs the resolution rule against clauses which are essential to the unsatisfiability of the entire set. For this reason (and based on experimental results), we consider this paper to be a robust answer to a research problem posed by Larry Wos [23]: *What strategy can be used to restrict paramodulation at the term level to the same degree that the set of support strategy restricts all inference rules at the clause level?*

2 Preliminaries

We present here a brief overview of the notation and preliminary definitions necessary for the paper; for a more thorough coverage, see the books [3, 21].

Equations, Clauses, and Closures A *multiset* is an unordered collection with possible duplicate elements; for a multiset M, we denote the number of occurrences of an object x by $M(x)$, and define the union of multisets $M \cup N$ as the multiset Q such that $Q(x) = M(x) + N(x)$ for every x. For multisets M, N, we say M is included in N, denoted $M \subseteq N$, if there exists a multiset Q such that $N = M \cup Q$. An *equation* is a binary multiset $\{s,t\}$, conventionally represented $s \approx t$, where s and t are first-order terms over a given signature. A *clause* is a pair of multisets of equations,[1] written $\Gamma \rightarrow \Delta$, where Γ is the *antecedent* and Δ the *succedent.* We usually write Γ_1, Γ_2 and A, Γ instead of $\Gamma_1 \cup \Gamma_2$ and $\{A\} \cup \Gamma$. A clause represents an implication $A_1 \wedge \cdots \wedge A_m \supset B_1 \vee \cdots \vee B_m$; the empty clause, a contradiction. Clauses of the form $\Gamma, A \rightarrow A, \Delta$ or $\Gamma \rightarrow \Delta, t \approx t$ are called *tautologies.*

By a *ground* expression (i.e., a term, equation, etc.) we mean an expression containing no variables. The application of a substitution σ to an object Φ, denoted $\Phi\sigma$, produces an *instance* of Φ (a *ground instance*, if $\Phi\sigma$ is ground). Composition of substitutions is denoted by juxtaposition; if τ and ρ are substitutions, then $x\tau\rho = (x\tau)\rho$, for all variables x. We define $dom(\sigma) = \{x \mid x\sigma \neq x\}$.

We shall distinguish terms of the original clauses from terms introduced by substitutions. A *closure* is a pair $C \cdot \sigma$ consisting of a clause C (the *skeleton*) and a substitution σ. A closure $C \cdot \sigma$ is a *representation* of a clause D if $D = C\sigma$. A closure $C \cdot \sigma$ is called a *ground closure* if $C\sigma$ is ground. The closure $C \cdot \sigma\rho$ is an *instance* of $C \cdot \sigma$ (by a substitution ρ). We identify a clause C with the closure $C \cdot \mathrm{id}$, where id is the identity substitution. Two closures $C \cdot \sigma$ and $C \cdot \tau$ with the same skeleton C need not be distinguished if σ and τ agree on each variable that occurs in C. We say that two closures $C \cdot \sigma$ and $D \cdot \tau$ *have disjoint variables* whenever $var(C) \cup var(C\sigma)$ and $var(D) \cup var(D\tau)$ are disjoint; and we can identify $C \cdot \sigma$ and $D \cdot \tau$ with $C \cdot \rho$ and $D \cdot \rho$, respectively, where $\rho = \sigma\tau$. It is assumed in what follows that all closures are formed from idempotent substitutions.

Equality Herbrand Interpretations By an (*equality Herbrand*) *interpretation* we mean a congruence on ground terms. A ground clause $C = \Gamma \rightarrow \Delta$ is *true in an interpretation* I, denoted $\models_I C$, if either $\Gamma \not\subseteq I$ or $\Delta \cap I \neq \emptyset$. For a non-ground clause C, $\models_I C$ iff $\models_I C\sigma$ for every ground instance $C\sigma$. Similarly, $\models_I C \cdot \sigma$ iff $\models_I C\sigma$. An interpretation I is called a (*equality Herbrand*) *model* of a set N of clauses or closures, respectively, if it satisfies all members of N. N is called *consistent* if it has a model; and *inconsistent* (or *unsatisfiable*), otherwise. We say that N *implies* C, denoted $N \models C$, if every model of N satisfies C.

We write $\Phi[s]$ to indicate that Φ contains s as a subexpression and (ambiguously) denote by $\Phi[t]$ the result of replacing a particular occurrence of s by t. A relation $\Rightarrow$ is a *rewrite relation* if $s \Rightarrow t$ implies $u[s\sigma] \Rightarrow u[t\sigma]$, for all terms s, t and u, and substitutions σ. Furthermore, we write $s \Downarrow t$ to indicate that s and t can be rewritten to a common form. A rewrite relation $\Rightarrow$ is said to be *Church-Rosser* if the two relations $\Leftrightarrow^*$ and $\Downarrow$ are the same. We assume the reader is familiar with

[1] An atomic formula $P(t_1, \ldots, t_n)$, where P is some predicate symbol other than equality, is considered an abbreviation for the equation $P(t_1, \ldots, t_n) \approx \top$, where $\top$ is a distinguished constant.

the notions of *normal forms* and *convergent* (i.e., Church-Rosser and well-founded) sets of rewrite rules.

Any ordering $\succ$ on a set S can be extended to an ordering $\succ_{mul}$ on finite multisets over S as follows: $M \succ_{mul} N$ if (i) $M \neq N$ and (ii) whenever $N(x) > M(x)$ then $M(y) > N(y)$, for some y such that $y \succ x$. If $\succ$ is a total [well-founded] ordering, so is $\succ_{mul}$. Given a set (or multiset) S and an ordering $\succ$ on S, we say that x is *maximal* relative to S if there is no $y \in S$ with $y \succ x$; and *strictly maximal* if there is no $y \in S \backslash \{y\}$ with $y \succeq x$. If $\succ$ is an ordering on terms, then the multiset ordering $\succ_{mul}$ is an ordering on equations, which we denote by $\succ^e$.

We identify an occurrence of an equation $s \approx t$ in the antecedent of a clause with the multiset (of multisets) $\{\{s, \bot\}, \{t, \bot\}\}$, and an occurrence in the succedent with the multiset $\{\{s\}, \{t\}\}$, where $\bot$ is a new symbol.[2] We identify clauses with finite multisets of occurrences of equations. By $\succ^o$ we denote the twofold multiset ordering $(\succ_{mul})_{mul}$ of $\succ$, which is an ordering on occurrences of equations; by $\succ^c$ we denote the multiset ordering $\succ^o_{mul}$, which is an ordering on clauses. If $\succ$ is a well-founded [total] ordering, so are $\succ^e$, $\succ^o$, and $\succ^c$.

3 Basic Inference Rules

We present two inference systems $\mathcal{S}$ and $\mathcal{P}$ on closures. The inference rules are defined with respect to a reduction ordering $\succ$ and we also assume that the premises of a binary inference rule have disjoint variables (the variables in one premise are renamed if necessary).

We say that a clause $C = \Gamma \rightarrow \Delta, s \approx t$ is *reductive* for $s \approx t$ if $t \not\succeq s$ and $s \approx t$ is a strictly maximal occurrence of an equation in C.

The first rule encodes the reflexivity of equality:

Equality Resolution: $$\frac{(\Lambda, u \approx v \rightarrow \Pi) \cdot \rho}{(\Lambda \rightarrow \Pi) \cdot \theta}$$

where $\theta = \rho\sigma$, with σ a most general unifier of $u\rho$ and $v\rho$, and $u\theta \approx v\theta$ is a maximal occurrence of an equation in $\Lambda\theta, u\theta \approx v\theta \rightarrow \Pi\theta$.

The next inference rule represents a variant of factoring, restricted to the succedent of clauses:

Positive Factoring: $$\frac{(\Gamma \rightarrow \Delta, A, B) \cdot \rho}{(\Gamma \rightarrow \Delta, A) \cdot \theta}$$

where $\theta = \rho\sigma$, with σ a most general unifier of $A\rho$ and $B\rho$,and $A\theta$ is a maximal occurrence of an equation in $\Gamma\theta \rightarrow \Delta\theta, A\theta, B\theta$.

The remaining rules are restrictions of paramodulation.

Basic Left Superposition: $$\frac{(\Gamma \rightarrow \Delta, s \approx t) \cdot \rho \quad (u[s'] \approx v, \Lambda \rightarrow \Pi) \cdot \rho}{(u[t] \approx v, \Gamma, \Lambda \rightarrow \Delta, \Pi) \cdot \theta}$$

[2] The symbol $\bot$ is not in the language and is minimal with respect to $\succ$, i.e., $t \succ \bot$, for all terms t.

where (i) $\theta = \rho\sigma$, with σ a most general unifier of $s\rho$ and $s'\rho$, (ii) the clause $\Gamma\theta \rightarrow \Delta\theta, s\theta \approx t\theta$ is reductive for $s\theta \approx t\theta$, (iii) $v\theta \not\succeq u\theta$ and $u\theta \approx v\theta$ is a maximal occurrence of an equation in $u\theta \approx v\theta, \Lambda\theta \rightarrow \Pi\theta$, and (iv) s' is not a variable.

Basic Right Superposition: $$\frac{(\Gamma \rightarrow \Delta, s \approx t) \cdot \rho \quad (\Lambda \rightarrow u[s'] \approx v, \Pi) \cdot \rho}{(\Gamma, \Lambda \rightarrow u[t] \approx v, \Delta, \Pi) \cdot \theta}$$

where (i) $\theta = \rho\sigma$, with σ a most general unifier of $s\rho$ and $s'\rho$, (ii) the clause $\Gamma\theta \rightarrow \Delta\theta, s\theta \approx t\theta$ is reductive for $s\theta \approx t\theta$, (iii) the clause $\Lambda\theta \rightarrow u\theta \approx v\theta, \Pi\theta$, is reductive for $u\theta \approx v\theta$, (iv) $s\theta \approx t\theta \not\succeq^{e} u\theta \approx v\theta$, and (v) s' is not a variable.

Basic Merging Paramodulation:
$$\frac{(\Gamma \rightarrow \Delta, s \approx t) \cdot \rho \quad (\Lambda \rightarrow u \approx v[s'], u' \approx v', \Pi) \cdot \rho}{(\Gamma, \Lambda \rightarrow z \approx v[t], z \approx v', \Delta, \Pi) \cdot \{z \mapsto u\}\theta}$$

where z is a new variable not occurring in either premise and (i) $\theta = \rho\sigma$, with σ the composition $\alpha\beta$ of a most general unifier α of $u\rho$ and $u'\rho$, and a most general unifier β of $s\rho$ and $s'\rho\alpha$, (ii) the clause $\Gamma\theta \rightarrow \Delta\theta, s\theta \approx t\theta$ is reductive for $s\theta \approx t\theta$, (iii) the clause $\Lambda\theta \rightarrow \Pi\theta, u\theta \approx v\theta, u'\theta \approx v'\theta$ is reductive for $u\theta \approx v\theta$, (iv) $u\theta \succ v\theta$ and $v'\theta \not\succeq v\theta$, and (v) s' is not a variable.

The above inference system, denoted $\mathcal{S}$, is called *basic superposition*. Note that in the last rule we have "imported" the terms u, u' into the substitution part of the conclusion.

If we weaken certain ordering constraints, such as $v\theta \not\succeq u\theta$ in the right superposition rule, we may import additional terms into the substitution part of the conclusion. *Basic ordered paramodulation*, denoted $\mathcal{P}$, consists of equality resolution, positive factoring, plus the following three rules.

Basic Left Paramodulation: $$\frac{(\Gamma \rightarrow \Delta, s \approx t) \cdot \rho \quad (u[s'] \approx v, \Lambda \rightarrow \Pi) \cdot \rho}{(u[z] \approx v, \Gamma, \Lambda \rightarrow \Delta, \Pi) \cdot \{z \mapsto t\}\theta}$$

for some new variable z not occurring in either premise, and where conditions (i)–(iv) are identical to those for Basic Left Superposition, plus: (v) $s\theta \approx t\theta \not\succ^{e} u\theta \approx v\theta$.

Basic Right Paramodulation I: $$\frac{(\Gamma \rightarrow \Delta, s \approx t) \cdot \rho \quad (\Lambda \rightarrow u[s'] \approx v, \Pi) \cdot \rho}{(\Gamma, \Lambda \rightarrow u[z] \approx v, \Delta, \Pi) \cdot \{z \mapsto t\}\theta}$$

where z is a new variable not occurring in the premises, and conditions (i)–(v) are the same as for Basic Right Superposition.

Basic Right Paramodulation II:
$$\frac{(\Gamma \rightarrow \Delta, s \approx t) \cdot \rho \quad (\Lambda \rightarrow u \approx v[s'], \Pi) \cdot \rho}{(\Gamma, \Lambda \rightarrow z_1 \approx v[z_2], \Delta, \Pi) \cdot \{z_1 \mapsto u, z_2 \mapsto t\}\theta}$$

where z_1 and z_2 are new variables not occurring in the premises, and (i) $\theta = \rho\sigma$, with σ a most general unifier of $s\rho$ and $s'\rho$, (ii) the clause $\Gamma\theta \rightarrow \Delta\theta, s\theta \approx t\theta$ is reductive for $s\theta \approx t\theta$, (iii) $u\theta \succ v\theta$ and $u\theta \approx v\theta$ is strictly maximal in $\Lambda\theta \rightarrow u\theta \approx v\theta, \Pi\theta$, and (iv) s' is not a variable.

Note that the basic strategy is enforced in our inference rules by the restriction "s is not a variable" in the context of closures. We may impose the following additional restrictions on $\mathcal{S}$ and $\mathcal{P}$: if $C \cdot \rho$ and $D \cdot \rho$ are the premises and $E \cdot \theta$ the conclusion of a binary inference, then $C\theta \not\succeq^c D\theta$ and $E\theta \not\succeq^c D\theta$.

The reader should notice that in the two calculi $\mathcal{P}$ and $\mathcal{S}$, there is a trade-off between the strength of the ordering restrictions and the strength of the "basic" restriction, viz., in $\mathcal{S}$, we may not perform inferences into the smaller sides of equations in an unrestricted way, but when this is relaxed, in $\mathcal{P}$, we may strengthen the "basic" restriction by importing right-hand sides of equations into the substitution part of the conclusion. In fact, if we weaken the ordering requirements even further, so that inferences are permitted into non-maximal equations, then we may import *every* term from the left premise into the substitution of the conclusion. This is called simply *ordered paramodulation* and is described in detail in the full paper [4].

4 Refutation Completeness

These calculi are refutationally complete in the sense that a contradiction (the empty clause) can be derived from any inconsistent set of clauses. The proof relies on reasoning about ground instances of clauses and inferences. For the following definition recall that we do not distinguish between a clause C and a closure $C \cdot \mathrm{id}$.

Definition 1 Let π be an inference in $\mathcal{I}$ with premises $C_1 \cdot \rho_1, \ldots, C_n \cdot \rho_n$ and conclusion $C \cdot \rho$. By a *ground instance* of π we mean any inference in $\mathcal{I}$ with premises $C_1\rho_1\sigma, \ldots, C_n\rho_n\sigma$ and conclusion $C\rho\sigma$, where all clauses $C_i\rho_i\sigma$ and $C\rho\sigma$ are ground.

The main problem in proving completeness for paramodulation systems is that the usual lifting lemma does not hold without qualifications, as the position of the inference may be lifted off with the substitution. The solution is to work with reduced substitutions; in our method we go further and require that clauses be "hereditarily reduced," so that no inference need be performed inside *any* substitution position.

Definition 2 We say that a ground closure $C \cdot \sigma$ is *reduced* with respect to a rewrite rule $u \approx v$ if every variable x, for which $x\sigma$ is reducible by $u \approx v$, occurs only in equations of the form $x \approx t$ in the succedent of C, where $x\sigma = u$ and $v \succeq t\sigma$. A closure is reduced with respect to a rewrite system R if it is reduced with respect to all rules in R.[3] By a reduced instance of N we mean any ground clause $D = C\sigma\tau$, where $C \cdot \sigma$ is a closure in N and $C \cdot \sigma\tau$ is reduced.

For example, if $a \succ b \succ c \succ d$, then the closure $(\rightarrow x \approx c, x \approx d)\cdot\{x/a\}$ is reduced with respect to $a \approx b$, but not with respect to $a \approx d$. Note that closures $C \cdot \mathrm{id}$ with an

[3] Alternately, a closure $C \cdot \sigma$ is reduced with respect to R if for any equation occurrence $s \approx t$ in C and any x in $s \approx t$, $x\sigma$ is irreducible by any $u \approx v$ in R such that $s\sigma \approx t\sigma \succ^o u \approx v$.

empty substitution part are reduced with respect to any rewrite system R. The next result shows under what conditions ground inferences are ground instances of basic inferences, and is easily proved using the the properties of most general unifiers.

Lemma 1 (Lifting Lemma) *Let C and D be clauses with no variables in common, and let $C\sigma = \Gamma \to \Delta, s \approx t$ and $D\sigma$ be ground instances, such that $D\sigma \succ^c C\sigma$, $s \approx t$ is a maximal occurrence of an equation in $C\sigma$, and $D\cdot\sigma$ is reduced with respect to $s \approx t$. Then any basic superposition or paramodulation inference with premises $C\sigma$ and $D\sigma$ is a ground instance of a similar inference from $C \cdot \tau$ and $D \cdot \tau$ for any τ which subsumes σ, i.e., $\sigma = \tau\rho$ for some ρ.*

Let N be a set of closures and $\succ$ be reduction ordering which is total on ground terms. We define an interpretation I by means of a convergent rewrite system R as follows. For any ground rewriting system R, let R^* be the least interpretation (i.e., ground congruence) containing R.

First, we use induction on the clause ordering $\succ^c$ to define sets of equations E_C, R_C, and I_C, for all ground instances C of closures of N. Let C be such a ground instance and suppose that $E_{C'}$, $R_{C'}$, and $I_{C'}$ have been defined for all ground instances C' of N for which $C \succ^c C'$. Then

$$R_C = \bigcup_{C \succ^c C'} E_{C'} \quad \text{and} \quad I_C = R_C^*.$$

Moreover

$$E_C = \{s \approx t\}$$

if $C = \Gamma \to s \approx t, \Delta$ is a reduced instance of N with respect to R_C such that (i) C is reductive for $s \approx t$, (ii) s is irreducible by R_C, (iii) $\Gamma \subseteq I_C$, and (iv) $\Delta \cap I_C = \emptyset$. In that case, we also say that C *produces* the equation (or rule) $s \approx t$. In all other cases, $E_C = \emptyset$. Finally, we define I to be the equality interpretation R^*, where $R = \bigcup_C E_C$ is the set of all equations produced by ground instances of clauses of N.

Clauses that produce equations are also called *productive*. Note that a productive clause C is false in $I_C = R_C^*$, but true in $(R_C \cup E_C)^*$. Moreover only clauses which are reduced ground instances of N can be productive. The sets R_C and R are constructed in such a way that they are convergent, and hence the truth value of an equation can be determined by rewriting: $u \approx v \in I$ if and only if $u \Downarrow_R v$.

The interpretation I is a model of N, provided N is consistent and saturated (i.e., closed under sufficiently many applications of the appropriate basic inference rules). We define saturation, which is one of the key notions in our approach, so as to take into account a notion of redundancy which will enable us to add simplification and deletion techniques to the basic calculi without impairing refutational completeness. It is well known that simplification and deletion are essential for the practical application of paramodulation and resolution systems.

Intuitively, a closure is redundant if it follows from smaller closures in N, where "smaller" refers to an ordering $\succ^s$ on closures which uses the notion of *basic subsumption*. Let $s \cdot \sigma$ and $t \cdot \theta$ be closures of terms. We say that $s \cdot \sigma$ is *η-dominated* by $t \cdot \theta$, for some substitution η, written $s \cdot \sigma \sqsubseteq_\eta t \cdot \theta$, iff $s\sigma\eta = t\theta$ and for each $x \in dom(\sigma)$ if x occurs in s at position p, then p is at or below a variable position in t. For equations, we say that $(s \approx t) \cdot \sigma \sqsubseteq_\eta (u \approx v) \cdot \theta$ iff either $s \cdot \sigma \sqsubseteq_\eta u \cdot \theta$ and $t \cdot \sigma \sqsubseteq_\eta v \cdot \theta$, or if $s \cdot \sigma \sqsubseteq_\eta v \cdot \theta$ and $t \cdot \sigma \sqsubseteq_\eta u \cdot \theta$. For closures of multisets of equations, we have $\Delta_1 \cdot \theta_1 \sqsubseteq_\eta \Delta_2 \cdot \theta_2$ iff there exists an injection φ from $\Delta_1 \cdot \theta_1$ into $\Delta_2 \cdot \theta_2$ such that if $\varphi(e_1 \cdot \theta_1) = e_2 \cdot \theta_2$, then $e_1 \cdot \theta_1 \sqsubseteq_\eta e_2 \cdot \theta_2$. For closures of clauses, we have $(\Delta \rightarrow \Gamma) \cdot \sigma \sqsubseteq_\eta (\Pi \rightarrow \Theta) \cdot \rho$ iff $\Delta \cdot \sigma \sqsubseteq_\eta \Pi \cdot \rho$ and $\Gamma \cdot \sigma \sqsubseteq_\eta \Theta \cdot \rho$. We write $\Phi \sqsubseteq \Psi$ to indicate that there exists some η such that $\Phi \sqsubseteq_\eta \Psi$; in the case of clauses, Φ is said to be a *basic subsumer* of Ψ. If $\Phi \sqsubseteq \Psi$ but $\Psi \not\sqsubseteq \Phi$, then Φ is a *proper basic subsumer* of Ψ. Now let $C \cdot \sigma\tau$ and $D \cdot \theta\rho$ be ground instances of closures $C \cdot \sigma$ and $D \cdot \theta$. We define $C \cdot \sigma\tau \succ^s D \cdot \theta\rho$ if and only if either (i) $C\sigma\tau \succ^c D\theta\rho$ or (ii) $C\sigma\tau = D\theta\rho$ and $D\theta$ is a proper basic subsumer of $C\sigma$.

The basic idea of the relation $\sqsubseteq$ is that all terms in the closure substitution on the left side must overlap the right side inside the closure substitution. This definition has been designed so that if $s \cdot \sigma$ and $t \cdot \theta$ are two closures of terms such that $s \cdot \sigma \sqsubseteq_\eta t \cdot \theta$, then $t \cdot \theta$ is reducible at a substitution position by a rule $l \approx r$ whenever $s \cdot \sigma\eta$ is.

Definition 3 Let D be a ground instance of a closure (not necessarily in N). We call D *redundant* with respect to N, if for any ground rewriting system R for which D is reduced, there exist ground instances $C_1 \cdot \sigma_1\rho_1, \ldots, C_k \cdot \sigma_k\rho_k$ of N such that (i) $D \succ^s C_i \cdot \sigma_i\rho_i$, for all i, $1 \leq i \leq k$, (ii) each $C_i \cdot \sigma_i\rho_i$ is reduced wrt R, and (iii) if each $C_i \cdot \sigma_i\rho_i$ is true in R^*, then D is true in R^*. A closure is redundant if all its ground instances are redundant. An inference with premises $C_1 \cdot \rho \cdots C_n \cdot \rho$ and conclusion $C \cdot \theta$ is said to be redundant if one of the closures $C_1 \cdot \theta \cdots C_n \cdot \theta$ is redundant (note carefully the use of θ instead of ρ). A set N is *saturated* if the conclusion of every non-redundant inference from N is in N or is redundant in N.

Tautologies are trivially redundant. A fundamental property of redundancy, which we shall use later, is that redundancy is preserved if additional closures are added or if redundant closures are deleted.

Lemma 2 *(i) If $N \subseteq N'$, then any closure which is redundant with respect to N is also redundant with respect to N'.*

(ii) If $N \subseteq N'$ and all closures in $N' \setminus N$ are redundant with respect to N', then any closure redundant with respect to N' is also redundant with respect to N.

We shall only sketch the main ideas of the completeness proof and refer the reader to the full paper [4] for details (there are also similarities to the completeness proofs in [1]). A saturated set of closures N satisfies various properties with respect to its associated interpretation I. For instance, if a reduced ground instance of N is redundant, then it is true in I. Moreover, if $C = \Gamma \rightarrow \Delta, s \approx t$ is a productive ground instance of N, then $\Gamma \subseteq I$ and $(\Delta \cap I) = \emptyset$. (We emphasize that these properties need not hold if N is not saturated.)

These properties can be used to show that all reduced ground instances of N are true in I, whenever N is consistent. For if some reduced instance is not true in I, one can choose a minimal counterexample $D = \Sigma \to \Pi$ (with respect to the given well-founded ordering on clauses), for which $\Sigma \subseteq I$ and $(\Pi \cap I) = \emptyset$. Furthermore, it is possible to show that there exists an inference with premise D, the conclusion D' of which is simpler than D and false in I, which contradicts that D is a minimal counterexample.

With this result one can easily prove the following completeness theorem:

Theorem 1 *Let K be a set of clauses and let N be a saturated set of closures such that $C \cdot \mathrm{id}$ is in N for any clause in K and such that any closure in N follows from K. Then K is consistent if and only if N does not contain the empty clause. In the latter case, I is a model of K and N.*

Proof. If N contains the empty clause, K is inconsistent. On the other hand, if N does not contain the empty clause, then I is a model of any reduced instance of N. Now let $C\sigma$ be a ground instance of K. We define a substitution τ by $x\tau = t_x$, where t_x is the normal form of $x\sigma$ by R. Then $C \cdot \tau$ is a reduced instance of the closure $C \cdot \mathrm{id}$ in N. Therefore $C\tau$, and also $C\sigma$, is true in I. □

In the next section we shall discuss specific instances of redundancy and describe ways of constructing saturated sets.

5 Theorem Proving with Simplification and Deletion

A (finite or countably infinite) sequence $N_0, N_1, N_2, \ldots$ of sets of closures is called a *theorem proving derivation* if the substitution part of every closure in N_0 is empty, and if each set N_{i+1} can be obtained from N_i by adding the conclusion of some inference from $\mathcal{S}$ $[\mathcal{P}]$, or by deletion of a redundant clause. The set $N_\infty = \bigcup_j \bigcap_{k \geq j} N_k$ is called the *limit* of the derivation.

Definition 4 A theorem proving derivation is called *fair* if the conclusion of every non-redundant inference from N_∞ is in $\bigcup_j N_j$.

A fair derivation can be constructed, for instance, by systematically adding conclusions of non-redundant inferences from persisting closures.

Lemma 3 *If $N_0, N_1, N_2, \ldots$ is a fair theorem proving derivation, then both $\bigcup_j N_j$ and N_∞ are saturated.*

Theorem 2 *Let $N_0, N_1, N_2, \ldots$ be a fair theorem proving derivation. If $\bigcup_j N_j$ does not contain the empty closure, then N_0 is consistent.*

Redundancy provides a very general syntactic criterion for proving the completeness of various deletion strategies. It covers not only tautology deletion, but also more complex kinds of deletion such as subsumption, simplification, and blocking.

Subsumption. If C and D are closures in N, such that $C \sqsubseteq D$, then D can be deleted from N. Formally, this is because D is redundant. The standard form of subsumption is not complete in the basic setting, as can be shown by the following example.

$$\begin{array}{rcl} P(x,y) & \rightarrow & P(x,b) \\ P(a,b) & \rightarrow & \\ & \rightarrow & a=c \\ & \rightarrow & P(c,b) \end{array}$$

Assume the constants are ordered $a \succ b \succ c$. By resolving the first two clauses, we obtain $P(\boxed{a},y) \rightarrow$ (where a is forbidden), which then triggers the deletion of the second clause above. But then the reader may verify that there is no refutation. When we have a subsumer in the standard, but not the basic sense, we must instantiate the substitution of C so as to form the *intersection* of the substitution parts of the atoms (alternately, the union of the skeletons) which are matched up during the subsumption step, so that the relation $C \sqsubseteq D$ holds; thus, any term which is forbidden in both clauses is forbidden in the subsumer. For example, if $P(f(\boxed{a}),\boxed{g(b)},c) \rightarrow$ is subsumed by $P(\boxed{f(a)},g(\boxed{b}),x) \rightarrow$, then we must instantiate this latter clause to $P(f(\boxed{a}),g(\boxed{b}),x) \rightarrow$ when the former is deleted.

Simplification. Let $C = (\rightarrow l \approx r)\cdot\sigma$ be a unit closure such that $Var(r) \subseteq Var(l)$ and $D[l']\cdot\theta$ be a closure containing a non-variable term l' such that $l\cdot\sigma \sqsubseteq_\rho l'\cdot\theta$ for some matching substitution ρ, where $l\sigma\rho \succ r\sigma\rho$, and $D\theta \succ^c C\rho$. Then we can simplify $D[l']\cdot\theta$ to $D[r\rho]\cdot\sigma\rho\theta$.[4] An example of the incompleteness of arbitrary simplification in the basic setting is as follows:

$$\begin{array}{rcl} & \rightarrow & P(f(x)),\ f(x)=b \\ P(f(a)) & \rightarrow & \\ Q & \rightarrow & a=c \\ f(c)=b & \rightarrow & \\ & \rightarrow & Q \end{array}$$

(We assume a lexicographic path ordering based on $P \succ f \succ a \succ b \succ c \succ Q$.) Let us assume that saturation begins with superposing the first onto the second clause, producing the new clause $\rightarrow f(\boxed{a}) = b$, which is then used to simplify the clause, yielding a new second clause $P(b) \rightarrow$. If a is forbidden in the simplifier, then it is impossible to derive the empty clause, as the calculus does not admit a superposition of $a \approx c$ into $f(\boxed{a}) \approx b$. This example illustrates the problem for both $\mathcal{S}$ and $\mathcal{P}$.

As with subsumption, if the conditions for simplification, but not basic simplification, are met, then we must partially instantiate the simplifier so that the

[4]It is possible to refine this rule somewhat by specifying which parts of the matching substitution ρ are from σ, and importing these from $r\rho$ into the substitution; see the full paper [4] for details. Note that the only variables instantiated in applying ρ to r are those in $var(r) - dom(\sigma)$.

conditions $Var(r) \subseteq Var(l)$ and $l \cdot \sigma \sqsubseteq_\rho l' \cdot \theta$ are satisfied. The latter, again, amounts to taking the "intersection" of the substitution parts of the two terms.

Basic Blocking We now discuss our last deletion rule, in which we can delete a clause which we know not to be reduced, due to the presence of a simplifier in the set. A closure $C \cdot \theta$ is *blocked* with respect to a set of closures N if there exists an occurrence of an equation e in C and a variable x in e such that $x\theta$ is reducible by an instance $l\sigma\rho \approx r\sigma\rho$ of a simplifier $(\rightarrow l \approx r) \cdot \sigma$ in N, where $Var(r) \subseteq Var(l)$, $l\sigma\rho \succ r\sigma\rho$, and $e\theta \succ^o l\sigma\rho \approx r\sigma\rho$. The justification for this is that a blocked clause is redundant. It is also possible to block inferences because certain atoms in the premises are simplifiable, as in [18]; for details, see the full paper [4].

6 Change of Closure Representation

We now briefly describe an additional technique for restricting $\mathcal{S}$ and $\mathcal{P}$, which is based on the fact that the boundary between the skeleton and substitution of a closure can be changed under certain conditions; for example it can be shown that we can always instantiate a closure by its substitution (since then we are removing a restriction). It is clearly more fruitful to consider techniques for importing skeleton terms into the substitution part, e.g., replacing a clause $C \cdot \sigma$ by $C' \cdot \eta\sigma$, where $C = C'\eta$, since this restricts the application of the inference rules. For example, we know we can replace $P(\boxed{a}, b), P(a, c) \rightarrow$ by $P(\boxed{a}, b), P(\boxed{a}, c) \rightarrow$ if $c \prec b$, since if there exists no rule smaller than $P(a, b)$ in R to reduce a, then clearly the same holds for $P(a, c)$. The most general form of this transformation may be stated as follows. Let $C[u \approx v][l \approx r]$ delineate two occurrences of equations (not necessarily distinct) in a clause C.

$$\textbf{Variable Abstraction:} \qquad \frac{C[u[t] \approx v][l[x] \approx r] \cdot \sigma}{C[u[z] \approx v][l[x] \approx r] \cdot \{z \mapsto t\}\sigma}$$

unless we have the pathological case that (i) $l \approx r$ occurs positively, (ii) $l = x$, (iii) $l\sigma \not\preceq r\sigma$, (iv) and $(u \approx v)\sigma \not\preceq^o (l \approx r)\sigma$. This rule defines a replacement.

In other words, a piece of the substitution can be "propagated" from one part of a clause to another where an identical term occurs, except in the rare case that the original substitution term occurs at the top of a maximal side of a positive equation, and the term propagated to occurs in a larger or incomparable equation. For example, terms can always be propagated from "predicate equations".

The reason that this rule is valid is that the conditions insure that for every ground instance $C\sigma\tau$ of the premise closure, there exists no possible ground equation $w \approx w'$ which reduces $t\sigma\tau$ and such that $(u \approx v)\sigma\tau \succ^o w \approx w' \succeq^o (l \approx r)\sigma\tau$. Thus the new closure is reduced relative to the old one, and hence in the completeness proof if the old closure has a reduced ground instance, then the corresponding ground instance of the new one is also reduced, and if this replacement is made, the contradiction upon which the completeness result depends still goes through. For obvious reasons, this transformation of representation should be applied *eagerly* to all clauses when they are generated.

7 Benchmarks

We have implemented these strategies in part in a modified version of Otter [11]. Code was added to "mark" substitution positions in clauses with a Boolean flag, and also to mark the right-hand sides of equations used in basic ordered paramodulation steps, and to mark the entire left premise (as in the basic paramodulation strategy). In addition, we created routines to take the intersection of the marked positions during subsumption and the intersection of the marked positions in the left-hand sides of demodulators during simplification (the marks in all right-hand sides were erased). These modifications do not affect completeness. We have also run trials without taking the intersections of the marks in subsumption and simplification.

A summary of our experimental data, in terms of the number of clauses generated, is given in the following chart. We used consistent flag settings throughout these experiments; our purpose was to compare the basic with the non-basic strategy and not to find the fastest possible proof.

	super	para	b-super	b-ord	b-para	best
s1	70	103	65	85	85	63
s2	17	36	15	22	22	15
p48	113	111	113	130	108	62
p49	161	213	149	114	77	69
p63	156	138	292	119	119	110
p64	45	69	42	63	63	42
p65	69	66	55	71	71	50
o1	160	237	145	201	201	130
o2	413	252	416	146	146	123
o3	75	74	71	90	90	54

The first two problems are, respectively, (s1) – "xx is a constant in the Boolean implication algebra," and (s2) – "$x(xy) = xy$ in Boolean algebra." The next five are from the numbered list in [14]. Problems o1–o3 are respectively kb_comm, tba_gg, and x2_quant in the test directory accompanying Otter (Release 2.0).

In these experiments paramodulation is allowed only from a maximal side of an equality and into any literal in a clause. The (super) column represents experiments where we only paramodulate into a maximal side of an equality. In (para) we paramodulated into both sides. We used the same Otter flag settings for the "basic" refinements of these strategies, with corresponding data listed in columns (b-super) – basic superposition, (b-ord) – basic ordered paramodulation, and (b-para) – basic paramodulation. Finally, in column (best) we list our attempt to produce a shorter refutation by picking the best trial among the basic trials, and then running Otter without intersecting the marks in subsumption and simplification. It turned out that this (incomplete) strategy produced a shortest refutation in each case.

Several points deserve mention. First, the modification of Otter necessary to implement the basic strategies here was accomplished in two weeks by someone not familiar with the Otter code; hence our claim that our strategies are "easy to implement" in any paramodulation theorem prover seems justifiable. Second, we find it interesting that although there is the expected variation, there is a rather consistent improvement across the board. The average savings (with respect to the number of clauses) of the fastest trial among b-super, b-ord, and b-para, compared with the fastest non-basic trial for each problem is 23%, whereas the average savings with respect to *bèst* was 34%. We should remark here that these problem sets are the only ones we have tried since finishing our modifications to Otter; this is *not* a selection of the best data for our system. Naturally, much work remains to be done, but our initial results seem extremely promising.

We also emphasize that these modifications to Otter do not reflect all the restrictions imposed by our calculi. For instance, they do not include ordering constraints or any form of blocking (although it seems to be relatively simple to add at least blocking to Otter). We have simultaneously worked on a prototype Common Lisp implementation, which unfortunately was not finished in time for this submission; however, our limited experience with this new prover (in which the exact inference systems given in this paper have been implemented, including ordering constraints and blocking) suggests that the further restrictions will speed up these trials even more. For example, we noticed that during completion problems under non-basic superposition, critical pairs which would normally be reduced to identities and removed are instead blocked in the basic strategy, and so the time normally spent in reduction is saved.

8 Summary

Due to a lack of space, we can not review here the full development of the paramodulation calculus, but only remark on the history of the basic strategy. *Narrowing* is a method for generating complete sets of R-unifiers with respect to a canonical set of rewrite rules R (see [6]). The basic strategy was introduced explicitly—as far as we know—for the first time by Hullot in this context [8], and further studied by [16, 9]; it was noticed here that the basic strategy conflicts to some degree with simplification, and a method for dealing with this was described. In the case of paramodulation, D. Plaisted has remarked to us that Brand's proof [5] in fact uses something similar by virtue of his clause transformation, and Plaisted's theorem prover [12], which uses an analogous transformation, thus also uses a basic-like strategy; also a critical pair criterion similar to the basic strategy is described in [19]. A related effort to restrict the addresses where paramodulation may be applied is discussed in [10]. The current project grew out of a lemma necessary in the proof of [20], and was presented in a preliminary form at the *4th Unification Workshop*, Barbizon, France (June 1991), without deletion or blocking rules, and using a very different style of proof. We have recently learned that R. Nieuwenhuis and A. Rubio have independently developed an inference system similar to basic superposition.

Our results, in addition to providing a means of making paramodulation theorem provers (and related systems, such as completion procedures) more efficient, show that substitutions, which are produced initially as most general unifiers which calculate the intersection of ground instances of universally quantified clauses, in fact play *only* this role in theorem proving, in the sense that they need not be subject to equational inferences themselves. We view these results as a robust answer to the question posed by L. Wos and cited in the introduction in the following sense. Essentially, our results depend on the fact that terms in clauses can be forbidden for paramodulation inferences when, at the ground level, they represent irreducible terms in the construction of the model described in Section 4. Specifying additional forbidden terms in the original set of clauses—which would be more in the spirit of set-of-support—seems to require that we can prove that these clauses are *reduced* to start with; since in general it is difficult or impossible to know what models could be constructed for a set of clauses being saturated (except in a limited sense when simplifiers arise), it seems difficult to extend the analogy of set-of-support to paramodulation in non-trivial ways. Rather, we have developed a class of restrictions to paramodulation based on the role of substitutions in the process of saturating a set of clauses.

Acknowledgments. We wish to thank Dennis Kfoury and Steve Homer of Boston University for graciously providing funds for the third author during the academic years 1990–92. We would also like to thank Michael Rusinowich, David Plaisted, Pierre Lescanne, Deepak Kapur, and H.-J. Ohlbach for much helpful discussion.

References

[1] L. Bachmair and H. Ganzinger. On restrictions of ordered paramodulation with simplification. In *Proc. 10th Int. Conf. on Automated Deduction*, Lect. Notes in Comput. Sci., vol. 449, pp. 427–441, Berlin, 1990. Springer-Verlag.

[2] L. Bachmair and H. Ganzinger. Rewrite-based equational theorem proving with selection and simplification. Technical Report MPI-I-91-208, Max-Planck-Institut für Informatik, Saarbrücken, Germany, 1991. To appear in *Journal of Logic and Computation*.

[3] L. Bachmair. *Canonical Equational Proofs*. Birkhauser Boston, Inc., Boston MA (1991).

[4] L. Bachmair, H. Ganzinger, C. Lynch, and W. Snyder. Basic Paramodulation and Superposition. Technical Report No. 92-001, Boston University, Boston, MA, 1992.

[5] D. Brand. Proving theorems with the modification method. *SIAM Journal of Computing 4*:4 (1975) pp. 412–430.

[6] M. Fay. First-order unification in an equational theory. In *Proc. 4th Workshop on Automated Deduction*, Austin, Texas (1979).

[7] J. Hsiang and M. Rusinowitch. Proving refutational completeness of theorem proving strategies: The transfinite semantic tree method. *J. ACM 38:3* (July 1991) pp. 559–587.

[8] J.-M. Hullot. Canonical forms and unification. In *Proc. 5th Int. Conf. on Automated Deduction*, Lect. Notes in Comput. Sci., vol. 87, pp. 318–334, Berlin, 1980. Springer-Verlag.

[9] W. Nutt, P. Réty, and G. Smolka. Basic Narrowing Revisited. Reprinted in *Unification*, C. Kirchner (ed.), Academic Press, London (1990).

[10] W. McCune. Skolem functions and equality in automated deduction. In *Proc. 8th Nat. Conf. on AI*, MIT Press, 1990, pp. 246–251.

[11] W. McCune. OTTER 2.0 users guide. *Technical Report ANL–90/9*, Argonne National Laboratory, Argonne IL, 1990.

[12] X. Nie and D. Plaisted. A complete semantic back-chaining proof system. In *Proc. 10th Int. Conf. on Automated Deduction*, Lect. Notes in Comput. Sci., vol. 449, pp. 16–27, Berlin, 1990. Springer-Verlag.

[13] J. Pais and G. Peterson. Using forcing to prove completeness of resolution and paramodulation. *J. Symbolic Computation 11* (1991) pp.3-19.

[14] F. Pelletier. Seventy-five problems for testing automatic theorem provers. *Journal of Automated Reasoning 2* (1986) pp. 191-216.

[15] G. Peterson. A technique for establishing completeness results in theorem proving with equality. *SIAM Journal of Computing 12* (1983) pp. 82–100.

[16] P. Rety. Improving basic narrowing techniques. In P. Lescanne, editor, *Proc. of 2nd Int. Conf. on Rewrite Techniques and Applications*, LNCS vol. 256, pp. 228–241, Berlin, 1987. Springer-Verlag.

[17] G.A. Robinson and L. T. Wos. Paramodulation and theorem proving in first order theories with equality. In B. Meltzer and D. Michie, editors, *Machine Intelligence 4* pp. 133–150. American Elsevier, New York, 1969.

[18] J. Slagle. Automated theorem proving with simplifiers, commutativity, and associativity. *J. ACM 21* (1974) pp. 622–642.

[19] M. Smith and D. Plaisted. Term-rewriting techniques for logic programming I: completion. Report TR88-019, Department of Computer Science, Univ. North Carolina (1988).

[20] W. Snyder and C. Lynch. Goal directed strategies for paramodulation. In *Proc. 4th Int. Conf. on Rewriteing Techniques and Applications*, Lect. Notes in Comput. Sci., vol. 488, pp. 150–161, Berlin, 1991. Springer-Verlag.

[21] W. Snyder. *A Proof Theory for General Unification*. Birkhauser Boston, Inc., Boston MA (1991).

[22] L. T. Wos, G. A. Robinson, D. F. Carson, and L. Shalla. The concept of demodulation in theorem proving. *Journal of the ACM*, Vol. 14, pp. 698–709, 1967.

[23] L. Wos. Automated Reasoning: 33 Basic Research Problems. Prentice Hall, Englewood Cliffs, New Jersey, 1988.

[24] H. Zhang. *Reduction, Superposition, and Induction: Automated Reasoning in an Equational Logic*. Ph.D. Thesis, Rensselaer Polytechnic Institute (1988).

Theorem Proving with Ordering Constrained Clauses

Robert Nieuwenhuis and Albert Rubio

Technical University of Catalonia, dept. Lenguajes y Sistemas Informáticos
Pau Gargallo 5, E-08028 Barcelona, Spain
E-mail: roberto@lsi.upc.es rubio@lsi.upc.es

Abstract: We use *clauses with ordering constraints* to reduce the search space in ordered inference systems for clauses with or without equality, such as ordered resolution or superposition. In our completion procedure for ordering constrained clauses *redundant inferences* can be ignored and *redundant clauses* can be deleted without loosing refutational completeness. Two new results needed for fast ordering constraint solving and *incrementality* of the set of function symbols are given. We discuss the use of our methods for reasoning about infinite sets of clauses defined by a finite number of ordering constrained ones.

1. Introduction

The constraint paradigm has been applied to theorem proving in e.g. [Hue 72, KKR 90, Pet 90, NR 92], being its main difficulty the falsity of the usual lifting lemmata in the constraint case. However, by means of *equality constraints* and using a new lifting technique, we have recently proved the completeness of *basic superposition* [NR 92], the expected counterpart of *basic narrowing* ([Hul 80]) in Knuth-Bendix-like completion (cf. independent work by Bachmair et al. in this conference on similar basic strategies). Basic superposition is the restriction of normal superposition to those occurrences of subterms that have *not* been originated in previous inferences. For example

$$\frac{f(g(a)) \simeq a \qquad h(f(x)) \simeq h(x)}{h(a) \simeq h(g(a))}$$

is an inference by (equational) superposition whose conclusion is the instantiation of $h(a) \simeq h(x)$ with the unifier of $f(x)$ and $f(g(a))$. Therefore, no further basic superposition steps have to be applied to its subterm $g(a)$. Our proof of completeness is based on a representation in which the accumulated unifiers are kept in equality constraints, i.e. the conclusion of the inference would be $h(a) \simeq h(x)\, [\![x = g(a)]\!]$. Further superpositions can then take place only upon non-variable subterms of $h(a) \simeq h(x)$.

In this paper we extend these techniques for much further restricting ordered inference systems by means of *ordering constrained clauses*. The interest of similar ordering constraints has been pointed out earlier by H. and C. Kirchner and Rusinowitch [KKR 90], although, as far as we know, no proofs had been found up to now. The basic idea is very simple: in ordered inference rules the search space is reduced by using only the *maximal* (terms and) literals to compute inferences with. Therefore, if a clause is obtained in an inference, in fact we want to keep only those ground instances of it for which the literal selected is really the biggest one. This information can be stored in its constraint.

Future choices of maximal literals that are incompatible with this constraint can then be shown to be unnecessary (by proving the unsatisfiability of ordering constraints, which is decidable [Com 90]). For example, if we denote by $t \simeq t' [\![T]\!]$ the ground instances of an equation $t \simeq t'$ satisfying the constraint T, then the inference rule of basic superposition with ordering constraints for the equational case is:

$$\frac{s \simeq s' [\![T']\!] \qquad t \simeq t' [\![T]\!]}{t[s']_u \simeq t' [\![T' \wedge T \wedge s \succ s' \wedge t \succ t' \wedge t|_u = s]\!]} \quad \text{where } t|_u \notin \mathcal{V}ars(t)$$

which, as we can see, is a powerful and also elegant representation for ordered inference rules, since information from the meta-level, such as the ordering restrictions and the accumulated unifiers, is included into the formulae and used later on. Obviously, this method reduces the search space importantly. Especially in ordered inference rules for full first-order clauses the constraints become quickly very restrictive, which cuts down the number of axioms retained drastically.

Here we have chosen to present our theorem prover for constrained clauses as a Knuth-Bendix-like completion procedure. Although traditionally completion techniques have been applied to equational logic (starting with [KB 70]), important results have also been obtained recently for completion of full first-order clauses with and without equality ([HR 89, BG 90,91, NO 91, NR 92]). The reason is that on one hand a completion procedure can be seen as an efficient refutationally complete theorem prover, but on the other hand, it also transforms consistent sets of axioms in such a way that very efficient proof strategies become complete by using the final *saturated* set. The powerful *redundancy* notions defined by Bachmair and Ganzinger (1991) make completion especially interesting for theorem proving in first-order logic.

To illustrate the difficulty of dealing with constraints, consider $f \succ a \succ b \succ c$ and

1. $f(x) \simeq c [\![x \succ b]\!]$
2. $a \simeq c$

In this example there are no critical pairs, but there is no rewrite proof for $f(c) = c$, which is a valid consequence, since $f(c) \leftrightarrow_2 f(a) \leftrightarrow_1 c$. The problem is caused by the falsity in the constraint case of lifting lemmata like the critical pair lemma, based on the existence of *all* ground instances of the axioms, which is not the case here, since $f(c) = c$ is *not* a ground instance of the first equation. For this reason, Peterson (CADE 1990) needed to additionally define the so-called *easy* critical pairs: for every equation $t \simeq t' [\![c]\!]$, the equation $t \simeq t' [\![\neg c]\!]$ had to be added. An important disadvantage is the large amount of equations generated by this inference rule, which is moreover sound only if the initial equations have no constraints.

In this paper we solve this disadvantage by using new lifting techniques. Our completion procedure for ordering constrained clauses allows to ignore *redundant inferences* and delete *redundant clauses* without loosing refutational completeness. It can handle clauses without equality having arbitrary ordering constraints. This allows to describe (and reason about) infinite sets of normal formulae. In the presence of clauses with equality, completeness is preserved only if the *initial* set of ordering constrained clauses fulfills a property called *purity*. In particular, sets of clauses without constraints are pure. Our work improves Peterson's completion techniques for ordering constrained equations since we can deal with full first-order clauses (with or without equality) using no additional inference rules like *easy* critical pairs, we allow initial axioms with constraints and we can combine our methods with basic superposition.

We are working on an implementation of these techniques (and expect to report on its results at this conference) within the TRIP system [NOR 90].

This paper is structured as follows. The basic definitions and notations are given in section 2. Section 3 states the refutational completeness of the inference system and the completion procedure, based on techniques for the case without constraints of [Zha 88, BG 90,91]. Section 4 is devoted to two new results needed for fast ordering constraint solving and *incrementality* of the set of function and constant symbols, and section 5 is on the use of ordering constraints for expressing infinite sets of formulae.

2. Basic notions and terminology

We adopt the standard notations and definitions for term rewriting given in [DJ 90]. Furthermore, by *(ordering and equality) constraints* we mean quantifier-free first-order formulae built over the binary predicate symbols '$\succ$' and '$=$' relating terms in $\mathcal{T}(\mathcal{F},\mathcal{X})$, where '$=$' denotes syntactic equality of terms, and '$\succ$' denotes a given simplification ordering that is total on ground terms. A constraint T is *satisfiable* if there exists some ground instance $T\sigma$ that is (equivalent to) true, denoted $T\sigma \equiv \mathit{true}$. Satisfiability is decidable if $\succ$ is interpreted as the lexicographic path ordering (LPO) [Com 90]. Ground constraints are always equivalent to *true* or to *false*, as $\succ$ is total.

By an *equation* we mean a multiset $\{s,t\}$, denoted by $s \simeq t$ (or equivalently by $t \simeq s$), where s and t are terms in $\mathcal{T}(\mathcal{F},\mathcal{X})$. A first-order clause $\Gamma \rightarrow \Delta$ is a pair of (finite) multisets of equations Γ and Δ, called resp. its *antecedent* and *succedent*. A constrained clause is a pair (C,T), denoted $C\,[\![T]\!]$, where C is a first-order clause and T is a constraint. Such a pair can be seen as a shorthand for the set of *ground instances* of $C\,[\![T]\!]$: those ground clauses $C\sigma$ such that $T\sigma \equiv \mathit{true}$. In this note, distinct clauses are supposed not to share variables.

An interpretation I is a congruence on ground terms. It satisfies a ground clause $\Gamma \rightarrow \Delta$, denoted $I \models \Gamma \rightarrow \Delta$, if $I \not\supseteq \Gamma$ or else $I \cap \Delta \neq \emptyset$. The empty clause is therefore satisfied by no interpretation. An interpretation I satisfies (is a model of) $C\,[\![T]\!]$, denoted $I \models C\,[\![T]\!]$, if it satisfies every ground instance of $C\,[\![T]\!]$, i.e. clauses with unsatisfiable constraints are tautologies. Therefore, $C\,[\![T]\!]$ is said to be an empty clause only if C is empty and T is satisfiable. I satisfies a set of clauses S, denoted by $I \models S$, if it satisfies every clause in S. A clause C *can be deduced* from a set of clauses S (denoted by $S \models C$), if C is satisfied by every model of S.

For dealing with non-equality predicates, we express atoms A by equations $A \simeq \mathit{true}$ where *true* is a special symbol, i.e. we treat atoms as terms. No significant changes have to be made for distinguishing between atoms and terms in e.g. inference rules or orderings. In the ordering $\succ$, the special symbol *true* is the smallest symbol. We use $\succ_{mul}$ ($\succ_{mul^n}$) to denote its (n-fold) multiset extension.

We use the definitions of [DJ 90] for rewriting-related notions like *normal form, confluence, convergence, reducibility,* etc. We denote *ground rewrite rules* (ground equations $t \simeq t'$ with $t \succ t'$) by $t \Rightarrow t'$. The congruence generated by a set of ground rewrite rules R (which is an interpretation) will be denoted by R^*.

3. Refutational Completeness and Completion

The inference system we use below (first proved complete in [NR 92]) includes only one inference rule for equality factoring, instead of, apart from "normal" factoring, inference rules for *merging paramodulation* [BG 90,91] or *equality factoring left* and *equality factoring right* [BG 90]. Our methods do not depend on the specific inference system we use here, since our lifting techniques can be easily adapted to other systems, such as calculi which consider only one arbitrary *marked* negative literal for superposition or resolution, as done in [NN 91] for Horn clauses with equality, and in [BG 91] by means of *selection functions* on negative literals.

In the following ordering $\succ_C$ on ground clauses, the terms appearing in antecedents of clauses are slightly more complex than the ones in succedents:

Definition 1: The *multiset expression* of an equation $t \simeq t'$ in a clause $\Gamma \to \Delta$ is

(i) $\{\{t,t\},\{t',t'\}\}$ if $t \simeq t'$ belongs to Γ
(ii) $\{\{t\},\{t'\}\}$ if $t \simeq t'$ belongs to Δ

The ordering $\succ_E$ on ground equations appearing in clauses is defined as the ordering $\succ_{mul^2}$ on their multiset expressions.
The ordering $\succ_C$ on ground clauses is defined as the ordering $\succ_{mul^3}$ on the multisets containing the multiset expressions of their equations.

Definition 2: A ground equation e is called *maximal* (resp. *strictly maximal*) in a ground clause C if $e \succeq_E e'$ (resp. $e \succ_E e'$), for every other equation e' in C.

In the following inference system $\mathcal{O}$ (here $\mathcal{O}$ stands for "Ordering constrained"), inferences take place only in equations of succedents that are strictly maximal and in equations of antecedents that are maximal, for some ground instance. Moreover, only the maximal terms in each equation are used. Therefore, in each ground inference, the conclusion is strictly smaller (wrt. $\succ_C$) than the maximal premise, i.e. $\mathcal{O}$ is *monotonic*, which is essential for computing *fair* theorem proving derivations.

For simplicity, we have chosen not to express the inference rules here in their "basic superposition" version, i.e. including the unifiers as an equality within the constraints, as done in the second example of the introduction. The completeness proofs for that case are a combination of the results below and [NR 92].

Definition 3: The inference rules of $\mathcal{O}$ are the following (we always consider maximality of equations in clauses wrt. $\succ_E$):

1) *superposition right:*

$$\frac{\Gamma' \to \Delta', s \simeq s' \,[\![T']\!] \qquad \Gamma \to \Delta, t \simeq t' \,[\![T]\!]}{(\,\Gamma', \Gamma \to \Delta', \Delta, t[s']_u \simeq t' \,[\![T \wedge T' \wedge T'']\!]\,)\sigma} \quad \text{where } t|_u \notin \mathcal{V}ars(t) \text{ and } \sigma = mgu(t|_u, s)$$

where T'' includes:

a) $t \succ t'$, $s \succ s'$, and $t \simeq t' \succ_E s \simeq s'$;
b) $s \simeq s'$ is strictly maximal in $\Gamma' \to \Delta', s \simeq s'$;
c) $t \simeq t'$ is strictly maximal in $\Gamma \to \Delta, t \simeq t'$.

2) *superposition left:*

$$\frac{\Gamma' \to \Delta', s \simeq s' \,[\![T']\!] \qquad \Gamma, t \simeq t' \to \Delta \,[\![T]\!]}{(\,\Gamma', \Gamma, t[s']_u \simeq t' \to \Delta', \Delta \,[\![T \wedge T' \wedge T'']\!]\,)\sigma} \quad \text{where } t|_u \notin \mathcal{V}ars(t) \text{ and } \sigma = mgu(t|_u, s)$$

where T'' includes:

a) $t \succ t'$ and $s \succ s'$;
b) $s \simeq s'$ is strictly maximal in $\Gamma' \to \Delta', s \simeq s'$;
c) $t \simeq t'$ is maximal in $\Gamma, t \simeq t' \to \Delta$.

3) *equality resolution:*

$$\frac{\Gamma, t \simeq t' \to \Delta \,[\![T]\!]}{(\,\Gamma \to \Delta \,[\![T \wedge T']\!]\,)\sigma} \quad \text{where } \sigma = mgu(t, t')$$

where T' includes: $t \simeq t'$ is maximal in $\Gamma, t \simeq t' \to \Delta$.

4) *factoring:*

$$\frac{\Gamma \to \Delta, t \simeq s, t' \simeq s' \,[\![T]\!]}{(\,\Gamma, s \simeq s' \to \Delta, t \simeq s \,[\![T \wedge T']\!]\,)\sigma} \quad \text{where } \sigma = mgu(t, t')$$

where T' includes:

a) $t \succ s$ and $t' \succ s'$;
b) $t \simeq s$ is maximal in $\Gamma \to \Delta, t \simeq s, t' \simeq s'$.

Note that our inference rule for factoring is a generalization to the equality case of "normal" factoring. For instance, if t and t' are atoms, then both s and s' are the symbol *true* and the equation $true \simeq true$ can be omitted in the antecedent. Superposition left is equivalent to ordered resolution when applied to atoms and superposition right is then unnecessary as it generates only tautologies. Also note that no factoring on negative literals is needed.

As we can see, the ordering constraints can become quickly very restrictive, since many conditions can be encoded within the part of the constraints generated in every inference. The important improvement wrt. ordered inference rules without constraints is that the ordering restrictions are inherited, which allows to prove the redundancy of many more inferences. Of course, there is a need for efficient techniques for proving the unsatisfiability of ordering constraints, since clauses with unsatisfiable constraints are tautologies and can be deleted, and also for simplification of constraints, i.e. keeping smaller equivalent ones. These problems are addressed in section 4.

The refutational completeness of our techniques is proved as follows. First we define a method for *generating* from a set S of clauses, a (convergent) set of ground rewrite rules R_S. After this, we describe how to effectively compute ***saturated*** sets, i.e. sets of clauses S that are closed (up to redundant inferences) under $\mathcal{O}$. Then we prove that saturated sets S not containing the empty clause have a model, namely the congruence generated by R_S, i.e. $R_S^* \models S$. This implies that the empty clause belongs to S whenever S is inconsistent.

As said, the difficulty with constrained superposition lies in the fact that the usual lifting lemmata (like the critical pair lemma in the equational case) do not hold. Let us look again at an example:

Example 4: Let S be the set of the following three clauses with $P \succ a \succ b \succ c$:

1. $\rightarrow P(x)\, [\![x \succ b]\!]$
2. $P(c) \rightarrow$
3. $\rightarrow a \simeq c$

Now S is inconsistent but no inferences can be made (since the only instance of the first clause is $\rightarrow P(a)$). As said, the problem is caused by the falsity in the constraint case of the usual lifting lemmata, whose proofs are based on the existence of *all* ground instances of the axioms.

Another consequence that we can draw from the previous example is that we cannot start with arbitrary sets of ordering constrained clauses with equality (at least, not without adding new inference rules). Therefore, for the moment we will suppose that the *initial* set of axioms contains only clauses $C\, [\![T]\!]$ where T is an empty (or trivially true) constraint, sometimes written $C\, [\![true]\!]$. In section 5 we will show how this requirement can be weakened.

Below we apply a new lifting technique, consisting in first proving our results for the particular set of instances with, in some sense, *irreducible substitutions* and, after that, generalizing these results to *all* instances.

Definition 5: A ground substitution σ is *irreducible* wrt. a set of ground rewrite rules R if $x\sigma$ is irreducible wrt. R, for every variable x in the domain of σ. A ***normal form*** of σ wrt. R is a substitution σ' with the same domain as σ, and such that $x\sigma'$ is a normal form wrt. R of $x\sigma$, for every variable x in the domain.

Definition 6: Let $C\sigma$ be a ground instance $\Gamma \rightarrow \Delta, t \simeq s$ of a clause $C\,[\![T]\!]$ in a set S. Then $C\sigma$ *generates* a rule $t \Rightarrow s$ if the following conditions hold:

(1) $R_C^* \not\models C\sigma$

(2) $t \simeq s$ is strictly maximal (wrt. $\succ_E$) in $C\sigma$ with $t \succ s$

(3) $R_C^* \not\models s \simeq s'$, for every $t \simeq s'$ in Δ

(4) t is irreducible by R_C

(5) σ is irreducible by R_C

where R_C is the set of rules generated by ground instances smaller than C (wrt. $\succ_C$) of clauses in S. R_S is the set of rules generated by *all* ground instances of clauses in S.

Definition 7: Let S be a set of constrained clauses, and let R be a set of ground rewrite rules. The set of ground instances of clauses in S with substitutions that are irreducible wrt. R is denoted by $irred_R(S)$, i.e.

$$irred_R(S) = \{\ C\sigma \mid C\,[\![T]\!] \in S,\ \ T\sigma \equiv true,\ \ \sigma\ ground,\ \ \sigma\ irreducible\ wrt.\ R\ \}$$

We do not pretend the definitions below to be constructive, e.g. the definition of *theorem proving derivations* does not provide (yet) a way to compute them (at least not if the redundancy notions are exploited). This point will be made clear below.

Definition 8: Let $S_0, S_1, \ldots$ be a sequence of sets of constrained clauses.

a) The set S_∞ of *persistent* clauses in $S_0, S_1, \ldots$ is defined as $\bigcup_j(\bigcap_{k\geq j} S_k)$.

b) A clause $C\,[\![T]\!]$ is *redundant* in a set S_j if for every ground instance $C\sigma$ of it with σ irreducible wrt. R_{S_∞}, there exist instances D_i in $irred_{R_{S_\infty}}(S_j)$, for $i = 1 \ldots m$, such that $C\sigma \succ_C D_i$ and $R_{S_\infty} \cup \{D_1, \ldots, D_m\} \models C\sigma$.

Definition 9: A *theorem proving derivation* is a sequence of sets of constrained clauses $S_0, S_1, \ldots$, such that

$$S_i = S_{i-1} \cup \{C\,[\![T]\!]\} \quad \text{where } S_{i-1} \models C\,[\![T]\!], \text{ or}$$
$$S_i = S_{i-1} \setminus \{C\,[\![T]\!]\} \quad \text{if } C\,[\![T]\!] \text{ is redundant in } S_{i-1}.$$

An inference is redundant in a set S_j of a theorem proving derivation if in every irreducible (wrt. R_{S_∞}) instance of it, the conclusion can be deduced from irreducible instances of clauses in S_j smaller than the maximal premise:

Definition 10: Let $S_0, S_1, \ldots$ be a theorem proving derivation and let π be an inference of $\mathcal{O}$ with premises $C_1\,[\![T_1]\!], \ldots, C_n\,[\![T_n]\!]$, and with conclusion $C\,[\![T]\!]$.

a) Every inference of $\mathcal{O}$ with premises $C_1\sigma, \ldots, C_n\sigma$, and conclusion $C\sigma$ for some ground substitution σ with $T\sigma \equiv true$, is a *ground instance* $\pi\sigma$ of π.

b) The inference π is *redundant* in a set of constrained clauses S if for every ground instance $\pi\sigma$ of π with σ irreducible wrt. R_{S_∞}, there exist instances D_i in $irred_{R_{S_\infty}}(S)$, for $i = 1 \ldots m$, such that $max(C_1\sigma, \ldots, C_n\sigma) \succ_C D_i$ and $R_{S_\infty} \cup \{D_1, \ldots, D_m\} \models C\sigma$.

c) A set S is *saturated* if every inference of $\mathcal{O}$ with premises in S is redundant in S.

Definition 11: A theorem proving derivation $S_0, S_1, \ldots$ is *fair* if every inference of the inference system $\mathcal{O}$ with premises in S_∞ is redundant in some S_j wrt. R_{S_∞}.

As we can see, in theorem proving derivations we consider instances with substitutions irreducible wrt. R_{S_∞}. For example, a *clause* is redundant if all its instances that are irreducible in that sense can be deduced from other smaller irreducible instances.

However, in practice, during the computation of a fair derivation, one cannot prove the redundancy of clauses or inferences in a set S_j, since at that point R_{S_∞} is unknown. Therefore, sufficient conditions for redundancy have to be used. We will define them in detail at the end of this section, and we suppose for the moment that we can indeed compute fair theorem proving derivations (by systematically adding conclusions from persistent clauses).

Below we sketch only the main ideas of the completeness proofs. For detailed proofs (included in the version refereed for this conference, but not here due to space limitations) we refer to the full version of this paper, and to similar techniques in [NR 92]. Proofs of results similar to the following classical lemma can be found in [BG 91] or [NR 92]:

Lemma 12: If $S_0, S_1, \ldots$ is a fair theorem proving derivation, then S_∞ is saturated.

Lemma 13: Let S be a saturated set of clauses not containing the empty clause. Then $R_S^* \models irred_{R_S}(S)$.

Lemma 13 is proved by deriving a contradiction from the existence of a minimal (wrt. $\succ_C$) instance $C\sigma$ in $irred_{R_S}(S)$ of a clause $C\,[\![T]\!]$ in S, such that $R_S^* \not\models C\sigma$.

E.g. in the case where $C\sigma$ is a clause $\Gamma\sigma, t\sigma \simeq t'\sigma \to \Delta\sigma$, where $t\sigma \simeq t'\sigma$ is maximal in $C\sigma$ and $t\sigma \succ t'\sigma$ we have $R_S^* \models t\sigma \simeq t'\sigma$, because $R_S^* \not\models C\sigma$, i.e. $t\sigma$ is reducible by a rule $s\sigma' \Rightarrow s'\sigma'$ in R_S (note that R_S is canonical) generated by a clause C' in S of the form $\Gamma' \to \Delta', s \simeq s'\,[\![T']\!]$, s.t. $t\sigma|_u = s\sigma'$. The following inference π by superposition left

$$\frac{\Gamma' \to \Delta', s \simeq s'\,[\![T']\!] \qquad \Gamma, t \simeq t' \to \Delta\,[\![T]\!]}{(\ \Gamma', \Gamma, t[s']_u \simeq t' \to \Delta', \Delta\,[\![T \wedge T' \wedge T'']\!]\ )\theta}$$

can be made: $s\sigma'$ does not reduce $t\sigma$ below a variable position of t because σ is irreducible wrt. R_S (no additional lifting is needed here!). The conclusion C'' of π has a ground instance D of the form $\Gamma'\sigma', \Gamma\sigma, t\sigma[s'\sigma']_u \simeq t'\sigma \to \Delta'\sigma', \Delta\sigma$ that is not satisfied by R_S^*.

Moreover, D is an instance of C'' with a ground substitution that is irreducible by R_S, because of the irreducibility wrt. R_S of (i) σ and (ii) σ' restricted to the domain $\mathcal{V}ars(C') \cap \mathcal{V}ars(C'')$. The fact that (ii) holds here is an interesting technical difference wrt. the equality constraint case of [NR 92]: the unifiers are applied to the conclusions, instead of being kept in constraints, which allows to prove $l \succ x\sigma'$ for all rules $l \Rightarrow r$ in $R_S \setminus R_{C'\sigma'}$ and for all x in $\mathcal{V}ars(C') \cap \mathcal{V}ars(C'')$.

Since S is saturated, π must be redundant in S, i.e. there exist instances $D_1, \ldots, D_m$in $irred_{R_S}(S)$ such that $R_S \cup \{D_1, \ldots, D_m\} \models D$ and $C\sigma \succ_C D_i$. The fact that R_S^* does not satisfy D implies that R_S^* does not satisfy at least one of the D_i either, which contradicts the minimality of $C\sigma$.

Theorem 14: Let $S_0, S_1, \ldots$ be a fair theorem proving derivation, where S_0 is a set of clauses with empty constraints. Then S_0 is inconsistent if, and only if, the empty clause belongs to some S_j.

Proof. (Sketch) By soundness of $\mathcal{O}$, if the empty clause is in S_j then S_0 is inconsistent. For the reverse implication, we prove that S_0 is consistent if the empty clause belongs to no S_j. In that case, it is not in S_∞ and $R^*_{S_\infty} \models irred_{R_{S_\infty}}(S_\infty)$ because S_∞ is saturated. Now, by the notion of redundant clause, it can be shown that $R_{S_\infty} \cup irred_{R_{S_\infty}}(S_\infty) \models irred_{R_{S_\infty}}(S_0)$, and therefore also $R^*_{S_\infty} \models irred_{R_{S_\infty}}(S_0)$. This implies $R^*_{S_\infty} \models S_0$: it suffices to show that $R^*_{S_\infty} \models C\sigma$ for every instance $C\sigma$ of a clause $C\,[\![true]\!]$ in S_0. Now let σ' be the normal form of σ wrt. R_{S_∞}. Then $R^*_{S_\infty} \models C\sigma'$, because $C\sigma'$ is in $irred_{R_{S_\infty}}(S_0)$ since it is an existing instance of $C\,[\![true]\!]$. From $R^*_{S_\infty} \models C\sigma'$ and $R_{S_\infty} \cup \{C\sigma'\} \models C\sigma$ it follows that $R^*_S \models C\sigma$. But, if $R^*_{S_\infty} \models S_0$ then S_0 is consistent, as we wanted to prove. ■

In a similar way, for saturated sets E of constrained *equations*, it holds that $R_{E_\infty} \models E$, which implies that E is ground confluent (cf. [NR 92]).

As said, the definition of fair theorem proving derivations does not provide a practical method for proving the redundancy of inferences and clauses at a given point of the derivation, because R_{S_∞} is not known then. Therefore, sufficient conditions for redundancy have to be used. The redundancy notions of [BG 91], which include most known simplification and redundancy criteria*, state, roughly speaking, that a clause is redundant in S if all its ground instances can be deduced from smaller instances of other clauses in S, and an *inference* is redundant if, for all its instances, the conclusion can be deduced from instances smaller than the maximal premise.

However, our definitions of redundancy require every instance *with an irreducible substitution* to be deducible from other smaller instances *with irreducible substitutions*, and also R_{S_∞} may be used. These different notions are really needed (although only for clauses with equality), i.e. quite natural simplification methods like the one used in the following example become incorrect:

Example 15: Let the following four equations be the initial set with $a \succ f \succ g \succ h \succ b$:

1. $f(x) \simeq b$
2. $f(x) \simeq g(x, h(a))$
3. $g(x, h(b)) \simeq h(x)$
4. $a \simeq b$

Now superposing 1 and 2 gives $g(x, h(a)) \simeq b\,[\![x \succeq a]\!]$ which is simplified with 4 into:

5. $g(x, h(b)) \simeq b\,[\![x \succeq a]\!]$. Superposition of 5 and 3 gives:
6. $h(x) \simeq b\,[\![x \succeq a]\!]$. Simplifying 2 with 6 and 1 (i.e. 2 is deleted) gives:
7. $g(x, b) \simeq b$.

Now the set $\{1, 3, 4, 5, 6, 7\}$ is a saturated set of constrained equations, but there is no rewrite proof for the equation $h(b) \simeq b$, which is a valid consequence of the initial set.

Definition 16: Let $C\,[\![T]\!]$ be a clause. Then a variable x in $Vars(C)$ is *lower bounded* by T if there exist ground substitutions σ and σ' such that $T\sigma \equiv true$, $T\sigma' \equiv false$, $x\sigma \succ x\sigma'$ and $y\sigma = y\sigma'$ for every other variable y in $Vars(C)$.

Below we prove that we can always apply the notions of [BG 91], although sometimes we have to slightly *weaken* some ordering constraints of clauses used in redundancy proofs. More precisely, a clause $C\,[\![T]\!]$ can be used in a redundancy proof with the notions of

* e.g. deletion of tautologies, demodulation, clausal rewriting, splitting, critical pair criteria, etc. Subsumption can easily be included by combining the ordering $\succ_C$ with the subsumption ordering.

[BG 91], if for every variable x in $\mathcal{V}ars(C)$, x is not lower bounded by T or else x takes the same value in the proof as some variable in the clause proved:

Lemma 17: Let $S_0, S_1, \ldots$ be a theorem proving derivation. The clause $C\,[\![T]\!]$ is redundant in a set S_j if

(i) for every ground instance $C\sigma$ of it, there are ground instances $D_i\sigma_i$ for $i = 1 \ldots m$ of clauses $D_i\,[\![T_i]\!]$ in S_j such that $\{D_1\sigma_1, \ldots, D_m\sigma_m\} \models C\sigma$ and $C\sigma \succ_C D_i\sigma_i$, and

(ii) for every i in $1 \ldots m$, and for every x in $\mathcal{V}ars(D_i)$, x is not lower bounded by T_i, or else $x\sigma_i = y\sigma$, for some variable y in C.

For a proof of this lemma, and the equivalent lemma for proving the redundancy of *inferences*, cf. [NR 92]. Might all the conditions of the previous lemma fail for some variable x, and we still want to carry out the redundancy proof, then we can always *weaken* T for x: we can eliminate the part of the ordering constraint that establishes the lower bound on x. Note that it is sound to do this only if we have started with an initial set of clauses without constraints.

The interest of ordering constrained completion of first-order clauses lies not only in the gain of efficiency as a consequence of the more reduced search space, but also in the higher probability of obtaining *saturated* sets of clauses S for a given theory. By using such saturated sets very efficient proof strategies become complete, e.g. the *rewrite proofs* in the equational and conditional equational cases, or the set-of-support strategy and in some cases the deducibility of clauses from saturated sets is even decidable.

4. Efficient constraint solving and incrementality

In our completion procedure the ordering constraints become quickly very restrictive, since in every inference many conditions are added to the constraints inherited from the premises. Therefore we need efficient techniques for proving the unsatisfiability of ordering constraints, since clauses with unsatisfiable constraints can be deleted. One known method for deciding the satisfiability of ordering constraints was defined by Comon [Com 90] for the case in which $\succ$ is interpreted as the lexicographic path ordering (LPO). Here we cannot use its extension to the recursive path ordering with status (RPOS), given by Jouannaud and Okada [JO 91], since RPO is not total on ground terms. However, since these methods are quite inefficient in practice, we have developed a new faster algorithm for the LPO case that has shown its practical value in our implementation. The report [RN 91] includes detailed proofs of all the results appearing in this section.

The LPO ordering $\succ_{lpo}$ on terms is defined as an extension of an ordering $\succ_{\mathcal{F}}$ on a set of function symbols $\mathcal{F}$: $s = f(s_1, \ldots, s_m) \succ_{lpo} g(t_1, \ldots, t_n) = t$ if

a) $s_i \succeq_{lpo} t$, for some i with $1 \leq i \leq m$ or

b) $f \succ_{\mathcal{F}} g$, and $s \succ_{lpo} t_j$, for all j with $1 \leq j \leq n$ or

c) $f = g$, $(s_1, \ldots, s_m) \succ\!\!\succ_{lpo} (t_1, \ldots, t_n)$, and $s \succ_{lpo} t_j$, for all j with $1 \leq j \leq n$

where $(s_1, \ldots, s_m) \succ\!\!\succ_{lpo} (t_1, \ldots, t_n)$ if $\exists j \leq n, \forall i < j, s_i = t_i$ and $s_j \succ_{lpo} t_j$.

In what follows we denote variables by $x, y \ldots$ and terms (with variables) by s and t. Every ordering constraint can be expressed by an equivalent disjunction of *solved forms* [Com 90], by keeping it in disjunctive normal form, eliminating negations with $(t \not\succ t') \equiv (t' \succ t \vee t = t')$ and $(t \neq t') \equiv (t' \succ t \vee t \succ t')$, and applying the definition of LPO:

Definition 18: A *solved form* I is either $\top$, $\perp$ or a formula

$$x_1 \succ t_1 \wedge \ldots \wedge x_n \succ t_n \quad \wedge \quad t'_1 \succ x'_1 \wedge \ldots \wedge t'_m \succ x'_m \quad \wedge \quad y_1 = s_1 \wedge \ldots \wedge y_r = s_r$$

where t'_j is not a variable and y_k appears only once in I for $j = 1 \ldots m$ and $k = 1 \ldots r$.

A constraint T is satisfiable, i.e. it has a solution (a ground substitution σ s.t. $T\sigma \equiv \mathit{true}$), iff one of its solved forms is. The equalities $y_i = s_i$ moreover can be ignored if we are not interested in obtaining an actual solution of the constraint but only in deciding whether it is satisfiable or not, i.e. deleting these equalities does not affect the satisfiability of a solved form.

For such a solved form I, we call the literals $x_i \succ t_i$ *right term* literals, and literals $t'_j \succ x'_j$ (where t'_j is not a variable) *left term* literals. We call $\{t_1, \ldots, t_n\}$ the *set of right terms* of I, and a right term is called *maximal* in a solved form I if it is maximal wrt. $\succ_{lpo}$ in $\{t_1\sigma, \ldots, t_n\sigma\}$, for some solution σ of I.

From now on, we consider a set of function (and predicate) symbols $\mathcal{F}$ (containing at least one constant) whose elements are totally ordered by $\succ_{\mathcal{F}}$. Let f and 0 be respectively the smallest (wrt. $\succ_{\mathcal{F}}$) non-constant and constant symbols in $\mathcal{F}$ and suppose moreover that there is no constant c with $c \neq 0$ such that $f \succ_{\mathcal{F}} c$ (note that $\succ_{\mathcal{F}}$ can be arbitrarily chosen). By $\succ_{lpo}$ we denote the extension of $\succ_{\mathcal{F}}$ to the lexicographic path ordering on terms. Now we define the *successor* of a ground term t as the smallest ground term greater than t:

Definition 19: Given a ground term t (i.e. $t \in \mathcal{T}(\mathcal{F})$), the *successor* of t denoted by $\mathit{succ}_{\mathcal{F}}(t)$, is a ground term s.t. there is no ground term t' with $\mathit{succ}_{\mathcal{F}}(t) \succ_{lpo} t' \succ_{lpo} t$.

Since f is the smallest non-constant symbol and there is at most one constant symbol smaller than f, it is easy to prove the following lemma:

Lemma 20: Let t be a term in $\mathcal{T}(\mathcal{F}, \mathcal{X})$. Then $\mathit{succ}_{\mathcal{F}}(t\sigma) = f(0, \ldots, 0, t\sigma)$ for every ground substitution σ.

Example 21: Assume $\mathcal{F} = \{g \succ f \succ 0\}$ where f is a unary function symbol. Intuitively, if the constraint $[\![x \succ y \ \wedge \ g(z) \succ x]\!]$ is satisfiable, then the constraint obtained by replacing x by $f(y)$, i.e. $[\![f(y) \succ y \ \wedge \ g(z) \succ f(y)]\!]$ is satisfiable too, since, by the previous lemma, $f(y)$ is the smallest term greater than y. Less clear is the situation when there are also (left term) literals of the form $t[x]_p \succ z$ with $p \neq \lambda$.

In the following two lemmas we show some properties of solved forms. The first one states that a solved form (different from $\perp$) with no right term literals is always satisfiable. The second one shows which right terms can be maximal.

Lemma 22: Let I be a solved form $t'_1 \succ x'_1 \wedge \ldots \wedge t'_m \succ x'_m$, where t'_j is not a variable (i.e. I has no right term literals). Then the substitution α defined by $\alpha(x) = 0 \ \ \forall x \in \mathit{Vars}(I)$ is a solution for I.

Lemma 23: Let I be a solved form $x_1 \succ t_1 \wedge \ldots \wedge x_n \succ t_n \ \wedge \ t'_1 \succ x'_1 \wedge \ldots \wedge t'_m \succ x'_m$. If (i) $I\sigma \equiv \mathit{true}$, (ii) $n > 0$ (i.e. there is some right term literal) and (iii) $t_1\sigma$ is maximal wrt. $\succ_{lpo}$ in $\{t_1\sigma, \ldots, t_n\sigma\}$, then $x_1 \notin \mathit{Vars}(t_i)$ for all i with $1 \leq i \leq n$.

A consequence of the previous lemma is that x_1 only can appear in the formula as:

(a) left hand side of a right term literal ($x_1 \succ t_i$),
(b) right hand side of a left term literal ($t'_j \succ x_1$) or
(c) a proper subterm of a term in a left term literal $t'_j[x_1]_p \succ x'_j$ with $p \neq \lambda$.

Lemma 24: Let I be a solved form $x_1 \succ t_1 \wedge \ldots \wedge x_n \succ t_n \;\wedge\; t'_1 \succ x'_1 \wedge \ldots \wedge t'_m \succ x'_m$ with $n > 0$ and let σ be a solution of I s.t. $t_1\sigma$ is maximal in $\{t_1\sigma, \ldots, t_n\sigma\}$.

Then there exists a solution σ' for I s.t. (i) $t_1\sigma'$ is maximal in $\{t_1\sigma', \ldots, t_n\sigma'\}$ and (ii) $f(0,\ldots,0,t_1\sigma') \succeq_{lpo} x\sigma'$ for every x in $\mathcal{V}ars(I)$.

Proof. (Sketch) The proof is by induction on the number k of variables in $\mathcal{V}ars(I)$. If $k = 1$ then $\sigma' = \{x \leftarrow f(0,\ldots,0,t_1)\}$ fulfills the conditions. The key to the proof of the induction step is the following construction of a solved form I':

$$I' \in \textit{Solved-forms}(\; [I \setminus \{t'_j[x_1] \succ y_j\} \cup \{t_1 \succeq t_2 \wedge \ldots \wedge t_1 \succeq t_n\}](x_1 \leftarrow f(0,\ldots,0,t_1))\;)$$

with $I'\sigma \equiv \textit{true}$. Note that at least one such an I' must exist, and that I' has at least one variable less than I. If I' has no right term literals then σ' can be the substitution defined by $x_1\sigma' = f(0,\ldots,0,t_1\alpha)$ and $x\sigma' = 0$ otherwise (plus the instantiations of variables produced when computing I' as a solved form).

If there is some right term literal in $I'\sigma$ and $t''\sigma$ is the maximal one, then by the induction hypothesis, there exists a solution σ'' of I' with a maximal right term $t''\sigma''$ in $I'\sigma''$ and $f(0,\ldots,0,t''\sigma'') \succeq_{lpo} \sigma''$ for every x in $\mathcal{V}ars(I')$.

Now it can be shown, by the way solved forms are obtained, that $t_1\sigma'' \succeq_{lpo} t''\sigma''$, and therefore, $f(0,\ldots,0,t_1\sigma'') \succeq_{lpo} x\sigma''$ for every x in $\mathcal{V}ars(I')$. Furthermore, due to the way I' is defined and by the previous lemma, a substitution σ' fulfilling the conditions is obtained by taking $x\sigma' = x\sigma''$ if x belongs to the domain of σ'', $x_1\sigma' = f(0,\ldots,0,t_1\sigma'')$ and $y\sigma' = 0$ otherwise (plus the instantiations of variables produced when computing I' as a solved form). ■

Theorem 25: Let I be a solved form $x_1 \succ t_1 \wedge \ldots \wedge x_n \succ t_n \wedge t'_1 \succ x'_1 \wedge \ldots \wedge t'_m \succ x'_m$.
Then I is satisfiable wrt. $\succ_{\mathcal{F}}$ iff $n = 0$ or else there exists a maximal right term in I such that $I[x_i \leftarrow f(0,\ldots,0,t_i)]$ is satisfiable.

Proof. If $n = 0$ then the substitution α is a solution. Otherwise, by the previous lemma, we know that for at least one such a maximal right term t_i the satisfiability is preserved when replacing x_i by $f(0,\ldots,0,t_i)$. ■

The previous theorem states the correctness of our constraint solving algorithm: in practice, to find a solution (or decide the satisfiability) of a given constraint, for each maximal right term t_i in I we can decide the satisfiability of $I[x_i \leftarrow f(0,\ldots,0,t_i)]$ by computing its set of solved forms and applying the theorem recursively. This process terminates because the number of variables decreases in every step.

Example 26: We want to solve $[\![x \succ y \;\wedge\; x \succ z \;\wedge\; z \succ y]\!]$ with $\mathcal{F} = \{f \succ 0\}$ where f is a unary function symbol.
First we can try replacing x by $f(y)$ obtaining $[\![f(y) \succ z \;\wedge\; z \succ y]\!]$, and then z by $f(y)$ obtaining $[\![f(y) \succ f(y)]\!]$, which has no solution. The only other possibility is to reconsider the first choice. This time we replace x by $f(z)$ getting $[\![f(z) \succ y \;\wedge\; z \succ y]\!]$, and z by $f(y)$ obtaining $[\![f(f(y)) \succ y]\!]$, which has α as a solution, since it has no right terms.

4.1. The incrementality problem

Another problem for which we report a solution here is the fact that the satisfiability of ordering constraints depends on the set of function symbols $\mathcal{F}$ and the ordering $\succ_{\mathcal{F}}$. Adding new symbols to $\mathcal{F}$ we obtain a so-called *extension* $\mathcal{F}'$ of $\mathcal{F}$ if the ordering $\succ_{\mathcal{F}}$ is preserved (i.e. for every g and h in $\mathcal{F}$ s.t. $g \succ_{\mathcal{F}} h$ we have $g \succ_{\mathcal{F}'} h$). Now it can happen that a set of axioms S over $\mathcal{F}$ that is saturated wrt. $\succ_{\mathcal{F}}$ becomes non-saturated wrt. $\succ_{\mathcal{F}'}$, i.e. what we call the *incrementality problem* appears. Due to this problem in principle we can loose the completeness of the set of support strategy for the refutation of clauses with new Skolem constants when using a saturated set of clauses. The incrementality problem also appears when we join saturated sets of clauses as the following example (for the equational case) shows:

Example 27: Assume $\mathcal{F} = \{g \succ a \succ f\}$ for the set **A** and $\mathcal{G} = \{g \succ b \succ f\}$ in **B**:

A	**B**
1) $a \simeq f(x)$	1) $b \simeq f(x)$
2) $g(a) \simeq a$	2) $g(b) \simeq b$

Both sets are saturated due to the fact that a and b are the smallest constants in $\mathcal{F}$ and $\mathcal{G}$ respectively. E.g. in **A** the equation $g(f(x)) \simeq a[\![a \succ x]\!]$ is a tautology, since $[\![a \succ x]\!]$ is unsatisfiable wrt. $\mathcal{F}$. Therefore, when considering $\mathcal{F} \cup \mathcal{G}$ as a new set of function symbols, at least one of the sets **A** and **B** must become non-saturated. This implies that, when working with the union of **A** and **B**, apart from computing inferences between equations form **A** and **B** as usual, it becomes also necessary to compute new inferences between equations from the same set. Theorem 28 provides a solution:

Theorem 28: Let $\mathcal{F}$ be a set of function symbols and let T be an ordering constraint defined over $\mathcal{F}$. Moreover, let $\mathcal{F}_0$ be the extension of $\mathcal{F}$ s.t. $\mathcal{F}_0$ is $\mathcal{F} \cup \{f, 0\}$ with $f \succ_{\mathcal{F}_0} 0$ and $g \succ_{\mathcal{F}_0} f$ for every symbol g in $\mathcal{F}$.

Then T is satisfiable wrt. $\succ_{\mathcal{F}_0}$ if, and only if, there exists some extension $\mathcal{F}'$ of $\mathcal{F}$ such that T is satisfiable wrt. $\succ_{\mathcal{F}'}$.

Proof. The left-to-right implication is trivial since $\mathcal{F}_0$ is an extension of $\mathcal{F}$. The reverse is proved as follows: if T is satisfiable wrt. $\succ_{\mathcal{F}'}$ then T is satisfiable wrt. $\succ_{\mathcal{F}'_0}$ (where $\mathcal{F}'_0$ is $\mathcal{F}' \cup \{f, 0\}$ with smallest symbols f and 0 as in $\mathcal{F}_0$). But by applying the algorithm proposed in theorem 25 to check the satisfiability of T wrt. $\succ_{\mathcal{F}'_0}$, we obtain a solution in which only symbols from $\mathcal{F} \cup \{f, 0\}$ appear, which obviously implies that T is also satisfiable wrt. $\succ_{\mathcal{F}_0}$. ■

The previous theorem states that sets of axioms S over $\mathcal{F}$ that are saturated wrt. $\mathcal{F}_0$ remain saturated under every extension of $\mathcal{F}$. Therefore, solving constraints during the completion process wrt. $\mathcal{F}_0$ is a method to overcome the *incrementality problem* mentioned above.

If we review example 26, but now working with $\mathcal{F}_0$ and $\mathcal{G}_0$ then we can check that e.g. for the set **A** now the constrained equation $g(f(x)) \simeq a[\![a \succ x]\!]$ is not a tautology, since the constraint has a solution. Therefore the set **A** is *not* saturated wrt. $\mathcal{F}_l$.

5. Using constraints to describe infinite sets of formulae

Here we address another advantage of our techniques for ordering constrained completion: we can also handle initial sets of axioms that do have non-trivial ordering constraints. This allows to describe and reason about infinite sets of normal formulae:

Example 29: The clause $\rightarrow P(x,y)\,[\![a \succ x]\!]$, under the ordering $a \succ f \succ b$ denotes the infinite set of clauses $\rightarrow P(b,y)$, $\rightarrow P(f(b),y)$, $\rightarrow P(f(f(b)),y), \ldots$

It has to be further investigated which languages can be expressed, and how, by this kind of constraints, e.g. regular tree languages. The fact that the clauses of the initial set S_0 have no constraints is only required in theorem 14 for proving that $R^*_{S_\infty} \models irred_{R_{S_\infty}}(S_0)$ implies $R^*_{S_\infty} \models S_0$. This is true because the instances $C\sigma$ with *reducible* substitutions of clauses $C\,[\![true]\!]$ in S_0 can be deduced from R_{S_∞} and their *irreducible* versions $C\sigma'$ where σ' is the normal form of σ wrt. R_{S_∞}. We can be sure that these instances $C\sigma'$ do really exist because the clauses in S_0 have no constraints.

Now first we show that if we deal with first-order clauses without equality, then *every* ground substitution σ is irreducible wrt. R_{S_∞}, since R_{S_∞} consists only of rewrite rules of the form $A \Rightarrow true$, where A is an atom, and therefore $R^*_{S_\infty} \models irred_{R_{S_\infty}}(S_0)$ trivially implies $R^*_{S_\infty} \models S_0$, because $irred_{R_{S_\infty}}(S_0)$ is equal to S_0. This means that the initial set of axioms can be *any* set of constrained first-order clauses.

Theorem 30: Let $S_0, S_1, \ldots$ be a fair theorem proving derivation, where S_0 is a set of ordering constrained clauses without equality literals (with arbitrary ordering constraints). Then S_0 is inconsistent iff the empty clause belongs to some S_j.

If there are clauses with equality literals and also clauses with ordering constraints in S_0 then S_0 must fulfil the *purity* property:

Definition 31: A set S_0 of constrained clauses is called *pure* if $R^*_{S_\infty} \models irred_{R_{S_\infty}}(S_0)$ implies $R^*_{S_\infty} \models S_0$ for every fair theorem proving derivation $S_0, S_1, \ldots$

Note that sets of ordering constrained clauses without equality literals are also pure. This definition reflects what is needed, but not how to check this property in practice. Below we give a simple sufficient condition for purity, although other more powerful ones can be found. A consequence of this lemma is the purity of sets S_0 containing only clauses without constraints. The final theorem summarizes the case of clauses with equality.

Lemma 32: A set of ordering constrained clauses S_0 is pure if for every clause $C\,[\![T]\!]$ in S_0 there are no variables in C that are lower bounded by T.

Theorem 33: Let $S_0, S_1, \ldots$ be a fair theorem proving derivation, where S_0 is a pure set of ordering constrained clauses. Then S_0 is inconsistent if, and only if, the empty clause belongs to some S_j.

Acknowledgements: We wish to thank Marianne Haberstrau, Fernando Orejas, Pilar Nivela, Harald Ganzinger and Leo Bachmair for their interest and advice on this work, and Hubert Comon for his comments on the methods described in section 4.

6. References

[BG 90] L. Bachmair, H. Ganzinger: On restrictions of ordered paramodulation with simplification. In Proc. 10th CADE, Kaiserslautern, LNCS 449, 1990.

[BG 91] L. Bachmair, H. Ganzinger: Rewrite-based equational theorem proving with selection and simplification, Report 208 Max-Planck Inst. Saarbrücken, 1991.

[Com 90] H. Comon: Solving Symbolic Ordering Constraints. In proc. 5th IEEE Symp. Logic in Comp. Sc. Philadelphia. (1990).

[DJ 90] N. Dershowitz, J-P. Jouannaud: Rewrite systems, in Handbook of Theoretical Comp. Sc. vol. B: Formal Methods and Semantics. Norht Holland, 1990.

[HR 89] J. Hsiang, M. Rusinowitch: Proving refutational completeness of theorem proving strategies: The transfinite semantic tree method. Submitted. (1989).

[Hue 72] G. Huet: Constrained Resolution: A complete method for higher-order logic. PhD. Thesis. Case Western Reserve University. 1972.

[Hul 80] J.M. Hullot: Compilation de Formes Canoniques dans les Teories Equationnelles, These de 3eme Cycle, Universite de Paris Sud, 1980.

[JO 91] J-P. Jouannaud, M. Okada: Satisfiability of systems of ordinal notations with the subterm property is decidable, ICALP 1991, LNCS 510, Madrid, 1991.

[KB 70] D.E. Knuth, P.B. Bendix: Simple word problems in universal algebras. J. Leech, editor, Computational Problems in Abstract Algebra, 263-297, Pergamon Press, Oxford, 1970.

[KKR 90] C. and H. Kirchner, M. Rusinowitch: Deduction with Symbolic Constraints, Revue Francaise d'Intelligence Artificielle, Vol 4, no. 3, pp. 9-52, 1990.

[NN 91] R. Nieuwenhuis, P. Nivela: Efficient deduction in equality Horn logic by Horn-completion, Information Processing Letters, no. 39, pp. 1-6, July 1991.

[NO 91] R. Nieuwenhuis, F. Orejas: Clausal Rewriting, 2nd Intl. Workshop on Conditional and Typed Term Rewriting, Montreal LNCS 516, pp. 246-261, (1991).

[NOR 90] R. Nieuwenhuis, F. Orejas, A. Rubio: TRIP: an implementation of clausal rewriting, Proc. 10th Int. Conf. on Automated Deduction, Kaiserslautern, LNCS 449, pp. 667-668 (1990).

[NR 92] R. Nieuwenhuis, A. Rubio: Basic Superposition is Complete, Proc. European Symposium On Programming, Rennes, France, LNCS, February 1992.

[Pet 90] G.E. Peterson: Complete Sets of Reductions with Constraints, Proc. 10th Int. Conf. on Automated Deduction, Kaiserslautern, LNCS 449, pp. 381-395 (1990).

[RN 91] A. Rubio, R. Nieuwenhuis: Handling ordering constraints of clauses in theorem proving, Research Report LSI-UPC, 1991.

[Zha 88] H. Zhang: Reduction, Superposition, and Induction: Automated Reasoning in an Equational Logic, PhD. Thesis, Renselaer Polytechnic Institute, (1988).

The Special-Relation Rules are Incomplete

Zohar Manna
Computer Science Department
Stanford University
and
Applied Mathematics Department
Weizmann Institute of Science

Richard Waldinger
Artificial Intelligence Center
SRI International
and
Computer Science Department
Stanford University

Abstract. The special-relation rules give accelerated treatment to transitivity, substitutivity, and other axioms classed as "monotonicity properties". These rules extend paramodulation and other equality rules to relations other than equality. In this paper, it is established that these rules are all logically incomplete. The incompleteness of the negative paramodulation rule of Wos and McCune is also demonstrated.

1 Introduction

A continuing trend in automated theorem-proving research has been the introduction of inference rules applicable only to particular theories. Such domain-dependent rules allow shorter and better motivated proofs of sentences valid in the theory. If completeness results for these rules are obtained, we are able to remove the corresponding axioms from our clause set, without losing the proof of any valid sentence. The search space may be reduced significantly.

For example, the paramodulation rule [15] applies to theories with equality and allows us to remove several equality axioms. Resolution with various forms of equational unification [16] allows us to remove axioms that have the form of equations $s = t$ by effectively building them into the unification algorithm. Theory resolution [18] allows us to remove many nonequational axioms and introduce decision procedures instead. Inference rules have been introduced especially for the real numbers [1] and for set theory [4].

The special-relation rules [8] were introduced to give special treatment to orderings, such as the less-than-or-equal-to relation $\leq$ or the subset relation $\subseteq$, and to allow us to discard the transitivity and monotonicity axioms for these relations, much as the paramodulation rule enables us to discard equality axioms. When the rules first appeared, it was speculated that they are complete; in this paper we establish that they are not.

We first review three rules that have been introduced for equality: paramodulation, the RUE-NRF rule [3], and negative paramodulation [19]. We then briefly

This research was supported in part by the National Science Foundation grants CCR-89-04809, CCR-89-11512, and CCR-89-13641, by the Defense Advanced Research Projects Agency under contract NAG2-703, and by the United States Air Force Office of Scientific Research under contract AFOSR-90-0057.

describe how each of these rules is extended to apply to an arbitrary binary relation $\prec$, obtaining three special-relation rules: relation replacement, relation matching, and negative relation replacement, respectively. For this purpose we define the notion of polarity of a subexpression (i.e., term or formula) with respect to a pair of relations. We then exhibit a counterexample that shows that these rules are incomplete. We also show that the negative paramodulation rule itself is incomplete.

To ensure that this paper is self-contained, we review some concepts from the original paper [9]; the reader may refer there for technical details and to *The Logical Basis for Computer Programming* [8,10] for logical prerequisites. Although the original paper presented all rules in a nonclausal context, the rules here are all phrased clausally. The relation-stripping rule, the negative relation-replacement rule, and all the incompleteness results are new to this paper.

2. Theories With Equality

In the absence of any equality rule, we may use a general inference rule for first-order logic, such as the resolution rule [14], and include in our clause set the clausal forms of the equality axioms: reflexivity, symmetry, transitivity, and functional and predicate substitutivity.

Except for the reflexivity axiom $x = x$, these axioms are troublesome to theorem provers because they spawn numerous consequences, most of them irrelevant to the theorem at hand. For example, the resolution rule can be applied to the transitivity axiom and any clause containing an equality $s = t$. That is why the need for an equality rule was recognized early.

2.1 The Paramodulation Rule

The paramodulation rule [15] allows us to replace equals with equals; its soundness follows from the substitutivity property of equality. Completeness results (e.g., [5,13]) establish that if the rule is employed in combination with the resolution rule, all of the equality axioms except for the harmless reflexivity axiom $x = x$ may be dropped from the clause set, without sacrificing the ability to prove any valid sentence.

Paramodulation has been a central part of the theorem-proving systems developed at the Argonne National Laboratory (e.g., OTTER [11]). *Narrowing*, a special case of paramodulation, has been employed in theorem provers based on term rewriting (e.g., RRL [6]).

2.2 Resolution with Unification and Equality

The RUE (resolution with unification and equality) rule [3], which resembles the earlier E-resolution rule [12], allows us to generate a negated equality clause when two nearly complementary literals fail to unify.

The RUE rule is used in conjunction with the NRF (*negative reflective function*) rule, which allows us to drop function symbols from a negated equality. The soundness of these rules is based on the substitutivity properties of equality. Completeness for this pair of rules has not been established in the general case, although Digricoli and Harrison take some steps in that direction. An implementation of the RUE-NRF rules by Digricoli performs better than paramodulation on some problems; the role of the two rules is complementary.

2.3 Negative Paramodulation

The paramodulation rule allows us to reason only from clauses containing positive equality atoms. The *negative paramodulation* rule [19] enables us to reason from clauses containing negative equality atoms. It may appear at first that reasoning from negative equalities is not nearly so important as reasoning from positive ones. However, if the theorem to be proved is itself a positive equality, it will be negated in the context of a refutation proof procedure. If the paramodulation rule alone is employed, the negation of the theorem may be able to play only a passive rule in the search for a refutation.

In experiments with the Argonne system, Wos and McCune found that the negative paramodulation rule allowed them to obtain easy proofs of some theorems that were difficult or impossible with paramodulation alone.

The negative paramodulation rule is intended to replace certain properties of the function and predicate symbols in our clauses. We shall show subsequently that the rule is relatively incomplete, in that it fails to allow us to discard these properties without losing the ability to prove some valid sentences.

Most of the programs that have been able to prove difficult results in theories with equality have used one or another of the equality rules. In the next section, we see how these rules may be extended to relations other than equality.

3. The Special-Relation Rules

Many relations other than equality have axioms and properties that are as unpleasant as the transitivity and substitutivity of equality but that are not dealt with by the equality rules. For example, in numerical reasoning, the less-than-or-equals relation $\leq$ exhibits transitivity and various monotonicity properties. In reasoning about sets, the subset relation $\subseteq$ also exhibits transitivity and monotonicity properties. If we represent such properties as axioms, we shall see the same proliferation of irrelevant clauses as from the equality axioms. Can we extend the equality rules to reason about these relations as well as equality? And will these extended rules have completeness properties, so that when using them we can discard the axioms, without running the risk of losing proofs? Well, yes and no.

In our earlier paper [9], we introduced two special-relation rules for reasoning about arbitrary binary relations with monotonicity properties, analogous to paramodulation and the RUE rule, respectively. In this paper we introduce special-relation rules analogous to the NRF rule and to negative paramodulation, respectively, also for reasoning about relations with monotonicity properties. Each special-relation rule resembles the corresponding equality rule, but whereas the equality rules allow arbitrary occurrences of the subterms s to be replaced, the new rules allow only particular subterm occurrences to be replaced. Precisely which occurrences s may be replaced depends on the monotonicity properties of the function and predicate symbols that surround s.

3.1 Monotonicity Properties

We now say what we mean by a monotonicity property. In what follows, an *expression* is a term or a formula. We regard a connective as a special sort of predicate symbol, denoting a relation over the truth-values. An *operator* is a function symbol or a predicate symbol.

A monotonicity property is a sentence of one of the following forms:

if $x \prec_1 y$
then $x \prec_2 y$

if $y \prec_1 x$
then $x \prec_2 y$

if $x \prec_1 y$
then $f(-, x, -) \prec_2 f(-, y, -)$

if $y \prec_1 x$
then $f(-, x, -) \prec_2 f(-, y, -)$.

Here $\prec_1$ and $\prec_2$ are binary predicate symbols and f is an n-ary operator. (By convention, u, v, w, x, y, and z, with or without subscripts or superscripts, are variables. All free variables are understood to have implicit universal quantification.) Also, $f(-, x, -)$ is an abbreviation for $f(z_1, \ldots, z_{i-1}, x, z_{i+1}, \ldots, z_n)$.

For example,
if $x \leq y$
then $x + z \leq y + z$

is a monotonicity property. The transitivity of a relation $\prec$ may also be regarded as a monotonicity property, because it may be expressed by the property
if $x \prec y$
then $z \prec x \rightarrow z \prec y$.

Here $\prec_1$, $\prec_2$, and f are taken to be $\prec$, $\rightarrow$, and $\prec$, respectively. Note that we use $\rightarrow$ interchangeably with *if-then* to denote implication. We are taking advantage of the fact that the connective $\rightarrow$ is regarded as a predicate symbol.

The symmetry of a relation $\prec$ is a monotonicity property
if $y \prec x$
then $x \prec y$.

In particular, the symmetry axiom for equality is a monotonicity property. The functional- and predicate-substitutivity axioms for equality are also monotonicity properties.

Monotonicity properties are used in the definition of relational polarity, our key notion.

3.2 Relational Polarity

We shall define the polarity of an occurrence of a subexpression s of an expression $e\langle s\rangle$, with respect to a pair of binary relations $\langle \prec_1, \prec_2\rangle$ and relative to a set of monotonicity properties $\mathcal{M}$. A subexpression occurrence may have positive (+) or negative (−) polarity, or both (±), or none. The notion will be defined so that the following property will hold.

Property (polarity substitutivity)

If an occurrence of x is positive in an expression $e\langle x^+\rangle$, or negative in an expression $e\langle x^-\rangle$, with respect to binary relations $\langle \prec_1, \prec_2\rangle$, relative to a set of monotonicity properties $\mathcal{M}$, then the properties $\mathcal{M}$ imply the sentences

if $x \prec_1 y$	*if* $y \prec_1 x$
then $e\langle x^+\rangle \prec_2 e\langle y^+\rangle$	*then* $e\langle x^-\rangle \prec_2 e\langle y^-\rangle$.

Here $e\langle y^+\rangle$ is the result of replacing the (positive) occurrence of x with y in $e\langle x^+\rangle$, and $e\langle y^-\rangle$ is the result of replacing the (negative) occurrence of x with y in $e\langle x^-\rangle$; only one occurrence is replaced in each case. ┘

The polarity-substitutivity property may be paraphrased by saying that making a positive subexpression occurrence bigger (with respect to $\prec_1$) makes the entire expression bigger (with respect to $\prec_2$), while making a negative subexpression occurrence smaller (with respect to $\prec_1$) makes the entire expression bigger (with respect to $\prec_2$).

In the case in which the second relation $\prec_2$ is transitive, more than one occurrence of x may be replaced with y in $e\langle x^+\rangle$ or $e\langle x^-\rangle$; this is particularly useful when $\prec_2$ is the implication connective.

The definition of relational polarity is inductive.

Definition (relational polarity)

Let $\prec_1$ and $\prec_2$ be binary relations and $\mathcal{M}$ be a set of monotonicity properties. Then, with respect to $\langle \prec_1, \prec_2\rangle$ and relative to $\mathcal{M}$:

- For any expression s, s has positive polarity in s itself iff $(\textit{if } x \prec_1 y \textit{ then } x \prec_2 y)$ belongs to $\mathcal{M}$.
- s has negative polarity in s iff $(\textit{if } y \prec_1 x \textit{ then } x \prec_2 y)$ belongs to $\mathcal{M}$.

For any operator f,

- s has positive polarity in $f(-, s, -)$ iff $(\textit{if } x \prec_1 y \textit{ then } f(-, x, -) \prec_2 f(-, y, -))$ belongs to $\mathcal{M}$.
- s has negative polarity in $f(-, s, -)$ iff $(\textit{if } y \prec_1 x \textit{ then } f(-, x, -) \prec_2 f(-, y, -))$ belongs to $\mathcal{M}$.

For any expression $e\langle s\rangle$ containing an occurrence of s,

- s has positive polarity in $f(-, e\langle s\rangle, -)$ iff there exists a binary relation $\prec$ such that the polarity of s in $e\langle s\rangle$ with respect to $\langle \prec_1, \prec\rangle$ is the same as the polarity of $e\langle s\rangle$ in $f(-, e\langle s\rangle, -)$ with respect to $\langle \prec, \prec_2\rangle$.
- s has negative polarity in $f(-, e\langle s\rangle, -)$ iff there exists a binary relation $\prec$ such that the polarity of s in $e\langle s\rangle$ with respect to $\langle \prec_1, \prec\rangle$ is opposite to the polarity of $e\langle s\rangle$ in $f(-, e\langle s\rangle, -)$ with respect to $\langle \prec, \prec_2\rangle$.

Recall that a subexpression occurrence may have both positive and negative polarities. If we say that a subexpression occurrence has positive [or negative] polarity, we do not exclude the possibility that it actually has both polarities. If two occurrences both have positive polarity or both have negative polarity, we say that they have the *same* polarity, even if one of them has both polarities and the other does not. Similarly, if one occurrence has positive polarity and the other has negative polarity, we say that they have *opposite* polarities, even if one of them has both polarities and the other does not. ┘

Example

Suppose $\mathcal{M}_0$ is the set of the following monotonicity properties:

1. *if* $y \to x$ *then* $\neg x \to \neg y$
2. *if* $y \leq x$ *then* $x \leq z \to y \leq z$
3. *if* $x \leq y$ *then* $z \cdot x \leq z \cdot y$
4. *if* $y \subseteq x$ *then* $z \sim x \subseteq z \sim y$
5. *if* $x \subseteq y$ *then* $card(x) \leq card(y)$.

Note that we allow (in 1) variables that range over truth-values. Also, $z \sim x$ is the set difference between sets z and x, and $card(x)$ is the number of elements in the (finite) set x.

Then, relative to $\mathcal{M}_0$,

s	is negative in	$a \sim s$	with respect to $\langle \subseteq, \subseteq\rangle$, by 4
$a \sim s$	is positive in	$card(a \sim s)$	with respect to $\langle \subseteq, \leq\rangle$, by 5
$card(a \sim s)$	is positive in	$2 \cdot card(a \sim s)$	with respect to $\langle \leq, \leq\rangle$, by 3
$2 \cdot card(a \sim s)$	is negative in	$2 \cdot card(a \sim s) \leq b$	with respect to $\langle \leq, \to\rangle$, by 2
$2 \cdot card(a \sim s) \leq b$	is negative in	$\neg(2 \cdot card(a \sim s) \leq b)$	with respect to $\langle \to, \to\rangle$, by 1.

Hence, by the definition,

s is negative in $\neg(2 \cdot card(a \sim s) \leq b)$ with respect to $\langle \subseteq, \to\rangle$.

Therefore, by the polarity-substitutivity property, the properties $\mathcal{M}$ imply the sentence

if $t \subseteq s$
then $\neg(2 \cdot card(a \sim s) \leq b) \to \neg(2 \cdot card(a \sim t) \leq b)$. ┘

By repeated application of the definition, we may compute the polarities of all the subexpressions of a given expression, with respect to given relations $\langle \prec_1, \prec_2\rangle$ and relative to a given set of monotonicity properties.

If the equality axioms (other than reflexivity) are included in a set of monotonicity properties $\mathcal{M}$, all subterms of a given term t will have both polarities in t with respect to $\langle =, = \rangle$, relative to $\mathcal{M}$; also, all subterms of a given formula $\mathcal{P}$ will have both polarities in $\mathcal{P}$ with respect to $\langle =, \rightarrow \rangle$, relative to $\mathcal{M}$.

The notion of relational polarity is being used by Douglas Smith in extending the KIDS system to formulate connections between specifications [17].

3.3 The Relation-Replacement Rule

The relation-replacement rule is the analogue for an arbitrary binary relation $\prec$ of the paramodulation rule for the equality relation $=$. It is expressed in the ground case as follows:

Rule (relation replacement)

Relative to a set $\mathcal{M}$ of monotonicity properties, the following deductions may be performed:

$$\frac{\begin{array}{l} s \prec t,\ \mathcal{Q} \\ \mathcal{P}\langle s^{+}\rangle \end{array}}{\mathcal{P}\langle t^{+}\rangle,\ \mathcal{Q}} \qquad \frac{\begin{array}{l} t \prec s,\ \mathcal{Q} \\ \mathcal{P}\langle s^{-}\rangle \end{array}}{\mathcal{P}\langle t^{-}\rangle,\ \mathcal{Q}}$$

Here $\mathcal{P}\langle t^{+}\rangle$ is the result of replacing one or more occurrences of the subterm s with the term t in the clause $\mathcal{P}\langle s^{+}\rangle$; the replaced occurrences must all be positive in $\mathcal{P}\langle s^{+}\rangle$ with respect to $\langle \prec, \rightarrow \rangle$, relative to $\mathcal{M}$. Similarly for $\mathcal{P}\langle t^{-}\rangle$; the replaced occurrences must all be negative in $\mathcal{P}\langle s^{-}\rangle$. ▟

Example

Relative to the set $\mathcal{M}_0$ of monotonicity properties given in the previous section, the following deduction may be performed by the relation-replacement rule:

$$\frac{\begin{array}{l} t \subseteq s \\ \neg(2 \cdot card(a \sim s) \leq b) \end{array}}{\neg(2 \cdot card(a \sim t) \leq b)}$$

This application of the rule is legal because, as we have seen, the replaced occurrence of s is negative in $\neg(2 \cdot card(a \sim s) \leq b)$ with respect to $\langle \subseteq, \rightarrow \rangle$, relative to $\mathcal{M}_0$. ▟

To extend the rule from the ground case to the general case, we first apply an appropriate most-general unifier θ, and then invoke the ground version of the rule.

The paramodulation rule is a special case of the relation-replacement rule in which we take the relation $\prec$ to be the equality relation $=$. This is because any subterm s has both positive and negative polarities in $\mathcal{P}\langle s\rangle$ with respect to $\langle =, \rightarrow \rangle$, relative to the equality axioms (other than reflexivity).

To discuss the soundness of the special-relation rules, it will be convenient to introduce a notion of relative soundness.

Definition (relative soundness)

A deduction rule is *sound relative* to a set $\mathcal{M}$ of properties if, whenever the rule allows us to deduce, from the clauses $\mathcal{C}_1, \ldots, \mathcal{C}_n$, the clause $\mathcal{D}$, we have that $\mathcal{M}$ and the given clauses $\mathcal{C}_1, \ldots, \mathcal{C}_n$ imply the deduced clause $\mathcal{D}$. ⌟

We can use this notion to establish the relative soundness (in the ground case) of the positive part of the relation-replacement rule.

Justification (relation replacement rule)

It suffices to show that the monotonicity properties and the given clauses imply the deduced clause. Suppose (under a given interpretation) that the properties $\mathcal{M}$ and the two given clauses are all true; it suffices to show that the deduced clause

$$\mathcal{P}\langle t^+\rangle,\ \mathcal{Q}$$

is also true. If its subclause $\mathcal{Q}$ is true, the deduced clause is surely true; so let us assume that $\mathcal{Q}$ is false. Then, because the given clause

$$s \prec t,\ \mathcal{Q}$$

is true, the atom $s \prec t$ is true. Because the replaced occurrences of s are positive in $\mathcal{P}\langle s^+\rangle$ with respect to $\langle \prec, \rightarrow\rangle$ relative to $\mathcal{M}$, it follows from the polarity-substitutivity property that $\mathcal{M}$ implies

$$\begin{array}{l} \textit{if } s \prec t \\ \textit{then } \mathcal{P}\langle s^+\rangle \rightarrow \mathcal{P}\langle t^+\rangle. \end{array}$$

(Because the implication connective $\rightarrow$ is transitive, more than one occurrence of s may be replaced.) Therefore, because $\mathcal{M}$, $s \prec t$, and the given clause

$$\mathcal{P}\langle s^+\rangle$$

are all true, it follows that the subclause $\mathcal{P}\langle t^+\rangle$ of the deduced clause is true. Hence the deduced clause itself is true. ⌟

The statement of the relation-replacement rule suggests the following implementation. For a given theory, the user identifies a set of monotonicity properties $\mathcal{M}$ whose validity has already been established. The system computes the polarity of every subterm of every clause relative to $\mathcal{M}$, with respect to $\langle \prec, \rightarrow\rangle$, for every relation $\prec$ for which a monotonicity property has been provided. It is then easy to identify possible applications of the relation-replacement rule.

A version of the relation-replacement rule is being introduced by J Moore into the NQTHM theorem prover [2].

3.4 The Relation-Matching Rule

The relation-matching rule is the analogue for an arbitrary binary relation $\prec$ of the RUE rule [3]. In the ground case, it is expressed as follows:

Rule (relation matching)

Relative to a set $\mathcal{M}$ of monotonicity properties, the following deductions may be performed:

$$\frac{\mathcal{P}\langle s^+\rangle,\ \mathcal{Q} \qquad \neg\mathcal{P}\langle t^+\rangle,\ \mathcal{R}}{\neg(s \prec t),\ \mathcal{Q},\ \mathcal{R}} \qquad\qquad \frac{\mathcal{P}\langle s^-\rangle,\ \mathcal{Q} \qquad \neg\mathcal{P}\langle t^-\rangle,\ \mathcal{R}}{\neg(t \prec s),\ \mathcal{Q},\ \mathcal{R}}$$

Here the atom $\mathcal{P}\langle t^+\rangle$ is obtained from the atom $\mathcal{P}\langle s^+\rangle$ by replacing one or more occurrences of the term s with the term t, where the replaced occurrences must have positive polarity in $\mathcal{P}\langle s^+\rangle$ with respect to $\langle \prec, \rightarrow\rangle$, relative to $\mathcal{M}$. Similarly for $\mathcal{P}\langle t^-\rangle$. ┘

The relative soundness of the rule follows from the polarity-substitutivity property. Like the RUE rule, the relation-matching rule can be extended to allow several distinct terms to be replaced.

Just as the RUE rule is used in conjunction with the NRF rule, the relation-matching rule may be used in conjunction with a *relation-stripping* rule, which, in the ground case, is expressed as follows:

Rule (relation stripping)

Relative to a set $\mathcal{M}$ of monotonicity properties, the following deductions may be performed:

$$\frac{\neg(e\langle s^+\rangle \prec_2 e\langle t^+\rangle),\ \mathcal{Q}}{\neg(s \prec_1 t),\ \mathcal{Q}} \qquad\qquad \frac{\neg(e\langle s^-\rangle \prec_2 e\langle t^-\rangle),\ \mathcal{Q}}{\neg(t \prec_1 s),\ \mathcal{Q}}$$

Here $e\langle t^+\rangle$ is obtained from $e\langle s^+\rangle$ by replacing precisely one occurrence of s with t, where the replaced occurrence is positive in $e\langle s^+\rangle$ with respect to $\langle \prec_1, \prec_2\rangle$, relative to $\mathcal{M}$. Similarly for $e\langle t^-\rangle$. ┘

The soundness of the rule relies on the following instances of the polarity-substitutivity property:

if $s \prec_1 t$
then $e\langle s^+\rangle \prec_2 e\langle t^+\rangle$

if $t \prec_1 s$
then $e\langle s^-\rangle \prec_2 e\langle t^-\rangle$.

The relation-matching and -stripping rules extend from the ground case to the general case; first we apply an appropriate substitution θ; then we invoke the ground case of the rule. What is meant by "appropriate" is clarified in the example.

Examples

Relative to the set $\mathcal{M}_0$ of monotonicity properties given in Section 3.3, we may perform the following deduction:

$$\frac{2 \cdot card(a \sim s) \le x,\ q(x) \qquad \neg(2 \cdot card(y \sim t) \le b),\ r(y)}{\neg(s \subseteq t),\ q(b),\ r(a)}$$

Here the substitution θ: $\{x \leftarrow b, y \leftarrow a\}$ is applied before the ground case of the rule is invoked. This application of the rule is legal because the replaced occurrence of s is positive in $(2 \cdot card(a \sim s) \leq b)$ with respect to $\langle \subseteq, \rightarrow \rangle$, relative to $\mathcal{M}_0$.

Relative to the set consisting of the single monotonicity property

if $sisters(x, y)$
then $father(x) = father(y)$,

the following deduction can be performed:

$$\frac{\neg[father(jean) = father(mary)]}{\neg sisters(jean, mary)}$$

This application of the rule is allowed because *jean* is positive in *father*(*jean*) with respect to $\langle sisters, = \rangle$, relative to the property. ▟

3.5 The Negative Relation-Replacement Rule

The negative relation-replacement rule is the analogue for an arbitrary binary relation $\prec$ of the negative paramodulation rule [19] discussed earlier. In the ground case, it may be expressed as follows:

Rule (negative relation-replacement rule)

Relative to a set of monotonicity properties $\mathcal{M}$, the following deductions may be performed:

$$\frac{\begin{array}{l} s \prec t,\ \mathcal{Q} \\ \mathcal{P}\langle s^+ \rangle,\ \mathcal{R} \end{array}}{\neg\mathcal{P}\langle t^+ \rangle,\ \mathcal{Q},\ \mathcal{R}} \qquad \frac{\begin{array}{l} t \prec s,\ \mathcal{Q} \\ \mathcal{P}\langle s^- \rangle,\ \mathcal{R} \end{array}}{\neg\mathcal{P}\langle t^- \rangle,\ \mathcal{Q},\ \mathcal{R}}$$

Here $\mathcal{P}\langle t^+ \rangle$ is obtained from $\mathcal{P}\langle s^+ \rangle$ by replacing a single occurrence of s with t, where the replaced occurrence must have positive polarity in $\mathcal{P}\langle s^+ \rangle$ with respect to $\langle \prec, \not\wedge \rangle$, relative to $\mathcal{M}$. Similarly for $\mathcal{P}\langle t^- \rangle$. The relation $\not\wedge$ is the *nand* connective, the negation of conjunction. ▟

The soundness of the rule relies on the following instances of the polarity-substitutivity property:

if $s \prec t$
then $\mathcal{P}\langle s^+ \rangle \not\wedge \mathcal{P}\langle t^+ \rangle$

if $t \prec s$
then $\mathcal{P}\langle s^- \rangle \not\wedge \mathcal{P}\langle t^- \rangle$

or, equivalently,

if $s \prec t$
then $\mathcal{P}\langle s^+ \rangle \rightarrow \neg\mathcal{P}\langle t^+ \rangle$

if $t \prec s$
then $\mathcal{P}\langle s^- \rangle \rightarrow \neg\mathcal{P}\langle t^- \rangle$.

The rule extends from the ground case to the general case; in the usual way.

The negative paramodulation rule is a special case of the negative relation-replacement rule.

Rule (negative paramodulation, ground case)

Relative to a set $\mathcal{M}$ of monotonicity properties, the following deduction may be performed:

$$\frac{\begin{array}{l} s \neq t,\ \mathcal{Q} \\ \mathcal{P}\langle s^{+}\rangle,\ \mathcal{R} \end{array}}{\neg\mathcal{P}\langle t^{+}\rangle,\ \mathcal{Q},\ \mathcal{R}}$$

Here $\mathcal{P}\langle t^{+}\rangle$ is obtained from the atom $\mathcal{P}\langle s^{+}\rangle$ by replacing precisely one occurrence of s with t, where the replaced occurrence is of positive polarity in $\mathcal{P}\langle s^{+}\rangle$ with respect to $\langle \neq, \not\prec \rangle$, relative to $\mathcal{M}$. ┘

Wos and McCune expressed the same rule with a slightly different vocabulary.

Note that the not-equals relation is expressed by a single predicate symbol $\neq$. The relation between the two symbols $\neq$ and $=$ must be described axiomatically.

Example

The negative paramodulation rule will enable us to perform the following deduction in the theory of integers:

$$\frac{\begin{array}{l} s \neq t \\ a + 2s = b \end{array}}{a + 2t \neq b}$$ ┘

Note that we may not replace more than one occurrence of s with t, e.g.

$$\frac{\begin{array}{l} s \neq t \\ s = s \end{array}}{t \neq t}$$

is not a sound deduction, even though both replaced occurrences of s are of positive polarity in $s = s$ with respect to $\langle \neq, \not\prec \rangle$, relative to the set of properties

if $x \neq y$ *then* $x = z \not\prec y = z$ *if* $x \neq y$ *then* $z = x \not\prec z = y$.

4. Completeness

We are now ready to discuss the completeness of the special-relation rules. We shall define completeness relative to a finite set of properties $\mathcal{M}$, usually taken to be a set of monotonicity properties. The properties in $\mathcal{M}$ will be assumed to be put into clausal form.

Definition (relative satisfiability)

A set of clauses $\mathcal{C}$ is *satisfiable relative* to a set of properties $\mathcal{M}$ if $\mathcal{C} \cup \mathcal{M}$ is satisfiable. Accordingly, $\mathcal{C}$ is *unsatisfiable relative to* $\mathcal{M}$ if $\mathcal{C} \cup \mathcal{M}$ is unsatisfiable. ┘

Definition (relative completeness)

A set of rules $\mathcal{R}$ is *complete relative to* $\mathcal{M}$ if there is a refutation, invoking only rules from $\mathcal{R}$, for any clause set $\mathcal{C}$ that is unsatisfiable relative to $\mathcal{M}$. ┘

If a set of rules is complete (and sound) relative to a finite set of monotonicity properties $\mathcal{M}$, then for any clause set $\mathcal{C}$, $\mathcal{C}$ can be refuted by the set of rules if and only if $\mathcal{C} \cup \mathcal{M}$ can be refuted using resolution alone. This follows from the soundness and completeness of the resolution rule.

4.1 The Special-Relation Rules are Incomplete

Now we have the vocabulary to discuss the completeness of the special relation rules. In the original paper we speculated that the relation-replacement and relation-matching rules were complete if used with the resolution rule, relative to a given set of monotonicity properties. The following example, devised in collaboration with Jonathan Traugott, shows that there are sets of monotonicity properties relative to which the relation-replacement, relation-matching, relation-strippping, and resolution rules together are not complete.

Counterexample (incompleteness of the special-relation rules)

Let $\mathcal{M}$ be (the clausal version of) the following set of monotonicity properties:

1. *if* $x \prec_1 y$
   then $g(x) \prec_2 g(y)$

2. *if* $x \prec_2 y$
   then $p(u, x, v) \rightarrow p(u, y, v)$.

Note that p has a monotonicity property only for its second argument x. Furthermore $\mathcal{M}$ does not contain the transitivity property for either $\prec_1$ or $\prec_2$.

Let $\mathcal{C}$ be the following set of clauses:

3. $a \prec_1 b$ 4. $b \prec_1 c$ 5. $p(u, u, g(c))$ 6. $\neg p(g(a), v, v)$.

We claim that the clause set $\mathcal{C}$ is unsatisfiable relative to $\mathcal{M}$, but that no refutation of $\mathcal{C}$ is possible using the resolution rule and the special-relation rules relative to $\mathcal{M}$.

To see that the clause set $\mathcal{C}$ is unsatisfiable relative to $\mathcal{M}$, we form the clause set $\mathcal{C}'$ obtained by replacing clauses 5 and 6 with their ground instances

5′. $p(g(a), g(a), g(c))$ 6′. $\neg p(g(a), g(c), g(c))$.

If a clause set is satisfiable, any set of instances of its clauses will be satisfiable too. Hence, to show that $\mathcal{C}$ is unsatisfiable, it suffices to show that $\mathcal{C}'$ is unsatisfiable. This is demonstrated by the following refutation of $\mathcal{C}'$:

7. Relation replacement applied to 3 and 5′:
   $$p(g(a), g(b), g(c)).$$
   This step is allowed because the annotated occurrence of a is positive in $p(g(a), g(a^+), g(c))$ with respect to $\langle \prec_1, \rightarrow \rangle$ relative to $\mathcal{M}$, by 1 and 2.

8. Relation replacement applied to 4 and 7:

$p(g(a), g(c), g(c))$.

The justification for this step is similar to that for the previous step.

9. Resolution applied to 8 and 6′:

$\Box$.

Note that the first step of this refutation is impossible if we attempt to use clause 5 instead of its instance 5′, because clause 5 has no subterm of positive polarity with respect to $\langle \prec_1, \rightarrow \rangle$. In fact, no refutation of C is possible, even if we may apply all the special-relation rules and the resolution rule. Only two inferences are possible:

10. Relation matching applied to 5 and 6, $\{u \leftarrow g(a), v \leftarrow g(c)\}$:

$\neg(g(a) \prec_2 g(c))$.

This step is allowed because (the annotated occurrence of) u is positive in $p(u, u^+, g(c))$ with respect to $\langle \prec_2, \rightarrow \rangle$, relative to $\mathcal{M}$.

11. Relation stripping applied to 10:

$\neg(a \prec_1 c)$.

This step is allowed because a is positive in $g(a)$ with respect to $\langle \prec_1, \prec_2 \rangle$, relative to $\mathcal{M}$. ┘

4.2 Negative Paramodulation is Incomplete

No completeness claim was made for the negative paramodulation rule [19]. In fact, an example similar to the above counterexample illustrates that, relative to certain sets of monotonicity properties, negative paramodulation with resolution, and even paramodulation is incomplete.

Counterexample (incompleteness of negative paramodulation)

Let $\mathcal{M}$ be the following set of monotonicity properties:

1. *if* $x \neq y$ *then* $g(x) \neq g(y)$

2. *if* $x \neq y$ *then* $p(u, x, v) \not\leftrightarrow p(u, y, v)$.

Let C be the following set of clauses:

3. $a \neq b$ 4. $p(u, u, g(b))$ 5. $p(g(a), v, v)$.

We claim that this clause set is unsatisfiable relative to $\mathcal{M}$ but that there does not exist a refutation of C using negative paramodulation with resolution.

To see that the clause set is unsatisfiable, we form the clause set C' obtained by replacing 4 and 5 with their ground instances

4′. $p(g(a), g(a), g(b))$ 5′. $p(g(a), g(b), g(b))$.

To show that C is unsatisfiable, it suffices to show that C' is unsatisfiable. We demonstrate the unsatisfiability of C' by the following refutation, using negative paramodulation and resolution.

6. Negative paramodulation applied to 3 and 4′:

$\neg p(g(a), g(b), g(b))$.

This step is allowed because the annotated occurrence of a is positive in the atom $p(g(a), g(a^+), g(b))$ with respect to $\langle \neq, \not\prec \rangle$, relative to $\mathcal{M}$, by 1 and 2.

7. Resolution applied to $5'$ and 6:
   $\square$.

No refutation of the original clause set C exists; only four inferences are possible. For example:

8. Negative paramodulation applied to 3 and 4, $\{u \leftarrow a\}$:
   $\neg p(a, b, g(b))$.
   This step is allowed because the annotated occurrence of u is positive in the atom $p(u, u^+, g(b))$ with respect to $\langle \neq, \not\prec \rangle$, relative to $\mathcal{M}$, by 2.

No further inferences are possible even if the paramodulation rule is allowed. ┘

5. Conclusions

There is the possibility that the special-relation rules may be found to be complete in more restricted situations, or that the rules may be extended to be complete in all situations.

An extended special-relation rule [7] unifies the relation-replacement, the negative relation-replacement, and (hence) the negative paramodulation rule into a single rule; however, the extended rule suffers from the same completeness problems as the original rules.

The results we have presented do not imply that the special-relation rules cannot be used in practice; after all, the counterexamples we have used are a bit arcane and involve situations that do not arise naturally. Nevertheless, our results suggest that the rules be used with some caution.

Acknowledgements

We would like to thank Jieh Hsiang, Doug Smith, Mark Stickel, Jon Traugott, and the referees for valuable suggestions and help. Anuchit Anuchitanukul did a detailed reading of the manuscript.

References

[1] W.W. Bledsoe and L. Hines: Variable elimination and chaining in a resolution-based prover for inequalities. *Proc. 5th Conf. Automated Deduction*, 281–292, 1980.

[2] R.S. Boyer and J S. Moore: *A Computational Logic.* Academic Press, New York, 1979.

[3] V.J. Digricoli and M.C. Harrison: Equality-based binary resolution. *J. ACM*, 33(2):253–289, 1986.

[4] L. Hines: Str$\dot{+}$ve $\subseteq$: The str$\dot{+}$ve-based subset prover, *Proc. 10th Int. Conf.*

Automated Deduction, 193–206, 1990.

[5] J. Hsiang and M. Rusinowitch: A new method for establishing refutational completeness in theorem proving. *Proc. 8th Int. Conf. Automated Deduction*, 141–152, 1986.

[6] D. Kapur and P. Narendran: An equational approach to theorem proving in the first-order predicate calculus. *Proc. 9th Int. Conf. Automated Deduction*, 1146–1153, 1985.

[7] Z. Manna, M. Stickel, and R. Waldinger: Monotonicity properties in automated deduction. In *Artificial Intelligence and Mathematical Theory of Computation: Papers in Honor of John McCarthy*, V. Lifschitz (editor), Academic Press, 261–279, 1991.

[8] Z. Manna and R. Waldinger: *Logical Basis for Computer Programming*. Volume 1: *Deductive Reasoning*. Addison–Wesley, 1985.

[9] Z. Manna and R. Waldinger: Special relations in automated deduction. *J. ACM*, 33(1):1–59, 1986.

[10] Z. Manna and R. Waldinger: *Logical Basis for Computer Programming*. Volume 2: *Deductive Systems*. Addison–Wesley, 1990.

[11] W.W. McCune: OTTER 2.0 Users' Guide. Mathematics and Computer Science Division, Argonne National Lab., 1990.

[12] J.B. Morris: E-resolution: An extension of resolution to include the equality relation. *Int. Conf. Artificial Intelligence*, 287–294, 1969.

[13] G.E. Peterson: A technique for establishing completeness results in theorem proving with equality. *SIAM J. Comput.*, 12(1):82–100, 1983.

[14] J.A. Robinson: A machine-oriented logic based on the resolution principle. *J. ACM*, 12(1):23–41, 1965.

[15] G. Robinson and L. Wos: Paramodulation and theorem-proving in first-order theories with equality. In *Machine Intelligence 4*, B. Meltzer and D. Michie (editors), American Elsevier, 135–1509, 1969.

[16] J.H. Siekmann: Unification theory. *J. of Symbolic Computation*, 7(3/4):207–274, 1989.

[17] D.R. Smith: Constructing specification morphisms. Technical Report, Kestrel Institute, 1992.

[18] M.E. Stickel: Automated deduction by theory resolution. *J. Automated Reasoning*, 1(4):333–335, 1985.

[19] L. Wos and W.W. McCune: Negative paramodulation. *Proc. 8th Conf. Automated Reasoning*, 229–239, 1986.

An Improved Method for Adding Equality to Free Variable Semantic Tableaux*

Bernhard Beckert Reiner Hähnle

Institute for Logic, Complexity and Deduction Systems
University of Karlsruhe, 7500 Karlsruhe, Germany
{beckert,haehnle}@ira.uka.de

Abstract. Tableau–Based theorem provers can be extended to cover many of the nonclassical logics currently used in AI research. For both, classical and nonclassical first–order logic, equality is a crucial feature to increase expressivity of the object language. Unfortunately, all so far existing attempts of adding equality to semantic tableaux have been more or less experimental and turn out to be useless in practice. In the present work we introduce an approach that leads much further and sets the stage for more advanced developments. We identify the problems that stem specifically from choosing semantic tableaux as a framework and state soundness and completeness results for our method.

Introduction

In this paper we present a theoretical basis for, as well as an actual implementation of equality handling in tableau–based theorem provers. We do not claim that it can compete with state–of–the–art equality reasoning systems; it is, however, to our best knowledge, the first equality extension of *semantic tableaux* that can grasp beyond textbook examples. Using tableaux as a logical basis for mechanical theorem proving has three major merits: First, they do not commit one to the usage of normal forms; second, they can be extended to cover many of the nonclassical logics (as has been done for example in [3, 6]) currently used in AI research; and third, counterexamples can be generated for non–tautologies. Moreover, while tableau–based systems may not be the most powerful provers being around, they have shown to be sophisticated enough to be interesting in real applications [9].

For both classical and nonclassical first–order logic equality is a crucial feature to increase expressivity of the object language. Unfortunately, all so far existing attempts of adding equality to semantic tableaux have been more or less experimental and turn out to be useless in practice. In the present work we introduce an approach that leads further and sets the stage for more advanced developments.

We assume that the reader is familiar with semantic tableaux and first–order logic with equality (if not, an excellent introduction can be found in [4]). To enhance the readability of our paper we give a short account of the version of the tableau system used by us in Section 1, together with some technical definitions that will be used later. In Section 2 we carry out a careful analysis of the shortcomings in previous approaches. As a result we can clearly identify the problems that stem specifically from choosing semantic tableaux as a framework. This analysis becomes the basis for our own treatment of equality in Section 3 which avoids the aforementioned drawbacks. We state soundness and completeness results for our method. In Section 4 the method is illustrated with an example. Our approach has been implemented as part of the many–valued theorem proving system $_3T^AP$ [5, 6] and we provide some empirical data from test runs of this implementation. Finally we

*This work has been partly supported by IBM Germany.

explain the limitations of the present work and suggest some directions where future research seems promising. Due to space constraints we give no proofs in this paper. All proofs may be found in [1].

1 Preliminaries

Syntax and Semantics

Let us fix a first-order language **L** which is built up from countable sets **R** of predicate symbols, **F** of function symbols, **C** of constant symbols and **Var** of object variables in the usual manner.[1] The quantifier symbols are $\forall, \exists$ and the binary logical connectives consist of $\wedge$ (conjunction), $\vee$ (disjunction) and $\supset$ (material implication), the only unary connective is $\neg$ (negation). Furthermore let us assume that **R** contains a binary predicate symbol for equality which we denote by $\approx$ such that no confusion with the meta-level equality predicate $=$ can arise. We stress that there is no restriction where equalities can occur in formulæ.

We are using the standard notions of free/bound variable, sentence, model, valuation, satisfiability and tautology.[2] Substitutions are mappings from variables to terms and are extended to formulæas usual. Since we will only be concerned with substitutions that are the identity mapping up to a finite number of variables, we will denote a substitution σ by $\{x_1 \leftarrow t_1, \ldots, x_n \leftarrow t_n\}$, where $\{x_1, \ldots, x_n\}$ are the variables that are changed by σ.

A model $\mathbf{M} = \langle \mathbf{D}, \mathbf{I} \rangle$ (with domain **D** and interpretation **I**) is called **normal** iff $\approx^{\mathbf{I}}$ is the identity relation on **D**. A model is called **canonical** if, moreover, for every $d \in \mathbf{D}$ there is a term t in **L** such that $t^{\mathbf{I}} = d$. The following theorem shows that canonical models are analogous to Herbrand models.

Theorem 1.1 *If a set S of sentences is satisfied by a normal model, then there is also a canonical model that satisfies S.*

Since in the tableau proofs it will be necessary to introduce Skolem terms, we extend our first order language to a language $\mathbf{L}^{\mathbf{Sko}}$ by adding countably many constant symbols and function symbols for each arity which do not appear already in **L**. From now on we are working in $\mathbf{L}^{\mathbf{Sko}}$ and we consider only canonical models.

Semantic Tableaux

Semantic (or analytic) tableaux have been introduced in the 1950s by E. W. Beth and K. J. J. Hintikka, its ancestors being Gentzen systems. R. Smullyan [13] gave a particularly elegant version of tableaux which increased their popularity largely and most tableau systems used today are based on his formulation. Tableau systems are available in two versions, namely signed and unsigned, from which we will be using the former.

For our purposes it is sufficient to visualize a tableau proof as a finite labelled binary tree. The node labels are first-order formulæwhich are prefixed with a sign, i.e. an element from $\{\mathsf{T}, \mathsf{F}\}$. To prove tautologyhood of a formula ϕ we begin with a tree whose single node is labelled by $\mathsf{F}\,\phi$, i.e. we assume that ϕ is false in some model. A tableau proof represents a systematic search for such a model. For every combination of sign/leading connective (resp. sign/leading quantifier) there exists a tableau expansion rule which reflects its semantics. We call a maximal path in a tableau proof tree *branch* and say that a branch is *closed* if it contains a pair of unifiable atomic formulæwith complementary signs. A tableau proof tree represents a proof of the root formula when all branches in the tree can be closed simultaneously

[1] For each arity greater than 0 there are countably many function and predicate symbols.

[2] If in doubt, the reader should consult [4] for the precise definitions.

$\dfrac{\alpha}{\alpha_1 \atop \alpha_2}$	$\dfrac{\beta}{\beta_1 \mid \beta_2}$	$\dfrac{\gamma}{\gamma(x)}$ where x is a free variable.	$\dfrac{\delta}{\delta(f(x_1,\ldots,x_n))}$ where $x_1,\ldots,x_n$ are the free variables occurring in δ and f is a new function symbol.

Table 1: Tableau rule schemes for different formula types.

(i.e. using the same unifier); in other words: when every attempt to construct a model that makes the root formula false leads to a contradiction.
Following Smullyan we divide the tableau expansion rules for signed formulæinto four classes: α–rules for propositional formulæof conjunctive type (e.g. $\mathsf{F}\, X \vee Y$), β–rules for propositional formulæof disjunctive type (e.g. $\mathsf{T}\, X \vee Y$), γ–rules for quantified formulæof universal type (e.g. $\mathsf{F}\,(\exists x)\phi(x)$), and finally δ–rules for quantified formulæof existential type (e.g. $\mathsf{T}\,(\exists x)\phi(x)$). The rule patterns are summarized in Table 1. It should be obvious how the α_i and β_i are computed from the semantical definitions. The quantifier rules have traditionally been working with ground terms, i.e. from a universal type formula like $\mathsf{T}\,(\forall x)\phi(x)$ the formula $\mathsf{T}\,\phi(t)$ may be inferred, where t is any ground term; and from an existential type formula like $\mathsf{T}\,(\exists x)\phi(x)$ the formula $\mathsf{T}\,\phi(c)$ may be inferred, where c is a Skolem constant not occurring on the current branch. Since a proof is only found when the right ground terms are "guessed", these rules are a source of much indeterminism, which in turn inevitably leads to expensive backtracking. Recent versions of tableau systems [4] therefore work with free variables that are instantiated *on demand*, i.e. when a branch is closed. We remark that the δ–rule we are using here is more liberal than that used in [4] and has recently been proposed and proved sound by Hähnle and Schmitt [7]. We will see in Section 3 that the availability of a liberal free version of tableaux is crucial for efficient equality handling.
For an example of a proof tree see Section 4. To achieve completeness some additional mechanism for handling equality must be provided. In the next section we review the most important approaches.

2 Analysis of Previous Approaches

Jeffrey's Approach

Jeffrey's approach is a very natural and straightforward way of adding equality to semantic tableaux. It is described in [8]; a summary can be found in [12]. The method is based on semantic tableaux without free variables. Therefore the ground versions of the quantifier rules are being used. As noted above it must be possible to add all the additional formulæ to a branch that are valid in canonical models. For this purpose Jeffrey introduced the following additional tableau expansion rules:
If a branch B has a formula $\phi(t)$ on it and an equality $\mathsf{T}\,(t \approx s)$ or $\mathsf{T}\,(s \approx t)$, then $\phi(s)$ may be added to B, where $\phi(s)$ is constructed by substituting one of the occurrences of t in $\phi(t)$ by s.

$$\frac{\mathsf{T}\,(t \approx s) \quad \phi(t)}{\phi(s)} \qquad \frac{\mathsf{T}\,(s \approx t) \quad \phi(t)}{\phi(s)}$$

Example 2.1 *Supposed the formulæ* $\mathsf{T}\,(a \approx b)$ *and* $\mathsf{T}\,P(a,a)$ *are on a branch B, by the application of Jeffrey's new expansion rule the new formulæ* $\mathsf{T}\,P(a,b)$, $\mathsf{T}\,P(b,a)$ *and* $\mathsf{T}\,P(b,b)$ *can be added to B. Note that it is not possible to derive* $\mathsf{T}\,P(b,b)$ *in a single step.*

Besides the expansion rules there is an additional closure rule. As before a branch B is closed if it contains complementary formulæ$\mathsf{T}\,G$ and $\mathsf{F}\,G$. But additionally it is also closed if it contains an inequality $\mathsf{F}\,(t \approx t)$, where t is any ground term.
The new expansion rules are symmetrical and their application is completely unrestricted. This leads to an enormous number of irrelevant formulæthat can be

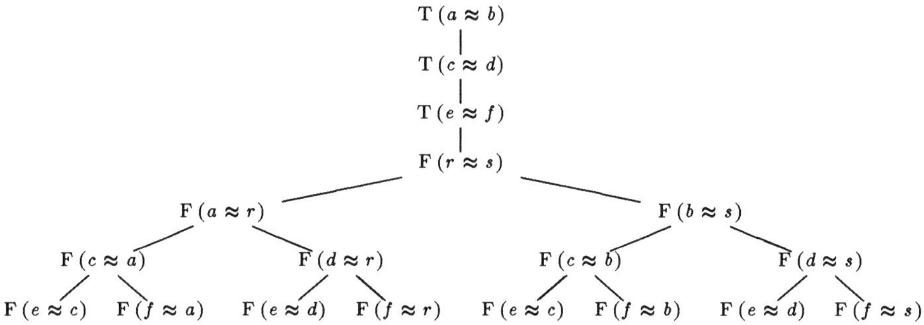

Figure 1: The disadvantage of Reeves's method: Three equalities and one inequality result in eight branches; and even more branches could be added to the tableau.

added. For example, if a branch contains formulæT $(f(a) \approx a)$ and T $P(a)$, then it can subsequently be expanded by all the formulæin $\{$T $P(f(a))$, T $P(f(f(a)))$, T $P(f(f(f(a))))$, $\ldots\}$. According to Reeves [12] Jeffrey's method is therefore "useless in practice".

Reeves's Approach

Reeves uses an expansion rule that is based on the fact that in a canonical model where T $P(a_1, \ldots, a_n)$ and F $P(b_1, \ldots, b_n)$ or F $(f(a_1, \ldots, a_n) \approx f(b_1, \ldots, b_n))$ is valid at least one of the formulæ F $(a_1 \approx b_1)$, ..., F $(b_1 \approx b_n)$ has to be valid, yielding the following equality expansion rules:

$$\frac{\mathsf{T}\ P(a_1, \ldots, a_n) \qquad \mathsf{F}\ P(b_1, \ldots, b_n)}{\mathsf{F}\ (a_1 \approx b_1 \wedge \ldots \wedge a_n \approx b_n)} \qquad \frac{\mathsf{F}\ (f(a_1, \ldots, a_n) \approx f(b_1, \ldots, b_n))}{\mathsf{F}\ (a_1 \approx b_1 \wedge \ldots \wedge a_n \approx b_n)}$$

Expanding the formula F $(a_1 \approx b_1 \wedge \ldots \wedge a_n \approx b_n)$ results in n new subbranches each containing one of the inequalities F $(a_i \approx b_i)$ $(1 \leq i \leq n)$.
Without doubt Reeves's approach has some advantages. The search for the closure of a branch is more directed. Only atomic formulæthat potentially close a branch are used for expansion.
But the method also has a big disadvantage: If a branch contains several equalities and an inequality, the problem shown in Figure 1 occurs. As the new expansion rule can as well be applied to pairs of equalities and inequalities, a large number of new branches is added to the tableau. It grows exponentially with the number of equalities on the branch. As a consequence, Reeves's approach is not suitable for implementation.
Nevertheless we adopted the idea of transforming pairs of potentially closing atoms into disjunctions of inequalities. In our approach, however, building disjunctions from pairs of equalities and inequalities is avoided.

Popplestone's Approach

Popplestone's approach [11] is based on Jeffrey's method. It uses the same tableau expansion rules. The only improvements made essentially regard implementation. The main idea is to attach a graph to each formula in a tableau. The nodes of this graph are labelled with terms that occur above the formula on the branch or that can be derived from these terms using equalities present on the branch. Two nodes in the graph are adjacent if one can be transformed into the other by a single equality application. Thus, two nodes in the same graph are known to be equal if and only if they are connected. Therefore, a branch is closed if it contains a formula F $(t \approx s)$ which has a graph attached to it proving t to be equal to s. A closure caused by complementary atoms can be found in a similar way.

This method has several advantages. Any heuristic can be used to search in the graphs and to expand them. Also, the information about the equality of certain terms can be reused in new formulægenerated by application of a tableau expansion rule.
But there is still a problem: information about the equality of all the terms in all the formulæin a tableau is generated. As will be shown later, this is not necessary. It suffices to take a closer look on some of the *atomic formulæ* and the terms they contain.

Fitting's Approach

Like Popplestone's method Fitting's is also an improvement of Jeffrey's approach. In [4] a complete implementation of a tableau-based theorem prover with equality in PROLOG is contained.
Fitting combines free variable tableaux with treatment of equality. The advantages of using free variables have already been discussed — only free variable substitutions that are necessary are performed. But with equalities present in a tableau, substitutions are also necessary at other points than closure of branches. There might be equality rule applications that require substitutions of free variables. Fortunately, these substitutions are easy to obtain, namely in a similar way as those substitutions that are needed to close a branch. If an equality $\mathsf{T}\ (t \approx s)$ is to be applied to a formula $\mathsf{T}\ Z(t')$, the application of an MGU σ of t and t' to the tableau is sufficient and necessary. Fitting's tableau expansion rules are thus the rules shown on the right. Fitting also addresses the problem of indeterminism embodied in the tableau expansion rules. As indeterminism is difficult to implement and inevitably leads to expensive backtracking, its elimination is an important improvement. To achieve this the application of equality rules is separated from the application of the standard tableau expansion rules. The application of the latter has to be exhausted before equalities are being applied to a tableau. In this process γ–rules may be applied to a certain formula only a finite number of times, say q. As a consequence, completeness has to be relativized: If ϕ is an unsatisfiable formula, then a closed tableau for ϕ will only be found when the limit on the number of γ–rule applications is sufficiently high. In practice one could implement an incremental prover which tries to find a proof by increasing the γ–limit after each try.

$$\frac{\begin{array}{c}\mathsf{T}\ (t \approx s)\\ \phi(t')\end{array}}{(\phi(s))\sigma} \qquad \frac{\begin{array}{c}\mathsf{T}\ (s \approx t)\\ \phi(t')\end{array}}{(\phi(s))\sigma}$$

where σ is an MGU of t and t'.

Fitting proves his method to be complete in the above sense if only the order of rule applications is "fair", i.e. if there is a standard tableau expansion rule that can be applied to a certain formula, this application will eventually happen.
Besides the elimination of indeterminism, there is another important point involved here. After the first stage of tableau expansion is finished and the tableau is exhausted (observing the limit q for γ–rule applications), it is sufficient to expand the tableau in the second stage solely by equality applications to *atomic formulæ*. That greatly decreases the number of possible equality applications.
In addition, the method has advantages for implementation, as it is not necessary to switch between the application of equalities and the application of other expansion rules. Thus an appropriate data structure for dealing with equalities can be chosen.
Indeterminism is only partly resolved yet, since there may be several ways to close a branch or to apply equalities that require different free variable substitutions. Fitting's system subsequently tries each of these possibilities by means of PROLOG's backtracking. The resulting inefficiency presumably is the main reason for his concluding remark "...But remember the resulting system, while complete, may take years to prove anything interesting. Good heuristics are vital now."
Fitting proposes something similar to a heuristic: an orientation is assigned to the equalities. They are applied only from left to right. This makes it possible to avoid those applications of equalities like $\mathsf{T}\ (f(a) \approx a)$ which yield a large number of new formulæ. But there is a drawback, of course. Completeness would be lost if

there were not the following new expansion rule: The formula $\mathsf{T}\,(\forall x)(x \approx x)$ may be added to every branch. Unfortunately, this enables one to reverse the orientation of equalities. If a branch contains an equality $\mathsf{T}\,(f(a) \approx a)$, one can add $\mathsf{T}\,(\forall x)(x \approx x)$ and deduce $\mathsf{T}\,(x \approx x)$ from the latter. Applying the first equality to the *left* side of $\mathsf{T}\,(x \approx x)$ yields $\mathsf{T}\,(a \approx f(a))$. Therefore, this technique can only be described as a heuristic that specifies a preferred ordering of equalities in which they are tried first.

Completion–Based Methods

An essential prerequisite for using methods that are based on the completion of an equality theory is that the equalities are universally closed. This condition does not hold for the equalities in tableaux with free variables; there has to be a *single* substitution which, when applied, allows to close all branches simultaneously.
A method of equality theory completion would be needed, where the resulting complete theory not only provides the information whether two terms are equal, but also the set of all free variable substitutions that allow to prove equality of these terms using the initial equalities. This problem has, to our best knowledge, not been addressed so far.
On the other hand, for adding equality to ground semantic tableaux rewrite systems may well be used. In [2] an example of such a system is given. The problem of course is that the success of the rewrite part depends on the right guesses of terms in γ–rule applications. This will inevitably lead to frequent backtracking over the whole tableau and is hopelessly inefficient in practice.

3 An Improved Equality Module for Tableau–Based Provers

Two Separate Tableau Expansion Stages

From Fitting's approach we adopted the idea of separating the tableau expansion into two stages. In the first stage the standard rules are applied until the tableau is exhausted (observing the limit for γ–rule applications). Thus, in the second stage, it is possible to restrict equality applications and to avoid the generation of useless new formulæ. Equality rule applications can be limited to inequalities $\mathsf{F}\,(t \approx s)$ and pairs of atomic formulæ$\langle \mathsf{T}\,P(t_1,\ldots,t_n), \mathsf{F}\,P(s_1,\ldots,s_n)\rangle$ that potentially close a branch, where P is not the equality predicate $\approx$. In particular, for preserving completeness it is not necessary to apply equalities to other equalities.
When a tableau is exhausted, most of its formulæare not needed any more. From then on only equalities, inequalities, and pairs of potentially closing atoms are of interest. Therefore the tableau can be left and equality reasoning is done on a more suitable (and efficient) data structure. For this optimization it is crucial that backtracking to the first building stage of the tableau does not occur too frequently.

Disjunctions of Inequalities

The new data structure mentioned in the previous section consists of two sets of certain formulæfor each branch B of an exhausted tableau. The first one is the set $\mathrm{Gl}(B)$ defined as

$$\mathrm{Gl}(B) := \{s \approx t \;:\; \mathsf{T}\,(s \approx t) \in B\},$$

i.e. the set of all equalities present on B. The second is the set $\mathrm{Dis}(B)$, which consists of disjunctions of inequalities. It embodies the two remaining types of important formulæ. For every pair $\langle \mathsf{T}\,P(t_1,\ldots,t_n), \mathsf{F}\,P(s_1,\ldots,s_n)\rangle$ of atoms that potentially close B, in $\mathrm{Dis}(B)$ there is the n–place disjunction $t_1 \not\approx s_1 \vee \ldots \vee t_n \not\approx s_n$; and for every inequality $\mathsf{F}\,(t \approx s)$ on B, in $\mathrm{Dis}(B)$ there is the (one–place) disjunction $t \not\approx s$.

Example 3.1 *Consider the branch shown in the left part of Figure 2. The corresponding set of equalities is $\{a \approx b,\ b \approx c\}$; the set of disjunctions of inequalities is $\{a \not\approx c\}$.*

A tableau T with branches $B_1, \ldots, B_k$ is closed if there is a substitution σ of free variables such that for each branch B_i there is a disjunction $D_i \in \mathrm{Dis}(B_i)$ which can be proved to be unsatisfiable using the equalities in $\mathrm{Gl}(B_i)$. Soundness of this new closure rule is justified by the following consideration that holds for each branch B_i: If a disjunction $D_i = (t \not\approx s)$ has emerged from an inequality $\mathsf{F}\,(t \approx s) \in B_i$ and D_i can be proved to be unsatisfiable, we have in fact proved that, when σ is applied, $\mathsf{F}\,(t \approx t)$ is valid in every canonical model of B_i, which is clearly a contradiction. If, on the other hand, $D_i = (t_1 \not\approx s_1 \vee \ldots \vee t_n \not\approx s_n)$ has emerged from a pair of potentially complementary atoms $\langle \mathsf{T}\,P(t_1, \ldots, t_n), \mathsf{F}\,P(s_1, \ldots, s_n)\rangle$, unsatisfiability of D_i implies that $\mathsf{F}\,P(t_1, \ldots, t_n)$ holds and the branch can be closed as usual.
A straightforward method to prove an inequality $t \not\approx s$ to be unsatisfiable is to calculate step by step the equivalence classes of the terms t and s using equalities in $\mathrm{Gl}(B_i)$ and look in these classes for common elements.
One has to take into account that different free variable substitutions may lead to different equivalence classes. The problem of finding a suitable substitution that allows to refute all inequalities in a disjunction simultaneously is discussed in the next section.
The transformation of pairs of potentially closing atoms into disjunctions of inequalities corresponds to the application of Reeves's expansion rule; but the disadvantage of Reeves's approach is avoided, as disjunctions of inequalities are no longer allowed to be built up from pairs of equalities and inequalities.
The gain of efficiency in closing a branch by calculating equivalence classes instead of adding new formulæis illustrated by the following example.

Example 3.2 *If the branch shown in the left part of Figure 2 is expanded and closed according to Jeffrey's method, the information that c can be derived from a has to be generated twice to derive the formula $\mathsf{T}\,P(c,c)$, that is used to close the branch. Irrelevant formulæ, such as $\mathsf{T}\,P(a,c)$, are repeatedly added to the branch. With no heuristic at hand to avoid useless equality applications, up to eight new formulæare added to close the branch.*
In Figure 2 the branch is closed based on the transformation into disjunctions of inequalities. There is only one disjunction, containing two identical inequalities, that are merged into one. It is possible to prove the inequality to be unsatisfiable after no more than two new terms have been derived.

Example 3.2 demonstrates that for Jeffrey's method the number of formulægenerated in order to close a branch grows exponentially with the arity n of the involved predicate symbols, even if there is only one pair of potentially closing atoms, if for each occurring term there is only one equality that can be applied to it, and if no free variable substitutions have to be applied. It is very difficult to avoid the worst case, because a heuristic would be needed to recognize irrelevant formulæ. Only in the best case the number of generated formulægrows linearly with n, whereas the number of equality applications, when the new method is used, *always* grows linearly, even in the worst case.
In addition it is possible to use any heuristic to direct the search for common elements in the equivalence classes.

Most General Substitutions

The problem of finding a free variable substitution that allows to close all inequalities in one of the disjunctions of each branch simultaneously will be addressed next. A possible but inefficient solution would be to subsequently try each grounding substitution.
Fortunately, there is a much better method, namely, not to restrict the search to grounding substitutions. Instead, a closer look is taken at those substitutions σ that are an MGU of one side of an equality and an already derived term t or one of its subterms, i.e. that allow to apply this equality to t. This proves to be sufficient for preserving completeness as the following holds: If the term s can be derived from t

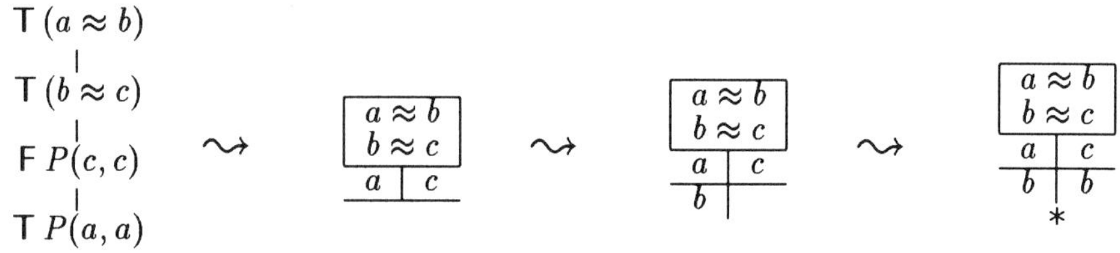

Figure 2: The branch of a tableau being closed using the method based on the transformation into disjunctions of inequalities and the calculation of equivalence classes. The new data structures are represented by a box containing the equalities having below it the two sides of the inequality with the terms in the equivalence classes of the left– and the right–hand term computed so far.

with the help of σ, then s subsumes every term s' that can be derived from t with the help of a grounding substitution τ more special than σ, i.e. there is a substitution σ' such that $s' = s\sigma'$. Therefore, an inequality that can be closed using the term s' can be closed using s as well.
Since equivalence classes may contain non–closed terms when they are built up using most general substitutions, an inequality is not only closed if the equivalence classes corresponding to its left– and right–hand side terms have a common element, but it is already closed if two *unifiable* terms occur in these classes.

Example 3.3 *$\sigma = \{y \leftarrow x\}$ is a most general substitution that allows to apply the equality* T *$(f(x) \approx x)$ to the term $g(f(y))$. The term $g(x)$ would be the result of this application. If the grounding substitution $\sigma' = \{y \leftarrow a,\ x \leftarrow a\}$, which is more special than σ, was used, the term $g(a)$ could be derived from $g(f(y))$.*
Now, if an inequality becomes closed, because $g(a)$ is a common element of the equivalence classes corresponding to its left– and right–hand sides, that inequality may as well be closed using $g(x)$, since $g(x)$ subsumes $g(a)$.

Breadth–First–Search for Substitutions

Using most general substitutions to build up equivalence classes leads to a search tree for each side of an inequality whose edges are coloured with most general substitutions and whose nodes are the sets of terms that can be derived from the present equalities using terms already on the branch and the substitution on its incoming edge.
The edge leading to the root node of a search tree is labelled with the most general of all, the empty substitution and the root node itself consists of the terms that can be derived in one step without applying any substitutions, in particular including the starting term itself. The tree branches when the application of equalities to a term on one of the leaf nodes requires additional substitutions that are more special than those associated with its incoming edge.
Each of the terms in a search tree is associated with the substitution σ on its incoming edge and as well with the substitutions that are more special than σ. Substitutions become more special towards the leaves. Figure 3 shows an example.
One may prove that while using these search trees all substitutions that allow to close an inequality can be found. A depth–first–search could easily be implemented based on PROLOG's backtracking.
In general, however, search trees have infinitely long branches. It is therefore very difficult to realize when another branch should be tried, the problem is in fact undecidable. In addition, it is often necessary to look for more than one closing substitution for each inequality, as a disjunction is only closed provided that *all* substitutions used to close its inequalities are compatible. Consequently, heuristics would have to be used in order to decide at which point the search in a branch should be abandoned and backtracking should be initiated. Obviously, either these

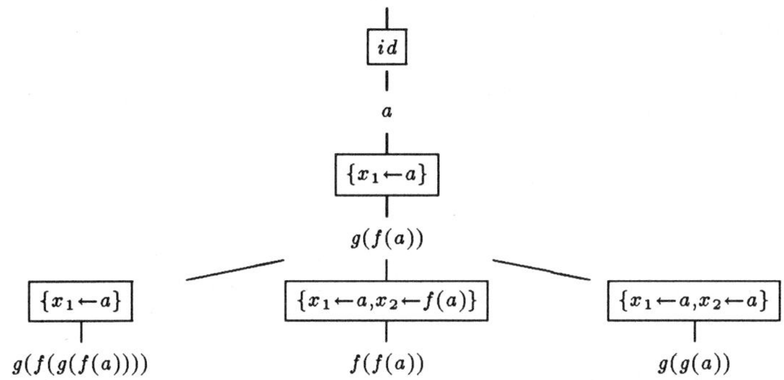

Figure 3: The search tree that is built from the right side of the inequality $g(g(a)) \not\approx a$ if $\mathrm{Gl}(B)$ contains the equalities $g(f(x_1)) \approx x_1$ and $g(x_2) \approx f(x_2)$.

heuristics react too quickly such that many possibilities for closing inequalities are being missed; or they react too slowly and one has to wait too long before the right branches are tried.
All these problems are avoided if breadth–first–search is used, i.e. all branches are searched simultaneously. Breadth–first–search is much more powerful, because any heuristic can be used to push ahead the search in some of the branches more quickly than in others. Therefore, virtually any other search method can easily be simulated. Since breadth–first–search cannot be based on PROLOG's backtracking, it is, however, slightly more difficult to implement in PROLOG.

Data Structures for Breadth–First–Search

Breadth–first–search as proposed in the previous section can be implemented using sets $\langle t\rangle_B$ that contain all the equivalence classes of a term t on a branch B associated with different substitutions. The elements s_σ of $\langle t\rangle_B$ are terms labelled with the substitution which is needed to derive them from t using equalities in $\mathrm{Gl}(B)$. If, for example, the application of $\{x_1 \leftarrow a\}$ leads to an equivalence class of the term a that contains $g(f(a))$, the element $g(f(a))_{\{x_1\leftarrow a\}}$ is in $\langle a\rangle_B$.
One can view the sets $\langle t\rangle_B$ as a representation of the search trees introduced in the previous section; $s_\sigma \in \langle t\rangle_B$ if and only if s occurs on some node of the search tree whose incoming edge is coloured with σ.
Thus all required equivalence classes for checking an inequality are handled simultaneously. $\langle t\rangle_B$ is in general an infinite set, but there is a sequence of approximations $(\langle t\rangle_B^n)_{n\geq 0}$ to $\langle t\rangle_B$ that can be computed with a deterministic algorithm.
The only element in the first set $\langle t\rangle_B^0$ is t_{id}. The additional elements in $\langle t\rangle_B^{n+1}$ are those that can be derived in one step from a certain element $t' \in \langle t\rangle_B^n$ using the equalities in $\mathrm{Gl}(B)$, where $t' = \mathcal{H}(\langle t\rangle_B^n)$ is chosen by a certain heuristic $\mathcal{H}$. Elements that are subsumed by others are not included, since an inequality that can be proved to be unsatisfiable using a term $s'_{\sigma'}$ can as well be proved to be unsatisfiable using any term s_σ that subsumes $s'_{\sigma'}$.

Definition 3.4 (Subsumption) *Suppose s, s' are terms and σ, σ' are substitutions. s_σ* **subsumes** *$s'_{\sigma'}$ if s and s' are unifiable with an MGU τ such that $s\tau = s'$ and both σ and τ are more general than σ'.*

Example 3.5 *$f(x)_{\{y\leftarrow a\}}$ subsumes $f(a)_{\{x\leftarrow a,y\leftarrow a\}}$; $a_{\{x\leftarrow y\}}$ subsumes $a_{\{x\leftarrow b,y\leftarrow b\}}$. On the other hand, $a_{\{x\leftarrow f(y)\}}$ does not subsume $a_{\{x\leftarrow f(b)\}}$; both of these terms do,* *however, subsume* $a_{\{x \leftarrow f(b), y \leftarrow b\}}$.

Definition 3.6 (Sequence of Sets $\langle t \rangle_B^n$) *The sets* $\langle t \rangle_B^n$ *are inductively defined as follows:*

- $\langle t \rangle_B^0 = \{t_{id}\}$
- *Let the set* Θ_n *contain all the elements from* $\langle t \rangle_B^n$ *and in addition the terms* r_τ *that can be derived in one step from* $s_\sigma = \mathcal{H}(\langle t \rangle_B^n)$, *where* r_τ *can be derived in one step from* s_σ *if*

  1. τ *is an MGU of a subterm of* s *and one side of an equality* $G \in \mathrm{Gl}(B)$;
  2. r *can be derived from* $s\tau$ *by application of* $G\tau$;
  3. σ *is more general than* τ.

  Then $\langle t \rangle_B^{n+1}$ *contains all elements from* Θ_n *that are not subsumed by another element in* Θ_n. *If there are several elements subsuming each other, an arbitrary one is chosen.*

Example 3.7 *Table 2 shows the computation of* $\langle a \rangle_B^n$ *(for* $n = 0, 1, 2$*) using the set of equalities* $\mathrm{Gl}(B) = \{g(f(x_1)) \approx x_1,\ g(x_2) \approx f(x_2)\}$.

n	$\langle a \rangle_B^n$	$\mathcal{H}(\langle a \rangle_B^n)$	Θ_n
0	a_{id}	a_{id}	a_{id} $g(f(a))_{\{x_1 \leftarrow a\}}$
1	a_{id} $g(f(a))_{\{x_1 \leftarrow a\}}$	$g(f(a))_{\{x_1 \leftarrow a\}}$	a_{id} $g(f(a))_{\{x_1 \leftarrow a\}}$ $a_{\{x_1 \leftarrow a\}}$ $g(f(g(f(a))))_{\{x_1 \leftarrow a\}}$ $g(g(a))_{\{x_1 \leftarrow a, x_2 \leftarrow a\}}$ $f(f(a))_{\{x_1 \leftarrow a, x_2 \leftarrow f(a)\}}$
2	a_{id} $g(f(a))_{\{x_1 \leftarrow a\}}$ $g(f(g(f(a))))_{\{x_1 \leftarrow a\}}$ $g(g(a))_{\{x_1 \leftarrow a, x_2 \leftarrow f(a)\}}$ $f(f(a))_{\{x_1 \leftarrow a, x_2 \leftarrow a\}}$		

Table 2: Computation of $\langle a \rangle_B^n$ ($n = 0, 1, 2$) (Example 3.7).

Definition 3.8 (Closed Tableau) *A tableau* T *with branches* $B_1, \ldots, B_k$ *is* **closed** *if for each branch* B_i $(1 \le i \le k)$ *there is a disjunction*

$$D_i = (t_{i1} \not\approx s_{i1} \vee \ldots \vee t_{in_i} \not\approx s_{in_i}) \in \mathrm{Dis}(B_i)$$

such that for $1 \le j \le n_i$ *elements*

$$r^{ij}_{\rho_{ij}} \in \langle t_{ij} \rangle_{B_i}^{l_{ij}} \quad \textit{and} \quad \overline{r}^{ij}_{\overline{\rho}_{ij}} \in \langle s_{ij} \rangle_{B_i}^{\overline{l}_{ij}}$$

can be found for some $l_{ij}, \overline{l}_{ij} \ge 0$, *where* r^{ij} *and* $\overline{r}^{ij}$ *are unifiable with an MGU* $\overline{\overline{\rho}}_{ij}$, *and there is a grounding substitution* σ *such that all of the substitutions* $\rho_{ij}, \overline{\rho}_{ij}, \overline{\overline{\rho}}_{ij}$ $(1 \le i \le k, 1 \le j \le n_i)$ *are more general than* σ.

If one shows a tableau to be closed, the substitution σ mentioned in the above definition allows to prove one disjunction of inequalities for each branch of the tableau to be unsatisfiable.
One can check whether a tableau T is closed according to Definition 3.8 by gradually computing the sets $\langle t\rangle^0_{B_i}$, $\langle t\rangle^1_{B_i}$, ... for every term t occurring in Dis(B_1), ..., Dis(B_k), where $B_1,\ldots,B_k$ are the branches of T. Definition 3.6 provides an effective way to do this.

Heuristics for Term Search

Before soundness and completeness of the method based on the sets $\langle t\rangle^n_B$ can be stated, we have to say something about the heuristics. The heuristic $\mathcal{H}$ for choosing an element from $\langle t\rangle^n_B$ to which equalities are applied has to be "fair" in the following sense:

Definition 3.9 (Fair Heuristic) *A heuristic $\mathcal{H}$ is* **fair** *if for each term t, each $n \geq 0$, and each element $s_\sigma \in \langle t\rangle^n_B$ there is an $m \geq 0$ such that $\mathcal{H}(\langle t\rangle^m_B)$ subsumes s_σ.*

For example, the heuristic that always chooses the syntactically shortest term which has not been chosen before is fair. We propose that the heuristic in Definition 3.6 is fair in the above sense. The heuristic we have been using for our implementation is described below.

Theorem 3.10 *Suppose T is a tableau with root formula ϕ.
If T is closed according to Definition 3.8, then ϕ is not satisfiable in a normal model.
If ϕ is not satisfiable in a normal model and if the limit q for γ-rule applications is sufficiently high, T is closed according to Definition 3.8.*

The proof is based on the following lemma which clarifies the connection between the sets $\langle t\rangle^n_B$ and the equivalence classes of the term t that are associated with different substitutions.

Lemma 3.11 *If $r_\sigma \in \langle t\rangle^n_B$ and σ is more general than the grounding substitution τ, then the equivalence class of t that is associated with τ contains the term $r\tau$.
If the equivalence class of t that is associated with the grounding substitution τ contains a term s, then for some $n \geq 0$ there is an element $r_\sigma \in \langle t\rangle^n_B$ that subsumes s_τ.*

For our implementation we used the following heuristic for choosing the next element from $\langle t\rangle^n_B$ to which equalities are applied:

Definition 3.12 (Implemented Heuristic) *The criteria for selection, ordered by their importance, are as follows:*

1. *Elements that have been chosen before are not considered again.*
2. *The term weight $G(s)$ and the distance $D(s) = G(s') - G(s)$ to the weight of the term s' from which s has been derived.*[3] *Terms s are preferred that*
   (a) *have a positive weight distance $D(s)$,*
   (b) *have a lower weight $G(s)$,*
   (c) *have a higher weight distance $D(s)$.*
3. *The number of steps necessary to derive a term. Terms that can be derived in fewer steps are preferred.*

We remark that this heuristic is fair in the sense of Definition 3.9. The term weight is by default given by the lexicographic length of the terms, but can easily be changed in order to prove theorems over specific domains.

[3] $D(s)$ is not defined for the single element t_{id} in $\langle t\rangle^0_B$, since it is not derived from a term. However, t_{id} is always chosen, as it is the only element in $\langle t\rangle^0_B$.

Universal Formulæ

An equality has often to be applied more than once in order to close a branch, each time with different substitutions for the variables occurring in it. A typical example is the associativity axiom $As = (\forall x)(\forall y)(\forall z)[(x \cdot y) \cdot z \approx x \cdot (y \cdot z)]$ from group theory. In most cases it has to be applied several times with different instantiations of x, y and z to prove even simple theorems of group theory.
In semantic tableaux the mechanism to do so usually is to apply the γ–rule more than once to As and thus generate several instances of As, each with different free variables substituted for x, y and z.
Consequently, to prove a theorem the γ–limit q has to be at least as high as the maximal number of necessary applications of the same equality with different substitutions for the free variables it contains. Before equalities can be applied, however, the tableau has to be exhausted, but the higher the limit q is, the more branches have to be closed and the bigger the tableau becomes. Moreover, it is quite difficult to choose the limit q appropriately, because one does not know how many equality applications will be needed.
The problem could be avoided if equations in the initial tableau were not allowed to occur nested in other formulæ, but appeared only on the top–level. We did, however, not want to employ this restriction in order to allow for a natural formulation of problems.[4] Nevertheless, the problem can at least partly be solved if one is able to recognize formulæand in particular equalities that are "universal", i.e. that can be used repeatedly with different substitutions for the variables they contain:

Definition 3.13 (Universal Formula) *Suppose ϕ is a formula on some tableau branch B. If $\phi = \mathsf{T}\, F$ for some F let $\phi_u = F$ else if $\phi = \mathsf{F}\, F$ let $\phi_u = \neg F$.
ϕ is* **universal** *with respect to x if the following holds for every normal model* **M** *and every grounding substitution σ:*

$$\textit{If } \mathbf{M} \models B\sigma, \textit{ then } \mathbf{M} \models ((\forall x)\phi_u)\sigma.$$

The problem of recognizing universal formulea is in general undecidable. However, a wide and important class of universal formulæcan be recognized easily:

Theorem 3.14 *A formula ϕ on a branch B is universal with respect to x if either*

1. *ϕ has been generated by applying a γ–rule to a γ–formula and x is the free variable that has been substituted for the variable bound by the leading quantifier in the γ–formula or*
2. *ϕ has been generated by applying an α– or a γ–rule to a formula which is universal with respect to x.*

Once formulæare recognized as being universal this knowledge can be taken advantage of in the following way:

- If an equality $\mathsf{T}\,(t \approx s)$ on a branch B is universal with respect to a variable x, the equality is implicitly universally quantified by marking all occurrences of x in the equality as being "universal".[5] Thus, later on, the variable x does not have to be instantiated to apply the equality.
- To make use of the universal validity of other formulæthe variables with respect to which they are universal are substituted by new ones that hitherto do not occur on the tableau. The occurrences of the same variable in different formulæare substituted by different new variables. This is done before the sets Dis(B) are being built. Therefore, it is less likely that a branch is prevented from being closed by a substitution which interferes with the closure of other branches.

[4] Cf. the example at the end of the next section.
[5] To mark an occurrence of a variable x as being "universal" it is given in the following a different typeface.

Theorem 3.10 still holds if one takes advantage of the universal validity of formulæ, i.e. soundness is preserved. Since elements subsuming each other are removed, the sets $\langle t\rangle_B^n$ are still finite, although $\mathrm{Gl}(B)$ may contain implicitly universally quantified formulæ.

4 Examples

In this section we are going through an example in detail which shows how the method works and which also demonstrates most (though not all) of its advantages. The formula we are going to establish as a tautology is

$$(\forall x)[g(f(x)) \approx c \wedge (f(x) \approx g(x) \vee \phi)] \supset g(g(a)) \approx g(f(b)),$$

where ϕ is an arbitrary formula that can be proved to be unsatisfiable without the use of equalities.
It will turn out that a γ–limit of 1 is sufficient, thus γ–formulæwill be removed after the first rule application to them.
The first stage of tableau expansion yields the tableau shown in Figure 4. According to Theorem 3.14 some of the formulæare universal with respect to x. Since the formulæmarked with an asterisk are no longer present on the tableau, the expansion using logical rules is finished, but the tableau cannot be closed.

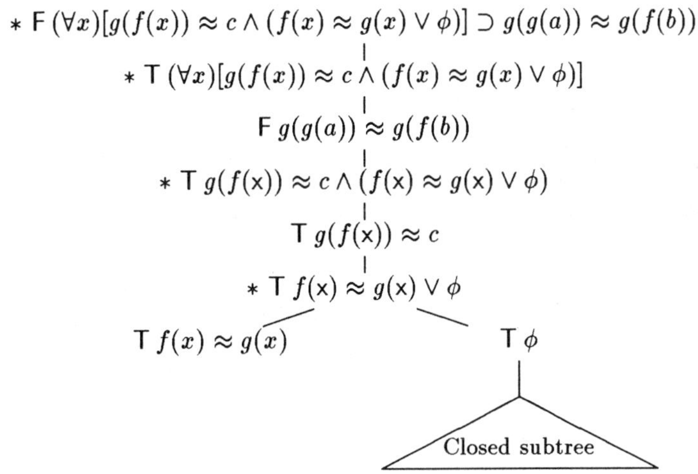

Figure 4: Expanded tableau after first stage.

At this stage the equality reasoning part is invoked. Let us refer with the letter B to the open branch. Then B is converted into the following data structure:

$$\begin{aligned} \mathrm{Gl}(B) &= \{g(f(\mathrm{x})) \approx c,\ f(x) \approx g(x)\} \\ \mathrm{Dis}(B) &= \{g(g(a)) \not\approx g(f(b))\} \end{aligned}$$

This can be conveniently represented as in the top left part of Figure 5. The algorithm now tries to find unifiable elements in the sets $\langle g(g(a))\rangle_B^n$ and $\langle g(f(b))\rangle_B^n$. In the first step by equation (2) the terms $g(f(a))$ and $f(g(a))$ are derived from $g(g(a))$. If depth–first– instead of a breadth–first–search was used, and the useless term $f(g(a))$ was derived first, this would prevent one from deriving $g(f(a))$, since the required substitutions $\{x \leftarrow a\}$ and $\{x \leftarrow g(a)\}$ are incompatible. Backtracking would become necessary. Instead, both $f(g(a))_{\{x \leftarrow g(a)\}}$ and $g(f(a))_{\{x \leftarrow a\}}$ are included in $\langle g(g(a))\rangle_B^1$.

In the second step, first, the useless elements $f(f(b))_{\{x \leftarrow f(b)\}}$, $g(g(b))_{\{x \leftarrow b\}}$ are added to $\langle g(f(b)) \rangle_B^1$ using equality (2). Then the term c is derived from $g(f(b))$ using equation (1). Since equation (1) is marked as being universal with respect to x, the required substitution $\{x \leftarrow b\}$ has not been actually performed. Thus the resulting term c is indexed with the empty substitution. In the next step we can see this is crucial, because the substitution $\{x \leftarrow a\}$ is needed to derive c from $g(f(a))$. If equality (1) had not been universal, the derivation would be stuck at this point with two incompatible substitutions. Backtracking to the first tableau expansion stage would be required to produce another copy of equation (1). Moreover, a γ–limit of 1 would not be sufficient, causing the fully expanded tableau and in particular the subtree for T ϕ to be much bigger. In addition two copies of the subtree would have to be closed. If ϕ is a complex formula, this is very expensive.
Now, the disjunction $g(g(a)) \not\approx g(f(b))$ and thus branch B can be closed using the compatible elements $c_{\{x \leftarrow a\}} \in \langle g(g(a)) \rangle_B^2$, $c_{id} \in \langle g(f(b)) \rangle_B^1$.

(1) $g(f(x)) \approx c$ (2) $f(x) \approx g(x)$	
$g(g(a))$	$g(f(b))$

$\rightsquigarrow$

(1) $g(f(x)) \approx c$ (2) $f(x) \approx g(x)$	
$g(g(a))$	$g(f(b))$
$f(g(a))_{\{x \leftarrow g(a)\}}$	
$g(f(a))_{\{x \leftarrow a\}}$	

(1) $g(f(x)) \approx c$ (2) $f(x) \approx g(x)$	
$g(g(a))$	$g(f(b))$
$f(g(a))_{\{x \leftarrow g(a)\}}$	$f(f(b))_{\{x \leftarrow f(b)\}}$
$g(f(a))_{\{x \leftarrow a\}}$	$g(g(b))_{\{x \leftarrow b\}}$
	c_{id}

$\rightsquigarrow$

(1) $g(f(x)) \approx c$ (2) $f(x) \approx g(x)$	
$g(g(a))$	$g(f(b))$
$f(g(a))_{\{x \leftarrow g(a)\}}$	$f(f(b))_{\{x \leftarrow f(b)\}}$
$g(f(a))_{\{x \leftarrow a\}}$	$g(g(b))_{\{x \leftarrow b\}}$
$c_{\{x \leftarrow a\}}$	c_{id}

Figure 5: Second stage of tableau expansion.

If the same formula was to be proven to be a tautology using Fitting's method, first of all, the γ–limit q would have to be raised and more branches would have to be closed in the subtree for T ϕ. In addition, more equality applications would be necessary to prove the inequality $g(g(a)) \not\approx g(f(b))$ to be unsatisfiable. If the same heuristic was used to select the terms and equalities that we used for our implementation, then about 15 equality applications would be needed. Several times free variable substitutions would be applied that prevent the branch from being closed resulting in backtracking.
If Jeffrey's approach were used, the number of equality rule applications would heavily depend on the order of the equalities (there would be four equalities on the branch). In the best case only two branches have to be closed; a slightly different order, however, can lead to several hundred branches.
Our implementation solves most of the problems stated by Pelletier [10] in a few seconds. Below Pelletier's 55th problem[6] is stated both in natural language and in first–orde logic:

[6]Although a γ–limit of $q = 1$ is sufficient, the resulting tableau is too large to be depicted here. Our implementation closes 33 branches, and succeeds in 4.18s in proving that the problem as it is stated above is a tautology (running on a SUN–4 SPARC SLC workstation).

$$
\begin{array}{c}
(\exists x)(L(x)\wedge K(x,a)) \\
L(a)\wedge L(b)\wedge L(c) \\
(\forall x)(L(x)\supset(x\approx a\vee x\approx b\vee x\approx c)) \\
(\forall x)(\forall y)(K(x,y)\supset H(x,y)) \\
(\forall x)(\forall y)(K(x,y)\supset\neg R(x,y)) \\
(\forall x)(H(a,x)\supset\neg H(c,x)) \\
(\forall x)(\neg(x\approx b)\supset H(a,x)) \\
(\forall x)(\neg R(x,a)\supset H(b,x)) \\
(\forall x)(H(a,x)\supset H(b,x)) \\
(\forall x)(\exists y)(\neg H(x,y)) \\
\neg(a\approx b) \\
\hline
K(a,a)
\end{array}
$$

Someone who lives in Dreadsbury Mansion killed Aunt Agatha. Agatha, the butler, and Charles live in Dreadsbury Mansion, and are the only people who live therein. A killer always hates his victim, and is never richer than his victim. Charles hates no one that Aunt Agatha hates. Agatha hates everyone except the butler. The butler hates everyone not richer than Aunt Agatha. The butler hates everyone Agatha hates. No one hates everyone. Agatha is not the butler.
Therefore: Agatha killed herself.

5 Conclusion

We have presented a method for handling equality within a first–order analytic tableaux framework that outperformances all previous approaches considerably. Recent tableau proving techniques, such as free variables, universal formulæ, and a liberalized δ–rule are made full use of. Although performance cannot compete with completion–based equality provers or resolution systems using paramodulation, we think that the effort is justified, since tableau–based systems offer superior possibilities for many non–classical logics. We stress that, in contrary to most other approaches, the equality theory needs not to be specified explicitly, but equalities may occur arbitrarily nested in formulæ.
Our considerations suggest that it might be better to abandon the depth–first branch–by–branch closure of tableaux in favour of a breadth-first approach which expands a tableau fully, before attempting to close all branches simultaneously.
Other topics for further investigations are the use of successful equality reasoning methods, such as completion–based algorithms (though one has to take into account the problem of non–universal variables), and implementation of many–valued "equality" predicates.

References

[1] B. Beckert. Konzeption und Implementierung von Gleichheit für einen tableau–basierten Theorembeweiser. Studienarbeit, Univ. Karlsruhe, July 1991.

[2] R. J. Browne. Ground term rewriting in semantic tableaux systems for first–order logic with equality. Technical Report UMIACS–TR–88–44, College Park, MD, 1988.

[3] M. Fitting. First-order modal tableaux. *JAR*, 4:191 – 213, 1988.

[4] M. Fitting. *First–Order Logic and Automated Theorem Proving.* Springer, 1990.

[5] R. Hähnle. Spezifikation eines Theorembeweisers für dreiwertige First–Order Logik. IWBS Report 136, Wiss. Zentrum, IWBS, IBM Deutschland, 1990.

[6] R. Hähnle. Towards an efficient tableau proof procedure for multiple–valued logics. In *Proc. WS Computer Science Logic, Heidelberg*, pp 248 – 260. Springer LNCS 533, 1990.

[7] R. Hähnle and P. H. Schmitt. The liberalized δ–rule in free variable semantic tableaux. *to appear*, 1991.

[8] R. C. Jeffrey. *Formal Logic. Its Scope and Limits.* McGraw–Hill, New York, 1967.

[9] F. Oppacher and E. Suen. HARP: A tableau-based theorem prover. *JAR*, 4:69 – 100, 1988.

[10] F. J. Pelletier. Seventy-five problems for testing automatic theorem provers. *JAR*, 2:191 – 216, 1986.

[11] R. J. Popplestone. Beth-tree methods in automatic theorem proving. In *Machine Intelligence 1*, pp 31 – 46. Oliver and Boyd, 1967.

[12] S. V. Reeves. Adding equality to semantic tableaux. *JAR*, 3:225 – 246, 1987.

[13] R. Smullyan. *First–Order Logic.* Springer, New York, second edition, 1968.

Proof Search in the Intuitionistic Sequent Calculus

N. Shankar*
SRI International Computer Science Laboratory
Menlo Park, CA 94025 USA
email: shankar@csl.sri.com

Abstract

The use of Herbrand functions (sometimes called *Skolemization*) plays an important role in classical theorem proving and logic programming. We define a notion of Herbrand functions for the full intuitionistic predicate calculus. This definition is based on the view that the proof-theoretic role of Herbrand functions (to replace universal quantifiers), and of unification (to find instances corresponding to existential quantifiers), is to ensure the eigenvariable conditions on a sequent proof. The propositional impermutabilities that arise in the intuitionistic but not the classical sequent calculus motivate a generalization of the classical notion of Herbrand functions. This generalization of Herbrand functions also applies to the sequent calculus formalizations of logics other than intuitionistic predicate calculus.

1 Introduction

Proof search is an effective way to do automated theorem proving since there is more meaningful information in a proof or a failed proof than there is in a theorem or a failed conjecture. Intuitionistic proofs are interesting because they contain more information than the corresponding classical proofs. Intuitionistic logic can be shown to contain classical logic via Gödel's double-negation interpretation [9]. However, proof search (and theorem proving) is significantly more difficult for the intuitionistic predicate calculus than it is for the classical predicate calculus since there are no convenient normal forms such as prenex, conjunctive, or disjunctive forms.

This paper describes an effective technique for carrying out automated proof search in the intuitionistic sequent calculus. The sequent calculus is useful as a foundational medium for studying the mathematics of proof search. There is a vast body of important proof-theoretic research on sequent calculi. Other proof search techniques such as the tableau and matrix methods can be viewed as optimizations of sequent calculus

*The main part of this work was performed during 1987-88 at Stanford University, where it was funded by NSF Grant No. CCR-8718605. The preparation of this paper has been supported by the SRI International Computer Science Laboratory.

proof search and the sequent calculus makes it possible to study these optimizations in their full generality. The technique described here is actually extremely general and can be applied to proof search in any sequent calculus with conventional quantifier rules and a cut-elimination theorem. The technique is described solely for the intuitionistic calculus in order to keep the exposition concrete, but we briefly discuss how it can be easily generalized to other sequent calculi.

Proof search techniques for classical sequent calculi are fairly well-known [3, 4, 11]. These techniques work by Herbrandizing[1] the goal sequent to eliminate quantifiers and then use the propositional rules along with unification to carry out the proof search on the resulting quantifier-free sequent. Consider the attempt to search for a proof of the following unprovable sequent:

$$(\forall y\colon (\exists x\colon p(x,y))) \vdash (\exists x\colon (\forall y\colon p(x,y))).$$

In its Herbrandized form, the quantiers of existential strength have been replaced by variables (to be instantiated during proof search) and the quantifiers of universal strength have been replaced by terms, $f(u)$ and $g(v)$ where f and g are newly chosen. The result is:

$$p(f(u),u) \vdash p(v,g(v)).$$

Here, f and g are the Herbrand functions, and u and v are called the Herbrand variables. The Herbrand terms $f(u)$ and $g(v)$ mention only the Herbrand variables corresponding to the surrounding quantifiers of existential strength. The propositional rules of Gentzen's sequent calculus are applied backwards, *i.e.*, from conclusion to premise(s). A branch of the proof is successfully terminated when a formula to the right of the sequent arrow unifies with a formula to the left of the sequent arrow. The resulting unifier is propagated to the next remaining branch and the process is repeated. Proof search on the example above fails, as it should, because $p(f(u),u)$ and $p(v,g(v))$ do not unify.[2]

The above system of proof search has several advantages. Proof search is transparent, since it closely follows the proof rules of the sequent calculus. User interaction becomes more manageable. The choice of the term to be existentially generalized is postponed to the axiom case where the term can be constructed by means of unification. The amount of backtracking is minimized. In fact for proof search in the classical sequent calculus, backtracking only occurs in the selection of unifiers and not in the order of the proof steps. It is worth emphasizing this point: *sequent calculus proof search without the use of Herbrand functions would be entirely impractical due to the amount of backtracking involved in selecting the order of the proof steps.*

The present paper demonstrates that the above classical sequent based method can be adapted for theorem proving in the intuitionistic predicate calculus. The modification requires a careful proof-theoretic analysis of the role of Herbrand functions

[1]Skolemization is the process of eliminating quantifiers from a sentence so that free variables replace (essentially) universally quantified variables, and Skolem terms of the form $f(u_1,\ldots,u_n)$ replace (essentially) existentially quantified variables that are within the scope of the universal quantifiers binding the u_i. A sentence is satisfiable iff the universal closure of its Skolemized form is satisfiable. The Herbrandized form of a sentence is the Skolemization of its negation.

[2]The actual explanation of proof search is a little more complicated since it needs to account for the relabelling of Herbrand variables, and the copying of formulas corresponding to the use of the contraction rule in a proof.

and variables, and of unification in proof search [10]. The crucial aspect of our proof-theoretic approach is the observation that Herbrand functions along with unification, serve to enforce the eigenvariable conditions on a sequent calculus derivation. The approach of Herbrandizing the goal formula prior to proof search does not work for nonclassical calculi such as the intuitionistic one. Herbrand functions have to be introduced during the proof search. One obvious approach to introducing a Herbrand function f corresponding to universal quantifier steps is to make the Herbrand term $f(u_1, \ldots, u_n)$ depend on all the Herbrand variables $u_1, \ldots, u_n$, introduced in the existential quantifier steps in the previous steps of the proof search. This approach[3] does take the guesswork out of discovering the term to be existentially generalized but has no other advantage over nondeterministically guessing the entire proof since one cannot ignore the ordering of the quantifier steps during the proof search. Our main observation is that a Herbrand term does not have to depend on all the currently active Herbrand variables but only on a minimal set of these variables that are computed using the impermutabilities between the sequent calculus rules. The use of Herbrandization to record the impermutabilities in the intuitionistic sequent calculus reduces (but clearly does not eliminate) the amount of backtracking since it ignores the relative order of the permutable inferences in the proof search. The unifier returned by a successful proof search can be used to derive the order of introduction of quantifier rules needed to construct a correct sequent calculus proof by permuting the propositional proof returned by the search procedure.

There are a number of results in the literature directed at mechanical theorem proving for intuitionistic logic. We list only a few of these below. Nuprl [5] is an interactive proof checking environment for an intuitionistic logic. Mints [15] describes a resolution-style proof method for intuitionistic logic and a general technique for deriving resolution-style systems from sequent calculi. Beeson [1] has implemented a Prolog proof search procedure called GENTZEN for an intuitionistic sequent calculus. Felty and Miller [6] have written a λ-*Prolog* procedure for proof search in the intuitionistic sequent calculus that maintains the eigenvariable conditions through the use of higher-order unification and hereditary Harrop program clauses [14, 17]. The procedure we describe has a significant efficiency advantage over the approaches of Beeson, and Felty and Miller, since it postpones commitments on the order of introduction of certain quantifiers until the terminal nodes of the search tree are reached. Wallen [20] uses Gödel's translation [8] of intuitionistic logic into the modal logic S4 to define a matrix proof procedure for intuitionistic logic that employs string unification to ensure that the introduction rules for S4 modalities are respected. While Wallen's method is an ingenious and effective one, the translation into S4, whether implicit or explicit, complicates the intuitive reading of the given intuitionistic formula and its proof. Pym and Wallen [18] present an approach to proof search in an intuitionistic type theory that is similar to our approach but they do not exploit Herbrand functions to encode impermutabilities during the proof search. Mints [16] and Bellin [2] studying variants of the Herbrand theorem for intuitionistic logic.

[3]The technique of introducing Herbrand functions in this manner is fairly well-known in automated reasoning. Fitting [7] describes this technique as applied to tableau systems.

2 Proof Theoretic Background

We first fix some of the syntax and list some well-known results in the proof theory of sequent calculi. The alphabet (along with the metavariable conventions) consists of the logical symbols: $\neg, \supset, \vee, \wedge, \forall, \exists$; the variable symbols represented by the metavariables u, v, x, y, z; the function symbols (with non-negative arity) represented by f, g, h; the predicate symbols (with non-negative arity) represented by p, q, r; and the parameters represented by a, b. The metavariables c and d range over the constants, *i.e.*, 0-ary functions applied to the empty argument list. The atomic formula got by applying a 0-ary predicate p to an empty argument list is simply represented as p.

A *term* is either a variable symbol, or of the form $f(t_1, \ldots, t_n)$, where f is an n-ary function symbol and $t_1, \ldots, t_n$ are terms. The metavariables s and t range over terms. An *atomic formula* is an n-ary predicate symbol followed by n terms. A *formula* is either an atomic formula or of one of the forms: $\neg A, A \supset B, A \vee B, A \wedge B, (\forall x{:}A), (\exists x{:}\, A)$, where A and B are smaller formulas and x is a variable. A *sequent* is of the form $\Gamma \vdash \Delta$, where Γ and Δ are multisets of formulas. In $\Gamma \vdash \Delta$, Γ contains the *antecedent* formulas and Δ contains the *succedent* formulas. If Γ is of the form $A_1, \ldots, A_m$ and Δ is of the form $B_1, \ldots, B_n$, then the sequent $\Gamma \vdash \Delta$ can be interpreted as

$$A_1 \wedge \ldots \wedge A_m \supset B_1 \vee \ldots \vee B_n$$

where the empty conjunction is taken to be logically true, and the empty disjunction is logically false.

The notions of positive and negative occurrences of subformulas in a formula and in a sequent should be clear. A positive occurrence of $\exists$ or a negative occurrence of $\forall$ is labelled an *existential* quantifier. Correspondingly, a positive $\forall$ occurrence or a negative $\exists$ occurrence is a *universal* quantifier. So existential and universal quantification refer to the strengths of the quantification and not to the quantifier symbol that is used.

Note that Γ and Δ range over (possibly empty) multisets of formulas rather than lists or sets. Both Γ, A and A, Γ represent the multiset union of $\{A\}$ and Γ, and Γ_1, Γ_2 represents the multiset union of Γ_1 and Γ_2. An additional piece of syntax is required for these rules: terms are extended to include the parameters a, b, used in the quantifier rules below. It is useful to present the sequent calculi for both classical logic, called LK, and for intuitionistic logic, called LJ [19]. The presentation of LK in Figure 1 is essentially the same as Kleene's G3 [12]. Figure 2 displays the rules for LJ and these are similar to those of LK with the important restriction that the succedent part of any sequent in a proof can contain at most one formula.

A proof using these rules of LK or LJ is a tree rooted from below by the conclusion sequent, where the leaf sequents are all axioms, and each non-leaf sequent is derived from its premise sequents by a rule application. The systems LK and LJ have some important properties [12, 19].

Proposition 2.1 (Cut Elimination) *The Cut rule is redundant in both LK and LJ:*

1. *If $\Gamma \vdash_{LK} \Delta$, then $\Gamma \vdash_{\{LK-Cut\}} \Delta$.*
2. *If $\Gamma \vdash_{LJ} \Delta$, then $\Gamma \vdash_{\{LJ-Cut\}} \Delta$.*

Left	Right
$\dfrac{}{\Gamma, A \vdash A, \Delta}(Ax)$	
$\dfrac{\Gamma, \neg A \vdash A, \Delta}{\Gamma, \neg A \vdash \Delta}(\neg\vdash)$	$\dfrac{\Gamma, A \vdash \neg A, \Delta}{\Gamma \vdash \neg A, \Delta}(\vdash\neg)$
$\dfrac{\Gamma, (A \supset B), B \vdash \Delta \qquad \Gamma \vdash A, \Delta}{\Gamma, (A \supset B) \vdash \Delta}(\supset\vdash)$	$\dfrac{\Gamma, A \vdash B, (A \supset B), \Delta}{\Gamma \vdash (A \supset B), \Delta}(\vdash\supset)$
$\dfrac{\Gamma, (A \vee B), A \vdash \Delta \qquad \Gamma, (A \vee B), B \vdash \Delta}{\Gamma, (A \vee B) \vdash \Delta}(\vee\vdash)$	$\dfrac{\Gamma \vdash A, B, (A \vee B), \Delta}{\Gamma \vdash (A \vee B), \Delta}(\vdash\vee)$
$\dfrac{\Gamma, (A \wedge B), A, B \vdash \Delta}{\Gamma, (A \wedge B) \vdash \Delta}(\wedge\vdash)$	$\dfrac{\Gamma \vdash A, (A \wedge B), \Delta \qquad \Gamma \vdash B, (A \wedge B), \Delta}{\Gamma \vdash (A \wedge B), \Delta}(\vdash\wedge)$
$\dfrac{\Gamma \vdash C, \Delta \qquad \Gamma, C \vdash \Delta}{\Gamma \vdash \Delta}Cut(C)$	
$\dfrac{\Gamma, (\forall x\colon A), A\{t/x\} \vdash \Delta}{\Gamma, (\forall x\colon A) \vdash \Delta}(\forall\vdash)$	$\dfrac{\Gamma \vdash A\{a/x\}, (\forall x\colon A), \Delta}{\Gamma \vdash (\forall x\colon A), \Delta}(\vdash\forall)^*$
$\dfrac{\Gamma, (\exists x\colon A), A\{a/x\} \vdash \Delta}{\Gamma, (\exists x\colon A) \vdash \Delta}(\exists\vdash)^*$	$\dfrac{\Gamma \vdash A\{t/x\}, (\exists x\colon A), \Delta}{\Gamma \vdash (\exists x\colon A), \Delta}(\vdash\exists)$

*: a not free in Γ, Δ (*eigenvariable* condition).

Figure 1: Rules for *LK*

Note that $|\Delta| \leq 1$.

Left	Right
$\dfrac{}{\Gamma, A \vdash A}(Ax)$	
$\dfrac{\Gamma, \neg A \vdash A}{\Gamma, \neg A \vdash \Delta}(\neg\vdash)$	$\dfrac{\Gamma, A \vdash \neg A}{\Gamma \vdash \neg A}(\vdash\neg)$
$\dfrac{\Gamma, (A \supset B), B \vdash \Delta \qquad \Gamma \vdash A}{\Gamma, (A \supset B) \vdash \Delta}(\supset\vdash)$	$\dfrac{\Gamma, A \vdash B}{\Gamma \vdash (A \supset B)}(\vdash\supset)$
$\dfrac{\Gamma, (A \vee B), A \vdash \Delta \qquad \Gamma, (A \vee B), B \vdash \Delta}{\Gamma, (A \vee B) \vdash \Delta}(\vee\vdash)$	$\dfrac{\Gamma \vdash A}{\Gamma \vdash (A \vee B)}(\vdash\vee_1) \qquad \dfrac{\Gamma \vdash B}{\Gamma \vdash (A \vee B)}(\vdash\vee_2)$
$\dfrac{\Gamma, (A \wedge B), A, B \vdash \Delta}{\Gamma, (A \wedge B) \vdash \Delta}(\wedge\vdash)$	$\dfrac{\Gamma \vdash A \qquad \Gamma \vdash B}{\Gamma \vdash (A \wedge B)}(\vdash\wedge)$
$\dfrac{\Gamma \vdash C \qquad \Gamma, C \vdash \Delta}{\Gamma \vdash \Delta}Cut(C)$	
$\dfrac{\Gamma, (\forall x\colon A), A\{t/x\} \vdash \Delta}{\Gamma, (\forall x\colon A) \vdash \Delta}(\forall\vdash)$	$\dfrac{\Gamma \vdash A\{a/x\}}{\Gamma \vdash (\forall x\colon A)}(\vdash\forall)^*$
$\dfrac{\Gamma, (\exists x\colon A), A\{a/x\} \vdash \Delta}{\Gamma, (\exists x\colon A) \vdash \Delta}(\exists\vdash)^*$	$\dfrac{\Gamma \vdash A\{t/x\}}{\Gamma \vdash (\exists x\colon A)}(\vdash\exists)$

*: a not free in Γ, Δ (*eigenvariable condition* condition).

Figure 2: Rules for *LJ*

The cut rule is the only rule in which there is a formula appearing in the premise sequent(s) that is not a subformula of its conclusion sequent. Substituting the term t for the free occurrences of x in A yields $A\{t/x\}$, where the bound variables in A are renamed to avoid any clashes. Note that $A\{t/x\}$ for any t is taken to be a subformula of both $(\forall x: A)$ and $(\exists x: A)$. Cut-free proofs therefore have the subformula property, *i.e.*, every formula occurrence in the proof is a subformula of all the sequents appearing below it, and in particular, of the conclusion sequent of the proof. Each of the non-cut rules introduces exactly one new formula occurrence into the conclusion sequent, namely the *principal formula* of the rule.

Proposition 2.2 (Subformula Property) *The formulas in a sequent in a cut-free LK or LJ proof are all subformulas of the conclusion sequent of the proof.*

Proposition 2.3 (Permutability) *The order in which the rules occur in a cut-free LK proof can be permuted without changing the conclusion of the proof, provided:*

1. *The eigenvariable conditions are not violated.*
2. *The subformula property is not violated.*

In particular, the order of the rules in a propositional proof is permutable so long as the subformula property is maintained. For instance, the rules in the proof

$$\dfrac{\dfrac{A, B \vdash B \quad A, B \vdash A}{A, B \vdash B \wedge A}(\vdash \wedge)}{A \wedge B \vdash B \wedge A}(\wedge \vdash)$$

can be permuted as

$$\dfrac{\dfrac{A, B \vdash B}{A \wedge B \vdash B}(\wedge \vdash) \quad \dfrac{A, B \vdash A}{A \wedge B \vdash A}(\wedge \vdash)}{A \wedge B \vdash B \wedge A}(\vdash \wedge)$$

There are certain impermutable pairs of inferences in LJ that are permutable in LK, as described in Section 3 below. The above properties of the sequent calculus play a significant role in justifying the proof search procedures.

3 Proof Search in LJ

In this section, we present and analyze a proof search predicate, *Search*, that searches for LJ proofs. To examine why intuitionistic proof search is different from classical proof search, we consider the following sentence that is classically provable but is not intuitionistically provable:

$$(\forall x: (p(x) \vee q)) \supset (q \vee (\forall x: p(x))).$$

A classical proof search would proceed by Herbrandizing this formula to form the sequent

$$\vdash (p(u) \vee q) \supset (q \vee p(c))$$

then applying the $(\vdash\supset)$ and $(\vdash\vee)$ rule to get

$$(p(u) \vee q) \vdash q, p(c)$$

which by $(\vee\vdash)$ reduces to the two subgoals

$$\begin{aligned} p(u) &\vdash q, p(c) \\ q &\vdash q, p(c). \end{aligned}$$

The second of these subgoals is a propositional axiom. The first subgoal is established by instantiating the Herbrand variable u with the Herbrand constant c.

In searching for a sequent calculus proof as shown above, each application of a proof rule transforming a sequent matching the conclusion of the rule to the corresponding premises of the rule, is termed a *reduction*.

We can attempt a similar proof search with the Herbrandized sentence using only the rules of LJ. Starting with

$$\vdash (p(u) \vee q) \supset (q \vee p(c))$$

we apply $(\vee\vdash)$ to get the subgoals

$$\begin{aligned} p(u) &\vdash q \vee p(c) \\ q &\vdash q \vee p(c). \end{aligned}$$

The first subgoal can be completed by applying $(\vdash\vee_1)$ instantiating u with c. Applying $(\vdash\vee_2)$ to the second subgoal reduces it to an axiom, and completes the proof search successfully.

Since we started with an intuitionistically unprovable sentence, there must be a flaw in the above approach to intuitionistic proof search. The flaw is highlighted when we attempt to reconstruct the proof found by the search procedure as shown below:

$$\dfrac{\dfrac{\dfrac{\dfrac{p(a) \vdash p(a)}{p(a) \vdash (\forall x\colon p(x))}(\vdash\forall)}{p(a) \vdash q \vee (\forall x\colon p(x))}(\vdash\vee) \qquad \dfrac{q \vdash q}{q \vdash q \vee (\forall x\colon p(x))}(\vdash\vee)}{\dfrac{(p(x) \vee q) \vdash (q \vee (\forall x\colon p(x)))}{\dfrac{(\forall x\colon (p(x) \vee q)) \vdash (q \vee (\forall x\colon p(x)))}{\vdash (\forall x\colon (p(x) \vee q)) \supset (q \vee (\forall x\colon p(x)))}(\vdash\supset)}(\forall\vdash)}(\vee\vdash)$$

The above "proof" obviously violates the eigenvariable condition in the $(\vdash\forall)$ step. There is no way to repair this violation of the eigenvariable condition by reordering the inferences since the only way to prove prove $p(c) \vee q \vdash q \vee p(c)$ in LJ is with the $(\vee\vdash)$ rule applied below the $(\vdash\vee)$ rule. In LK it would be possible to change the order of the rules as required since $(\vee\vdash)$ rule permutes above the $(\vdash\vee)$ rule, and the resulting LK proof would satisfy the eigenvariable condition.

We thus observe that the impermutability of certain pairs of inferences in LJ makes it incorrect to directly use Herbrandization for proof search. Our solution to this problem is to dynamically introduce Herbrand functions to replace universal

	R1/R2	Example
1.	$(\forall\vdash)/(\vdash\forall)$	$(\forall x\colon A(x)) \vdash (\forall x\colon (A(x) \vee B))$
2.	$(\forall\vdash)/(\exists\vdash)$	$(\forall x\colon A(x)), (\exists x\colon \neg A(x)) \vdash$
3.	$(\vdash\exists)/(\exists\vdash)$	$(\exists x\colon (A(x) \wedge B)) \vdash (\exists x\colon A(x))$
4.	$(\supset\vdash)/(\vdash\supset)$	$(A \supset \neg A) \vdash (A \supset B)$
5.	$(\neg\vdash)/(\vdash\supset)$	$\neg A \vdash (A \supset B)$
6.	$(\supset\vdash)/(\vdash\neg)$	$(A \supset \neg A) \vdash \neg A$
7.	$(\neg\vdash)/(\vdash\neg)$	$\neg A \vdash \neg(A \wedge A)$
8.	$(\supset\vdash)/(\vee\vdash)$	$A \vee B, A \supset B \vdash B$
9.	$(\vdash\vee)/(\vee\vdash)$	$(A \vee B) \vdash (B \vee A)$
10.	$(\neg\vdash)/(\vee\vdash)$	$(A \vee B), \neg A \vdash B$
11.	$(\vdash\exists)/(\vee\vdash)$	$A(a) \vee A(b) \vdash (\exists x\colon A(x))$

Figure 3: Impermutabilities in *LJ*

quantifiers during the proof search so that the variables $u_1, \ldots, u_n$ in the Herbrand term $h(u_1, \ldots, u_n)$ can encode the propositional impermutabilities that arise in *LJ* proofs. With this change, the $(\vdash \forall)$ step in the above proof search must replace the universal quantified formula $(\forall x\colon p(x))$ by $p(h(u))$, where the Herbrand term $h(u)$ is used instead of c to encode the constraint that u goveerns the $(\vee \vdash)$ rule which cannot be permuted above the $(\vdash \vee)$ rule. In the above proof search, this yields the subgoal

$$p(u) \vdash p(h(u)),$$

where occurs-checking prevents u from being unified with $h(u)$. There are of course other directions in which the above proof search could have proceeded but they all fail rather more easily.

The restriction on the succedent of *LJ* sequents leads to a number of impermutabilities that are not present in *LK*. Kleene [13] lists the basic impermutabilities along with examples of theorems whose *LJ* proofs can only be constructed with R1 above R2 (*i.e.*, R1/R2). These are displayed in Figure 3. In each instance of R1/R2, the rule R2 cannot be permuted above the rule R1 in any proof of the corresponding example. It can be checked that for any pair of rules R1 and R2 not listed below, an occurrence of R2 can always be moved above an occurrence of R1 in an *LJ* proof.

We call the impermutabilities 4–11 in Figure 3 the *LJ* impermutabilities since they are peculiar to *LJ*. We first observe certain uniformities in the *LJ* impermutabilities listed in Figure 3:

- In any *LJ* impermutability R1/R2 above, the only Left rule in the R2 position is $(\vee \vdash)$.
- The only Right rules are $(\vdash\supset)$ and $(\vdash \neg)$.
- When R2 is either $(\vdash\supset)$ or $(\vdash \neg)$, R1 is either $(\supset\vdash)$ or $(\neg \vdash)$.

These uniformities are exploited in the proof search procedure below. Since we employ a partially Herbrandized sequent to represent the goals in proof search, we define a *form* to be a pair $[X\colon A]$ consisting of a formula A and the set X of governing Herbrand variables. The initial sequent to the proof search should not contain any free variables, and the set of governing variables in each form is empty. The proof search procedure is presented as a predicate that takes five arguments: a (partially Herbrandized) sequent consisting of a set of zero or more antecedent forms and at most one succedent form, a set of Herbrand variables $lvars$, set of Herbrand variables $rvars$, an input unifier, and an output unifier. The set $lvars$ contains those Herbrand variables governing any form to which the $(\vee \vdash)$ rule has been applied in the course of the proof search so far. The set $rvars$ contains those Herbrand variables governing any form to which either $(\vdash \supset)$ or $(\vdash \neg)$ has been applied in the course of the proof search so far. The input unifier is initially empty.

Multisets of forms are represented by Σ, Θ, etc. Sets of variables are represented by X, Y, L, R. Substitutions are represented by the letter U, V, W, etc. The search sequent is represented as $\Sigma \vdash \Theta$. We omit the definition of $Unify(A; B; U; V)$ while noting that V is required to be a unifier for A and B that extends U. That is, V is of the form $U' \circ U$ for some substitution U'. If such a unifier does not exist, then the result V is **No**. The parameter U is not used in any essential way in the definition below, but does permit a variation of *Search* in which the unifier returned from one successful branch of the search can be given as the input to the other branch. The parameter U also turns out to be useful for the proof of completeness in Theorem 4.1. The relation *Search* can be defined inductively by the following scheme:

Unifier: $Search(\Sigma \vdash \Theta;\ L;\ R;\ \mathbf{No};\ \mathbf{No})$.

Axiom: $Search([X\colon A], \Sigma \vdash [Y\colon B];\ L;\ R;\ U;\ V)$ if $Unify(A; B; U; V)$ and $V \neq \mathbf{No}$.

$(\neg \vdash)$: $Search([X\colon \neg A], \Sigma \vdash \Theta;\ L;\ R;\ U;\ V)$ if $Search([X\colon \neg A], \Sigma \vdash [X \cup L \cup R\colon A];\ L;\ R;\ U;\ V)$.
X is augmented with $L \cup R$ to encode the impermutabilities of the form $(\neg \vdash)/\mathrm{R2}$.

$(\vdash \neg)$: $Search(\Sigma \vdash [X\colon \neg B];\ L;\ R;\ U;\ V)$ if $Search([X;\ B], \Sigma \vdash\ ;\ L;\ X \cup R;\ U;\ V)$.
R is augmented with X to encode the impermutabilities of the form $\mathrm{R1}/(\vdash \neg)$.

$(\supset\vdash)$: $Search([X\colon (A \supset B)], \Sigma \vdash \Theta;\ L;\ R;\ U;\ V)$ if $Search([X\colon (A \supset B)], \Sigma \vdash [X \cup L \cup R\colon A];\ L;\ R;\ U;\ V)$, and $Search([X \cup L \cup R\colon B], [X\colon (A \supset B)], \Sigma \vdash \Theta;\ L;\ R;\ U;\ V)$.
X is augmented with $L \cup R$ to encode the impermutabilities of the form $(\supset\vdash)/\mathrm{R2}$.[4]

[4]The construction of the unifier V for the left-implication rule can be carried out in stages by recording the output of the first branch as W and using W as the input unifier to the second branch. The same applies to the $(\vee \vdash)$ and $(\vdash \vee)$ steps.

($\vdash\supset$): $Search(\Sigma \vdash [X\colon (A \supset B)];\ L;\ R;\ U;\ V)$ if
$Search([X\colon A], \Sigma \vdash [X\colon B];\ L;\ X \cup R;\ U;\ V)$.
R is augmented with X to encode the impermutabilities of the form R1/($\vdash\supset$).

($\vee\vdash$): $Search([X\colon (A \vee B)], \Sigma \vdash \Theta;\ L;\ R;\ U;\ V)$ if
$Search([X\colon B], [X\colon (A \vee B)], \Sigma \vdash \Theta;\ X \cup L;\ R;\ U;\ V)$,
and $Search([X\colon A], [X\colon (A \vee B)], \Sigma \vdash \Theta;\ X \cup L;\ R;\ U;\ V)$.
L is augmented with X to encode the impermutabilities of the form R1/($\vee\vdash$).

($\vdash\vee$): $Search(\Sigma \vdash [X\colon (A \vee B)];\ L;\ R;\ U;\ V)$ if
$Search(\Sigma \vdash [L \cup X\colon A];\ L;\ R;\ U;\ V)$ and $V \neq$ **No**
or $Search(\Sigma \vdash [L \cup X\colon B];\ L;\ R;\ U;\ V)$ and $V \neq$ **No**
otherwise $V =$ **No**.
X is augmented with L to encode the impermutability ($\vdash\vee$)/($\vee\vdash$).

($\wedge\vdash$): $Search([X\colon (A \wedge B)], \Sigma \vdash \Theta;\ L;\ R;\ U;\ V)$ if
$Search([X\colon A], [X\colon B], [X\colon (A \wedge B)], \Sigma \vdash \Theta;\ L;\ R;\ U;\ V)$.

($\vdash\wedge$): $Search(\Sigma \vdash [X\colon (A \wedge B)];\ L;\ R;\ U;\ V)$ if
$Search(\Sigma \vdash [X\colon B];\ L;\ R;\ U;\ V)$
and $Search(\Sigma \vdash [X\colon A];\ L;\ R;\ U;\ V)$.

($\exists\vdash$): $Search([X\colon (\exists x\colon A)], \Sigma \vdash \Theta;\ L;\ R;\ U;\ V)$ if
$Search([X\colon A\{h(X)/x\}], \Sigma \vdash \Theta;\ L;\ R;\ U;\ V)$,
where h is a new n-ary function symbol, and if X is $\{u_1, \ldots, u_n\}$ then $f(X)$ represents $f(u_1, \ldots, u_n)$.

($\vdash\exists$): $Search(\Sigma \vdash [X\colon (\exists x\colon A)];\ L;\ R;\ U;\ V)$ if
$Search(\Sigma \vdash [\{u\} \cup L \cup X\colon A\{u/x\}];\ L;\ R;\ U;\ V)$,
where u is a new Herbrand variable. X is augmented with L to encode the impermutability ($\vdash\exists$)/($\vee\vdash$).

($\forall\vdash$):

$Search([X\colon (\forall x\colon A)], \Sigma \vdash \Theta;\ L;\ R;\ U;\ V)$ if
$Search([\{y\} \cup X\colon A\{u/x\}], [X\colon (\forall x\colon A)], \Sigma \vdash \Theta;\ L;\ R;\ U;\ V)$,
where u is a new Herbrand variable.

($\vdash\forall$): $Search(\Sigma \vdash [X\colon (\forall x\colon A)];\ L;\ R;\ U;\ V)$ if
$Search(\Sigma \vdash [X\colon A\{h(X)/x\}];\ L;\ R;\ U;\ V)$,
where h, X, and $h(X)$ are as in ($\exists\vdash$).

The reader might wish to apply the procedure *Search* to the following sequents that are classically valid, but not intuitionistically.

1. $(p \supset (\exists x\colon q(x))) \vdash (\exists x\colon p \supset q(x))$.

2. $\neg(\forall x\colon q(x)) \vdash (\exists x\colon \neg q(x))$.

3. $((\forall x\colon q(x)) \supset p) \vdash (\exists x\colon q(x) \supset p)$.

The next step is to argue that *Search* is sound and complete.

4 Soundness and Completeness

Search is *sound* if whenever it returns a unifier other than **No** on a sequent, there is an LJ proof of that sequent. *Search* is *complete* if when given any sequent provable in LJ, *Search* nondeterministically returns a unifier (other than **No**). It is easy to see that *Search* is both sound and complete on the propositional part of LJ since it is a nondeterministic search procedure for cut-free proofs in this fragment.

For demonstrating completeness, we restrict the Herbrand functions to be taken from $h_1, h_2, \ldots$, and to each h_i we associate the parameter a_i. We restrict LJ proofs to only contain parameters a_i. If Σ is a multiset of forms, let $\underline{\Sigma}$ represent the corresponding multiset of formulas, but where any terms of the form $h_i(\ldots)$ in Σ have been replaced by the corresponding parameter a_i. Applying the substitution U to Σ yields the multiset of forms $U(\Sigma)$.

Lemma 4.1 *If $\Gamma \vdash \Delta$ is provable in LJ, then for any two multisets of forms Σ and Θ, and unifier U such that Γ is $\underline{U(\Sigma)}$ and Δ is $\underline{U(\Theta)}$, we can find a V distinct from* **No** *where $Search(\Sigma \vdash \Theta;\ L;\ R;\ U;\ V)$.*

Proof. By induction on cut-free LJ proofs of $\Gamma \vdash \Delta$. Note that L and R do not play any role in this proof. We consider only a single case of the proof.

In the ($\vdash \exists$) case, the given proof derives $\Gamma \vdash (\exists x\colon A)$ from $\Gamma \vdash A\{t/x\}$. The corresponding step in the search replaces $(\exists x\colon A)$ with $A\{u/x\}$. Let U in the induction hypothesis be replaced with $U\{t/u\}$ so that $\underline{U\{t/u\}([\{y\} \cup L \cup X\colon A\{u/x\}])}$ is just $A\{t/x\}$ and it can be shown that the required V is provided by the induction hypothesis.

The other cases are similar ∎

When all the forms in Σ and Θ are of the form $[\{\}\colon A]$, and U is the empty substitution $\emptyset$ (which, of course, is idempotent), then $\underline{U(\Sigma)} \vdash \underline{U(\Theta)}$ is just the original sequent, and we get the completeness theorem below, where $\{\}\colon\Gamma$ represents the multiset of forms got by replacing each A in Γ with $[\{\}\colon A]$.

Theorem 4.2 (Completeness) *If $\Gamma \vdash \Delta$ is provable in LJ, then there is a V distinct from* **No** *such that $Search(\{\}\colon\Gamma \vdash \{\}\colon\Delta;\ L;\ R;\ \emptyset;\ V)$ holds.*

The soundness argument is more delicate since it sheds light on the use of Herbrand functions. The goal is to show that if proof search succeeds on a sequent, then that sequent has a proof in LJ. The proof proceeds by constructing a proof search tree from a successful proof search and extracting an ordering on the quantifiers from the output unifier. We then show that the proof search tree can be permuted so as to respect the ordering on the quantifiers and that the resulting proof obeys the eigenvariable condition. Call Π an *LJ proof structure* if it is an LJ proof that possibly violates the eigenvariable condition. A proof structure Π is *hygienic* in its use of parameter names if each parameter name a_i is associated with at most one quantifier step in Π.

Lemma 4.3 *If $Search(\Sigma \vdash \Theta;\ L;\ R;\ U;\ V)$ succeeds with $V \neq$* **No**, *we can construct a corresponding hygienic LJ proof structure Π of $\underline{V(\Sigma)} \vdash \underline{V(\Theta)}$.*

Proof. By induction on the computation corresponding to *Search*. We only deal with a single case.

Consider the $(\vdash \forall)$ case when the induction hypothesis yields a hygienic proof structure Π_1 for $\underline{V(\Sigma)} \vdash \underline{V([X\colon A\{h_i(X)/x\}])}$ corresponding to $Search(\Sigma \vdash [X\colon A\{h_i(X)/x\}];\ L;\ R;\ U;\ V)$. Note that $\underline{V([X\colon A\{h_i(X)/x\}])}$ is just $\underline{V([X\colon A\{a_i/x\}])}$. It is, however, possible for $\underline{V(\Sigma)}$ to contain occurrences of a_i so that the proof tree below could violate the eigenvariable condition.

$$\cfrac{\begin{array}{c}\Pi_1\\ \vdots\\ \underline{V(\Sigma)} \vdash \underline{V([X\colon A\{h_i(X)/x\}])}\end{array}}{\underline{V(\Sigma)} \vdash \underline{V([X\colon (\forall x\colon A)])}}\,(\vdash \forall)$$

The other cases are similar. ■

Given a proof structure Π, if a rule occurrence R1 occurs above a rule occurrence R2 in Π, then R1>R2 in Π iff either

- R1/R2 is one of the *LJ* impermutabilities 1–11
- the principal formula of R1 is a proper subformula of the principal formula of R2
- or there is a rule occurrence R3 such that R1>R3>R2

Lemma 4.4 *If for any n, rule* R1 *occurs n steps above* R2 *in proof structure Π and* R1$\not>$R2 *in Π, then the order of rules in Π can be permuted to yield a new proof structure Π' with the same conclusion as Π, so that* R2 *occurs above* R1 *in Π'. If Π is hygienic, then there are no new pairs of quantifier rule occurrences that violate the eigenvariable condition in Π'.*

Proof. By induction on n. The details are omitted.

In a hygienic proof structure, the only way to introduce new violations of the eigenvariable condition would be to invert the order of an existential quantifier step R1 that was above a universal quantifier step R2 in Π. This never happens since we would have R1>R2 in this case. ■

Lemma 4.5 *Let U be an idempotent substitution such that for any i, the search sequent $U(\Sigma) \vdash U(\Theta)$ contains (zero or more occurrences of) at most one term of the form $h_i(X)$. If $Search(\Sigma \vdash \Theta;\ L;\ R;\ U;\ V)$ returns an idempotent substitution V such that $V \neq$ No, then we can find an idempotent V' such that $Search(\Sigma \vdash \Theta;\ L;\ R;\ U;\ V')$ and V' contains at most one term of the form $h_i(\ldots)$ for any Herbrand function h_i, namely $h_i(V'(X))$.*

Proof. By induction on the computation of *Search* taking care to return the most general idempotent unifier at every stage. ■

Denote the quantifier step corresponding to the Herbrand term $h_i(X)$ as $\forall_i$, and the quantifer step corresponding to the Herbrand variables u_j as $\exists_j$.

Lemma 4.6 *If $\forall_i > \exists_j$ in a proof structure Π corresponding to a computation of* Search, *then $u_j \in X$ for the Herbrand term $h_i(X)$ introduced in the computation of* Search.

Proof. By induction according to the definition of R1>R2. ∎

Lemma 4.7 *Given that U satisfies the conditions of Lemma 4.5, Search($\Sigma \vdash \Theta$; L; R; U; V) holds, and V is an idempotent unifier, if $u_j \in X$ for the Herbrand term $h_i(X)$ in the computation of* Search, *then h_i does not occur in $V(u_j)$.*

Proof. Since by Lemma 4.5, we can ensure that V is such that h_i only occurs in the form $h_i(V(X))$ in V. If h_i occurs in $V(u_j)$, then either V is not idempotent or contains an infinite term since no finite term can be a proper subterm of itself. Both these possibilities are ruled out. ∎

Theorem 4.8 (Soundness) *If $\underline{U(\Sigma)} \vdash \underline{U(\Theta)}$ contains no parameters and Search($\Sigma \vdash \Theta$; L; R; U; V) succeeds with $V \neq$ **No**, then $\underline{V(\Sigma)} \vdash \underline{V(\Theta)}$ is provable in LJ.*

Proof. By Lemma 4.3, we can derive a proof structure Π corresponding to *Search*($\Sigma \vdash \Theta$; L; R; U; V). The proof of Theorem 4.8 is by induction on the number of pairs of quantifier rules that violate the eigenvariable condition in the proof structure Π. By Lemma 4.6, for any rule occurrences $\forall_i$ and $\exists_j$ in Π, if $\forall_i > \exists_j$ then $u_j \in h_i(X)$. Here $h_i(X)$ is the Herbrand term introduced in the quantifier step $\forall_i$. By Lemma 4.7, h_i does not occur in $V(u_j)$. Therefore, if h_i does occur in $V(u_j)$, then $\forall_i \not> \exists_j$ and by Lemma 4.4, $\forall_i$ can be permuted below $\exists_j$ in Π to yield the proof structure Π' with the same conclusion as Π.

Note that by Lemma 4.4, no new violations of the eigenvariable condition are introduced into Π'. We have thus succeeded in reducing the number of violations of eigenvariable condition and the induction hypothesis can be applied to the resulting proof structure Π'. ∎

5 Conclusions

We have described a search procedure for proofs in the intuitionistic sequent calculus and proved it to be sound and complete. The search procedure uses a generalization of Herbrand functions so that the Herbrand term introduced for a universal quantifier depends only on those Herbrand variables corresponding to existential quantifier steps that cannot always be permuted above the universal quantifier step. This makes it possible to reconstruct the proof from the search structure and the unifier, by permuting the search structure to reorder the quantifier steps to satisfy the order induced by the unifier.

Both the search algorithm and proof can be generalized to any sequent calculus with a cut elimination theorem and conventional quantifier rules. The generalized procedure is to first identify the impermutabilities other than those given by the eigenvariable conditions in the given sequent calculus, and derive a search procedure similar to *Search* that records these impermutabilities when forming Herbrand terms corresponding to universal quantifier steps. The above proof sketch would essentially apply without change. Note that in *Search*, we have taken advantage of certain patterns in the impermutability pairs to reduce the number of auxiliary parameters for accumulating Herbrand variables, and such optimizations may not always be possible.

It is not possible to improve upon this form of Herbrandization without examining the bodies of the quantified formulas themselves. In other words, if any Herbrand variables that are included in Herbrand terms in the search procedure above are dropped, then the procedure is unsound. It is possible to construct a counterexample of an unprovable sequent on which the search procedure would succeed. This does not preclude optimizations where the internal structure of the quantified formula is examined in order to rule out such counterexamples. We can also identify certain classes of formulas where it is always sound to Herbrandize the initial sequent. These are classes of formulas that are defined solely by syntactic restrictions on the positive and negative occurrences of various connectives and quantifiers. Herbrandization will always work for a class of formulas that does not admit an *LJ* impermutability R1/R2, where an existential quantifier can govern the principal formula of R2, and the principal formula of R1 contains a universally quantified subformula. For example, Herbrandization fails for the class of hereditary Harrop formulas [14] since the unprovable formula

$$((\forall x{:}\, p(x)) \supset q) \supset (\exists y{:}\, (p(y) \supset q))$$

has the Herbrandized form

$$(p(c) \supset q) \supset (p(u) \supset q)$$

which is easily seen to be intuitionistically provable with c instantiating Herbrand variable u. This counterexample arises from the $(\supset\vdash)/(\vdash\supset)$ impermutability permitted by the class of hereditary Harrop formulas.

Acknowledgements: The ideas and comments of Michael Beeson, Jussi Ketonen, Lincoln Wallen, Gianluigi Bellin, Grigori Mints, Dale Miller, Patrick Lincoln, Sam Owre, and Roy Dyckhoff have contributed to this presentation, as have the comments of the anonymous referees. Sol Feferman's course on proof theory provided the initial stimulus for this work.

References

[1] M. J. Beeson. Some applications of Gentzen's proof theory in automated deduction. In P. Schroeder-Heister, editor, *Extensions of Logic Programming, Lecture Notes in Computer Science 475*, pages 101–156. Springer-Verlag, 1991.

[2] G. L. Bellin. Herbrand's theorem for calculi of sequents LK and LJ. In D. Prawitz, editor, *Proceedings of the Third Scandinavian Logic Symposium*, 1980.

[3] W.W. Bledsoe and P. Bruell. A man-machine theorem-proving system. In *Advance Papers of Third International Joint Conference on Artificial Intelligence.* W.W. Bledsoe, 1974.

[4] K. A. Bowen. Programming with full first-order logic. In J. E. Hayes, D. Michie, and Y.-H. Pao, editors, *Machine Intelligence 10*, pages 421–440. Halsted Press, 1982.

[5] R. L. Constable, *et al.*. *Implementing Mathematics with the Nuprl.* Prentice-Hall, New Jersey, 1986.

[6] A. Felty and D. Miller. Specifying theorem provers in a higher-order logic programming language. In E. Lusk and R. Overbeek, editors, *Ninth International Conference on Automated Deduction, Lecture Notes in Computer Science 310*, pages 61–80, Argonne, Illinois, May 1988.

[7] M. Fitting. *First-Order Logic and Automated Theorem Proving.* Springer-Verlag, 1990.

[8] K. Gödel. An interpretation of the intuitionistic propositional calculus. In S. Feferman, editor, *Kurt Gödel: Collected Works Vol. 1*, pages 301–303. Oxford University Press, 1986.

[9] K. Gödel. On intuitionistic arithmetic and number theory. In S. Feferman, editor, *Kurt Gödel: Collected Works Vol. 1*, pages 287–295. Oxford University Press, 1986.

[10] J. Herbrand. Investigations in proof theory. In J. van Heijenoort, editor, *From Frege to Gödel: A Source Book of Mathematical Logic*, pages 525–581. Harvard University Press, Cambridge, Mass., 1967.

[11] J. Ketonen and R. Weyhrauch. A decidable fragment of predicate calculus. *The Journal of Theoretical Computer Science*, 32:297–307, 1984.

[12] S. C. Kleene. *Introduction to Metamathematics.* North-Holland, Amsterdam, 1952.

[13] S. C. Kleene. Permutability of inferences in Gentzen's calculi LK and LJ. *Memoirs of the AMS*, 10, 1952.

[14] D. Miller, G. Nadathur, F. Pfenning, and A. Scedrov. Uniform proofs as a foundation for logic programming. *Annals of Pure and Applied Logic*, 51:125–157, 1991.

[15] G. Mints. Gentzen-type systems and resolution rules, Part I: Propositional logic. In P. Martin-Löf and G. Mints, editors, *COLOG-88, Lecture Notes in Computer Science 417*, pages 198–231. Springer-Verlag, 1988.

[16] G. E. Mints. Analog of Herbrand's theorem for non-prenex formulas of constructive predicate calculus. In *Studies in constructive mathematics and mathematical logic, I.* Steklov Mathematical Institute Seminars in Mathematics, 1969.

[17] G. Nadathur and D. Miller. Higher-order Horn clauses. *Journal of the ACM*, 37(4):777–814, 1990.

[18] D. J. Pym and L. A. Wallen. Investigations into proof-search in a system of first-order dependent function types. In M. E. Stickel, editor, *Tenth Conference on Automated Deduction, Lecture Notes in Computer Science 449*, pages 236–250. Springer-Verlag, 1990.

[19] M. E. Szabo, editor. *The Collected Papers of Gerhard Gentzen.* North-Holland, 1969.

[20] L. A. Wallen. *Automated Proof Search in Non-Classical Logics.* MIT Press, 1990.

Implementing the Meta-Theory of Deductive Systems

Frank Pfenning and Ekkehard Rohwedder
School of Computer Science
Carnegie Mellon University
Pittsburgh, Pennsylvania 15213-3890, U.S.A.
fp+@cs.cmu.edu and er+@cs.cmu.edu

Abstract. We exhibit a methodology for formulating and verifying meta-theorems about deductive systems in the Elf language, an implementation of the LF Logical Framework with an operational semantics in the spirit of logic programming. It is based on the mechanical verification of properties of transformations between deductions, which relies on type reconstruction and *schema-checking*. The latter is justified by induction principles for closed LF objects, which can be constructed over a given signature. We illustrate our technique through several examples, the most extensive of which is an interpretation of classical logic in minimal logic through a continuation-passing-style transformation on proofs.

1 Introduction

Formal deductive systems have become an important tool in computer science. They are used to specify logics, type systems, operational semantics and other aspects of languages. The role of such specifications is three-fold. Firstly, inference rules serve as a high-level notation which helps to explain the meaning of the language under consideration. This was one of Gentzen's original motivations for his calculus of natural deduction [10]. Secondly, they can form the basis for an implementation of a deductive system. For example, it is not difficult to translate an operational semantics presented in the style of natural deduction [18, 13, 4] into an implementation of an interpreter. Thirdly, deductive systems help in developing the meta-theory of a language. For example, the soundness of a type system with respect to an operational semantics is most easily expressed as a property of two inference systems.

The LF Logical Framework [16] has been designed to provide an appropriate language for the high-level specification of deductive systems. In LF, *judgments* are represented as *types* and *deductions* as *objects*. The validity of a deduction is reduced to the well-typedness of the representing object. Since type-checking in the LF type theory is decidable, purported deductions can be checked automatically for validity.

However, LF is a powerful basis for much more comprehensive tasks than mere proof-checking. Unification and proof search algorithms have been developed [7, 27, 28, 24] and it has been amenable to an operational interpretation which is realized in the Elf programming language [21, 23]. A wide range of deductive systems have been specified in LF and implemented in Elf [1, 15, 19, 20].

In this paper we investigate the use of Elf to implement the *meta-theory* of deductive systems, thus addressing the third of the principal applications listed above. Our approach is based on three observations. Firstly, in LF deductions are represented as objects and can thus be part of higher-level judgments. For example, it is easy to write down rules defining the judgment of "normal form" as it applies to natural deductions. Secondly, proofs of properties of deductive systems (henceforth called meta-proofs) often rely on transformations between deductive systems. Such transformations can be represented in LF (and implemented in Elf) as judgments relating deductions. Thirdly, due to the rich type structure of LF, it is often possible to check certain properties of such judgments purely mechanically.

Thus our methodology for the verification of meta-theorems presents itself as follows. Stage 1 (**Syntax**) is the representation of the object languages under consideration. Stage 2 (**Judgments**) is the definition of the deductive systems following the LF methodology. Stage 3 (**Deduction Transformations**) is the formulation of transformations between the deductive systems as higher-level judgments. Stage 4 (**Schema-Checking**) is the mechanical verification of a property of the transformations axiomatized in Stage 3. We have carried out this methodology for a number of examples, the most intricate of which is a verification of the subject reduction property for Mini-ML [5, 20]. Currently, Stage 4 is done mostly by hand, as we have not yet implemented a general schema-checker. As we will illustrate, in the current Elf implementation schema-checking can be directly achieved through a set of queries which can be constructed from a signature on a case-by-case basis.

It is unclear if the methodology and implementation (when it is complete) as described so far deserves the label "automated theorem prover". Clearly, it can verify many theorems far beyond the scope of current theorem provers. Moreover, a significant part of the verification is carried out automatically during term reconstruction, type-checking, and schema-checking. On the other hand, for difficult meta-theorems, the transformations constructed in Stage 3 must be carefully engineered so as to be amenable to schema-checking.

The difficulty of constructing the crucial transformations at Stage 3 varies greatly from problem to problem. In some cases —exemplified in Section 3— a routine automatic construction of the transformation from the inference systems given appears feasible as well as sufficient for complete verification. In other cases it is difficult— either because of the sheer intricacy and size of the systems involved, or because of the inherent difficulty of the meta-proof. An example of the latter is the normalization theorem for the polymorphic λ-calculus, which is beyond the scope of our techniques as we have developed them so far. We have been able to verify:

- evaluation of a Mini-ML expression results in a value if it terminates,
- equivalence of an algorithmic and a more declarative operational semantics for Mini-ML,
- type-soundness of Mini-ML (sometimes called the subject reduction theorem), including polymorphism,
- correctness of a compiler from Mini-ML to a variant of the Categorical Abstract Machine [14],

- the deduction theorem for an axiomatic formulation of propositional logic in the style of Hilbert,
- equivalence of natural deduction and Hilbert's calculus,
- soundness of two theorem provers, one using Prolog-style depth-first search and one employing bounded search,
- equivalence of two formulations of the Lambek-calculus,
- correctness of 8 different logic interpretations in the propositional calculus,
- soundness of an algorithm for deciding equality in the simply-typed λ-calculus.

In each of these cases the meta-proofs turned out to be very natural, compact, and relatively easy to construct, since they are operationally meaningful (as translations between deductions). They are also very close to an informal argument one might give to prove the corresponding meta-theorems and, with an appropriate interface, could be used to *explain* the meta-theorems and their proofs.

Schema-checking as presented in this paper verifies properties of signatures. Therefore, our work draws upon a calculus for LF signatures [17]. In that paper, it is also shown how some simple meta-theoretic properties can be witnessed directly by *realizations* (functions between signatures). The limitations of this alternative approach are also discussed, but more work is required to understand the precise relationship to schema-checking. Another line of investigation is followed by Basin and Constable [2], who propose using inductively defined types in the NuPrl type theory in order to represent deductive systems. However, their approach is not especially tailored towards developing the meta-theory of deductive systems, but applies an already existing apparatus to a new and more difficult problem. We feel that one can gain significantly by moving to a meta-language such as LF which has been specifically designed for the task of formalizing deductive systems. One can then take advantage of built-in support for such ubiquitous concepts as free and bound variables, substitution, hypothetical reasoning, or schematic judgments.

The primary difficulty in applying our methodology is to construct the transformations between deductions. Due to the strong constraints imposed by the dependent types, we believe that in many cases such transformations could be constructed automatically. Other work in inductive theorem proving and logic programming such as, for example [3, 12, 26], should be applicable in our setting to aid in the construction of such transformations. Closely related to the ideas presented here is work by Fribourg [9] in the simpler setting of Horn clauses. Again, his ideas could add to the degree of automation available within our methodology.

The remainder of this paper is organized as follows: In Section 2 we present some of the basic ideas behind Elf, schema-checking, and its connection to proof by induction, using the natural numbers as a very simple example. In Section 3 we illustrate the use of transformations between deductions and introduce the important notion of *unit refinement*. As an example we demonstrate that evaluation of an expression in a simple functional language always returns a value. In Section 4 we sketch a more comprehensive example, verifying an interpretation of classical logic in minimal logic by way of a continuation-passing-style (CPS) transformation on proofs. We end with a summary and some speculation about future directions in Section 5.

2 LF Signatures and Induction

Syntax. The first stage in the representation of a deductive system is to declare the underlying languages. We begin with a particularly simple and familiar example: the natural numbers.

```
nat : type.
z   : nat.
s   : nat -> nat.
```

`nat` is declared as a type and `z` and `s` as constructors for data of this type. Valid (that is, well-typed), closed objects of type `nat` represent natural numbers. We refer to a list of declarations such as the ones above as a *signature*. A calculus of signatures for Elf is described in [17]. As this module calculus has not yet been implemented, we only use the Elf core language in this exposition.

Judgments. The calculus of functions underlying LF is not very rich: it permits only λ-abstraction and application, and functions cannot be defined by primitive recursion, for example. This is an important requirement and cannot easily be relaxed, because it would destroy the adequacy of encodings of deductive systems in LF (see [16] and [17] for further discussion). Thus, non-trivial operations on objects must be defined as relations. Fortunately, such relations are operationally adequate within the Elf programming language, as Elf gives them an operational reading in the style of Prolog. We consider a simple `double` predicate.

```
double : nat -> nat -> type.
dbl_z  : double z z.
dbl_s  : double (s N) (s (s M)) <- double N M.
```

The relation `double` is realized as a so-called *type family* indexed by two objects which are natural numbers. For readers familiar with the Curry-Howard isomorphism between propositions and types, it should come as no great surprise that we can represent relations this way. The left-arrow is mere syntactic sugar, and `B <- A` and `A -> B` are parsed into the same internal form. The constants `dbl_z` and `dbl_s` construct objects which represent deductions. For example,

```
dbl_s (dbl_s (dbl_z)) : double (s (s z)) (s (s (s (s z)))).
```

represents a deduction which establishes that 4 is the double of 2. This object would be constructed by the Elf interpreter when executing the query

```
?- double (s (s z)) M.
```

Here `M` is a free variable which is treated as a logic variable. That is, we simultaneously search for objects `M : nat` and `P : double (s (s z)) M`.

Free variables in declarations are implicitly quantified as in Prolog. In Elf, such implicit quantifiers are inferred during parsing. The explicit form for the last declaration above would be

```
dbl_s  : {N:nat} {M:nat} double N M -> double (s N) (s (s M)).
```

`{x:A} B` is Elf's concrete syntax for $\Pi x{:}A.\ B$, where Π is the dependent function type constructor. Thus, `dbl_s` is really a function of three arguments: it accepts a natural number `N`, a natural number `M`, and then a deduction establishing that `M` is the double of `N`. It constructs a deduction showing that `(s (s M))` is the double of `(s N)`. Implicit quantifiers give rise to implicit arguments, and in its full form the proof object from the example above would be

```
(dbl_s (s z) (s (s z)) (dbl_s z z (dbl_z)))
  :  double (s (s z)) (s (s (s (s z)))).
```

The gaps in the first form shown above are filled during term reconstruction in Elf, which employs an algorithm for solving constraints between types, described in [24]. For a further discussion on term reconstruction and the operational semantics of Elf, the reader is referred to [23].

Schema-Checking — Induction. In this first example, we will directly verify a property of `double`, so there is no need to formulate a translation between deductions as they arise in Sections 3 and 4. The meta-theorem we verify states that `double` is total in its first argument.

> For every valid, closed object `n` of type `nat`, there exists a valid, closed object `M` of type `nat` and a valid, closed object `P` of type `double n M`.

Implicit in this statement is the signature from which the constants in `n`, `M`, and `P` can be drawn, which consists of all the declarations we have considered so far. Henceforth we will omit the adjective "valid" and only consider valid objects. This meta-theorem can be proven by an induction over the canonical forms of LF types and objects constructible from constants in the given signature. This induction argument cannot be internalized. That is, there are no closed objects `M` and `P` such that `P: {n:nat} double n (M n)`.

Now we are at an important crossroads. One choice is to formalize LF and build meta-theorem proving tools so that statements about signatures of the form above can be expressed and verified. This option appears prohibitively complex. The second choice is to identify general, decidable criteria for the totality of relations. A desired theorem is then verified if we can show that it is equivalent to one which satisfies such a schematic criterion.

These alternatives have a connection to the ideas behind primitive recursion. If we would like to show that a function is total, we can either reason about the function within an appropriate logic, or we can present its definition in the form of a primitive recursion. Interestingly, neither choice is *a priori* weaker than the other. For example, the functions provably total in second-order arithmetic are exactly the functions which can be defined using a schema of primitive recursion at higher types [8].

As we hope to illustrate in this paper, the latter choice is a very natural one, and many examples can be treated very elegantly. Moreover, it does not preclude the application of automated theorem proving methods, as they can be used to synthesize schematic relations. This is one of Fribourg's basic observations [9] and illustrated in Section 3. But even without any automatic assistance beyond term reconstruction it is feasible to demonstrate non-trivial meta-theorems.

We now return to the example. The induction principle for objects constructed over the given signature is just the familiar induction principle for natural numbers.

> For any property P, if P holds for `z`, and, whenever P holds for a closed object `n` of type `nat` then it also holds for `(s n)`, then P holds for every closed object of type `nat`.

Interestingly, even though this principle cannot be internalized, instances of this schema for a certain P can be checked by formulating an appropriate Elf query. In this simple case, we can prove that `double` must be total in its first argument if the queries

```
?- double z Qz.
?- {n:nat} {m:nat} double n m -> double (s n) (Qs n m).
```

both succeed. Note that the substitution terms for `Qz` and `Qs`, namely

```
Qz = z,     Qs = [n:nat] [m:nat] s (s m)
```

can be used to synthesize a schema of primitive recursion for the functional version `dbl` of `double`:

```
dbl z = Qz = z
dbl (s n) = Qs n (dbl n) = dbl (s (s n))
```

Here, `[x:A] B` is Elf concrete syntax for $\lambda x{:}A.\ B$.

Many types, such as the type of `exp` defined below, are not inductive in the usual sense (see, for example, [6]). However, we can still derive a form of an induction principle for those types, as we limit ourselves to closed LF terms constructed over a fixed given signature.

3 A Functional Language Fragment

To illustrate our technique further, consider a fragment of some functional language (here $\langle\,\rangle$ represents a unit constant):

$$\textit{Expressions} \quad e \quad ::= \quad x \mid \lambda x.e \mid e\ e \mid \langle\,\rangle \mid \langle e, e\rangle \mid \pi_1(e) \mid \pi_2(e)$$

Syntax. The above syntax can be formulated directly with the following declarations:

```
exp : type.                    unit: exp.
lam : (exp -> exp) -> exp.     pair: exp -> exp -> exp.
app : exp -> exp -> exp.       pi1 : exp -> exp.          pi2 : exp -> exp.
```

The binding construct $\lambda x.e$ is represented using *higher-order abstract syntax*, whereby meta-language abstraction represents object-level binding. This also means that Elf variables of type `exp` serve as variables of our object language.

Judgments. The following is an inference system defining (call-by-value) evaluation of expressions:

$$\frac{}{\lambda x.e \hookrightarrow \lambda x.e}\,\text{lam} \qquad \frac{e_1 \hookrightarrow \lambda x.\, e_1' \qquad e_2 \hookrightarrow v_2 \qquad [v_2/x]e_1' \hookrightarrow v}{e_1\, e_2 \hookrightarrow v}\,\text{app}$$

$$\frac{}{\langle\,\rangle \hookrightarrow \langle\,\rangle}\,\text{unit} \qquad \frac{e_1 \hookrightarrow v_1 \qquad e_2 \hookrightarrow v_2}{\langle e_1, e_2\rangle \hookrightarrow \langle v_1, v_2\rangle}\,\text{pair} \qquad \frac{e \hookrightarrow \langle v_1, v_2\rangle}{\pi_i e \hookrightarrow v_i}\,\text{pi}_i$$

Three more rules determine which expressions are considered to be values:

$$\frac{}{\downarrow \langle\,\rangle}\,\text{unit} \qquad \frac{}{\downarrow \lambda x.\, e}\,\text{lam} \qquad \frac{\downarrow e_1 \qquad \downarrow e_2}{\downarrow \langle e_1, e_2\rangle}\,\text{pair}$$

These inference rules can be transcribed directly into Elf.

```
eval          : exp -> exp -> type.
eval_unit     : eval unit unit.
eval_pair     : eval (pair E1 E2) (pair V1 V2) <- eval E1 V1 <- eval E2 V2.
eval_pi1      : eval (pi1 E) V1               <- eval E (pair V1 V2).
eval_pi2      : eval (pi2 E) V2               <- eval E (pair V1 V2).
eval_lam      : eval (lam E) (lam E).
eval_app_lam : eval (app E1 E2) V
                  <- eval E1 (lam E1')
                  <- eval E2 V2
                  <- eval (E1' V2) V.

value         : exp -> type.
val_unit      : value unit.
val_pair      : value E1 -> value E2 -> value (pair E1 E2).
val_lam       : value (lam E).
```

Deduction Transformations. We would now like to state and verify a simple property relating evaluation and values, namely that the evaluation of an expression always yields a value. This is accomplished via the relation:

```
vr : eval E V -> value V -> type.
```

We have to write this relation in such a way that it can be used to establish the following:

> For every closed object `p` of type `eval e v` (where `e`, `v` are closed objects of type `exp`), there exists a closed object `VP` of type `value v` and a closed object `R` of type `vr p VP`.

To substantiate the above claim we need to define the transformation `vr` by covering the possible cases for `eval`, *i.e.*, for each proof that some expression `E` evaluates to `V` we have to supply a deduction showing that `V` is indeed a value. This is accomplished with the following clauses:

```
vr_unit     : vr (eval_unit) (val_unit).
vr_pair     : vr (eval_pair P2 P1) (val_pair VP1 VP2)
                <- vr P1 VP1
                <- vr P2 VP2.
vr_pi1      : vr (eval_pi1 P) VP1 <- vr P (val_pair VP1 VP2).
vr_pi2      : vr (eval_pi2 P) VP2 <- vr P (val_pair VP1 VP2).
vr_lam      : vr (eval_lam) (val_lam).
vr_app_lam  : vr (eval_app_lam P3 P2 P1) VP3 <- vr P3 VP3.
```

Schema-Checking As in the previous section, schema-checking can be performed by formulating a sequence of queries which check the various cases of the induction proof.

```
?- vr (eval_unit) Qunit.
?- vr (eval_lam) Qlam.
?- {e1:exp} {e2:exp} {v1:exp} {v2:exp} {m:eval e2 v2} {n: eval e1 v1}
   {q: value v2} {q':value v1}
     vr m q -> vr n q'
       -> vr (eval_pair m n) (Qpair e1 e2 v1 v2 m n q q').
?- {e:exp} {v1:exp} {v2:exp} {m:eval e (pair v1 v2)} {q: value (pair v1 v2)}
     vr m q -> vr (eval_pi1 m) (Qpi1 e v1 v2 m q).
?- {e1:exp} {e1':exp -> exp} {e2:exp} {v:exp} {v2:exp} {m:eval (e1' v2) v}
   {n: eval e2 v2} {o: eval e1 (lam e1')} {q: value v} {q': value v2}
   {q'': value (lam e1')}
     vr m q -> vr n q' -> vr o q''
       -> vr (eval_app_lam m n o) (Qapp_lam e1 e1' e2 v v2 m n o q q' q'').
```

Unfortunately, the cases for `eval_pi1` and `eval_pi2` fail, while the others succeed. When analyzing the rule

```
vr_pi1 : vr (eval_pi1 P) VP1 <- vr P (val_pair VP1 VP2).
```

the reason becomes clear: in order to show totality of `vr` in its first argument, we need to know that, in this case, `vr P (val_pair VP1 VP2)` always succeeds! But this relies on a subtle point: `P` has type `eval E (pair V1 V2)` for some `E`, `V1`, and `V2`, and therefore the second argument of `vr P VP` must have type `value (pair V1 V2)`. But there is only one rule constructing a deduction with this conclusion, and therefore any closed `VP` of this type *must* have the form `(val_pair VP1 VP2)` for some `VP1` and `VP2`. This observation gives rise to the following *unit refinement* principle:

> For any property P such that P holds of `q : value (pair v1 v2)` there exist `q1 : value v1` and `q2 : value v2` such that P holds of `value_pair q1 q2`.

Note that such a principle holds whenever the principal constructor of deductions of a given judgment is uniquely determined from the judgment. This kind of reasoning arises quite often in informal meta-proofs, sometimes in a more general form using an auxiliary induction. It is closely related to the notion of *iff*-completion of logic programs.

If we Skolemize this unit refinement principle and refine our induction accordingly, the queries can now be executed successfully. The additional assumptions appear as constants `q1` and `q2`.

```
?- {e:exp} {v1:exp} {v2:exp} {m:eval e (pair v1 v2)} {q:value (pair v1 v2)}
   {q1:value (pair v1 v2) -> value v1} {q2:value (pair v1 v2) -> value v2}
     ({Q:value (pair v1 v2)} vr m Q -> vr m (val_pair (q1 Q) (q2 Q)))
     -> vr m q
     -> vr (eval_pi1 m) (Qpi1 e v1 v2 m q q1 q2).
```

This succeeds with substitution `Qpi1 = [e][v1][v2][m][q][q1][q2] q1 q`.

4 Logic Interpretations and CPS Transform

In [11], Griffin presents a number of interpretations between logics and shows how they can be viewed as computational simulations. We have transcribed and verified 8 of these interpretations. In this section we will verify the type-soundness of the *continuation-passing-style* (CPS) transform of Plotkin [25], which is one of Griffin's examples.

Syntax. Once more, the first task will be to represent the logics under consideration. Here we deal with a propositional logic, which allows their direct interpretation as types of a programming language (using the Curry-Howard isomorphism). We use α and β to range over formulas.

$$\textit{Formulas} \quad \alpha \quad ::= \quad p \mid \bot \mid \alpha \to \alpha \mid \alpha \wedge \alpha \mid \alpha \vee \alpha$$

The representation of these in Elf is straightforward. Propositional variables p are not directly represented, but become meta-variables, that is, variables in the meta-language Elf.

```
form : type.
bot : form.                   imp : form -> form -> form.
and : form -> form -> form.   or  : form -> form -> form.
```

Judgments. Now we would like to formulate the necessary judgments. These are provability in minimal and classical propositional logic. Instead of separating expressions in a programming language and proofs in the propositional calculi, we can think of the deductions of formulas in these logics directly as functional programs.

$$\begin{array}{lll}\textit{Proofs} \quad M & ::= & x \mid \lambda x.M \mid MM \mid \langle M, M\rangle \mid \pi_1(M) \mid \pi_2(M) \\ & & \mid \mathrm{inj}_1(M) \mid \mathrm{inj}_2(M) \mid \mathrm{case}(M, \lambda x.M, \lambda x.M)\end{array}$$

Not all expressions which follow this grammar are actually meaningful. For example, $\pi_1(\lambda x.x)$ does not make sense as a proof. In the Elf representation below we directly capture the conditions under which a proof is meaningful or *valid.* Essentially, a proof is indexed by the formula that it establishes. This can also be thought of as a refinement of the representation of untyped programs in Section 3, by indexing expressions by their type, thus dividing and restricting the space of legal programs. For a further discussion of such issues of representation, the reader is referred to [16].

```
pf : form -> type.
```

```
%% minimal logic
lam  : (pf A -> pf B) -> pf (imp A B).
app  : pf (imp A B) -> pf A -> pf B.
pair : pf A -> pf B -> pf (and A B).
pi1  : pf (and A B) -> pf A.
pi2  : pf (and A B) -> pf B.
inj1 : pf A -> pf (or A B).
inj2 : pf B -> pf (or A B).
case : pf (or A B) -> (pf A -> pf C) -> (pf B -> pf C) -> pf C.
aa   : pf bot -> pf A.                     %% intuitionistic logic
kk   : pf (imp (imp A bot) A) -> pf A.    %% classical logic
```

For example, **app** takes a proof of $\alpha \to \beta$ and a proof of α and returns a proof of β.[1] This representation is *adequate* in the sense that every natural deduction of a formula α can be represented by an object of type **pf A** and vice versa. Here **A** is the representation of α.

The proof constructors **aa** (representing the intuitionistic rule of absurdity) corresponds to an aborting operator, and **kk** corresponds to Scheme's `call/cc`, though this correspondence is not explored here.

Deduction Transformations. We now present the interpretation $\overline{()}$ which maps classically provable formulas into formulas provable in minimal logic.

$$\overline{\alpha} = \neg\neg\alpha^*$$

$$\begin{aligned} \bot^* &= \bot \\ p^* &= p \\ (\alpha \wedge \beta)^* &= \alpha^* \wedge \beta^* \\ (\alpha \vee \beta)^* &= \alpha^* \vee \beta^* \\ (\alpha \to \beta)^* &= \alpha^* \to \overline{\beta} \end{aligned}$$

The mutually recursive definitions of translations $\overline{()}$ and $()^*$ are easily represented relationally in Elf.[2] This representation is operationally adequate.

```
cps- : form -> form -> type.
cps* : form -> form -> type.

dblneg : cps- A (imp (imp A* bot) bot) <- cps* A A*.

cps*_bot : cps* bot bot.
cps*_and : cps* (and A B) (and A* B*) <- cps* A A* <- cps* B B*.
cps*_or  : cps* (or A B)  (or A* B*)  <- cps* A A* <- cps* B B*.
cps*_imp : cps* (imp A B) (imp A* B-) <- cps* A A* <- cps- B B-.
```

This represents the translations above in the following sense: $\alpha^* = \beta$ holds iff `cps* A B`, where **A** is the representation of α, and **B** is the representation of β. Similary, translation $\overline{()}$ is represented by `cps-`. We will verify the following theorem:

[1] The familiar rules for natural deduction are easily recognized: **lam** $\simeq$ $\supset$-Intro, **app** $\simeq$ $\supset$-Elim, **pair** $\simeq$ $\wedge$-Intro, **pi**$_i$ $\simeq$ $\wedge$-Elim, **inj**$_i$ $\simeq$ $\vee$-Intro, and **case** $\simeq$ $\vee$-Elim.

[2] Note that **A***, **B-**, *etc.* are Elf variables.

If α is provable in classical logic, then $\overline{\alpha}$ is provable in minimal logic.

This is shown by a translation on the deductions: every classical deduction of α is transformed into an minimal deduction of $\overline{\alpha}$.

$$
\begin{array}{rcl}
\overline{x} & = & \lambda k.kx \\
\overline{\lambda x.M} & = & \lambda k.k(\lambda x.\overline{M}) \\
\overline{MN} & = & \lambda k.\overline{M}(\lambda m.\overline{N}(\lambda n.mnk)) \\
\overline{\langle M, N\rangle} & = & \lambda k.\overline{M}(\lambda m.\overline{N}(\lambda n.k\langle m, n\rangle)) \\
\overline{\pi_i(M)} & = & \lambda k.\overline{M}(\lambda m.k(\pi_i(m))) \\
\overline{\mathrm{inj}_i} & = & \lambda k.\overline{M}(\lambda m.k(\mathrm{inj}_i(m))) \\
\overline{\mathrm{case}(M, \lambda x.N, \lambda y.Q)} & = & \lambda k.\overline{M}(\lambda m.\mathrm{case}(m, \lambda x.\overline{N}k, \lambda y.\overline{Q}k)) \\
\overline{\mathcal{A}(M)} & = & \lambda k.\overline{M}(\lambda m.m) \\
\overline{\mathcal{K}(M)} & = & \lambda k.\overline{M}(\lambda m.m(\lambda z.\lambda d.kz)k)
\end{array}
$$

The representation of this translation in Elf is very direct and reproduced below. It is worth noting that the higher-level judgment connecting deductions of α and $\overline{\alpha}$ must explicitly refer to this translation. It would not suffice to specify

```
cps : pf A -> pf A- -> type.
```

since we need to show how to relate a proof of `A` to a proof of `A-`, where `A-` is the translation of `A` under $\overline{(\ )}$. Thus the main judgment is

```
cps : pf A -> cps- A A- -> pf A- -> type.
```

where `A` and `A-` are implicitly quantified. This judgment can be specified with the following rules (omitting the symmetric cases for π_2 and inj_2):

```
cps_lam: cps (lam M) (dblneg (cps*_imp CpsB- CpsA*)) (lam [k] app k (lam M-))
        <- ({x} {x-} cps x (dblneg CpsA*) (lam [k] app k x-)
                     -> cps (M x) CpsB- (M- x-)).
cps_app: cps (app M N) CpsB-
           (lam [k] app M- (lam [m-] app N- (lam [n-] app (app m- n-) k)))
        <- cps M (dblneg (cps*_imp CpsB- CpsA*)) M-
        <- cps N (dblneg CpsA*) N-.
cps_pair: cps (pair M N) (dblneg (cps*_and CpsB* CpsA*))
            (lam [k] app M- (lam [m-] app N- (lam [n-]app k (pair m- n-))))
         <- cps M (dblneg CpsA*) M-
         <- cps N (dblneg CpsB*) N-.
cps_pi1: cps (pi1 M) (dblneg CpsA*)
           (lam [k] app M- (lam [m-] app k (pi1 m-)))
        <- cps M (dblneg (cps*_and CpsB* CpsA*)) M-.
cps_inj1: cps (inj1 M) (dblneg (cps*_or CpsB* CpsA*))
            (lam [k] app M- (lam [m-] app k (inj1 m-)))
         <- cps M (dblneg CpsA*) M-.
cps_case: cps (case M N Q) (dblneg CpsC*)
            (lam [k] app M- (lam [m-] case m- ([x-] app (N- x-) k)
                                              ([y-] app (Q- y-) k)))
         <- cps M (dblneg (cps*_or CpsB* CpsA*)) M-
         <- ({x} {x-} cps x (dblneg CpsA*) (lam [k] app k x-)
```

```
                    -> cps (N x) (dblneg CpsC*) (N- x-))
        <- ({y} {y-} cps y (dblneg CpsB*) (lam [k] app k y-)
                    -> cps (Q y) (dblneg CpsC*) (Q- y-)).
cps_aa: cps (aa M) (dblneg CpsA*)
           (lam [k] app M- (lam [m-] m-))
       <- cps M (dblneg cps*_bot) M-.
cps_kk: cps (kk M) (dblneg CpsA*)
           (lam [k] app M- (lam [m-]app (app m- (lam [z]lam [d]app k z)) k))
       <- cps M (dblneg (cps*_imp (dblneg CpsA*)
                            (cps*_imp (dblneg cps*_bot) CpsA*)))
              M-.
```

What does type-checking of this signature guarantee for us here? Reexamining the type of cps

```
cps : pf A -> cps- A A- -> pf A- -> type.
```

shows us that if a query `?- cps M CpsA N.` succeeds for some `M`, `CpsA`, and `N`, then `M` will be a proof of some formula `A`, `N` will be a proof of `A-`, and `CpsA` will be a deduction showing that `A-` is the translation of `A`. This is an important property, and many obviously incorrect translations will be caught at this stage because of type-checking errors, but it does not guarantee our theorem in any way—this is where additional *schema-checking* is required.

Schema-Checking. In order to demonstrate the theorem, we need to show that cps- is total in its first argument, and we also need to show that cps is total in its first two arguments.

Showing the totality of cps- demonstrates that for a given `A` we can construct an `A-` and a deduction showing that `cps- A A-`. This amounts to showing that $\overline{(\,)}$ is a total function on formulas.

Showing that cps is total in its first two arguments means that for a given proof `M` of type `pf A` and translation `cps- A A-` there exists a proof `M-` of type `pf A-`. That such translations always exist almost verifies the claimed theorem: It remains to be shown that the translated proof lies within the minimal fragment of the logic under consideration. This can be guaranteed through proper use of the module system for Elf, which is beyond the scope of this paper and can be found in [17].

Translations $\overline{(\,)}$ and $(\,)^*$ are mutually recursive. We add the induction hypotheses for both functions in constructing the query for this particular case (the other cases can be proven similarly):

```
?- {m:form} {n:form} {p:form} {q:form} {r:form} {s:form}
     cps* m p -> cps- m q -> cps* n r -> cps- n s
       -> cps* (imp m n) (Qand m n p q r).

Qand = [m] [n] [p] [q] [r] imp p (imp (imp r bot) bot)
```

In the induction proof for the totality of cps we will have to employ unit refinement again: Any proof `q : cps- A A-` will necessarily use the clause dblneg. In Skolemized form, this means that there is a total function which maps any deduction of `q : cps- A A-` to a deduction `q' : cps* A A*` such that q has the form

dblneg q'. A special case of this can be expressed by the first two of the following six declarations—the others express the totality of cps- and cps*.[3]

```
cps-_refine: cps- A A- -> cps* A A*.
cps-_refine_lemma: cps X Q Z -> cps X (dblneg (cps-_refine Q)) Z.
cps*_tot: form -> form.          cps*_tot_lemma: cps* M (cps*_tot M).
cps-_tot: form -> form.          cps-_tot_lemma: cps- N (cps-_tot N).
```

Representative for schema-checking we display the query for the case of pairs and its result:

```
?- {a:form} {b:form} {a*: form} {b*: form} {x:pf a} {y:pf b}
   {m:cps- a (imp (imp a* bot) bot)} {p:pf (imp (imp a* bot) bot)}
   {n:cps- b (imp (imp b* bot) bot)} {q:pf (imp (imp b* bot) bot)}
     cps x m p -> cps y n q
       -> cps (pair x y) (Qpair a b a* b* x y m p n q)
                         (Rpair a b a* b* x y m p n q).
Rpair = [a] [b] [a*] [b*] [x] [y] [m] [p] [n] [q]
          lam ([k:pf (imp (and a* b*) bot)]
                 app p
                     (lam ([m-:pf a*]
                              app q (lam ([n-:pf b*] app k (pair m- n-)))))),
Qpair = [a] [b] [a*] [b*] [x] [y] [m] [p] [n] [q]
          dblneg (cps*_and (cps-_refine n) (cps-_refine m)).
```

5 Conclusion

We have outlined a practical methodology for the implementation and verification of the meta-theory of deductive systems. This methodology employs the LF logical framework as a basis and consists of four stages: (1) representation of syntax, (2) specification of judgments, (3) implementation of transformations between deductions, and (4) checking that the transformations are total in some of their arguments. This last stage is called *schema-checking* and relies on induction over the closed valid terms which can be constructed over a signature in the LF type theory. While a significant part of the verification is automatic through term reconstruction, type-checking, and schema-checking, much of the work is still mechanical and, we hope, amenable to methods from the field of inductive theorem proving. This would mean that an Elf theorem prover would try to synthesize an appropriate deduction transformation, such as tr or cps above.

We currently have a small prototype implementation of schema-checking as an extension of the current Elf core language [22]. Several further verifications of standard meta-theorems in logic and computer science using the methods described here are subject of current work.

[3] We omit from this discussion the unit refinements for the different cases of cps* that are needed in other parts of the proof.

References

[1] Arnon Avron, Furio A. Honsell, and Ian A. Mason. Using typed lambda calculus to implement formal systems on a machine. Technical Report ECS-LFCS-87-31, Laboratory for Foundations of Computer Science, University of Edinburgh, Edinburgh, Scotland, June 1987.

[2] David A. Basin and Robert L. Constable. Metalogical frameworks. Submitted, July 1991.

[3] Alan Bundy, Frank van Harmelen, Alan Smaill, and Andrew Ireland. Extensions to the rippling-out tactic for guiding inductive proofs. In M. E. Stickel, editor, *10th International Conference on Automated Deduction*, pages 132–146. Springer-Verlag LNAI 449, 1990.

[4] Rod Burstall and Furio Honsell. Operational semantics in a natural deduction setting. In Gérard Huet and Gordon Plotkin, editors, *Logical Frameworks*, pages 185–214. Cambridge University Press, 1991.

[5] Dominique Clément, Joëlle Despeyroux, Thierry Despeyroux, and Gilles Kahn. A simple applicative language: Mini-ML. In *Proceedings of the 1986 Conference on LISP and Functional Programming*, pages 13–27. ACM Press, 1986.

[6] Thierry Coquand and Christine Paulin. Inductively defined types. In P. Martin-Löf and G. Mints, editors, *COLOG-88*, pages 50–66. Springer-Verlag LNCS 417, December 1988.

[7] Conal M. Elliott. *Extensions and Applications of Higher-Order Unification*. PhD thesis, School of Computer Science, Carnegie Mellon University, May 1990. Available as Technical Report CMU-CS-90-134.

[8] Steven Fortune, Daniel Leivant, and Michael O'Donnell. The expressiveness of simple and second-order type structures. *Journal of the ACM*, 30:151–185, 1983.

[9] Laurent Fribourg. Extracting logic programs from proofs that use extended Prolog execution and induction. In David H.D. Warren and Peter Szeredi, editors, *Proceedings of the Seventh International Conference on Logic Programming*, pages 685–699. MIT Press, June 1990.

[10] Gerhard Gentzen. Untersuchungen über das logische Schließen. *Mathematische Zeitschrift*, 39:176–210, 405–431, 1935.

[11] Timothy G. Griffin. Logical interpretations as computational simulations. Draft paper. Talk given at the North American Jumelage, AT&T Bell Laboratories, Murray Hill, New Jersey 1991, October 1991.

[12] Lars Hallnäs and Peter Schroeder-Heister. A proof-theoretic approach to logic programming. *Journal of Logic and Computation*, 1(5):635–660, 1991.

[13] John Hannan and Dale Miller. From operational semantics to abstract machines: Preliminary results. In M. Wand, editor, *ACM Conference on Lisp and Functional Programming*, pages 323–332. ACM Press, 1990.

[14] John Hannan and Frank Pfenning. Compiler verification in LF. In *Seventh Annual IEEE Symposium on Logic in Computer Science*, Santa Cruz, California, June 1992. IEEE Computer Society Press. To appear. Preliminary version available as POP Report 91–003, School of Computer Science, Carnegie Mellon University.

[15] Robert Harper. Systems of polymorphic type assignment in LF. Technical Report CMU-CS-90-144, Carnegie Mellon University, Pittsburgh, Pennsylvania, June 1990.

[16] Robert Harper, Furio Honsell, and Gordon Plotkin. A framework for defining logics. *Journal of the ACM*, To appear. A preliminary version appeared in *Symposium on Logic in Computer Science*, pages 194–204, June 1987.

[17] Robert Harper and Frank Pfenning. Modularity in the LF logical framework. Submitted. Available as POP Report 91–001, School of Computer Science, Carnegie Mellon University, November 1991.

[18] G. Kahn. Natural semantics. In *Proceedings of the Symposium on Theoretical Aspects of Computer Science*, pages 22–39. Springer-Verlag LNCS 247, 1987.

[19] Ian A. Mason. Hoare's logic in the LF. Technical Report ECS-LFCS-87-32, Laboratory for Foundations of Computer Science, University of Edinburgh, Edinburgh, Scotland, June 1987.

[20] Spiro Michaylov and Frank Pfenning. Natural semantics and some of its meta-theory in Elf. In Lars Hallnäs, editor, *Extensions of Logic Programming*. Springer-Verlag LNCS. To appear. A preliminary version is available as Technical Report MPI–I–91–211, Max-Planck-Institute for Computer Science, Saarbrücken, Germany, August 1991.

[21] Frank Pfenning. Elf: A language for logic definition and verified meta-programming. In *Fourth Annual Symposium on Logic in Computer Science*, pages 313–322. IEEE, June 1989.

[22] Frank Pfenning. An implementation of the Elf core language in Standard ML. Available via ftp over the Internet, September 1991. Send mail to elf-request@cs.cmu.edu for further information.

[23] Frank Pfenning. Logic programming in the LF logical framework. In Gérard Huet and Gordon Plotkin, editors, *Logical Frameworks*, pages 149–181. Cambridge University Press, 1991.

[24] Frank Pfenning. Unification and anti-unification in the Calculus of Constructions. In *Sixth Annual IEEE Symposium on Logic in Computer Science*, pages 74–85. IEEE Computer Society Press, July 1991.

[25] G. D. Plotkin. Call-by-name, call-by-value and the λ-calculus. *Theoretical Computer Science*, 1:125–159, 1975.

[26] Lutz Plümer. *Termination Proofs for Logic Programs*. Springer-Verlag LNAI 446, 1991.

[27] David Pym. *Proofs, Search and Computation in General Logic*. PhD thesis, University of Edinburgh, 1990. Available as CST-69-90, also published as ECS-LFCS-90-125.

[28] David Pym and Lincoln Wallen. Investigations into proof-search in a system of first-order dependent function types. In M.E. Stickel, editor, *10th International Conference on Automated Deduction, Kaiserslautern, Germany*, pages 236–250. Springer-Verlag LNCS 449, July 1990.

Tactic-based Theorem Proving and Knowledge-based Forward Chaining: an Experiment with Nuprl and Ontic

Wilfred Z. Chen*

Department of Computer Science
Cornell University
Ithaca, NY 14853 USA
chen@cs.cornell.edu

Abstract. We explore a new approach to interactive theorem proving which combines a knowledge-based notion of obvious reasoning with a tactic-based notion of obvious reasoning. We study the interplay of two particular systems and apply our approach to a proof of the Fundamental Theorem of Arithmetic. We achieve both shorter and more robust proofs. It is our opinion that the kind of control information contained in interactive proofs is a more important issue than their mere size. We analyze our proof scripts in terms of the control information they contain and suggest that stronger knowledge-based notions of obviousness and declarative representations of tactics are needed to further reduce low-level control information.

1 Introduction

Tactic-based theorem proving has received a lot of attention lately [13, 9, 12, 11, 27]. Forward chaining theorem proving has also been studied [8, 19]. In this paper, we explore the potential for combining these two approaches. In particular, we study the combination of **Nuprl** and **Ontic**.

The motivation for studying the combination of these two different approaches is the hope that their strengths and weaknesses will compensate each other and produce proofs that are both shorter and more robust.

Although the current collection of tactics in **Nuprl** has provided indispensable assistance in developing proofs [14, 23, 1], these tactics are often sensitive to minor variations in the formulation of the problem, to changes in the lemma library and to improvements in some standard tactics – the Autotactic, for example. This problem is acute – strengthening a tactic or the lemma collection can damage previously checked proof scripts. It is also general – other **LCF** descendants, such as **HOL**, suffer similarly: providing low-level control information to the system is not only extremely tedious, it also makes proofs harder to reuse. By applying knowledge in the form of a large lemma collection, we hope to reduce the amount of detail.

In contrast, **Ontic** proof scripts replay under any strengthening of the lemma collection, modulo practical feasibility considerations. But there is no mechanism to

* The author would like to acknowledge support for this work under NSF grant CCR-9108062 and ONR grant N000014-88K-0409.

encode simple patterns of inference, so similar subproofs are duplicated. So we hope to find regular patterns of inferences – clichés – that might be captured using tactics.

We expect fruitful cooperation between the tactic-style theorem proving and the forward chaining technique because the former is suitable for capturing and combining dynamic patterns in inferences while the latter does the same to static patterns, in the form of lemmas. Both forms of knowledge are necessary in problem solving.

How are these (dynamic) patterns discovered? Our first step is to implement a knowledge-based forward chaining autotactic and gain some experience by applying it to compress some known proofs in **Nuprl**.

This paper contains an introduction to the main ideas of the knowledge-intensive forward chaining technique and a report on a case study (the fundamental theorem of arithmetic). We obtain shorter and more robust proof scripts by combining the tactical approach and the forward chaining approach.

The paper is organized as follows: Section 2 lays down the general background of this work, Section 3 is an introduction to the forward chaining approach, to prepare the reader for a case study of integrating forward chaining into a tactic-based environment in Section 4, finally we summarize and expand in Section 5.

2 The Theorem Proving Context

2.1 Tactical Theorem Proving

Tactic-based theorem proving systems (**Nuprl**, **HOL**, **Isabelle**) are descendants of **LCF**. Although their object languages differ, a common feature is the meta-language ML [2]. Tactics are programs written in ML to refine a (sequent style) goal into (presumably) simpler subgoals. The simplest tactics are the primitive refinement rules. The higher-order programming style of ML facilitates the combination of simple tactics into complex ones.

In these systems, a collection of tactics defines a notion of obvious inferences.

In our work, we use a subset of the (object) language of **Nuprl**. The **Nuprl** subset we use is essentially an order-sorted first order logic with subset, pairing and lists. Equality reasoning in this language is simpler than in the full language of **Nuprl**. In particular, it permits the use of a single global equality [16].

2.2 Knowledge-Intensive Forward Chaining

There are two important issues in applying a large number of lemmas to a reasoning task: selecting the relevant lemmas and ensuring termination. **Ontic** addresses both issues with the concept of focusing. The design objectives of **Ontic** are: (1) to obtain a practically effective proof verification system and (2) to model low-level human mathematical reasoning. It has been used to verify the Stone representation theorem for boolean lattices. Most existing interactive proof development systems do not share the second objective of **Ontic**.

[2] **Oyster**, the Edinburgh version of **Nuprl**, uses PROLOG as its meta-language.

Ontic has a knowledge-based (declarative) counterpart to the tactic-based (procedural) notion of obvious inference in the systems mentioned above.

We explain **Ontic** in more detail below.

3 The Forward Chaining Tactic

This section serves two purposes: to explain the basic ideas of **Ontic** and to present, at a high level, a particular **Ontic**-like system used in our work. Our version of **Ontic** removes some of the features of **Ontic** that are not necessary for our experiment – among them, automatic universal generalization [3].

3.1 Ontic in a Nutshell

The decision procedure **Ontic** is defined by a set of inference rules and a set of lemmas. Basically, inference rules are applied under a *mention restriction*, where only those instances containing subterms in the subterm closure of all input formulas are allowed; lemmas are applied under *focus restriction*, where all members in a certain closure of the set of lemmas are instantiated on a given set of terms, the *focus set*.

The reader is referred to [19] for a complete description of **Ontic** and to [20] for a general theory of tractable inferences.

3.2 Forward Chaining Inference Relations

We present the forward chaining autotactic as a family of tractable inference relations indexed by a finite set of terms, called *focus terms*.

The inference relation presented below is not new. It is different from those in the original **Ontic** mainly in the constructive validity of the inference rules and the use of (almost) standard first order syntax (**Ontic** takes advantage of non-standard syntax to obtain a larger tractable fragment [18]).

Let $\vdash$ denote the usual logical derivability relation in first order logic, specialized to our particular language. Let $\vdash\!\!\circ_{\mathcal{F}}$ denote the parameterized tractable approximations to $\vdash$ that we shall define. Let $\vdash\!\!\circ$ abbreviate $\vdash\!\!\circ_{\{\}}$.

For each $\mathcal{F}$, $\vdash\!\!\circ_{\mathcal{F}}$ is quickly (polynomial time) decidable fragment of $\vdash$. A *high level proof* consists of steps of single inference rules that define $\vdash$ mixed with steps of $\vdash\!\!\circ_{\mathcal{F}}$.

$\vdash\!\!\circ_{\mathcal{F}}$ will be presented in two steps: first the inference rules that make up $\vdash\!\!\circ_{\mathcal{F}}$, and second, the restrictions on when these rules are applicable. For the algorithms that implement these rules and restrictions, we refer the reader to the original papers of McAllester.

[3] There are two reasons for this: (1) tactics already perform introduction of universal quantifiers and, more subtly, (2) adding universal generalization reduces the efficiency of forward chaining considerably.

3.3 The Inference Rules

Here we give a set of inference rules that are constructively valid, so that forward chaining proofs using them can be converted to **Nuprl** proofs.

The inference rules that go into $\vdash\!\circ_{\mathcal{F}}$ shall be presented in sequent calculus style. So a typical rule would have the form

$$\frac{\Sigma \vdash\!\circ_{\mathcal{F}} \Phi \quad \Sigma' \vdash\!\circ_{\mathcal{F}'} \Psi}{\Sigma'' \vdash\!\circ_{\mathcal{F}''} \Theta}.$$

However, when the Σ's and $\mathcal{F}$'s are identical in all places, the simpler form $\frac{\Phi \quad \Psi}{\Theta}$ is used.

Propositional Rules Following are the so-called Boolean Constraint Propagation rules:

$$\frac{\Phi \quad \Psi}{\Phi \& \Psi} \quad \frac{\Phi \& \Psi}{\Phi} \quad \frac{\Phi \& \Psi}{\Psi} \quad \frac{\Phi}{\Phi \vee \Psi} \quad \frac{\Psi}{\Phi \vee \Psi} \quad \frac{\Phi \quad \Phi \Rightarrow \Psi}{\Psi} \quad \frac{\bot}{\Phi}$$

These rules do not define a complete inference relation for intuitionistic propositional logic because $\vee$–elim and $\Rightarrow$–intro are absent. These rules require searching or guessing hence are not used automatically in forward chaining.

Equality Rules The forward chaining rules for equality reasoning are reflexivity, symmetry, transitivity and congruence (substitution of equals for equals). A complete decision procedure for these rules is congruence closure [17, 26, 10] which is used by all implementations of **Ontic**. Congruence closure is valid for the fragment of **Nuprl** used here [16].

Quantifier Rules

$$\frac{\Sigma \vdash\!\circ_{\mathcal{F}} \forall x : T.\Phi(x) \qquad \Sigma \vdash\!\circ_{\mathcal{F}} (a \,\mathbf{in}\, T)}{\Sigma \vdash\!\circ_{\mathcal{F} \cup \{a\}} \Phi(a)} \ (\forall\text{-E}) \qquad \frac{\Sigma \vdash\!\circ_{\mathcal{F}} \Phi(a) \qquad \Sigma \vdash\!\circ_{\mathcal{F}} (a \,\mathbf{in}\, T)}{\Sigma \vdash\!\circ_{\mathcal{F} \cup \{a\}} \exists x : T.\,\Phi(x)} \ (\exists\text{-I})$$

$\forall$-E is much more important than $\exists$-I in practice, and it will receive more attention in the following explanations.

Note the absence of $\forall$-I and $\exists$-E. $\forall$-I is usually applied by tactics in the beginning of a refinement style proof. $\exists$-E introduces new constants that the rest of the proof somehow refers to. If this is done by the forward chaining tactic, the rest of the proofs would have no access to these constants.

Type Inference Rules The following rules allow the inference of types of focus terms in the presence of set types:

$$\frac{}{\forall x : T.\, x \,\mathbf{in}\, T} \qquad \frac{\Sigma \vdash\!\circ_{\mathcal{F}} \Phi(a) \qquad \Sigma \vdash\!\circ_{\mathcal{F}} (a \,\mathbf{in}\, T)}{\Sigma \vdash\!\circ_{\mathcal{F} \cup \{a\}} a \,\mathbf{in}\, \{x : T | \Phi(x)\}}$$

Note that the first rule cannot be a lemma schema, because lemmas are only applied to focus terms and the application requires that 'x **in** T' be true.

3.4 Restrictions on Rule Applicability

In addition to the *focus restriction* which is implicit in the quantifier rules, all rules are subjected to the so-called *mention restriction.* These two restrictions together ensure that $\vdash\!\circ_{\mathcal{F}}$ is quickly decidable. Focus restriction can be understood by inspecting the quantifier rules carefully; mention restriction is the requirement that for each application of a forward chaining rule, the Σs, Φs, Ψs and Θs must all be subexpressions of the rule set, lemma library, current goal sequent or focus set.

3.5 Examples

Example 1. Let $\Sigma = \{\forall x, y : int.\, x + y = y + x, \forall x : int.\, x + 0 = x\}$. Then we can prove $\Sigma \vdash \forall x : int.\, 0 + x = x$ in two 'high-level' steps:

$$\cfrac{\dfrac{\forall x, y : int.\, x + y = y + x \quad \forall x : int.\, x + 0 = x}{0 + x = x}\;(\vdash\!\circ_{\{x,0\}})}{\forall x : int.\, 0 + x = x}\;(\forall\text{-I})$$

Note that in the first step, all that the user has to supply is the focus set $\{x, 0\}$; applications of both lemmas and some equality inferences are done by the system.

In practice, due to the high complexity in terms of the focus set size (see [19]), we often limit the focus set size to that of the deepest quantifier nesting. The next example shows that a larger focus set means fewer steps.

Example 2. Let $f : S \to T$ and let $\Sigma = \{\forall x : T.\, \Phi(x), \forall x : S.\, \Phi(f(x)) \Rightarrow \Psi(x)\}$. Suppose we want to prove $\forall x : S.\, \Psi(x)$. Let a be a new constant of type S, then

$$\frac{\forall x : T.\, \Phi(x)}{\Phi(f(a))}\;(\vdash\!\circ_{\{f(a)\}}) \qquad \frac{\forall x : S.\, \Phi(f(x)) \Rightarrow \Psi(x)}{\Phi(f(a)) \Rightarrow \Psi(a)}\;(\vdash\!\circ_{\{a\}})$$

and $\Psi(a)$ follows by boolean constraint propagation.

Using the focus set $\mathcal{F} = \{a, f(a)\}$, the above can be done in a single step of $\forall$-E (boolean constraint propagation does not count as a *visible* step).

3.6 Some Properties of Ontic

What is the difference between **Ontic**'s use of lemmas and a naive backchaining search through the entire collection of lemmas? **Ontic** is polynomially bounded in some sense, hence can be regarded as forward chaining rather than searching. However, we wish to be a little more precise about the complexity of **Ontic**. The space complexity of **Ontic** is $\mathcal{O}(n^d)$, where n is the size of the focus set and d is the deepest quantifier nesting. In general, n is at least as large as d. In practice, d is about 3 or 4.

Subformulas Property The focused instantiation rule ∀-E lacks subformulas property in a rather strong sense.

Example 3. Let

$$\Sigma = \left\{ \begin{array}{l} \forall x, y, z : \mathcal{N}.\, max3(x,y,z) = max(max(x,y),z), \\ \forall x, y : \mathcal{N}.\, max(x,y) = max(y,x), \\ \forall x, y, z : \mathcal{N}.\, max(max(x,y),z) = max(x, max(y,z)) \end{array} \right\}$$

Then $max3(a,b,c) = max3(a,c,b)$ follows from Σ, with a one step proof which requires focusing on $\{a,b,c\}$. But $max3(a,b,c) = max3(c,a,b)$ requires, in addition, $max(a,b)$ in the focus set for a one step proof. Note that $max(a,b)$ does not occur as a subterm of the goal or any lemmas in the library.

Repeated Use of Lemmas Let $\Sigma = \{\forall x : int.\, x+1 > x\}$. It has the obvious consequence $(\ldots(a+1)+\ldots+1) > a$, but **Ontic** cannot deduce this in a single step because it needs to focus on the intermediate terms. Interestingly, we can infer the same from $a+1 > a$, where a lemma about the monotonicity of $+$ is used multiple times, implicitly.

Consider again one of the lemmas in Example 2. Let $\Sigma = \{\forall x : S.\, \Phi(f(x)) \Rightarrow \Phi(x)\}$. Then

$$\frac{\forall x : S.\, \Phi(f(x)) \Rightarrow \Phi(x) \qquad \Phi(f^n(a))}{\Phi(a)} \; (\vdash\!\!\circ_{\{a, f(a), \ldots, f^{n-1}(a)\}})$$

The conclusion is equally obvious for any n, but the focus set size grows linearly with it. Induction and rewriting are two possible solutions to this.

Sequencing of Focus Sets Because of the space complexity associated with large focus sets, simulating a large focus set with a sequence of smaller focus sets is appealing. Let $\vdash\!\!\circ_{<\mathcal{F}_1,\ldots,\mathcal{F}_n>}$ denote the inference relation based on the sequence of focus sets.

In order for the above to be different from $\vdash\!\!\circ_{\mathcal{F}_n}$, facts derived from the previous focus sets must be saved. The definition of $\vdash\!\!\circ_{<\mathcal{F}_1,\ldots,\mathcal{F}_n>}$ depends on what facts are saved between successive focus sets. A reasonable choice is to use the mention restriction again. Under this notion of *focus sequencing*, $\vdash\!\!\circ_{<\mathcal{F}_1,\mathcal{F}_2>}$ is not equivalent to $\vdash\!\!\circ_{<\mathcal{F}_2,\mathcal{F}_1>}$ in general. Moreover, there exists a focus set $\mathcal{F}$ such that no focus sequencing over proper subsets of $\mathcal{F}$ is equivalent to focusing on $\mathcal{F}$ itself [6] .

Relationship to Unification-based Systems Because ∀-E and ∃-I are restricted by the focus set, matching and unification can be avoided, and lemma instantiation in **Ontic** is implemented via pre-instantiation on generic constants, focus binding with these constants and congruence closure. This approach gives both the tractability of **Ontic** and some of the weaknesses mentioned above, because lemma application on intermediate results is limited.

What is a better trade-off? This question can only be answered in the context of a theorem proving environment: when several tools are available, the virtue of each may not be its individual power, but rather how well it works with other tools. We cannot give a full answer to this question here.

4 The Fundamental Theorem of Arithmetic: a case study

Here we wish to evaluate the effectiveness of applying mathematical knowledge in the form of a lemma library via focused forward chaining similar to **Ontic**. Our choice of number theory is motivated by the need to improve arithmetical reasoning in **Nuprl**. We want to see whether the knowledge-based approach gives a significant boost to the tactical approach. Furthermore, **Nuprl** has two tactics for arithmetical reasoning: **Arith** and **Mono**. Their presence offers possibilities of cooperation and a limited form was exploited in this study.

Arith is a decision procedure for a fragment of number theory (see [8] for details). It is part of Autotactic. **Mono** is a fixed collection of derived rules of inference involving inequalities.

Below we discuss various aspects of the experiment. We refer to our implementation of the forward chaining tactic as **OnticTac** below.

4.1 The Library

The lemma collection determines the effectiveness of **OnticTac**. Adding more lemmas can make **OnticTac** arbitrarily stronger, but also slower, because of the memory performance of **OnticTac**. Furthermore, adding arbitrary lemmas to the library can also invalidate the result of the experiment. For both of these reasons, we have limited the lemma library to a moderate size and only included relatively general facts.

There are 109 lemmas expressing basic facts of number theory: including type inclusion, basic facts about constants, and basic facts about primitive and defined operators. The rather large total number is due to the number of redundant lemmas included.

The redundancy can be classified by form as follows:

- specializations of some lemmas on constants
- overlappings of certain frequently used lemmas
- obvious consequences of **Arith** and **Mono**, in a sense, **OnticTac** acted as a cache for **Arith** and **Mono**

The justifications for the redundancies can be seen by classifying them according to function:

- Obvious facts not provable by Autotactic.
- Obvious facts that, though provable by Autotactic, need to be chained together with other forward chaining proof steps. This is a form of internalization of some of the power of Autotactic by **OnticTac**. This can also be thought of as using **OnticTac** to cache some of the results of Autotactic. Many simple rewriting or monotonicity lemmas are included for this reason.
- Obvious facts that, though establishable by **Arith**, are not established because the simplification procedure in **Arith** goes further. Using **Arith** to prove these lemmas and then using these lemmas with a particular focus binding can be thought of as forcing **Arith** to retain particular intermediate results.

- Some special version or overlaps of more general lemmas that have frequent applications. These are included to reduce the sizes of focus sets.

As an example of when not to include redundant lemmas, imagine adding the instantiations of all lemmas on a particular ground term t (a form of partial evaluation). Although this reduces the focus set in the sense that now we do not ever have to focus on t, the space cost of doing this is at least as high as always adding an extra focus term t, which is a rather inflexible way of using the focusing mechanism.

4.2 Examples, Issues and Observations

All examples shown in this section appear as they do in actual **Nuprl** sessions. `AFC`, `SFC` and `ASFC` are interface tactics to **OnticTac**. `AFC` selects the focus set from the variable declarations among the hypotheses, `SFC` takes an explicitly given list of focus terms with their typing as an argument, `ASFC` does both. Also, '`--*`' and '`--`' are used to indicate an Autotactic wrapper – meaning that the Autotactic is applied to unproven subgoals. `Seq` (a generalized version of the cut rule) sequences in a list of formulas as intermediate goals, producing subgoals with each of these and the original conclusion as conclusions and with previous formulas in the list as additional hypotheses. `THENL` matches a list of tactics to the list of subgoals.

Proofs are shorter

Example 4.

```
>> ∀i,j,p:N+.prime(p) & p|i*j=>p|i ∨ p|j
  BY -- Repeat Intro
  | 1. i:N+
  | 2. j:N+
  | 3. p:N+
  | 4. p prime
  | 5. p|i*j
  |->> p|i ∨ p|j
    BY -- Cases ['p|i';'¬p|i']
           THENL [AFC;ILeft;IRight THEN AFC]
```

A proof without **OnticTac** [15] has a size of 43 top level refinement steps and 828 primitive refinement steps (abbreviated as 43/828); it is too large to be shown here.

The new proof has size (2/68). Even including the proofs of the lemmas

$$\forall p,q:\mathcal{N}^+.\,prime(p)\&\neg p|q => rel\_prime(p,q) \quad (5/135)$$
$$\forall a,b,c:\mathcal{N}^+.\,a|b*c\&rel\_prime(a,b) => a|c \quad (4/95)$$
$$\forall i,n:int.\,i|n \vee \neg i|n \quad (13/109)$$

the new proof is significantly shorter than the old one. Two of the lemmas mentioned here are applied by **OnticTac** in a single step for the second case of the case analysis. The proofs of these lemmas have a combined size of (9/230). In comparison, the old proof has a size of (36/661) for this case.

On the whole, when we add up all the significant lemmas leading up to the prime factor existence lemma, the new proof is shorter than the previous proof by a factor of about 2-3, both in terms of top level (visible) refinement steps and the primitive refinement steps [4].

This reduction in proof length is a result of several factors: (1) **Mono** was not available at the time when old proof was completed, (2) **OnticTac** chains together applications of several lemmas in a single step and (3) redundant lemmas increase the effect of (2). The absence of the monotonicity rule turns out not to affect the comparison significantly because in Howe's proof[15], many monotonicity lemma had been used instead.

Proofs are less brittle Using **OnticTac** produces more stable proofs because user directives contain no explicit references to lemma names or hypothesis numbering . Furthermore, provability is monotonic in the set of lemmas and hypotheses – adding or strengthening lemmas or hypotheses does not invalidate a proof step.

Cooperation of OnticTac and Autotactic Rewriting of moderate length plus congruence reasoning with small number of ground equations seems quite obvious to human users, but poses great difficulty to automation. Roughly, Autotactic is weak because it only uses a fixed set of rewrite rules, does not make use of (ground) equational hypotheses in rewriting, and cannot incorporate equational lemmas. Congruence closure based forward chaining uses ground equations very effectively, but is weak in reasoning about equality and inequalities because multiple instances of lemmas are often needed, requiring large focus sets.

It seems that **OnticTac** and Autotactic might be able to compensate for each other's deficiencies through some mechanism of cooperation. We explored one particular form of cooperation: dividing proof burdens through the use of the cut rule (called 'seq' in **Nuprl**) and the 'THENL' construct of the meta-language.

As a first step, mediation is done by the human user by explicitly distributing proof obligations between Autotactic and **OnticTac**. Automation of this turned out to be difficult, for reasons given below.

The human mediation introduced a lot of low level control information into proof scripts. This is undesirable in general.

We illustrate these issues and the user decisions by examples.

Example 5. This is one subgoal in the proof of a lemma that says two numbers are relatively prime iff the greatest-common-divisor of them equals to 1.

```
1. l:int
2. 2<l
3. ∀a,b:N+.a+b<(l-1)&rel_prime(a,b)=>(a,b)=1
4. a:N+
```

[4] The number of top level refinement steps is the more important measure of the amount of user interaction. The primitive refinement steps are included to rule out the possible biasing effect of collapsing multiple steps into a single step using tacticals.

```
5. b:N+
6. a+b<l
7. rel_prime(a,b)
8. a<b
>> a+(b-a)<l-1
  BY -- Seq ['b<l-a';'1≤a';'l-a≤l-1';'b<l-1']
        THENL [AFC;AFC;ASFC [('1','int')];Idtac;Idtac]
```

Idtac leaves its goal unchanged, hence passing it to the Autotactic wrapper, hence to **Arith**. As a result, $b < l - a, 1 \leq a$ and $l - a \leq l - 1$ are proved by **OnticTac**; $b < l - 1$ and $a + (b - a) < l - 1$ are proved by **Arith**.

Although this proof is robust in the sense of being monotonic in the lemma collection and hypotheses, it still requires a lot of detailed user knowledge about both **OnticTac** and **Arith** in breaking up the focus set and distributing subgoals.

The above goal can also be proved by a single step of **OnticTac**, with the focus set $\{a, b, -a, l, 1, a - 1, l - 1, a + (b - a)\}$, assuming the standard lemmas about commutativity and associativity of $+$, cancellation with $-$ and transitivity of $<$ and $=$. A focus set of such a size would have taken significantly longer running time than one of size 4 and, even more critically, more space. Moreover, the choice of such a focus set still requires a lot of knowledge about **OnticTac**.

As another example of the kind of knowledge required to guide this kind of proofs by choosing cut formulas and focus sets, here is the complete proof of a lemma mentioned earlier.

Example 6.

```
>> ∀a,b,c:N+.a|b*c&rel_prime(a,b)=>a|c
  BY -- Repeat I
  | 1. a:N+
  | 2. b:N+
  | 3. c:N+
  | 4. a|b*c
  | 5. rel_prime(a,b)
  |->> a|c
    BY -- Seq ['(a,b)=1'] THENL [AFC; OnLastHyp (RPE ['r';'s'])]
    | 6. r:int
    | 7. s:int
    | 8. r*a+s*b=1
    |->> a|c
      BY -- OnLastHyp (Times '0<c')
      | 9. c*(r*a+s*b) = c*1 in int
      |->> a|c
        BY -- Seq ['c=(c*r)*a+s*(b*c)'; 'a|(c*r)*a'; 'a|s*(b*c)']
              THENL
              [Idtac;
               SFC [('a','int');('c*r','int')];
               SFC [('a','int');('s','int');('b*c','int')];
               SFC [('a','int');('(c*r)*a','int');('s*(b*c)','int')]]
```

Let us focus on the last step in the above example and examine: (1) the choice of the cut goals, and (2) the choice of the focus terms for each subgoal.

Again this step is possible by **OnticTac** in a single step with a large focus set (exercise for the reader). However, due to the high cost of large focus sets, we chose to divide the goal into smaller pieces and feed them to **OnticTac** and Autotactic according to their respective strengths.

Selecting cut formulas and focus terms requires detailed knowledge of how the proof will proceed. Take the first cut formulas, for example. $c = (c*r)*a+s*(b*c)$ is a rewrite of hypothesis 9. It was so chosen because **Arith** is capable of proving this by rewrite (if given this formula explicitly) and it matches the two divisibility subgoals better. Without this cut formula, the two divisibility subgoals would require larger focus sets in order to prove the dividends equal to respective subterms in hypothesis 9 by commutativity and associativity.

The choice of these cut goals requires

- knowledge of the workings of the Autotactic (**Arith** and **Mono**)
- knowledge of **OnticTac**
- knowledge of how the proof will proceed

Although requiring some knowledge about the capabilities of these tactics is reasonable, one would like to minimize this requirement as much as possible. And detailed knowledge about how the proof will proceed really subverts the idea of having the machine filling in tedium. At present, we know of no general purpose interactive theorem proving systems that do not require the user to have fairly detailed knowledge of the hidden subproofs that these systems supposedly 'fill-in'. Rewriting steps may be an exception, but simplifying rewrite only plays a limited role in these systems.

These examples raise serious questions about the relationship between **OnticTac** and **Arith/Mono**. One way to pose these questions is as a choice among:

1. Putting enough lemmas into the library so that, with very large focus sets, **OnticTac** can supersede **Arith/Mono**.
2. Using a library similar to the above but dividing proof burdens among different incarnations (*i.e.*, different focus sets) of **OnticTac** to reduce the focus set size
3. Dividing proof burdens among **OnticTac**, **Arith** and **Mono**, as we have done in this experiment
4. Integrating **Arith**, **Mono** and other useful decision procedures directly into **Ontic**.

(1), besides having high space cost, also has limitations: **Arith** applies a few facts deeply while **Ontic** applies many facts shallowly. It would be a test of the hypothesis that results of deep applications of some important facts can be packaged into lemmas and applied shallowly. This approach requires extending the object language with meta-constructs. Moreover, it is unlikely that either style of reasoning can subsume the other.

(2) is different from focus sequencing because the different incarnations of **OnticTac** can exchange facts among themselves, thus there is no dependence on the ordering as in focus sequencing. This may not require focus sets as large as for (1), but is not expected to overcome the limitations of (1).

(3) requires intelligent decisions, either by the user or some tactics. These are difficult options: while machine intelligence is difficult to achieve, applying human intelligence in making nitty-gritty decisions often reduces proof reusability.

At the least, (4) requires combining rewrite with congruence closure, an interesting research direction in its own right [7, 28, 22]. Whether integrating rewrite and 'obvious' induction [21] into **Ontic** will add enough deep applications of facts to provide a natural notion (approximating human judgement) of obviousness remains to be seen. In general, it also raises the question of how to integrate decision procedures. Cooperation among disjoint and convex decision procedures is studied in [25]

Focus Set Selection Large focus sets are impractical for currently known algorithms implementing the tractable inference relations defined by **Ontic**. Lacking the subformulas property implies that heuristics for automatically choosing focus sets are likely to be inaccurate. These two facts make automatic selection of focus terms difficult: poor performance of large focus sets rules out over-estimates, lack of subformulas property makes over-estimates highly likely. On the hopeful side, automatic focus set selection opens up the possibility of simulating large focus sets with a sequences of small focus sets.

Let us observe that each lemma defines a 'redex' for 'reduction' and focus sets that do not contain redices are useless. For example, applying the commutativity lemma requires two focus terms that are operands of a binary operator. Such redices usually generate small focus sets. By preprocessing the lemma library, we can obtain the redices and use them to select small focus sets. Because of the order dependence of focus set sequencing, it is necessary to either choose a particular order to sequence or to try all orderings. Not all pairs of focus sets are order sensitive, many are independent of each other. It seems possible, at the time of pre-processing of the lemma library, to identify those 'redices' that have order dependence.

A crude heuristic for selecting small focus set can be found by noting that most 'redices' are 'close relatives' in the term structure tree. [5] A crude heuristic for ordering is to sequence the focus sets in a bottom-up order. The combination of these simple heuristics works relatively well for many cases, but is not always the right order to sequence focus sets. These simple heuristics have been tested to cut down large focus sets selected by the user and the experimental result is somewhat encouraging: focus sets with up to 10 terms have been specified and the sequencing of them takes several minutes of real time. [6] In contrast, a single small focus step typically takes on the order of a few seconds to half a minute, and a focus set of 10 terms would have taken on the order of an hour.

In particular, examples 5 and 6, among others of similar complexity, are proved in a single step with focus sequencing on large focus sets chosen by the user. However, focus sequencing requires additional terms be mentioned in order to save enough facts

[5] If the logical connective '&' appears in a lemma, then members of a redex are not necessarily close neighbors and are not easily recognized. So this approximation picks out those cases that are efficiently recognizable.

[6] Real time is a better measure because it includes the impact of paging, a significant factor in this experiment. Time is measured on a Sparcstation 1.

between each subset focusing. These additional terms correspond to the subterms in the cut formulas of the proof shown earlier, plus others needed to permit focusing to simulate rewriting. Moreover, choosing the focus terms and the additional mention terms require knowledge of the proof, to about the same degree of detail as before.

Recognition versus Indexing Whether a tactic is applicable in a given situation is the *recognition problem*. What tactics are applicable and what parameters to apply them with (in the case of parameterized tactics, such as focusing) is the *indexing problem*. Search can reduce indexing to that of recognition, but the search space can be quite large, sometimes infinite, when tactics take parameters.

A minimum requirement to a solution of the recognition problem is some kind of declarative representation of tactics. In [3, 2, 4], a refinement of the tactic-based approach is taken, where tactics are given partial specifications in the form of pre-conditions and post-effects. In contrast to McAllester's notion of 'obvious by known facts', Bundy's approach can be characterized as defining a notion of 'obvious by known methods'. The relationship between these two notions is extremely intricate and does not seem to be explored in the literature.

The recognition problem itself requires theorem proving. A sensible solution might be to use another notion of 'obviousness', to perform limited inference in recognizing pre-conditions of tactics. Moreover, the internal data-structure used by an **Ontic**-like tactic might also provide a solution to the indexing problem.

Tactics that are like decision procedures tend to have difficult recognition problems. In this sense, **Arith** and **OnticTac** are both too powerful. The focus set parameter of **OnticTac** also makes the indexing problem difficult. In contrast, Bundy advocates synthesizing decision procedures out of simple 'proof plans' in [5]. Perhaps some of our focus selection and sequencing heuristics can be formulated as simple 'proof plans' with pre-conditions.

Focus sequencing heuristics blur the distinction between mechanisms to achieve a basic notion of obviousness and proof plans based on that. It is not clear whether proof plans should be used to repair defects in a minimum obviousness decision procedure or should rely on one. Perhaps the distinction is not real.

There seems to be a parallel between the relationship of facts and methods on the one hand, and forward chaining and backward chaining as evaluation/inference mechanisms on the other. Combinations of forward and backward chaining at the same syntactic level have been studied in the context of deductive databases [30, 24] and of theorem proving [29]. However, lemmas and plans do not seem to belong to the same class of objects.

5 Conclusions and Future Work

We have seen that applying a knowledge-based tactic can significantly shorten interactive proofs and increase robustness in the sense of replayability.

Restricting the size of focus sets keeps the response time acceptably low, but forces us to resort to either: (1) cooperation with other tactics, such as **Arith** or (2) sequencing small subsets of a larger focus set. Currently, both approaches require a fair amount of control information.

We have applied some simple heuristics for selecting and ordering focus sets and had some success. We are currently exploring heuristics that will identify 'relevant' terms to mention during sequencing. This is a less daunting task because the complexity of **Ontic** only depends on the mentioned terms polynomially, hence over-estimating mention terms is not as big a problem as over-estimating focus sets.

Focus sequencing and refinements via lemma library pre-processing seems to be a promising direction to extend our work.

Improving **OnticTac** by incorporating rewriting and/or induction are important research problems just beginning to be explored. We expect progress in these directions to complement our approach.

Acknowledgement

The author is grateful to Robert Constable, Charles Elkan, Douglas Howe, David McAllester and the referees for helpful comments on the paper and to Bill Aitken, James Allan and Keith Gasser for reading earlier drafts.

References

1. D. Basin. An environment for automated reasoning about partial functions. In R. Lusk and R. Overbeek, editors, *Ninth International Conference on Automated Deduction*, volume 310 of *Lecture Notes in Computer Science*, pages 101–110. Springer Verlag, May 1988. Also as Cornell CS TR 87-884.
2. A. Bundy, F. van Harmelen, J. Hesketh, and A. Smaill. Experiments with proof plans for induction. *Journal of Automated Reasoning*, page (in press) Earlier version available from Edinburgh as Research Paper No 413, 1988.
3. Alan Bundy. The use of explicit plans to guide inductive proofs. In R. Lusk and R. Overbeek, editors, *Ninth International Conference on Automated Deduction*, volume 310 of *Lecture Notes in Computer Science*, pages 111–120. Springer Verlag, 1988. Longer version available from Edinburgh as Research Paper No. 349.
4. Alan Bundy. A science of reasoning. In J.-L. Lassez and G. Plotkin, editors, *Computational Logic: Essays in Honor of Alan Robinson*, pages 178–198. MIT Press, Cambridge, MA, 1991.
5. Alan Bundy. The use of proof plans for normalization. Report, Dept. of Artificial Intelligence, University of Edinburgh, 1991.
6. Wilfred Chen. Tactic-based theorem proving and knowledge-based forward chaining: an experiment with Nuprl and Ontic. Technical report, Department of Computer Science, Cornell University, March 1992.
7. Leslie Paul Chew. An improved algorithm for computing with equations. In *the 21st Annual Symposium of Foundations of Computer Science*, pages 108–117. IEEE Computer Society Press, 1980.
8. Robert L. Constable, Scott D. Johnson, and Carl D. Eichenlaub. *Introduction to the PL/CV2 Programming Logic*, volume 135 of *Lecture Notes in Computer Science*. Springer-Verlag, New York, 1982.
9. Robert L. Constable, et al. *Implementing Mathematics with the Nuprl Proof Development System*. Prentice-Hall, Englewood Cliffs, New Jersey, 1986.
10. Peter J. Downey, Ravi Sethi, and Robert E. Tarjan. Variations on the common subexpression problem. *JACM*, 27(4):758–771, October 1980.

11. A. Felty and D. Miller. Specifying theorem provers in a higher-order logic programming language. In R. Lusk and R. Overbeek, editors, *Ninth International Conference on Automated Deduction*, volume 310 of *Lecture Notes in Computer Science*, pages 61–80. Springer Verlag, 1988.
12. M. Gordon. A proof generating system for higher-order logic. In *Proceedings of the Hardware Verification Workshop*, 1989.
13. Michael J. Gordon, Robin Milner, and Christopher P. Wadsworth. *Edinburgh LCF: A Mechanized Logic of Computation*, volume 78 of *Lecture Notes in Computer Science*. Springer-Verlag, 1979.
14. D.J. Howe. *Automating Reasoning in an Implementation of Constructive Type Theory*. PhD thesis, Cornell University, Ithaca, NY, April 1988.
15. Douglas J. Howe. Implementing number theory, an experiment with Nuprl. In *Eighth International Conference on Automated Deduction*, volume 230 of *Lecture Notes in Computer Science*, pages 404–415. Springer Verlag, 1986.
16. Douglas J. Howe. Equality in lazy computation systems. In *Proceedings of the Fourth Annual IEEE Symposium on Logic in Computer Science*, pages 198–203. IEEE Computer Society, June 1989.
17. Dexter C. Kozen. Complexity of finitely presented algebras. In *Proceedings of the Ninth Annual ACM Symposium on the Theory of Compututation*, pages 164–177, 1977.
18. D. McAllester, R. Givan, and T. Fatima. Taxonomic syntax for first order inference. In *Proceedings of the First International Conference on Principles of Knowledge Representation and Reasoning*, pages 289–300, 1989. To Appear in JACM.
19. David McAllester. *Ontic: A Knowledge Representation System For Mathematics*. MIT Press, Cambridge, Massachusetts, 1989.
20. David McAllester. Automatic recognition of tractability in inference relations. Memo 1215, MIT Artificial Intelligence Laboratory, February 1990. To appear in JACM.
21. David McAllester. Some observations on cognitive judgements. In *AAAI-91*, pages 910–915. Morgan Kaufmann Publishers, July 1991.
22. David A. McAllester. Grammar rewriting. In *Eleventh International Conference on Automated Deduction*. Springer Verlag, 1992.
23. C. Murthy. An evaluation semantics for classical proofs. In *Proceedings of the Sixth Annual IEEE Symposium on Logic in Computer Science*, pages 96–107, Amsterdam, The Netherlands, July 1991.
24. Jeffery F. Naughton and Taghu Ramakrishnan. Bottom-up evaluation of logic programs. In J.-L. Lassez and G. Plotkin, editors, *Computational Logic: Essays in Honor of Alan Robinson*, pages 640–700. MIT Press, Cambridge, MA, 1991.
25. Greg Nelson. Combining satisfiability procedures by equality-sharing. In *Automated Theorem Proving: After 25 Years*, pages 201–211. American Mathematical Society, 1983.
26. Greg Nelson and Derek Oppen. Fast decision procedures based on congruence closure. *JACM*, 27(2):356–364, April 1980.
27. L. Paulson. Isabelle: The next 700 theorem provers. In P. Odifreddi, editor, *Logic and Computer Science*, pages 361–385. Academic Press, 1990.
28. David J. Sherman. Lazy directed congruence closure. Tech report 90-028, University of Chicago, September 1990.
29. Mark E. Stickel. Upside-down meta-interpretation of the model elimination theorem proving procedure for deduction and abduction. Technical Report TR-664, Institute for New Generation Computer Technology, Tokyo, Japan, May 1991.
30. Jeffery D. Ullman. *Principles of Database and Knowledge-base Systems*. Computer Science Press, 1989. Chapter 13.

Little Theories*

William M. Farmer, Joshua D. Guttman, F. Javier Thayer

The MITRE Corporation

Abstract

In the "little theories" version of the axiomatic method, different portions of mathematics are developed in various different formal axiomatic theories. Axiomatic theories may be related by inclusion or by theory interpretation. We argue that the little theories approach is a desirable way to formalize mathematics, and we describe how IMPS, an Interactive Mathematical Proof System, supports it.

1 Introduction

In this paper, we will argue in favor of implementing a particular version of the axiomatic method in mechanical theorem provers. By the axiomatic method we mean the practice of reasoning logically from a set of sentences in a formal language. Such a set of sentences is called an axiomatic theory.[1] For instance, Peano arithmetic and the theory of an ordered field are familiar axiomatic theories.

There are two contrasting ways of using the axiomatic method, both well established in modern mathematical practice. We will refer to them as the "big theory" version and the "little theories" version. In the big theory version, we select a powerful and highly expressive set of axioms, with the property that any model of these axioms will contain all the objects that will be of interest to us. Logical derivation from these powerful axioms will allow us to prove our theorems about these objects, so all of the reasoning can be carried out within this single theory. Zermelo-Fraenkel set theory is frequently used by mathematicians for this purpose.

---

*Supported by the MITRE-Sponsored Research Program. Authors' address: The MITRE Corporation, 202 Burlington Rd, Bedford MA 01730 USA; Telephone: 617-271-2749; Email: `farmer, guttman, jt@mitre.org`.

[1] Thus, in our usage, a theory amounts to a set of axioms in a specified formal language. The word "theory" is also frequently used to mean a set of sentences closed under logical consequence. The latter usage allows one to speak, for instance, of different axiomatizations of the same theory. However, for our purposes it is convenient to focus on the axioms, and to express the idea of alternative axiomatizations by saying that two theories have the same consequences.

In the little theories version of the axiomatic method, a number of theories will be used in the course of developing a portion of mathematics. Different theorems will be proved in different theories, depending on the amount of structure required. For instance, one theorem may be true in any arbitrary topological space, while another may hold only in a metric space. Theorems are proved in a particular theory by logical derivation from the axioms available in that theory.

The goal of this paper is to demonstrate the usefulness of the little theories approach in mechanized reasoning. In addition, we will indicate how IMPS, an Interactive Mathematical Proof System [9], supports the approach.

We will not focus on logical issues related to the little theories approach, including its use of theory interpretations. On the contrary, the logic of theory interpretations is well understood, and a version for the particular logic we use is available in [8]. Interpretations have been effectively used in the logical literature since at least the 1950's [27], and in mathematics for much longer. Indeed, this makes interpretations especially attractive to us. Our overall goal in IMPS is to mechanize traditional tools of classical mathematical reasoning, partly because they are understood by a wide range of potential users, and partly because their intellectual power has been demonstrated over a long period of time.

1.1 Little Theories in Mathematics: An Example

The utility of the little theories approach in mathematics is largely due to the power that theory interpretations provide. A theory interpretation is a syntactic translation between the languages of two theories which preserves theorems. That is, if a formula is a theorem of the source theory, then its image is a theorem of the target theory. To establish that a translation is a theory interpretation, one must show that the source theory *axioms* are translated to target theory theorems, together with some additional obligations that depend on the details of the logical context. Theory interpretations are used in two crucial ways in the little theories approach:

- To reuse theorems of the source theory in the target theory;
- To establish the consistency of the source theory relative to the target theory. As a special case, to establish that a formula φ is independent of a theory $\mathcal{T}$, one may show that both $\mathcal{T} \cup \{\varphi\}$ and $\mathcal{T} \cup \{\neg\varphi\}$ are consistent.

A theory interpretation from a source theory $\mathcal{T}$ to a target theory $\mathcal{T}'$ also allows us to infer a relation between models of the theories. Namely, given an arbitrary model of $\mathcal{T}'$, the interpretation tells us how to select a subdomain and distinguished values that furnish a model of $\mathcal{T}$.

Partly because of its usefulness for establishing consistency and independence, the little theories approach has become a deeply entrenched way of organizing mathematical knowledge. For instance, in the introduction to *The Foundations of Geometry* [19], Hilbert wrote that one of his aims was:

> to bring out as clearly as possible the significance of the different groups of axioms and the scope of the conclusions to be derived from the individual axioms.

This expresses one of the major themes of modern mathematics, which aims to determine not just which mathematical statements are true, but also which assumptions are needed to deduce them.

The Foundations of Geometry illustrates how the axiomatic method can serve as an organizing principle in examining an axiomatization for Euclidean geometry. The goal of the book is to study relations of independence among subsets of the axioms, in combination with two crucial theorems, namely Pascal's theorem and Desargues's theorem. As a tool, Hilbert also considers a number of algebraic theories, primarily the theory of an ordered field.

A simple example of a theory interpretation is given in §§15, 17, where certain line segments are shown to form an ordered field with appropriately chosen operations. In this particular case, we are not concerned with an independence proof; instead, Hilbert introduces the theory interpretation so that he can use the familiar algebraic theorems to carry out computations in the course of giving geometrical proofs. The interpretation guarantees that the results of the algebraic computation will be sound in the geometrical context. This illustrates theorem reuse, in the helpful case of equations and conditional equations.[2]

In these cases, Hilbert makes no reference to any particular model of either theory (in the usual semantic sense). In some other cases, his language is more ambiguous, and he appears to be working within a specific model (for instance, the reals). However, he is explicit about the algebraic axioms his construction actually relies on, and emphasizes that the theory has denumerable models (see, for instance, §9). Thus, these other cases could be easily recast into a terminology in which we are unambiguously concerned with theory interpretations.

1.2 Little Theories in Mechanized Reasoning

Quite apart from the intellectual appeal of fine-grained knowledge about the logical power of particular axioms, and the relations among them, the little theories approach has two important practical advantages for mechanized reasoning.

- It allows the use of minimal axiomatizations for specific groups of theorems. This ensures maximal generality of reuse for the results, through interpretation into other theories, or through direct inclusion in larger theories.

  When equipped with procedures that can construct the great majority of the interpretations needed, an interactive theorem prover can offer the user great power in applying previously proved theorems across a wide range of new theories.

- The hand-crafted framework of a little theory allows theorems to be written in simpler forms. Cues suggesting the applicability of particular lemmas or other techniques are easier to identify.

[2] See Section 2.3 for a mechanism in IMPS implementing a similar kind of reuse. For a more elaborate use of theory interpretations in Hilbert, see for instance [19, §§24–26, 28–29], where he exploits a whole sequence of theory interpretations.

> Moreover, the work of establishing the applicability of a whole group of theorems can be carried out once and for all, and then encoded in a theory interpretation. The theory interpretation then serves as a "license" authorizing us to use them repeatedly in the future.

The first of these advantages will be discussed primarily in Section 2. The second will be the focus of Section 3. Section 4 discusses the big theory approach. We will argue that the little theory approach does not create a disadvantage when we may want to use a powerful theory like **ZF**. Moreover, we will point out some cases in which this is desirable. In Section 5, we will present a substantial piece of mathematics as it appears in the little theory approach; this example has been developed with the interactive theorem prover IMPS.

1.3 Previous Work; IMPS

In spite of its utility, the little theories version of the axiomatic method lies off the main path of previous work in mechanized theorem proving.

The little theories idea is, however, a familiar ingredient in work on specification languages, probably first introduced by Burstall and Goguen [2]. It was a central tenet of work on Clear [3, 4], and it was also a motivating idea in Larch [16]. The idea is also an ingredient in more recent work on logical frameworks [18]. In the logical frameworks context, however, it appears in an unusual guise in which not just *theories* but also *logics* may be combined. It is not clear whether this additional generality will prove to be a benefit in practical use, as it may impede the process by which users develop strong and reliable semantic intuitions. Moreover, a single familiar logic, such as simple type theory, suffices to formalize conveniently a wide range of problems.

There has also been some work on supporting little theories in mechanized theorem proving. Sannella and Burstall undertook to implement some of the ideas of Clear in an extended version of LCF [24]. However, according to the the published description, theory interpretation and parameterized theories had not yet been implemented [24, pp. 384, 389]. Although OBJ [14] incorporates a translation ("view") mechanism, the user is responsible for deciding whether the translation is in fact a theory interpretation: OBJ itself makes no attempt to prove the images of the source theory axioms. Moreover, because of OBJ's equational logic, its usefulness as a theorem prover is, in our opinion, highly restricted. Curiously, the Larch Prover LP [12] does not give strong support for the little theories approach: there is only one theory available in the prover at a time, and thus theory interpretations cannot be used in proofs. E. Gunter [15] has made a start on implementing little theories within HOL.

So far as we know, IMPS, an Interactive Mathematical Proof System, is the first interactive theorem prover to have been designed from the start to support little theories. Moreover, IMPS implements a strong logic—rather than the weak systems generally used in algebraic specification languages—that is suited to expressing and proving sophisticated mathematical statements. One example, a simple inverse function theorem for Banach spaces, is presented in Section 5.

For a general description of IMPS, see [10, 9]. Examples of IMPS proofs are found in [10, 11]. All of the examples given below (except where explicitly noted) represent material we have developed using IMPS.

All concept formulation, calculation, and inference in IMPS is performed with respect to a formal logic that is a version of simple type theory. The logic, called LUTINS[3], provides strong support for specifying and reasoning about partial (and total) functions, and is equipped with a system of types and subtypes. Types and subtypes are collectively called *sorts*.

The treatment of partial functions in LUTINS is studied in [7], while the treatment of sorts and interpretations is the subject of [8]. For a detailed presentation of the syntax and semantics of LUTINS, see [17].

A language is built in IMPS from a signature—a list of sort and constant declarations. A theory consists of a language $\mathcal{L}$ plus a set of axioms (i.e., sentences of $\mathcal{L}$). Theories are the basic units in IMPS for specifying mathematical objects and concepts and for organizing automated deduction.

2 Theory Interpretations and Theorem Reuse

The notion of an interpretation of one axiomatic theory in another is a fundamental concept of mathematics and logic. As we mentioned in the introduction, this notion is formalized using certain syntactic translations that are known in logic as "theory interpretations" [6, 25]. Intuitively, a theory interpretation from $\mathcal{T}$ to $\mathcal{T}'$ specifies one of the (possibly many) ways of embedding $\mathcal{T}$ in $\mathcal{T}'$, while preserving theorems.

Logicians have used theory interpretations to prove metamathematical properties about theories, particularly consistency, decidability, and undecidability. The classic work of Tarski, Mostowski, and Robinson [27], for example, illustrates how theories can be proved undecidable by means of theory interpretation. References on theory interpretations include [23, 26, 28, 29].

2.1 Theory Interpretations in IMPS

The notion of a theory interpretation in LUTINS is similar to the standard notion in first-order logic (see [6, 25]). However, since LUTINS admits partial functions and subtypes, the notion in LUTINS is necessarily more complicated. A precise definition of a LUTINS theory interpretation is given in [8].

In brief, a LUTINS *translation* from a theory $\mathcal{T}$ to a theory $\mathcal{T}'$ is specified by a pair (μ, ν) of functions—where μ maps the sorts of $\mathcal{T}$ to sorts, sets, or unary predicates of $\mathcal{T}'$ and ν maps the constants of $\mathcal{T}$ to expressions of $\mathcal{T}'$—which satisfy certain syntactic conditions. The translation is a homomorphism I from the expressions of the source theory to the expressions of the target theory, i.e., $I(c(e_1, \ldots, e_n)) = c(I(e_1), \ldots, I(e_n))$, where $c(e_1, \ldots, e_n)$ is a compound expression composed of a logical constant c and n subexpressions $e_1, \ldots, e_n$.[4]

[3] Pronounced as the word in French.

[4] When μ maps sorts to sets or unary predicates, expressions beginning with variable binders such as $\forall$ or λ must be "relativized." For example, if I maps a sort α to a unary predicate φ on a sort β, then $I(\forall x{:}\alpha.\psi) = \forall x{:}\beta.\varphi(x) \supset I(\psi)$.

Every translation I from $\mathcal{T}$ to $\mathcal{T}'$ determines a set of formulas in the target theory called *obligations*, which includes $I(\theta)$ for each axiom θ of $\mathcal{T}$. By the Interpretation Theorem for LUTINS [8], if each obligation of I is a theorem of $\mathcal{T}'$, then I translates each theorem of $\mathcal{T}$ to a theorem of $\mathcal{T}'$, i.e., I is a theory interpretation.

2.2 Theorem Reuse in Mathematics

Mathematicians commonly use a kind of informal theory interpretation for reusing theorems. A result about an abstract theory, such as a group, will be applied to a more concrete theory such as a field, by (for instance) observing that multiplication over the nonzero elements has the structure of a group.

In the the little theories version of the axiomatic method, mathematical reasoning is distributed over a network of theories linked to one another via theory interpretations. These theory interpretations provide the means to "transport" a theorem from the theory it was proved in to any number of other theories. For instance, the theorem that a sequence of points converges to at most one limit can be proved once in a theory of an abstract metric space and then applied to a sequence of reals, to a sequence of points in a normed space, and so on.

Similarly, a theory of an abstract monoid (that is, an associative operator $\oplus$ with an identity e), taken together with the integers as an additional type, allows one to define an iterated monoid summation operator $\sum$ such that $\sum_{k=i}^{j} f = e$ for $j < i$, and $\sum_{k=i}^{j} f = (\sum_{k=i}^{j-1} f) \oplus f(j)$ otherwise. Facts derived in this theory can then be applied to a large class of iterated operators, including the normal numerical sum and product operators. Many interesting facts about algorithms can be developed in this general framework [1].

Another very useful technique is to interpret a theory into itself or one of its subtheories. For example, many similar theorems about algebraic groups are just different instantiations of a particular abstract theorem about group actions [20, pp. 71–79]. The abstract theorem φ can be proved in a theory $\mathcal{A}$ of an abstract group action, which includes a theory $\mathcal{B}$ of an abstract group as a subtheory, and then the various instantiations can be obtained by transporting φ from $\mathcal{A}$ to $\mathcal{B}$ via appropriate interpretations.

Sometimes, this second technique provides proofs by symmetry or duality. For instance, the duality of points and lines in projective geometry can be formalized by observing that the translation that interchanges them is an interpretation. Having done so, we can prove a theorem in one form and then immediately infer that its dual is also a theorem.[5]

2.3 Theorem Reuse in IMPS

Although theory interpretations can be used in IMPS in several different ways, theorem reuse is certainly the most important application. An IMPS user can build translations, verify that they are theory interpretations, and transport

[5]This example has not been done in IMPS, but several other examples of proof by symmetry have been formalized in IMPS using interpretations.

theorems with them as desired. The development of a portion of mathematics in IMPS will usually involve several theories and a great many theory interpretations, most of which are quite simple. However, building theory interpretations can be distracting or burdensome, especially in the midst of a proof. If the user were responsible for explicitly creating all these theory interpretations, it would be very difficult to concentrate on the major task at hand.

Consequently, the great majority of the theory interpretations needed by the IMPS user are built by software without user assistance. In addition, when handcrafted interpretations have been added by the user, the system can retrieve the right one in most situations.

There are about a half dozen mechanisms in IMPS that automatically find, extend, or create theory interpretations. Of these the most useful is a kind of polymorphic matching called *translation matching*, which allows for inter-theory matching of expressions. An expression e in a theory $\mathcal{T}'$ *translation matches* a pattern expression p in a theory $\mathcal{T}$ if there is a theory interpretation I from $\mathcal{T}$ to $\mathcal{T}'$ and a substitution σ such that $\sigma(I(p)) = e$. The translation matching algorithm works in IMPS as follows. First, the variables and constants in p are matched with subexpressions of e yielding a "sort association list" and a "constant association list." If the entries of these lists are compatible with each other (e.g., no constant is associated with two distinct expressions), IMPS will try to find a theory interpretation defined by (μ, ν) such that the sort and constant association lists specify subfunctions of μ and ν, respectively. If that fails, IMPS will try to build a theory interpretation directly from the two association lists. If a theory interpretation I is obtained and e matches $I(p)$ in the ordinary sense, then σ is just the substitution which matches the expressions.

Translation matching is employed in several kinds of machinery in IMPS for theorem reuse. For example, when theorems about generic objects like sets and sequences are of appropriate form to be used as rewrite rules, IMPS uses translation matching to automatically create the theory interpretations and substitutions needed to apply them whenever relevant in the course of simplification.

A collection of theorems can be applied as conditional rewrite rules in an organized way in IMPS using extremely simple procedures called *macetes*[6]. The exact behavior of an *elementary macete* (built from a particular theorem) depends on the syntactic form of the theorem. In the central case of a theorem of the form $\forall \vec{x} \,.\, \varphi_1 \wedge \ldots \wedge \varphi_n \supset s = t$, the elementary macete replaces matches to s by the corresponding matches to t, if the instances of the φ_i can be recognized to be true. Compound macetes may be constructed from elementary and other kinds of atomic macetes using a small number of very simple combining forms. For instance, the `repeat` form takes a list of macetes and applies them repeatedly until they are no longer applicable. A small number of special atomic macetes allow the user to intersperse beta-reduction, definition expansion, and calls to the simplifier [10].

A *transportable macete*, like an elementary macete, is another kind of atomic macete that is built from a theorem. They are used, when working in a theory $\mathcal{T}$, to apply theorems that reside outside of $\mathcal{T}$. The mechanism is similar to

[6] *Macete*, in Portuguese, means a chisel, or in informal usage, a clever trick.

that for elementary macetes except that translation matching is used instead of ordinary matching.

3 Controlling Theorem Provers

The axiomatic theory is a natural unit to use in organizing and guiding the behavior of a theorem prover. In this section we will discuss several ways that information is gathered around an axiomatic theory in order to control the behavior of IMPS.

3.1 Expressing Theorems in Simple Form

There is frequently an advantage to expressing facts in a simple form, as they can then be used by a theorem prover in some special way in the course of computing with expressions. For instance, equations and biconditionals can be used as (unconditional) rewrite rules in the course of simplification. Within the logical framework we use, there are two ways that the little theories approach aids in recasting facts in the simplest syntactic form.

First, a theorem that in pure logic would take the form $\bigwedge_{i<n} \varphi_i \supset \psi$ may be written in the form ψ in any axiomatic theory whose axioms include (or entail) the formulas φ_i. Naturally, in order to *apply* the theorem in a new theory under some translation (possibly the identity translation), one must check that the translations of the assumptions φ_i are in fact satisfied. This requirement amounts to proving that the translation is an interpretation.

Although the work of discharging these obligations must still be done, there is often an advantage to this approach. For, the interpretation itself can be treated in the theorem proving program as a data object, so that once it has been certified, all the theorems of the source theory may be made available. Similarly, the interpretation as a data object makes it easy to avoid proving the same conditions repeatedly when a theorem (or group of related theorems) are to be applied many times.

Second, the LUTINS logic supports subtypes called sorts. For instance, in our theory of arithmetic, the natural numbers and integers are subtypes of the reals. In an axiomatic theory that involves a subtype σ of some type τ, $\forall x : \sigma \,.\, \psi(x)$ can replace $\forall y : \tau \,.\, \varphi(y) \supset \psi(y)$, where φ is the unary predicate corresponding to membership in the subtype σ. As before, this is no magic way to eliminate work: given a term t, we must ensure that t is defined with a value in the sort σ. Although this is semantically similar to discharging the assumption $\varphi(t)$, a theorem prover may be equipped with algorithms to resolve questions of this form effectively.

IMPS gains considerably from a range of algorithms for these "sort definedness" assertions. They use information extracted from theorems, such as that σ is closed under particular functions. They frequently reduce the assertion that a complicated term t is defined with a value in σ to a few assertions about small subterms. If these in turn are not discharged by the simplifier, they may be presented to the user for proof.

3.2 Simplification and Decision Procedures

Many theorem provers use efficient algorithms, exploiting facts about particular operators, either to simplify expressions or to decide formulas in some syntactic class. These hand-coded procedures, which we will call *processors*, may be far more efficient than applying basic inferences to derive the same conclusion.

However, the same algorithm is often sound for a range of theories, so long as they satisfy a number of presuppositions. Suppose a number of formal symbols, such as $+$ and $*$, or $\oplus$ and $\otimes$, represent operations the algorithm might be applied to. The processor can then generate a number of concrete formulas representing presuppositions for the manipulations that it will perform. For instance, these presuppositions might assert that the operators are associative and satisfy a distributive property. The simplifier for that theory can soundly call the processor if these presuppositions are theorems.

Moreover, a processor may be installed with subsets of the possible operations, so that a processor for simplifying polynomials in a field can be installed in a commutative ring (with unit) simply by not supplying a division operator. Alternatively, some properties of an operator may be optional: a processor designed for a commutative ring may be applied to a non-commutative ring simply by stipulating that the multiplication may not commute. The code of the processor must of course be written so that these optional choices make sense.

IMPS allows a user to tailor such a processor for his theory with no need to write code. For instance, if a theory of bitstrings is needed, it is trivial to instantiate an existing algebraic processor to provide efficient simplification for expressions involving the bitstring operations. The theory is the natural unit for this sort of theorem proving aid.

4 Relation to the Big Theory Approach

It might be thought that there is also a disadvantage to the little theories approach. Namely, in some situations we may want to use a "big theory" like **ZF** as a context to reason about structures satisfying different properties. Naturally, set theories can also be treated within IMPS. But, if we do so, can we make use of theorems that we have proved using the little axiomatic theories that characterize these structures?

In fact, we do not sacrifice the opportunity to use a "big theory" in connection with results proved using the little theories approach. For instance, suppose that we have a collection of theorems we have proved in a theory $\mathcal{M}$ of monoids. $\mathcal{M}$ asserts of the constants e and $\oplus$ that e is an identity for $\oplus$, and that $\oplus$ is associative. Within the chosen big theory $\mathcal{B}$, define a ternary relation *monoid* between a set, an object, and a binary function by the condition:

$$
\begin{aligned}
& \mathit{monoid}(s,x,f) =_{df} \\
& \qquad x \in s \\
& \wedge \quad \forall z \,.\, z \in s \supset f(z,x) = f(x,z) = z \\
& \wedge \quad \forall z_0, z_1, z_2 \,.\, z_0, z_1, z_2 \in s \supset f(z_0, f(z_1, z_2)) = f(f(z_0, z_1), z_2)
\end{aligned}
$$

Suppose we can establish a formula of the form $monoid(s_0, x_0, f_0)$, possibly relative to some assumptions $\{\varphi_1, \ldots, \varphi_n\}$. Then we will want to apply the theorems of $\mathcal{M}$ (under these assumptions) to the structure that $\langle s_0, x_0, f_0\rangle$ represents.

Context Theory Interpretations. To see how this can be accomplished, consider the syntactic translation I that sends the domain to s_0, the constant e to x_0, and the operator $\oplus$ to f_0. I is unlike the most usual notion of interpretation in two ways:

- The expressions s_0, x_0 and f_0 may involve free variables, so that the image of a closed expression under I need not be closed;
- If the set of assumptions $\{\varphi_1, \ldots, \varphi_n\}$ is non-empty, then the images of the monoid axioms under I may not be theorems of $\mathcal{B}$.

Although I is not properly a theory interpretation, it is what we call a *context theory interpretation* from $\mathcal{M}$ to $\mathcal{B}$ relative to the "context" [22, 10] containing the assumptions $\{\varphi_1, \ldots, \varphi_n\}$. Context theory interpretations are used in IMPS in much the same as way as ordinary theory interpretations, so long as our position in a proof licenses us to use the assumptions $\{\varphi_1, \ldots, \varphi_n\}$. In the case at hand, they justify citing the theorems of $\mathcal{M}$ in reasoning about s_0, x_0 and f_0.

Context theory interpretations greatly extend the power of the little theories approach. As a consequence, a commitment to the little theories approach need entail no reduplication of effort when structures within **ZF** are of interest.[7]

Proving Consistency. An objection is often raised to the free use of axiomatic theories in the formal verification of software or hardware. All but the most sophisticated users will introduce axioms that are wrong, and sometimes even inconsistent.

Naturally, no formal method can prevent a user from writing a specification that does not accurately reflect his intent. For instance, if the specification is to be built up by means of a sequence of definitions within a fixed big theory, the user may select the wrong definitions, and may sometimes even define unsatisfiable predicates. Then, although his theorems will be truths of the formal system he is working within, their content will not correspond to his intuitive understanding.

While the little theories approach does not prevent a user from doing stupid things, it is by no means worse off than the alternative. An interpretation I from $\mathcal{T}$ to $\mathcal{T}'$ will allow us to assure ourselves that a theory $\mathcal{T}$ is consistent, if $\mathcal{T}'$ is a well-established theory trusted to be consistent. Moreover, a theory

[7] Indeed, the big theories approach is particularly useful when a theorem concerns an arbitrary family of structures rather than a collection of a fixed number of structures. Since, within **ZF**, any family of structures is simply represented as another set, **ZF** is an advantageous framework for the reasoning.

By contrast, in case a theorem concerns some *fixed* number of structures of some kind, such as three arbitrary metric spaces, IMPS provides a convenient mechanism for synthesizing a suitable theory.

interpretation also conveys more information. The user may have a conception of what sort of structures ought to satisfy $\mathcal{T}$. The interpretation I gives him a way of isolating, within any structure satisfying $\mathcal{T}'$, a structure satisfying $\mathcal{T}$. If these structures can not be made to look as he expects, then he has reason to think he has got the formulation of $\mathcal{T}$ wrong.

This method is by no means limited to formal methods, as it is also commonly used in giving relative consistency proofs in standard mathematical practice. For instance, the consistency of the Generalized Continuum Hypothesis with **ZF** is sometimes proved in essentially this way (see, for instance, [21]).[8] In this case, both $\mathcal{T}$ and $\mathcal{T}'$ happen to be the same theory, namely **ZF**.

5 Example

In this section we discuss an IMPS proof which exemplifies a number of advantages of the little theories approach. Our example is a well-known "inverse function theorem" (Theorem 1 below) for a mapping from a Banach space into itself which is near the identity (see [5, §10.1]). We begin by describing the sophisticated network of interrelated theories used in the proof.

5.1 The Network of Theories

The theories are constructed step-by-step using theory extension and theory instantiation (as recommended in [13]), beginning from a few general mathematical theories:

- A theory of an abstract ring.
- A theory of an abstract module over a ring.
- A theory of an abstract metric space (denoted $\mathcal{M}$).
- A theory of real arithmetic (formalized as a complete ordered field).

From these theories we build a number of other theories and interpretations:

- A theory of an abstract real vector space is obtained by instantiating the ring of the module theory with the field of real numbers via the obvious interpretation from the module theory to the theory of real arithmetic.
- A theory of an abstract normed space is built as an extension of the real vector space theory by adding a new constant (denoted $||\cdot||$) together with axioms which characterize this constant as a norm.
- A theory interpretation (denoted Φ) from the metric space theory $\mathcal{M}$ to the normed space theory is specified by interpreting the distance function of $\mathcal{M}$ as the function $(x, y) \mapsto ||x - y||$ and the sort of points of $\mathcal{M}$ as the sort of points of the normed space.

[8] This example has not been done in IMPS.

- Using Φ, the definitions in $\mathcal{M}$ of constants such as "open", "connected", and "complete" are transported to the normed space theory.
- A theory of an abstract Banach space (denoted $\mathcal{B}$) is formed by adding an axiom that says the propositional constant complete-normed-space holds.

Thus, this one theory $\mathcal{B}$, in which the theorem will be stated and proved, illustrates a variety of ways that theories can be built in IMPS.

5.2 The Theorem and Its Proof

Before stating the theorem we make some preliminary definitions.

We will use the symbols $\mathbf{P}$ and d to denote the sort of points and distance function in the metric space theory $\mathcal{M}$, and $\mathbf{V}$ and $\| \ \|$ to denote the sort of vectors and norm in the Banach space theory $\mathcal{B}$. A partial function $\varphi : \mathbf{P} \to \mathbf{P}$ is a *contraction with modulus* $c < 1$ if, for all x, y in the domain of φ, $d(\varphi(x), \varphi(y)) < c \cdot d(x, y)$. For $a \in \mathbf{P}$ and $r \in \mathbf{R}$, $\mathrm{B}(a, r)$ and $\bar{\mathrm{B}}(a, r)$ denote, respectively, the open and closed balls with center a and radius r. As illustrated above, these definitions can be readily transported via Φ to the normed space theory and thus to $\mathcal{B}$. Given a partial function $\varphi : \mathbf{V} \to \mathbf{V}$, $f_\varphi : \mathbf{V} \to \mathbf{V}$ is the partial function $x \mapsto x + \varphi(x)$, provided $\varphi(x)$ is defined.

Theorem 1 *Let $\varphi : \mathbf{V} \to \mathbf{V}$ be a contraction whose domain is open. Then the range of f_φ is itself open.*[9]

The key idea behind the proof of this result is the following classical theorem of abstract analysis attributed to Banach:

Theorem 2 (Contraction Principle) *A contraction on a complete metric space has a unique fixed point.*

The Contraction Principle is proved in IMPS in the theory $\mathcal{M}$. The proof is very similar to the standard textbook proof. From the perspective of machine-aided deduction, this proof is noteworthy for two reasons: First it involves reasoning at various levels of abstraction; secondly, it uses widely different proof techniques. Specifically, the proof uses definitions and properties naturally stated in a theory of an abstract metric space at the same time that it uses assorted facts about the real numbers. Some of these facts, such as the convergence to zero of r^n as $n \to \infty$ (which follows from Bernoulli's inequality) involve the order, metric, and algebraic structure of the real numbers. Other pertinent facts have a purely algebraic nature, such as the geometric series formula,

$$\sum_{j=p}^{q} r^j = \frac{r^p \cdot (1 - r^{q-p+1})}{1 - r},$$

[9] From this conclusion, it is easy to show that the mapping is also a homeomorphism (i.e., that it is continuous with a continuous inverse).

which holds provided $0 \leq p \leq q$ and $r \neq 0, \neq 1$. Bernoulli's inequality and the geometric series formula are in turn proved using induction and algebraic and order properties of the reals.

The proof of Theorem 1 requires the following two lemmas:

Lemma 1 *Suppose the metric space* $(\mathbf{P}, d)$ *is complete;* $a \in \mathbf{P}$*;* $r \in \mathbf{R}$ *with* $0 \leq r$*;* $\varphi : \mathbf{P} \rightarrow \mathbf{P}$ *is a contraction with modulus* c *that is defined on* $\bar{\mathrm{B}}(a, r)$*; and* $d(a, \varphi(a)) < r \cdot (1 - c)$*. Then* φ *has a fixed point in* $\bar{\mathrm{B}}(a, r)$*.*

Lemma 2 *Suppose* $a \in \mathbf{V}$*,* $r \in \mathbf{R}$ *with* $0 \leq r$*, and* $\varphi : \mathbf{V} \rightarrow \mathbf{V}$ *is a contraction with modulus* c *that is defined on* $\bar{\mathrm{B}}(a, r)$*. Then*

$$\mathrm{B}(f_\varphi(a), r \cdot (1 - c)) \subseteq f_\varphi(\bar{\mathrm{B}}(a, r)).$$

Proof of Theorem 1 in IMPS. First, Lemma 1 is stated and proved in $\mathcal{M}$ as follows. For technical convenience, a new theory $\mathcal{M}'$ is constructed by adding individual constants a and r of sort $\mathbf{P}$ and $\mathbf{R}$, respectively, and the axiom $0 \leq r$ to $\mathcal{M}$. A version θ of Lemma 1 is formulated in $\mathcal{M}'$ by taking a and r in the statement of the lemma to be individual constants instead of universally quantified variables. θ is derived in $\mathcal{M}'$ from an appropriate instantiation of the Contraction Principle. The instantiation is obtained by transporting the Contraction Principle from $\mathcal{M}$ to $\mathcal{M}'$ via the interpretation which interprets the sort $\mathbf{P}$ as $\bar{\mathrm{B}}(a, r)$ and the distance function d as the restriction of d to $\bar{\mathrm{B}}(a, r)$. To get a version of Lemma 1 in $\mathcal{M}$, in which a and r are universally quantified variables, θ is transported from $\mathcal{M}'$ to $\mathcal{M}$ via a context interpretation that is created automatically by IMPS.

Next, Lemma 1 is transported to $\mathcal{B}$ via the interpretation Φ defined above. Lemma 2 is derived in $\mathcal{B}$ from the translation of Lemma 1 by unfolding the definition of open set and then performing, in a straightforward manner, some user-directed logical reasoning.

Finally, Theorem 1 follows from Lemma 2 by a straightforward proof in $\mathcal{B}$.□

Another theorem which can be proved with a technique similar to the one we have outlined here is the Picard-Lindelöf existence theorem for ordinary differential equations (see [5]). Its proof involves an application of the Contraction Principle to a space of continuous functions on an interval. Much of this proof (which involves the construction of a large network of theories) has already been carried out by Robert Givan using the IMPS system.

6 Conclusion

Mathematics of any complexity requires a mixture of strikingly different kinds of reasoning. For instance, the proof of the Contraction Principle depends on the numerical properties of the reals, such as the geometric series formula, as well as using induction on integers and the much more abstract properties of a metric space. A sequence of points must be built up by applying a second order iteration operator to the contractive mapping, and an ϵ/δ argument must be given to show that the sequence is Cauchy. Algebra, ordering properties, and many lemmas must be used to compute with the various subgoals that arise. The proof illustrates what Wittgenstein called the "motley of mathematics."

Little theories provide a way of organizing this variety both at the implementation level and at the conceptual level.

References

[1] R. S. Bird. An introduction to the theory of lists. Technical Report PRG-56, Oxford University Computing Laboratory, 1986.

[2] R. Burstall and J. Goguen. Putting theories together to make specifications. In *Fifth International Joint Conference on Artificial Intelligence*, 1977.

[3] R. Burstall and J. Goguen. The semantics of Clear, a specification language. In *Advanced Course on Abstract Software Specifications*. Springer Verlag, 1980.

[4] R. Burstall and J. Goguen. An informal introduction to specifications using Clear. In R. Boyer and J Moore, editors, *The Correctness Problem in Computer Science*, pages 185–213. Academic Press, 1981.

[5] J. Dieudonné. *Foundations of Modern Analysis*. Academic Press, 1960.

[6] H. B. Enderton. *A Mathematical Introduction to Logic*. Academic Press, 1972.

[7] W. M. Farmer. A partial functions version of Church's simple theory of types. *Journal of Symbolic Logic*, 55:1269–91, 1990.

[8] W. M. Farmer. A simple type theory with partial functions and subtypes. Technical report, The MITRE Corporation, 1991. Submitted to *Annals of Pure and Applied Logic*.

[9] W. M. Farmer, J. D. Guttman, and F. J. Thayer. IMPS: an Interactive Mathematical Proof System (system description). In these proceedings.

[10] W. M. Farmer, J. D. Guttman, and F. J. Thayer. IMPS: an Interactive Mathematical Proof System. Technical Report M90-19, The MITRE Corporation, 1991. Submitted to *Journal of Automated Reasoning*.

[11] W. M. Farmer and F. J. Thayer. Two computer-supported proofs in metric space topology. *Notices of the American Mathematical Society*, 38:1133–1138, 1991.

[12] S. J. Garland and J. V. Guttag. A guide to LP, the Larch prover. Technical report, MIT, 1991. Available by ftp from larch.lcs.mit.edu.

[13] J. A. Goguen. Principles of parameterized programming. Technical report, SRI International, 1987.

[14] J. A. Goguen and T. Winkler. Introducing OBJ3. Technical report, SRI International, August 1988.

[15] E. Gunter. The implementation and use of abstract theories in HOL. In *Third HOL Users Meeting.* Computer Science Department, Aarhus University, 1990.

[16] J. V. Guttag, J. J. Horning, and J. M. Wing. Larch in five easy pieces. Technical Report 5, Digital Systems Research Center, 1985.

[17] J. D. Guttman. A proposed interface logic for verification environments. Technical Report M91-19, The MITRE Corporation, 1991.

[18] R. Harper, D. Sannella, and A. Tarlecki. Structure and representation in LF. In *Fourth Annual Symposium on Logic in Computer Science*, pages 226–37. IEEE Computer Society, 1989.

[19] D. Hilbert. *The Foundations of Geometry.* Open Court, Chicago, 1902.

[20] N. Jacobson. *Basic Algebra I.* W. H. Freeman, second edition, 1985.

[21] J. Krivine. *Introduction to Axiomatic Set Theory.* Reidel, Dordrecht, 1971.

[22] L. G. Monk. Inference rules using local contexts. *Journal of Automated Reasoning*, 4:445–462, 1988.

[23] J. Mycielski. A lattice of interpretability types of theories. *Journal of Symbolic Logic*, 42, 1977.

[24] D. Sannella and R. Burstall. Structured theories in LCF. In *CAAP'83: Trees in Algebra and Programming, 8th Colloquium*, volume 159 of *Lecture Notes in Computer Science*, pages 377–91. Springer-Verlag, 1983.

[25] J. R. Shoenfield. *Mathematical Logic.* Addison-Wesley, 1967.

[26] L. W. Szczerba. Interpretability of elementary theories. In R. E. Butts and J. Hintikka, editors, *Logic, Foundations of Mathematics, and Computability Theory*, pages 129–145. Reidel, 1977.

[27] A. Tarski, A. Mostowski, and R. M. Robinson. *Undecidable theories.* North-Holland, 1953.

[28] J. van Bentham and D. Pearce. A mathematical characterization of interpretation between theories. *Logica Studia*, 43:295–303, 1984.

[29] A. Visser. The formalization of interpretability. *Studia Logica*, 50:81–105, 1991.

Some Termination Criteria for Narrowing and E-Narrowing

Jim Christian*
Kyoto University
and
ASTEM Research Institute of Kyoto

Abstract

Given a set of first-order equations T, we describe some restrictions on T which ensure that there are no infinite narrowing derivations using T. The criteria are also applicable when T is used to perform E-narrowing modulo a set of simple linear permutative equations E.

1 Introduction

Narrowing [5, 9] is a powerful technique with a number of applications including equational unification, functional programming, and logic programming. We are interested in using it to extend the second-order matching algorithm of Huet and Lang [6], to perform second-order matching modulo a set of first-order rewrite rules defining "built-in" function symbols. In this application, we very much desire that the resulting algorithm also be a decision procedure, since we plan to incorporate it into a compiler for a programming language. Unfortunately, the use of narrowing generally results in a semi-decision procedure for the unification problem, even when the set of rewrite rules constitutes a decision procedure for the associated term equality problem. To a first approximation, the problem is that the (ordinary) unification operation used to match the left-hand side of a rewrite rule with a subterm to be narrowed can introduce new structure into the subterm, possibly with the result that the narrowed term does not decrease according to any well-founded term ordering.

*Research Institute for Mathematical Sciences, Kyoto University, Kitashirakawa, Sakyo-ku, Kyoto 606, Japan. email: jimc@astem.or.jp

Because of this, we have developed a simple restriction which will ensure that narrowing must eventually terminate. The restriction is quite strong – roughly, the left-hand side of a rewrite rule must be of the form $f(t_1, \ldots, t_n)$ where each t_i is either ground or a variable – but it happens to be adequate for the program matching and transformation problems in which we are interested. Also, it is localized to individual equations, making it very easy to verify. In addition, it is possible to modify the restriction to work even when E-narrowing is used, providing E consists only of simple linear permutative equations. We provide some examples which illustrate difficulties involved in weakening the restriction.

2 Narrowing

Given a set of rewrite rules T the basic operation of narrowing is as follows: Let t be a nonvariable term such that $\sigma(t[u]) = \sigma(l)$ for some substitution σ, subterm occurrence u, and rewrite rules $l \rightarrow r \in T$. Then $\sigma(t[u \leftarrow r])$ is the result of narrowing t by $l \rightarrow r$ at u. (Note that, in general, a single subterm occurrence might unify with the left sides of multiple rules, and so several narrowings are possible.) We will write $t \mapsto t'$ to denote narrowing of t to produce t'.

More generally, let E be a set of equations for which an E-unification algorithm is available. Then E-narrowing is defined as above, except that σ is an E-unifier of $t[u]$ and l. Remember that there may be multiple E-unifiers for a pair of terms.

Narrowing is typically used to match or unify pairs of terms. For instance, we can use narrowing to find a substitution by which $U+(V+W)$ and W are equal modulo the set $\{0+X \rightarrow X\}$. First, we can unify $U+(V+W)$ with $0+X$ by the substitution $\{U \leftarrow 0, X \leftarrow V+W\}$ and then narrow to produce $V+W$. Narrowing again by the substitution $\{V \leftarrow 0, W \leftarrow X\}$, we obtain W.

To see why narrowing can be problematic, consider the rewrite rule $R = f(f(X)) \rightarrow X$. Clearly, R is terminating when used only for rewriting. However, there is an infinite sequence of narrowings for the term $f(U)$: $f(U) \mapsto f(U') \mapsto f(U'') \mapsto \ldots$. On the other hand, for the rule $f(X) \rightarrow g(h(X))$, there is only a one-step narrowing sequence possible for $f(U)$: $f(U) \mapsto g(h(U)$. Naturally, the situation can be much more complicated when the set of rewrite rules contains multiple rules. Since rewrite rules have the equivalent power of Turing machines, the termination problem for rewriting and narrowing is undecidable. The best one can hope is to identify large classes of rewrite systems for which termination can be guaranteed.

3 Criteria for Termination

Our approach is straightforward. First, we define a global property called *n-reduction* (for "narrowing reduction") which ensures termination of narrowing. Then, we show that conventional termination orderings [3, 8, 4] can be coupled with a simple restriction on the syntax of left sides of rewrite rules, to construct a (conservative) test for the presence of the n-reduction property.

Define a *termination ordering* to be any well-founded ordering $<$ on ground terms such that if $s < t$, then $\sigma(s) < \sigma(t)$ for any substitution σ.

Given a termination ordering $<$ and a set of rewrite rules T, we say that T is *n-reducing* with respect to $<$ if and only if whenever any term t narrows to t' using a rule in T, then $t' < t$.

Lemma 1 *If T is n-reducing with respect to some termination ordering, then there are no infinite narrowing sequences using only rules from T.*

> **Proof:** If T is n-reducing with respect to $<$, then any sequence $t_1 \mapsto t_2 \mapsto \ldots$ must be such that $t_i < t_{i-1}$. By the well-foundedness of $<$, any such sequence must be finite.□

We say that a term $f(s_1, \ldots, s_n)$ is *flat* if $n = 0$ (i.e. f is a constant), or s_i is either a variable or a ground term for $1 \leq i \leq n$.

Given a termination ordering $<$, we define $<_{\mathcal{L}}$ to be the extension of $<$ such that $s <_{\mathcal{L}} t$ if and only if either

- the number of distinct variables in s is less than the number in t; or
- the number of distinct variables in s is the same as the number in t, and s and t are identical everywhere (up to renaming of the variables in s) except at some occurrence u, where $s[u] < t[u]$.

Clearly, $<_{\mathcal{L}}$ is well-founded if and only if $<$ is.

The following lemma is our main result:

Lemma 2 *Let T be a set of rewrite rules all of whose left-hand sides are flat, and let $<$ be a termination ordering. If for every rule $l \rightarrow r$ in T, $r < l$, then T is n-reducing with respect to $<_{\mathcal{L}}$.*

Proof: We show that if $r < l$ and $\sigma(t[u]) = \sigma(l)$, then $\sigma(t[u \leftarrow r]) <_{\mathcal{L}} t$; hence, T is n-reducing with respect to $<_{\mathcal{L}}$.

By hypothesis, l is flat. Because of this, and because narrowing is never performed at a variable occurrence (i.e. $t[u]$ is a nonvariable), σ cannot increase the number of variables in $t[u]$. (Any variable in $t[u]$ must either be bound to a ground term, in which case all occurrences of the variable are eliminated by σ; or bound to a variable in l, merely causing all occurrences of the variable to be uniformly renamed; or not bound at all.) Next, each variable in r must have occurred in l; otherwise, $<$ could not be a termination ordering. (This is a well-known property of termination orderings.) Hence, narrowing by $l \longrightarrow r$ cannot increase the number of distinct variables in $t[u]$, and hence in t.

So, if narrowing results in no change in the number of distinct variables, then each variable in t is either unbound by σ, or renamed to another variable. So $\sigma(t[u \leftarrow r])$ is identical (up to renaming of variables) to t everywhere except at occurrence u; and since $<$ is a termination ordering, we have $\sigma(t[u \leftarrow r])[u] < \sigma(t[u])(= \sigma(l))$.

Thus, $\sigma(t[u \leftarrow r]) <_{\mathcal{L}} t$.□

So, we can use any conventional term rewriting ordering – for instance, the recursive path ordering, or the Knuth-Bendix ordering – to check for n-reducibility, providing that the left side of each rewrite rule is flat. See [4] for a survey of rewriting termination orderings.

3.1 Narrowing Modulo Simple Linear Permutative Equations

We can easily extend our results to accomodate E-narrowing when E is a set of simple linear permutative equations. An equation is a simple linear permuter when it is of the form $f(x_1, \ldots, x_n) = f(x_{\pi(1)}, \ldots, x_{\pi(n)})$, where the x_i are distinct variables and π is a permutation function ranging over $\{1 \ldots n\}$.

Let $[t]_E$ be the equivalence class of t in E. Then a set of rewrite rules T is *n-reducing modulo* E with respect to a well-founded ordering $<$ if whenever t narrows to t' using a rule in T, then $u < v$ for all $u \in [t']_E$ and $v \in [t]_E$. It is easy to see that the presence of this property assures that all E-narrowing derivations using T are finite.

The following fact generalizes Lemma 2:

Lemma 3 *Let T be such that l' is flat and $r' < l'$ for each $r' \in [r]_E$ and $l' \in [l]_E$, for all $l \rightarrow r \in T$, according to a termination ordering $<$. Then T is n-reducing modulo E with respect to $<$.*

Proof: The proof is a straightforward generalization of the proof of Lemma 2.

Notice that if E contains only simple linear permuters, then l' is flat for any $l' \in [l]_E$, where l is flat. Also, if $<$ is the recursive path ordering [3], then $r < l$ implies $r' < l'$ for $r' \in [r]_E$ and $l' \in [l]_E$; this is because the multiset comparison used by the recursive path ordering is insensitive to positions of subterms. Hence, it isn't necessary to physically test each possible permutation. However, for other orderings, like the Knuth Bendix ordering, it is necessary to compare each possible permutation of l and r.

4 Example

Of course, the flatness requirement is a strong restriction, but it covers some interesting sets of rewrite rules. For instance, the motivation for the work presented here was to show that we can safely narrow with rules like the following, all of which have flat left hand sides and which are reductions according to the recursive path ordering. Notice that E contains a number of commutativity equations. According to Lemma 3, the following theory yields no infinite narrowing sequences:

```
E:
  or(X, Y) = or(Y, X).
  and(X, Y) = and(Y, X).
  *(X, Y) = *(Y, X).
  +(X, Y) = +(Y, X).

T:
  and(t, X) --> X.
  and(f, X) --> f.
  or(X, f) --> X.
  or(X, t) --> t.
  *(X, 0) --> 0.
  *(X, 1) --> X.
  +(X, 0) --> X.
  -(X, 0) --> X.
  /(0, X) --> 0.
  if(t, X, Y) --> X.
  if(f, X, Y) --> Y.
  if(X, Y, f) --> and(X, Y).
  if(X, t, Y) --> or(X, Y).
```

```
(not t) --> f.
(not f) --> t.
```

5 Closing Remarks

It is tempting to try to find a weaker restriction than that of flatness of left-hand sides, but this seems difficult to do without losing the ability to test individual rewrite rules in isolation. Also, it would be nice to identify more general classes of equations that are compatible with our criteria for termination of E-narrowing. But this, too, seems nontrivial.

For example, the rule $f(f(x)) \to x$ is "safe" when used to narrow a linear term like $g(f(u), v)$. It produces the term $g(x, v)$, which can't be further narrowed. But the nonlinear term $g(f(u), u)$ can be narrowed indefinitely: $g(f(u), u) \mapsto g(x, f(x)) \mapsto g(f(x'), x') \mapsto \ldots$. In general, whenever the left side of a rewrite rule is not flat, aliasing due to repeated variables can cause troublesome propagation of structure.

Next, consider E-narrowing when E is an associative-commutative theory. Let $+$ be associative-commutative, and narrow the term $u + v$ using $x + 0 \to x$. One unifier of $u + v$ and $x + 0$ is $\{u \leftarrow 0 + z, x \leftarrow z + v\}$. So, $u + v \mapsto z + v \mapsto z' + v \ldots$. Hence, associative-commutative theories lie outside the scope of our techniques.

There are a few other approaches to improving the termination characteristics of ordinary narrowing, but they typically require inspection of the term being narrowed as well as of the rewrite rules being used for narrowing, and they don't consider E-narrowing. For example, Hullot [7] has shown that ordinary narrowing can be restricted to "basic" positions without affecting completeness. Intuitively, a basic position is one which wasn't introduced by a narrowing substitution. Basic narrowing can handle the equation $f(f(x)) \to x$ mentioned above; but it also allows an infinite narrowing derivation from $h(x)$ using the rule $h(f(g(y))) \to h(f(y))$ (example from [1]).

Likewise, Chabin and Réty [1] have developed a graph-based approach to improving termination of narrowing, but their graph structure is built using information about the term being narrowed. Moreover, much time is apparently required to build the graph.

Also, we note that Comon, Haberstrau, and Jouannaud [2] have been investigating some ground decidability properties of *shallow* terms; these appear to be the same as what we call flat terms. (This was brought to our attention by a referee; we haven't actually obtained the research report yet.)

Despite the strength of the flatness restriction, the results presented here are useful. The class of theories for which narrowing is decidable according to our criteria contains a number of interesting sets. For instance, the set of simplifiers presented in the previous section is especially appropriate for the matching application which we are exploring; there, a general template must be specialized at compile-time using second-order matching and narrowing to attempt to fit it to a particular piece of code. Knowing that the narrowing process is decidable makes us less concerned about adding the matcher to a compiler, and ensures that we needn't sacrifice completeness of matching to guarantee a bounded running time. An important property of our criteria is that termination may be checked syntactically, without inspection of the term to be narrowed.

References

[1] Jacques Chabin and Pierre Réty. Narrowing directed by a graph of terms. Springer-Verlag LNCS 488, *Rewriting Techniques and Applications 1991*, 112–123.

[2] H. Comon, M. Haberstrau, and J.-P. Jouannaud. Decidable problems in shallow equational theories. Research report, LRI, December 1991.

[3] Nachum Dershowitz. Orderings for term rewriting systems. *Theoretical Computer Science*, 17:279–301, 1982.

[4] Nachum Dershowitz. Termination of rewriting. *Journal of Symbolic Computation*, 1987.

[5] Michael Fay. First-order unification in an equational theory. In *Proceedings of the Fourth Workshop on Automated Deduction*, pp161–167, Austin, Texas 1979.

[6] G. Huet and B. Lang. Proving and applying program transformations expressed with second-order logic. *Acta Informatica* 11, 31–55, 1978.

[7] J.-M. Hullot. Canonical forms and unification. *Proceedings of the Fifth Conference on Automated Deduction*, Les Arcs, France. pp318–334, July 1980.

[8] Dallas Lankford. On proving term rewriting systems are noetherian. Technical Report MTP-3, Dept. of Mathematics, Louisiana Tech University, Ruston, Louisiana, 1979.

[9] P. Rety, C. Kirchner, H. Kirchner, and P. Lescanne. Narrower: a new algorithm for unification and its application to logic programming. In *Lecture Notes in Computer Science 202*, Springer-Verlag, 1985.

Decidable Matching for Convergent Systems *

— Preliminary Version —

Nachum Dershowitz, Subrata Mitra
Department of Computer Science
University of Illinois
Urbana, IL 61801, U.S.A.
{nachum, mitra}@cs.uiuc.edu

G. Sivakumar
Department of Computer Science
Indian Institute of Technology
Bombay 400076, India.
siva@cse.iitb.ernet.in

Abstract

We provide a simple system, based on transformation rules, which is complete for certain classes of semantic matching problems, where the equational theory with respect to which the semantic matching is performed has a convergent rewrite system. We also use this transformation system to describe decision procedures for semantic matching problems. We give counterexamples to show that semantic matching becomes undecidable (as it generally is) when the conditions we give are weakened. Our main result pertains to convergent systems with variable preserving rules, with some particular patterns of defined functions on the right hand sides.

1 Introduction

Equation solving is the process of finding a substitution which makes two terms equal in a given theory, while *semantic unification* is the process that generates a basis set of such unifiers. A simpler version of this problem, *semantic matching*, restricts the substitution to apply only to the term on the left, say (the *pattern*). Semantic matching has potential applications in pattern-directed languages. For example, if we could match with respect to addition, the function definition

$$\begin{aligned} \mathit{half}(x+x) &= x \\ \mathit{half}(x+x+1) &= x \end{aligned}$$

could be applied to a term like *half*(17), by finding that the pattern in the second definition matches the term when $x = 8$.

It is well-known that any strategy for finding a complete set of unifiers (or matchings) for two terms, with respect to a given theory, may not terminate, even when the theory is presented as a finite and *convergent* (terminating and confluent) set of rewrite rules [HH87, Bo87]. But, for some special classes of theories—associativity and commutativity, for instance—semantic unification is decidable.

It is, therefore, of interest to find suitable cases for which a particular equation-solving procedure is provably terminating, thus implying that the semantic unification or semantic matching problems in the corresponding theories are decidable. In this paper, we consider only

*This research was supported in part by the U. S. National Science Foundation under Grants CCR-90-07195 and CCR-90-24271.

equational theories for which there is a finite convergent rewrite system. We specialize the unification procedure given in [DS87, Mit90, JK91] and study the effect of some syntactic and semantic restrictions on the rewrite system presenting a theory, which result in decidability.

In the remainder of this section, we briefly review the relevant basic notions, terminology and results for equational theories and rewrite systems. For surveys of this area, see [DJ90] and [JK91].

Terms are constructed from a given set of function symbols and variables. We normally use ϱ, ρ, l, r, s, and t for arbitrary terms, and x, y, and z for variables. A *ground* term is one containing no variables (such as, $0+0$). A term t is said to be *linear* in a variable x if x occurs only once in t. For example, the term $x+s(y)*z$ is linear in all three variables. The *size* of a ground term is the number of function symbols it has, whereas its *depth* is the length of the longest path in its tree representation. A *substitution* is a mapping from variables to terms. We use lower case Greek letters θ, σ and μ to denote substitutions, and write them out as $\{x_1 \mapsto s_1, \ldots, x_m \mapsto s_m\}$.

A (ground) term t *matches* a pattern (term) s in a theory E if $E \models s\sigma = t$ for some substitution σ. We also say that t is an *instance* of s in this case. For example, $0+s(0)$ matches $y+x$ with the substitution $\{x \mapsto 0, y \mapsto s(0)\}$ in the theory $\{x+y=y+x\}$. A term s *unifies* with a term t in a theory E if $E \models s\sigma = t\sigma$ for some substitution σ. We say that a solution σ is *at least as general* as a solution ρ if there exists a substitution τ such that ρ and the composition of σ and τ give equal terms (equal, in E), for each variable in the problem. For example, a most general unifier of $x+y$ and $u+v$ is the substitution $\{x \mapsto u, y \mapsto v\}$. Semantic unification is the process of finding all such substitutions.

An *equation* is an *unordered* pair of terms, written in the form $s=t$. Either or both of s and t may contain variables; which are understood as being universally quantified. A *rewrite rule* is an oriented equation between terms, written $l \rightarrow r$; a *rewrite system* is a set of such rules. A rewrite rule is *left linear* if its left hand side is linear for all the variables, for example $s(x)*y \rightarrow y+(x*y)$. A rewrite rule is said to be *variable-preserving* if all the variables in its left hand side also appear in its right hand side term. A function symbol f is said to be a *defined function* with respect to a rewrite system R if there exists a rule in R with f as the top-most symbol of its left hand side; if there is no such rule, then f is called a *constructor*. We will use $\equiv$ for identity of terms, to distinguish it from other forms of equality.

For a given system R, the rewrite relation $\rightarrow$ replaces an instance $l\sigma$ of a left-hand side l by the corresponding instance $r\sigma$ of the right-hand side r. Unlike equations, replacements are not allowed in the reverse direction. We write $s \rightarrow t$, if s rewrites to t in one step; $s \rightarrow^* t$, if t is *derivable* from s, that is, if s rewrites to t in zero or more steps; $s \downarrow t$, if s and t *join*, that is, if $s \rightarrow^* w$ and $t \rightarrow^* w$ for some term w. A term s is said to be *irreducible*, or in *normal form*, if there is no term t such that $s \rightarrow t$. We write $s \rightarrow^! t$ if $s \rightarrow^* t$ and t is in normal form. All the matching problems we consider are of the form $s \rightarrow^? N$, meaning: find a substitution σ such that $s\sigma$ has normal form N. (We will frequently use N to stand for a term in normal form.) A solution is *irreducible* if each of the terms substituted for the variables is irreducible.

A rewrite relation is *terminating* if there is no infinite chain of rewrites: $t_1 \rightarrow t_2 \rightarrow \cdots \rightarrow t_k \rightarrow \cdots$. A rewrite relation is *ground confluent* if whenever two ground terms, s and t, are derivable from a term u, then a term v is derivable from both s and t. That is, if $u \rightarrow^* s$ and $u \rightarrow^* t$, then $s \rightarrow^* v$ and $t \rightarrow^* v$ for some term v. A rewrite system that is both ground confluent and terminating is said to be *ground convergent*; whenever we say "convergent" in this paper, we mean "ground convergent". Convergent rewrite systems are important for the following reason:

If R is a convergent rewrite system and E is the underlying equational theory (E is R with each rule taken as an equation), then $E \models s = t$ (for ground terms s and t) iff $s \downarrow t$ in R.

Example 1.1. *The following convergent system has an undecidable semantic unification problem.*

$$\begin{array}{rcl} 1 + x & \rightarrow & s(x) \\ s(x) + y & \rightarrow & s(x + y) \end{array}$$

$$\begin{array}{rcl} 1 * x & \rightarrow & x \\ s(x) * 1 & \rightarrow & s(x) \\ s(x) * s(y) & \rightarrow & s(y + (x * s(y))) \end{array}$$

The system defines addition ($+$) and multiplication ($*$) over positive integers, which are represented in unary notation, using the constant 1 and successor function s.

It can be shown that in general it is undecidable if an equation has a solution with respect to the rewrite system given above, since were there a decision procedure for this, then it would solve Hilbert's undecidable Tenth Problem. We will prove later that the semantic matching problem is, nevertheless, decidable for this theory. ([Bo87, DJ90] use similar examples to show that in general semantic matching and unification are undecidable even for convergent systems.)

In the most general case, however, semantic matching can be as difficult as full semantic unification: For example, adding a new rule $eq(x, x) \rightarrow true$ to the above example makes unifying two terms s and t the same as matching $eq(s, t)$ to $true$ in the augmented theory.

2 The Matching Procedure

We describe a method for semantic matching that is complete for the special cases of matchings that we will consider in Section 3, and later in Section 4. This is a simplified version of the generally complete system for unification appearing in [DS87, Mit90, JK91], which is a refinement of *narrowing*, as studied in [Fay79, Hul80, NRS89, Ret87], and others.

We consider equational theories that are given as finite convergent rewrite systems. Convergent systems allow one to ignore reducible solutions to semantic unification and matching problems. For an equational goal like $s(0) + x \rightarrow^? s(s(0))$, in the theory of addition $(\{0 + x = x, s(x) + y = s(x + y)\})$, the only solution of interest is $\{x \mapsto s(0)\}$. Reducible solutions, like $\{x \mapsto 0 + s(0)\}$, are redundant if we collect all irreducible ones. We will, therefore, be interested only in finding solutions at least as general as all the irreducible ones. In the decidable cases we describe, there are only finitely many such solutions.

We always begin with a goal of the form $s \rightarrow^? N$, where N is a ground normal form, since instead of matching s with an arbitrary t, we can take N to be its normal form. The transformation rules keep track of the current list of subgoals to be solved. A matching is found when all the subgoals are of the form $x \mapsto N$, where x is a variable and N is a normal form, provided that whenever the same variable appears on the left in more than one subgoal, the identical term appears on the right. To get a complete set of solutions we need to consider different ways of applying the following (non-deterministic) transformation rules:

Eliminate	$\{x \rightarrow^? t\} \cup G$ $\rightsquigarrow$ $\{x \mapsto t\} \cup G\{x \mapsto t\}$

Decompose	$\{f(s_1, \ldots, s_n) \rightarrow^? f(t_1, \ldots, t_n)\} \cup G$ $\rightsquigarrow$ $\{s_1 \rightarrow^? t_1, \ldots, s_n \rightarrow^? t_n\} \cup G$

Mutate	$\{f(s_1, \ldots, s_n) \rightarrow^? t\} \cup G$ $\rightsquigarrow$ $\{s_1 \rightarrow^? l_1, \ldots, s_n \rightarrow^? l_n, r \rightarrow^? t\} \cup G$ where $f(l_1, \ldots, l_n) \rightarrow r$ is a renamed rule in R

We need not try all transformations on all goals, as shown in the proof of the following completeness result:

Theorem 2.1 (Completeness). *Let R be either a variable-preserving or a left-linear convergent rewrite system. If the goal $s \rightarrow^? N$ has a solution θ (that is, $s\theta \rightarrow^! N$), then there is a derivation of the form*

$$\{s \xrightarrow{?} N\} \overset{!}{\rightsquigarrow} \mu,$$

such that μ is a substitution at least as general as θ.

Proof. The first observation is that if $s \equiv f(s_1, \ldots, s_n)$, and we consider an innermost rewriting strategy, then

$$f(s_1\theta, \ldots, s_n\theta) \rightarrow^* f(N_1, \ldots, N_n),$$

where $s_i\theta \rightarrow^! N_i$. (If s is a variable, then $s \mapsto N$ is the solution we're looking for.) Thus, at this stage there are two possibilities for the topmost position of $f(N_1, \ldots, N_n)$:

- No rule applies at this position, and thus $N \equiv f(N_1, \ldots, N_n)$. This situation is simulated by the **Decompose** transformation rule, which generates the subgoals corresponding to $s_i\theta \rightarrow^! N_i$.

- Some rule applies at this position. This is handled by the **Mutate** transformation rule.

After a finite number of decompositions, the mutation corresponding to the next rewrite step in the derivation of N from $s\theta$ becomes possible, making progress towards the desired solution.

We show next that, since R is variable-preserving, we need only deal with subgoals which have ground normal forms on the right. This guarantees that whenever we have a subgoal with a variable on the left, no further work remains. Clearly, **Decompose** preserves this property of right hand sides. **Mutate** does not, since the l_i may have variables in them. But, if we solve $r \rightarrow^? t$ first to get a partial solution σ, then we can apply (using **Eliminate**) the solutions we get to *each* of the variables in the l_i terms. (We get ground substitutions for all of them, on account of the system's being variable-preserving.) Since we need only look at innermost rewriting, the $l_i\sigma$ must be in normal form.

We now consider the case when the rewrite system is left-linear. The selection strategy that we use in this case is identical to the one mentioned above, that is, solve the $r \rightarrow^? t$ subgoal first, and then apply these solutions to the l_i terms and so forth. However, the major difference is that in this case, since there could be variable dropping rules, such partial solutions may be non-ground. Thus, in general, we have to solve equations of the form $s \rightarrow^? t(\bar{x})$, where $\bar{x}$ denotes the set of variables that t may contain.

Now, note that for any goal of the form $s \rightarrow^? t(\bar{x})$, all the variables ($\bar{x}$) must have either come from the right hand side of the initial goal (which in this case was a ground term N) or from the left hand side of some rule which was used for mutation. The rewrite systems that we are considering are left-linear, and furthermore, every application of **Mutate** renames variables of the rule uniformly using variables that do not appear elsewhere. Thus, under this situation, a variable can occur at most once in the right hand side of some goal. Therefore, while using the selection strategy outlined before, if we encounter a subgoal of the form $s \rightarrow^? x$, we do not need to solve this goal any further, since this goal will be trivially satisfiable for any solution to the variables of s. (This observation is important, since it means that such subgoals does not constrain the solutions to the original goal in any way. Note that for a system which has non left-linear rules, this argument would not work if such a non-linear variable happened to be in s, and in such cases we would require new transformation rules to handle such goals.) In any other case, at least one of the transformation rules mentioned before must apply to this goal.

Finally, we have to show that the computed answer (μ) is at least as general as the solution θ. This can be done by induction on the well-founded multiset extension of $\rightarrow^+$, which compares multisets of the left hand side terms of a list of goals, with the solution θ applied to each such term, along any suitable derivation sequence. □

The termination proofs in later sections assume particular strategies for selecting subgoals or discarding subgoals. These strategies are instances of the selection strategy used in the above completeness proof, namely, always find solutions to the last subgoal of **Mutate** first, and eliminate goals whenever possible. Of course, **Decompose** and **Mutate** may both apply to the same subgoal, and there may be many ways of mutating, one for each rule of the rewrite system with the same outermost symbol as the left side of the subgoal.

3 Variable-Preserving Rules

In looking for decidable matching problems, we started with the following result (a special case of Theorem 3.6 which we prove later): *If R is a variable-preserving convergent term rewriting system for which:*

- *all right hand sides of rules are either variables, or have a constructor at the top-level, and*
- *there are no nested defined functions in any right hand side,*

then the semantic matching problem is decidable for R.

Example 3.1. *By this result, the following system has a decidable semantic matching problem.*

$$\begin{array}{rcl} app(nil, x) & \rightarrow & x \\ app(cons(a, x), y) & \rightarrow & cons(a, app(x, y)) \end{array}$$

In [HH87], there is an example of a system with a single defined function in every right hand side, which has an undecidable semantic matching problem. There, the defined function on the right hand side of rules does not appear below a constructor, but it obeys the other restrictions. This shows that defined functions must appear below at least one constructor.

Next, we tried to allow some nested function symbols on the right hand side of the rewrite rules. We require the following definitions:

Definition 3.2 (Suitable Property). A *suitable property* is a measure (like *depth*, *size*, etc.) associated with ground terms, along with a well-founded total ordering $>$ which compares values of P, such that P is strictly larger, under $>$, for terms than for its subterms.

Definition 3.3 (Non-Decreasing). Let P be a suitable property. A function symbol f is defined to be *non-decreasing* (with respect to P) if whenever $f(\widehat{s_1},\ldots,\widehat{s_n}) \rightarrow^! N$, where each $\widehat{s_i}$ and N is in ground normal form, $P(\widehat{s_i}) \leq P(N)$. Any function which does not have this property is said to be a *potentially decreasing* function (with respect to P).

We can similarly define the notion of strict increasingness of a function.

It is not possible to always decide whether a function defined by a given convergent rewrite system is non-decreasing with respect to a property P, even for a simple suitable property like depth.

Lemma 3.4. *It is undecidable if a function symbol is depth non-decreasing.*

Proof. Consider

$$\begin{array}{rcl} g(x) & \rightarrow & h(f(S_1 \circ \$, S_2 \circ \$, x), x) \\ h(f(\$, \$, \$), x) & \rightarrow & \$ \end{array}$$

where f is as detailed in [HH87]. If S_1 and S_2 are respectively the start symbols for two context free grammars G_1 and G_2, we have

$$f(S_1 \circ \$, S_2 \circ \$, x) \rightarrow^! f(\$, \$, \$) \text{ iff } x \in G_1 \text{ and } x \in G_2.$$

By the above construction, g can be depth non-decreasing if and only if

$$\forall x . x \notin (G_1 \cap G_2).$$

Thus, a decision procedure for this problem could be used to decide if the intersection of two arbitrary context free grammars is empty, which is impossible. □

This lemma shows that in general it is not possible to decide if a function is depth (increasing) non-decreasing, even for convergent systems. However, certain decidable subclasses with the property are easy to identify. For example, any function which has a variable dropping rule, with the dropped variable appearing immediately below the top-level function on the left hand side, cannot be depth (increasing) non-decreasing. Again, for each rule $l \rightarrow r$ which defines a function f, if $depth(l) \leq depth(r)$ then f is depth non-decreasing. We can also have similar sufficient conditions using the depth of each variable in the rule. For example, if every variable occurs below at least the same number of constructors on the right side, as on the left, then the corresponding function is depth non-decreasing. We can use the last criterion to show that $+$, as defined in Example 1.1, is depth non-decreasing.

Unfortunately, if the right hand sides in rewrite rules have defined functions nested below a potentially (depth) decreasing function, then the resulting system may have undecidable semantic matching problems. This we show by considering the rules shown below, together with the definitions of $+$ and $*$ given in Example 1.1.

Example 3.5.

$$\begin{array}{rcl} half(s(1)) & \rightarrow & 1 \\ half(s(s(x)) & \rightarrow & s(half(x)) \end{array}$$

$$\begin{array}{rcl} f(1,1) & \rightarrow & s(1) \\ f(s(x),s(y)) & \rightarrow & s(half(f(x,y))) \end{array}$$

Here *half* is a potentially (depth) decreasing function. We have the following property for f:

$$f(x,y) = \begin{cases} s(1) & \text{if } x = y \\ \textit{undefined} & \text{otherwise} \end{cases}$$

We can now try to solve the goal $f(t_1,t_2) \rightarrow^? s(1)$, where t_1 and t_2 are terms involving $+$ and $*$. This goal has a solution σ iff $t_1\sigma$ and $t_2\sigma$ have the same ground normal form (because of the observation about f). Thus, if this problem has a decision procedure, then we could use the same for deciding the semantic unification problem mentioned in Example 1.1. Therefore, no such decision procedure can exist.

Based on the counterexample above, it can be seen that a function definition in terms of some potentially decreasing functions is not suitable for our purpose, and we therefore restrict the right hand sides of rules to have potentially decreasing functions only at the lowest level (that is, no other defined function symbol can be nested below them). We have:

Theorem 3.6. *Let R be a convergent variable-preserving term rewriting system, and P be some suitable property. If*

- *all right hand sides for rules in R are either variables, or have a constructor at the top-level, and*
- *all right hand sides are such that no defined function is nested below any function decreasing with respect to P,*

then the semantic matching problem is decidable for R.

Proof. Let $\succ$ be a well-founded ordering on goals such that $s_1 \rightarrow^? N_1 \succ s_2 \rightarrow^? N_2$ iff either $P(N_1) > P(N_2)$ or $P(N_1) = P(N_2)$ and s_2 is a subterm of s_1.

We prove that it is possible to find all solutions (in finite time) to any goal of the form $\varrho \rightarrow^? N$, where ϱ is a term which has no defined function nested below any decreasing function and N is a ground normal form. This we do by induction with respect to the ordering $\succ$.

The interesting case is the one in which $\varrho \equiv f(\varrho_1,\ldots,\varrho_n)$, and f is a defined function. It is therefore possible to use the transformation rule **Mutate** on this goal, applying some rule $f(l_1,\ldots,l_n) \rightarrow \rho$. The essential steps are:

$$\begin{array}{lll} \{\varrho \rightarrow^? N\} & \leadsto_{\textbf{Mutate}} & \{\varrho_i \rightarrow^? l_i,\ldots,\varrho_n \rightarrow^? l_n, \rho \rightarrow^? N\} \\ & \leadsto_{\textbf{Decompose}} & \{\varrho_i \rightarrow^? l_i,\ldots,\varrho_n \rightarrow^? l_n, \rho_1 \rightarrow^? N_1,\ldots,\rho_m \rightarrow^? N_m\} \end{array}$$

Since every right hand side, by assumption, has constructors at the top, we have shown the decomposition step which may be applied to $\rho \rightarrow^? N$, assuming that the top-level constructor of ρ has m immediate subterms.

The subgoals $\{\rho_j \rightarrow^? N_j\}$ produced after the decomposition step are smaller than the original goal, that is, $\{\varrho \rightarrow^? N\} \succ \{\rho_j \rightarrow^? N_j\}$, for each j. Thus, by applying the inductive hypothesis we can assume that all the solutions to each of the goals in $\{\rho_j \rightarrow^? N_j\}$ (and therefore also for their collection, that is, $\rho \rightarrow^? N$) can be found in finite time. Let σ be the solution obtained along one feasible branch for the goal $\rho \rightarrow^? N$. Now, since all the rules are variable-preserving, ρ contains all the variables that are in any of the l_i terms. Furthermore, because of the variable-preserving nature of all rules, any such σ must be a ground solution (if not, we will have a situation where a non-ground term will rewrite using only variable-preserving rules to a ground term, which is not possible). Thus, for any such σ, each $l_i\sigma$ term must be ground.

There are now two different cases to be considered.

- Function f is *potentially decreasing*. By assumption there is no defined function below it, that is, no ϱ_i has a defined function, and therefore all the $\varrho_i \rightarrow^? l_i$ subgoals can be decomposed immediately to solved forms ($x \mapsto N'$), or to unsolvable goals with different constructors at the top. This shows that all the solutions for $\varrho \rightarrow^? N$ can be found in finite time in this case.

- Function f is *non-decreasing*. The important point to note is that each left hand side in the list of subgoals has the property that no defined function is nested below a potentially decreasing function. Let us now consider the ground solution σ as described above. Since f is known to be non-decreasing with respect to P, each of the $l_i\sigma$ terms must be such that $P(l_i\sigma) \leq P(N)$, or else the partial solution σ violates the condition that f is non-decreasing, and can be ignored. (In this case, the goal $\varrho \rightarrow^? N$ has no solution using this rule for **Mutate**.)

  Thus, for all feasible paths, we have that $P(l_i\sigma) \leq P(N)$, and therefore, we get

  $$\{\varrho \xrightarrow{?} N\} \succ \{\varrho_i \xrightarrow{?} l_i\sigma\},$$

  for each i, since each ϱ_i is a subterm of ϱ. Thus, by the induction hypothesis, each subgoal $\varrho_i \rightarrow^? l_i\sigma$ can be solved, and therefore the goal $\varrho \rightarrow^? N$ itself can also be solved.

Using this result, it is easy to show (by induction on the size of the left-sides of goals) that for any term s (even without the restrictions imposed on ϱ), and ground normal form $\widehat{N}$, the goal $s \rightarrow^? \widehat{N}$ is solvable. The idea is that for every application of **Mutate** with the goal $s \rightarrow^? \widehat{N}$, the subgoal $r \rightarrow^? N$ is solvable (by the above argument). Thus, we can replace the multiset of subgoals generated by **Mutate**, by a finite number of such multisets (without $r \rightarrow^? N$) corresponding to each of the solutions of $r \rightarrow^? N$. □

A few comments about the restrictions used for the proof are in order:

- Theorem 3.6 uses semantic restrictions on the right hand sides of rewrite rules of the system, by requiring that certain defined functions can appear only below non-decreasing functions. Although there can be no decision procedure to check if a function is increasing in general, certain simple sufficient restrictions are easy to check, given the corresponding set of rewrite rules. For instance, the special case, mentioned at the beginning of this section and applied in Example 3.1, uses only syntactic restrictions.

- The fact that we do not have defined functions nested below potentially decreasing function(s) is important because of the counterexample given in Example 3.5.

In certain special cases it is possible to relax the requirement that all right hand sides with defined functions must have a constructor at the top-level. For example, if we assume that the top-level function on the right hand side is strictly increasing, and that it eventually generates a constructor in a finite number of steps, then the above theorem holds.

The following example illustrates this point:

Example 3.7.

$$\begin{array}{rcl} 1+x & \to & s(x) \\ s(x)+y & \to & s(x+y) \end{array}$$

$$\begin{array}{rcl} fib(1) & \to & 1 \\ fib(s(1)) & \to & 1 \\ fib(s(s(x))) & \to & fib(s(x)) + fib(x) \end{array}$$

Here $+$ is a strictly *(depth) increasing* function, and fib defines the Fibonacci numbers, both being defined over positive integers. Furthermore, both rules for $+$ have constructors at the top-level on the right hand side, and the remaining rules have the properties required by Theorem 3.6. The semantic matching problem is decidable for this system.

The essential idea is that if any sequence of applications of **Mutate** for increasing functions generate a constructor eventually, then the matching problem is decidable. Here is an outline of the proof.

If the top-level symbol (of ϱ in Theorem 3.6) is an increasing function, the applicable transformation rule generates the following derivation.

$$\{\varrho \to^? N\} \quad \leadsto_{\textbf{Mutate}} \quad \{\varrho_{i_1} \to^? l_{i_1}, \rho_1 \to^? N\}$$

The goal $\rho \to^? N$ is not decreasing as such. However, since we assumed that all such derivations eventually generate a constructor at the top, we must have at least one step of decomposition if we continue to mutate this goal. The derivation therefore would look like:

$$\begin{array}{rll} \{\varrho \to^? N\} & \leadsto_{\textbf{Mutate}^*} & \{\varrho_{i_1} \to^? l_{i_1}, \ldots, \varrho_{i_m} \to^? l_{i_m}, \rho_m \to^? N\} \\ & \leadsto_{\textbf{Decompose}} & \{\varrho_{i_1} \to^? l_{i_1}, \ldots, \varrho_{i_m} \to^? l_{i_m}, \bar{\rho_m} \to^? \bar{N}\} \end{array}$$

We assume that we only mutate the $\rho_m \to^? N$ subgoal at every stage, in keeping with the selection strategy mentioned in the completeness proof of Section 2.

Now, we can show that the subgoals are decreasing with respect to the ordering $\succ$, as in Theorem 3.6.

Also, the system described in Example 1.1 obeys all the restrictions of Theorem 3.6, and thus has a decidable semantic matching problem.

4 Variable Dropping Rules

In this section we deal with the possibility of incorporating variable dropping rules into the rewrite system, and we will try to extend Theorem 3.6 suitably to handle such cases. However, before we do so, we point out some cases which cause problems, by way of counterexamples.

Example 4.1. *Consider the rule given below, together with the definitions of + and * from Example 1.1*

$$eq(x,x) \rightarrow true$$

The new rule is the only one which is variable dropping. For this set of rules, the problem of $\{eq(s,t) \rightarrow^? true\}$ is not solvable in general, because once again a solution of this problem would mean a decision procedure for some variation of Hilbert's Tenth Problem.

Example 4.2. *This time we consider two rules and the definitions of + and * as before*

$$\begin{aligned} f(s(x),0) &\rightarrow 0 \\ eq(x,x) &\rightarrow x \end{aligned}$$

Here the only variable dropping rule is the one for f. Consider a goal of the form $\{f(eq(t_1,t_2),y) \rightarrow^? 0\}$, where t_1 and t_2 are terms involving + and *. The possible solution steps for this goal are shown below:

$$\begin{aligned} \{f(eq(t_1,t_2),y) \rightarrow^? 0\} &\rightsquigarrow_{\textbf{Mutate}} \{eq(t_1,t_2) \rightarrow^? s(x_1)\}, \sigma = \{y \mapsto 0\} \\ &\rightsquigarrow_{\textbf{Mutate}} \{t_1 \rightarrow^? x, t_2 \rightarrow^? x\}, \sigma = \{y \mapsto 0, x \mapsto s(x_1)\} \end{aligned}$$

Thus, if this goal has a solution, then we can also solve the semantic unification problem with respect to + and *, which is not possible. This example illustrates the fact that a system with a single (left linear) variable dropping rule may admit undecidable matching problems, even when the other rules are variable preserving.

It is important to note that a single dropped variable may interact with non-linear variables of the left hand side of some other rule in a way which may lead to undecidability. Here we give an example which has a single non left-linear rule, and one of the rules is variable dropping. Once again we assume that we have the definitions of + and * as before.

Example 4.3.

$$\begin{aligned} eq(x,x,y) &\rightarrow s(g(x,y)) \\ g(x,0) &\rightarrow true \end{aligned}$$

With this system of rules, and the goal $eq(t_1,t_2,z) \rightarrow^? s(true)$, where t_1 and t_2 are terms involving + and *, we have the following solution steps:

$$\begin{aligned} \{eq(t_1,t_2,z) \rightarrow^? s(true)\} &\rightsquigarrow_{\textbf{Mutate}} \{t_1 \rightarrow^? x, t_2 \rightarrow^? x, z \rightarrow^? y, s(g(x,y)) \rightarrow^? s(true)\} \\ &\rightsquigarrow_{*} \{t_1 \rightarrow^? x, t_2 \rightarrow^? x\}, \sigma = \{y, z \mapsto 0, x \mapsto x_1, \ldots,\} \end{aligned}$$

Thus, for reasons similar to Example 4.2, this problem is undecidable. The last example shows that even a linear variable occurring immediately below the top-level symbol on the left hand side of a rule may result in the elimination of a variable that is non-linear in another rule. There is a subtle difference between Examples 4.2 and 4.3. In the former, the dropped variable x appears below some constructor (in the first rule), which implies that the subgoal $eq(t_1,t_2) \rightarrow^? s(x_1)$ is forced to be further mutated, thus causing the problem. Note that if this rule had a variable (say z) instead of the $s(x)$ term, then the corresponding subgoal would have been trivially solvable. The last example is a variation of Example 4.1; only in this case eq drops variables one at a time. This situation is simpler to check.

It is possible to have variable dropping rules and still have a decidable semantic matching algorithm:

Theorem 4.4. *Let R be a left linear rewrite system and P be a suitable property. If*

- *all right hand sides of rules are either variables, or have a constructor at the top-level, and*
- *there are no nested defined functions in the right hand side,*

then the semantic matching problem is decidable for R.

Proof.
Since we now have a left-linear system, we only need to solve for goals of the form $s \rightarrow^? t$, where any variable x in t has the property that it is linear in t and does not occur in the right hand side of any other subgoal. Thus, we can apply a proof quite similar to that in Theorem 3.6, and show that the procedure is terminating for this case.

We use a well-founded ordering (like $\succ$ in Theorem 3.6), which compares goals using a suitable property for right hand sides and subterm property for the left hand sides. Let us consider a goal of the form $\varrho \rightarrow^? N(\bar{x})$, with ϱ is a term without any nested defined functions. We use $N(\bar{x})$ to denote a term which has some variables $\bar{x}$, such that N is linear with respect to each of them, and furthermore, no other subgoal in the multiset of goals being solved has any of these variables in any right hand side. We show by induction that all solutions to such goals can be finitely generated. As before, consider an application of **Mutate**, with the rule $f(l_1, \ldots, l_n) \rightarrow \rho$ (assume that $\varrho \equiv f(\varrho_1, \ldots, \varrho_n)$). We have the following derivation (like in Theorem 3.6):

$$\begin{array}{lll} \{\varrho \rightarrow^? N(\bar{x})\} & \leadsto_{\textbf{Mutate}} & \{\varrho_i \rightarrow^? l_i, \rho \rightarrow^? N(\bar{x})\}, 1 \leq i \leq n \\ & \leadsto_{\textbf{Decompose}} & \{\varrho_i \rightarrow^? l_i, \rho_j \rightarrow^? N(\bar{x})_j\} \end{array}$$

We can now apply the inductive hypothesis on the $\rho_j \rightarrow^? N(\bar{x})_j$ subgoals, which implies that $\rho \rightarrow^? N(\bar{x})$ itself is solvable. Let σ be a solution to this subgoal. By assumption, there are no nested defined functions in ϱ. Therefore, each of the remaining goals can be solved using decomposition alone. □

We next attempted to introduce nested defined functions on the right hand side. The main difficulty is with the ordering using a suitable property. For the general case, like in Theorem 3.6, we have to show that each of the $l_i\sigma$ terms for goals of the form $\varrho_i \rightarrow^? l_i\sigma$ are less than the original goal, and this may not be possible for terms with variables. Thus, some further restrictions are required. We restrict the system such that if ρ be the right hand side of a rule where any function which has variable dropping rule occurs, then any goal of the form $\rho \rightarrow^? N$ has only ground solutions. In effect, we are trying to combine Theorems 3.6 and 4.4. Here we provide examples of two systems which have the required property:

Example 4.5. *In this example $+$ is depth non-decreasing, while $*$ is potentially depth decreasing, and has variable dropping rules. Furthermore, rules for $+$ do not use any function which has variable dropping rules. Finally, the last rule (which is the only one which has the variable dropping function $*$ on the right hand side) satisfies the condition mentioned above.*

$$\begin{array}{rcl} 0 + x & \rightarrow & x \\ s(x) + y & \rightarrow & s(x + y) \end{array}$$

$$\begin{array}{rcl} 0 * x & \rightarrow & 0 \\ s(x) * 0 & \rightarrow & 0 \\ s(x) * s(y) & \rightarrow & s(y + (x * s(y))) \end{array}$$

In the example which follows, *insert* is a strictly depth increasing function, which uses *min* (a variable dropping function) in its right hand side (the last rule). However, since both the variables which can potentially be dropped (x and y of the last rule) also appear in the second subterm of *cons*, under a non-variable dropping function *max*, the entire rule can be treated as variable preserving.

Example 4.6.

$$\begin{array}{rcl} min(x,0) & \rightarrow & 0 \\ min(0,x) & \rightarrow & 0 \\ min(s(x),s(y)) & \rightarrow & s(min(x,y)) \end{array}$$

$$\begin{array}{rcl} max(x,0) & \rightarrow & x \\ max(0,x) & \rightarrow & x \\ max(s(x),s(y)) & \rightarrow & s(max(x,y)) \end{array}$$

$$\begin{array}{rcl} sort(nil) & \rightarrow & nil \\ sort(cons(x,y)) & \rightarrow & insert(x,sort(y)) \end{array}$$

$$\begin{array}{rcl} insert(x,nil) & \rightarrow & cons(x,nil) \\ insert(x,cons(y,z)) & \rightarrow & cons(min(x,y),insert(max(x,y),z)) \end{array}$$

Thus, in order to introduce nested defined functions on the right hand side, we have to essentially make sure that whenever there is a possibility of a variable being dropped, there must be another subgoal which instantiates that variable.

5 Related Results

Some results similar to those given here have been reported in [Hul80, KN87], where they are interested in the more general problem of semantic unification. Hullot [Hul80] shows that the *narrowing* procedure terminates when all right hand sides are either variables or ground terms. Furthermore, it has been demonstrated by Kapur and Narendran [KN87] that if each right hand side is either ground or a subterm of the left hand side, the unification problem for the corresponding theory is NP-complete.

Using techniques similar to those in this paper, and the *full* set of transformation rules in [Mit90], it is not hard to show that there is a strategy which is terminating for Hullot's case. The full system essentially gives additional decomposition rules necessary to handle cases when the right hand side of a goal $s \rightarrow^? x$ happens to be a variable. For the systems that we have considered so far, such a goal can either be ignored (in some left linear cases), or can be replaced by a different goal which doesn't have this property (for the variable preserving case). However, for the general semantic unification problem, such simplifications are not possible, and thus we require new transformation rules to handle these cases. Since both sides of a goal may contain variables the orderings we used in Section 3 do not decrease, having no a priori bound on the measure of the right side of goals.

It is possible to extend Hullot's system somewhat:

Theorem 5.1. *If all right hand side of rules are either constructor terms or ground terms, then semantic unification is decidable.*

Proof. Consider r in the mutation rule. In the ground case, the goal $r \rightarrow^? t$ has a solution only if the normal form of r syntactically matches t. In the constructor case, r is free of defined functions, and $r \rightarrow^? t$ has a solution only if r and t can be syntactically unified. □

Applying similar techniques, we can also prove the following theorem from [KN87]:

Theorem 5.2. *Let R be a rewrite system in which every right hand side is a subterm of the corresponding left hand side. Then, the unification problem is decidable for R.*

Proof. The basic idea is that, since the full set of transformation rules ([Mit90]) simulates *innermost* rewriting in reverse (see [DS87, Mit90] for details), at any stage the right hand side t of a goal of the form $s \rightarrow^? t$ must be irreducible for a solution to be feasible. In other words, if at any stage we have to mutate at a position which comes from the left hand side of a previous rule application, then the corresponding solution is reducible and can be ignored (the procedure will enumerate another solution which is equivalent but irreducible). Now, every right hand side of a rule in R is a subterm of the corresponding left hand side. Thus, based on the previous argument, we never need to further mutate the $r \rightarrow^? t$ subgoal generated from any application of **Mutate**, which implies that the solution tree generated from this subgoal is finite. □

6 Future Work

It would be interesting to develop the ideas of the previous section. Further restrictions on the system, such as having *completely defined* functions (a function f is said to be *completely defined* by a rewrite system R, if the normal form of any ground term containing f is a constructor-only term), may help because of the following result:

Lemma 6.1. *If R consists of only completely defined function symbols, then innermost narrowing is complete with respect to ground solutions.*

Proof. Observe that, in the final normal form, all the defined functions must be removed. Thus, we can apply rules at any position where a defined function occurs, and, in particular, to the innermost position. □

With this understanding, in certain cases it will be possible to have a decision procedure for semantic unification. The idea is to apply a minimal substitution to a subset of the variables in the goal, so that the terms are reducible and compare the top-level constructor generated along the two sides. If the subterms generated after decomposition happen to be "smaller" than the original goal terms in a well-founded ordering, then the system will be solvable.

It may also be possible to extend Theorem 4.4 somewhat, to the general case of left-linear systems with proper function nesting on the right hand side. Furthermore, it is possible that we could accommodate non left-linear systems, after demonstrating that these non-linear variables are never eliminated by the procedure at a later stage, when using a variable dropping rule in **Mutate**. It may be possible to give sufficient conditions to check such properties by analyzing a graph of the terms of the rewrite system. Example 4.3 illustrates such a possibility.

Acknowledgement

We thank Deepak Kapur and the referees for their helpful comments.

References

[Bo87] Alexander Bockmayr. A Note on a Canonical Theory with Undecidable Unification and Matching Problem. In *Journal of Automated Reasoning*, Vol 3, pages 379–381, 1987.

[DS87] Nachum Dershowitz and G. Sivakumar. Solving Goals in Equational Languages. In *Proceedings of the First International Workshop Conditional Term Rewriting System*, Orsay, France, July 1987. Vol. 308, pages 45–55, of *Lecture Notes in Computer Science*, Springer Verlag (1987).

[DJ90] Nachum Dershowitz and Jean-Pierre Jouannaud. Rewrite systems. In J. van Leeuwen, editor, *Handbook of Theoretical Computer Science*, chapter 6, pages 243–320, North-Holland, Amsterdam, 1990.

[Fay79] M. Fay. First-order unification in an equational theory. In *Proceedings of the Fourth Workshop on Automated Deduction*, pages 161–167, Austin, TX, February 1979.

[Hul80] Jean-Marie Hullot. Canonical forms and unification. In R. Kowalski, editor, *Proceedings of the Fifth International Conference on Automated Deduction*, pages 318–334, Les Arcs, France, July 1980. Vol. 87 of *Lecture Notes in Computer Science*, Springer, Berlin.

[HH87] Stephan Heilbrunner and Steffen Holldobler. The Undecidability of the Unification and Matching Problem for Canonical Theories. In *Acta Informatica*, Vol 24, pages 157–171, 1987.

[JK91] Jean-Pierre Jouannaud and Claude Kirchner. Solving Equations in Abstract Algebras: A Rule-Based Survey of Unification. In J.-L. Lassez and G. Plotkin, editors, *Computational Logic: Essays in Honor of Alan Robinson*, MIT Press, Cambridge, MA, 1991.

[KN87] Deepak Kapur and Paliath Narendran. Matching, Unification and Complexity. In *ACM SIGSAM Bulletin*, (1987) Vol. 21, Number 4, pages 6–9.

[Mit90] Subrata Mitra. Top-Down Equation Solving and Extensions to Associative and Commutative Theories. Master's thesis, Department of Computer and Information Sciences, University of Delaware, Newark, DE, 1990.

[NRS89] Werner Nutt, Pierre Réty and Gert Smolka. Basic Narrowing Revisited. In *J. of Symbolic Computation*, (1989) Vol. 7, pages 295–317.

[Ret87] Pierre Réty. Improving basic narrowing techniques. In P. Lescanne, editor, *Proceedings of the Second International Conference on Rewriting Techniques and Applications*, pages 228–241, Bordeaux, France, May 1987. Vol. 256 of *Lecture Notes in Computer Science*, Springer, Berlin.

Free Sequentiality in Orthogonal Order-Sorted Rewriting Systems with Constructors

Delia Kesner

INRIA Rocquencourt
Domaine de Voluceau, BP 105
78153 Le Chesnay Cedex, France
kesner@inria.inria.fr

CNRS and LRI
Bât 490, Université de Paris-Sud
91405 ORSAY Cedex, France
kesner@lri.lri.fr

Abstract

We introduce the notions of sequentiality and strong sequentiality in order-sorted rewriting systems, both closely related to the subsort order and to the form of declarations of the signature. We define *free sequentiality* for the class of orthogonal systems with constructors, a notion which does not impose conditions over the signature. We provide an effective decision procedure for free sequentiality that gives at the same time a simple construction of a non-deterministic pattern matching tree. These trees describe how the refinement of sorts and structures has to be done along the reduction sequence in such a way that wasteful computations are avoided.

1 Introduction

Term rewriting systems provide an elegant framework for programming in equational logic. *Sequentiality* is a property related to the possibility of *systematically* expanding any term step-by-step in order to get its normal form. This means that for every term not in normal form, there exists a redex occurrence which we necessarily need to reduce in every sequence that normalizes the term and this occurrence can be found without looking ahead. Unfortunately, there is no algorithm to compute these *needed redexes* , and there is also no algorithm to decide if a given system is sequential. These results of undecidability motivate the search for more realistic and effective approaches. Our purpose is to present a notion of sequentiality in orthogonal order-sorted rewriting systems with constructors. This notion is related to the possibility of expanding a term by refining either its sort or its structure. We obtain a characterization of a class of order-sorted systems, for which there is an efficient implementation that avoids useless reductions. We discuss some related work in order to place our results in context.

Orthogonal term rewriting systems are defined to have *linear* and *non-ambiguous* left hand sides of rules (LHS). In this class of systems, the parallel-outermost strategy, a normalization strategy which reduces all outermost redexes in each step, has been shown to be complete. Huet and Levy [HL79] showed that a restricted notion of sequentiality, called *strong sequentiality*, that only considers the LHS's, seems to be much more useful in order to perform efficient implementations of orthogonal rewriting systems. They developed a normalization algorithm which eliminates useless computations by only reducing needed redexes.

In term rewriting systems with *constructors*, the outermost symbols of LHS's are only allowed to appear at a root position. All other symbols are called *constructors* and LHS's are called *patterns*. Thus, subterms of patterns do not have reducible instances, and so reducing a term requires essentially the head normal form of some of its subterms in order to match the reduced term with some LHS rule. A *compilation* scheme will transform a set of LHS rules into a case-expression presented as a *discrimination tree*. However, the resulting tree can fail to terminate for some terms that can be matched by a pattern. This is due to wasteful reductions that try to normalize terms which do not have normal forms. A refined notion of strong sequentiality for term rewriting systems with constructors can be found in [Tha87].

Order-sorted algebras provide a substantially improved expressiveness over many sorted algebras. Sorts are partially ordered and terms are constrained with subsorts. In order-sorted rewriting systems sort information is not only useful for runtime type-checking, but also for pattern matching verifications. *Sort decreasing* rewriting systems refine the sort of the terms during the reduction process. As a consequence, a non-terminating reduction always has a minimal sort, and even if this sort is not always known, we can use some *partial information* in order to perform pattern matching. Thus, terms will need to be evaluated only as far as necessary in order to make either a structure or a subsort verification possible. In general, a few computation steps could be sufficient without the need to reduce terms to full normal forms. For example, suppose σ is a sort greater than η, the variable x is constrained by η, denoted as x^η, f is a constructor and t is a term of sort σ. Suppose that t reduces in one step to a term not in normal form whose sort is a subsort of η. If t has to be matched with the term $f(y)$, its head-normal form will be required, whereas if it has to be matched with the term x^η, only one step of reduction will be sufficient. We really take advantage in this way of a *lazy evaluation* mechanism.

Puel and Suárez [PS90] devised a compilation scheme to statically generate pattern matching algorithms in languages with lazy evaluation. They define a notion of strong sequentiality for sets of terms restricted by *structure constraints*[1]. This sequentiality notion has been generalized in [Kes91] to sets of terms restricted not only by structure, but also by *subsort constraints*. However, only some particular order-sorted signatures, called *unitary signatures*, were considered. Unitary signatures are defined to be regular and to verify some properties as for example minimality of the sort that appears in the codomain of every declaration of a function symbol. However, since Comon [CD91] has noticed that order-sorted signatures can also be seen as *finite bottom-up tree automata*, order-sorted rewriting systems can be studied from a more effective point of view. The main advantage consists of the elimination of restrictions to the function declarations and relation subsort orders of the signature. As an example, unification of order-sorted terms can be expressed in this framework with a finite set of constraints, even if there is an infinite set of most general unifiers. In this paper we will use Comon's formalism in order to study sequentiality in a general case, with no restrictions to the signatures.

[1]For example, a term $f(x)$ restricted by "x different from a" is written as $f(x)|x\Diamond a$.

Section 2 presents order sorted signatures and orthogonal order-sorted rewriting systems with constructors. In section 3 we show how to encode order-sorted signatures with bottom-up tree automata and we give the semantics of terms in the tree language recognized by its automaton. Section 4 introduces sequentiality of monotonic predicates over order-sorted partial terms. Our notion of sequentiality not only takes into account the structure of terms, but also their sorts. The idea is that the expansion of a term consists now of the refinement of either its sort or its structure. In section 5 we define *strong sequentiality*, a notion which is based only on left-hand sides of rules. It turns out that terms are not always forced to be reduced to head normal forms because they are evaluated only as far as required to make either a structure or a subsort verification possible. We provide in section 6 a construction method of non-deterministic *pattern matching trees* (PMT's), also called *pattern matching automata* or *pattern matching algorithms*, having transitions labeled with either constructors or subsort symbols. In section 7 we point out the main problems that arise when using such algorithms with order-sorted systems. Finally, we define *free sequential orthogonal constructor* systems, which allow to construct *optimal* PMT's for the evaluation of *well-sorted* terms, even if they are build with "strange" signatures. Optimal PMT's evaluates only the part of the term which is strictly necessary in order to satisfy the matching predicate, taking advantage in this way of the lazy reduction mechanism. The reduction of a well-sorted term t using an optimal PMT causes either a subsort or a structure verification and avoids wasteful computations. Our restrictions becomes in this way a sufficient condition in order to construct optimal order-sorted pattern matching trees. In reading the paper, familiarity with order-sorted algebras and sequentiality would be helpful, but the treatment is self-contained enough to be meaningful on its own.

2 Order-Sorted Rewriting Systems

An *order-sorted signature* $\Sigma = \langle \mathcal{S}, \leq, \mathcal{F}, \mathcal{V}, \mathcal{D} \rangle$ consists of a finite set of sort symbols $\mathcal{S} = \{\top, \bot, \sigma, \eta, \delta, \ldots\}$, a partial order $\leq$ on $\mathcal{S}$ such that $\forall \sigma \in \mathcal{S}, \bot \leq \sigma, \sigma \leq \top$, a set of function symbols $\mathcal{F} = \{f, g, h, \ldots\}$, a set of $\mathcal{S}$-indexed variables $\mathcal{V}^\sigma = \{x^\sigma, y^\sigma, z^\sigma, \ldots\}$ with a countably infinite number of variables for each sort symbol σ, and a set of declarations $\mathcal{D}$ of the form $f : \sigma_1 \ldots \sigma_n \rightarrow \sigma$ where $f \in \mathcal{F}$, $\sigma_1 \ldots \sigma_n$ is the *domain* of f, σ its *codomain* and n its *arity*. The sets $\mathcal{S}$, $\mathcal{F}$ and $\mathcal{V}$ are mutually disjoint. For brevity, we will write $s \in \Sigma$ for any symbol s in $\mathcal{S}$, $\mathcal{F}$, $\mathcal{V}$ or $\mathcal{D}$. From now on, we will use $\overline{\sigma_n}$ to denote a string $\sigma_1 \ldots \sigma_n$ and $f : \overline{\sigma_n} \rightarrow \sigma$ to denote a declaration $f : \sigma_1 \ldots \sigma_n \rightarrow \sigma$.

Terms are constructed as usual: a variable $x^\sigma \in \Sigma$ is a term and $f(t_1 \ldots t_n)$ is a term if and only if n is an arity of f and for all $1 \leq i \leq n$, t_i is a term. Σ-terms are *well-sorted* terms constructed in the following way: a variable $x^\sigma \in \Sigma$ is a Σ-term of sort η if $\sigma \leq \eta$, and $f(t_1 \ldots t_n)$ is a Σ-term of sort η if and only if there is $f : \sigma_1 \ldots \sigma_n \rightarrow \sigma \in \Sigma$ such that $\sigma \leq \eta$ and for all $1 \leq i \leq n$, t_i is a Σ-term of sort σ_i. A signature Σ is regular iff every Σ-term has a least sort. In the following, we will drop the prefix Σ when it is clear from the context.

A substitution θ is a function from terms to terms such that $\mathcal{D}\theta = \{x \mid \theta x \neq x\}$ is finite, $\theta f(t_1 \ldots t_n) = f(\theta t_1 \ldots \theta t_n)$ and when s is a term of sort σ, then θs is also of sort σ. A term is called *linear* if its variables occur at most once and *ground* if it contains no variables. We say that t' is an *instance* of t or t' *matches* t if there is a substitution θ such that $\theta t = t'$. Two terms t and t' are said to be *unifiable* or *compatible*, denoted as $t \uparrow t'$ iff there exists a substitution θ such that $\theta t = \theta t'$. Syntactic unification in finite order-sorted signatures is decidable [SNGM89].

A *rewriting rule* is an ordered pair of terms written as $t \rightarrow r$ such that t is not a variable, and every variable occurring in r occurs in t. A rewriting rule $t \rightarrow r$ is *sort decreasing* if, for every substitution θ and every sort symbol σ, θr has sort σ if θt has sort σ. It is decidable whether a finite rewriting system is sort decreasing [SNGM89]. The set of *positions* or *occurrences* of a term t, denoted $\mathcal{O}(t)$, is defined as usual as finite sequences of positive integers such that $\epsilon \in \mathcal{O}(t)$ (the root of t) and $k.u \in \mathcal{O}(f(t_1 \ldots t_n))$ if $u \in t_k$. The subterm of t at position u, denoted t/u, is defined as $t/\epsilon = t$ and $f(t_1 \ldots t_n)/k.u = t_k/u$. The replacement of the subterm of t at position u by a term p is denoted as $t[u \leftarrow p]$ or $t[s]_u$. When we don't care about the term where the replacement was done we will also write $C[s]_u$. The function symbol f appearing in t at position u is denoted $t|_u$.

A *base* $\mathcal{L}$ for a rewriting system is a finite set of linear terms such that when l_1 and l_2 are two distinct terms in $\mathcal{L}$, $u \in \mathcal{O}(l_1)$ and l_1/u is not a variable, then l_1 and l_2 are not unifiable. A *rewriting system* $\mathcal{R}$ is a finite set $\{l_i \rightarrow r_i\}_{i=1}^n$ of rewriting rules, and we say that it is *orthogonal* when $\{l_i\}_{i=1}^n$ is a base.

A term is a *redex* if it is an instance of some $l \in \mathcal{R}$. If t/u is a redex, u is said to be a *redex occurrence* in t. We say that t *reduces to* h, denoted $t \rightarrow_{\mathcal{R}} h$, iff there is a substitution θ and $u \in \mathcal{O}(t)$ such that $t/u = \theta l$ for some $l \rightarrow r \in \mathcal{R}$, where $h = t[u \leftarrow \theta r]$. Note that when $\mathcal{R}$ is a sort decreasing rewriting system the replacement of a subterm of a term t at position u always yields a well-sorted term. We denoted as $\rightarrow_{\mathcal{R}}^*$ the transitive reflexive closure of $\rightarrow_{\mathcal{R}}$. A term t is in *normal form* iff there is no redex occurrence in t. A term t is in *head normal form* if there is no redex s such that $t \rightarrow_{\mathcal{R}}^* s$.

If every $l_i \in \mathcal{L}$ has the form $f_i(t_1^i, \ldots, t_{m_i}^i)$, $\mathcal{F} = \{f_1 \ldots f_n\}$ is called the set of *head* symbols of $\mathcal{L}$ and $\mathcal{C} = \mathcal{F} - \{f_1 \ldots f_n\}$ the set of *constructors* symbols of $\mathcal{L}$. A term $f(t_1 \ldots, t_m)$ is said to be a *pattern* iff f is a head symbol and each t_i is a term without head symbols. A base $\mathcal{L}$ is a *constructor base* iff every $l_i \in \mathcal{L}$ is a pattern and a system based on such $\mathcal{L}$ is called a *constructor system*. For the rest of the paper we will only deal with sort decreasing systems that follow this constructor discipline. Sort decreasing systems guarantee that the sort of a term may only become smaller, and this is precisely what allows an efficient reduction strategy.

3 Signatures as Bottom-Up Tree Automata

Comon [CD91] has noticed that an order-sorted signature Σ is a finite bottom-up tree automaton (Q, A, F, T), where the set of states Q is the set of sorts $\mathcal{S}$, the set of terminal symbols A is $\mathcal{F}$, the set of final states F is also $\mathcal{S}$ and the set of transitions T is defined by the following rules:

- If $\sigma \leq \eta \in \Sigma$, then there is a ϵ-transition rule $\sigma \Rightarrow \eta$ in T
- If $f : \overline{\sigma_n} \rightarrow \sigma \in \Sigma$, then there is an f-transition rule $f(\overline{\sigma_n}) \Rightarrow \sigma$ in T

We will denote by $T(\Sigma)$ the set of ground terms constructed with Σ and by $T(\Sigma|_\sigma)$ the regular tree language recognized by Σ when replacing the set of final states F by $\{\sigma\}$. It turns out that the set of Σ-terms is $T(\Sigma|_\top)$. We define the set of *sort expressions*, denoted by $\mathcal{E}_S$, in the following way:

- $\sigma \in \mathcal{E}_S$, for any σ in Σ
- If $\sigma, \eta \in \mathcal{E}_S$, then $\sigma \vee \eta, \sigma \wedge \eta \in \mathcal{E}_S$
- If $\sigma_1 \ldots \sigma_n \in \mathcal{E}_S$ and $f \in \mathcal{F}$, $f(\sigma_1 \ldots \sigma_n) \in \mathcal{E}_S$

Given an automaton Σ, the semantics of a given sort expression σ in $T(\Sigma)$, denoted by $[\![\sigma]\!]^\Sigma$ or simply by $[\![\sigma]\!]$, is a set of ground terms defined in the following way:

- $[\![\bot]\!]^\Sigma = \emptyset$
- $[\![\sigma]\!]^\Sigma = T(\Sigma|_\sigma)$
- $[\![\sigma_1 \vee \sigma_2]\!]^\Sigma = [\![\sigma_1]\!]^\Sigma \cup [\![\sigma_2]\!]^\Sigma$
- $[\![\sigma_1 \wedge \sigma_2]\!]^\Sigma = [\![\sigma_1]\!]^\Sigma \cap [\![\sigma_2]\!]^\Sigma$
- $[\![f(\sigma_1 \ldots \sigma_n)]\!]^\Sigma = \{f(\eta_1 \ldots \eta_n) \in T(\Sigma) \mid \eta_i \in [\![\sigma_i]\!]^\Sigma, \text{ for } i = 1 \ldots n\}$

Intuitively, there is no well-sorted term of sort $\bot$, t is of sort expression $\sigma_1 \wedge \sigma_2$ iff it is of σ_1 and σ_2, t is of sort expression $\sigma_1 \vee \sigma_2$ iff it is of σ_1 or σ_2 and, t is of sort expression $f(\sigma_1 \ldots \sigma_n)$ iff $t = f(h_1 \ldots h_n)$ and each h_i is of sort expression σ_i. The meaning of a sort expression of the form $f(\sigma_1 \ldots \sigma_n)$ is to approximate the sort of a given term by describing its skeleton form, even if this skeleton is not well-sorted. We refer to [GS84] in order to show that intersection and union of regular tree languages (languages recognized by bottom-up tree automata) are also regular tree languages.

We extend the order $\leq$ over sort expressions in such a way that $\eta_1 \preceq \eta_2$ iff $[\![\eta_1]\!] \subseteq [\![\eta_2]\!]$. Note that $\eta \leq \sigma$ implies $\eta \preceq \sigma$ but the converse is false. It turns out that the greatest lower bound (resp. least upper bound) of σ_1, σ_2 is $\sigma_1 \wedge \sigma_2$ (resp. $\sigma_1 \vee \sigma_2$) and that $\sigma_1 \preceq \sigma_2$ and $\sigma_2 \preceq \sigma_2$ does not necessary imply that σ_1 and σ_2 are the same sort expression. We write $\sigma_1 \uparrow \sigma_2$ if $\sigma_1 \wedge \sigma_2 \neq \bot$ and $\sigma_1 \sharp \sigma_2$ if $\sigma_1 \wedge \sigma_2 = \bot$. Finally $\sigma \uparrow \Gamma$ if $\exists \delta \in \Gamma$ such that $\sigma \uparrow \delta$.

We are now able to decorate variables with sort expressions as for example in $x^{\sigma \vee \eta}$. Analogously, Σ-terms are constructed replacing $\leq$ by $\preceq$ in the definition of section 2 and all the notions presented there are naturally extended. It turns out that every term has a least regular tree language containing it, even if it has many sorts or it has no least sort. Order-sorted signatures as bottom-up tree automata allow to associate a natural notion of "minimality" to each term, and therefore regularity [SNGM89] becomes a redundant property of signatures.

4 Sequentiality of monotonic predicates

New constant symbols Ω^σ will be added to the signature Σ in order denote terms that contain no information about their structures and sorts. Intuitively, we know nothing about a term $\Omega^\top$ and we only know the sort of a term Ω^σ, with $\sigma \in \mathcal{E}_S$ and $\sigma \neq \top, \bot$. Formally, the set of Ω-terms is defined as follows: Ω^σ is an Ω-term if $\sigma \in \mathcal{E}_S$ and $f(t_1 \ldots t_n)$ is an Ω-term iff f has arity n and each t_i is an Ω-term.

Well-sorted Ω-terms are constructed as in section 2: Ω^σ is an Ω-term of sort η if and only if $\sigma, \eta \in \mathcal{E}_S$ and $\sigma \preceq \eta$ and $f(t_1 \ldots t_n)$ is an Ω-term of sort η if and only if $\eta \in \mathcal{E}_S$, there is $f : \sigma_1 \ldots \sigma_n \to \sigma \in \Sigma$ such that $\sigma \preceq \eta$ and for all $1 \leq i \leq n$, t_i is a well-sorted Ω-term of sort σ_i. All the conventional notions regarding substitutions, assignments and unification are naturally extended to Ω-terms. An information ordering on Ω-terms is defined by:

- $\Omega^\sigma \sqsubseteq t$ iff t is of sort η and $\eta \preceq \sigma$.
- $f(t_1 \ldots t_n) \sqsubseteq f(u_1 \ldots u_n)$ iff $\forall i = 1 \ldots n,\ t_i \sqsubseteq u_i$.

From now on, $t \sqcup t'$ denotes the least upper bound of the Ω-terms t and t'. It is important to recall that the formalism of order-sorted signatures as bottom-up tree automata assures that $t \sqcup t'$ is also an Ω-term. For example, the least upper bound of Ω^σ and Ω^η is the Ω-term $\Omega^{\sigma \wedge \eta}$. See [CD91] for details. We will also use $max(S)$ and $min(S)$ to denote the maximal and minimal (w.r.t $\sqsubseteq$) Ω-terms of S respectively.

Let $\mathcal{L}$ be a base and t a term. t_Ω indicates the Ω-term obtaining from t by replacing each variable in t with Ω and $\mathcal{L}_\Omega = \{l_\Omega \mid l \in \mathcal{L}\}$. From now on, we will abuse the notation by writing $\mathcal{L}$ instead of $\mathcal{L}_\Omega$ in order to make the paper more readable. This is possible because $\mathcal{L}$ has linear terms. An *Ω-normal form* is an Ω-term without redexes, containing at least one occurrence of Ω. We reserve the phrase *normal form* for terms containing neither redexes nor Ω's. Every ground term can be translated to an Ω-term by the following *context Ω-rule*:

$$C[s] \to_\Omega C[\Omega^\top], \text{ where } s \uparrow \mathcal{L} \text{ and } s \neq \Omega$$

The $\to_\Omega$ reduction is confluent and terminating as it is shown in [KM91], and the generalization of these results for order-sorted systems is straightforward.

Let the truth values be ordered by $false \ll true$. Let P be a monotonic predicate on Ω-terms. Increasing information about a given term (in the sense of $\sqsubseteq$) can only change the value of the predicate to a more favorable one.

In order to define the notion of sequentiality of a monotonic predicate, let's introduce some definitions that we will use for that purpose. The *acceptable set* of Ω-terms associated to a predicate P and a term t, denoted by $Accept(P, t)$ is defined to be the set of Ω-terms h such that $t \sqsubseteq h$ and $P(h) = true$. For each $u \in \mathcal{O}(t)$, we identify two subsets of $Accept(P, t)$ in such a way that the first one has only those terms with constructors at position u and the second one has only those terms with Ω at position u:

$$StrucAccept(P, t, u) = Accept(P, t) \cap \{h[f(t_1 \ldots t_n)]_u\}$$
$$SortAccept(P, t, u) = Accept(P, t) \cap \{h[\Omega^\eta]_u\}$$

We characterized positions of a term t where expansions may be performed, *i.e,* occurrences where we need to expand either the structure or the sort of t in order for the predicate P to become true on the expanded term. Formally, a position $u \in \mathcal{O}(t)$ is said to be a *potential direction* of P in $t[\Omega^\sigma]_u$ if and only if $\forall h \in Accept(P, t)$, $t/u \sqsubset h/u$. A position $u \in \mathcal{O}(t)$ is a *structure direction* of P in $t[\Omega^\sigma]_u$ if and only if $Accept(P, t) = StrucAccept(P, t, u)$ (*i.e,* $SortAccept(P, t, u) = \emptyset$). In the untyped case, a position is a potential direction if and only if it is a structure one and the definition coincides with that of direction in [HL79]. Finally, a position

$u \in \mathcal{O}(t)$ is a *sort direction* when it is a potential but not a structure direction, *i.e*, $t/u = \Omega^{\sigma}$, u is a potential direction and $SortAccept(P, t, u) \neq \emptyset$.

> *(Sequentiality)* The predicate P is said to be *sequential at* t if and only if, whenever $P(t) = false$ and there exists h such that $t \sqsubseteq h$ and $P(h) = true$, then there is either a structure or a sort direction of P in t. Finally, P is *sequential* (resp. *sort sequential*) iff it is sequential at every Ω-normal form (resp. at every well-sorted Ω-normal form).

This notion turns out to be the same as that in [HL79] when the subsort order is empty, and sort directions do not make sense.

Consider for example the predicate $\mathcal{NF}_{\mathcal{R}}$ over Ω-terms, where $\mathcal{R}$ is a term rewriting system and $\mathcal{NF}_{\mathcal{R}}(t) = true$ iff $t \rightarrow^{*}_{\mathcal{R}} u$ and u is in normal form (without redexes nor Ω's). The *system* $\mathcal{R}$ is said to be *sequential* iff $\mathcal{NF}_{\mathcal{R}}$ is a sequential predicate. This kind of sequentiality is undecidable [HL79], as the corresponding directions cannot in general be found effectively even if a given system is known to be sequential. Sequentiality is therefore a too general property for practical applications.

5 Strong Sequentiality

Huet-Levy showed in [HL79] that a restricted notion of sequentiality, called *strong sequentiality* was more appropriate. The technical formulation is based only on the left-hand sides of rules by using the idea of arbitrary reduction sequences, *i.e*, sequences in which a redex is replaced at each step by an arbitrary ground term.

Let t be an Ω-term, $\mathcal{L}$ a constructor base and $Match_{\mathcal{L}}$ a monotonic predicate defined as $Match_{\mathcal{L}}(t) = true$ if and only if there exists $l \in \mathcal{L}$ such that $l \sqsubseteq t$. Huet-Levy define a *base* $\mathcal{L}$ *to be match sequential* iff $Match_{\mathcal{L}}$ is sort-sequential at every Ω-normal form. Match sequentiality is a stronger notion than strong sequentiality as defined in [HL79] in the general case of orthogonal systems, although it coincides with the latter in the case of constructor systems as it is shown in [Tha87]. Since in this paper we restrict our attention to constructor systems and the formulation of match sequentiality is much simpler than that of strong sequentiality, we define directly a *base* $\mathcal{L}$ *to be strongly sequential* (resp. *strongly sort sequential*) iff $Match_{\mathcal{L}}$ is sort-sequential at every Ω-normal form (resp. at every well-sorted Ω-normal form).

We define also $Comp_{\mathcal{L}}(t) = \{l \sqcup t \mid l \in \mathcal{L} \text{ and } l \uparrow t\}$. For every $u \in \mathcal{O}(t)$, the set $Comp_{\mathcal{L}}(t, u)$ denotes the subset of $Comp_{\mathcal{L}}(t)$ such that $k \in Comp_{\mathcal{L}}(t, u)$ if $t/u \sqsubset k/u$. In order to distinguish the terms appearing in $Comp_{\mathcal{L}}(t, u)$ we define:

$$StrucComp_{\mathcal{L}}(t, u) = Comp_{\mathcal{L}}(t, u) \cap \{l[f(t_1 \ldots t_n)]_u\}$$
$$SortComp_{\mathcal{L}}(t, u) = Comp_{\mathcal{L}}(t, u) \cap \{l[\Omega^{\eta}]_u\}$$

Example 1 *Let* $\mathcal{L} = \{f(\Omega^{\eta}, \Omega^{\sigma})\ f(\Omega^{\rho}, a)\ f(\Omega^{\phi}, b)\ f(a, \Omega^{\sigma})\ f(\Omega^{\omega}, a)\}$, t *the* Ω*-term* $f(\Omega^{\top}, \Omega^{\sigma})$, *and* Σ *the following signature:*

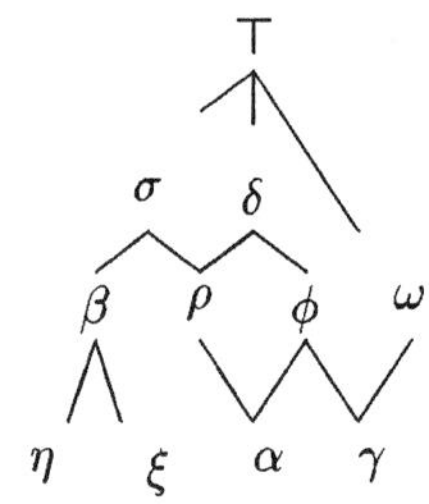

$c :\rightarrow \gamma$
$b :\rightarrow \eta$
$a :\rightarrow \xi$

$p : \eta \times \rho \rightarrow \rho$
$q : \xi \times \rho \rightarrow \rho$
$f : \sigma \times \sigma \rightarrow \phi$
$f : \beta \times \rho \rightarrow \alpha$
$h : \top \times \top \times \top \rightarrow \gamma$

$$\begin{array}{l} Comp_{\mathcal{L}}(t) = Comp_{\mathcal{L}}(t,1) = \mathcal{L} \\ StrucComp_{\mathcal{L}}(t,1) = \{f(a,\Omega^{\sigma})\} \\ SortComp_{\mathcal{L}}(t,1) = \{f(\Omega^{\eta},\Omega^{\sigma}),\ f(\Omega^{\rho},a),\ f(\Omega^{\phi},b)\ f(\Omega^{\omega},a)\} \\ Comp_{\mathcal{L}}(t,2) = StrucComp_{\mathcal{L}}(t,2) = \{f(\Omega^{\rho},a),\ f(\Omega^{\phi},b),\ f(\Omega^{\omega},a)\} \\ SortComp_{\mathcal{L}}(t,2) = \emptyset \end{array}$$

Remark 1 $Comp_{\mathcal{L}}(t) \subseteq Accept(Match_{\mathcal{L}},t)$.

Lemma 1 *$\forall h \in Accept(Match_{\mathcal{L}},t)$, $\exists k \in Comp_{\mathcal{L}}(t)$ such that $k \sqsubseteq h$.*

Proof. Take any $h \in Accept(Match_{\mathcal{L}},t)$. Then, $Match_{\mathcal{L}}(h) = true$ (*i.e.* $\exists l \in \mathcal{L}$ such that $l \sqsubseteq h$) and since $t \sqsubseteq h$ we have $t \sqcup l \in Comp_{\mathcal{L}}(t)$ and $t \sqcup l \sqsubseteq h$.

Lemma 2 *A position u is a potential direction of $Match_{\mathcal{L}}$ in $t[\Omega^{\sigma}]_u$ if and only if $Comp_{\mathcal{L}}(t,u) = Comp_{\mathcal{L}}(t)$.*

Proof. Let u be a potential direction of $Match_{\mathcal{L}}$ in $t[\Omega^{\sigma}]_u$ and let $l \in \mathcal{L}$ such that $l \uparrow t$. Then $l \sqcup t$ is in $Comp_{\mathcal{L}}(t)$ and so in $Accept(Match_{\mathcal{L}},t)$. Since u is a potential direction, $t/u \sqsubset (l \sqcup t)/u$ and then $l \sqcup t$ is also in $Comp_{\mathcal{L}}(t,u)$.

Conversely, let $h \in Accept(Match_{\mathcal{L}},t)$. By lemma 1, there is $k \in Comp_{\mathcal{L}}(t)$ such that $k \sqsubseteq h$ and hypothesis k is also in $Comp_{\mathcal{L}}(t,u)$. Therefore $t/u \sqsubset k/u$, by transitivity $t/u \sqsubset h/u$, and thus u is a potential direction.

Lemma 3 *A position u is a structure direction of $Match_{\mathcal{L}}$ in $t[\Omega^{\sigma}]_u$ if and only if $StrucComp_{\mathcal{L}}(t,u) = Comp_{\mathcal{L}}(t)$.*

Proof. Let u be a structure direction and let $l \in \mathcal{L}$ such that $l \uparrow t$. Since $l \sqcup t$ belongs to $Comp_{\mathcal{L}}(t)$ it also belongs to $Accept(Match_{\mathcal{L}},t)$, by hypothesis $Accept(Match_{\mathcal{L}},t) = StrucAccept(Match_{\mathcal{L}},t,u)$ and therefore $(l \sqcup t)|_u = f$ and $l \sqcup t$ belongs also to $StrucComp_{\mathcal{L}}(t,u)$.

Conversely, let $h \in Accept(Match_{\mathcal{L}},t)$. By lemma 1, $\exists k \in Comp_{\mathcal{L}}(t)$ such that $k \sqsubseteq h$ and by hypothesis k is also in $StrucComp_{\mathcal{L}}(t,u)$ and therefore $(l \sqcup t)|_u = f$. As $k \sqsubseteq h$, $h|_u = f$, and we conclude that $h \in StrucAccept(Match_{\mathcal{L}},t,u)$.

Lemma 4 *A position u is a subsort direction of $Match_{\mathcal{L}}$ in t if and only if $t/u = \Omega^{\sigma}$, $Comp_{\mathcal{L}}(t,u) = Comp_{\mathcal{L}}(t)$ and $SortComp_{\mathcal{L}}(t,u) \neq \emptyset$.*

Proof. Let u be a subsort direction of $Match_{\mathcal{L}}$ in t. By hypothesis $t/u = \Omega^{\sigma}$, and by lemma 2, $Comp_{\mathcal{L}}(t,u) = Comp_{\mathcal{L}}(t)$. Since u is a subosrt direction, $\exists h \in SortAccept(Match_{\mathcal{L}},t,u)$ and in particular h is in $Accept(Match_{\mathcal{L}},t)$. By lemma 1, there is $k \in Comp_{\mathcal{L}}(t)$ such that $k \sqsubseteq h$. Since u is a potential direction, k is also in $Comp_{\mathcal{L}}(t,u)$ (*i.e.* $t/u \sqsubset k/u$) and since h/u is of the form Ω, k/u is also an Ω.

Conversely, by hypothesis $t/u = \Omega^{\sigma}$ and by lemma 2, u is a potential direction of $Match_{\mathcal{L}}$ in t. By hypothesis $\exists k \in SortComp_{\mathcal{L}}(t,u)$. In particular k belongs to $Comp_{\mathcal{L}}(t)$ and so it belongs to $Accept(Match_{\mathcal{L}},t)$. Since k/u is an Ω and $t/u \sqsubset k/u$, then $k \in SortAccept(Match_{\mathcal{L}},t,u)$.

Example 2 *Let $\mathcal{L}$ be the base of example 1. We have that* 1 *is a sort direction of $Match_{\mathcal{L}}$ in $f(\Omega^{\sigma},\Omega^{\sigma})$, $f(\Omega^{\rho},b)$ and $f(\Omega^{\phi},a)$, while* 2 *is a structure direction of $Match_{\mathcal{L}}$ in $f(\Omega^{\rho},\Omega^{\sigma})$, $f(\Omega^{\phi},\Omega^{\sigma})$ and $f(\Omega^{\omega},\Omega^{\sigma})$.*

On the other hand, 2 *is not a direction of $Match_{\mathcal{L}}$ in $f(\Omega^{\sigma},\Omega^{\sigma})$ and* 1 *is neither a direction of $Match_{\mathcal{L}}$ in $f(\Omega^{\rho},\Omega^{\sigma})$ nor in $f(\Omega^{\phi},\Omega^{\sigma})$ nor in $f(\Omega^{\omega},\Omega^{\sigma})$.*

6 Pattern Matching Trees

The decision procedure of strong sequentiality in untyped constructor systems can also be used to build pattern matching algorithms [Tha87]. However, this is not the case in the order-sorted framework: the difficulty lies in the definition of the order-signature. In this paper we do not discuss the properties that a signature has to verify in order to allow a really sequential pattern matching process; this notion can be found in [Kes91]. Our interest is to discuss strongly sorted sequential systems, and to characterize those algorithms that do not take into account any property of the signature. We assume there is an orthogonal mechanism of type checking that takes place before the pattern matching process does and therefore, we study those algorithms that allow to perform sequential searching of directions at every *well-typed* Ω-normal form. We first define pattern matching trees and we give an intuitive outline of its behavior.

A *pattern matching tree (PMT)* associated to a base $\mathcal{L}$ is defined as:

- Each node is an Ω-pattern.
- If h is an internal node, it has a distinguished position u (denoted $h\ \{u\}$).
- $\Omega^{\top}\{\epsilon\}$ is the root
- If $\mathcal{L}'$ is the set of leaves, $\mathcal{L} \subseteq \mathcal{L}'$ and $\forall l' \in \mathcal{L}'\ \exists l \in \mathcal{L}$ such that $l \sqsubseteq l'$.
- If p_i is a child of $p\ \{u\}$, then p_i and p only differs at position u and $p/u \sqsubset p_i/u$.

PMT's as defined here may be non-deterministic, since different children of a given node are not necessary incompatible. This non-determinism only happens because of sort directions, and it is one of the main differences with classical pattern matching trees [PS90]. Since we are interested in strong sort sequentiality, nodes of a PMT are also required to be *well-sorted.* A PMT associated to a base $\mathcal{L}$ can also be seen as a non-deterministic case-expression, where each step of the tree is seen as a constraint to be satisfied by the term to be matched. The evaluation of a term m is performed by $Match_{\mathcal{L}}(m,\Omega^{\top},\epsilon)$, where the procedure $Match_{\mathcal{L}}$ is defined as follows:

When p is a leaf, the procedure terminates with a succesful result:

$Match_{\mathcal{L}}(t, p, \epsilon) = true$

When $p_1\ \{u_1\}, \ldots, p_k\ \{u_k\}$ are the children of a node $p\ \{u\}$, the reduction of the subterm at position u of the term to be matched is produced:

$Match_{\mathcal{L}}(t, p, u) =$
    case t/u **of:**
    $p_1/u \sqsubseteq t/u :\Rightarrow Match_{\mathcal{L}}(t, p_1, u_1)$
    $\vdots$
    $p_k/u \sqsubseteq t/u :\Rightarrow Match_{\mathcal{L}}(t, p_k, u_k)$
    otherwise: if $Normal\_Form(t/u)$
        then $false$
        else $Match_{\mathcal{L}}(t[u \leftarrow Reduction(t, u)], p, u)$

$Reduction(t, u) =$
    case u **of:**
    u is a sort direction: One_Step_Reduction(t/u)
    u is a structure direction: Head_Normal_Form(t/u)

In sort decreasing systems, if a term t reduces to another term t', the sort of the second is smaller or equal than that of the first. Therefore, a one-step reduction may possibly refine the sort information of a term in order to satisfy a subsort constraint (the sort direction case). On the other hand, since only constructor function symbols are allowed inside left-hand sides of rules, the head-normal form will be always required in order to satisfy a structure constraint (the structure direction case).

In the untyped case, the well known example of Berry, consisting of the non-strongly sequential base $\mathcal{L} = \{f(a, a, \Omega), f(b, \Omega, a), f(\Omega, b, b)\}$, shows how unnecessary verifications lead to non optimal PMT's. Let t_1, t_2 and t_3 be three redexes in $t = f(t_1, t_2, t_3)$. The choice of normalizing t_1 before t_2 and t_3 fails to reach a matching for the term t, if t_1 has no normal form and t_2 and t_3 have both normal forms b. The same reasoning applies if t_2 or t_3 are chosen to be normalized first. As [HL79] have shown, there is no optimal pattern matching algorithm associated to $\mathcal{L}$.

In the order-sorted case, the following set illustrates a non strongly sequential base for which there is no optimal pattern matching algorithm. Let us consider the set $\mathcal{L}_1 = \{h(\Omega^{\top}, \Omega^{\eta}, \Omega^{\xi}), h(\Omega^{\xi}, \Omega^{\top}, \Omega^{\eta}), h(\Omega^{\eta}, \Omega^{\xi}, \Omega^{\top})\}$, a slight modification of Berry's example, where the subsort order is the same of example 1. Let t_1, t_2 and t_3 be three redexes in $t = f(t_1, t_2, t_3)$. The choice of normalizing t_1 before t_2 and t_3 fails to reach a matching for the term t, if t_1 has no normal form and t_2 and t_3 have normal forms b and a respectively. The same reasoning applies if t_2 or t_3 are chosen to be normalized first.

The algorithm below can be seen as an implicit method of construction of PMT's associated to $\mathcal{L}$. The main idea is to find directions at some well-sorted terms that are prefixes of $\mathcal{L}$. $SortSeq(\mathcal{L}, t)$ proceeds as follows: if there is $l \in \mathcal{L}$ such that $l \sqsubseteq t$, return true. Otherwise, try to find a direction u of $\mathcal{L}$ in t. If such a direction cannot be found, $\mathcal{L}$ is not strongly sequential at t and then fail. Otherwise, a set of terms called *STEPS* is computed and strong sequentiality is verified recursively at the well-sorted terms expanded by substituting the subterm at position u of t by a well-sorted term in STEPS. We denote as $g(\overline{\eta_k})$ the term $g(\Omega^{\eta_1}\ldots\Omega^{\eta_k})$.

$SortSeq(\mathcal{L},t) =$
if $\exists l \in \mathcal{L}$ such that $l \sqsubseteq t$,
then $true$
else if $\exists u \in \mathcal{O}(t)$ such that
 $StrucDirec(\mathcal{L},t,u) = [true, \text{STEPS}]$ **or** $SortDirec(\mathcal{L},t,u) = [true, \text{STEPS}]$
 then $\bigwedge_{p \in STEPS} SortSeq(Comp_{\mathcal{L}}(t[u \leftarrow p]), t[u \leftarrow p])$
 else $false$

$StrucDirec(\mathcal{M},t,u) =$
if $(t/u = \Omega^{\sigma})$ **and** $(\forall l \in \mathcal{M},\ l|_u$ is a constructor$)$
 then $[true, min\{f(\overline{\eta_k}) \mid f : \overline{\eta_k} \to \eta \in \Sigma$ and $\eta \preceq \sigma$ and $\exists l \in \mathcal{M},\ l|_u = f\}]$
 else $[false, \emptyset]$

$SortDirec(\mathcal{M},t,u) =$
if $(t/u = \Omega^{\sigma})$ **and** $(\forall l \in \mathcal{M},\ t/u \sqsubset l/u)$ **and** $(\exists l \in \mathcal{M},\ l/u = \Omega^{\eta})$
 then $[true, min\{l/u \mid l \in \mathcal{M}\}]$
 else $[false, \emptyset]$

Example 3 *Let* $\mathcal{L}_1 = \{f(\Omega^{\beta}, p(b,\Omega^{\rho})),\ f(a, q(a,\Omega^{\rho})),\ f(\Omega^{\rho},\Omega^{\sigma})\}$, *where* Σ_1 *is the signature of example 1. Figure 1 shows a PMT for* $\mathcal{L}_1$.

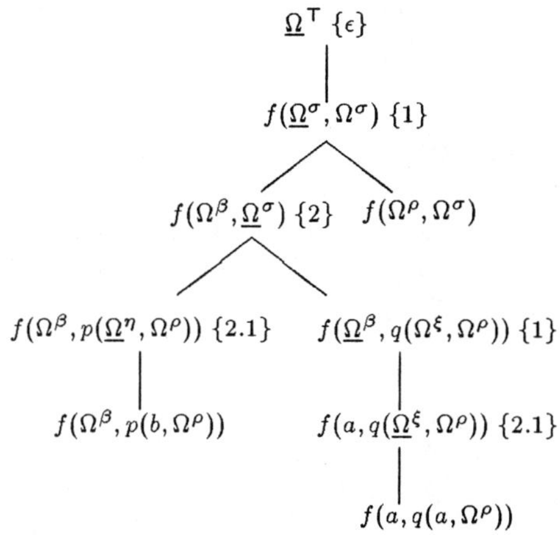

Figure 1: Pattern Matching Tree for $\mathcal{L}_1$

Example 4 *Let* Σ_2 *be signature containing the subsort order of example 1 and the following set of declarations:*

$f : \delta \times \sigma \times \top \to \delta$	$f : \beta \times \sigma \times \top \to \delta$	$m :\to \gamma$	$k :\to \eta$
$f : \delta \times \sigma \times \top \to \phi$	$f : \beta \times \sigma \times \top \to \rho$	$l :\to \xi$	

Let $\mathcal{L}_2 = \{f(k,\Omega^\sigma,\Omega^\top),\ f(l,\Omega^\sigma,\Omega^\top),\ f(\Omega^\rho,k,\Omega^\top)\ f(\Omega^\rho,l,m)\ f(\Omega^\phi,l,k)\}$. *We show a PMT for* $\mathcal{L}_2$ *in figure 2.*

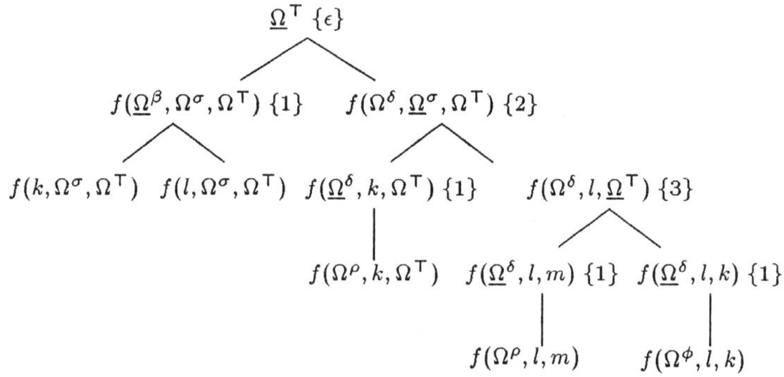

Figure 2: Pattern Matching Tree for $\mathcal{L}_2$

Remark 2 *If* t *is a node of a PMT associated to* $\mathcal{L}$ *and* $t_1 \ldots t_m$ *its children, then* $\forall l \in \mathcal{L}$ *such that* $l \uparrow t$, $\exists 1 \leq i \leq m$, $t_i \sqsubseteq l$.

Theorem 1 *If* $\mathcal{L}$ *is strongly sequential, then* $SortSeq(\mathcal{L},\Omega^\top) = true$.

Proof. If $\mathcal{L}$ is strongly sequential, $Match_\mathcal{L}$ has a structure or a sort direction at every term in Ω-normal form. In particular, every node constructed by $SortSeq(\mathcal{L},\Omega^\top)$ is a well-sorted Ω-normal form and thus $SortSeq(\mathcal{L},\Omega^\top) = true$.

7 Free Sequentiality

In order-sorted rewriting systems, sort sequentiality is related to the definition of order-sorted signatures, which introduces complex situations that are well illustrated by the following example. Let $\sigma, \eta, \delta, \beta$ be four pairwise disjoint sorts and g a constructor symbol declared as:

$$\begin{array}{ll} g : \top \times \sigma \times \eta \to \delta & g : \top \times \top \times \top \to \top \\ g : \eta \times \top \times \sigma \to \delta & h : \top \to \top \\ g : \sigma \times \eta \times \top \to \delta & h : \delta \to \top \end{array}$$

Let $\mathcal{L}_3 = \{h(\Omega^\delta)\}$ be a base and $t = h(g(\Omega^\top,\Omega^\top,\Omega^\top))$. We have $Match_{\mathcal{L}_3}(t) = false$ but there exists an instance of t such that $Match_{\mathcal{L}_3}$ becomes true. If $\mathcal{L}_3$ is strongly sequential a direction of $Match_{\mathcal{L}_3}$ in t must be found, and this direction does not depend only on the predicate, but on the signature Σ. In the example, choosing a direction of $Match_{\mathcal{L}_3}$ at $h(g(\Omega^\top,\Omega^\top,\Omega^\top))$ is equivalent to choosing a direction of $WellSorted_\delta$ at $g(\Omega^\top,\Omega^\top,\Omega^\top)$, where the last predicate is defined as $WellSorted_\delta(t) = true$ if and only if t is of sort δ. But the set of domains of the three last declarations of g is just Berry's example on signatures and this set is known to be not strongly sequential [Ber78], so we cannot find a direction. Thus, we need a notion of "sequential order-sorted signatures" and sequential type-checking, and moreover, such notion must be independent of any predicate. In this paper we want to get rid of the property of sequentiality of signatures, and so instead of restricting the class of signatures to sequential ones, we forbid the declarations of constructors that allow the expansion of terms that do not lead to an immediate realization of the match predicate. A reader interested in sequential signatures can be referred to [Kes92].

We need a stronger notion of sort direction in some terms of the PMT in order to avoid such problems. Let us recall that $g(\overline{\eta_k})$ denotes $g(\Omega^{\eta_1}\ldots\Omega^{\eta_k})$. We say that t has the *free sort expression (FSE)* property at position u if and only if $\forall g : \overline{\eta_k} \to \eta \in \Sigma$ such that $t[g(\overline{\eta_k})]_u$ is well-sorted and $t[g(\overline{\eta_k})]_u \uparrow \mathcal{L}$, $\exists h \in Comp_{\mathcal{L}}(t)$ such that $h/u \sqsubseteq g(\overline{\eta_k})$. By constructionof the tree, every internal node with a structure direction as distinguished position has the FSE property, but this is not true in general for every internal node of the tree. The FSE property is clearly decidable because the set of declarations of the signature is finite.

> *(Free Sequentiality)* A base $\mathcal{L}$ is *free sequential* iff $SortSeq(\mathcal{L},\Omega^{\top}) = true$ and every node of the PMT constructed by $SortSeq(\mathcal{L},\Omega^{\top})$ has the FSE property.

Note that $\mathcal{L}_1$ of example 3 is not free sequential since for example $f(\Omega^{\sigma},\Omega^{\sigma})$ has not the FSE property because of the declaration $f : \sigma \times \sigma \to \phi$. If we replaced $f : \sigma \times \sigma \to \phi$ by $f : \sigma \times \sigma \to \alpha$ in the signature Σ of example 1, every node has the FSE property and thus $\mathcal{L}_1$ is free sequential. The base $\mathcal{L}_2$ of example 4 is free sequential and the base $\mathcal{L}_3$ presented at the beginning of this section does not: the node $h(\Omega^{\delta})$ has not the FSE property because of the declaration $g : \top \times \top \times \top \to \top$.

Theorem 2 *If $\mathcal{L}$ is free sequential, then $\mathcal{L}$ is strongly sort sequential.*

Proof. Let m be a well-sorted Ω-normal form such that $Match_{\mathcal{L}}(m) = false$ and $m \uparrow \mathcal{L}$. As $\Omega^{\top}$ is the root of the PMT but m does not match any pattern of $\mathcal{L}$, there exists a node t with children $t_1 \ldots t_k$ and distinguished position u such that $t \sqsubseteq m$ and for each $1 \le i \le k$, $t_i \not\sqsubseteq m$. Since t differs only at position u which each t_i, then

$$\forall 1 \le i \le k,\ t_i/u \not\sqsubseteq m/u \tag{1}$$

We will first prove that u is a potential direction of $Match_{\mathcal{L}}$ in m:

a) Suppose $m/u = g(l_1 \ldots l_n)$. Since m/u is well-sorted and $t/u = \Omega^{\sigma} \sqsubseteq m/u$, there is a declaration $g : \overline{\eta_k} \to \eta$ such that $g(\overline{\eta_k}) \sqsubseteq m/u$ and $\eta \preceq \sigma$. Then $\Omega^{\sigma} \sqsubseteq g(\overline{\eta_k})$ and $t[g(\overline{\eta_k})]_u$ is well-sorted. Therefore, $t[g(\overline{\eta_k})]_u \sqsubseteq m[g(\overline{\eta_k})]_u \sqsubseteq m$ and since $m \uparrow \mathcal{L}$, also $t[g(\overline{\eta_k})]_u \uparrow \mathcal{L}$. By hypothesis, t has the FSE property at position u and so $\exists l \in Comp_{\mathcal{L}}(t)$ such that $l/u \sqsubseteq g(\overline{\eta_k})$. By construction of the tree, $\exists i$ such that $t_i \sqsubseteq l$, and thus $t_i/u \sqsubseteq g(l_1 \ldots l_n) = m/u$ holds by transitivity which leads to a contradiction with equation 1. Then m/u has the form Ω^{δ} with $\delta \preceq \sigma$.

b) Let $l \uparrow m$ and suppose $m/u = (l \sqcup m)/u$. As $t \sqsubseteq m$, also $l \uparrow t$ and by construction of the tree, $\exists i$ such that $t_i/u \sqsubseteq (l \sqcup t)/u$. Therefore $t_i/u \sqsubseteq$

$(l \sqcup t)/u \sqsubseteq (l \sqcup m)/u = m/u$ which leads to a contradiction with equation 1. Thus $m/u \sqsubset (l \sqcup m)/u$ and u is a potential direction of $Match_{\mathcal{L}}$ in m.

By definition, if u is a potential direction, it is either a structure or a sort direction and this is sufficient to show that $\mathcal{L}$ is strongly sort sequential. But in particular we can add that, whenever u is a structure direction of $Match_{\mathcal{L}}$ in t, it is also a structure direction of $Match_{\mathcal{L}}$ in m, and whenever u is a sort direction of $Match_{\mathcal{L}}$ in t, it may be either a sort or a structure direction of $Match_{\mathcal{L}}$ in m.

The intuitive idea of this result is that whenever a system is free sequential and we supply an orthogonal mechanism of type checking, pattern matching can be efficiently performed, by avoiding unnecessary reductions to head normal forms.

As we have noticed above, the FSE property always holds in nodes having a structure direction and, if the system is untyped, sort directions do not make sense. Therefore our result of theorem 2 can be seen as a generalization of that in [Tha87], that in terms of our formalism can be stated as follows:

Corollary 1 *For every untyped base such that $SortSeq(\mathcal{L}, \Omega^{\top}) = true$, $\mathcal{L}$ is strongly sequential.*

The problem we have encountered with order-sorted systems is that there is a "hidden" concept of sequentiality in the presentation of their signatures. Each time there is a sort constraint to be satisfied, the type checker needs a sequential mechanism to verify such constraint and this leads to the problem of defining some kind of "sequential type checking" which is also related to the declarations of function symbols. Thus, we can have a sequential signature but a non strongly sequential set of rules, or a strongly sequential set of rules but a sequential signature. While strong sequentiality is equivalent to optimal computations in orthogonal untyped rewriting systems with constructors, a combination of sequential matching (strongly sequential bases) and sequential type checking will be necessary in the order-sorted case in order to have a complete result of effective computations.

Conclusions

We have introduced the notion of sequentiality and strong sequentiality in order-sorted rewriting systems and we have characterized via a decidable property those orthogonal constructor systems which admit an easy and simple implementation. Thanks to the formalism of order-sorted signatures as bottom-up tree automata, we simplify and avoid many restrictions that often appear in the treatment of order-sorted signatures, such as regularity, downward completeness and coregularity. We think that this is a first step towards efficient implementation of order-sorted rewriting systems. In this paper we suppose that an orthogonal mechanism of type verification is given, *i.e*, terms to be evaluated are always well-sorted. Future work will focus on sequential order-sorted signatures and sequential type checking.

Acknowledgements

I would like to thank Jean-Pierre Jouannaud and Huber Comon for stimulating suggestions that motivate many extensions of my previous work. I am very grateful to Laurence Puel for fruitful discussions and helpful improvements in the final version of the paper. I would also like to thank Roberto Di Cosmo and Chet Murthy for their help in the presentation of the work.

References

[Ber78] Gérard Berry. Séquentialité de l'évaluation formelle des lambdas-expressions. *Proc. 3rd International Colloquium on Programming*, Paris, France, 1978.

[CD91] Hubert Comon and Catherine Delor. Equational Formulas with Membership Constraints. Technical Report 649, Laboratoire de Recherche en Informatique, 1991. To appear in Information and Computation.

[GS84] Ferenc Gécseg and Magnus Steinby. Tree Automata. Akademiai Kiado, Budapest, 1984.

[HL79] Gérard Huet and Jean-Jacques Lévy. Call by need computations in non ambiguous linear term rewriting systems. Technical Report IRIA Laboria 359, INRIA, Le Chesnay, France, 1979.

[Kes91] Delia Kesner. Pattern Matching in Order-Sorted Languages. In *Mathematical Foundations of Computer Science*. LNCS 520, Springer-Verlag. Also in PRL Research Report Number 10, Digital Equipment Corporation, Paris Research Laboratory, 1991.

[Kes92] Delia Kesner. Sequential signatures. Draft, 1992.

[KM91] Jan Willem Klop and Aart Middeldorp. Sequentiality in orthogonal term rewriting Systems. In *J. Symbolic Computation*. Academic Press, 1991.

[PS90] Laurence Puel and Ascánder Suárez. Compiling Pattern Matching by Term Decomposition. In *1990 ACM Conference on Lisp and Functional Programming*. ACM Press. Also in PRL Research Report Number 4, Digital Equipment Corporation, Paris Research Laboratory, 1990.

[SNGM89] Gert Smolka, Werner Nutt, Joseph Goguen, and José Meseguer. Order Sorted Equational Computation. In Hassan Aït-Kaci and Maurice Nivat, editors, *Resolution of Equations in Algebraic Structures. Volume 2: Rewriting Techniques*, pages 297–367. Academic Press, 1989.

[Tha87] Satish Thatte. A Refinement of Strong Sequentiality for Term Rewriting systems with Constructors. *Information and Computation*, Number 72, 1987.

Programming with Equations: A Framework for Lazy Parallel Evaluation*

R.C. Sekar† and I.V. Ramakrishnan
Dept. of Computer Science, SUNY at Stony Brook, NY 11794.
E-mail: rcs@sbcs.sunysb.edu, ram@sbcs.sunysb.edu

Abstract

Huet and Levy pioneered lazy sequential evaluation of equational programs based on the concepts of *strong-sequentiality* and *needed redexes*. Natural extensions of their strategy are not well-suited for parallel evaluation since they do not support independent searches for needed redexes along different paths in the input term. Furthermore, the size of compiled code can be exponential in program size. We therefore propose a different notion of sequentiality called *path-sequentiality* that overcomes these drawbacks and thus provides a natural framework for lazy parallel evaluation. We present a sound and complete algorithm for lazy parallel normalization of path-sequential systems. We show that our algorithm is optimal in the sense that its time complexity is bounded only by the time required to perform the needed reductions. The results presented in this paper are applicable to functional languages as well through the transformation of Laville.

1 Introduction

Programming with equations is a style of declarative programming in which systems of oriented equations (called *rewrite rules*) of the form $l \rightarrow r$ (l and r are terms with variables in them universally quantified implicitly) constitute the set of logical assertions describing a program. An implementation of an equational language takes such a program and a term t as input and produces as output a "simpler" term s called the *normal form* of t (i.e., s contains no instances of the left-hand side (lhs) of any rule). Such an implementation is said to be *complete* if it finds the normal form s whenever $s = t$ follows logically from the input rules.

An equational program is evaluated by *term rewriting*, i.e., by repeatedly replacing *redexes* (instances of lhs) in the input term by the corresponding right-hand sides (rhs) until the input term is normalized. An implementation based on term rewriting can be complete if the input programs are guaranteed to be *confluent* or Church-Rosser. For programming purposes it is natural to do this by requiring input

*Partially supported by NSF grants CCR-8805734, 9102159 and NYS S&T grant RDG 90173.
†Current address: Bellcore, 445 South Street, Morristown, NJ 07962.

programs to be *orthogonal*[1] [DJ90].

For orthogonal systems (without overlaps at the root) Huet and Levy [HL79] developed a theory of optimal reduction strategies based on the concept of *needed redexes.* (A needed redex for a term t is one that is reduced in every reduction sequence that normalizes t.) They showed that in such systems, every term that is not in normal form will have a needed redex and that a normalization strategy that reduces a needed redex at every step is complete. Any such strategy that reduces only needed redexes is said to be *lazy.* Unfortunately, needed redexes are not computable for all orthogonal programs[2]. Therefore they developed the class of *strongly-sequential systems* for which needed redexes can be computed based on the lhs alone. Subsequently, Hoffman and O'Donnell [HO84] developed an equational interpreter for *strongly-left-sequential systems*, which are a subclass of strongly-sequential systems.

The normalization algorithms mentioned above are all sequential (since they search for and reduce only one needed redex at a time) and the design of an efficient algorithm for lazy parallel evaluation of equational programs has remained an interesting open problem. In this paper we address the design of such a strategy. Our high-level model of a parallel computation consists of (evaluator) processes interconnected by links. (The links may be implemented either through shared-memory or a network.) In [SSR89] we had proposed a notion of sequentiality called *path-sequentiality* that facilitates parallel evaluation in a natural way on such a model. But the focus in [SSR89] was only to formally demonstrate that the expressive power of path-sequential systems is the same as that of strongly-sequential systems. We show that in [SSR89] by syntactically transforming any strongly-sequential system into an equivalent path-sequential system. But *all* questions regarding parallelization of path-sequential and strongly-sequential systems were left unanswered, e.g., what are the inherent difficulties in parallelizing strongly-sequential systems on such a model? Do path-sequential systems overcome such problems? Furthermore, the important problem of developing a sound and complete normalization strategy for lazy parallel evaluation of path-sequential systems was not addressed. We address all these questions in this paper and present a simple and efficient normalizing algorithm and study its computational and communication complexities. The results in this paper complement the results on expressive power of path-sequential systems in [SSR89]. Together they complete a thorough study of problems, mechanisms and solutions to lazy parallel evaluation of equational programs. The following is a summary of results.

1.1 Main Results

• A lazy parallel normalization algorithm must search for and reduce several needed redexes in parallel. A natural way to do this is to search (in parallel) along different paths in a term and reduce needed redexes. To avoid excessive communication and synchronization overheads, the search on different paths should proceed inde-

[1] In such systems, the lhs of rules are linear and no two lhs unify except possibly at the root. If two lhs unify then their most general unifier must also unify the corresponding rhs. Until recently [DJ90] the term *regular* was used in the literature in place of *orthogonal.*

[2] Although Kennaway [Ken89] has presented a *sequential* algorithm for all orthogonal programs, this algorithm does not compute needed redexes.

pendently. In section 2, we show that natural extensions of Huet-Levy strategy for parallel evaluation do not support this independent search and incur (potentially) large communication overheads. Another serious problem with such extensions is that they cause an *exponential blow-up* in the size of compiled code.

• We propose a new framework for lazy parallel evaluation of equational programs, based on the concept of *path-sequentiality*. For such systems, we outline a strategy (described in detail in section 3) that searches for needed redexes *independently* along different paths with low communication and synchronization overheads. More importantly, the size of the compiled code (with this strategy) is linear in the size of input program. In contrast, natural extensions of the Huet-Levy strategy will result in exponential blow-up of code size *even for path-sequential systems*.

• In section 3 we devise a parallel lazy normalization algorithm (called *Normalize*) for path-sequential systems. Our algorithm is simple and efficient and demonstrates the appropriateness of path sequential systems as a basis for lazy parallel evaluation.

• We establish the soundness and completeness of *Normalize* in section 4. In spite of the simplicity of *Normalize*, these proofs are quite tricky and involved.

• Although it may appear that *Normalize* may spend a lot of time searching for needed redexes, we show in section 5 that the time complexity of *Normalize* is bounded only by the time required to perform the needed reductions. Thus the search for needed redexes does not add to the complexity. We also show that matching (i.e., checking for the presence of a redex) is done with minimal communication.

• In section 6 we show that the parallelism exploited by *Normalize* cannot be enhanced without looking at the rhs. We then remark how parallelism can be enhanced by analyzing the rhs using a technique called *ee*-analysis [SSR90]. Finally, we describe modifications to *Normalize* to enhance parallelism using this information.

For the rest of the paper, we will be dealing only with programs that follow *constructor discipline*, i.e., the outermost symbols of the lhs will be distinct from any other symbol appearing inside. It may appear that path-sequential constructor-based systems are restrictive. Therefore we now show how our results can be applied to functional or equational programming languages.

• Most functional languages such as ML [HMT88] and Hope [BMS80] permit overlapping rewrite rules, but make use of priorities to disambiguate among them. Laville [Lav87] has shown how these rules can be transformed into equivalent constructor-based orthogonal system. The transformed programs must be strongly-sequential (as they most often are) in order to have a complete lazy normalization procedure. We can now transform this system into a path-sequential system using our transformation presented in [SSR89]. Thus our results are applicable to functional programming languages as well.

• For languages based on orthogonal systems such as O'Donnell's equational language [OD85], programs that violate the constructor discipline can be transformed into constructor-based systems using Thatte's transformation procedure [That85][3].

We would like to remark that the above transformation procedures are often not necessary since most programs in practice (e.g., almost all the programs in [OD85]) appear to be already path-sequential and constructor-based.

---

[3] A caveat here is that Thatte's transformation does not always preserve strong-sequentiality.

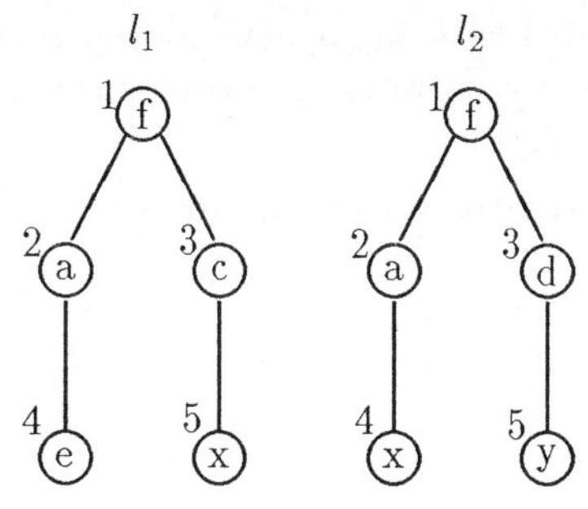

Figure 1

1.2 Notational Preliminaries

We use the obvious tree representation for terms, and use *terms* and *trees* interchangeably. The notion of a *path* (also called *occurrence*) is used to refer to subterms in a term as follows. A path is either the empty string Λ that reaches the root or $p.i$ (p is a path and i an integer) which reaches the ith argument of the root of the subterm reached by p. We use t/p to refer to the subterm of t reached by p. We say $p \leq q$ whenever $\exists r\ p.r = q$. For a path $p = i_1.i_2....i_k$, let $s_0, s_1, ..., s_k$ be the symbols on this path in a term t (listed in the order from root to the node at p). The path string for a term t and a path p, denoted by $pathstr(t,p)$, refers to the string $s_0.i_1.s_1.i_2....i_k.s_k$. In fig.1, the paths 2.1 and 1 reach nodes 5 and 2 respectively in l_1, and $pathstr(l_1, 1.1) = f.1.a.1.e$. We use $root(t)$ to denote the root symbol of t.

A term rewriting system (TRS) R is a set of *rewrite rules* of the form $l_i \longrightarrow r_i$ such that the variables in r_i are a subset of those in l_i. Let $\mathcal{F}$ be the set of outermost symbols of all lhs (called function symbols) and $\mathcal{C}$ be the set of non-variable symbols (called constructors) that appear inside left-hand sides. A TRS is said to be a *constructor system* if $\mathcal{F}$ and $\mathcal{C}$ are disjoint, the lhs are linear (i.e., no variable in an lhs occurs more than once in it) and no two lhs unify. Throughout the rest of this paper we will be using the language of constructor systems and denote them by $\mathcal{R}$. $\mathcal{L}$ denotes the corresponding set of lhs.

A *substitution* maps variables to terms. An *instance* $t\beta$ of a term t is obtained by substituting $\beta(x)$ for every variable x in t. If t is an instance of u then we say $u \leq t$ and call u a *prefix* of t. A term u is a *redex* if $u \geq l$ for some lhs l. A reduction $t \longrightarrow_l u$ means that u is obtained by replacing a redex $l\beta$ in t by $r\beta$, where $l \longrightarrow r$ is a rewrite rule. (Sometimes we omit the subscript on $\longrightarrow$.) The reflexive and transitive closure of $\longrightarrow$ is denoted $\longrightarrow^*$. A term t is in *normal form* (NF) if it contains no redexes. It is in *head normal form* (HNF) if there is no term t' such that $t \longrightarrow^* t'$ contains a redex at the root. A *needed redex* is one that is rewritten in every reduction of t to NF.

2 Issues in Lazy Parallel Evaluation

In this section we motivate the use of path-sequential systems as the basis for lazy parallel evaluation of equational programs. We first describe the problems that arise when extending the Huet-Levy algorithm for parallel evaluation. We then show how these problems can be avoided by using a new normalizing strategy based on path-sequential systems. We start by outlining the Huet-Levy strategy.

2.1 Huet-Levy Normalization Strategy

Suppose the TRS $\{f(a(e), c(x)) \rightarrow x, f(a(x), d(y)) \rightarrow x\}$ (the lhs of this TRS are shown in fig.1) is used to normalize the term $t = f(a(t_1), t_2)$, where $t_1 = f(a(e), c(e))$ and $t_2 = f(a(c(e)), d(e))$. In order for t to become a redex, certain reductions must be performed within its subterms. In particular, t_2 must be reduced into a term with a constructor root (c or d). More generally, let u be a *constructor prefix* of t, i.e., all nonvariable symbols in u, except the one at its root, are constructors. If all the lhs that unify with u have a nonvariable symbol at p (where u/p is a variable) then it is clear that the subterm t/p must be reduced to a term with constructor root (i.e., t/p must be in HNF) before t can become a redex. Such a p is called an *index* for u. In the above example $u = f(a(x), y)$ and $p = 2$. In a strongly-sequential system every constructor prefix that unifies with a nonempty set of lhs has an index.

The Huet-Levy normalization algorithm constructs such a prefix u, identifies its index p and reduces t/p to a term in HNF. If $root(t/p)$ is now a constructor (otherwise t can never become a redex) then u can be "extended" into a prefix u' by adding this constructor to it. This process is repeated until either t becomes a redex or we find that t can never become a redex (i.e., t is in HNF). In order to head normalize t/p the same process is used to identify and reduce redexes at the root of t/p until it is in HNF. In the above example t_2 is a redex and can be reduced to $c(e)$. The enlarged prefix $f(a(x), c(e))$ has 1.1 as an index. Now $t/1.1 = t_1$ is reduced to e, thereby yielding a redex at the root. Reducing this redex yields the normal form e.

Thus matching in the above normalization algorithm involves identifying a prefix u, computing its index p, extending u into u' (after reducing the subterm at t/p into HNF) and repeating this on u' and so on till t becomes a redex. Since identification of a prefix and computation of its index are very time consuming, language implementations compile the lhs into an automata (which may either be represented by tables or case statements) that drives the matching process. A state v of the automaton remembers the prefix u_v inspected in reaching this state from the start state. We (arbitrarily) choose one particular index of u_v and have transitions from v on pairs (p, c), where c is any symbol that could appear at p in any lhs l that unifies with u. Fig.2 is such an automaton for the above example.

2.2 Extending Huet-Levy Strategy for Parallel Evaluation

Observe that in the Huet-Levy strategy, each call to head normalize a subterm may reduce several redexes before returning. Thus a significant amount of parallelism can arise if we head normalize in parallel all subterms t/p where p is an index for u. In a direct extension of the above lazy normalization algorithm for exploiting this parallelism, the processor P initiating a match at the root of t would compute all indexes $p_1, ..., p_k$ and then initiate computations to head normalize $t/p_1, ..., t/p_k$ in parallel. Note that although head normalization of different subterms could proceed in parallel on different processors, the computation of the prefix and its indexes are all done *centrally* in P because of the nature of matching in strongly-sequential systems.

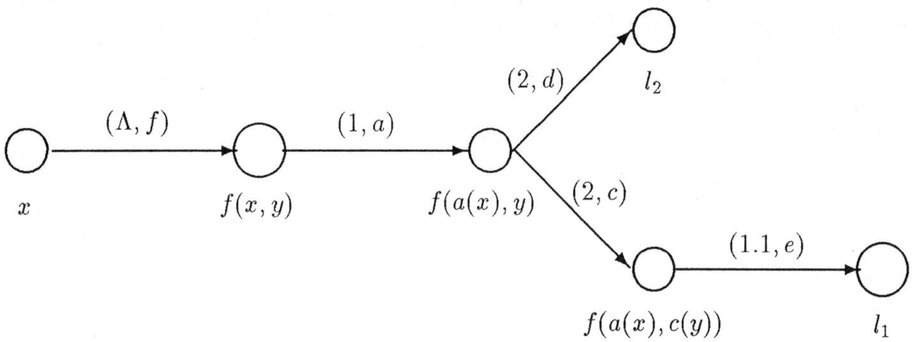

Figure 2: Automaton for patterns in fig.1

Such a centralized strategy suffers from two serious drawbacks. Firstly, P may have to successively copy (potentially) a large number of nodes in order to extend the initial prefix $f(x_1, ..., x_n)$ (where f is the root symbol of t) into a redex. This may result in a significant amount of communication ($O(|l|)$ messages, where $|l|$ is the size of the lhs for which a redex is ultimately discovered). The second problem relates to the size of the automaton that drives such a matching process. Note that as soon as any one t/p_i is head normalized, the prefix u can be extended at p_i into an u' which may have additional indexes $q_1, ..., q_m$. In order to maximize parallelism, processor P must initiate computations to head normalize the subterms $t/q_1, ..., t/q_m$. However, the computations to head normalize all t/p_i's may complete in any order. Therefore P must be capable of extending the prefix at any p_i. This means that the state of the matching automaton that corresponds to having inspected the prefix u must have transitions on (p_i, c) for *all* $1 \leq i \leq k$ and c such that for some lhs l that unifies with u has $root(l/p_i) = c$. This contrasts sharply with the sequential case wherein there are transitions (p, c) for *only one* index p.

We will now show that the requirement to have transitions on all p_i's may result in the automaton having one state corresponding to each possible prefix of an lhs. Consider a TRS $\{l \to r\}$ where l is a complete binary tree of height h with only variables as leaves. Note that for a prefix u of l, any path p whose length is less than h (and for which u/p is a variable) is an index for u. It can be readily shown (by induction on the number of nonvariable symbols in a prefix) that the automaton will have states corresponding to every prefix of l.

We now show that the number of such prefixes is exponentially large. Let $T(h)$ denote the number of possible prefixes of a tree of height h. Note that a prefix of l is either a variable (i.e., the prefix is "empty") or $f(t_1, t_2)$ where f is the root of l and t_1 and t_2 are any prefixes of the left and right subtrees of l respectively. There are $T(h-1)$ prefixes for t_1 and t_2 and so

$$T(h) = 1 + (T(h-1))^2$$

with $T(1) = 2$. It can be readily established by induction on h that $T(h) \geq O(2^{2^{h-1}}) \geq 2^n$, where n is the number of nonvariables in the tree.

2.3 Path-Sequential Systems

The exponential blow-up discussed in the previous section occurs because the index computation is based on the entire prefix. It is possible to avoid this blow-up if indexes can be computed based only on a part of the prefix. We therefore propose a new framework for lazy parallel evaluation based on *path-sequential systems* wherein we can decide whether a path p is an index based only on the symbols on this path in the prefix. This enables matching (and consequently the search for needed redexes) to be performed independently along different paths in a distributed fashion. In particular, let $t = f(t_1, ..., t_k)$ be split among processors $P, P_1, ..., P_k$ such that P holds the root of t and the subterm t_i is in processor P_i. In our framework, P will initiate a match at the root and ask $P_1, ..., P_k$ to continue the match. P_i attempts to match t_i with the ith subterm of all lhs whose outermost symbol is f and reports back to P the subset of these lhs that matched. Now P decides whether there is a redex at t based on the results of the submatches thus returned. Note that copying of nodes to processor P from P_i's is completely avoided.

To avoid excessive communication, the matching process within a P_i must take place without any information about the outcomes of the submatches proceeding in other processors. If this happens then observe that communication takes place only between those processors across which the portion of the term to be matched is split. For matching within a P_i to proceed independently of other submatches we require that P_i be able to determine, based only on the information sent to it by its parent P, whether t_i is a variable substitution or is to be head normalized to match a nonvariable symbol of an lhs. This is possible for the class of *path-sequential systems*:

Definition 1 *A constructor system $\mathcal{R}$ is* path-sequential *iff there are no two lhs l_1 and l_2, a path p and an integer i such that $pathstr(l_1, p) = pathstr(l_2, p)$, $l_1/p.i$ is a variable and $l_2/p.i$ is a nonvariable.*

In other words, path-sequential systems are a subclass of strongly-sequential systems wherein, given any prefix u and a position p in u, we can determine whether p is an index for u based only upon the symbols on the path p. The lhs in fig.1 are strongly-sequential, but not path-sequential since $pathstr(l_1, 1) = pathstr(l_2, 1) = f.1.a$, but $l_1/1.1$ is a variable whereas $l_2/1.1$ is a nonvariable. In contrast, the set $\{f(b(e), c(x)), f(a(x), c(y))\}$ of lhs is path-sequential.

Since we can determine whether p is an index based only on the path string from root to the parent of the node at p, matching can be based on path strings. Therefore the matching automaton will contain one state corresponding to each prefix of a *path string*. Observe that each such prefix starts at the root of an lhs and ends at a unique nonvariable node of this lhs. Consequently, the number of such prefixes is bounded by the total number of distinct nonvariable nodes in different lhs and is thus linear in the sum of the sizes of the lhs in the TRS.

In summary, a natural extension of the Huet-Levy strategy for parallel evaluation makes use of a centralized strategy that incurs significant amount of communication and synchronization overheads. More importantly, this strategy results in an exponential blow-up in the size of the matching automaton[4] *even for programs that are*

[4] Although this was established with a TRS consisting of a single rule, note that for TRS consisting of many rules, the automaton can have one state corresponding to each possible prefix of every lhs.

path-sequential. In contrast path-sequential systems provide a natural and elegant framework for efficient lazy parallel evaluation of equational programs. They permit us to search and reduce needed redexes independently along different paths. More importantly, the size of the automaton is linear in size of the rules.

3 Parallel Normalization Algorithm

The normalizing algorithm makes use of three procedures: *Normalize*, *Hnorm* and *Match*. All these procedures take a term v, passed by reference, as an argument. *Normalize* normalizes its argument. It makes use of *Hnorm* to find the HNF v_h of v. Since the root symbol of v_h appears in its normal form, the normal form of v can be computed by normalizing the subterms rooted at the children of v_h.

Hnorm uses the procedure *Match* to check if there is any redex at the root of its argument v. If a redex is found then it is rewritten and this process is repeated until v is reduced to a term v' in HNF. We refer to v' as the "term returned" by *Hnorm*. The details of *Normalize* are as follows. In this algorithm, let $nvsons(\mathcal{L}, p)$ denote the set of integers

$$\{i \mid \exists l \in \mathcal{L}\ root(l/p.i) \text{ is a nonvariable }\}$$

Note that if $pathstr(l_1, p) = pathstr(l_2, p)$ then by definition of path-sequentiality, $l_1/p.i$ is a nonvariable iff $l_2/p.i$ is a nonvariable.

procedure $Normalize(v)$ /* v is the term to be normalized */
1. $Hnorm(v)$;
2. $Normalize(u \mid u$ is an immediate subterm of $v)$

procedure $Hnorm(v)$ /* v is the term whose HNF is to be found */
3. /* If $root(v)$ is a constructor then we can return immediately. */
4. while true do
5.   if $Match(\Lambda, v, \mathcal{L}) \neq \phi$
6.   then rewrite the redex v
7.   else return; /* Now v must be in HNF */
8. end

function $Match(p, v, \mathcal{S})$ /* p is a path, v a term and $\mathcal{S}$ is a set of lhs. */
9. $\mathcal{S}' := \{l \in \mathcal{S} \mid root(l/p) = root(v)\}$
10. if $nvsons(\mathcal{S}', p) = \phi$ then return $\mathcal{S}'$
11. else return $\bigcap_{j \in nvsons(\mathcal{S}')} Match(p.j, Hnorm(v/j), \mathcal{S}')$.

Note that procedure *Match* performs matching based on path strings. Let u be the innermost subterm above v where $Match(\Lambda, u, \mathcal{L})$ was invoked. Then the argument $\mathcal{S}$ in a call $Match(p, v, \mathcal{S})$ consists of all lhs whose path strings matched the path string from the root of u to the parent of v. $\mathcal{S}'$ computed at line 9 consists of lhs in $\mathcal{S}$ whose symbols at p match that at u/p and thus $\mathcal{S}' = \{l \in \mathcal{L} \mid pathstr(l, p) =$

Thus exponential blow-ups can occur in practice.

$pathstr(u,p)\}$. Since the symbols on the inside of all lhs are constructors, *Hnorm* (which would reduce any term with a function symbol root into one with a constructor root whenever possible) is called at line 11 before *Match* is called recursively. Note that these calls to *Hnorm* implicitly change v to v' by reducing the subterms below v and we refer to v' as "the term returned" by *Match* at line 11. The set returned by $Match(p, v, \mathcal{S})$ consists of all lhs $l \in \mathcal{S}$ such that l/p matches v'.

Note that in order to perform the rewrite at step 6, we need the substitutions for the variables in the lhs that matched. This are passed back to the processor P holding the root of the match in the following way. Each processor holding a substitution communicates directly with P to pass a pointer to the root of the substitution. To do this, the identity of P will be propagated downwards as a fourth argument to the recursive calls of *Match*.

- There are two sources of parallelism in this algorithm. First, the normalization of the different u's at line 2 can proceed in parallel. Second, at line 11, $Hnorm(v/j)$ and the subsequent matching on the paths $p.j$, for different j's can be done in parallel. The parallelism in the latter case is more significant in practice, since each invocation of *Hnorm* may in turn result in several parallel invocations of *Hnorm* on subterms below and so on, thereby yielding (potentially) a large degree of parallelism.

4 Soundness and Completeness

We now establish the soundness and completeness of *Normalize*. First we need the following facts.

Fact 1 When $Match(p, v, \mathcal{S})$ is invoked, let u be the innermost subterm above v where *Match* has been invoked with Λ as the first argument. Then $\mathcal{S}$ is the set of all lhs l such that $pathstr(u,p)$ and $pathstr(l,p)$ are identical except for $root(v)$.

This is easily proved by induction on length of p (and is omitted). Now note that $Match(p, v, \mathcal{S})$ is invoked at p only if some lhs in $l \in \mathcal{S}$ has a nonvariable at p. By path-sequentiality and fact 1, we have

Fact 2 All $l \in \mathcal{S}$ have a nonvariable term at p whenever $Match(p, v, \mathcal{S})$ is invoked.

This ensures that the comparison made at line 9 involves only nonvariable symbols. Now we are ready to prove correctness of *Match*.

Lemma 2 (Correctness of ***Match*****)** *If* $Match(\Lambda, v, \mathcal{L})$ *returns a set* $\mathcal{S}$ *and a term* v' *then* $\mathcal{S} = \{l \in \mathcal{L} \mid u \geq l\}$.

Proof: Is immediate from the proof of the following stronger assertion:

If $Match(p, v, \mathcal{S})$ *returns with the set* $\mathcal{S}''$ *and a term* v' *then* $\mathcal{S}'' = \{l \in \mathcal{S} \mid v' \geq l/p\}$.

We prove this assertion by induction on the maximum height of l/p for any $l \in \mathcal{S}$. The base case is when this height is zero, in which case l/p must be a unary constructor for every $l \in \mathcal{S}$. In this case $\mathcal{S}'$ computed at line 9 is either empty or consists of all $l \in \mathcal{S}$ such that $v' = l/p$. In either case $\mathcal{S}'$ is the desired set and it is returned by *Match* at line 10. For the induction step, note that *Match* may either return from line 10 or line 11. In the former case, the proof is similar to the basis step. In the latter case, note that the maximum height of any $l/p.j$ for $l \in \mathcal{S}' \subseteq \mathcal{S}$ is less than the maximum height of any l/p for $l \in \mathcal{S}$. Therefore we can inductively assume that each (recursive) invocation $Match(p.j, Hnorm(v/j), \mathcal{S}')$ returns with $\mathcal{S}_j$ and term v_j such that $\mathcal{S}_j = \{l \in \mathcal{S}' \mid v_j \geq l/p.j\}$. Now it is easy to see that the intersection of $\mathcal{S}_j$'s yields $\mathcal{S}''$. ∎

Lemma 3 (Correctness of ***Hnorm*****)** *If $Hnorm(v)$ returns with a term v' then v' is in HNF.*

Proof: Note that *Hnorm* returns only when *Match* returns ϕ at line 7. Therefore it suffices to show that whenever $Match(\Lambda, v, \mathcal{L})$ returns ϕ and a term v' then v' is in HNF. This is immediate from the proof of the following stronger assertion:

If $Match(p, v, \mathcal{S})$ returns with a set $\mathcal{S}''$ and a term v' that is not a redex then v' cannot be reduced to a $v'' \geq l/p$ for any $l \in \mathcal{S} - \mathcal{S}''$.

This is established by induction on height of v'. For the base case, height of v' is 0. Since v' is not a redex it must be a constructor and cannot be further reduced. Thus $\mathcal{S}'$ is the set of *all* lhs $l \in \mathcal{S}$ with $v' \geq l/p$ and *Match* correctly returns this set at line 10. For the induction step there are three cases to consider:

Case 1. *Match* returns from line 11: Let $\mathcal{S}_j$ be the set and v_j the term returned by a recursive call $Match(p.j, Hnorm(v/j), \mathcal{S}')$ at line 11. Then $\mathcal{S}'' = \bigcap \mathcal{S}_j$ and $v'/j = v_j$. Since the heights of v_j are less than that of v' we can inductively assume that v_j cannot be reduced to an instance of $l/p.j$ for any $l \in \mathcal{S}' - \mathcal{S}_j$. Now $root(v')$ can be either a constructor or a function symbol. In the former case it is clear that the lemma holds. In the latter case, $p = \Lambda$ and $\mathcal{S} = \mathcal{L}$ (or otherwise $\mathcal{S}'$ will be empty and *Match* will return from line 10). Also note that $\mathcal{S}'' = \phi$ since v' is not a redex. This means that for every lhs l with the outermost symbol $root(v')$ there exists a j such that v_j cannot be reduced into an instance of l/j. Therefore v' cannot be reduced into an instance of any l and hence the lemma holds.

Case 2. $root(v')$ is constructor and *Match* returns from line 10: Any v'' to which v' can be reduced has $root(v'') = root(v')$ and so $v'' \geq l/p$ holds for an $l \in \mathcal{S}$ only if $root(l/p) = root(v')$. All such lhs are in the set $\mathcal{S}'' = \mathcal{S}'$ and hence the lemma.

Case 3. $root(v')$ is a function symbol and *Match* returns from line 10: First note that $v = v'$. In this case if $\mathcal{S}' \neq \phi$ then $v = v'$ will match a nonvariable subterm of some lhs l (by correctness of *Match*). This subterm must be l itself since $root(v)$ is a function symbol and so $v = v'$ will be a redex, which is a contradiction. Therefore $\mathcal{S}' = \phi$. Furthermore, this invocation of *Match* must have been made from line 11 and *not* line 5, since in the latter case $p = \Lambda$ and $\mathcal{S} = \mathcal{L}$, which implies that $\mathcal{S}'$ cannot be empty when $root(v) \in \mathcal{F}$. Therefore v must have been returned by an invocation of $Hnorm(u)$ at line 11. This call of *Hnorm* would have performed a series of invocations of *Match* followed by rewrites (in the loop at lines 4–8) till a call $Match(\Lambda, u', \mathcal{L})$ returned ϕ and v. For this invocation of *Match*, $root(u') = root(v)$ is a function symbol and $\mathcal{S}'$ computed at line 9 will be nonempty. Therefore this invocation can return ϕ only from line 11. In such a case, by case 1, $v = v'$ can never be reduced into a redex and so any v'' to which v' can be reduced will have $root(v'') = root(v') \in \mathcal{F}$ and so will not match l/p (Note $p \neq \Lambda$) for any $l \in \mathcal{S}$. ∎

Theorem 4 (Soundness) *$Normalize(t)$ terminates with the normal form of t.*

Proof: We need to show that if *Normalize*(t) terminates with t_n then t_n is in NF. This is done by induction on height of t_n. The base case is when height of t_n is zero. Then *Normalize*(t) calls *Hnorm* which returns with t_n. By correctness of *Hnorm*, t_n is in HNF and hence in NF. For the induction step, note that *Normalize* calls *Hnorm* and then proceeds to normalize the immediate subterms of the term returned by it. The heights of the terms returned by these recursive invocations are less than that of t_n and so we can assume, by inductive hypothesis, that they are in NF. Then t_n will be in NF since it is in HNF and all its subterms are in NF. ∎

Theorem 5 (Completeness) *Normalize(t) terminates if t has a normal form.*

Proof: Note that any procedure that repeatedly rewrites needed redexes is complete for constructor systems [HL79]. Therefore we need only show that *Normalize* rewrites only needed redexes. To do this, we first show that *Hnorm*(v) reduces only those redexes that will remain outermost in v irrespective of other reductions inside v. Observe that whenever *Normalize* invokes *Hnorm*(v), every subterm (of the term t being normalized) that is above v is in HNF. Therefore the redexes reduced by *Hnorm*(v) will remain outermost in t irrespective of other reductions and hence are needed.

Now we need to show that any redex selected at line 6 by *Hnorm* will remain outermost in v. This is implied by the following stronger assertion:

For every p such that Hnorm(v/p) is invoked, no subterm v/q for $q < p$ can be reduced into a redex unless v/p is reduced into a term with a constructor root.

We prove this by induction on length of p. For the base case $p = \Lambda$ in which case there is no $q < p$ and hence the above assertion holds vacuously. For the induction step, *Hnorm*(v/p) is called from line 11. It can be easily seen that there exists a longest path $q < p$ such that

1. *Hnorm*(v/q) was invoked.
2. This resulted in a call of *Match* at line 5.
3. This call of *Match* resulted in further calls to *Match* at line 11 until *Match*$(p', Hnorm(v/p), \mathcal{S}')$ was called.

By fact 1 and path-sequentiality, every lhs that could potentially match at q has a nonvariable (constructor) at p and thus no redex can be found at q without reducing v/p to a term with constructor root. By inductive hypothesis, no redex can be found at any $r < q$ without reducing v/q into a term with constructor root. These two facts imply that no subterm at any $r < p$ can be reduced into a redex without reducing v/p to a term with constructor root. ∎

5 Complexity Issues

We modify *Normalize* as follows before we proceed to establish its time complexity. First, observe that *Normalize* uses a tree representation of terms. Because of this, reducing an outer redex can result in duplicating inner redexes. So the number of reductions performed by it may not be minimal although it performs no wasteful reductions. We can avoid duplication by keeping the rhs as fully compact dags and thus sharing multiple occurrences of these redexes. The details are straight-forward and so we omit them here.

Second, note that *Normalize* operates by repeatedly searching for redexes and replacing them. The search may often involve repeated traversals of the term and so a lot of time may be spent in such traversals rather than in performing reductions. Therefore we associate a flag with each node to indicate whether the term rooted at the node is in HNF. This flag is set on return from and checked before entry into (and the call skipped if set) *Hnorm*.

With the above modifications we now show that the time spent by *Normalize* is bounded by the time required to perform the needed reductions. To establish this result, we assume that the number of lhs with the same function symbol (typically 2–3) is smaller than the number of bits in a word (typically 32). We also assume that the input term is *legal*, i.e., its normal form will not contain function symbols. In languages that follow the constructor discipline (such as ML and HOPE), it is an error attempting to normalize a term that is not legal and so we ignore such terms for the time complexity result.

Theorem 6 *Let $t \rightarrow_{l_1} t_1 \rightarrow \cdots \rightarrow_{l_n} t_n$ be the sequence of needed reductions performed by Normalize, $E = \sum_{i=1}^{n}(|l_i| + |r_i|)$ and N be the size of the normal form. Then the normalization procedure executes in $O(N+E)$ time.*

Proof: We split the total time spent into times spent inside the body of *Match*, *Hnorm* and *Normalize* respectively. 1. Time spent in *Normalize*: Note that it takes $O(1)$ time to invoke *Hnorm*. Therefore each invocation of normalize spends constant amount of time. Furthermore, it can be seen that there will be one invocation of *Normalize* for each edge in the dag representation of the normal form and thus the time spent is bounded by $O(N)$.

2. Time spent in *Hnorm*, i.e., lines 3 – 8. If *Hnorm* returns without performing any rewrites, then $O(1)$ time is spent inside *Hnorm* and we "charge" this to the call of *Hnorm* (made at line 2 or 11). Otherwise, it is clear that the cost is bounded by the time taken for rewrites performed at line 6 and hence is bounded by $O(E)$.

3. Time spent inside *Match*. We claim that this is bounded by $O(E)$. We first show that each invocation of *Match* spends constant amount of time. Observe that line 9 can be executed in $O(1)$ time using a matching automaton for path strings, wherein the state reached by a path string s stores the set of all lhs that have s in them. If $|nvsons(\mathcal{S}')| = k$ at 11 then it can be computed in time proportional to k and the set intersection operation can also be done in $O(k)$ time with the bitvector representation. We distribute this cost as 1 unit each for the k recursive invocations of *Match* on the k subterms. Thus each execution of *Match* takes $O(1)$ time.

For an invocation of $Match(p, u', \mathcal{S})$, let u be the innermost subterm above u' with $root(u) \in \mathcal{F}$ such that the *Match* was invoked on u and this resulted in recursive invocations of *Match* on subterms of u till $Match(p, u', \mathcal{S})$ was called. If the match at u succeeded with a redex for l then a reduction will be performed, whose cost $|l| + |r|$ is accounted for in E. Note that $O(1)$ cost of $Match(p, u', \mathcal{S})$ is included in $|l|$. If the match at u fails then there must exist a path q such that $u = t/q$ and all subterms t/r for $r \leq q$ are in HNF. Otherwise, by proof of completeness, $Hnorm(u)$ and hence $Match(\Lambda, u, \mathcal{L})$ will never have been invoked. Thus the function symbol at the root of u will appear in the NF and hence the input term is not legal. ∎

We remark that any algorithm based on term rewriting must spend at least $O(N+E)$ time. This is because the redexes reduced by *Normalize* must all be reduced by any other algorithm and furthermore, it takes at least $O(N)$ time to check if the output term is in normal form. Thus *Normalize* is optimal.

5.1 Communication Complexity of *Match*

As mentioned earlier (see section 2), processors need to communicate during the match phase if parts of a subterm being matched are split across different processors. If these split matches are handled naively (e.g., by first copying all the nodes to be matched to the processor holding the node at which the match is initiated and then performing the match locally) then it can lead to a large volume of communication. Ideally, we would like handle each split using $O(1)$ communication steps and this in fact is achieved by *Normalize*. The invocation of *Match* at line 11 results in one communication step if $root(v/j)$ is in a processor different from the one in which $root(v)$ is. The return value of the recursive call is transmitted back to the processor holding $root(v)$ in another communication step.

Finally, note that the communication required to pass the substitutions to the processor P holding the root of the match can also be minimized: one communication step for each substitution that resides on a processor different from P.

6 Enhancing Parallelism

Normalization algorithms based on lhs alone will never reduce a needed redex u if it is possible to create an outer redex v by rewriting redexes other than u. This is because an algorithm based on lhs cannot rule out the possibility that the rewrite of the redex v may "throw away" the redex u and hence u may not be needed. For instance, consider the TRS $\{f(a(x)) \rightarrow b\}$ and the term $f(a(f(a(b))))$. Here, the inner redex $f(a(b))$ is discarded when the redex at the root is rewritten. Therefore such algorithms postpone reducing a redex until it is clear that the redex will remain outermost. Consequently, by not reducing such redexes early, we encounter significant loss of parallelism even in path-sequential systems. We later show how this problem can be alleviated by analyzing the rhs. First we show that based on lhs alone, it is not possible to enhance the parallelism detected by *Normalize*.

Observe that an invocation of $Match(\Lambda, v, \mathcal{L})$ results in a downward traversal of all paths from the root of v. All redexes encountered on these traversals are reduced. The traversal on a path halts only on encountering a node that corresponds to a variable substitution for all potentially matching lhs. Therefore the parallelism can be enhanced only if the algorithm searches for and replaces redexes inside variable substitutions. But this cannot be done by any lazy algorithm that does not inspect the rhs since it is possible that the corresponding variable does not appear on the rhs. In such a case, all the redexes below this variable substitution are discarded and are therefore not needed.

6.1 Enhancing Parallelism by Analyzing rhs

By analyzing the rhs, we can often rule out the possibility of some redexes being "thrown away". For instance, consider the TRS $\{f(a(x), y) \rightarrow y\}$ and the term

$f(a(b), f(a(c), b))$. Analysis of the rhs indicates that the substitution for the second argument of f is not discarded and therefore any redex inside this argument can be reduced. Therefore we can reduce the root redex and the redex $f(a(c), b)$ *simultaneously*. Observe that this would not be possible if we had only considered the lhs since we would then be unable to determine if the redex $f(a(c), b)$ is needed.

We have developed a technique called *ee*-analysis that analyzed the rhs of rewrite rules and specified which arguments of a function f must possess normal forms in order for an application of f to possess a normal form. The function f is said to be *ee*-strict in such arguments. Observe that by this definition, all constructors are *ee*-strict in all their arguments. The details of this analysis have appeared in [SSR90]. Our technique is conceptually simple and very effective in practice.

We now modify *Normalize* to identify and reduce needed redexes as early as possible using the information provided by *ee*-analysis. With this modification (given below), *Normalize* detects all available parallelism in many example programs such as matrix multiplication and sorting.

procedure $Normalize(v)$ /* v is the term to be normalized */
0. for $i := 1$ to $arity(root(v))$ do
   if $root(v)$ is ee-strict in ith argument then $Normalize(v/i)$
1. $Hnorm(v)$;
2. $Normalize(u \mid u$ is an immediate subterm of $v)$

7 Concluding Remarks

In this paper we introduced a new class of strongly-sequential systems called path-sequential systems. We showed that such systems provide an elegant framework for lazy parallel evaluation of equational programs. In these systems the search for needed redexes can proceed independently along different paths of the input term with minimal communication. Furthermore, matching in such systems can be driven by an automaton whose size is linear in input program size. We then presented an algorithm for lazy parallel normalization of such systems based on the lhs alone and established its soundness and completeness. We also showed that it is optimal in the sense that its time complexity is bounded only by the time required to perform the needed reductions; the search for redexes does not add to this complexity. We then showed how to enhance the parallelism exploited using information obtained by analyzing rhs.

We have been continuing our work on parallel evaluation of equational programs and recently completed implementing EQUALS [OPRRS92], a system to support equational programs on the Sequent shared-memory multiprocessor. Our initial results were obtained on a six-processor Sequent Symmetry. On many example programs such as matrix multiplication and sorting our implementation achieves substantial speedups over existing sequential implementations of functional languages such as Standard ML of New Jersey. Preliminary results on a larger Sequent indicate that we obtain almost linear speedups on many programs even upto 16 processors.

References

[BMS80] R.Burstall, D.MacQueen and D.Sanella, HOPE: An Experimental Applicative Language, *Proc. of 1st Int. LISP Conf., 1980.*

[DJ90] N. Dershowitz and J.P. Jouannaud, Rewrite Systems, *Handbook of Theoretical Computer Science, Ch. 6, Vol II, North-Holland, 1990.*

[FGJM85] K. Futatsugi, J. Goguen, J.P. Jouannaud and J. Meseguer, Principles of OBJ2, *POPL 1985, pp. 52-66.*

[HMT88] R.Harper, R.Milner and M.Tofte, The Definition of Standard ML, *Report ECS-LFCS-88-62, LFCS, U. of Edinburgh, 1988.*

[HO84] C. Hoffmann, and M.J. O'Donnell , Implementation of an Interpreter for Abstract Equations, *POPL 1984, pp. 111 – 120.*

[HL79] G. Huet and J.J. Levy, Computations in Nonambiguous Linear Term Rewriting Systems, *TR No. 359(1979), INRIA, France.*

[Ken89] J.R. Kennaway, Sequential Evaluation Strategies for Parallel-Or and Related Reduction Systems, *Annals of Pure and Applied Logic 43, pp. 31–56, 1989.*

[Lav87] A. Laville, Lazy Pattern Matching in the ML Language, *Proc. of FST&TCS, Springer-Verlag LNCS, 1987.*

[OD85] M.J. O'Donnell, Equational Logic as a Programming Language, *MIT Press (1985).*

[OPRRS92] K. Owen, S. Pawagi, C.R. Ramakrishnan, I.V. Ramakrishnan and R.C. Sekar, Fast Parallel Implementation of Lazy Languages – The EQUALS Experience, *Lisp and Functional Programming, 1992.*

[SSR89] R.C. Sekar, S. Pawagi and I.V. Ramakrishnan, Transforming Strongly Sequential Rewrite Systems with Constructors for Efficient Parallel Execution, *RTA 1989.*

[SSR90] Sekar, R.C., Shaunak Pawagi and Ramakrishnan, I.V., Small Domains Spell Fast Strictness Analysis, *POPL 1990.*

[Str88] Robert I. Strandh, Compiling Equational Programs into Efficient Machine Code, *PhD Thesis, Johns Hopkins University, 1988.*

[That85] Satish Thatte, On the correspondence between two classes of Reduction systems, *Information Processing Letters 20 (2), pp. 83-85.*

A Many Sorted Logic with Possibly Empty Sorts

A G Cohn
Division of Artificial Intelligence, School of Computer Studies
University of Leeds, LS2 9JT, England
agc@ai.leeds.ac.uk

Abstract

Normally the interpretation of sort symbols in sorted logics are required to be non empty subsets of the universe of discourse. In this paper we explore the consequences of relaxing this restriction whilst maintaining a separation between the specialised sort theory and the main theory. A sound and complete inference system for a resolution based logic is presented.

1. Introduction

A many sorted logic is one in which the universe of discourse is divided up into subsets called *sorts*. A sorted logic has a set of *sort symbols*[1] S which denote sorts in any interpretation. For example, one might subdivide the universe of discourse into different kinds (sorts) of animals and plants (cf Walther (1985), Cohn (1985)). There are also mechanisms for attaching sort constraints to variables (restricting their domain of quantification) and describing the sortal behaviour of the non logical symbols. These declarations comprise the *signature* of a sorted logic and affect the semantics by restricting the possible interpretations of the non logical symbols and allowing function symbols to be interpreted as partial functions. A sorted logic is therefore a logic which can treat *taxonomic* information specially.

AI's main interest in sorted logics has been their computational efficiency: the search space may be much reduced compared to a corresponding formulation in unsorted logic because it is easy to detect that certain inferences cannot lead to a successful proof; these inferences are those that produce *ill sorted formulae,* i.e. where the sorts of terms do not match the sorts of the arguments positions they occupy or those generated by the application of *ill sorted substitutions,* i.e. where the sort of a term does not fit the sort constraints of the variable it is being substituted for. Another advantage claimed for sorted logics is that the ontology of the domain is made explicit and the sortal behaviour of the non logical symbols explicitly declared. This can be valuable not only for documentation purposes but also for knowledge base verification and integrity checking through well sortedness checking.

However, in a sorted logic, the sorts must be non empty. This is for the same reason that in standard unsorted logic the universe of discourse must be non empty: from $\forall x\, P(x)$ we expect to be able to infer $\exists x\, P(x)$ (soundly); similarly in a many sorted logic from $\forall x{:}\tau\, P(x)$ we expect to be able to infer $\exists x{:}\tau\, P(x)$, where the notation '$x{:}\tau$' indicates that x is restricted to range over sort τ; if τ could be empty then this inference would be invalid.

---

1 Where no confusion will arise, we may sometimes be lax and refer to 'sorts' where strictly we should have said 'sort symbols'.

This restriction on empty sorts means that not every monadic predicate can be considered as a sort, only those known *a priori* to be non empty: if τ is a sort symbol, then by definition τ is interpreted as a non empty set in every interpretation; thus to assert $\neg\exists x{:}\tau\ \tau(x)$ is contradictory.[2] Of course, just because it may not be known explicitly that a predicate τ is non empty, does not mean that $\exists x\ \tau(x)$ does not follow from a particular axiomatisation. For example, consider the unsorted theory:

Motorcycle(vehicle42)
Has(vehicle42,sidecar)
$\forall x$ [[Motorcycle(x) $\wedge$ Has(x,sidecar)] $\rightarrow$ Threewheeledvehicle(x)]
Fivewheeledvehicle(a)$\vee$Fivewheeledvehicle(b)

It is a simple database retrieval to prove that Motorcycle is non empty and thus could be a sort if we were to convert this axiomatisation to a sorted one. However it requires a certain amount of inference to deduce that Threewheeledvehicle(vehicle42) and thus that Threewheeledvehicle is nonempty. Obviously the deduction might have been much more difficult and in fact since first order logic is only semi-decidable, proving the non emptiness of all monadic predicates *a priori* is not feasible in general. The predicate Fivewheeledvehicle is also provably non empty, even though there is no named individual for which it is provably true.

The goal of this paper therefore is to develop a many sorted logic which allows possibly empty sorts. When we say 'possibly empty sorts' what we mean is that a symbol may be a sort without it being non empty in every interpretation of the *signature*; however it may be required to prove that such a sort is actually non empty *dynamically* (i.e. during the proof) in order to preserve soundness.

Another advantage of allowing possibly empty sorts is that the structure of the *sort hierarchy* can be very flexible; most many sorted logics allow an ordering '$\sqsubseteq$' on the set of sort symbols such that if $\tau_1 \sqsubseteq \tau_2$ then τ_1 is interpreted as a subset of τ_2 in any interpretation.[3] Suppose that $\tau_3 \sqsubseteq \tau_1$ and $\tau_3 \sqsubseteq \tau_2$; if τ_3 is actually empty and there is no other common subsort of τ_1 and τ_2 then τ_1 and τ_2 are effectively disjoint. In some many sorted logics the sort hierarchy is interpreted in such a way that if $\tau_3 = \tau_1 \sqcap \tau_2$ (i.e. τ_3 is the greatest lower bound (glb) of τ_1 and τ_2), then the interpretation of τ_3 is *equal* to the intersection of the interpretations of τ_1 and τ_2 (and similarly for the other lattice operators,'$\sqcup$' and '\') – this kind of sort structure has been called *lattice theoretic*. In contrast, the former kind where the interpretation of τ_3 may be a strict subset of the intersection of the interpretations of τ_1 and τ_2 has been called an *order theoretic* sort structure. Suppose the sort structure is lattice theoretic, and is as illustrated in figure 1. Consider the base sorts, just above $\perp$, marked with an asterisk: from left to right, if only the first is empty then τ_1 is (semantically) a subsort of τ_2, if only the second is empty then τ_3 is (semantically) the intersection of τ_2 and τ_3 and if only the third is empty then τ_2 is (semantically) a subsort of τ_1. Thus with possibly empty sorts, we can express all kinds of indefiniteness in the sort hierarchy whilst retaining the extra expressive power that a

---

[2] Note that a sort symbol appears here as a predicate symbol. Not all many sorted logics allow this. See Cohn (1989) for a discussion on the appearance of characteristic literals (i.e. literals whose predicate name is a sort symbol) in many sorted logics. As we shall see below, sort literals will be required by the inference rules of a many sorted logic with possibly empty sorts.

[3] Such logics are sometimes called *order sorted logics* to emphasise this feature.

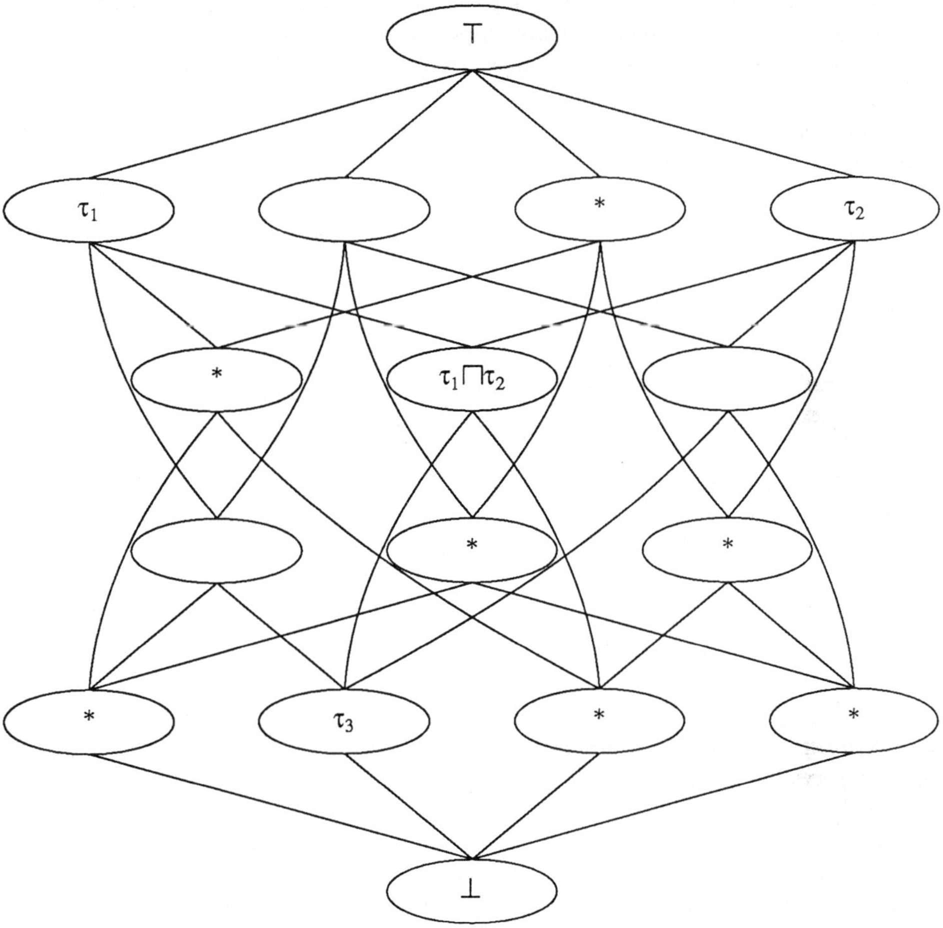

Figure 1: A Lattice Theoretic sort structure in the form of a complete Boolean lattice. Nodes marked with an asterisk are *possibly empty* sorts.

lattice theoretic interpretation gives.

In the following sections we will first of all consider the language of a many sorted logic with possibly empty sorts and then the inference mechanism required; we will then compare the logic to other proposals before going on to sketch the completeness proof.

2. The Language

In this section we will formally present a many sorted logic which allows possibly empty sorts and its semantics. In order to keep the presentation as general as possible, we will assume that the signature is specified axiomatically[4], in the style of Frisch (1989), though

[4] Of course, it is likely that for the purposes of an implementation, the sort theory would be stored and processed in some special way in order to take exploit the restricted nature of syntax allowed.

we choose to restrict the possible set of interpretations by the signature, whilst for Frisch this does not happen. This representation does mean that it is not possible to describe the sortal behaviour of predicate symbols (apart from sort symbols themselves of course) in the signature, because the only predicate symbols allowed are the sort symbols. This could be allowed but it would complicate the logic considerably (for example necessitating further inference rules) and we wish to keep the presentation simple and brief in this paper.

The language has an associated fixed set of sort symbols S which always includes the symbols $\top$ and $\bot$. A formula is either an *S-formula* or an *A-formula*. S-formulae are written in standard first order predicate calculus with the restriction that every predicate symbol is a member of S. A-formulae are written in standard first order predicate caclulus except that all occurrences of variables[5] are annotated with a sort symbol;[6] for example $x{:}\tau$ indicates that x is of sort τ. All occurrences of a variable governed by the same quantifier must have identical annotations. By convention, only one instance of any given variable need have an annotation, the rest implicitly have the same annotation. Also, if a particular variable has no annotations then this is equivalent to an annotation of $\top$. We will refer to the set of S-formulae as Σ (or alternatively as the *SBOX*) and the set of A-formulae as Π (or alternatively the *ABOX*).

The only allowable interpretations of A-formulae are those which satisfy all the S-formulae, Σ, i.e. $\{\sigma| \sigma \models \Sigma\}$; such interpretations will be called Σ-interpretations. In order to make the signature non trivial, we will require it be be satisfiable (i.e. there should be at least one Σ-interpretation). S is divided into three disjoint subsets, S^{NE}, S^{PE} and S^{E} which are the *non empty*, *potentially empty* and *definitely empty* sorts respectively:

$$S^{NE} = \{\tau \in S| \text{ for all } \Sigma\text{–interpretations } \sigma,\ \sigma(\tau)\neq\emptyset\}$$
$$S^{E} = \{\tau \in S| \text{ for all } \Sigma\text{–interpretations } \sigma,\ \sigma(\tau)=\emptyset\}$$
$$S^{PE} = S-(S^{NE}\cup S^{E}).$$

We will also define the *potentially inhabited* sorts, $S^{PI} = S-S^{E} = S^{NE}\cup S^{PE}$. We can now also impose a syntactic restriction on A-formulae – that all variable annotations are in S^{PI} – in order to eliminate empty quantifications. Further restrictions on Σ-interpretations are that $\bot \in S^{E}$ and that $\Sigma \models \forall x\ \top(x)$ from which it follows that $\top \in S^{NE}$.

We will need the notion of the *relativisation* $\hat{A}$ of a quantifier free formula A (i.e. the semantically equivalent unsorted formula). In a many sorted logic the relativisation of a quantifier free formula A whose free variables (which are implicitly universally quantified) are $x_1{:}\tau_1 \cdots x_n{:}\tau_n$ is defined to be $(\tau_1(x_1)\wedge \cdots \wedge\tau_n(x_n))\rightarrow A'$ where A' is obtained from A by dropping all the sort annotations from the variables (eg (Walther 1987)).

A Σ-interpretation σ satisfies an A-formula A, in an environment ϕ (i.e. a mapping of variables to individuals), written $\sigma\phi \models_\Sigma A$, iff $\sigma\phi \models \hat{A}$,[7] i.e. if it satisifes the relativisation of A. This means that $\forall x{:}\tau\Phi$ does not imply $\exists x{:}\tau\Phi$ unless τ is non empty, unlike the corresponding situation in unsorted logic (where the domain is of course

5 We will use italic font to indicate variables.

6 It is possible to allow more general annotations, e.g. lattice theoretic expressions, such as $\tau_1 \sqcap \tau_2$ or $\tau_1 \sqcup \tau_2$, but for simplicity we will not do this here.

7 We can choose to say that $\sigma\phi \models_\Sigma A$, iff $\sigma\phi \models \hat{A}$, rather than $\sigma\phi \models_\Sigma A$, iff $\sigma\phi \models (\Sigma\rightarrow\hat{A})$, because σ is a Σ interpretation and thus satisifes Σ.

guaranteed to be non empty and thus $\forall$ always implies $\exists$). We could alter the semantics so that $\forall x{:}\tau\Phi$ could not be true unless τ is non empty, but this considerably complicates the logic (particularly in the quantifier free clausal form form, because of a desire to insist only that "just enough" sorts are non empty) to no great practical effect: in both cases there will be an identical obligation in the proof system to only infer $\exists x{:}\tau\Phi$ from $\forall x{:}\tau\Phi$ if τ is non empty.

Although we are specifying the signature axiomatically, we will still allow ourselves the liberty of using lattice theoretic operators on sort symbols as shorthand; so when we, for example, write $\tau_1 \sqsubseteq \tau_2$, we mean, $\Sigma \models \forall x\, \tau_1(x) \rightarrow \tau_2(x)$; similarly[8] $\tau_1 \sqcap \tau_2 = \tau_3$ is shorthand for $\Sigma \models \forall x\, (\tau_1(x) \wedge \tau_2(x)) \leftrightarrow \tau_3(x)$ and $\tau_2 \sqsubseteq \top \backslash \tau_1$ means $\Sigma \models \forall x\, \tau_2(x) \rightarrow \neg \tau_1(x)$.

A *well sorted formula* is normally defined to be one where the sort of every term (as determined by Σ) matches the sort of the argument positions they occur in. The concept of a well sorted formulae is a little difficult to define since we are not allowing arbitrary predicates to appear in S-sentences and therefore the sortal behaviour of predicates cannot be described. The concept of a term being well sorted can however be expressed: a term β is of sort τ if $\Sigma \models \hat{\tau}(\beta)$, and it is well sorted if it is of some sort $\tau \in \mathbf{S}^{PI}$. Note that this definition ensures that a variable $x{:}\tau$ is of sort τ. We can then simply define a well sorted formula as one which only contains well sorted terms.[9] Also note that if a ground term is of sort τ then $\tau \in \mathbf{S}^{NE}$.

A substitution $\langle \beta_1/\alpha_1{:}\tau_1, \cdots, \beta_n/\alpha_n{:}\tau_n \rangle$ is *well sorted* iff each β_i is of sort $\tau_i \in \mathbf{S}^{PI}$. Unfortunately, as we shall see below, well sortedness is sometimes too strong a concept. We therefore introduce the notion of a substitution being *sort consistent:* each β_i must have some instance $\beta_i\theta$ which is of some sort τ'_i such that $\Sigma\ \hat{\tau}_i(\beta_i\theta)\ \hat{\tau}'_i(\beta_i\theta) \nvDash$ False; i.e. a sort of an instance of β_i must be consistent with the sort of the variable being substituted for.[10] Well sorted substitutions are sort consistent but not vice versa in general. Substitutions which are not sort consistent are called *ill sorted.* For example, if $\Sigma = \{S1(x) \rightarrow S2(f(x))\}$,[11] then $\langle f(x{:}\top)/y{:}S2 \rangle$ is sort consistent but not well sorted, whilst $\langle f(x{:}S1)/y{:}S2 \rangle$ is well sorted.

It has been pointed out that it may be possible to create a well sorted substitution from one which is not by composing it with a further a substitution; this is called *weakening* (Walther 1987, Frisch 1989). For example in the first case in the example above, one could weaken the subsitution by composing it with $\langle z{:}S1/x{:}\top \rangle$. However, in our logic we might lose completeness if we only considered this weakened substitution, because there might be a way of proving that $f(x{:}\top)$ is of sort S2 without having to weaken x by using Π as well as just Σ; this was pointed out by Beierle et al (1989). Also note that weakening could never turn an illsorted substitution into a sort consistent one. However to maintain soundness we will have to add extra literals during the inference process as we will see below. In a sorted logic which does not allow characteristic literals (such as those of Walther (1987), Frisch (1989) or Schmidt-Schauss (1987)) then there is

8 Note that we do not necessarily require that the lattice 'operators' such as $\sqcap$ are unique.

9 This is as though the sort associated with every predicate argument position is $\top$.

10 This notion was first defined in Cohn (1983) though it was not called sort consistency there. According to Kowalski and Hayes (1971), Gordon Plotkin first had the idea of allowing what we call sort consistent substitutions.

11 Note that we assume that $\forall x\, \top(x)$ and $\forall x\, \neg\bot(x)$ are always in Σ; if these are the only wffs in Σ then we say that Σ is empty.

no need to add explicit negative information to Σ (e.g. to say that two sorts are disjoint) because a reduction in the search space will be achieved in any case simply because two sorts are incompatible (i.e. cannot be proved to have a lower bound not equal to $\perp$). However in our logic, it is of benefit to add explicit negative information to Σ because this might allow us to detect that we could never prove a term is of a particular sort, even using Π. For example, if Σ is empty then $\langle z{:}S2/x{:}S1\rangle$ is sort consistent even though Π might not contain any sort literals (and thus there would be no way to show that the intersection of S1 and S2 was non empty), but if we added $\neg(S1(x)\wedge S2(x))$ to Σ then $\langle z{:}S2/x{:}S1\rangle$ would be ill sorted rather than sort consistent. In the logic LLAMA (Cohn 1983, 1987) the addition of negative information was enforced by requiring the (lattice theoretic) sort hierarchy to be in the form of a complete Boolean lattice and requiring sorting functions (i.e. the functions describing the sortal behaviour of function symbols) to be total on the set of sorts.

3. Inference

In this section we will present an inference system for the A-sentences. We will assume that all A-sentences are represented in clausal form and that we wish to develop a resolution calculus. This will in turn require proofs of certain theorems from the SBOX. We will not consider here how deduction in the SBOX is performed, but will treat the SBOX as a 'black box' and assume there is some oracle which can answer certain questions. First, we want to be able to determine whether a substitution component $t/x{:}\tau$ is well sorted. In this case we will call on the oracle to determine whether $\Sigma \models \hat{\tau}(t)$. Note that this allows the sorts annotating the free variables of t to be possibly empty; e.g. consider $f(x{:}\tau_1)/y{:}\tau_2$ where $\Sigma \models \{\tau_1(x) \rightarrow \tau_2(f(x))\}$ and $\tau_2 \in S^{PE}$ then $\Sigma \models \hat{\tau}_2(f(x{:}\tau_1))$.

Secondly, we want to be able to determine whether a substitution component $t/x{:}\tau$ is sort consistent. In this case we will call on the oracle to determine whether there exists a sort τ' and a substitution θ such that $t\theta$ is of sort τ' and $\Sigma \cup \{\hat{\tau}(t\theta), \hat{\tau}'(t\theta)\} \not\models$ False.

Frisch (1989) postulates a similar oracle[12] but does not allow for empty sorts, i.e. the wellsortedness of his substitutions do not depend on the non emptiness of sorts, whilst our notion of well sorted substitutions does allow sorts to be possibly empty. However his logic has no mechanism for proving non emptiness outwith the SBOX since he does not allow sort literals in the ABOX. Also his oracle does not need to take account of sort consistent rather than well sorted substitutions.

Later, (in the rule of *characteristic resolution* and the rule of *evaluation*) we will require other computations of the oracle which will be specified below.

We will take the view that a resolution comprises two parts, *instantiation* and *clashing*. (Factoring can be seen as just instantiation). In unsorted logic, instantiation simply comprises a textual substitution: if θ is a substitution $\langle \beta_1/\alpha_1, \cdots, \beta_n/\alpha_n\rangle$, then $A\theta$ is obtained by substituting each β_i for the corresponding variable α_i in turn. In sorted logic the sortal constraints on variables have to be checked.

[12] Abrams and Frisch (1991) have also shown that even if the oracle uses ordinary deduction, rather than some special purpose computational mechanism, then computation in a many sorted logic can be still more efficient in most cases.

If a substitution component is well sorted, then instantiation is standard; however if it is only sort consistent, then as mentioned earlier, extra literals need to be added to the instance; following Cohn (1983) we call these *prosthetic* literals. For example, suppose Σ is empty, then the substitution $t/x{:}S1$ is sort consistent but not well sorted, and applying this substitution to the clause $P(x{:}\ S1)$, yields the instance $S1(t) \rightarrow P(t)$. We thus define the notion of applying a sort consistent substitution to a clause C.[13]

Definition (Instantiation)

Let C be a clause and θ a sort consistent substitution. Then

$$C\downarrow\theta =_{def} C\theta \cup \{\neg\tau(\beta) \mid \beta/\alpha{:}\tau \text{ is a substitution component of } \theta \text{ and } \beta/\alpha{:}\tau \text{ is not well sorted}\}.$$

The soundness of this instantiation rule is easy to see and follows straightforwardly from a similar definition and proof of soundness in Cohn (1983).

The problem with empty sorts makes itself felt in a resolution based calculus in the clashing rule when two clauses (assuming we are just considering binary resolution) are combined to form a resolvent which consists of all the literals in the two parent clauses except the resolved upon literals: if the resolved upon literals contained the only occurrence of a variable which was of a possibly empty sort, then forming the resolvent in this standard way is unsound.

For example, suppose we are resolving $P(x{:}\tau_1) \vee \Phi$ (where Φ is some further formula not containing x) with $\neg P(y{:}\tau_1)$, then the standard resolvent would be just the formula Φ. If $\tau_1 \in S^{PE}$ then this inference is unsound. Unless there is an object of sort τ_1, both $P(x{:}\tau_1)$ and $\neg P(y{:}\tau_1)$ are true and the inference is thus unsound since any interpretation in which τ_1 is empty can satisfy both the parents but not the resolvent. Of course if Φ contains x as well, then x will still occur in the resolvent and we can delay worrying about the possible emptiness of τ_1 since x will still be labelled with τ_1; if this resolvent forms part of the refutation, then the literal(s) in which x occurs will have to be resolved away and either x will be unified with a non variable term which will have been proven to be of sort τ_1 thus ensuring the non emptiness of of τ_1 (because of the restriction to sort consistent substitutions) or it will be unified with another variable which means the case we are currently considering applies. If x does not occur in Φ, then we must ensure that a literal $\neg\tau_1(z{:}\tau_1)$ is added to the resolvent. The following definition captures the above notion.

Definition (Instantiation with respect to variables with possibly empty sorts)

Let C be a clause, θ a sort consistent substitution and Γ a set of variables, then

$$C\underset{\Gamma}{\downarrow}\theta =_{def}$$

$$C\downarrow\theta \cup \{\neg\tau(\beta) \mid \alpha \in \Gamma \text{ and } \alpha\theta \text{ is a variable of the form } \beta{:}\tau \text{ not occurring in } C\downarrow\theta \text{ and } \tau \in S^{PE}\}$$

---

13 Note that we treat a clause either as an explicit disjunction or implication or as an implicitly disjoined set of literals as convenient.

We can now define the rule of binary resolution.

Definition (Binary Resolution)

Let C_1 and C_2 be two clauses containing literals $L_1 \in C_1$ and $L_2 \in C_2$ and let θ be a sort consistent substitution such that $L_1\theta = \neg L_2\theta$. Then $((C_1 \setminus L_1) \cup (C_2 \setminus L_2)) \downarrow_{\Gamma} \theta$ is a binary resolvent of C_1 and C_2, where Γ is the set of variables occurring in $C_1 \cup C_2$.

As can be seen from the above definition, new sort literals may be added to a resolvent for one of two reasons: because a variable with a possibly empty sort constraint is disappearing or because a sort consistent rather than a well sorted subsitution is being used. As has been remarked elsewhere (Cohn 1989) the explicit appearance of sort literals can have a potentially rather damaging effect on the search space. The first kind of literal is not worrysome in this regard because they always predicate variables and there is only one such literal per variable, and such literals predicating variables will only be present when that is the only occurrence of the variable in the clause. Thus the potentially bad computational effects are virtually non existent here since we do not have to worry about choosing an inappropriate instantiation for x when resolving $\neg\tau(x{:}\tau_1)$ away, because, since there are no other occurrences of x, any instantiation will do.

On the other hand, sort literals introduced because of substitutions being sort consistent rather than well sorted are less welcome, but are necessary and the search space is no worse than would have been the case in unsorted logic, as argued in (Cohn 1989). As mentioned above, we cannot use weakening in this logic because of the possibility of losing completeness because Π could actually be used to prove the substitution 'well sorted'. However, we could add a further simplification rule to the logic which could exploit additional information in Σ. For example if $\Sigma = \{(S1(x) \wedge S2(x)) \rightarrow S3(x)\}$ and we are trying to resolve $P(x{:}S1)$ with $\neg P(y{:}S2)$, then the subsitution under consideration is $\langle x{:}S1/y{:}S2\rangle$, which is sort consistent but not well sorted. Thus we must add the literal $\neg S2(x{:}S1)$ to the resolvent. We cannot just weaken x to S3 because it might be the case that from Π we can prove $S1(x) \wedge S2(x) \rightarrow S4(x)$ and S3 and S4 might be disjoint. However if Σ actually contained the further information that $S3(x) \rightarrow (S1(x) \wedge S2(x))$ then we could avoid adding the prosthetic literal and simply change the sort annotation on the variable. Such a rule would allow one to infer Φ' from $\Phi \vee L$ where L is a sort literal predicating a variable $\alpha{:}\tau$ which occurs in Φ and $\Sigma \models \neg\tau'(\alpha) \leftrightarrow \hat{L}$ for some $\tau \in S^{PI}$, where Φ' is like Φ but with all occurrences of $\alpha{:}\tau$ replaced by $\alpha{:}\tau'$. However since this rule is not needed for completeness we will not prolong its discussion here.

As an example which shows the requirement for sort consistent substitutions rather just well sorted ones, consider the following axiomatisation:[14] $\Sigma = \{B(r) \vee D(r)\}$ and $\Pi = \{A(x{:}B), A(x{:}D), \neg A(r)\}$. $\Sigma \cup \Pi$ is unsatisifiable, but there are only two possible resolutions of clauses in Π, yielding the resolvents: $\neg B(r)$ and $\neg D(r)$. In both these cases the unifier required a substitution which was only sort consistent rather than well sorted because the sort of the variable for which r was substituted was not a supersort of the sort of r. This example also shows the need for an inference rule to allow $\neg B(r)$ and $\neg D(r)$ to

[14] This example is derived from one in Frisch (1989). Note that both B and D are possibly empty sorts in our formulation.

be resolved even though they are not complementary in the standard sense. The required rule is a form of Stickel's *Theory Resolution* (Stickel 1985). In order to keep the rule binary (i.e. considering just two clauses), we need *partial wide theory resolution.*[15]

Definition (Characteristic Resolution)

Let C_1 and C_2 be two clauses containing characteristic literals L_1 and L_2 respectively, and θ be a sort consistent substitution such that $\Sigma \models ((L_1 \wedge L_2)\theta \rightarrow \Phi)$ where Φ is a disjunction (possibly null) of sort literals, then

$C_3 = \Phi \vee ((C_1 \backslash L_1) \cup (C_2 \backslash L_2)) \downarrow_{\Gamma} \theta$ is a *binary characteristic resolvent* of C_1 and C_2,

where Γ is the set of variables occurring in $\{C_1 \cup C_2\}$.

The soundness of this rule follows straightforwardly from the argument above. A single clause may contain a characteristic literal which is false by virtue of the sort theory alone. In this case the literal could be deleted; for example, trivially both $\bot(\alpha)$ and $\neg\top(\alpha)$ may be deleted from any clause without affecting satisfiability. We will now formally define such a delete rule, which we call *evaluation.*

Definition (Evaluation)

If L is a sort literal and Φ_1 and Φ' are formulae, then $L \vee \Phi_1 \equiv \Phi'$ if $\Sigma \models \hat{L}' \vee \Phi_2$, where $L' = \neg L$ and Φ_2 is a (possibly null) disjunction of characteristic literals, and $\Phi' = ((\Phi_1 \cup \Phi_2) \downarrow_{\Gamma} \{\}$ where Γ is the set of variables occurring in L.

If Φ_2 is null, then we will call the rule *elementary evaluation* following a similar rule in Cohn (1987). If Φ_2 is not null, then we will call the rule *partial evaluation* (by analogy with with Stickel's partial theory resolution). The former always has a beneficial effect on the search space (because a literal is evaluated to false and can thus be deleted) and might be performed 'eagerly' whilst the latter may increase the search space and could be performed only when determined to be necessary.

The soundness of the rule is immediate from its form.

As examples of this rule, each of the following equivalences are valid under the conditions given.

$\neg\tau_1(\alpha{:}\tau_2) \equiv$ False if $\tau_2 \sqsubseteq \tau_1$.

$\neg\tau_1(x{:}\tau_2) \vee P(c) \equiv P(c)$ if $\tau_1 \sqcap \tau_2$ is provably (from the SBOX) non empty.

$\neg\tau_1(c) \equiv$ False if c is of sort τ_1.

$\tau_1(c) \equiv \tau_2(d)$ if $\Sigma = \{\tau_1(x) \rightarrow \tau_2(x)$.

In every case, whenever a literal is deleted which contains the last occurrence of a potentially empty sort, then another sort literal predicating the non emptiness of that sort must be added as discussed above in the case of standard resolution (this is ensured by the definition of Φ'). E.g. if $f(x{:}\tau_1)$ is of sort τ_2 and x is not in Φ, then $\neg\tau_2(f(x{:}\tau_1)) \vee \Phi$ can be simplified to Φ if $\tau_1 \in S^{NE}$ but to $\neg\tau_1(x{:}\tau_1) \vee \Phi$ if $\tau_1 \in S^{PE}$.

15 Actually, I believe that *total wide theory resolution* would be complete, but I specify partial wide theory resolution here since this may enable shorter proofs (which use fewer instances of *partial evaluation* – see below).

To see the necessity for partial evaluation rather than simply elementary evalutation consider the theory specified by Σ = {S1(a)∨S2(a), ¬S1(a)∨S2(a), S1(a)∨¬S2(a)} and Π = {¬S1(a)∨¬S2(a)}. No other rule apart from partial evaluation is applicable initially, but $\Sigma \cup \Pi$ is unsatisfiable.

Obviously, a rule in which sort literals can be evaluated to True can be formulated in an equivalent manner allowing the detection of clauses which are tautologous by virtue of sort information but we omit this definition here for reasons of space and the fact that tautology detection does not affect the completeness of refutation proof.

4. Related Work

The only previous resolution many sorted calculus with possibly empty sorts appears to be that of Weidenbach and Ohlbach (1990). Their language is very different to ours. Rather than having a separate sort theory or signature they effectively just have the ABOX, but with two distinguished predicates: '∈' and '⊑'. Sorts appear as constant symbols occurring as arguments to these two predicates and as constraints on variables. They point out that their logic can handle 'conditioned function declarations'; for example in their axiomatisation

1) $\forall x{:}x{:}$rat ∈ real (or equivalently: rat⊑real)
2) $\forall x{:}$rat,$y{:}$rat $y{:}$rat→div($x{:}$rat,$y{:}$rat) ∈ rat
3) a ∈ rat
4) b ∈ rat
5) b≠0

axiom (2) is a conditioned function declaration. Of course, since we allow sort literals in the ABOX, we too can handle conditioned function declarations, as can any similar logic such as LLAMA (Cohn 1987) or that of Beierle et al (1989). It is also interesting to note that in order to prove completeness, Weidenbach and Ohlbach have to impose a certain ordering on deductions which would appear to be an analogue of the enumeration restriction which appears in the completeness proof below.

Their calculus also has to add sort literals in certain circumstances to inferred clauses, but in their case they always add a literal of the form $\alpha \notin \tau$ whenever a variable of sort τ is instantiated to a non variable term α, whereas we have the possibility of determining that the substitution is ill sorted which will restrict the search space.

To sum up the difference between the two logics, Weidenbach and Ohlbach's logic is homogenous and not *hybrid* (in the sense discussed in Frisch and Cohn (1990)). This division does mean that we require additional inference rules to ensure the necessary communicatin between the ABOX and SBOX: in particular we require the evaluation rule of inference. However, because we distinguish between well sorted, sort consistent and ill sorted substitutions, we have the possibility of reducing the search space, by not inserting prosthetic literals unless really necessary and forbidding some inferences (which use ill sorted substitutions) altogether. In fairness it must be pointed out that the computation of well sortedness and sort consistency must be taken into account when making the comparison. However, we believe that much can be gained, both computationally (because specialised inference engines can be used for specialised theories and because of reductions in the search space caused by partitioning, as observed by Abrams and Frisch (1989) even if a standard inference engine is used for the specialised theory), and from a pragmatic viewpoint (because knowledge is better structured if it can be separated into separate 'chunks' with well defined interactions).

Gent (1991) presents a first order framework with restricted quantification and a tableau based proof system for it. This framework can implement a sorted logic with possibly empty sorts, but since Gent has full first order logic (with existential quantifiers), he does not need to include sort literals in Π explicitly: the existential quantifier elimination rule effectively adds new literals to Σ which can be used for proving non emptiness dynamically. Thus his logic is much simpler since he does not need a rule corresponding to evaluation nor does he have to worry about variables constrained by possibly empty sorts 'disappearing' as we have to. His framework also applies to modal logics and will handle non serial modal logics which is the modal equivalent of a many sorted logic with possibly empty sorts. Scherl (1991) and Ohlbach (1988) also present logics for non serial modal logics.

Finally, it is worth pointing out that this work is related to that presented in Cohn (1992). There, we developed a way of relaxing the strict complete Boolean lattice required by the many sorted logic LLAMA (Cohn 1987) by embedding a partially specified sort hierarchy in a complete Boolean lattice where some sorts could be possibly empty in the original interpretation; however the user of the logic had no direct access to these possibly empty sorts as is possible in the logic here.

5. Completeness of the Resolution Calculus

In this section we sketch a completeness proof for the resolution based calculus outlined above. Our proof is founded on the semantic tree based proof of Kowalski and Hayes (1969).

The notions of a Herbrand universe **H**, Herbrand base $\tilde{\mathbf{H}}$ and Herbrand interpretation are all standard. Note that the Herbrand base contains all ground instances of literals in $\Sigma \cup \Pi$, whether well sorted, sort consistent or ill sorted.

The notion of a semantic tree differs somewhat. We have an enumeration restriction (essentially the same as that to be found in Cohn (1983)) which ensures that if a literal L contains a term t, then any sort literal predicating t occurs as an ancestor of the node labelled by t and not as a descendant.

We say that a clause C *fails* at N iff $\vec{N} \models_{\Sigma} \neg C$, where $\vec{N}$ denotes the set of labels from the root of the semantic tree upto and including N. We say that $C \in \Sigma \cup \Pi$ *fails* at N (a *failure point*) iff C is some ground instance of C whose literals are all negated by some node at N or above; i.e. $\exists \theta \exists C' \in C{\downarrow}\theta$ s.t. $\overline{C'} \subseteq \vec{N}$, where $\overline{C'} = \{\neg\alpha \mid \alpha \in C'\}$. Note that the ground instance may contain prosthetic literals but these must all be negated as well. A tree is *closed* if every branch contains a failure point. A node N with daughters D1 and D2 is an *inference node* iff D1 is failed by some clause in Π and either $\Sigma \models \neg\vec{D2}$ or D2 is also failed by some clause in Π. Thus an inference node has one daughter failed by a clause in the ABOX and the other daughter is either also failed by some clause in the ABOX or is inconsistent with the SBOX. The appropriate version of Herbrand's theorem is easy to prove.

To see the need for the enumeration restriction and to illustrate the various definitions above consider the following axiomatisation; let Π be the following set of clauses:

1) $\neg$ P(x:S1) $\vee$ Q
2) P(x:S1)
3) S1(a)

4) $\neg$Q

and let Σ just be

5) $\top(x)$.

6) $\neg\bot(x)$.

The Herbrand base is {P(a), S1(a), Q, $\top$(a), $\bot$(a)}. Consider the two semantic trees depicted in figures 2 and 3.

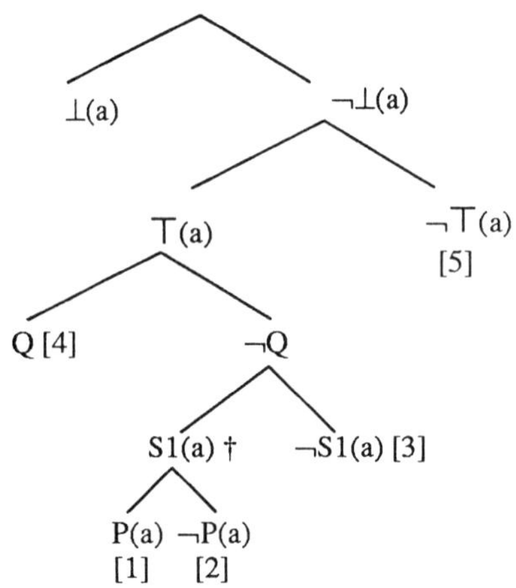

Figure 2: A Semantic Tree obeying the enumeration restriction.

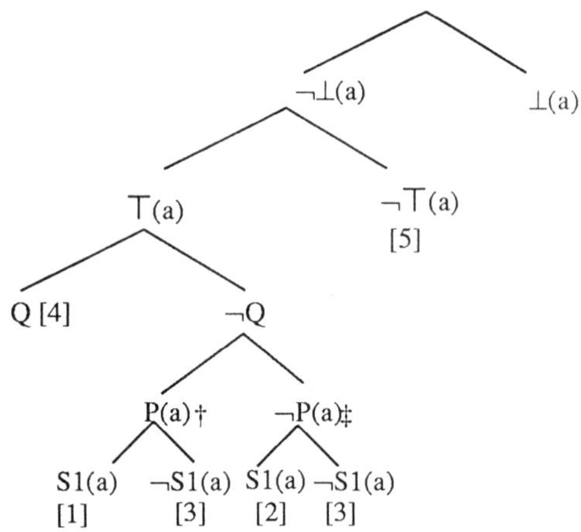

Figure 3: A Semantic Tree not obeying the enumeration restriction.

The nodes where each of the clauses (1-6) fail are marked within square brackets. In the first tree there is one inference node (marked with †) which has two daughters where (1) and (2) fail. The resolvent of (1) and (2) is $\neg$S1(x:S1)$\vee$Q which fails at the inference node in figure 2. However in figure 3 (which violates the enumeration restriction) there are two inference nodes (marked † and ‡). Unfortunately there is no inference possible from either (1) and (3) or (2) and (3) i.e. from the clauses failing just below † and ‡. Thus we see the need for the enumeration restriction.

We can now turn to the completeness proof itself. The crux of a semantic tree based completeness proof relies on showing that given an unsatisfiable clause set S not containing the empty clause, then an inference can be constructed at any inference node such that the inferred clause fails at the inference node (an inference node is bound to exist since S is unsatisfiable and does not contain the empty clause). Let N be an inference node. The proof now is a case analysis depending on the descendants of N.

1) N has two daughters labelled by L and $\neg L$, which by definition are failure points for clauses $C_1 \in \Pi$ and $C_2 \in \Pi$ respectively.

1A) Both daughters are non characteristic literals. Since C_1 and C_2 don't fail at N, they contain literals L_1 and L_2 such that $\neg L$ is an instance of L_1 and L is an instance of L_2. Thus C_1 and C_2 (or any factors of C_1 and C_2) resolve on L_1 and L_2 to produce some clause C. To show that C fails at N is standard except for the possiblity of C containing newly introduced characteristic literals because of the restriction on eliminating variables annotated by possibly empty sorts, or because of literals introduced by sort consistent substitutions. However, in the latter case clearly the definition of failure point ensures that such literals are failed by $\vec{N}$ and the former case reduces to the latter since a ground substitution for a variable constrained by a possibly empty sort will not be well sorted and will thus produce a prosthetic literal.

1B) Both the daughters are characteristic literals. In order to show that an inference (a characteristic resolution) is possible we need to show that $\neg L$ is an instance of a literal which actually occurs in C_1 (ie $\neg L$ is not a prosthetic literal in the ground instance of C_1 which is falsified by the label of a daughter of N). Assume the contrary: i.e. $\neg L$ is not an instance of a literal in C_1, but then some literal L' in C_1 contains a variable $\alpha{:}\tau$ where τ is the predicate symbol of L. $\neg L$ must be a characteristic literal since otherwise the enumeration restriction would force L' to occur lower down the semantic tree than L and thus C_1 would not fail here. However, we can reorder the levels of the semantic tree (whilst still obeying the enumeration restriction) so that the level corresponding to L' comes just below that of L. Consider any clause which failed at the level associated with L or above: if its failure was independent of the L' level, then it will still fail at the same node; otherwise it will now fail two levels below N. However, providing we always work on the highest level inference node, no inference nodes wil be moved down the semantic tree by this reordering (indeed some might be moved up), so such a reordering cannot take place infinitely many times, so we can be sure that we will still 'make progress'. This thus shows that we can ensure that $\neg L$ is indeed an instance of some literal in C_1.

The proof now is very similar to (1A) except that the literals in C_1 and C_2 which are not falsified at N are not necessarily of the form $\alpha(...)$ and $\neg\alpha(...)$. However, it is easy to see that $\Sigma \models (L \rightarrow \neg L_1)$ and $\Sigma \models (L \rightarrow \neg L_2)$ and therefore that a characteristic resolution of C_1 and C_2 is possible and that this fails at N.

2) N has two daughters D1 and D2, labelled by characteristic literals L and $\neg L$ and one of these daughters (say, D1) is failure point for some clause in Σ but no clause in Π (they cannot both be failure points for clauses in Σ, since Σ is guarantee be consistent because of the definition of inference node). An analagous argument to (1B) shows that $\neg L$ and L are both instances of actual literals occurring in the clauses failing at D1 or D2. However it is easy to see that the clause in Π failing at D2 can be evaluated and that the inferred clause fails at N or above.

6. Final Comments

As Weidenbach and Ohlbach (1990) point out, the transformation to clausal form in a logic with possibly empty sorts is non standard and they give an appropriately revised algorithm.

It might appear that coping with possibly empty sorts is very easy – as deductions are performed just collect all the sorts which must be non empty to preserve soundness and check for non emptyness at the end of the proof. This is unsatisfactory for two reasons. Firstly, doing all the checking at the end may be a bad idea computationally because if a sort is empty (and this might be very easy to show) then the current branch of the search space could be pruned earlier rather than later. Secondly, if sort literals occur in A-formulae along with non sort literals, so proving the non emptyness of a sort may involve arbitrary reasoning, so it is not 'done all at the end' after all. So, our concern here to build a many sorted logic with the apparatus required to handle possibly empty sorts fully integrated into the deductive machinery (and the semantics) is justified. Although the actual trick to extend resolution to handle possibly empty sorts is simple, the requirement to allow characteristic literals into the ABOX adds considerable complexity and care is required to ensure completeness. However we believe the distinction between definitely empty, possibly empty and non empty sorts, and similarly between ill sorted, sort consistent and well sorted substitutions are distinctions worth making because they can be compuationally exploited.

In order to be general we have made very few assumptions or restrictions on the SBOX. One assumption we have made is that it is consistent. If no restriction on the appearance of function symbols in the SBOX is made then deciding consistency is not possible in general of course. However restrictions can be placed on the language[16] of the SBOX to ensure that it is consistent or to ensure that its consistency can be decided. For example if the SBOX only contains definite clauses then it is consistent; however in the case since it contains no negative information, so all substitutions are sort consistent so much of the utility of a many sorted logic is lost in this case. Alternatively, we might insist that any SBOX formula containing a function symbol is a definite clause, although other wffs (where all terms are universally quantified variables) need not be definite clauses. In this case it is clear that the SBOX's consistency is decidable (providing it is finite of course) but not all substitutions need be sort consistent. Such a language would allow negative information about the sort hierarchy to be expressed, but not about any function symbols.

Acknowledgements

I wish to thank everyone with whom I have ever discussed many sorted logic, and Alan Frisch in particular. The anonymous comments of several referees are also gratefully acknowledged. The financial assistance of the SERC under grants GR/G/36852 and GR/G/24231 is gratefully acknowledged.

References

A M Frisch and A G Cohn, "Thoughts and Afterthoughts on the 1988 AAAI Workshop on Hybrid Reasoning," *AI Magazine*, vol. 11 (5), pp. 77-83 , 1990.

C Abrams and A Frisch, *An examination of the increased efficiency of sorted deduction*, University of Illinois at Urbana Campaign, 1991.

C Beierle, U Hedstueck, U Pletat, and J Siekmann, "An Order Sorted Predicate Logic with Closely Coupled Taxonomic Information," LILOG Report 86, IBM Germany, Stuttgart, 1989.

A G Cohn, "Mechanising a Particularly Expressive Many Sorted Logic," PhD Thesis, University of Essex, 1983.

A G Cohn, "On the Solution of Schubert's Steamroller in Many Sorted Logic," *Proc IJCAI 9*, pp. 1169-1174, Morgan Kaufmann, Los Altos, 1985.

[16] There are of course other reasons why we might want to restrict the expressive power of the SBOX. In particular we would expect that reasoning in the SBOX is somehow efficient; however if the language of the SBOX is not restricted then any first order problem can be encoded by resort to the SBOX oracle!

A G Cohn, ''A More Expressive Formulation of Many Sorted Logic,'' *J Automated Reasoning*, vol. 3, no. 2, pp. 113-200, 1987.

A G Cohn, ''On the Appearance of Sortal Literals: a Non Substitutional Framework for Hybrid Reasoning,'' in *Principles of Representation and Reasoning*, ed. R J Brachman, H J Levesque & R Reiter, Morgan Kaufmann, Los Altos, 1989.

A G Cohn, ''Completing Sort Hierarchies,'' *Computers and Mathematics with Applications*, vol. 23, pp. 477-491, 6-9, 1992.

A M Frisch, ''A General Framework for Sorted Deduction: Fundamental Results for Hybrid Reasoning,'' in *Principles of Representation and Reasoning*, ed. R Brachman, H Levesque & R Reiter, Morgan Kaufmann, Los Altos, 1989.

I P Gent, *Analytic Proof Systems for Classical and Modal Logics of Restricted Quantification.*, University of Warwick, 1992.

P J Hayes and R Kowalski, ''Semantic Trees in Automatic Theorem Proving,'' in *Machine Intelligence 4*, Edinburgh University Press, 1969.

R Kowalski and P J Hayes, ''Lecture Notes on Automatic Theorem Proving,'' DCL Memo 40, University of Edinburgh, 1971.

H J Ohlbach, ''A resolution calculus for modal logic,'' in *Proc CADE 9, LCNS 310*, Springer Verlag, Berlin, 1988.

R B Scherl, ''Hybrid Reasoning and Modal Logic Theorem Proving,'' in *Presented at the AAAI Fall Symposium*, Monterey, 1991.

M E Stickel, ''Automated Deduction by Theory Resolution,'' *J Automated Reasoning*, vol. 1, pp. 333-355, Kluwer, 1985.

C Walther, ''A Mechanical Solution of Schubert's Steamroller by Many-Sorted Resolution,'' *Artificial Intelligence*, vol. 26, pp. 217-224, 1985.

C Walther, *A Many Sorted Calculus Based on Resolution and Paramodulation,* Pitman, London, 1987.

C Weidenbach and H J Ohlbach, ''A Resolution Calculus with Dynamic Sort Structures and Partial Functions,'' in *Proc ECAI*, ed. L C Aiello, pp. 688-693, Pitman, London, 1990.

Theorem proving in non-standard logics based on the inverse method

Andrei Voronkov[1]

ECRC, Arabellastr.17, 8000 Munich 81, Germany (voronkov@ecrc.de)*

Abstract. We present a general approach to constructing efficient theorem provers which is applicable to classical and some non-classical logics. To define machine-oriented versions of these logics we use Maslov's inverse method. To make the proof-search more efficient one needs to use strategies for reducing search space. Instead of studying concrete systems like the classical predicate calculus we first present an abstract definition of a deductive system and then prove completeness of several strategies for deductive systems. Afterwards we project these theorems from the abstract level of deductive systems to concrete systems, obtaining efficient strategies for them.

1 Introduction

This paper has two main aims. The first is to show how to construct machine-oriented calculi for a wide class of logics based on the ideas of Maslov's inverse method [Mas67, Mas87]. The second is to give a general way to characterize strategies which reduce search space for the machine-oriented calculi.

The title of this paper is somehow misleading because our approach is applicable to classical logic as well. As examples of application of the general techniques from our paper we consider classical logic **Cl** and modal logic **S4**.

In Section 2 we outline the general ideas of the inverse method and construct machine-oriented calculi for classical predicate logic and for **S4**. However the naive construction gives us inefficient calculi which have many redundancies. To make the proof search more efficient one needs to introduce strategies for reducing the search space. As we show, it is possible to introduce strategies which are similar to the subsumption strategy and the tautology deletion strategy of the resolution method, as well as some other strategies.

To this end in Section 3 we define an abstract notion of deductive systems. This notion intuitively captures the general idea of what a theorem proving system is. Then we introduce several strategies for the deductive systems and prove their completeness in the general case.

In Section 4 we show how these abstract definitions can be projected to particular logical systems thus giving efficient machine-oriented versions for them. We also briefly sketch some aspects of the implementation of the strategies in the theorem proving system LISS [Vor90a]. The techniques used here is quite abstract and applicable to many particular systems. So during the presentation we shall sometimes skip from a more concrete level of particular logical systems to a more abstract level of general notions and back.

* On leave from the International Laboratory of Intelligent Systems (SINTEL), 630090, Universitetski Prospect 4, Novosibirsk 90, Russia.

2 Maslov's Inverse Method

Usually proof-search in theorem proving systems is organized either in the form of backward reasoning or in the form of forward reasoning. The advantages and disadvantages of both approaches are well known. Usually backward reasoning is used because it is more goal-directed. Maslov's inverse method uses a version of forward reasoning strategy which is however more goal-oriented than the usual forward reasoning methods. In this section we will informally explain the main idea of the inverse method and apply this method to two particular logical systems: the classical predicate calculus and modal logic **S4**.

The first publications on Maslov's inverse method refer to 1964. Later some of Maslov's paper were published in a more accessible literature (see e.g. [Mas67, Mas69, Mas87]). Lifschitz, Zamov and Mints made another steps towards the explanation of the ideas of the inverse method [Lif86, Zam89, Min90]

The main idea of the inverse method is the following. Suppose that we have a logical system consisting of axioms **Ax** and inference rules **IR**. Suppose then that we want to find a proof of a (goal) formula γ. The (naive) backward procedure starts from γ and tries to apply all possible rules that can be used to prove it, reducing this goal to subgoals. Then this procedure applies to subgoals until (in case of success) we come to axioms.

A normal forward reasoning procedure uses a different approach: starting from axioms it generates new formulas (or sequents) until the goal is reached.

It is well known that the main disadvantage of the forward reasoning that it is not goal-directed. Suppose however that we have a criterion which allows us to select only some relevant initial axioms $\mathbf{Ax}_\gamma$ and inference rules $\mathbf{IR}_\gamma$ which could be used to prove the goal γ. In this case we can specialize the forward procedure to make it a procedure oriented to prove only γ.

It allows one to exploit all properties of a forward reasoning procedure but to make it more goal oriented at the same time. This is the key idea of the inverse method.

One of the problems in applying the inverse method is that of specializing the calculus. For example if we take cut-free sequent calculi for classical predicate logic or for **S4** which are quite friendly for the backward proof search, then the problem is the infiniteness of the sets of axioms and of inference rules.

However it is possible to make the specialized system finite using the notions of unification and the most general unifier. Let us consider as an example the classical predicate calculus **Cl**.

Below $\Gamma, \Delta, \Sigma, \Xi$ are *sets* of formulas. Sequent is an expression of the form $\Gamma \to \Delta$. In all calculi from this paper provable objects are sequents.

System Cl

- Axioms:

$$\varphi \to \varphi$$

- Rules:

$$(\to\wedge)\quad \frac{\Gamma \to \Delta, \varphi \qquad \Sigma \to \Xi, \psi}{\Gamma, \Sigma \to \Delta, \Xi, \varphi\wedge\psi}$$

$$(\wedge\to)\quad \frac{\varphi, \Gamma \to \Delta}{\varphi\wedge\psi, \Gamma \to \Delta} \qquad \frac{\psi, \Gamma \to \Delta}{\varphi\wedge\psi, \Gamma \to \Delta}$$

$$(\rightarrow \vee) \quad \frac{\Gamma \rightarrow \Delta, \varphi}{\Gamma \rightarrow \Delta, \varphi \vee \psi} \qquad \frac{\Gamma \rightarrow \Delta, \psi}{\Gamma \rightarrow \Delta, \varphi \vee \psi}$$

$$(\vee \rightarrow) \quad \frac{\varphi, \Gamma \rightarrow \Delta \qquad \psi, \Sigma \rightarrow \Xi}{\varphi \vee \psi, \Gamma, \Sigma \rightarrow \Delta, \Xi}$$

$$(\rightarrow \supset) \quad \frac{\Gamma \rightarrow \Delta, \psi}{\Gamma \rightarrow \Delta, \varphi \supset \psi} \qquad \frac{\varphi, \Gamma \rightarrow \Delta}{\Gamma \rightarrow \Delta, \varphi \supset \psi}$$

$$(\supset \rightarrow) \quad \frac{\Gamma \rightarrow \Delta, \varphi \qquad \psi, \Sigma \rightarrow \Xi}{\varphi \supset \psi, \Gamma, \Sigma \rightarrow \Delta, \Xi}$$

$$(\neg \rightarrow) \quad \frac{\Gamma \rightarrow \Delta, \varphi}{\neg \varphi, \Gamma \rightarrow \Delta} \qquad (\rightarrow \neg) \quad \frac{\varphi, \Gamma \rightarrow \Delta}{\Gamma \rightarrow \Delta, \neg \varphi}$$

$$(\rightarrow \forall) \quad \frac{\Gamma \rightarrow \Delta, \varphi}{\Gamma \rightarrow \Delta, \forall x \varphi} \qquad (\forall \rightarrow) \quad \frac{\varphi[x \leftarrow t], \Gamma \rightarrow \Delta}{\forall x \varphi, \Gamma \rightarrow \Delta}$$

$$(\rightarrow \exists) \quad \frac{\Gamma \rightarrow \Delta, \varphi[x \leftarrow t]}{\Gamma \rightarrow \Delta, \exists x \varphi} \qquad (\exists \rightarrow) \quad \frac{\varphi, \Gamma \rightarrow \Delta}{\exists x \varphi, \Gamma \rightarrow \Delta}$$

Note: In the rules $(\rightarrow \forall)$ and $(\exists \rightarrow)$ the variable x has no free occurences in Γ, Δ.

To make the system finite one needs to make finite the set of axioms. To this end we can use the subformula property of the classical predicate calculus: all formulas occuring in a cut-free proof of a formula φ are subformulas of φ. If φ contains quantifiers then the number of its subformulas is infinite, but we can consider only the most general axioms using the notion of unification. To make this idea precise we shall introduce some notions.

The notions of (<u>free</u>) (<u>immediate</u>) <u>subformula</u> are defined as in [Wal90].

1. φ is an immediate (and a free immediate) subformula of $\neg\varphi$, both φ and ψ are immediate (and free immediate) subformulas of $\varphi \wedge \psi$, $\varphi \supset \psi$ and $\varphi \vee \psi$.
2. For any term t, the formula $\varphi[x \leftarrow t]$ is an immediate subformula and φ is the free immediate subformula of $\forall x \varphi$ and $\exists x \varphi$.
3. If φ is a (free) immediate subformula of ψ or identical to ψ, then φ is a (free) subformula of ψ.
4. If φ is a (free) subformula of ψ and ψ is a (free) subformula of χ then φ is a (free) subformula of ψ.

   For modal logics we add to this definition the following:
5. φ is the free immediate subformula of the formulas $\Box\varphi$ and $\Diamond\varphi$.

We make two agreements to be held throughout the paper:

From now on we assume that γ is a fixed ground formula to be proved.
In what follows "subformula" means "subformula with the fixed occurrence". (In other words, we consider different occurences of the same subformula as different subformulas.)

The notions of negative and positive occurrence of a subformula to a formula are defined as usual. To stress that a subformula φ occurs in the formula γ positively (negatively) we shall use the notation φ^+ (φ^-). The notion of a (most general) unifier is defined as usual. Let E_1, E_2 be terms or formulas, θ_1, θ_2 substitutions. Then the pair $\langle\theta_1, \theta_2\rangle$ is called r-unifier of E_1, E_2 iff $E_1\theta_1$ is identical to $E_2\theta_2$. Such a unifier is a most general r-unifier if for any r-unifier $\langle\theta_1', \theta_2'\rangle$ of E_1, E_2 there is a substitution θ such that $\theta_i'\theta$ is equal to θ_i, $i = 1, 2$.

Now we are ready to define the specialized system $\mathbf{Cl}_\gamma$. The set of sequents of $\mathbf{Cl}_\gamma$ consists of sequents of the form $\varphi_1, \ldots, \varphi_n \rightarrow \psi_1, \ldots, \psi_m$ such that φ_i are positive and ψ_j are negative subformulas of γ. It is easy to see that any subformula of γ can be obtained from a free subformula by applying some substitution. For this reason in the formulation of $\mathbf{Cl}_\gamma$ below we shall represent formulas of $\mathbf{Cl}_\gamma$ in the form $\varphi\theta$ where φ is a free subformula of γ, θ a substitution[2].

System $\mathbf{Cl}_\gamma$.

In all the rules given below φ, ψ are free subformulas of γ.

- Axioms:

$$\varphi_1\theta_1 \rightarrow \varphi_2\theta_2$$

where φ_1 is a negative and φ_2 a positive free subformulas of γ, $\langle\theta_1, \theta_2\rangle$ the most general r-unifier of φ_1, φ_2.[3]

- One-premise rules[4]:

$$(\wedge\rightarrow)\quad \frac{\varphi[\bar{x}\leftarrow\bar{t}], \Gamma \rightarrow \Delta}{(\varphi\wedge\psi)[\bar{x}\leftarrow\bar{t}], \Gamma \rightarrow \Delta} \qquad \frac{\psi[\bar{y}\leftarrow\bar{s}], \Gamma \rightarrow \Delta}{(\varphi\wedge\psi)[\bar{y}\leftarrow\bar{s}], \Gamma \rightarrow \Delta}$$

$$(\rightarrow\vee)\quad \frac{\Gamma \rightarrow \Delta, \varphi[\bar{x}\leftarrow\bar{t}]}{\Gamma \rightarrow \Delta, (\varphi\vee\psi)[\bar{x}\leftarrow\bar{t}]} \qquad \frac{\Gamma \rightarrow \Delta, \psi[\bar{y}\leftarrow\bar{s}]}{\Gamma \rightarrow \Delta, (\varphi\vee\psi)[\bar{y}\leftarrow\bar{s}]}$$

$$(\rightarrow\supset)\quad \frac{\Gamma \rightarrow \Delta, \psi[\bar{y}\leftarrow\bar{s}]}{\Gamma \rightarrow \Delta, (\varphi\supset\psi)[\bar{y}\leftarrow\bar{s}]} \qquad \frac{\varphi[\bar{x}\leftarrow\bar{t}], \Gamma \rightarrow \Delta}{\Gamma \rightarrow \Delta, (\varphi\supset\psi)[\bar{x}\leftarrow\bar{t}]}$$

$$(\neg\rightarrow)\quad \frac{\Gamma \rightarrow \Delta, \varphi[\bar{x}\leftarrow\bar{t}]}{\neg\varphi[\bar{x}\leftarrow\bar{t}], \Gamma \rightarrow \Delta} \qquad (\rightarrow\neg)\quad \frac{\varphi[\bar{x}\leftarrow\bar{t}], \Gamma \rightarrow \Delta}{\Gamma \rightarrow \Delta, \neg\varphi[\bar{x}\leftarrow\bar{t}]}$$

Note: in all one-premise rules φ, ψ are free subformulas of γ, $\bar{x}$ are all free variables of φ, $\bar{y}$ are all variables of ψ.

- Quantifier rules:

$$(\rightarrow\forall)\quad \frac{\Gamma \rightarrow \Delta, \varphi[\bar{x}\leftarrow\bar{t}, y\leftarrow v]}{\Gamma \rightarrow \Delta, \forall y\varphi[\bar{x}\leftarrow\bar{t}]} \qquad (\forall\rightarrow)\quad \frac{\varphi[\bar{x}\leftarrow\bar{t}, y\leftarrow s], \Gamma \rightarrow \Delta}{\forall y\varphi[\bar{x}\leftarrow\bar{t}], \Gamma \rightarrow \Delta}$$

$$(\rightarrow\exists)\quad \frac{\Gamma \rightarrow \Delta, \varphi[\bar{x}\leftarrow\bar{t}, y\leftarrow s]}{\Gamma \rightarrow \Delta, \exists y\varphi[\bar{x}\leftarrow\bar{t}]} \qquad (\exists\rightarrow)\quad \frac{\varphi[\bar{x}\leftarrow\bar{t}, y\leftarrow v], \Gamma \rightarrow \Delta}{\exists y\varphi[\bar{x}\leftarrow\bar{t}], \Gamma \rightarrow \Delta}$$

---

[2] Another reason for such representation is that it is actually used in the implementation [Vor90a].

[3] It is sufficient to allow only atomic formulae in the axioms.

[4] We shall freely use the overbar notion for the tuples — whenever $\bar{x}$ is met in the text it denotes the tuple $x_1, \ldots, x_n$.

Note: In all quantifier rules φ, ψ are free subformulas of γ, $\bar{x}, y$ are all free variables of φ and the variable v has no free occurences in $\Gamma, \Delta, \varphi[\bar{x}\leftarrow\bar{t}]$.

- Two-premise-rules:

$$(\rightarrow\wedge)\quad \frac{\Gamma\rightarrow\Delta,\varphi[\bar{x}\leftarrow\bar{t}] \qquad \Sigma\rightarrow\Xi,\psi[\bar{x}\leftarrow\bar{s}]}{\Gamma\theta,\Sigma\theta\rightarrow\Delta\theta,\Xi\theta,(\varphi\wedge\psi)[\bar{x}\leftarrow\bar{t}\theta]}$$

$$(\vee\rightarrow)\quad \frac{\varphi[\bar{x}\leftarrow\bar{t}],\Gamma\rightarrow\Delta \qquad \psi[\bar{x}\leftarrow\bar{s}],\Sigma\rightarrow\Xi}{(\varphi\vee\psi)[\bar{x}\leftarrow\bar{t}\theta],\Gamma\theta,\Sigma\theta\rightarrow\Delta\theta,\Xi\theta}$$

$$(\supset\rightarrow)\quad \frac{\Gamma\rightarrow\Delta,\varphi[\bar{x}\leftarrow\bar{t}] \qquad \psi[\bar{x}\leftarrow\bar{s}],\Sigma\rightarrow\Xi}{(\varphi\supset\psi)[\bar{x}\leftarrow\bar{t}\theta],\Gamma\theta,\Sigma\theta\rightarrow\Delta\theta,\Xi\theta}$$

Note: in all two-premise rules φ, ψ are free subformulas of γ, $\bar{x}$ are all free variables of φ, ψ, θ is a most general unifier of $\bar{s}$ and $\bar{t}$.

- Factoring rules:

$$(f\rightarrow)\quad \frac{\varphi[\bar{x}\leftarrow\bar{t}],\varphi[\bar{x}\leftarrow\bar{s}],\Gamma\rightarrow\Delta}{\varphi[\bar{x}\leftarrow\bar{t}\theta],\Gamma\theta\rightarrow\Delta\theta} \qquad (\rightarrow f)\quad \frac{\Gamma\rightarrow\Delta,\varphi[\bar{x}\leftarrow\bar{t}],\varphi[\bar{x}\leftarrow\bar{s}]}{\Gamma\theta\rightarrow\Delta\theta,\varphi[\bar{x}\leftarrow\bar{t}\theta]}$$

Note: in the factoring rules φ is a free subformula of γ, $\bar{x}$ are all free variables of φ, θ is a most general unifier of $\bar{s}$ and $\bar{t}$.

Factoring rules are needed for capturing contraction rules. Without them the system is incomplete. The techniques for proving completeness may be found in [Vor90b]:

Theorem 1. *Let γ be a ground formula. Then the sequent $\rightarrow\gamma$ is provable in* **Cl** *iff it is provable in* $\mathbf{Cl}_\gamma$. □

The main difference between the system $\mathbf{Cl}_\gamma$ and the original system **Cl** is that instead of arbitrary substitutions the most general unifiers are used. This makes the system finite in the following sense:

1. The set of axioms of $\mathbf{Cl}_\gamma$ is finite up to renaming variables.
2. For each sequent (pair of sequents) of $\mathbf{Cl}_\gamma$ there is only a finite number of rules applicable to this sequent (these pair of sequents). For each applied rule the conclusion is unique up to renaming variables.

However in spite of the finiteness the system $\mathbf{Cl}_\gamma$ has a lot of redundancies which will be studied later.

Another example of the specialization of a system is that of **S4**. The set of sequents of **S4** is defined similar to **Cl**. The axioms and rules of **S4** are all axioms and rules of **Cl** plus the rules

$$(\rightarrow\Box)\quad \frac{\Box\Gamma\rightarrow\Diamond\Delta,\varphi}{\Box\Gamma\rightarrow\Diamond\Delta,\Box\varphi} \qquad (\Box\rightarrow)\quad \frac{\varphi,\Gamma\rightarrow\Delta}{\Box\varphi,\Gamma\rightarrow\Delta}$$

$$(\rightarrow\Diamond)\quad \frac{\Gamma\rightarrow\Delta,\varphi}{\Gamma\rightarrow\Delta,\Diamond\varphi} \qquad (\Diamond\rightarrow)\quad \frac{\varphi,\Box\Gamma\rightarrow\Diamond\Delta}{\Diamond\varphi,\Box\Gamma\rightarrow\Diamond\Delta}$$

where $\Box\Gamma=\{\Box\psi\mid\psi\in\Gamma\}$ and $\Diamond\Delta=\{\Diamond\psi\mid\psi\in\Delta\}$. The specialized system $\mathbf{S4}_\gamma$ has the same rules as $\mathbf{Cl}_\gamma$ plus the above four rules restricted to the sequents of $\mathbf{S4}_\gamma$.

The technique of specializing described above is applicable at least to the following calculi:

1. the classical predicate calculus;
2. the intuitionistic predicate calculus;
3. several logics without the contraction rule: the direct logic of Ketonen and Weychrauch [KW84], Grishin's logic without the contraction rule [Gri81], Mey's logic without the contraction rule [Mey89], linear logic [GLT89];
4. Some modal logics: **S4**, **K**.

It can be also extended to some other logics using the technique of hiding special axioms into the unification algorithm [OH91] but we shall not dwell on it here.

Not all the logics are so easy to handle as **S4** or **Cl**. To illustrate rules which are more difficult to specialize we give the example of the rule ($\vee\rightarrow$) of the intuitionistic sequent calculus **Int**. The difference between **Int** and **Cl** is that there can be at most one formula in the right part (succedent) of a sequent. This seemingly inessential difference produces the rule

$$\frac{\varphi[\bar{x}\leftarrow\bar{s}_1], \Gamma \rightarrow \chi[\bar{x}\leftarrow\bar{t}_1] \qquad \psi[\bar{x}\leftarrow\bar{s}_2], \Delta \rightarrow \chi[\bar{x}\leftarrow\bar{t}_2]}{\varphi\vee\psi[\bar{x}\leftarrow\bar{s}_1\theta], \Gamma\theta, \Delta\theta \rightarrow \chi[\bar{x}\leftarrow\bar{t}_1\theta]}$$

where θ is a most general unifier of $\bar{s}_1, \bar{t}_1$ with $\bar{s}_2, \bar{t}_2$. As one can see, we have to unify not only $\bar{s}_1$ with $\bar{s}_2$, as in the case of classical logic, but also $\bar{t}_1$ with $\bar{t}_2$.

3 Deductive Systems and Strategies

The next problem with an efficient implementation of the inverse method is to introduce strategies reducing the search space. But instead of looking for the strategies for concrete systems we shall introduce a generalization called deductive systems. Intuitively, deductive system is a set of axioms and inference rules operating on a set of clauses[5]. Some clauses are distinguished and called goals. Below we shall formulate a notion of subsumption ordering for the deductive systems which is later used for proving completeness of several strategies for the deductive systems in general.

In the next section we show how to develop strategies for classical logic and for **S4** based on the results from this section. Classical logic and **S4** were chosen only as examples. We do not directly consider here other interesting logical systems like the intuitionistic predicate calculus due to the lack of space. However the results of this section are applicable at least to all logics mentioned in Section 2.

Now we shall proceed to the formal definitions. This part of paper is rather technical but the intuition behind definitions is usually clear and in most of cases we omit explanations.

A triple $\langle \mathbf{C}, \mathbf{G}, \mathbf{R} \rangle$ is called a deductive system iff

1. $\mathbf{C}$ is a set;
2. $\mathbf{G} \subseteq \mathbf{C}$;
3. $\mathbf{R} \subseteq \mathbf{C}^\star \times \mathbf{C}$.

where $\mathbf{C}^\star$ denotes the set of all finite subsets of $\mathbf{C}$.

In what follows the elements of $\mathbf{C}$ will be called clauses, $\mathbf{G}$ is the subset of goal clauses, $\mathbf{R}$ the the set of inference rules of the deductive system. The rule $\langle\{c_1, \ldots, c_n\}, c\rangle$ will usually be written in the form $c_1, \ldots, c_n/c$.

Axiom is an inference rule with $n = 0$. We impose one restriction to deductive systems: if $c_1, \ldots, c_n/c \in \mathbf{R}$ then for every $i \in \{1, \ldots, n\}$, we have $c_i \notin \mathbf{G}$.[6]

[5] Clauses should not necessarily be as in the resolution method. They may be e.g. sequents.

[6] This condition is not essential, but it simplifies many proofs.

The notion of provability is defined as usual. A clause c is provable from the clauses $c_0, \ldots, c_m$ iff there exists a sequence $d_0, \ldots, d_n$ of clauses such that

1. $d_n = c$;
2. for every $i \in \{0, \ldots, n\}$ either $d_i \in \{c_0, \ldots, c_m\}$ or there is a rule $\bar{e}/d_i \in \mathbf{R}$ such that $\bar{e} \subseteq \{d_0, \ldots, d_{i-1}\}$.

Such a sequence $d_0, \ldots, d_n$ is called a proof of c from $c_0, \ldots, c_m$. The number $n+1$ is the length of the proof $d_0, \ldots, d_n$. We shall also say that c is provable from $c_1, \ldots, c_m$ in $n+1$ steps.

A clause c is provable (in n steps) iff it is provable (in n steps) from the empty set of clauses. A set of clauses is successful (in n steps) iff some goal clause $g \in \mathbf{G}$ is provable from this set (in n steps).

The following definition is related to the subsumption strategy of resolution (as well as that of the other methods defined below). Let $\prec$ be some partial ordering on $\mathbf{C}$. It is called a subsumption ordering iff the following two condition hold:

1. Let $R \in \mathbf{R}$ be an inference rule of the form $c_1, \ldots, c_n/c$ and $c_1 \prec c'_1, \ldots, c_n \prec c'_n$. Then either $c \prec c'_i$ for some $i \in \{1, \ldots, n\}$, or there is a rule $R' \in \mathbf{R}$ such that R' takes the form $\bar{e}/c'$ with $e \subseteq \{c'_1, \ldots, c'_n\}$ and $c \prec c'$;
2. If $c \in \mathbf{G}$ and $c \prec c'$ then $c' \in \mathbf{G}$.[7]

Let $\prec$ be a binary relation on clauses. We expand this relation to the sets of clauses in the following way: for $\bar{c}, \bar{c}'$ sets of clauses $\bar{c} \prec \bar{c}'$ iff for every $c \in \bar{c}$ there is a $c' \in \bar{c}'$ such that $c \prec c'$.

It is interesting that some known strategies of the resolution method are in a special way based on the notion of subsumption. It will be clear from the definitions and theorems given below.

In the following we shall use the following notation for provability: $c \vdash^n_{\mathbf{D}} c$ means that c is provable from $\bar{c}$ in n steps in the deductive system $\mathbf{D}$. We usually omit subscripts $\mathbf{D}$ and n when they are clear from the context or unimportant.

Theorem 2. *Let $\bar{c}$ be a successful set of clauses and $\bar{c} \prec \bar{c}'$. Then $\bar{c}'$ is successful. Moreover if for a goal clause g, $\bar{c} \vdash^n g$ then there is a goal clause g' such that $\bar{c} \vdash^m g'$ for some $m \leq n$.*

Proof. Let $d_0, \ldots, d_n$ be a proof of some goal clause c. We construct step by step sequences of clauses $\bar{e}_0, \ldots, \bar{e}_n$ such that for every $i \in \{0, \ldots, n\}$, $\{d_0, \ldots, d_i\} \prec \bar{e}_i$. Let $\bar{e}_{-1}$ be the empty sequence. We suppose that the sequence $\bar{e}_{i-1}$ is already constructed. Consider d_i. If $d_i \in \mathbf{C}$ then there exists a $d \in \bar{c}'$ such that $d_i \prec d$. In this case we let $\bar{e}_i = \bar{e}_{i-1}, d$. Otherwise by the definition of a proof there is $R \in \mathbf{R}$ of the form $d_j, \ldots, d_k/d_i$ with $j, \ldots, k < i$. Let $d'_j, \ldots, d'_k$ be elements of $\bar{e}_{i-1}$ with $d_j \prec d'_j, \ldots, d_k \prec d'_k$. Since $\prec$ is a subsumption ordering then either $d_i \prec d'_l$ for some $l \in \{j, \ldots, k\}$ or there is a rule $R' \in \mathbf{R}$ of the form $\bar{d}/d'_i$ such that $\bar{d} \subseteq \{d'_j, \ldots, d'_k\}$ and $d_i \prec d'_i$. In the former case we let $\bar{e} = \bar{e}_{i-1}$, in the latter $\bar{e}_i = \bar{e}_{i-1}, d'_i$.

It is easy to verify that $\{d_0, \ldots, d_i\} \prec \bar{e}_i$ and that $\bar{e}_i$ is a proof. Consider $\bar{e}_n$. Since $\{d_0, \ldots, d_n\} \prec \bar{e}_n$, then there is $e \in \bar{e}_n$ such that $d_n \prec e$. By the definition of subsumption orderings $e \in \mathbf{G}$. Deleting from $\bar{e}_n$ the part following e we obtain the needed proof of the goal clause e from $\bar{c}'$. □

[7] If $\prec$ is a subsumption ordering and $c \prec d$, then we say that d subsumes c. Sometimes in the literature the notation for "d subsumes c" is $d \leq c$, but we prefer the other way, since subsumption intuitively means that the subsuming clause is better than the subsumed one.

From this theorem the correctness and completeness of the subsumption strategies follows: one can delete from a set of clauses a clause subsumed by another clause from this set:

Theorem 3. *Let* $\prec$ *be a subsumption ordering,* $c_0 \prec c_1$. *Then* $\{c_0, \ldots, c_n\}$ *is successful iff* $\{c_1, \ldots, c_n\}$ *is successful.*

Proof. Immediately follows from Theorem 2 and from the fact that $\{c_0, \ldots, c_n\} \prec \{c_1, \ldots, c_n\}$ and $\{c_1, \ldots, c_n\} \prec \{c_0, \ldots, c_n\}$. □

Below we introduce some relationships between deductive systems called homomorphisms. If there is a homomorphism from a deductive system to another deductive system then to know whether some set of clauses is successful in the first system it is sufficient to know whether the image of this set under homomorphism is successful in the second system. If the second system is in some way simpler then the first one, it is reasonable to try to construct such a homomorphism. There may be examples of homomorphisms in logical calculi. For example, different refinements of the same calculus can be homomorphic.

This definition of homomorphism is somehow restrictive and we then introduce a modification of the notion of homomorphism called $\prec$-homomorphism, which is based on some subsumption ordering. $\prec$-homomorphisms have the same property as the usual ones: a set of clauses is successful iff the image of this set under homomorphism is successful. In Theorem 6 we investigate some particular example when such a homomorphism exists. Surprisingly this theorem is applicable to many logics. For example, in [Vor90b] we in fact introduced the $\prec$-homomorphic image of a variant of the classical predicate calculus. Examples of application of Theorem 6 are given in the next section.

Let $\mathbf{D}_0 = \langle \mathbf{C}_0, \mathbf{G}_0, \mathbf{R}_0 \rangle$ and $\mathbf{D}_1 = \langle \mathbf{C}_1, \mathbf{G}_1, \mathbf{R}_1 \rangle$ be two deductive systems. The provability in $\mathbf{D}_0$ and $\mathbf{D}_1$ will be denoted by $\vdash_0$ and $\vdash_1$ respectively. A homomorphism from $\mathbf{D}_0$ to $\mathbf{D}_1$ is any mapping $f : \mathbf{C}_0 \rightarrow \mathbf{C}_1$ such that

1. If $\bar{c} \vdash_0 d$ then $f(\bar{c}) \vdash_1 f(d)$;
2. If $f(\bar{c}) \vdash_1 f(d)$ then there exists $e \in \mathbf{C}_0$ with $f(e) = f(d)$ such that $\bar{c} \vdash_0 e$.
3. $\bar{c}$ is successful in $\mathbf{D}_0$ iff $f(\bar{c})$ is successful in $\mathbf{D}_1$.

Theorem 4. *Let* $\mathbf{D}_0 = \langle \mathbf{C}_0, \mathbf{G}_0, \mathbf{R}_0 \rangle$ *and* $\mathbf{D}_1 = \langle \mathbf{C}_1, \mathbf{G}_1, \mathbf{R}_1 \rangle$ *be two deductive systems. Let* $f : \mathbf{C}_0 \rightarrow \mathbf{C}_1$ *be such that* $\mathbf{G}_1 \subseteq f(\mathbf{C}_0)$. *Then* f *is a homomorphism iff the following conditions hold:*

1. If $\bar{c}/d \in \mathbf{R}_0$, *then* $f(\bar{c}) \vdash_1 f(d)$;
2. If $f(\bar{c})/f(d) \in \mathbf{R}_1$ *then for some* e *with* $f(e) = f(d)$, $\bar{c} \vdash_0 e$.
3. (a) If $g \in \mathbf{G}_0$ *then* $\{f(g)\}$ *is successful in* $\mathbf{D}_1$.
   (b) If $f(c) \in \mathbf{G}_1$ *then* $\{c\}$ *is successful in* $\mathbf{D}_0$.

The definition of a shortening homomorphism is similar to the definition of a homomorphism. A mapping f is a shortening homomorphism iff:

1. If $\bar{c} \vdash_0^n d$ then $f(\bar{c}) \vdash_1^m f(d)$ for some $m \leq n$.
2. If $f(\bar{c}) \vdash_1 f(d)$ then there exists $e \in \mathbf{C}_0$ with $f(e) = f(d)$ such that $\bar{c} \vdash_0 e$.
3. $\bar{c}$ is successful in $\mathbf{D}_0$ iff $f(\bar{c})$ is successful in $\mathbf{D}_1$, and if for a goal clause g we have $\bar{c} \vdash_0^n g$ then there is a goal clause $g' \in \mathbf{G}_1$ such that $f(\bar{c}) \vdash_1^m g'$ for some $m \leq n$.

As we noted above the notion of homomorphism is somewhat restricted and below we introduce a similar notion of $\prec$-homomorphism.

Let $\prec_0$ and $\prec_1$ be subsumption orderings for $\mathbf{D}_0$ and $\mathbf{D}_1$ respectively. A $\langle\prec_0,\prec_1\rangle$-homomorphism (or in the following simply $\prec$-homomorphism) from $\mathbf{D}_0$ to $\mathbf{D}_1$ is any mapping $f : \mathbf{C}_0 \to \mathbf{C}_1$ such that

1. If $\bar{c}\vdash_0 d$ then $f(\bar{c})\vdash_1 e$ for some $e\in\mathbf{C}_1$ such that $f(d)\prec_1 e$.
2. If $f(\bar{c})\vdash_1 f(d)$ then there exist $e,h\in\mathbf{C}_0$ such that $f(e)=f(d)$, $e\prec_0 h$, $\bar{c}\vdash_0 h$.
3. $\bar{c}$ is successful in $\mathbf{D}_0$ iff $f(\bar{c})$ is successful in $\mathbf{D}_1$.

There are characterizations of shortening homomorphisms and $\prec$-homomorphisms similar to that of Theorem 4, which are omitted here.

A shortening $\prec$-homomorphism from $\mathbf{D}_0$ to $\mathbf{D}_1$ is any mapping $f : \mathbf{C}_0 \to \mathbf{C}_1$ such that

1. If $\bar{c}\vdash_0^n d$ then $f(\bar{c})\vdash_1^m e$ for some $m \leq n$ and $e\in\mathbf{C}_1$ such that $f(d)\prec_1 e$.
2. If $f(\bar{c})\vdash_1 f(d)$ then there exist $e,h\in\mathbf{C}_0$ such that $f(e)=f(d)$, $e\prec_0 h$, $\bar{c}\vdash_0 h$.
3. $\bar{c}$ is successful in $\mathbf{D}_0$ iff $f(\bar{c})$ is successful in $\mathbf{D}_1$, and if for a goal clause $g\in\mathbf{G}_0$ we have $\bar{c}\vdash_0^n$ then there is a goal clause $g'\in\mathbf{G}_1$ such that $f(\bar{c})\vdash_1^m g$ for some $m \geq n$.

Theorem 5. *Let $\mathbf{D}_0 = \langle\mathbf{C}_0,\mathbf{G}_0,\mathbf{R}_0\rangle$ and $\mathbf{D}_1 = \langle\mathbf{C}_1,\mathbf{G}_1,\mathbf{R}_1\rangle$ be two deductive systems. Let $f : \mathbf{C}_0 \to \mathbf{C}_1$ is such that $\mathbf{G}_1 \subseteq f(\mathbf{C}_0)$. Then f is a shortening $\prec$-homomorphism iff the following conditions hold:*

1. *If $\bar{c}/d\in\mathbf{R}_0$ then either $\{f(d)\}\prec_1 f(\bar{c})$ or for some $e\in\mathbf{C}_1$ such that $f(d)\prec_1 e$ we have $f(\bar{c})/e\in\mathbf{R}_1$.*
2. *If $f(\bar{c})/f(d)\in\mathbf{R}_1$ then for some $e,h\in\mathbf{C}_0$ we have $f(e)=f(d)$, $e\prec_0 h$, $\bar{c}\vdash_0 h$.*
3. *(a) If $g\in\mathbf{G}_0$ then $f(g)\in\mathbf{G}_1$.*
   (b) If $f(c)\in\mathbf{G}_1$ then $\{c\}$ is successful in $\mathbf{D}_0$.

Note. The notions of a homomorphism and a shortening homomorphism are instances of the notions of a $\prec$-homomorphism and a shortening $\prec$-homomorphism when $\prec_0$ and $\prec_1$ stand for the equality.

Here we will show some general construction of $\prec$-homomorphisms. Let $\mathbf{D}_0 = \langle\mathbf{C}_0,\mathbf{G}_0,\mathbf{R}_0\rangle$ be a deductive system. We construct its homomorphic image $\mathbf{D}_0 = \langle\mathbf{C}_0,\mathbf{G}_0,\mathbf{R}_0\rangle$ in the following way. Let $f : \mathbf{C}_0 \to \mathbf{C}_1$ be any function giving by any $c_0\in\mathbf{C}_0$ some $c_1\in\mathbf{C}_1$ such that $c\vdash c_1$ and $c\prec c_1$. We define $\mathbf{G}_1$ as $\mathbf{G}$ and $\mathbf{C}_1$ as $f(\mathbf{C})$. Define $\mathbf{R}_1$ in the following way. Let $c_1,\ldots,c_n/c\in\mathbf{R}$, $d_1 = f(c_1),\ldots,d_n = f(c_n)$ and d be a clause from $\mathbf{C}_1$ with $f(c)\vdash d$. Then $d_1,\ldots,d_n/d\in\mathbf{R}_1$ and all rules from $\mathbf{R}_1$ are obtained in this way. Finally let $\mathbf{D}_1 = \langle\mathbf{C}_1,\mathbf{G}_1,\mathbf{R}_1\rangle$.

Note. The construction of $\mathbf{D}_1$ from $\mathbf{D}_0$ is the most simple in the case when there is exactly one clause from $\mathbf{C}_1$ deducible from the given clause. This is true for all particular deductive systems considered in the next section.

Finally we let $\prec_0 = \prec_1 = \prec$. For this construction $\mathbf{D}_1 = \langle\mathbf{C}_1,\mathbf{G}_1,\mathbf{R}_1\rangle$ the following theorem holds.

Theorem 6. *The function f is a shortening $\prec$-homomorphism from $\mathbf{D}_0 = \langle\mathbf{C}_0,\mathbf{G}_0,\mathbf{R}_0\rangle$ to $\mathbf{D}_1$.*

Proof. We apply Theorem 5 and prove 1–3b.

1. Obvious.

2. Let $f(\bar{c})/(d)\in\mathbf{R}_1$. Then there is a rule $\bar{a}/e\in\mathbf{R}_0$ such that $f(\bar{a}) = f(\bar{c})$ and $f(e) = f(d)$. For a set of clauses $\bar{b}$ we denote by $\bar{c}\vdash\bar{b}$ the fact that for every $b\in\bar{b}$ $\bar{c}\vdash b$. Since $\bar{a}\prec f(\bar{c})$ and $a\vdash_0 e$ then there is e_1 such that $e\prec e_1$ and $f(\bar{c})\vdash e_1$. Since $e\prec e_1$ and $e\vdash_0 f(d)$ then there is a clause h such that $f(d)\prec h$ and $f(\bar{c})\vdash_0 h$. Applying $\bar{c}\vdash_0 f(\bar{c})$ and $e\prec f(d)$ we obtain $\bar{c}\vdash_0 h$ and $e\prec h$.
3. (a) It is easy to see that if $g\in\mathbf{G}_0$ then $f(g) = g$.
   (b) Trivially follows from $\mathbf{G}_1 = \mathbf{G}_0$. □

How to apply Theorem 6? For this we have to find a function f with the above properties. Theorem 6 does not help very much in the case of resolution. The only non-trivial examples of clauses c, d with $c\vdash d$ and $c\prec d$ in the resolution system are subsuming factors. Fortunately in the logical systems obtained by the inverse method the examples are numerous.

A clause c is called maximal iff there is no clause d with $c\vdash d$ and $c\prec d$. If there is a set of clauses which are obviously non-maximal, then one can try to refine the deductive system so as not to consider these clauses, constructing the homomorphic image. It could be especially efficient if one can discover such sets of clauses during preprocessing. In the subsequent sections we show that examples of non-maximal clauses are numerous in $\mathbf{Cl}_\gamma$ and $\mathbf{Mod}_\gamma$. The strength of the notion of non-maximal clauses is also demonstrated by the example from Section 5.

It is generally known that we may not to generate tautologies in the resolution method. We can say that the set of all tautologies is useless in resolution systems. Below we shall formulate a rather general theorem which gives some sufficient conditions for a set of clauses to be useless. It is interesting that in particular this theorem applies directly to the tautologies in the resolution method[8].

A set of clauses U is called useless iff for any clauses $c_1, \ldots, c_n\in\mathbf{C}$, if there exists a proof of a goal $g\in\mathbf{G}$ from $c_1, \ldots, c_n$ then there exists a proof of some $g_1\in\mathbf{G}$ from $c_1, \ldots, c_n$ in which no clause from U occurs. It is easy to note that the set of all tautologies of resolution (i.e. clauses containing a contrary pair of literals) is useless. The following definition and Theorem 7 below allow one to prove uselessness of a set of clauses.

Let $\mathbf{D} = \langle\mathbf{C}, \mathbf{G}, \mathbf{R}\rangle$ be a deductive system and $\prec$ a subsumption ordering on $\mathbf{D}$. A set of clauses $U \subseteq \mathbf{C}$ is called $\prec$-decreasing iff

1. $U\bigcap\mathbf{G} = \emptyset$;
2. For any inference rule $\bar{c}/c_0$ such that $\bar{c}\bigcap U \neq \emptyset$, either $c_0\in U$ or $\{c_0\}\prec\bar{c}$.

Theorem 7. *If U is $\prec$-decreasing then it is useless.*

Proof. Let U be a $\prec$-decreasing set of clauses. We call a proof in $\mathbf{D}$ clear iff no clause from U occurs in this proof. First we prove the following statement:

> Let the clause c is provable from $c_1, \ldots, c_n$. Then either there is a clause $d\in U$ with $c\prec d$ or there is a clear proof from $c_1, \ldots, c_n$ of a clause d such that $c\prec d$.

We use induction on the length l of the proof of c from $c_1, \ldots, c_n$.

1. $l = 1$. In this case the proof consists of one clause c. If $c\in U$ then the statement is trivially satisfied. If $c\notin U$ then the proof is clear.

[8] We had discovered some useless classes of sequents during the first implementation of LISS [Vor90a] in 1987. Only later we had found out that the useless classes of sequents relate to the notion of subsumption in the same way as tautologies to subsumption in the resolution method.

2. $l \neq 1$. The proof takes the form $d_1, \ldots, d_l$ where $d_1 = c$. If $c \in \{c_1, \ldots, c_n\}$ then we prove as in case 1. Let $c \notin \{c_1, \ldots, c_n\}$. Then c is obtained by applying a rule $e_1, \ldots, e_k/c$ where $\{e_1, \ldots, e_k\} \subseteq \{d_1, \ldots, d_{l-1}\}$.By inductive hypothesis $e_1, \ldots, e_k$ satisfy the statement. Consider two cases.
   (a) For all e_i there exists a clear proof of some f_i such that $e_i \prec f_i$. By the definition of subsumption orderings either $c \prec e_i$ for some i or there is a rule $f_1, \ldots, f_m/f \in \mathbf{R}$ with $c \prec f$. In the former case take f as d, in the latter take e_i as d.
   (b) There is some e_i, say e_1, and $f \in U$ such that $e_i \prec f$. Thus $\{e_1, \ldots, e_k\} \prec \{f, e_2, \ldots, e_k\}$. By the definition of a subsumption ordering either $c \prec \{f, e_2, \ldots, e_k\}$, from which the theorem trivially follows, or there exists a rule $\bar{a}/h$ with $\bar{a} \subseteq \{f, e_2, \ldots, e_k\}$ and $c \prec h$. Let $f \in \bar{e}$. Since $f \in U$ and U is $\prec$-decreasing, then either $h \in U$ or $h \prec \{f, e_2, \ldots, e_k\}$. In both cases c satisfies the statement. Now let $\bar{a} \subseteq \{e_2, \ldots, e_k\}$. If all $e_2, \ldots, e_k$ have clear proofs, then the statement of the theorem is satisfied, otherwise we continue as before for $e_1, \ldots, e_k$.

To complete the proof we note that if a goal clause g is provable from $c_1, \ldots, c_n$ then either $g \prec u$ for some $u \in U$ (which is impossible since $U \bigcap \mathbf{G} = \emptyset$) or there is a clear proof of some g_1 from $c_1, \ldots, c_n$ such that $g \prec g_1$ (and hence $g_1 \in \mathbf{G}$). □

4 Applications to Particular Logics

In this section we apply the general techniques from the previous section to refine the specialized systems $\mathbf{Cl}_\gamma$ and $\mathbf{S4}_\gamma$ from Section 2. In both cases we shall denote the corresponding deductive systems by the same notations.

We start from $\mathbf{Cl}_\gamma$. Let $\mathbf{Cl}_\gamma$ be the following deductive system $\langle \mathbf{C}, \mathbf{G}, \mathbf{R} \rangle$:

1. The set $\mathbf{C}$ of clauses is the set of all sequents of $\mathbf{Cl}_\gamma$;
2. The set $\mathbf{R}$ of inference rules is that of the axioms and inference rules of $\mathbf{Cl}_\gamma$;
3. $G = \{\rightarrow \gamma\}$.

Now we define a binary relation $\prec_{\mathrm{Cl}}$ on $\mathbf{C}$ in the following way: $c \prec_{\mathrm{Cl}} d$ iff there is a substitution θ such that for every formula $\psi \in d\theta$ there exists a formula $\varphi \in c$ such that φ is a free subformula of ψ.

Theorem 8. *The relation $\prec_{\mathrm{Cl}}$ is a subsumption ordering.*

Note. There is a close relation between proofs in Gentzen-type systems and resolution proofs pointed out by Maslov for classical logic [Mas69] and considered in more details by Mints [Min90]. In principle we could easily modify the subsumption strategy of resolution to formulate a similar strategy for the Gentzen-type predicate calculus. But in this case we obtain a weaker notion of subsumption. This weaker relation $\prec$ may be obtained from our definition of the ordering $\prec_{\mathrm{Cl}}$ by replacing the words "is free subformula" to the words "is identical to". It is weaker e.g. because under our subsumption ordering a sequent consisting of a free subformula φ of γ can be subsumed by any free subformula ψ of γ such that φ is a subformula of ψ. But under the usual definition of subsumption φ cannot be subsumed by any non-goal sequent different from φ.

Now we apply Theorem 6 for the case of $\mathbf{Cl}_\gamma$.

Theorem 9. *A sequent $c \in \mathbf{C}$ is not maximal if there is a formula $\varphi \in c$ such that φ is an immediate subformula of a subformula of γ of one of the following forms: $(\varphi_1 \vee \varphi_2)^+$, $(\varphi_1 \supset \varphi_2)^+$, $\neg\varphi$, $(\varphi_1 \wedge \varphi_2)^-$, $(\exists x \varphi_1)^+$, $(\forall x \varphi_1)^-$.*

The immediate application of this theorem is that we can exclude these non-maximal formulas from the considerations. It is implemented in the system LISS [Vor90a] in the following way: on the preprocessing phase we construct the calculus in which the formulas which are not maximal according to Theorem 9 may not occur.

Now consider Theorem 7 applied to $\mathbf{Cl}_\gamma$. Let the set $U_{\text{Cl}} = U_1 \bigcup U_2$ where

$U_1 = \{c \mid \exists\varphi_1, \varphi_2 \in c$ such that there is a free subformula $(\psi_1 \wedge \psi_2)^+$ (or $(\psi_1 \vee \psi_2)^-$, or $(\psi_1 \supset \psi_2)^-$), $\bar{x}$ are all free variables of ψ_1, ψ_2, and there exists a tuple of terms $\bar{t}$ such that φ_1 is a subformula of $\psi_1[\bar{x} \leftarrow \bar{t}]$, φ_2 is a subformula of $\psi_2[\bar{x} \leftarrow \bar{t}]\}$;

$U_2 = \{c \mid \forall\varphi \in c$, φ is an immediate subformula of $(\forall x \varphi_1)^+$ or $(\exists x \varphi_1)^-$.

Theorem 10. *The set* U_{Cl} *is* $\prec$*-decreasing.*

Note. There is an algorithm to verify whether $c \in U_{\text{Cl}}$ which is not much more complicated than the algorithm verifying whether a clause in the resolution method is a tautology.

The strategy of deleting the sequents belonging to U_{Cl} from the search space proved to be very strong on many tests. The reason for it is that usually (as well as for the resolution tautologies) many rules are applicable to clauses from U_{Cl} and that sometimes clauses from U_{Cl} are hard to subsume.

Now we shall consider modal logic **S4**.

If φ is a subformula of ψ then path from φ to ψ is the sequence of formulae $\chi_1, \ldots, \chi_n$ such that $\varphi = \chi_1$, $\psi = \chi_n$ and for every $i \in \{1, \ldots, n-1\}$ χ_i is an immediate subformula of χ_{i+1}.

The specialized system $\mathbf{Mod}_\gamma$ is defined like $\mathbf{Cl}_\gamma$. To define a subsumption ordering for $\mathbf{Mod}_\gamma$ let us first introduce a relation $\leq_{\text{Mod}}$ between subformulas of γ. Let φ, ψ be subformulas of γ, φ is a subformula of ψ and $\pi = (\varphi_0, \ldots, \varphi_n)$ is the path from φ to ψ. Then $\varphi \leq_{\text{Mod}} \psi$ iff

1. φ is a free subformula of ψ;
2. ψ takes the form $(\Box\psi_1)^-$ or $(\Diamond\psi_1)^+$ or there is no subpath in π of the form $(\Box\chi_1)^-, \chi_2$ or $(\Diamond\chi_1)^+, \chi_2$.

Now we let $c \prec_{\text{Mod}} d$ iff there is a substitution θ such that for every formula $\psi \in d\theta$ there exists a formula $\varphi \in c$ such that $\varphi \leq_{\text{Mod}} \psi$.

Theorem 11. *The relation* $\prec_{\text{Mod}}$ *is a subsumption ordering.*

The following theorem gives us examples of non-maximal sequents for $\mathbf{Mod}_\gamma$:

Theorem 12. *A sequent* $c \in \mathbf{C}$ *is not maximal if there is a formula* $\varphi \in c$ *such that one of the following conditions hold:*

1. *φ is an immediate subformula of one of the following subformulas of γ: $(\varphi_1 \vee \varphi_2)^+$, $(\varphi_1 \supset \varphi_2)^+$, $\neg\varphi$, $(\varphi_1 \wedge \varphi_2)^-$, $(\exists x \varphi_1)^+$, $(\forall x \varphi_1)^-$ and φ does not take the form $(\Box\chi)^-$ or $(\Diamond\chi)^+$;*
2. *φ is the immediate subformula of a subformula of γ of the form $(\Box\chi)^-$ or $(\Diamond\chi)^+$.*

Now let $U_{\text{Mod}} = \{c \mid \exists\varphi_1, \varphi_2 \in c$ such that there is a free subformula of γ of the form $(\psi_1 \wedge \psi_2)^+$ (or $(\psi_1 \vee \psi_2)^-$, or $(\psi_1 \supset \psi_2)^-$), and there is a substitution θ such that $\varphi_1 \leq_{\text{Mod}} \psi_1\theta$, $\varphi_2 \leq_{\text{Mod}} \psi_2\theta\}$;

Theorem 13. *The set* U_{Mod} *is* $\prec$*-decreasing (in* $\mathbf{Mod}_\gamma$*).*

Note. In the case of **S4** it is possible to find some other natural $\prec$-decreasing sets. For example, we can add to U the set $\{c \mid \exists\varphi, \psi \in c$ such that φ is the immediate subformula of a subformula of γ of the form $\forall x\varphi$, ψ is the immediate subformula of $\Box\psi$ and x has a free occurrence to $\psi\}$.

There are other applications of the techniques introduced in this paper. For example in the case of logics without the contraction rule the notion of $\prec$-decreasing set of clauses allows one to prove decidability of some of the systems mentioned in the introduction:

Theorem 14. *Let* **D** *be a deductiove system based on any of the following logics:*

1. *Direct logic of Ketonen and Weyhrauch [KW84];*
2. *The logic of Grishin [Gri81];*
3. *The logic of Mey [Mey89];*
4. *The linear fragment of linear logic [GLT89].*

Let U be the set

$\{c \mid$ *There are $\varphi_1, \varphi_2 \in c$ and free subformulae ψ_1, ψ_2 of γ such that*
(a) ψ_1 is a subformula of ψ_2;
(b) $\varphi_1, \varphi_2 \in c$ have different occurences in c $\}$.

Then

1. *U is $\prec$-decreasing;*
2. *For every $\bar{c} \in$ **C**, the set $\{d \mid \bar{c} \vdash d$ and $d \notin U\}$ is finite.*

Corollary 15. *All logics from Theorem 14 are decidable.*

5 An example

All the techniques described here are implemented in the theorem proving system LISS [Vor90a]. It may seem from the definition that the algorithms for implementing the strategies are very complicated. However in practice they are not so different from those of resolution. The algorithms for applying inference rules of the specialized systems, for checking subsumption and for $\prec$-decreasing sets are very similar to the application of the resolution rule, checking for subsumption and checking for tautologies in the resolution method. All the work connected with the non-maximal clauses is done during the preprocessing phase.

Consider a simple example from the paper of Mints [Min90]. Let γ be the following formula to be proved in **S4**:

$$\gamma = \Box\neg\Box\neg(\Box a \vee \Box\neg\Box a)$$

It has the following free subformulae:

$$\begin{array}{llll}
\varphi_1^+ & = \neg\Box\neg(\Box a \vee \Box\neg\Box a) & \varphi_6^+ & = a \\
\varphi_2^- & = \Box\neg(\Box a \vee \Box\neg\Box a) & \varphi_7^+ & = \Box\neg\Box a \\
\varphi_3^- & = \neg(\Box a \vee \Box\neg\Box a) & \varphi_8^+ & = \neg\Box a \\
\varphi_4^+ & = \Box a \vee \Box\neg\Box a & \varphi_9^- & = \Box a \\
\varphi_5^+ & = \Box a & \varphi_{10}^- & = a
\end{array}$$

According to Theorem 12 any sequent containing one of the formulae $\varphi_3, \varphi_4, \varphi_5, \varphi_7, \varphi_{10}$ is not maximal, so we may omit them. Thus refined system $\mathbf{Mod}_\gamma$ have only one axiom $\varphi_9 \rightarrow \varphi_6$ and five inference rules:

$$(r1)\ \frac{\Box\Gamma \to \Diamond\Delta, \varphi_1}{\Box\Gamma \to \Diamond\Delta, \gamma} \qquad (r2)\ \frac{\varphi_2, \Gamma \to \Delta}{\Gamma \to \Delta, \varphi_1} \qquad (r3)\ \frac{\Box\Gamma \to \Diamond\Delta, \varphi_6}{\varphi_2, \Box\Gamma \to \Diamond\Delta}$$

$$(r4)\ \frac{\Box\Gamma \to \Diamond\Delta, \varphi_8}{\varphi_2, \Box\Gamma \to \Diamond\Delta} \qquad (r5)\ \frac{\varphi_9, \Gamma \to \Delta}{\Gamma \to \Delta, \varphi_8}$$

The proof consists of 5 steps:

1. $\varphi_9 \to \varphi_6$	[axiom]	4. $\varphi_2 \to$	[rule $(r4)$ from 3]
2. $\varphi_2, \varphi_9 \to$	[rule $(r3)$ from 1]	5. $\to \varphi_1$	[rule $(r2)$ from 4]
3. $\varphi_2 \to \varphi_8$	[rule $(r5)$ from 2]	6. $\to \gamma$	[rule $(r1)$ from 5]

The proof search in our system was deterministic. The only generated clauses that were not immediately subsumed are exactly the ones shown in the proof. The maximal number of clauses kept in the memory during the proof search is 2, because clause 1 is subsumed by clause 2 and clause 4 subsumes all previously generated clauses. In the prover implemented by Basylev, which is based on the presentation from [Min90] there are 9 initial clauses, the number of clauses actually used in the proof is 15 and the total number of generated clauses is 50 (the statistics is taken from [Min90]).

6 Related Work

In our opinion the notion of strategy for automated deduction is of a great importance and needs more research. At the time we know only a few papers which consider the relation between strategies for different proof systems.

In the paper of Maslov [Mas69] the relation between strategies of resolution and those of inverse method are considered. In [AEH90] some strategies of resolution were generalized to modal logics. However in both papers the definitions are much more concrete than in our paper. We do not know any other paper in which definitions are as abstract and applicable to different kinds of logics as in our paper.

The possibility of applying the inverse method to non-classical logics as well was noted originally by Maslov. However this application was not intensively studied. A general method of theorem proving in non-classical logics was studied in [Wal90]. He uses the connection method [Bib82] instead of the inverse method. Some techniques used in the proofs in [Wal90] are very similar to our techniques. But in our opinion the inverse method applied to non-classical logics is much simpler than the connection method. The efficiency of the two methods is difficult to compare. To the best of our knowledge no strategies for the connection method (like our subsumption strategy) have been developed.

A very similar treatment of the inverse method had been proposed by Mints [Min90]. He views the inverse method as a transformation of the formula to be proved to a set of clauses in a way similar to ours. The main difference is that we make more efforts to the classification of rules,while the clauses are very simple. Mints has more different types of clauses and less different types of rules. In our opinion, the stress on rules is more efficient, because the rules can be *compiled.* Mints does not consider strategies like ours.

Proof search in the intuitionistic sequent calculus was recently considered by Shankar [Sha92]. His system is based on a variant of tableau method. To decrease the search space he considers impermutabilities [Kle52] among the rules of intuitionistic predicate logic. We achieve almost the same effect by using subsumption and the notion of non-maximal clauses. It is also possible to use the information about permutabilities in our approach. It is interesting to see if this information can improve performance. Shankar uses top-down linear strategy of derivation.

Our techniques concerned with with non-maximal clauses can also be used in Shankar's system. Intuitively, if a clause is not maximal, then there is a rule of the sequent calculus, to which one can commit, not considering other possible rules. What is important, it is possible to find a lot of non-maximal clauses during preprocessing, thus compiling more efficient rules. In Shankar's system it could possibly be achieved by a kind of abstract interpretation of the set of his Prolog-like rules.

Acknowledgements

Norbert Eisinger had made many comments which helped to improve this paper. Mark Wallace had corrected my English in the preliminary version of the paper.

References

[AEH90] I. Auffray, P. Enjalbert, and J-J. Hebrard. Strategies for modal resolution: Results and problems. *Journal of Automated Reasoning*, 6:1–38, 1990.

[Bib82] W. Bibel. *Automated theorem proving.* Vieweg Verlag, 1982.

[GLT89] J-Y. Girard, Y. Lafont, and P. Taylor. *Proofs and Types.* Cambridge University Press, 1989.

[Gri81] V.N. Grishin. Predicate and set-theoretic calculi based on logic without the contraction rule (in Russian). *Izvestiya Akad. Nauk SSSR*, 45(1):47–68, 1981.

[KW84] J. Ketonen and R. Weyhrauch. A decidable fragment of predicate calculus. *Theoretical Computer Science*, 32(3):297–309, 1984.

[Kle52] S.C. Kleene. Permutability of inferences in Gentzen's calculi LK and LJ. *Memoirs of the American Mathematical Society*, 10, 1952.

[Lif86] V. Lifschitz. What is the inverse method? *Journal of Automated Reasoning*, 5(1):1–24, 1986.

[Mas67] S.Yu. Maslov. An inverse method for establishing deducibility of nonprenex formulas of the predicate calculus. In J.Siekmann and G.Wrightson, editors, *Automation of Reasoning (Classical papers on Computational Logic)*, volume 2, pages 48–54. Springer Verlag, 1983.

[Mas69] S.Yu. Maslov. Relationship between tactics of the inverse method and the resolution method. In J.Siekmann and G.Wrightson, editors, *Automation of Reasoning (Classical papers on Computational Logic)*, volume 2, pages 264–272. Springer Verlag, 1983.

[Mas87] S.Yu. Maslov. *Theory of Deductive Systems and its Applications.* MIT Press, 1987.

[Mey89] D. Mey. *Private communications.*

[Min90] G. Mints. Gentzen-type systems and resolution rules. Part I. Propositional logic. In P. Martin-Löf and G. Mints, editors, *COLOG-88*, volume 417 of *Lecture Notes in Computer Science*, pages 198–231. Springer Verlag, 1990.

[OH91] H.-Yü. Ohlbach and A. Herzig. Parameter structures for parameterized modal operators. In *Proc. IJCAI'91*, pages 512–517, 1991.

[Sha92] N. Shankar. Proof search in the intuitionistic sequent calculus. In *12th International Conference on Automated Deduction*, Lecture Notes in Computer Science, 1992.

[Vor90a] A.A. Voronkov. LISS - the logic inference search system. In Mark Stickel, editor, *Proc. Int. Conf. on Automated Deduction*, volume 449 of *Lecture Notes in Computer Science*, pages 677–678, Kaiserslautern, Germany, 1990. Springer Verlag.

[Vor90b] A.A. Voronkov. A proof-search method for the first order logic. In P.Martin-Lof and G.Mintz, editors, *COLOG'88*, volume 417 of *Lecture Notes in Computer Science*, pages 327–340. Springer Verlag, 1990.

[Wal90] L.A. Wallen. *Automated Deduction in Nonclassical Logics.* MIT Press, 1990.

[Zam89] N.K. Zamov. Maslov's inverse method and decidable classes. *Annals of Pure and Applied Logic*, 42(2):165–194, 1989.

One more logic with uncertainty and resolution principle for it

Konstantine Vershinin and *Igor Romanenko*
Institute for Cybernetics
Glushkov av.20, Kiev 252207, Ukraine, USSR

Abstract

The paper presents a three-valued logic aimed at the formalization of the "algorithmic-style" and expert reasoning. The "third" truth value is treated as "incomprehensible" rather than "unknown" or "undefined". The corresponding language and its semantics are given, and the existence of various normal forms is demonstrated. We introduce several reasonable notions of logical consequence, distinguish two "basic" among them and present analogs of the resolution method for establishing each of the above mentioned consequences. The completness of both variants of resolution can be shown by means of semantic-tree arguments. Several simple examples, solved by the implemented prover, are given.

A tendency to reason formally about programs and algorithms leads to the necessity to represent formally various sentences containing uncertain (unknown, undefined) terms and to handle the sentences of the kind automatically. Several formalisms were suggested for this purpose but only two of them are known to the author as to be aimed at subsequent automation of the proof search in corresponding calculi [LK87, Smi87]. Both these interesting attempts contain several results, concerning an extension of the resolution method for some three-valued logics. However, there are errors in [LK87] and only a particular case is considered in [Smi87].

In what follows we suggest one more (three-valued) logic with "uncertainty". We also formulate and ground some "resolution-like" principles for the logic.

Consider a first order language with usual notions of terms and atoms, with logical constants *true*, *false*, propositional connectives & , $\vee$, $\rightarrow$, $\neg$, $\sim$ and quantifiers $\forall$ and $\exists$. Formulae of the language are defined in the usual way.

Let $\Omega = \{\bot, \Lambda, \top\}$ be the set of the logical values, linearly ordered in the following way: $\bot < \Lambda < \top$. Let & and $\vee$ be evaluated correspondingly as usual minimum and maximum on the Ω and $\rightarrow$, $\neg$ and $\sim$ be defined on the Ω by the following truth-tables:

x	$\bot$	$\bot$	$\bot$	Λ	Λ	Λ	$\top$	$\top$	$\top$
y	$\bot$	Λ	$\top$	$\bot$	Λ	$\top$	$\bot$	Λ	$\top$
$x \rightarrow y$	$\top$	$\top$	$\top$	$\bot$	$\top$	$\top$	$\bot$	Λ	$\top$

x	$\bot$	Λ	$\top$
$\neg x$	$\top$	$\bot$	$\bot$

x	$\bot$	Λ	$\top$
$\sim x$	$\top$	Λ	$\bot$

Quantifiers may be then defined as infinite conjunction and disjunction correspondingly.

Why do we choose the above mentioned connectives among other possible primitives? First of all, we would like to understand "uncertain" as "incomprehencible" rather than "undefined" or "unknown". A similar situation we usually face when encountering a word or a phrase which is not included in any dictionary (say, "mock-turtle soup") but, nevertheless, it should participate in a reasoning. Furthermore, as it was mentioned at the very beginning, we are going to deal with "algorithmic style" reasoning, therefore we need an implication which is as "constructive" as possible (particularly, we wish to preserve the validity of $\varphi \to \varphi$). Moreover, our experience shows that the most adequate formal analog of natural "if...then..." connective in the above mentioned context is the relative pseudocomplement on the corresponding lattice.

As to unary connectives, the choice was free and we could fix "truth" operators instead of the pseudocomplement or any appropriate combination of them. Our choice was not principal and, maybe, it was done according to an old dissident custom "to deny" rather than "to confirm". The problem of finding an adequate formalization is itself very deep and interesting one, but further discussion may lead us far beyond the scope of this paper.

Let $\Gamma \models \varphi$ abbreviate "φ is a logical consequence of Γ". Since the underlying logic is three-valued, this concept can be defined in a more precise form in several ways. We shall mention four of them below:

(#1) $\Gamma \models_1 \varphi$ iff for every evaluation **ev** $\min\{\mathbf{ev}(\psi) \mid \psi \in \Gamma\} \leq \mathbf{ev}(\varphi)$;

(#2) $\Gamma \models_2 \varphi$ iff for every evaluation **ev** if $\min\{\mathbf{ev}(\psi) \mid \psi \in \Gamma\} = \top$ then $\mathbf{ev}(\varphi) = \top$;

(#3) $\Gamma \models_3 \varphi$ iff for every evaluation **ev** if $\min\{\mathbf{ev}(\psi) \mid \psi \in \Gamma\} \neq \bot$ then $\mathbf{ev}(\varphi) \neq \bot$.

(#4) $\Gamma \models_4 \varphi$ iff for every evaluation **ev** either no formula from Γ is evaluated to $\top$ or $\mathbf{ev}(\varphi) \neq \bot$.

Below $\models_i \varphi$ will stand for *true* $\models_i \varphi$ and $\Gamma \models_i$ — for $\Gamma \models_i$ *false*. We shall call two formulae *equivalent* iff every evaluation transforms them into the same element of Ω. It is easy to show that:

($1) both distributive laws hold for & and $\vee$;

($2) $\neg(\varphi \vee \psi)$ is equivalent to $\neg\varphi \,\&\, \neg\psi$,
$\sim(\varphi \vee \psi)$ is equivalent to $\sim\varphi \,\&\, \sim\psi$,
$\neg(\varphi \,\&\, \psi)$ is equivalent to $\neg\varphi \vee \neg\psi$,
$\sim(\varphi \,\&\, \psi)$ is equivalent to $\sim\varphi \vee \sim\psi$,
(these are analogs of de Morgan laws);

($3) $\varphi \to \psi$ is equivalent to $\neg\varphi \vee \psi \vee (\neg\neg\sim\varphi \,\&\, \neg\neg\psi)$;

($4) quantifiers in a formula may be moved into its prefix;

($5) $\Gamma \models \varphi$ iff $sk^-(\Gamma) \models sk^+(\varphi)$ where $sk^-(\Gamma)$ consists of skolemized w.r.t. satisfyability elements of Γ, $sk^+(\varphi)$ is skolemized w.r.t. validity φ.

The properties **($1)**–**($5)** immediately give us the opportunity to transform a given formula into various (equivalent to it) normal forms. The following properties of $\models_1$, $\models_2$, $\models_3$ and $\models_4$ can also be easily proved:

($6) $\Gamma \models_2 \varphi$ iff $\Gamma, \neg\neg\sim\varphi \models_2$;

($7) $\Gamma \models_3 \varphi$ iff $\Gamma, \neg\varphi \models_3$;

($8) there exists no unary propositional form $f(.)$ in the given signature such that $\Gamma \models_1 \varphi$ iff $\Gamma, f(\varphi) \models_1$;

($9) $\Gamma \models_1 \varphi$ iff $\Gamma \models_2 \varphi$ and $\Gamma \models_3 \varphi$;

($10) $\Gamma \models_4 \varphi$ iff $\Gamma \models_2 \varphi$ or $\Gamma \models_3 \varphi$.

The implication and the consequence relations are connected in the following way:

($11) $\Gamma, \varphi \models_2 \psi$ iff $\Gamma, \neg\varphi \vee \neg\neg\sim\varphi \models_2 \varphi \rightarrow \psi$.

Note that $\neg\varphi \vee \neg\neg\sim\varphi$ intuitively expresses the fact that φ is "certain".

($12) $\Gamma, \varphi \models_3 \psi$ iff $\Gamma \models_3 \varphi \rightarrow \psi$.

Let a set Γ of formulae be called $\models_i$ inconsistent iff $\Gamma \models_i$. Then **($6)**–**($10)** mean particularly that the problem of establishing logical consequence may be reduced to the problem of checking inconsistency (with respect to either $\models_2$ or $\models_3$) for appropriate set of formulae. Moreover we can restrict ourselves to formulae of special canonical form, e.g. of clausal form. As usual, a clause is a (possible empty which is denoted as $\Box$) set of literals, where literal is either an atom or has one of the following forms: $\neg P$, $\sim P$, $\neg\neg P$, $\neg\sim P$, $\neg\neg\sim P$ (here P is an atom, which is called base of the literal). We will show that for solving the last mentioned problem it is sufficient to change only the notion of contrary pair in usual resolution rule, while other mechanisms (e.g. unification, factoring, subsumption etc.) can be preserved.

Let P_1, P_2 be unifiable (in the usual sense) atoms. The following pairs of literals are called contrary w.r.t $\models_2$: $\langle P_1, \neg P_2\rangle$, $\langle P_1, \sim P_2\rangle$, $\langle P_1, \neg\neg\sim P_2\rangle$, $\langle \sim P_1, \neg\sim P_2\rangle$, $\langle \sim P_1, \neg\neg P_2\rangle$, $\langle \neg P_1, \neg\neg P_2\rangle$, $\langle \neg P_1, \neg\sim P_2\rangle$, $\langle \neg\sim P_1, \neg\neg\sim P_2\rangle$. The following pairs of literals are called contrary w.r.t. $\models_3$: $\langle P_1, \neg P_2\rangle$, $\langle \sim P_1, \neg\sim P_2\rangle$, $\langle \neg P_1, \neg\neg P_2\rangle$, $\langle \neg P_1, \neg\sim P_2\rangle$, $\langle \neg\sim P_1, \neg\neg\sim P_2\rangle$.

Moreover, it can be easily shown that w.r.t. $\models_2$ the literal $\neg\sim P$ is satisfied by those and only those evaluations which satisfy the literal P. Therefore we can substitute the literal P for all occurrences of the literal $\neg\sim P$ in the set of clauses, preserving it's (un)satisfiability. Another pair of interchangable w.r.t. $\models_2$ literals is $\sim P$ and $\neg P$. For $\models_3$ corresponding pairs are $(P, \neg\neg P)$ and $(\sim P, \neg\neg\sim P)$. Reducing the set of clauses this way often results in shortening both the individual clauses and the set as a whole (the latter due to the subsumption, see below).

With regard to notions of contrary pairs introduced above, we can formulate two versions of resolution rules. Let C be a set of clauses, $D_1 = D_1{}' \cup \{L\}$, $D_2 = D_2{}' \cup \{L\}$ be either elements of C or their factors with selected literals L_1, L_2.

Rule $\mathbf{R_2}$: if L_1, L_2 are contrary with respect to $\models_2$ and θ is the most general unifier of their bases, then it is permitted to generate from D_1, D_2 their $\mathbf{R_2}$-resolvent $(D_1' \cup D_2')\theta$ (denoted by $R_2(D_1, D_2)$) and add it to C.

Rule $\mathbf{R_3}$: is analogous to $\mathbf{R_2}$ with the only difference, that "contrary with respect to $\models_3$" stands instead of "contrary with respect to $\models_2$".

Let $R_i(C)$, $i = 2, 3$ be the set of all possible R_i-resolvents of elements of C. Let $R_i^0(C) = C$, $R_i^{k+1}(C) = R_i^k(C) \cup R_i(R_i^k(C))$, $R_i^*(C)$ be the union of all $R_i^k(C)$, $k \in N$.

Theorem: *C is $\models_i$-inconsistent iff $\Box \in R_i^*(C)$.*

The proof of the theorem is carried out with the help of semantic-tree arguments and follows the usual scheme. The proof method shows that some strategies (e.g. semantic clash resolution) can be easily defined and their completness can be established.

Analogs of factoring and subsumption can be easily introduced. But in addition to the usual factoring (i.e. factoring of literals with identical signs and unifiable atoms), we can factorize literals with different signs. Namely, w.r.t. $\models_2$ the clause $P \vee \neg\neg P' \vee C$ with P and P' being unifiable atoms with m.g.u. σ can be factorized to $\neg\neg P\sigma \vee C\sigma$ (but not $P\sigma \vee C\sigma$). We can also factorize ${\sim}P \vee \neg\neg{\sim}P' \vee C$ to $\neg\neg{\sim}P\sigma \vee C\sigma$. With respect to $\models_3$ we can factorize P and $\neg{\sim}P'$ (keeping $P\sigma$) and ${\sim}P$ and $\neg P'$ (keeping ${\sim}P\sigma$).

The resolution method just discussed is implemented for IBM/PC family of computers under MS-DOS operating system. The implementation includes some useful features:

- many-sorted signatures are allowed with function overloading;
- unification and matching are performed modulo AC;
- term rewriting is available for computations over abstract terms;
- several frequently used sorts (like numbers, strings, etc.) are built-in.

We conclude the paper by presenting two simple examples. First, we will show that

$$\models_2 (P \to Q) \vee (Q \to P)$$

According to (**$6**) we must prove that

$$\neg\neg{\sim}((P \to Q) \vee (Q \to P)) \models_2$$

Converting this problem to clauses we get

$$\neg\neg{\sim}Q \quad (1)$$

$$\neg\neg P \quad (2)$$

$$\neg\neg{\sim}P \quad (3)$$

$$\neg\neg Q \tag{4}$$

$$\neg{\sim}Q \vee \neg P \tag{5}$$

$$\neg{\sim}P \vee \neg Q \tag{6}$$

Resolving (1) and (5) we get the clause

$$\neg P$$

which in combination with (2) produces an empty clause.

Now let us consider the formula

$$((P \rightarrow (Q_1 \rightarrow Q_2)) \,\&\, (\neg\neg Q_1 \vee \neg\neg{\sim}Q_1)) \rightarrow ((P \,\&\, Q_1) \rightarrow Q_2)$$

with the clausal form

$$\neg\neg Q_1 \tag{1}$$

$$\neg\neg P \tag{2}$$

$$\neg\neg{\sim}Q_2 \tag{3}$$

$$\neg Q_2 \vee \neg Q_1 \vee \neg P \vee \neg{\sim}Q_2 \vee \neg\neg{\sim}Q_1 \vee \neg\neg{\sim}P \tag{4}$$

$$\neg Q_2 \vee \neg{\sim}P \tag{5}$$

$$\neg Q_2 \vee \neg{\sim}Q_1 \tag{6}$$

$$\neg Q_1 \vee \neg P \vee \neg\neg Q_2 \tag{7}$$

The following proof was found in less than 1 second by the prover running on a 20 MHz IBM PC/AT

$$(1),(7) \vdash \neg P \vee \neg\neg Q_2 \tag{8}$$

$$(2),(8) \vdash \neg\neg Q_2 \tag{9}$$

$$(3),(4) \vdash \neg Q_2 \vee \neg Q_1 \vee \neg P \vee \neg\neg{\sim}Q_1 \vee \neg\neg{\sim}P \tag{10}$$

$$(5),(10) \vdash \neg Q_2 \vee \neg Q_1 \vee \neg P \vee \neg\neg{\sim}Q_1 \tag{11}$$

$$(2),(11) \vdash \neg Q_2 \vee \neg Q_1 \vee \neg\neg{\sim}Q_1 \tag{12}$$

$$(6),(12) \vdash \neg Q_2 \vee \neg Q_1 \tag{13}$$

$$(1),(13) \vdash \neg Q_2 \tag{14}$$

$$(9),(14) \vdash \square$$

References

[LK87] A. Letichevsky and J. Kapitonova. On constructive mathematical description of problem domains. *Kibernetika*, (4), 1987. (in Russian).

[Smi87] P.H. Smitt. Computational aspects of three-valued logic. *Lecture Notes in Computer Science*, 230:191–198, 1987.

A Natural Deduction Automated Theorem Proving System

Li Dafa
Department of Applied Mathmatics
Tsinghua University
Beijing, CHINA 100084

Abstract Recently many people are researching the natural deduction system adapted from Gentzen system, it need not transform formulas into other normal form. Robinson's unification algorithm only uses the substitution rule, therefore it can't be used to handle formulas with quantifiers. we concluded an algorithm which can treat any wffs with quantifiers. that is , given two wffs, the algorithm tries to apply the rules for quantifiers to them to test if they become equal. The automatic natural deduction proving system based on our algorithm has been implemented, the Andrews Challeges, Bledsoe Challenges and Pelletier Challenges were proved by our system.

1. Introduction

The natural deduction system for the first-order logic predicate calculus was adapted from Gentzen's work in 1934-1935 [6]. It can give fairly natural and well structured proofs, and express the forms of arguments which arise in mathematical practice. Therefore the proofs constructed in natural deduction style are reasonably readable and comprehensible. And the proofs of theorems in resolution style are not readable, many people want to transform their proofs in other style into proofs in antural deduction style[2]. Recently many peolpe reseach non-resolution theorem proving [2,3,5,9,10], and specially natural deduction [5,9,10]. We have implemented an automatic natural deduction proving system (we call it ANDP system) by programming our unification algorithm. We take $\sim, \vee, \wedge, \rightarrow, \leftrightarrow$ as primitive connectives and $\forall$ and $\exists$ as primitive quantifiers for our natural deduction system. We use expressions of the form $\Gamma \vdash B$, where Γ is any finite set of zero or more wffs called premises or hypotheses, and B is wff. A natural deduction system consists of axioms and rules of inference. A proof of a wff A from the set Γ of premises or hypotheses is a sequence of wffs such that A is the last wff in the sequence and each wff in the sequence is an axiom or a member of Γ or derived from previous wffs in the sequence by one of the rules of inference. We write $\Gamma \vdash A$ to express that there is a proof of A from Γ. The axioms and rules of inference in our ANDP system are as follows.

Rules of inference for natural deduction

The axiom and basic rules

It is omitted.

Rules for the connectives

(a) MP (modus ponens) MT (modus tollens) (b) LDS (left disjunctive syllogism)

* The project supported by NSFC

RDS (right disjunctive syllogism) (c) CP (conditional proof) (d) ADD (addition) (e) CONJUN (conjunction) SIMP (simplication) (f)IP (indirect proof) and $\sim$ elimination (g)EQUIV (equivalence) (h)DILEMMA (cases) (i)IMP (implication)
Rules for the quantifiers

(a) UG (universal generalization)(b)US (universal specialization)

(c) EG (existential generalization)

(d) $\frac{\Gamma\vdash\exists x A(x) \text{ and } \Gamma, A(a)\vdash C}{\Gamma\vdash C}$, ES (or EE) (eliminating existential quantifier)

where a is an individual constant not accuring in any member of Γ, in $\exists$xA(x), or in C.

It is difficult to construct proofs of theorems in the natural deduction style. We will give an example to show what the natural deduction proof is like. The example is Socrates argument which is given by all men are mortal, Socrates is a man, therefore Socrates is mortal. Here Hx means that x is human, Mx means that x is mortal. The following is its proof in natural deduction style.

(1)	$\{1\}$	$\vdash (\forall x)[Hx \rightarrow Mx]$	Premise
(2)	$\{1,2\}$	$\vdash Hs$	Premise
(3)	$\{1,2\}$	$\vdash Hs \rightarrow Ms$	US 1
(4)	$\{1,2\}$	$\vdash Ms$	MP 2,3

2. Main Software Packages and Their Functions

In resolution method we have to decide whether or not two literals L_1, and L_2 of parent clauses C_1 and C_2 are equal after applying substitution rule to them. For natural deduction system the similar things will happen. The package 1 for the rule MP, when we have $\Gamma \vdash A$ and $\Gamma \vdash A' \rightarrow B'$, we want to apply the rule MP to them, we have to test if A and A' become equal after applying quantifier rules UG, US, EG and ES to them. The package 2 for the rule MT. The package 3 for the rule LDS. The pakage 4 for the rule RDS. The package 5 for the rule IP, in the case we want to find an indirect proof of a theorem, when we have $\Gamma, \sim A \vdash B$ and $\Gamma, \sim A \vdash \sim B'$, we have to test if B and B' can become equal after applying UG,US,EG and ES to them. The package 6 for rule $\sim$ elimination. The package 7 for testing if the antecedence line $\Gamma \vdash A$ is equal to the goal line $\Gamma \vdash A'$ in any time, it is necessary to test if A can become equal to A' after applying UG, US, EG and ES to $\Gamma \vdash A$. The package 8 for the rule US, when we apply US to $\Gamma \vdash (\forall x)A(x)$ we have to decide what term t is substituted for variable x in $A(x)$ to obtain $\Gamma \vdash A(t)$. The package 9 for the rule EG, when we apply the rule EG to $\Gamma \vdash A(t)$, we have to decide what term t and which occurences of the term are existentially generalized upon to obtain $\Gamma \vdash (\exists x)A(x)$. In natural deduction it is obvious that we have to have an algorithm to decide how to apply quantifier rules UG, US, EG and ES to two wffs to test if they become equal. Clearly Robinson's algorithm only uses the substitution rule,it can only handle clauses, our algorithm can handle any wffs with quantifiers[7], it is used in each package above. For example the package MP is used to search the two lines being of the forms A and $A' \rightarrow B$ from antecedence lines, if A and A' can become equal after applying UG, US, EG, ES to the lines, then the result lines are infered by the rules UG, US, EG, ES and MP, it returns; if it can not find the two lines, it returns and tries other rules. We apply the rules to antecedence lines in the following order: CP, CONJUN, SIMP, MP, MT, LDS, RDS, UG, ES, US, CASES.

The main procedure of ANDP system is given in the following flow chart.

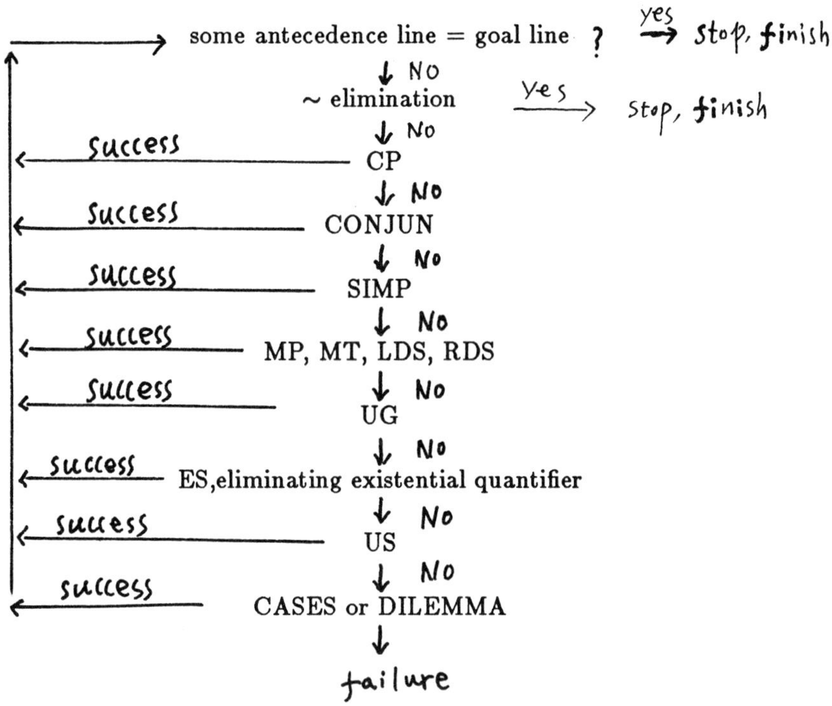

3. How Does ANDP System Work?

Next we show how our system proves the following example. The premises are $(\exists z)(\forall u)(\forall v)Pzuv$ and $(\forall x)[(\forall y)Pxyy \rightarrow Qx]$, we want to prove $(\exists x)Qx$ from the two premises.

The natural deduction proof of the example above constructed by using our unfication algorithm[7] is as follows.

1.	$\{1\} \vdash (\exists z)(\forall u)(\forall v)Pzuv$	Premise
2.	$\{1,2\} \vdash (\forall x)[(\forall y)Pxyy \rightarrow Qx]$	Premise
3.	$\{1,3\} \vdash (\forall u)(\forall v)Pauv$	HYPO
4.	$\{1,2\} \vdash (\forall y)Payy \rightarrow Qa$	US 2
5.	$\{1,3\} \vdash (\forall v)Pauv$	US 3
6.	$\{1,3\} \vdash Pauu$	US 5
7.	$\{1,3\} \vdash (\forall u)Pauu$	UG 6
8.	$\{1,2,3\} \vdash Qa$	MP 7, 4
9.	$\{1,2,3\} \vdash (\exists x)Qx$	EG 8
10.	$\{1,2\} \vdash (\exists x)Qx$	EE 1, 9

When we apply the rule US to line 2, it must be assured that the rule MP can be applied to line 3 and the result line obtained from line 2 by using the rule US, our unification algorithm can decide to substitute the term a for the individual variable x in wf $(\forall y)Pxyy \rightarrow Qx$ in line 2 for this purpose, thus line 4 is infered. Now

$(\forall u)(\forall v)Pauv$ in line 3 is not equal to antecedence $(\forall y)Payy$ of $(\forall y)Payy \rightarrow Qa$ in line 4, we can not apply MP to line 3 and line 4, and we can not eliminate the quantifier in line 4 by the reason above. We can only apply the rule US to line 3,then line 5 is obtained. $(\forall v)Pauv$ is not equal to the antecedence $(\forall y)Payy$ of $(\forall y)Payy \rightarrow Qa$ in line 4, we have to apply the rule US to line 5. It must be assured that the rule MP can be applied to line 4 and the result line obtained from line 5 by applying the rule US to line 5, for the purpose our algorithm can decide to substitute the term u for the variable v in wf $Pauv$ in line 5, then line 6 is infered. Now $Pauu$ in line 6 is not equal to the antecedence $(\forall y)Payy$ of $(\forall y)Payy \rightarrow Qa$ in line 4, in order to apply the rule MP to line 6 and 4 our algorithm can decide to apply the rule UG to line 6, where the individual variable u will be universally generalized upon, then line 7 is infered. After line 8 is infered from line 4 and 7 by using MP,it becomes into the case in which we test whether or not an antecedence line 8 is equal to the goal line 10, our algorithm can decide to apply the rule EG to line 8 for this purpose, the line 9 is obtained. Now we finished the proof of the example in natural deduction style.

It is not hard to see that Robinson's algorithm [12] and other algorithms can not used in natural deduction system. Our unification algorithm can decide how to apply quantifier rules UG,US,EG and ES to wffs to find their unifier.

4. Formats for Logic Formulas

We use (Ax), (Ex) to stand for ($\forall$ x), and ($\exists$ x), respectively. When entering wff, &, or, $->$, iff and $\sim$ are used to express conjunction, disjunction, implication, equivalence and negation. The other formats are like those in [1].

5. Testing Logic Formulas Entered

After the formulas are entered, the system checks whether they are wff, the definition of wff is same as that in any logic book. If they are not wff, system will give the error information and ask to enter them again.

6. Direct Proof and Indirect Proof in Natural Deduction Style

The system can construct not only direct natural deduction proofs but also indirect natural deduction proofs of logic theorems.

7. Top-down and Bottom-up

It is convenient to construct proof of theorems first up from bottom by using the rules CP, CONJUN, UG, IMP then down from top by using other rules.

8. The Proof of Andrews' Challenge and Proofs of Other Challenges

The problem is $[(\exists x)(\forall y)[Px \leftrightarrow Py] \leftrightarrow [(\exists x)Qx \leftrightarrow (\forall y)Qy]] \leftrightarrow [(\exists x)(\forall y)[Qx \leftrightarrow Qy] \leftrightarrow [(\exists x)Px \leftrightarrow (\forall y)Py]]$ [11]. It will produce 1600 clauses [11]. The size makes it impossible to prove it by resolution.

We constructed a direct proof in the natural deduction style of the challenge with our ANDP system, the system is loaded in SUN 4 Work-Station, it is written in Common LISP, has 7000 lines, occupies about 250k bytes store spaces. The programme for ANDP is not compiled. The length of the proof of the challenge constructed by ANDP is 522 steps. The CPU time is 8 seconds.

Other Andrews challenges, Bledsoe challenges, Pelletier challenges [1, 11] are proved by our ANDP system.

Acknowledgement

Thank Intelligence Technology and System Lab. and College of Science of Tsinghua University for supporting me to do the research.

References

1. Andrews, P.B., An Introduction To Mathematical Logic And Type Theory: To Truth Trough Proof, Orlando, Academic Pr., Inc., 1986.
2. Andrews, P.B., Transforming Mating Into Natural Deduction Proofs. 5th Conference On Automated Deduction, Les Arcs, France, edited by G.Goos and J. Hartmanis, Lecture Notes in Computer Science 138, Spring-Verlag, July 8-11,1980.
3. Bledsoe, W.W.,Non-resolution theorem proving, Artificial Intelligence 9 (1977) 1-35.
4. Chang, C.L. and Lee, R.C.T., Symbolic Logic and Mechanical Theorem Proving, Academic Press, New York, 1973.
5. Dan Sahlin, Torkel Franzen and Seif Haridi, An Intuitionistic Predicate Logic Theorem Prover. SICS Research Report R89001, ISSN 0283-3638. Swedish Institute Of Computer Science.
6. Gentzen, G. Investications into Logical Deductions. The Collected Papres of Gerhard Gentzen, M.E.Szabo,ED., North-Holland Publishing CO., Amsterdam, 1969, pp.68-131.
7. Li Dafa, Unification Algorithm with Quantifiers in the First-Order Logic, Science Report 89005, Dept. of Applied Mathematics of Tsinghua University, Nov. 1989.
8. Manna, Z. Mathematical Theory of Computation, McGraw Hill, New york, 1974.
9. Pastre, Dominigue, MUS(ADET: An Automatic Theorem Proving System Using Knowledge and Metaknowledge in Mathematics, *Artificial Intelligence 38 (1989) 257-318.*
10. Pelletier, J.,Further Developments in THINKER, an Automated Theorem prover: Technical Report TR-ARP-16/87, Dept. of Computer Science, University of Alberta, CANADA.
11. Pelletier, J., Seventy-Five Problems for Testing Automatic Theorem Provers, J. of Automated Reasoning 2 (1986) 191-216.
12. Robinson,J.A., A Machine-orienteal Logic Based on the Resolution Principle, *JACM, 12, 1(Jan.1965),23-41.*

Isabelle-91 *

Tobias Nipkow[†] and Lawrence C. Paulson[‡]

Isabelle is a generic theorem prover. Object-logics are formalized within higher-order logic, which is Isabelle's meta-logic. Proofs are performed by a generalization of resolution, using higher-order unification. The latest incarnation of Isabelle, *Isabelle-91*, features a type system based on order-sorted unification; this supports polymorphism and overloading in logic definitions.

1 Defining logics

Isabelle's meta-logic is intuitionistic higher-order logic with implication ($\Longrightarrow$), universal quantifiers ($\bigwedge$), and equality ($\equiv$) [5, 6]. The presentation of an object-logic consists of a signature introducing the types and constants of the logic, i.e. its abstract syntax, and axioms describing the inference rules. As a tiny example, consider the following definition of minimal logic.

$$\begin{array}{rcl} \mathcal{M}in & = & \textbf{types}\ \textit{form} \\ & & \textbf{consts}\ \_ \longrightarrow \_ : \textit{form} \to (\textit{form} \to \textit{form}) \\ & & \qquad\qquad [\![\_]\!] : \textit{form} \to \textit{prop} \\ & & \textbf{rules}\ ([\![P]\!] \Longrightarrow [\![Q]\!]) \Longrightarrow [\![P \longrightarrow Q]\!] \\ & & \qquad\quad [\![P \longrightarrow Q]\!] \Longrightarrow [\![P]\!] \Longrightarrow [\![Q]\!] \end{array}$$

$\mathcal{M}in$ introduces the type *form* (of object-formulae) and the infix constant $\longrightarrow$ (for object-implication). The constant $[\![\_]\!]$ maps *form* to the predefined type *prop* of meta-logic propositions. The proposition $[\![P]\!]$ should be read as "the formula P is true"; we usually distinguish object-level formulae (*form*) from meta-level formulae (*prop*). In practice the brackets can be dropped; parser and pretty-printer take care of such matters.

The two rules for minimal logic are typical natural deduction rules for implication introduction and elimination, which are usually written as follows:

$$\frac{\begin{array}{c}[P]\\ \vdots \\ Q\end{array}}{P \longrightarrow Q} \qquad\qquad \frac{P \longrightarrow Q \quad P}{Q}$$

Logics can be combined (taking their union) and can be extended with new types, constants and rules. A minimal predicate logic can be defined as an extension of $\mathcal{M}in$ by adding a type of terms and a quantifier:

$$\begin{array}{rcl} \mathcal{P}red & = & \mathcal{M}in + \textbf{types}\ \textit{term} \\ & & \qquad\quad \textbf{consts}\ \forall : (\textit{term} \to \textit{form}) \to \textit{form} \\ & & \qquad\quad \textbf{rules}\ (\bigwedge x.[\![P(x)]\!]) \Longrightarrow [\![\forall(P)]\!] \\ & & \qquad\qquad\qquad [\![\forall(P)]\!] \Longrightarrow [\![P(t)]\!] \end{array}$$

---

*Research supported by ESPRIT BRA 3245, *Logical Frameworks*.

[†]Author's address: Institut für Informatik, TU München, Postfach 20 24 20, 8000 München 2, Germany. E-mail: Tobias.Nipkow@Informatik.TU-Muenchen.De.

[‡]Author's address: University of Cambridge, Computer Laboratory, Pembroke Street, Cambridge CB2 3QG, England. E-mail: Larry.Paulson@cl.cam.ac.uk.

Because Isabelle is based on higher-order logic, its expressions are simply typed λ-terms. The λ-calculus notions of free and bound variables handle quantifiers. The formula $\forall x.P(x)$, where P is of type $term \to form$, is internally represented as $\forall(\lambda x.P(x))$. The parser and pretty-printer translate between the concrete syntax and the internal form.

The two inference rules formalize the usual rules of quantifier introduction and elimination:

$$\frac{P}{\forall x.P} \qquad \frac{P[t/x]}{\forall x.P}$$

The introduction rule is subject to the proviso that x is not free in the assumptions. In Isabelle, this proviso is automatically enforced because the premise $[\![P(x)]\!]$ is in the scope of a local $\bigwedge x$ [5]. Similar techniques handle existential quantifiers and expressions such as $\Sigma_{i=0}^{n} k_i$ (summations), $\Pi_{x\in A} B(x)$ (dependent types) and $\bigcup_{\alpha<\gamma} H_\alpha$ (large unions).

2 Order-sorted polymorphism

$\mathcal{P}red$ formalizes a single sorted predicate logic. To support many-sorted and polymorphic logics, Isabelle-91 introduces *order-sorted polymorphism* [3]. This is ML-polymorphism where the algebra of types is order-sorted — there is a new level of partially ordered *sorts* classifying the types. Type variables are qualified by sorts, thus restricting the set of types they range over. This is a generalization of Standard ML's equality types [2] and is closely related to Haskell's type classes [4]. As an example consider the following definition of a polymorphic first-order logic:

$$\begin{aligned} \mathcal{FOL} \;=\; & \textbf{sorts}\ i < \top \\ & \textbf{types}\ form : \top \\ & \textbf{consts}\ \_ \longrightarrow \_ : form \to form \to form \\ & \qquad\qquad \forall : (\alpha_i \to form) \to form \end{aligned}$$

The first line introduces a sort i ("individuals") which is a subsort of the predefined sort $\top$ of all types. The type of formulae is classified as being of sort $\top$. Implication is as before, whereas $\forall$ has acquired a polymorphic type: the type variable α_i ranges over all types of sort i. In particular α_i does not range over *form* and function types, because neither are of sort i. This rules out quantification over predicates and functions, thus ensuring that FOL is indeed a first-order logic and not a higher-order one in disguise.

The sort i is initially empty but further extensions may change this:

$$\begin{aligned} \mathcal{N}at \;=\; \mathcal{FOL} + \; & \textbf{types}\ nat : i \\ & \textbf{consts}\quad 0 : nat \\ & \qquad\qquad succ : nat \to nat \end{aligned}$$

The formula $\forall x.P(x) \longrightarrow P(succ(x))$ is legal, with x having the inferred type *nat* which is of sort i.

A polymorphic equality operator would have type $\alpha_i \to (\alpha_i \to form)$. For higher-order logic, *form* would be declared to have sort i, to permit quantification over formulae [3].

2.1 Overloading

Order-sorted polymorphism can also be used to specify *ad-hoc* polymorphism or *overloading.* Suppose we would like to use the symbol + at more than one type, for example both natural numbers and strings (where + might denote concatenation). Isabelle's type system does not allow the simultaneous declaration

$$\begin{aligned} \textbf{consts } \_+\_ &: nat \to nat \to nat \\ \_+\_ &: string \to string \to string \end{aligned}$$

However, + can be declared once and for all as a polymorphic operator:

$$\begin{aligned} \mathcal{FOL}_+ \quad = \quad \mathcal{FOL} \; + \; & \textbf{sorts } a < i \\ & \textbf{consts } \_+\_ : \alpha_a \to \alpha_a \to \alpha_a \end{aligned}$$

The sort a is meant to represent those types which provide addition (+). More precisely, any type τ of sort a automatically gives rise to a constant $+ : \tau \to \tau \to \tau$. So far there is only the generic operation but there are no instances. The latter are created by asserting that some type is of sort a:

$$\begin{aligned} \mathcal{N}at_+ \quad = \quad \mathcal{FOL}_+ \; + \; \mathcal{N}at \; + \; & \textbf{types } nat : a \\ & \textbf{rules } 0 + n = n \\ & \qquad succ(m) + n = succ(m + n) \end{aligned}$$

3 Theorem proving in Isabelle

Object-logic proofs can be performed by applying one rule at a time, or in large tactic steps.

3.1 Resolution

The central notion in Isabelle is that of a theorem. All axioms in a logic definition are available as theorems; new theorems can only be derived by combining existing ones. Ignoring $\bigwedge$ and $\equiv$, the general form of a theorem is $\phi_1 \Longrightarrow \ldots \Longrightarrow \phi_m \Longrightarrow \phi$ which we abbreviate as $[\phi_1, \ldots, \phi_m] \Longrightarrow \phi$. Theorems are combined by resolution. Given theorems

$$[\phi_1, \ldots, \phi_m] \Longrightarrow \phi \quad \text{and} \quad [\psi_1, \ldots, \psi_n] \Longrightarrow \psi$$

and a substitution s such that $s(\phi) = s(\psi_k)$, we can infer the new theorem

$$s([\psi_1, \ldots, \psi_{k-1}, \phi_1, \ldots, \phi_n, \psi_{k+1}, \ldots, \psi_n] \Longrightarrow \psi)$$

Isabelle computes the substitution s by *higher-order unification* [1].

If theorems are viewed as derived rules, resolution corresponds to forward proof. Backward proof is obtained by a change of perspective: theorem $[\psi_1, \ldots, \psi_n] \Longrightarrow \psi$ can be read as an intermediate state in a backward proof where ψ is the overall goal and the ψ_i are the subgoals yet to be proved. In this case, resolution with rule $[\phi_1, \ldots, \phi_m] \Longrightarrow \phi$ corresponds to Prolog-style backward chaining. The initial state of a backward proof is the trivially valid implication $\psi \Longrightarrow \psi$. Identifying proof states with theorems means that intermediate proof states are also valid theorems, albeit hypothetical ones.

Due to the presence of universal quantifiers and the possibility of natural deduction style proofs, resolution may also have to "lift" one rule over the quantifiers and local assumptions of the other one [5].

3.2 Tactics

Tactics are the functional programmer's answer to the tedium of single-step theorem proving. They combine arbitrary algorithmic sequences of proof steps (and searches) into a single function. Since Isabelle identifies proof states with theorems, tactics are simply functions over theorems. To allow for backtracking, tactics are in fact functions from theorems to streams of theorems. Backtracking can occur over the choice of rule, like in Prolog, and also over the choice of unifier (because of higher-order unification).

Tactics can be written from scratch or can be assembled from existing tactics with tacticals like THEN, ORELSE, REPEAT, DEPTH_FIRST, BEST_FIRST, etc. The tactic writer is supplied with a rich language of control structures, including search strategies.

Isabelle comes with several generic packages for writing tactics. There are two packages to support rewriting with user-defined object-level equalities; more generally, they can rewrite with equivalence relations and reduction systems. Another package supports classical predicate calculus reasoning, and can prove many of the problems in Pelletier [8]; this same package supports automated reasoning in set theory.

4 General information

Isabelle is written in Standard ML and should run on any Standard ML system. Interaction with Isabelle is through ML. Logics, theorems and tactics are ML values; they are manipulated by calling ML functions for extending a logic, resolving two theorems, applying a tactic, or whatever.

Isabelle comes with 8 different logics, including LCF and some modal logics. The most substantially developed logics are first-order logic, set theory, and higher-order logic. Many non-trivial theorems have been proved in them, including the Schröder-Bernstein theorem, the well-founded recursion theorem, and soundness and completeness of propositional logic.

Isabelle can be obtained by anonymous ftp from `ftp.cl.cam.ac.uk`. Using binary mode, get the file `isabelle.tar.Z` from directory `ml`. The LaTeX-sources for the manual [7] are included.

References

[1] G. Huet. A unification algorithm for typed λ-calculus. *Theoretical Computer Science*, 1:27–57, 1975.

[2] R. Milner, M. Tofte, and R. Harper. *The Definition of Standard ML*. MIT Press, 1990.

[3] T. Nipkow. Order-sorted polymorphism in isabelle. In G. Huet, G. Plotkin, and C. Jones, editors, *Proc. 2nd Workshop on Logical Frameworks*, pages 307–321, 1991.

[4] T. Nipkow and G. Snelting. Type classes and overloading resolution via order-sorted unification. In *Proc. 5th ACM Conf. Functional Programming Languages and Computer Architecture*, pages 1–14. LNCS 523, 1991.

[5] L. C. Paulson. The foundation of a generic theorem prover. *J. Automated Reasoning*, 5:363–397, 1989.

[6] L. C. Paulson. Isabelle: The next 700 theorem provers. In P. Odifreddi, editor, *Logic and Computer Science*, pages 361–385. Academic Press, 1990.

[7] L. C. Paulson and T. Nipkow. Isabelle tutorial and user's manual. Technical Report 189, University of Cambridge, Computer Laboratory, 1990.

[8] F. J. Pelletier. Seventy-five problems for testing automatic theorem provers. *J. Automated Reasoning*, 2:191–216, 1986. Errata, JAR 4 (1988), 236–236.

The Semantically Guided Linear Deduction System

Geoff Sutcliffe
Department of Computer Science, The University of Western Australia
Nedlands, Western Australia, 6009. Email : geoff@cs.uwa.oz.au

1. Introduction

The Semantically Guided Linear Deduction System (SGLD) is the successor of GCTP [Sutcliffe, 1990]. SGLD has been constructed by imposing semantic guidance onto a chain format linear deduction system called Guided Linear Deduction (GLD). GLD is broadly based on the Graph Construction (GC) procedure [Shostak, 1976], but has added features which improve performance.

2. Deduction Operations

GLD's eight deduction operations consist of six inference operations - extension, unit subsumed extension, A-reduction, identical A-reduction, C-reduction, and identical C-reduction; and two bookkeeping operations - A-truncation and C-truncation. The extension operations are, in combination, equivalent to the extension operations of other chain format systems. GLD's reduction operations combine features from Model Elimination (ME) [Loveland, 1969] and the GC procedure. A significant feature of GLD's non-compulsory reduction operations is the use of a selection rule, which provides a search guidance point not available in most other chain format systems. The reduction operations work in tandem with A-truncation to implement re-use of deduced information.

Orthogonal to the operational divisions, the deduction operations may be split into two groups based on whether or not alternative successor center chains need be considered once the operation has been completed. The latter group are the *compulsory* operations, being unit subsumed extension, identical A-reduction and identical C-reduction. An important difference between the non-compulsory and the compulsory inference operations is that the non-compulsory inference operations operate on a selected B-literal in the rightmost cell of a center chain, while the compulsory inference operations may eventually use any B-literal in a center chain.

3. Deduction Chunks

The desirable feature of using coarse grain deduction steps in a deduction system has been approached in GLD by combining multiple deduction operations into indivisible deduction chunks. The philosophy underlying GLD's operation chunking is that no center chain is stored whilst it contains a literal that can be removed by a compulsory operation. After each non-compulsory operation, a maximal sequence of compulsory operations is performed before the resulting center chain is stored. The intermediate chains deduced whilst building the maximal sequence are discarded. The initial non-compulsory operation and the sequence of compulsory operations form a deduction chunk.

4. Search Strategy

For each chunk of a GLD deduction, two choices have to be made. The first is to select a B-literal for the non-compulsory operation; the second is to choose an order in which alternative successor center chains are considered. GLD uses two methods for selecting a B-literal and two methods for ordering alternative successors. These in combination provide four possible search styles - literal-selected, literal-ordered, cell-selected and cell-ordered.

The literal-selected and literal-ordered search styles use a trivial B-literal selection method, simply taking the rightmost B-literal. This provides no search guidance. In the cell-selected and cell-ordered styles the B-literal selection aims for the selection that is most likely to lead to failure. For each possible selection of B-literal the set of successor center chains is deduced, the heuristic value of each successor in each set calculated, and the best value in each set noted as the heuristic value of the set. The B-literal whose successor set has the worst value, is selected.

The literal-selected and cell-selected search styles use a default ordering of alternative successor center chains. The default ordering is guided by (i) a 'fewest-literals preference' maxim, and (ii) by avoiding changing scope values (see below) if possible. The literal-ordered and cell-ordered search styles order the alternative successor center chains in order of worsening heuristic value.

The overall search strategy of GLD is a modified consecutively bounded depth first search which places a bound on the number of center chain A- and B-literals.

5. Re-use of Deduced Information

GLD uses a combination of ME's lemma mechanism and the GC procedure's C-literal mechanism, to re-use deduced information. Because of the advantages of the lemma mechanism over the C-literal mechanism (principally, lemmas are persistent), preference is given to the creation of lemmas. The addition of lemma chains to the input set is limited by the use of forward subsumption. Further, non-unit lemmas are only added to the input set if they backward subsume at least one existing input chain. If no existing input chains are backward subsumed then a C-literal is inserted into the center chain. In this manner the number of non-unit input chains does not increase. Unit lemmas are always added to the input set if they are not forward subsumed. To implement the combined mechanism, A-literals maintain a scope value. The C-point of an A-truncation is determined from the scope values.

6. Admissibility Restrictions

The admissibility restrictions in GLD are based on those of the GC procedure. The GC procedure specifies that no two non-B-literals in any center chain may have identical atoms. GLD imposes these and five new restrictions. (i) No A-literal may be to the left of an identical B-literal. (ii) No C-literal may be to the left of an identical B-literal. (iii) No B-literal may be in the same cell as a complementarily identical B-literal. (iv) No B-literal may be in the cell immediately to the left of a complementarily identical A-literal. (v) No A-literal formed in a non-unit extension may be complementarily subsumed by a unit input chain.

7. Semantic Guidance

SGLD is GLD using two forms of semantic guidance : (i) Truth value deletion in some parts of deductions, and (ii) A heuristic function that has a preference for FALSE chains.

7.1 Linear Input Analysis

There are syntactically identifiable situations in which A- and C-reduction does not occur in GLD, i.e. situations in which linear-input subdeductions are performed. Three methods of analysing sets of input chains have been developed for detecting these situations. The first method (Horn subset analysis) focuses on Horn input chains while the second (LISS analysis) and third (LISL analysis) are successive generalisations of the first method. The start of a linear-input subdeduction is identified by the selection of a linear-input B-literal for an extension operation. This literal is called the top literal of the subdeduction. A linear-input subdeduction ends when its top literal is truncated.

In a linear-input subdeduction the reduction operations can be explicitly ignored, so that no effort is expended attempting to find A-literals or C-literals to reduce against. If Horn subset analysis is used then only the positive literals of Horn subset input chains need ever be considered when searching for a suitable input chain in an extension operation. A more significant benefit that may be derived from linear-input analysis is the completeness of a truth value deletion strategy in linear-input subdeductions.

> **Def'n** : The *rightwards subchain* of a literal in a center chain consists of the literal and all literals to its right.

A truth value deletion system which requires all rightwards subchains of the top literal in a linear-input subdeduction to be interpreted as FALSE, in all models of the side parent chains used in the subdeduction, is complete. Such a system is integrated into SGLD's search guidance system : the heuristic function fails to return a value if a center chain has no ground instances which are acceptable to the deletion system, and the center chain under consideration is rejected. Linear-input analysis and the truth value deletion system are described in [Sutcliffe, 1992].

7.2 FALSE Preference

Truth value deletion rigidly expects chains to be interpreted as FALSE in an interpretation. By softening this expectation to having a preference for parent chains that are interpreted as FALSE, a new strain of truth value guidance systems is formed. These guidance systems are called *truth value preference systems*. A truth value preference system is used in SGLD by using a heuristic function which calculates the FALSity level of center chains. The FALSity level of a center chain is a non-worsening function of the FALSity levels of its ground universe instances; the FALSity level of a ground universe instance of a center chain is a non-worsening function of the FALSity levels of its literals; and the FALSity level of a ground literal is FalseScore if the literal is interpreted as FALSE, TrueScore if the literal is interpreted as TRUE. The values of FalseScore and TrueScore are parameters to the preference system. FalseScore is a better FALSity level than TrueScore. The effect of the FALSE preference is to minimise the number of reduction operations in a deduction.

8. Conclusion

SGLD has been implemented in Prolog, and performance testing shows the efficacy of using semantic information to guide search. As well the features described above, SGLD has facilities for imposing sort value deletion and for embedding equality. Sort value deletion is imposed via the same mechanism as the truth value deletion, but throughout deductions. Equality is embedded via an extension of the RUE and NRF inference rules [Digricoli, 1979].

References

Digricoli V.J., Automatic Deduction and Equality, In Martin A.L. (Ed.), *Proceedings of the Annual Conference of the ACM* (Detroit, MI, 1979), ACM Press, New York, NY, (1979), 240-250.

Loveland D.W., A Simplified Format for the Model Elimination Theorem-Proving Procedure, In *Journal of the ACM* 16(3), ACM Press, New York, NY, (1969), 349-363.

Shostak R.E., Refutation Graphs, In *Artificial Intelligence* 7, Elsevier Science, Amsterdam, The Netherlands, (1976), 51-64.

Sutcliffe G., A General Clause Theorem Prover, In Stickel M.E. (Ed.), *Proceedings of the 10th International Conference on Automated Deduction* (Kaiserslautern, Germany, 1990), (Siekmann J.H. (Ed.), *Lecture Notes in Artificial Intelligence 449*), Springer-Verlag, New York, NY, (1990), 675-676.

Sutcliffe G., Linear-Input Subset Analysis, In Kapur D. (Ed.), *Proceedings of the 11th International Conference on Automated Deduction* (Saratoga Springs, NY, 1992), Springer-Verlag, New York, NY, (1992).

The SHUNYATA System*

Kurt Ammon†
Windmühlenweg 27, W-2000 Hamburg 52
Germany

SHUNYATA contains heuristics which model reasoning processes of mathematicians in the discovery of proofs. For example, a contradiction heuristic proves the negation of a proposition. It introduces the assumption that the proposition holds and attempts to derive a contradiction. A case heuristic applies the method of proof by cases. It first constructs a proposition, which represents a case, and derives a proof on the assumption that either the case or its negation hold. A composition heuristic composes predicates and functions which represent elementary concepts. These compositions, which represent formulas or sets, form the central "ideas" of proofs. Another heuristic changes the representation of a proposition to be proved and attempts to prove the changed proposition. Some heuristics control the application of other heuristics, for example by time limits which interrupt a heuristic if it achieves no result. SHUNYATA selects the definitions and lemmas needed in a proof from a knowledge base of definitions and lemmas and produces readable proofs. In experiments, it generated proofs of Banach's fixed point theorem (Ammon, 1988a), the Schröder-Bernstein theorem, Gödel's incompleteness theorem, the Bolzano-Weierstrass theorem, and Heine's theorem. The Schröder-Bernstein theorem says that two sets M and N are equinumerous if M is equinumerous with a subset of N and N is equinumerous with a subset of M.[1] Gödel's incompleteness theorem says that every formal number theory contains an undecidable formula, i.e., neither this formula nor its negation can be proved in the theory.[2] In the proof of this theorem, the composition heuristic produced an undecidable formula. The Bolzano-Weierstrass theorem says that every infinite bounded set of real numbers has an accumulation point (see Apostol, 1957, p. 43). In the proof of this theorem, the composition heuristic produced a set whose supremum is an accumulation point of an infinite bounded set of real numbers. Heine's theorem says that every continuous function on a compact set of real numbers is uniformly continuous (see Apostol, 1957, p. 75). The generation of each proof of these theorems took several hours on an IBM PC AT. To my knowledge, SHUNYATA is the first program that automatically constructed proofs of very significant theorems in mathematical logic and standard analysis (see Bledsoe, 1977, p. 31).[3] The proofs were generated without any human intervention.

---

*This work, in whole or in part, describes components of machines or processes protected by one or more patents or patent applications in Europe, the United States of America, or elsewhere. Further information is available from the author.

†This work was supported in part by the German Science Foundation (DFG).

[1] Kaufmann (1989) checked a proof of the Schröder-Bernstein theorem.

[2] Shankar (1987) checked a detailed proof of Gödel's incompleteness theorem.

[3] Ballantyne and Bledsoe (1977) generated proofs of the Bolzano-Weierstrass theorem and Heine's theorem in nonstandard analysis which simplifies these proofs considerably.

1.	*assume* $f(function(f, \mathbf{N}) \wedge$ $\forall n(n \in \mathbf{N} \rightarrow f(n) \subseteq \mathbf{N}) \wedge$ $\forall S(S \subseteq \mathbf{N} \rightarrow \exists n(n \in \mathbf{N} \wedge S = f(n))))$	Contradiction
2.	*assume* $S(S \subseteq \mathbf{N} \wedge$ $\forall n(n \in \mathbf{N} \rightarrow (n \in S \leftrightarrow \neg n \in f(n))))$	Composition
3.	$\exists n(n \in \mathbf{N} \wedge S = f(n))$	Composition, 1, 2
4.	*assume* $n(n \in \mathbf{N} \wedge S = f(n))$	Composition, 3
5.	$\forall x(x \in S \rightarrow x \in f(n))$	Case, 4, Definition
6.	$\forall x(x \in f(n) \rightarrow x \in S)$	Case, 4, Definition
7.	*assume* $(n \in S)$	Case
8.	$\neg n \in f(n)$	Forward, 2, 4, 7
9.	$n \in f(n)$	Forward, 5, 7
10.	$\neg n \in S$	Case, 7, 8, 9
11.	$\neg n \in f(n)$	Forward, 6, 10
12.	$n \in S$	Forward, 2, 4, 11
13.	$\neg \exists f(function(f, \mathbf{N}) \wedge$ $\forall n(n \in \mathbf{N} \rightarrow f(n) \subseteq \mathbf{N}) \wedge$ $\forall S(S \subseteq \mathbf{N} \rightarrow \exists n(n \in \mathbf{N} \wedge S = f(n))))$	Contradiction, 1, 10, 12

Table 1: SHUNYATA's proof that the set of all sets of natural numbers is not enumerable

We illustrate the architecture of SHUNYATA by its proof of the theorem that the set of all sets of natural numbers is not enumerable, i.e., there is no function f on the set of natural numbers $\mathbf{N}$ into the subsets of $\mathbf{N}$ such that for all subsets $S \subseteq \mathbf{N}$, there is $n \in \mathbf{N}$ with $S = f(n)$. A formalization of the theorem is:

$$\begin{aligned} &\neg \exists f(function(f, \mathbf{N}) \wedge \\ &\quad \forall n(n \in \mathbf{N} \rightarrow f(n) \subseteq \mathbf{N}) \wedge \\ &\quad \forall S(S \subseteq \mathbf{N} \rightarrow \exists n(n \in \mathbf{N} \wedge S = f(n)))) \end{aligned}$$

Table 1 contains a proof of this theorem which was generated by SHUNYATA. Each proof step consists of a number, a proposition or an assumption, and an explanation which gives the heuristic, preceding proof steps, and definitions that were used to derive the proposition. For example, step 3 was generated by the composition heuristic which applied the universal proposition in the last line of step 1 to the relation $S \subseteq \mathbf{N}$ in step 2. The proof in Table 1 was generated by SHUNYATA as follows: Because the theorem to be proved is the negation of a proposition, the contradiction heuristic introduces the assumption in step 1 that this proposition holds. In order to apply the antecedent $S \subseteq \mathbf{N}$ of the universal proposition in the third line of step 1, the composition heuristic constructs examples of such sets $S \subseteq \mathbf{N}$. Because $f(n) \subseteq \mathbf{N}$ for all $n \in \mathbf{N}$, the application of the element relation "$\in$" and the negation "$\neg$" to an element $n \in \mathbf{N}$ yields the predicate $\neg n \in f(n)$ which defines a subset $S \subseteq \mathbf{N}$ in step 2. Therefore, the universal proposition in the third line of step 1 implies the existence of an $n \in \mathbf{N}$ such that $S = f(n)$ in steps 3 and 4. The application of the definition of the equality relation between sets to the equality relation $S = f(n)$ in step 4 yields steps 5 and 6. Analogously to the composition

$\{ (t_1 = t_2, a, s) :$	$\textit{is-a-proof-step}(t_1 = t_2, a, s) \wedge$ $t_1 = \textit{left-side}(\textit{THEOREM}) \vee$ $t_1 \in \textit{right-sides}(\textit{equations}(\textit{PREC.-STEPS})) \wedge$ $\textit{substituents}(s) \subseteq \textit{subterms}(\textit{left-side}(\textit{THEOREM})) \}$
$\{ (t_1 = t_2, a, s) :$	$\textit{is-a-proof-step}(t_1 = t_2, a, s) \wedge$ $t_1 \in \textit{right-sides}(\textit{equations}(\textit{PREC.-STEPS})) \wedge$ $t_2 \neq \textit{left-side}(\textit{THEOREM}) \wedge$ $\textit{variables}(\textit{left-side}(a)) \supseteq \textit{variables}(\textit{right-side}(a)) \wedge$ $\textit{number-of-symbols}(\textit{left-side}(a)) \geq$ $\textit{number-of-symbols}(\textit{right-side}(a)) \}$

Table 2: A theorem prover which was generated by SHUNYATA

heuristic, the case heuristic constructs the assumption $n \in S$ in step 7. Because the forward heuristic, which applies propositions in preceding proof steps, produces a contradiction in steps 8 and 9, the case heuristic generates the negation of the assumption $n \in S$ in step 10. Now, another application of the forward heuristic produces another contradiction in steps 11 and 12 which yields the theorem to be proved in step 13.

The composition heuristic in SHUNYATA was originally developed as a learning procedure which constructs mathematical conjectures, heuristics, and even theorem provers. In an experiment, the procedure constructed Goldbach's conjecture which says that every even number is the sum of two prime numbers. For example, the even number 8 is the sum of the prime numbers 3 and 5. Another conjecture in this experiment said that for every prime number p, there is a prime number q and a natural number n such that $p + q = 3n$ holds. This conjecture came as a surprise to me because I could not easily find a proof or a refutation of this conjecture. Even a specialist in number theory of the mathematics department of the University of Hamburg could not immediately tell whether the conjecture was known or provable. After some days he told me that it was true and related to the theory of arithmetic progressions. A proof of this conjecture, which was discovered with the assistance of the SHUNYATA program, is: For $p = 3$, you can choose $q = 3$ and $n = 2$. For any prime number $p \neq 3$, you can choose $q = 2$ or $q = 7$. Any sequence of three consecutive natural numbers such as 7, 8, and 9 contains one number that is divisible by 3 such as 9. In particular, the sequence p, $p + 1$, and $p + 2$ contains a number that is divisible by 3. Therefore, either $p + 1$ or $p + 2$ is divisible by 3 because p is a prime number. If $p + 1$ is divisible by 3, $p + 7$ is also divisible by 3 because the difference 6 between $p + 1$ and $p + 7$ is divisible by 3. It follows that either $p + 7$ or $p + 2$ is divisible by 3, i.e., either $p + 7 = 3n$ or $p + 2 = 3n$ holds with a natural number n. This completes the proof.

In another experiment, the learning procedure produced the theorem prover in Table 2 by analyzing a proof of the simple theorem that $(x^{-1})^{-1} = 1$ holds for all elements x of a group. Ammon (1988b) describes an earlier comparable experiment which applied the learning procedure to another proof in group theory. A group is a set with a binary associative operation such that the axioms $1x = x = x1$ and $x^{-1}x = 1 = xx^{-1}$ hold for all elements x of the group. A very simple theorem in

Equation ($t_1 = t_2$)	Axiom (a)	Substitution (s)
$1^{-1} = 1^{-1}1$	$x = x1$	$1^{-1}/x$
$1^{-1}1 = 1$	$x^{-1}x = 1$	$1/x$

Table 3: A proof of the *THEOREM* $1^{-1} = 1$

group theory is $1^{-1} = 1$, i.e., the inverse of the identity 1 of a group is equal to the identity 1 itself. Table 3 gives a proof of this theorem which was generated by the theorem prover in Table 2. The proof consists of two proof steps. The first proof step substitutes the term 1^{-1} for the variable x in the axiom $x = x1$ which yields the equation $1^{-1} = 1^{-1}1$. The second proof step substitutes the identity 1 for the variable x in the axiom $x^{-1}x = 1$ which yields the equation $1^{-1}1 = 1$. The first proof step is generated by the first set constructor in Table 2. The first three lines in this set constructor say that the left side t_1 of an equation $t_1 = t_2$ is the left side of the *THEOREM* or the right side of an equation in a preceding proof step: The left side 1^{-1} of the equation in the first proof step is the left side of the *THEOREM* $1^{-1} = 1$. The last line of the first set constructor says that the substituents of the substitution s are subterms of the left side of the *THEOREM*: The substituent 1^{-1} in the first proof step in Table 3 is a subterm of the left side 1^{-1} of the *THEOREM* $1^{-1} = 1$. The second set constructor in Table 2 produces the second proof step. The last three lines of the second set constructor say that the variables in the right side of an axiom are contained in its left side and that the number of symbols in the right side of an axiom is less than or equal to the number of symbols in its left side. This entails a simplification of right side $1^{-1}1$ of the first equation which results in the right side 1 of the *THEOREM* $1^{-1} = 1$ in the second proof step.

An optimizing compiler translates the theorem prover in Table 2, which is represented by two set constructors, into efficient procedural code. In an experiment, the translated theorem prover produced proofs of nine theorems in group theory without any human intervention. The first five theorems said that the identity element of a group is unique, the inverse of each element is unique, a left identity is a right identity, a left inverse is a right inverse, and the inverse of the identity is the identity itself (see Table 3). The sixth theorem said that $x^{-1}y^{-1}yx = 1$ holds for all elements x and y of a group, the seventh theorem that $x^2 = x$ implies $x = 1$, the eighth theorem that linear equation solvability implies the existence of an identity, and the ninth theorem that $x^2 = 1$ implies group commutativity. Table 4 compares the run times of the theorem prover in Table 2 and McCune's (1990, 1991) theorem prover OTTER 2.2 for proving these theorems on an IBM PC AT. It shows that the theorem prover in Table 2, which was generated by SHUNYATA on the basis of an analysis of a simple proof, is faster than OTTER 2.2 in all theorems but the last one. OTTER 2.2 takes 8 minutes and 22.03 seconds to generate a proof of SAM's Lemma on an IBM PC AT from the axioms for a modular lattice in McCharen *et al.* (1976). This set of axioms is not complete because it does not contain the ordinary lattice operations "join" and "meet". A variant of the theorem prover in Table 2 generated a proof of SAM's Lemma from a complete set of axioms in ordinary representation in 3 minutes and 43.66 seconds. Because the complexity of the heuristics in SHUNYATA, which produce proofs of very significant theorems in mathematics

Theorem	Theorem prover in Table 2	OTTER 2.2
1	.66 sec	3.46 sec
2	9.06 sec	10.88 sec
3	2.69 sec	6.26 sec
4	8.01 sec	9.51 sec
5	1.93 sec	3.41 sec
6	9.45 sec	31.92 sec
7	6.70 sec	9.62 sec
8	11.46 sec	98.41 sec
9	29.27 sec	10.16 sec

Table 4: A run time comparison between the theorem prover in Table 2 and OTTER 2.2

such as Heine's theorem and Gödel's incompleteness theorem, is comparable with the complexity of the theorem prover in Table 2, the experiments with the learning procedure suggest that SHUNYATA's heuristics and thus SHUNYATA itself can be developed automatically.

SHUNYATA is written in COMMON LISP. Its source code is some 600,000 characters long. This number includes comments and an optimizing compiler for the language CL in which the theorem prover in Table 2 is represented.

References

Ammon, K. 1988a. Discovering a proof for the fixed point theorem: A case study. *Proceedings of the Eighth European Conference on Artificial Intelligence*, August 1–5, Munich, West Germany, pp. 613–618.

Ammon, K. 1988b. The automatic acquisition of proof methods. *Proceedings of the Seventh National Conference on Artificial Intelligence*, August 21–26, St. Paul, Minnesota, pp. 558–563.

Apostol, T. M. 1957. *Mathematical Analysis.* Reading, Mass.: Addison-Wesley.

Ballantyne, A. M., and Bledsoe, W. W. 1977. Automatic proofs of theorems in analysis using nonstandard techniques. *Journal of the Association of Computing Machinery* 24, pp. 353–374.

Bledsoe, W. W. 1977. Non-resolution theorem proving. Artificial Intelligence 9, pp. 1–35.

Kaufmann, M. 1989. DEFN-SK: An extension of the Boyer-Moore Theorem Prover to handle first-order quantifiers. Technical Report 43, Computational Logic, Inc., Austin, Texas.

McCharen, J. D., Overbeek, R. A., and Wos, L. A. 1976. Problems and experiments for and with automated theorem-proving programs. *IEEE Transactions on Computers*, Vol. C-25, No. 8, pp. 773–782.

McCune, W. W. 1990. OTTER 2.0 Users Guide. Report ANL-90/9, Argonne National Laboratory, Argonne, Illinois.

McCune, W. W. 1991. What's new in OTTER 2.2. Report ANL/MCS-TM-153, Argonne National Laboratory, Argonne, Illinois.

Shankar, N. 1987. Proof checking metamathematics. Technical Report 9, Computational Logic, Inc., Austin, Texas.

A Geometry Theorem Prover for Macintoshes*

Shang-Ching Chou

Department of Computer Science
The Wichita State University, Wichita, KS 67208, USA
e-mail: chou@wiz.wsu.ukans.edu

This geometry prover for Macintoshes is based on a prover developed on Symbolics Lisp Machines which is in turn based on the work by Wu [6] [7]. It can prove a subset of theorems that the original prover proved. It addresses a class of geometry statements called class C ([1] or [3]) and is complete for a subclass of class C, called class C_E [1]. It is powerful enough to prove many hard theorems such as Pappus' theorem, Simson's theorem, Pascal's theorem, the nine-point theorem, Feuerbach's theorem, Steiner's theorem, etc. With a further extension, it is expected to prove over 90% of the theorems that the original prover proved, including Morley's trisector theorem. this prover is mainly based on the method described in [1] or [3].

A disk of Mac plus or SE version can be obtained by writing to me. In the disk, there are a system fold, the program, a file (named "sample.lisp") containing 12 sample examples, and a documentation [2] (serving as a kind of manual) in the TEX file form. The disk can run on the mac plus and SE(SE/30) without using a hard disk.

1. Input (or Specification of a Geometry Statement)

The first step to convert a geometry statement into its algebraic form is to assign coordinates to the points in the statement. On the prover this can be done either manually or automatically. Here we show how to specify the input (the geometry statement) for the program to assign coordinates automatically. *The statement must be specified in a constructive way.*

Most theorems (involving only equalities) in elementary geometry can be stated in a constructive way, i.e., given a set of arbitrarily chosen points, new points are added in a constructive way by introducing one or two geometry conditions (constructions) for each point, such as taking an arbitrary point on a line or on a circle (one condition), or taking intersections of two lines or one line and one circle, or two circles (two conditions). The conclusion is a (equality) relation among these points.

The user *must* have a clear idea about the order in which the points are introduced.

The input is in the Lisp list form (though the present program was written in Pascal and has nothing to do with Lisp; we use this input format for consistency with the input of our original prover in Lisp).

* The work reported here was supported in part by the NSF Grants CCR-9002362 and CCR–9117870

The first element in the input is a list beginning with the keyword "point-order", followed by the points in the constructive order (arbitrarily chosen points are first). The following elements (but the last) of the input are the geometry conditions to construct the points. They are the equality part of the hypotheses of the geometry statement. The order of these geometry conditions is not important. They can be arranged in any order. But it is recommended that they be written according to the point order, which is important. The last element is the geometry condition representing the conclusion to be proved. Following are two examples [3].

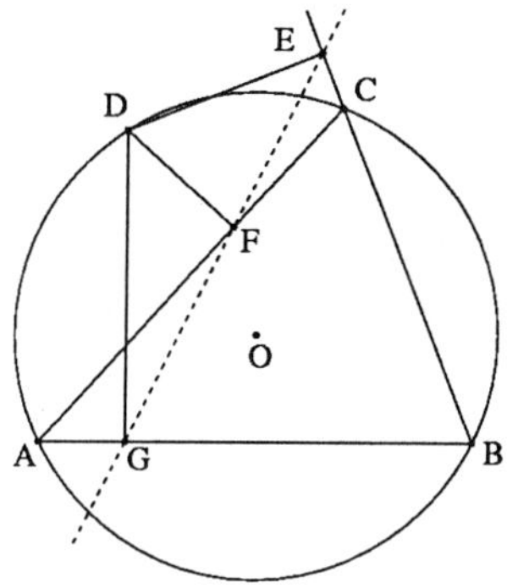

Figure 1

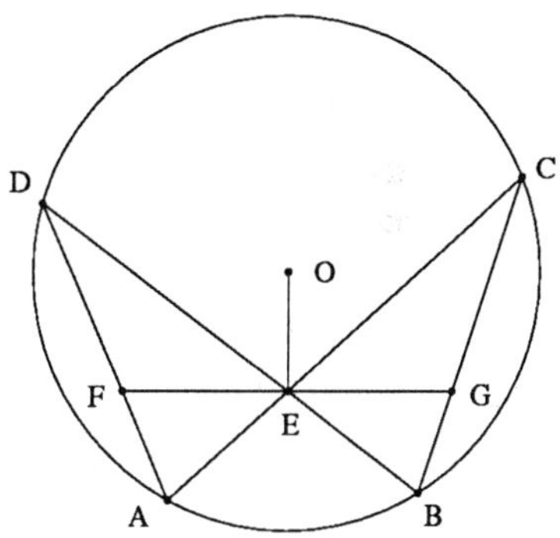

Figure 2

Example 1. (Simson's theorem) Let D be a point on the circumscribed circle (O) of triangle ABC. From D three perpendiculars are drawn to the three sides BC, CA and AB of $\triangle ABC$. Let E, F and G be the three feet respectively. Show that E, F and G are collinear (Figure 2).

The input is

```
(setq simson
  '((point-order A B C O D E F G)
    (collinear A B G)
    (collinear B C E)
    (collinear A C F)
    (perpendicular E D B C)
    (perpendicular F D A C)
    (perpendicular G D A B)
    (eqdistance A O B O)
    (eqdistance A O C O)
    (eqdistance A O D O)
    (collinear E F G)))
```

Here we deliberately write the hypothesis conditions not in the order in which the points are introduced. Permuting those conditions will not affect the constructions (hence the selection of coordinates in any way). But if you change the point order, then the constructions will be changed. E.g., if you put point O first, then the program can figure out you first have a circle, then have three points on the circle, etc. For the above input, the constructions are (You can select the Item "Show Thm in Constructions" in the Info Menu to show them on the screen):

Points A, B, C, are arbitrarily chosen
O is on B(A B) and B(A C)
D is on R(O, A O)
E is on L(B C) and T(D, B C)
F is on L(A C) and T(D, A C)
G is on L(A B) and T(D, A B).

For method for generating construction sequences from the input, see [1] or [2]. The most common lines which can be constructed from the preceding points are one of the following four types:

L(A B): the line passing through points A and B.

P(A, B C): the line passing through A and parallel to line BC.

T(A, B C): the line passing through A and perpendicular to line BC.

B(A B): the perpendicular bisector of the segment AB.

The most common circles which can be constructed from the preceding points are one of the following two types:

R(O, A B): the circle with the center O and radius AB.

C(A B C): the circle passing through the three points A, B, and C.

If you change the first line of input to (point-order $O\ A\ B\ C\ D\ E\ F\ G$), then the construction sequence is

Points O, A, are arbitrarily chosen
B is on R(O, A O)
C is on R(O, A O)
D is on R(O, A O)
E is on L(B C) and T(D, B C)
F is on L(A C) and T(D, A C)
G is on L(A B) and T(D, A B).

The program generates non-degenerate conditions from these construction sequences according to the method described in [3]. Select the Item "Show Degenerate Conds" in the Info Menu to see the differences of non-degenerate between these two point orders (Also see Section 2).

Example 2. (The Butterfly Theorem). A, B, C and D are four points on circle (O). E is the intersection of AC and BD. Through E draw a line perpendicular to OE, meeting AD at F and BC at G. Show that $FE \equiv GE$ (Figure 2).

The input is

(setq butterfly
  '((point-order $E\ O\ A\ B\ C\ D\ F\ G$)
    (eqdistance $O\ A\ O\ B$)
    (eqdistance $O\ A\ O\ C$)

(eqdistance O A O D)
(perpendicular O E E F)
(collinear A E C)
(collinear B E D)
(collinear A D F)
(collinear F E G)
(collinear B C G)
(midpoint F E G)))

In the file "sample.lisp", there are another 12 theorems. These 12 theorems can be used to test a new geometry theorem prover based on similar algebraic methods. For a list of geometry conditions for the input, see [2].

2. On Non-Degenerate Conditions

The geometric conditions of input (or their algebraic equivalents) usually include degenerate cases. For example, the exact meaning of (parallel A B C D) is $[(A = B) \vee (C = D) \vee (AB \parallel CD)]$. It is not instructive to exclude the cases when $A = B$ and $C = D$. By doing so, we still cannot exclude all degenerate conditions necessary for a geometry statement to be valid.

In the current research, there are two Approaches or Formulations to prove geometry statements [4] and [5].

Formulation F1 is to prove the above statement to be generally or generically true, at the same time giving sufficiently many non-degenerate conditions. Those conditions usually are in algebraic form. However, for a class of geometry statements of constructive type specified in [1] and [3], we can generate sufficiently many non-degenerate conditions in *geometric form*. In [1]or [3], we have proved a theorem stating that if a statement is still not valid under these machine generated non-degenerate conditions, then the statement cannot be valid no matter how many more reasonable non-degenerate conditions are added.

After selecting a theorem from the Menu "Theorems", click the Item "Show degenerate Conds" in the Info Menu, you will get those degenerate conditions.

Acknowledgment. The author wishes to thank R. S. Boyer for his advice, help, and encouragement to write a Mac version of my geometry theorem prover and thank N. McPhee for his help and suggestions in writing this Mac program.

References

[1] S.C. Chou and X.S. Gao, "A Class of Geometry Statements of Constructive Type and Geometry Theorem Proving", TR-89-37, Computer Sciences Department, The University of Texas at Austin, November, 1989.

[2] S.C. Chou, "A Geometry Theorem Prover for Macintoshes", TR-91-8, Computer Sciences Department, The University of Texas at Austin, March, 1991.

[3] S.C. Chou and X. S. Gao, " Proving Geometry Statements of Constructive type", submitted to CADE–11.

[4] S.C. Chou and G.J. Yang, "On the Algebraic Formulation of Certain Geometry Statements and Mechanical Geometry Theorem Proving", *Algorithmica*, Vol. **4**, 1989, 237–262.

[5] D. Kapur and Hoi K. Wan, "Refutational Proofs of Geometry Theorems via Characteristic Set Computation", in *Proceedings of International Symposium on Symbolic and Algebraic Computation*, ACM Press, 1990, 277–284.

[6] Wu Wen-tsün, "On the Decision Problem and the Mechanization of Theorem Proving in Elementary Geometry", *Scientia Sinica* **21** (1978), 157-179.

[7] Wu Wen-tsün, "Basic Principles of Mechanical Theorem Proving in Geometries", *J. of Sys. Sci. and Math. Sci.* **4(3)**, 1984, 207-235, republished in *Journal of Automated Reasoning* **2(4)** (1986), 221-252.

FRI: Failure-Resistant Induction in RRL*

Xin Hua, Hantao Zhang
Department of Computer Science
The University of Iowa
Iowa City, IA 52242, U.S.A.
{xin, hzhang}@cs.uiowa.edu

1 Introduction

Inductive reasoning is known to play a critical role in analyzing and designing specifications as well as reasoning about software and hardware. Four years ago, a method called the cover set induction was implemented in *RRL* and was presented at CADE-88 [10]. The method is based on the concept of the ***cover set*** of a data type for mechanizing proofs by explicit induction for equational specifications. The method is a generalization of Burstall's structural induction principle and is closely related to Boyer and Moore's approach [1]. During the past year, we have applied the method to problems in number theory [6], graph theory [8], software verification [7] and hardware verification [3].

In the beginning of this project, we made a great effort to completely automate the method. However, as the method is applied to larger problems, the search space becomes intractable. It is impractical to design a fully inductive theorem prover capable of doing what *RRL* can do today. Moreover, a drawback of the first implementation of the method is that when it fails, it is often not clear why: — the conjecture being proved might be false, or a proof of the conjecture needs either a different sequence of proof steps or additional lemmas.

Since there are no general inductive heuristics, when a proof fails we need facilities for human intervention. That is the motivation for FRI (Failure-Resistant Induction), a proof manager for supporting the cover set method. When a failure occurs, the user is provided with a set of tools which are useful in completing the proof. Through this manager, we obtain a failure-resistant implementation of the cover set method.

Basically, FRI dynamically maintains a proof tree of the current task. Each node of the tree is marked by a subgoal (the root is marked by the goal) and the link to its parent is marked by the name of the inference rule which derived that subgoal from its parent. If a proof is completed successfully, the user may refer the proof tree to see how the proof is obtained. When a proof fails, the user may backtrack along some branch of the proof tree to any node and retry that step with hints from the user or with some special techniques provided by FRI. Thus, the user can choose the extent of automation from a manual proof to a non-stop automatic proof.

Our work has benefited from the AFFIRM project [2]. In the AFFIRM theorem prover, proofs are presented by trees to facilitate the management of large proofs. We were also inspired by

*Partially supported by the National Science Foundation Grants no. CCR-9009414 and INT-9016100.

Kaufmann's work [5]: In addition to the hint facilities provided by the Boyer-Moore Theorem Prover (BMTP) [1], Kaufmann has successfully built an interactive interface on the top of BMTP.

2 Major Facilities of FRI

FRI is written in *RRL* [4], a theorem proving environment for experimenting and developing reasoning methods based on rewrite techniques and equational logics. FRI is an integral part of *RRL* supporting the cover set induction method [10]. That is, the codes of FRI and *RRL* are indistinguishable.

There are two ways to start FRI: Invoke the option **manager** provided by the *RRL* main menu; the other occurs automatically when a failure in an induction proof takes place. In the later case, when the user is using the cover set induction method in *RRL*, FRI works together with *RRL* to maintain the proof tree, but is not visible to the user. Only when a failure occurs, FRI will show up to offer help. The help includes manipulations of the proof tree and various manual processing tools.

In the following, we describe briefly the facilities of FRI offered when an induction proof fails; these facilities are similar to those offered by FRI when the user enters FRI by invoking the option **manager** from the main menu of *RRL*.

2.1 Manipulation of Proof Trees

When the cover set induction method fails at some step, *RRL* will report the failed subgoal to the user and enter FRI's Failure-Handler menu.

```
      **********************************************
      *          Failure Handler of FRI            *
      **********************************************
Type Nodes, One-node, Tree, Cursor, Restart, Exit, Abort, Help -->
```

The above menu provides tools to the user for manipulating proof trees.

- **Nodes:** List the equations associated with each node in the proof tree.
- **One-node:** Show all the details of the current node pointed to by the cursor, including the source, state, status and so on.
- **Tree:** Display the proof tree.
- **Cursor:** Enter the Cursor Control menu which allows the user to move the cursor around the proof tree and cut subtrees from the proof tree if necessary. (The description of the Cursor Control menu is omitted here for brevity.)
- **Restart:** The user is asked to specify a node in the proof tree. Then FRI will backtrack to that node and enter the Manual Proof Control menu; the proof will be restarted with the control of the user.

2.2 Manual Proof Control

After the user has chosen **restart** in the Failure Handler's menu, FRI jumps to the node indicated by the user and shows the Manual Proof Control menu as follows.

```
          +--------------------------------------------+
          |          Manual Proof Control of FRI        |
          +--------------------------------------------+
Type New-rule, Old-rule, Reverse, Continue, Induction, Hypothesis, Structure
     Generalize, Add-premise, Split, Assumed, Abort, Exit, Help -->
```

We describe briefly some of the options provided in the menu:

- **New-rule:** Let the user enter a lemma to assist the proof. FRI converts the new lemma into a rewrite rule and try to rewrite the current goal using this rule. When the whole proof is finished, FRI will check whether there exist unproved lemmas and prompt the user to prove them (if any) to ensure the soundness of the entire proof.

- **Old-rule:** Let the user choose one rule from the existing rule set to rewrite the current goal once.

- **Reverse:** The same as "Old-rule" except that the chosen rule is used in reverse. With this option, the user can obtain a proof even though the current rewrite system is not *Church-Roser*, because this option allows any equation to be used in both directions without having the terminating problem.

- **Induction**: Let the user choose a different induction scheme by providing a function name; this function may not exist in the current subgoal, but must exist in the system and must be complete with respect to constructors [6]. Actually, the definition of that function decides the chosen induction scheme. The user may define a *dummy* function to generate the desired induction scheme.

- **Hypothesis:** Convert the induction hypothesis into a general rewrite rule so that the user can simulate the proofs in the approach of inductive completion. This option and the above option allow the user to find the best induction scheme and the best way of using induction hypotheses.

- **Add-premise:** Let the user add a valid premise into the premises of the current subgoal. For instance, if $prime(x)$ is in the premises of the current subgoal, then $(x \neq 0)$ is a valid premise which can be added by this option, assuming the equation $x \neq 0$ **if** $prime(x)$ is already proved before.

- **Assumed:** Assume the current subgoal is proved, and continue the proofs of other subgoals; when other subgoals are all proved, the user has to prove this subgoal again.

2.3 An Example

We illustrate the usage of FRI by a small example. Suppose $+$ and $<$ are defined as usual. Let the remainder function *rem* be defined as follows:

```
rem(x, 0) := x
rem(x, y) := x if x < y
rem(y + x, y) := rem(x, y)
```

We input the equation **rem(x, suc(0)) == 0** to *RRL* and *RRL* prints the following message:

```
Let P(x) be [main] rem(x, suc(0)) == 0

The induction will be done on x and will follow the scheme:
    [#1] P(0)
    [#2] P(suc(x)) if  { P(x) }
 ... ...
    The proof failed at node (3):
    [#2] rem(suc(x), suc(0)) == 0 if (rem(x, suc(0)) = 0)
```

RRL then enters FRI. With the assistance of FRI, we finish the proof by introducing a simple lemma.

```
    WELCOME to FRI (Failure-Resistance Induction) !!!

     ***********************************************
     *          Failure Handler of FRI             *
     ***********************************************
Type Nodes, One-node, Tree, Cursor, Restart, Exit, Abort, Help --> restart
     Please indicate the node number: 3

      +-------------------------------------------+
      |          Manual Proof Control of FRI      |
      +-------------------------------------------+
The current equation is:  [#2] rem(suc(x), suc(0)) == 0 if (rem(x, suc(0)) = 0)

Type New-rule, Old-rule, Reverse, Continue, Induction, Hypothesis,
     Generalize, Add-premise, Split, Assumed, Abort, Exit, Help --> New-rule
Enter your equation: suc(x) == (suc(0) + x)
 ... ...
 By the rule:  [10] suc(x) ---> (suc(0) + x)
    rem(suc(x), suc(0)) == 0  if  { (rem(x, suc(0)) = 0) }
    is reduced to: rem((suc(0) + x), suc(0)) == 0  if   { (rem(x, suc(0)) = 0) }

 By the rule:   [9] rem(y + x, y) ---> rem(x, y)
    rem((suc(0) + x), suc(0)) == 0  if   { (rem(x, suc(0)) = 0) }
    is reduced to:  rem(x, suc(0)) == 0  if   { (rem(x, suc(0)) = 0) }
 ... ...
All subgoals of [main] are proven, hence
    [main] rem(x, suc(0)) == 0
    is an inductive theorem.
```

Before the end of the proof, FRI reminds us to prove the added lemma, which can be easily proved from the definition of $+$. Note that in this example, even $suc(x) = (suc(0)+x)$ is proved in advance, FRI's assistance is still necessary, because *RRL* can orient this equation only as $(suc(0) + x) \rightarrow suc(x)$.

3 Summary

We have described briefly a proof manager called FRI, which supports the cover set induction method of *RRL*. We believe that FRI makes the cover set method more powerful and applicable to a wider class of equational theories for automating inductive proofs. Without FRI, we could not obtain a proof of the Chinese Remainder theorem [9].

Any successful inductive prover must be based on heuristics, which involve at least the following operations: (a) choice of induction schemes, (b) use of induction hypotheses, (c) conjecturing lemmas, (d) generalization of induction formulas, (e) simplification strategies. In addition to the heuristics provided by ***RRL***, FRI provides the facilities to perform any of the operations listed above.

FRI can also help us to develop new heuristics for further automating cover set induction. In fact, when we started to work on Ramsey's theorem [8], ***RRL*** could not prove each lemma of Ramsey's theorem automatically; they were proved semi-automatically with the assistance of FRI. The semi-automatically generated proof inspired us to add new techniques to increase the automatic power of ***RRL***. Finally, we were able to obtain an automatic proof of Ramsey's theorem.

In the near future, we will further improve FRI and plan to integrate FRI with the TECTON system developed by Kapur *et al.* so that proof trees created by FRI can be displayed graphically in a hypertext environment.

References

[1] Boyer, R.S., Moore, J.S.: (1979) A computational logic. New York: Academic Press.

[2] Erickson, R.W., Musser, D.R.: (1980) The AFFIRM theorem prover: proof forests and management of large proofs. *Proc. 7th Principles of Programming Languages.*

[3] Hua, X., Zhang, H.: (1991) Axiomatic semantics of a hardware specification language. Submitted to the Second IEEE Great Lakes Symposium on VLSI, Kalamazoo, MI, February 1992.

[4] Kapur, D., Zhang, H.: (1989) An overview of RRL: Rewrite Rule Laboratory. *Proc. of the third International Conference on Rewriting Techniques and its Applications* (RTA-89), April 1989, Chapel Hill, NC, Springer Verlag LNCS 355, 513-529.

[5] Kaufmann, M.: (1988) An interactive enhancement to the Boyer-Moore theorem prover. Proc. of *Ninth International Conference on Automated Deduction (CADE-9)*, Argonne, IL, May 1988. Springer-Verlag LNCS 310, pp. 735-736.

[6] Zhang, H.: (1988) Reduction, superposition and induction: automated reasoning in an equational logic. Ph.D. Thesis, Department of Computer Science, Rensselaer Polytechnic Institute, Troy, NY.

[7] Zhang, H., Guha, A., Hua, X.: (1991) Using algebraic specifications in Floyd-Hoare assertions. In: Rus, T. (ed.): Proc. of Second International Conference on Algebraic Methodology and Software Technology, Iowa City, Iowa.

[8] Zhang, H., Hua, X.: (1991) Proving Ramsey theorem by cover-set induction: a case and comparison study. Accepted for presentation at Second International Symposium on Artificial Intelligence and Mathematics. Fort Lauderdale, Florida.

[9] Zhang, H., Hua, X.: (1991) A computer proof of the Chinese remainder theorem. Submitted to CADE-92.

[10] Zhang, H., Kapur, D., Krishnamoorthy, M.S.: (1988) A mechanizable induction principle for equational specifications. Proc. of *Ninth International Conference on Automated Deduction (CADE-9)*, Argonne, IL, May 1988. Springer-Verlag LNCS 310, pp. 250-265.

Herky: High Performance Rewriting in RRL*

Hantao Zhang
Department of Computer Science
The University of Iowa
Iowa City, IA 52242, U.S.A.
hzhang@cs.uiowa.edu

1 Introduction

Herky[1] is a program of *RRL*[2] in which several high-performance operations are implemented to support the rewriting/completion based methods for equational reasoning.

The two key operations in the extensions of the well-known Knuth-Bendix completion procedure are ***rewriting*** and ***superposition***. To implement these two operations efficiently, we have adopted in Herky the new data structures for equations and rewrite rules, the term-indexing technique of discrimination trees for fast rewriting and term-sharing, the constraint rewriting for handling nonterminating equations [7], the cost-free techniques for detecting unnecessary critical pairs [10], the special completion procedures for Ablian group theories and the distributivity laws. The adoption of these techniques results in a state-of-the-art of a theorem prover for equational reasoning.

In the following, some examples are presented to show the performance of Herky and the use of the discrimination trees and the special completion procedures.

2 Overbeek's Competition Problems

We think that the high performance of Herky may be manifested by the proofs of Overbeek's competition problems for theorem provers. Overbeek's competition consists of two sets of problems [6]: One for the clause-based theorem provers and the other for the equation-based theorem provers. Herky can solve the first nine of the 10 problems for the equation-based theorem provers (currently, Herky cannot solve the 10th problem).

In Table 1, we listed the statistics of Herky for solving the nine problems for equational resonning. Column 2 lists the number of critical pairs (or new inferences) for each problem. Columns 3 and 4 list the numbers of rewrite rules made and retained by Herky at the time a proof is found. Column 5 gives the number of rewrite rules (including the input equations) in the proof found by Herky. The last two columns give, respectively, the total computing time and the time for normalization. The times (in seconds) are measured in

---

*Partially supported by the National Science Foundation Grants no. CCR-9009414 and INT-9016100.

[1] Herky stands for "High-performance key operations".

[2] *RRL* (*Rewrite Rule Laboratory*) is a theorem proving environment for experimenting and developing reasoning methods based on rewriting techniques and equational logics [4].

Problem	Critical pairs	Generated rules	Retained rules	Proof length	Total time	Norm. time
1	379	51	32	25	3.93	2.11
2	388	80	17	47	10.80	6.23
3	1484	440	391	33	12.26	5.25
4	528	391	26	211	6.43	1.44
5	368493	8465	8463	1660	1495.97	668.29
6	171	73	73	9	0.98	0.18
7	248	20	17	19	8.13	5.34
8	3664	2807	2807	26	214.45	14.01
9	199	61	22	36	13.97	8.77

Table 1: Statistics of Herky on Overbeek's competition problems

Sun Common Lisp on a Sun Sparc station 2 (with 64 megabyte main memory); they are not exactly the CPU time.

3 Term Indexing Techniques

The key operation of interest in theorem provers based on completion is to compute a normal form of a term with respect to a set of rewrite rules. A normal form of a term is computed by repeatedly applying an applicable rewrite rule to the term. By the use of discrimination nets for subterm indexing, the efficiency of normalization process is dramatically improved. The benchmark examples of Jim Christian [2] can be used to manifest this. These examples are represented by the following schema:

$$\begin{array}{rcll} A_n: \; f(f(x,y),z) & = & f(x,f(y,z)). & \\ f(e_j,x) & = & x & \text{for } 1 \le j \le n. \\ f(x,i_j(x)) & = & e_j & \text{for } 1 \le j \le n. \end{array}$$

That is, each A_n defines a binary operator f which is associative, and has n left-units and n right-inverses.

We have tried to complete A_n, for $n = 2i$, $1 \le i \le 12$ and for $n = 10j$, $3 \le j \le 10$, in three theorem provers: Otter [5], Hiper [2] and Herky. Table 2 presents the statistics of these runs. All the run times are collected by choosing the best of three identical runs on a Sun Sparc station 1 (16 megabyte memory) and are measured in seconds.

In all these examples Dershowitz' recursive path ordering is used with the following precedence relation:

$$e_1 < e_2 < \cdots < e_n < f < i_1 < i_2 < \cdots < i_n$$

In addition, f has the left-to-right status. As a result, all the three theorem provers produce the same canonical rewrite system for each A_n (the number of the rules in each canonical system is given in the second column of the table).

Otter is a resolution-based theorem prover implemented in C by Bill McCune at Argonne National Laboratory [5]. Otter supports a rich set of inference rules, including the

Prob.	Final rules	Otter pairs	Otter time	Hiper pairs	Hiper time	Herky pairs	Herky time
A_2	12	491	0.4	149	0.5	112	0.7
A_4	18	1261	1.6	421	2.0	224	1.5
A_6	24	2655	4.1	957	4.6	376	2.7
A_8	30	4865	9.6	1853	9.9	568	4.2
A_{10}	36	8083	20.9	3205	17.8	800	6.8
A_{12}	42	12501	41.8	5109	32.7	1072	8.6
A_{14}	48	18311	79.6	7661	51.3	1384	11.7
A_{16}	54	25705	140.5	10957	81.6	1436	15.3
A_{18}	60	34875	238.3	15093	119.9	2128	19.9
A_{20}	66	46013	381.8	20165	175.0	2560	25.0
A_{22}	72	59311	594.4	26269	244.9	3032	31.5
A_{24}	78	74961	895.0	33501	328.7	3544	38.5
A_{30}	96	224543	2850.0	62925	688.3	6374	79.9
A_{40}	126	307873	10976.7	143485	2251.6	10884	178.9
A_{50}	156	579803	35384.8	–	–	16594	323.3
A_{60}	186	–	>10 hr.	–	–	23504	622.5
A_{70}	216	–	–	–	–	31614	1030.5
A_{80}	246	–	–	–	–	40924	1609.9
A_{90}	276	–	–	–	–	59534	2238.4
A_{100}	306	–	–	–	–	73144	2963.9

Table 2: Statistics of Otter, Hiper and Herky on a Sun Sparc station

Knuth-Bendix completion procedure. In spite of the fact that Otter is the most efficient resolution-based theorem prover, from the table, we can see that the performance of Otter degrades quickly because of rapid growing of critical pairs.

Hiper was claimed by its creator Jim Christian as "the fastest completion procedure in the world" [2]. We used the Lisp version (the C version is even faster) in our experiment.[3] The run times include the time for parsing the input file, printing the initial equations and completing the set. For A_{50}, Hiper fails to produce a complete system because of running out of memory. This is not a surprise, because, according to Christian, Hiper's speedup comes at the price of extra memory consumption — typically five to six times that required for the ordinary version.

Both Otter and Hiper were implemented with the performance as the first consideration. As a result, in terms of the number of equations processed per second, Otter is the fastest and Herky is the slowest of the three. However, Herky takes less time than both Otter and Hiper to complete A_n for $n \geq 6$, because Herky generates far less critical pairs. For instance, Herky can easily complete A_{80} (7,366 rewrite rules; the 85% of the time is spent on unification); while both Otter and Hiper have difficulty to complete A_{60}.

[3]We compiled the code in Sun Common Lisp and set the variable `*benchmark*` to `t` to collect the statistics.

Problem	SBR2 time	SBR2 pairs	SBR2 rules	Herky time	Herky pairs	Herky rules
1. $(x*y)*x = x*(y*x)$	32	67	15	0.58	19	9
2. $a(x,y,z) + a(y,x,z) = 0$	48	127	19	0.48	12	8
3. $x*(y*(x*z)) = ((x*y)*x)*z$	2:30:49	452	41	16.08	172	48
4. $((z*x)*y)*x = z*((x*y)*x)$	1:56:36	427	39	15.91	172	48
5. $(x*y)*(z*x) = (x*(y*z))*x$	1:56:09	638	47	16.03	172	48
6. $f(x,x,y,z) = 0$	2:07:58	463	39	22.10	190	59

Table 3: Statistics of SBR2 and Herky on alternative ring problems

4 Special Completion Procedures

We implemented in Herky special procedures for handling the Abelian group theory and the distributivity laws. We did not use any special unification or mathing algorithms for these theories. Instead, we used the procedures to simulate superpositions and rewritings related to the rewrite rules of these theories. These theories are practically useful in commutative Thue systems and non-associative ring theories. In the following, we give some experimental results of Herky, in comparison with that of Anantharaman and Hsiang, on problems in alternative ring theories.

An alternative ring is a special case of nonassociative rings (the associativity of the multiplication $*$ is not present) in which both $(x*y)*y = x*(y*y)$ and $(x*x)*y = x*(x*y)$ hold. Rick Stevens was the first to attack problems in nonassociative rings by a computer program [8]. In [1], automatic proofs of some alternative ring problems are reported by Anantharaman and Hsiang using SBR2. In Table 3, we listed the statistics of SBR2 and Herky.[4] The definitions of a and f are as follows:

$$\begin{aligned} a(x,y,z) &= (x*y)*z + -(x*(y*z)) \\ f(w,x,y,z) &= a(w*x,y,z) + -(x*a(w,y,z)) + -(a(x,y,z)*w) \end{aligned}$$

Proofs reported in [1] are obtained in a cumulative way in the sense that when a theorem is proved, it is added into the system for proving subsequent theorems. In Herky, every theorem is proved directly from the original axiom set. Equations 3-5 are the famous Moufang Identities. In Herky, all the three identities can be proved when Rule [47] is generated (that is why they have the same proof statistics):

$$[47]\ \ (y*x)*(z*y) \rightarrow ((y*x)*z)*y + (z*(y*x))*y + -(((z*y)*x)*y)$$

Proof of equation 6 is a part of the proof that the function f is skew-symmetry. It is said in [1] that Anantharaman and Hsiang's proof was the first such proof which did not any auxiliary functions. In fact, the proof reported in [1] relies on the use of Skolem functions and the Moufang identities (in addition to the alternative ring axioms). In contrast, our proof is obtained directly from the alternative ring axioms and uses neither Skolem functions (i.e., it is pure forward search) nor auxiliary functions (the functions a and f serve only as a shorthand when inputing the theorems into Herky).

[4]The time of SBR2 was measured on a Sun 3/60 and that of Herky is on a Sun Sparc station 1 (in seconds). We are told by Hsiang that the new version of SBR2 is much more faster. The number of generated rules for SBR2 does not count the number of non-orientable equations and unequalities.

Extensions of the Knuth-Bendix procedure for special equational theories have been proposed over years [9], [3]. Most of these extensions aim to extend the scope of problems accepted by the Knuth-Bendix procedure by building non-terminating or non-convergent equations into the procedure. We believe that our approach is one of the few attempts of building the theories into the completion procedure with the sole objective of improving the performance of the completion procedure, even though there exist canonical rewrite systems for such theories. Our experimental results show that this approach is promising.

References

[1] Anantharaman, S., Hsiang, J.: (1990) Automated proofs of the Moufang identities in alternative rings. *J. of Automated Reasoning* **6**: 79-109.

[2] Christian, J.: (1989) Fast Knuth-Bendix completion: summary. ***Proc. of Third Intl. Conf. of Rewriting Techniques and Applications*** (RTA-89). Springer-Verlag LNCS 355, 548-510.

[3] Jouannaud, J.P., Kirchner, H.: (1986) Completion of a set of rules modulo a set of equations. *SIAM J. on Computing*, 15:1155–1194.

[4] Kapur, D., Zhang, H.: (1989) An overview of RRL: Rewrite Rule Laboratory. ***Proc. of the third International Conference on Rewriting Techniques and its Applications*** (RTA-89), April 1989, Chapel Hill, NC, Springer Verlag LNCS 355, 513-529.

[5] McCune, B.: (1990) OTTER 2.0 users guide. ANL-90/9, Argonne National Laboratory, Argonne, IL.

[6] Overbeek, R.: (1990) A proposal for a competition. Argonne National Laboratory, Argonne, IL.

[7] Peterson, G.L.: (1990) Complete sets of reductions with constraints. Proc. of 10th International Conference on Automated Deduction. Lecture Notes in Artificial Intelligence Vol. 449. Springer-Verlag, Berlin, pp. 381-395

[8] Stevens, R.L.: (1987) Some experimental in nonassociative ring theorem with an automated theorem prover. *J. of Automated Reasoning* **3**: 221-221

[9] Stickel, M.E.: (1985) Automated deduction by theory resolution. *J. of Automated Reasoning* **1**: 333-355

[10] Zhang, H.: (1991) Criteria of critical pair criteria: a practical approach and a comparative study. Submitted to J. of Automated Reasoning.

IMPS: System Description*

William M. Farmer, Joshua D. Guttman, F. Javier Thayer

The MITRE Corporation

IMPS, an Interactive Mathematical Proof System [5], aims at computational support for traditional techniques of mathematics. It is based on three observations about rigorous mathematics:

- Mathematics emphasizes the axiomatic method. Characteristics of mathematical structures are summarized in axioms. Theorems are derived for all structures satisfying the axioms. Frequently, a clever change of perspective shows that a structure is an instance of another theory, thus also bringing its theorems to bear.

- Many branches of mathematics emphasize functions, including partial functions. Moreover, the classes of objects studied may be nested, as are the integers and the reals; or overlapping, as are the bounded functions and the continuous functions.

- Proof proceeds by a blend of computation and formal inference.

Support for the Axiomatic Method. IMPS supports the "little theories" version of the axiomatic method, as opposed to the "big theory" version. In the big theory approach, all reasoning is carried out within one theory—usually some highly expressive theory, such as the Zermelo-Fraenkel set theory. In the little theories approach, reasoning is distributed over a network of theories. Results are typically proved in compact, abstract theories, and then transported as needed to more concrete theories, or indeed to other equally abstract theories. Theory interpretations provide the mechanism for transporting theorems. The little theories style of the axiomatic method is employed extensively in mathematical practice; in [4], we discuss its benefits for mechanical theorem provers, and how the approach is used in IMPS.

Logic. Standard mathematical reasoning in many areas focuses on functions and their properties, together with operations on functions. For this reason,

*Supported by the MITRE-Sponsored Research Program. Authors' address: The MITRE Corporation, 202 Burlington Rd, Bedford MA 01730 USA; Telephone: 617-271-2907; Email: **farmer, guttman, jt@mitre.org**.

IMPS is based on a version of simple type theory.[1] However, we have adopted a version, called LUTINS,[2] containing partial functions, because they are ubiquitous in both mathematics and computer science. Although terms, such as 2/0, may be nondenoting, the logic is bivalent and formulas always have a truth value. In particular, an atomic formula is false if any of its constituents is nondenoting. This convention follows an approach common in traditional rigorous mathematics, and it entails only small changes in the axioms and rules of classical simple type theory [2].

Moreover, LUTINS allows subtypes to be included within types. Thus, for instance, the natural numbers form a subtype of the reals, and the continuous (real) functions a subtype of the functions from reals to reals. The subtyping mechanism facilitates machine deduction, because the subtype of an expression gives some immediate information about the expression's value, if it is defined at all. Moreover, many theorems have restrictions that can be stated in terms of the subtype of a value, and the theorem prover can be programmed to handle these assertions using special algorithms [4].

This subtyping mechanism interacts well with the type theory only because functions may be partial. If σ_0 is a subtype of τ_0, while σ_1 is a subtype of τ_1, then $\sigma_0 \rightarrow \sigma_1$ is a subtype of $\tau_0 \rightarrow \tau_1$. In particular, it contains just those partial functions that are never defined for arguments outside σ_0, and which never yield values outside σ_1.

Computation and Proof. One problem in understanding and controlling the behavior of theorem provers is that a derivation in predicate logic contains too much information.

The mathematician devotes considerable effort to proving lemmas that justify computational procedures. Although these are frequently equations or conditional equations, they are sometimes more complicated quantified expressions, and sometimes they involve other relations, such as ordering relations. Once the lemmas are available, they are used repeatedly to transform expressions of interest. Algorithms justified by the lemmas may also be used; the algorithm for differentiating polynomials, for example. The mathematician has no interest in those parts of a formal derivation that "implement" these processes within predicate logic.

On the other hand, to understand the structure of a proof (or especially, a partial proof attempt), the mathematician wants to see the premises and conclusions of the informative formal inferences.

Thus, the right sort of proof is broader than the logician's notion of a formal derivation in, say, a Gentzen-style formal system. In addition to purely formal inferences, IMPS allows also inferences based on sound computations. They are treated as atomic inferences, even though a pure formalization might require

[1]This version is *many-sorted*, in that there may be several types of basic individuals. Moreover, it is *multivariate*, in that a function may take more than one argument. Currying is not required. However, taking (possibly n-ary) functions is the only type-forming operation.

[2]Pronounced as in French. See [2, 3] for studies of logical issues associated with LUTINS; see [7] for a detailed description of its syntax and semantics.

hundreds of Gentzen-style formal inference steps. We believe that this more inclusive conception makes IMPS proofs more informative to its users.

The System

The IMPS system consists of four components.

Core. All the basic logical and deductive machinery of IMPS on which the soundness of the system depends is included in the *core* of IMPS. The core is the specification and inference engine of IMPS. The other components of the system provide the means for harnessing the power of the engine.

The organizing unit of the core is the IMPS *theory*. Mathematically, a theory consists of a LUTINS language plus a set of axioms. A theory is implemented, however, as a data structure to which a variety of information is associated. Some of this information procedurally encodes logical consequences of the theory that are especially relevant to low-level reasoning within the theory. For example, the great majority of questions about the definedness of expressions are answered automatically by IMPS using tabulated information about the domain and range of the functions in a theory. Theories may be enriched by the definition of new sorts and constants and by the installation of theorems.

Proofs in a theory are constructed interactively using a natural style of inference based on sequence calculus. IMPS builds a data structure which records all the actions and results of a proof attempt. This object, called a *deduction graph*, allows the user to survey the proof and to choose the order in which he works on different subgoals. Alternative approaches may be tried on the same subgoal. Deduction graphs also are suitable for analysis by software.

The user is only allowed to modify a deduction through the application of *primitive inferences*, which are akin to rules of inference. Most primitive inferences formalize basic laws of predicate logic and higher-order functions. Others implement computational steps in proofs. For example, one class of primitive inferences performs expression simplification, which uses the logical structure of the expression [8], together with algebraic simplification, deciding linear inequalities, and applying rewrite rules.

Another special class of primitive inferences "compute with theorems" by applying extremely simple procedures called macetes [9].[3] An *elementary macete*, which is built by IMPS whenever a theorem is installed in a theory, applies the theorem in a manner determined by its syntax (e.g., as a conditional rewrite rule). Compound macetes are constructed from elementary and other kinds of atomic macetes (including macetes that beta-reduce, unfold defined constants, and perform expression simplification). They apply a collection of theorems in an organized way.

In addition to the machinery for building theories and reasoning within them, the core contains machinery for relating one theory to another via theory interpretations. A theory interpretation can be used to "transport" a theorem from

[3] *Macete*, in Portuguese, means a chisel, or in informal usage, a clever trick.

the theory it is proved in to any number of other theories. Theory interpretations are also used in IMPS to show relative consistency of theorems, to formalize symmetry and duality arguments, and to prove universal facts about polymorphic operators [5, 4, 1]. The great majority of the theory interpretations needed by the IMPS user are built by software without user assistance. For example, when a theorem is applied outside of its home theory via a *transportable macete*, IMPS automatically builds the required theory interpretation if needed.

Supporting Machinery. We have built an extension of the core machinery with the following goals in mind:

- To facilitate building and printing of expressions by providing various parsing and printing procedures.
- To make the inference mechanism more flexible and more autonomous. In particular, the set of commands available to users includes an extensible set of inference procedures called *strategies*. Strategies affect the deduction graph only by using the primitive inference procedures of the core machinery, but are otherwise unrestricted.
- To facilitate construction of theories and interpretations between them.

User Interface. From the outset IMPS has been designed to provide users with facilities to easily direct and monitor proofs. This is accomplished by a user interface which controls three critical elements of an interactive system:

- The display of the state of the proof. This includes graphical displays of the deduction graph as a tree, TEX typesetting of the proof history, and TEX typesetting of individual subgoals in the deduction graph. The graphical display of the deduction graph permits the user to visually determine the set of unproven subgoals and to select a suitable continuation point for the proof. On the other hand, the TEX typesetting facilities allow the user to examine the proof in a mathematically more appealing notation than is possible by raw textual presentation alone.
- The presentation of options for new proof steps. For any particular subgoal, the interface presents the user with a well-pruned list of commands and macetes to apply. This list is obtained by using syntactic and semantic information which is made available to the interface by the IMPS supporting machinery. For example, in situations where over 400 theorems are available to the user, there are rarely more than 10 macetes presented to the user as options.
- Processing of user commands and submitting them to the inference software, requesting from the user whenever necessary, additional arguments required to execute the command.

The user interface, which is a completely detachable component of IMPS, is primarily written in Emacs; the other components are written in T, a sophisticated version of Scheme.

Theory Library. IMPS is equipped with a library of basic theories, which can augmented as desired by the user. The theory of the reals, the central theory in the library, specifies the real numbers as a complete ordered field. (The completeness principle is formalized as a second-order axiom, and the rationals and integers are specified as the usual substructures of the reals.) The library also contains various "generic" theories that contain no nonlogical axioms (except possibly the axioms of the theory of the reals). These theories are used for reasoning about objects such as sets, unary functions, and sequences.

Applications

We have formulated a variety of different kinds of mathematics within IMPS, with emphasis on mathematical analysis and abstract algebra. In our theory of the reals, we have proved a large number of basic facts such as the combinatory identity, Bernoulli's inequality, and the Archimedean property of the reals. In analysis, we have proved Banach's Contraction Principle in a theory of an abstract metric space, and the theorem that the continuous image of a connected set is itself connected in a theory of two metric spaces (see [6]). The proof in IMPS of a simple "inverse function theorem" in a theory of an Banach space is described in [4]. In algebra, we have proved a version of the binomial theorem for commutative rings, and various facts about an abstract iterated product operator in a theory of an abstract monoid, which can be transported to the reals as theorems about the Σ and Π operators. The Schröder-Bernstein Theorem and Fundamental Counting Theorem of group theory, of which Lagrange's theorem is an immediate corollary, have also been proved in IMPS.

References

[1] W. M. Farmer. Abstract data types in many-sorted second-order logic. Technical Report M87-64, The MITRE Corporation, 1987.

[2] W. M. Farmer. A partial functions version of Church's simple theory of types. *Journal of Symbolic Logic*, 55:1269–91, 1990.

[3] W. M. Farmer. A simple type theory with partial functions and subtypes. Technical report, The MITRE Corporation, 1991. Submitted to *Annals of Pure and Applied Logic*.

[4] W. M. Farmer, J. D. Guttman, and F. J. Thayer. Little theories. In these proceedings.

[5] W. M. Farmer, J. D. Guttman, and F. J. Thayer. IMPS: an Interactive Mathematical Proof System. Technical Report M90-19, The MITRE Corporation, 1991. Submitted to *Journal of Automated Reasoning*.

[6] W. M. Farmer and F. J. Thayer. Two computer-supported proofs in metric space topology. *Notices of the American Mathematical Society*, 38:1133–1138, 1991.

[7] J. D. Guttman. A proposed interface logic for verification environments. Technical Report M91-19, The MITRE Corporation, 1991.

[8] L. G. Monk. Inference rules using local contexts. *Journal of Automated Reasoning*, 4:445–462, 1988.

[9] F. J. Thayer. Obligated term replacements. Technical Report MTR-10301, The MITRE Corporation, 1987.

Proving Equality Theorems with Hyper-Linking

Geoffrey D. Alexander
IBM Corporation
P. O. Box 12195
Research Triangle Park, NC 27709

David A. Plaisted
Department of Computer Science
University of North Carolina
Chapel Hill, NC 27599-3175

Abstract

Hyper-linking is an instance-based refutation theorem proving strategy for first-order logic, recently developed by Lee and Plaisted[4]. An implementation of hyper-linking is the theorem prover CLIN[3]. We extend hyper-linking to support equality by using Brand's E-modification[1]. We discuss how term rewriting is added to improve performance. We describe a number of other improvements, both in support of equality theorems and non-equality theorems.

1 Hyper-Linking

Hyper-linking is an instanced-based refutation theorem proving strategy. It determines if a set of clauses is unsatisfiable by iteratively generating instances of the clauses and performing a propositional unsatisfiability test on a ground set of the instances.

Instances are generated from a set of clauses by the hyper-link operation. Let S be a set of clauses, $\{ M_1, \ldots, M_m \}$ the set of literals in S, and $\{ L_1, \ldots, L_n \}$ a clause in S. We say that a literal L_i *links* with a literal M_j if there exists a substitution θ such that $L_i\theta$ and $M_j\theta$ are complementary. The hyper-link operation maps the clause $\{ L_1, \ldots, L_n \}$ into an instance $\{ L_1\theta, \ldots, L_n\theta \}$ where θ is a most general unifier which links each L_i. We refer to $\{ L_1, \ldots, L_n \}$ as the *nucleus* of the hyper-link, θ as the *substitution* of the hyper-link, each M_i as an *electron* of the hyper-link, and $\{ L_1\theta, \ldots, L_n\theta \}$ as the *instance* of the hyper-link. For example, consider the following set of clauses $\{ \{ \sim p(y), \sim s(z), \sim r(y,z) \}, \{ \sim q(x), r(a,x) \}, \{ p(a) \}, \{ s(b) \} \}$. An instance of $\{ \sim p(y), \sim s(z), \sim r(y,z) \}$ generated by the hyper-link operation is $\{ \sim p(a), \sim s(b), \sim r(a,b) \}$.

2 Brand's E-Modification

We add equality support to hyper-linking by using Brand's E-modification[1]. The E-modification of a clause is based on the idea of pulling out subterms and predicate arguments. For example, consider the set of clauses $\{ \{ p(a) \}, \{ \sim p(b) \}, \{ a = b \} \}$ which is unsatisfiable with respect to equality. Pulling out the argument a from the clause $\{ p(a) \}$ gives $\{ \{ a \neq x, p(x) \}, \{ \sim p(b) \}, \{ a = b \} \}$ which is unsatisfiable under all interpretations.

The E-modification of a clause C is defined inductively as follows. If all nonvariable terms of C occur as arguments of equality, the E-modification of

C is C. Otherwise, let s be a nonvariable term occurring as the argument of a non-equality predicate or as the argument of a function which is the argument of equality. Let C' be $\{ s \neq w \} \cup C[w]$ where w is a variable not occurring in C and $C[w]$ is C with the chosen occurrence of s replaced by w. The E-modification of C is the E-modification of C'. For example, the E-modification of the clause $\{ p(f(a, g(x))), f(x, y) = g(a) \}$ is $\{ a \neq w_4, g(x) \neq w_3, a \neq w_2, f(w_2, w_3) \neq w_1, p(w_1), f(x, y) = g(w_4) \}$. The E-modification of a set of clauses is the E-modification of each clause in the set.

A set of clauses is unsatisfiable with respect to equality if its E-modification along with the reflexivity and transitivity axioms is unsatisfiable under all interpretations. Thus, Brand's E-modification offers a simple method of adding equality support to clause-based refutation proof procedures. For hyper-linking, we apply Brand's E-modification after converting the input closed formulas to clause form.

3 Term Rewriting

Hyper-linking with Brand's E-modification generates many clauses on all but the most trivial equality theorems. One method for reducing the number of clauses is term rewriting. An overview of term rewriting can be found in Dershowitz and Jouannaud[2] .

We add term rewriting to hyper-linking with Brand's E-modification by using the lexicographic path ordering to order equations into rewrite rules. Clauses are rewritten to normal form prior to the clause grounding phase of hyper-linking. Note that in adding term rewriting to hyper-linking with Brand's E-modification, it becomes necessary to apply the E-modification to clauses produced by the hyper-link operation as well as to input clauses in order to preserve completeness.

Brand's E-modification has the potential of producing clauses with many literals. This is especially true when applying the E-modification to clauses generated by the hyper-link operation. Clauses with many literals present a serious problem since the hyper-link operation is exponential in number of literals.

To address this problem, we restrict term rewriting so that the left hand side of equations and inequations are not rewritten at the outer level. With this restriction, it is no longer necessary to apply the E-modification to clauses generated by the hyper-link operation. Hyper-linking with Brand's E-modification and restricted term rewriting is illustrated in Figure 1.

4 Implementation

We have implemented hyper-linking with Brand's E-modification and term rewriting by extending the theorem prover CLIN. The addition of Brand's E-modification was fairly straightforward. However, the addition of term rewriting

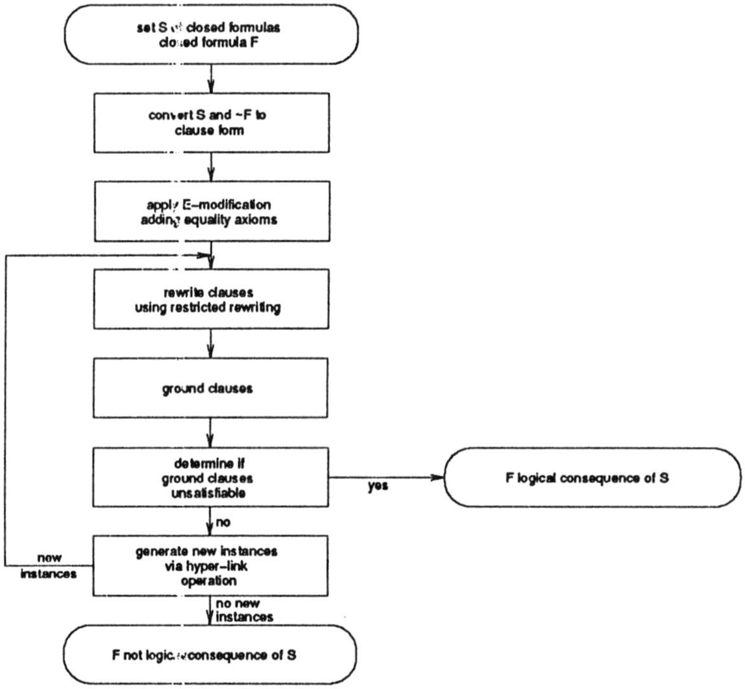

Figure 1: Hyper-Linking with Brand's E-modification and Restricted Term Rewriting

was more challenging. We encountered two major performance issues when adding term rewriting.

The first performance issue was which reduction strategy to use in rewriting clauses to normal form. After trying a number of methods, we decided upon a hybrid method in which we first rewrite a term fully at the outer level and then recursively rewrite the term's arguments. This process is repeated until no change occurs.

The second performance issue was how to prevent the generation of huge terms. At times, term rewriting generated clauses of over 150,000 characters. Many well-known theorem provers address this problem by setting a bound on the number of reductions allowed on a single clause. What we do is to rewrite clauses during instance generation using only rewrite rules in which no variable occurs more often in the right hand side. Later, when full rewriting is performed, we stop rewriting whenever the resulting term size exceeds a predetermined priority bound. This works well as CLIN automatically sets a priority bound which decreases as more and more instances are generated.

5 Refinements

In addition to adding equality support to CLIN, we also introduced a number of other improvements. These improvements include a modification to sliding priority, early termination of the hyper-link operation, automated clause splitting, automated group detection, and ordered rewriting.

Lee[3] developed a priority method, called *sliding priority*, in which CLIN automatically sets a priority bound based on the amount of work performed. Clauses whose size exceed the priority bound are deleted. We modified sliding priority so that it better reflects the work performed.

In CLIN, clauses are deleted for a number of reasons. For example, a clause is deleted if it is subsumed by a unit clause, it is a tautology, or its size exceeds the priority bound set by sliding priority. Also, duplicate clauses are not retained in CLIN. We noted that these properties hold for clauses in which some but not all literals are linked. We refer to a clause in which some but not all literals are linked as a *partial instance.* For example, consider a partial instance which is unit subsumed. All instances coming from the partial instance will also be unit subsumed and thus deleted. This means that we can terminate the hyper-link operation as soon as we determine that a partial instance is unit subsumed. Similar results hold for tautology detection, priority testing, and duplicate detection.

Lee[3] describes a method in which a clause can be split into two or more clauses having fewer literals. For example, the clause $\{ L_1, \ldots, L_m, L_{m+1}, \ldots, L_n \}$ can be split into the clauses $\{ L_1, \ldots, L_m, p(\overline{x}) \}$ and $\{ \sim p(\overline{x}), L_{m+1}, \ldots, L_n \}$ where p is a new predicate symbol and $\overline{x}$ are the variables common to $\{ L_1, \ldots, L_m \}$ and $\{ L_{m+1}, \ldots, L_n \}$. Brand's E-modification tends to produce clauses with many literals. We automated clause splitting to address the problem with long clauses.

Group theory is decidable by a ten-rule term rewriting system. We developed a procedure which automatically detects groups by looking for associativity, left inverse, and left identity axioms or associativity, right inverse, and right identity axioms. Our detection is completely generic in that we allow the group operation, inverse, and identity to be arbitrary terms containing two, one, and zero variables respectively. Once a group is detected, we assert the ten rewrite rules for the group. We also assert six replace rules to simulate the left and right cancellation laws, uniqueness of the left and right inverse, and uniqueness of the left and right identity.

Ordered rewriting is an extension to term rewriting which allows unorderable equations to be used as rewrite rules whenever their use simplifies the term. We have extended our term rewriting implementation with ordered rewriting.

6 Results

Lusk and Overbeek[5] give the following six equality problems.

Problem 1: In a group, if $x^2 = e$ for all x in the group, then the group is Abelian (for all x and y, $xy = yx$).

Problem 2: In a group, $(x^{-1})^{-1} = x$ for all x in the group.

Problem	Extended CLIN*	OTTER 2.2*
1	0.95	0.44
2	1.15	0.21
3	119.90	20.42
4	81.12	7.75
5	1,141.68	958.23
6	———	———

***Note that extended CLIN is written in Prolog, while OTTER 2.2 is written in C. We feel that we could get an order a magnitude improvement in extended CLIN performance by rewriting extended CLIN in C or Lisp.**

Figure 2: Equality Results

Problem 3: In a ring, if $x^2 = x$ for all x in the ring, then $xy = yx$ for all x, y in the ring.

Problem 4: In a group, if $x^3 = e$ for all x in the group, then the commutator $h(h(x,y),y) = e$ for all x and y. The commutator $h(x,y)$ is defined as $xyx^{-1}y^{-1}$.

Problem 5: In a ternary Boolean algebra with the third axiom removed, it is true that $f(x,g(x),y) = y$ for all x and y.

Problem 6: In a ring, if $x^3 = x$ for all x in the ring, then $xy = yx$ for all x and y in the ring.

We have attempted to prove all six of the above theorems with both extended CLIN and with OTTER 2.2. Results are given in Figure 2 with times in CPU seconds on a DECsystem 5500. In addition, since CLIN works well for non-Horn problems[4], we expect our approach to excel on problems combining equality and highly non-Horn clauses.

References

[1] D. Brand. Proving theorems with the modification method. *SIAM J. Comput.*, 4(4):412–430, December 1975.

[2] Nachum Dershowitz and Jean-Pierre Jouannaud. Rewriting systems. In Jan van Leeuwen, editor, *Handbook of Theoretical Computer Science*, volume B, pages 243–320. Elsevier Science Publishers, Amsterdam, 1990.

[3] Shie-Jue Lee. *CLIN: An Automated Reasoning System Using Clause Linking.* PhD Dissertation, University of North Carolina at Chapel Hill, 1990.

[4] Shie-Jue Lee and David A. Plaisted. Eliminating duplication with the hyper-linking strategy. *J. Automated Reasoning*, To appear.

[5] E. L. Lusk and R. A. Overbeek. Reasoning about equality. *J. Automated Reasoning*, 1:209–228, 1985.

Xpnet: A Graphical Interface to Proof Nets with an Efficient Proof Checker

Jawahar Chirimar Carl A. Gunter Myra VanInwegen[1]

Department of Computer and Information Science
University of Pennsylvania, Philadelphia, PA 19104

We describe an automated proof-checker for linear logic proof nets based on efficient algorithms for deciding necessary and sufficient *path conditions.* The system, which is called *Xpnet,* is an experiment in the graphical representation of mathematical proof objects.

Linear logic is a new logical system introduced by Jean-Yves Girard [Gir87]. The main difference between classical and linear logic is that in the latter structural rules for copying or deleting assumptions are disallowed. Full linear logic is ordinarily presented as a sequent calculus in the style of Gentzen. Proof nets, which are the linear logic analogs of natural deduction, can be formed from a subset of the logic known as the *multiplicative fragment.*

Proof nets are represented as graphs with two types of edges (see Figure 1). Their most distinctive feature is the condition required for *proof-checking.* In most logics proof-checking involves verifying that each step of a proposed proof object is an instance of an axiom or rule of the system. This is insufficient for proof nets: checking the steps by which the graph is constructed only ensures that it is a *proof structure*, a type of graph that is a proof net exactly when it satisfies a special connectivity property. Stated mathematically, this property sounds as if it would take exponential time to check. Girard asserted that his path condition for proof nets could be checked in time proportional to a polynomial in the number of edges. We have developed an algorithm with this efficiency using an equivalent condition given by Danos and Regnier [DR89].

Checking even moderately-sized proof structures by hand can be very tedious. The motivation behind Xpnet (for X Proof NET) was to use an implementation of our algorithm to automate this process. Since proof nets can be described pictorially it is natural to have a graphical user interface to aid in the sketching of proof structures to be checked.

There are relatively few systems for the graphical manipulation of mathematical objects that can be said to 'understand' the objects they manipulate. An example of a system that does is LatticeDraw, designed for the MacIntosh by Alan Day. This program allows the user to construct an ordered collection of elements (a lattice) by drawing its Hasse diagram. It aids in the creation of a well-formed Hasse diagram by guiding legitimate uses of a mouse to describe elements and relationships between them. It is then capable of providing information about the lattice by responding to questions selected from a menu such as 'show me all of the primes'. It is also capable of calculating and displaying related lattices. For example, the program can calculate the lattice of lower sets of a lattice and display its Hasse diagram. This

[1]The authors' email addresses are `chirimar`, `gunter`, and `myra @saul.cis.upenn.edu`. Gunter's research is partially supported by NSF grant CCR-8912778 and an ONR Young Investigator award. Chirimar's research is supported by NSF grant CCR-8912778. VanInwegen's work is supported by ARO grant DAAL03-89-C-0031.

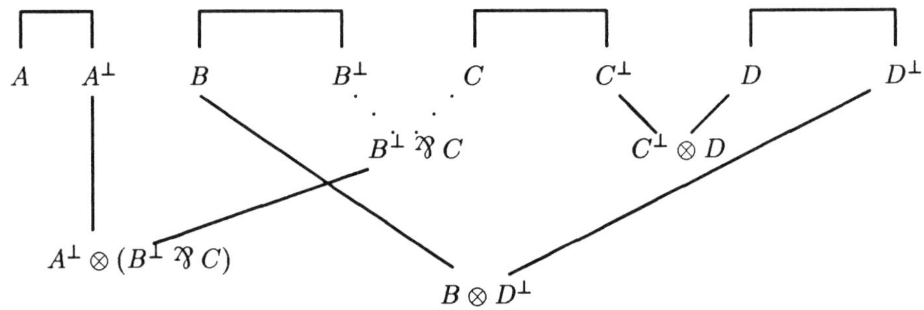

Figure 1: A proof net

might take hours of work to do by hand, even for fairly small examples.

Other examples of systems with a graphical interface to a mathematical system are the programs Turing's World and Tarski's World of Jon Barwise. The first of these aids the user in providing a mouse sketch of a Turing machine which can then be run as a program. The latter determines the truth of logical propositions about a configuration of blocks which the user can create with a mouse.

The 'look and feel' of Xpnet is partly inspired by LatticeDraw. Axioms and rules are placed by mouse clicks, and line displays guide the user in the legitimate placement of his nodes. Diagnostics are carried out by clicking on a button. Unlike LatticeDraw, Xpnet is implemented using the X Toolkit and runs under the X Window System. Its internal structure is similar to that of Xfig. The system is portable and robust as well as easy and fun to use.

Below we describe how to use Xpnet to build and manipulate proof structures, and then we discuss the basis and implementation of our proof-checking algorithm.

Description of the Program. Figure 2 shows Xpnet in operation. There are three main windows in its display area. The canvas is the large window in which proof structures are drawn. The message window is below the canvas; messages to the user are printed there. The panel (for control panel) is to the left of the canvas; users initiate commands by clicking on the buttons it contains.

A proof structure is built of components called *clusters*. The cluster types are axiom, pax (short for proper axiom), tensor, par, and cut, and they are placed in the structure using command buttons in the panel. Xpnet employs user-selected ASCII strings to represent the symbols for tensor ($\otimes$), par ($⅋$), and the linear negation perp ($^\perp$); here we use the defaults "tensor", "par", and "~".

To build a proof structure one starts off with axiom and pax clusters. Axiom clusters have as formulae A and $A^\perp$, where A is an atomic formula entered by the user. Pax clusters may have any formulae; the user types in the name of a file that contains them. In both cases, the user selects the position of the formulae with the mouse and then determines the height of the solid edge connecting the formulae.

After placing axioms or paxes the user can add tensor, par, or cut clusters. Each requires the selection of two premises. A formula may be used at most once as a premise. To place a tensor or par cluster, the user selects the premises (by clicking on the formulae) and then places the resulting formula. The position of the formula

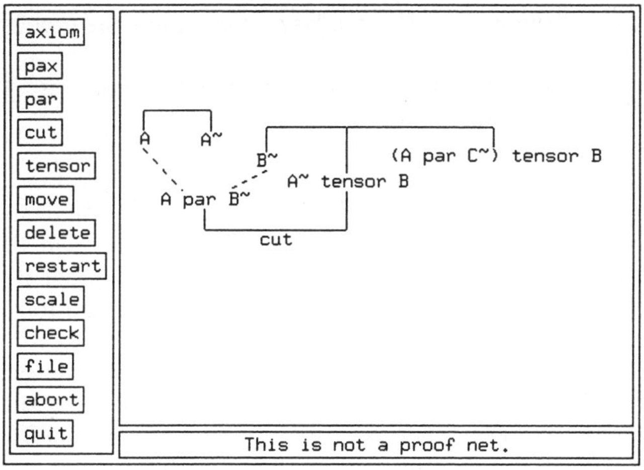

Figure 2: Xpnet with a proof structure that is not a proof net.

is forced by the program to be below the premises. Edges (solid for a tensor, dotted for a par) connect the formula to its premises. Since these formulae can grow to be rather large, the user has the option of displaying only the type of the cluster. The premises for a cut cluster must be inverses. After selecting the premises for a cut, the user selects the position of the solid edge below the premises which connects them. No formula is generated by the cut cluster. If the resulting proof structure is a proof net, then the conclusions of the proof are the formulae that are not used as premises. For example, the conclusions of Figure 1 are A, $A^{\perp} \otimes (B^{\perp} ⅋ C)$, $B \otimes D^{\perp}$, and $C^{\perp} \otimes D$.

Examples of axiom, pax, par, and cut clusters appear in Figure 2. The axiom consists of the nodes labeled **A** and **A~** connected above with a bar. The pax consists of the nodes labeled **B~**, **A~ tensor B**, and **(A par C~) tensor B** connected above with a bar. The premises of the par are **A** and **B~**, and the resulting formula is **A par B~**. The premises of the cut are **A par B~** and **A~ tensor B**, which are connected below with a bar. Figure 1 contains three tensors.

Editing is supported by the *move, delete, restart,* and *scale* commands. The *move* command moves a connected component of the graph. The *delete* command deletes a cluster if none of its formulae are used as premises. The entire proof structure can be erased using *restart* or scaled by an arbitrary amount using *scale*.

Once a proof structure has been drawn, the user can obtain information about it using the *expand* and *check* commands. The *expand* command causes the entire formula corresponding to a node to be printed out in the message window and is only useful when the user chooses to have only the cluster type displayed at tensor and par clusters. The *check* command invokes the proof net verification algorithm, and the result (**Net is ok.** or **This is not a proof net.**) is printed to the message window. In Figure 2, the *check* command has just been applied.

The *file* command brings up a menu with several options on it. The *save* and *read* options allow the user to save a textual representation of a proof net to a file and to retrieve one of these files. The *xpn2ps* option dumps the proof structure in a file in PostScript format.

The *abort* command terminates a command already in progress. The *quit* command terminates the program.

Proof-Checking. A naive implementation of proof-checking based on the condition for proof structures to be proof nets takes exponential time. We define a *switching* of a par cluster to be the solidification of one of the edges from the formula to its premises and the deletion of the other. A switching of a proof structure is the selection of a switching for each par in the graph. The Danos and Regnier condition for a proof structure to be a proof net is that each graph resulting from a switching of the proof structure is a tree. As there are two possible switchings for each par, checking this directly takes time exponential in the number of pars in the structure.

The idea for our proof-checker comes from the *Sequentialization Theorem*, which states that for any proof net there is a sequent proof with the same conclusions. The proof of this theorem provides a method by which a sequent proof is constructed from a proof net. This method, which is shown to be correct using the Danos and Regnier condition, can be used to determine whether or not a given proof structure is a proof net: we check if it is possible to construct an equivalent sequent proof, and if so, the structure is a proof net. Structures consisting of just an axiom or pax cluster are the base case. Converting these to sequent proofs is trivial, so such structures are proof nets. To convert a structure with a par conclusion we remove the par, convert the rest of the structure, and then add a par rule to the resulting sequent proof. To check such a structure, we remove the par cluster and check the rest of the structure. The only cases left are structures with only tensor conclusions and cuts, and structures consisting of two or more axioms or paxes. The latter are not connected and so are not proof nets.

The difficulty is in structures that have only tensor conclusions and cuts. To convert these we would like to remove a tensor or cut, convert the two resulting structures, and apply the appropriate rule to the conversions. However, removing an arbitrary cluster may not split the proof structure into two separate pieces: in Figure 1 we can split on only the leftmost tensor. The *Tensor Splitting Lemma* states that we can split on at least one cluster in a proof net. Thus to check such structures, we search for a cluster to split on, at the same time checking that the graph is connected. If we succeed, then we split and check the two resulting structures. To determine if we can split on a cluster D with premises A and B, we first mark D with the number 3 and all the other formulae in the structure with 0. Then we mark with 1 both A and all formulae connected to it by graph edges except the one between A and D. If in doing so we encounter D, then we have come to D through B, so removing D will not split the structure. If we succeed, we mark with 2 both B and all formulae connected to it by edges except the one between B and D. If there are any formulae still marked 0, then the structure is not connected.

The algorithm above is order of E^3, where E is the number of edges in the graph. To check a structure we remove a cluster and then verify what is left. Removing an axiom, pax, or par takes time proportional to the number of nodes in the cluster. Removing a tensor or cut takes $O(NE)$, where N is the number of nodes: we test $O(N)$ clusters to find one to split on, and this test requires $O(E)$ time. Thus checking

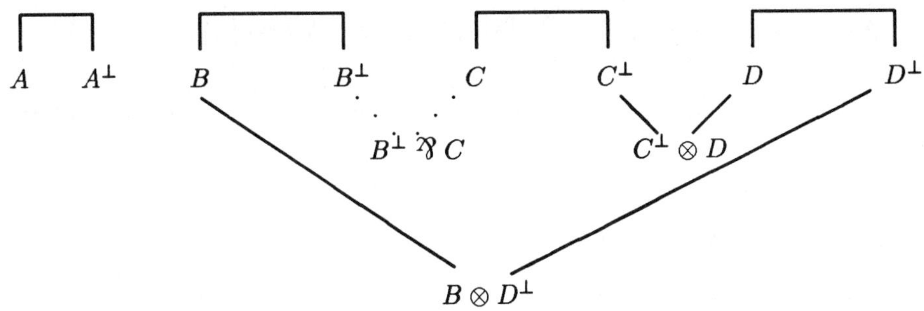

Figure 3: Verification: after splitting

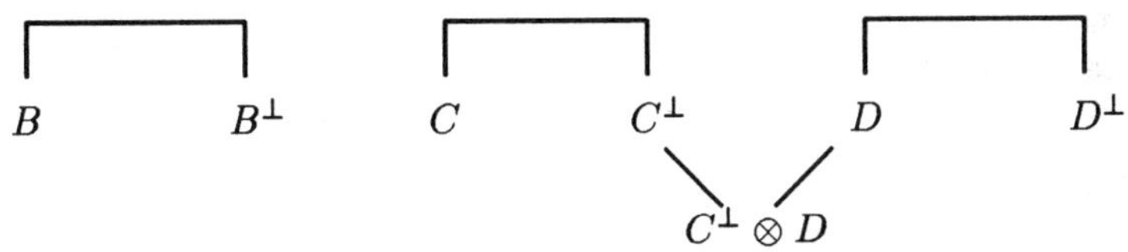

Figure 4: Verification: after removing par and splitting again

the entire graph takes $O(N^2E)$ time. Since in a proof structure the number of edges is proportional to the number of nodes, this is $O(E^3)$.

To demonstrate the algorithm, we will go through the steps involved in checking the structure in Figure 1. After splitting on the leftmost tensor, we get the proof structure in Figure 3. The structure on the left is immediately a proof net since it is an axiom. To check the other one, we remove the par and can then split on either tensor. Say the bottom one is chosen, resulting in Figure 4. Again the left hand structure is immediately a proof net, and we split the right one on the remaining tensor, getting two structures which are axioms.

Obtaining the program. To get an executable that runs on the Sun4, FTP to ftp.cis.upenn.edu and log in as anonymous (give your email address as password). The file is in the pub directory and is called xpnet.tar.Z. To use it you must uncompress it and extract the contents from the resulting tar file. One way to do this is "`uncompress < xpnet.tar.Z | tar -xf -`" which will place the extracted files in the current directory.

We would like to acknowledge assistance and encouragement from Jean Gallier, Vijay Gehlot, Andre Scedrov, Hay Tran, and Loan Dinh in our efforts on this project.

References

[DR89] V. Danos and L. Regnier. The structure of multiplicatives. *Archive for Mathematical Logic*, 28:181–203, 1989.

[Gir87] J. Y. Girard. Linear logic. *Theoretical Computer Science*, 50:1–102, 1987.

&: Automated Natural Deduction[1]

Dave Barker-Plummer[2] Sidney C. Bailin[3] Andrew S. Merrill[4]

Abstract

In this paper we describe a sequent calculus-based theorem prover called &[5]. The underlying logic of & is that of Zermelo set theory. In addition to the usual rules of first-order sequent based systems, the logic contains inference rules to handle set abstraction terms, including the ability to unify formulae involving such terms, and the ability to introduce new abstraction terms as instantiations. & also has novel derived rules and heuristics for making choices in the course of the proof of a theorem.

1 Introduction

In this paper we describe the & theorem proving system. & is based on a cut-free sequent calculus for the normalizable fragment of Zermelo set theory. The underlying logic is described in [Bai].

The usual sequent calculus rules for first-order logic are the basis of &. These are not discussed in this paper. The main novelties of &, which we report here, are new inference rules and a unification algorithm which allow & to handle set abstraction terms (sections 3 and 4). & is written in the Quintus dialect of PROLOG. We hope to make a version of the system generally available in the near future.

We have used & to prove automatically some non-trivial theorems in set theory, including the correctness of transfinite induction and Cantor's theorem. These proofs are completely automatic, in that & itself finds all necessary term instantiations. In its role as the underlying prover for our graphical theorem proving system, GROVER, & has automatically proved the Diamond Lemma, a highly non-trivial proof in the theory of well-founded relations.

2 The Logic

We do not describe the underlying logic in detail in this paper. The interested reader is referred to [Bai] for the full details. The handling of logical connectives has much in common with previous work on automating natural deduction, particularly [BT75]. The main novelties of the system are the development of new inference rules for set theory, and new search control heuristics.

One feature of &'s logic is that the role of variables is played by *schematic* terms, and variables are treated as most provers treat constants. This is a consequence of skolemizing as in formal logic (i.e., skolem functions replace essentially existentially

[1] This paper is based on work supported by the National Science Foundation under award number ISI-8701133, and by Swarthmore College through a summer research award to Andrew Merrill and through the college's Research Support Fund.

[2] Swarthmore, College, Swarthmore, PA, 19081. `plummer@cs.swarthmore.edu`

[3] CTA, INCORPORATED, 6116 Executive Blvd, Rockville, Maryland, 20852. `sbailin@cta.com`

[4] Swarthmore, College, Swarthmore, PA, 19081. `merrill@cs.swarthmore.edu`

[5] & is an abbreviation of an acronym: Automated Natural Deduction.

quantified variables). Universal quantifiers on the right-hand side of a sequent, and existential quantifiers on the left-hand side, are essentially universal and are kept as variables when the quantifiers are removed. Existential quantifiers on the right-hand side of a sequent, and universal quantifiers on the left-hand side, are essentially existential and are replaced by schematic terms (skolem functions) when the quantifiers are removed. This is the dual of skolemization as treated by most automated theorem provers.

The order in which quantifiers have been removed is implicitly represented in the arguments of the schematic terms. When a schematic term is introduced into a sequent, it takes as its arguments all of the variables that are currently free in the sequent. Schematic terms may be bound only to terms whose variables are among the schematic's argument variables. This is the dual of the "occurs-check" required in other systems.

The rules implemented for & are grouped into classes pertaining to logical connectives, quantifiers, equality, definitions, set abstractions, and instantiation.

Set abstractions are handled by inference rules which replace formulae of the form $t \in \{x|P[x]\}$ by the formula $P[t]$. We call these rules {} *-left* and {} *-right* depending on the side of the turnstile on which the reduced formula appears. & has the ability to unify formulae containing set abstraction terms using the λ-unify algorithm outlined in section 3. In addition, the $\mathcal{Z}$-match inference rule is able to introduce new set abstraction terms to complete the proof of some sequents. $\mathcal{Z}$-match is described in section 4.

The system may be used in either automatic or interactive modes. Additional rules are available to users of the system when it is in interactive mode.

3 λ-Unification

The *λ-unify* rule in & implements the algorithm described in [Bai88] with some optimizations. The role of free variables in the original algorithm is played by schematic terms in &.

Given a sequent $A_1 \cdots A_m \vdash C_1 \cdots C_n$, the rule attempts to λ-unify some A_i with some C_j. The main difference between λ-unify and the usual first-order unification algorithm is that λ-unify recognizes the equivalence of the forms $t \in \{x|P[x]\}$ and $P[t]$. The algorithm opportunistically performs such rewriting steps, and to provide such opportunities it will generate abstraction terms for schematics appearing as the second argument to the $\in$ predicate. Thus λ-unify will succeed in unifying the formulae $t \in f(x_1 \ldots x_n)$ and $P[t]$ by instantiating $f(x_1 \ldots x_n)$ to the term $\{x|P[t]\}$, provided this is a legal set expression.

The λ-unification algorithm is similar to Huet's algorithm [Hue] for unification in the typed λ-calculus (as the name makes clear). Huet's algorithm does not apply to set theory because variables in set theory are not typed—they range over the entire universe of sets. A thorough comparison of the algorithms can be found in [Bai88], but the main differences are:

1. Since set-theoretic formulae cannot be normalized in advance, the λ-unification algorithm has to perform set contractions (replacing $t \in \{x|P[x]\}$ with $P[t]$) "on the fly" as the algorithm executes.

2. Huet's *projection* operation is inappropriate in the context of set theory.

3. The choice of formulae that can be used to form set abstractions is restricted by the particular set theory under discussion. These restrictions are necessary in order to avoid the familiar set-theoretic paradoxes. No such restriction is necessary in the typed λ-calculus.

λ-unify is a semi-decision algorithm. It will find a unifier if one exists, but if one does not exist then the algorithm may not terminate. We therefore place an arbitrary depth bound on the search for a unifier. The depth bound guarantees termination at the expense of completeness: some λ-unifiable formulae will not be unifiable through the bounded search. The depth bound for λ-*unify* is specified as a defined constant, compiled into the system (but this can be overridden in interactive mode). The depth bound specifies the maximum allowable number of calls to the *match* procedure along any single branch of the unification search tree. The *match* procedure generates instantiations by matching a schematic term or formula with another term or formula.

4 $\mathcal{Z}$-match: Introducing Set Abstractions

We give only a brief description of the $\mathcal{Z}$-match rule here. The interested reader is referred to [BBP] for the full details of the rule and its justification. That paper also contains a discussion of the relation between $\mathcal{Z}$-match and Bledsoe's work on set instantiation [BF90,Ble79].

Simple $\mathcal{Z}$-match. The simplest form of $\mathcal{Z}$-match looks for a formula $t \in f(x_1 \ldots x_n)$, where f is schematic (we call this the *schematic formula* and t the *member term*), and any other formula from the sequent, P, called the *matching formula*. Subject to some conditions outlined below, the formula P may be arbitrarily chosen.

The result of $\mathcal{Z}$-match is that the schematic term $f(x_1 \ldots x_n)$ is instantiated with a set abstraction term. The particular term is chosen on the basis of the relative polarities of the schematic and matching formulae. The simple forms of the instantiations are generated by the following rules:

1. If the schematic formula is an assumption, and the matching formula is a conclusion, then $f(x_1 \ldots x_n)$ is bound to $\{y | y \in g(x_1 \ldots x_n) \wedge P\}$, where g is a new schematic symbol.

2. If the schematic and matching formulae are both assumptions, then $f(x_1 \ldots x_n)$ is bound to $\{y | y \in g(x_1 \ldots x_n) \wedge \neg P\}$.

3. If the schematic formula is a conclusion, and the matching formula is an assumption, then $f(x_1 \ldots x_n)$ is bound to $\{y | y \in \{t\} \cup g(x_1 \ldots x_n) \wedge P\}$.

4. If the schematic and matching formulae are both conclusions, then $f(x_1 \ldots x_n)$ is bound to $\{y | y \in \{t\} \cup g(x_1 \ldots x_n) \wedge \neg P\}$.

These rules are subject to conditions on the free variables that occur in the formula P and the term t. These conditions are analogous to the "occurs check". In the context of our logic (the dual of the usual formulation), a schematic term such as $f(x_1 \ldots x_n)$ may not be bound to *any* term, but only to those terms whose free variables are a subset of the x_i. If either P or t contain variables that are not among the arguments to the schematic being bound (we call such variables "bad") then we cannot include P or t, respectively, as a subterm of the new abstraction term, so the rule will fail unless $\mathcal{Z}$-match can find a workaround. Our implementation of $\mathcal{Z}$-match has several such workarounds available, for example introducing new quantifiers so that the "bad" variables in the formula P are no longer free.

Extended $\mathcal{Z}$-match. The solutions returned by the simple form of $\mathcal{Z}$-match described above are, in a sense, trivial. If $\{y|y \in g(x_1 \ldots x_n) \wedge P\}$ is a solution generated by that method, then P does not contain y free; thus, depending on whether P is true or not, the solution is equivalent to either $g(x_1 \ldots x_n)$ or $\emptyset$. These simple solutions are often inadequate to prove the remaining subgoals. We have devised extended forms of $\mathcal{Z}$-match to provide solutions in which the abstraction variable y does occur free in P; the extended forms are *required* when P contains bad free variables. In such cases, we build an abstraction term from a derivative of P that "has the same effect" as P in this context.

1. The *first extended* case occurs when all of the "bad" variables of P are localized within a single subterm, which happens to be t. That is, P is of the form $P_1[t/z]$ such that the free variables of P_1 are among $z, x_1 \ldots x_n$. $\mathcal{Z}$-match then works as in the simple case, described above, except that we use $P_1[y/z]$ in place of P.

2. The *second extended* form of $\mathcal{Z}$-match is required when the "bad" variables of P are all localized within a term s (that is, P is of the form $P_1[s/z]$), and s is different from t. In this case, we require that t be schematic, and that the free variables of s are among those of t. The solution is to bind t to s and then proceed as in the first extended case.

3. We allow the matching formula P to be constructed from several different atomic formulae in the sequent, conjoined or disjoined together depending on whether they come from the hypothesis or conclusion of the sequent. To avoid considering all possible subsets of formulae in the sequent, & compares these possible matching formulae with the right-hand side of the definitional rewrite rules it has stored. It then selects those sets of formulae that can be reverse-rewritten into a defined predicate, with the justification that those formulae are more likely to be useful to the proof.

Our implementation of $\mathcal{Z}$-match returns all of the instantiations generated by these extensions, even if P contains no "bad" variables. This is an essential feature of the rule, as the extended solutions are usually more meaningful than those generated by the simple method.

Backtracking and Ordering Solutions. For any sequent containing a formula of the form $t \in f(x_1 \ldots x_n)$, it is likely that the $\mathcal{Z}$-match rule may be used in any one of a number of forms. Different matching formulae may be chosen, different forms of the rule may be used, and there may even be multiple schematic formulae within the sequent. Each of these possibilities may yield a different substitution of abstraction terms for schematic terms. In order to handle this, backtracking is used: if the proof fails using one instantiation of variables, it will return and try another. The order in which & tries the instantiations produced by $\mathcal{Z}$-match is determined by the form of $\mathcal{Z}$-match used (extended solutions precede simple ones); in the case of compound matching formulae, we have introduced the notion of *contraction position* to help determine the order in which to attempt solutions.

The idea of *contraction position* applies to the schematic $f(x_1 \ldots x_n)$ that has just be bound to the abstraction term, and its appearances in the remaining sequents to be proved. If a formula of the form $s \in f(x_1 \ldots x_n)$ appears with negative polarity in subsequent sequents, then the $\{\}$ *-left* rule will (eventually) cause each of the sub-formulae in the instantiation of f to become hypotheses, so having more formulae will help the prover. Likewise, if $s \in f(x_1 \ldots x_n)$ appears with positive polarity, then the embedded sub-formulae will become conclusions, so having more formulae will produce more work for the prover. & uses compound matching formula instantiations first if there are more hypothesis contraction positions in the remaining subgoals, and nearly last if there are more conclusion contraction positions.

5 Conclusions

In this paper we have described the & theorem prover. & is a natural deduction based prover for Zermelo set theory, with the usual rules for handling logical connectives, and additional, novel rules for handling set abstraction terms. & has been used to prove non-trivial theorems in set theory and in the theory of well-founded relations.

References

[Bai] S.C. Bailin and D. Barker-Plummer. Automated Deduction in Set Theory. Technical Report NSF/ISI-90006.

[Bai88] S.C. Bailin. A λ unifiability test for set theory. *Journal of Automated Reasoning*, 4(3), September 1988.

[BBP] S.C. Bailin and D. Barker-Plummer. Z-match: An inference rule for incrementally finding set instantiations. Submitted to the Journal of Automated Reasoning.

[BF90] W.W. Bledsoe and G. Feng. Completeness of set-var. Technical Report ATP104, University of Texas at Austin, Department of Computer Sciences, Austin, Texas, 78712, December 1990.

[Ble79] W.W. Bledsoe. A maximal method for set variables in automatic theorem proving. In *Machine Intelligence 9*, pages 53–100. Ellis Harwood, Ltd., Chichester, 1979.

[BT75] W.W. Bledsoe and M. Tyson. The UT interactive prover. Memo ATP-17, Mathematics Department, University of Texas, May 1975.

[Hue] G.P. Huet. A unification algorithm for typed λ-calculus. *Theoretical Computer Science 1*, pages 27–57, 1974.

An Overview of FRAPPS 2.0: A Framework for Resolution-based Automated Proof Procedure Systems

Tomás E. Uribe,* Alan M. Frisch, and Michael K. Mitchell**

Dept. of Computer Science and Beckman Institute, University of Illinois,
405 North Mathews Avenue, Urbana, Illinois, 61801, USA

1 Introduction

The Framework for Resolution-based Automated Proof Procedure Systems (FRAPPS) is a set of functions written in Common Lisp that facilitate the construction of a wide variety of resolution-based deductive systems. FRAPPS offers the basic functionality necessary to build such systems, freeing users from low-level implementation concerns. It is not intended for use in the construction of high-performance theorem provers, but rather to provide a modular and customizable system useful for rapid prototyping and experimentation in teaching and research.

In addition to data structures that represent clauses and procedures that implement inference rules, FRAPPS provides a *derivation graph* that records all the inferences that have been performed and a *priority queue* that keeps track of potential future inferences. A number of common deletion and search strategies are predefined, and many more can be implemented with a minimum amount of effort. Proof procedures can be conducted automatically or interactively; FRAPPS offers a wide range of functions to access the derivation graph, so that users can examine and guide the progress of a particular search.

An extension to FRAPPS enables the introduction of specialized unification procedures and constraints, thus supporting the implementation of substitutional hybrid reasoners as defined by Frisch [1991]. The latest version of FRAPPS is fully documented in the FRAPPS 2.0 Users' Guide [Mitchell *et al.*, 1991], and a companion document [Uribe, 1991] describes the extension to constraint unification. Mitchell [1989] provides an in-depth description of the original design of FRAPPS.

2 Proof Procedures Using FRAPPS

A proof procedure is a combination of inference rules and deletion strategies, together with a search strategy that controls the order in which these operations are performed. Proof procedures using FRAPPS grow a derivation graph, where nodes correspond to retained clauses and directed arcs link nodes with their parents and children. New nodes are formed by applying inference rules to previously existing

* Current address: Apartado Aéreo 100677, Bogotá, Colombia.

** Current address: IBM Corp., Internal Zip 9151, 11400 Burnet Rd., Autstin, Texas, 78758, USA.

nodes and then *integrating* the resulting clauses into the derivation graph. The inference rules provided by FRAPPS are factoring, binary and general resolution, and common resolution refinements such as SLD-resolution. Answer extraction is supported through the use of *answer literals*.

The derivation graph is organized by levels: The initial set of clauses is integrated at level 0, and each derived clause is integrated at the level following the maximum level of its parent(s). Nodes are assigned unique integer *node-ids*, reflecting the order in which they are created, and are indexed according to their level as well as the number and type of literals that their clauses contain. A set of functions allow for efficient retrieval of nodes according to various properties they might posses. Particularly useful are those functions that return all nodes whose clause contains a literal equal to, more general than, or an instance of, any given literal. Besides the clause itself, nodes contain additional information, including a *user field* that can be used to associate arbitrary extra information with each node.

FRAPPS offers a number of common deletion strategies, including length, complexity and tautology deletion. Back and forward subsumption procedures are also provided; these can be used to their full power or restricted to consider only instance or variant tests. Users can also decide whether answer literals, if present, should participate in subsumption tests. In general, the user can label nodes as *active* or *inactive*; inactive nodes cannot be used to generate new clauses and are not considered when updating the priority queue.

The priority queue maintained by FRAPPS contains pairs of nodes ordered by a user-defined *cost-function*, and provides a general mechanism for implementing search strategies. A typical proof procedure is conducted as follows:

1. Set the derivation graph and the priority queue to be empty, and let $\mathcal{S}$ be the initial set of clauses.
2. For each clause $c \in \mathcal{S}$, do:
   If c is not deleted by any of the deletion strategies in effect, then integrate c as a new node in the derivation graph. If back subsumption is in effect, label as *inactive* all other nodes whose clause is subsumed by c. Retrieve all active nodes whose clause can be resolved with c, and add to the priority queue the set of pairs that contain one of these nodes and the new node for c, ordered by the user-defined *cost function* currently in effect.
3. If the priority queue is empty or a termination condition is satisfied (for example, a given number of answer or empty clauses is found or a maximum time or space limit is reached), stop.
4. Pop the priority queue, removing the "cheapest" node pair, and repeat, if necessary, until the queue is empty or a pair is obtained such that both nodes are active. If a pair of active nodes is found, let $\mathcal{S}$ be the result of applying the chosen inference rule(s) to the clauses in the two nodes,[3] and go to Step 2.

This basic proof procedure can be implemented in less than a single page of Lisp code, using the higher-level functions provided by FRAPPS. The default dele-

[3] Note that, in general, more than one group of literals can be resolved upon at a given resolution step. Thus, a *set* of resolvents is obtained from a pair of nodes. Given two clauses, users can control which literals are factored or resolved upon.

tion, inference and control strategies can be easily overridden to obtain new proof procedures.

The node-pair cost function is defined as the weighted sum of an arbitrary number of individual components. In this way, a large number of complex search strategies can be implemented, starting out from a small set of primitive cost function components. New components are simple to write; each should simply return a numerical value given a pair of nodes. Node-pairs exceeding a maximum cost set by the user are not added to the queue.

Note that when the above procedure adds a new node to the graph, all possible resolvent pairs that can be formed using the new node are added to the priority queue. Efficiency can be improved by delaying the formation of these node pairs; if an underestimate of their cost were available, these pairs should be added to the queue only when the cost of all other pairs exceeds the given estimated cost. FRAPPS implements this scheme if given a user-defined function of single nodes that computes the cost-function estimate.

To further improve the efficiency of a search, users can redefine the node-pair creation process. By default, a node is paired up with *all* the previously created nodes that can resolve with it. Modifying this procedure to suit a particular resolution refinement such as unit, input or SLD resolution, for example, can exclude a large number of pairs from consideration based on indexed properties of nodes.

3 Integrating Constraint-solving Unification

FRAPPS has been extended to support the introduction of unification procedures that incorporate constraint-processing mechanisms [Uribe, 1991]. The user-defined unification procedure can take into account, and generate, constraint fields that are associated with each clause.

This extension of FRAPPS can be used to construct *hybrid reasoners* in accordance to the *substitutional framework* [Frisch, 1991]. The extended system could, for example, incorporate specialized equational reasoning through E-unification, taxonomic reasoning by using sorted logic and sorted unification, or even modal deduction, using constraints on terms that denote possible worlds [Frisch and Scherl, 1991].

Users are responsible for implementing unification of pairs of terms, used in binary resolution, and, optionally, unification of sets of terms, used in factoring and general resolution. (This distinction is due to efficiency considerations. If the more general procedure is not provided, the unification procedure for pairs of terms is used instead.) The new unification procedure can produce a finite number of unifiers; each should be identified by a substitution and a new set of constraints. If back and forward subsumption are to be done, users should also provide functions that implement subsumption tests under the new constraints: In general, there is no way of implementing a subsumption algorithm given only an arbitrary unification procedure.

Other than the points mentioned above, all of the functionality present in FRAPPS 2.0 is included in this extension, which we call *Hooked-on-*FRAPPS.

4 Strengths and Weaknesses of FRAPPS

The major strengths of FRAPPS are its flexibility and ease of use. Being written in Common Lisp, FRAPPS is highly portable[4] and relatively simple to use and modify, thereby providing a practical tool for quickly implementing and experimenting with resolution-based systems. As such, it has proved to be particularly useful to students and researchers in automated deduction. Starting with Version 1.0 in 1988, FRAPPS has been used on three separate occasions in a graduate-level introductory course on automated deduction at the University of Illinois. Students are usually able to design their own proof procedures within a few of days of exposure to the system. This hands-on experience provides a better understanding of the ideas covered in the course.

The very same virtues mentioned above are also the source of FRAPPS' main weakness, namely, its relative inefficiency. Even though FRAPPS includes features that make proof procedures more efficient (such as structure sharing and literal indexing), it is *not* intended for use in building high-performance systems. Though its speed could certainly be improved, it appears that some efficiency must always be sacrificed if the flexibility and modularity currently provided by FRAPPS are to be retained. Theorem provers achieve high performance in part by tightly integrating their components; in FRAPPS, these are only "loosely" coupled.[5] On the other hand, added flexibility can enhance overall performance by allowing the implementation of special-purpose proof procedures that are not offered by other systems; thus, raw performance comparisons may not always be a good indicator of a system's success.

FRAPPS' overall efficiency can be improved by adding more support for specialized node-retrieval functions. Ideally, derivation graph nodes would be cross-indexed according to an arbitrary number of pre-determined properties (including the usual level, literal and length indexing), so that search procedures could *efficiently* identify all the nodes that satisfied any arbitrary combination of such properties. Currently, nodes can only be retrieved according to one property at a time, so set intersection must be used to answer conjunctive queries.

A related problem is that specialized inference rules to handle equality, such as paramodulation and demodulation, are currently not available in FRAPPS; its indexing capabilities would have to be further extended in order to support the efficient implementation of these rules.

There are many other ways in which FRAPPS can be extended. The most outstanding one is to generalize the priority queue mechanism, so that not only pairs but arbitrary *tuples* of nodes can participate in the queue. This capability would be used together with inference rules that are a function of more than two clauses, such

[4] Identical code has been run under Allegro, Kyoto and Lucid Common Lisps.

[5] Given the many proof procedures that can be implemented using FRAPPS as a starting point, as well as the different ways of implementing each, how to determine its real "speed" is unclear. On an IBM RISC 6000 machine running Allegro Common Lisp, the compiled code can generate (and integrate) on the order of 10-20 clauses per second using binary resolution in a breadth-first search, while a theorem prover like OTTER [McCune, 1990] under equivalent circumstances can generate around 200. FRAPPS' relative performance degrades as the derivation graph grows and when subsumption tests are used.

as *UR-resolution* and *hyperresolution*. Currently, the priority queue mechanism can be simply turned off if a particular proof procedure does not use it.

5 System Acquisition

The entire FRAPPS and Hooked-on-FRAPPS systems, including commented source code, demonstration examples, and documentation in postscript, is publicly available. It can all be found in the tar file `/pub/frapps` obtained by anonymous FTP to `a.cs.uiuc.edu` (128.174.252.1).

6 Acknowledgments

This work was partially supported by NASA, under grant NAG 1-613, and by the University of Illinois. We thank the many students who have used the system and provided valuable feedback.

References

[Frisch, 1991] Alan M. Frisch. The substitutional framework for sorted deduction: Fundamental results on hybrid reasoning. *Artificial Intelligence*, 49:161–198, 1991.

[Frisch and Scherl, 1991] Alan M. Frisch and Richard B. Scherl. A general framework for modal deduction. In James Allen, Richard Fikes, and Erik Sandewall, editors, *Principles of Knowledge Representation and Reasoning: Proceedings of the Second International Conference*, pages 196–207, Morgan Kaufman, San Mateo, CA, 1991.

[McCune, 1990] William W. McCune. OTTER *2.0 Users' Guide.* Technical Report ANL-90/9, Argonne National Laboratory, Argonne, Illinois, March 1990.

[Mitchell, 1989] Michael K. Mitchell. *The Anatomy of a Framework for Resolution-Based Automated Proof Procedure Systems.* Master's thesis, University of Illinois, Urbana, Illinois, July 1989.

[Mitchell *et al.*, 1991] Michael K. Mitchell, Tomás E. Uribe, and Alan M. Frisch. *The Framework for Resolution-based Automated Proof Procedure Systems –* **FRAPPS** *2.0 Users' Guide.* University of Illinois, Urbana, Illinois, September 1991.

[Uribe, 1991] Tomás E. Uribe. *Constraints in* **FRAPPS** *2.0.* University of Illinois, Urbana, Illinois, September 1991.

This article was processed using the LaTeX macro package with LLNCS style

The GAZER Theorem Prover[1]

Dave Barker-Plummer
Computer Science Program,
Swarthmore College,
Swarthmore, Pennsylvania, 19081

plummer@cs.swarthmore.edu

Alex Rothenberg
Computer Science Program,
Swarthmore College,
Swarthmore, Pennsylvania, 19081

alex@cs.swarthmore.edu

Abstract

We describe GAZER theorem proving system which was designed as a testbed for developing ideas about choosing which definitions and lemmas to use in the search for a proof. GAZER is a sequent calculus based system for first-order logic. The main novelty is the use of a new inference rule, *gazing*, which enables the system to determine which of a possibly large number of definitions and lemmas to use at any point in the proof. This decision is made on the basis of multiple abstractions of the current conjecture and of the database of definitions and lemmas. We have used gazer to prove theorems from naïve set theory, and a detailed comparison of the success of the *gazing* inference rule in cutting down the search for proofs of these theorems is presented in [BP89b].

1 Overview

GAZER is a first-order logic, sequent-based, automatic reasoning system which is able to prove theorems in formally axiomatized theories. GAZER draws on many existing systems and original research to form a novel contribution to automated reasoning research [BT75, Ble83, BM79, Bro78, BP89b].

GAZER is used for two types of activity: *theory building* and *attempting proofs*. By theory building, we mean stating axioms and definitions to the system. GAZER allows one to add definitions and axioms to the theory, and later retract them in favor of alternatives if that is required. The system therefore provides a flexible tool for exploring the consequences of different choices in the development of an axiomatization. In other words, GAZER is designed to be a general purpose prover, rather than as a theorem prover tailored to a particular domain.

Proofs are attempted to determine whether a given formula, a *conjecture*, follows from the current theory. If it does, the conjecture is called a *theorem* of the theory, if not it is a non-theorem. The system carries out the proof using a sequent-based calculus which the system utilizes in a subgoaling manner. Repeated application of the inference rules may cause the theorem prover to

[1] This paper is based on work supported in part by Swarthmore College, through a summer research award to Alex Rothenberg, and through the college Research Support Fund.

produce goals that are instances of basic theorems, in which case the original formula is a theorem. The process of repeatedly applying the inference rules will eventually terminate, whether or not the conjecture is a theorem. It follows that GAZER is incomplete (i.e. there are some theorems that cannot be proved by GAZER), and the failure to find a proof does not indicate that the conjecture is not a theorem.

The interface to the system is "user-friendly". For example formulae are presented to the system in a language which differs only from standard logical notation because of the limitations of a standard computer keyboard. No knowledge of any programming language is required to use the system.

2 Relation to Existing Systems

The GAZER theorem prover was developed at Duke University between January and August 1989 and extended at Swarthmore College in July and August 1991. GAZER is based on earlier work on the prototype VOYEUR system which was implemented at the University of Edinburgh and the University of Texas at Austin between 1984 and 1987.

GAZER was developed to provide a test-bed for carrying out research in automated theorem proving. In particular we wish to investigate ways of controlling the use of definitions, axioms and previously proved lemmas within the development of a proof. The *gazing* inference rule was the product of this research, and the GAZER prover provides proof of concept. See [BP89b, Plu88] for a detailed description of *gazing*.

Apart from the *gazing* inference rule, much of the logic of GAZER is derived from earlier work in natural deduction theorem proving. In particular, the logic is quite similar to that of the UT theorem provers [BT75, Ble83]. The main differences between GAZER and these systems are in the handling of quantifiers and in the sequent representation of goals to prove.

GAZER differs from the UT provers in its representation of goals. The earlier UT systems [BT75] represented goals as an implication of two formulae which led to extra search when, for example, only some conjunct of the hypothesis was required in a proof. Later versions of the UT prover [Ble83] retained the formula representation of the consequent but represented the hypotheses as a set of (implicitly conjoined) formulae, allowing random access to the members of this set as required. In GAZER we have completed the move to a fully sequent based system representing the consequent as a set of (implicitly disjoined) formulae. This last move means that GAZER implements a version of Gentzen's $\mathcal{NK}$ system, while the UT prover implements Gentzen's $\mathcal{NJ}$, [Gen69].

In the UT provers, quantifiers are handled by skolemizing the conjecture to be proved. This leads to problems caused by the inability to copy universally (existentially) quantified antecedents (consequents). In GAZER quantifiers are not skolemized away, rather we use a technique for controlled copying of quantified formulae due to Frank M. Brown, [Bro78].

3 The *Gazing* Inference Rule

The GAZER system is unique in its ability to control access to a database of definitions, axioms, and preproved theorems. The use of lemmas and definitions allows the user to state the conjecture in a more natural manner, in terms of complex concepts, and can encapsulate many inferences into a single step and can therefore be beneficial in reducing the search for a proof. However giving the prover access to such formulae is costly. Providing the prover with all formulae that might be relevant rapidly swamps the prover with unnecessary information, and presenting the theorem prover with only the formulae that are necessary for the proof might be viewed as "cheating" and requires that the user prove the theorem ahead of time [Wos87]. In real-world applications of theorem proving, this is almost certain to be infeasible. The *gazing* inference rule is a heuristic designed to allow the prover to select intelligently the appropriate rule to apply.

The *gazing* inference rule is an abstraction based heuristic motivated by the belief that, unlike machine provers, humans remember formulae by what they are *about* – what they *say* or *mean* – rather than as a string of variables, connectives, functions and predicates, [Bun83]. Gazing attempts to capture this semantic information, but with an operational view of the meaning of a formula. The view corresponds to "what can this formula do" in the proof. Rather than relating different concepts, we view the formulae as means by which we can trade concepts in the conjecture for other concepts. For example we may want to exchange one predicate symbol for another that will be "more useful".

The gazing rule selects formulae by examining representations of the database at varying levels of abstraction. Each representation of the theory is produced by applying some abstraction mapping to each of the formulae in the database. An abstraction mapping is one which ignores certain *details* of the formula.

The abstraction spaces are ordered into a hierarchy where each "less abstract" space contains all the information from the preceding spaces with the addition of some extra detail. The gazing inference rule uses the hierarchy as a refinement mechanism. Gazing searches for appropriate formulae in each of the abstraction spaces in turn, starting with the most abstract space. Each space accepts an abstraction of the conjecture to be proved and a representation of the formulae whose use is suggested by the conjecture in the previous abstraction space (called a plan). The given plan is critiqued by considering whether each of these formulae are suggested in the light of the new information present in this space. Some formulae previously under consideration may be dropped as a consequence of this new information, and similarly new formulae may be suggested. Once the existing plan has been refined in a given abstraction space, the result is passed to the next abstraction space for further refinement. When all abstraction spaces have been considered, the original conjecture is rewritten as indicated by the plan, and the result is a new subgoal for the prover.

4 Abstraction Hierarchies

If a given abstraction hierarchy of the theory is to provide useful search control the abstract search spaces should be smaller than the search space produced by the formula representation of the theory. This can occur by mapping different formulae into the same abstract representation, or by reducing some non-trivial formula to a trivial abstraction. It is unlikely that any single abstraction hierarchy will provide useful guidance for *all* possible theories, since different theories will have different, but characteristic, syntactic forms. For example, in an equational theory where = is the only predicate, any abstraction that "remembers" only the predicate symbol which appears in a formula is going to provide no guidance in the search for an appropriate formula to use: they will all look the same in the abstraction space.

In view of this observation, we have implemented GAZER so that different abstraction hierarchies can coexist. This enables us to run experiments to determine which of a given suite of abstraction hierarchies provide more useful guidance when given a particularly theory. We hope that these experiments will enable us to extend GAZER so that it can eventually choose for itself which of a number of hierarchies to use in a particular context. Currently we have three different hierarchies implemented.

One of these hierarchies is described in detail in [BP89b]. The second hierarchy that we have implemented is based on the work of Sacerdoti, [Sac74]. Although Sacerdoti implemented his abstraction hierarchy for a function-free calculus, we have been able to adapt it to the more general context of first-order logic. The third hierarchy is closely related to the second. We are as yet unable to present results comparing the properties of these abstraction hierarchies with respect to the guidance gained by their use. We hope to be able to do so in a future paper.

5 Conclusion

Expanding machine provers so that they are able to utilize a theory of known facts is a necessary advance before we can expect them to prove complex theorems in reasonable time. They should be able to cite known facts and avoid the work of repeating the proof of the fact.

We have built the GAZER system to allow us to explore the use of the gazing inference rule with different abstraction hierarchies. Our eventual aim is to isolate features of theories which indicate that one particular abstraction hierarchy should be used in preference to another. If this is possible, we will be able to exploit the presence of large databases of formulae more efficiently within the search for a proof.

Although GAZER is a general purpose prover, we have only tested it on approximately forty theorems of naive set theory. Those theorems are listed in [BP89b]. In the experiment reported in that paper, we were not so much interested in the ability of the prover to prove difficult theorems, but rather in the success of the *gazing* inference rule in limiting the search for proofs.

A user manual for GAZER is available as [BP89a]. GAZER is written in the Quintus dialect of PROLOG and is available from the authors on request.

References

[Ble83] W.W. Bledsoe. The UT interactive prover. Memo ATP-17B, Mathematics Department, University of Texas, April 1983.

[BM79] R.S. Boyer and J S. Moore. *A Computational Logic.* ACM Monograph Series. Academic Press, 1979.

[BP89a] D. Barker-Plummer. The gazer theorem prover: Description and users guide. Technical Report CS-1989-15, Department of Computer Science, Duke University, Durham, North Carolina, 27706, USA, 1989.

[BP89b] D. Barker-Plummer. Gazing: An approach to the problem of definition expansion. Technical Report CS-1989-13, Department of Computer Science, Duke University, Durham, North Carolina, 27706, USA, 1989. To Appear in the *Journal of Automated Reasoning.*

[Bro78] F.M. Brown. Towards the automation of set theory and its logic. *Artificial Intelligence,* 10:281–316, 1978.

[BT75] W.W. Bledsoe and M. Tyson. The UT interactive prover. Memo ATP-17, Mathematics Department, University of Texas, May 1975.

[Bun83] A. Bundy. Finding a common currency - a new proof plan. Internal Note 159, Department of Artificial Intelligence, University of Edinburgh., January 1983.

[Gen69] G. Gentzen. Untersuchen über das logische schliessen [investigations into logical deduction]. In M.E. Szabo, editor, *The Collected Papers of Gerhard Gentzen,* chapter 3, pages 68–131. North-Holland, 1969. First appeared in 1935 as an article in *Mathematische Zeitschrift* 39. This work was accepted as Gentzen's Inaugural Dissertation at the University of Göttingen.

[Plu88] D. Plummer. *Gazing: A Technique for Controlling the Use of Rewrite Rules.* PhD thesis, Department of Artificial Intelligence, University of Edinburgh, May 1988.

[Sac74] E.D. Sacerdoti. Planning in a hierarchy of abstraction spaces. *Artificial Intelligence,* 5:115–135, 1974.

[Wos87] L. Wos. The problem of definition expansion and contraction. *Journal of Automated Reasoning,* 3(4), December 1987.

Roo: A Parallel Theorem Prover*

Ewing L. Lusk
William W. McCune
Mathematics and Computer Science Division
Argonne National Laboratory
U.S.A.
{lusk,mccune}@mcs.anl.gov

John Slaney
Automated Reasoning Project
Australian National University
Canberra, ACT
Australia
slaney@arp.anu.edu.au

Background. Roo is a parallel theorem prover for shared-memory multiprocessors. Its basic algorithm is based on a parallel algorithm for closure computations reported in [4]. Since the fundamental algorithm of the Otter sequential theorem prover is a closure computation, we can apply the parallelizing approach of [4] to get a parallel theorem proving algorithm. This algorithm is written in C using the **p4** parallel programming library developed at Argonne National Laboratory by Ralph Butler and Ewing Lusk. It provides a high-level parallel programming abstraction (monitors) well suited to this algorithm.

The Parallel Algorithm. The basic problem to be solved is the use by multiple processes of a shared database. A variety of uses are made of the database of clauses (multiple inference rules, forward and backward subsumption, forward and backward demodulation), and these must be allowed to proceed in parallel, while updates to the database are serialized to maintain integrity. The flow of data is shown in Figure 1.

The ellipses in the this diagram represent potential work units, not processes, Each process is in a perpetual loop requesting a work unit and then processing it. The priority of tasks is update, back subsumption, back demodulation, and then inference with the one restriction that only one process can be working on the update task at a time. The inference processes do most of the forward demodulation and forward subsumption. These are redone by the update task as a final check before updating the database.

Using Roo. Roo is compatible with Otter in the sense that all Otter input is acceptable to Roo, and all but a very few features of Otter are present in Roo. In fact, almost all source code is shared. A useful feature of Roo is its ability to generate log files displayable by `upshot`[1]. The display of details of the parallel execution has enabled us to identify and remove bottlenecks in Roo. Roo is in the

*This work was partially supported by the Applied Mathematical Sciences subprogram of the Office of Energy Research, U.S. Department of Energy, under Contract W-31-109-Eng-38.

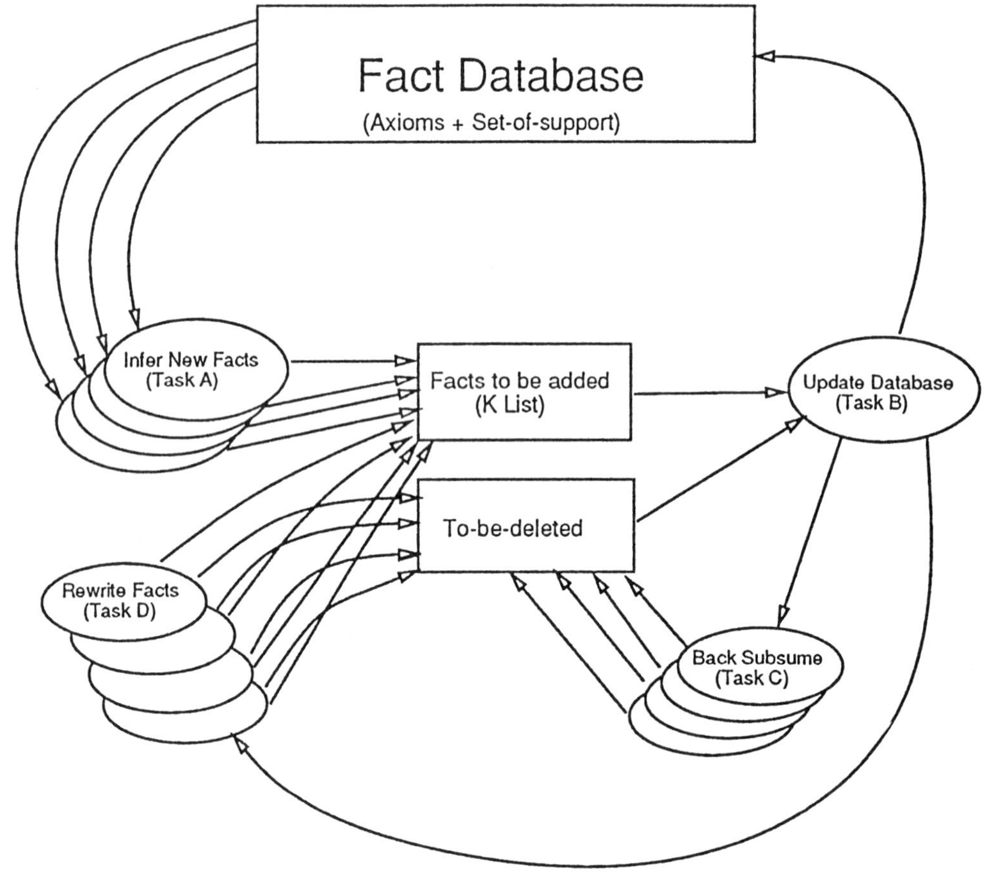

Figure 1: Flow of Data in Roo

public domain and the Sequent Symmetry version can be obtained by anonymous ftp from `info.mcs.anl.gov`. We show in Figure 2 a portion of the trace from the proof of SAM's Lemma.

Performance. Roo's performance is in general good, with regular and near-linear speedups. On some problems it performs badly, and on others superlinear speedups are obtained, due to the reordering of the search space introduced by parallelism. We show examples, one small and one large, of good performance in Tables 1 and 2. The first is the problem known as SAM's Lemma, the other is the IMP-4 problem of implicational propositional calculus [3]. The clauses for these problems are given in [2]. These runs were done on a 386-based Sequent Symmetry.

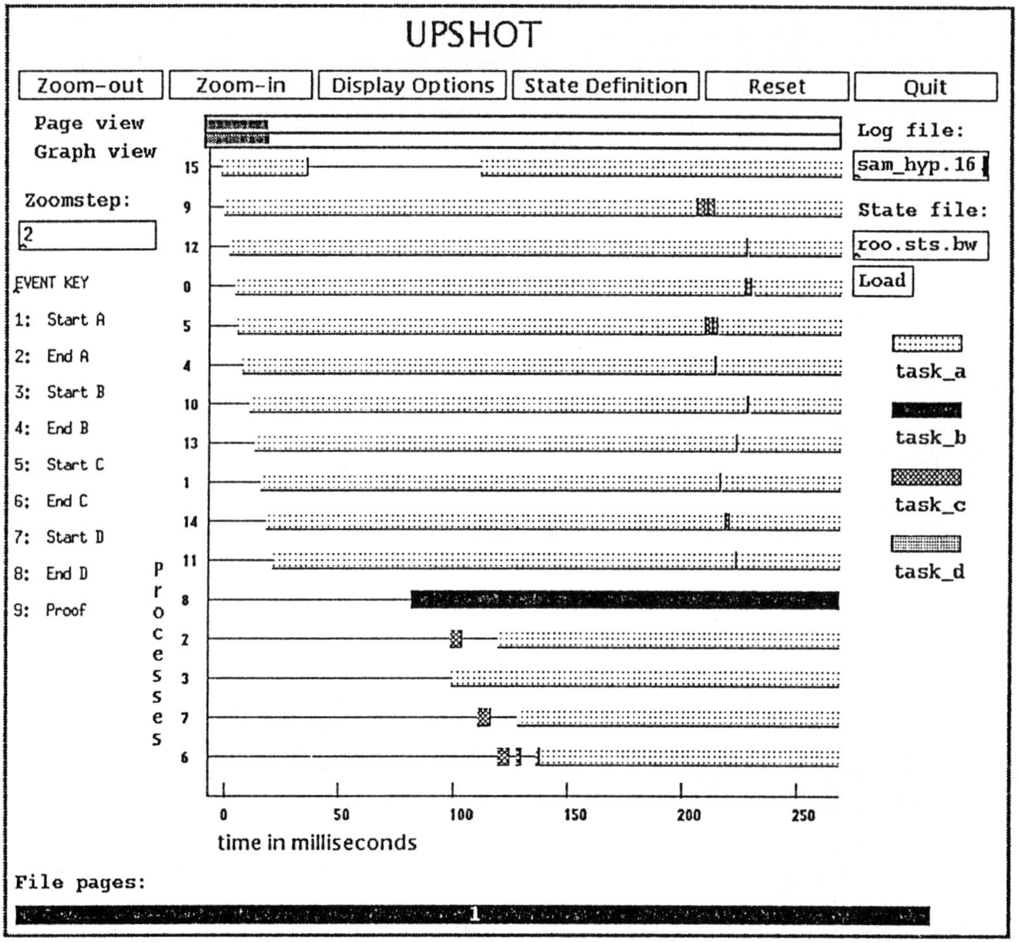

Figure 2: Trace of SAM's Lemma

	OTTER	ROO-1	ROO-4	ROO-8	ROO-12	ROO-16	ROO-20	ROO-24
Run time	30.70	32.04	7.91	4.37	2.57	2.85	1.98	1.62
Generated	5924	5981	5977	6079	6022	6208	6143	5837
Kept	134	134	131	131	130	130	129	124
Memory (K)	95	159	220	344	468	592	716	840
Gen/sec	192	186	755	1391	2343	2178	3102	3603
Speedups:								
Proof	1.0	1.0	3.9	7.0	11.9	10.8	15.5	19.0
Gen/sec	1.0	1.0	3.9	7.2	12.2	11.3	16.2	18.8

Table 1: SAM's Lemma

	OTTER	ROO-1	ROO-4	ROO-8	ROO-12	ROO-16	ROO-20	ROO-24
Run time	29098.32	29984.67	7180.29	3440.89	2462.52	1844.03	1492.35	1269.28
Generated	6706380	6668046	6413924	6208570	6649996	6662397	6826296	7108289
Kept	20410	20342	20309	20242	20438	18281	19534	18759
Memory (K)	7185	7791	8653	9737	13119	14235	17109	19498
Gen/sec	230	222	893	1804	2700	3612	4574	5600
Speedups:								
Proof	1.0	1.0	4.1	8.5	11.8	15.8	19.5	22.9
Gen/sec	1.0	1.0	3.9	7.8	11.7	15.7	19.9	24.3

Table 2: IMP-4

A survey of ROO's performance on a wide variety of problems is given in [2].

References

[1] Virginia Herrarte and Ewing Lusk. Studying parallel program behavior with upshot. Technical Report ANL–91/15, Argonne National Laboratory, Argonne, IL 60439, 1991.

[2] Ewing L. Lusk and William W. McCune. Experiments with ROO, a parallel automated deduction system. (To appear in Springer Lecture Notes in Artificial Intelligence), 1992.

[3] F. Pfenning. Single axioms in the implicational propositional calculus. In E. Lusk and R. Overbeek, editors, *Proceedings of the 9th International Conference on Automated Deduction, Lecture Notes in Computer Science 310*, pages 710–713. Springer-Verlag, 1988.

[4] John K. Slaney and Ewing L. Lusk. Parallelizing the closure operation in automated deduction. In M. E. Stickel, editor, *Proceedings of the 10th International Conference on Automated Deduction, Lecture Notes in Artificial Intelligence 449*, pages 28–39. Springer-Verlag, 1990.

RVF: An Automated Formal Verification System

T. C. Wang and Allen Goldberg
Kestrel Institute
3260 Hillview Avenue
Palo Alto, CA 94304

Overview. RVF (Reacto-VeriFier) is a mechanical verifier developed for supporting the design of reliable reactive systems. In order to make the formal verification of large and/or complex systems tractable, the verifier has been enhanced by a proof manager, a knowledge-base manager, and a dependency maintenance procedure. The proof manager extracts unproven verification conditions, and supports off-line development of proofs for them. The knowledge-base manager supports a flexible use of a large set of axioms and rules derived from the domain theory or user-defined abstract data types of the specification language. The dependency maintenance procedure permits the user to trace the history of a derivation, and supports efficient addition and/or retraction of assumptions. The theorem prover is based on a goal-oriented proof procedure, hierarchical deduction, incorporated with term-rewriting, partial-evaluation, and forward-inference procedures. The prover can be used as an automated system, or as an interactive proof checker. Fig. 1 depicts the main components of RVF and a portion of the user interface.

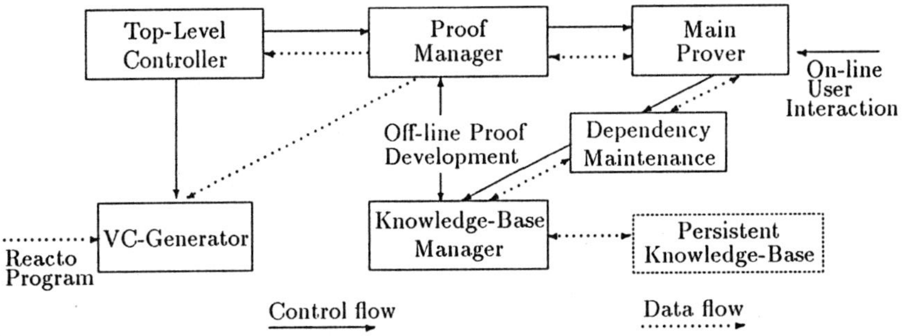

Fig. 1: The Architecture of Reacto-VeriFier

RVF is the consistency checker for Reacto: a system for the specification, verification and transformational implementation of reliable reactive systems. Reacto uses hierarchically-structured finite state machines to model reactive systems [4]. RVF is designed to prove the consistency of the assertions (typically, of safety properties) associated with the state of a Reacto program with its operational behavior [7]. It is based on an extension of Floyd's inductive assertion method. RVF can be used both for batch-style verification and for supporting incremental development of specifications and proofs. For a simplest application of the verifier, the user needs only to apply it to the root state of the program or subprogram that is to be verified. The verifier will attempt to verify the correctness of each transition contained in the state according to the assertions associated with its originating and destination states. For each transition that can not be verified automatically, the proof manager will create a *proof-object* associated with it and store it in a proof directory. Upon termination, the verifier will give a summary of the unproven transitions and the associated unproven conjectures. For each unproven conjecture, the user can determine the reason of proof failure by examining the corresponding proof-object that has been saved. If the conjecture is indeed a theorem, then the user can help the prover to finally prove it by editing the proof-object with additional definitions, rules, intermediate lemmas, proving strategies, control parameters, etc.

1 A Typed KR&R Environment.

Reacto uses a strongly-typed functional language to specify the conditions and actions associated with transitions. In accordance with the strong typing of the language, RVF has been developed in a *typed* knowledge representation and reasoning (KR&R) environment. This environment includes a typed first order language [8], a typed knowledge base, and a typed theorem prover. The typed language permits one to specialize the knowledge representation for a rich set of data types and type relations, including those associated with overloading and (parametric) polymorphism. The language contains two kinds of normalized forms: typing rules and typed clauses. Typing rules are used for specifying type relations. Typed clauses are clauses over general predicates attached with type restrictions. RVF contains a library of typing rules, called the basic typing theory, that is formulated for specifying the typing system of the programming language. The knowledge base of the theorem prover contains mainly theories of various of data types. Each theory is a collection of axioms and rules written in the typed language. The theorem prover employs typed resolution and typed paramodulation to make deductions on typed clauses. The prover employs special procedures for efficiently deducing types for terms generated by the the prover, and for checking the type consistency for newly generated resolvents.

This typed KR&R environment has demonstrated three advantages over the untyped one. One is the clarity of the representation. There is a clear distinction between type descriptions and kernel logical formulations. Another is the efficiency of the deduction. The inference on type relations can be done more efficiently, and it is used mainly for filtering out certain redundant consequences that may be created by the deduction on kernel logical formulations. The last is the soundness of the system. The typed discipline provides a system support to enforce the type consistency of all logical expressions.

2 The Verification Condition Generator

Due to the explicit notion of states and transitions, the verification of a Reacto program is decomposed naturally into the verification of the correctness of each transition.

The task of the verification condition generator (VCG) is to deduce, for each transition, a *verification condition* (VC) from the assertions associated with the originating state, the terminating state and containing states of the transition, as well as the guard condition and the action of the transition. VCG is designed by using existing techniques based on Hoare-style rules [2]. According to state hierarchy semantics, the task of verifying a transition is transformed into the proof of a basic statement, "if an assertion p holds and an action α is taken then an assertion q holds after α terminates." The VCG is called which first deduces a weakest liberal precondition $wlp(\alpha, q)$ from α and q. Then it produces $p \Rightarrow wlp(\alpha, q)$ as a VC of α with respect to p and q. The actual design of VCG is complicated by a number of pragmatic features of the language which extend the basic finite state machine model. These features include hierarchical states, substate calls (which are similar to procedure calls of conventional programs), and specially interpreted self-transitions. Of course, the VC deduced by VCG for a transition tr itself is usually not a theorem. But if tr is correct with respect to the associated assertions, then VC should be a logical consequence of the underlying domain theory of the language.

3 Knowledge-Base Management

A large set of axioms exist from domain theory and axiom schemata induced by induction principles for recursively constructed data types. This is a distinguishing feature of mechanical theorem proving associated with program verification. In order to help the theorem prover avoid the search space explosion that can result from the existence of such a large KB, we have transformed as many axioms and lemmas as possible into term-rewriting rules. In addition, a knowledge-base manager is designed to support a flexible use of the KB rules. We have defined a number of distinct rule types. Associated with each rule type is a set of restrictions and control strategies. The user can control the use of each axiom or lemma by assigning it a rule type and/or by defining specific constraints. By the help of KB manager, the user can freely choose the subset of rules and define special constraints that he/she believes to be useful for the prover in solving a non-simple problem.

RVF maintains three distinct KBs: a *persistent knowledge-base* (PKB), a *global knowledge-base* (GKB), and a *working knowledge-base* (WKB). The PKB is used to store axioms from the domain theory and those extensions of the theory which have been approved as reusable. The WKB is the knowledge base used by the prover in proving a particular conjecture. While the PKB is relatively stable, the WKB is usually modified dramatically during a proof process. GKB is the initial state of WKB, which contains some rules loaded from PKB and those given by the user. However, the content of GKB is not fixed: it is determined by RVF or by the user according to the given theorem proving task. For instance, in proving a theorem which requires a deep reasoning, we may limit the GKB to the set of term-rewriting rules of the PKB plus some user-defined rules. A user-defined rule can be attached with special constraint. For example, one may add a rule to GKB, and define backward-implication as the rule-type, then the rule can only be used by a backward-inference procedure.

4 Proof Management

The proof manager of RVF is designed for handling verification failure and for supporting incremental development of proofs. More precisely, it is designed to do the following:

1. Decompose and simplify each VC into a set of conjectures. Create a proof-object for each conjecture with proving-strategies and control parameters provided by default or by the user, and then call the main prover to prove it.
2. Provide a working environment for the user to help RVF prove those conjectures that can not be proved by RVF in its initial proof attempts.
3. Provide a working environment for proving additional rules and lemmas, and maintain persistent records for those that are allowed to be reused.
4. Keep track of the entire verification process and maintain the validity and consistency of the verification results when changes occur in the specifications or in the knowledge base.

The proof manager contains a *natural deduction controller*. Since the verification condition is usually a fairly large and complex formula, the natural deduction controller is used to first decompose the formula into a set of conjectures, and name each conjecture in such a way, that the same name will be assigned to the corresponding conjecture when RVF is invoked again to verify the same program. The name is useful for the proof-object manager to identify the proof-file which was created by previously running the RVF for the same component VC, and which may have been edited by the user. The decomposition is based on the IMPLY algorithm of Bledsoe's UT Interactive Prover [2]. A *proof-object* is a record of information about the proof of a main conjecture, such as proof status, problem origin, lemmas to be exported, theories to be used, additional rules, supplementary conjectures, etc. Besides the logical formula, each conjecture may contain specific values of control parameters given by the user to help the prover.

5 The Theorem Prover

The theorem prover consists of a simplifier, a forward-inference procedure (FIP) and a backward-inference procedure (BIP). Given a conjecture to be proved, it will be simplified immediately by the simplifier. If the resulting formula is *true*, then a proof has been found. Otherwise the formula is negated and transformed into a set S of clauses (or a set of sets of clauses, if the prover is directed to do case-analysis). Then the prover initializes the WKB with GKB, and calls the FIP with S. The FIP will add S to the WKB, and deduce additional consequences from S and rules contained in the WKB, and add some or all of these consequences to the WKB. The prover will terminate with a proof if a refutation is found. Otherwise the BIP is invoked to deduce a contradiction with the WKB and with goal clauses chosen from S.

The *simplifier* is designed to replace a term or a subterm by a simpler, but semantically equivalent term or subterm. The simplification is carried out by *term-rewriting,* and *partial evaluation.* Term-rewriting is the main component of the simplifier, whose behavior is determined by the set of term-rewriting rules contained in the WKB. Partial evaluation is used to deduce canonical forms for those terms or subterms, that are computable.

The FIP accepts a set of clauses as the set-of-support, and deduces a set of clauses by unit resolution and paramodulation, using rules stored in the WKB.

The BIP is the most powerful proof procedure of RVF, which is based on a goal-oriented proving procedure, typed hierarchical deduction. The BIP proves a theorem by traversing a tree of nodes. Each node contains a different set of rule clauses. All candidate goal clauses are contained in a goal-list. Each literal of a goal is indexed by a node name, through which a set of nodes can be located to obtain rule clauses for the resolution and paramodulation upon that literal. The root node of this tree is the WKB. The "legal" resolvents of BIP are produced under a set of constraints or narrowing strategies, such as local subsumption, constraints on common tails, proper reduction, global subsumption, subgoal reordering, partial set of support, and semantic guidance, etc. The reader interested in a discussion of these constraints and strategies may refer to [5].

6 Implementation of RVF

The VCG and the proof manager are written in Refine [1]. The theorem prover is written in Common-lisp. Terms maintained by the prover are implemented with a term-integration technique [6]. This technique helps improve the efficiency of the KB manager and the dependency maintenance procedure.

References

1. Abraido-Fandino, L. "An overview of REFINE^{TM} 2.0." *Proceedings of Second Int. Symp. Knowledge Engineering*, Spain (1987).
2. Apt, K. R. "Ten years of Hoare's logic: a survey — part 1." *ACM Transaction on Programming Language and Systems*, Vol. 3, No. 4 (1981).
3. Bledsoe, W. W. "The UT interactive prover." Tech. Report ATP-17B, Department of Mathematics, The University of Texas at Austin (1983).
4. Gilham Li-mei, Goldberg, A., Wang, T. C., "Toward reliable reactive systems." in *Proceedings of 5th International Workshop on Software Specification and Design*, 68-74, IEEE Computer Society Press, (May, 1989).
5. Wang, T. C., and Bledsoe, W. W. 'Hierarchical Deduction', *Journal of Automated Reasoning,* 3(1), 35-71 (1987).
6. Wang, T. C. 'Term integration: an indexing method for storage and retrieval of large number of terms' (abstract), Presented in *1989 AAAI Spring Symposium: Representation and Compilation in High Performance Theorem Proving* (1989).
7. Wang, T. C., and Goldberg G. A. 'A mechanical verifier for supporting the design of reliable reactive systems', In *Proceedings of International Symposium on Software Reliability Engineering*, Austin, IEEE Press, 131-138 (1991).
8. Wang, T.C. 'A typed resolution principle for reasoning about programs', Tech. Report KES.U.92.2, Kestrel Institute (1992).

KPROP — An AND-parallel Theorem Prover for Propositional Logic implemented in KL1 System Abstract

Johann M. Ph. Schumann*
Intellektik
Institut für Informatik
Technische Universität München
Augustenstr. 46
W-8000 München 2
Germany
email: schumannlan.informatik.tu-muenchen.de

KPROP is a theorem prover for first order propositional logic with *Model Elimination* [Lov78] as its underlying calculus. The given input clauses in conjunctive normal form are fanned out into *contrapositives* (see e.g. citeSti88,LSBB89). After that, they are compiled into KL1 clauses. KL1 is the parallel logic language, based on Guarded Horn Clauses which has been developed at ICOT [Sus89].

An implementation of the Model Elimination proof procedure (i.e. the extension and reduction steps) can be accomplished quite easily in that language, since each contrapositive can be compiled into one KL1 rule. With this approach, only very little code has to be added to yield the entire theorem prover.

The exploitation of AND-parallelism is done by the KL1 language itsself. As defined in the language, all subgoals of a clause (i.e. all literals which follow the guard) are tried in parallel. No additional synchronisation has to be performed, since we are working with propositional logic where the problem of shared variables (see e.g. [KR88]) does not occur.

The performance of this theorem prover could be increased substantially by encorporating a rather powerful method for pruning the search space, called *Equal Predecessor Fail* (see e.g. ([LSBB90], [AL91], and [Sti88]). In the Model Elimination tableau this means that no node may be dominated by a node in the same branch which is marked with the same literal. If such a situation occurs, this branch can considered to be failing, and a backtracking step can be performed. With this pruning method the

*This work has been done during a stay of the author at ICOT, Tokyo.

calculus remains complete, but the search space becomes considerably smaller. Especially in the case of propositional calculus, this method has two additional advantages:

- The check whether two nodes of the tableau are equal can be accomplished in constant time.
- The number of different labels for the nodes in the tableau is finite, namely twice the number of variables. Therefore, the length of branches of the tableau is restricted with the given pruning method being applied. This allows an unrestricted depth first search to be performed. In contrast to this, theorem provers for predicate calculus have to perform iterative deepening.

First measurements with this theorem prover have been made on a simulation of KL1 running on a PSI-machine at ICOT. The following table shows the run-time and the number of inferences of the proof for several examples. For comparison, the last column gives the run-time obtained by SETHEO [LSBB90] on a sun 3/260. The pigeon-hole examples are from [Pel86], the problem salt and mustard by L. Carroll is from [McC88], and $full_i$ is a formula constructed out of all permutations of i variables[1] Such a formula consists of 2^i clauses.

Example	Inferences	run-time [s]	run-time [s] SETHEO[2]
pigeon3	67	0.083	0.045
pigeon4	393	0.254	0.312
pigeon5	2611	1.452	1.910
salt	1518	0.777	0.933
$full_2$	7	0.065	<0.016
$full_3$	34	0.068	<0.016
$full_4$	197	0.141	0.083
$full_5$	1306	0.993	0.900

References

[AL91] O. L. Astrachan and D. W. Loveland. METEORs: High performance theorem provers using model elimination. Technical Report CS-1991-08, Dept. of CS, Duke University, Durham, North Carolina, 1991.

[KR88] J. Chassin de Kergommeaux and P. Robert. An abstract machine to implement efficiently OR–AND parallel Prolog. Technical report, ECRC, Munich, 1988.

[1]These formulae have been given to me by R. Letz.

[2]The run-times of SETHEO have been measured in 16ms intervals. A mean value over 5 runs has been taken.

[Lov78] D. W. Loveland. *Automated Theorem Proving: a Logical Basis.* North–Holland, 1978.

[LSBB90] R. Letz, J. Schumann, S. Bayerl, and W. Bibel. SETHEO: A High-Performance Theorem Prover. Technical report, Technische Universität München, 1990. To appear in Journal of Automated Reasoning.

[McC88] W. McCune. OTTER users' guide. Technical report, Mathematics and Computer Sci. Division, Argonne National Laboratory, Argonne, Ill., USA, 1988.

[Pel86] F. J. Pelletier. Seventy–five problems for testing Automated Theorem Provers. *Journal of Automated Reasoning*, 2:191–216, 1986.

[Sti88] M. E. Stickel. A Prolog Technology Theorem Prover: Implementation by an Extended Prolog Compiler. *Journal of Automated Reasoning*, 4:353–380, 1988.

[Sus89] Kasumi Susaki. KL1 Programming. Technical Report TM-949, ICOT, Tokyo, Japan, 1989.

A Report on ICL HOL

K.Blackburn, ICL Associated Services Division,
Eskdale Road, Winnersh, Wokingham, Berks, UK, RG11 5TT.
Phone +44-734-693131, E-mail K.Blackburn@win0109.uucp

Abstract This paper reports on research and development work which has been going on in the High Assurance Team at ICL Associated Services Division concerned with re-engineering the HOL system to industrial high-assurance standards, while improving the productivity.

1 Introduction

Over the last five years ICL have been using the public domain HOL proof tool developed by Mike Gordon and others at Cambridge University. Several large proofs relating to real-life security applications have been delivered to customers. In most of our applications work the customers have mandated the use of the specification language Z [3] and so a significant amount of effort has been devoted to extending and customising the HOL system to support reasoning about Z specifications. The Cambridge system has evolved over the years from work of several, mainly academic, contributors. Good documentation for the Cambridge HOL system has been available since late in 1989, see [5].

Since the beginning of 1990, the High Assurance Team at ICL Associated Services Division, the Universities of Cambridge and Kent and Program Validation Ltd. have been involved in a collaborative programme of research and development entitled "Foundations and Tools for Formal Verification". One of ICL's main contributions to this programme is a re-engineering of the HOL system to meet more fully industrial requirements for specification and proof support. The re-engineering activity has as its objectives:

- product quality implementation and documentation;
- high assurance in the integrity of the proof system, i.e., a high degree of confidence in the validity of the theorems it is used to prove;
- improved ease of use, for the novice user, for the expert and for the implementer of extensions to the system;
- good facilities for extending the system to provide proof support for formalisms other than HOL, including other specification and programming languages.

For the product development, we wished to use a more widely available and standardised meta-language and Standard ML was the obvious candidate. To gain experience with this language, our development work began in the spring of 1990 with the production of a prototype reimplementation of HOL. At the time of writing the product system is being readied for external field trials.

2 System Structure

The structure of the new system is shown schematically as follows, in which we view the main subsystems of the proof development system as layers of functionality, each layer being supported by the layers below it:

Library Theories	theories such as of arithmetic, lists, and sets
Specification Tools	interfaces for loose specification and deferral of consistency proofs
Elementary Proof Tools	general purpose derived rules and tactics, the subgoal package
Basic Theories	theories which are required by the logic
User Interface	parser, type checker, pretty printer
Logical Kernel	the abstract data type of theorems and its support
Utilities	A library of functional programming tools

3 Comparisons with Cambridge HOL

We can divide the changes of the re-engineered system from the original into three kinds:

- the further use of good engineering practice;
- giving many system features user-modifiable contexts;
- and other changes that come from taking the opportunities offered during a complete re-implementation.

3.1 Good Engineering Practice

Of particular importance in the industrial use of a proof development system is its *integrity*, i.e., the level of confidence that users, customers and evaluators can have in the correctness of the proof tool itself. Production of formal proofs is a difficult and time-consuming activity and it is important that all concerned can be confident that the tools which are used do prove theorems, and not fallacies.

Though Cambridge HOL comes from a varied academic background, in practice it has been found to already have a high degree of integrity. However, just developing the product to meet the industrial quality standards should improve the level of confidence in the correctness of the tool over that of a product of academic research.

A key piece of good engineering is already present in the LCF paradigm, a paradigm which Cambridge HOL follows. This is using a logical kernel of code, which contains all the critical material, upon which the bulk of the non-critical code depends. Systems following the paradigm only create theorems by using the primitive inference rules, with an extensive set of derived inference rules and tactics being provided as interfaces to these primitives. Thus the correctness of the system depends only upon the correctness of the primitive rules.

The re-engineered HOL also follows the LCF paradigm. The primary extension over Cambridge HOL is to use an abstract data type covering both theorems and theories (which is the critical code), so that only safe interfaces to the critical code are available. Such safe interfaces were not a goal of the Cambridge HOL system, where the user has access to all the system builders tools, including the means to create arbitrary theorems. Using such tools in a proof would be intentional, and the Cambridge goal is only to prevent accidental misuse of the system, not deliberate fraud. However, preventing fraud becomes important in a commercial environment. The identified logical kernel is only some 11% of the current system's code, and there are opportunities to shrink this further, by extracting non-critical code.

Other changes in our development are:

Formal Specification The critical requirements of the system were formally specified before the product development started. These requirements are roughly that the logical kernel behaves properly, and that the rest of the system only uses (often indirectly) the logical kernel to form theorems. It would not be feasible to formally specify the entire tool within the project, and as the logical kernel is by design the critical part of the tool, it is therefore the correct place to apply most of the formal effort. Formal specification and other development rules followed in this project are required for the critical portions of tools providing formal methods support (such as ICL HOL) in, for instance, the UK INTERIM Defence Standard 00-55 [4].

Namespace One aim of the new system is to organise and present the many functions available in the metalanguage. The intent is to provide, and then help the user to find, the functions they need. Apart from the provision of tutorial and reference documentation three other methods have been used:

- the use of naming conventions, covering suffixes, prefixes and case of letters;
- comprehensive coverage - for instance the inference rules provided include all the rules given in a standard presentation of the predicate calculus;
- consistent coverage - if a pattern of coverage emerges then it is followed to its conclusion.

Using Standard ML Structures The norm within the project is for there to be a one to one correspondence between detailed design and implementation documents. A detailed design document contains a "signature", which is an ML object used in the system build which defines the visibility of objects within an ML structure, and places type constraints on visible objects. Each object's type constraint is associated with a narrative description which serves as its individual detailed design. Each implementation document then contains a structure, constrained by the signature provided by the associated detailed design.

3.2 Contexts

A key aim of the product is to be able to support multiple object languages. Though the language we are most interested in supporting is Z, Cambridge HOL has been used in an ad hoc way to support reasoning in many languages (e.g. Hardware Description Languages), and we wish to simplify this support. This has led to many areas being designed to have modifiable contextual data. This has covered areas such as parsing and pretty-printing single- and mixed-language terms, as well as good provision for document preparation. However, further contextual features have been built into the proof tools.

"Stripping" is a process of generating subgoals from a goal or breaking a theorem into simpler parts by considering its outermost connective, or top level "form". "Rewriting" in backward proof is a process of traversing a goal instantiating members of a set of equational theorems to match sub-terms encountered, and using the instantiations to infer a new goal. Similarly, theorems can be rewritten directly, producing new theorems, and equational theorems formed by rewriting terms.

These two processes, rewriting and stripping, are very common in the HOL proof style. Any language or subject domain is likely to have its own rewriting theorems and ways of stripping goals, and we enable the user to incorporate them into the standard set of proof tools. We have also enhanced the rewriting facilities in ICL HOL to apply what are referred to in HOL as "conversions" along with using equational theorems. A conversion is a function capable of deriving an equational theorem, which is given as an argument the LHS that the resulting equation must have. Perhaps the best example of them is a function that given a term which is a β-redex, returns the theorem that the redex is equal to its reduced form. The resulting collection of equations, one for each possible β-redex cannot be captured in a single theorem.

3.3 Other Changes

Subgoal Package The Cambridge HOL subgoal package uses the requests for the applications of tactics to build up a "pending" inference rule, that was only executed when subgoals were "proven", not when they were just replaced by a set of new subgoals. The correctness of the inference rules is checked, though the method chosen to do this does introduce a loophole that would allow a fraudulent user to prove an untruth as a theorem. Without such a check a proof could apparently have succeeded so far, but on finishing the last step the pending inference rules would fail.

In the new product we address this problem as follows. The subgoal package maintains a "goal state theorem", that contains the effect of all the inference work to date. Instead of checking the result of a tactic's application, we actually use the result of the application immediately to modify this goal state theorem: this avoids any pending inference rules in a safe manner. A completed proof leaves the goal state theorem with no assumptions, stating the desired result in an encoded form. When the user requests the result of the proof a theorem matching the original goal is derived from the goal state theorem. In addition, a theorem can be inferred from the goal state theorem when the proof is incomplete, whose conclusion is the original goal, but which depends on more assumptions than the original goal requested. This "partial proof" of a goal can be used to resume the development of the proof, but more importantly can be offered to a customer as a partial and intelligible result if budgets are limited.

Built-in Knowledge We have consistently added a few kinds of "built-in knowledge" to functions supplied to the user. For instance, all but the most primitive functions "understand" α-convertibilty, and functions intended for the user "understand" handling constructs that bind pairs of variables. This avoids some drudgery for the user.

Theory Management We have provided a simple scoping facility, that allows for theories to go out of scope, and for definitions, axioms and types to be deleted, without prejudicing the integrity of the primitive inference rules. A deletion places any theorem produced after the introduction of the deleted item (whether logically dependent or not), but before the deletion, out of scope. Even this limited solution provides the ability to try, and discard, experiments with definitions, without having to restart the system and rebuild the theory, as would be required in Cambridge HOL.

Given that the scope of theories can change the system also stores context data, such as operator fixity, in theories.

4 Prospects

4.1 Support for Other Languages

The transfer of the Z support in the prototype onto the new HOL system is in progress at the time of writing. This is seen as an important pipe-cleaning exercise for many of the new facilities we have provided, and should result both in a valuable tool in its own right and in a HOL system which we can confidently claim permits ready extension to provide proof support for multiple formalisms.

The FST Project also includes a work package to provide proof support for another language, SPARK, which is a subset of Ada[1]. SPARK was developed, and supported with program analysis and verification tools, by Program Validation Ltd. In particular we are interested in the verification of SPARK programs against Z specifications.

4.2 Proof Automation

The requirement for ICL HOL is that it offers significant productivity improvements over previous practice. Though the re-engineering offers some benefits, we also expect to make significant gains in proof automation (also an area of active research at Cambridge). One approach we will take is to provide decision procedures and other "black box" provers as tactics and inference rules, to be slotted into the "normal" HOL proof approach. When the user recognises a task that can be solved by such tools the appropriate tool may be invoked. It will either prove the

[1]Ada is a registered trademark of the US Government, Ada Joint Program Office.

goal, or fail. We intend to do little more than port the best (or a synthesis of the best) of current approaches to our system, considering such areas as existence proofs, resolution proofs, arithmetic decision procedures, and term normalisation.

When choosing which tools to develop we are particularly interested in determining what will assist proofs whose goals are expressed in terms of Z, rather than HOL. The underlying representation of such Z goals is such that there is a lot of "clutter", e.g. type information, whose handling would be best automated, as it is "uninteresting" to the user.

4.3 Improved User Interface

The current system works with a SUN [2] command tool and textedit windows: a simple environment, but still quite a productive one. We are working on augmenting this environment, providing window and mouse-based features, such as a theory browser, while not denying the user the power and flexibility the metalanguage currently offers. This is being done via a high level interface to X Windows.

5 Conclusions

An industrial quality implementation of a proof development system is being produced. The design of the new system offers a good degree of assurance of its correctness, and responds to various new requirements arising from commercial use of the Cambridge HOL system and from new directions in research into HOL. We are in particular interested in its use to support multiple logical formalisms. Within ICL these capabilities are being exploited in the production of a proof support system for the Z specification language. The formal specification of the tool is further discussed in [1], and the incidental use of the tool in the verification of critical properties is described in [2].

By following the LCF approach, and by formal treatment of the critical components (i.e. the abstract data type of theorems), we hope to achieve a level of assurance for the ICL HOL system comparable with that required of the high-assurance applications which it is used to verify.

Acknowledgements

The FST project (IED project 1563), under which the work described here is being carried out, is jointly funded by ICL Computers Limited, Program Validation Limited, the UK Science and Engineering Council and by the Information Engineering Directorate of the UK Department of Trade and Industry.

References

[1] R.D. Arthan. Formal Specification of a Proof Tool. In S.Prehn and W.J.Toetenel, editors, *VDM '91, Formal Software Development Methods, LNCS 551*, volume 551, pages 356–370. Springer-Verlag, 1991.

[2] R.B. Jones. Methods and Tools for the Verification of Critical Properties. In R.Shaw, editor, *proc. 5th BCS-FACS refinement workshop*. Springer-Verlag (to appear), 1992.

[3] J.M. Spivey. *The Z Notation: A Reference Manual.* Prentice-Hall, 1989.

[4] INTERIM Defence Standard 00-55 Issue 1. *The Procurement of Safety Critical Software in Defence Equipment.* Ministry of Defence, 5th April 1991.

[5] *The HOL System: Description.* SRI International, 4 December 1989.

[2] Trademark of Sun Microsystems, Inc

PVS: A Prototype Verification System

S. Owre, J. M. Rushby, and N. Shankar
SRI International Computer Science Laboratory
Menlo Park, CA 94025 USA
email: {owre, rushby, shankar}@csl.sri.com

1 Introduction

PVS is a prototype system for writing specifications and constructing proofs. Its development has been shaped by our experiences studying or using several other systems[1] and performing a number of rather substantial formal verifications (e.g., [5, 6, 8]). PVS is fully implemented and freely available. It has been used to construct proofs of nontrivial difficulty with relatively modest amounts of human effort. Here, we describe some of the motivation behind PVS and provide some details of the system.

Automated reasoning systems typically fall in one of two classes: those that provide powerful automation for an impoverished logic, and others that feature expressive logics but only limited automation. PVS attempts to tread the middle ground between these two classes by providing mechanical assistance to support clear and abstract specifications, and readable yet sound proofs for difficult theorems. Our goal is to provide mechanically-checked specifications and proofs that contribute to the social process by which purported theorems come to be discarded or accepted, and designs for critical systems get certified.

PVS combines an expressive logic with a powerful but highly interactive proof checker that supports top-down proof exploration and construction. In addition to its proof checker, the PVS system includes a parser, prettyprinter, and typechecker. We describe the PVS logic and proof checker in the following sections.

2 The PVS Specification Logic

The philosophy behind the PVS logic has been to exploit mechanization in order to augment the expressiveness of the logic. PVS features a strongly typed higher-order logic with a rich type system. Higher-order logic was chosen since we and others have found it conducive to the construction of compact and perspicuous specifications. Strong typing is needed to keep higher-order logic consistent, but typechecking is also a simple and effective way to discover very many errors in specifications.

[1] Lack of space prevents us from discussing or explicitly referencing the many systems and notations that have influenced us in one way or another. These include Affirm, Automath, EHDM, EKL, EVES, FDM, Gypsy, HOL, IMPLY, IMPS, LCF, LP, Muse, Nqthm, Nuprl, OBJ, Ontic, PC-Nqthm, RAISE, RRL, STP, TPS, Tecton, VDM, Veritas, and Z among others. Most of our ideas can be traced to one or other of these earlier efforts.

PVS specifications are structured into parameterized theories that can have constraints attached to the parameters. Constraints can also be attached to the types in a PVS specification. These choices make it possible to be very explicit about allowed instantiations of theories, and about the domains and ranges of functions, thereby contributing to the clarity of the specification. The price paid is that typechecking in PVS is not algorithmically decidable: it can require theorem proving to establish that expressions satisfy the constraints attached to parameters and types. However, the inference mechanisms of PVS perform most of the necessary theorem proving automatically, and thereby allow an enriched specification language at little cost.

For instance, the division operation is typed so that it is constrained to nonzero divisors. Constraints are attached to types in PVS using *predicate subtypes*, so that the signature for division can be given as:

```
nonzero : TYPE = {x : rational | x /= 0}
/ :  [rational, nonzero -> rational]
```

where `rational` is the (built-in) type of rational numbers, and `nonzero` is defined here to be the subtype of the rational numbers different from zero. When the PVS typechecker is invoked on the formula:

```
x /= y IMPLIES (y - x)/(x - y) < 0
```

it recognizes that the divisor expression `(x - y)` must be shown to be nonzero and generates a proof obligation (known as a *type correctness condition*) of the form:

```
(FORALL (y, x: rational): x /= y IMPLIES (x - y) /= 0)
```

Notice that the logical context in which `(x - y)` occurred appears as part of the hypothesis to the proof obligation. The PVS decision procedures are powerful enough to automatically discharge the large majority of proof obligations, such as the one above; more difficult ones must be proved under user-guidance, and can be postponed until convenient.

PVS also has a mechanism for automatically generating theories for abstract datatypes that generalizes the *shell* principle of the Boyer-Moore prover [1]. The PVS type system includes numbers, enumerations and uninterpreted types, together with tuple, record, and function types, and also dependent forms of these type constructions. Numerous other features of the language are omitted from this brief description.

Just as the use of powerful inference procedures during typechecking allows the specification language to be enriched, so, conversely, do several of the features of the specification language contribute to the effectiveness of the proof checker: constructions such as abstract datatype definitions, predicate subtypes, and dependent types supply constraints that can be used effectively by the inference mechanisms. Thus we find a synergistic interaction between the language and inference capabilities.

3 The PVS Proof Checker

Our experience with mechanical verification of complex designs and algorithms has led us to conclude that, just as with software, there is a lifecycle to a mechanically-checked proof. In the initial *exploratory* phase of proof development, we are mainly interested in debugging the specification and putative theorems, and in testing and revising the key, high-level ideas in the proof. An important requirement in this phase is early and useful feedback when a purported theorem is, in fact, false. Once the basic intuitions have been acquired and the formalization is stable, the proof checking enters a *development* phase where we take care of the details and construct the proof in larger leaps. Efficiency of proof development is a key requirement here. In the third, *presentation* phase, the proof is honed and polished for presentation in order to be scrutinized by the social process. Readability and intellectual perspicuity of the output is the goal here. The final phase is *generalization* where we carefully analyze the finished proof, weaken and generalize the assumptions, extract the key insights and proof techniques, and make it easier to carry out subsequent verifications of a similar nature. *Maintenance*, is a special application of generalization, where we adapt a verification to slightly changed assumptions or requirements. Robustness of the proof procedure is a useful attribute here.

The goal of the PVS proof checker is to support the efficient development of readable proofs in all stages of the proof development lifecycle. The PVS proof checker implements a small set of powerful primitive inference rules, a mechanism for composing these into proof strategies, a facility for rerunning proofs, and another to check that all secondary proof obligations (such as type correctness conditions) have been discharged. The first two of these are described below.

Representation. A sequent representation is used for proof goals since it nicely encapsulates all the information that is relevant to a branch of the proof for presentation to a user as well as the machine.

Goal-directed Proof Search. A proof is constructed by starting from the conclusion sequent and progressively applying the inference steps to generate subgoals until the subgoals are trivially provable. This makes it easy to present the proof as it is being developed.

Primitive Inferences. The inference steps in PVS were chosen to be powerful in comparison with the simple rules given in textbook introductions to logic. Powerful primitive inferences make the composed inference steps correspondingly more powerful, and allow the proof to be represented in a manner that is robust and can be rerun efficiently. A small and carefully chosen set of primitive inferences makes the system easier to learn and use. Each inference step is flexible, so it can be used in a variety of related ways, and takes optional parameters that adjust its behavior. For example, the beta reduction rule eliminates all redexes (and for flexibility many things are regarded as redexes) from a set of sequent formulas specified by a parameter (the default is all formulas).

The primitive inferences in PVS include the propositional and quantifier steps, beta-reduction, equality replacement, and the use of decision procedures and lemmas. The propositional rules of inference include a propositional axiom rule, a disjunctive simplification rule, a conjunctive splitting rule, the Cut rule for introducing case splits, a rule for lifting IF-conditionals to the top level of a formula (to enable the corresponding case splits), and a rule for deleting formulas from a goal sequent (the weakening rule). The quantifier rules consist of a rule for replacing universally quantified variables with Skolem constants, and a rule for instantiating existentially quantified variables with terms. The equality rules include a rule for beta-reducing redexes, and one for replacing one side of an equality premise by another.

In addition to the rules above, there are rules for introducing an instance of a lemma as a premise formula in a goal sequent, for introducing an extensionality axiom for a given type as a premise formula, for introducing the type constraints of a given expression as premise formulas, for invoking the decision procedures on a goal sequent, and for enabling and disabling the automatic use of rewrite rules.

Decision Procedures. Formal proofs of even trivial facts can be quite difficult to construct, and the typical user is seldom curious about the trivial details. Decision procedures help to automatically discharge such trivial subgoals. PVS uses decision procedures for ground equalities and linear inequalities (based on the work of Shostak [9]) in order to simplify IF-expressions, datatype expressions, function definitions, and conditions of conditional rewrite rules. As a result, the user usually only sees the relevant case of a large definition and often never has to deal with the conditions of a conditional rewrite rule. Through this use of decision procedures, there are fewer cases to a proof and the sequents themselves are kept to manageable size.

Proof Strategies. It is useful to be able to compose frequently used patterns of proofs into single steps. We call these *strategies* in PVS. Typical PVS strategies are propositional simplification, which applies all the proposition proof rules and returns with the remaining subgoals. A more powerful version of this strategy employs the decision procedures as well. Rewriting with a definition or lemma is also described by a strategy. We have implemented a simple language with two basic constructs for describing strategies. The `(IF condition strategy1 strategy2)` construct evaluates the `condition` against the current proof goal and applies `strategy1` or `strategy2` accordingly. The `(TRY strategy1 strategy2 strategy3)` applies `strategy1` to the current goal and if that succeeds, then `strategy2` is applied to the resulting subgoals, otherwise `strategy3` is applied to the current proof goal. These simple constructs can be combined to achieve the effects of *tactics* and *tacticals* in LCF-style systems [3]. For example, the propositional simplification strategy `(prop*)` has the form

```
(prop*) = (TRY (propax) nil (TRY (dsimp) (prop*) (TRY (split) (prop*) nil)))
```

where `nil` indicates that there are no further steps and the current goal should be postponed, `(propax)` checks if the current goal is propositional axiom, and `(dsimp)` and `(split)` represent the primitive inference steps for disjunctive simplification and conjunctive splitting, respectively.

4 Experience and Prospects

Only a few modest-sized example verifications have so far been carried out using PVS; these include the correctness of the Boyer-Moore majority algorithm [2], the proof that insertion into an ordered binary tree preserves order, the *Oral Messages* Algorithm for Byzantine Agreement [4], Cantor's theorem, the Schröder-Bernstein theorem, and the equivalence of a pipelined and an unpipelined microprocessor design [7]. All of these examples took on the order of a day or less of human effort to verify on the first attempt using PVS. The time taken to rerun finished proofs is on the order of minutes.

Future Work. Although we are reasonably satisfied that PVS is an effective tool for developing readable specifications and formal proofs with considerable human efficiency, we are still significantly short of our goal of employing mechanization to produce proofs that humans find truly perspicuous. We would also like to extract robust and reusable proof outlines from individual proofs. We plan to enhance the expressive power of the specification language by introducing structural subtypes, inductive definitions, and refinement mappings between theories. We also plan to define powerful higher-level proof strategies to further mechanize proof construction, and enhance the user-interface to the system.

Acknowledgements. Friedrich von Henke (SRI, currently at U. of Ulm, Germany), David Cyrluk (Stanford), Judy Crow (SRI), Steven Phillips (Stanford), Carl Witty (currently at MIT), contributed to the design, implementation, and testing of PVS. We also thank Mark Moriconi, Director of the SRI Computer Science Laboratory, for his support. Development of PVS was funded entirely by SRI International.

References

[1] R. S. Boyer and J S. Moore. *A Computational Logic.* Academic Press, New York, 1979.

[2] R. S. Boyer and J S. Moore. MJRTY—a fast majority vote algorithm. In Robert S. Boyer, editor, *Automated Reasoning: Essays in Honor of Woody Bledsoe*, volume 1 of *Automated Reasoning Series*, pages 105–117. Kluwer Academic Publishers, Dordrecht, The Netherlands, 1991.

[3] M. Gordon, R. Milner, and C. Wadsworth. *Edinburgh LCF: A Mechanized Logic of Computation*, volume 78 of *Lecture Notes in Computer Science.* Springer Verlag, 1979.

[4] L. Lamport, R. E. Shostak, and M. Pease. The Byzantine generals problem. *ACM TOPLAS*, 4(3):382–401, July 1982.

[5] J. M. Rushby. Formal specification and verification of a fault-masking and transient-recovery model for digital flight-control systems. In Vytopil [10], pages 237–257.

[6] J. M. Rushby and F. W. von Henke. Formal verification of algorithms for critical systems. In *SIGSOFT '91: Software for Critical Systems*, New Orleans, LA, December 1991. Published as ACM SIGSOFT Engineering Notes, 16(5):1–15.

[7] J. B. Saxe, S. J. Garland, J. V. Guttag, and J. J. Horning. Using transformations and verification in circuit design. Technical Report 78, DEC Systems Research Center, Palo Alto, CA, September 1991.

[8] N. Shankar. Mechanical verification of a generalized protocol for Byzantine fault-tolerant clock synchronization. In Vytopil [10], pages 217–236.

[9] R. E. Shostak. Deciding combinations of theories. *Journal of the ACM*, 31(1):1–12, 1984.

[10] J. Vytopil, editor. *Formal Techniques in Real-Time and Fault-Tolerant Systems*, volume 571 of *Lecture Notes in Computer Science*, Nijmegen, The Netherlands, January 1992. Springer Verlag.

The KIV System: Systematic Construction of Verified Software

Wolfgang Reif *
University of Karlsruhe
reif@ira.uka.de

Summary

In this abstract we give a brief overview over the Karlsruhe Interactive Verifier (KIV) and sketch one of its applications: The design and verification of large modular systems.

1 The KIV System

In the KIV system the paradigm of *tactical theorem proving* is applied to realize a deduction-based programming environment for the systematic development of verified software. Basically, it provides a functional programming language PPL (Proof Programming Language), which can be used to implement both the formal and the informal aspects of rigorous program development. Examples are the design, the refinement and the administration of formal specifications, the generation of proof obligations for verification and synthesis, and the design of strategies for proof search, proof management and reuse. Potential users might be interested in implementing their own verification or synthesis strategies in PPL, in combining or comparing these with strategies that are already available, or just in applying the verification and synthesis strategies implemented by the KIV group over the last six years. Most of the verification and synthesis strategies found in the literature, are implemented in the KIV system as well as a new strategy for verifying modular systems. The KIV system is described in [HRS 88], [HRS 90], [HRS 91].

1.1 The System Architecture

The KIV system is a tactical theorem prover in the tradition of the Edinburgh LCF system, [GMW 79], or other systems like Nuprl, [Co 86], Oyster, [Bun 89], Isabelle, [Pau 86] etc. A main characteristic of tactical theorem provers is that they do not only support the

* Author's address: Institut für Logik, Komplexität und Deduktionssysteme, Universität Karlsruhe, Postfach 6980, W-7500 Karlsruhe, FRG, Tel. +721-608-4245. The KIV system is a joint development of the KIV group.

automation of deduction but are also designed for interactive proof engineering if the proof search gets stuck. Furthermore, tactical theorem provers support the definition, extension, and integration of proof methods. Due to these architectural features, tactical theorem provers are very successful in large applications, which cannot be tackled fully automatically. The KIV system exhibits the typical system structure: a logical formalism (Dynamic Logic, see 1.2) is embedded in a functional metalanguage (PPL, see 1.3). In this framework proof methods are represented as PPL programs constructing proofs in the underlying logical formalism. A proof method is implemented in terms of *tactics* and *strategies.* Tactics are used to define the elementary proof steps of the method, whereas strategies reflect its pragmatics. Tactics reduce goals to subgoals. Strategies control the proof search, decide how to select and to combine tactics, keep track of the still unproved subgoals, and are responsible for the interaction with the user. By adding heuristic information to the strategies, the degree of automation may be increased gradually.

1.2 The Logical Basis

Currently, the KIV system is tuned for the construction and verification of Pascal-like, imperative programs and modular systems. The specification language is first-order logic. Since correctness proofs involve both the programs and their specifications, a logic is required, where complex interrelations between programming- and specification language are expressible. Therefore, the KIV system is based on Dynamic Logic (DL, [Ha 79], [Go 82], [HRS 89], [Ste 89]) which is tailor-made for that purpose. DL extends ordinary predicate logic by formulas $[\pi]\varphi$ ("box π φ") and $\langle\pi\rangle\varphi$ ("diamond π φ"), where π is a program, and φ is again a DL-formula. The intuitive meaning of $[\pi]\varphi$ is: "if π terminates, φ holds after execution of π ". The formula $\langle\pi\rangle\varphi$ has to be read as: "π terminates and φ holds after execution of π ". The imperative programs that may occur in such program formulas are built up from skip, abort (the never halting program), assignments, conditionals, while loops, local variables and mutually recursive procedures (allowing value-, reference-, and procedural parameters). In DL many interesting properties of programs are expressible: Examples are partial correctness $\varphi \rightarrow [\pi]\psi$, total correctness $\varphi \rightarrow \langle\pi\rangle\psi$, termination $\langle\pi\rangle$true, non-termination $\neg\langle\pi\rangle$true, the equivalence of two programs π and π' with respect to a program variable x $\langle\pi\rangle x=x' \leftrightarrow \langle\pi'\rangle x=x'$, more general relations between programs $\langle\pi\rangle\varphi \rightarrow \langle\pi'\rangle\psi$, and the correctness of generic modules, [Re 91], [Re 92].

1.3 The Metalanguage

The metalanguage PPL is a typical functional language enriched by operations to construct formal proofs in DL. Proofs are represented explicitly as proof trees, and may be manipulated by forward- and backward reasoning. The basic rules of the DL calculus can be enriched by arbitrary *user-defined rules*, to support the definition of high-level, application-specific proof steps. However, *validations* have to be provided for these rules in order to guarantee soundness. Validations are proof constructing PPL programs. When called, they try to prove the applications of a user-defined rule in a proof to be sound. The explicite representation of proof trees is a very important prerequisite for proof search optimizations using learning techniques or reusing proof experience. A special feature of the KIV system are the *metavariables*. These act as placeholders for formulas, programs, terms etc., and may be instantiated later. This feature allows to specify and to prove schematic statements, to postpone decisions, and to specify synthesis problems. For example, the specification of a synthesis problem may be expressed as a total correctness assertion $\varphi \rightarrow \langle \$C \rangle \psi$ with known precondition φ and postcondition ψ but a yet unknown program \$C. \$C is a metavariable which gets successively instantiated by applying synthesis tactics to the goal.

1.4 The Tool Box of Verification- and Development Strategies

Currently the KIV system offers a number of strategies for classical verification and synthesis. Examples for verification methods are Hoare's invariant assertion method [Ho 69], and Burstall's intermittent assertion method [Bu 74], [HRS 87]. For the construction of programs the KIV system offers an integration of various methods due to Gries, Dijkstra, Smith, Dershowitz, Manna and Waldinger, [HRS 88], [HRS 91], [He 91]. The most elaborate strategy of the KIV system deals with the verification of large modular systems, [Sc 89], [St 90], [Re 91]. This application is sketched in the following section. The KIV system is implemented in Lisp and runs on Sun workstations under Unix. Most of the strategies have a graphical, window oriented interface.

2 Verifying Large Software Systems with the KIV System

The systematic construction and verification of large software systems is a good example where the success of a strategy not only depends on the performance of the automated deduction system used to verify the proof obligations arising during the development. It is also extremely important to consequently pursue a strict decompositional discipline of specifying and programming, in order to bound the number and the complexity of the arising deduction problems. In the sequel we sketch the design discipline supported by the KIV system, as well as the correctness and deduction issues, [Re 91], [Re 92].

2.1 The Discipline of Specifying and Programming

Software systems are described using loose algebraic specifications. The specification language is first-order logic, and is not restricted to universal Horn clauses or equations like in [GTW 78], [EM 85]. This increases the syntactic flexibility in practical applications. The main specification problem is, that the description of a large system is large as well. Therefore, the specification cannot be designed as a monolithic block. Structuring operations are needed to break the specification down into smaller and tractable pieces, starting with the overall structure, and proceeding towards the details. This kind of structuring is usually called *horizontal structuring*. The corresponding operations are disjoint union, enlargement and actualization of parameterized specifications.

After the development of a well-structured specification, the aim is to implement the abstract notions of the specification in a conventional programming language. However, in general, a direct implementation is not possible, since the abstract notions are too far away from those available in the target programming language. In this situation it is useful to design intermediate specifications which are closer to the implementation level, but still abstract enough to facilitate an implementation of the original specification in terms of the intermediate one. By recursively applying this technique to the intermediate specification, the "vertical" distance between the original specification and the target level can be bridged. This process of successively implementing one specification by another is called *vertical refinement*, and a single step a *vertical refinement step* or a *module*. In the KIV system additional constraints are imposed on the use of vertical refinement steps. These restrictions guarantee the following three design properties for the resulting modular systems:

- *Compatibility of the vertical- and the horizontal structure*: Vertical refinements respect the horizontal structure in the sense that different parts of a specification are implemented by seperate refinements. In this case the refinements can be developed independently.
- *Compositionality of correctness*: The correctness of a modular system can be reduced to the correctness of the single refinements. Consequently, the verification of a large system can be reduced to the verification of smaller parts. This is the key property to control the complexity of the verification task.
- *Substitutivity of correct refinements*: The correctness of a system is not affected if the implementation of one correct refinement is replaced by another.

2.2 Correctness and Automated Verification of Modular Systems

Due to the aforementioned compositionality property, the correctness of large modular systems can be reduced to the correctness of single refinements (modules). This is the key idea to make verification tractable. The correctness of single refinements is translated to a set of proof obligations in DL, which are necessary and sufficient for refinement correctness [Re 91], [Re 92]. For these proof obligations a special proof strategy has been designed and implemented in the KIV system. The strategy is based on symbolic execution and induction, and has been successfully tested in a number of case studies. It carries out the major part (80-95%) of the proof steps on its own, communicates with the user in "critical" situations, keeps track of the yet unsolved subgoals, and produces readable proof transscripts. Currently, this and other strategies are used in the national VSE project (Verification Support Environment). In this project KIV is combined with a CASE tool (EPOS), a formal specification system (SL, von Henke et al.), first-order-, and induction provers (MKRP, INKA, Siekmann et al., [EO 86], [BHHW 86]). It is applied in an industrial context (Dornier / Mercedes Benz), to produce verified software for a national radio network and verified access control software for nuclear power plants.

References

[BHHW 86] Biundo, S., Hummel, B., Hutter, D., Walther, C., The Karlsruhe Induction Theorem Proving System, 8th Int. Conference on Automated Deduction, Springer LNCS 230, 1986

[Bu 74] R. Burstall, Program Proving as Hand Simulation with a little Induction, Information Processing 74, North-Holland Publishing Company (1974)

[Bun 89] Bundy, A. Automatic Guidance of Program Synthesis Proofs. Proc. Workshop on Automating Software Design, IJCAI 89. Kestrel Institute, Palo Alto (1989), pp. 57-59

[Co 86] Constable, R. et al. Implementing Mathematics with the Nuprl Proof Development System. Prentice Hall, Englewood Cliffs (1986)

[EO 86] Eisinger, N., Ohlbach H.-J., The Markgraf Karl Refutation Procedure (MKRP), 8th International Conference on Automated Deduction, J. Siekmann (ed), Springer LNCS 230 (1986), pp. 681-682

[EM 85] Ehrig, H., Mahr, B., Fundamentals of Algebraic Specification 1, Equations and Initial Semantics, EATCS Monographs on Theoretical Computer Science, Vol. 6, Springer 1985

[GMW 79] Gordon,M/Milner,R./Wadsworth,C. Edinburgh LCF. Springer LNCS 78 (1979)

[GTW 78] Goguen, J., Thatcher, J., Wagner, E., An Initial Algebra Approach to the Specification, Correctness and Implementation of Abstract Data Types, Current Trends in Programming Methodology IV, Yeh, R. (Ed.), Prentice-Hall, Englewood Cliffs, 1978, pp. 80-149

[Go 82] Goldblatt, R., Axiomatising the Logic of Computer Programming, Springer LNCS 130

[Ha 79] Harel, D., First-Order Dynamic Logic, Springer, LNCS 68, 1979

[Ho 69] Hoare, C.A.R., An Axiomatic Basis for Computer Programming, Comm. ACM 12 (1969)

[He 91] Heisel, M., Formal Program Development Using Dynamic Logic, Dissertation, University of Karlsruhe, 1991 (in German)

[HRS 87] Heisel, M., Reif, W., Stephan, W., Program Verification by Symbolic Execution and Induction, Proc. 11th German Workshop on AI, K. Morik (eds), Springer Informatik Fachbericht 152, 1987, 201-210.

[HRS 88] Heisel, M., Reif, W., Stephan, W., Implementing Verification Strategies in the KIV system, Proc. 9th International Conference on Automated Deduction, E. Lusk, R. Overbeek (eds), Springer LNCS 310 (1988), pp. 131-140

[HRS 89] Heisel, M., Reif, W., Stephan, W., A Dynamic Logic for Program Verification, Meyer, A., Taitslin, M., Logic at Botik 1989, Pereslavl-Zalessky, USSR, Springer LNCS 363

[HRS 90] Heisel, M., Reif, W., Stephan, W., Tactical Theorem Proving in Program Verification, 10th International Conference on Automated Deduction, Kaiserslautern, FRG, July 1990, Springer LNCS 449, pp. 117-131

[HRS 91] Heisel, M., Reif, W., Stephan, W., Formal Software Development in the KIV System, in Automating Software Design, Lowry McCartney (eds), AAAI press 1991, and Proc. Workshop on Automating Software Design, IJCAI-89, Kestrel Institute, Palo Alto (1989),

[Pau 86] Paulson, L. C., Natural Deduction as Higher-Order Resolution, Journal of Logic Programming, 1986, 3, 237-258.

[Re 91] Reif, W., Correctness of Specifications and Generic Modules, Dissertation, University of Karlsruhe, 1991 (in German)

[Re 92] Reif, W., Correctness of Generic Modules, Symposium on Logical Foundations of Computer Science, Tver, USSR, Springer LNCS, 1992 (to appear)

[Sc 89] Schellhorn, G., Examples for the Verification of Modules in Dynamic Logic, Institut für Logik, Komplexität und Dedutktionssysteme, University of Karlsruhe 1989, (in German)

[St 90] Stenzel, K., Design and Implementation of a Proof Strategy for Module Verification in the KIV System, University of Karlsruhe 1990, (in German)

[Ste 89] Stephan, W., Axiomatising Recursive Procedures in Dynamic Logic, habil. thesis, University of Karlsruhe, 1998 (in German)

The Tableau–Based Theorem Prover $_{3}T^{A}P$ for Multiple–Valued Logics*

Bernhard Beckert Stefan Gerberding
Reiner Hähnle Werner Kernig

Institute for Logic, Complexity and Deduction Systems
University of Karlsruhe, 7500 Karlsruhe, Germany
{beckert,gerberding,haehnle,kernig}@ira.uka.de

$_{3}T^{A}P$ is an acronym for **3**–valued **ta**bleau–based theorem **p**rover. It is based on the method of analytic tableaux. $_{3}T^{A}P$ has been developed at the University of Karlsruhe in cooperation with the Institute for Knowledge Based Systems of IBM Germany in Heidelberg. Despite its name $_{3}T^{A}P$ is able to deal with "classical" — i.e. two–valued — first–order predicate logic as well as with any finite–valued first–order logic, provided the semantics is specified by truth–tables. Currently implemented versions are working for two–valued and for a certain three–valued first–order predicate logic, which is a variant of the strong Kleene logic, see [3]. The multiple–valued version implements the concept of *generalized signs*. These may be seen as sets of ordinary tableau signs or prefixes, see [6] and [7] for details. Without generalized signs one has to build a separate tableau for each non–designated sign to refute a formula. $_{3}T^{A}P$ needs to close only one tableau using generalized signs. The system has been implemented in Quintus Prolog and is running on SUN and IBM PS/2. The use of Prolog and the modular design makes it easy to extend or modify the prover.

$_{3}T^{A}P$'s input is given by a set of axioms and theorems contained in a database file, which can be precompiled. The user may specify a theorem to be proved or (s)he proves the consistency of the database. The formulae need not to be in normal form, hence the database remains readable. The use of a sort hierarchy in the database to distinguish terms of different nature is supported. For inspection of proofs some output formatting utilities allow the indentation of a proof or the conversion to LaTeX–syntax. During the compilation of a database static links are being built. These links are used to decide whether a formula can possibly lead to a branch closure and needs to be used in the tableau or not.

The free variable tableau calculus used by $_{3}T^{A}P$ is basically the one described in [10] or [4] with a few modifications concerning the handling of free variables and the extension to the multiple–valued case.

Through branch closures free variables become instantiated and they are used up. Since a formula is often needed with a different substitution of free variables, a mechanism has been employed that allows to mark formulae as *universal* with respect to a variable occurring freely in it. Basically, this is the case when the formula can

*This work has been supported by IBM Germany.

be deduced from the γ–formula that created the free variable without causing any branching of the tableau. Variables with respect to which formulae are marked as universal do not become instantiated when they are used in a branch closure. This feature allows $_3T^AP$ to find shorter proofs for many theorems.

The quantified variable x in a δ–formula is substituted by a new Skolem function, which depends on exactly the free variables in that formula except x. This version of the δ–rule generalizes the one in [4], where the newly introduced Skolem function depends on all free variables introduced so far on the current branch, i.e. our Skolem function depends on fewer variables and — most important for implementation — one needs not to keep track of the variables introduced on the branch in focus. That the liberalized δ–rule preserves completeness is obvious, soundness is proved in [8].

Another feature of $_3T^AP$ may also shorten proofs: The generation of lemmata. Lemmata may be generated for tableau rules with more than one extension in their conclusions. If the extensions $E_1, \ldots, E_n$ are not logically disjoint and some of the corresponding branches have already been closed, one can retrieve information from the intersection of the already closed branches corresponding to, say, $E_{i_1}, \ldots, E_{i_k}$. From this information one may extract a formula, which may be added to the next branch, the one corresponding to $E_{i_{k+1}}$ say, which possibly rules out some models for that branch and thus shortens the proof. The generation of lemmata can also be seen as a kind of analytic cut rule. It lifts proof length complexity in the propositional case, as is proved for a similar system in [2].

$_3T^AP$ is able to deal with equality [1], which is treated as a two–valued predicate in the multiple–valued case. Many–valued similarity predicates are not yet supported.

For efficiency reasons not the whole tableau is kept in memory, but only the branch in focus is stored. If a branch has been closed it is discarded.

The classical version of $_3T^AP$ is able to solve most of the problems stated in [9] in less than one second on a SUN–4.

References

[1] Bernhard Beckert. Konzeption und Implementierung von Gleichheit für einen tableau–basierten Theorembeweiser. Studienarbeit, Fakultät für Informatik, Universität Karlsruhe, July 1991.

[2] Marcello D'Agostino. *Investigations into the Complexity of some Propositional Calculi.* PhD thesis, Oxford University Computing Laboratory, Programming Research Group, November 1990. Also Technical Monograph PRG–88, Oxford University Computing Laboratory.

[3] Jens Erik Fenstad, Per-Kristian Halvorsen, Tore Langholm, and Johan F.A.K. van Benthem. Equations, schemata and situations: A framework for linguistic semantics. Technical Report CSLI–85–29, Center for the Studies of Language and Information Stanford, 1985.

[4] Melvin C. Fitting. *First–Order Logic and Automated Theorem Proving.* Springer, New York, 1990.

[5] Reiner Hähnle. Spezifikation eines Theorembeweisers für dreiwertige First–Order Logik. IWBS Report 136, Wissenschaftliches Zentrum, IWBS, IBM Deutschland, 1990.

[6] Reiner Hähnle. Towards an efficient tableau proof procedure for multiple–valued logics. In *Proceedings Workshop on Computer Science Logic, Heidelberg.* Springer, LNCS 533, 1990.

[7] Reiner Hähnle. Uniform notation of tableaux rules for multiple–valued logics. In *Proc. International Symposium on Multiple–Valued Logic, Victoria.* IEEE Press, 1991.

[8] Reiner Hähnle and Peter H. Schmitt. The liberalized δ–rule in free variable semantic tableaux. *To appear*, 1991.

[9] Francis Jeffry Pelletier. Seventy-five problems for testing automatic theorem provers. *Journal of Automated Reasoning*, 2:191 – 216, 1986.

[10] Peter H. Schmitt. The THOT theorem prover. Technical Report TR–87.09.007, IBM Heidelberg Scientific Center, 1987.

Analytica — A Theorem Prover in Mathematica

Edmund Clarke Xudong Zhao
School of Computer Science
Carnegie Mellon
Pittsburgh, PA 15213

Current automatic theorem provers, particularly those based on some variant of resolution, have concentrated on obtaining ever higher inference rates by using clever programming techniques, parallelism, etc. We believe that this approach is unlikely to lead to a useful system for actually doing mathematics. The main problem is the large amount of domain knowledge that is required for even the simplest proofs. In this paper, we describe an alternative approach that involves combining an automatic theorem prover with a symbolic computation system. The theorem prover, which we call *Analytica*, is able to exploit the mathematical knowledge that is built into this symbolic computation system. In addition, it can guarantee the correctness of certain steps that are made by the symbolic computation system and, therefore, prevent common errors like division by an expression that may be zero.

Analytica is written in the Mathematica programming language and runs in the interactive environment provided by this system [4]. Since we wanted to generate proofs that were similar to proofs constructed by humans, we have used a variant of the sequent calculus in the inference phase of our theorem prover. However, quantifiers are handled by skolemization instead of explicit quantifier introduction and elimination rules. Although inequalities play a key role in all of analysis, Mathematica is only able to handle very simple numeric inequalities. We have developed a technique that is complete for linear inequalities and is able to handle a large class of non-linear inequalities as well. This technique is more closely related to the BOUNDER system developed at MIT [3] than to the traditional SUP-INF method. Another important component of Analytica deals with expressions involving summation and product operators. A large number of rules are devoted to the basic properties of these operators. We have also integrated Gosper's algorithm [2] for hypergeometric sums with the other summation rules, since it can be used to find closed form representations for a wide class of summations that occur in practice.

In a related project that we plan to describe in a forthcoming paper, we have managed to prove all of the theorems and examples in Chapter 2 of Ramanujan's Collected Works[1] completely automatically. The techniques that we use are similar to those described in this paper. We believe that the examples that we have been able to prove provide convincing justification for combining powerful symbolic computation techniques with theorem provers.

Analytica consists of four different phases: skolemization, simplification, inference, and rewriting. When a new formula is submitted to Analytica for proof, it

This research was sponsored in part by the National Science Foundation under Contract Number CCR-8722633, and also by the Defense Advanced Research Projects Agency (DOD), ARPA Order No. 4976, Amendment 20, under Contract Number F33615-87-C-1499.

is first skolemized to a quantifier free form. Then it is simplified using a collection of algebraic and logical reduction rules. A formula is simplified with respect to its *proof context*. The proof context consists of the formulas that may be assumed true when the formula is encountered in the proof. If the formula reduces to true, the current branch of the inference tree terminates with success. If not, the theorem prover checks to see if the formula matches the conclusion of some inference rule. If a match is found, Analytica will try to establish the hypothesis of the rule. If the hypothesis consists of a single formula, then it will try to prove that formula. If the hypothesis consists of a series of formulas, then Analytica will attempt to prove each of the formulas in sequential order. Special tactics are also included in the inference phase for handling inequalities and constructing inductive proofs. The inequality tactic is complete for linear inequalities and can handle many nonlinear inequalities as well. The induction tactic enables Analytica to select a suitable induction scheme for the formula to be proved and attempts to establish the basis and induction steps. If no inference rule is applicable, then various rewrite rules are used attempting to convert the formula to another equivalent form. If the rewriting phase is unsuccessful, the search terminates in failure; otherwise the simplification, inference and rewriting phases will repeat with the new formula. Backtracking will cause the entire inference tree to be searched before the proof of the original goal formula terminates with failure.

In the appendix, we give a few examples that Analytica can prove. Due to lack of space, we omit the proofs in all but the first example. Mathematica automatically generates LaTeXcommands to typeset formulas involving algebraic expressions. In addition to these examples, Analytica has also proved some other non-trivial theorems including the Bernstein approximation theorem for continuous functions and main parts of the Fundamental Theorem of Algebra.

References

[1] B.C.Berndt, *Ramanujan's Notebooks, Part I*, Springer-Verlag, 1985, pp 25-43.

[2] R.W.Gosper, *Indefinite Hypergeometric sums in MACSYMA*,

[3] E.Sacks, *Hierarchical Inequality Reasoning*, Technique Report, MIT Laboratory for Computer Science, 1987.

[4] S. Wolfram. *Mathematica: A System for Doing Mathematics by Computer*, Wolfram Research Inc., 1988.

Appendix. Examples proved by Analytica

A Closed form for a summation

Theorem :

$$(integer(n) \wedge 0 \leq n \wedge m \neq 1 \Rightarrow \sum_{k=0}^{n} \frac{2^k}{1+m^{2^k}} = \frac{1}{m-1} + \frac{2^{n+1}}{1-m^{2^{n+1}}})$$

Proof :

prove

$$\sum_{k=0}^{n} \frac{2^k}{1+m^{2^k}} = \frac{1}{-1+m} + \frac{2 \cdot 2^n}{1-m^{2 \cdot 2^n}}$$

use induction on n

base case with n = 0

$$m = 1 \vee \frac{1}{1+m} = \frac{1}{-1+m} + \frac{2}{1-m^2}$$

reduces to

$$True$$

induction step

$$integer(n) \wedge 0 \leq n \wedge \sum_{k=0}^{n} \frac{2^k}{1+m^{2^k}} = \frac{1}{-1+m} + \frac{2 \cdot 2^n}{1-m^{2 \cdot 2^n}} \Longrightarrow$$
$$m = 1 \vee \sum_{k=0}^{1+n} \frac{2^k}{1+m^{2^k}} = \frac{1}{-1+m} + \frac{4 \cdot 2^n}{1-m^{4 \cdot 2^n}}$$

calculate summations

$$integer(n) \wedge 0 \leq n \wedge \sum_{k=0}^{n} \frac{2^k}{1+m^{2^k}} = \frac{1}{-1+m} + \frac{2 \cdot 2^n}{1-m^{2 \cdot 2^n}} \Longrightarrow$$
$$m = 1 \vee \frac{2 \cdot 2^n}{1+m^{2 \cdot 2^n}} + \left(\sum_{k=0}^{n} \frac{2^k}{1+m^{2^k}}\right) = \frac{1}{-1+m} + \frac{4 \cdot 2^n}{1-m^{4 \cdot 2^n}}$$

substitute using equation

$$integer(n) \wedge 0 \leq n \wedge \sum_{k=0}^{n} \frac{2^k}{1+m^{2^k}} = \frac{1}{-1+m} + \frac{2 \cdot 2^n}{1-m^{2 \cdot 2^n}} \Longrightarrow$$
$$m = 1 \vee \frac{2 \cdot 2^n}{1+m^{2 \cdot 2^n}} + \frac{1}{-1+m} + \frac{2 \cdot 2^n}{1-m^{2 \cdot 2^n}} = \frac{1}{-1+m} + \frac{4 \cdot 2^n}{1-m^{4 \cdot 2^n}}$$

reduces to

$$True$$

□

B Stereographical projection

Consider the function that maps each point (x_1, x_2, x_3) in 3-space to the complex plane C:

$$sp(x_1, x_2, x_3) = \frac{x_1 + ix_2}{1 - x_3}$$

We will use Analytica to prove that this mapping is bijection between the unit sphere S whose equation is $x_1^2 + x_2^2 + x_3^2 = 1$ and the complex plane. We will also prove that this mapping is a projection, i.e. if $sp(x_1, x_2, x_3) = a + bi$, then the north pole(0,0,1), (x_1, x_2, x_3) and $(a, b, 0)$ are all collinear.

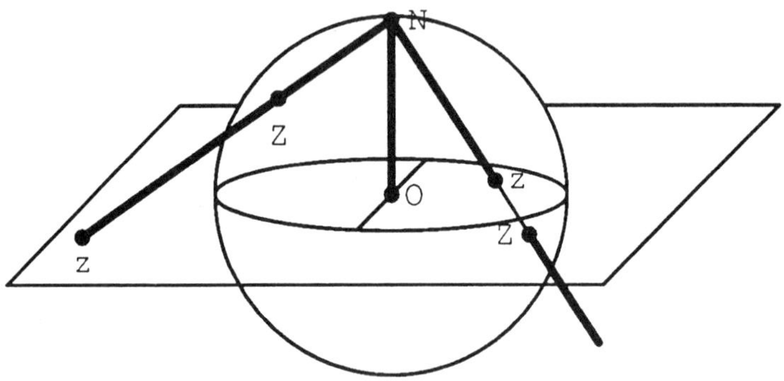

With the definitions that:

$unit(x_1, x_2, x_3) = T$ if and only if $x_1^2 + x_2^2 + x_3^2 = 1$

$collinear(\{x_1, x_2, x_3\}, \{y_1, y_2, y_3\}, \{z_1, z_2, z_3\}) = T$ if and only if

$$\begin{vmatrix} x_1 & x_2 & 1 \\ y_1 & y_2 & 1 \\ z_1 & z_2 & 1 \end{vmatrix} = 0 \bigwedge \begin{vmatrix} x_1 & x_3 & 1 \\ y_1 & y_3 & 1 \\ z_1 & z_3 & 1 \end{vmatrix} = 0 \bigwedge \begin{vmatrix} x_2 & x_3 & 1 \\ y_2 & y_3 & 1 \\ z_2 & z_3 & 1 \end{vmatrix} = 0$$

The following theorems show the properties of the Stereographical projection:

$$(unit(x_1, x_2, x_3) \wedge unit(y_1, y_2, y_3) \wedge sp(x_1, x_2, x_3) = sp(y_1, y_2, y_3)$$

$$\Rightarrow x_3 = y_3 \wedge x_1 = y_1 \wedge x_2 = y_2)$$

$$\exists\{x_1, x_2, x_3\}[sp(x_1, x_2, x_3) = z_1 + z_2 i \wedge unit(x_1, x_2, x_3)]$$

$$z = sp(x_1, x_2, x_3) \Rightarrow collinear(\{R(z), I(z), 0\}, \{x_1, x_2, x_3\}, \{0, 0, 1\})$$

All three of these theorems are proved completely automatically. In second theorem, Analytica is able to find the appropriate instantiation for the existentially quantified variables. Analytica avoids division by zero in this example by making sure that x_3 in $sp(x_1, x_2, x_3)$ is not equal to 1.

C Weierstrass' everywhere continuous but nowhere differentiable function

Weierstrass's non-differentiable function is defined be the series

$$f(x) = \sum_{n=0}^{\infty} b^n \cos(a^n \pi x)$$

where $0 < b < 1$, and a is a odd positive integer. When $ab > 1 + \frac{3}{2}\pi$, the derivative of the function does not exist for any value of x.

With

$$\mathit{diff}(h, f, x) = \frac{-f(x) + f(h + x)}{h}$$

$$h = \frac{1 - a^m x + \mathit{round}(a^m x)}{a^m}$$

It is sufficient to prove that

$$\mathit{Continuous}(f(x), \{x, x_0\})$$

$$\forall \epsilon [\left(\epsilon > 0 \Rightarrow \lim_{m \to \infty} |h| < \epsilon\right)]$$

$$\forall M [\lim_{m \to \infty} |\mathit{diff}(h, f, x)| > M]$$

The first two theorems can be proved without help from the user, however, the last one cannot be proved fully automatically. Three lemmas are added:

$$\mathit{diff}(h, f, x) = R(m) + S(m)$$

$$|R(m)| \geq \frac{2 a^m b^m}{3}$$

$$|S(m)| < \frac{\pi a^m b^m}{ab - 1}$$

where

$$S(m) = \frac{a^m \left(\sum_{n=0}^{-1+m} b^n \left(-\cos(\pi a^n x) + \cos(\pi a^{-m+n} \left(1 + \mathit{round}(a^m x)\right))\right)\right)}{1 - (a^m x - \mathit{round}(a^m x))}$$

$$R(m) = -\frac{(-1)^{\mathit{round}(a^m x)} a^m \left(\sum_{n=m}^{\infty} b^n \left(1 + \cos(\pi a^{-m+n} \xi(m))\right)\right)}{1 - (a^m x - \mathit{round}(a^m x))}$$

All three lemmas are proved automatically. Using these lemmas, Analytica can prove the remaining theorem without further help.

The FAUST - Prover

Klaus Schneider, Ramayya Kumar and Thomas Kropf

Institute of Computer Design and Fault Tolerance
Prof. Dr. D. Schmid
P.O. Box 6980
W-7500 Karlsruhe
Germany
schneide,kumar,kropf @ira.uka.de

1 Introduction

Unfortunately first-order logic has certain limitations which are felt in many applications such as in hardware verification. The use of proof assistants like the HOL system ([Gord88]) is therefore resorted to. However many theorems of higher order logic can also be proven by methods of first order logic as well. Being aware of this situation, we have implemented a prover based on Sequent Calculus within the HOL system, which can be used to mechanize proofs of necessary, but tedious lemmata required for a large proof in HOL. In order to find an efficient implementation, we have introduced the concept of unification in our prover. These modifications have resulted in a calculus called the "restricted sequent calculus "($\mathcal{RSEQ}$) and an automatic prover based on it called $\mathcal{FAUST}$[1] [2].

The well known sequent calculus[3] ($\mathcal{SEQ}$) introduced by Gentzen [Gent35] has a major disadvantage as far as the so called γ-rules are concerned. The application of a γ-rule extends a sequent by an instance of a quantified formula of the sequent on which the rule is applied to. Unfortunately the γ-rule cannot be easily automated as the 'right' choice of the term for instantiation cannot be easily computed at the stage of rule application. The rule itself allows the use of any term, but usually only special terms lead to the desired proof. To overcome this deficiency we have introduced the concept of metavariables described in the next section.

2 The Concepts in $\mathcal{RSEQ}$

Metavariable: γ-rule applications in $\mathcal{RSEQ}$ do not introduce arbitrary terms, but special place-holders called *metavariables*[4] , which do not belong to the language of our logic. For syntactical convenience Metavariables can be treated like usual variables, but it should be noted that input formulae must not contain any metavariables, since they are available only for the proof process.

[1] First-Order Automation using Unification in a modified Sequent calculus Technique

[2] A public domain version of $\mathcal{FAUST}$ can be obtained from the authors or along with the HOL system sources.

[3] For details of syntax and semantics of $\mathcal{SEQ}$ the reader is referred to standard textbooks on logic, e.g. [Gall86].

[4] A similar concept was developed independently by Reeves [Reev87].

In figure 2, an example illustrating the application of $\gamma-, \alpha-$ and $\delta-$rules are shown. It can be observed that before the $\gamma-$rule on Φ is applied, one could substitute $[m_1 \leftarrow y_1][m_2 \leftarrow f(a, y_1)]$ in order to obtain non-disjoint antecedent and succedent. However, this substitution is forbidden, since $y_1 \in fs_{m_1}$. Therefore the tree cannot be closed at this stage and further rule applications are necessary.

$$
\begin{gathered}
\vdash \underbrace{\exists x.\forall y.\exists z.P(x,z) \rightarrow P(y, f(a,y))}_{\Phi} \parallel \{\} \\
\downarrow \gamma \\
\vdash \forall y.\exists z.P(m_1,z) \rightarrow P(y, f(a,y)), \Phi \parallel \{(m_1, \{\})\} \\
\downarrow \delta \\
\vdash \underbrace{\exists z.P(m_1,z) \rightarrow P(y_1, f(a,y_1))}_{\Psi}, \Phi \parallel \{(m_1, \{y_1\})\} \\
\downarrow \gamma \\
\vdash P(m_1,m_2) \rightarrow P(y_1, f(a,y_1)), \Psi, \Phi \parallel \{(m_1, \{y_1\}), (m_2, \{\})\} \\
\downarrow \alpha \\
P(m_1,m_2) \vdash P(y_1, f(a,y_1)), \Psi, \Phi \parallel \{(m_1, \{y_1\}), (m_2, \{\})\} \\
\downarrow \gamma \text{ applied on } \Phi \\
P(m_1,m_2) \vdash P(y_1, f(a,y_1)), \Psi, \forall y.\exists z.P(m_3,z) \rightarrow P(y, f(a,y)), \Phi \parallel \\
\{(m_1, \{y_1\}), (m_2, \{\}), (m_3, \{\})\} \\
\downarrow \delta \\
P(m_1,m_2) \vdash P(y_1, f(a,y_1)), \Psi, \exists z.P(m_3,z) \rightarrow P(y_2, f(a,y_2)), \Phi \parallel \\
\{(m_1, \{y_1, y_2\}), (m_2, \{y_2\}), (m_3, \{y_2\})\} \\
\downarrow \gamma \\
P(m_1,m_2) \vdash P(y_1, f(a,y_1)), \Psi, P(m_3,m_4) \rightarrow P(y_2, f(a,y_2)), \Phi \parallel \\
\{(m_1, \{y_1, y_2\}), (m_2, \{y_2\}), (m_3, \{y_2\}), (m_4, \{\})\} \\
\downarrow \alpha \\
P(m_1,m_2), P(m_3,m_4) \vdash P(y_1, f(a,y_1)), \Psi, P(y_2, f(a,y_2)), \Phi \parallel \\
\{(m_1, \{y_1, y_2\}), (m_2, \{y_2\}), (m_3, \{y_2\}), (m_4, \{\})\} \\
\downarrow \\
[m_3 \leftarrow y_1][m_4 \leftarrow f(a, y_1)]
\end{gathered}
$$

Figure 2: Example Proof Tree

3 Implementational Details of $\mathcal{FAUST}$

Since the proof construction process generates a tree, it is possible to either have a depth-first or a breadth-first construction procedure. The depth-first procedure continues to apply rules on the leftmost branch until it can be closed by computing a set of unifiers. These unifiers can then be refined by the unifiers found in the next leftmost branch so that the next branch is also closed. This process is repeated until all branches are closed. Details of the depth-first algorithm are given in [ScKK91a]. The breadth-first procedure on the other hand applies the rules on all the branches until a closing substitution can be obtained. If none can be found, each branch of the tree is extended by further rule applications.

The depth-first search is much faster than the breadth-first approach, since the breadth-first approach extends each branch even if it can already be closed by some unifier. On the other hand, the depth-first approach is inherently incomplete, while the breadth-first search can prove each valid formula (except for complexity constraints).

Optimizations. If the above-mentioned technique for depth-first search is implemented naively, then the number of unifiers to be manipulated grows in an uncontrolled manner. In order to minimize the number of unifiers and to speed up the process of proof tree construction, the following enhancements have been made (see [Schn91] for details):

- Metavariables are managed locally within the branches. Thus if two unifiers only differ with respect to the instantiation of metavariables local to a branch, the unifiers become identical after leaving this branch.
- According to the generality of the unifiers only the most general ones need to be refined and all others can be removed.
- Antecedents and succedents are split up into four sets: $\alpha-$ and $\delta-, \beta-, \gamma-$ and for atomic formulae. For example, looking for an $\alpha-$rule now avoids searching through the whole sequent.
- All forbidden sets are stored in a global list outside the proof tree.
- There is no need to use proof trees at all. All the information to extend a branch or to search for a closing substitution is kept in the leaves. Hence, we work on a list of sequents rather than on a binary proof tree.

4 Experimental Results

The set of problems given by Pelletier [Pell86] has been proven and their runtimes (in seconds) on a SUN 4/330 are given in the table below. The first column contains the runtimes for a simple depth-first search, the second column contains the runtimes of a prover which skolemizes its input before the depth-first search and in the third column the runtimes of a breadth-first prover are given.

P_1	0.1	0.1	0.3	P_{17}	0.2	0.2	1.6	P_{33}	0.7	0.7	2.4
P_2	0.0	0.0	0.1	P_{18}	0.2	0.2	0.3	P_{34}	17.6	31.2	***
P_3	0.1	0.1	0.2	P_{19}	0.3	0.3	0.7	P_{35}	0.4	0.4	0.4
P_4	0.0	0.0	0.3	P_{20}	0.4	0.4	0.8	P_{36}	1.6	0.7	1.0
P_5	0.1	0.1	0.2	P_{21}	0.4	0.3	2.5	P_{37}	1.6	1.6	9.6
P_6	0.0	0.0	0.0	P_{22}	0.2	0.3	0.4	P_{38}	13.0	9.9	***
P_7	0.0	0.0	0.1	P_{23}	0.2	0.2	0.4	P_{39}	0.4	0.5	0.2
P_8	0.1	0.1	0.1	P_{24}	1.3	1.2	5.4	P_{40}	9.0	0.7	0.5
P_9	0.1	0.1	0.9	P_{25}	0.4	0.4	3.8	P_{41}	2.7	0.7	1.4
P_{10}	0.1	0.1	1.2	P_{26}	0.9	1.3	5.6	P_{42}	0.7	1.0	8.4
P_{11}	0.1	0.1	0.1	P_{27}	0.7	0.6	2.5	P_{43}	**	**	147.5
P_{12}	0.2	0.2	2.5	P_{28}	0.7	0.7	2.0	P_{44}	1.2	0.8	1.2
P_{13}	0.1	0.1	0.7	P_{29}	0.9	0.9	2.4	P_{45}	5.7	2.5	24.4
P_{14}	0.1	0.1	0.7	P_{30}	0.4	0.5	0.5	P_{46}	184.0	170.4	9.9
P_{15}	0.1	0.1	0.2	P_{31}	0.5	0.5	1.0				
P_{16}	0.1	0.1	0.1	P_{32}	0.7	0.7	2.8				

Restricted Sequents: The introduction of metavariables introduces new problems as far as the application of the δ-rules are concerned. A δ-rule application requires that the variable introduced for the quantifier elimination is *new*, i.e. the variable must not occur in the sequent on which the δ-rule is applied to. As the sequent may contain several metavariables which are not instantiated at this stage, one can not check whether a variable occurs after instantiation of the metavariables in the given sequent or not. The use of a δ-rule restricts therefore the domain of the metavariables occuring in the sequent, because only terms which do not contain the variable introduced by the δ-rule are allowed for instantiation of the metavariables. Hence, for each metavariable m of the sequent all variables introduced by further δ-rule applications are stored in its so-called forbidden set fs_m. The forbidden sets of all metavariables occuring in a sequent are kept in a restriction list which contains an ordered pair (m, fs_m) for each metavariable m. Therefore, a restricted sequent has the form $\Gamma \vdash \Delta \parallel R$.

Metasubstitution: As in $\mathcal{SEQ}$, the axiom scheme is a sequent with non-disjoint antecedent and succedent. Hence, given a restricted sequent $\Gamma \vdash \Delta \parallel R$, we try to find a substitution σ of the metavariables such that $\sigma(\Gamma) \cap \sigma(\Delta) \neq \{\}$ holds. Of course, the restriction has to be observed, in other words, for each metavariable m no variable of $\sigma(m)$ is an element of fs_m.
Given a restricted sequent $\Gamma \vdash \Delta \parallel R$ one can compute a "closing substitution" by unifying each formula of the antecedent Γ with each formula of the succedent Δ. The underlying unification algorithm must therefore consider only metavariables as replaceable subterms and other variables are treated like constants. A closing substitution is a unifier which does not violate the restrictions.

Rules and Proof Trees in $\mathcal{RSEQ}$: Using the concepts of metavariables and restricted sequents the rules of $\mathcal{SEQ}$ are modified as follows: $\alpha-$ and $\beta-$rules are quite the same as in $\mathcal{SEQ}$, except that the restrictions in the premise are copied to each conclusion of the rule. In the $\gamma-$ and $\delta-$ rules given in figure 1, both the variable y and the metavariable m have to be *new*, i.e. they do not appear in the sequent on which the rule is applied to. The function ϱ_y updates the restrictions of the existing metavariables by adding y to each forbidden set.

$$\frac{\forall x.\phi, \Gamma \vdash \Delta \parallel R}{[\phi]^m_x, \forall x.\phi, \Gamma \vdash \Delta \parallel \{(m, \{\})\} \cup R} \qquad \frac{\Gamma \vdash \forall x.\phi, \Delta \parallel R}{\Gamma \vdash [\phi]^y_x, \Delta \parallel \varrho_y(R)}$$

$$\frac{\exists x.\phi, \Gamma \vdash \Delta \parallel R}{[\phi]^y_x, \Gamma \vdash \Delta \parallel \varrho_y(R)} \qquad \frac{\Gamma \vdash \exists x.\phi, \Delta \parallel R}{\Gamma \vdash [\phi]^m_x, \exists x.\phi, \Delta \parallel \{(m, \{\})\} \cup R}$$

Figure 1: Quantifier rules of $\mathcal{RSEQ}$

All the benchmarks except Schubert's Steamroller Problem and the first-order formulae with equality (since our application domain does not require them) have been proven in acceptable times, although the implementation was done in a very naive way to obtain a prototype and the ML used for implementation is itself bootstrapped using an underlying Lisp. This implementation language has been chosen since our prover has been primarily developed for proving tedious lemmata in large HOL proofs.

5 Summary and Conclusions

We have shown how the integration of unification within sequent calculus can greatly improve its efficiency. This has been achieved by the introduction of the metavariables and its associated restrictions. The calculus thus developed ($\mathcal{RSEQ}$) has been proven to be sound and complete and has been implemented in the HOL environment. The resulting prover $\mathcal{FAUST}$ is not only capable of solving first-order formulae, but also certain second-order formulae which occur in the domain of hardware verification. We have also implemented a function, which generates a HOL-tactic representing a proof by natural deduction. $\mathcal{FAUST}$ has also been embedded in an environment for proving Hardware called $\mathcal{MEPHISTO}$ [ScKK91b].
We are currently reimplementing the prover for use within the new HOL system based on Standard ML. We are also adding some new concepts which make the proof-process faster and are also going to incorporate some hardware domain specific rules within the calculus. Experimentation will also been undertaken with linear unification algorithms.

References

[Gall86] Gallier J.H.,*Logic for Computer Science—Foundations of Automated Theorem Proving*, Harper & Row Publishers, New York, 1986

[Gord88] M.J.C. Gordon, *HOL: A Proof Generating System for Higher–Order Logic*, in *VLSI Specification, Verification and Synthesis*, G. Birtwistle, P.A. Subrahmanyam (eds.), Kluwer Academic Press, 1988

[Gent35] Gentzen G., *Untersuchungen über das logische Schließen*, Mathematische Zeitschrift, Vol. 39, 1935

[Pell86] F.J. Pelletier, *Seventy-Five Problems for Testing Automatic Theorem Provers*, Journal of Automated Reasoning, Vol. 2, pp 191-216,1986

[Reev87] S. Reeves, *Semantic tableaux as a framework for automated theorem-proving*, Departement of Computer Science and Statistics, Queen Mary College, Univ. of London, 1987

[Schn91] Schneider K., *Ein Sequenzenkalkül für die Hardware-Verifikation in HOL*, Diploma Thesis, Institute of Computer Design and Fault-Tolerance, University of Karlsruhe, 1991

[ScKK91a] Schneider K., Kumar R., Kropf T.,*Automating most parts of hardware proofs in HOL*, Proc. Workshop on Computer Aided Verification, Aalborg, July 1991

[ScKK91b] Schneider K., Kumar R., Kropf T.,*Structuring Hardware Proofs: First Steps towards Automation in a Higher-Order Environment*, Proc. VLSI 1991, Edinburgh, P.B. Denyer, A. Halaas (Eds.), North-Holland, 1991

This article was processed using the LaTeX macro package with LLNCS style

Eves System Description

Dan Craigen, Sentot Kromodimoeljo, Irwin Meisels, Bill Pase, and Mark Saaltink

ORA Canada
265 Carling Avenue, Suite 506
Ottawa, Ontario K1S 2E1
CANADA

Internet: eves@ora.on.ca
Phone: (613) 238 7900

1 Introduction

The primary purpose of the Eves project[1] was to develop a verification system, integrating state-of-the-art techniques from automated deduction, mathematics, language design, and formal methods, so that the system is both sound and useful.

The development of Eves has followed two principal directions:

- The design of a specification and implementation language called Verdi. This direction included the development of the formal semantics and the proof system with a demonstration of soundness.
- The implementation of a verification environment. The main component is a theorem prover called Never.

We will only briefly describe the Verdi language, and concentrate instead on the Eves verification environment that has been implemented, particularly the theorem prover component. Additional introductory information may be found in [CKMPS 91] and [CS 90]. [Cra 91] is the reference manual for Verdi, and [Saa 90a] and [Saa 90b] present the mathematical foundations of Eves.

2 Verdi

Verdi is a formal notation that may be used to write programs that are to be verified formally using the Eves verification environment. Consequently, Verdi consists of syntactic forms for expressing specifications (what effect a program is to have), implementation (how a program is to cause an effect), and proofs (justification that a program meets its specification).

Fundamentally, Verdi is a formal notation based on a version of untyped set theory [Fra 68], which can be used to express rigorous mathematical concepts [Saa 91a, Saa 91b].

[1] The development of Eves was sponsored by the Canadian Department of National Defence through DSS contract W2207-0-AF09 and various other contracts. The ORA Canada reference number for this conference paper is CP–91–6017–44, November 1991.

For example, Verdi can be used to prove theorems of set theory (e.g., Schroeder-Bernstein, Cantor), to prove functional correctness of hardware designs (e.g., n-bit adder), or to prove security-critical properties (e.g., versions of non-interference).

A Verdi theory consists of a current vocabulary and a set of axioms specifying relationships among the names of the vocabulary. Verdi determines an initial theory which introduces names and axioms that can be used by any Verdi program. The Eves database contains a representation of a current theory, heuristic information associated with the theory, and records of complete and partial proofs.

The Verdi syntax is similar to the s-expressions of Lisp. While Verdi is, in general, an untyped language, typing information is necessary for syntactic forms that are to be executed.

3 Never

An integral component of the Eves system is a theorem prover called Never. It is an interactive theorem prover that is capable of performing large proof steps. Features of Never include reduction commands, Boyer-Moore induction, low-level inference commands, and command modifiers that perform variations on reduction commands.

Never is neither a fully automatic nor an entirely manual theorem prover. Although Never provides powerful deductive techniques for the automatic proof of theorems, it also includes simple user steps which permit its use as a system more akin to a proof checker than a theorem prover. Combining the manual and automatic functions within a single system creates the possibility of a synergy between abilities of the system (fast and accurate) and the user (the necessary insight).

A user of Eves starts with the initial theory of Verdi, and extends it with Verdi declarations. When a user enters a declaration into the system, the system may generate a proof obligation for the declaration.[2] The user may then use Never to discharge the proof obligation. The proof may be worked on immediately or it may be deferred. In addition, Eves has a library facility (described in Section 7 below) that allows users to write specifications without proofs and to provide the models with proofs separately.

4 Reduction

The workhorse of Never is the reducer. The reducer reduces a formula to its normal form by an innermost application of simplification, rewriting, and function invocation to each of the subexpressions. A user may tell the reducer to use its full power, only simplification and rewriting, or only simplification, using the `reduce`, `rewrite`, and `simplify` commands respectively. Using command modifiers, a user may tell the reducer to perform variations on reduction commands, such as reducing only certain subexpressions, reducing without variable instantiations, and reducing with specific declarations disabled (e.g., not applying specific rewrite rules). Command modifiers provide tight control over reduction commands, without increasing the complexity of the command set.

---

[2] Generally, the proof obligations are such that, if successfully proven, the declaration results in a conservative extension of the Verdi theory.

The simplifier part of the reducer recognizes propositional tautologies and reasons about integers and equalities. The equality and integer reasoning mechanisms are based on the work of Nelson and Oppen [Nel 81]. A user may extend the simplifier by declaring forward chaining rules (called frules and grules). Using these rules, decision procedures may be added for abstract data types that can be presented as constructor-accessor theories. The system uses forward chaining rules to reason about types used for executable programs. The simplifier tries to instantiate quantified variables when trying to affirm or refute a subformula.

The rewriter applies rewrite rules in addition to performing simplification. A rewrite rule may be conditional, in which case it is only applied if the rewriter shows that the condition is satisfied. Currently, Never does not force the user to enter only terminating sets of rewrite rules although it does detect some cases where rules may loop. Having Never perform checks that warn the user if rewrite rules do not obey some system-defined reduction ordering (such as those used by the conditional term rewriting community) is under consideration. In addition, there is a prototype of a Knuth-Bendix procedure based on Ganzinger's extension [Gan 87] to handle conditional rewrite rules. The prototype is still being used for experiments to determine its suitability for incorporation into Eves.

The full power of the reducer uses function definitions in addition to performing simplification and rewriting. Non-recursive definitions are always expanded unless they are disabled. For recursive definitions, Never uses heuristics based on those used by Boyer-Moore.

5 Low-Level Inference Commands

Never has commands to perform low-level inference steps. The low level inference commands include commands to instantiate quantified variables, to prenex a formula, to apply axioms manually (axioms include rewrite and forward chaining rules, and those rules associated with function definitions), to perform equality substitution, and to perform a case split.

A user may need to use low-level inference commands to complete a proof. If, for example, the reducer can not find instantiations for quantified variables automatically, the user can give the instantiations explicitly using the `instantiate` command.

Low-level inference commands may be used to search for a proof that the prover was unable to complete automatically. Alternatively, they may be used to step through a proof that was completed automatically, but that the user does not fully understand. The low-level inferences are also useful for debugging conditional rewrite rules that were expected to complete a proof but did not apply. A user may explicitly apply specific conditional rules and then use reduction commands—perhaps with the subexpression command modifier to restrict the reductions to the subexpressions that correspond to the applied rules—to study the prover's effect upon the result of the application.

6 Induction

Never uses the Boyer-Moore induction technique, extended to handle quantified variables. The heuristics of Boyer-Moore automatically select an induction scheme. A user may override the heuristics by specifying a recursive function term to generate the induction scheme. In addition, Never allows induction in the presence of quantifiers both in the conjecture to be proved and in the recursive function that generates the induction scheme (however, the induction variables must be free in the conjecture). The latter allows one to perform strong induction, by quantifying the recursive call in the function definition.

7 Libraries

The Eves environment supports abstraction and modularity by means of its library facility. The facility allows the user to `make`, `load`, `edit`, and `delete` library objects. The kinds of objects that may be stored in a library include *spec*, *model*, and *freeze* objects.

A freeze object contains a representation of the Eves system's state. This can be "thawed" at a later time to restore Eves to that state.

A spec object contains an axiomatic specification. In this specification, it is not necessary to provide complete definitions for any types, functions, or procedures that are declared; instead, axioms can be used to describe their properties. Axioms in a specification do not need to be proved (they are proved in the corresponding model).

A model object contains complete definitions of the types, functions, and procedures declared in the corresponding specification, and contains proofs of all the axioms of the specification. It may contain additional definitions and lemmas that might be used in the main definitions and proofs.

Spec objects may be used as specifications of program modules, in which case the corresponding model defines the implementation of the module; or as a presentation of a mathematical theory, in which case the model demonstrates the soundness of the theory.

The system performs a consistency check between a spec object and its corresponding model object. All symbols declared in the spec object must have a corresponding declaration in the model object. The consistency check is basically syntactic. However, the system allows corresponding declarations to have different heuristics, and it allows stub declarations in spec units.

The library facility is quite flexible: it does not force the user to make spec objects before model objects or vice versa, and it does not force the user to make the corresponding model (spec) object immediately after making a spec (model) object. The user may develop the model objects in a top-down, bottom-up, or mixed fashion.

8 General Remarks

With portability in mind, we wrote the Eves system in a carefully chosen subset of Common Lisp. We have successfully compiled the system under various Common Lisp systems on a number of hardware platforms, including Symbolics 3600, Sun-3, Sun SPARCstation, Apple Mac II, Data General AViiON, and Digital Equipment DecStation.[3]

The Eves environment has been used on a number of examples, the biggest being a PICO interpreter (about 13,000 lines source) [KP 90]. We translated the specification of a PICO interpreter (from [BHK 89]) into Verdi in the form of specification objects,

[3] These systems are the trademarks of their respective manufacturers.

provided models for them, wrote an imperative program for the interpreter, and verified that the program meets the specification.

We have also built library objects [Saa 91a, Saa 91b] that include most of the "Mathematical Toolkit" as described in Spivey's Z reference manual [Spi 87]. Many additional laws (beyond those explicitly mentioned in the Toolkit) were needed to complete the proofs of the laws in the Toolkit. During this work, minor errors in the laws were found.

References

[BHK 89] J.A. Bergstra, J. Heering, and P. Klint. *Algebraic Specification.* ACM Press, New York, New York, 1989.

[BM 79] R.S. Boyer, J S. Moore. *A Computational Logic.* Academic Press, 1979.

[CKMPS 91] Dan Craigen, Sentot Kromodimoeljo, Irwin Meisels, Bill Pase and Mark Saaltink. EVES: An Overview. In *Proceedings of VDM '91.* Noordwijkerhout, The Netherlands, October 1991.

[CS 90] Dan Craigen and Mark Saaltink. Simple Type Theory in EVES. In *Proceedings of the Fourth Banff Higher Order Workshop.* Graham Birtwistle (editor), Springer-Verlag, New York, September 1990.

[Cra 91] Dan Craigen. Reference Manual for the Language Verdi. TR-91-5429-09a, ORA Canada, Ottawa, September 1991.

[Fra 68] Abraham Fraenkel. *Abstract Set Theory.* North-Holland, 1968.

[Gan 87] Harald Ganzinger. A Completion Procedure for Conditional Equations. In *Conditional Term Rewriting System, 1st International Workshop*, LNCS 308, Springer-Verlag, Berlin, July 1987.

[KP 90] Sentot Kromodimoeljo, Bill Pase. Using the Eves Library Facility: A PICO Interpreter. FR-90-5444-02, ORA Canada, Ottawa, February 1990.

[Nel 81] G. Nelson. Techniques for Program Verification. CSL-81-10, Xerox Palo Alto Research Center, CA, June 1981.

[Saa 90a] Mark Saaltink. A Formal Description of Verdi. TR-90-5429-10a, ORA Canada, Ottawa, November 1990.

[Saa 90b] Mark Saaltink. Alternative Semantics for Verdi. TR-90-5446-02, ORA Canada, November 1990.

[Saa 91a] Mark Saaltink. Z and Eves. TR-91-5449-02, ORA Canada, Ottawa, October 1991.

[Saa 91b] Mark Saaltink. The Eves Library. TR-91-5449-03, ORA Canada, Ottawa, August 1991.

[Saa 92] Mark Saaltink. Z and Eves: A summary. In *Proceedings of the 6th Annual Z User Meeting*, held December 1991, Workshop in Computing Series, Springer-Verlag, 1992.

[Spi 87] J.M. Spivey. *The Z Notation: A Reference Manual.* Prentice-Hall, 1987.

This article was processed using the LaTeX macro package with LMAMULT style

MGTP: A Parallel Theorem Prover Based on Lazy Model Generation

Ryuzo HASEGAWA
Institute for New Generation Computer Technology
1-4-28 Mita, Minato-ku, Tokyo 108, Japan

Miyuki KOSHIMURA
Toshiba Information Systems (Japan)

Hiroshi FUJITA
Mitsubishi Electric Corporation

hasegawa@icot.or.jp, koshi@icot.or.jp, fujita@sys.crl.melco.co.jp

Abstract. We have implemented a model-generation based parallel theorem prover in KL1 on a parallel inference machine, PIM. We have developed several techniques to improve the efficiency of forward reasoning theorem provers based on lazy model generation. The tasks of the model-generation based prover are the generation and testing of atoms to be the elements of a model for the given theorem. The problem with this method is the explosion in the number of generated atoms and in the computational cost in time and space, incurred by the generation processes. Lazy model generation is a new method that avoids the generation of unnecessary atoms that are irrelevant to obtaining proofs, and to provide flexible control for the efficient use of resources in a parallel environment. With this method we have achieved a more than one-hundred-fold speedup on a PIM consisting of 128 PEs.

1 Introduction

The aim of this research is to make high performance theorem provers for first-order logic by using the programming techniques of the parallel logic programming language, KL1. As a first step in developing KL1-technology theorem provers, we adopted the model generation method of SATCHMO as the basis. Our reasons were as follows: 1) SATCHMO has a good property that full unification is not necessary and that matching suffices for range-restricted problems. This makes it very convenient for us to implement provers in KL1 since KL1, as a committed choice language, provides very fast one-way unification. 2) It is easier to incorporate mechanisms for lemmatization, subsumption tests, and the other deletion strategies that are indispensable in solving difficult problems such as condensed detachment problems [Overbeek 90].

Two types of model generation provers were built: one for ground models (MGTP/G) and the other for nonground models (MGTP/N). Utilizing merit 1), above, MGTP/G becomes very simple and efficient [Fujita 91] where a new algorithm, called RAMS, is implemented to avoid redundancy in conjunctive matching. MGTP/G can prove non-Horn problems very efficiently on a distributed memory multi-processor, the Multi-PSI, by exploiting OR parallelism.

MGTP/N, on the other hand, aims at proving hard Horn problems by exploiting AND parallelism. For MGTP/N, we developed a new parallel algorithm that runs with optimal load balancing on a distributed memory architecture, and with the minimal amount of computation and memory to obtain proofs. This article provides a short introduction to MGTP/N and its main features.

2 Model Generation Method

A clause is represented in an implicational form:

$$A_1, A_2, \ldots, A_n \rightarrow C_1; C_2; \ldots; C_m$$

where $A_i(1 \leq i \leq n)$ and $C_j(1 \leq j \leq m)$ are atoms; the antecedent is a conjunction of $A_1, A_2, \ldots, A_n$; and the consequent is a disjunction of $C_1, C_2, \ldots, C_m$. A clause is said to be *tester* if its consequent is $false(m = 0)$, otherwise it is called *generator*.

The model generation method incorporates the following two rules:

- Model extension rule: If there is a generator clause, $A \rightarrow C$, and a substitution σ such that $A\sigma$ is satisfied in a model candidate M and $C\sigma$ is not satisfied in M, then extend M by adding $C\sigma$ into M.
- Model rejection rule: If a tester clause has an antecedent $A\sigma$ that is satisfied in a model candidate M, then reject M.

We call the process of obtaining $A\sigma$, a *conjunctive matching (CJM)* of the antecedent literals against the elements in a model candidate.

To achieve high performance in model-generation based provers, it is important to avoid redundant computation in conjunctive matching[1]. For this, we developed RAMS [Fujita 91], MERC [Hasegawa 91], and Δ-M [Hasegawa 92] algorithms.

3 Lazy Model Generation

A more important issue with regard to the efficiency of model-generation based provers is how to reduce the total computation amount and memory space required for proof processes. This problem becomes more critical when dealing with harder problems that require deeper inferences (longer proofs). To solve this problem, it is important to recognize that proving processes are viewed as *generation-and-test* processes and that generation should be performed only when required by the testing. We achieved high performance by introducing a new method called *lazy model generation*, based on the idea of "generate-only-at-test", designed for a demand-driven style of computation.

Figure 1 shows the lazy algorithm. In this algorithm, it is assumed that two processes, an algorithm for generator clauses and the other for tester clauses, run in parallel and communicate with each other.

Tester process 1) requests a set of atoms, Δ, to the generator process, 2) performs a subsumption test on Δ against $M \cup D$ where D represents a set of atoms to be added to M, and 3) performs a rejection test on $\Delta(CJM_T)$. For the generator process, 1) if a buffer, Buf, that stores a set of atoms that are the results of an application of the model extension rule, is empty, select an atom, e out of D and set a code for model

[1]The operation may be performed for identical combinations of atoms more than once.

process tester:
  repeat forever
    $request(generator, \Delta)$;
    $\Delta' := subsumption(\Delta, M \cup D)$;
    if $CJM_T(\Delta', M \cup D) \ni \bot$ **then return**(success);
    $D := D \cup \Delta'$.

process generator:
  repeat forever
    while $Buf = \phi$ **do begin**
      $D := D - \{e\}; Buf :=$ **delay**$CJM_G(e, M); M := M \cup \{e\}$ **end**;
    $wait(tester)$;
    $\Delta :=$**force**Buf;
  until $D = \phi$ **and** $Buf = \phi$.

Figure 1: Lazy algorithm

Table 1: Comparison of complexities (for unit tester clause)

Algorithm	T	S	G	M
Basic	ρm^2	$\mu\rho^2 m^{4\alpha}$	$\rho^2 m^4$	$\rho^3 m^4$
Full-test/Lazy	m^2	$\mu m^{2\alpha}$	ρm^2	ρm^2
Lazy & Lookahead	m^2	$(\mu/\rho)m^\alpha$	m/ρ	m

† m is the number of elements in a model candidate when $false$ is detected in the basic algorithm.
‡ ρ is the survival rate of a generated atom, μ is the rate of successful conjunctive matchings ($\rho \leq \mu$), $\alpha(1 \leq \alpha \leq 2)$ is the efficiency factor of a subsumption test.

extension (**delay** CJM_G) for e and M onto Buf. 2) waits for a request of Δ from the tester process, and 3) forces the buffer, Buf, to generate Δ.

The lazy mechanism can be used to control the difference in speed between the generator and tester processes, thereby avoiding the unnecessary consumption of time and space caused by over generation.

From a simple analysis, it is estimated that the time complexity of the model extension and subsumption test decreases from $O(m^4)$ in the algorithms[2] without lazy control to $O(m)$ in the lazy one. Table 1 compares the complexities of the algorithms, where T(S/G) represents the number of rejection tests(subsumption tests/model extensions), and M represents the number of atoms stored. For details, refer to [Hasegawa 92].

[2] The basic algorithm taken by OTTER[McCune 90] generates a bunch of new atoms before completing rejection tests for previously generated atoms. The full-test algorithm completes the tests before the next generation cycle, but still generates a bunch of atoms each time. Lookahead is an optimization method for testing wider spaces than in Full-test/Lazy.

4 AND Parallel Implementation

We have several choices when parallelizing model-generation based theorem provers: 1) proof changing or unchanging according to the number of PEs, 2) model sharing (copying in a distributed memory architecture) or model distribution, and 3) master-slave or master-less.

A proof changing prover may achieve super-linear speedup while a proof unchanging prover can achieve at most linear speedup.

The merit of model sharing is that time consuming subsumption testing and conjunctive matching can be performed at each PE independently with minimal inter-PE communication. On the other hand, the merit of model distribution is that we can obtain "memory scalability". The communication cost, however, increases as the number of PEs increases, since generated atoms need to be flown to all PEs for subsumption testing.

The master-slave configuration makes it easy to build a parallel system by simply connecting a sequential version of MGTP/N on a slave PE to the master PE. However, its devices must be designed to minimize the load on the master process. On the other hand, a master-less configuration such as ring connection allows us to achieve pipeline effects with better load balancing, whereas it becomes harder to implement suitable control to manage collaborative work among PEs.

Our policy in developing parallel theorem provers is that we should distinguish between the speedup effect caused by parallelization and the search-pruning effect caused by strategies. In the proof changing parallelization, changing the number of PEs is merely betting, and may cause the strategy to be changed badly even though it results in the finding of a shorter proof.

Given the above, we implemented a proof unchanging version of MGTP/N in a master-slave configuration based on the lazy model generation. In this system, generator and subsumption processes run in a demand-driven mode, while tester processes run in a data-driven mode. The main features of this system are as follows: 1) Proof unchanging allows us to obtain greater speedup as the number of PEs increases; 2) By utilizing the synchronization mechanism supported by KL1, sequentiality in subsumption testing is minimized; 3) Since slave processes spontaneously obtain tasks from the master and the size of each task is well equalized, good load balancing is achieved; 4) By utilizing the stream data type of KL1, demand driven control is easily and efficiently implemented.

By using demand-driven control, we can not only suppress unnecessary model extensions and subsumption tests but also maintain a high running rate, that is the key to achieving linear speedup.

Table 2 shows performance obtained by running the MGTP/N prover for Theorems 5 and 7 [Overbeek 90] on Multi-PSI with 64 PEs. We did not use heuristics such as sorting, but merely limited term size and eliminated tautologies. Full unification is written in KL1, which is thirty to hundred times slower than that written in C on SUN/3 and SPARC. Note that the average running rate per PE for 64 PEs is even a little bit higher than that for 16 PEs. With this and other results, we were able to obtain almost linear speedup.

Recently we obtained a proof of Theorem 5 on PIM/m with 127 PEs in 2870.62 sec and 43939240329 reductions (thus 120 KRPS/PE). Taking into account that the CPU of PIM/m is about two times faster than that of Multi-PSI, we found that almost

Table 2: Performance of MGTP/N (Theorem 5 and Theorem 7)

Problem		16 PEs	64 PEs
Theorem 5	time(sec)	41725.98	11056.12
	Reductions	38070940558	40759689419
	KRPS/pe	57.03	57.60
	speedup	1.00	3.77
Theorem 7	time(sec)	48629.93	13514.47
	Reductions	31281211417	37407531427
	KRPS/pe	40.20	43.25
	speedup	1.00	3.60

linear speedup can be achieved at least up to 128 PEs.

5 Conclusion

It is important to avoid explosive increases in the amount of time and space consumed when proving hard theorems that require deep inferences. For this we proposed the lazy model generation method and techniques to improve its performance. Experimental results show that a significant saving in the computation and memory amounts can be realized by using the lazy algorithm. The lazy model generation method can be easily extended from Horn clauses to non-Horn clauses.

We developed AND parallelization methods to solve Horn problems. The recent results of our experiments show that we could achieve linear speedup up to 128 PEs. The key technique is laziness in model generation that avoids unnecessary computation and memory space while maintaining a high running rate.

REFERENCES

[Fujita 91] H. Fujita and R. Hasegawa, A Model-Generation Theorem Prover in KL1 Using Ramified Stack Algorithm, In *Proc. of the Eighth International Conference on Logic Programming,* The MIT Press, 1991.

[Hasegawa 91] R. Hasegawa, A Parallel Model Generation Theorem Prover: MGTP and Further Research Plan, In Proc. of the Joint American-Japanese Workshop on Theorem Proving, Argonne, Illinois, 1991.

[Hasegawa 92] R. Hasegawa, M. Koshimura and H. Fujita, Lazy Model Generation for Improving the Efficiency of Forward Reasoning Theorem Provers, ICOT TR-751, 1992.

[Manthey 88] R. Manthey and F. Bry, SATCHMO: a theorem prover implemented in Prolog, In *Proc. of CADE 88, Argonne, Illinois,* 1988.

[McCune 90] W. W. McCune, OTTER 2.0 Users Guide, Argonne National Laboratory, 1990.

[Overbeek 90] R. Overbeek, Challenge Problems, (private communication) 1990.

Benchmark Problems in Which Equality Plays the Major Role[1]

E. Lusk and L. Wos

Mathematics and Computer Science Division
Argonne National Laboratory
Argonne, Illinois 60439-4801

Abstract. We have recently heard rumors that researchers are again studying paramodulation [7] in the context of strategy for its control. In part to facilitate such research, and in part to provide test problems for evaluating other approaches to equality-oriented reasoning, we offer in this article a set of benchmark problems in which equality plays the dominant role. The test problems are taken from group theory, ring theory, Robbins algebra, combinatory logic, and other areas. For each problem, we include appropriate clauses and comment as to its status with regard to provability by an unaided automated reasoning program. To complement the test problems, we offer various open questions.

1. Group Theory

A group is a nonempty set G in which multiplication is associative such that a two-sided identity e exists whose product with x is x and for which the two-sided inverse of x exists. To study group theory, one can use the following clauses; throughout the remainder of this paper, we use the notation of William McCune's program OTTER [3,4], where " | " means **or** and " - " means **not.**

```
EQ(prod(e,x),x).          EQ(prod(x,inv(x)),e).
EQ(prod(x,e),x).          EQ(prod(prod(x,y),z),prod(x,prod(y,z))).
EQ(prod(inv(x),x),e).     EQ(x,x).
```

The last clause (for reflexivity) is included, for its presence is required when paramodulation is used. When attempting to prove that some set of equalities is an axiom system for group theory, one proves each of these axioms, except for the clause for reflexivity, one at a time.

Problem GT1, simple. If the square of every x is the identity, the group is commutative.

```
EQ(prod(x,x),e).
```

Problem GT2, moderate. Prove that the following equality (taken from Meredith [6] is a single axiom for groups in which the square of every x is the identity. In particular, using the single axiom, derive the axioms for groups (given earlier) and the axiom that asserts that the square of every element is the identity. In this problem e is represented by the term $prod(x,x)$, once one proves that $prod(x,x) = prod(y,y)$.

[1]This work was supported by the Applied Mathematical Sciences subprogram of the Office of Energy Research, U.S. Department of Energy, under Contract W-31-109-Eng-38.

EQ(prod(prod(prod(x,y),z),prod(x,z)),y).

Problem GT3, moderate. Prove that $[[x,y],y] = e$ when the cube of every x is e, where $[x,y]$ is the product of x, y, the inverse of x, and the inverse of y.

EQ(prod(x,prod(x,x)),e). EQ(com(x,y),prod(x,prod(y,prod(inv(x),inv(y))))).

Problem GT4, moderate. Prove that the following equality axiomatizes group theory; the corresponding theorem, proved by McCune using OTTER [5], is a new contribution to the literature. In this problem and the next two, e is prepresented by the term $prod(x, inv(x))$, once one has proved that $prod(x, inv(x)) = prod(y, inv(y))$.

EQ(prod(x,inv(prod(y,prod(prod(prod(z,inv(z)),inv(prod(u,y))),x)))),u).

Problem GT5, moderate. Prove that the following equality axiomatizes commutative group theory; this new single axiom was found by William McCune using OTTER [5].

EQ(prod(prod(prod(x,y),z),inv(prod(x,z))),y).

Problem GT6, never proved in single runs for each axiom. Prove that the following equality axiomatizes commutative group theory; the result was verified by Ken Kunen [private correspondence] using OTTER as an assistant.

EQ(prod(inv(prod(inv(prod(inv(prod(x1,x2)),prod(x2,x1))),prod(inv(prod(z,y)), prod(z,inv(prod(prod(v,inv(x)),inv(y))))))),x),v).

2. Robbins Algebra

A Robbins algebra is a nonempty set A satisfying the following four axioms, expressed in clause notation, in which the function o can be interpreted as plus and the function n as negation.

(R1) EQ(o(x,y),o(y,x)). (R3) EQ(n(o(n(o(x,y)),n(o(x,n(y))))),x).
(R2) EQ(o(o(x,y),z),o(x,o(y,z))). (R4) EQ(x,x).

A Boolean algebra is a nonempty set S with two operations, plus and times, and a 1 and a 0. Each operation is commutative, and each distributes over the other. The 1 is a multiplicative identity, and the 0 is an additive identity. In addition, for every x, the negation of x exists with x plus its negation equal to 1 and x times its negation equal to 0. An alternative axiomatization of Boolean algebra consists of (R1), (R2), and Huntington's axiom (H3) [2]: EQ(o(n(o(n(x),y)),n(o(n(x),n(y)))),x). Whether Robbins implies Boolean is still an open question. What is known is that the addition to the three Robbins axioms of any one of a number of properties of a Boolean algebra suffices to yield Boolean.

Problem RA1, simple. Prove that, if the axiom EQ(o(x,0),x). is adjoined to the axioms for a Robbins algebra, the resulting algebra is Boolean. We recommend trying to prove Huntington's axiom (H3).

Problem RA2, moderate. Prove that the addition of the equality EQ(o(x,x),x). to Robbins yields Boolean.

Problem RA3, hard. Prove that, where c is a constant, the addition of the equality EQ(o(c,c),c). to Robbins yields Boolean.

Problem RA4, never proved in a single run unaided. Prove that, where c and d are constants, the addition of the equality EQ(o(c,d),d). to Robbins yields Boolean.

Problem RA5, never proved in a single run unaided. Prove that, where c and d are constants, the addition of the equality EQ(n(o(c,d)),n(d)). to Robbins yields Boolean.

3. Combinatory Logic

Barendregt [1] defines combinatory logic as an equational system satisfying the combinators S and K with $((Sx)y)z = (xz)(yz)$ and $(Kx)y = x$. Rather than studying this logic in its entirety, one finds challenging test problems by replacing one or both of S and K by one or more combinators and focusing on questions concerning the possible presence of the strong fixed point property. The set consisting of the combinators under study is called a *basis*, and the set of combinators generated by a basis is called a *fragment*. Where **A** is a given fragment with basis **B**, the strong fixed point property holds for **A** if and only if there exists a combinator y such that, for all combinators x, $yx = x(yx)$, where y is expressed purely in terms of elements of **B**. The problems we offer focus on various subsets of the following combinators. We use $a(S,x)$ to represent Sx.

EQ(a(a(a(B,x),y),z),a(x,a(y,z))).
EQ(a(a(a(C,x),y),z),a(a(x,z),y)).
EQ(a(a(a(H,x),y),z),a(a(a(x,y),z),y)).
EQ(a(I,x),x).
EQ(a(M,x),a(x,x)).
EQ(a(a(a(N,x),y),z),a(a(a(x,z),y),z)).
EQ(a(a(a(S,x),y),z),a(a(x,z),a(y,z))).
EQ(a(a(W,x),y),a(a(x,y),y)).

For each of the following problems, the object is to use the combinators specified in the problem and no others and prove that the strong fixed point property holds by finding an appropriate object.

Problem CL1, simple. The set consists of B, M, and W.

Problem CL2, hard. The set consists of B and W.

Problem CL3, hard. The set consists of B and N.

Problem CL4, hard. The set consists of B and H.

Problem CL5, hard. The set consists of B, C, I, and S.

To complement the preceding test problems, we offer two open questions. Does the set consisting of B and M alone permit the construction of an object that proves that the strong fixed point property holds? Does the set consisting of B and S alone permit the construction of an object that proves that the strong fixed point property holds?

4. Many-Valued Sentential Calculus

The axioms are the following, where the constant T can be interpreted as "true" and the functions i and n as implication and negation, respectively.

EQ(i(T,x),x).	EQ(i(i(x,y),y),i(i(y,x),x)).
EQ(i(i(x,y),i(i(y,z),i(x,z))),T).	EQ(i(i(n(x),n(y)),i(y,x)),T).

Problem MV1, simple. Prove that each of the two equalities EQ(i(n(n(x)),x),T). and EQ(i(x,n(n(x))),T). holds in many-valued sentential calculus.

Problem MV2, moderate. Prove that EQ(i(i(x,y),i(i(z,x),i(z,y))),T). holds in the calculus.

Problem MV3, moderate. Prove that EQ(i(i(x,y),i(n(y),n(x))),T). holds in the calculus.

Problem MV4, hard. Prove that EQ(i(i(i(x,y),i(y,x)),i(y,x)),T). holds in the calculus.

5. Nonunit Problems

With the intention of spurring research focusing on paramodulation in which nonunit clauses occur, we offer the following problems.

Problem NU1, moderate. The problem asks one to prove the following identity in modular lattices in which a 0 and 1 exist, where $\wedge$ is *meet*, $\vee$ is *join*, and $'$ is *complement*. This is the equational form of Sam's Lemma.

$$((A \vee B)' \vee ((A \wedge B)' \wedge B)) \wedge ((A \vee B)' \vee ((A \wedge B)' \wedge A)) = (A \vee B)'$$

The following clauses can be used, the first 18 of which capture the properties of a modular lattice.

EQ(meet(0,x),0).	EQ(meet(meet(x,y),z),meet(x,meet(y,z))).
EQ(meet(x,0),0).	EQ(join(join(x,y),z),join(x,join(y,z))).
EQ(join(0,x),x).	EQ(meet(x,join(x,y)),x).
EQ(join(x,0),x).	EQ(join(x,meet(x,y)),x).
EQ(meet(1,x),x).	-EQ(meet(x,z),x) | EQ(meet(z,join(x,y)),join(x,meet(y,z))).
EQ(meet(x,1),x).	EQ(x,x).
EQ(join(1,x),1).	EQ(join(r2,meet(a,b)),1).
EQ(join(x,1),1).	EQ(meet(r2,meet(a,b)),0).
EQ(meet(x,x),x).	EQ(join(r1,join(a,b)),1).
EQ(join(x,x)x).	EQ(meet(r1,join(a,b)),0).
EQ(meet(x,y),meet(y,x)).	EQ(join(r1,meet(a,r2)),b2).
EQ(join(x,y),join(y,x)).	EQ(join(r1,meet(b,r2)),a2).
	-EQ(meet(a2,b2),r1).

Problem NU2, moderate. Prove that subgroups of index 2 are normal, using the clauses given earlier to study group theory. O(x) means that x is in the subgroup, j(x,y) is an element of the subgroup that exists if x and y are not in the subgroup (for giving the definition of index 2).

```
O(e).                                 O(x) | O(y) | EQ(prod(x,j(x,y)),y).
-O(x) | O(inv(x)).                    O(b).
-O(x) | -O(y) | O(prod(x,y)).         -O(prod(a,prod(b,g(a)))).
O(x) | O(y) | O(i(x,y)).
```

Problem NU3, hard. In set theory, prove that if two ordered pairs are equal, then they are equal componentwise. In the following clauses, op(x,y) means the ordered pair, up(x,y) means the unordered pair, IN is set membership, and sing is the singleton set consisting of x.

```
EQ(x,x).                         -IN(x,up(y,z)) | EQ(x,y) | EQ(x,z).
IN(x,sing(x)).                   EQ(op(x,y),up(sing(x),up(x,y))).
-IN(x,sing(y)) | EQ(x,y).        EQ(op(m1,r1),op(m2,r2)).
IN(x,up(x,y)).                   -EQ(m1,m2) | -EQ(r1,r2).
IN(y,up(x,y)).                   EQ(op(x,y),up(sing(x),up(x,y))).
```

References

1. Barendregt, H. P. *The Lambda Calculus: Its Syntax and Semantics,* North-Holland, Amsterdam (1981).

2. Huntington, E., "New sets of independent postulates for the algebra of logic, with special reference to Whitehead and Russell's Principia Mathematica", *Trans. of AMS,* **35**, pp. 274–304 (1933).

3. McCune, W., *OTTER 2.0 Users Guide,* Technical Report ANL-90/9, Argonne National Laboratory, Argonne, Illinois, 1990.

4. McCune, W., *What's New in OTTER 2.2,* Mathematics and Computer Science Division Technical Report ANL/MCS-TM-153, Argonne National Laboratory, 1991.

5. McCune, W., *Single Axioms for Groups and Abelian Groups with Various Operations,* Mathematics and Computer Science Division Preprint MCS-P270-1091, Argonne National Laboratory, Argonne, Illinois, 1991.

6. Meredith, C. A., and Prior, A. N., "Equational logic", *Notre Dame J. Formal Logic,* **9**, pp. 212–226 (1968).

7. Wos, L., *Automated Reasoning: 33 Basic Research Problems,* Prentice-Hall, Englewood Cliffs, N.J., 1987.

Computing Transitivity Tables: A Challenge For Automated Theorem Provers *

D A Randell, A G Cohn and Z Cui
Division of Artificial Intelligence
School of Computer Studies
University of Leeds, Leeds LS2 9JT, England
{dr, agc, cui}@dcs.leeds.ac.uk

Abstract. Implementations of Allen's interval-based temporal logic and a recently developed simulation system for reasoning about space and time, both require the use of transitivity tables. Although strategies exist to construct such tables, the proofs which underly the entries in the table are both tedious to do and in some cases difficult to secure. Often a difficult proof is only obtained via lemmas; moreover, finding models for satisfiable sets of dyadic relations in the theory introduces its own difficulties. This paper presents the problems. Any automated theorem prover which can effectively generate the entries for such transitivity tables would mark significant progress in automated theorem proving.

1 Introduction

A transitivity table is defined as follows. Given a particular theory Σ supporting a set of mutually exhaustive and pairwise disjoint dyadic relations, three individuals, a, b and c and a pair of dyadic relations R_1 and R_2 selected from Σ such that $R_1(a,b)$ and $R_2(b,c)$, the transitive closure $R_3(a,c)$ represents a disjunction of all the possible dyadic relations holding between a and c in Σ. Each $R_3(a,c)$ result can be represented as one entry of a matrix for each $R_1(a,b)$ and $R_2(b,c)$ ordered pair. If there are n dyadic relations supported by Σ, then there will be $n \times n$ entries in the matrix. This matrix is called a transitivity table.

A well known example of the use of a transitivity table arises in an implementation of Allen's temporal logic[1]. Allen's theory assumes an ontology of (convex) intervals upon which are defined a set of 13 mutually exclusive and exhaustive dyadic relations, e.g. $Before(x,y)$, $Meets(x,y)$, and $Overlaps(x,y)$.

In contrast to Allen, Randell and Cohn[9, 11] develop a formal theory for reasoning about both space and time based on earlier work by Clarke. A subset of the defined dyadic relations are used to construct a transitivity table. Initially, only 9 mutually inclusive and exhaustive defined relations are used to generate the matrix, but other spatial relations are subsequently defined which extend this set to 23. These tables are used in an implementation of a variant of Theory Resolution[10], and for consistency checking in an envisionment-based simulation program[4]. We

*The support of the SERC under grant no. GR/G36852 is gratefully acknowledged.

have constructed the complete transitivity table for the basic part of the theory see Table 1[1], but not for the extended theory yet [2]. The transitivity table in this extension of the theory is very large with 529 theorems to be proved and verified for minimality.

Proving such a large number of theorems by hand or with machine assistance is both tedious to do and is a computationally intensive problem. However, given the difficulties we have encountered constructing such transitivity tables, we believe the task of constructing transitivity tables presents an interesting problem for automated theorem provers in its own right. Moreover, we believe this is the first challenge which has been posed that has arisen from a genuine AI application, rather than being a problem out of some branch of mathematics or an artificial problem such as 'Schubert's Steamroller', or the 'Lion and the Unicorn'.

R2(b,c) / R1(a,b)	DC	EC	PO	TPP	NTPP	TPP^{-1}	$NTPP^{-1}$	TPI	NTPI
DC	no.info	DR,PO,PP	DR,PO,PP	DR,PO,PP	DR,PO,PP	DC	DC	DC	DC
EC	DR,PO,PP^{-1}	DR,PO TPP,TP^{-1}	DR,PO,PP	EC,PO,PP	PO,PP	DR	DC	EC	×
PO	DR,PO,PP^{-1}	DR,PO,PP^{-1}	no.info	PO,PP	PO,PP	DR,PO,PP^{-1}	DR,PO PP^{-1}	PO	PO
TPP	DC	DR	DR,PO,PP	PP	NTPP	DR,PO TPP,TP^{-1}	DR,PO PP^{-1}	TPP	×
NTPP	DC	DC	DR,PO,PP	NTPP	NTPP	DR,PO,PP	no.info	NTPP	NTPP
TPP^{-1}	DR,PO,PP^{-1}	EC,PO,PP^{-1}	PO,PP^{-1}	PO,TPP,TP^{-1}	PO,PP	PP^{-1}	$NTPP^{-1}$	TPP^{-1}	×
$NTPP^{-1}$	DR,PO,PP^{-1}	PO,PP^{-1}	PO,PP^{-1}	PO,PP^{-1}	O	$NTPP^{-1}$	$NTPP^{-1}$	$NTPP^{-1}$	$NTPP^{-1}$
TPI	DC	EC	PO	TPP	NTPP	TPP^{-1}	$NTPP^{-1}$	TPI	×
NTPI	DC	×	PO	×	NTPP	×	$NTPP^{-1}$	×	NTPI

Table 1: Transitivity table for the 9 basic relations. If R_1(a,b) and R_2(b,c), it follows that R_3(a,c) where R_3 is looked up in the table. "×" entries mean that the corresponding conjunction R_1(a,b) and R_2(b,c) cannot be simultaneously satisfied, and "no info." that no base relation is excluded. Multiple entries in a cell are interpreted as disjunctions. Note that DR stands for DC and EC, PP for TPP and NTPP, PP^{-1} for TPP^{-1} and $NTPP^{-1}$, TP^{-1} for TPP^{-1} and TPI, and O for PO, TPP, NTPP, TPP^{-1}, $NTPP^{-1}$, TPI and NTPI.

2 Constructing transitivity tables

There are two aspects to be considered when constructing transitivity tables. The first is to prove that $R_1(a,b) \wedge R_2(b,c) \vdash R_3(a,c)$ and the second to show that the disjunction of entries for each $R_3(a,c)$ relation is the strongest possible result, i.e., given $R_1(a,b)$ and $R_2(b,c)$, a model also exists for each atomic disjunct in $R_3(a,c)$. If a single entry for $R_3(a,c)$ exists, then this is the strongest possible result, but if

[1] This was initially constructed by hand but has been subsequently verified using OTTER 2.0[6]. Some machine proofs were assisted by the use of hand chosen lemmas, and by restricting the set of axioms and definitions given to the system. Listings of mechanical proofs can be obtained by e-mailing the authors.

[2] Very recently, we have constructed an extended table by writing a program which constructively enumerates and tests configurations in a diagrammatic representation of space. However we have no proof of the correctness of this table with respect to the underlying formal theory though a few entries have been verified.

$R_3(a,c)$ consists of at least two disjuncts, we need to ensure that no strict subset of $R_3(a,c)$ is also a logical consequence of $R_1(a,b)$ and $R_2(b,c)$. Sometimes $R_3(a,c)$ excludes no base relation; in this case $R_1(a,b) \wedge R_2(b,c) \vdash R_3(a,c)$, is vacuously true.

Allen's logic benefits from having a relatively small number of base relations, together with a one-dimensional model. This provides a useful constraint when constructing models satisfying each $R_3(a,c)$ entry in his table. However, in our case we have (in the extended theory) substantially more base relations to consider, plus the fact that we cannot use a one-dimensional model. Another complication arises from the fact that in our theory, individuals need not be (topologically) connected (in one-piece). Other complications arise when it is not known whether or not Σ has a finite model - if this cannot be guaranteed a model-building program may not terminate. If, on the other hand, an infinite model exists, a finite model may also exist, although constructing one may require much ingenuity.

3 The Axiomatisation

The basic part of the formalism assumes one primitive dyadic relation: '$\mathrm{C}(x,y)$' read as 'x connects with y', and that the individuals are interpreted as (either spatial or temporal) regions. In terms of points incident in regions, $\mathrm{C}(x,y)$ holds when region x and y share a common point. Using C, a basic set of dyadic relations are defined: DC (is disconnected from), P (is a part of), PP (is a proper part of), = (is identical with), O (overlaps), DR (is discrete from), EC (is externally connected with), TP (is a tangential part of), NTP (is a nontangential part), TPP (is a nontangential proper part of), NTPP (is a nontangential proper part of), TPI (is the identity tangential part of), and NTPI (is the identity nontangential part of). The relations: P, PP, TP, NTP, TPP and NTPP support inverses[3]. DC, EC, PO, TPP, NTPP, TPI, NTPI, TPP^{-1} and NTPP^{-1} are mutually exhaustive and pairwise disjoint.

$\forall x \mathrm{C}(x,x)$ | $\forall x \forall y[\mathrm{C}(x,y) \rightarrow \mathrm{C}(y,x)]$

$\mathrm{DC}(x,y) \equiv_{def} \neg\mathrm{C}(x,y)$ | $\mathrm{P}(x,y) \equiv_{def} \forall z[\mathrm{C}(z,x) \rightarrow \mathrm{C}(z,y)]$

$\mathrm{PP}(x,y) \equiv_{def} \mathrm{P}(x,y) \wedge \neg\mathrm{P}(y,x)$ | $x = y \equiv_{def} \mathrm{P}(x,y) \wedge \mathrm{P}(y,x)$

$\mathrm{O}(x,y) \equiv_{def} \exists z[\mathrm{P}(z,x) \wedge \mathrm{P}(z,y)]$ | $\mathrm{PO}(x,y) \equiv_{def} \mathrm{O}(x,y) \wedge \neg\mathrm{P}(x,y) \wedge \neg\mathrm{P}(y,x)$

$\mathrm{DR}(x,y) \equiv_{def} \neg\mathrm{O}(x,y)$ | $\mathrm{TP}(x,y) \equiv_{def} \mathrm{P}(x,y) \wedge \exists z[\mathrm{EC}(z,x) \wedge \mathrm{EC}(z,y)]$

$\mathrm{EC}(x,y) \equiv_{def} \mathrm{C}(x,y) \wedge \neg\mathrm{O}(x,y)$ | $\mathrm{NTP}(x,y) \equiv_{def} \mathrm{P}(x,y) \wedge \neg\exists z[\mathrm{EC}(z,x) \wedge \mathrm{EC}(z,y)]$

$\mathrm{TPP}(x,y) \equiv_{def} \mathrm{TP}(x,y) \wedge \neg\mathrm{P}(y,x)$ | $\mathrm{NTPP}(x,y) \equiv_{def} \mathrm{NTP}(x,y) \wedge \neg\mathrm{P}(y,x)$

$\mathrm{TPI}(x,y) \equiv_{def} \mathrm{TP}(x,y) \wedge \mathrm{P}(y,x)$ | $\mathrm{NTPI}(x,y) \equiv_{def} \mathrm{NTP}(x,y) \wedge \mathrm{P}(y,x)$

For the extension to the theory, a primitive function 'conv(x)' ('the convex-hull of x') is added and axiomatised as follows:

$\forall x \mathrm{P}(x, \mathrm{conv}(x))$ | $\forall x \forall y \forall z[[\mathrm{P}(x,\mathrm{conv}(y) \wedge \mathrm{P}(y,\mathrm{conv}(z))] \rightarrow \mathrm{P}(x,\mathrm{conv}(z))]$

$\forall x \mathrm{P}(\mathrm{conv}(\mathrm{conv}(x)), \mathrm{conv}(x))$ | $\forall x \forall y[[\mathrm{P}(x,\mathrm{conv}(y)) \wedge \mathrm{P}(y,\mathrm{conv}(x))] \rightarrow \mathrm{O}(x,y)]$

$\forall x \forall y[[\mathrm{DR}(x,\mathrm{conv}(y)) \wedge \mathrm{DR}(y,\mathrm{conv}(x))] \rightarrow \mathrm{DR}(\mathrm{conv}(x),\mathrm{conv}(y))]$

We use this function to define a further set of dyadic relations which describe regions being inside, partially inside and outside others, e.g. INSIDE (is inside), P-INSIDE (is partially inside) and OUTSIDE (is outside):

[3]We omit formal definitions for the inverse relations (e.g. $\mathrm{P}^{-1}(x,y)$) supported by the theory.

$\text{INSIDE}(x,y) \equiv_{def} \text{DR}(x,y) \wedge \text{P}(x,\text{conv}(y))$ $\text{OUTSIDE}(x,y) \equiv_{def} \text{DR}(x,conv(y))$
$\text{P-INSIDE}(x,y) \equiv_{def} \text{DR}(x,y) \wedge \text{PO}(x,\text{conv}(y))$

Each of these relations has an inverse. This particular set of relations specialises DR in the basic theory. The developed theory actually supports many more specialisations of these particular relations, with, for example, one region being wholly outside, or just outside, or just inside, or wholly inside another - see [11]. However, here we restrict the set of defined relations to the three specialisations defined above, their inverses, and the set of relations that result from their non-empty intersections, in each case defining an EC and a DC variant. This set of base relations which are pairwise disjoint and which mutually exhaust DR, are constructed according to the following schema:

$\alpha\_\beta^{\delta}(x,y) \equiv_{def} \alpha(x,y) \wedge \beta(x,y) \wedge \delta(x,y)$

where: $\alpha \in$ {INSIDE, P-INSIDE, OUTSIDE}, $\beta \in$ {INSIDE^{-1}, P-INSIDE^{-1}, OUTSIDE^{-1}}, and $\delta \in$ {EC, DC}, excepting where α = INSIDE and β =INSIDE^{-1}. Half these basic relations specialise EC and the other half specialise DC. There are now a total of 23 basic relations, the 9 original ones except DC and EC and 16 new ones.

4 Tactics

All the relations have been embedded in a relational lattice[10] . Lemmas reflecting this structure can be effectively used in securing proofs. For example, given the pair <DC(a,b),EC(b,c)>, the lemma: $\forall x \forall y[\text{PP}(x,y) \leftrightarrow [\text{TPP}(x,y) \vee \text{NTPP}(x,y)]]$, 'collapses' a subset of the entries in the $R_3(a,c)$ cell to a more general equivalent disjunction[4]. Other useful lemmas are those that express transitivity properties of individual relations, e.g. for the relation P, and lemmas that map 'across' different relations, e.g. $\forall x, \forall y, \forall z[[\text{NTP}(x,y) \wedge \text{C}(z,x)] \rightarrow \text{O}(z,y)]]$. This presupposes that one already has a good idea what the $R_3(a,c)$ result will be, but the real challenge would be not to assume this. This is particularly relevant in the extended theory, where constructing models is much more difficult.

It is also interesting to note that when constructing the transitivity table for the basic part of the theory, we found that a duality existed between the scale of difficulty in constructing models, and securing mechanical proofs of the $R_3(a,c)$ result using a resolution proof procedure. The less constrained the models for an entry, the easier it seemed to be to secure the mechanical proof, while the more constrained the model, the more difficult it was.

In [9, 8], the set of base relations for the basic part of the theory is expressed in terms of a graph. Arcs in the graph corresponds to transitions which preserve continuity [5]. Thus, for example, if DC(a,b) is true, then over time this could change directly to EC(a,b), then to PO(a,b), and then either to TPP(a,b), TPP$^{-1}(a,b)$ or to TPI(a,b). Entries in each $R_3(a,c)$ cell in the transitivity table reflect this structure for no disjunction contains an 'isolated' predicate. For example, nowhere

[4] We have used such lemmas in the generation of table 1 in order to reduce the number of entries in the table and the space in the page it occupies

[5] Nökel[7] and Freksa[5] using Allen's logic also interpret direct transitions between pairs of individuals in terms of a graph.

do we find DC and PO appearing without EC. This suggests the following strategy. When constructing models by machine, we work inward from, say DC and TPI using the graph as the model. If, for example, we find a model for DC and TPI, then the results so far suggest models will also exist for EC and PO too.

5 The Challenges

There are several challenges that can be readily formulated for automated theorem provers from the automated construction of transitivity tables. The first is to generate the basic transitivity table given in Table 1 with no assistance; that is to say without the ad hoc introduction of lemmas which we used, or by restricting the set of axioms and definitions given to the theorem prover. The second challenge is to construct the transitivity table using the extended set of definitions. An interesting sub-problem can also be formulated, which is to prove which subset of the named relations are pairwise disjoint, or to prove the properties of the relational lattice they can be embedded in, simply from the set of axioms and definitions given, i.e. to construct the lattice specified in [9, 10]. The final challenge would be to construct a model-building program that can prove that logical models exist for each disjunct of each $R_3(a, c)$ cell in the transitivity tables generated.

References

[1] Allen, J. F. *Maintaining Knowledge about Temporal Intervals.* Comm. ACM 26(11), 1983.

[2] Clarke, B. L. [1981] *A Calculus of Individuals based on 'Connection',* Notre Dame Journal of Formal Logic, 22(3).

[3] Clarke, B. L. [1985] *Individuals and Points,* Notre Dame Journal of Formal Logic, 26(1).

[4] Cui, Z., Randell, D. A. and Cohn, A. G. *Qualitative Simulation based on Logical Formalism of Space and Time,* to appear in Proceedings of AAAI–92, AAAI Press, 1992.

[5] Freksa, C. *Qualitative Spatial Reasoning,* in Proceedings of Workshop RAUM, University of Koblenz, 1990.

[6] McCune W. *OTTER 2.0 Users Guide.* Arg. Nat. Lab, Argonne, Illinois, 1990.

[7] Nökel, K. *Convex Relations between Time Intervals,* SEKI Report SR-88-17, Universität Kaiserslautern, 1988.

[8] Randell, D. A. *Analysing the Familiar: Reasoning about Space and Time in the Everyday World,* Ph.D thesis, University of Warwick, UK, 1991.

[9] Randell, D. A. & Cohn, A. G. *Modeling Topological and Metrical Properties in Physical Processes.* in Proc. of 1st Int. Conf. on Princ. of Know. Rep. and Reasoning, ed. R.J. Brachman, et al, Morgan Kaufmann, San Mateo, Ca., 1989

[10] Randell, D. A. & Cohn, A G *Exploiting Lattices in a Theory of Space and Time.* in Computers and Mathematics, with Applications, Vol 23, No 6–9, 1992.

[11] Randell, D. A., Cohn, A. G. and Cui, Z. *Naive Topology: modeling the force pump.* to appear in P. Struss and B. Faltings (eds) Recent Advances in Qualitative Physics, MIT press, 1992.

Author Index

Lecture Notes in Artificial Intelligence (LNAI)

Vol. 383: K. Furukawa, H. Tanaka, T. Fujisaki (Eds.), Logic Programming '88. Proceedings, 1988. IX, 251 pages. 1989.

Vol. 390: J.P. Martins, E.M. Morgado (Eds.), EPIA 89. Proceedings. 1989. XII, 400 pages. 1989.

Vol. 395: M. Schmidt-Schauß, Computational Aspects of an Order-Sorted Logic with Term Declarations. VIII, 171 pages. 1989.

Vol. 397: K.P. Jantke (Ed.), Analogical and Inductive Inference. Proceedings, 1989. IX, 338 pages. 1989.

Vol. 406: C.J. Barter, M.J. Brooks (Eds.), AI '88. Proceedings, 1988. VIII, 463 pages. 1990.

Vol. 418: K.H. Bläsius, U. Hedtstück, C.-R. Rollinger (Eds.), Sorts and Types in Artificial Intelligence. Proceedings, 1989. VIII, 307 pages. 1990.

Vol. 419: K. Weichselberger, S. Pöhlmann, A Methodology for Uncertainty in Knowledge-Based Systems. VIII, 132 pages. 1990.

Vol. 422: B. Nebel, Reasoning and Revision in Hybrid Representation Systems. XII, 270 pages. 1990.

Vol. 437: D. Kumar (Ed.), Current Trends in SNePS – Semantic Network Processing System. Proceedings, 1989. VII, 162 pages. 1990.

Vol. 444: S. Ramani, R. Chandrasekar, K.S.R. Anjaneyulu (Eds.), Knowledge Based Computer Systems. Proceedings, 1989. X, 546 pages. 1990.

Vol. 446: L. Plümer, Termination Proofs for Logic Programs. VIII, 142 pages. 1990.

Vol. 449: M.E. Stickel (Ed.), 10th International Conference on Automated Deduction. Proceedings, 1990. XVI, 688 pages. 1990.

Vol. 451: V. Marík, O. Stepánková, Z. Zdráhal (Eds.), Artificial Intelligence in Higher Education. Proceedings, 1989. IX, 247 pages. 1990.

Vol. 459: R. Studer (Ed.), Natural Language and Logic. Proceedings, 1989. VII, 252 pages. 1990.

Vol. 462: G. Gottlob, W. Nejdl (Eds.), Expert Systems in Engineering. Proceedings, 1990. IX, 260 pages. 1990.

Vol. 465: A. Fuhrmann, M. Morreau (Eds.), The Logic of Theory Change. Proceedings, 1989. X, 334 pages. 1991.

Vol. 475: P. Schroeder-Heister (Ed.), Extensions of Logic Programming. Proceedings, 1989. VIII, 364 pages. 1991.

Vol. 476: M. Filgueiras, L. Damas, N. Moreira, A.P. Tomás (Eds.), Natural Language Processing. Proceedings, 1990. VII, 253 pages. 1991.

Vol. 478: J. van Eijck (Ed.), Logics in AI. Proceedings. 1990. IX, 562 pages. 1991.

Vol. 481: E. Lang, K.-U. Carstensen, G. Simmons, Modelling Spatial Knowledge on a Linguistic Basis. IX, 138 pages. 1991.

Vol. 482: Y. Kodratoff (Ed.), Machine Learning – EWSL-91. Proceedings, 1991. XI, 537 pages. 1991.

Vol. 513: N. M. Mattos, An Approach to Knowledge Base Management. IX, 247 pages. 1991.

Vol. 515: J. P. Martins, M. Reinfrank (Eds.), Truth Maintenance Systems. Proceedings, 1990. VII, 177 pages. 1991.

Vol. 517: K. Nökel, Temporally Distributed Symptoms in Technical Diagnosis. IX, 164 pages. 1991.

Vol. 518: J. G. Williams, Instantiation Theory. VIII, 133 pages. 1991.

Vol. 522: J. Hertzberg (Ed.), European Workshop on Planning. Proceedings, 1991. VII, 121 pages. 1991.

Vol. 535: P. Jorrand, J. Kelemen (Eds.), Fundamentals of Artificial Intelligence Research. Proceedings, 1991. VIII, 255 pages. 1991.

Vol. 541: P. Barahona, L. Moniz Pereira, A. Porto (Eds.), EPIA '91. Proceedings, 1991. VIII, 292 pages. 1991.

Vol. 542: Z. W. Ras, M. Zemankova (Eds.), Methodologies for Intelligent Systems. Proceedings, 1991. X, 644 pages. 1991.

Vol. 543: J. Dix, K. P. Jantke, P. H. Schmitt (Eds.), Nonmonotonic and Inductive Logic. Proceedings, 1990. X, 243 pages. 1991.

Vol. 546: O. Herzog, C.-R. Rollinger (Eds.), Text Understanding in LILOG. XI, 738 pages. 1991.

Vol. 549: E. Ardizzone, S. Gaglio, F. Sorbello (Eds.), Trends in Artificial Intelligence. Proceedings, 1991. XIV, 479 pages. 1991.

Vol. 565: J. D. Becker, I. Eisele, F. W. Mündemann (Eds.), Parallelism, Learning, Evolution. Proceedings, 1989. VIII, 525 pages. 1991.

Vol. 567: H. Boley, M. M. Richter (Eds.), Processing Declarative Kowledge. Proceedings, 1991. XII, 427 pages. 1991.

Vol. 568: H.-J. Bürckert, A Resolution Principle for a Logic with Restricted Quantifiers. X, 116 pages. 1991.

Vol. 587: R. Dale, E. Hovy, D. Rösner, O. Stock (Eds.), Aspects of Automated Natural Language Generation. Proceedings, 1992. VIII, 311 pages. 1992.

Vol. 590: B. Fronhöfer, G. Wrightson (Eds.), Parallelization in Inference Systems. Proceedings, 1990. VIII, 372 pages. 1992.

Vol. 592: A. Voronkov (Ed.), Logic Programming. Proceedings, 1991. IX, 514 pages. 1992.

Vol. 596: L.-H. Eriksson, L. Hallnäs, P. Schroeder-Heister (Eds.), Extensions of Logic Programming. Proceedings, 1991. VII, 369 pages. 1992.

Vol. 597: H. W. Guesgen, J. Hertzberg, A Perspective of Constraint-Based Reasoning. VIII, 123 pages. 1992.

Vol. 599: Th. Wetter, K.-D. Althoff, J. Boose, B. R. Gaines, M. Linster, F. Schmalhofer (Eds.), Current Developments in Knowledge Acquisition - EKAW '92. Proceedings. XIII, 444 pages. 1992.

Vol. 604: F. Belli, F. J. Radermacher (Eds.), Industrial and Engineering Applications of Artificial Intelligence and Expert Systems. Proceedings, 1992. XV, 702 pages. 1992.

Vol. 607: D. Kapur (Ed.), Automated Deduction – CADE-11. Proceedings, 1992. XV, 793 pages. 1992.

Lecture Notes in Computer Science

Vol. 570: R. Berghammer, G. Schmidt (Eds.), Graph-Theoretic Concepts in Computer Science. Proceedings, 1991. VIII, 253 pages. 1992.

Vol. 571: J. Vytopil (Ed.), Formal Techniques in Real-Time and Fault-Tolerant Systems. Proceedings, 1992. IX, 620 pages. 1991.

Vol. 572: K. U. Schulz (Ed.), Word Equations and Related Topics. Proceedings, 1990. VII, 256 pages. 1992.

Vol. 573: G. Cohen, S. N. Litsyn, A. Lobstein, G. Zémor (Eds.), Algebraic Coding. Proceedings, 1991. X, 158 pages. 1992.

Vol. 574: J. P. Banâtre, D. Le Métayer (Eds.), Research Directions in High-Level Parallel Programming Languages. Proceedings, 1991. VIII, 387 pages. 1992.

Vol. 575: K. G. Larsen, A. Skou (Eds.), Computer Aided Verification. Proceedings, 1991. X, 487 pages. 1992.

Vol. 576: J. Feigenbaum (Ed.), Advances in Cryptology - CRYPTO '91. Proceedings. X, 485 pages. 1992.

Vol. 577: A. Finkel, M. Jantzen (Eds.), STACS 92. Proceedings, 1992. XIV, 621 pages. 1992.

Vol. 578: Th. Beth, M. Frisch, G. J. Simmons (Eds.), Public-Key Cryptography: State of the Art and Future Directions. XI, 97 pages. 1992.

Vol. 579: S. Toueg, P. G. Spirakis, L. Kirousis (Eds.), Distributed Algorithms. Proceedings, 1991. X, 319 pages. 1992.

Vol. 580: A. Pirotte, C. Delobel, G. Gottlob (Eds.), Advances in Database Technology – EDBT '92. Proceedings. XII, 551 pages. 1992.

Vol. 581: J.-C. Raoult (Ed.), CAAP '92. Proceedings. VIII, 361 pages. 1992.

Vol. 582: B. Krieg-Brückner (Ed.), ESOP '92. Proceedings. VIII, 491 pages. 1992.

Vol. 583: I. Simon (Ed.), LATIN '92. Proceedings. IX, 545 pages. 1992.

Vol. 584: R. E. Zippel (Ed.), Computer Algebra and Parallelism. Proceedings, 1990. IX, 114 pages. 1992.

Vol. 585: F. Pichler, R. Moreno Díaz (Eds.), Computer Aided System Theory – EUROCAST '91. Proceedings. X, 761 pages. 1992.

Vol. 586: A. Cheese, Parallel Execution of Parlog. IX, 184 pages. 1992.

Vol. 587: R. Dale, E. Hovy, D. Rösner, O. Stock (Eds.), Aspects of Automated Natural Language Generation. Proceedings, 1992. VIII, 311 pages. 1992. (Subseries LNAI).

Vol. 588: G. Sandini (Ed.), Computer Vision – ECCV '92. Proceedings. XV, 909 pages. 1992.

Vol. 589: U. Banerjee, D. Gelernter, A. Nicolau, D. Padua (Eds.), Languages and Compilers for Parallel Computing. Proceedings, 1991. IX, 419 pages. 1992.

Vol. 590: B. Fronhöfer, G. Wrightson (Eds.), Parallelization in Inference Systems. Proceedings, 1990. VIII, 372 pages. 1992. (Subseries LNAI).

Vol. 591: H. P. Zima (Ed.), Parallel Computation. Proceedings, 1991. IX, 451 pages. 1992.

Vol. 592: A. Voronkov (Ed.), Logic Programming. Proceedings, 1991. IX, 514 pages. 1992. (Subseries LNAI).

Vol. 593: P. Loucopoulos (Ed.), Advanced Information Systems Engineering. Proceedings. XI, 650 pages. 1992.

Vol. 594: B. Monien, Th. Ottmann (Eds.), Data Structures and Efficient Algorithms. VIII, 389 pages. 1992.

Vol. 595: M. Levene, The Nested Universal Relation Database Model. X, 177 pages. 1992.

Vol. 596: L.-H. Eriksson, L. Hallnäs, P. Schroeder-Heister (Eds.), Extensions of Logic Programming. Proceedings, 1991. VII, 369 pages. 1992. (Subseries LNAI).

Vol. 597: H. W. Guesgen, J. Hertzberg, A Perspective of Constraint-Based Reasoning. VIII, 123 pages. 1992. (Subseries LNAI).

Vol. 598: S. Brookes, M. Main, A. Melton, M. Mislove, D. Schmidt (Eds.), Mathematical Foundations of Programming Semantics. Proceedings, 1991. VIII, 506 pages. 1992.

Vol. 599: Th. Wetter, K.-D. Althoff, J. Boose, B. R. Gaines, M. Linster, F. Schmalhofer (Eds.), Current Developments in Knowledge Acquisition - EKAW '92. Proceedings. XIII, 444 pages. 1992. (Subseries LNAI).

Vol. 600: J. W. de Bakker, K. Huizing, W. P. de Roever, G. Rozenberg (Eds.), Real-Time: Theory in Practice. Proceedings, 1991. VIII, 723 pages. 1992.

Vol. 601: D. Dolev, Z. Galil, M. Rodeh (Eds.), Theory of Computing and Systems. Proceedings, 1992. VIII, 220 pages. 1992.

Vol. 602: I. Tomek (Ed.), Computer Assisted Learning. Proceedigs, 1992. X, 615 pages. 1992.

Vol. 603: J. van Katwijk (Ed.), Ada: Moving Towards 2000. Proceedings, 1992. VIII, 324 pages. 1992.

Vol. 604: F. Belli, F. J. Radermacher (Eds.), Industrial and Engineering Applications of Artificial Intelligence and Expert Systems. Proceedings, 1992. XV, 702 pages. 1992. (Subseries LNAI).

Vol. 605: D. Etiemble, J.-C. Syre (Eds.), PARLE '92. Parallel Architectures and Languages Europe. Proceedings, 1992. XVII, 984 pages. 1992.

Vol. 606: D. E. Knuth, Axioms and Hulls. IX, 109 pages. 1992.

Vol. 607: D. Kapur (Ed.), Automated Deduction – CADE-11. Proceedings, 1992. XV, 793 pages. 1992. (Subseries LNAI).